C0-CCC-335

INTERNATIONAL CELL BIOLOGY · 1976-1977

INTERNATIONAL

B. R. BRINKLEY and

KEITH R. PORTER, Editors

PAPERS PRESENTED AT THE

FIRST INTERNATIONAL CONGRESS

ON CELL BIOLOGY

BOSTON, MASSACHUSETTS, 1976

CELL BIOLOGY

1976-1977

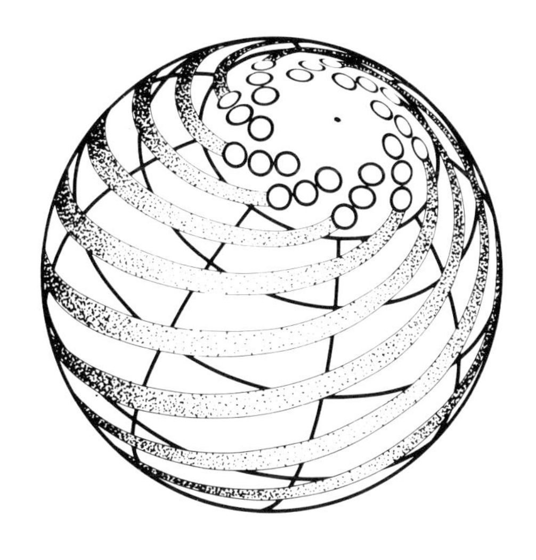

THE ROCKEFELLER UNIVERSITY PRESS

IN COOPERATION WITH THE AMERICAN

SOCIETY FOR CELL BIOLOGY AND

THE JOURNAL OF CELL BIOLOGY • 1977

LIBRARY OF CONGRESS CATALOGUE NUMBER 77-79991

ISBN 87470-027-2

PRINTED IN THE UNITED STATES OF AMERICA

SECOND PRINTING 1977

PREFACE

B. R. BRINKLEY and KEITH R. PORTER, *Editors*

This volume brings together most of the papers presented in twenty-two of the symposia at the First International Congress on Cell Biology, convened in Boston, September 5th to 10th, 1976. Because the topics selected for the symposia are among those most actively discussed and studied at the present time, this book becomes a historic document of cell biology in 1976. Future scholars should find it a valuable source of information from a period during which cells became the objects of intensive investigation. Naturally, we hope that subsequent Congresses will find in this publication a precedent which they will choose to follow, so that in the decades to come we shall have a continuous summary of our progress toward understanding the cell.

The various symposia topics were selected by the Program Committee many months before the meeting and after thorough discussion of what would be important and interesting in September, 1976. They also selected the symposia chairpersons, who, in their turn, invited speakers and then marshaled them and their papers before the Congress at the scheduled time and place. Great meetings depend on the efforts of great people, and we were fortunate to find so many among our Committee members who were willing and eager to help in organizing this meeting. We are especially grateful to these men and women, for without their cooperation the Congress would have been less successful and production of this volume would have been quite impossible.

We wish that we might have published in book form many of the other papers presented during the scientific sessions and workshops, but the magnitude of the task and the expense would have been prohibitive. Fortunately, abstracts of contributed papers and posters are preserved in the August, 1976, issue of *The Journal of Cell Biology*.

The opening ceremony of this Congress included speeches by distinguished guests and the presidents of the American Society for Cell Biology, which organized the meetings, and the International Federation for Cell Biology, which sponsored the Congress. George Palade and Daniel Mazia have generously provided us with the texts of their remarks for inclusion in this volume. They speak to the organization of cell biologists on a worldwide scale and remind us that cell biology has come of age and looks toward new horizons for exploration in the domains of structural organization and functional integration within cells.

Congresses, including published records of this kind, require an immense effort on the part of a few people. To these, the Organizing Committee and Program Committee extend sincerest thanks. For the assembly and editing of this book, we are especially grateful to Raymond B. Griffiths, Executive Editor of *The Journal of Cell Biology,* and to Helene Jordan and William A. Bayless of The Rockefeller University Press.

Though anticipating an acceleration in the tempo of discovery, we hope that this volume on cell biology in 1976 will remain relatively current until the next Congress in 1980. As we go to press, we learn that that meeting will be sponsored by the European Cell Biology Organization (ECBO) and will be held in Berlin.

Like the Boston Congress, it will acknowledge support from both public (government) and private agencies and foundations. Without the help and interest of those individuals who are instrumental in awarding the benefactions of such institutions, we would not enjoy the freedom to sense these things as they are and to record our findings in volumes such as this.

GREETINGS TO THE CONGRESS

From The American Society for Cell Biology

GEORGE E. PALADE, *President*

Distinguished guests, fellow cell biologists, ladies and gentlemen:

On behalf of the Council and membership of The American Society for Cell Biology I would like to welcome you to the First International Congress on Cell Biology, which is also the 17th Annual Meeting of our Society.

We are greatly honored to be the hosts of our colleagues from abroad on this unique occasion, which stands a good chance of being remembered for many years to come. We feel particularly honored, as we recognize among our guests so many pioneers and so many outstanding scientists.

We are also happy to meet again many of our American colleagues and to see that the frequency of young, fresh faces is increasing.

With the help of many of you, our program committee has endeavored to provide a copious scientific fare — four full days of science — which, I hope, you will find interesting and exciting, although perhaps exhausting.

Our local arrangements committee has worked out a program that gives you a chance to see some of the memorable sites of this city so deeply involved in the history — political and intellectual (scientific included) — of this country. The committee has also succeeded in introducing into the program a few divertisements and a few social and political intermezzos, destined to show you how we live on festive occasions, and how we converse with our political leaders in an election year.

Speaking to our foreign guests, I am sure that many members of our Society would be happy to act as your guides through the intricacies of this city and the complications of our daily life, and to serve as translators, if need be, from the English of Boston to that of dictionaries, or vice versa.

In the vast domain of life sciences, cell biology is a field reborn. It came back to life about 25 years ago after half a century of quiescence. In this relatively short time, it has become a central field in basic biological sciences, thanks to the work done by you, your friends, your collaborators, and your competitors.

Modern cell biology deserves credit not only for the mountain of new information it has produced in its short history, but also — and especially — for two major accomplishments of broad and lasting impact. First, modern cell biology has established that the common basis of biological organization extends far beyond the molecules of biochemistry through the levels of macromolecular assemblies and cell organs up to the cell itself. The cell theory has been with us for more than 130 years, and the universality of the biochemical basis of life has emerged as a basic concept during the last three or four decades. But the universality of whatever exists in biological organization between molecules and cells was neither assumed nor predicted. It emerged directly and progressively from the evidence collected over the last 25 years. This applies to the universality of ribosomes in all cells as a counterpart to the universality of proteins and protein synthesis, and it also applies to the universality of mitochondria in eukaryotic cells as an expression of the generality of oxidative phosphorylation and ATP utilization. The list can be easily extended: it includes, in fact, all important cell organs and practically all types of macromolecular assemblies. At the level of subcellular components, all basic bioengineering patterns are surprisingly similar throughout the biosphere — an impressive example of biological conservatism or, in other words, a clear expression of the stringency of the conditions that prevailed at the beginning of biological evolution. Each cell organ, as well as the cell itself (in its two main variants), appears today as the epitome of a unique formula, fashioned, tested, and perfected in early evolution and perpetuated forever after with only minor variations.

It is this extension of the common basis of biological organization from biochemistry all the way up to cells that has made cell biology, as well as its close neighbor molecular biology, the common ground and the starting base for all the other branches of biology that deal with either organisms or disciplines.

The other major achievement has been bridging the gap that, until recently, separated structural studies from biochemical or physiological inquiries. During the last two decades we have witnessed a progressive merging of once-separate disciplines, we have broadened our horizons, and have arrived at a syncretic, comprehensive, reasonably well-integrated view of biological organization. Historically, syncretic movements are periods of great advance, times when disparate pieces of information or isolated concepts merge into enlightenment. We have been involved successfully in such a movement; in fact, we are still moving with it. We can look back at what we have succeeded in adding to human knowledge and we can be reasonably pleased with our performance, but we should forget neither our predecessors who set many of the premises from which we started, nor our colleagues from other fields of science who provided us with most of the technology and many of the concepts we have used in our explorations.

Perhaps I should mention that the bridging of the gap and the disappearance of traditional boundaries have also cost us some losses. Like the old kingdom of Spain, we have lost, to some enterprising late-comers, quite a number of attractive provinces on which our standards were originally planted. The mitochondria have been virtually annexed by the biochemists, and the ribosomes have been taken over almost entirely by molecular biologists. But, as long as we are still experiencing a period of relative scientific affluence, it gives us a good feeling to know that we have contributed to the prosperity of other branches of biology, irrespective of their own state of development.

If we consider the future, we should realize that the differences between us and our neighbors — the molecular biologists and the biochemists — will further decrease, because our tendency is to become more molecular and their tendency, in turn, is to become more cellular. This is simply an indication of the character of the interesting problems of the foreseeable future, and of the means by which they will be solved.

We should also realize that the period of discovery and initial exploration in cell biology is practically over, and that the old idea of integrating structure, biochemistry, and function for each subcellular component considered in isolation, though still valid, is no longer sufficient. The interest is already shifting toward regulatory mechanisms operating over the whole domain of cellular organization and integrating the activity of each cell organ within the overall activity of the cell.

Recent investigations on the regulation of cholesterol synthesis in human cells suggests that the mechanisms at work involve different molecules in a number of different cell compartments, which means that control in this case has to be studied and understood as a problem in cell biology, rather than in traditional biochemistry.

Work on regulatory mechanisms, especially in eukaryotic cells, promises to be hard, intellectually exciting, and socially useful. It may well provide the key for a better understanding, and eventual control, of some of the major health problems of our times, such as cancer, vascular disease, and mental diseases, because in each of them some basic regulatory mechanisms seem to be at fault.

When we arrive at a reasonable understanding of cell-wide regulatory mechanisms, cell biology will be directly and immediately relevant to a number of major targets in disease control and health improvement. But we are not yet there, and neither our sincere desire to be useful nor the repeated proddings of our target-minded institutions will bring us there in the absence of a critical amount of knowledge about the nature of these regulatory mechanisms and of the ways in which they operate.

Like many other active fields of science, the cell biology of today is the result of the endeavors of many scientists from many countries; it has been and it will remain a truly international or supranational venture. We have, of course, national and regional societies which are used as a means of communication within smaller groups. But it would be misleading and unrealistic to put much stress, or any stress, on national contributions. To take the example of our own society, with which I am quite familiar, our membership is already multinational by law or by extraction and, in developing our work, we have learned from colleagues from at least four other continents and too many countries to allow precise counting. In turn, I am sure, they have learned from us.

The Federation of Societies for Cell Biology is a move in the right direction. Perhaps we can improve our international channels of communication by making available to the members of the federated societies as many national meetings as possible. On behalf of The American Society for Cell Biology I would like to propose that such a move be considered.

This First International Congress has some special human connotations which deserve to be brought forward. Our history is short. It started with a few centers that have generated a relatively small number of scientific genealogies. Welcoming you here is more than welcoming a bevy of scientific colleagues; it is welcoming some special entities, somewhere between friends and bench relatives, who have succeeded in overcoming all those barriers imposed by great distances and small travel allocations and have arrived in Boston for a conclave.

In human terms, this conclave may well be unique. The time of a major change of the guard is rapidly approaching. Perhaps this is the last time when the old guard — that medley of hard-working pioneers, wise founders, and demanding bosses — will parade in strength. The performance may well amount to an interesting piece of history.

Again, it is my privilege to welcome you to whatever the Congress offers: science, human experience, entertaining history, and — who knows? — even mistakes.

From the International Federation for Cell Biology

DANIEL MAZIA, *President*

I welcome you in the name of the International Federation for Cell Biology, whose only contribution to this Congress was to nucleate it, leaving the assembly entirely to The American Society for Cell Biology. The Federation itself is a modest body, which exists only to advance the international activities of our international company of cell biologists. Its modesty can be measured by the most common standard of value. What respect can I claim for an organization that has spent only 488 pounds and 16 pence in three years?

The Federation is constituted according to strict federal principles; its members are societies and regional organizations; its decisions are voted by delegates of these autonomous associations. It now includes the Japan Society, the American Society, and the European Cell Biology Organization, which contains the individuals and societies of eastern and western Europe, including societies in Britain, Belgium, France, Hungary, Poland, the German-speaking countries, and the Scandinavian countries. Tomorrow we expect the new Indian Society for Cell Biology to join, and we are making efforts toward worldwide participation.

The first responsibility of the Federation is to beget our international Congresses. Each Congress is put completely in the hands of the Federation member which is the host. As we open the first Congress in Boston tonight, I claim the right to be the first of the many who will thank the American Society for this superb scientific feast. We announce the second Congress, to be held in Berlin in 1980, organized by the European Cell Biology Organization through the German Society for Cell Biology.

The Federation announces its founding of a new journal, *Cell Biology: International Reports,* which will make its appearance at the beginning of 1977. It will publish very short papers with a short publication time, enlisting a large and distinguished international board of editors in the selection and editing of the contributions. The chief editor is Dr. Sam Franks of the Imperial Cancer Research Fund Laboratories in London. Not all of you will feel that your greatest need is to keep up with more literature, but the sincere purpose of this journal of the International Federation is to enhance the flow of discoveries between authors all over the world and readers all over the world in all fields of cell biology.

One should not have to plead for the advancement of the international life of our very international science. That life is natural. The obstacles are intolerable, whether they arise in tyranny or in the foolish misconception that travel is a touristic indulgence for scientists. We send our regrets to those who should have been with us and cannot be here.

Still, we have come together in our thousands, from all six continents of this planet—and Washington. To communicate, we agree to speak in the language of a medium-sized island off the coast of France. We will disagree, but we will not quarrel. We will be exhausted by Friday. We may remember good company and forget boring speeches, but most will recognize that our future work will gain from what was learned and discussed here.

This is not, of course, the first Congress of cell biology; it is only the first brought about by the new confederation. Before it were Congresses organized by the parent of this Federation—the International Society for Cell Biology—and there were others before that Congress in Stockholm in 1947, at which the Society was founded. Some here were present then. Scarcely two years after a long and awful war, during which cell biologists were doing anything but cell biology, the necessity of talking about cells with colleagues from other countries brought them together. And what they talked about was—what we are talking about. The difference is that many *questions* in the 120 papers presented in 1947 have become *assumptions* in the 1,200 papers to be presented here.

Then the time came when a single society could no longer contain the company of cell biologists, so the International Society for Cell Biologists created this Federation.

We want to thank and honor our precursors and especially those of them who are with us tonight.

CONGRESS COMMITTEES

ORGANIZING COMMITTEE

Keith Porter, *Chairman**
Joyce Albersheim, *Congress Secretary*
Michael Abercrombie
Olav Behnke
Wilhelm Bernhard
H. G. Callan
Albert Claude
J. F. David-Ferreira
Eduardo De Robertis
Lars Ernster
Pierre Favard
Werner Franke
Sam Franks
Pieter J. Gaillard
Ian Gibbons*
Gonzalo Giménez-Martín
Luiz-Carlos Junqueira
Charles Leblond
Adolfo Martínez-Palomo
Daniel Mazia*
Ernest A. McCulloch
Georg Melchers
George Palade*
Ennio Pannese
Mary Lou Pardue*
W. James Peacock
Jean-Paul Revel*
R. Robineaux
Satimaru Seno
B. R. Seshachar
J. Roberto Sotelo
Olga Stein
Herbert Stern*
Hewson Swift*
Jury Vasiliev
Ewald Weibel
Eichi Yamada

*Executive Committee

LOCAL COMMITTEE

Elizabeth Hay, *Chairman*
Chantal Fujiwara, *Assistant Chairman*
Eugene Bell

LOCAL COMMITTEE, CONTINUED

Chandler Fulton
Elinor O'Brien
Thomas Pollard
Michael Shelanski
David Shepro
William Sullivan
Robert Trelstad

FINANCE COMMITTEE

George Pappas, *Chairman*
Sydney S. Breese, Jr., *Congress Treasurer*
Don Fawcett
Gordon Kaye
Cyrus Lindgren
Dorothy Skinner

PROGRAM COMMITTEE

Bill Brinkley, *Chairman*
Betti Ledlie,
Administrative Assistant
Michelle Wiktorowicz,
Administrative Assistant
Robert Allen
Sam Barranco
Donald Brown
Walter Eckhart
Virginia Evans
T. C. Hsu
James Jamieson
William Jensen
Morris Karnovsky
Jean Lafontaine
Vincent Marchesi
David Prescott
David Robinson
Joel Rosenbaum
Russell Ross
Frank Ruddle
Emma Shelton
Richard Sidman
Philip Siekevitz
Igor Tamm
Jonathan Warner

CONTENTS

Cell Surface and Neoplasia

Cells and Hormone Action

Organization and Assembly of Chloroplasts

Somatic Plant Hybridization by Fusion of Protoplasts

Biogenesis of Mitochondria

Endoplasmic Reticulum-Golgi Apparatus and Cell Secretion (Plant Cells)

Endoplasmic Reticulum-Golgi Apparatus and Cell Secretion (Animal Cells)

Microtubules and Flagella

Molecular Basis of Motility

Eukaryotic Cell Cycle

Cytoplasmic Control of Nuclear Expression

Chromatin Structure and Function

Functional Organization of Chromosomes

Molecular Cytogenetics of Eukaryotes

Viral Gene Function in Cell Transformation

Morphogenesis of Gametes

Differentiation and Regulation of Photoreceptors

Cells of the Artery Wall and Atherosclerosis

Plasma Membrane Organization

INTRODUCTORY REMARKS

D. BRANTON

In this symposium, we are concerned with the molecular organization of biological membranes. The problems of molecular organization refer both to the location and arrangement of component molecules and to the shapes, interactions, and movements of those molecules.

Danielli and Davson depicted the membrane as a protein covered lipid bilayer. By the 1950s, it appeared that this organization could be directly visualized in the electron microscope. However, it also became apparent that this model provided only a limited insight into the molecular interactions in membrane. Physical probes and thermodynamic reasoning led J. Singer and G. Nicolson to suggest that lipid-protein interactions are crucial in many membranes, and they summarized the results of work in many laboratories by suggesting a fluid mosaic model in which proteins were embedded into or traversed a fluid lipid bilayer.

However, it is obvious that conjectures which postulate how molecules are organized in membranes must be bolstered by knowledge of the specific composition, modes of attachment, and function of individual components. This symposium will show how biochemical perturbations and analyses can indicate the disposition and function of membrane proteins; how spectroscopic measurements can analyze molecular motion and interactions; and how high-resolution electron microscopy can relate the structure of individual molecules to the overall architecture of cellular membranes.

D. BRANTON The Biological Laboratories, Harvard University, Cambridge, Massachusetts

PROTEIN ENSEMBLES IN THE HUMAN RED-CELL MEMBRANE

THEODORE L. STECK and JAMES F. HAINFELD

We shall briefly summarize our current understanding of the disposition of the predominant protein species in the isolated human erythrocyte membrane, and then consider in more detail two interesting supramolecular ensembles. Recent research on red-cell membrane proteins has been repeatedly reviewed; for example, by Juliano (1973), Steck (1974), Marchesi et al. (1976), and Kirkpatrick (1976).

Proteins contribute approximately 60% of the mass of the human red-cell membrane (ghost), with roughly 10% of the protein fraction being bound carbohydrate. The remainder of the membrane is comprised chiefly of phospholipids, cholesterol, and a small amount of glycolipid. Eighty per cent or more of the protein fraction derives from perhaps a dozen major polypeptides and glycoproteins, which have been resolved and enumerated by electrophoresis in sodium dodecyl sulfate on gels of polyacrylamide (cf. Steck, 1974; Marchesi et al., 1976). These predominant species have received detailed study in many laboratories, even though their functions are, in many cases, unknown or uncertain.

We can now list several interesting aspects of the organization of these polypeptides that may be taken as a set of hypotheses worth testing in other membrane systems. Fig. 1 is offered as a guide to this discussion.

1. There appears to be an absolute asymmetry in the distribution of proteins between the two surfaces of the membrane. That is, no protein domain has thus far been shown to occupy both membrane surfaces. Even those polypeptides which span the membrane do so anisotropically, presenting distinctly different aspects at each surface (Fig. 1; cf. Steck, 1974; Marchesi et al., 1976). This profound asymmetry may be a reflection of the degree to which the activities of the membrane differ at its two surfaces, but even isotropic transmembrane functions (such as the facilitated diffusion of solutes) may be mediated by anisotropic structures.

The carbohydrate groups of both the glycoproteins and glycolipids are confined exclusively to the outer (external) membrane surface. All of the polypeptides thus far identified at the external surface are in fact glycosylated (cf. Steck and Dawson, 1974). The functions of the oligosaccharide moieties are not now evident. (It is interesting in this regard that the red cell membranes of individuals with the rare blood type En[a-] are apparently deficient in the major sialoglycoproteins stained by the periodic acid-Schiff [PAS] reagent, including glycophorin [PAS-1], without associated clinical disability [Tanner and Anstee, 1976].)

2. There are two distinct modes of association of the major polypeptides with the membrane. A discrete set of polypeptides, comprising approximately half of the protein mass, is selectively solubilized by treating isolated ghosts with any of several agents which denature or covalently modify protein structure (e.g., 6 M guanidine hydrochloride, 0.1 N NaOH, or vigorous succinylation). The complementary set of polypeptides (and all of the carbohydrate and lipid) remain in a membranous form. It appears that the latter proteins (which are more hydrophobic in behavior and in amino acid composition) are integrated into the apolar core of the lipid stratum. Harsh perturbation of their tertiary and quarternary structure does not unseat them (i.e., dissociate hydrophobic lipid-protein contacts). In contrast, the solubilized polypeptides apparently bind to specific sites at

THEODORE L. STECK and JAMES F. HAINFELD Departments of Biochemistry and Medicine and the Enrico Fermi Institute, the University of Chicago, Chicago, Illinois

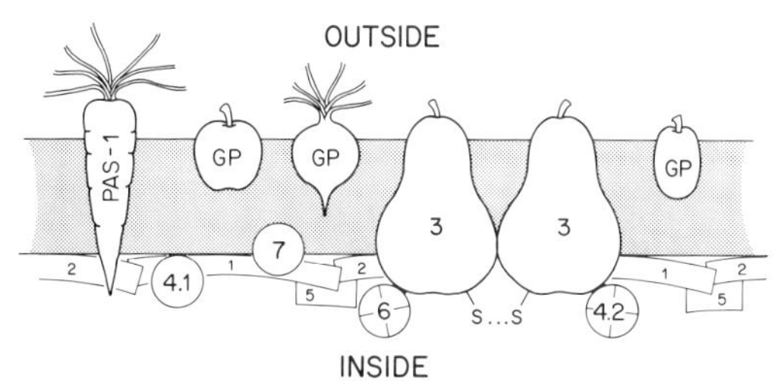

FIGURE 1 A speculative representation of the organization of the principal erythrocyte membrane polypeptides. This diagram is intended to convey (*a*) the "sidedness" of the constituents; (*b*) their modes of association with the membrane; (*c*) certain homo-oligomeric associations; and (*d*) two supramolecular ensembles, as described in the text. The numbered shapes shown correspond to polypeptides enumerated according to their electrophoretic mobilities on polyacrylamide gels. Reproduced from Steck (1974), with the permission of The Rockefeller University Press.

the membrane surface from which they are released if their structure is perturbed (cf. Steck, 1974).

This behavior relates to the classifications introduced by Singer and Nicolson (1972) and by Capaldi and Vanderkooi (1972), placing membrane proteins into integral (intrinsic) and peripheral (extrinsic) categories. However, Singer and Nicolson (1972), and more recently, Marchesi *et al.* (1976), supposed that denaturants liberate hydrophobically anchored proteins from their lipid environment, while this does not seem to be the case here (cf. Steck, 1974).

In some cases, specific nondenaturing treatments suffice to solubilize certain proteins. Polypeptide bands 1 and 2 (spectrin) and 5 (actin) are liberated from the membrane by incubation in low ionic strength, alkaline media. Gel band 6 (the protomer of glyceraldehyde 3-P dehydrogenase, G3PD) is selectively eluted by raising the ionic strength or, more specifically, by exposure to certain of its specific ligands, such as NADH. Other membrane-associated glycolytic enzymes are also extracted at high ionic strength. Aldolase is specifically released from the membrane by its substrate, fructose diphosphate, but not by NADH (Strapazon and Steck, 1976).

Although the hydrophobically anchored proteins are quite resistant to solubilization by strong protein perturbants, they dissolve readily in a variety of detergents, both nondenaturing (e.g., Triton X-100 or sodium cholate) and denaturing (e.g., sodium dodecyl sulfate). Presumably, protein solubilization involves displacement of hydrophobic contacts with an extended (hence, insoluble) lipid bilayer by small detergent monomers or micelles, yielding water-soluble (and often functional), detergent-protein complexes.

It is significant that all of the readily eluted polypeptides are confined to the cytoplasmic aspect of this membrane, whereas the external surface bears only the hydrophobic, tightly anchored (glycoprotein) species (cf. Fig. 1).

3. Some polypeptides are organized into homo-oligomers. In favorable cases, water-soluble proteins can be isolated from the membrane and shown to be oligomeric by classic techniques; for example, band 6 is the polypeptide subunit of the water-soluble tetramer, G3PD (Fig. 1).

The quaternary structure of the more tenaciously bound or insoluble membrane proteins is more difficult to establish. Their solubilization may be attended by the disruption of interpolypeptide associations, as occurs in sodium dodecyl sulfate. However, cross-linking reagents can be used to fix such contacts covalently before the dissolution of the membrane (e.g., Steck, 1972; Wang and Richards, 1974). By this means, certain of the major polypeptides have been found to exist as oligomers in the membrane; Fig. 1 illustrates this feature for gel bands 1 and 2, band 3, and band 4.2.

A second approach to this problem is to solubilize integral membrane constituents in a nondenaturing detergent in which their state of association can be assessed with physical techniques, such as ultracentrifugation (Clarke, 1975; Yu and Steck, 1975). These studies support the conclusions depicted in Fig. 1. It now seems that both intrinsic and extrinsic polypeptides are frequently organized into simple oligomers comprised of identical or closely related chains, both in the red cell and in other membranes.

4. Some membrane proteins are associated into supramolecular ensembles. Two examples of this phenomenon will be considered; the first involves a "loosely bound" system (spectrin plus actin) and the second, an integral protein (band 3).

Marchesi and his colleagues and several other investigators have identified filamentous polymers at the cytoplasmic surface of isolated erythrocyte membranes. The polypeptides associated with these structures appear to be spectrin (a doublet of two very high molecular weight polypeptides, gel bands 1 and 2) and band 5 (which is thus far indistinguishable from actin). These components have been selectively solubilized and have been induced to form filaments in isolation by the addition of electrolytes, divalent cations in particular (cf. Kirkpatrick, 1976; Marchesi et al., 1976). It is not now evident, however, how spectrin and actin interact nor is the molecular architecture of these filaments understood.

A central question is the disposition of these filaments *in situ*. Originally, Marchesi and Palade (1967) induced structures resembling actin filaments by treating ghosts with trypsin. Subsequently, patches of fibrillar material were observed issuing from the cytoplasmic surface in electron micrographs of thin sections of isolated ghosts (cf. Kirkpatrick, 1976). More recently it was shown that extraction of membranes with Triton X-100 solubilized the bulk of the lipid layer and its integrated proteins, but left an insoluble ghost-shaped reticulum of filaments, enriched in spectrin and actin polypeptides (Yu et al., 1973). It thus appeared that a matrix of filaments lay under the membrane proper, maintained not by adherence to the lipid stratum but by extensive self-associations among the proteins.

We have recently attempted to examine this network more directly in the scanning electron microscope (SEM). As shown in Fig. 2, our preparations of ghosts generally appeared globoid and wrinkled, in contrast to the intact red cell which has a smooth surface and the shape of a biconcave disk. The ghosts also bore dark patches which may represent holes induced by hemolysis, since they were absent from intact erythrocytes.

On occasion, ghosts (for example, that shown in Fig. 3) appeared to tear so as to reveal a submembrane meshwork extending across the defect. At higher magnification (not shown), the reticulum appeared to be comprised of cylindrical processes approximately 20 nm in diameter, which formed both elongated filaments and closely spaced annular figures bearing holes 20 nm or greater in diameter.

Ghosts were also prepared on supports coated with polylysine, to which the membranes adhered

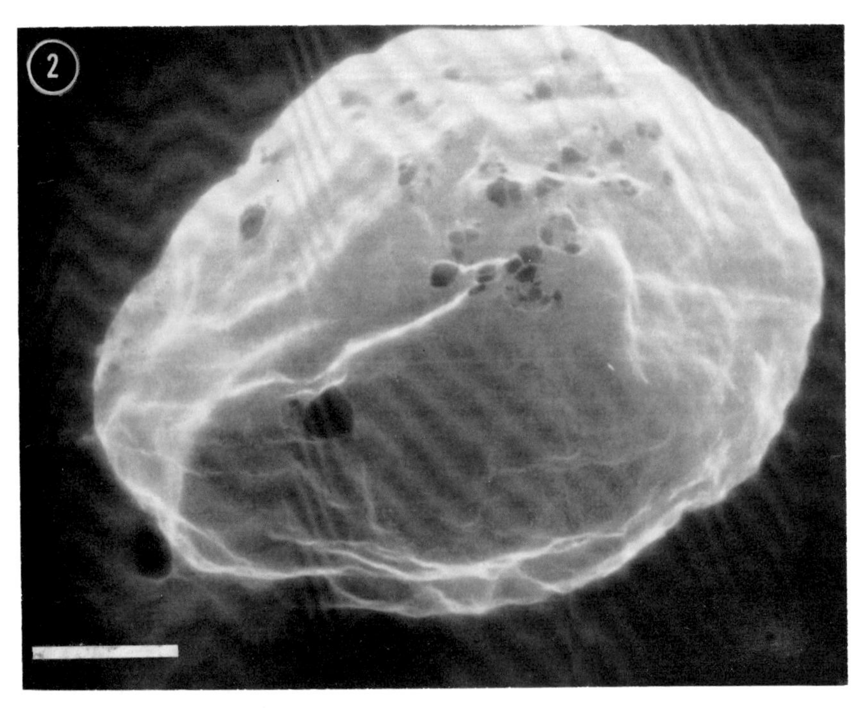

FIGURE 2 A SEM of a nonflattened ghost. Human erythrocytes were washed, applied to an aluminum disc without polylysine, and hemolysed by extensive washing with 5 mM Na phosphate, pH 8.0. After fixation in glutaraldehyde, the disc was rinsed thoroughly with deionized water, freeze-dried, and coated with 5–10 nm of palladium/gold [Pd/Au] (40:60). A field emission Hitachi HFS-2 SEM was used to examine the specimen. Calibration bar equals 1 μm. (Details of these procedures are contained in a manuscript by J. F. Hanifeld and T. L. Steck, submitted for publication.)

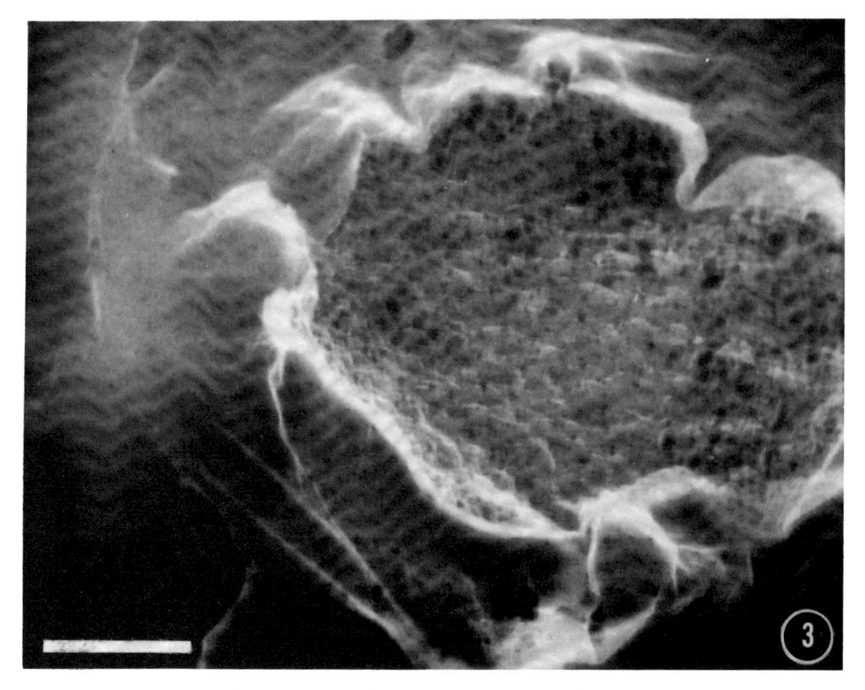

FIGURE 3 A torn, nonflattened ghost. The preparation and calibration are as described in Fig. 2.

quite strongly. A typical image is shown in Fig. 4. The ghost has flattened into an oval disk approximately 7 μm in its greatest diameter and has split open to reveal a reticulum extending continuously from the margins of the tear across the floor of the ghost. The network is clearly demarcated near the raised, torn margins but is less readily discerned in the flattened central region.

We believe that the network draped between the torn edge and the floor of the ghosts has, in fact, pulled away from the membrane. An extreme example of this phenomenon is seen in Fig. 5, where an entire filamentous matrix appears to have everted onto the support disk through a hole in a flattened ghost. Another untethered reticulum is shown at higher magnification in Fig. 6. These networks clearly do not depend on contacts with the membrane—but rather on self-association—for their integrity.

The reticulum flattened on polylysine-coated supports is comprised of filaments as small as 5–10 nm in diameter and interstices of up to 100 nm (e.g., in the floor region in Fig. 4). In general, the networks shown in Figs. 4 and 5 assumed a more extended structure than the relatively condensed forms seen in Figs. 3 and 6. It thus appears that the ultrastructure of this webbing can vary, being stretched out when drawn flat onto polylysine and relaxed when free. This behavior suggests that the network has elastic properties, a concept previously proposed for the spectrin-actin system on other grounds (cf. Kirkpatrick, 1976).

To establish the relationship of this network to the spectrin-actin system, we extracted membranes adherent to support disks with Triton X-100, a treatment known to yield ghostlike residues rich in bands 1, 2, and 5 (Yu et al., 1973). The SEM image of these residues (Fig. 7) revealed a reticulum of filaments which closely resembled those seen in Figs. 4 and 5. While lacking the membrane proper, Triton X-100 residues still conformed to the size and shape of ghosts. As an additional control, we have examined membranes from which polypeptide bands 1, 2, and 5 had been selectively extracted; no networks or filaments were observed on these inside-out and right-side-out vesicles. We infer that spectrin and actin are major contributors to the reticulum visualized by SEM.

We have also examined the surfaces of *intact* erythrocytes at high magnification for evidence of an underlying reticulum. In this case (Fig. 8), a fine network composed of ridges, 5–10 nm in width, was observed in low relief. We have speculated that the membrane layer reflects the contours of an underlying meshwork. If so, these data

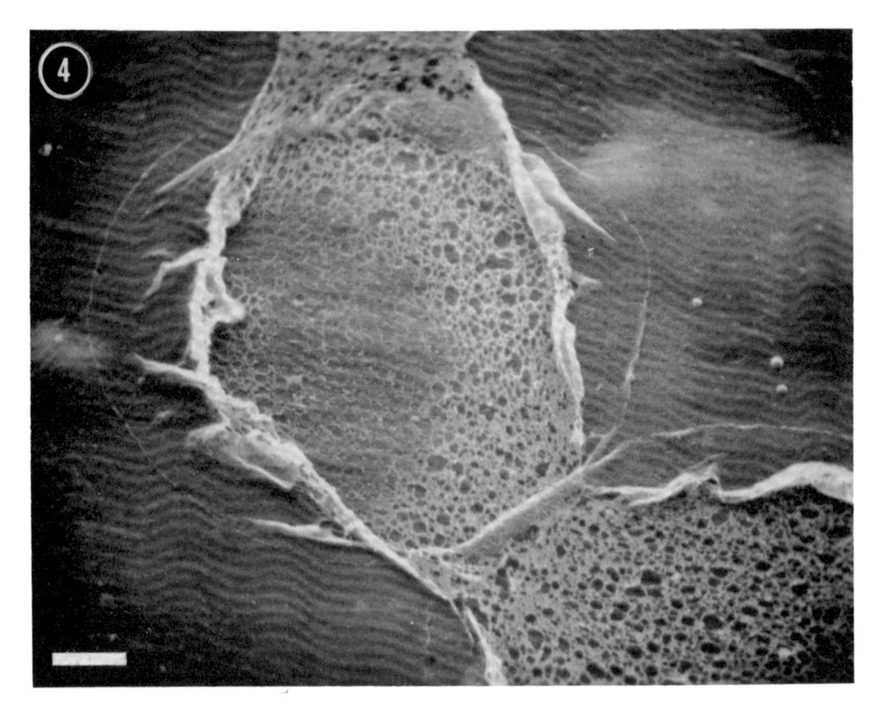

FIGURE 4 A torn and flattened ghost. The preparation and calibration are as described in Fig. 2, except that the aluminum disc was precoated with polylysine according to J. F. Hainfeld and T. L. Steck (manuscript submitted for publication).

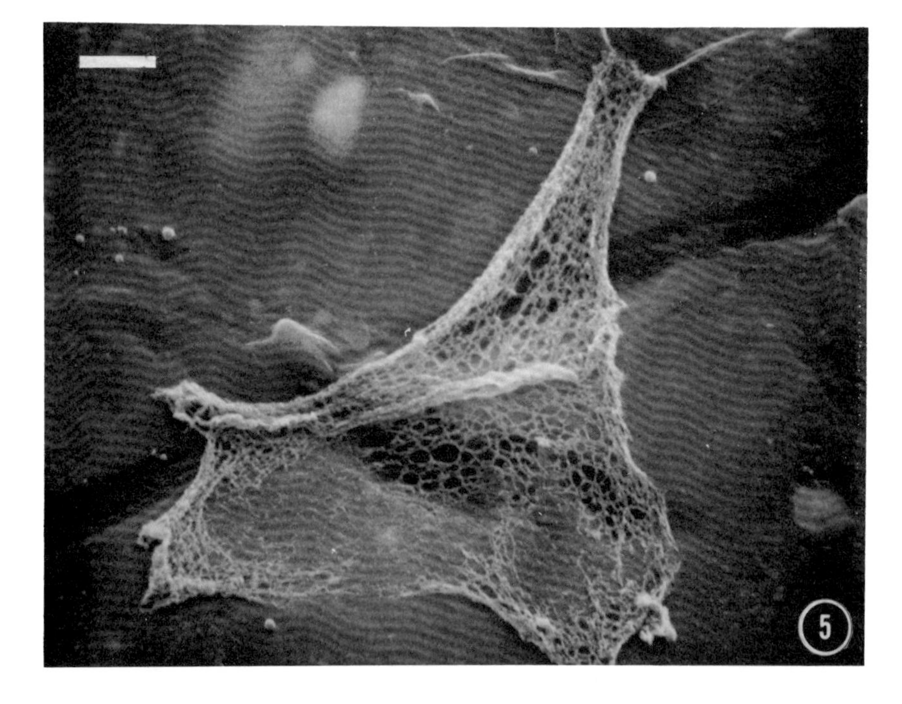

FIGURE 5 An everted reticulum. The preparation and calibration are as described in Fig. 4.

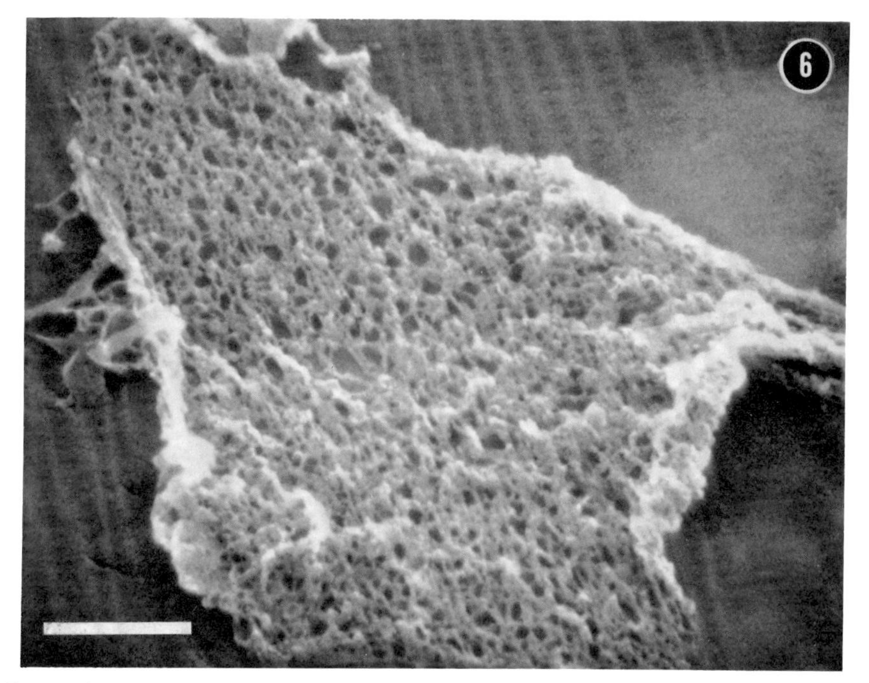

FIGURE 6 An untethered reticulum. The preparation and calibration are as described in Fig. 4.

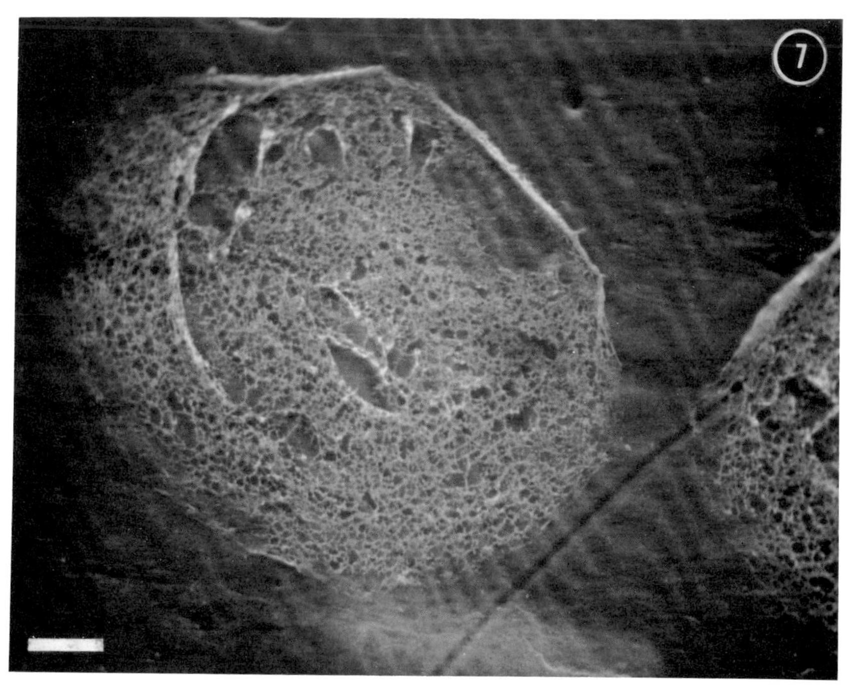

FIGURE 7 A ghost residue extracted with Triton X-100. Erythrocytes were affixed to an aluminum disc coated with polylysine, lysed to form ghosts as described in Fig. 2, and then washed well with 0.2% Triton X-100. The disc was rinsed, fixed, freeze-dried, and coated as described in Fig. 2. Calibration bar equals 1 μm.

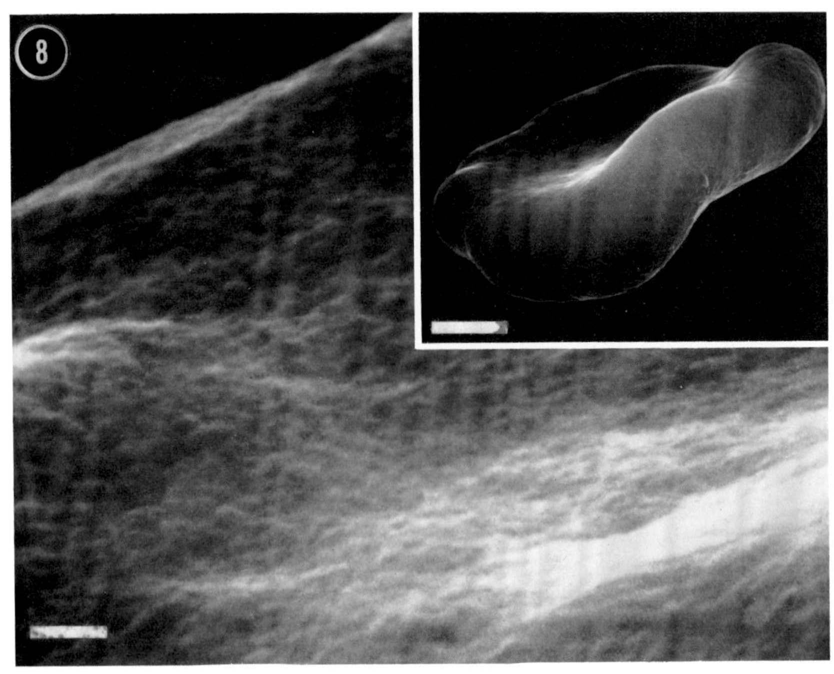

FIGURE 8 The surface of an intact erythrocyte. Thoroughly washed human erythrocytes were applied to an aluminum disc coated with polylysine. The sample was rinsed, fixed with glutaraldehyde, washed in deionized water, and processed as in Fig. 2. The *inset* shows a single cell; the calibration bar equals 1 μm. The figure itself shows a portion of that cell; the calibration bar equals 0.1 μm.

suggest that a reticulum is indeed present at the cytoplasmic surface of intact red cells and that its disposition can be studied by these means in unbroken cells in varied physiologic states. Perhaps a clearer understanding of the role of the reticulum in red-cell shape changes and deformability may be obtained thereby (cf. Kirkpatrick, 1976).

Finally, let us consider ensembles formed from gel band 3 (Fig. 1). This 90,000 dalton membrane-spanning polypeptide is the predominant protein constituent in the human red-cell membrane, comprising roughly 25% of the protein mass. It bears a small amount of carbohydrate at its outer surface pole, but very little sialic acid; thus, it is distinctly different from glycophorin (PAS-1 and PAS-2 in Fig. 1; cf. Steck, 1974; and Marchesi et al., 1976).

Band 3 polypeptides are found as noncovalent dimers both in the membrane (Steck, 1972; Wang and Richards, 1974) and when solubilized in nondenaturing detergents (Yu and Steck, 1975). There are enough dimers of band 3 (approximately 500,000–600,000/cell) to account for all of the ~8-nm particles seen in freeze-fracture electron micrographs of this membrane. Indeed, recent reconstitution studies strongly suggest that the intramembrane particles are comprised of band 3 dimers and not glycophorin (PAS-1 and 2; Yu and Branton, 1976).

There is impressive, but as yet indirect, evidence suggesting that the band 3 polypeptide is involved in the facilitated diffusion of anions, and thus mediates a primary function of the erythrocyte membrane, the exchange of chloride and bicarbonate ions (cf. Marchesi et al., 1976). One can readily envision that a membrane-spanning dimer could play such a role, the interface between protomers serving as the pathway for solute transfer. Nevertheless, it remains to be directly demonstrated that band 3 performs this function; furthermore, the molecular basis of the transport process is unknown. The fact that band 3 is fixed and asymmetrical in its membrane orientation places constraints on the interpretation of data on the kinetics of anion flux (e.g., the classic, symmetrical mobile carrier model is excluded).

We have recently attempted to correlate known features of band 3 structure with anion transport kinetics. We developed a spectrophotometric assay for the transport of the organic anion, pyruvate, using lactic dehydrogenase and NADH trapped within sealed ghosts or inside-out vesicles as the detection system (Rice and Steck, 1976). We found that pyruvate transport is indeed asymmetrical, in that (*a*) the apparent kinetic constants differed for flux in the two directions; (*b*) induction of a disulfide cross-link in the band 3 dimer altered pyruvate flux differently in the two directions; and (*c*) probenecid, an inhibitor of anion transport, exerted its effect only from the cytoplasmic surface (W. R. Rice and T. L. Steck, manuscript in preparation).

The cytoplasmic pole of band 3 has been shown to be the exclusive binding site for at least three peripheral proteins in the isolated membrane. These are band 4.2 (a 72,000 dalton polypeptide of unknown function), G3PD, and aldolase (Fig. 1; Yu and Steck, 1975; Strapazon and Steck, 1976). Each band 3 polypeptide can bind one enzyme molecule. The catalytic activity of aldolase is reversibly inhibited when bound to band 3 in ghosts or in solution (E. Strapazon and T. L. Steck, manuscript in preparation). A central question now under study is whether these in vitro associations actually occur in intact cells and what their physiologic role might be.

Band 3 has been selectively cleaved by proteases *in situ* so as to generate major fragments representing the outer surface, transmembrane, and inner surface regions of the polypeptide (Steck et al., 1976). The former two pieces are distinctly hydrophobic in their behavior and remain in the membrane after proteolysis. Cytoplasmic surface fragments are water-soluble and are released from the membrane upon cleavage. The latter soluble fragments apparently remain dimeric and carry the binding site(s) for G3PD and aldolase. It is interesting that proteolytic excision of the cytoplasmic domain of band 3 does not markedly alter pyruvate transport.

The intramembrane freeze-fracture particles of the erythrocyte membrane (hence, band 3 molecules) appear to be influenced in their lateral disposition by the spectrin-actin system (cf. Steck, 1974; Marchesi et al., 1976). This phenomenon has been demonstrated directly, using lipid vesicles reconstituted to include the relevant purified protein constituents (Yu and Branton, 1976). It may be that the two supramolecular ensembles discussed here are actually linked and that an even greater integration exists among the proteins of erythrocyte membrane that can presently be defined.

ACKNOWLEDGMENTS

Original studies presented here were supported by a National Institutes of Health fellowship to Dr. Hainfeld (1F 32 GM 01797) and an American Cancer Society grant (BC-95C and D). Dr. Steck is a Faculty Research Awardee of the American Cancer Society.

REFERENCES

Capaldi, R. A., and G. Vanderkooi. 1972. The low polarity of many membrane proteins. *Proc. Natl. Acad. Sci. U. S. A.* **69:**930–932.

Clarke, S. 1975. The size and detergent binding of membrane proteins. *J. Biol. Chem.* **250:**5459–5469.

Juliano, R. L. 1973. The proteins of the erythrocyte membrane. *Biochim. Biophys. Acta.* **300:**341–378.

Kirkpatrick, F.H. 1976. Spectrin: current understanding of its physical, biochemical, and functional properties. *Life Sci.* **19:**1–18.

Marchesi, V. T., H. Furthmayr, and M. Tomita. 1976. The red cell membrane. *Annu. Rev. Biochem.* **45:**667–698.

Marchesi, V. T., and G. E. Palade. 1967. Inactivation of adenosine triphosphatase and disruption of red cell membranes by trypsin: protective effect of adenosine triphosphate. *Proc. Natl. Acad. Sci. U. S. A.* **58:**991–995.

Rice, W. R., and T. L. Steck. 1976. Pyruvate flux into resealed ghosts from human erythrocytes. *Biochim. Biophys. Acta.* **433:**39–53.

Singer, S. J., and G. L. Nicolson. 1972. The fluid mosaic model of the structure of cell membranes. *Science (Wash. D.C.).* **175:**720–731.

Steck, T. L. 1972. Cross-linking the major proteins of the isolated erythrocyte membrane. *J. Mol. Biol.* **66:**295–305.

Steck, T. L. 1974. The organization of proteins in the human red blood cell membrane. *J. Cell Biol.* **62:**1–19.

Steck, T. L., and G. Dawson. 1974. Topographical distribution of complex carbohydrates in the erythrocyte membrane. *J. Biol. Chem.* **249:**2135–2142.

Steck, T. L., B. Ramos, and E. Strapazon. 1976. Proteolytic dissection of band 3, the predominant transmembrane polypeptide of the human erythrocyte membrane. *Biochemistry.* **15:**1154–1161.

Strapazon, E., and T. L. Steck. 1976. Binding of rabbit muscle aldolase to band 3, the predominant polypeptide of the human erythrocyte membrane. *Biochemistry.* **15:**1421–1424.

Tanner, M. J. A., and D. J. Anstee. 1976. The membrane change in En(a-) human erythrocytes. *Biochem. J.* **153:**271–277.

Wang, K., and F. M. Richards. 1974. An approach to nearest neighbor analysis of membrane proteins. *J. Biol. Chem.* **249:**8005–8018.

Yu, J., and D. Branton. 1976. Reconstitution of intramembrane particles in erythrocyte band 3-lipid recombinants: effects of spectrin-actin association. *Proc. Natl. Acad. Sci. U. S. A.* **73**:3891–3895.

YU, J., D. A. FISCHMAN, and T. L. STECK. 1973. Selective solubilization of proteins and phospholipids from red blood cell membranes by nonionic detergents. *J. Supramol. Struct.* **1**:233–248.

YU, J., and T. L. STECK. 1975. Associations of band 3, the predominant polypeptide of the human erythrocyte membrane. *J. Biol. Chem.* **250**:9176–9184.

LIPID-PROTEIN INTERACTIONS IN A RECONSTITUTED CALCIUM PUMP

J. C. METCALFE and G. B. WARREN

The ATP-dependent accumulation of calcium is almost the only membrane function carried out by isolated sarcoplasmic reticulum vesicles, and this singularity of function is matched by the simplicity of the protein composition. About 85% of the protein in the membrane is a Ca^{2+}, Mg^{2+}-dependent ATPase, which stoichiometrically couples the uptake of two Ca^{2+} ions to the hydrolysis of one molecule of ATP. If the nonspecific efflux of Ca^{2+} from the vesicles is negligible, the initial rate of Ca^{2+} accumulation will be the same as the rate of Ca^{2+} uptake through the protein, and the efficiency of Ca^{2+} accumulation as measured by the mole ratio of initial Ca^{2+} accumulation to ATP hydrolysis (Ca^{2+}/ATP) will have the maximum value of 2.0. Any increase in the nonspecific permeability of the vesicles to Ca^{2+} will lower this value, because the initial rate of calcium accumulation is determined by the difference between Ca^{2+} uptake through the protein and the nonspecific efflux of calcium. In this discussion, we regard the stoichiometric coupling of ATP hydrolysis to Ca^{2+} uptake as an intrinsic property of the functional transport system, which is preserved, for example, when the protein is transporting calcium into a disrupted vesicle, from which it is immediately lost. In particular, we assume that the intrinsic stoichiometry of the protein function is unaffected by any of the reconstitution conditions to be described, so that a low efficiency of Ca^{2+} accumulation ($Ca^{2+}/ATP \ll 2$) is not the result of a change in the stoichiometry of coupling but is due to an increase in the nonspecific efflux of Ca^{2+} from the vesicles.

With this assumption, it is clear that the lipids in the vesicles must serve at least two functions. They must maintain the calcium transport protein in the appropriate conformation to support the specific uptake of calcium and, if the calcium is to be retained by the vesicle, the lipids must also restrict its nonspecific efflux from the vesicle. There are more than 50 species of lipids in the membrane of sarcoplasmic reticulum, in marked contrast to the simplified protein composition. To determine the extent to which this heterogeneity of lipids is involved both in supporting Ca^{2+} uptake through the protein and in restricting the nonspecific efflux of Ca^{2+}, we have asked whether the 50 species of lipids are individually required for efficient Ca^{2+} accumulation or whether certain averaged physical and chemical properties of the lipid mixture provide a suitable two-dimensional environment for the calcium transport protein. Our data show that the high efficiency of the native membrane vesicles can be retained when the 50 lipid species are replaced by a simple mixture of two synthetic phospholipids, and that the efficiency of calcium accumulation is mainly determined by the interaction between the lipids and the calcium transport protein. We also suggest that it is the zwitterionic phospholipids that support the activity of the transport protein and that the minor lipids with net negative charges may be involved in membrane functions not directly related to the protein.

To determine the effect of lipid composition on the function of the penetrant calcium transport protein, we replaced the endogenous lipids of sarcoplasmic reticulum membranes with defined synthetic phospholipids, by using a technique (Warren et al., 1974*a, b*) that we have applied to a variety of membrane structures (Houslay et al., 1975, 1976).

Reconstitution of a Calcium Pump by the Use of a Single Species of Synthetic Phospholipid

The technique of reconstitution consists of two main steps. The endogenous lipid of the mem-

J. C. METCALFE and G. B. WARREN Department of Biochemistry, University of Cambridge, Cambridge, England

brane is replaced by synthetic phospholipid, and the resulting lipid-protein complexes are then assembled into closed vesicular structures which can be tested for their ability to accumulate Ca^{2+}.

Sarcoplasmic reticulum vesicles are pretreated with cholate at a concentration that solvates the calcium transport protein together with the endogenous lipid that is interacting with the protein. This mixture is then treated with an excess of synthetic phospholipid in the same detergent. Cholate catalyzes a lipid exchange between the endogenous and the synthetic lipid pool until the composition of the lipid in contact with the protein is the same as that of the total lipid pool (Fig. 1). We used 100 times more synthetic phospholipid than endogenous lipid in these experiments and, by removing excess lipid and detergent from the lipid-protein complex, have been able to demonstrate that more than 98% of the lipid in contact with the isolated protein is the synthetic phospholipid. Once the lipid pools have equilibrated, cholate is slowly removed by dialysis so that closed vesicular structures reform. These are able to retain Ca^{2+} taken up by the protein to an extent dependent on the nature of the synthetic phospholipid used for reconstitution.

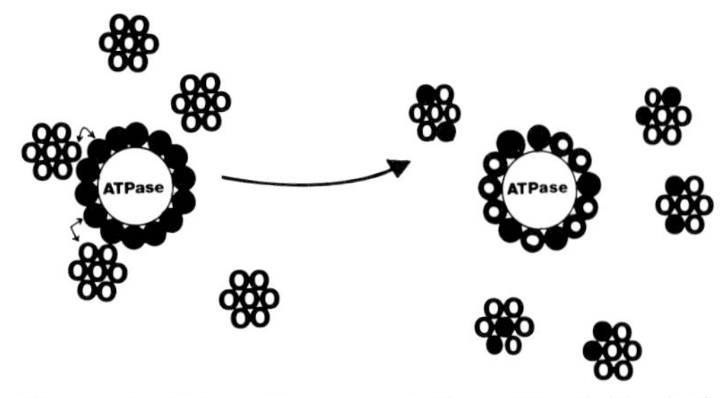

FIGURE 1 Schematic representation of the lipid substitution procedure. The protein, together with lipid in contact with it, is solvated by cholate (omitted from the figure for clarity) and treated with excess synthetic phospholipid. Cholate catalyzes an exchange of lipid between the endogenous and synthetic lipid pools.

TABLE I
The Fatty Acid Chain Specificity of a Reconstituted Ca^{2+} Pump

Phospholipid	Phase-transition temperature	ATPase activity (IU/mg at 25°C)	Ca^{2+}/ATP
DPL	41°C	0.05	0
DML	23°C	0.1	0
DEL	11°C	1.3	0.1
DOL	−20°C	1.4	0.53

Abbreviations used are: DEL, dielaidoyllecithin; DML, dimyristoyllecithin; DOL, dioleoyllecithin, DPL, dipalmitoyllecithin.

The Effect of Phospholipid Structure on the ATPase Activity of the Protein in Reconstituted Vesicles

The ATPase activity of the protein depends on the rigidity of the phospholipid fatty-acid chains in contact with the protein. Dipalmitoyllecithin (DPL) and dimyristoyllecithin (DML) have rigid fatty-acid chains at 25°C and when reconstituted complexes of the protein with either of these phospholipids are assayed below 25°C, the vesicles exhibit very low ATPase activity (Warren et al., 1974*c*), and hence a low rate of Ca^{2+} uptake by protein into the vesicles. Complexes of the protein with either dioleoyllecithin (DOL) or dielaidoyllecithin (DEL) exhibit high ATPase activity at 25°C because both these phospholipids are well above their phase-transition temperatures and their fatty acid chains are fluid (Table I).

The charge on the phosphoryl headgroup of the phospholipid also has a marked effect on the ATPase activity of the protein in reconstituted vesicles (Fig. 2). Zwitterionic phospholipids such as DOL and dioleoylphosphatidylethanolamine (DOPE) support high ATPase activities whereas phospholipids with a single net negative charge support

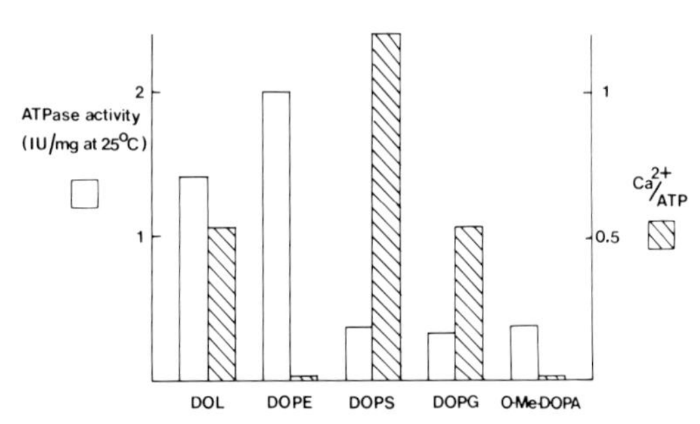

FIGURE 2 The headgroup specificity of a reconstituted Ca^{2+} pump. DOPA, dioleoylphosphatidic acid; other abbreviations are given in the text.

lower activity which is relatively unaffected by variation in the precise chemical structure of the negatively charged headgroup. Dioleoylphosphatidic acid with two net negative charges supports very low ATPase activity (Warren et al., 1975*a*).

In these studies on the effect of phospholipid structure on the ATPase activity of the protein, we assume that we are determining the extent to which a phospholipid will support the uptake of calcium through the protein into the vesicle. We can now ask whether a phospholipid that supports a high ATPase activity also restricts to a low level the nonspecific efflux of Ca^{2+} from reconstituted vesicles.

The Effect of Phospholipid Structure on Ca^{2+} Accumulation in Reconstituted Vesicles

No Ca^{2+} is accumulated by reconstituted complexes of the protein with either DPL or DML, because neither phospholipid supports a significant uptake of Ca^{2+} through the protein at 25°C. Complexes of the protein with DEL have a low Ca^{2+}/ATP ratio despite the high rate at which Ca^{2+} is taken up by the protein into the vesicle. This suggests that DEL cannot restrict the nonspecific efflux of Ca^{2+} from reconstituted vesicles. DOL, like DEL, supports high ATPase activity, but reconstituted complexes of the protein with DOL support reasonably efficient Ca^{2+} accumulation (Table I). DOL and DEL differ only in the configuration of the double bond in the fatty acid chains so it is clear that the precise stereochemistry of these chains can markedly affect the extent to which a phospholipid can restrict the nonspecific efflux of Ca^{2+}.

We have also determined the efficiency of Ca^{2+} accumulation in reconstituted complexes of the protein with a series of dioleoylphospholipids (Fig. 2). DOPE supports a high ATPase activity but no Ca^{2+} is accumulated. Dioleoylphosphatidylserine (DOPS) and dioleoylphosphatidylglycerol (DOPG) support lower ATPase activity but Ca^{2+} accumulation is reasonably efficient. O-Methyldioleoylphosphatidic acid supports an ATPase activity similar to that supported by DOPS and DOPG but no Ca^{2+} accumulates. We conclude that there is no simple correlation between the ability of a phospholipid to support Ca^{2+} uptake by the protein and its ability to restrict nonspecific efflux of Ca^{2+}. We must, however, insert a word of caution regarding the interpretation of the properties of reconstituted vesicles. In these experiments, the same conditions were used for reconstitution, whatever the nature of the phospholipid, and the vesicles differ somewhat in their size and in the amount of protein incorporated into each vesicle. Furthermore, by changing the conditions of reconstitution, we can vary widely the efficiency of Ca^{2+} accumulation in reconstituted vesicles. Nevertheless, we believe our conclusion above to be valid qualitatively, because such phospholipids as DEL, which support high ATPase activity and hence Ca^{2+} uptake into the vesicle, do not restrict the nonspecific leakage of Ca^{2+} from reconstituted vesicles, despite wide variation in the experimental conditions used for reconstitution. It is only under special conditions, in which the effects of nonspecific leakage are minimized by making the ratio of the internal volume of the vesicles large compared with the number of calcium pump units, that we have been able to obtain efficient calcium accumulation using any single species of phospholipid (e.g., DOL). We suggest that, at normal lipid to protein ratios, a single species of phospholipid may be unable to satisfy the two requirements of maintaining high ATPase activity and sealing the protein tightly into the bilayer. For example, the most impermeable vesicles with a single lecithin species are those consisting of a large excess of crystalline DML or DPL bilayer, but the rigidity of the chains is sufficient to inhibit Ca^{2+} uptake through the protein almost completely. Rapid Ca^{2+} uptake through the protein requires fluid fatty acid chains, but the increase in fluidity apparently leads to an increase in nonspecific efflux of Ca^{2+}. In contrast, native sarcoplasmic reticulum vesicles, with their heterogeneous lipid composition, support both high ATPase activity and restrict the initial rate of nonspecific efflux of Ca^{2+} to a negligible level. We find that when we reconstitute complexes of the ATPase with a mixture of two synthetic phospholipids, the reconstituted vesicles can exhibit similar efficiencies of calcium accumulation to the native membrane. We have already emphasized that such results might reflect variation in the requirements for effective reconstitution with different lipids rather than intrinsic differences in their ability to support calcium accumulation. However, our best guess is that the data are indicating significant features of lipid requirements for efficient calcium accumulation.

Reconstitution of a Calcium Pump with Mixtures of Synthetic Phospholipids

The most efficient reconstituted vesicles are those comprising a mixture of two synthetic phospholipids, particularly if one of the phospholipids is DOL. A good example is provided by a mixture of DOL and DOPE, because reconstituted complexes of the protein with these two phospholipids accumulate Ca^{2+} with an efficiency measured by the Ca^{2+}/ATP ratio of 1.9, close to the maximum

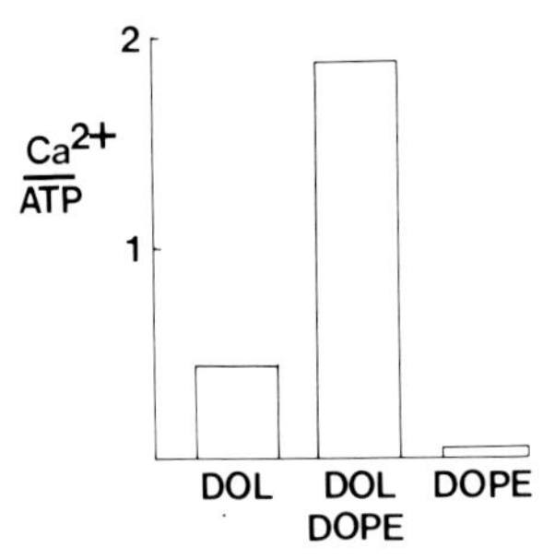

FIGURE 3 The effect of phospholipid mixtures on the efficiency of a reconstituted Ca^{2+} pump.

value of 2.0 (Fig. 3). The overall effect of this phospholipid mixture is synergistic, as the efficiency of the mixture is greater than the efficiency of either phospholipid alone. Approximately 85% of the phospholipid headgroups in native sarcoplasmic reticulum are lecithins and phosphatidylethanolamines, and the fluidity of the average fatty acid chain is probably similar to that of a mixture of DOL and DOPE. This synthetic mixture approximates the chemical and physical properties of the sarcoplasmic reticulum lipid bilayer, and it is quite clear that such a mixture is sufficient for efficient ATP-dependent accumulation of Ca^{2+}. The 50 species of lipids in sarcoplasmic reticulum membranes can be replaced by two synthetic phospholipids, at least with respect to the calcium transport function of these membranes. As we shall see, minor sarcoplasmic reticulum lipids with headgroups other than lecithin or phosphatidylethanolamine may have functions unrelated to Ca^{2+} uptake and accumulation.

The Phospholipids in Contact with the Protein Determine the Efficiency of Calcium Accumulation

We would now like to develop the hypothesis that the phospholipids in immediate contact with the penetrant calcium transport protein are responsible for the efficient accumulation of Ca^{2+} in reconstituted vesicles. These phospholipids are sufficient to support the ATPase activity of the protein and also to restrict the nonspecific Ca^{2+}, which we suggest occurs mainly at a boundary between the protein and the lipid bilayer (Fig. 4). Thus, by restricting Ca^{2+} efflux at this boundary, the phospholipids are, in effect, sealing the protein into the membrane. In contrast, the lipid bilayer outside this boundary is almost unperturbed by the presence of the protein and retains its characteristic impermeability to Ca^{2+} (Warren et al., 1975*b*).

To test this hypothesis we have tried to distinguish between those phospholipids that are in contact with the protein and those that make up the rest of the sarcoplasmic reticulum lipid bilayer. This can be done by using phospholipase D, which, in aqueous solution, breaks down only the outer monolayer of single-shelled phospholipid vesicles to phosphatidic acid. In sarcoplasmic reticulum vesicles, there are about 50 molecules of phospholipid in the outer monolayer associated with each molecule of the calcium transport protein and of these, about 12–14 molecules of phospholipid are inaccessible to phospholipase-D digestion. These 12–14 molecules are not intrinsically resistant to digestion by phospholipase D, and presumably they are inaccessible because they are in contact with the calcium transport protein. When native sarcoplasmic reticulum vesicles are disrupted, phospholipase D can digest both monolayers and about 30 molecules of phospholipid per ATPase are now resistant to attack by phospholipase D (Bennett et al., unpublished results).

We can now picture the penetrant calcium transport protein surrounded by a single bilayer shell of phospholipid, which is, to a first approximation, symmetrically disposed in the halves of the membrane bilayer. The number of 30 molecules in the annulus is found to be invariant in complexes of the protein with proportions of synthetic phospholipids, such as DOL, ranging from 30–150 molecules phospholipid/molecule of protein. These vesicular complexes can be made very leaky, so that phospholipase D will digest both monolayers. For all complexes, digestion stops with 30 molecules of DOL remaining intact per molecule of protein, and this annulus can only be

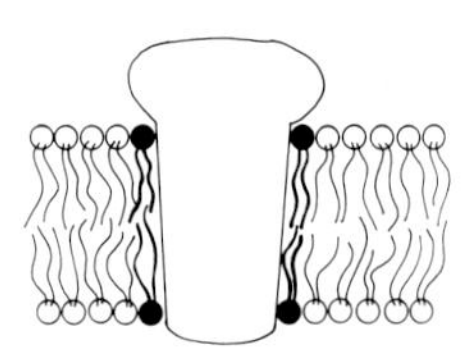

FIGURE 4 A schematic cross-sectional view of the calcium transport protein embedded in lipid bilayer. The first bilayer shell of phospholipid surrounding the penetrant part of the protein is the annulus which we consider to be sufficient to support active Ca^{2+} uptake through the protein and to be responsible for restricting the nonspecific efflux of Ca^{2+} which occurs at the boundary between protein and lipid bilayer.

degraded further by treatment with phospholipase D at elevated temperatures and in the presence of fluidizing agents, such as benzyl alcohol or cholate (Bennett et al., unpublished results). As we shall see, phospholipids in contact with the protein normally exchange freely with those in the extra-annular lipid bilayer, so it is not immediately clear why the phosphatidic acid molecules produced by phospholipase-D digestion do not simply exchange with the phospholipids in the annulus so that all the phospholipid in the outer monolayer is eventually degraded to phosphatidic acid. The reason is that phospholipase D will digest phospholipids only in the presence of a high concentration of Ca^{2+}, and as phosphatidic acid molecules are produced, they are chelated by Ca^{2+} to form a rigid structure that prevents their exchange with the phospholipids in the annulus. Unfortunately, we cannot assess the response of the ATPase activity of the protein to the changing phospholipid composition during digestion with phospholipase D because the high concentration of Ca^{2+} inhibits the protein. However, if Ca^{2+} is removed subsequently, the phosphatidic acid molecules are free to exchange with the DOL molecules in the annulus and, as the exchange occurs, there is a rapid fall in the ATPase activity of the protein.

To assess the response of the ATPase activity of the protein to varying amounts of lipid, we can remove the lipids from sarcoplasmic reticulum vesicles by using cholate (Warren et al., 1974*c*). Of the 100 molecules of phospholipid associated with each molecule of the calcium transport protein in a native sarcoplasmic reticulum vesicle, about 70 can be removed without affecting the ATPase activity of the protein (Fig. 5). Progressive removal of the remaining 30 molecules leads to an irreversible loss of ATPase activity. At least 30 phospholipid molecules are needed for maximum ATPase activity, and we suggest that these are the same phospholipid molecules that are inaccessible to digestion by phospholipase D. If we prepare complexes of the protein with varying amounts of a synthetic phospholipid such as DPL, we find that the maximum ATPase activity is considerably reduced but 30–35 molecules of the synthetic phospholipid are still needed to support the maximum activity of 1 molecule of the protein (Hesketh et al., 1976).

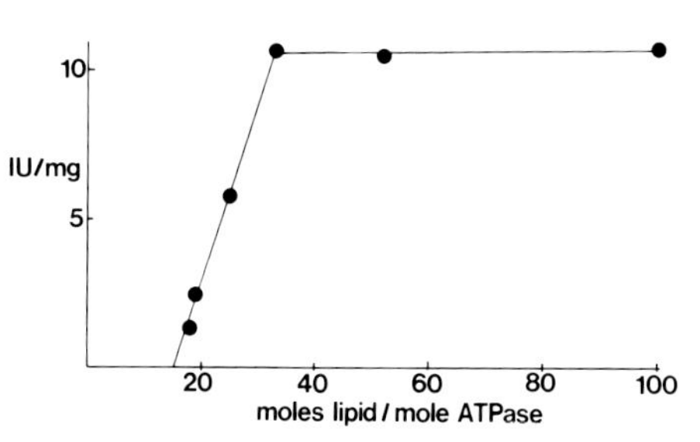

FIGURE 5 The effect of phospholipid depletion on the ATPase activity of the calcium transport protein. Sarcoplasmic reticulum vesicles were solvated by varying amounts of cholate and the lipid-protein complexes isolated by sucrose density gradient centrifugation. Each molecule of the protein needs 30 molecules of phospholipid to maintain maximal ATPase activity.

The Phospholipid Annulus Determines the ATPase Activity of the Protein

Complexes of the protein with DPL show unusual temperature-activity profiles, in that they retain significant ATPase activity down to 30°C, well below the phase-transition temperature of DPL at 41°C. Apparently, in the temperature range from 30–40°C, the 30 molecules of DPL in the annulus retain sufficient fluidity to support significant ATPase activity under conditions where the extra-annular DPL can be shown to undergo a normal phase transition at 41°C (Hesketh et al., 1976). Inasmuch as the temperature profiles of ATPase activity are very similar irrespective of the lipid to protein ratio, we conclude that it is the annular phospholipids that determine the activity profile, rather than the extra-annular lipid bilayer.

The Site of Nonspecific Ca^{2+} Efflux

From the above results we have indicated in Fig. 6 that there is a significant discontinuity in the lipid phase at the boundary between the annular and the extra-annular lipid. Discontinuities are known to occur at the liquid-crystalline to crystalline phase-transition temperature of pure lipids when solid regions of lipid are in equilibrium with lipid in the liquid state. The nonspecific permeability of a pure lipid bilayer to ions shows a sharp optimum at the phase-transition temperature when both solid and liquid phases coexist, which suggests that the site of nonspecific leakage may be localized at the boundary between the liquid and solid lipid bilayer (Marsh et al., 1976). By analogy, we suggest that the site of nonspecific leakage for Ca^{2+} in native and reconstituted vesicles is at the boundary of the annular and the extra-annular lipid. Normally, this boundary will

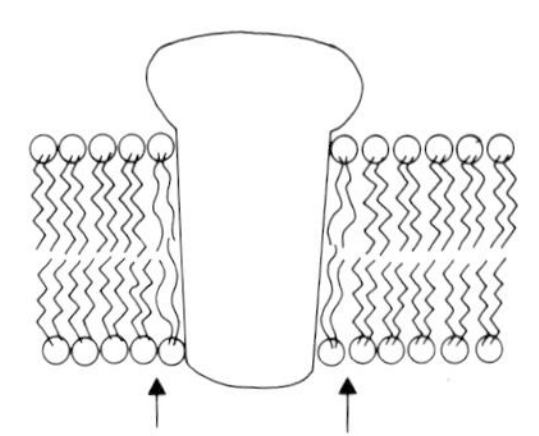

FIGURE 6 A schematic cross-sectional view of the calcium transport protein embedded in a DPL bilayer between 30 and 40°C. The DPL annulus is sufficiently fluid in this temperature range to support ATPase activity, but the DPL in the extra-annular bilayer is in the crystalline state. The arrows indicate the proposed site at which Ca^{2+} leaks through the membrane.

not be between annular phospholipid and crystalline extra-annular lipid, because the fatty acid composition of eukaryotic membranes is adjusted so that the bilayer is fluid at the growth temperature. If our analogy is valid, we have to demonstrate a physical distinction between the phospholipids in the annulus and those lipids in the extra-annular bilayer and show that it is this discontinuity in physical properties that gives rise to the leakage of ions at the boundary.

We can show that the phospholipids in the annulus have physical properties distinct from those in the extra-annular bilayer by preparing complexes of the protein with varying amounts of DOL containing 1 mol percent of spin-labeled phospholipid (Montecucco et al., 1977). The mobility of the fatty acid chains was estimated from $2T_{11}$, and this was plotted as a function of the lipid to protein ratio. As shown in Fig. 7, there is a clear discontinuity at about 28 mol phospholipid/mol protein. Below this lipid to protein ratio, we are titrating the binding sites on the protein surface and all bound phospholipid molecules show similar restricted mobilities. It is only in complexes with more than 28 mol phospholipid/mol protein that the phospholipid is free to distribute in the extra-annular bilayer, as well, where the freedom for chain movement is considerably higher. This accounts for the monotonic increase in $2T_{11}$ in complexes with more than 28 mol phospholipid/mol protein. The motional freedom of phospholipid in the annulus is physically restricted when compared to the phospholipids in the extra-annular bilayer. This results in a physical discontinuity in the lipid phase at the boundary of the annular and extra-annular phospholipid and it is this boundary which we consider to be the main site at which Ca^{2+} is lost from the vesicle. If this is correct, then the annular phospholipid that determines the ATPase activity of the protein must also be responsible for restricting the nonspecific efflux of Ca^{2+}.

The Rate at Which Phospholipids in the Annulus Exchange with Those in the Extra-Annular Bilayer

The number of phospholipid molecules physically restricted on the surface of the protein has been confirmed by ^{13}C nuclear magnetic resonance studies of complexes of the protein with varying amounts of ^{13}C-enriched DPL. In addition, such studies have given information on the rate at which phospholipids in the annulus exchange with those in the extra-annular bilayer (Montecucco et al., 1977). Phospholipids in the extra-annular bilayer have a calculated residence time at any location of the order of 10^{-7} s. The ^{13}C nuclear magnetic resonance data gives a residence time for phospholipids in the annulus of at least 0.5 ms, or 5,000 times longer than the residence time in phospholipid bilayer. The calcium transport protein will also undergo rotational diffusion in the plane of the membrane, with an estimated correlation time of about 20 μs calculated by analogy with rhodopsin in rod outer segment membranes (Poo and Cone, 1974) so that when the ATPase diffuses it will carry with it an annulus

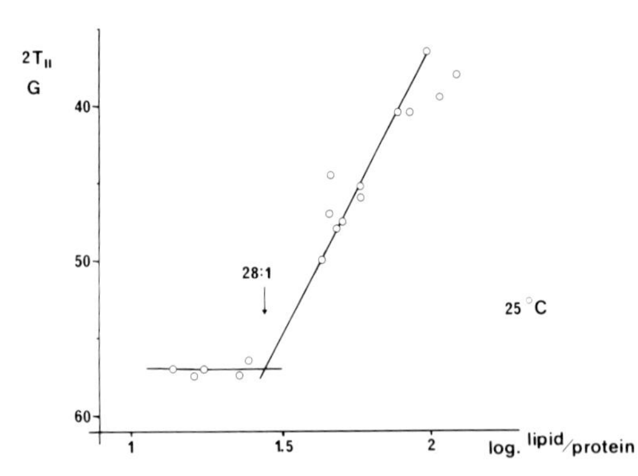

FIGURE 7 The number of phospholipid molecules that are physically restricted on the surface of each molecule of the calcium transport protein. The mobility (as measured by $2T_{11}$) of the phospholipid in complexes of the protein with DOL is restricted and constant until all the lipid binding sites on the protein surface have been filled. Phospholipid that is not in contact with protein has considerably more freedom of movement, and this accounts for the increase in mobility above 28 mol DOL/mol protein.

of phospholipid molecules. From this evidence for a long-lived complex of the ATPase with annular lipids, we suggest that another function of the annulus is to solvate the membrane protein. We speculate that this annulus is able to prevent protein-protein contacts in the hydrophobic phase of the bilayer, which would lead to two-dimensional aggregation and consequent exclusion of lipid molecules. Most penetrant membrane proteins are irreversibly inactivated when lipid is removed and they are forced to interact through protein-protein contacts, and we suggest that this solvating effect of lipids in the annulus may be a general feature of membrane architecture.

The Protein Can Segregate Phospholipids in the Plane of the Membrane

Having distinguished the phospholipids in the annulus from those in the extra-annular bilayer by differences in their physical properties, we can now ask whether they can also be distinguished by differences in their chemical composition. We show that the calcium transport protein can segregate phospholipids in the plane of the membrane and select for its annulus those lipids that support the highest activity.

The protein cannot discriminate among phosphatidylcholines with different fatty acid chains (Warren et al., 1975*c*) but it can discriminate between dioleoylphospholipids whose headgroups differ in their charge structure. One of the clearest examples is the response of the ATPase activity of the protein to a mixture of dioleoylphospholipids comprising DOL and dioleoylphosphatidic acid (DOPA) in varying proportions (Fig. 8). When the protein is complexed with DOPA alone, the ATPase activity is very low. If, however, we start with a complex of the protein with DOL and gradually increase the content of DOPA, the protein maintains its activity at a level characteristic of the protein with an annulus of DOL molecules. When the content of DOPA in the complexes is 70%, we can show that the protein is surrounded by an annulus of which only a minor proportion is DOPA, and more than 70% of the phospholipids in this annulus are DOL. Quite clearly, the protein is able to segregate phospholipids with different headgroup charges, in the plane of the membrane.

Although we have shown that a mixture of DOL and DOPE gives reconstituted vesicles which support very efficient Ca^{2+} accumulation,

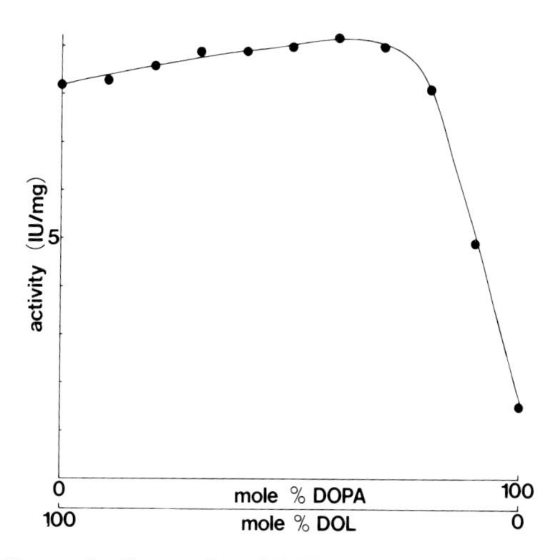

FIGURE 8 Segregation of DOPA and DOL by the calcium transport proteins. The protein can maintain ATPase activity as DOPA replaces DOL with the lipid-protein complexes by segregating these two phospholipids in the plane of the membrane and selecting DOL for the lipid annulus.

there is no evidence to suggest that the calcium transport proteins can segregate these two zwitterionic phospholipids in the plane of the membrane. It is possible that they could be distributed asymmetrically across the membrane, rather than in its plane, one phospholipid comprising the outer monolayer of the phospholipid annulus and the other, the inner monolayer. Such transverse segregation does not require that the chemical composition of the annulus be different from that of the extra-annular bilayer. At present, there is no evidence to suggest that lecithins and phosphatidylethanolamines have such a disposition in native or reconstituted membranes. Indeed, preliminary evidence suggests that the lecithins, at least, are symmetrically disposed in the halves of the native sarcoplasmic reticulum bilayer. If phospholipid segregation occurs in the native membrane, the data suggest that the protein will select the zwitterionic phospholipids for its annulus and exclude those phospholipids with net negative charges, which comprise about 15% of the total phospholipid in the membrane. We speculate that phospholipid segregation may be more than a means by which membrane proteins might select those phospholipids from the bilayer which best support their function. All penetrant proteins we have examined are active in the presence of zwitterionic phospholipids, which comprise the majority of phospholipids in most biological membranes.

Hence, even in the absence of phospholipid segregation, the membrane proteins would exhibit considerable activity. Some of the minor phospholipid components of biological membranes are thought to be involved in specific membrane functions not directly connected with the penetrant proteins. For example, phosphatidylinositol may be involved generally in the process of membrane fusion (Michell, 1975). Most of these minor phospholipid components have one thing in common: they all possess a net negative charge, which we expect will exclude them from the annulus of penetrant membrane proteins. Phospholipid segregation would then be the means by which these important minor phospholipid species would be free to reside in the extra-annular bilayer; they would not be adsorbed onto penetrant membrane proteins and removed from the bilayer where they are required for their functions.

To maintain optimal membrane function, the fatty acid chain composition of many prokaryotic membranes is adjusted in response to changes in such environmental conditions as temperature, and it is clear that these changes in composition directly affect the activity of the penetrant membrane proteins. If these proteins were able to segregate lipids according to their fatty acid chain composition, their response to changes in the fatty acid chain composition presumably would be minimized. However, if they behave like the calcium transport protein, their activity will respond directly to changes in fatty acid chain composition, because the protein is unable to select between lecithins of different chain structure.

SUMMARY

We conclude by summarizing what we believe to be the functions of the annulus of phospholipids that is physically restricted on the penetrant surface of the calcium transport protein, and on other membrane proteins such as cytochrome oxidase (Jost et al., 1973). The phospholipids in the annulus support the functioning of the calcium transport protein, and phospholipid headgroups that do not support function can be excluded. The annulus seals the membrane protein into the bilayer by restricting the nonspecific efflux of accumulated Ca^{2+}. Lastly, the annulus solvates this, and probably other membrane proteins, preventing their two-dimensional crystallization and consequent inactivation. The phospholipid annulus is the structural feature of biological membranes that provides the bridge between the impermeability of the lipid bilayer and the specific permeability of the penetrant transport proteins.

REFERENCES

Hesketh, T. R., G. A. Smith, M. D. Houslay, K. A. McGill, N. J. M. Birdsall, J. C. Metcalfe, and G. B. Warren. 1976. Annular lipids determine the ATPase activity of a calcium transport protein complexed with dipalmitoyllecithin. *Biochemistry.* **15:**4145–4151.

Houslay, M. D., T. R. Hesketh, G. A. Smith, G. B. Warren, and J. C. Metcalfe. 1976. The lipid environment of the glucagon receptor regulates adenylate cyclase activity. *Biochim. Biophys. Acta.* **436:**495–504.

Houslay, M. D., G. B. Warren, N. J. M. Birdsall, and J. C. Metcalfe. 1975. Lipid phase transitions control β-hydroxybutyrate dehydrogenase activity in defined-lipid protein complexes. *FEBS (Fed. Eur. Biochem. Soc.) Lett.* **51:**146–151.

Jost, P. C., O. H. Griffith, R. A. Capaldi, and G. Vanderooi. 1973. Evidence for boundary lipid in membranes. *Proc. Natl. Acad. Sci. U. S. A.* **70:**480–484.

Marsh, D., A. Watts, and P. F. Knowles. 1976. Evidence of phase boundary lipid-permeability of tempo-choline into dimyristoylphosphatidylcholine vesicles at the phase transition. *Biochemistry.* **15:**3570–3578.

Michell, R. H. 1975. Inositol phospholipids and cell-surface receptor function. *Biochim. Biophys. Acta.* **415:**81–147.

Montecucco, C., G. A. Smith, N. J. M. Birdsall, G. B. Warren, and J. C. Metcalfe. 1977. Structural and dynamic properties of the lipids interacting with a calcium transport protein. *Biochemistry.* In press.

Poo, M. -M., and R. A. Cone. 1974. Lateral diffusion of rhodopsin in the photoreceptor membrane. *Nature (Lond.).* **247:**438–441.

Warren, G. B., P. A. Toon, N. J. M. Birdsall, A. G. Lee, and J. C. Metcalfe. 1974*a*. Reconstitution of a calcium pump using defined membrane components. *Proc. Natl. Acad. Sci. U. S. A.* **71:**622–626.

Warren, G. B., P. A. Toon, N. J. M. Birdsall, A. G. Lee, and J. C. Metcalfe. 1974*b*. Complete control of the lipid environment of membrane-bound proteins: application to a calcium transport system. *FEBS (Fed. Eur. Biochem. Soc.) Lett.* **41:**122–124.

Warren, G. B., P. A. Toon, N. J. M. Birdsall, A. G. Lee, and J. C. Metcalfe. 1974*c*. Reversible lipid titrations of the activity of pure adenosine triphosphatase-lipid complexes. *Biochemistry.* **13:**5501–5506.

Warren, G. B., J. C. Metcalfe, A. G. Lee, and N. J. M. Birdsall. 1975*a*. Mg^{2+} regulates the ATPase activity of calcium transport protein by interacting with bound phosphatidic acid. *FEBS (Fed. Eur. Biochem. Soc.) Lett.* **50:**261–264.

WARREN, G. B., J. C. METCALFE, A. G. LEE, and N. J. M. BIRDSALL. 1975*b*. The lipids surrounding a calcium transport protein: their role in calcium transport and accumulation. Proc. 10th FEBS Meeting. American Elsevier, New York. **41:**3–15.

WARREN, G. B., M. D. HOUSLAY, N. J. M. BIRDSALL, and J. C. METCALFE. 1975*c*. Cholesterol is excluded from the phospholipid annulus surrounding an active calcium transport protein. *Nature* (*Lond.*). **255:**684–687.

MEMBRANE PROTEINS: STRUCTURE ANALYSIS BY ELECTRON MICROSCOPY

P. N. T. UNWIN and R. HENDERSON

A good deal is now known about membrane proteins in terms of their biochemical properties and their organization in the lipid bilayer, but not much has yet been found out about their three-dimensional molecular structure. This information is required from each of them, or from each class of membrane protein, to obtain a full understanding of the mechanisms by which they function. Knowledge of the molecular structure of at least a few membrane proteins involved in different processes should provide a valuable insight into the characteristics of membrane proteins as a whole.

The electron microscope, in principle, provides a direct means of high resolution structure analysis. Its value is restricted by the conventional staining and embedding techniques for preserving specimens, which are at present inadequate in resolution and accuracy. However, recent improvements in preparation procedures have been made and there are now a number of ways in which proteins, sensitive to dehydration, and lipid bilayers can be maintained and observed in the microscope vacuum, essentially in their native state (Matricardi et al., 1972; Hui and Parsons, 1974; Taylor and Glaeser, 1974, 1976; Unwin and Henderson, 1975; Unwin, 1975) and without the complications associated with the use of fixatives and stains. These improvements promise to make the electron microscope a much more powerful tool for investigation of membrane protein structure in the future.

Radiation damage limits high resolution electron microscope structure analysis of unstained specimens to well-ordered arrays of molecules (see below). Membranes containing ordered arrays of protein molecules do not occur in abundance in nature. However, it is reasonable to hope that many membrane proteins when purified could, with some effort, be induced to form crystalline sheets, just as many soluble proteins can be induced to form large three-dimensional crystals suitable for X-ray analysis. Cytochrome oxidase provides an example in which such sheets have been produced artificially (Vanderkooi et al., 1972).

A brief account is given here of the principles involved in electron microscope structure analysis, and of how they have been used to investigate the structure of the purple membrane from *Halobacterium halobium*. The aim is to give an idea of the scope of the method—its likely value and limitations in relation to membrane proteins in general.

Electrons as a Source of Radiation

High resolution structure analysis with X-rays is a well-established technique. What advantages do electrons offer over X-rays for investigating membrane structure? There are two obvious ones. First, electrons are able to form images; these can be measured to give the relative displacements of the sine waves from which the diffracting matter in the object is built up; images, therefore, dispense with the need for phase determination by the less direct methods used in X-ray crystallography. Second, electrons interact much more strongly with matter and therefore are an appropriate form of radiation for investigating material that is only a few tens of Ångstroms thick (the ratio of elastic scattering cross sections, electrons/X-rays is $\sim 10^8$ for biological matter at medium resolutions).

Substantial radiation damage occurs in both cases, but manifests itself more strikingly in the case of electrons. This is because the smaller the crystal the more the energy has to be concentrated on each unit cell to provide a picture of the aver-

P. N. T. UNWIN and R. HENDERSON Medical Research Council, Laboratory of Molecular Biology, Cambridge, England

age unit cell of given statistical significance. It can be shown that if the energy is concentrated over a single unit cell or isolated molecule, a dose of at least 500 electrons/Å^2 would be required to define secondary structure with reasonable statistical accuracy. Unfortunately, the energy absorbed with this dose, as a result of inelastic scattering, is not trivial. It is equivalent to ~ 4 × 10^{11} rads (Grubb, 1974; ~ 10^9 times the lethal dose for a human being) and is enough to transform the native conformation of a protein molecule into something quite unrecognizable.

Five hundred electrons/Å^2 is about the dose that would normally be given to a specimen in recording its picture at some intermediate magnification (say × 50,000). However, 0.5 electrons/Å^2 seems to be about the maximum a molecule can tolerate without gross structural alteration, according to electron diffraction results (Unwin and Henderson, 1975). There is, therefore, a discrepancy of roughly 1,000 (depending on the specimen involved and the method of preservation) between the electron dose required to map out secondary structure in an isolated molecule and the dose it can tolerate without much destruction.

In theory, one could circumvent these difficulties caused by electron damage and obtain a useful map of an isolated molecule by averaging 1,000 or more identical images of it, each recorded at 0.5 electrons/Å^2. But there would be formidable practical difficulties in carrying out such a task. If, instead, the molecules form a crystalline array, an altogether more favorable opportunity for averaging presents itself, because in a crystal the positions and orientations of the constituent molecules are related by fixed vectors. The same number of copies of the same thing is still required, so the minimum size of crystal appropriate for analysis is one containing about 1,000 unit cells. Therefore, any *quantitative high resolution* method of structure analysis using electrons requires fairly extensive ordered arrays of molecules.

Diffraction and Microscopy

An electron micrograph recorded with a "safe" dose of only about 0.5 electrons/Å^2 shows statistical fluctuations in the number of electrons from one image element to the next. At 10 Å resolution, these are about 15% of the mean number in each element. They are large in comparison to the genuine periodic fluctuations ($\lesssim 1\%$) which would arise from a membrane sheet containing an ordered array of molecules. Therefore, at high resolution the micrograph would appear featureless.

Electron diffraction patterns or optical or Fourier transforms of micrographs, on the other hand, compress the periodic fluctuations into a relatively few peaks arranged on a lattice, and these peaks are easily observed. Diffraction peaks express the average information about the sheet directly, and it therefore makes sense to study diffraction patterns and Fourier transforms of micrographs, rather than the micrographs themselves, in carrying out a structure analysis of a radiation-sensitive object.

Electron diffraction intensities give information about the strengths of the sine waves from which the diffracting matter in the object is built up. Transforms of micrographs not only give this information but also show how they are displaced relative to each other. Both types of information, diffraction intensities (or rather, the square roots of the intensities, the amplitudes) and phases are, of course, required for determining a structure.

A crystalline sheet of membrane is periodic only in the plane of the sheet and does not repeat in the direction perpendicular to this plane. Its diffraction pattern is, therefore, continuous in this direction. It is not a three-dimensional lattice of peaks, as would be the case with a true three-dimensional crystal, but a two-dimensional lattice of continuous lines of intensity pointing in the direction perpendicular to the membrane plane (Fig. 1).

In recording an electron diffraction pattern, a two-dimensional lattice of spots is observed. This is the central section, perpendicular to the electron beam, through the line lattice. The Fourier transform of a micrograph of the specimen oriented in the same way to the electron beam would contain the phases and amplitudes for the same central section. With the specimen oriented normal to the beam, the central section would be as in Fig. 1 and would correspond to the normal projection of the structure.

By tilting membranes through different angles to obtain different views, the same lattice can be sampled along different sections (Fig. 2). In this way, by using data from diffraction patterns and micrographs of many separate membranes (a single membrane can only be used for one record because of radiation damage), a picture of the continuous amplitude distributions (electron diffraction patterns) and phase distributions (Fourier transforms of micrographs) along each of the lattice lines is built up. Eventually, when a complete set of data has been collected, a three-dimensional

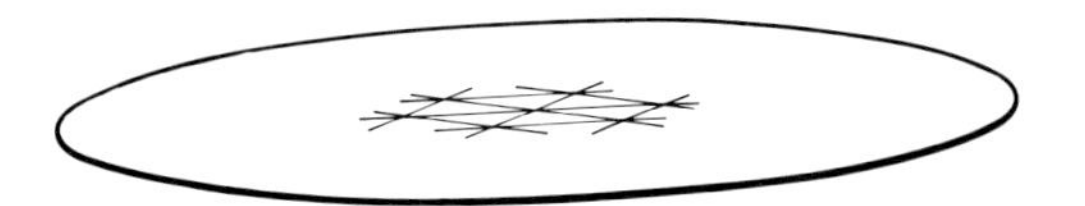

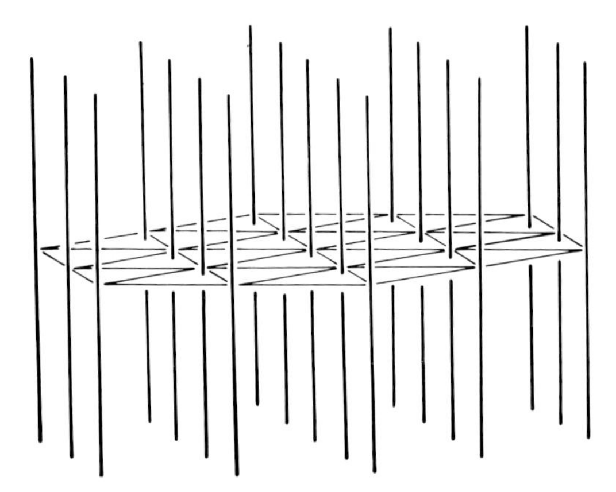

FIGURE 1 Schematic representation of the purple membrane and its three-dimensional diffraction pattern. The protein and lipid molecules are arranged on an hexagonal lattice and form a diffraction pattern which consists of continuous, hexagonally spaced, lines of intensity pointing in the direction perpendicular to the membrane plane. The central section normal to these lines provides the data on which the projection map, Fig. 3, is based.

Fourier synthesis is carried out to give a three-dimensional map.

The general principle of tilting specimens to obtain a number of different views and combining these views, by making use of Fourier transforms, to obtain a three-dimensional map was originally put forward by De Rosier and Klug in 1968. A similar approach was outlined by Hoppe et al. in the same year. Electron diffraction is not an essential part of the procedure, but with large crystalline sheets it can be expected to provide a more accurate set of amplitudes, particularly at high resolutions, than the micrographs.

Unfortunately, micrographs (and hence their Fourier transforms) suffer from a number of defects and restrictions not apparent with electron diffraction patterns. Their detailed appearance depends critically on the focus level at which they are taken; they are extremely sensitive to electron-optical instabilities, vibrations and any other slight disturbance, and, at low magnification, to limitations of photographic emulsions. The influence of the focus level can be corrected for in the transform of the micrograph, although several micrographs may be required to give a complete set of data for one particular view. However, the remaining effects result in loss of information. Thus it may be difficult to obtain phases for the high resolution peaks visible in electron diffraction patterns, and it is the quality of the micrographs that finally dictate the resolution to which the structure analysis can be taken.

The Purple Membrane Protein

The purple membrane is a specialized part of the cell membrane of *Halobacterium halobium* that functions in vivo, under the influence of light, as a hydrogen ion pump (Oesterhelt and Stoeckenius, 1971). It contains a single species of protein molecule, bacteriorhodopsin (Oesterhelt and Stoeckenius, 1971), which is arranged in the lipid bilayer on a P3 space group lattice (a = 62 Å; Henderson, 1975). The membrane thickness is 45 Å and it extends up to about 1 μm in diameter (Blaurock and Stoeckenius, 1971; Henderson,

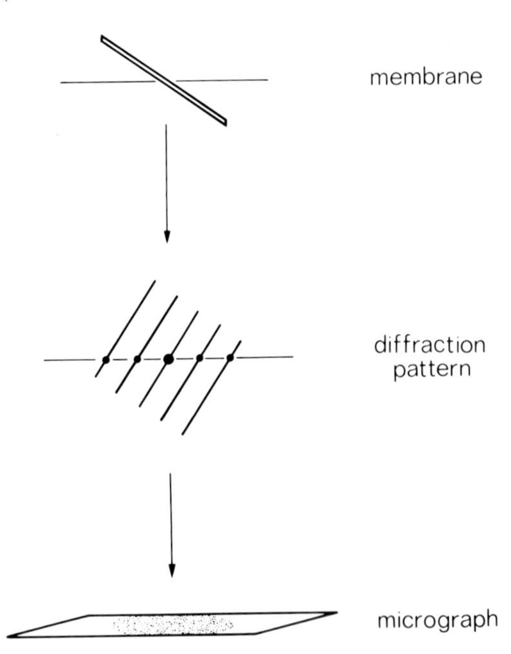

FIGURE 2 Microscope geometry with the membrane tilted, as in collecting data for a three-dimensional map. Diffraction patterns recorded with different angles of tilt differ according to the way the central section, perpendicular to the electron beam, samples the line lattice. Micrographs record the projected views of the membrane corresponding to these diffraction patterns.

1975), so that the maximum number of unit cells within a sheet is about 20,000 – or about 20 times the minimum required number estimated earlier. Its three-dimensional structure was determined, by the methods outlined in the preceding paragraphs, to a resolution of 7 Å (Henderson and Unwin, 1975).

A projection map of the membrane, Fig. 3, shows that each protein molecule is comprised of at least three rodlike segments aligned roughly perpendicular to the membrane plane. But a fairly continuous region of higher than average density indicates that there may be several more rods tilted sufficiently to overlap and which therefore contribute a smear in projection. The dimensions and spacings of these rods identify them as being closely packed α-helices, in agreement with X-ray studies (Henderson, 1975; Blaurock, 1975). The more uniform spaces between and within the protein clusters are composed mainly of lipid bilayer.

Figure 3 is relatively informative in comparison with projections obtained from true three-dimensional crystals, where the detail is normally confused by the superposition of molecules from adjacent layers. This property applies generally to projections of isolated membrane sheets and also to more complicated aggregates, such as collapsed vesicles or multilayer assemblies, unless, of course, the separate layers interact specifically so that their contributions to the diffraction pattern cannot be separated.

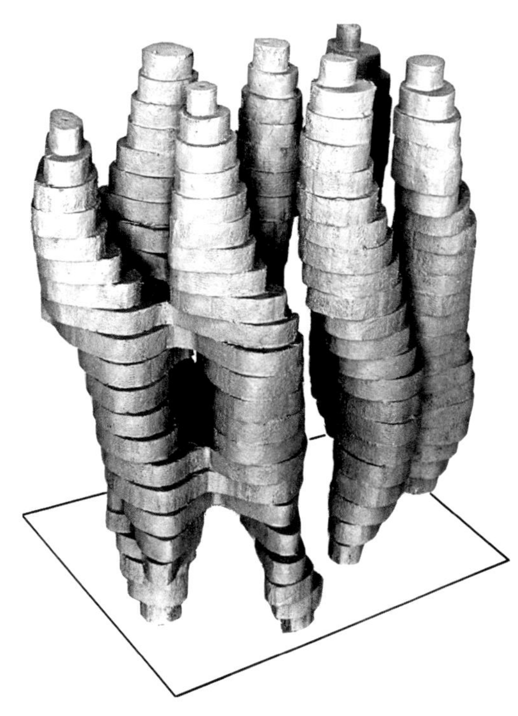

FIGURE 4 Model of the purple membrane protein constructed from a 7 Å three-dimensional map (Henderson and Unwin, 1975). The view is that of the molecule outlined in Fig. 3 seen from below. Thus the four most strongly inclined α-helices are in the foreground, and the other three, which are approximately perpendicular to the membrane surfaces, are at the back.

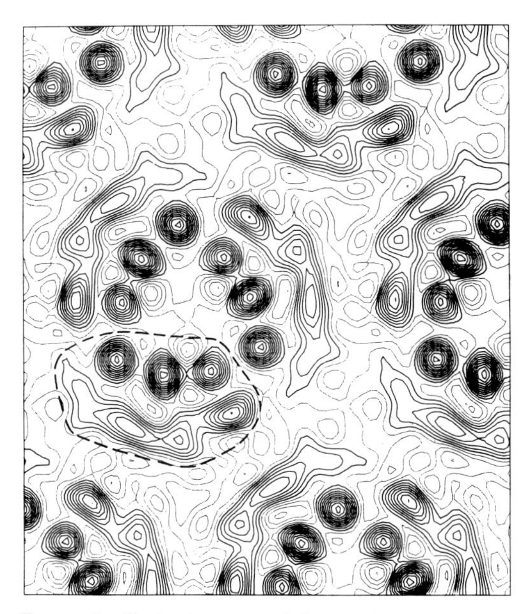

FIGURE 3 Projection map of the purple membrane at 7 Å resolution (Unwin and Henderson, 1975). The protein molecules (heavier contours) pack tightly together around threefold axes in clusters of three. One of the molecules is outlined.

The three-dimensional map of the purple membrane shows that the continuous region of higher than average density in Fig. 3 is indeed made up from α-helices which are slightly tilted. There are seven close-packed α-helical segments in the protein altogether and each is 30–40 Å in length. The molecule is about 45 Å long in the direction perpendicular to the bilayer, and so almost certainly extends through it to be exposed to the environment on either side.

A photograph of a model constructed from the map, and illustrating these features, is given in Fig. 4. The resolution is insufficient to show regions of unfolded polypeptide chain, but because the molecular weight is only 26,000 (Oesterhelt and Stoeckenius, 1971), these regions can make up only 20–30% of the total polypeptide chain and therefore are accounted for largely by the connecting links between the α-helical segments.

The retinal molecule, which is covalently bound to the protein (Oesterhelt and Stoeckenius, 1971), is also invisible at this resolution.

The study of the purple membrane emphasizes the power of the electron microscope as a tool for structure analysis of membrane proteins generally. But it also draws attention to limitations inherent in the technique which make even higher levels of resolution difficult to attain. Problems arise, especially in directions away from the plane of the membrane sheet, where the lattice lines tend to become increasingly broad unless the sheets can be made perfectly flat and where there is also a technical difficulty in recording data from objects tilted more than about 60° to the incident beam direction. The resolution is, therefore, always likely to be better in the plane of the membrane than away from it.

With the purple membrane, anisotropy in resolution from these sources is to some extent disguised by the fact that the arrangement of α-helices is such as to concentrate the diffraction, at least to medium resolution, toward the plane of the membrane. Fortunately, the effect of this anisotropy will only be very pronounced in the unlikely event of much of the diffraction being concentrated in a direction perpendicular to the membrane sheet, as, would be the case, say, with α-helices stacked predominantly parallel to the membrane surface.

Closer to the plane of the membrane, resolutions approaching the capability of a modern electron microscope, 2–3 Å, should be attainable. But technical problems increase substantially as the microscope is taken toward its theoretical limit. Useful data from micrographs at this resolution require that they be corrected for electron optical distortions before Fourier analysis and that careful attention be paid to such factors as the modulation transfer characteristics of the photographic emulsion, chromatic aberration, and partial coherence of the electron beam—all of which now become critically important.

Clearly, then, the structures of membrane proteins will be difficult to solve by electron microscopy alone at quite the resolutions we have come to expect from X-ray diffraction of soluble proteins; we can hardly expect to be able to determine directly the primary conformation of the polypeptide chain in this way.

However, for obtaining information on a slightly coarser scale—the organization of the polypeptide chain into conformationally discrete units such as α-helices and β-pleated sheet—the effectiveness of electron microscope structure analysis is unquestionable. Molecular structure at this level of resolution is of considerable value, because it provides a framework upon which, with the aid of additional data, a more detailed picture can be built up.

REFERENCES

BLAUROCK, A. E. 1975. Bacteriorhodopsin: a trans-membrane pump containing α-helix. *J. Mol. Biol.* **93**:139–158.

BLAUROCK, A. E., and W. STOECKENIUS. 1971. Structure of the purple membrane. *Nat. New Biol.* **233**:152–155.

DE ROSIER, D. J., and A. KLUG. 1968. Reconstruction of three-dimensional structures from electron micrographs. *Nature (Lond.).* **217**:130–134.

GRUBB, D. T. 1974. Radiation damage and electron microscopy of organic polymers. *J. Mat. Sci.* **9**:1715–1736.

HENDERSON, R. 1975. The structure of the purple membrane from *Halobacterium halobium*: analysis of the X-ray diffraction pattern. *J. Mol. Biol.* **93**:123–138.

HENDERSON, R., and P. N. T. UNWIN. 1975. Three-dimensional model of purple membrane obtained by electron microscopy. *Nature (Lond.).* **257**:28–32.

HOPPE, W., R. LANGER, G. KNESCH, and CH. L. POPPE. 1968. Protein-kristallstruktur analyse mit elektronen strahlen. *Naturwissenschaften.* **1**:333–336.

HUI, S. W., and D. F. PARSONS. 1974. Electron diffraction of wet biological membranes. *Science (Wash. D.C.).* **184**:77–78.

MATRICARDI, V. R., R. C. MORETZ, and D. F. PARSONS. 1972. Electron diffraction of wet proteins: catalase. *Science (Wash. D.C.).* **177**:268–269.

OESTERHELT, D., and W. STOECKENIUS. 1971. Rhodopsin-like protein from the purple membrane of *Halobacterium halobium. Nat. New Biol.* **233**:149–152.

TAYLOR, K., and R. M. GLAESER. 1974. Electron diffraction of frozen, hydrated protein crystals. *Science (Wash. D.C.).* **186**:1036–1037.

TAYLOR, K., and R. M. GLAESER. 1976. Electron microscopy of frozen, hydrated biological specimens. *J. Ultrastruct. Res.* **55**:448–456.

UNWIN, P. N. T. 1975. Beef liver catalase structure: interpretation of electron micrographs. *J. Mol. Biol.* **98**:235–242.

UNWIN, P. N. T., and R. HENDERSON. 1975. Molecular structure determination by electron microscopy of unstained crystalline specimens. *J. Mol. Biol.* **94**:425–440.

VANDERKOOI, G., A. E. SENIOR, R. A. CAPALDI, and H. HAYASHI. 1972. Biological membrane structure. III. The lattice structure of membranous cytochrome oxidase. *Biochim. Biophys. Acta.* **274**:38–48.

Cell-to-Cell Interactions

INTRODUCTORY REMARKS

L. WOLPERT

Cell-to-cell interactions refer to the processes taking place, particularly those involving communication, when cells are relatively close to each other, or touch one another. It refers to interactions that occur over distances of about 1 μm: thus hormones, for example, which enable cells to interact with each other over long distances, are excluded. On the whole, this is a field in which we are still uncertain about what is happening, and our ignorance is much greater than our knowledge. In this symposium, we will consider mainly developmental aspects of the problem.

When cells interact with one another, the cell surface membrane has to be involved. It is helpful to classify interactions into three main classes, according to the role played by the membrane (Wolpert and Gingell, 1969). In the first, the cell membrane acts as a mechanical sensor, and enables the cell to respond, for example, to differences in adhesiveness in the environment. In the second, it behaves as a channel whereby a signal from one cell can pass directly into another cell, and the formation of gap junctions exemplifies such an interaction. In the third class, it acts as a transducer: that is, some message or some effect from an adjacent cell has an effect on the membrane, whereby the information in that message, or the nature of that message, is transformed or transduced by the membrane to something else which occurs in the interior of the cell. Examples of this would be an extracellular stimulus that either activates adenyl cyclase in the membrane and therefore causes more cyclic AMP to be made inside, or causes a change in permeability of the membrane (Gingell, 1971). The following papers will, to some extent, illustrate aspects of these three classes of interaction. Dr. Gerisch will discuss the nature of cell adhesion, which can largely determine the membrane's role as a mechanical sensor; Dr. Pitts will discuss gap junctions where the cell is acting as a channel; and Dr. Hay will discuss the situations in which cells are not really in direct contact and where whatever interaction there is between the cells probably involves the intercellular matrix. The matrix most likely exerts its influence by the cell membrane acting as a transducer.

A main area of interest in cell-to-cell interactions relates to the complexity and specificity of the interaction. Emphasis is often placed on the nature of the communication involved and, to put it more anthropomorphically, we want to know what sort of conversations cells have with each other. It is a widespread idea that the nature of such conversations is complex. In this view, conversations between cells are thought to be very interesting, and this implies that there are all sorts of qualitative messages passing between cells. On the other hand, there is a view, and one to which I adhere, that conversations between cells are very simple and really rather uninteresting, and the important thing is the cell's response (Wolpert and Lewis 1975). In terms of the three classes of membrane-mediated interactions, it is important to realize that the specificity of cell-to-cell interaction may reside at various sites. It could be, for example, in the membrane, but might equally well reside in the interior of the cell, especially when a channel is formed or when the membrane acts as a transducer. It has been insufficiently appreciated that the widespread occurrence of gap junctions provides a channel whereby the internal contents of cells become accessible to interactions, and that this is where the specificity of response may reside, rather than in the membrane itself.

L. WOLPERT Department of Biology as Applied to Medicine, The Middlesex Hospital Medical School, London, England

Contact Inhibition

As an example of the problems that one has to face when dealing with cell-to-cell interaction, I want to draw attention to an experiment of Dunn (1971) relating to contact inhibition. Contact inhibition is the phenomenon described by Abercrombie (1967) for fibroblasts in culture, in which the locomotion of a cell is inhibited when it makes contact with another cell. Dunn has examined nerve cells in culture and made the important observation that at the growth cone, where there are large numbers of filopods, when one filopod touches another nerve cell, but not when it touches, for example, some solid object or a fibroblast, it and other filopods withdraw and the cell tends to move off in another direction. Not only is this a very nice example of contact inhibition; it also raises problems as to what the mechanism is. There have been suggestions, for example, that contact inhibition can be explained in terms of relative adhesiveness (Martz and Steinberg, 1973). However, in this case, such a mechanism is most unlikely, as it could not explain why, when one filopod touches the adjacent nerve cell, adjacent filopods also withdraw. Rather, it seems, one requires a mechanism whereby the effect of contacting another nerve cell is propagated throughout the growth cone, and this could involve the effect of contact being transduced by the membrane into, for example, a permeability change, or a channel formed at the site of contact, allowing communication between the internal contents of the cells. It is not known which mechanism is involved, but it is clear that the specificity of the interaction may lie in the initial contact or in the response to the changes it brings about.

Cell Adhesion

It is still not clear, but the popular current view suggests that cells are held together by specific molecules acting as ligands (Moscona, 1974; Marchase et al., 1975). On the other hand, one should be aware that all cells are subject to physical forces when they come close to each other: because of their negative surface charge there is a repulsive force which prevents cells coming together, and there is also now good evidence for attractive forces of the long-range van der Waals-London type (Parsegian and Gingell, 1973; Gingell and Fornés, 1976). The relative importance of these two types of mechanisms for cell adhesion remains unclear. One should be careful at this stage not to make too strong an assumption about the specificity of cell adhesion. For example, to my mind there is still little evidence on the biological side for the type of specificity that is found in antigen/antibody or enzyme substrate specificity, which are the paradigms we have in mind when we talk about specificity. Even in the nervous system, where it is generally believed there is a high degee of adhesive specificity determining neural connections, at this stage, the biological evidence remains rather weak. I would emphasize that the requirement for a high degree of biological specificity in adhesiveness still remains to be proved. Gaze and Hope (1976), for example, looking at the development of retinotectal connections, show that one can develop models that require rather a low degree of specificity of interaction between retinal and tectal cells.

Signaling in Hydra

It may be useful to draw attention to work on the biological side that provides some evidence for interaction over quite long distances, and in which cell contact is required and diffusion might be involved. This is just the sort of situation in which gap junctions might be expected to provide the channel for communication. The system we have studied is the regeneration of the head in hydra, where there is good evidence (Hicklin et al., 1973) that the head end of hydra exerts an inhibitory influence that prevents the formation of other heads.

Our experiments are based on the observation of Wilby and Webster (1970) that one head can inhibit formation of another head end over relatively long distances (0.5 mm), and that this inhibition can be propagated in a proximodistal direction, that is, from the foot end toward the head end. In a typical experiment, a head is grafted onto the proximal end of the gastric region of another hydra and, at a later time, the host head is removed, and one simply observes whether the grafted head can inhibit head regeneration (Fig. 1). (The head and gastric region of hydra can be represented by H*1234*). When an additional head is grafted onto the *3* region to give H*123*/H then (at 18°C) it is necessary to wait for about 8 hr before removing the host head if regeneration is to be inhibited. We explored the time/distance relationship, and the results showed that the time between grafting the additional head and removing the host head so that inhibition occurred was highly sensitive to distance (Wolpert et al., 1972). For example, for a short length corresponding to region *1*, the additional head could be grafted 6 hr after removal of the host head. Thus, there is a 14-hr difference between the times of grafting the

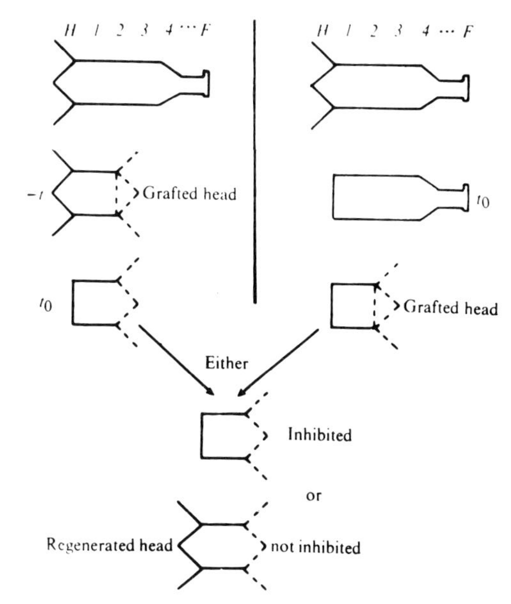

FIGURE 1 Scheme for experiments to test the time/distance relationship for an inhibitory signal from the head end of hydra. On the left, the head is grafted on before the host head is removed, and this is necessary for inhibition of 123/H. Host head removal is at t_0, and in this case the graft is at $-t$. On the right, the head is removed from the host before grafting of the head and this can lead to inhibition for shorter distances.

additional head when the distance is region *1* or region *123*. From our results, together with a theoretical analysis, we concluded that diffusion of a molecule with an effective diffusion constant of 2×10^{-7} cm²/s could provide the mechanism whereby the inhibitory signal was transmitted. We have also obtained evidence that signaling will not occur unless cell contact has been established (Hicklin et al., 1973). Taken together, these results suggest that the long-range inhibitory signals might be mediated by a diffusible, intracellular signal. This requires a channel for cell-to-cell signaling, such as a gap junction. Hydra have gap junctions.

Cell-to-cell Interactions Via Extracellular Matrix

There is a variety of cell-to-cell interactions that appear to be mediated by the extracellular matrix. Among the most dramatic are those involving epithelial mesenchymal interactions in relation to the epidermis. Sengel (1976) and others have shown that a graft of embryonic mesoderm, made during development of a region of the chick embryo that would normally form scales to a region in which feathers would normally form, can result in the overlying epidermis forming scales instead of feathers. In such interactions, the mesoderm and the epidermis are separated by a basement membrane, and there is no direct cell-to-cell contact. Any signaling must be through the extracellular matrix, and the cell membrane very likely acts as a transducer. Dr. Hay will consider such interaction.

Positional Signaling in the Developing Limb

The basic idea of positional information is that it provides a mechanism for pattern formation in development in terms of a two-step process: first, a coordinate system is set up, whereby the cells have their positions specified with respect to boundaries; second, the cells interpret this positional information with the appropriate cytodifferentiation (Wolpert, 1969). Such a mechanism implies that quantitative signals can provide the basis for positional information and that the differences between patterns of cellular differentiation arise from the interpretation of the signal, rather than from the signal itself. In the case of the chick wing, we have suggested that positional information probably is specified along three axes; proximodistal, anteroposterior, and dorsoventral (Wolpert, et al., 1975). Here I will confine myself to the anteroposterior axis, for which there is evidence that positional information is specified by a signal from a region at the posterior margin of the limb, called the ZPA (zone of polarizing activity), discovered by Saunders and Gasseling (1968). The basic observation is that when this region is grafted to an anterior position in the early limb bud, the limb is duplicated in a mirror image along the anteroposterior axis. This region may set up a gradient of positional information, possibly in the form of a diffusible morphogen, which enables cells to know their position along the anteroposterior axis. This may be investigated by looking at the specification of digits in the chick limb that are clearly distinguishable one from another: the anteroposterior sequence is digits *2*, *3*, and *4*. The hypothesis is that the digits are specified with respect to their distance from ZPA: when they are close to the ZPA, they will form digit *4*, further away, digit *3*, and still further away digit *2* (Fig. 2). We investigated this hypothesis by grafting an additional ZPA to different positions along the

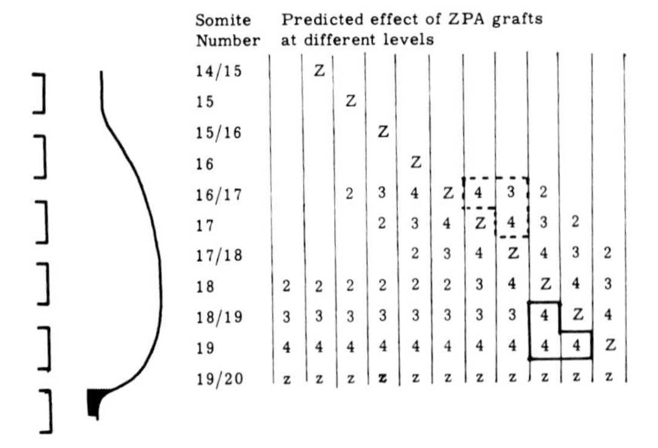

FIGURE 2 Scheme to show the expected pattern of digits when an additional 2PA (Z) is grafted to the early limb bud at different positions along the anteroposterior axis. Each column shows the graft at a different position and the expected result in terms of the digits is specified by their distance from the ZPA. The host ZPA (*z*) is always present and its position in the early bud is shown by the shaded region on the left. On the whole the predicted results were obtained, but the dotted and solid lines show where some deviation was observed (Tickle et al., 1975).

anteroposterior axis and, on the whole, our predictions have been confirmed (Tickle et al., 1975). These results can be interpreted in terms of a gradient in a diffusible morphogen arising from the ZPA—perhaps in a manner similar to the inhibitory signal in hydra—and from the cells responding to various thresholds in the concentration of this morphogen (Lewis et al., 1977). It must however, be emphasized that we do not really know that it is a diffusible signal, but merely that our results are consistent with it. Two other observations also are consistent with it. Whenever there is reason to believe that the signal is attenuated, digit *2* forms, and this is consistent with the idea that we are dealing with a diffusible gradient, because, of course, the first threshold that the cells will read is that for digit *2*. Second, grafting a second ZPA to the site of the existing ZPA has no effect: this is to be expected if the morphogen is held at constant concentration at the source.

Further evidence relates to the possibility of the response being the important feature, rather than the signal. We have grafted the region of the mouse embryonic limb, which we thought corresponded to the ZPA, into the chick limb bud, and have found that it induced chick digits (Tickle et al., 1975). This seems to be a good basis for suggesting that the signal is the same in all vertebrate limbs and what differs in different animals is the response.

ACKNOWLEDGMENT

This work is supported by the Medical Research Council.

REFERENCES

ABERCROMBIE, M. 1967. Contact inhibition: the phenomenon and its biological implications. *Natl. Cancer Inst. Monogr.* **26:**249–277.

DUNN, G. A. 1971. Mutual contact inhibition of chick sensory nerve fibers *in vitro*. *J. Comp. Neurol.* **14:**491–508.

GAZE, R. M., and A. R. HOPE. 1976. The formation of continuously ordered mappings. *Prog. Brain Res.* **45:**327–357.

GINGELL, D. 1971. Cell Membrane Surface Potential as a Transducer. *In* Membranes and Ion Transport. Vol. 3. E. E. Bittar, editor. Wiley-Interscience, London. 317–357.

GINGELL, D., and J. A. FORNÉS. 1976. Interaction of red blood cells with a polarized electrode: evidence of long-range intermolecular forces. *Biophys. J.* **16:**1131–1153.

HICKLIN, J., A. HORNBRUCH, L. WOLPERT, and M. R. B. CLARKE. 1973. Positional information and pattern regulation in hydra: formation of boundary regions following axial grafts. *J. Embryol. Exp. Morphol.* **30:**727–740.

LEWIS, J., J. M. W. SLACK, and L. WOLPERT. 1977. Thresholds in development. *J. Theor. Biol.* In press.

MARCHASE, R. M., R. J. BARBERA, and S. ROTH. 1975. A molecular approach to retinotectal specificity. *Ciba Found. Symp.* **29:**315–326.

MARTZ, E., and M. S. STEINBERG. 1973. Contact inhibition of what? An analytical review. *J. Cell Physiol.* **81:**25–38.

MOSCONA, A. A. 1974. Surface specification on embryonic cells: lectin receptors, cell recognition, and specific cell ligands. *In* The Cell Surface in Development. A. A. Moscona, editor. John Wiley & Sons, Inc., New York. 67–99.

PARSEGIAN, V. A., and D. GINGELL. 1973. A physical force model of biological membrane interaction. *In* Recent Advances in Adhesion. L. Lee, editor. Gordon & Beech, London. 153–190.

SAUNDERS, J. W., and M. T. GASSELING. 1968. Ectomesodermal—mesenchymal interacts in the origin of limb symmetry. *In* Epithelial Mesenchymal Interactions. R. Fleischmajer and R. E. Billingham, editors. The Williams & Wilkins Company, Baltimore, Md. 78–97.

SENGEL, P. 1976. Morphogenesis of Skin. Cambridge University Press, London, England. 277.

TICKLE, C., D. SUMMERBELL, and L. WOLPERT. 1975. Positional signalling and specification of digits in chick limb morphogenesis. *Nature* (*Lond.*). **254:**199–202.

WILBY, O. K., and G. WEBSTER. 1970. Experimental studies on axial polarity in hydra. *J. Embryol. Exp. Morphol.* **24:**595–613.

Wolpert, L. 1969. Positional information and the spatial pattern of cellular differentiation. *J. Theor. Biol.* **25:**1–47.

Wolpert, L., M. R. B. Clarke, and A. Hornbruch. 1972. Positional signalling along hydra. *Nature* (*Lond.*). **239:**101–105.

Wolpert, L., and D. Gingell. 1969. The Cell Membrane and Contact Control. *In* Ciba Foundation Symposium on Homeostatic Regulator. C. E. W. Wolstenholme and J. Knight, editors. Churchill Ltd., London. 241–259.

Wolpert, L., and J. H. Lewis. 1975. Towards a theory of development. *Fed. Proc.* **34:**14–20.

Wolpert, L., J. H. Lewis, and D. Summerbell. 1975. Morphogenesis of the vertebrate limb. *Ciba Found. Symp.* **29:**99–119.

MEMBRANE SITES IMPLICATED IN CELL ADHESION: THEIR DEVELOPMENTAL CONTROL IN DICTYOSTELIUM DISCOIDEUM

G. GERISCH

Dictyostelium *or the Easy Way to Study Cell Adhesion*

To a certain extent, cell aggregation in the cellular slime mold *Dictyostelium discoideum* can be taken as a model for tissue reconstruction from dissociated cells. Advantages of *D. discoideum* are the growth of the organism in the form of single cells, which renders trypsinization unnecessary; the sharp separation in time of growth and cell differentiation; the possibility of cultivating the cells up to the aggregation phase in a uniform suspension, in which they develop morphogenetic capabilities; and the availability of mutants, which grow as single cells but are blocked at different steps of differentiation. Sorting out of aggregating cells from interspecies mixtures indicates specificity of either one or both aggregation mechanisms: chemotaxis and cell adhesion (Konijn, 1972; Raper and Thom, 1941; Bonner and Adams, 1958). One of the signals that controls cell differentiation from the growth phase to the aggregation-competent state has been identified as cyclic AMP, which, in order to be effective, has to be administered in the form of reiterated pulses (Gerisch et al., 1975; Darmon et al., 1975).

Candidates for Cell Adhesion Sites

The search for membrane sites specifically involved in cell adhesion and recognition can be based on the following criteria: (1) cell surface location of the sites and their extension into the intercellular space; (2) species specificity; (3) developmental regulation in cells differentiating from the growth-phase stage to aggregation competence; (4) absence in nonadhesive mutants; (5) binding to the cell surface; (6) either promotion or inhibition of cell adhesion by the addition of a solubilized factor to aggregation-competent cells; (7) blockage of cell adhesion by univalent antibody fragments binding to specific cell-surface sites.

On the basis of various of these criteria, several cell-surface sites have been suggested as participating in cell adhesion: (1) target sites for aggregation-blocking, univalent, antibody fragments (Fab), (two types of such sites, contact sites A and B, have been distinguished, one of which is developmentally regulated [Beug et al., 1973*a*]); (2) a carbohydrate-binding protein, discoidin I, first identified as a lectin that agglutinates sheep erythrocytes (Rosen et al., 1973); (3) discoidin receptors, which are believed to have carbohydrate moieties complementary to the carbohydrate-recognition site of discoidin (Reitherman et al., 1975; Siu et al., 1976); (4) a concanavalin A-binding glycoprotein of 150-kilodalton molecular weight (Geltosky et al., 1976).

Some Critical Remarks on Criteria for Adhesion Sites

The developmental regulation of various cell-surface constituents is listed in Table I, other characteristics of these structures in Table II. The increase in either quantity or activity from the growth phase to the aggregation stage is neither a sufficient nor a necessary criterion for surface sites to participate in cell adhesion. Two sites involved in cyclic-AMP recognition and regulation show a similar behavior: cyclic-AMP receptors and cyclic-AMP phosphodiesterase. If, on the other hand, cell adhesion is mediated by a multiple component

G. GERISCH Biozentrum der Universität Basel, Basel, Switzerland

TABLE I
Developmental Regulation of Cell Surface Proteins and Related Components

Factor	Reference	Developmental stages: Growth phase	Aggregation competence	Aggregated cells
Factors implicated in cell adhesion				
Contact sites A	Beug et al., 1973*a*	Almost absent	High	Probably high
Glycoprotein antigen II	Wilhelms et al., 1974	Low	High	?
Glycoprotein 150	Geltosky et al., 1976	Low	?	High
Protein band no. 4	Smart and Hynes, 1974	Low	Low	High
Discoidin I				
soluble	Rosen et al., 1973	Low in strain NC-4	High	High
	Malchow, pers. comm.	High in strains Ax-2 and v-12		
surface bound	Frazier, 1976	Low in strain NC-4	High	High
	Siu et al., 1976	Always present in axenically grown cells of strain Ax-3		
"Discoidin receptor"	Reitherman et al., 1975	Low affinity	High affinity	?
"Discoidin receptor"	Siu et al., 1976	Low	High	?
Other cell surface sites				
Cyclic-AMP receptors	Gerisch and Malchow, 1976	Low	High	?
Cyclic-AMP phosphodiesterase	Malchow and Gerisch, 1972	Low	High	?

system, certain constituents may be present all the time, whereas others, while completing the system, render the cells able to aggregate.

Inasmuch as *D. discoideum* cells are able to differentiate into aggregation-competent cells in strongly agitated suspensions, they can be kept single up to this stage. Transfer onto a supporting surface induces these cells to aggregate instantaneously (Gerisch, 1968). This makes a clear distinction possible between sites that are formed in response to cell contact and those present in cells which are still single but are ready to aggregate. Smart and Hynes (1974) have shown that their band no. 4 protein is of the first type. The exact developmental regulation of glycoprotein 150 remains to be clarified. This glycoprotein increases strongly during the aggregation process (Geltosky et al., 1976) and might therefore be a product, rather than a prerequisite, of aggregation.

Developmental regulation of a membrane site loses its value as a criterion if it differs from strain to strain. Discoidin is strongly regulated in one strain, but in others it is present in considerable quantities in growth-phase cells, which nevertheless do not aggregate (Table I).

The absence of certain cell-surface constituents from nonaggregating mutants is another criterion of limited value because of the pleiotropism of most of the mutants; those blocked at early steps of differentiation miss all the surface sites that would appear after the block (Gerisch et al., 1974).

Is Cell-to-Cell Adhesion the Result of Carbohydrate-Protein Interaction?

The possibility that glycoproteins mediate cell adhesion by protein-carbohydrate interaction between contiguous cells has focused attention on lectin-binding proteins. The chemistry of one developmentally regulated concanavalin A binding protein of *D. discoideum* is partially known (Wilhelms et al., 1974). However, the cell-surface location of this glycoprotein has not been established, and attempts to demonstrate its participation in cell adhesion have been unsuccessful: Fab directed against its carbohydrate moiety did not block cell adhesion. Contact sites A are the only concanavalin A-binding glycoproteins of the cell surface for which evidence for a function in cell adhesion has yet been obtained (Huesgen, 1975).

Table II

Properties of Cell Surface Proteins and Related Components

Factor	Major references	Cell surface	Cyto-sol	Species specificity	Defec-tive in mutants	mol wt (1,000)	Con A bind-ing	Blockage by Fab
Factors implicated in cell adhesion								
Contact sites A	Beug et al., 1973*a, b*; Huesgen and Gerisch, 1975	<3 × 10^5/cell	–	+*	+	120–130	+	+
Glycoprotein antigen II	Wilhelms et al., 1974	?	?	+*	+	?	+	–(Anticarbohydrate Fab)
Glycoprotein 150	Geltosky et al., 1976	+	?	?	+	150	+	?
Protein band no. 4	Smart and Hynes, 1974	+	?	?	?	130	?	?
Discoidin I	Frazier et al., 1976	+	+	+	+	4 × 26	–	?
"Discoidin receptor"	Reitherman et al., 1975	5 × 10^5/cell	?	affinity K = 10^9 vs. 10^8	?	?	?	?
"Discoidin receptor"	Siu et al., 1976	+	?	?	+	56	?	?
Other developmentally regulated cell surface sites								
Cyclic-AMP receptors	Gerisch and Malchow, 1976	10^5–10^6/cell	?	+	+	?	?	?
Cyclic-AMP phosphodiesterase	Malchow and Gerisch, 1972	+	+	?	+	?	+	?

* The species specificity indicated is immunological specificity.

Cyclic-AMP phosphodiesterase, for which a role in cell adhesion seems unlikely, indicates that developmental regulation and concanavalin A binding together do not identify a protein as an adhesion site.

The interesting finding that *D. discoideum* produces a lectin that binds preferentially *N*-acetylgalactosamine, and the discovery of lectins of different specificities in other cellular slime molds, has prompted a series of investigations on the possible function of these carbohydrate-binding proteins in cell-to-cell adhesion (Frazier, 1976). Behind these studies is the idea that the lectin acts as a multivalent ligand that connects carbohydrate residues of adjacent cell surfaces, similar to aggregation factors in sponges (Humphreys, 1963; Weinbaum and Burger, 1973; Moscona, 1974).

Binding of labeled discoidin to glutaraldehyde-fixed cells has been employed for the demonstration of discoidin receptors.Discoidin binds with high affinity to *D. discoideum* cells, but simultaneously strong cross-reactivity of the lectins of different species with heterologous cell surfaces has been obtained. For the *D. discoideum* × *Polysphondylium pallidum* pair association, constants of 1 to 5 × 10^8 have been found for the binding of lectins to heterologous receptors, compared with 1 to 4 × 10^9 in the homologous combinations (Reitherman et al., 1975). It is hard to imagine, in the light of the high cross-species affinity, how the cells are able to sort out on the basis of the lectin system.

Another possibility to identify discoidin receptors is the search for proteins that coprecipitate with antidiscoidin antibodies (Siu et al., 1976). A 56-kilodalton cell-surface protein did coprecipitate, indicating an affinity of discoidin to this 56-kilodalton moiety, provided that the antibody itself did not cross-react with the latter.

In spite of the accumulated evidence in favor of a function of discoidin and the lectins of related species in cell adhesion, several points remain to be clarified. Discoidin is also a cytoplasmatic protein, making up 1% of the total soluble proteins in aggregating cells (Siu et al., 1976). What is its function in the cytoplasm: is it a precursor accumulated for subsequent incorporation into the cell membrane? After the aggregation stage, a galactosamine containing mucopolysaccharide is synthesized and finally surrounds the spore mass of the fruiting body (White and Sussman, 1963). Does discoidin interact with this material, attaching it to the cells?

Two pieces of direct evidence for lectin-mediated cell adhesion have been obtained in cellular slime molds. Glutaraldehyde-fixed *D. discoideum* cells are agglutinated by discoidin. Spontaneous agglutination of *P. pallidum* cells is inhibited by Fab directed against their endogenous lectin, pallidin (Rosen et al., 1976). In both cases, impairment of the natural adhesion of the cells seems to be required for a detectable effect. The *P. palli-*

dum cells were incubated in glucose of high ionic strength, which by itself inhibits cell adhesion. What, then, is the reason for the weak aggregation-blocking power of antilectin Fab, in contrast to Fab against other cell-surface sites which completely blocks adhesion of cells aggregating under optimal conditions?

Fab that Blocks Cell Adhesion: What Are the Target Sites?

Fab directed against specific cell-surface sites inhibits cell adhesion, whereas Fab binding to other surface constituents does not. Fab directed against the membrane fraction of aggregation-competent cells completely blocks adhesion. After absorption with either growth-phase cells or their membranes, the blockage is no longer complete, but can be restituted by supplementation with antigrowth-phase Fab. Antigrowth-phase Fab alone blocks only the adhesion of growth-phase cells completely. EDTA has the same effect and, in fact, can replace antigrowth-phase Fab (Beug et al., 1973*a*). These results demonstrate the presence of two classes of contact sites on the surface of aggregation-competent cells: contact sites A are characteristic for the aggregation phase, as contact sites B are already present in growth-phase cells (Fig. 1). Each site functions independently.

Aggregating cells assemble into streams; typically they are elongated and adhere to each other preferentially at their ends, but also side-by-side (Fig. 1). This pattern of cell assembly changes into pure end-to-end association upon blockage of contact sites B by either Fab or EDTA. Blockage of contact sites A by Fab still allows the cells to form loose, irregular assemblies, where they adhere to each other, often side-by-side. Activity of contact sites B in growth-phase cells is indicated by their EDTA-sensitive agglutination, which is also present in nondifferentiating mutants (Gerisch et al., 1974).

Purification of contact sites can be based on the neutralization of the aggregation-blocking activity of Fab. Contact sites A constitute a small fraction of the membrane proteins of aggregation-competent cells. After solubilization with deoxycholate, they appear in a single peak on Sephadex G 200 in the 130-kilodalton range (Huesgen and Gerisch, 1975). Periodate and pronase sensitivity is in accord with the glycoprotein nature of contact sites A. As shown in Table III, contact sites A are not identical with discoidin.

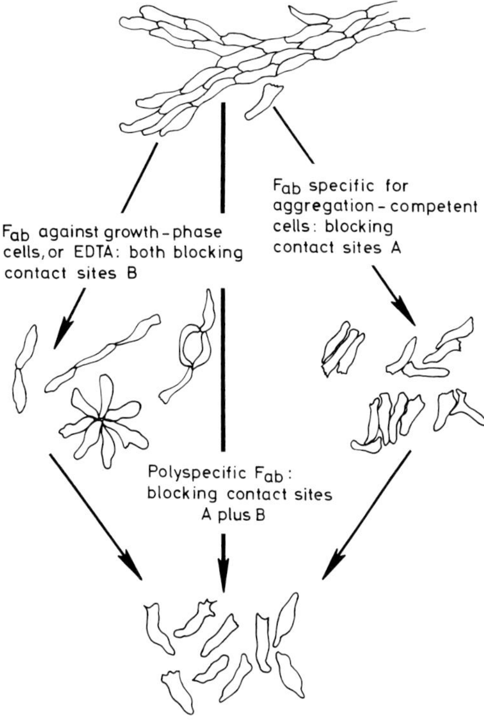

FIGURE 1 Selective blockage of either contact sites A or B, and complete inhibition of cell adhesion by blocking both. In the absence of a blocking agent, the cells associate into streams which move towards aggregation centers (top). Blockage of the B sites by either Fab or EDTA still allows the cells to assemble into chains or rosettes (left). After blockage of the A sites, only loose assemblage into irregular groups remains possible (right). Inhibition of both A and B sites leads to dissociation into single cells (bottom) which still move actively and are capable of chemotactic orientation (from Gerisch et al., 1974).

The Fab-blocking test for the identification of contact sites does not enable one to specify their function in cell adhesion. One possibility is, of course, that contact sites are the molecules which actually form the bridges between adhering cells. Contact sites A would be able to form transmembrane homodimers by protein-carbohydrate or protein-protein interaction. However, the possibility remains that contact sites are regulatory sites for cell adhesion. The molecules directly involved in adhesion might be different, and perhaps not amenable to blockage by Fab, because of their high affinity to each other.

Quantitative and Topological Aspects

Fab directed against the carbohydrate moieties

TABLE III

Differences between Contact Sites A and Discoidin I

Contact sites A	Discoidin
No agglutination of formalinized sheep erythrocytes	Agglutination of these erythrocytes
Not detectable in the soluble cytoplasmic fraction (100,000 *g* supernate)	1% of the soluble cytoplasmic protein in aggregating cells is discoidin
Absence from growth-phase cells of strains Ax-2 and v-12 (membrane fractions)	present in growth-phase cells of these strains; in aggregation-competent cells only about twofold higher titers
A concanavalin A binding glycoprotein of about 130 kilodaltons	A tetrameric protein of 25 kilodaltons subunit size

of certain cell-surface antigens does not influence cell aggregation, even if 2×10^6 Fab molecules are bound per living cell (Beug et al., 1973*b*). Fluorescent-labeled Fab of this specificity is detectable within the space between contiguous cell surfaces, indicating that molecules of $60 \times 35 \times 35$ Å size can fit between adjacent membranes without impairing adhesion. Ferritin-labeling of antibodies of the same specificity reveals a uniform distribution of the corresponding antigens over the whole cell surface (Fig. 2). Molecules that form bridges between the cell membranes must be intercalated between these antigens and, consequently, cannot be associated into larger patches.

Contact sites A are completely inactivated by binding of not more than 3×10^5 Fab molecules per cell (Beug et al, 1973*b*). These antibody fragments cover less than 2% of the surface area of the cells, establishing that cell adhesion can be traced to specific loci on the cell surface.

The end-to-end assembly of the cells indicates spatial heterogeneity of contact site A activities along the surface of a cell. Preliminary evidence has been obtained that contact sites A are nevertheless homogeneously distributed (Gerisch et al., 1974). Antibody labeling of those cell-surface antigens that are specific for aggregation-competent cells did not reveal differences between different areas of a cell surface. Although contact sites A are certainly not the only antigens labeled under these conditions, they are believed to constitute a considerable fraction of the labeled antigens. A cell can form a new tip within a few seconds, simultaneously losing its adhesiveness at the original ends. These results suggest that the activity of contact sites A is spatially controlled, depending on their position relative to the actual ends of a cell.

Do the molecules, which represent the active spots for cell adhesion, extend far enough into the intercellular space to bridge the distance of about 150 Å between adjacent cell membranes? Figure

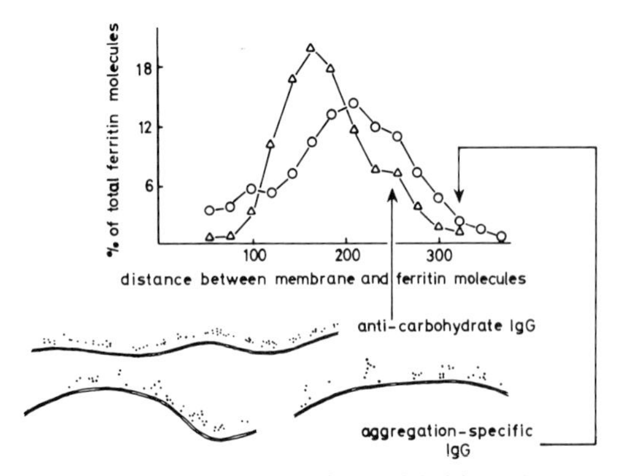

FIGURE 2 Distances of the ferritin label from the outer electron-dense layer of the plasma membrane. For labeling, either anticarbohydrate IgG (△) or IgG against aggregation-specific antigens (○) was used as the first layer, and in both cases the same ferritin-conjugated antirabbit IgG goat IgG as the second layer. The maximal possible distance contributed by the IgG is about 2×120 Å. Examples for the pattern of labeling corresponding to the histograms on top are given for both the anticarbohydrate IgG (middle) and the IgG directed against those antigens which are present on aggregation-competent cells and absent from growth-phase cells (bottom) (from Schwarz, 1973; Gerisch et al., 1974).

2 shows carbohydrate residues that form an array of uniform distances from the cell surface. These are the target antigens of nonblocking Fab. In contrast, contact sites A, together with other antigens which are specific for aggregation-competent cells, form a more heterogeneous pattern perpendicular to the membrane (Fig. 2, bottom). On the histogram of distances between membrane and ferritin, antigenic sites are seen that extend 30–40 Å beyond the carbohydrate layer (Fig. 2, top). With an estimated thickness of about 30 Å for this layer, the total length would be sufficient for a molecule to reach the middle line between adjacent membranes, and to interact with molecules that extend a similar distance beyond the opposite cell surface.

Cyclic-AMP Receptors and the Developmental Control of Contact Sites A

In contrast to contact sites B, which remain almost constant during cell differentiation from the growth-phase stage to aggregation competence, contact sites A undergo a drastic change from nondetectability to a maximal number in aggregation-competent cells (Fig. 3). This change parallels the ability of the cells to form EDTA-stable contacts, indicating that, during normal development, the appearance of contact sites A on the cell membrane is immediately followed by the detectability of their function.

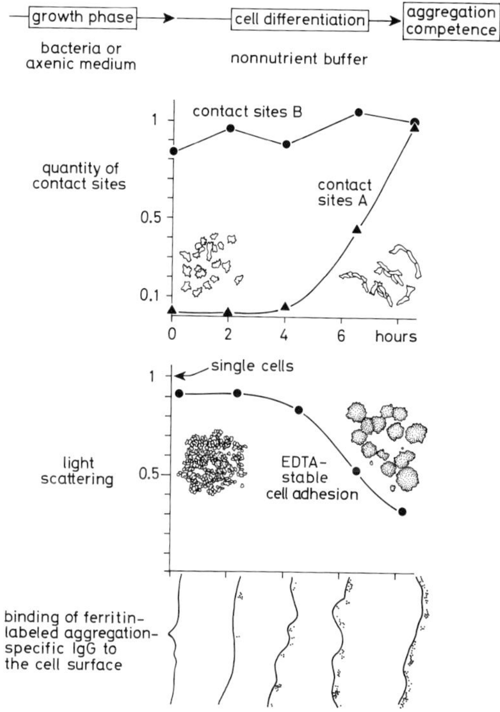

FIGURE 3 Developmental regulation of contact sites A and EDTA-stable cell adhesion. In the absence of nutrients, the cells differentiate within a period of about 8 hr into aggregation-competent cells which readily associate into chains and streams. Immunological evidence indicates that contact sites A are strongly regulated during this period in contrast to contact sites B (top). Simultaneously, cells develop the ability to adhere to each other in the presence of EDTA, as evidenced by their agglutination into large groups. The agglutination can be quantitated by measuring light scattering in cell suspensions which, in the presence of EDTA, stays near to the single-cell value in growth-phase cells, and turns into lower values as characteristic for agglutinates in aggregation-competent cells (middle). During the differentiation period, new cell surface antigens become successively detectable (bottom) (from Beug et al., 1973*a*; Schwarz, 1973).

Their clear-cut regulation, together with the availability of convenient assays, predestines contact sites A as markers for studies on the control of membrane differentiation. One control mechanism is known: cyclic-AMP pulses strongly accelerate the formation of contact sites A in the wild type (Gerisch et al., 1975), and induce contact site-A activity in certain types of nonaggregating mutants (Darmon et al., 1975). During the preaggregation stage, wild-type cells produce cyclic-AMP pulses rhythmically every 3–9 min (Gerisch and Malchow, 1976). Contact sites A become detectable after about 20 pulses.

The cells recognize the signals by means of cyclic-AMP receptors at their surface. A fast increase of the extracellular cyclic-AMP concentration results in a transient activation of adenylate cyclase and, consequently, in the amplification of the signals (Roos and Gerisch, 1976). The temporal behavior of the response system makes cell differentiation in this case dependent on the temporal pattern of the signals. Folic acid has a similar effect on the expression of contact sites A, and again the administration of pulses is crucial (Wurster and Schubiger, 1977, Oscillations and cell development in *Dictyostelium discoideum* stimulated by folic acid pulses, submitted for publication).

The continuation of the signal-processing pathway presumably involves protein phosphorylation by cyclic-nucleotide-dependent proteinkinases and regulation of gene transcription. The elucidation of this part of the control pathway for contact sites A is an intriguing problem for future research.

ACKNOWLEDGMENTS

Our work was supported by the Stiftung Volkswagenwerk, the Deutsche Forschungsgemeinschaft, and the Schweizerischer Nationalfonds.

REFERENCES

BEUG, H., F. E. KATZ, and G. GERISCH. 1973*a*. Dynamics of antigenic membrane sites relating to cell aggregation in *Dictyostelium discoideum*. *J. Cell Biol.* **56:**647–658.

BEUG, H., F. E. KATZ, A. STEIN, and G. GERISCH. 1973*b*. Quantitation of membrane sites in aggregating *Dictyostelium* cells by use of tritiated univalent anti-

body. *Proc. Natl. Acad. Sci. U. S. A.* **70:**3150–3154.
BONNER, J. T., and M. S. ADAMS. 1958. Cell mixtures of different species and strains of cellular slime molds. *J. Embryol. Exp. Morphol.* **6:**346–356.
DARMON, M., P. BRACHET, and L. H. PEREIRA DA SILVA. 1975. Chemotactic signals induce cell differentiation in *Dictyostelium discoideum. Proc. Natl. Acad. Sci. U. S. A.* **72:**3163–3166.
FRAZIER, W. A. 1976. The role of cell surface components in the morphogenesis of the cellular slime molds. *Trends in Biochem. Sci.* **1:**130–133.
GELTOSKY, J. E., C.-H. SIU, and R. A. LERNER. 1976. Glycoproteins of the plasma membrane of *Dictyostelium discoideum* during development. *Cell.* **8:**391–396.
GERISCH, G. 1968. Cell aggregation and differentiation in *Dictyostelium. Curr. Top. Dev. Biol.* **3:**157–197.
GERISCH, G., H. BEUG, D. MALCHOW, H. SCHWARZ, and A. STEIN. 1974. Receptors for intercellular signals in aggregating cells of the slime mold, *Dictyostelium discoideum. In* Biology and Chemistry of Eucaryotic Cell Surfaces. Miami Winter Symposia. Vol. VII. Academic Press Inc., New York. 49–66.
GERISCH, G., H. FROMM, A. HUESGEN, and U. WICK. 1975. Control of cell-contact sites by cyclic AMP pulses in differentiating *Dictyostelium* cells. *Nature (Lond.).* **255:**547–549.
GERISCH, G., and D. MALCHOW. 1976. Cyclic AMP receptors and the control of cell aggregation in *Dictyostelium. Adv. Cyclic Nucleotide Res.* **7:**49–68.
HUESGEN, A. 1975. Biochemische Untersuchungen deaggregations-spezifischen Zellkontakts bei *Dictyostelium discoideum.* Ph.D. Thesis, University of Tübingen.
HUESGEN, A., and G. GERISCH. 1975. Solubilized contact sites A from cell membranes of *Dictyostelium discoideum. FEBS (Fed. Eur. Biochem. Soc.) Lett.* **56:**46–49.
HUMPHREYS, T. 1963. Chemical dissolution and in vitro reconstruction of sponge cell adhesions. I. Isolation and functional demonstration of the components involved. *Dev. Biol.* **8:**27–47.
KONIJN, T. M. 1972. Cyclic AMP as a first messenger. *Adv. Cyclic Nucleotide Res.* **1:**17–31.
MALCHOW, D., and G. GERISCH. 1972. Membrane-bound cyclic AMP phosphodiesterase in chemotactically responding cells of *Dictyostelium discoideum. Eur. J. Biochem.* **28:**136–142.
MOSCONA, A. A. 1974. Surface specification of embryonic cells: lectin receptors, cell recognition, and specific cell ligands. *In* The Cell Surface in Development. A. A. Moscona, editor. John Wiley & Sons, Inc., New York. 67–99.
RAPER, K. B., and C. THOM. 1941. Interspecific mixtures in the *Dictyosteliaceae. Am. J. Bot.* **28:**69–78.
REITHERMAN, R. W., S. D. ROSEN, W. A. FRAZIER, S. H. BARONDES. 1975. Cell surface species-specific high affinity receptors for discoidin: developmental regulation in *Dictyostelium discoideum. Proc. Natl. Acad. Sci. U. S. A.* **72:**3541–3545.
ROOS, W., and G. GERISCH. 1976. Receptor-mediated adenylate-cyclase activation in *Dictyostelium discoideum. FEBS (Fed. Eur. Biochem. Soc.) Lett.* **68:** 170–172.
ROSEN, S. D., P. L. HAYWOOD, and S. H. BARONDES. 1976. Inhibition of intercellular adhesion in a cellular slime mould by univalent antibody against a cell-surface lectin. *Nature (Lond).* **263:**425–427.
ROSEN, S. D., J. A. KAFKA, D. L. SIMPSON, and S. H. BARONDES. 1973. Developmentally regulated, carbohydrate-binding protein in *Dictyostelium discoideum. Proc. Natl. Acad. Sci. U. S. A.* **70:**2554–2557.
SCHWARZ, H. 1973. Immunelektronenmikroskopische Untersuchungen über Zelloberflächenstrukturen aggregierender Amöben von *Dictyostelium discoideum.* Ph.D. Thesis, University of Tübingen.
SIU, C.-H., R. A. LERNER, G. MA, R. A. FIRTEL, and W. F. LOOMIS. 1976. Developmentally regulated proteins of the plasma membrane of *Dictyostelium discoideum*: the carbohydrate-binding protein. *J. Mol. Biol.* **100:**157–178.
SMART, J. E., and R. O. HYNES. 1974. Developmentally regulated cell surface alterations in *Dictyostelium discoideum. Nature (Lond.).* **251:**319–321.
WEINBAUM, G., and M. M. BURGER. 1973. Two component system for surface guided reassociation of animal cells. *Nature (Lond.).* **244:**510–512.
WHITE, G. J., and M. SUSSMAN. 1963. Polysaccharides involved in slime-mold development. II. Water-soluble acid mucopolysaccharide(s). *Biochim. Biophys. Acta.* **74:**179–187.
WILHELMS, O.-H., O. LÜDERITZ, O. WESTPHAL, and G. GERISCH. 1974. Glycosphingolipids and glycoproteins in the wild-type and in a non-aggregating mutant of *Dictyostelium discoideum. Eur. J. Biochem.* **48:**89–101.

DIRECT COMMUNICATION BETWEEN ANIMAL CELLS

JOHN D. PITTS

The concept of the cell as a unit of all living things has provided a basis for much of our understanding of biological systems. It highlights a fundamental similarity between all living organisms from the simple unicellular prokaryotes to the complex multicellular eukaryotes. However, the simplicity and universality of the theory have tended to obscure one of the basic differences between unicellular and multicellular organisms. The individuality and independence of a bacterial cell, for example, make it strikingly different from a differentiated cell in an animal tissue. Multicellular organisms are populations of interacting and interdependent cells with many properties that cannot be ascribed to individual cells. Coordination of cellular activity and cellular proliferation through cell-cell interactions is necessary for the propagation and sharing of growth signals and for the organization of levels of cellular (or tissue) activity that are intercompatible.

Such cell interactions can be indirect, between cells in different parts of the tissue or organism, or direct, between cells in contact. Hormonal control is an example of indirect cell interactions, and intercellular junction formation is an example of direct interactions.

Intercellular junctions are a characteristic of multicellular organization. Three general types are found in higher animals, each having a specific function. Desmosomes (Kelly, 1966; Skerrow and Matoltsy, 1974) form firm points of attachment between epithelial cells. They also provide anchor sites for cytoplasmic filaments and thus form part of a cytoskeletal system which gives mechanical strength to a tissue. Tight junctions (Farquhar and Palade, 1963; Hudspeth, 1975), or septate junctions in invertebrates (Staehelin, 1974), form seals between epithelial cells lining cavity organs and thus prevent luminal contents from penetrating the interstitial spaces.

The third type of junction is permeable to ions and small molecules, and so allows the exchange of small molecular weight cell components between coupled cells (cells are said to be coupled if they are joined by permeable junctions). The physical basis of the permeable junction is thought to be the gap junction (see Gilula, this volume).

Electrophysiological and microinjection techniques (Loewenstein, 1966; Furshpan and Potter, 1968; and see Loewenstein, this volume) have shown that these junctions are permeable to ions and synthetic fluorescent probes with a molecular weight less than 1,000–2,000. Much of this work has been done with the large cells of the salivary glands of *Drosophila* and *Chironomus*. However, more limited studies with vertebrate cells suggest that the permeability properties of invertebrate and vertebrate junctions are similar.

This paper describes other approaches with biochemical techniques which have shown that the junctions between vertebrate cells in tissue culture are permeable to many (probably all) small molecular weight cellular metabolites but are impermeable to cellular macromolecules.

The Permeability of Intercellular Junctions

The first indications of junctional communication by these methods came from a chance observation of metabolic cooperation (Subak-Sharpe et al., 1966, 1969) between mutant and wild-type strains of the hamster fibroblast cell line BHK21/13 (BHK cells; Macpherson and Stoker, 1962). Cells lacking the enzyme hypoxanthine:guanine phosphoribosyltransferase (HGPRT; E.C. 2.4.2.8.) are unable to convert the base hypoxanthine to the nucleotide inosine monophosphate (IMP) and, accordingly (unlike the wild-type cells), do not incorporate exogenous [^{3}H]-

JOHN D. PITTS Department of Biochemistry, University of Glasgow, Glasgow, Scotland

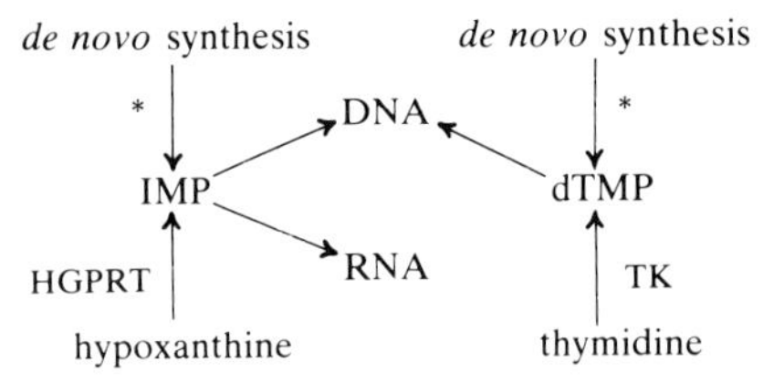

FIGURE 1 Pathways of nucleotide metabolism. HGPRT, hypoxanthine:guanine phosphoribosyltransferase; TK, thymidine kinase; asterisks mark pathways blocked by aminopterin.

hypoxanthine into their nucleic acid (see Fig. 1). However, when mutant and wild-type cells are grown in mixed cultures in the presence of [^{3}H]hypoxanthine, autoradiographic analysis shows that mutant cells are labeled if they are in contact (either directly or through other mutant cells) with wild-type cells. This phenotypic modification of mutant cells in contact with wild-type cells is so extensive that in confluent 1:1 mixed cultures, the two cell types are equally labeled and cannot be distinguished (Bürk et al., 1968; Pitts, 1971). However, if the experiment is repeated with wild-type and mutant L cells (a mouse cell line; Sanford et al., 1948), half the cells are labeled (phenotypically wild-type) and half are unlabeled (phenotypically mutant). Metabolic cooperation is a property of BHK cells but not of L cells (Pitts, 1971).

The phenotypically modified mutant BHK cells have not gained HGPRT activity, that is, neither the enzyme nor information to make the enzyme, is transferred from the wild-type to the mutant cells (Cox et al., 1970; Pitts, 1971), and it was concluded that the explanation of metabolic cooperation lies in the intercellular transfer of labeled purine nucleotides which are beyond the enzyme block and can be incorporated into mutant cell nucleic acid. The alternative explanation of labeled nucleic acid transfer was ruled out because metabolic cooperation also occurs between mutant BHK cells lacking the enzyme thymidine kinase (TK; E.C. 2.7.1.21) and wild-type BHK cells. Complete equilibration of the DNA between these two cell types in mixed culture (which would be required to explain metabolic cooperation in terms of nucleic acid transfer) is incompatible with the genetic stability of these cells.

Nucleotides, like most other intermediate metabolites, are effectively retained in the cell by the permeability barrier of the cytoplasmic membrane, so the equilibration of nucleotide pools during metabolic cooperation requires intercellular junctions that are permeable to nucleotides. BHK cells form these junctions and L cells do not. This has been confirmed by subsequent work (Gilula et al., 1972) which also showed the absence of electrical coupling and gap junctions between L cells but not between other fibroblasts which, like BHK cells, show metabolic cooperation (thus providing a correlation between permeable junctions and gap junctions).

Many different cell types in culture will form junctions that can be detected by metabolic cooperation and other methods (see below). Of more than 30 cell types tested, only 3 have been found not to form junctions. Of these, the L cells is the most thoroughly studied.

L cells do not form junctions with BHK cells (i.e., inability to form junctions is dominant in mixed cultures; Pitts, 1972), but BHK-L cell hybrids do form junctions (McCargow and Pitts, 1971; Pitts, 1972). These observations suggest that L cells lack an active gene product necessary for junction formation, but the nature of this product, and whether it is the result of a defect developed during the many years in culture or whether the cells were originally derived from some non-junction-forming tissue cell is not known.

A more direct demonstration of intercellular nucleotide transfer can be obtained with a different approach (Pitts, 1976; Pitts and Simms, 1977). After animal cells in culture have been labeled with [^{3}H]uridine, the cells contain labeled uridine, labeled uridine nucleotides, and labeled RNA. Washing such labeled cells with unlabeled medium removes all the [^{3}H]uridine but leaves the labeled nucleotides and labeled RNA inside the cells. If such prelabeled and washed BHK cells are cocultured with unlabeled BHK cells, subsequent autoradiography shows that labeled material rapidly spreads from the labeled donor cells to the recipient cells in contact, but not to unlabeled cells not in contact (Fig. 2*a*). As in the interpretation of metabolic cooperation, the transfer could represent either intercellular nucleotide movement or intercellular RNA movement.

In this system, nucleotide transfer and RNA transfer can be distinguished experimentally. If the labeled donor cells are cultured in unlabeled medium, the labeled nucleotide pools are chased into RNA, and after 24 h are almost completely depleted. When such chased donor cells are cocultured with unlabeled cells, there is very little transfer of labeled material to recipient cells in contact

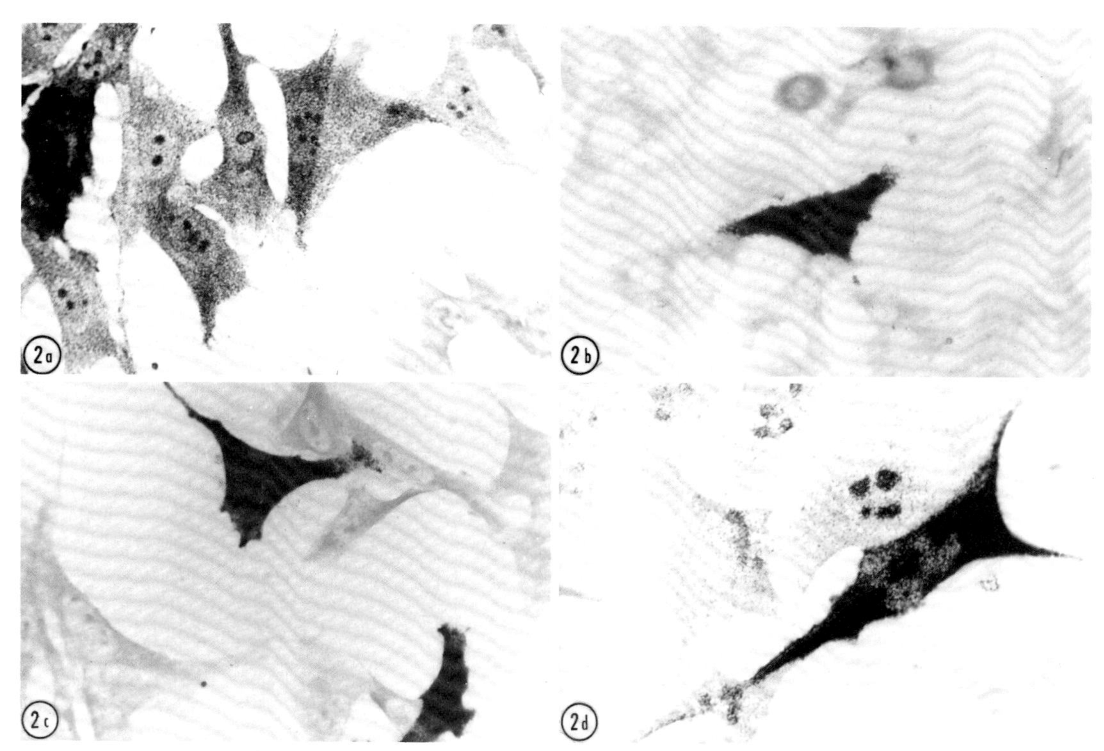

FIGURE 2 Junctional communication between cells in culture. (*a*) Transfer of uridine nucleotides between BHK cells; (*b*) no transfer of RNA between BHK cells; (*c*) no transfer of uridine nucleotides from L cells to BHK cells; (*d*) transfer of uridine nucleotides from *Xenopus* cells to BHK cells. The donor cells (black cells) in *a, c,* and *d* were prepared by labeling with [^{3}H]uridine for 3 h and then washing with unlabeled medium. Such donor cells contain labeled uridine nucleotides and labeled RNA (in about equal amounts). Unlabeled cells were added to the washed donor cells and the mixtures cultured for 3 h. The cultures were then fixed, acid washed, and processed for autoradiography. In *a* and *d* labeled nucleotides were transferred by intercellular junctions to recipient cells in contact (lightly labeled cells) but not to cells not in contact (unlabeled cells). In *c*, because L cells do not form junctions, nucleotide transfer did not occur. The "chased" donor cells in *b* were prepared by culturing the prelabled donor cells in unlabeled medium for 24 h before adding the unlabeled cells. Such "chased" donor cells contain labeled RNA but almost no labeled nucleotides. The labeled RNA is not transferred to the added cells. For further details see text and Pitts and Simms (1977).

(Fig. 2*b*). The remaining amount of transfer to primary recipient cells (i.e., recipient cells in direct contact with donor cells) can be quantitated by grain counting. Such analyses show that the amount of transfer is always directly proportional to the activity in the donor nucleotide pools and is quite unrelated to the activity of the donor RNA. This and other approaches (Pitts and Simms, 1977) show clearly that uridine nucleotides but not RNA are transferred between coupled cells. If L cells are used either as donors or recipients, uridine nucleotide transfer does not occur (Pitts and Simms, 1977, and Fig. 2*c*).

This method of examining junctional permeability is more versatile than metabolic cooperation. Mutant cells are not required and precursors other than [^{3}H]uridine can be used. A number of labeled precursors (tritiated hypoxanthine, thymidine, glucose, fucose, glucosamine, 2-deoxyglucose, choline, and amino acids) have been used to show that other nucleotides, sugar phosphates, and choline phosphate or CDP-choline are all transferred between junction-forming cells in contact (but not between L cells) and that the macromolecular products (RNA, DNA, protein, macromolecular carbohydrate) derived from these intermediate metabolites are not transferred (Pitts and Finbow, 1977; Finbow and Pitts, manuscript in preparation). Nor is phospholipid transferred, but this may be on account of its special position in the cell rather than its molecular size (Finbow and Pitts, manuscript in preparation). The absence of phospholipid exchange between cells, even those joined by junctions, suggests that the membranes of cells in contact do not form any kind of continuum, and the mixing of membrane components of different cells requires some form of cell fusion.

Other methods, designed for specific purposes and not of general application, have been used to show the transfer of proline and the vitamin-derived cofactor tetrahydrofolate between coupled cells (Finbow and Pitts, manuscript in preparation). Interestingly, the active form of tetrahydrofolate, the tetraglutamate derivative, appears not to be transferred, whereas the forms containing fewer glutamate residues are transferred. This suggests that the junctions (which have already been shown to be permeable to small metabolites but not to macromolecules) have a permeability size limit of about 960 mol wt for substituted peptides such as these. This cutoff point is similar to that suggested by Rose and Loewenstein for fluorescent-labeled synthetic peptides (see Loewenstein, this volume).

Intercellular junctions are permeable to all the small molecular weight cellular molecules so far examined. If inorganic ions and the small molecular weight synthetic tracers shown to pass through junctions by electrophysiological and microinjection techniques are added to the list, it seems reasonable to conclude that transfer mechanism is nonselective (except against large molecules). A transfer mechanism that is selective only in terms of molecular size is consistent with the concept that the 20 Å diameter aqueous channels penetrating gap junctional subunits are the pathways of intercellular communication (see Gilula, this volume).

Specificity of Junction Formation

Some other properties of intercellular junctions have been characterized in the course of this work. Junctions form rapidly (within minutes) between most cell types that grow in tissue culture. There appears to be no species specificity—cells from *Xenopus*, chick, and man will all form junctions with each other as efficiently as with themselves (Fig. 2*d*). Similarly, there are examples of cells from different tissues that will form junctions, but there are some instances of specificity. The most fully characterized of these examples is a system (Pitts and Bürk, 1976) using liver epithelial cells and fibroblasts (BHK cells). The epithelial cells form junctions rapidly and extensively among themselves and the fibroblasts do likewise. In mixed cultures, however, junction formation between an epithelial cell and a fibroblast is rare, although once junctions have been established between a heterologous cell pair, the rate of uridine nucleotide transfer appears to be similar to that through junctions between homologous cells. These two cell types tend to sort out in culture, and it has been suggested (Pitts and Bürk, 1976) that the specificity of interaction is not the result of different types of junctional proteins but is a function of the frequency with which the cytoplasmic membranes of two cells come close enough together to allow the necessary interaction and formation of the junctional channels.

The absence of specificity between most cells in culture may not be representative of specificity in vivo. Many of the cells that survive and grow in culture are fibroblasts or fibroblastlike, and many differentiated cells fail to grow in culture. It may therefore be dangerous to generalize about specificity (or lack of it) until patterns of communication have been mapped in vivo.

The Rate and Extent of Junctional Transfer

An estimate of the rate of movement of molecules through intercellular junctions can be obtained from a simple model system in culture. Aminopterin blocks the *de novo* synthesis of IMP and thymidine monophosphate (dTMP; see Fig. 1) but wild-type cells can grow in the presence of the drug if the medium is supplemented with hypoxanthine and thymidine (HAT medium). Mutant cells lacking either HGPRT or TK cannot grow in HAT medium because they are unable to make IMP and dTMP, respectively. However, they do grow when cocultured with wild-type cells if both cell types are able to form permeable junctions. Such rescue of mutant cells by wild-type cells has been called the "kiss of life" (Fujimoto et al., 1971). Furthermore, because each mutant is wild type with respect to the defect of the other, mixed cultures of HGPRT and TK-deficient BHK cells will also grow in HAT medium (Pitts, 1971). One cell type (TK-deficient) produces all the purine nucleotides for both cell types and the other (HGPRT-deficient) produces all the dTMP. The mixed cultures can grow at the wild-type rate with a division time of 14 h. Under such conditions, enough purine nucleotides must enter each HGPRT-deficient cell in 14 h by junctional transfer to make a full cellular complement of nucleic acid. This rate of movement is approximately 10^6 nucleotides/cell per second and presumably all other intermediate metabolites are passing between cells at similar rates. Junctional communication is not trivial leakage from one cell to the next, but massive exchange and equilibration of small ions and molecules between all cells in a coupled population.

In other model systems, [^{3}H]hypoxanthine incorporation can be followed at an edge between two contiguous monolayers, one of wild-type cells and the other of HGPRT-deficient cells, to show that nucleotides can be transferred through cell sheets over distances of more than 30 cell diameters (about 1 mm) in a few hours (Michalke, 1977; Pitts, unpublished data; and see Pitts, 1976). The concentrations necessary to allow autoradiographic detection in these systems may be orders of magnitude greater than those necessary to trigger some biological response by a nucleotide-sized signal molecule. This means that small signal molecules traveling through intercellular junctions could act over distances of 1 mm and perhaps considerably more.

Junctions between Cells In Vivo

All the different junction-forming cells examined in tissue culture make junctions that have similar permeability properties; several observations suggest that junctions in vivo are the same.

Tumors derived from HGPRT-deficient cells incorporate [^{3}H]hypoxanthine by metabolic cooperation with the surrounding wild-type cells of the host (Pitts, 1972). Electrophysiological and microinjection techniques applied to cells in organ culture and in vivo show that these cells behave like those in tissue culture. HGPRT deficiency in man (Lesch-Nyhan syndrome) is an X-linked recessive condition. In heterozygotes, half the cells are mutant and half are wild type as a result of random inactivation of the X chromosome, and such females are phenotypically normal (Migeon et al., 1968). This would be expected if the biochemical defect in the mutant cells was masked by metabolic cooperation (although the complete story is likely to be more complex). In allophenic mice made from early embryos of two strains with electrophoretically distinct isozymes of isocitrate dehydrogenase, the hybrid enzyme is found in muscle (where the different cell types fuse) but not in liver and other organs (Mintz and Baker, 1967). This shows that junctions are impermeable to the enzyme monomers because they do not mix in tissues (like liver) where gap junctions abound.

Finally, the morphology of gap junctions characterized by electron microscopy is the same in tissues in vivo and in culture.

Coordinate Control in Coupled Cell Populations

Different cell types in coupled populations will have different macromolecules but they will share common metabolite pools. In this way every cell benefits from, or is affected by, the metabolic activities of the other cells, and a population will have properties that cannot be specifically associated with individual cells but will be characteristic of the particular mixture of cells (whether the mixture is an artificial one in culture or is a tissue or an organism).

The sharing of intermediate metabolites can lead directly to the intercellular control of enzyme activity and cell proliferation, and such coordinate control has been studied in model systems in culture.

Wild-type cells respond to increased concentrations of exogenous hypoxanthine by an increased activity of the HGPRT pathway and a decreased rate of *de novo* IMP synthesis (see Fig. 1). Mutant cells lacking HGPRT are unaffected by exogenous hypoxanthine but in mixed culture (if both wild-type and mutant cells form intercellular junctions), the metabolism of both cell types changes. The HGPRT activity in the wild-type cells is several times higher than that in the wild-type cells cultured alone, and the *de novo* pathway is inhibited in both cell types (Sheridan et al., 1975, 1977). The possible explanations of these changes are somewhat complex and have been discussed elsewhere (Sheridan, 1977), but whatever the mechanisms, it is clear that all the enzymic pathways for making IMP in both cell types are subject to coordinate control.

This is not a surprising result because if the activities of metabolic pathways are controlled by the cellular concentrations of small molecules, it seems inevitable that these pathways will be jointly controlled in all the cells in a coupled population, both in culture and in vivo.

In experiments described earlier, two mutant BHK cell strains (one lacking HGPRT and the other TK) were shown to grow in HAT medium by mutual nucleotide exchange. A further property inherent in this system is revealed if unequal proportions (1:20 or 20:1) of the two cell types are cultured in HAT medium. Initially, there is cell death but this is followed by growth until the culture is confluent again. Examination of the final mixtures shows that there are approximately equal proportions of the two cell types. Because the cells are interdependent, the cultures tend to stabilize at a 1:1 mixture. Such metabolic interdependence, resulting in the coordinate control of cell proliferation, is a very simple way of regulating the proportions of different cells in a population. As yet there is no evidence to show whether or not such systems operate in vivo, but the difficulties experienced when attempting to grow differentiated cells in culture and their frequent requirement for other cells or feeder layers (which can form junctions) could mean that some cells are deficient in essential metabolic pathways and depend for maintenance and proliferation on metabolic cooperation with other cell types.

The Role of Junctional Communication in Development

Intercellular junctions are widespread in embryonic tissues in all but the earliest stages of development (de Laat et al., 1976; Sheridan, 1977). They could provide pathways for developmental signaling, but as yet there is little evidence to show that they do.

Junctions could provide an extra mechanism for cell-cell recognition which has previously been thought to involve only surface-surface interactions. Junction formation would allow cells to sample each other's contents before commitment to some developmentally significant interaction. It is not known if such sampling is important or significant but, if junctions are there, it must occur and it should be kept in mind as a possible addition to the repertoire of processes which might be involved in the spatial organization of cells in embryos.

Developmental signaling between adjacent cell populations is required for embryonic induction (Saxen, 1975), and there is some evidence that gap junctions may be formed between the interacting cells. In one induction system, mesoderm was prelabeled with [^{3}H]uridine, and labeled material (either uridine nucleotides or RNA) was shown to be transferred to added ectoderm (Kelley, 1968). Later studies, with different techniques, were unable to detect RNA transfer (Grainger and Wessells, 1974). The system is therefore very analogous to the studies described earlier in this paper showing junction formation and nucleotide transfer between cells in culture. However, even if cell-cell contact is required and intercellular junctions are formed, junctional communication need not be the route of inductive signaling. There may be a number of different mechanisms operating either at the same time or in different systems (see Hay, this volume) but in some systems at least, junctions offer a possible explanation.

In some organisms, development appears to proceed by a process of sequential subcompartmentation (Garcia-Bellido, 1975). All cells in the same compartment remain associated and form a specific region of the embryo and, eventually, the adult. Differences between differentiated cells within a compartment may be specified by positional information in the form of concentration gradients of diffusible morphogens (Wolpert, 1969, and this volume). There is evidence for gradient discontinuities and abrupt changes of concentration at compartment boundaries (Lawrence et al., 1972), so if morphogens are small molecules which pass through the cells in a compartment by junctional transfer, it might be expected that junctions would not be formed between cells on opposite sides of a compartment boundary. This expectation has been shown to be incorrect, as electrical coupling (Caveney, 1974), and morphologically identifiable gap junctions (Lawrence and Green, 1975) are found between cells from different compartments. Some other property must define the boundaries.

Intercellular junctions provide a communication system during development and in adult tissues which could carry signals fast enough and far enough to account for many developmental processes (see Wolpert, this volume). However, the system appears to be rather nonspecific, and the signals it carries are not informational (in the sense that nucleic acids are informational) but only trigger molecules for the activation and inhibition of preexisting responses.

REFERENCES

Bürk, R. R., J. D. Pitts, and J. H. Subak-Sharpe. 1968. Exchange between hamster cells in tissue culture. *Exp. Cell. Res.* **53**:297–301.

Caveney, S. 1974. Intercellular communication in a positional field: movement of small ions between insect epidermal cells. *Dev. Biol.* **40**:311–322.

Cox, R. P., M. R. Krauss, M. E. Balis, and J. Dancis. 1970. Evidence for transfer of enzyme product as the basis of metabolic cooperation between tissue culture fibroblasts of Lesch-Nyhan disease and normal cells. *Proc. Natl. Acad. Sci. U. S. A.* **67:**1573–1579.

FARQUHAR, M. G., and G. E. PALADE. 1963. Junctional complexes in various epithelia. *J. Cell Biol.* **17:**375–412.

FUJIMOTO, W. Y., J. H. SUBAK-SHARPE, and J. E. SEEGMILLER. 1971. *Proc. Natl. Acad. Sci. U. S. A.* **68:**1516–1519.

FURSHPAN, E. J., and D. D. POTTER. 1968. Low resistance junctions between cells in embryos and tissue culture. *Curr. Top. Dev. Biol.* **3:**95–127.

GARCIA-BELLIDO, A. 1975. Genetic control of wing disc development in *Drosophila. Ciba Found. Symp.* **29:**161–182.

GILULA, N. B., O. R. REEVES, and A. STEINBACH. 1972. Metabolic coupling, ionic coupling and cell contacts. *Nature (Lond.).* **235:**262–265.

GRAINGER, R. M., and N. K. WESSELLS. 1974. Does RNA pass from mesenchyme to epithelium during an embryonic tissue interaction? *Proc. Natl. Acad. Sci. U. S. A.* **71:**4747–4751.

HUDSPETH, A. J. 1975. Establishment of tight junctions between epithelial cells. *Proc. Natl. Acad. Sci. U. S. A.* **72:**2711–2713.

KELLEY, R. O. 1968. An electron microscopic study of chordamesoderm-neurectoderm association in gastrulae of a toad *Xenopus laevis. J. Exp. Zool.* **172:**153–180.

KELLY, D. E. 1966. Fine structure of desmosomes. *J. Cell Biol.* **28:**51–72.

LAAT, S. W. DE, P. W. J. A. BARTS, and M. I. BAKKER. 1976. New membrane formation and intercellular communication in the early *Xenopus* embryo. *J. Membr. Biol.* **27:**109–129.

LAWRENCE, P. A., F. H. C. CRICK, and M. MUNRO. 1972. A gradient of positional information in an insect, *Rhodnius, J. Cell Sci.* **11:**815–853.

LAWRENCE, P. A., and S. M. GREEN. 1975. The anatomy of a compartment border. *J. Cell Biol.* **65:**373–382.

LOEWENSTEIN, W. R. 1966. Permeability of membrane junctions. *Ann. N. Y. Acad. Sci.* **137:**441–472.

MACPHERSON, I., and M. STOKER. 1962. Polyoma transformation of hamster cell clones—an investigation of genetic factors affecting cell competence. *Virology.* **16:**147–151.

MCCARGOW, J., and J. D. PITTS. 1971. Interaction properties of cell hybrids formed by fusion of interacting and non-interacting mammalian cells. *Biochem. J.* **124:**48P.

MICHALKE, W. 1977. A gradient of diffusible substance in a monolayer of cultured cells. *J. Membr. Biol.* In press.

MIGEON, B. R., V. M. KALOUSTIAN, and W. L. NYHAN. 1968. X-linked hypoxanthine:guanine phosphoribosyl transferase heterozygote has two clonal populations. *Science (Wash. D. C.).* **160:**425–427.

MINTZ, B., and W. W. BAKER. 1967. Normal mammalian muscle differentiation and gene control of isocitrate dehydrogenase synthesis. *Proc. Natl. Acad. Sci. U. S. A.* **38:**592–598.

PITTS, J. D. 1971. Molecular exchange and growth control in tissue culture. *Ciba Found. Symp.: Growth Control in Cell Cultures.* 89–105.

PITTS, J. D. 1972. Direct interactions between animal cells. *In* Le petit Colloquium on Cell Interactions. L. G. Silvestri, editor. North-Holland Publishing Co., Amsterdam. 277–285.

PITTS, J. D. 1976. Junctions as channels of direct communication between cells. *In* Developmental Biology of Plants and Animals. C. F. Graham, and P. F. Wareing, editors. Blackwell Scientific Publications Ltd., Oxford. 96–110.

PITTS, J. D., and R. R. BÜRK. 1976. Specificity of junctional communication between animal cells. *Nature (Lond.).* **264:**762–764.

PITTS, J. D., and M. E. FINBOW. 1977. Junctional permeability and its consequences. *In* Intercellular communication. W. C. DeMello, editor. Plenum Publishing Corp., New York. 61–86.

PITTS, J. D., and J. W. SIMMS. 1977. Permeability of junctions between animal cells: intercellular transfer of nucleotides but not macromolecules. *Exp. Cell Res.* **104:**153–163.

SANFORD, K. K., W. R. EARLE, and G. D. LIKELY. 1948. The growth *in vitro* of single isolated tissue cells. *J. Natl. Cancer Inst.* **9:**229–246.

SAXEN, L. 1975. Transmission and spread of kidney tubule induction. *In* Extracellular matrix influences on gene expression. H. C. Slavkin, and L. A. Bavetta, editors. Academic Press, Inc., New York. 523–529.

SHERIDAN, J. D. 1977. Cell coupling and cell communication during embryogenesis. *In* The Cell Surface in Animal Embryogenesis and Development. G. Poste and G. L. Nicholson, editors. Elsevier North-Holland Publishing Co., New York. In press.

SHERIDAN, J. D., M. E. FINBOW, and J. D. PITTS. 1975. Metabolic cooperation in culture: possible involvement of junction transfer in regulation of enzyme activities. *J. Cell Biol.* **67:**396a(Abstr.).

SKERROW, C. J., and A. G. MATOLTSY. 1974. Chemical characterization of isolated epidermal desmosomes. *J. Cell Biol.* **63:**524–530.

STAEHELIN, L. A. 1974. Structure and function of intercellular junctions. *Int. Rev. Cytol.* **39:**191–283.

SUBAK-SHARPE, J. H., R. R. BÜRK, and J. D. PITTS. 1966. Metabolic cooperation by cell-cell transfer between genetically different mammalian cells in tissue culture. *Heredity.* **21:**342.

SUBAK-SHARPE, J. H., R. R. BÜRK, and J. D. PITTS. 1969. Metabolic cooperation between biochemically marked mammalian cells in tissue culture. *J. Cell Sci.* **4:**353–367.

WOLPERT, L. 1969. Positional information and the spatial pattern of cell differentiation. *J. Theor. Biol.* **25:**1–47.

CELL-MATRIX INTERACTION IN EMBRYONIC INDUCTION

ELIZABETH D. HAY

Perhaps the classic example of so-called cell-cell interaction in the embryo is the phenomenon termed embryonic induction. The concept of embryonic induction originated in the early part of this century, largely as a result of the work of Spemann (1938) and his collaborators. They showed that the dorsal lip of the amphibian blastopore induced a neural tube to form when transplanted to a competent embryo. By the end of the 1940s, investigators had discovered that even killed calves' liver could mimic the effect of the dorsal lip or chorамesoderm in causing a competent ectoderm to form neural folds. Primary induction, that is, the interaction between chordamesoderm and presumptive neural ectoderm, seemed completely nonspecific, at least as far as the evocator was concerned (Needham, 1942).

It remained for the tissue-culture approach, introduced by Grobstein in 1953, to reawaken widespread interest in embryonic induction, or tissue interaction, as it is now commonly called (Grobstein, 1955). As we shall see, this approach has served to focus attention on the role of cell-matrix interaction in second-order inductions (tissue interactions subsequent to primary induction). In retrospect, it was inevitable that this should be so, because in the Grobstein approach one removes the matrix between two tissues, such as a gland epithelium and its mesenchyme, and then recombines the tissues in culture across a filter (Fig. 1) through which, it turns out, new matrix secreted by the mesenchyme can pass over to the epithelium (Grobstein, 1955; 1967). The idea that extracellular matrix (ECM) is an informative and vital component of the embryo captivated many earlier workers (see Baitsell, 1925, and Weiss, 1933), but the possibility that the phenomenon called embryonic induction is a matrix-mediated one in some cases, emerged directly from the Grobstein approach.

Grobstein published his matrix interaction theory of embryonic induction (Fig. 2) only a few years after he first began the work, but he had in mind the probability "that there is no single inductive mechanism" (Grobstein, 1955, p. 235). The possible morphogenetic effects of cells on each other and of hormones and metabolites on development were acknowledged (Fig. 2), but these do not appear to be involved in the kind of transfilter in vitro induction usually measured by the Grobstein approach. Rather, the evidence that has accumulated supports the idea that ECM is the missing component which mesenchymal cells furnish enzyme-isolated epithelial tissues in vitro.

We will review briefly the evidence that ECM is involved in second-order induction of this kind, and then we will discuss the relevance of our recent in vitro studies of the cornea to the subject of the role of ECM in tissue interaction. In a longer review on embryonic induction (Hay and Meier, 1977), the question of the initial state of differentiation of the "responding" tissue is discussed in detail, and we conclude that the starting tissue is already partially differentiated before the so-called second-order induction measured by the Grobstein approach. Grobstein (1955, p. 234) stated his working definition quite clearly:

> For present purposes, I would define embryonic induction as developmentally significant interaction between closely associated but dissimilarly derived tissue masses. "Developmentally significant" is used to imply a change *in* one or both masses, which is progressive in terms of the life cycle of the organism, and which is sufficiently stable to persist in some degree if the two masses are separated. Under this general definition, which deliberately avoids any reference to cellular determination or differentiation, we leave our minds quite open as to mechanism, and we are free to pick cases for analysis in terms of their experimental advantages.

What advantages would we like to have? First, an

ELIZABETH D. HAY Department of Anatomy, Harvard Medical School, Boston, Massachusetts

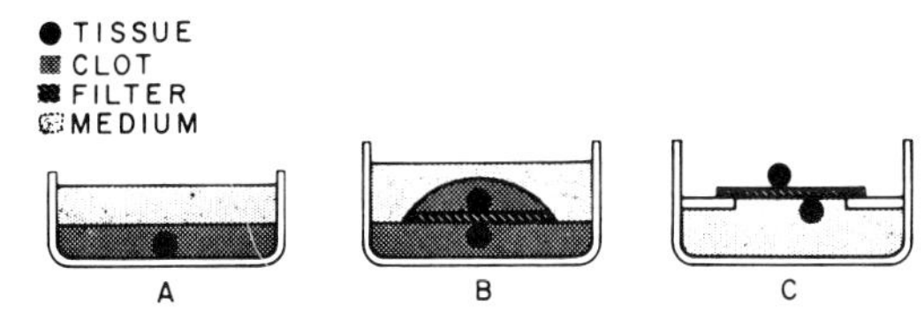

FIGURE 1 Diagrams of the types of organ cultures used in the Grobstein approach to the study of embryonic induction. In A, the tissue is placed at the interface between the bottom of the glass dish and a clot of chicken plasma. The culture medium consists of chick embryo extract, horse serum, and glucose in a balanced salt solution (after Grobstein, 1955). In B, one tissue is placed on a previously formed clot; then the Millipore filter is put on, and the second tissue positioned in a new clot layered over the first (after Grobstein, 1955). In C, one tissue is placed on the filter; then the filter is turned over and supported on the surface of the medium by lens paper. The second tissue is placed on the free surface of the filter, either directly above the first tissue as in B, or to one side of it as shown in C (after Lash et al., 1957).

unambiguous response, simple, yet clearly recognizable and characterizable. Second, a standardizable response. . . . Third, a response accessible to interruption of the reaction at any time. . . . Fourth, an isolable response occurring under controlled conditions, free of complicating variables including interactions with the rest of the organism. Fifth, a response which can be directly observed throughout the inductive course. Sixth, a response, if the nature of the mechanisms involved does not exclude it, occurring across an intervening space or interzone within which the effective agents can be visualized, and possibly analyzed, independently of the cellular masses.

To specify these advantages is to point strongly in the direction of an *in vitro* approach.

Studies of Second-Order Induction In Vitro

The salivary gland epithelium, when isolated by trypsin treatment and grown in a clot by itself, does not differentiate, but when recombined with its mesenchyme directly or across a filter (Fig. 1 B), it branches to become a gland. Grobstein, Parker, and Holtzer have shown that neural tube induces somite mesenchyme to become cartilage under similar in vitro conditions (Grobstein, 1955). Two interacting tissues can be separated across a Millipore (cellulose acetate) filter in the manner shown at the far right in Fig. 1 (Lash et al., 1957). If they are not directly across from each other, no induction occurs. Thus, a freely diffusible factor is probably not involved. Indeed, attempts to isolate a low molecular weight "inducer" (Lash et al., 1962) have not been repeatable. Nor does cell-to-cell contact seem to be required. Cell processes probably do not usually traverse these thick Millipore filters in the 10- to 40-h period (Grobstein, 1967; Lash et al., 1957) during which induction is believed to occur. Rather, the inducer seems to secrete a substance into the filter that brings about the induction. By staining the filter, Grobstein (1955) obtained evidence that the stuff in the filter had the characteristics of ECM; from this evidence he derived his ECM hypothesis (Fig. 2).

A whole new line of thinking was introduced in the mid-1960s by the work of Konigsberg and Hauschka (1965). Heretofore, it was supposed that high molecular weight substances of some specificity might be the active matrix ingredients. Konigsberg and Hauschka showed that the "inductive" action of mesenchyme on muscle differentiation could be mimicked, at least in part, by collagen alone. A number of studies immediately followed in various laboratories to rule in (or out) an inductive effect of so mundane a structural molecule as collagen; these studies were largely indirect, employing enzymes to digest collagen and other structural components of the ECM (see Grobstein, 1967; Bernfield et al., 1972). Kallman and Grobstein (1965) published autoradiographic evidence that mesenchyme secretes collagenlike proteins that polymerize in the Millipore filter and under the transfilter epithelium. Dodson (1963) discovered that frozen-killed dermis is as effective an inducer of epidermal differentiation as living mesenchyme, but attempts to influence gland differentiation directly by collagen were unsuccessful (Wessells and Cohen, 1966), possibly because polymerization of collagen into a substratum is im-

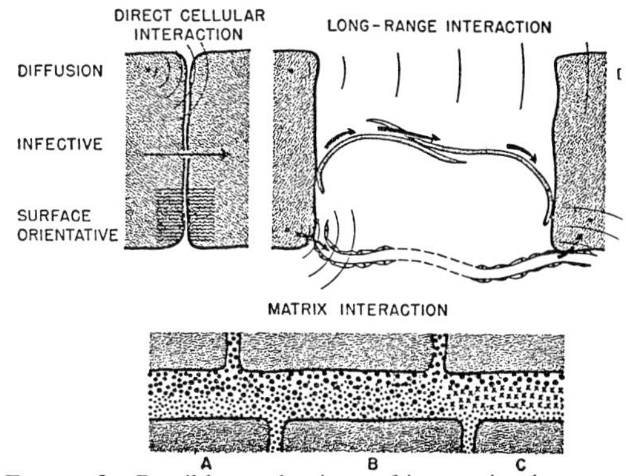

FIGURE 2 Possible mechanisms of interaction between embryonic tissues (from Grobstein, 1955).

portant if it is to affect epithelial differentiation (Meier and Hay, 1974*a*).

At first, it seemed that even if collagen were a nonspecific "mesenchyme common factor" stimulating epithelial differentiation, "mesenchyme specific factors" must also be present (Grobstein, 1967). Salivary gland epithelium seemed to demand its own mesenchyme, whereas pancreas did not. Recent evidence indicates that the salivary gland requirements may not be that specific (Cunha, 1972). Rather, the idea now emerging is that gland branching is under a rather subtle control by ECM, especially its glycosaminoglycan (GAG) component (Bernfield et al., 1972), whereas cytodifferentiation may be regulated by completely separate mechanisms (Spooner, 1974). A noncollagenous growth factor has also been implicated in gland cytodifferentiation (Levine et al., 1973).

The inductive interaction studied most extensively in recent years is the effect of neural tube and notochord on somite differentiation. As in the corneal system which we will discuss shortly, the end product of differentiation usually measured is the production of ECM by the tissue; the level of synthesis of cartilage matrix by somite mesenchyme is usually monitored by radioactive sulfate incorporation (Lash, 1968; Lash and Vasan, 1977). The effect of neural tube on the somite, however, may begin while the tissues are still in epithelial configuration (Fig. 3, class 1*b*) before the sclerotome disperses (Fig. 3, class 2) to give rise to the chondroblasts that form the axial skeleton. It is easy to believe that mesenchymal cells are producing collagen and GAG (Fig. 3, class 3), but the first conclusive evidence that an epithelium such as neural tube might be producing such products was not published until 1971 (Cohen and Hay, 1971); subsequently notochord was also shown to be a source of GAG and collagen (for

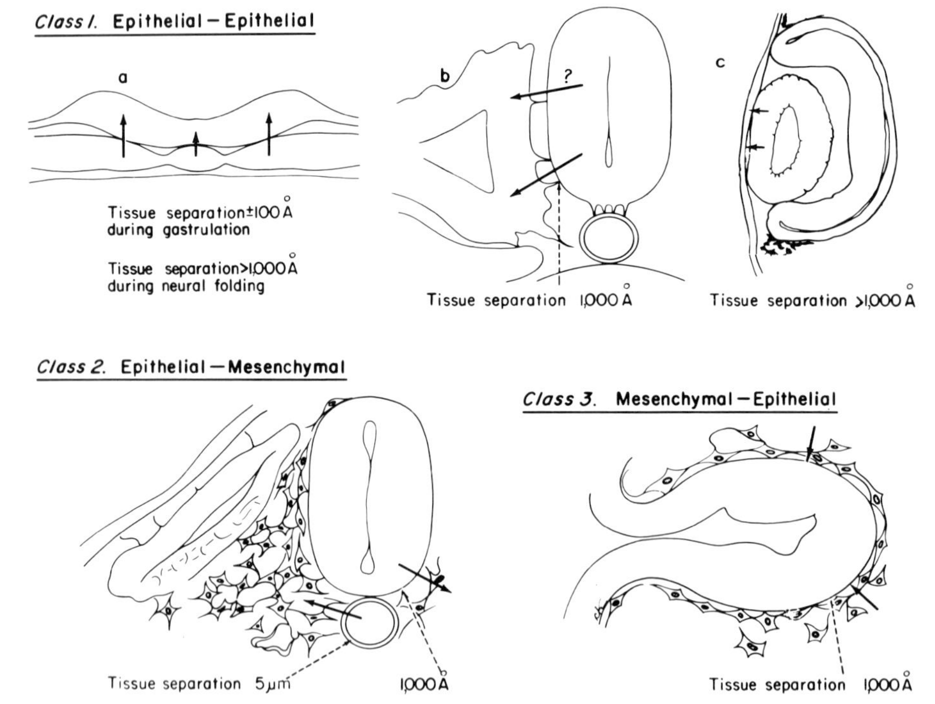

FIGURE 3 Diagram summarizing the main classes of tissue interactions in the developing embryo. Primary induction is an interaction between two epithelial tissues, the chordamesoderm and ectoderm, in which the neural tube is induced to form in the ectoderm (*a*). Second-order interactions between epithelia are shown at *b* and *c*. Neural tube interacts with somite mesoderm while it is still in the form of an epithelium (*b*) and also later on after the somite partly disperses as mesenchyme (class 2). Mesenchyme can also act as inducer, as for example in the differentiation of salivary gland (class 3). The minimal tissue separation is shown for each system depicted; the extracellular space between the tissues may be even greater (from Hay, 1973).

review see Hay and Meier, 1974; Lash and Vasan, 1977). Further impetus to the idea that GAG as well as collagen might stimulate chondrogenesis was published by Nevo and Dorfman in 1972; they were the first investigators to show that purified chondroitin sulfate and chondromucoprotein stimulate GAG synthesis by cells, in this case, chondrocytes isolated from long bones of 13-day-old chicks.

In 1972, O'Hare reported that collagenase and hyaluronidase collectively destroy the ability of irradiated spinal cord to induce somite chondrogenesis. Shortly thereafter, Kosher et al. (1973) published evidence that chondromucoprotein extracted from sternal and vertebral cartilage promotes [^{35}S]sulfate accumulation and visible cartilage formation in somites after 2 day in culture. Kosher and Church (1975) reported a mild but definite stimulatory effect of collagen and procollagen on collagen synthesis by somite mesenchyme, an effect which seems to be cyclic AMP-sensitive (Kosher, 1976). The stimulatory effect of various collagens on chondrogenesis in vitro has been confirmed by Lash, Vasan, and Kosher (see Lash and Vasan, 1977, for further review).

Cell Surface-Matrix Interaction in Corneal Morphogenesis

In the mid-1960s, Jean Paul Revel and I become interested in the development of the cornea of the chick embryo because its morphology indicated that it would be ideal for the study of what was then a heresy, the production of true collagen by epithelium. The corneal epithelium seemed to produce the primary corneal stroma under the influence of the adjacent lens (Fig. 3, class 1*c*) long before any mesenchymal cells entered the area (Hay and Revel, 1969). Dodson and Hay (1971, 1974) demonstrated that isolated corneal epithelium grown on killed lens capsule (a type IV collagen containing GAG) produces a collagenous stroma in vitro reminiscent of that formed in vivo. Subsequently, Meier and Hay (1974*a*) reported that any type of collagenous substratum tested (even one of pure type II collagen) supports the efforts of the corneal epithelium to produce a stroma (Fig. 4). On glass, plastic, and substrates of noncollagenous proteins, however, the corneal epithelium fails to produce a stroma.

Stroma production can be monitored morphologically by the appearance of orthogonal layers of collagen fibrils in vitro (there are 25 such layers in vivo) or it can be measured by incorporation of [^{3}H]proline into collagen (hot TCA-extractable protein) and [^{35}S]O_4 into GAG (Meier and Hay, 1974*a*). Collagens enhance the ability of the epithelium to produce *both* collagen and GAG. On the other hand, GAG (chondroitin and heparan sulfates) stimulates only GAG synthesis (Meier and Hay, 1974*b*). In preliminary experiments, Pamela Hartzband in our laboratory has found that hemocyanin and periodate do not block the ability of lens capsule collagen to stimulate corneal epithelial productions of ECM. It thus seems unlikely that the sugar component of collagen is involved in its stimulatory effect. Moreover, collagen fragments in solution have no measurable effect on the epithelium.

The question on which we have concentrated recently is one of interest to us as cell biologists: Where does the collagen-cell interaction take place? There are many reasons for suspecting that direct cell contact with the collagenous substratum is needed. Collagen in the form of a lens capsule or collagen gel (Fig. 4) is known not to be soluble under physiological conditions. If the corneal epithelium is blocked from contact with the underlying ECM by a Millipore filter, it fails to produce stroma (Fig. 5, left). Cell processes do not cross

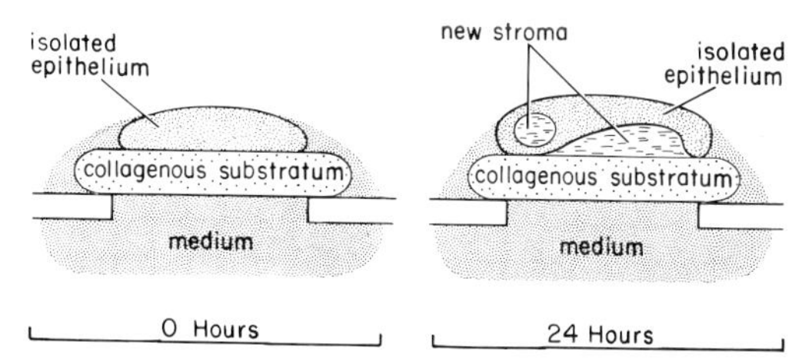

FIGURE 4 Diagrams of the culture method. Corneal epithelium isolated by trypsin-collagenase is grown on a collagenous substratum (such as lens capsule or collagen gels). Within 24 h, the epithelium has secreted a facsimile of the removed corneal stroma.

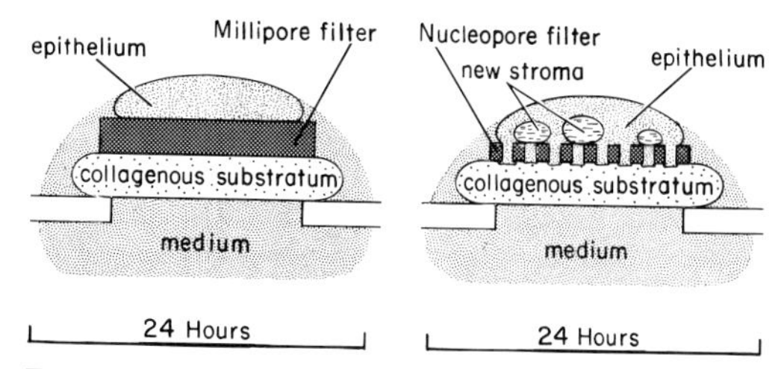

FIGURE 5 Diagrams of cultures employing filters to separate corneal epithelium from the collagenous substratum. Cell processes do not traverse Millipore filter (left) in this experiment, but they readily cross Nucleopore filter (right).

the filter under these circumstances (Fig. 6, left). On the other hand, a Nucleopore (polycarbonate) filter of a pore size (0.8 μm) that readily admits cell processes (see Saxen et al., 1976, for review) permits "induction" to take place across it (Fig. 5, right). Transmission electron microscopy (Fig. 7) and scanning electron microscopy (Fig. 6, right) reveal that cell processes completely cross Nucleopore filters to contact the substratum (Meier and Hay, 1975; Hay and Meier, 1976).

By using Nucleopore filters of differing pore size (0.1 μm–0.8 μm), it can be shown that the stimulatory effect of the collagenous substratum on the epithelium is related to the number of cell processes that traverse the filter (Meier and Hay, 1975). The level of the stimulatory effect is directly proportional to the total contact area calculated from scanning electron micrographs (Fig. 8). By using autoradiography and biochemical techniques, Hay and Meier (1976) showed that no detectable radioactivity passes from [^{3}H]proline-labeled lens capsule to the epithelium. Thus, the interaction between collagen and the epithelium is very likely to be at the cell surface rather than within the cytoplasm.

Relevance to Embryonic Induction

In many ways, the corneal epithelial system we have just described serves admirably as a model for understanding the kind of embryonic induction measured by the Grobstein approach. The work clearly calls attention to an interaction of the cell with the ECM that it contacts. *No cell to cell contact* exists between the epithelium and the "inducer," for the "inducer" is free of cells. The lens ECM (capsule) has the same effect as living lens in enhancing the ability of the corneal epithelium to produce a stroma in vitro (Hay, 1973; Dodson and Hay, 1974). Saxen and his co-workers (see Saxen et al., 1976, for review) have shown that spinal cord and metanephric mesenchyme can contact each other by cell processes across Nucleopore filters during induction of metanephros by spinal cord in vitro, but they did not examine the possibility that ECM was also transferred across the filters. They have shown that in some cases cell processes can, in time, even cross Millipore filters. Cell to cell contact was envisioned by Grobstein (Fig. 2) as playing a role in morphogenesis. What our experiments suggest, however, is that the emphasis Grobstein (1955; 1967) placed on the role of ECM in transfilter induction was in fact justified. In the model we have presented (Figs. 4 and 5), the only interaction that can take place is between cells and ECM. No living inducer is present.

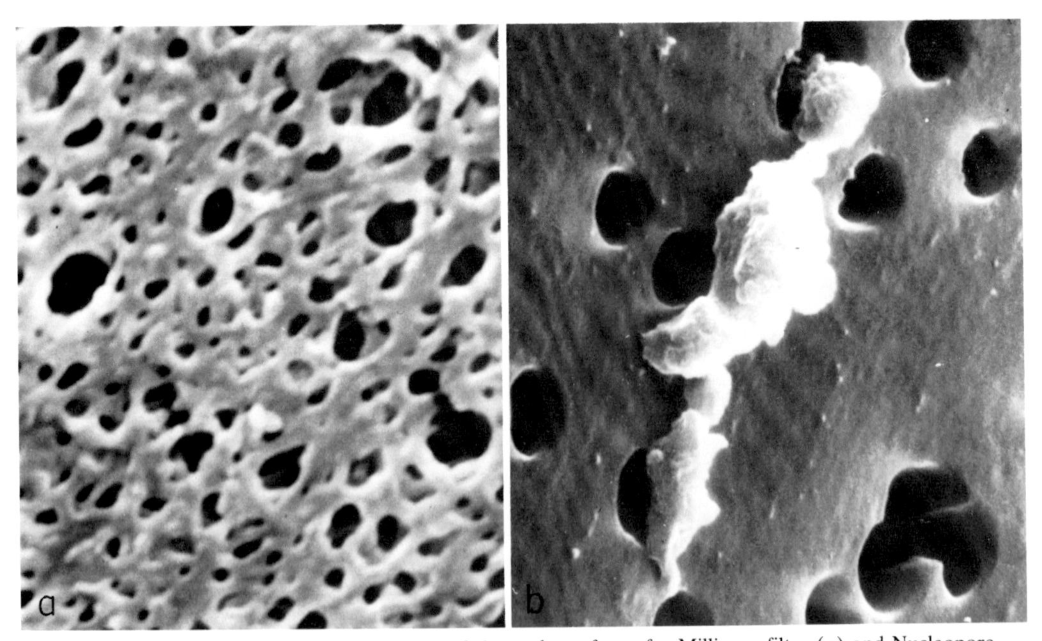

FIGURE 6 Scanning electron micrographs of the undersurface of a Millipore filter (*a*) and Nucleopore filter (*b*), each of which had corneal epithelium growing on the upper surface for 24 h. Cell processes traverse the Nucleopore but not the Millipore filter (from Meier and Hay, 1975). × 8,800.

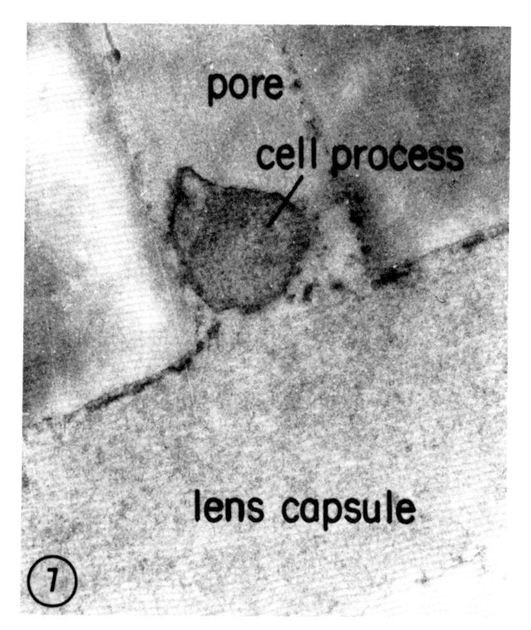

FIGURE 7 Transmission electron micrograph showing an epithelial cell process which traversed a Nucleopore filter to contact an underlying collagenous substratum (lens capsule) (from Hay and Meier, 1976). × 30,000.

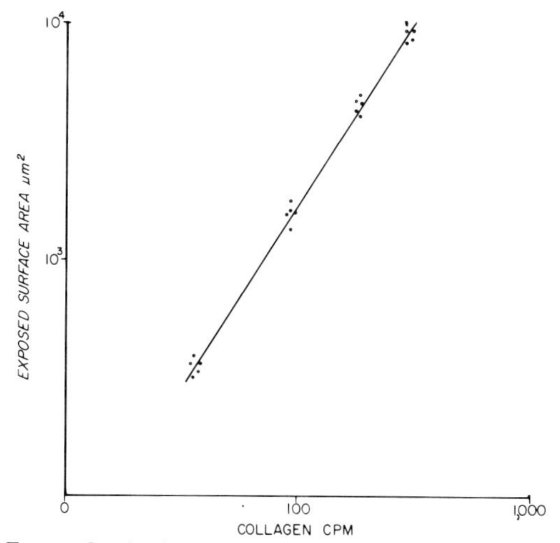

FIGURE 8 A plot on a log scale of epithelial surface area exposed under each Nucleopore filter compared with collagen synthesized per epithelium in 24 h. The 0.1-μm pore size allowed only a few cell processes to contact the underlying substratum, whereas the 0.8-μm pore size permitted a large surface area to be exposed to the substratum under the filter (from Meier and Hay, 1975).

How does an interaction between cells and ECM take place? Konigsberg and Hauschka (1965) suggested that the matrix may trap factors needed by cells for their continued differentiation. We found, however, that serum factors are not required for the interaction between corneal epithelium and lens capsule and that lens capsule does not bind factors of epithelial origin that condition its subsequent activity (Hay and Meier, 1976). In vivo, the function of the lens capsule may be supplanted in large part by the basal lamina of the epithelium itself. It is tempting to think that the collagen of the lamina is attached to receptors on the plasmalemma and that this attachment somehow stabilizes the underlying epithelial cytoplasm, but we do not know that this is so (see Hay and Meier, 1976; Kosher, 1976; Lash and Vasan, 1977, for further speculation).

What we do know is that enzymatic removal of the underlying ECM disrupts the morphogenesis of the epithelium. Likewise, in the somite-neural tube system we discussed, matrix is removed by enzyme treatment when the tissues are isolated for culture. The somite mesenchyme secretes collagen and GAG and, interestingly, is capable of some self-differentiation in confined quarters (Ellison and Lash, 1971). Does the epithelium also respond to the ECM it manufactures? Yes, its ECM seems to be autocatalytic. After 48 h of culture transfilter to lens capsule (Fig. 5, right), the epithelium and the stroma it deposited on the Nucleopore filter can be moved together to a noncollagenous substratum; the epithelium continues to synthesize collagen at about the same level as before (Hay and Meier, 1976).

Some years ago, Grobstein showed that the mesenchyme transfilter to salivary gland epithelium can be removed after 40 h in vitro without disrupting epithelial morphogenesis (see Grobstein, 1967). Lash et al. (1957) removed the transfilter neural tube after 10 h in vitro and found that somite chondrogenesis continued. The implication was that the gland epithelium and somites were now "determined" to follow a given course of differentiation. A genetic change was implied. The idea that we would like the reader to consider, however, is that the cell surface of these enzyme-isolated tissues was stabilized by the ECM secreted by the transfilter tissue and by ECM they may have produced themselves. The living "inducer" can be removed after 10–40 h by scraping the cells off the undersurface of the filter, but the transfilter ECM remains and probably continues to interact with the responding tissue.

This is a new way of looking at induction because it says that the so-called inducer must be more stable after isolation than the so-called responding tissue and a better producer of ECM. There is evidence for both these points—neural tube, notochord, and lens are able to grow and secrete ECM independently of other tissues in culture (Hay and Meier, 1974), and fibroblasts are notorious for their ability to produce collagen after isolation in vitro (Konigsberg and Hauschka, 1965).

If what we are measuring in our in vitro systems is the relative dependence of a particular cell function, such as cell elongation or collagen synthesis, on cell surface-ECM, then is it fair to continue to call these phenomena embryonic induction? We may well abandon the concept in the future. Nevertheless, I think it is fair to say that the systems discussed here do meet the Grobstein definition we quoted earlier: cell-ECM interaction is progressive, is stable, does occur across an interzone, and is accessible to interruption by the experimenter. However, it is not a single event in time. I think Grobstein sensed this point, for in the conclusion of his 1955 symposium talk he raised the possibility that "embryonic induction is not simply an enigmatic *event* peculiar to early development, but the beginning of a *process* which establishes and stabilizes relations among groups of cells and their matrices-relations which become increasingly fixed in what we recognize as structure, persist into the adult and are important in the maintenance of its tissue architecture" (p. 254).

It is to this wider concept, the role of the ECM *as well as* cells in structuring the developing organism, that I think we now want to direct our renewed attention.

REFERENCES

BAITSELL, G. A. 1925. On the origin of the connective-tissue ground-substance in the chick embryo. *Q. J. Microsc. Sci.* **69:**571–589.

BERNFIELD, M. R., S. D. BANERJEE, and R. H. COHN. 1972. Dependence of salivary epithelial morphology and branching morphogenesis upon acid mucopolysaccharide-protein (proteoglycan) at the epithelial surface. *J. Cell Biol.* **52:**674–689.

COHEN, A. M., and E. D. HAY. 1971. Secretion of collagen by embryonic neuroepithelium at the time of spinal cord-somite interaction. *Dev. Biol.* **26:**578–605.

CUNHA, G. R. 1972. Support of normal salivary gland morphogenesis by mesenchyme derived from accessory sexual glands of embryonic mice. *Anat. Rec.* **173:**205–212.

DODSON, J. W. 1963. On the nature of tissue interactions in embryonic skin. *Exp. Cell Res.* **31:**233–235.

DODSON, J. W., and E. D. HAY. 1971. Secretion of collagenous stroma by isolated epithelium grown *in vitro. Exp. Cell Res.* **65:**215–220.

DODSON, J. W., and E. D. HAY. 1974. Secretion of collagen by corneal epithelium. II. Effect of the underlying substratum on secretion and polymerization of epithelial products. *J. Exp. Zool.* **189:**51–72.

ELLISON, M. L., and J. W. LASH. 1971. Environmental enhancement of *in vitro* chondrogenesis. *Dev. Biol.* **26:**486–496.

GROBSTEIN, C. 1955. Tissue interaction in the morphogenesis of mouse embryonic rudiments *in vitro. In* Aspects of Synthesis and Order in Growth. D. Rudnick, editor. Princeton University Press, Princeton. 233–256.

GROBSTEIN, C. 1967. Mechanisms of organogenetic tissue interaction. *Natl. Cancer Inst. Monogr.* **26:**279–295.

HAY, E. D. 1973. Origin and role of collagen in the embryo. *Am. Zool.* **13:**1085–1107.

HAY, E. D., and S. MEIER. 1974. Glycosaminoglycan synthesis by embryonic inductors: neural tube, notochord, and lens. *J. Cell Biol.* **62:**889–898.

HAY, E. D., and S. MEIER. 1976. Stimulation of corneal differentiation by interaction between cell surface and extracellular matrix. II. Further studies on the nature and site of transfilter "induction." *Dev. Biol.* **52:** 141–157.

HAY, E. D., and S. MEIER. 1977. Concept of embryonic induction; inductive mechanisms in orofacial morphogenesis. *In* Textbook of Oral Biology. J. Shaw and S. Meller, editors. W. B. Saunders Company, Philadelphia. In press.

HAY, E. D., and J. P. REVEL. 1969. Fine structure of the developing avian cornea. *Monogr. Dev. Biol.* **1:**1–144.

KALLMAN, F., and C. GROBSTEIN. 1965. Source of collagen at epitheliomesenchymal interfaces during inductive interaction. *Dev. Biol.* **11:**169–183.

KONIGSBERG, I. R., and S. D. HAUSCHKA. 1965. Cell and tissue interactions in the reproduction of cell type. *In* Reproduction: Molecular, Subcellular and Cellular. M. Locke, editor. Academic Press, Inc., New York. 243–290.

KOSHER, R. A. 1976. Inhibition of "spontaneous," notochord-induced, and collagen-induced *in vitro* somite chondrogenesis by cyclic AMP derivatives and theophylline. *Dev. Biol.* **53:**265–276.

KOSHER, R. A., and R. L. CHURCH. 1975. Stimulation of *in vitro* somite chondrogenesis by procollagen and collagen. *Nature (Lond.)* **258:**327–330.

KOSHER, R. A., J. W. LASH, and R. R. MINOR. 1973. Environmental enhancement of *in vitro* chondrogenesis. IV. Stimulation of somite chondrogenesis by exogenous chondromucoprotein. *Dev. Biol.* **35:**210–220.

LASH, J. W. 1968. Somitic mesenchyme and its response

to cartilage induction. *In* Epithelial-Mesenchymal Interactions. R. Fleischmajer and R. E. Billingham, editors. The Williams and Wilkins Co., Baltimore. 165–172.

LASH, J., S. HOLTZER, and H. HOLTZER. 1957. Experimental analysis of development of spinal column. VI. Aspects of cartilage induction. *Exp. Cell Res.* **13:**292–303.

LASH, J. W., F. A. HOMMES, and F. ZILLIKEN. 1962. Induction of cell differentiation: the *in vitro* induction of vertebral cartilage with a low-molecular weight tissue component. *Biochim. Biophys. Acta.* **56:**313–319.

LASH, J., and N. S. VASAN. 1977. Tissue interactions and extracellular matrix components. *In* Cell and Tissue Interactions. M. M. Burger and J. W. Lash, editors. Raven Press, New York. In press.

LEVINE, S., R. PICTET, and W. J. RUTTER. 1973. Control of cell proliferation and cytodifferentiation by a factor of reacting with the cell surface. *Nat. New Biol.* **246:**49–52.

MEIER, S., and E. D. HAY. 1974*a*. Control of corneal differentiation by extracellular materials: collagen as a promoter and stabilizer of epithelial stroma production. *Dev. Biol.* **38:**249–270.

MEIER, S., and E. D. HAY. 1974*b*. Stimulation of extracellular matrix synthesis in the developing cornea by glycosaminoglycans. *Proc. Natl. Acad. Sci. U. S. A.* **71:**2310–2313.

MEIER, S., and E. D. HAY. 1975. Stimulation of corneal differentiation by interaction between cell surface and extracellular matrix. I. Morphometric analysis of transfilter induction. *J. Cell Biol.* **66:**275–291.

NEEDHAM, J. 1942. Biochemistry and Morphogenesis. Cambridge University Press, London. 165–188.

NEVO, A., and A. DORFMAN. 1972. Stimulation of chondromucoprotein synthesis in chondrocytes by extracellular chondromucoprotein. *Proc. Natl. Acad. Sci. U. S. A.* **69:**2069–2072.

O'HARE, M. J. 1972. Aspects of spinal cord induction of chondrogenesis in chick embryo somites. *J. Embryol. Exp. Morphol.* **27:**235–243.

SAXEN, L., E. LEHTONEN, M. KARKINEN-JAAS KELAINEN, S. NORDLING, and J. WARTOIVAARA. 1976. Are morphogenetic tissue interactions mediated by transmissible signal substances or through cell contacts? *Nature (Lond.).* **259:**662–663.

SPEMANN, H. 1938. Embryonic Development and Induction. Yale University Press, New Haven. 141–169.

SPOONER, B. S. 1974. Morphogenesis of vertebrate organs. *In* Concepts of Development. J. Lash and J. R. Whittaker, editors. Sinauer Associates, Inc., Stamford. 213–240.

WEISS, P. 1933. Functional adaptation and the role of ground substances in development. *Am. Nat.* **67:**322–340.

WESSELLS, N. K., and J. H. COHEN. 1966. The influence of collagen and embryo extract on the development of pancreatic epithelium. *Exp. Cell Res.* **43:**680–684.

Cell-to-Cell Communication

GAP JUNCTIONS AND CELL COMMUNICATION

NORTON B. GILULA

Over the past 20 years cell-to-cell communication has been established as a common property in both excitable and nonexcitable tissues. The device responsible for communication is referred to physiologically as an electrical synapse, an electrotonic synapse, or a low-resistance synapse. For communication to exist, the interacting cells must: (1) be in direct physical contact; (2) be joined by a low-resistance pathway that permits the passage of current (in the form of inorganic ions) with little voltage attenuation; and (3) have some selectivity or restriction on the nature of the molecules that are transferred. The latter characteristic may be important in distinguishing between communicating cells and fusing cells. In most systems that have been examined, a specific cell contact, the gap junction or nexus, appears to serve as the pathway for cell-to-cell transmission of small metabolites and inorganic ions.

In this brief review, I will attempt to synthesize the current available information on the properties of gap junctions, the biological role of gap junctions, and some of the future problems in this area of cell biology.

Gap Junctions and Cell Communication

Gap junctions are a class of cell contacts that, like cell-to-cell communication, have a wide distribution throughout the animal kingdom. These structures have been referred to as quintuple-layered, synaptic disks, the nexus, gap junctions, and, most recently, communicating junctions (Robertson, 1963; Dewey and Barr, 1962; Revel and Karnovsky, 1967; Simionescu et al., 1975). From studies on the most extensively characterized systems, it is clear that the gap junction can serve as the pathway for cell-to-cell communication (Dreifuss et al., 1966; Johnson and Sheridan, 1971; Gilula et al., 1972; Bennett, 1973).

NORTON B. GILULA The Rockefeller University, New York, New York

Experimentally, it is possible to define communication as the intercellular transfer of ions, the transfer of metabolites, or both. Thus, communicating cells may be referred to as ionically coupled or metabolically coupled.

Ionically coupled cells were first described between invertebrate neurons by Furshpan and Potter (1959). Since then, ionic coupling has been detected in a variety of nonexcitable and excitable tissues both in vivo and in culture with microelectrode impalements (Lowenstein, 1966; Furshpan and Potter, 1968; Bennett, 1973). The current transferred between coupled cells is presumably in the form of small ions, primarily K^+, Na^+, Cl^-. On the basis of the physiological observations, the pathway for ionic transfer should have polar or hydrophilic properties. Based on the hydrated ion size for molecules such as K^+, Na^+, Cl^-, the ionic coupling pathway should contain low-resistance channels that are at least 1–1.5 nm in diameter. In addition to small inorganic ions, the ionic coupling channels may also be permeable to a variety of iontophoretically injected dyes (Loewenstein, 1966; Furshpan and Potter, 1968; Bennett, 1973). In general, dyes below 1,000 daltons can permeate the low-resistance channels, whereas larger molecules are not successfully transferred from cell to cell. In certain developing embryos, ions can be transferred but dyes can not (for review, see Bennett, 1973).

Metabolically coupled cells were first described by Subak-Sharpe et al. (1969). In their initial studies, they described a metabolic exchange between cells and termed this phenomenon "metabolic cooperation between cells." Recent studies by John Pitts (this volume) and others (Rieske et al., 1975) indicate that a variety of small, metabolically significant molecules such as amino acids, sugars, phosphorylated sugars, and nucleotides, may be transferred between cells that are in direct physical contact. Thus far, macromolecular components, such as proteins, nucleic acids, etc., have not been demonstrated to take part in this metabolic exchange phenomenon.

In two different studies (Gilula et al., 1972; Azarnia et al., 1972), it has been demonstrated that metabolically coupled cells are also ionically coupled. This was further strengthened by the observation in those studies that communication-incompetent cells lack both ionic and metabolic coupling. Thus, the present evidence indicates that a communication-competent phenotype possesses the ability to transfer both ions and metabolites. Therefore, the communication pathway is probably the same for both metabolic and ionic coupling, with some qualitative restrictions. In one of these studies (Gilula et al., 1972), the gap junction was characterized as the structural pathway for the cell-to-cell communication. The communication-incompetent cells in this study did not express the gap junctional phenotype.

Gap Junctional Structure

A large number of pleiomorphic forms of gap junctions have been described (for reviews, see McNutt and Weinstein, 1973; Gilula, 1974*a*; Staehelin, 1974). With the exception of gap junctions in arthropods (Gilula, 1974*a*), the gap junctions in most organisms share basic structural features.

The gap junction, in its present form, was first described by Revel and Karnovsky in 1967. In thin-section electron microscopy, the gap junction appears as a complex of two adjacent plasma membranes separated by a small space or gap (Fig. 1). The gap is about 2–4 nm and the entire width of the junction is 15–19 nm. Electron-dense materials, such as lanthanum and ruthenium red, can penetrate the gap region of the junction to reveal the presence of a polygonal lattice of 8- to 9-nm subunits (Revel and Karnovsky, 1967). Gap junctions exist as plaquelike contacts between cells, and the size and number of junctions can vary considerably between different cells. The gap-junctional membranes can also be characterized with freeze-fracture electron microscopy (Fig. 2). Two complementary fracture faces are exposed at the junctional specialization. Most of the gap junctions, with the exception of those found in arthropods, contain these fracture face components. The arthropod gap junction is characterized by a variable arrangement of 10- to 30-nm particles on the outer membrane half (E fracture face), and a complementary arrangement of pits or depressions on the inner membrane half (P fracture face; Flower, 1972; Johnson et al., 1973; Peracchia, 1973; Gilula, 1974*a*). In the arthropod structure, the particles are not normally homogeneous in size, and they are associated with the opposite fracture face.

Gap junctions are normally present on the surface of cells that are in contact. However, in certain tissues, particularly those under hormonal influence, gap junctions are frequently present as intracellular vacuoles. These structures have been referred to as annular gap junctions or annular nexuses (Merk et al., 1973; Albertini and Anderson, 1974; Albertini et al., 1975). The intracellular gap junctions appear virtually identical in structural detail to those on the cell surface, and the annular structures often enclose cytoplasmic components and organelles. In some instances, deteriorating gap junctions are present in these vacuoles (Fig. 3). Thus, the internalization of gap junctions from the cell surface into the cytoplasm may represent a viable turnover mechanism for gap junctions in vivo. To date, no lysosomal activities have been demonstrated in these vacuoles.

Gap junctions also provide an important structural element in cell adhesion. The junctional complexes are extremely resistant to a variety of such treatments as proteolysis, removal of divalent cations, and mechanical or physical disruption. In fact, when cells are separated with these treatments, the gap junctions are retained as an intact complex by one of the dissociated cells (Fig. 4; Berry and Friend, 1969). At present, only one reported procedure, treatment with hypertonic sucrose, will effectively split or unzip the gap junctional complex (Barr et al., 1965; Dreifuss et al., 1966; Goodenough and Gilula, 1974), and this procedure may only be effective on intact tissues.

The formation of gap junctions between cells in vivo and in culture has recently been the focus of a number of studies (Revel et al., 1973; Johnson et al., 1974; Decker and Friend, 1974; Benedetti et al., 1974; Albertini and Anderson, 1974; Decker, 1976). In general, the formation process as examined by the freeze-fracture technique consists of the following stages: (1) the appearance of formation plaques; (2) the appearance of large "precursor" particles with a reduction of the intercellular space; (3) the appearance of smaller "junctional" particles in polygonal arrangements; and (4) the enlargement of junctions (Fig. 5).

Isolation and Biochemical Characterization of Gap Junctions

Gap junctions have been isolated as enriched subcellular fractions, primarily from rat and

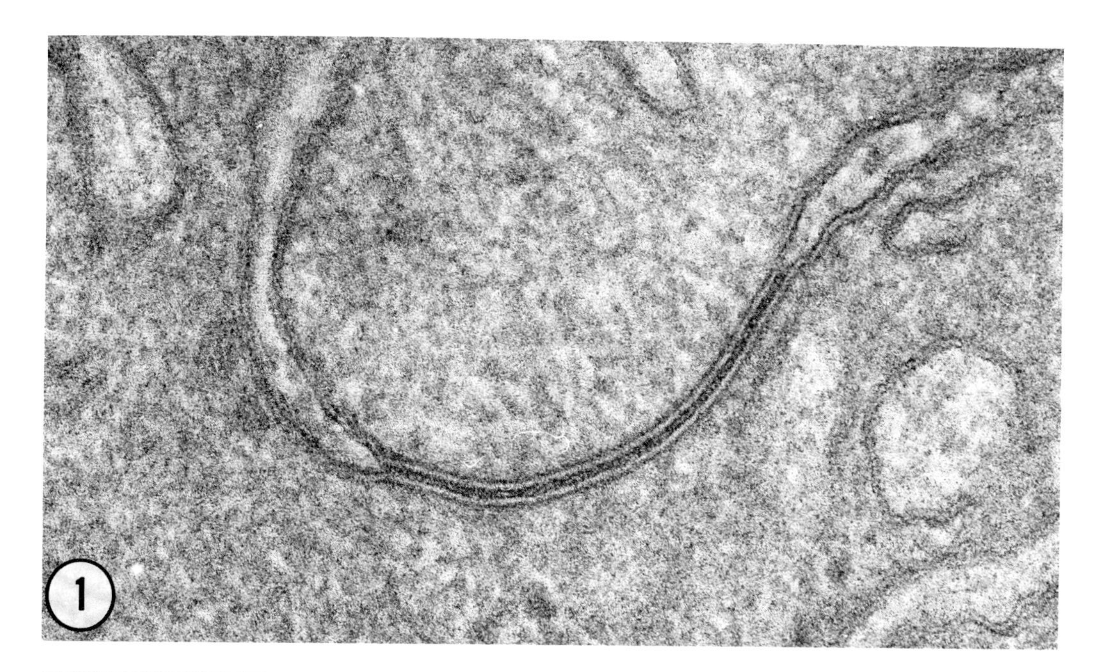

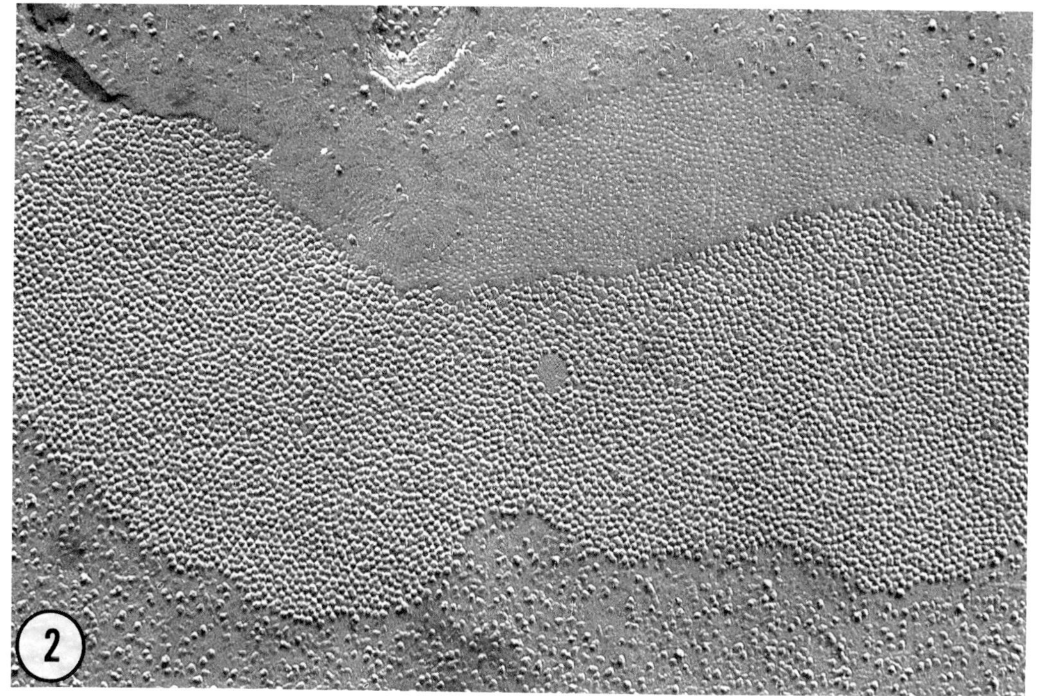

FIGURE 1 Thin-section appearance of a gap junction between rat hepatocytes. The surface membranes are separated by a 2- to 4-nm extracellular space at the site of the gap junction. This preparation hs been treated with procion brown to "penetrate" the extracellular space at the gap junction. × 220,000.

FIGURE 2 Freeze-fracture image of a gap-junctional plaque between mouse hepatocytes. The membranes have been split to expose the internal specializations at the site of the gap junction. The inner membrane half (P fracture face) contains a polygonal lattice of homogeneous 8- to 9-nm particles, whereas the outer membrane half (E fracture face) contains a complementary arrangement of pits or depressions (Chalcroft and Bullivant, 1970; McNutt and Weinstein, 1970). × 108,000.

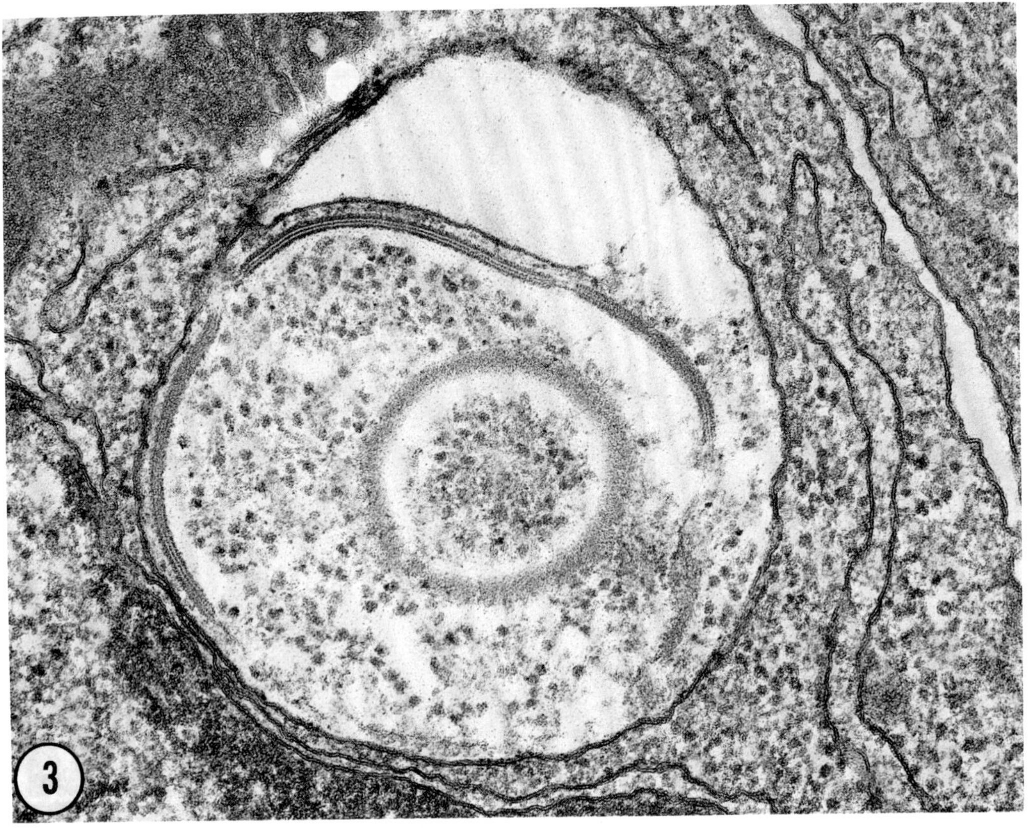

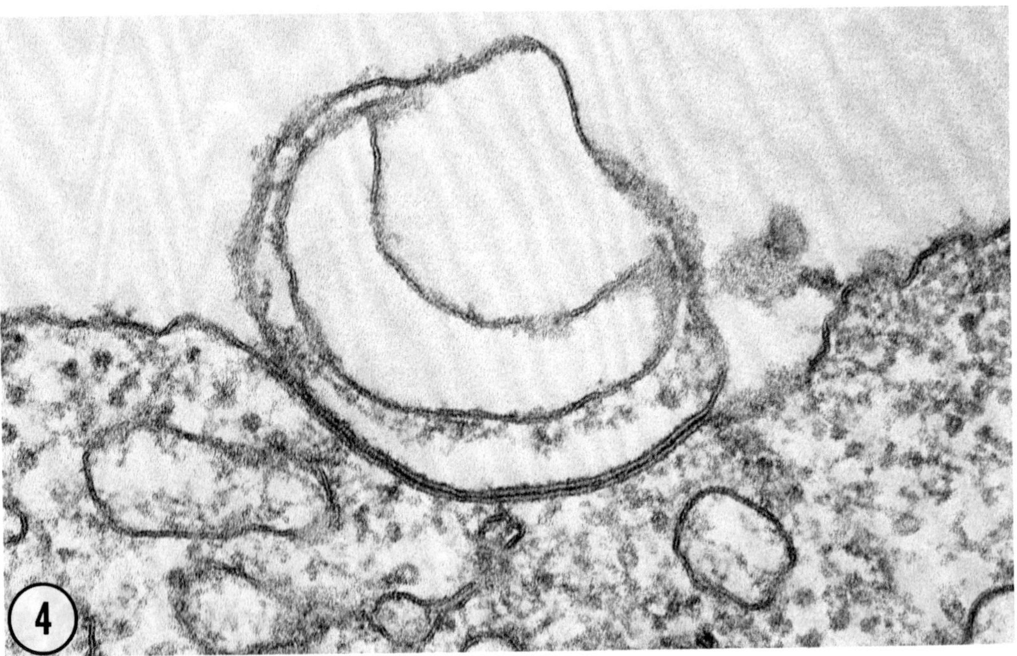

FIGURE 3 A vacuole inside a rat ovarian granulosa cell. The vacuole contains both intact and disintegrating regions of gap junctions that were previously located on the cell surface. × 72,100.

FIGURE 4 The remnants of a gap-junctional contact between rat ovarian granulosa cells after mechanical dissociation. The gap junction is retained as an intact complex on the surface of the dissociated cell. Note the nonjunctional plasma membrane and the trapped endoplasmic reticulum that are attached to the junction as a "bleb." × 100,000.

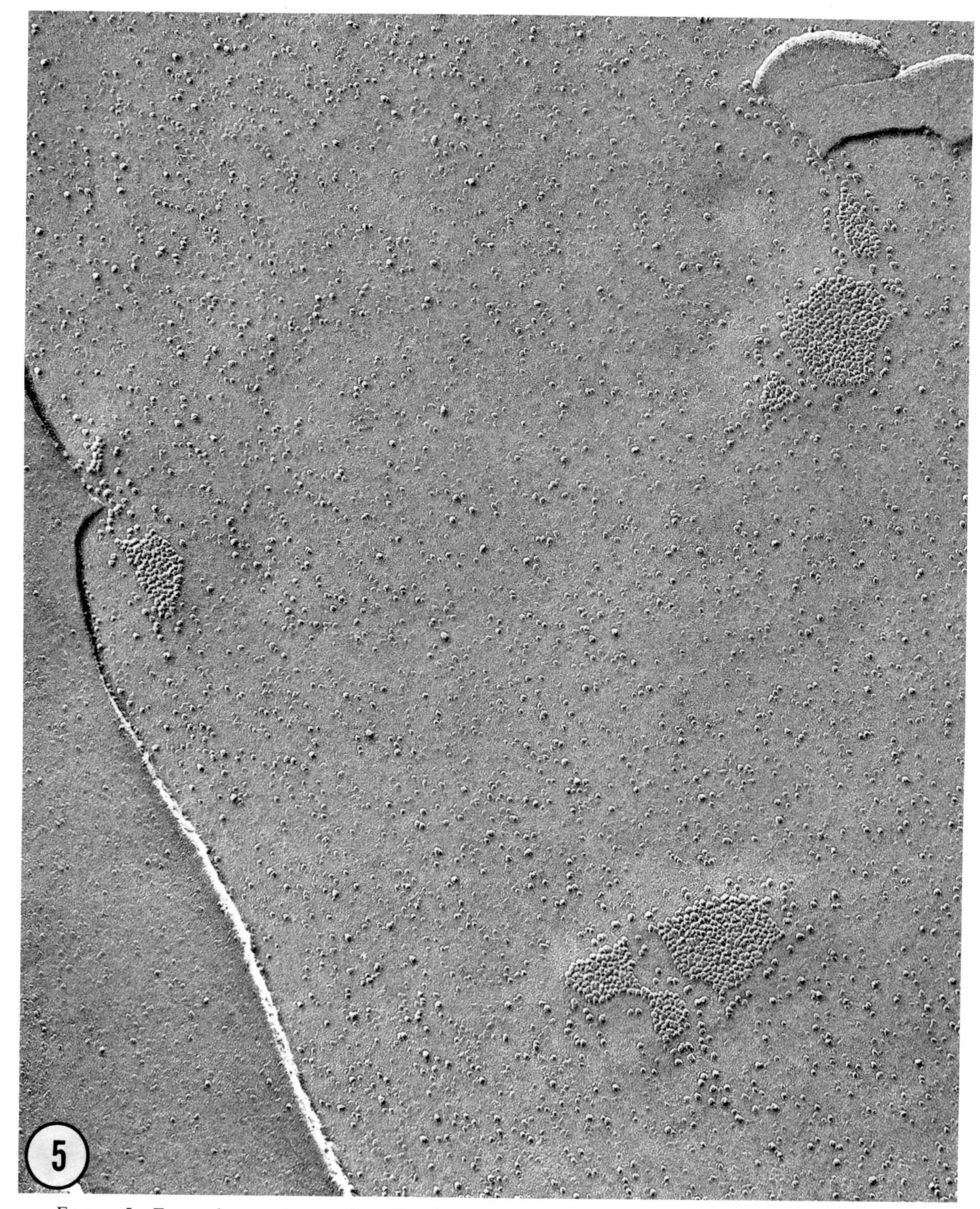

FIGURE 5 Freeze-fracture image of gap junction "formation" between rat ovarian granulosa cells. Large 10-nm particles are frequently detectable on the fracture faces before the appearance of polygonal aggregates of smaller 8- to 9-nm particles. This image reflects an intermediate stage in the formation process where the large particles are loosely arranged in close proximity to the polygonal aggregates of smaller gap-junctional particles. × 65,610.

mouse liver (Benedetti and Emmelot, 1968; Goodenough and Stoeckenius, 1972; Evans and Gurd, 1972; Gilula, 1974*b*; Dunia et al., 1974; Duguid and Revel, 1975). At present, no endogenous activity has been detected in these fractions, so the only criterion for purity is ultrastructural analysis.

Negative stain studies on isolated gap junctions

have revealed a polygonal lattice of 8- to 9-nm particles (Fig. 6) that is similar to the lattices revealed by lanthanum treatment in thin sections or freeze-fracturing. In certain preparations, a small 1.5- to 2-nm electron-dense dot occupies the central region of the 8- to 8.5-nm particles. This

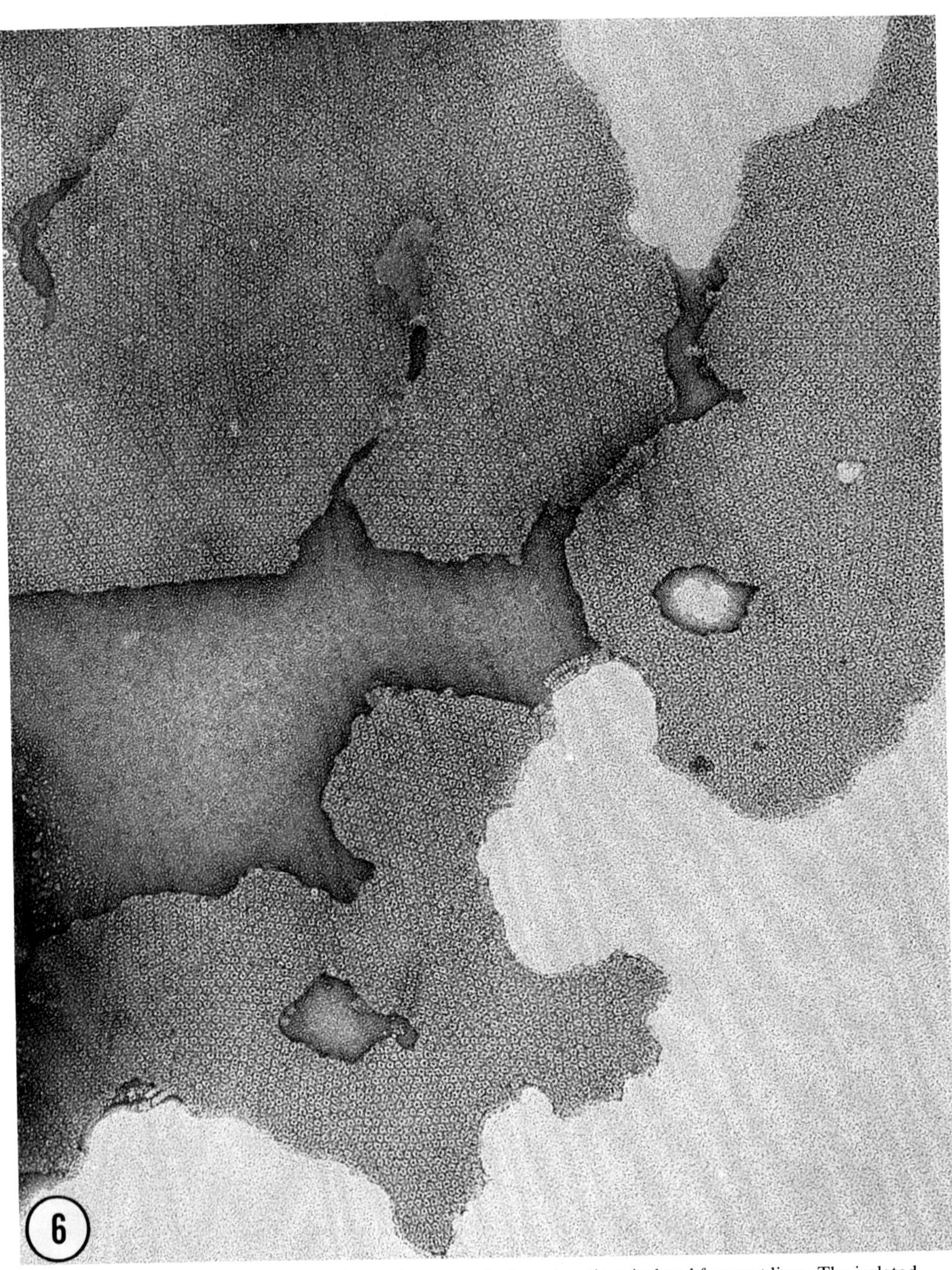

FIGURE 6 Negative stain (uranyl formate) treatment of gap junctions isolated from rat liver. The isolated junctions are plaques of a polygonal arrangement of 8- to 9-nm particles. The particles contain a central 1.5- to 2-nm electron-dense region that is the probable location of the polar channel for cell-to-cell communication. × 179,550.

1.5- to 2-nm region has been interpreted in thin sections and freeze-fracturing to represent the hydrophilic channel for cell-to-cell communication (Payton et al., 1969; McNutt and Weinstein, 1970).

In the isolated gap junction fractions, both protein and lipid are present, but no carbohydrate has been reported. In all of the isolated gap junction fractions from liver that have been reported, the junctional polypeptides have been affected by a proteolytic treatment that is used to reduce the collagen contamination. Nonetheless, it appears that a 25,000-dalton polypeptide can be detected, together with other polypeptides, as a prominent component in both mouse and rat liver preparations. Dunia et al. (1974) have isolated a fraction of junctions from bovine lens fibers, and they report the presence of a prominent 34,000-dalton polypeptide in this fraction.

Gap Junctions and Differentiation

Inasmuch as gap-junctional communication exists in a variety of developing tissues, it has been attractive to consider the potential regulatory role of this type of cellular interaction during differentiation. Unfortunately, it has been very difficult to demonstrate that gap-junctional communication is indeed directly or indirectly involved in any differentiation process. Recently, there have been several encouraging studies that may provide us with an excellent opportunity to explore the role of gap junctions in differentiation. I have subjectively selected two of these reports to illustrate this progress.

Blackshaw and Warner (1976) have found that, during somite formation in the amphibian embryos *Bombina* and *Xenopus*, the myotome cells in the segmented regions are coupled (communicating), the cells in the segmenting somite are poorly coupled to the cells in the segmented region, and the mesodermal cells in the unsegmented region are completely uncoupled from the cells in the segmented region. Thus, they have suggested that this coupling pattern potentially reflects the number of cells that have completed their morphogenetic movements. In these embryos, coupling is reestablished between cells of adjacent somites after segmentation is completed. Therefore, during this differentiation process, communication is initially present, lost, and then reestablished after the morphogenetic movements have ceased.

We have recently found that cell-to-cell communication is intimately associated with the follicular differentiation process in the mammalian ovary (Epstein et al., 1976). Gap-junctional communication, including fluorescent dry transfer, is present between the oocyte and cumulus oophorus (granulosa) cells in immature follicles. This observation is compatible with the morphological observations from other studies (Amsterdam et al., 1976; Anderson and Albertini, 1976). During follicular development, the communication gradually decreases as the time of ovulation approaches. After ovulation, the communication is not detectable between the cumulus cells and oocytes obtained from the oviduct. Hence, this pattern of communication between the cumulus and oocyte is perhaps closely related to the pattern of oocyte maturation that occurs during the same stages of development.

In the future, a clear definition of the relationship of cell communication to differentiation will depend on: (1) selecting a differentiation system that has a well-defined pattern of communication; and (2) developing procedures that will facilitate the experimental manipulation of communication either before it has been established or to disrupt communication that has been previously established.

Problems for the Future

In conclusion, I would like to list a few of the obvious problems that must be resolved in the near future in order to significantly amplify our understanding of the exciting sociology that exists between cells.

(1) Are gap junctions the only pathway for cell-to-cell communication? What about other junctional elements (tight junction, septate junction) and nonjunctional mechanisms?

(2) Are other (nonjunctional) mechanisms available for generating short, transient communication interactions?

(3) What is the relationship of gap-junctional size and frequency to gap-junctional permeability or coupling efficiency?

(4) How can we best study the biogenesis of cell communication and relate this to the biogenesis (synthesis and assembly) of gap junctions?

(5) What is the biological relevance of cell-to-cell communication in nonexcitable cell systems?

REFERENCES

Albertini, D. F., and E. Anderson. 1974. The appearance and structure of intercellular connections during the ontogeny of the rabbit ovarian follicle with particular reference to gap junctions. *J. Cell Biol.* **63**:234–250.

Albertini, D. F., D. W. Fawcett, and P. J. Olds. 1975. Morphological variations in gap junctions of ovarian granulosa cells. *Tissue Cell.* **7**:389–405.

Amsterdam, A., R. Josephs, M. E. Lieberman, and H. R. Lindner. 1976. Organization of intramembrane particles in freeze-cleaved gap junctions of rat graafian follicles: optical-diffraction analysis. *J. Cell Sci.* **21**:93–105.

Anderson, E., and D. F. Albertini. 1976. Gap junctions between the oocyte and companion follicle cells in the mammalian ovary. *J. Cell Biol.* **71**:680–686.

Azarnia, R., W. Michalke, and W. R. Loewenstein. 1972. Intercellular communication and tissue growth. VI. Failure of exchange of endogenous molecules between cancer cells with defective junctions and noncancerous cells. *J. Membr. Biol.* **10**:247–258.

Barr, L., M. M. Dewey, and W. Berger. 1965. Propagation of action potentials and the structure of the nexus in cardiac muscle. *J. Gen. Physiol.* **48**:797–823.

Benedetti, E. L., I. Dunia, and H. Bloemendal. 1974. Development of junctions during differentiation of lens fiber. *Proc. Natl. Acad. Sci. U. S. A.* **71**:5073–5077.

Benedetti, E. L., and P. Emmelot. 1968. Hexagonal array of subunits in tight junctions separated from isolated rat liver plasma membranes. *J. Cell Biol.* **38**:15–24.

Bennett, M. V. L. 1973. Function of electrotonic junctions in embryonic and adult tissues. *Fed. Proc.* **32**:65–75.

Berry, M. N., and D. S. Friend. 1969. High-yield preparation of isolated rat liver parenchymal cells: a biochemical and fine structure study. *J. Cell Biol.* **43**:506–520.

Blackshaw, S. E., and A. E. Warner. 1976. Low resistance junctions between mesoderm cells during development of trunk muscles. *J. Physiol.* (*Lond.*). **255**:209–230.

Chalcroft, J. P., and S. Bullivant. 1970. An interpretation of liver cell membrane and junction structure based on observations of freeze-fracture replicas of both sides of the fracture. *J. Cell Biol.* **47**:49–60.

Decker, R. S. 1976. Hormonal regulation of gap junction differentiation. *J. Cell Biol.* **69**:669–685.

Decker, R. S., and D. S. Friend. 1974. Assembly of gap junctions during amphibian neurulation. *J. Cell Biol.* **62**:32–47.

Dewey, M. M., and L. Barr. 1962. Intercellular connection between smooth muscle cells: the nexus. *Science* (*Wash. D. C.*). **137**:670–672.

Dreifuss, J. J., L. Girardier, and W. G. Forssman. 1966. Etude de la propagation de l'excitation dans le ventricule de rat du moyen de solutions hypertoniques. *Pflügers Arch. Eur. J. Physiol.* **292**:13–33.

Duguid, J. R., and J. P. Revel. 1975. The protein components of the gap junction. *Cold Spring Harbor Symp. Quant. Biol.* **40**:45–47.

Dunia, I., K. Sen, E. L. Benedetti, A. Zweers, and H. Bloemendal. 1974. Isolation and protein of eye lens fiber junctions. *FEBS* (*Fed. Eur. Biochem. Soc.*) *Lett.* **45**:139–144.

Epstein, M. L., W. H. Beers, and N. B. Gilula. 1976. Cell communication between the rat cumulus oophorus and the oocyte. *J. Cell Biol.* **70**:302a (Abstr.).

Evans, W. H., and J. W. Gurd. 1972. Preparation and properties of nexuses and lipid-enriched vesicles from mouse liver plasma membranes. *Biochem. J.* **128**:691–700.

Flower, N. E. 1972. A new junctional structure in the epithelia of insects of the order *Dictyoptera*. *J. Cell Sci.* **10**:683–691.

Furshpan, E. J., and D. D. Potter. 1959. Transmission at giant motor synapses of the crayfish. *J. Physiol.* (*Lond.*). **143**:289–325.

Furshpan, E. J., and D. D. Potter. 1968. Low-resistance junctions between cells in embryos and tissue culture. *Curr. Top. Dev. Biol.* **3**:95–127.

Gilula, N. B. 1974*a*. Junctions between cells. *In* Cell Communication. R. P. Cox, editor. John Wiley & Sons, Inc., New York. 1–29.

Gilula, N. B. 1974*b*. Isolation of rat liver gap junctions and characterization of the polypeptides. *J. Cell Biol.* **63**:111a (Abstr.).

Gilula, N. B., O. R. Reeves, and A. Steinbach. 1972. Metabolic coupling, ionic coupling, and cell contacts. *Nature* (*Lond.*). **235**:262–265.

Goodenough, D. A., and N. B. Gilula. 1974. The splitting of hepatocyte gap junctions and zonulae occludentes with hypertonic disaccharides. *J. Cell Biol.* **61**:575–590.

Goodenough, D. A., and W. Stoeckenius. 1972. The isolation of mouse hepatocyte gap junctions: preliminary chemical characterization and X-ray diffraction. *J. Cell Biol.* **54**:646–656.

Johnson, R. G., M. Hammer, J. Sheridan, and J. P. Revel. 1974. Gap junction formation between reaggregated Novikoff hepatoma cells. *Proc. Natl. Acad. Sci. U. S. A.* **71**:4536–4540.

Johnson, R. G., W. S. Herman, and D. M. Preus. 1973. Homocellular and heterocellular gap junctions in Limulus: a thin-section and freeze-fracture study. *J. Ultrastruct. Res.* **43**:298–312.

Johnson, R. G., and J. D. Sheridan. 1971. Junctions between cancer cells in culture: ultrastructure and permeability. *Science* (*Wash. D. C.*). **174**:717–719.

Loewenstein, W. R. 1966. Permeability of membrane junctions. *Ann. N. Y. Acad. Sci.* **137**:441–472.

McNutt, N. S., and R. S. Weinstein. 1970. The ultrastructure of the nexus: a correlated thin-section and freeze-cleave study. *J. Cell Biol.* **47**:666–687.

McNutt, N. S., and R. S. Weinstein. 1973. Membrane ultrastructure at mammalian intercellular junctions. *Prog. Biophys. Mol. Biol.* **26:**45–101.

MERK, F. B., J. T. ALBRIGHT, and C. R. BOTTICELLI. 1973. The fine structure of granulosa cell nexuses in rat ovarian follicles. *Anat. Rec.* **175:**107–125.

PAYTON, B. W., M. V. L. BENNETT, and G. D. PAPPAS. 1969. Permeability and structure of junctional membranes at an electrotonic synapse. *Science (Wash. D. C.).* **166:**1641–1643.

PERACCHIA, C. 1973. Low resistance junctions in crayfish. II. Structural details and further evidence for intercellular channels by freeze-fracture and negative staining. *J. Cell Biol.* **57:**66–76.

REVEL, J. P., L. CHANG, and P. YIP. 1973. Cell junctions in the early chick embryo: a freeze-etch study. *Dev. Biol.* **35:**302–317.

REVEL, J. P., and M. J. KARNOVSKY. 1967. Hexagonal array of subunits in intercellular junctions of the mouse heart and liver. *J. Cell Biol.* **33:**C7–C12.

RIESKE, E., P. SCHUBERT, and G. W. KREUTZBERG. 1975. Transfer of radioactive material between electrically coupled neurons of the leech central nervous system. *Brain Res.* **84:**365–382.

ROBERTSON, J. D. 1963. The occurrence of a subunit pattern in the unit membranes of club endings in Mauthner cell synapses in goldfish brains. *J. Cell Biol.* **19:**201–221.

SIMIONESCU, M., N. SIMIONESCU, and G. E. PALADE. 1975. Segmental differentiations of cell junctions in the vascular endothelium: the microvasculature. *J. Cell Biol.* **67:**863–886.

STAEHELIN, L. A. 1974. Structure and function of intercellular junctions. *Int. Rev. Cytol.* **39:**191–283.

SUBAK-SHARPE, J. H., R. R. BURK, and J. D. PITTS. 1969. Metabolic cooperation between biochemically marked mammalian cells in tissue culture. *J. Cell Sci.* **4:**353–367.

PERMEABILITY OF THE JUNCTIONAL MEMBRANE CHANNEL

WERNER R. LOEWENSTEIN

It has become clear over the past 12 years that most cells in organized tissues are interconnected by membrane channels built into the cell junction (Loewenstein, 1966, 1975; Furshpan and Potter, 1968). Many molecules can pass freely from one cell to another through these junctional channels; at variance with classic cell theory, the connected cell ensemble, rather than the single cell, is the unit in many functional respects. I shall deal here with the permeability of the channels and its regulation. In my talk at the Congress, I touched also on the formation and genetics of the channels. These aspects I have already reviewed elsewhere (Loewenstein, 1975; 1977).

The Permeability and Size of the Channel

The concept of the junctional channel unit was originally formulated on the basis of electrical measurements of high spatial resolution and of studies of diffusion of fluorescent and colorant tracers injected into large epithelial cells (Loewenstein, 1966). The junctional unit was thus defined as consisting of three elements: a pair of membrane elements of high permeability, the junctional membrane channels, matched on either side of the joined membranes; and an element of insulation that makes the channel pair leakproof at its junction (Fig. 1, top). The present-day general concept of a membrane channel is that of a protein spanning the lipid bilayer with the hydrophilic amino acid residues lining a continuous water channel (Singer, 1974). Thus, the junctional unit may now be envisioned as consisting of a pair of such protein channels on either membrane, tightly joined, where the insulation would be provided by the hydrophobic amino acid residue portions and their junction (Fig. 1, bottom; Loewenstein, 1974). The prevailing notion is that the channels are contained in the intramembranous particles seen in freeze-fracture electron microscopy aligned on the two sides of the membrane junction, forming closely packed aggregates (gap junction). The morphology of these particles and their aggregates is dealt with by N. B. Gilula in the preceding article.

An estimate of the electrical conductance of the junctional membrane channel may be obtained from limits of the junctional conductance and of the number of channels per unit junctional membrane area. A recent measurement across a minute junctional area (newt embryo cells) gave a lower limit of junctional conductance of unit membrane area of 10^2 mho/cm^2 (Ito et al., 1974*b*). The number of channels of unit membrane area may be estimated from electron micrographs of freeze-fractured junction on the assumption that the intramembranous particles of "gap" junction contain the channels. For particles with a spacing of $\sim$100 Å, approximately hexagonally arrayed as found in gap junction of many cell types (e.g., Goodenough and Revel, 1970), this amounts to 10^{12} channels/cm^2. Hence, from the aforegoing limit of membrane conductance, we obtain a lower limit of conductance for the single junctional channel, of 10^{-10} mho. For a 200-Å long cylindrical channel (two-membranes thick) with a cytoplasmic resistivity of 50 Ω cm (including the resistive component resulting from electrostatic channel interaction), this corresponds to a channel with a bore of the order of 10 Å (Loewenstein, 1975).

A closer estimate of the bore of the channel can be obtained with the aid of molecular probes. Here one may take advantage of the fact that the junctional channels are permeated also by mole-

---

WERNER R. LOEWENSTEIN Department of Physiology and Biophysics, University of Miami School of Medicine, Miami, Florida

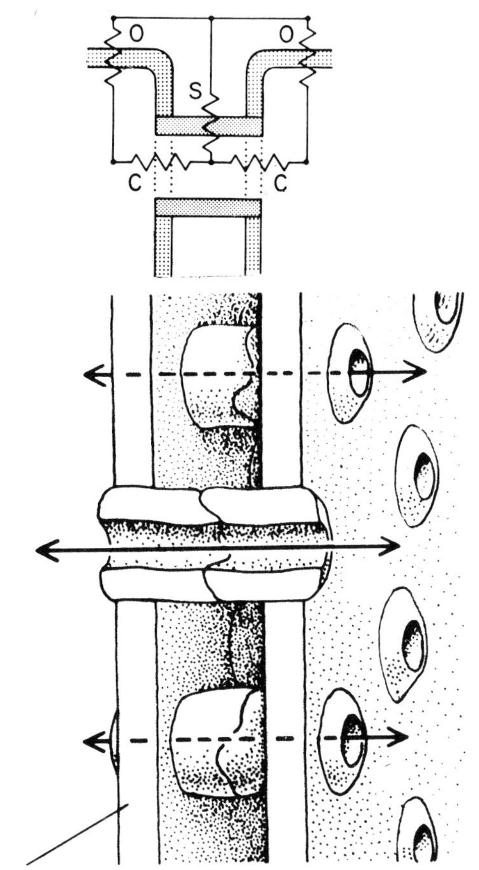

FIGURE 1 (Top) The *junctional unit* as inferred from electrical measurements and intracellular tracer diffusion. *O*, nonjunctional cell membrane; *C*, junctional membrane channels, matched on the two sites of the membrane junction; *S*, junctional insulation. The unit is represented entirely in terms of permeability properties (from Loewenstein, 1966). (Bottom) The unit now endowed with somewhat more specific structural attributes: two aligned protein channels where the junctional insulation, *S*, is given by the bonding between the hydrophobic outer portions of the proteins (reprinted with permission from Loewenstein, 1974). For further morphological specifications, see McNutt and Weinstein, 1973; Goodenough, 1975; Peracchia and Férnandez-Jaimovich, 1975; Gilula and Epstein, 1977.

cules larger than the inorganic ions, the carriers of electrical current. This was originally found in a salivary cell junction, using the strongly fluorescent fluorescein (300 daltons) as a tracer (Loewenstein and Kanno, 1964) and subsequently confirmed for a variety of cells (Pappas and Bennett, 1966; Furshpan and Potter, 1968; Rose, 1971; Sheridan, 1971; Azarnia and Loewenstein, 1971; Azarnia et al., 1974; Pollack, 1976). A number of tracer molecules that do not significantly permeate nonjunctional membrane have since been shown to pass through junction: several colorant molecules ranging from 300 to 990 daltons (Kanno and Loewenstein, 1966; Potter et al., 1966), the 550-dalton fluorescent Procion Yellow (Payton et al., 1969; Rose, 1971; Johnson and Sheridan, 1971), the fluorescent 370-dalton dansyl-DL-aspartate, and the 380-dalton dansyl-L-glutamate (Johnson and Sheridan, 1971). (See also Rieske et al., 1975, and Pitts, 1977, for demonstrations of cell-to-cell passage of molecules by autoradiographic techniques.) Ian Simpson, Birgit Rose, and I have recently constructed a series of fluorescent tracers, synthetic and natural peptide molecules, with the aim of determining the size of the junctional channels.

We set out to make fluorescent conjugates which incorporate some of the desirable features of the tracer fluorescein, such as water solubility, nontoxicity, low cytoplasmic binding, high fluorescent yield. To obtain conjugates of well-defined structure, we sought, for the nonfluorescent backbone, not only a molecule of known structure but one with few reactive sites, preferably only one. Thus, the primary amine group of the peptides listed in Table I was coupled with the fluorescent dyes fluorescein isothiocyanate (FITC), dansylchloride (DANS), or lissamine rhodamine B (LRB). These peptides make good, selective junctional membrane probes; none of them permeates nonjunctional cell membrane. They are also useful for simultaneous permeability probing with molecules of different size; their fluorescent color labels are easily distinguishable inside the cells (FITC and DANS fluoresce in the yellow-green; LRB, in the red). The probes were microinjected into *Chironomus* salivary cells, and the spread of the fluorescence inside the cells was observed in a microscope dark field; or for determination of diffusion velocities, the spread was scanned with an image intensifier-television system (Simpson et al., 1977).

The results obtained with these probes are summarized in Table I. The amino acids and peptides with sizes $\leq$1,158 daltons pass through the channels. These molecules move through the junction with velocities inversely related to their size. The peptides $\geq$1,926 daltons do not pass. The latter reflects an actual exclusion of the molecules from the channels, and not simply a channel alteration caused by these peptides: the smaller molecules injected together (or in succession) with the nonpermeants continue to pass through. Thus, in the example of Fig. 2, the 1,158-dalton LRB $(Leu)_3(Glu)_2$ OH (the largest permeant of the series) traverses the junction, whereas the 1,926-dalton FITC fibrinopeptide 'A' does not.

The results give a cutoff limit for permeation by peptides of about 1,200–1,900 daltons. An earlier estimate of a limit based on the junctional passage of FITC serum albumin (Kanno and Loewenstein, 1966) is thus shown to be wrong. The protein was probably enzymatically degraded in the cells, and an FITC-labeled fragment (with the antigenicity of the original molecule) went through the junction. The possibility of such a degradation of the probes in Table I is reasonably excluded by tests in which the permeant molecules are incubated for 2–12 h

TABLE I*

Cell-to-Cell Passage of Fluorescent Probes

Molecule	*mol wt*	Cell-cell spread cases†	Molecule	*mol wt*	Cell-cell spread cases†	Control molecule§
$DANS(SO_3H)$	251	2 (2)	FITC fibrinopeptide 'A'	1,926	0 (6)	LRB(Glu)OH
DANS(Glu)OH	380	12 (12)				$LRB(Leu)_3(Glu)_2OH$
$LRB(SO_3H)$	559	35 (35)	FITC microperoxidase	2,268	0 (1)	LRB(Glu)OH
$DANS(Gly)_6OH$	593	7 (7)	FITC insulin 'A' chain	2,921	0 (10)	LRB(Glu)OH
$DANS(Glu)_3OH$	640	12 (12)	DANS insulin 'A' chain	3,232	0 (3)	‖
LRB(Glu)OH	688	13 (13)	FITC insulin 'B' chain	3,897	0 (7)	$LRB(Glu)_3OH$
$FITC(Gly)_6OH$	749	2 (2)	LRB insulin 'A' chain	4,158	0 (7)	Fluorescein (330 mol wt)
$FITC(Glu)_3OH$	794	45 (45)				
$DANS(Leu)_3(Glu)_2OH$	849	12 (14)				
$LRB(Gly)_6OH$	901	13 (13)				
$LRB(Glu)_3OH$	950	13 (15)				
LRB(Glu-Tyr-Glu)OH	982	2 (2)				
$FITC(Leu)_3(Glu)_2OH$	1,004	2 (23)¶				
$LRB(Leu)_3(Glu)_2OH$	1,158	50 (56)				

(Glu)OH = glutamic acid; (Gly)OH = glycine; (Leu)OH = leucine; (Tyr)OH = tyrosine.

* From Simpson et al., 1977.

† The number of trials are in parentheses.

§ Small molecule (spreading) used as control for same junction.

‖ Not tested with control molecule, but junction was coupled electrically.

¶ In seven cases of no spread of this tracer, $LRB(Glu)_3OH$ or $LRB(Leu)_3(Glu)_2OH$ was used as control molecule and found to spread through the same junction.

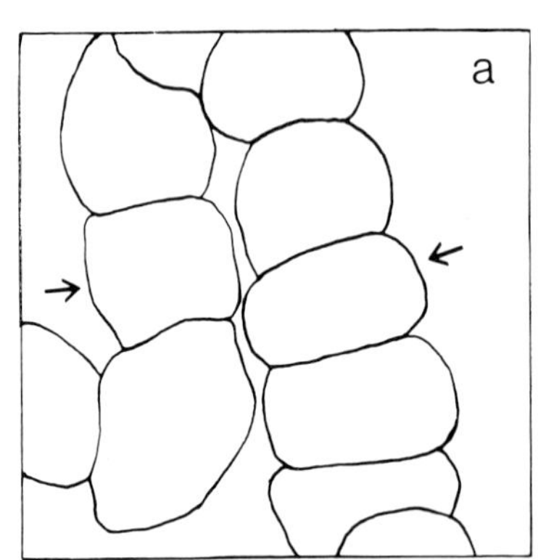

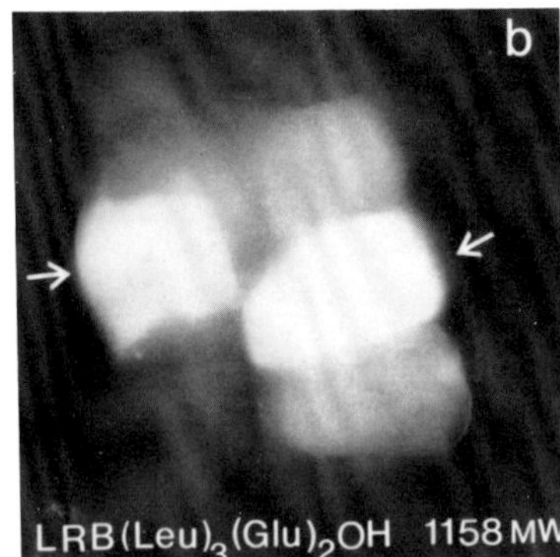

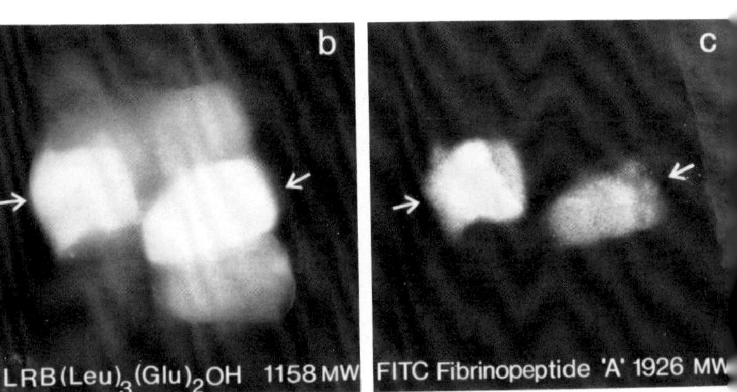

FIGURE 2 Probing junctional membrane channels with two molecules close to the size limit of permeation. The red fluorescent tracer $LRB(Leu)_3(Glu)_2OH$ of 1,158 daltons, the largest permeant molecule of the present series, is injected into the cells marked with an arrow, together with the yellow-green fluorescent tracer FITC-fibrinopeptide 'A' of 1,926 daltons. *b* shows the distribution of the red tracer (1,158 daltons) photographed (dark field) in black and white; and *c*, that of the yellow-green tracer (1,926 daltons). (The two fluorescences are set apart by the use of different excitation wavelengths and barrier filters.) The red tracer spread from the injected cells to several neighbors; the yellow-green tracer stayed within the injected cells. *a*, tracing of the cells from a bright-field photograph. Calibration, 100 μm. (Reprinted with permission from Simpson et al., 1977.)

with the cytoplasm of mashed cells, yielding single fluorescent spots with the mobility of the original labeled peptide.

We may now estimate from the molecular weight limit of junctional permeation the approximate bore size of the channel. The permeant probes are all short, simple peptide chains. Thus, we can bracket the channel bore between the sizes of two limiting shapes of the largest permeant: a sphere, providing the largest channel cross section; and a prolate spheroid with a minor diameter providing the smallest cross section and a major diameter representing the upper limit of molecular extension. The largest cross section (2 r) may then be approximated from

$$r = \left(\frac{3 \text{ mol wt } \bar{v}}{4\pi N}\right)^{1/3},$$

where N is Avogadro's number and $\bar{v}$ the specific volume assumed to be 0.7. The major diameter for the most extended molecular shape, as determined from molecular models, is 30 Å. Thus we obtain a diameter of the junctional membrane channel lying approximately between 14 and 10 Å, in satisfying agreement with the above order estimate based on electrical measurements. The actual channel bore probably lies closer to the upper value; for the molecules labeled with LRB and FITC, the small diameter of a realistic axiosymmetric equivalent is set at about this value by the size of the labels themselves.

Regulation of Channel Permeability by Ca^{2+}

CHANNEL CLOSURE: The channel permeability depends on Ca^{2+}. It is high in the low Ca^{2+} concentrations normally prevailing in cytoplasm ($\leq 10^{-7}$ M). It falls when the cytoplasmic Ca^{2+} rises or when the junctional insulation is broken and the channels are exposed to the high Ca^{2+} concentrations of the cell exterior. That the permeability depends on the cytoplasmic Ca^{2+} concentration is shown most simply by experiments in which a hole is made in the cell membrane and the cell interior is equilibrated with known concentrations of Ca^{2+} in the exterior. The conductance of junctional membrane and its permeability for molecules of the size of fluorescein fall drastically at Ca^{2+} concentrations above 5–8 × 10^{-5} M; the junctional channels then virtually close off (Oliveira-Castro and Loewenstein, 1971).

A more direct and analytically more powerful way of demonstrating the effect of Ca^{2+} is to inject the ion into the cell while monitoring the intracellular free Ca^{2+} concentration ($[Ca^{2+}]_i$). This was done by the use of the luminescent protein aequorin as a $[Ca^{2+}]_i$ indicator (Rose and Loewenstein, 1975*a*, 1976). The aequorin is injected into a pair of cells, and its light emission, approximately proportional to $[Ca^{2+}]^2$ (Shimomura and Johnson, 1969), is scanned with an image intensifier-television system (Fig. 3). This allows us to see where inside the cell the $[Ca^{2+}]$ is changing and by how much. The method has a spatial resolution of 1 μm and a sensitivity for 5 × 10^{-7} M $[Ca^{2+}]_i$ over the average volume of *Chironomus* salivary cells. In the experiment illustrated in Fig. 4, buffered Ca^{2+} is microinjected into a cell. A brief puff of Ca^{2+} (5 × 10^{-5} free

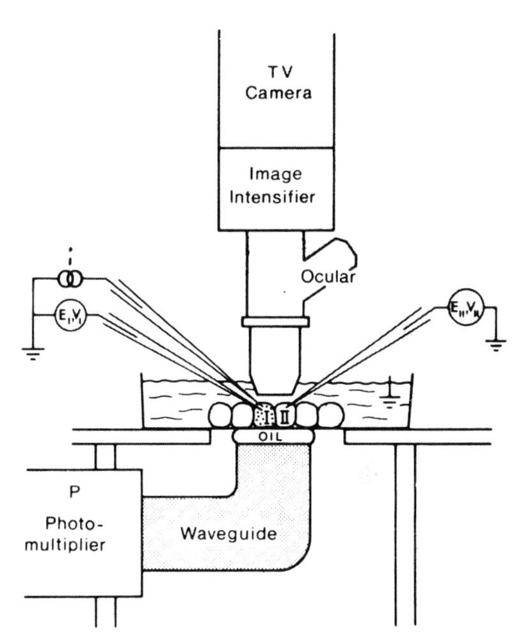

FIGURE 3 Image intensifier coupled to a television camera scans the aequorin luminescence in the cells through a microscope (dark field); luminescence is also measured by photomultiplier. Electrical coupling is measured by pulsing current (i) between interior and exterior of cell *I* and the resulting steady-state changes (*V*) in membrane potential (*E*) are measured in Cells *I* and *II*. Photomultiplier current *P* and the coupling parameters i, E_I, E_{II}, V_I, V_{II} are displayed on a chart recorder and a storage oscilloscope onto which a second television camera is focused. The two camera outputs are displayed simultaneously on a monitor and videotaped. (Reprinted with permission from Rose and Loewenstein, 1975*a*).

Ca^{2+}) is seen as an aequorin glow that is confined to the immediate vicinity of the injection micropipette (Fig. 4 A, *ii*). The confinement is mainly the result of fast energized Ca^{2+} sequestering by intracellular Ca^{2+} sinks (Rose and Loewenstein, 1975*b*). Such a local Ca^{2+} elevation, some 50 μm away from junction, does not affect the channels. However, when the injection saturates the Ca^{2+} sinks, and the $[Ca^{2+}]_i$ elevation reaches the junction, the channels close promptly (Fig. 4 A *iv*, and B). They open up again after the cell rids itself of the excess Ca^{2+} (Fig. 4 B).

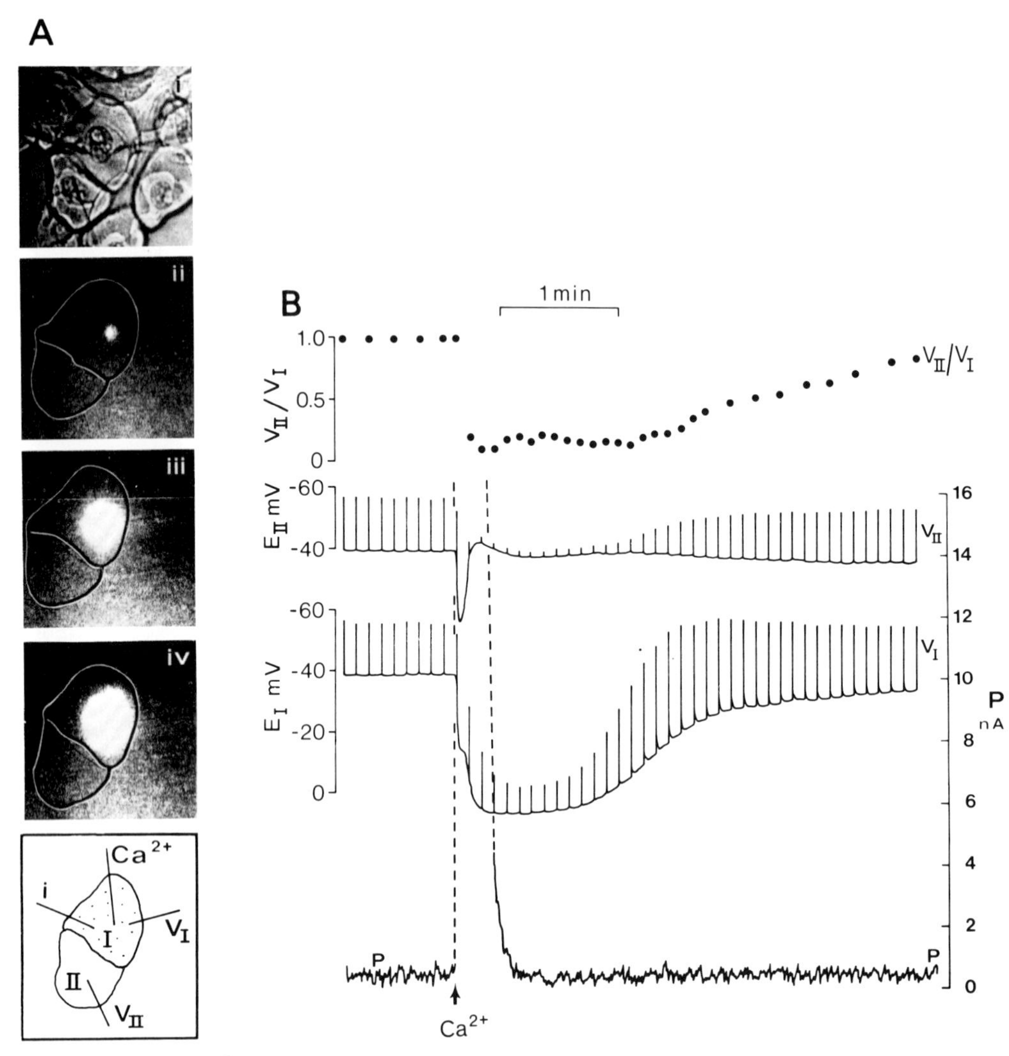

FIGURE 4 Channel closure by Ca^{2+}. Ca^{2+} is microinjected into cell *I*, while monitoring electrical coupling between cells *I* and *II*. (A) Dark-field television pictures (*ii–iv*) of aequorin luminescence produced by three puffs of 5×10^{-5} M free Ca^{++} (buffered with EGTA) of increasing magnitude delivered to about the center of the basal region of a *Chironomus* salivary gland cell. The pictures were each taken at the time of maximum luminescence spread. Cell diameter ca 100 μm. Puffs in *ii* and *iii* do not reach the junction of the cell and do not affect coupling. The puff in *iv* reaches one junction causing transient uncoupling, as shown by the electrical measurements in B: chart records of *P*, E_I, E_{II}, V_I, V_{II} ($i = 4 \times 10^{-8}$ A) and plot of coupling coefficient V_{II}/V_I. Note the rapid recoupling upon restoration of normal Ca^{++} activity. (*i*) Bright-field television picture of the cells. (A, *bottom*) Cell diagram showing location of microelectrodes and of Ca injection pipette (hydraulic); dotted cell preinjected with aequorin. (Reprinted with permission from Rose and Loewenstein, 1975*a*.)

A further informative demonstration of the action of Ca^{2+} is provided by experiments in which the cells are poisoned with cyanide or dinitrophenol. This leads to $[Ca^{2+}]_i$ elevation detectable as an aequorin glow that eventually extends to all junctions of the cell. When the $[Ca^{2+}]_i$ elevation reaches a junction, the channels invariably close (Fig. 5). Similar results and an equally good correlation between rising $[Ca^{2+}]_i$ and channel closure are obtained when the cells are treated with the Ca ionophores A23187 or X537A (Rose and Loewenstein, 1976).

In the above conditions, the $[Ca^{2+}]_i$ elevation generally produces depolarization of nonjunc-

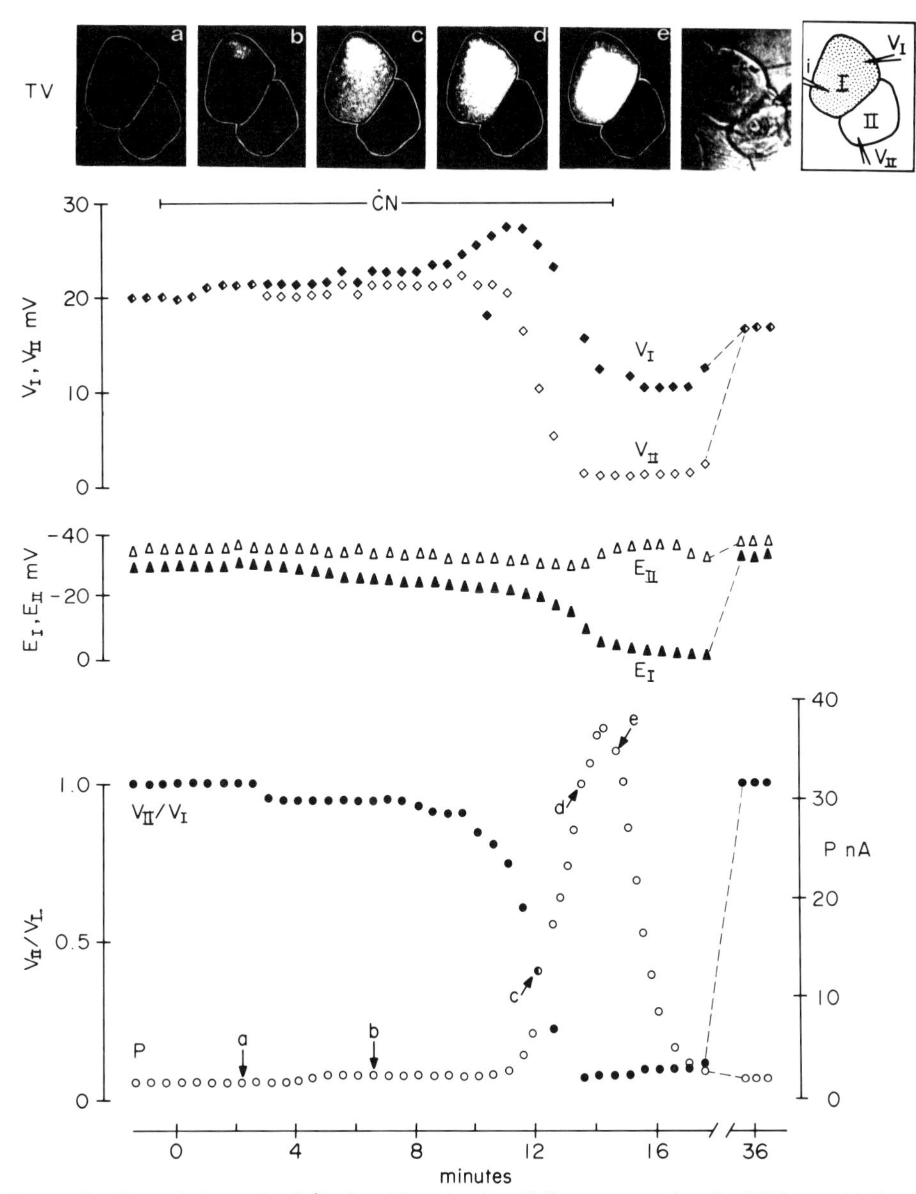

FIGURE 5 Channel closure by Ca^{2+}. Cyanide poisoning. Cells are exposed to 5 mM Na cyanide for the period indicated by the bar. They are in Ca-free medium throughout the experiment. $[Ca^{2+}]_i$ rises at first locally (television picture *b*), and then diffusely associated with uncoupling (*c–e*). $[Ca^{2+}]_i$ falls upon washout of cyanide and recoupling ensues. $i = 4 \times 10^{-8}$ A. (Reprinted with permission from Rose and Loewenstein, 1976.)

tional cell membrane. However, depolarization is neither sufficient nor necessary for junctional channel closure. When Ca^{2+} is injected into voltage-clamped cells, or it is injected deeply into the cells close to junction, avoiding the effects of Ca^{2+} on nonjunctional membrane, the channels close in the absence of significant depolarization. Conversely, when the cells are exposed to high [K], they depolarize completely within 2–5 min, but they stay coupled much longer, often for several hours; eventually, and rather abruptly, the channels close, and this is invariably associated with an abrupt $[Ca^{2+}]_i$ elevation (Rose and Loewenstein, 1976).

The channel reacts rapidly to Ca^{2+}. The channel closure appears to be so fast that it critically limits the flow of this ion through the channel. This is suggested by the results of experiments in which we tested for transjunctional flow by Ca^{2+} injections close to a junction while monitoring $[Ca^{2+}]_i$ on both sides of it. Ca^{2+} was not detectable transjunctionally although the chemical gradients across junction were steep; on the injected side of the junction, $[Ca^{2+}]_i$ was three orders of magnitude above detection threshold (Fig. 6). By contrast, fluorescein, although larger, traversed the junction in fractions of a second.

GRADED PERMEABILITY CHANGES: The preceding section dealt with $[Ca^{2+}]_i$ elevations to above 5×10^{-5} M, and at this level we have seen that the channels are effectively closed off for all molecular species down to the smallest inorganic ions. Junctional permeability can also undergo less radical changes. In the range of 10^{-7}–10^{-5} M, it changes selectively in respect to the size of the permeant molecules. The first suggestions to this effect came from work in which the junctional passage of fluorescein was found to be blocked at low $[Ca^{2+}]_i$ elevations where the cells still remained electrically coupled (Délèze and Loewenstein, 1976). Because measurements of electrical coupling are relatively insensitive to changes in junctional conductance at high levels of coupling, this result admitted two interpretations: (1) the channel permeability falls for the fluorescein molecule (blocking its passage) and not, or much less, for the smaller inorganic molecules carrying the current; or (2) a fraction of the channels close unselectively (i.e., completely).

The development of the junctional permeability probes (Table I) permitted us to distinguish between these possibilities. We probed the junction simultaneously with two molecules of different size and fluorescent color while raising $[Ca^{2+}]_i$ to different levels in the range of $10^{-7} - 10^{-5}$ M. It turned out that the permeability change is, indeed, selective in respect to molecular size: the size limit of channel permeation diminishes gradually with rising $[Ca^{2+}]_i$ (B. Rose, I. Simpson, and W. R. Loewenstein, to be published). Figure 7 illustrates an example of an experiment in which, in separate trials, three different tracer pairs are injected with Ca buffered to yield three different levels of free Ca^{2+}. The transit through junction by the molecules is blocked or slowed in order of their size, as the $[Ca^{2+}]$ in the injected solution rises. In general, measurements of junctional transit velocities in experiments of this kind with 10 molecular pairs (Table II) showed that junctional transit is retarded more for the larger molecule in each pair than it is for the smaller one and in some cases (as in Fig. 7, II, III, IV) is even sensibly blocked while the small molecule is passing. With the available repertoire of 10 probe pairs, we can distinguish at least two gradations in the size limit of permeation, one between 901 and 642 daltons (or between 794 and 559) and another between 559 and 330 daltons. If we extrapolate now to the earlier result showing a dissociation between the junctional passage of the 330-dalton fluorescein and electrical coupling (Délèze and Loewenstein, 1976), two further steps seem likely, one between 330 daltons and the size of the smallest inorganic ion, and a fourth where the latter is barred from passage, as the $[Ca^{2+}]_i$ rises above about 5×10^{-5} M. The actual grading may, of course, be finer.

Viewed in terms of the idea that the permeable membrane junction is made of junctional channel units (Loewenstein, 1966), the gradual decrease of the molecular size limit may simply be attributed to a gradual decrease of the effective channel size with rising $[Ca^{2+}]_i$. The junction would behave like a sieve in which the unit mesh is controlled by Ca^{2+}. Alternatively, the junction might contain channels of different size and Ca sensitivity; the result would be accounted for by an all-or-none closure of channels with Ca sensitivity directly related to channel size.

ON THE MECHANISM OF CA ACTION: We do not know the mechanism by which the calcium ion changes the channel permeability. A plausible mechanism is that the ion binds to junctional membrane (Loewenstein, 1967), causing a change in the channel's fixed charge or molecular conformation that reduces its effective size.

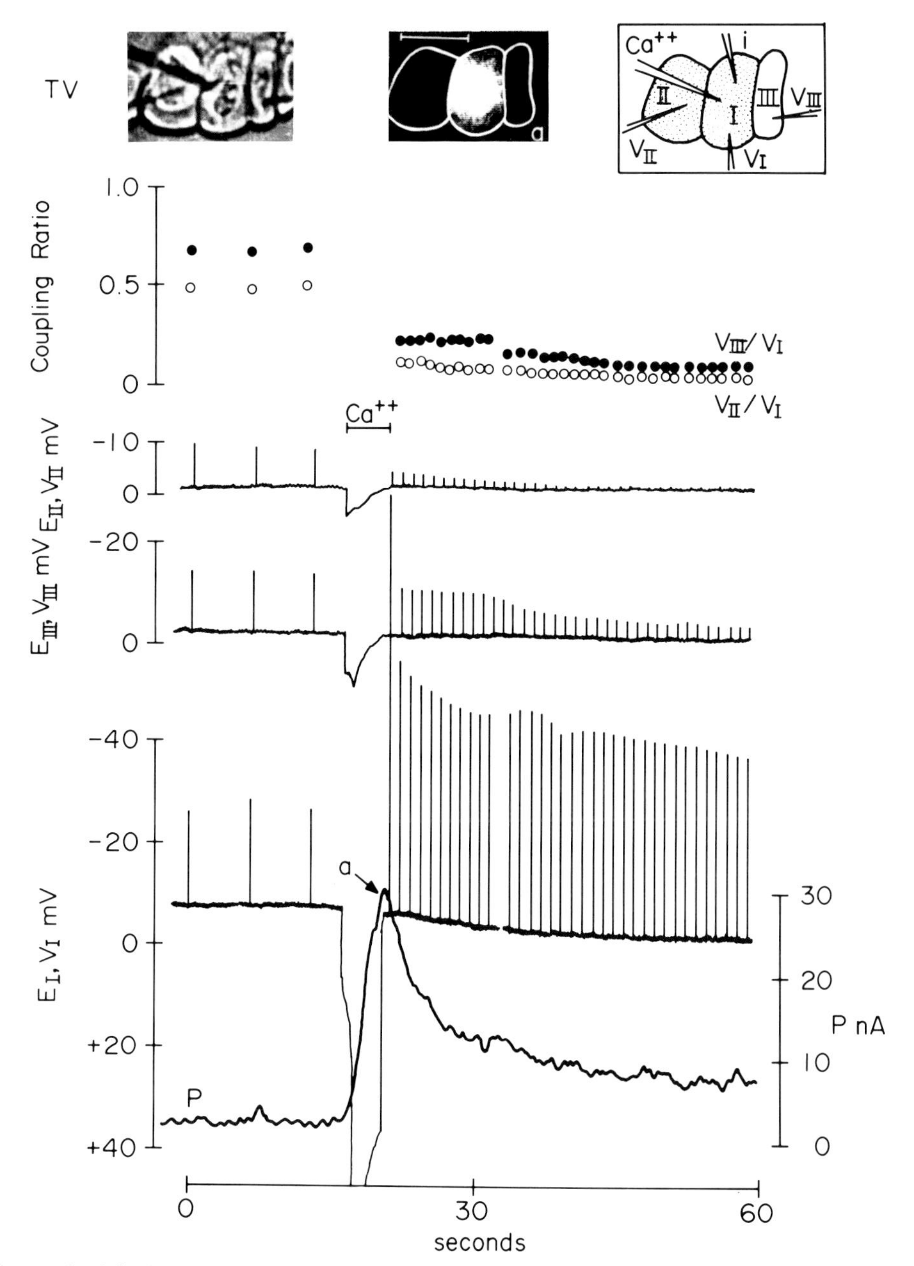

FIGURE 6 Ca^{2+} does not detectably flow through the junctional channel. Cells *I* and *II* are preinjected with aequorin. Ca^{2+} injection (iontophoresis) into cell *I* produces elevation of $[Ca^{2+}]_i$ along the entire length of junction 1:2 on the injected cell side but not on the other side (television picture *a*; calibration 100 μm). The junctional boundary here is seen as a sharply defined border of the glow sphere. The $[Ca^{2+}]_i$ elevation reaches also junction 1:3. Both junctions uncouple, with marked rise in V_I. $i = 4 \times 10^{-8}$ A. (Reprinted with permission from Rose and Loewenstein, 1976.)

I favor this simple notion because the permeability change is readily reversed when the normal $[Ca^{2+}]_i$ is restored in cells with normal Ca sequestering and Ca pumping. In this light, the fact that the reopening of the channels typically lags behind the decline in $[Ca^{2+}]_i$ (see Fig. 4B) suggests a slow release of junction-bound Ca^{2+}. This would of course in any event be expected if the membrane

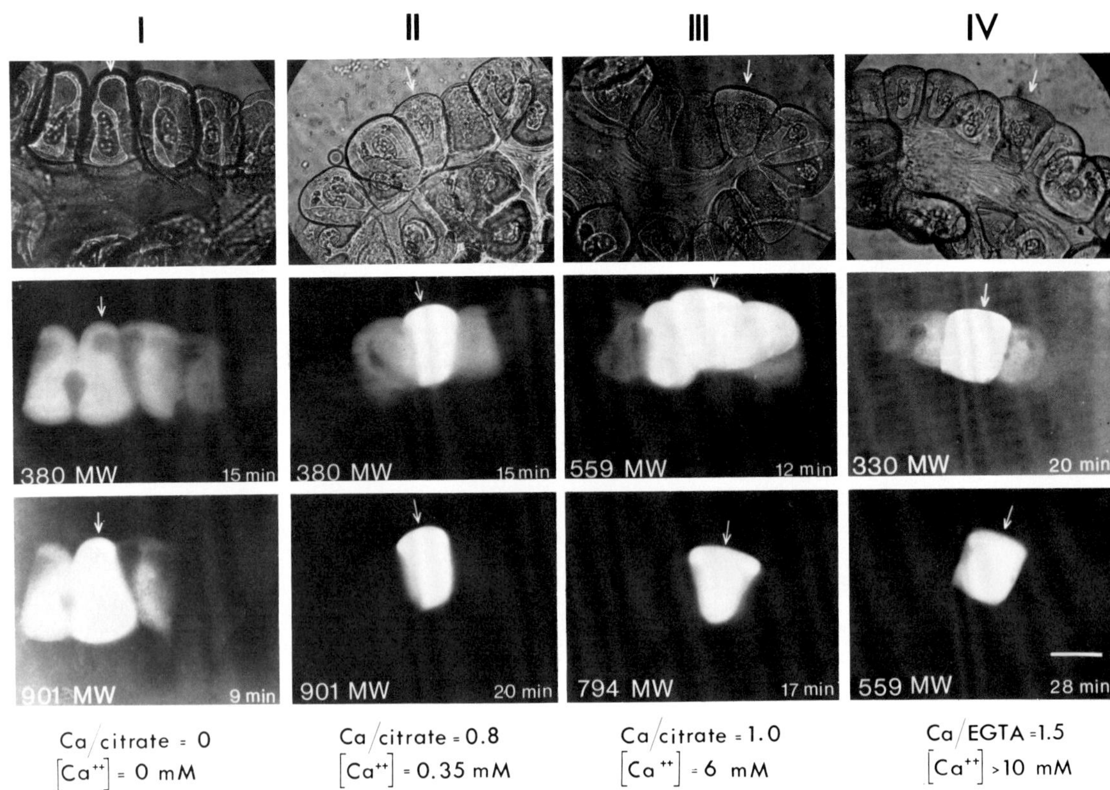

FIGURE 7 Graded changes in channel permeability by Ca^{2+}. Permeability of junctional membrane channels is probed with pairs of fluorescent-labeled tracers. The fluorescence of one tracer in each pair is yellow-green and the other is red. The pairs are microinjected into a cell (arrow) together with Ca^{2+} buffered with citrate or mixed with EGTA, and the tracer fluorescence is viewed and photographed through barrier filters. Shown are the permeability tests with four different tracer molecules at different Ca^{2+} concentrations. Each column (I–IV) corresponds to one test. In I, the cytoplasmic Ca^{2+} concentration is normal; the injected solution contains zero Ca. In II–IV, the solution contains free Ca^{2+} concentrations increasing from II to IV. On top of each column are the bright-field photomicrographs of the cell systems, and in the center and bottom, the dark-field photomicrographs of the color fluorescences of the paired tracers, taken in black-and-white through the respective barrier filters. The corresponding tracer molecule and the time the photographs were taken are indicated in each case, and the Ca/citrate or Ca/EGTA ratios and the free Ca^{2+} of the injected solution are given on the bottom of each column. In I, both tracer molecules of the pair pass from cell-to-cell; in II, III, and IV, the larger molecule in each pair is blocked. Calibration, 100 μm. (From B. Rose, I. Simpson, and W. R. Loewenstein, unpublished data.)

affinity for Ca^{2+} is high, but the lag could be magnified if the binding sites were in a narrow cleft or the release involved cooperativity (see, for instance, the binding and release of divalent cations by actin or phosphatidic acid; Loscalzo et al., 1975; Haynes, 1977).

As to the mode of channel closure by Ca^{2+}, two general mechanisms are readily envisaged: (1) the effective bore of the membrane channel is reduced (and here I include possible mechanical as well as electrostatic stricture); or (2) the channels in the two joined membranes are misaligned (without loss of junctional insulation). Structural alterations of this sort could conceivably be brought about by Ca binding to the channels themselves or, somewhat less directly, by Ca binding to other constituents of the junctional membrane whose change in conformation is transmitted to the channels. In this connection it is important to note the electron microscope findings of Peracchia and Dulhunty (1976) in uncoupled crayfish electrical synapses: the gap-junctional membrane particles become smaller and more closely packed after uncoupling by dinitrophenol or other agents producing elevation of $[Ca^{2+}]_i$.

ON THE PHYSIOLOGICAL ROLE OF CA^{2+} IN

TABLE II*

Tracer Pairs			
	mol wt		mol wt
$LRB(Leu)_3(Glu)_2OH$	1,158 -	$DANS(Glu)OH$	380
$LRB(Leu)_3(Glu)_2OH$	1,158 -	$DANS(Glu)_3OH$	640
$LRB(Leu)_3(Glu)_2OH$	1,158 -	$FITC(Glu)_3OH$	794
$LRB(Leu)_3(Glu)_2OH$	1,158 -	$DANS(Leu)_3(Glu)_2OH$	849
$LRB(Gly)_6OH$	901 -	$DANS(Glu)OH$	380
$LRB(Gly)_6OH$	901 -	$DANS(Glu)_3OH$	640
$DANS(Leu)_3(Glu)_2OH$	849 -	$LRB\ SO_3H$	559
$FITC(Glu)_3OH$	794 -	$LRB\ SO_3H$	559
$LRB\ SO_3H$	559 -	Fluorescein	330

LRB = lissamine rhodamine B; DANS = dansyl (dimethylaminonaphthalene sulfonyl); FITC = fluorescein isothiocyanate; Glu OH = glutamate; Leu OH = leucine; Gly OH = glycine.

* From Rose, Simpson, and Loewenstein, unpublished data.

THE REGULATION OF INTERCELLULAR COMMUNICATION: Among the possible functional adaptations of the channel closure by high $[Ca^{2+}]_i$, the most obvious one is the capacity of the connected cell ensemble to seal itself off from an unhealthy member. All elements of such a sealing reaction are built into the cell system and are criticially poised: the steep chemical and electrical gradients tend to drive Ca^{2+} inward, and the junctional channels are capable of closing rapidly in the presence of high $[Ca^{2+}]_i$. All that is required to set the reaction into motion is a discontinuity in the cell membrane or a depression of cellular energy metabolism on which intracellular sequestering and outward pumping of Ca depends. It is thus easy to see here two possible functional roles of Ca^{2+}: (1) that of uncoupling a cell community from a cell member with defective Ca pumping or sequestering mechanisms, and (2) that of uncoupling it from a cell member with a damaged membrane. The first is likely to apply to cells reaching the end of their life span; and the second, in fact, is known to apply to tissue injury (Loewenstein and Penn, 1967). Were it not for this fast mechanism of channel closure, tissues, such as skin, with widely interconnected cells, would not survive cell injury; or a liver, in which most cells are interconnected (Penn, 1966), would not survive the death of even a single cell.

As to the graded control of junctional permeability in the lower range of $[Ca^{2+}]_i$, this offers, in principle, a powerful mechanism for physiological regulation of intercellular communication. It provides a means for selective transmission of intercellular molecular signals if $[Ca^{2+}]_i$ changed in different physiological states. An interesting possibility is that such a selective signal transmission plays a role in embryonic development, namely, in the determination and fixation of cellular differentiations. I have discussed this in detail elsewhere (Loewenstein, 1968*a*, *b*). Here I shall add only that molecular sieving by junction, which then seemed remote, emerges now in the light of the finding of permeability gradation as a serious possibility. Also pointing in this direction is the discovery by Slack and Palmer (1969) that in the early embryo, where there is extensive electrical coupling between cells (cf. Furshpan and Potter, 1968; Loewenstein, 1968*b*; Warner, 1973), fluorescein may not pass between electrically coupled cells. It is not clear, because of the aforementioned limitations of the electrical method, whether this phenomenon reflects a molecular sieving or simply a nonselective reduction in the number of channels. If the former turns out to be the case, it will be interesting to see whether $[Ca^{2+}]_i$ is elevated at the early embryo stage.

INTERCELLULAR COMMUNICATION DURING PHYSIOLOGICAL ELEVATIONS OF $[Ca^{2+}]_i$: In many types of coupled cells, such as heart muscle, smooth muscle, nerve, visual-receptor, and gland cells, Ca^{2+} serves as an intracellular messenger, and $[Ca^{2+}]_i$ rises during physiological activity. This, however, does not necessarily mean that the cells become transiently uncoupled. As shown by the above experiments, what matters here is the $[Ca^{2+}]_i$ in the junctional region, not the overall $[Ca^{2+}]_i$. The Ca ion is quite unique in that it is rapidly sequestered by energized cellular organelles of high capacity (cf. Lehninger et al., 1967; Moore et al., 1975), limiting severely its domain of free diffusion in the cytosol (Baker and Crawford, 1972; Rose and Loewenstein, 1975*b*). Hence, it is entirely possible, in cells with a sufficiently dense Ca^{2+} sequestering machinery, for the junctional channels not to be affected by $[Ca^{2+}]_i$ elevations occurring away from junction. (For instance, in the case illustrated in Fig. 4, *ii*, the $[Ca^{2+}]_i$ in the junctional region stays below 5×10^{-7} M in the face of a more than 1,000-fold elevation in a central cell region, about 50 μm away.)

ON THE GENERALITY OF THE CA MECHANISM: Our analysis of the Ca mechanism (Nakas et al., 1966; Loewenstein et al., 1967; Politoff et al., 1969; Oliveira-Castro and Loewenstein, 1971; Rose and Loewenstein, 1971, 1975*a*, 1976) was carried out on *Chironomus* salivary cells which, because of their giant size and transparency, offer many experimental advantages. The mechanism has since been reported to operate in sheep and dog heart cells (Délèze, 1970, 1975; De Mello, 1975), in mammalian lymphocytes (Oliveira-Castro and Barcinsky, 1974), in newt embryo cells (Ito et al., 1974*a*), in crayfish electrical synapse (Peracchia and Dulhunty, 1976), and in insect TN cells in culture (Gilula and Epstein, 1977). Efforts to demonstrate the mechanism in mammalian cells in culture by means of treatments with metabolic inhibitors or ionophores have been inconclusive: the electrical coupling between 3T3 cells treated with cyanide, dinitrophenol (D. Garrison and W. R. Loewenstein, unpublished data), or ionophore A23187 (Gilula and Epstein, 1977; Garrison and Loewenstein, unpublished data) was affected only when the cells showed obvious signs of deterioration, and the effects were generally irreversible; and the electrical coupling between cultured mouse heart cells treated with the ionophore was largely unaffected (these cells also showed no morphological signs of ionophore effect; Gilula and Epstein, 1977). This, of course, does not necessarily imply a basic difference in the junctional mechanism of these cells. It may, for example, reflect differences in rates of intracellular Ca release, in junctional proximity to intracellular Ca depots, or in local intracellular Ca^{2+} sequestering power; in the case of the cultured heart cell, even insensitivity to the ionophore treatment is not excluded. Experiments of injection have not been done with any of the cultured cells.

SUMMARY AND CONCLUSIONS

Most cells of organized tissues have channels built into their membrane junctions that communicate the cell interiors with each other. These membrane channels have a high electrical conductance and a high permeability to hydrophilic molecules of a broad range of size. The estimated conductance of the single channel is 10^{-10} mho (lower limit). The molecular weight limit for channel permeation, as determined with short planar fluorescent peptide probes injected into *Chironomus* salivary gland cells, is 1,200–1,900 daltons. From the limiting geometries of the largest permeant probe, the channel bore is estimated at about 14 Å. The permeability of the channel is controlled by Ca^{2+}. This is shown by experiments in which Ca^{2+} is injected into a cell or the cytoplasmic Ca^{2+} concentration is elevated by treatments with metabolic inhibitors or Ca^{2+}-transporting ionophores, while the cytoplasmic free Ca^{2+} concentration in the junctional region is monitored with the luminescent protein aequorin as an indicator. As the cytoplasmic Ca^{2+} concentration is elevated in the range of 10^{-7} – 10^{-5} M, the molecular weight limit for channel permeation decreases gradually; at concentrations above about 5×10^{-5} M, the channel closure is virtually complete. The control of the junctional membrane channels by Ca^{2+} provides a powerful means for regulation of intracellular communication; some physiological implications are discussed.

REFERENCES

Azarnia, R., W. J. Larsen, and W. R. Loewenstein. 1974. The membrane junctions in communicating and non-communicating cells, their hybrids and segregants. *Proc. Natl. Acad. Sci. U. S. A.* **71:**880–884.

Azarnia, R., and W. R. Loewenstein. 1977. Intercellular communication and tissue growth. VIII. A genetic analysis of junctional communication and cancerous growth. *J. Membr. Biol.* In press.

Baker, P. F., and A. C. Crawford. 1972. Mobility and transport of magnesium in squid giant axons. *J. Physiol. (Lond.).* **227:**855–871.

Délèze, J. 1970. The recovery of resting potential and input resistance in sheep heart injured by knife or laser. *J. Physiol. (Lond.).* **208:**547–564.

Délèze, J. 1975. The site of healing over after local injury in the heart. *In* Recent Advances in Studies on Cardiac Structure and Metabolism. Vol. 5. A. Fleckenstein and N. S. Dhalla, editors. University Park Press, Baltimore. 223–235.

Délèze, J., and W. R. Loewenstein. 1976. Permeability of a cell junction during intracellular injection of divalent cations. *J. Membr. Biol.* **28:**71–86.

De Mello, W. C. 1975. Effect of intracellular injection of calcium and strontium on cell communication in heart. *J. Physiol. (Lond.).* **250:**231–245.

Furshpan, E. J., and D. D. Potter. 1968. Low resistance junctions between cells in embryos and tissue culture. *Curr. Top. Dev. Biol.* **3:**95–116.

Gilula, N. B., and M. L. Epstein. 1977. Cell-to-cell communication, gap junction and calcium. *Symp. Soc. Exp. Biol.* In press.

Goodenough, D. A. 1975. The structure and permeability of isolated hepatocyte gap junctions. *Cold Spring Harbor Symp. Quant. Biol.* **40:**37–45.

Goodenough, D. A., and J. P. Revel. 1970. A fine structural analysis of intercellular junctions in the mouse liver. *J. Cell Biol.* **45:**272–288.

Haynes, D. 1977. Divalent cation-ligand interactions of phospholipid membranes. *In* Metal-Ligand Interactions in Organic and Biochemistry. B. Pullman, editor. Ninth Jerusalem Symposium. Amsterdam. In press.

Ito, S., E. Sato, and W. R. Loewenstein. 1974*a*. Studies on the formation of a permeable cell membrane junction. I. Coupling under various conditions of membrane contact. Colchicine, Cytochalasin B, dinitrophenol. *J. Membr. Biol.* **19:**305–337.

Ito, S., E. Sato, and W. R. Loewenstein. 1974*b*. Studies on the formation of a permeable cell membrane junction. II. Evolving junctional conductance and junctional insulation. *J. Membr. Biol.* **19:**339–355.

Johnson, R., and J. D. Sheridan. 1971. Junctions between cancer cells in culture: ultrastructure and permeability. *Science (Wash. D. C.).* **174:**717–734.

Kanno, Y., and W. R. Loewenstein. 1966. Cell-to-cell passage of large molecules. *Nature (Lond.).* **212:**629–630.

Lehninger, A. L., E. Carafoli, and C. S. Rossi. 1967. Energy-linked ion movements in mitochondrial systems. *Adv. Enzymol. Relat. Areas Mol. Biol.* **29:**259–320.

Loewenstein, W. R. 1966. Permeability of membrane junctions. *Ann. N.Y. Acad. Sci.* **137:**441–472.

Loewenstein, W. R. 1967. Cell surface membranes in close contact: role of calcium and magnesium ions. *J. Colloid Interface Sci.* **25:**34–46.

Loewenstein, W. R. 1968*a*. Some reflections on growth and differentiation. *Perspect. Biol. Med.* **11:**260–272.

Loewenstein, W. R. 1968*b*. Communication through cell junctions: Implications in growth control and differentiation. *Dev. Biol.* **19:**(suppl. 2):151–183.

Loewenstein, W. R. 1974. Cellular communication by permeable junctions. *In* Cell Membranes: Biochemistry, Cell Biology and Pathology. G. Weissmann and R. Claiborne, editors. HP Publishing Co., Inc., New York. 105–114.

Loewenstein, W. R. 1975. Permeable junctions. *Cold Spring Harbor Symp. Quant. Biol.* **40:**49–63.

Loewenstein, W. R. 1977. Cell-to-cell membrane channels. Permeability, formation, genetics and functions. *In* The Physiological Basis for Disorders in Biomembranes. T. Andreoli, J. F. Hoffman, and D. D. Fanestil, editors. Plenum Press, New York. In press.

Loewenstein, W. R., and Y. Kanno. 1964. Studies on an epithelial (gland) cell junction. I. Modifications of surface membrane permeability. *J. Cell Biol.* **22:**565–586.

Loewenstein, W. R., M. Nakas, and S. J. Socolar. 1967. Junctional membrane uncoupling: permeability transformations at a cell membrane junction. *J. Gen. Physiol.* **50:**1865–1891.

Loewenstein, W. R., and R. D. Penn. 1967. Intercellular communication and tissue growth. II. Tissue regeneration. *J. Cell Biol.* **33:**235–242.

Loscalzo, J., G. H. Reed, and A. Weber. 1975. Conformational change and cooperativity in actin filaments free of tropomyosin. *Proc. Natl. Acad. Sci. U. S. A.* **72:**3412–3428.

McNutt, N. S., and R. S. Weinstein. 1973. Membrane ultrastructure and mammalian intercellular junction. *Prog. Biophys. Mol. Biol.* **26:**45–62.

Moore, L., T. Chen, H. R. Knapp, and E. J. Landon. 1975. Energy-dependent calcium sequestration activity in rat liver microsomes. *J. Biol. Chem.* **250:**4562–4568.

Nakas, M., S. Higashino, and W. R. Loewenstein. 1966. Uncoupling of an epithelial cell membrane junction by calcium ion removal. *Science (Wash. D. C.).* **151:**89–91.

Oliveira-Castro, G. M., and M. A. Barcinsky. 1974. Calcium-induced uncoupling in communicating human lymphocytes. *Biochem. Biophys. Acta.* **352:**338–343.

Oliveira-Castro, G. M., and W. R. Loewenstein. 1971. Junctional membrane permeability: effects of divalent cations. *J. Membr. Biol.* **5:**51–77.

Pappas, G. D., and M. V. L. Bennett. 1966. Specialized junctions involved in electrical transmission between neurons. *Ann. N. Y. Acad. Sci.* **137:**495–511.

Payton, B. W., M. V. L. Bennett, and G. D. Pappas. 1969. Permeability and structure of junctional membranes at an electrotonic synapse. *Science (Wash. D. C.).* **166:**1641–1656.

Penn, R. D. 1966. Ionic communication between liver cells. *J. Cell Biol.* **29:**171–173.

Peracchia, C., and A. Dulhunty. 1976. Low resistance junctions in crayfish: structural changes with functional uncoupling. *J. Cell Biol.* **70:**419–439.

Peracchia, C., and M. E. Fernández-Jaimovich. 1975. Isolation of intramembranous particles from gap junction. *J. Cell Biol.* **67:**330*a*. (Abstr.).

Pitts, J. (this volume)

Politoff, A. L., S. J. Socolar, and W. R. Loewenstein. 1969. Permeability of a cell membrane junction: dependence on energy metabolism. *J. Gen. Physiol.* **53:**498–515.

Pollack, G. H. 1976. Intercellular coupling in the atrioventricular node and other tissues of the rabbit heart. *J. Physiol. (Lond).* **255:**275–298.

Potter, D. D., E. J. Furshpan, and E. S. Lennox. 1966. Connections between cells of the developing squid as revealed by electrophysiological methods. *Proc. Natl. Acad. Sci. U. S. A.* **55:**328–344.

Rieske, E., P. Schubert, and G. W. Kreutzberg. 1975. Transfer of radioactive material between electrically coupled neurons of the leech central nervous system. *Brain Res.* **84:**365–382.

Rose, B. 1971. Intercellular communication and some structural aspects of cell junctions in a simple cell system. *J. Membr. Biol.* **5:**1–19.

Rose, B., and W. R. Loewenstein. 1971. Junctional membrane permeability. Depression by substitution of Li for extracellular Na, and by long-term lack of Ca and Mg: restoration by cell repolarization. *J. Membr. Biol.* **5:**20–50.

Rose, B., and W. R. Loewenstein. 1975*a*. Permeability of cell junction depends on local cytoplasmic calcium activity. *Nature (Lond.).* **254:**250–252.

Rose, B., and W. R. Loewenstein. 1975*b*. Calcium ion distribution in cytoplasm visualized by aequorin: diffusion in the cytosol is restricted due to energized

sequestering. *Science* (*Wash. D. C.*). **190:**1204–1206.
ROSE, B., and W. R. LOEWENSTEIN. 1976. Permeability of a cell junction and the local cytoplasmic free ionized calcium concentration: a study with aequorin. *J. Membr. Biol.* **28:**87–119.
SHERIDAN, J. 1971. Dye movement and low resistance junctions between reaggregated embryonic cells. *Dev. Biol.* **26:**627–643.
SHIMOMURA, O., and F. H. JOHNSON. 1969. Properties of the bioluminescent protein aequorin. *Biochemistry.* **8:**3991–4008.
SIMPSON, I., B. ROSE, and W. R. LOEWENSTEIN. 1977. Size limit of molecules permeating the junctional membrane channels. *Science* (*Wash. D. C.*). **195:**294–296.
SINGER, S. J. 1974. Architecture and topography of biological membranes. *In* Cell Membranes: Biochemistry, Cell Biology, and Pathology. G. Weismann and R. Claiborne, editors. H. P. Publishing Co., Inc., New York. 23–39.
SLACK, C., and J. P. PALMER. 1969. The permeability of intercellular junctions in the early embryo of *Xenopus laevis*, studied with fluorescent tracer. *Exp. Cell Res.* **55:**416–431.
WARNER, A. E. 1973. Electrical properties of the ectoderm in the amphibian embryo during induction and early development of the nervous system. *J. Physiol.* (*Lond.*). **235:**267–286.

ELECTRICAL AND CHEMICAL COMMUNICATION IN THE CENTRAL NERVOUS SYSTEM

CONSTANTINO SOTELO

Fast- and short-term interactions between neurons occur at specific junctional complexes, synapses. The burning question in neurobiology two decades ago was whether this kind of neuronal communication was chemically or electrically mediated. Today it is well established that both ways of synaptic transmission exist (Bennett, 1972). In addition, nonsynaptic neuronal interactions have also been described that can similarly be of a chemical or of an electrical nature. In the former, they are mainly related to the active bidirectional transneuronal molecular transport, as it has been suggested to occur for specific molecules between dendrites (see references in Schmitt et al., 1976). There is also electrophysiological evidence in favor of some electrical neuronal interactions in the absence of specific junctions, as a result of variations in the extracellular electric field, as is the case, for instance, in electrical inhibition (Furukawa and Furshpan, 1963).

This paper will first summarize the most important morphological features which underlie the electrical and chemical synaptic transmission, and will then discuss some problems correlated with the formation and the permanence of chemical synapses under abnormal conditions.

Electrical Synapses

Electrical synapses are characterized by the direct spread of depolarizing current, which can occur bidirectionally between the coupled neurons. This kind of transmission therefore takes place without synaptic delay, a property which is specific to chemical synapses. These two properties, fast transmission and reciprocity, underlie the functional significance of excitatory electrotonic transmission (Bennett, 1972). Direct electrophysiological evidence of the presence of electrotonic coupling between neurons has been obtained in invertebrates, as well as in some lower vertebrates (see references in Bennett, 1972). The fact that, in most of these instances, electrical transmission was found to take place by way of morphologically distinct junctions, which correspond to the gap junctions described between nonexcitable cells by Revel and Karnovsky (1967), has been considered as an indirect proof to correlate the gap junctions with low resistance pathways. A more direct proof to support this correlation has been provided by Pappas et al. (1971) in the septal synapses of the lateral giant axons of the crayfish, in which functional uncoupling of these synapses, obtained by different experimental manipulations, was accompanied by a morphological disruption of the gap junctions.

Our knowledge of the structural organization of gap junctions between nonexcitable cells has been extended in recent years to the molecular level (Gilula, this volume). Although similar studies on neuronal gap junctions are still lacking, it is reasonable to consider all gap junctions as belonging to a similar entity because they share the same permeability properties and the same morphological features.

MORPHOLOGY OF NEURO-NEURONAL GAP JUNCTIONS: One sees in the CNS of vertebrates, after uranyl acetate staining, regions of variable distance in which the membranes of the coupled cells come together, narrowing the extracellular space to a minor gap of about 2 nm in thickness (Fig. 1). Therefore, with these preparatory conditions, and when the plane of the section is perpendicular to the cell surfaces, the gap junction shows a heptalaminar configuration. A unique feature, almost invariably present in all observed electrical synapses, is the presence of a cytoplasmic semidense material which undercoats

---

CONSTANTINO SOTELO Laboratoire de Neuromorphologie, U-106 INSERM, Paris, France

the whole length of the inner surfaces of the junctional plasma membranes (Fig. 1). This cytoplasmic differentiation is generally absent in gap junctions between nonexcitable cells, even those between glial cells.

A closer examination at high magnification discloses, in favorable sections, that the narrow central gap is not uniform because there are spot-contacts bridging the gap between the outer leaflets of the plasma membranes. These bridges have a periodicity of about 9 nm. In some of the glial gap junctions we have succeeded in infiltrating the extracellular space with lanthanum (Revel and Karnovsky, 1967), and observe a complementary image of the gap. In these instances, the filled extracellular space exhibits a beaded appearance, in which neighboring opaque knobs are separated by narrow light spots (Fig. 2), which correspond to intermembranous bridges. "En face" views of gap junctions, in which the extracellular space is filled with lanthanum, reproduce the picture described by Robertson (1963) of the club endings on the Mauthner cells prepared by primary $KMnO_4$ fixation. The gap exhibits a honeycomb pattern composed of a system of lines disposed in a hexagonal network, in which the center of each hexagon is occupied by an electron-opaque spot.

The best analysis of the membrane organization at the level of gap junctions has been obtained by freeze-fracturing. By use of this technique, several examples of electrical synapses have been described in the CNS of vertebrates (Landis et al., 1974; Raviola and Gilula, 1975). In all of them, as in the one we studied, the electrical synapses of the chicken cerebellar glomeruli (Cantino and Sotelo, unpublished observations), the gap junctions appear as polygonal lattices of particles protruding on the protoplasmic face (P face) of the junctional membrane and complementary lattices of pits on the extracellular face (E face; Fig. 3). The intramembrane particles measure about 8 nm in diameter and, with the fixation technique that we used, they are arranged in a hexagonal array with a center-to-center spacing of about 9 nm. Therefore, these electrical synapses exhibit the same intramembrane organization as do the gap junctions between nonexcitable cells (Gilula, this volume). The composition of both junctional membranes is symmetrical, in such a way that the particles in one P face are in register with the particles of the other P face. These particles represent proteins that traverse the lipid bilayer completely. Each of the opposing particles, protruding into the extracellular space, spans the 2-nm gap which separates the two junctional membranes. Since the centers of these proteins are considered to be hydrophilic pores, the structural disposition described above confers on each couple of opposing particles the role of a permeability channel, which allows a direct cytoplasmic communication between the two junctional cells. A corollary of this symmetrical arrangement is that the current can flow in both directions. However, there are three examples of electrical synapses (Furshpan and Potter, 1959; see other references in Bennett, 1972) in which the current flow only in one direction. The fine morphology of these rectifying synapses has not been done.

ELECTROTONIC COUPLING BY WAY OF GAP JUNCTIONS IN THE MAMMALIAN BRAIN: Because electrotonic neurotransmission has been for years considered to be almost exclusively present in primitive forms of phylogeny, this review, by contrast, will be concerned only with the electrically transmitting synapses observed in the mammalian brain. If the anatomical demonstration of neuro-neuronal gap junctions is a sufficient criterion for electrotonic coupling, then the number of such synapses is considerable, including those described in primates (see references in Sotelo, 1975*a*). However, there are still few examples in which anatomical and correlative electrophysiological work have demonstrated the reality of electrotonic coupling in mammals. These are the rat mesencephalic trigeminal nucleus, the rat lateral vestibular nucleus, and the cat inferior olive (see references in Llinás, 1975, and in Sotelo, 1975*a*).

In the mesencephalic trigeminal nucleus, the neurons are coupled by direct somato-somatic gap junctions. However, in the lateral vestibular nucleus, the gap junctions occur between axon terminals and the perikarya of the giant cells of Deiters. The actual coupling can occur only through the presynaptic fibers. Several other examples of electrotonic coupling by way of presynaptic fibers are known to occur in the vertebrate brain (see references in Sotelo et al., 1975). In the inferior olive, the gap junctions have been mainly observed between dendritic appendages in a glomerular formation. This strategic location, in which the dendritic profiles involved in the neuronal coupling are surrounded by numerous axon terminals (Fig. 4), has been tentatively considered to be a functional device to modulate the electrotonic interactions by the shunting effects of the chemical synapses, in this way providing a variable electrotonic coupling (Llinás et al., 1974). A similar modulatory mechanism has been suggested by Spira and Bennett (1972) in the CNS of the mollusk *Navanax*. Gap junctions are rare between mesencephalic trigeminal neurons; they are more frequent in the lateral vestibular nucleus and in the inferior olive. This frequency correlates well with that found by electrophysiological measurements—the number of coupled neurons was low in the first case and was greater in the other two cases—emphasizing again that gap junctions are the sites for electrotonic coupling between neurons.

Chemical Synapses

The membranes involved in chemical transmission are separated by a high-resistance extracellular space, the synaptic cleft. In the presynaptic element, electrical events generate the release of transmitter, which diffuses into the synaptic cleft to affect the postsynaptic receptive membrane, the whole process taking place in a limited time.

SYNAPTIC MEMBRANES AND SPECIALIZATIONS: Several reviews dealing with the ultrastructure of chemical synaptic junctions have been published recently (Gray and Guillery, 1966; Peters et al., 1970; Sotelo, 1971; Pappas and Waxman, 1972). Therefore, I shall be concerned only with some recent findings mainly on synaptic membranes, which allow a close correlation between structure and function.

Differences in the permeability properties of various regions of neural membranes have been known for many years (see references in Katz, 1966). However, this heterogeneity was not correlated with specific ultrastructural features. A homogeneous unit-membrane structure characterizes neuronal membranes, even at the synaptic junctions. The presence of cytoplasmic differentiations that undercoat short segments of neural membranes at specific regions has been, until recently, the only indirect way to correlate functional heterogeneity with morphology. At the synaptic junctions, the cytoplasmic differentiations are asymmetrical (Fig. 5). The presynaptic membrane is covered by the vesicular grid (see references in Akert et al., 1972), whereas the postsynaptic one is marked by a continuous density. This asymmetry illustrates the unidirectional action of the chemical synapses. The signals can only cross from the side concerned with the secretory process, the presynaptic, to the membrane specialized for the reception, the postsynaptic.

With the application of the freeze-fracture technique to the study of synaptic junctions (Akert et al., 1972), it has been possible to disclose a conspicuous specialization in the internal organization of the synaptic membranes. The presynaptic membrane, in all studied synapses, contains large intramembrane particles and a varying number of small protuberances in its E face. The complementary picture in the P face of these protuberances appears as small depressions, which are in direct contact with the synaptic vesicles (Pfenninger et al., 1972). They are considered as transitory modulations of the presynaptic membrane taking place during the process of exocytosis (Heuser et al., 1974). The study of postsynaptic membranes in well-identified synapses in the cerebellar cortex and in the olfactory bulb, whose excitatory or inhibitory function are known electrophysiologically, has revealed that, at excitatory contacts, postsynaptic membranes contain an aggregate of homogeneous particles associated with their E face (Fig. 6). However, at inhibitory contacts, the particles in the P and E faces do not exhibit a specific organization, resembling those at nonsynaptic membranes (Landis and Reese, 1974; Landis et al., 1974).

All these results corroborate the old concept that there are specialized patches of the membranes at the synaptic interfaces which are the sites at which the chemical transmission takes place. These patches, or "synaptic complexes" (Palay, 1956), identified in thin sections by their cytoplasmic differentiation, correspond to the membrane zones that show the remarkable internal organization described above. Because of the fibrillary composition of the postsynaptic densities, it has been suggested that they play an important role in the anchorage of receptor proteins and other macromolecules at the postjunctional patches of the membrane. The presence of presynaptic vesicular grids and of postsynaptic densities will be the criteria used in the following sections to identify pre- and postsynaptic membranes.

FORMATION OF POSTSYNAPTIC DENSITIES IN THE ABSENCE OF PRESYNAPTIC ELEMENTS: If postsynaptic densities can be used as markers for identified receptor surfaces, it will be interesting to study such densities in neurons that have been almost completely devoid of synaptic inputs from their origin. Because of the relative simplicity of the corticocerebellar connections, and the fact that the pivotal element in this neuronal network, the Purkinje cells, receives over 95% of their inputs from one single origin, the granule cells, the cerebellum offers favorable material to analyze this question. Furthermore, two nonallelic mouse mutations, the *weaver* (wv) and the *reeler* (rl), provide Purkinje cells which have grown and differentiated in the absence of granule cells (Sidman, 1968). The postsynaptic densities are similar for Purkinje cells in wv/wv and for heterotopic Purkinje cells in rl/rl. All these cells develop innumerable dendritic spines with their corresponding postsynaptic densities (Fig. 7), similar to those in normal cerebellum. The important difference is that in the mutants the spines are free of innervation (Fig. 7), since the parallel fibers, the axons of granule cells, have not developed. Moreover, postsynapticlike densities are also present, in a lesser amount, at the dendritic smooth surfaces, where, in normal conditions, parallel fibers never contact these dendrites. The densities display a morphology similar to those characterizing postsynaptic membrane in Gray type 1 synapses (Fig. 8). They have a variable size, but often they are much larger than postsynaptic densities in any normal neuron (Fig. 8). Such dendritic segments can face glial processes, naked spines, and/or axonal terminals of the inhibitory interneurons. In the latter case, the terminals do not develop presynaptic vesicular grids (see references in Sotelo, 1975*b*).

These results point out that Purkinje cells are able to develop postsynaptic densities by an intrinsic mechanism, completely independent from interactions with parallel fibers. The majority of the large number of the autonomously developed densities exhibit the same morphology and localization as those of their normal counterparts. This is the case for dendritic spines. A recent freeze-fracturing study of these noninnervated spines in wv/wv (Hanna et al., 1976) has shown that the membrane corresponding to the segment bearing the postsynaptic density contains in its E face the same aggregate of homogeneous particles as do the innervated spines in normal cerebellum. This corroborates that, even in abnormal conditions, cytoplasmic differentiations underlie specific segments of neuronal membrane organization.

The presence of abnormal postsynaptic densities, gigantic in size and unusually located, has been tentatively explained as a reflection of the absence of a hypothetical feed-back mechanism, which would accompany the establishment of normal synaptic transmission, and would function to regulate the autonomous synthesis of the specific receptor protein in the postsynaptic neuron according to the firing of the presynaptic fiber (Sotelo, 1975*b*).

A completely different hypothesis to explain the presence of these noninnervated postsynaptic sites has been proposed by Hámori (1973). This author postulates that axon terminals of the climbing fibers, which synapse on Purkinje cells with a higher density in the agranular than in the normal cerebellum, are responsible for the indirect induction of the dendritic spines. By this interpretation, the

FIGURE 1 Gap junction between a mossy fiber (*MF*) and a granule cell dendrite (*GD*) in the cerebellum of the viper (*Vipera aspis*). Note the heptalaminar configuration of the gap junction and the cytoplasmic differentiations (arrows) undercoating the junctional membranes. The gap junction is surrounded by attachment plates (*AP*). × 184,000.

FIGURE 2 Interglial gap junction in which the extracellular space is filled with lanthanum. The arrow points to a zone in which the beaded appearance of this extracellular space may be seen. × 300,000.

FIGURE 3 Freeze-fractured membrane of a mossy fiber in the chicken cerebellum. Two gap junctions are present. In the one at the top only the *P* face is visible, whereas in the one at the bottom, both the *P* and the *E* faces may be seen. × 72,000.

FIGURE 4 Inferior olivary glomerulus of the cat. The central dendritic core is covered by axon terminals which establish chemical synapses (arrow heads) on them. Two of the central dendritic appendages (DA_1 and DA_2) are linked together by way of a gap junction (*GJ*). Note the strategic position of one axon terminal (*AT*) containing a pleomorphic population of synaptic vesicles, indicative of its possible inhibitory nature, and simultaneously synapsing on both dendrites coupled by the gap junction. The firing of this terminal may hypothetically uncouple the inferior olivary appendages as a result of synaptic conductance (from Sotelo et al., 1974). × 66,000.

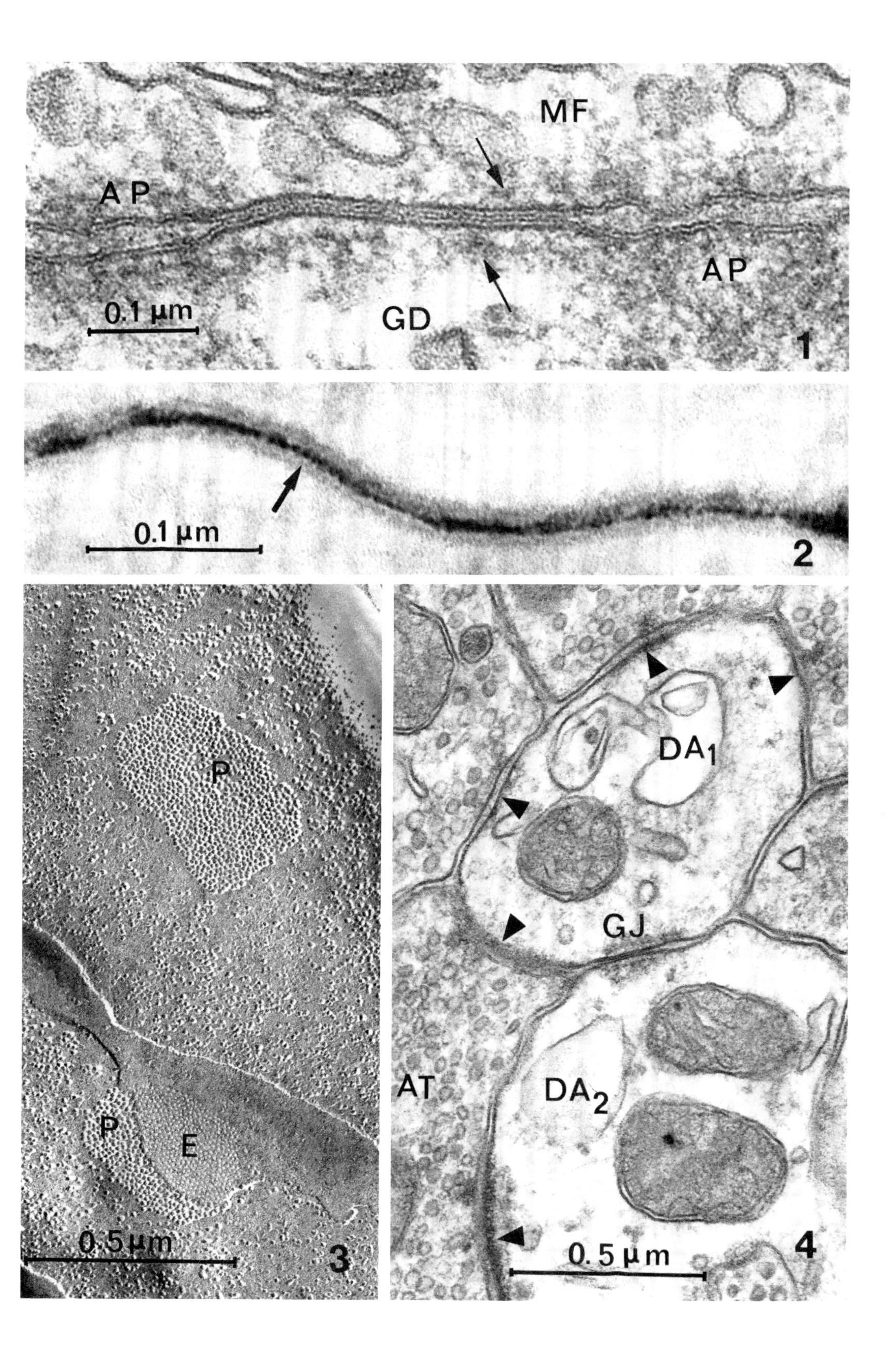
MF
AP
AP
0.1 µm
GD
1
0.1 µm
2
P
P
E
0.5 µm
3
DA1
GJ
AT
DA2
0.5 µm
4

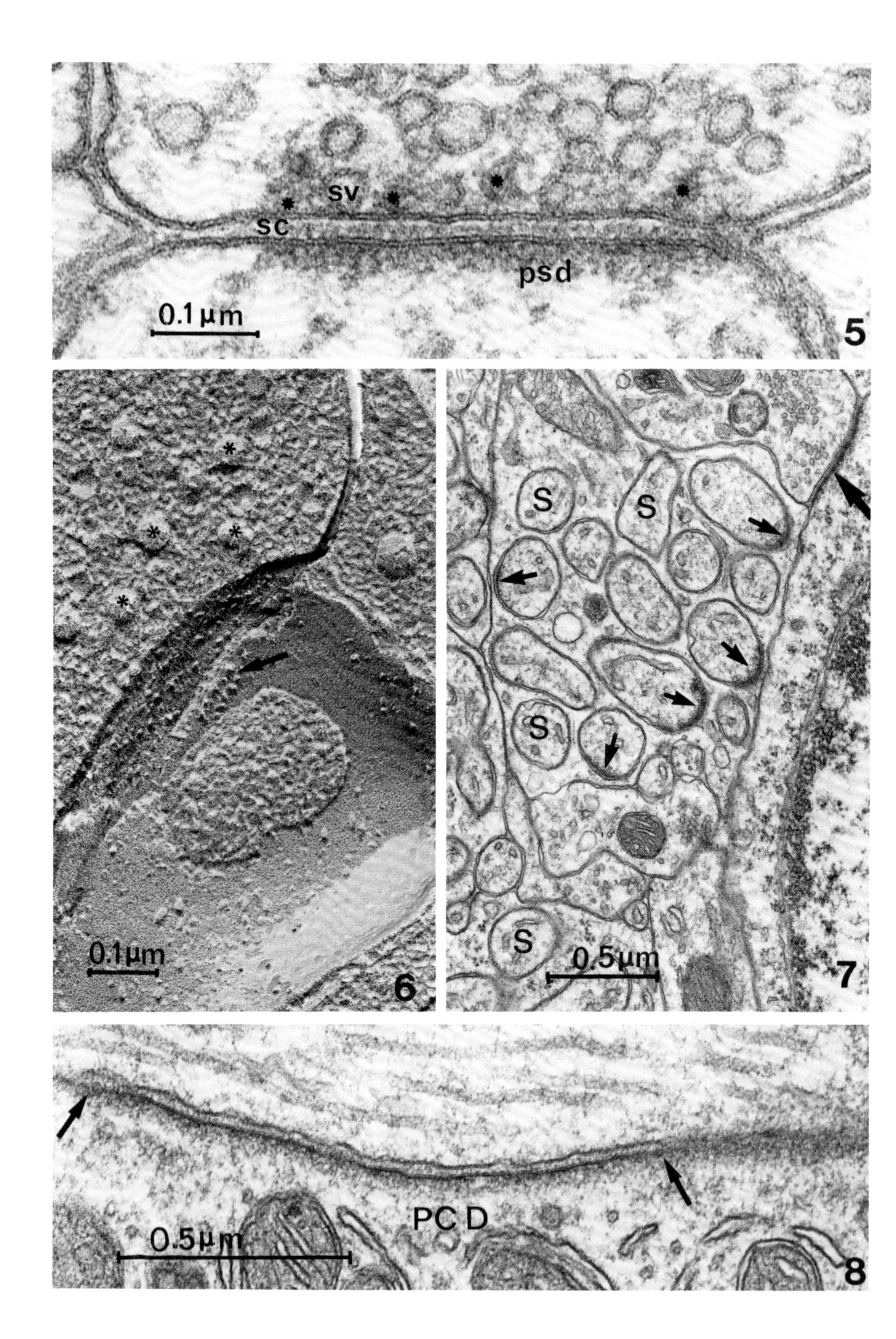
sv
sc
psd
0.1 µm
5
0.1µm
6
S
S
S
S
0.5µm
7
PC D
0.5µm
8

postsynaptic densities are not the result of an autonomous process, but to a heterotopic induction. In our recent work (Sotelo and Arsenio-Nunes, 1976), in which Purkinje cells develop in the absence of climbing fibers, we have failed to sustain the hypothesis of indirect induction. Under these conditions, Purkinje cells are able to develop spines; moreover, the number of these spines seems to be higher than in the presence of climbing fibers.

DEVELOPMENT AND STABILIZATION OF PRESYNAPTIC VESICULAR GRIDS: By using other neurological mutations affecting the cerebellum of the mouse, the question of the intrinsic development of presynaptic organelles can also be analyzed. In fact, there are two mutants, *staggerer* (sg) and *nervous* (nv), in which Purkinje cells seem to be directly affected. In the first mutant, Purkinje cells are unable to develop, at the appropriate time, the dendritic spines which are the receptive surfaces of parallel fibers (Sidman, 1968). Therefore, the parallel fibers must grow and evolve in the absence of postsynaptic sites. In the *nervous* mice, after a synaptogenic period in which parallel fibers establish normal chemical synapses with Purkinje cell spines, the Purkinje cells degenerate (Landis, 1973). Thus, parallel fibers lose their postsynaptic partners after the establishment of functional connections.

During the synaptogenic period of the *staggerer* cerebellum, numerous parallel fibers surround the abnormally smooth dendrites of the Purkinje cells. Under these conditions, parallel fibers form small varicosities filled with synaptic vesicles and, on some rare occasions, bearing vesicular grids, facing thin glial processes (Fig. 9). These axon terminals do not survive for a long time and they degenerate, producing a retrograde reaction in their perikarya of origin, which results in an extensive granule cell death (Sotelo and Changeux, 1974). This study shows that parallel fibers are capable of an intrinsic process of differentiation, which produces transient "boutons en passant" with normal presynaptic organelles. Thus, there is the possibility of an autonomous development of both postsynaptic densities and of presynaptic vesicular grids. The main difference, at least in cerebellum, is that the postsynaptic densities can remain during the whole life of the animal, whereas the presynaptic elements have only transient life.

Staggerer mutants, and especially *nervous* mutants, offer the possibility of studying the fate of presynaptic vesicular grids in axons devoid of their postsynaptic partners after the establishment of functional connections. In sg/sg, as a consequence of the progressive degeneration of granule cells, the mossy fibers slowly become denuded of their postsynaptic targets and are surrounded by astrocytic processes (Fig. 10). At this stage, normal vesicular grids can persist facing the glial cytoplasm. The number of these denuded mossy fibers increases with age. Some of these mossy fibers also degenerate (Sotelo and Changeux, 1974); however, the rate of this transsynaptic degenerative process occurs at a much slower rate than does the parallel fiber degeneration.

FIGURE 5 Synaptic complex, characterizing chemical transmission between a parallel fiber and a Purkinje cell spine in the cerebellar cortex of the cat. Note the high resistance extracellular space or synaptic cleft (*sc*) separating both neurons. In the parallel fiber, the presynaptic dense projections (asterisks) and their associate synaptic vesicles (*sv*) constitute the presynaptic vesicular grid. In the postsynaptic side, and facing the presynaptic vesicular grid, the membrane is undercoated by the postsynaptic density (*psd*). × 176,000.

FIGURE 6 Freeze-fractured parallel fiber-Purkinje spine synapse in the cerebellum of the rat. The axoplasm of the parallel fiber contains synaptic vesicles (asterisks). An aggregate of particles (arrow) is found on the E face of the postsynaptic membrane in the spine. × 120,000.

FIGURE 7 Cerebellar cortex of a 29-day *weaver* mouse. Numerous Purkinje cell dendritic spines (*S*) are free of innervation. Some of them exhibit a normal postsynaptic density (small arrows). The large arrow points to a postsynaptic density, on the perikaryon of a stellate cell, facing an axon terminal. Note the absence of vesicular grid at this terminal. × 39,000.

FIGURE 8 Cerebellar cortex of a 7-mo *reeler* mouse. Dendritic profile of a heterotopic Purkinje cell (*PCD*). The arrows mark the length of a gigantic postsynaptic density, undercoating a large segment of the dendritic smooth surface. × 80,000.

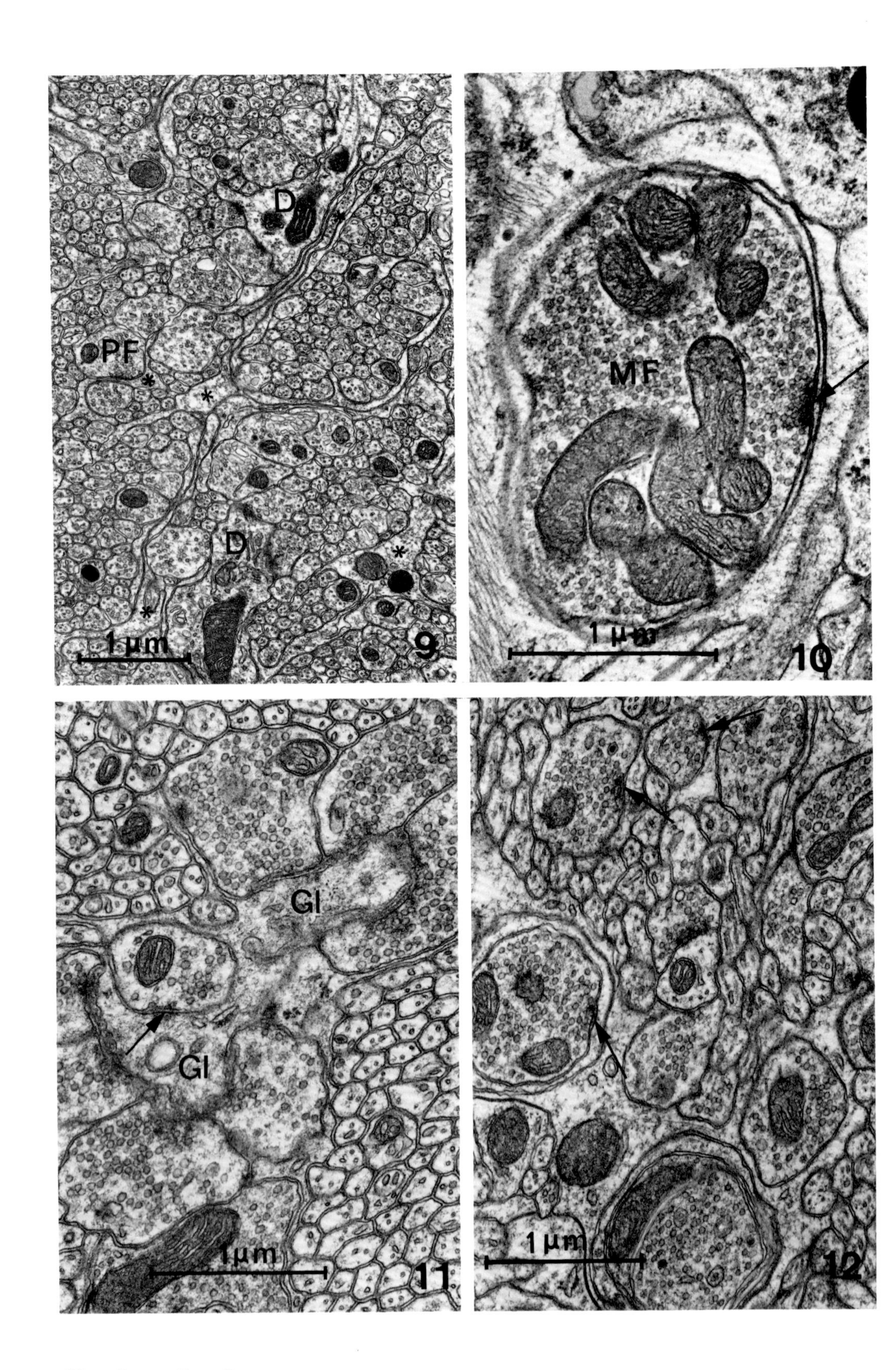
D
PF
D
1 μm
9
MF
1 μm
10
Gl
Gl
1μm
11
1 μm
12

In *nervous* mutants, 90% of the Purkinje cells follow a selective degenerative process between postnatal days 23 and 50 (Landis, 1973). I have studied this cerebellum in 6-mo to 1-yr old mice (Sotelo, unpublished observations) to investigate the fate of parallel fibers at long intervals after Purkinje cells have degenerated. In these old animals, clusters of parallel fibers keep their normal synaptic relationships with dark debris from Purkinje cell dendrites. Astroglial processes, filled with gliofilaments, invade the degenerative debris to transform it into discontinuous bands of sinuous contours, to which most of the parallel fibers remain attached (Fig. 11). During this progressive removal of the necrotic material, the glial processes take the place previously occupied by the Purkinje dendrite. The parallel fibers remain directly opposed to these glial profiles, retaining their presynaptic differentiation (Fig. 12). However, since in this cerebellum there is also a process of granule-cell death, although much slower and less complete than that in the *staggerer* mutant, it can be assumed that the parallel fibers are capable of a long survival period after they have lost their postsynaptic targets.

The most likely interpretation of all these results is that presynaptic fibers in the cerebellum are capable of an intrinsic growth and development, but they are incapable of independent survival, undergoing a rapid degenerative process when they fail to establish synaptic connection. However, once these fibers have been stabilized by function, they are able to survive for long periods even if they lose their postsynaptic partners. A retrograde signal, probably concomitant with functional activity, therefore becomes necessary for the permanent stabilization of presynaptic differentiation. This interpretation fits well with the mechanism postulated by Changeux et al. (1973) in their hypothesis of "selective stabilization of synapses."

REFERENCES

AKERT, K., K. PFENNINGER, C. SANDRI, and H. MOOR. 1972. Freeze-etching and cytochemistry of vesicles and membrane complexes in synapses of the central nervous system. *In* Structure and Function of Synapses. G. D. Pappas and D. P. Purpura, editors. Raven Press, New York. 67–86.

BENNETT, M. V. L. 1972. A comparison of electrically and chemically mediated transmission. *In* Structure and Function of Synapses. G. D. Pappas and D. P. Purpura, editors. Raven Press, New York. 221–256.

CHANGEUX, J. P., P. COURREGE, and A. DANCHIN. 1973. A theory of the epigenesis of neuronal networks by selective stabilization of synapses. *Proc. Natl. Acad. Sci. U. S. A.* **70:**2974–2978.

FURSHPAN, E. J., and D. D. POTTER. 1959. Transmission at the giant motor synapses of the crayfish. *J. Physiol. (Lond.).* **145:**289–325.

FURUKAWA, T., and E. J. FURSHPAN. 1963. Two inhibitory mechanisms in the Mauthner neurons of goldfish. *J. Neurophysiol.* **26:**140–176.

GRAY, E. G., and R. W. GUILLERY. 1966. Synaptic morphology in the normal and degenerating nervous system. *Int. Rev. Cytol.* **19:**111–182.

HÁMORI, J. 1973. Developmental morphology of dendritic postsynaptic specializations. *Rec. Dev. Neurobiol.* **4:**9–31.

HANNA, R. B., A. HIRANO, and G. D. PAPPAS. 1976. Membrane specializations of dendritic spines and glia in the weaver mouse cerebellum: a freeze-fracture study. *J. Cell Biol.* **68:**403–410.

HEUSER, J. E., T. S. REESE, and D. M. D. LANDIS. 1974. Functional changes in frog neuromuscular junctions studied with freeze-fracture. *J. Neurocytol.* **3:**109–131.

FIGURE 9 Cerebellar molecular layer of a 26-day *staggerer* mouse. The dendritic profiles (*D*) have a smooth contour. Dendritic spines are absent. Bundles of parallel fibers are segregated by glial processes (asterisks). Some parallel fiber (*PF*) varicosities are present, facing the thin glial processes. × 19,000.

FIGURE 10 Cerebellar granular layer of a 26-day *staggerer*. A denuded mossy fiber (*MF*) is completely surrounded by glia. The arrow points to a presynaptic vesicular grid facing the glia (from Sotelo and Changeux, 1974). × 35,000.

FIGURE 11 Cerebellar molecular layer of a 9-mo *nervous* mouse. A cluster of parallel fibers keep their synaptic relationships with a thin band of necrotic debris of a degenerating Purkinje dendrite. The astroglial cytoplasm (*Gl*) has invaded the necrotic debris. One bouton of a parallel fiber is already facing the glial process (arrow). × 30,000.

FIGURE 12 Same material as in Fig. 11. The arrows point to presynaptic vesicular grids on deafferented parallel fibers which are directly opposed to thin glial processes. × 27,000.

Katz, B. 1966. Nerve, Muscle and Synapse. McGraw-Hill Book Co., New York. 193.

Landis, S. C. 1973. Ultrastructural changes in the mitochondria of cerebellar Purkinje cells of nervous mutant mice. *J. Cell Biol.* **57**:782–797.

Landis, D. M. D., and T. S. Reese. 1974. Differences in membrane structure between excitatory and inhibitory synapses in cerebellar cortex. *J. Comp. Neurol.* **155**:93–126.

Landis, D. M. D., T. S. Reese, and E. Raviola. 1974. Differences in membrane structure between excitatory and inhibitory components of the reciprocal synapse in the olfactory bulb. *J. Comp. Neurol.* **155**:67–92.

Llinás, R. 1975. Electrical synaptic transmission in the mammalian central nervous system. *In* Golgi Centennial Symposium Proceedings. M. Santini, editor. Raven Press, New York. 379–385.

Llinás, R., R. Baker, and C. Sotelo. 1974. Electrotonic coupling between neurons in cat inferior olive. *J. Neurophysiol.* **37**:560–571.

Palay, S. L. 1956. Synapses in the central nervous system. *J. Biophys. Biochem. Cytol.* **2** (suppl.):193–202.

Pappas, G. D., Y. Asada, and M. V. L. Bennett. 1971. Morphological correlates of increased coupling resistance at an electrotonic synapse. *J. Cell Biol.* **49**:173–188.

Pappas, G. D., and S. G. Waxman. 1972. Synaptic fine structure-morphological correlates of chemical and electrotonic transmission. *In* Structure and Function of Synapses. G. D. Pappas and D. P. Purpura, editors. Raven Press, New York. 1–43.

Peters, A., S. L. Palay, and H. de F. Webster. 1970. The Fine Structure of the Nervous System: The Cells and Their Processes. Harper & Row, Publishers, New York. 198.

Pfenninger, K., K. Akert, H. Moor, and C. Sandri. 1972. The fine structure of freeze-fractured presynaptic membranes. *J. Neurocytol.* **1**:129–149.

Raviola, E., and N. B. Gilula. 1975. Intramembrane organization of specialized contacts in the outer plexiform layer of the retina: a freeze-fracture study in monkeys and rabbits. *J. Cell Biol.* **65**:192–222.

Revel, J. P., and M. J. Karnovsky. 1967. Hexagonal array of subunits in intercellular junctions of the mouse heart and liver. *J. Cell Biol.* **33**:c7–c12.

Robertson, J. D. 1963. The occurrence of a subunit pattern in the unit membranes of club endings in Mauthner cell synapses in goldfish brains. *J. Cell Biol.* **19**:201–221.

Schmitt, F. O., P. Dev, and B. H. Smith.. 1976. Electrotonic processing of information by brain cell. *Science (Wash. D.C.)*. **193**:114–120.

Sidman, R. L. 1968. Development of interneuronal connections in brains of mutant mice. *In* Physiological and Biochemical Aspects of Nervous Integration. F. D. Carlson, editor. Prentice-Hall Inc., Englewood Cliffs, N.J. 163–193.

Sotelo, C. 1971. General features of the synaptic organization in the central nervous system. *In* Chemistry and Brain Development. R. Paoletti and A. N. Davison, editors. Plenum Publishing Corp., New York. 239–279.

Sotelo, C. 1975*a*. Morphological correlates of electrotonic coupling between neurons in mammalian nervous system. *In* Golgi Centennial Symposium Proceedings. M. Santini, editor. Raven Press, New York. 355–365.

Sotelo, C. 1975*b*. Anatomical, physiological and biochemical studies of the cerebellum from mutant mice. II. Morphological study of cerebellar cortical neurons and circuits in the weaver mouse. *Brain Res.* **94**:19–44.

Sotelo, C., and M. L. Arsenio-Nunes. 1976. Development of Purkinje cells in absence of climbing fibers. *Brain Res.* **111**:389–395.

Sotelo, C., and J. P. Changeux. 1974. Transsynaptic degeneration "en cascade" in the cerebellar cortex of staggerer mutant mice. *Brain Res.* **67**:519–526.

Sotelo, C., R. Llinás, and R. Baker. 1974. Structural study of inferior olivary nucleus of the cat: morphological correlates of electrotonic coupling. *J. Neurophysiol.* **37**:541–559.

Sotelo, C., M. Rethelyi, and T. Szabo. 1975. Morphological correlates of electrotonic coupling in the magnocellular mesencephalic nucleus of the weakly electric fish *Gymnotus carapo*. *J. Neurocytol.* **4**:587–607.

Spira, M. E., and M. V. L. Bennett. 1972. Synaptic control of electrotonic coupling between neurons. *Brain Res.* **37**:294–300.

EXCITATORY AND INHIBITORY SYNAPTIC RESPONSES MEDIATED BY A DECREASE OF THE POSTJUNCTIONAL MEMBRANE PERMEABILITY

H. M GERSCHENFELD

In the last two decades, the ultrastructural and microphysiological studies on chemical synapses have revealed the subtlety and rich variety of the mechanisms operating at these intercellular junctions. In spite of the apparent simplicity of the message that chemical synapses transmit, either excitation or inhibition, the mechanisms operating to insure the specificity of the communication are amazingly complex. These include: (1) a variety of known (and probably many still unknown) transmitter molecules; (2) the possibility that a transmitter molecule can act on more than one specific receptor; and (3) the operation of different postsynaptic mechanisms. In this paper, we will discuss recent evidence regarding some "unusual" or "atypical" postsynaptic mechanisms.

"Classic" and "Atypical" Actions of Chemical Transmitters

The classic theory of synaptic transmission as advanced by Fatt and Katz (1951, 1953) postulates that synaptic transmitters cause excitation or inhibition of the postsynaptic cells by increasing the ionic permeability of their membranes, i.e., by "turning on" ionic pathways (channels) for either anions or cations in the membrane. Such a classic explanation has been shown to account for the function of the great majority of synapses studied to date (see review of Ginsborg, 1973; Gerschenfeld, 1973).

It is useful here to review briefly the behavior of excitatory and inhibitory classic synaptic responses when the membrane potential of the postsynaptic neuron is altered by passing current across it. In the case of the excitatory responses, known to be the result either of an increase in Na^+ permeability or of an increase in both Na^+ and K^+ permeabilities, the amplitude of the excitatory postsynaptic potentials (EPSP) increases gradually when the cell is hyperpolarized and decreases when the cell is depolarized, i.e., the EPSP grows in amplitude when the driving force for Na^+ ions is increased and diminishes when the Na^+ driving force is reduced. Depolarization of the postsynaptic membrane may cause a reversal of the polarity of the EPSP in cases where the transmitter increases both Na^+ and K^+ permeabilities, reversal being observed at $-15/-20$ mV. In the case of inhibitory synaptic responses, which are the result of an increase in either Cl^- or K^+ permeability, the inhibitory postsynaptic potential (IPSP) grows in amplitude when the cell is depolarized, i.e., when the driving force for either Cl^- or K^+ ions is increased. In contrast, when the cell is hyperpolarized, the amplitude of the classic IPSP decreases gradually until it reaches a null value. If the hyperpolarization of the neuron is pursued beyond this level, the inhibitory responses reverse their polarity. It has been demonstrated that, in many cases, the reversal potential of the IPSP corresponds to the equilibrium potential for either Cl^- ions (E_{Cl}) or K^+ ions (E_K).

An increasing number of synaptic responses that do not follow this behavior vis à vis the membrane potential have been reported recently. Fig. 1 shows an example of one of such unusual or atypical response. In this case, the ionophoretic application of 5-hydroxytryptamine (5-HT), a synaptic transmitter in the molluscan central nervous system (Gerschenfeld and Paupardin-Tritsch, 1974*b*), depolarizes a snail central neuron. When

H. M. GERSCHENFELD Laboratoire de Neurobiologie, Ecole Normale Supérieure, Paris, France

the cell is artificially hyperpolarized by passing inward current across its membrane, instead of showing, as do the classic depolarizing synaptic responses, a gradual increase of amplitude, the 5-HT depolarization decreases in amplitude and becomes null at −78 mV (Fig. 1*f*). Further hyperpolarization of the cell causes a reversal of the response (Fig. 1*g*).

It is evident that no mechanism involving an increase in the ionic permeability of the postsynaptic membrane can explain such behavior. On the contrary, such atypical behavior of the 5-HT response is well accounted for in a model first proposed by Weight and Votava (1970) to explain a similar behavior of the slow EPSP evoked by preganglionic stimulation in sympathetic ganglion neurons of frog. We interpreted these responses as the result of a *decrease* in the ionic permeability of the postsynaptic membrane. The experiment of Fig. 1 demonstrates that such is the case for the 5-HT atypical depolarization: (1) the input resistance of the neurons showing these responses increases when 5-HT is present in the extracellular medium (Fig. 1*a–c*); (2) an increase in the extracellular K^+ concentration causes a shift of the reversal potential of the 5-HT depolarization, which coincides well with the displacement of E_K predicted by the Nernst relation (Fig. 1*h–k*).

Synaptic transmitters also can evoke atypical inhibitory responses. Fig. 2 gives an example of such a response also evoked by 5-HT on another group of molluscan neurons. In this case, the ionophoretic application of 5-HT hyperpolarizes and inhibits the neurons' firing (Fig. 2*a*). When the cell is artificially hyperpolarized by injecting inward current, thus first decreasing and later reversing the driving force for Cl^- or K^+ ions, the atypical hyperpolarization, instead of decreasing and reversing as does the classic one, shows a gradual increase of its amplitude (Fig. 2 *a–e*). Furthermore, when the neuron is depolarized by passing outward current across its membrane, the atypical hyperpolarization decreases in amplitude and sometimes an inversion of the response can be obtained. The value of the reversal potential of these responses, measured directly or calculated by extrapolation (Fig. 2, graph), ranges between −20/−30 mV. Changes in the extracellular Cl^- concentration do not affect these 5-HT hyperpolarizations, whereas changes in either Na^+ and/or K^+ external concentrations cause a shift in their reversal potential (Fig. 2*f–j* and graph).

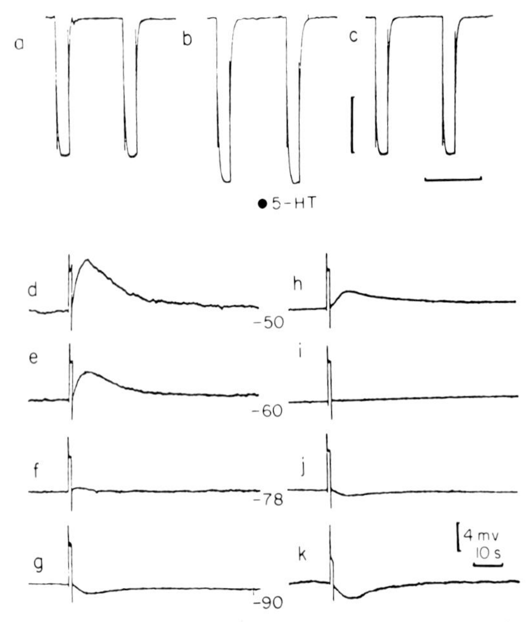

FIGURE 1 "Atypical" depolarizing response of a molluscan neuron to 5-hydroxytryptamine (5-HT). (*a–c*) Effects of 5-HT on the membrane resistance. In *a* the neuron is hyperpolarized by inward current to −50 mV, and square pulses are passed across the membrane to measure the input-resistance. In *b*, the bath application of a 5×10^{-5} M solution of 5-HT causes an evident increase in the input resistance. In *c*, the effects of 5-HT are reversed by removal of the amine. (*d–g*) Ionophoretic applications of 5-HT on the membrane of the same neuron results in an atypical depolarizing response which in a normal medium reverses at −78 mV (*f*). (*h–k*) When the external K^+ concentration is increased twice, the reversal potential of the response shifts to −60 mV (*i*). (Gerschenfeld and Paupardin-Tritsch, 1974*a*.)

These briefly described examples illustrate well the behavior of the atypical responses of neurons to chemical transmitters. From the analysis of these and other examples reported in recent years, it has been possible to conclude that the atypical synaptic depolarizations are associated with a decrease in the K^+ permeability of neuronal membranes (Weight and Votava, 1970; Paupardin-Tritsch and Gerschenfeld, 1973; Gerschenfeld and Paupardin-Tritsch, 1974*a*, *b*; Kehoe, 1975; Carew and Kandel, 1976) and that atypical hyperpolarizations are the result of a decrease in both Na^+ and K^+ permeabilities (Paupardin-Tritsch and Gerschenfeld, 1973; Weight and Padjen, 1973*a*, *b*; Gerschenfeld and Paupardin-Tritsch, 1974*a*, *b*; Delaleu, 1976).

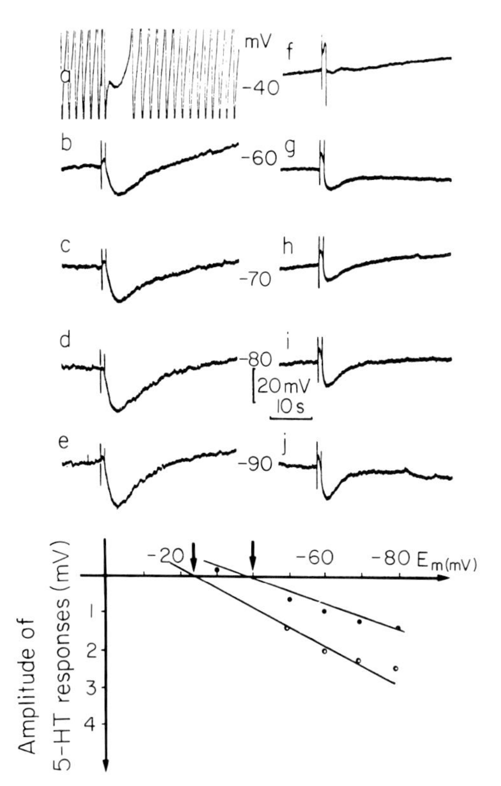

FIGURE 2 Molluscan neuron responding to 5-HT by an atypical hyperpolarizing response. (*a*) Ionophoretic application of 5-HT causes a hyperpolarization and an inhibition of the spontaneous spike firing. (*b–e*) Gradual hyperpolarization to −60, −70, −80, and −90 mV causes a gradual increase in the amplitude of the hyperpolarizing 5-HT response. In *f–j*, half of the NaCl content of the external environment is replaced by Tris Cl, causing a shift of the inversion potential of the atypical 5-HT response, which becomes directly observable at −40 mV (*f*). The graph at the bottom relates the amplitude of the atypical responses with the membrane potential values at which they were recorded. The open circles correspond to the experiment in normal medium and the closed circles, to the values recorded in the low Na^+ medium. Notice that the reversal potential shifts from the calculated value of −23 mV to the observed value of −40 mV. (Gerschenfeld and Paupardin-Tritsch, 1974*a*.)

Nature of the Ionic Permeability Decrease in Atypical Synaptic Responses

Figure 3 compares the simplified equivalent circuits of the cell membranes involved in classic (*a*) and atypical (*b*) responses to synaptic transmitters. In Fig. 3*a*, the closing of either switch A and/or switch A′ after the activation of specific receptors by the transmitter will introduce a shunt, Δg_{Na}, in the membrane circuit and thus cause an increase of the permeability to Na^+ ions. In an analogous manner, the closing of switch B or switch C by the activation of other different transmitter receptors will cause an increase in either K^+ or Cl^- permeability. In *b*, where the equivalent circuit of the membrane involved in an atypical response is represented, the cell membrane permeabilities to both Na^+ and K^+ are represented by two conductances in parallel (g_{Na} and Δg_{Na}, g_K and Δg_K). The transmitter produces an atypical depolarization, such as in Fig. 1, by opening the switch α, thus removing Δg_K from the circuit. The atypical hyperpolarization of Fig. 2 implies a displacement of the lever β, thus opening, at the same time, both switches, and therefore removing Δg_{Na} and Δg_K from the circuit (Fig. 3*b*).

The nature of the ionic pathways represented in the equivalent circuit diagram by Δg_{Na} and Δg_K raises some important questions: are these channels located on restricted spots of the neuronal membrane or are they present on the whole surface? Are they related to the ionic channels involved in the generation of the action potential?

Experiments on the already described atypical responses of molluscan neurons to 5-HT give some answers to those questions. Very few neurons in the molluscan CNS show such responses, and the atypical responses are only observed when the transmitter is applied to a precise and very restricted area of the sensitive neurons (Gerschenfeld and Paupardin-Tritsch, 1974*a*; see Gerschenfeld, 1976). Other experiments show that tetrodotoxin, which blocks selectively the Na^+ channels involved in the action potential, does not affect the atypical hyperpolarizing response to 5-HT (Paupardin-Tritsch and Gerschenfeld, unpublished data). Tetrodotoxin also has been shown previously to be totally ineffective on the Na^+ channels involved in the classic synaptic responses.

Some light is thrown on the nature of the turned-off channels by the experiments of Kehoe (1975), who observed that the firing of some cholinergic and noncholinergic interneurons in the pleural ganglion of *Aplysia* evokes in some other neurons a composite synaptic potential, whose late phase, a slow EPSP, shows an atypical behavior similar to that of the response in Fig. 1. In this case, the slow EPSP reverses at −80 mV and is also the result of a decrease in K^+ permeability.

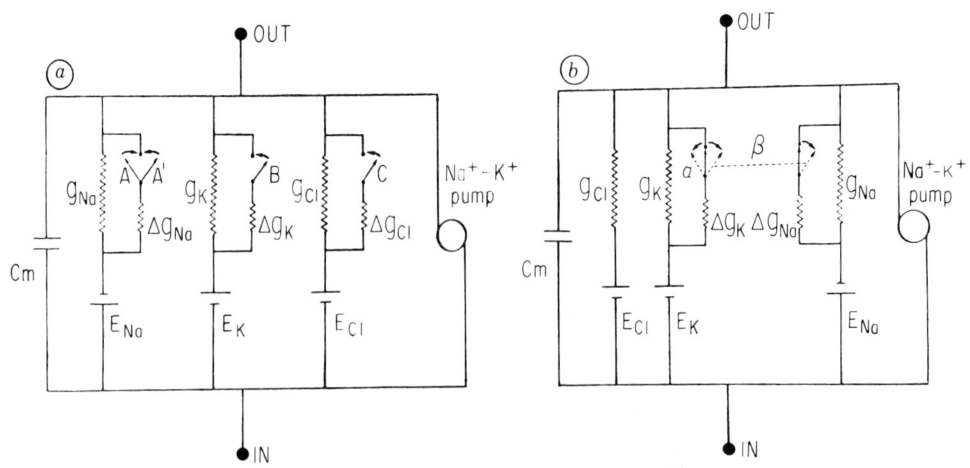

FIGURE 3 Simplified membrane circuits representing the conductance changes induced by the activation of different 5-HT receptors in molluscan neurons. (*a*) 5-HT responses involving an *increase* in membrane conductance. The closing of the A and/or the A′ switch will introduce the shunt Δg_{Na} in the circuit, increasing the Na^+ conductance. In the same way, closing either the B switch or the C switch will increase, respectively, the membrane conductances to K^+ or Cl^- ions. (*b*) 5-HT responses caused by a *decrease* in membrane conductances. The resting conductances to Na^+ and K^+ are represented by two conductances in parallel. 5-HT produces an atypical response by opening the switch α which removes the Δg_K conductance from the circuit, whereas the hyperpolarizing response is brought out by opening of the coupled switch (marked β) which results in the removal of both Δg_{Na} and Δg_K from the circuit. (Gerschenfeld and Paupardin-Tritsch, 1974*a*.)

That the K^+ channels inactivated during this response are somewhat "specific" is suggested by the observation of Kehoe (1975) that the amplitude of this slow EPSP is augmented by arecoline, an agonist of acetylcholine (ACh), known to act in *Aplysia* CNS only on a cholinergic receptor associated with the activation of K^+ channels (see Kehoe, 1972). Moreover, ACh antagonists known to block this receptor, such as methylxylocholine, depress the atypical slow EPSP. These observations suggest that the channels turned off during these atypical slow EPSPs likely are associated in some way with ACh receptors.

This result also raises an important question: have the channels inactivated in the atypical responses been previously turned on by the action of another transmitter, or do they exist in an "open" conformation in the membrane, closing randomly, but showing a predominance of turned-on states? Even if it is not yet possible to give an answer to this question, it is tempting to accept the first mechanism, as it would explain the atypical synaptic responses simply as a special case of a classic mechanism, in which the observed permeability decrease results from the closing of ionic channels previously turned on by a synaptic transmitter.

Recent microphysiological studies on the vertebrate retina show that this mechanism actually operates in some synapses. However, it does not appear to explain all the observed synaptic actions on retinal cells involving a permeability decrease.

The Evidence from Retinal Synapses

It must be remembered first that the responses of cones of lower vertebrates to light consist in a hyperpolarization that results from a decrease in permeability, probably mainly to Na^+ ions (Tomita, 1965; Kaneko and Hashimoto, 1967; Toyoda et al., 1969; Baylor and Fuortes, 1970; Lasansky and Marchiafava, 1974). The equivalent circuit of the vertebrate cone membrane would not differ much from the picture of Fig. 3*b*, the action of light simply removing Δg_{Na} from the circuit. Δg_{Na} would represent in the cone membrane the ionic channels kept turned-on in the dark. The "membrane noise" recently observed by Simon et al. (1975) in turtle cones in the dark has been interpreted as the result of the random turning-off of ionic channels in the membrane; the illumination of retina causing a marked reduction of such a "dark noise."

At the level of the outer plexiform layer, we find an example of a synaptic membrane, the ionic channels of which are kept activated by the persistent action of a tonically released synaptic transmitter; here, the phasic suppression of such release causes a decrease of the postsynaptic membrane permeability. On the basis of experiments analyzing the effects of transretinal extrinsic currents, Trifonov (1968) proposed that the synaptic transmitter released by photoreceptors is liberated continuously and that light stimulation causes the suppression of transmitter release. The horizontal cells, which receive direct input from the photoreceptors, are hyperpolarized by light stimulation (Tomita, 1965; Werblin and Dowling, 1969; Kaneko, 1970; Nelson, 1973) and this hyperpolarization involves a decrease in membrane permeability (Toyoda et al., 1969). Recent experiments show that, by adding to the extracellular medium either an excess of Mg^{++} ions or Co^{++} ions, the blockade of synaptic transmission in the retina hyperpolarizes the horizontal cells and depresses their responses to light without affecting the light responses of the photoreceptors (Dowling and Ripps, 1973; Cervetto and Piccolino, 1974; Trifonov et al., 1974; Kaneko and Shimazaki, 1975, 1976; Dacheux and Miller, 1976). Therefore, the hyperpolarization of the horizontal cells by light is a good example of how a physiological interruption of a continuous transmitter release results in a permeability decrease at the postjunctional membrane.

However, as pointed out before, other observations on the synapses made by the photoreceptors with second-order cells suggest the possibility that some retinal postsynaptic membranes may be endowed with "spontaneously" turned-on channels, which would become inactivated by the action of the photoreceptor transmitter.

Bipolar cells of lower-vertebrate retinas generally show an organization of their receptor fields characterized by a center-surround antagonism (Werblin and Dowling, 1969; Kaneko, 1970; Schwartz, 1974). Two types of bipolar cells have been described: the first, "depolarizing" bipolars, becomes depolarized when stimulated in the center of its receptive field by a light spot and becomes hyperpolarized when it is stimulated by an illuminated annulus, which activates the periphery of the receptive field. The second type, "hyperpolarizing" bipolars, responds to light and shows an opposite pattern, i.e., it is hyperpolarized by a central spot and depolarized by a peripheral light annulus. In the carp, Kaneko (1970) demonstrated that both types of bipolar cells receive direct input from photoreceptors when they are stimulated by a central light spot and from interneurons, probably horizontal cells, when they are stimulated by a peripheral light annulus (see also Richter and Simon, 1975). This would signify, at least, that the same photoreceptor synaptic transmitter exerts different and opposite actions on the depolarizing bipolar cells and on the horizontal cells. At variance with what happens in horizontal cells, the continuously released photoreceptor transmitter keeps the depolarizing bipolar cells hyperpolarized, and when the transmitter's release is depressed by illumination, the depolarizing bipolars become depolarized. This was recently confirmed by Kaneko and Shimazaki (1976) in the carp and by Dacheux and Miller (1976) in the mudpuppy. These authors observed that the perfusion of the retina with a medium containing Co^{++} ions, which blocks all transmitter release, depolarizes the depolarizing bipolars. Because previous results of Toyoda (1973) and Nelson (1973) suggest that light stimulation causes an increase in the membrane permeability of the depolarizing bipolars, it is likely that the membrane permeability of the depolarizing bipolar is decreased in the dark under the continuous action of the photoreceptor transmitter.

On the contrary, the effects of the continuously released photoreceptor transmitter in the dark and of its suppression, on the hyperpolarizing bipolar cells, do not appear to differ from those observed in the horizontal cells (Kaneko and Shimazaki, 1976; Dacheux and Miller, 1976). This is also supported by the recent observation by Simon et al. (1975) that, in the dark, the hyperpolarizing bipolar cells of the turtle retina are "noisy" and that spot central illumination suppresses their membrane noise.

It is still difficult to interpret the chain of synaptic events leading to the responses of both types of bipolar cells to the stimulation of the periphery of their receptive fields, except that they probably are mediated by horizontal cells (Richter and Simon, 1975; Kaneko and Shimazaki, 1976). Nevertheless, it can be supposed that the horizontal-cell synapses operate in a similar way to those of the photoreceptor synapses; they probably continuously release their synaptic transmitter, and this release is suppressed by light stimulation. If such is the case, since the stimulation of the periphery of the receptor field also has an opposite effect on the depolarizing and the hyperpolarizing bipolar cells, the primary effect of the transmitter released by the horizontal cells must also have opposite effects on both bipolar cell types.

Figure 4 attempts to summarize our present knowledge of the operation of the synapses between the photoreceptor and the second-order neurons. On the left, the direct connection between cones and both horizontal and hyperpolarizing bipolars has been schematized. In the dark, the continuous release of transmitter keeps the synaptic switch closed and the conductance (i.e., the permeability) $1/R_s$ shunts the membrane conductance $1/R_m$. When the cone is hyperpolarized by light, the synaptic-transmitter release stops or diminishes, the synaptic switch is open, and the postsynaptic cell-membrane permeability decreases because the synaptic shunt is removed from the circuit.

On the right, the case of the depolarizing bipolars is represented. The continuously released photoreceptor transmitter keeps the synaptic switch open in the dark, thus turning off the channels represented by the conductance $1/R_s$. When the cone hyperpolarization by light suppresses the transmitter release, the synaptic switch closes and the channels represented by $1/R_s$ are turned on and the membrane permeability increases. The nature of E_s and $1/R_s$ in the diagram has not yet been clarified, but it is likely that they could be equal to E_{Na} and g_{Na}.

Concluding Remarks

The atypical or unusual actions of chemical transmitters involve a decrease in the membrane permeability of the postsynaptic cells to specific ions. In some cases, these responses result from the turning off of ionic channels previously activated by the continuous action of a tonically released transmitter (horizontal and hyperpolarizing bipolar cells of the lower vertebrate retina, atypical slow EPSP in *Aplysia* pleural neurons). In other cases (slow EPSP and IPSP of sympathetic ganglion neurons, atypical serotoninergic IPSP in *Aplysia* buccal neurons, atypical slow EPSP of the inking reflex motor neurons of *Aplysia*, depolarizing bipolar cells of the retina), such a mechanism involving a continuous action of a transmitter in activating the channels which become inactivated during the atypical synaptic action remains to be proved. The possibility exists that, in some cases,

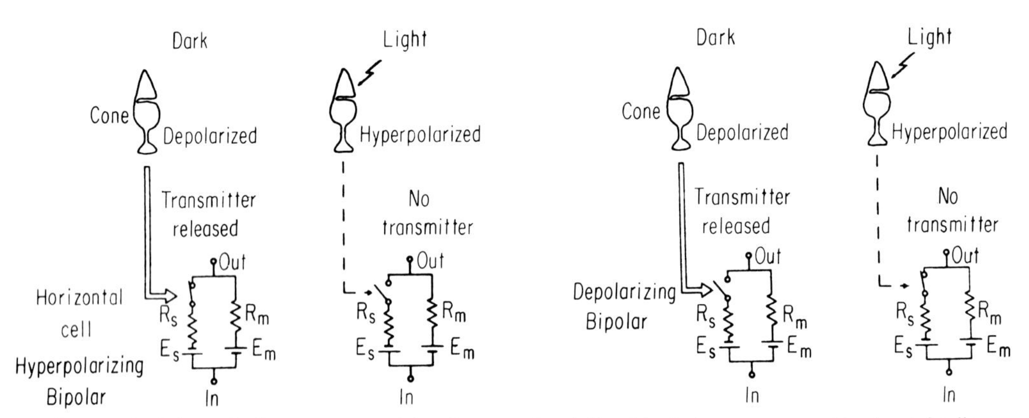

FIGURE 4 Schematic diagram summarizing the synaptic relationships between cones and horizontal cells, on the left, and between cones and "depolarizing bipolars," on the right. In the simplified membrane circuits represented in the lower part of the figure, R_m and E_m, respectively, represent the membrane resistance and the membrane potential of the cells, R_s represents the resistance of the synaptic membrane, and the battery E_s is equal to the equilibrium potential of the ions involved in the synaptic response. In the case of the horizontal cells, the cone continuously releases synaptic transmitter in the dark, keeping the switch closed and causing a sustained increase in the horizontal cell membrane conductance. When illumination hyperpolarizes the cones, the release of transmitter is suppressed, the switch is thus open, and the total membrane conductance of the horizontal cell is decreased. (This could be also the case for the "off-center" bipolars.) On the right, in the dark, the continuous release of transmitter from the cone would maintain the switch open, removing the conductance $1/R_s$ from the "on-center" bipolar membrane circuit, the membrane conductance of the bipolar cell would thus be kept low in the dark. The opening of the switch induced by the suppression of the cone transmitter release by light stimulation would cause an increase of the bipolar membrane conductance. If the case is already well substantiated for the cone-horizontal cell synapses, it is not fully proved for the synapses relating cones to "on-center" bipolars (modified from Kaneko and Shimazaki, 1975).

the postjunctional membrane could be endowed with spontaneously turned-on channels, as they actually exist in the membrane of vertebrate retinal cones.

Whatever may be their intrinsic ionic mechanism, the atypical synaptic actions probably play a role in insuring the specificity of connections in a complex circuitry. It has already been shown that they can intervene, modulating the electrical coupling between neurons (Carew and Kandel, 1976).

REFERENCES

Baylor, D. A., and M. G. F. Fuortes. 1970. Electrical responses of single cones in the retina of the turtle. *J. Physiol. (Lond.).* **207:**77–92.

Carew, T. J., and E. R. Kandel. 1976. Two functional effects of decreased conductance EPSP's: synaptic augmentation and increased electrotonic coupling. *Science (Wash. D.C.).* **192:**150–153.

Cervetto, L., and M. Piccolino. 1974. Synaptic transmission between photoreceptors and horizontal cells in the turtle retina. *Science (Wash. D.C.).* **183:**417–419.

Dacheux, R. F., and R. F. Miller. 1976. Photoreceptor-bipolar cell transmission in the perfused eyecup retina of the mudpuppy. *Science (Wash. D.C.).* **191:**963–964.

Delaleu, J. C. 1976. Evidence for a synaptically mediated decrease in conductance in a crustacean myocardium. *J. Exp. Biol.* **65,** 117–129.

Dowling, J. E., and J. H. Ripps. 1973. Effect of magnesium on horizontal cell activity in the skate retina. *Nature (Lond.).* **242:**101–103.

Fatt, P., and B. Katz. 1951. An analysis of the end-plate potential recorded with an intracellular electrode. *J. Physiol. (Lond.).* **115:**320–369.

Fatt, P., and B. Katz. 1953. The effect of inhibition nerve impulses on a crustacean muscle fiber. *J. Physiol. (Lond.).* **121:**374–389.

Gerschenfeld, H. M. 1973. Chemical transmission in invertebrate central nervous systems and neuromuscular junctions. *Physiol. Rev.* **53:**1–119.

Gerschenfeld, H. M. 1976. Multiple receptor activation by single transmitters. *In* The Synapse. G. A. Cottrell and P. N. R. Usherwood, editors. Blackie & Son Ltd., Glasgow. 157–176.

Gerschenfeld, H. M., and D. Paupardin-Tritsch. 1974*a*. Ionic mechanisms and receptor properties underlying the responses of molluscan neurons to 5-hydroxytryptamine. *J. Physiol. (Lond.).* **243:**427–456.

Gerschenfeld, H. M., and D. Paupardin-Tritsch. 1974*b*. On the transmitter role of 5-hydroxytryptamine at excitatory and inhibitory monosynaptic junctions. *J. Physiol. (Lond.).* **243:**457–481.

Ginsborg, B. L. 1973. Electrical changes in the membrane in the junctional transmission. *Biochim. Biophys. Acta.* **300:**289–318.

Kaneko, A. 1970. Physiological and morphological identification of horizontal, bipolar and amacrine cells in the goldfish retina. *J. Physiol. (Lond.).* **207:**623–633.

Kaneko, A., and H. Hashimoto. 1967. Recording site of the single cone response determined by an electrode marking technique. *Vision Res.* **7:**847–851.

Kaneko, A., and H. Shimazaki. 1975. Effect of external ions on the synaptic transmission from photoreceptors to horizontal cells in the carp retina. *J. Physiol. (Lond.).* **252:**509–522.

Kaneko, A., and H. Shimazaki. 1976. Synaptic transmission from photoreceptors to bipolar and horizontal cells in the carp retina. *Cold Spring Harbor Quant. Biol.* **45:**537–546.

Kehoe, J. S. 1972. Three acetylcholine receptors in *Aplysia* neurons. *J. Physiol. (Lond.).* **225:**115–146.

Kehoe, J. S. 1975. Analysis of a "resting" synaptic permeability that can be synaptically reduced. *J. Physiol. (Lond.).* **244:**23–24P.

Lasansky, A., and P. L. Marchiafava. 1974. Light induced resistance changes in retinal rods and cones of the Tiger salamander. *J. Physiol. (Lond.).* **236:**171–191.

Nelson, R. 1973. A comparison of electrical properties of neurons in *Necturus* retina. *J. Neurophysiol.* **36:**519–535.

Paupardin-Tritsch, D., and H. M. Gerschenfeld. 1973. Neuronal responses to 5-hydroxytryptamine resulting from membrane permeability decreases. *Nat. New Biol.* **244:**171–173.

Richter, A., and E. J. Simon. 1975. Properties of centre hyperpolarizing red sensitive bipolar cells in the turtle retina. *J. Physiol. (Lond.).* **248:**317–334.

Schwartz, E. 1974. Responses of bipolar cells in the retina of the turtle. *J. Physiol. (Lond.).* **236:**211–224.

Simon, E. J., T. D. Lamb, and A. L. Hodgkin. 1975. Spontaneous voltage fluctuations in retinal cones and bipolar cells. *Nature (Lond.).* **256:**661–662.

Tomita, T. 1965. Electrophysiological study of the mechanisms subserving color coding in the fish retina. *Cold Spring Harbor Symp. Quant. Biol.* **30:**559–566.

Toyoda, J. I. 1973. Membrane resistance changes underlying the bipolar cell response in the carp retina. *Vision Res.* **13:**295–306.

Toyoda, J., H. Nosaki, and T. Tomita. 1969. Light induced resistance changes in single photoreceptors of *Necturus* and *Gekko*. *Vision Res.* **9:**453–463.

Trifonov, Y. 1968. Study of synaptic transmission between photoreceptors and horizontal cells by means of electrical stimulation of the retina. (in Russian). *Biofizika.* **13:**809–817.

Trifonov, Y., A. L. Byzov, and L. M. Chilahian. 1974. Electrical properties of subsynaptic and nonsynaptic membranes of horizontal cells in the fish retina. *Vision Res.* **14:**229–241.

WEIGHT, F. F., and A. PADJEN. 1973*a*. Slow synaptic inhibition: evidence for synaptic inactivation of sodium conductance in sympathetic ganglion cells. *Brain Res.* **55:**219–224.

WEIGHT, F. F., and A. PADJEN. 1973*b*. Acetylcholine and slow synaptic inhibition in frog sympathetic ganglia. *Brain Res.* **55:**225–228.

WEIGHT, F. F., and J. VOTAVA. 1970. Slow synaptic excitation in sympathetic ganglion cells: evidence for synaptic inactivation of potassium conductance. *Science (Wash. D.C.)* **170:**755–758.

WERBLIN, S., and J. E. DOWLING. 1969. Organization of the retina of *Necturus maculosus*. II. Intracellular recording. *J. Neurophysiol.* **32:**339–355.

Cell Surface Immunoreceptors

B-LYMPHOCYTE RECEPTORS AND LYMPHOCYTE ACTIVATION

G. J. V. NOSSAL

There are a number of reasons why the lymphocyte is of special interest to cell biologists, and it presents both operational and theoretical advantages for many diverse studies. For example, lymphocytes are easy to obtain in single cell suspensions of high viability, and as such they permit ready study of the plasma membrane and its receptors by techniques such as immunofluorescence, immunoelectron microscopy, or cell surface macromolecule radioiodination. They represent, in the main, cells in the G_0 state of differentiation, but can be thrown into mitotic cycle and clonal proliferation with fair efficiency in vitro using a variety of stimuli. Though relatively uniform in morphology, lymphocytes fall into subsets with different properties and functions, thus presenting challenges for biophysical separation technology. Finally, B lymphocytes, when appropriately activated, secrete a highly characteristic product, antibody, which can be identified and measured with very sensitive techniques. For these reasons, lymphocytes have become a fashionable tool for studies that transcend the confines of immunology.

In this symposium, it will be our aim to present an up-to-date review of the two main categories of lymphocytes, known as T, or thymus-derived, and B, or bone marrow-derived lymphocytes, with particular reference to the nature and functions of the cell surface receptors on them, and to the mechanisms whereby antigens or artificial mitogens (which mimic antigen action) trigger the complex events leading not only to division but also to functional activation of the cells. Our plan is to begin with the B lymphocyte, the nature of which is better understood; then to consider the T lymphocyte; and then to explore possible molecular models of activation, using mitogens as probes. In so doing, we hope both to expose some of the issues lying at the moving edge of immunology, and to transmit enough information about the lymphocyte and its receptors to encourage cell biologists to explore its mysteries further.

G. J. V. NOSSAL The Walter and Eliza Hall Institute of Medical Research, Victoria, Australia

Key Questions in Cellular Immunology

GENERATION AND EXPRESSION OF IMMUNOLOGICAL DIVERSITY: A central feature of the immune response is its specificity for the antigen used to evoke it, and as the number of antigens that can be recognized and discriminated is very large, the repertoire of molecular recognition units able to be generated by an animal must also be very large. It is now generally agreed (Jerne, 1977) that the immunological specificity of both B and T lymphocytes rests in receptors possessing the variable domains of immunoglobulin (Ig) polypeptide chains. A recent estimate (Klinman et al., 1977) places the number of different antibodies able to be made by an adult mouse at 10^7. In the present paper, I shall describe in detail experiments that show that antibody formation works by a process of clonal selection (Burnet, 1957); that is, by the stimulation of that subset of cells that possesses preformed cell surface receptors capable of recognizing the antigen. Different unstimulated lymphocytes possess different receptor variable (V) regions, and the origin of this diversity amongst lymphocytes is the subject of much experimentation and speculation. Many hold the view that the number of different V genes in the germ line of an animal is relatively small, perhaps 30–100 for heavy and light chains, allowing (by random pairing of light and heavy chain V regions) 10^3–10^4 antibody types to reflect germ line genes. Thus, some process of somatic diversification (mutation, partial enzymic degradation and incorrect repair of DNA, or episomal insertion) may enlarge the available repertoire.

LYMPHOCYTE SUBSETS AND THEIR FUNCTIONS: Circulating antibody molecules are secreted by the differentiated progeny of antigenically activated B lymphocytes (Nossal et al., 1968). Several behavioral and biophysical characteristics distinguish "virgin" B lymphocytes, i.e., cells that have not yet encountered antigen, from "memory" B cells, i.e., cells that are the progeny of antigenically stimulated cells, and which are responsible for the heightened response capacity of preimmunized animals. While the same is true of T lymphocytes, the complete and unequivocal separation of virgin and memory cells from each other has not yet been achieved.

T lymphocytes are derived from the thymus and perform a variety of functions. They possess the capacity to initiate an inflammatory response around antigen depots, which reaches its maximum 24–48 h after injection of the test antigen, and is thus called delayed hypersensitivity. They can exert a direct cytotoxic action (which is not dependent on the classic complement pathway) on antigenic cells including grafts and tumors. They are also involved in complex but profoundly important interactions with B lymphocytes, which can either "help" the B cell (T cell help being essential for the generation of high affinity IgG antibodies) or suppress B-cell function. These different effector functions are mediated by different subsets of T cells, which can be distinguished from one another by the possession of different surface macromolecules. The present state of knowledge about T-cell receptors and lymphocyte interactions will be reviewed by Dr. D. H. Katz in the next paper of this symposium.

A very important subset of lymphocytes, considered later in this paper, are cells possessing the morphologic characteristics of lymphocytes but not yet displaying the surface receptors conferring the power of antigen recognition. These cells represent the penultimate result of the lymphopoietic process occurring in primary lymphoid organs, requiring only a nonmitotic maturation phase to complete this differentiation (Osmond and Nossal, 1974).

CELLULAR AND MOLECULAR MECHANISMS OF LYMPHOCYTE ACTIVATION: This key area of research will be the chief topic of all three papers, the present one focusing on the B cell. As an overall introductory comment, it may be fair to point out that the cellular phenomena are being more rapidly clarified than is the intracellular biochemistry, perhaps because the process of clonal lymphocyte differentiation is a complex and protracted one requiring not just a single signal, but a continuing exposure to antigen or mitogen (Pike and Nossal, 1976).

REGULATORY MECHANISMS IN THE IMMUNE RESPONSE, INCLUDING TOLERANCE, NEGATIVE FEEDBACK, ACTIVE SUPPRESSION, IMMUNE NETWORKS, AND GENETIC REGULATION: Two requirements of the immune response have imposed particularly tricky tasks on evolution. The first is the need for "self-not self" discrimination, to allow the formation of antibodies against molecules pulsed in unexpectedly from the outside world but not against the myriad potentially antigenic molecules of the body itself. The second is the need to maintain control over the cascade of clonal expansion that follows exposure to antigen, and to insure limits to the proliferation even if the antigen is not eliminated and chronic infection results. To these ends, an elaborate range of regulatory mechanisms has emerged, which we will not consider in depth in this symposium, but which must be mentioned in view of their great current importance.

Produced antibody exerts a negative feedback role, both by aiding the elimination of antigen and thus reducing the driving force for continued clonal expansion, and because antigen-antibody complexes can act as powerful blockaders of lymphocyte receptors, causing a failure of lymphocyte responsiveness (Stocker, 1976).

Active suppression, a recently discovered phenomenon, involves the genesis, through antigenic stimulation, of a population of T lymphocytes that produce soluble factors capable of reducing antibody formation. Controversy surrounds the question of whether the target for such suppressor factors is the B cell or the "helper" T cell needed to allow B cells to multiply optimally.

Immune network function is a concept generated by Jerne (1974), centered on the idea that antibody V regions can themselves act as antigens. An immune response generates an antibody, and if this antibody acts antigenically in the body, an anti-antibody (or a corresponding activated anti-antibody T lymphocyte) may be formed. This could act on lymphocytes bearing the original antibody as a surface receptor. Cytotoxic T lymphocytes could cause suppression (negative feedback loop), and, under certain circumstances, the anti-antibody, through its similarity to antigen, could cause stimulation (positive feedback loop).

A further area of complexity is that the immune

responses to many antigenic determinants are under genetic control, not only because Ig germ line genes influence the available recognition repertoire, but also, and perhaps more importantly, because specific immune response genes are involved in T cell-B cell interactions. This major facet of regulation is discussed in Dr. Katz's paper.

Nature of B-Cell Receptors

ONTOGENY OF Ig RECEPTORS: With this broad introduction as background, we can now consider the B lymphocyte in more detail.

In the mouse, B lymphocytes are generated in a multifocal fashion, first in the fetal liver in embryonic life and then in the bone marrow and spleen. Lymphoneogenesis persists in these organs throughout adult life. The process of B-lymphocyte formation is, in principle, no different from that involved in the formation of other blood cells. A pool of multipotential stem cells exists. Through a series of unclassified differentiation steps, large, rapidly cycling lymphocytes are generated which give rise to medium and small lymphocytes by a series of divisions. In the B-lymphocyte series, some Ig synthesis takes place in the large precursor cells and can be detected by staining the cytoplasm by immunofluorescent techniques. The display of these Ig molecules on the cell membrane as receptors comes as a later event in differentiation (Raff et al., 1976). In fact, the small lymphocyte (or pre-B cell) that is the end result of the process initially lacks surface receptors, at least by immunofluorescent or immunoautoradiographic techniques. These surface Ig molecules are inserted into the membrane in progressively larger numbers over the 48 h after the last division.

In the mouse embryo, the first cells with intracytoplasmic Ig appear in the fetal liver around day 12 of gestation, but the first mature B cells with surface Ig only around day 16.5 (Raff et al., 1976).

CLASS OF B-CELL Ig RECEPTOR: It has recently become apparent (Rowe et al., 1973; Vitetta and Uhr, 1975; Abney et al., 1976; Goding and Layton, 1976) that many B lymphocytes possess not one but two different classes of Ig as surface receptors, namely, IgM and IgD. This striking finding was not made sooner because (IgD being only a very minor constituent of serum) anti-δ sera have not been available. The recognition of a genetic polymorphism (allotype) in murine IgD, and the demonstration of antibodies to this molecule in many mouse antisera made against lymphoid cells of unrelated strains (Goding et al., 1976) will provide a powerful new tool for the study of B-cell differentiation.

The earliest-formed B lymphocytes, e.g., those in newborn mouse spleen, possess only IgM on their surface. In contrast to secreted, serum IgM, which has 10 heavy and 10 light chains and a sedimentation coefficient of 19S, the cell surface receptor is 7–8S and consists of two heavy and two light chains, therein resembling secreted IgG. In the adult mouse, newly formed B cells as found in the bone marrow also display only IgM. However, by about 1 wk of age, cells bearing IgM and IgD are apparent, particularly in lymph nodes. In adult tissues the picture is complex, cells positive for both IgM and IgD being the most frequent type of B cell, but both IgM positive, IgD negative and IgD positive, IgM negative cells constituting significant percentages (Goding and Layton, 1976).

It is tempting to speculate on the implications of these observations (Vitetta and Uhr, 1975). One possibility is that the B cell bearing only IgM is still immature, and acquires its capacity to be triggered by antigen only when it acquires surface IgD. This would involve IgD in a special way in immune triggering. A second possibility is that IgM^+ IgD^+ cells represent a distinct subset, for example, memory cells. In opposition to this, evidence exists to show that memory cells exhibit on their surfaces the class of antibody that they will synthesize after restimulation (e.g., IgA, IgG_1, etc.; Okumura et al., 1976; Mason, 1976). Evidence is currently being gathered (Goding and Layton, personal communication) that IgM^+ IgD^+ cells selectively lose surface IgD after antigenic stimulation. Despite all these hints for the importance of IgD in immune triggering, it is clear that anti-μ sera act as very powerful inhibitors of antigenic stimulation in vitro, whereas anti-δ sera appear to lack this capacity, at least as far as in vitro IgM responses are concerned. A further puzzle is why IgD, so prominent as a cell receptor, is not detectable in mouse serum (being secreted in only very small amounts if at all). These uncertainties are currently under active investigation in several laboratories.

SPECIFICITY OF B-CELL SURFACE RECEPTORS: The clonal selection theory (Burnet, 1957) predicts that a given B cell will display receptor molecules of only one antigen-combining specificity and that antigenic stimulation merely involves selecting, out of the total lymphocyte population, those cells with receptors for the antigen in question. Thus, if it were possible to select out of a random population of spleen cells from an unimmunized mouse that small fraction possessing receptors for a given antigen, then, in principle, *each* cell in the selected population should be able to be stimulated to produce an antibody-forming clone.

It is almost 20 yr since the enunciation of the rule that one cell can form only one antibody (Nossal and Lederberg, 1958), yet the above conceptually simple experiment has proven frustratingly difficult to perform. The chief practical hurdles have included difficulties in coping with the small numbers of B lymphocytes in vitro, nonspecific adhesion of lymphocytes to antigen-containing affinity chromatography columns, and poor cloning efficiency of the fractionated cells in vitro. It is pleasing to report that all of these have recently been overcome, and that it is now possible to produce a formal proof of clonal selection (Nossal et al., 1977).

In normal adult mouse spleen, about one cell in 2×10^4 can be triggered to form antibody to a typical haptenic antigen such as 4-hydroxy-3-iodo-5-nitrophenylacetic acid (NIP). One convenient method involves placing small numbers of spleen cells in microcultures in the presence of nonantibody-producing thymus lymphocytes acting as a "feeder" layer (Nossal and Pike, 1976), and stimulating them with the hapten conjugated onto a polymeric carrier antigen such as polymerized *Salmonella* flagellin. This provides a powerful in vitro stimulus, encouraging the formation of clones of anti-NIP antibody forming cells.

A major step forward came with the development of the hapten-gelatin fractionation technique of Haas and Layton (1975). This ingenious, simple method depends on the melting properties of gelatin. Instead of an antigen affinity column as tried by others, a thin layer of haptenated gelatin on the bottom of a Petri dish acted as the specific immunoabsorbant. At 4°C, 10^8 normal spleen cells were added to each dish and gently rocked. After 15 min, about 1 cell in 2,000 was firmly adherent to the haptenated monolayer and the rest were removed by extensive washing. Then the adherent cells were collected by the addition of tissue culture medium at 37°C (instantly melting the thin gel layer) and centrifugation. Next, the NIP-gelatin adherent to the receptors of NIP-specific cells was removed by brief treatment with collagenase. It was found that about 30% of these cells were indeed specific in that they fluoresced when stained with a haptenated, fluorescent antigen (Nossal and Layton, 1976). Of these, some 10% (i.e., 3% of the fractionated population) could form a clone of antibody-producing cells in the microculture system (Nossal and Pike, 1976).

Even though it might be unrealistic to expect 100% cloning efficiency in cultured B cells, some in vitro death being inevitable, it appeared that the fractionation procedure might be selecting cells with receptors of too low a median affinity for the hapten. Accordingly, we tried a more rigorous fractionation technique in the hope of selecting only those cells with the best-fitting receptors. The first cycle of NIP-gelatin fractionation was followed by a second, identical cycle, which resulted in a doubling of the proportion of cells staining with fluorescent antigen. Finally, the twice-fractionated cells were mixed with sheep erythrocytes to which NIP had been coupled. A proportion formed rosettes, i.e., attracted clusters of haptenated red cells to their surface. These could then be separated from the unrosetted cells by gradient centrifugation. The resulting population of rosetted cells was treated with hypotonic shock and trypsin to remove adherent red cells and debris.

This somewhat heroic series of treatments involves serious yield losses, and fewer than 1 cell in 10^5 from the original starting population is recovered. However, the population is indeed active in antibody formation. In three successive experiments, the mean proportion of cells capable of yielding an anti-NIP antibody-forming clone in vitro was 1 in 3.5. Two reasons suggest that this remarkably high figure may nevertheless be an underestimate. First, it was found that the initiation of clonal proliferation amongst B cells was asynchronous, newly arising clones still being noted on day 3 of culture. As this was the routine day of harvesting, the possibility could not be excluded that clonal precursors commencing division later than that were missed. Second, the three fractionation cycles involved a large amount of mechanical handling and centrifugation of the cells, and some impairment of cloning efficiency might be expected as a result.

The profound degree of specific enrichment in the capacity to form antibody achieved purely on the basis of affinity of cell surface receptors for the chosen antigen seems close to the ultimate vindication of clonal selection. Nevertheless, it remains to show that *all* of the receptor Ig on the surface of those hapten-specific lymphocytes was of uniform specificity, i.e., that all of it was anti-NIP in character. This we achieved by immunofluorescent techniques (Nossal and Layton, 1976). Using saturating concentrations of rhodamine-conjugated polyvalent antigen, hapten-fractionated cells were stained at 37°C, so that all the antigen-specific receptors were swept into a "cap" at one pole of the B cell. The cells were then fixed with paraformaldehyde to prevent further receptor redistribution and counterstained with a fluoresceinated antiglobulin. In all but a few exceptional cases, it was found that *all* the Ig had moved into the cap. In other words, *no* linearly distributed green fluorescence could be observed, as would have been the case if a second Ig with a different specificity had been present on the cell. Furthermore, when chain-specific anti-μ and anti-δ sera were used in experiments of this sort, it was found that the antigen had caused capping of *both* IgM and IgD (Goding and Layton, 1976). Thus it appears that one B cell possesses two Ig classes, each with an identical antigen-combining site, and thus presumably with identical Ig V regions. This leaves us with the question of how one V gene is simultaneously translocated to two different C genes in the one cell.

NUMBER OF Ig RECEPTORS ON B CELLS: It is estimated that a typical B cell has 3×10^4–10^5 Ig molecules on its surface, this representing 1–2% of total cell-surface protein. In contrast, a fully activated antibody forming cell contains around 10^7 Ig molecules in its cytoplasm, and secretes this quantity or more per hour.

NON-Ig RECEPTORS ON B LYMPHOCYTES: It is established that B lymphocytes have a number of other macromolecules on their surface that can function as receptors and may have great immunological importance. Not all B lymphocytes possess all the types of receptors, and moreover the physiological purpose of most of these is quite uncertain. In practical terms, the currently most useful aspect of these other cell surface macromolecules is that they aid in the definition and separation of different B-lymphocyte subsets.

Fc Receptors

These are receptors for a site on the Fc or "handle" portion of the heavy chains of certain Ig classes, the site being exposed when Ig is aggregated, as in antigen-antibody complexes, or heat-aggregated globulins. Fc receptors may function as an aid to antigen transport. There is evidence that foreign globulins function as particularly strong immunogens when attached to B cells via the Fc receptors, and, physiologically, many soluble antigens could be attached to B cells when either natural or immune antibody is present in the extracellular fluids.

C_3 Receptors

A fraction of B lymphocytes have a receptor for the third component of complement, and there has been some speculation that the complement pathway may be involved in B-lymphocyte activation.

Ia Antigens

A set of polymorphic genes in the I region of the major histocompatibility complex controls the production of glycoproteins which are expressed preferentially on B lymphocytes (though present in lower concentration on T cells as well; Goding et al., 1975*a, b*). It is possible that these molecules are receptors for regulatory molecules derived from T lymphocytes, and this area of work is covered in Dr. Katz's review.

Mitogen Receptors

A considerable number of compounds, prominently including polymers such as bacterial lipopolysaccharides, polymerized flagellin, polyvinyl pyrrolidine and dextran, possess the capacity to act as B-cell mitogens, i.e., to stimulate Ig synthesis and cell division even in lymphocytes not possessing Ig receptors specific for the mitogen. It is presumed that these molecules must make contact with some lymphocyte surface molecule in order to cause triggering. Some of these substances function as good specific immunogens in vitro and in vivo, in concentrations much lower than those required to produce nonspecific mitogenesis. Möller (1975) has speculated that this is because cells possessing Ig receptors specific for the mitogen focus large amounts of mitogen onto the cell even when the molar concentration of mitogen is low. He proposes that it is really the mitogen receptor that is the cell's trigger molecule, not the Ig receptor.

T lymphocytes can also be stimulated with mitogens, though a different set. Some of these, such as concanavalin A, adhere to practically every glycoprotein on the lymphocyte surface. The question of mitogen receptors and mitogen responsiveness is taken up in detail by Dr. M. J. Crumpton in the third paper in this symposium.

Hormone and Drug Receptors

In common with other cells, lymphocytes possess receptors for certain hormones and drugs.

These receptors and the biochemical changes consequent on interaction with the ligand, including changes in the differentiation state of the lymphocyte, are now being investigated in detail.

Mechanisms of B Lymphocyte Activation

In the absence of a consensus about mechanisms of B-lymphocyte activation, theories abound. Three major sets can be identified.

TWO SIGNAL THEORIES: Bretscher and Cohn (1968) initiated the idea that B-lymphocyte activation required two separate signals, namely, signal 1, union of Ig receptor with antigen, plus signal 2, being some short-range chemical inducer secreted by a T cell or perhaps a macrophage stimulated by a T cell. If signal 1 alone operates, the result would be inactivation or tolerance. Signal 1 is seen as raising intracellular cyclic AMP, and signal 2, cyclic GMP. The theory in its simplest form has been criticized for making the role of the T cell too obligatory. For example, B lymphocytes in agar dishes, separated by long distances from any other cell, can divide and partially differentiate (though not to full antibody-secreting status) under the influence of nonspecific mitogens (Metcalf et al., 1975). Schrader (1973) introduced the variant notion that a B cell mitogen might itself provide signal 2, achieving its triggering capacity by mimicking the physiological signal 2 substance derived from T cells or macrophages. His view was that signal 2 alone could trigger memory B cells, but signal 1 plus signal 2 were obligatory for triggering of virgin B cells. A substantial body of work is directed at determining what the stimulatory T-cell factor might be. Evidence that B cells, in the absence of T cells, can be stimulated by a factor derived from antigenically activated T cells has been presented (Taussig and Munro, 1975; Mozes, 1976). This factor is particularly interesting, in that it binds antigen and appears to be an *I* region gene product. In a different system (Tada et al., 1975), T cell-derived material with very similar characteristics was found to be not stimulatory but suppressive of B cell function. As well as these antigen-specific factors, other T cell-derived factors have been described which are not specific with respect to antigen, but act synergistically with antigen to increase the number of antibody-forming cells produced (Schimpl and Wecker, 1972; Armerding and Katz, 1974). We are still a long way from understanding how these various factors relate to each other, and how they might act biochemically.

CROSS-LINKING THEORIES: Lymphocyte Ig receptors, on union with appropriate cross-linking agents such as antiglobulin antibodies or specific polymeric antigens, move in the plane of the membrane to form patches and then caps over one pole of the cell. This observation, together with the fact that many of the antigens that can trigger B lymphocytes without T cell help are multivalent, and thus cause patching and capping of Ig receptors, has prompted the speculation that triggering is a consequence of receptor aggregation and contingent changes in submembrane assemblies (Yahara and Edelman, 1972). This theory in its simplest form is not acceptable. Many substances cause patching and capping of B-cell Ig receptors, such as NIP-gelatin or NIP-HGG. When tested on NIP-specific cells, these substances do not trigger but in fact can blockade B-cell receptors and prevent triggering by an authentic immunogen. Conversely, model nonspecific mitogens, such as lipopolysaccharides, do not aggregate the cells' Ig receptors to a detectable extent. However, receptor rearrangement is such a striking phenomenon with such a variety of fascinating features that it would be premature to consider it of no physiological consequence.

ONE NONSPECIFIC SIGNAL THEORY: The theory of Möller's group has already been mentioned. In its present form it has relatively little to say about T cell-dependent B-cell activation, which is clearly of great physiological importance. Nor does it propose specific biochemical mechanisms by which the diverse series of B-cell mitogens might act. Nevertheless, for the phenomenon of T-independent B-cell triggering, which is simpler to study than T-dependent events, it presents many ingenious and internally consistent arguments, and is therefore well worth pursuing further.

Needless to say, this listing by no means exhausts the field which, through the efforts of many workers, is moving ahead rapidly. Investigations on ion (particularly calcium) flux; allosteric changes in Ig; intracellular cyclic nucleotides; activation of complement and other enzymes; macrophage-lymphocyte interactions, and many other aspects of potential relevance are being reported, and should, together with the major efforts mentioned above, gradually move the field out of its present, rather unsatisfactory phase.

Clonal Abortion in Immunological Tolerance

Even a brief review of B-lymphocyte activation

must make some mention of the opposite phenomenon, namely, immunological tolerance or paralysis. Under appropriate circumstances, antigens can influence lymphocyte populations so as specifically to reduce or abolish their capacity to respond to the same antigen presented at a later time in immunogenic form. It is now clear that this process of immunological tolerance can be achieved by a variety of cellular mechanisms, which have been reviewed elsewhere (Nossal, 1974). Here, I wish only to mention one recently discovered mechanism which may be particularly relevant to self-tolerance, i.e., the nonreactivity each normal individual has to freely circulating bodily constituents. This is a process termed clonal abortion (Nossal and Pike, 1975).

One elegant way of insuring self-tolerance would be to arrange lymphocyte differentiation in such a manner that each lymphocyte, before maturing to capacity to react with antigen, had to pass through a phase where any contact with antigen destroyed or permanently paralyzed it. This would have the effect of "nipping in the bud" any potentially self-reactive cell. The ever-present self antigen would always "catch" the maturing cell during this phase. The unexpectedly pulsed in foreign antigen might catch a few cells in this stage, but would encounter many more that had passed through this stage before the antigen had entered the body. These cells would form antibody, rapidly eliminate the antigen, and thus prevent any clonal abortion in further differentiating cells.

A necessary postulate of this clonal abortion theory is that there should be distinct behavioral peculiarities in B cells that have just acquired their first Ig receptors. Indeed, such can readily be found. When such cells are isolated from mouse bone marrow, fetal liver, or neonatal spleen, and are treated with anti-Ig reagents, the Ig molecules on their surfaces patch and cap quite normally, after which they are pinocytosed. However, in contrast to normal mature B cells, which resynthesize and redisplay their receptor coat within 6–8 h, these immature cells remain bereft of receptors for the duration of the experiment (Raff et al., 1975; Sidman and Unanue, 1975). In fact, the addition of small quantities of anti-Ig serum to cultures of fetal liver taken from the embryo before any B cells have appeared can prevent completely the development of B cells which normally occurs in vitro. This suggests that the immature cells are extremely sensitive to anything that affects their Ig receptors.

We have studied both bone marrow (Nossal and Pike, 1975) and neonatal spleen (Stocker, 1977) B cells to determine their reactivity to antigen rather than to antiglobulin. With either tissue, it is possible to obtain a substantial *rise* in immune competence if the cells are cultured for 3 days in the absence of antigen. This reflects B-cell maturation. If antigen is added to the cultures, this rise can be specifically and completely abrogated, and with antigen concentrations in the low nanogram per milliliter range. In contrast, the immune potential of mature B cells is in no way affected. The phenomenon is entirely independent of T cells and appears to conform to all the predictions of the clonal abortion theory. Thus, the theory may well explain B-cell tolerance with respect to "self" antigens present at 10^{-9} M or higher concentrations. A number of other "fail-safe" regulatory mechanisms may exist as well, and the phenomenon of suppressor T cells may well be one of the most important ones.

CONCLUSIONS

Remarkable progress has been made in the technology applied to the problem of B-lymphocyte receptors and B-cell activation. Our knowledge of the B-cell membrane has reached a degree of detail and insight that would not have been dreamed of a decade ago, and the glycoproteins identified serologically and immunogenetically are beginning to be studied chemically. The tedious but vital process of categorizing B cells into subsets has begun. A number of ingenious cloning techniques have opened up possibilities of much more quantitative approaches to both activation and tolerance induction. The important regulatory role of the T cell, discovered only a decade ago, dominates much of current thinking. The avid search for biologically active T-cell factors is made more intriguing by the realization that the key to two deep puzzles, namely, the genetic regulation of the immune response and the function of the highly polymorphic gene cluster of the major histocompatibility complex, may be contained within them. There is a long way to go before activation, which is a shorthand for a highly complex series of proliferative and differentiative events, is reasonably well understood. The field thus stands at an interesting midpoint: well established, but with major challenges still ahead.

ACKNOWLEDGMENTS

This is publication number 2280 from the Walter and Eliza Hall Institute of Medical Research. This work was supported by the National Health and Medical Research Council, Canberra, Australia, the National Institutes of Health grant AI-O-3958, and was in part pursuant to contract number NIH-NCI-7-3889 with the National Cancer Institute.

REFERENCES

ABNEY, E. R., I. R. HUNTER, and R. M. E. PARKHOUSE. 1976. Preparation and characterization of an antiserum to the mouse candidate for immunoglobulin D. *Nature (Lond.)*. **259**:404.

ARMERDING, D., and D. H. KATZ. 1974. Activation of T and B lymphocytes in vitro. II: Biological and biochemical properties of an allogenic effect factor (AEF) active in triggering specific B lymphocytes. *J. Exp. Med.* **140**:19–34.

BRETSCHER, P. A., and M. COHN. 1968. Minimal model for the mechanism of antibody induction and paralysis by antigen. *Nature (Lond.)*. **220**:444.

BURNET, F. M. 1957. A modification of Jerne's theory of antibody production using the concept of clonal selection. *Aust. J. Sci.* **20**:67.

GODING, J. W., and J. E. LAYTON. 1976. Antigen-induced co-capping of IgM and IgD-like receptors on murine B lymphocytes. *J. Exp. Med.* **144**:852–857.

GODING, J. W., G. J. V. NOSSAL, D. C. SHREFFLER, and J. J. MARCHALONIS. 1975*a*. Ia antigens on murine lymphoid cells: distribution, surface movement and partial characterization. *J. Immunogenet.* **2**:9.

GODING, J. W., E. WHITE, and J. J. MARCHALONIS. 1975*b*. Partial characterization of Ia antigens on murine thymocytes. *Nature (Lond.)*. **257**:230.

GODING, J. W., G. W. WARR, and N. L. WARNER. 1976. Genetic polymorphism of IgD-like cell surface immunoglobulin in the mouse. *Proc. Natl. Acad. Sci. U. S. A.* **73**:1305.

HAAS, W., and J. E. LAYTON. 1975. Separation of antigen-specific lymphocytes. I. Enrichment of antigen-binding cells. *J. Exp. Med.* **141**:1004.

JERNE, N. K. 1974. Towards a network theory of the immune system. *Ann. Immunol. (Paris)*. **125 C**:373–389.

JERNE, N. K. 1977. Opening address. *Cold Spring Harbor Symp. Quant. Biol.* **41**:1–4.

KLINMAN, N. R., N. H. SIGAL, E. S. METCALF, P. J. GEARHART, and S. K. PIERCE. 1977. The interplay of evolution and environment in B cell diversification. *Cold Spring Harbor Symp. Quant. Biol.* **41**:165–173.

MASON, D. W. 1976. The class of surface immunoglobulin on cells carrying IgG memory in rat thoracic duct lymph: the size of the subpopulation mediating IgG memory. *J. Exp. Med.* **143**:1122.

METCALF, D., G. J. V. NOSSAL, N. L. WARNER, J. F. A. P. MILLER, T. E. MANDEL, J. E. LAYTON, and G. A. GUTMAN. 1975. B lymphocyte colonies in vitro. *J. Exp. Med.* **142**:1534.

MÖLLER, G. 1975. One non-specific signal triggers B lymphocytes. *Transplant. Rev.* **23**:127–137.

MOZES, E. 1976. The nature of antigen specific T cell factors involved in the genetic regulation of immune responses. *In* The Role of Products of the Histocompatibility Gene Complex in Immune Responses. D. H. Katz and B. Benacerraf, editors. Academic Press, Inc., New York. 485–505.

NOSSAL, G. J. V. 1974. Principles of immunological tolerance and immunocyte receptor blockade. *Adv. Cancer Res.* **20**:93–130.

NOSSAL, G. J. V., A. CUNNINGHAM, G. F. MITCHELL, and J. F. A. P. MILLER. 1968. Cell-to-cell interaction in the immune response. III. Chromosomal marker analysis of single antibody-forming cells in reconstituted, irradiated, or thymectomized mice. *J. Exp. Med.* **128**:839–853.

NOSSAL, G. J. V., and J. E. LAYTON. 1976. Antigen-induced aggregation and modulation of receptors on hapten-specific B lymphocytes. *J. Exp. Med.* **143**:155.

NOSSAL, G. J. V., and J. LEDERBERG. 1958. Antibody production by single cells. *Nature (Lond.)*. **181**:1419.

NOSSAL, G. J. V., and B. L. PIKE. 1975. Evidence for the clonal abortion theory of B lymphocyte tolerance. *J. Exp. Med.* **141**:904.

NOSSAL, G. J. V., and B. L. PIKE. 1976. Single cell studies on the antibody-forming potential of fractionated, hapten-specific B lymphocytes. *Immunology.* **30**:189.

NOSSAL, G. J. V., B. L. PIKE, J. W. STOCKER, J. E. LAYTON, and J. W. GODING. 1977. Hapten-specific B lymphocytes: enrichment, cloning, receptor analysis and tolerance induction. *Cold Spring Harbor Symp. Quant. Biol.* **41**:237–243.

OKUMURA, K., M. H. JULIUS, T. TSU, L. A. HERZENBERG, and L. A. HERZENBERG. 1976. Demonstration that IgG memory is carried by IgG bearing cells. *Eur. J. Immunol.* **6**:467.

OSMOND, D. G., and G. J. V. NOSSAL. 1974. Differentiation of lymphocytes in mouse bone marrow. II. Kinetics of maturation and renewal of antiglobulin-binding cells studied by double labeling. *Cell. Immunol.* **13**:132.

PIKE, B. L., and G. J. V. NOSSAL. 1976. Requirement for persistent extracellular antigen in cultures of antigen-binding B lymphocytes. *J. Exp. Med.* **144**:568.

RAFF, M. C., M. MEGSON, J. J. T. OWEN, and M. D. COOPER. 1976. Early production of intracellular IgM by B-lymphocyte precursors in the mouse. *Nature (Lond.)*. **259**:224.

RAFF, M. C., J. J. T. OWEN, M. D. COOPER, A. R. LAWTON, M. MEGSON, and W. E. GATHINGS. 1975. Differences in susceptibility of mature and immature mouse B lymphocytes to anti-immunoglobulin suppression in vitro. *J. Exp. Med.* **142**:1052.

ROWE, D. S., K. HUG, L. FORNI, and B. PERNIS. 1973. Immunoglobulin D as a lymphocyte receptor. *J. Exp. Med.* **138**:181.

SCHIMPL, A., and E. WECKER. 1972. Replacement of T-cell function by a T-cell product. *Nat. New Biol.* **237**:15–17.

SCHRADER, J. W. 1973. Specific activation of the bone marrow-derived lymphocyte by antigen presented in a non-multivalent form: evidence for a two-signal mechanism of triggering. *J. Exp. Med.* **137:**844–849.

SIDMAN, C. L., and E. R. UNANUE. 1975. Receptor-mediated inactivation of early B lymphocytes. *Nature (Lond.).* **257:**149.

STOCKER, J. W. 1976. Estimation of hapten-specific antibody-forming cell precursors in microcultures. *Immunology.* **30:**181.

STOCKER, J. W. 1977. Tolerance induction in maturing B cells. *Immunology.* In press.

TADA, T., M. TANIGUCHI, and T. TAKEMORI. 1975. Properties of primed suppressor T cells and their products. *Transplant. Rev.* **26:**106–129.

TAUSSIG, M. H., and A. MUNRO. 1975. Antigen-specific T cell factor in cell co-operation and genetic control of the immune response. *Proc. Leucocyte Cult. Conf.* **9:**791–803.

VITETTA, E. S., and J. W. UHR. 1975. Immunoglobulin receptors revisited. *Science (Wash. D. C.).* **189:**964.

YAHARA, I., and G. M. EDELMAN. 1972. Restriction of the mobility of lymphocyte immunoglobulin receptors by concanavalin A. *Proc. Natl. Acad. Sci.* U. S. A. **69:**608–612.

T-LYMPHOCYTE RECEPTORS AND CELL INTERACTIONS IN THE IMMUNE SYSTEM

DAVID H. KATZ

The immune system is one of the most intricate of all bodily systems, paralleling in many respects the endocrine system in terms of the multiplicity of functions required of it for maintaining homeostasis and integrity of each individual's health. Both systems exert control over discrete functions at great distances within the body by virtue of circulating components capable of performing their role(s) at sites quite removed from their point of origin, and in this sense they display a level of versatility not found in most other multicellular organ systems. They differ, however, in one important respect: the endocrine system encompasses several distinct endocrine organs, each endowed with specific, and limited, functional capabilities; each organ is comprised of distinctive cell types and architecture, and the complexity of the system itself stems from the intricate communications network existing among these organs (and their respective target tissues) mediated in large part by the hormonal products generally unique to each of them. In contrast, the immune system consists of relatively few distinct organs—i.e., thymus, spleen, bone marrow, and lymph nodes—which are composed of relatively few distinct cell types. Although it is true that there are numerous lymph nodes throughout the body, each one is generally structured in much the same manner as all the rest and appears to owe its multiplicity more to the strategic nature of the different locations than to any differences in general function.

Accordingly, the complexity of the immune system has evolved from an intriguing communications network established between the components of the system, designed in such a manner as to permit a multiplicity of effects to arise from relatively few distinct cell types. This has been accomplished by development of sophisticated regulatory mechanisms allowing either enormous amplification or contraction of a given response depending on the needs of the individual. Under normal circumstances, the system functions remarkably well in maintaining effective defenses against foreign agents and aberrant native cells which may have undergone neoplastic transformation either as a consequence of normal random mutational events or secondary to exogenous oncogenic influence. Many other circumstances exist, however, in which abnormalities in one or more components of the system result in some form of breakdown in the network, clinically manifested in various ways and levels of severity.

In the opening paper of this symposium, Dr. Nossal presented a thorough overview of what is known about the antigen-specific receptors on B lymphocytes and triggering signals involved in the activation and inactivation of that lymphocyte class. In this paper, the present state of our knowledge about immunoreceptors on thymus-derived (T) lymphocytes and the functions of subpopulations of this particular lymphocyte class will be discussed. In addition, I will discuss what we currently understand about regulatory cell-cell interactions involved in responses developed by the immune system, with particular emphasis on the genetic basis and control of such regulatory interactions. At the outset, it is worth noting that whereas we have been concentrating our efforts on delineating the basis of cell-cell communication between cells of the immune system, there are reasons to believe that what we are learning may have broader relevance to other biological systems as well. Moreover, it should be emphasized that one of the remarkable features of the immune system is that its cellular and molecular components are so enormously complex that the system

DAVID H. KATZ Department of Cellular and Developmental Immunology, Scripps Clinic and Research Foundation, La Jolla, California has evolved with an incredible degree of flexibility—rarely, does it seem, has the system created a single pathway to an end with no alternative avenue to take when a biological detour becomes advantageous.

Cells of the Immune System

The major cellular components of the immune system are the macrophages and lymphocytes. Macrophages are themselves very versatile in the functions they perform in a variety of immune responses (reviewed in Unanue, 1972). Although not themselves specific for any given antigen, they perform a crucial role in concentrating and presenting antigens to lymphocytes; in particular, they appear to determine, in some way, whether and which T lymphocytes will be induced to stimulation and function by various antigens. Moreover, macrophages secrete several biologically active mediators capable of regulating the type and magnitude of lymphocyte responses by either enhancing or suppressing cell division and/or differentiation (reviewed in Unanue, 1976).

The lymphocytes represent the *specific* cellular component of the system, specificity being conferred upon such cells by virtue of the existence of antigen-specific receptors on the surface membrane of each immunocompetent cell (Paul, 1970). The nature of receptor specificity is highly specialized in that each different clone of lymphocytes expresses its own unique specificity; the origin of such specialization—i.e., whether genetically inherited or induced by somatic mutation—is not defined as yet, and remains a subject of debate. Moreover, the nature of antigen receptors on the two major classes of lymphocytes may differ. Thus, as Dr. Nossal pointed out, it is well established that surface immunoglobulin (Ig) molecules serve as the antigen receptors for B lymphocytes (reviewed in Warner, 1974). The molecular nature of the antigen receptors of T lymphocytes has been a subject of intense debate for the past 6 yr for reasons that will be discussed in more detail below.

The two classes of T and B lymphocytes have very distinct functional capabilities. T lymphocytes do not themselves produce circulating antibodies, nor do they give rise to antibody-secreting cells. They can be subdivided into two major functional categories, based upon studies in the mouse in which the most extensive investigations have been made (reviewed in Katz and Benacerraf, 1972, and Katz, 1977*b*). (1) Regulatory T lymphocytes are those cells functioning either to facilitate or amplify ("helper" cells) or suppress ("suppressor" cells) the responses of either T lymphocytes or of B lymphocytes. These functions are mediated by distinct subpopulations of T cells inasmuch as helper and suppressor T cells have been found to differ in their phenotypic expression of certain cell surface antigenic markers (Cantor and Boyse, 1976). (2) Effector T lymphocytes are those cells responsible for cell-mediated immune reactions such as delayed hypersensitivity responses, rejection of foreign tissue grafts and tumors, and elimination of virus-infected cells. The latter two responses involve the participation of cytotoxic T lymphocytes (CTL), commonly referred to as "killer" cells; also involved in responses to foreign tissues are T cells which undergo rapid proliferation in mixed lymphocyte reactions (MLR). The cells responsible for MLR and cytotoxicity, respectively, can be distinguished from one another by the existence of different surface antigen phenotypes; likewise, CTL can be distinguished from delayed hypersensitivity (DH) cells on the basis of distinct surface markers (Cantor and Boyse, 1976).

However, DH, MLR, and helper T cells possess similar surface markers, and it remains to be established by other criteria whether or not these are functions performed by the same or distinct T-cell subpopulations. The same is true for CTL and suppressor T cells, which are indistinguishable in their surface antigen phenotype.

Categorization of functional subpopulations of B lymphocytes is most readily done on the basis of different Ig classes synthesized. B lymphocytes give rise to cells synthesizing and secreting all classes of circulating Ig—i.e., IgM, IgG, IgA, and IgE—and the respective B-cell precursors for these antibody-forming cells are $B\mu$, $B\gamma$, $B\alpha$, and $B\epsilon$. It is important to note that, in most instances, the successful differentiation from precursor to antibody-forming cells (AFC) requires the cooperative participation of T cells (see below). Memory B cells are functionally important for the development of rapid secondary antibody responses upon subsequent antigenic exposure; these cells can be distinguished from unprimed B lymphocytes by several features, including tissue distribution, size, migratory properties, and certain surface antigen differences (reviewed in Strober, 1975). There is no hard evidence for the existence of regulatory B lymphocytes analogous in function to regulatory T lymphocytes, although the discovery of such cells in the future would not be entirely surprising. The capacity of antibody molecules themselves to specifically regulate responses by the process of "antibody feedback" is well documented.

Antigen Receptors on T Lymphocytes

Now let us return to the question of the molecular nature of the receptors for antigens present on the surface of T lymphocytes, and examine the basis for the controversy on this point. First of all, it is relatively simple to find Ig molecules on the surface membranes of B lymphocytes by conventional techniques of immunohistochemistry, notably fluorescent antibody or immunoautoradiography (see G. J. V. Nossal, this volume). By appropriate experimental approaches, it was, moreover, shown unequivocally that these readily detectable surface Ig molecules on B lymphocytes are synthesized by these cells, serve as antigen-specific receptors, and parallel quite closely, in the structural sense, the ultimate antibody secreted by mature plasma cells (Warner, 1974).

When similar attempts were made to identify Ig molecules on the surface of T lymphocytes by the use of conventional immunohistochemical approaches, such techniques failed to reveal the presence of such molecules. It is important to understand, however, that the anti-Ig antibody reagents used as the probe in such studies react to a great extent with antigenic determinants on the Fc portion of the Ig molecule—hence, it was conceivable that failure to detect surface Ig on T cells merely reflected an inadequate probe, since it could well be that the tail of the Ig molecule is buried within the T-cell surface membrane and is, therefore, inaccessible to reaction with anti-Ig antibodies (reviewed in Marchalonis, 1975).

However, to complicate matters further, studies of another type suggested the possibility that T-cell receptors may be encoded by genes in a system completely unrelated and unlinked to immunoglobulin genes. These studies were those initially conducted by McDevitt and Benacerraf (1972) and their associates in which it was found that the capacity of inbred mice and guinea pigs to develop effective immune responses to certain well-defined antigenic determinants was controlled by genes located in the major histocompatibility gene complex of the species. These genes are known as immune response or *Ir* genes, and the regions of the histocompatibility complex where such loci exist are called *I* regions. Such *Ir* genes have subsequently been discovered in every animal species studied, including man (reviewed in Benacerraf and Katz, 1975). The relationship of *Ir* genes and their products to T-cell receptors stemmed from the fact that responses governed by such genes always involved the types of responses in which T lymphocytes were required (Benacerraf and McDevitt, 1972).

Thus, the phenotypic expression of the presence or absence of an *Ir* gene controlling the response to a given antigen is the ability or inability of the individual to develop both T cell-mediated responses, such as delayed hypersensitivity and antibody responses to the same antigen. It is important to reiterate here that the successful differentiation of precursor B cells to secretory antibody-forming cells in most instances requires the cooperative participation of regulatory T lymphocytes (Katz and Benacerraf, 1972).

As a consequence of (1) the difficulties in detecting surface Ig on T cells, and (2) the strong relationship of *Ir* gene function to T-cell responses, the hypothesis was developed that perhaps *Ir* genes coded for the molecular entity serving as the antigen receptor on T lymphocytes. This, of course, implied a distinctly *different* genetic origin of antigen receptors on T and B cells, respectively. In support of this possibility there have been experimental results in the last 2 yr indicating the existence of antigen-specific molecules that appear to be produced by T lymphocytes and which are capable of exerting biological regulatory effects on other lymphocytes very similar to those performed normally by T cells themselves. For example, Taussig et al. (1976) and Mozes (1976) have found that short-term stimulation of primed T lymphocytes with antigen under appropriate conditions in vitro results in elaboration of antigen-specific molecules capable of replacing the normal requirement for helper T cells in the development of antibody responses by B lymphocytes exposed concomitantly to antigen plus such molecules. Tada and his colleagues (reviewed in Tada and Taniguchi, 1976) have mechanically disrupted populations of antigen-primed T cells and recovered a soluble antigen-specific factor from such cells capable of replacing suppressor T cells in inhibiting antibody responses under appropriate conditions.

Interestingly, in both of the aforementioned systems the molecules exerting these contrasting biological effects do not possess antigenic determinants present on conventional Ig molecules, but do bear antigenic markers corresponding to known gene products of the histocompatibility gene complex and, more specifically, of genes located in the *I* region of the *H-2* complex of the mouse. These observations are therefore consistent, in principle, with the hypothesis proposing that histocompatibility-linked *Ir* genes may code for antigen-specific molecules serving as T-cell receptors (Benacerraf and McDevitt, 1972).

However, also during the past 2 yr some very compelling data have been obtained in two different experimental systems which imply that at least part of the genetic information determining the specificity of the T-cell receptor is identical to that determining the specificity of the Ig receptor on B cells (Binz and Wigzell, 1976; Rajewsky et al., 1976). Before discussing these studies, it is important to understand that the control of Ig synthesis is genetically unique in that the Ig gene system is the only one known to involve the participation of two discrete structural genes in the synthesis of a single polypeptide chain (reviewed in Gally and Edelman, 1972). Thus, a structural gene for the variable or *V* region – in which is located the specific antigen-combining site of the molecule – integrates somehow with another structural gene for the constant or *C* region to form a single chain comprised of *V* and *C* regions. There are antigenic determinants associated with each region, and the ones most pertinent to this discussion are those determinants unique to the area of the *V*-region combining site – such determinants are known as idiotypes, and antibodies reacting with idiotypic determinants are known as anti-idiotypic antibodies (reviewed in Nisonoff and Bangasser, 1976).

The experimental systems used by Binz and Wigzell (1976) and by Rajewsky et al. (1976) have been quite different in approach, but remarkably similar in principle. The idea, in essence, has been to utilize anti-idiotypic antibodies prepared in such a way as to display specific reactivity with idiotypic determinants present on Ig antibody molecules, and therefore encoded by *V*-region genes of conventional Ig, and to use such anti-idiotypic antibodies as a probe to analyze T-cell receptors for the presence or absence of identical idiotypic antigen determinants. A reciprocal approach has been to prepare antibodies against presumed idiotypiclike determinants of T-cell receptors and analyze such antibody preparations for reactivity with idiotypic determinants on conventional Ig molecules corresponding in their specificity to that of the T cells used to prepare such anti-idiotypic antibodies. It is not possible here to explain in any detail the nature of the two major systems utilized in such studies as they are highly complex in design; interested readers should therefore refer to Binz and Wigzell (1976) and Rajewsky et al. (1976) for recent reviews on the subject.

The crucial observations from such studies can be summarized as follows: anti-idiotypic antibodies directed against *V*-region determinants of conventional Ig molecules can, under appropriate experimental conditions, be shown to react with T lymphocytes whose receptor specificities correspond to the combining site specificities of the Ig molecules against which the anti-idiotypic antibodies are directed, and vice versa. Furthermore, it has been possible to specifically isolate molecules, which display antigen-binding capabilities, from T lymphocytes by utilizing appropriate anti-idiotypic antibody preparations. Biochemical and immunochemical analyses of molecules that can be isolated from T cells by use of such anti-idiotypic antibodies have indicated that T-cell receptors of this type consist of a single chain possessing an antigen-combining site; it is possible that this chain may exist on the cell membrane as a dimer. Although the molecular size of the isolated chain is similar to that of the heavy chain of conventional Ig, such T cell-derived chains do not possess any detectable antigenic determinants known to be present on conventional Ig. Nevertheless, inheritance of the relevant idiotype is clearly linked to Ig heavy chain genes. Moreover, such molecules do not possess antigenic markers of the histocompatibility gene complex.

Therefore, at the present time such data strongly imply that the *V* region of the T-cell receptor is encoded by the same genes encoding *V* regions of conventional Ig – however, the remainder of the polypeptide chain and its genetic origin remain a mystery. Furthermore, we are now in somewhat of a dilemma to understand the derivation and physiologic importance of the other non-immunoglobulin antigen-specific molecules, discussed above, that are believed to come from T cells and which bear antigenic determinants encoded by histocompatibility genes.

Genetic Basis of Cell Communication and Differentiation in the Immune System

Another area to be addressed in this paper concerns the nature of the mechanism(s) by which different populations and subpopulations of lymphocytes interact with one another as well as with macrophages. As stated above, it is now well documented that such cellular interactions underlie much of the regulatory control over the immune system (Katz and Benacerraf, 1972). T lymphocytes, B lymphocytes, and macrophages are capable of intercommunicating with one another by various means. At the moment, it appears that T lymphocytes exert the most sophisticated regulatory effects on the system in general, because it is now clear that subpopulations within this class communicate with one another as well as with macrophages and B lymphocytes (Katz, 1977*b*). The precise mechanism of cell-cell communication, namely whether this involves cell-cell contact, activity of secreted (or released) molecules, or a combination of cell contact and mediator release has not yet been ascertained.

However, in recent years it has been established that products of genes located in the major histocompatibility gene complex play a very important and integral role in regulating the processes of cell-cell communication as well as differentiation in the immune system (reviewed in Katz and Benacerraf, 1976; Katz, 1977*a*). We know this to be the case as a result of the discovery that in various cell interaction systems, the most efficient interactions tend to occur between cells derived from individuals sharing in common certain crucial genetic loci in the histocompatibility gene complex. For example, in development of antibody responses T and B lymphocytes interact very well when the donors of the cells are of the same histocompatibility type and not very well when their histocompatibility genotypes are different. Importantly, the critical genetic locus or loci involved in controlling T- and B-cell interactions map in the *I* region of the histocompatibility complex (Katz et al., 1975) which, as mentioned above, is precisely where the *Ir* genes are located. This may be merely coincidental, or it may be that the genes controlling cell interactions, which have been termed cell-interaction or *CI* genes, and those known as *Ir* genes are one and the same – this has yet to be sorted out.

Another example of involvement of histocompatibility gene products in cell-cell interactions of an entirely different type concerns the ability of CTL to effectively lyse virus-infected or chemically modified target cells. In such circumstances, it has been found that CTL are most efficient in lysing target cells derived from a similar histocompatibility genotype (reviewed in Zinkernagel and Doherty, 1976, and Shearer et al., 1976). The critical genetic locus or loci involved in controlling interactions between CTL and target cells map in the *K* and *D* regions of the histocompatibility complex, and differ from those involved in T- and B-cell interactions which are located in the *I* region.

Although it was initially believed that the mechanisms underlying genetic control of T- and B-cell interactions and interactions between CTL and target cells may be different, recent thinking has been toward considering the two phenomena along similar lines (Katz, 1977*a*; Zinkernagel, 1976). Hence, it is attractive to consider that cell interaction structures are present on the surface membranes of lymphoid cells and macrophages and it is via some type of molecular interaction at these sites that the communication takes place. This, in turn, results either in a signal for induction of differentiation in the case of helper T-cell interactions with B cells or a signal to initiate cytolysis in the case of CTL interactions with target cells. Whether this type of postulated molecular interaction between the respective cells occurs between homologous or complementary structures has not been elucidated.

The realization of the role played by the major histocompatibility gene complex in cell interactions in the immune system allows us to begin to question whether this family of highly polymorphic genes, whose products are displayed on the surface membranes of the majority of nucleated cells in multicellular systems, plays an essential role in governing cell communication and differentiation in many, or perhaps all, other organ systems in addition to the immune system. The question, in other words, is whether the products of histocompatibility genes represent a large family of cell interaction molecules which serve essentially as the doorway to effective cell communication in complex multicellular systems. Some of us believe this to be the case, but it is equally clear that an enormous task lies ahead before definitive answers will be obtained.

Complex Cell-Cell Interactions in the Development and Regulation of an Immune Response

As pointed out in a previous section, we have come to realize that the normal development of most immune responses involves, in addition to the inducing effects resulting from introduction of antigen into the system, a complex array of cell interactions which are necessary both to promote and to regulate the differentiation of cells participating in any given response. Although a precise delineation of the manner in which lymphocytes are activated and the mechanisms of cell-cell interactions has yet to be made, considerable information is now available which permits us to construct a reasonable picture of certain of the probable events involved.

One such picture of the various cell interactions occurring in the development of an antibody response is schematically illustrated in Fig. 1. The antigen depicted here is a hapten-carrier conjugate in which the carrier portion comprises those determinants recognized most readily by T cells, whereas the haptenic determinants are those recognized by the B-cell precursors of antibody-secreting cells. On the left, one notes that macrophages have bound antigen (nonspecifically) on their surface membranes and are displaying the relevant carrier determinants to an unprimed T cell possessing carrier-specific surface receptors. There is a second interaction between these two cells involving cell surface cell interaction (CI) molecules which, as stated above, are products of genes in the major histocompatibility complex. This depicts, therefore, the role of macrophage-associated CI molecules, in addition to antigen, in favoring the induction of carrier-specific helper T cells.

Once induced, the helper T cell is then capable of interacting with hapten-specific B cells which, as shown at the top right of Fig. 1, have interacted with haptenic determinants via their surface Ig receptors. This interaction between carrier-specific helper T cells and hapten-specific B cells also occurs via surface membrane-associated CI molecules although, as illustrated, it has not been defined as to whether direct cell-cell contact, soluble factor activity, or both constitute the primary means of interaction. Nevertheless, the consequence of such interactions between helper T cell and B cell is to drive the latter to differentiate into mature antibody-secreting plasma cells and/or to memory B cells. Suppressor T cells constitute a separate subpopulation of regulatory cells which are similarly specific for carrier determinants but serve as an opposing force to the effects of helper T cells. The suppressor T cell has been found to interfere with the aforementioned events at one or more of three possible points as indicated by the dashed arrows in Fig. 1: first, by inhibiting the activation of helper T cells; second, by interfering with the actual function of the helper T cell in terms of its facilitating interactions with B cells; or third, by directly inhibiting B-cell differentiation either before or after interaction of the precursors with helper. Hence, the overall response occurring at a given time appears to reflect the net sum of the opposing regulatory forces exerted by helper and suppressor T cells, respectively.

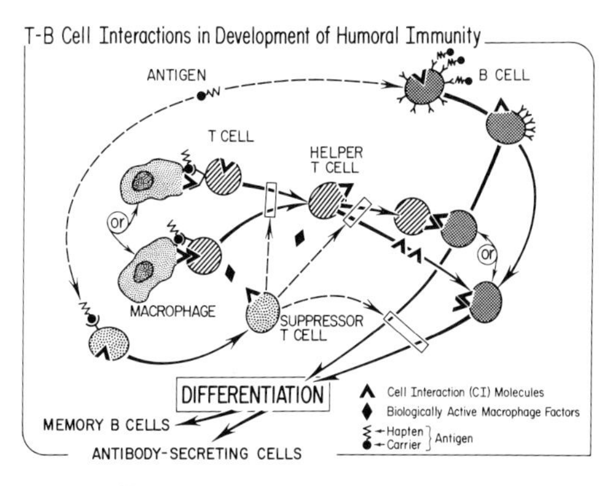

FIGURE 1 See text for explanation.

Concluding Remarks

We are still in the early stages of finding answers to the many questions concerning the genes, cells, and molecules of the immune system. It is clear that the active interest and participation of investigators with expertise in many other areas of cell biology will be increasingly required before certain of these questions will be answered. Moreoever, although most of the work discussed in this paper has been focused on lymphocytes and macrophages, it is not difficult to envisage that what is learned about cell communication and control of differentiation among these cells may provide important insights into comparable events in other tissues and cell types of multicellular organisms.

ACKNOWLEDGMENTS

I thank Candice LaMar for secretarial assistance in preparation of the manuscript.

Work conducted in my own laboratory is supported by National Institutes of Health grant AI-13874. This is publication number 1227 from the Immunology Departments and publication number 4 from the Department of Cellular and Developmental Immunology, Scripps Clinic and Research Foundation, La Jolla, California.

REFERENCES

BENACERRAF, B., and D. H. KATZ. 1975. The histocompatibility-linked immune response genes. *Adv. Cancer Res.* **21:**121–174.

Benacerraf, B., and H. O. McDevitt. 1972. Histocompatibility-linked immune response genes. *Science (Wash, D. C.)* **175**:273–280.

Binz, H., and H. Wigzell. 1976. Antigen-binding, idiotypic receptors from T lymphocytes: an analysis as to their biochemistry genetics and use as immunogens to produce specific immune tolerance. *Cold Spring Harbor Symp. Quant. Biol.* **41**:275–284.

Cantor, H., and E. A. Boyse. 1976. Regulation of cellular and humoral immune responses by T cell subclasses. *Cold Spring Harbor Symp. Quant. Biol.* **41**:22–32.

Gally, J. A., and G. M. Edelman. 1972. The genetic control of immunoglobulin synthesis. *Annu. Rev. Gen.* **6**:1–42.

Katz, D. H. 1977a. The role of the histocompatibility gene complex in lymphocyte diversity. *Cold Spring Harbor Symp. Quant. Biol.* **41**:611–624.

Katz, D. H. 1977b. Lymphocyte Differentiation, Recognition and Regulation. Academic Press, Inc., New York. In press.

Katz, D. H., and B. Benacerraf. 1972. The regulatory influence of activated T cells on B cell responses to antigen. *Adv. Immunol.* **15**:1–94.

Katz, D. H., and B. Benacerraf. 1976. Genetic control of lymphocypte interactions and differentiation. *In* The Role of Products of the Histocompatibility Gene Complex in Immune Responses. D. H. Katz and B. Benacerraf, editors. Academic Press, Inc., New York. 355–375.

Katz, D. H., M. Graves, M. E. Dorf, H. DiMuzio, and B. Benacerraf. 1975. Cell interactions between histoincompatible T and B lymphocytes. VII. Cooperative responses between lymphocytes are controlled by genes in the *I* region of the *H-2* complex. *J. Exp. Med.* **141**:263–271.

Marchalonis, J. J. 1975. Lymphocyte surface immunoglobulins. *Science (Wash. D. C.)*. **190**:20–26.

Mozes, E. 1976. The nature of antigen-specific T cell factors involved in the genetic regulation of immune responses. *In* The Role of Products of the Histocompatibility Gene Complex in Immune Responses. D. H. Katz and B. Benacerraf, editors. Academic Press, Inc., New York. 485–498.

Nisonoff, A., and S. A. Bangasser. 1976. Immunological suppression of idiotypic specificities. *Transplant. Rev.* **27**:100–122.

Paul, W. E. 1970. Functional specificity of antigen-binding receptors of lymphocytes. *Transplant. Rev.* **5**:130.

Rajewsky, K., G. J. Hämmerling, S. J. Black, C. Berek, and K. Eichmann. 1976. *In* The Role of Products of the Histocompatibility Gene Complex in Immune Responses. D. H. Katz and B. Benacerraf, editors. Academic Press, Inc., New York. 445–455.

Shearer, G. M., A.-M. Schmitt-Verhulst, and T. G. Rehn. 1976. Bifunctional histocompatibility-linked regulation of cell mediated lympholysis to modified autologous cell surface components. *In* The Role of Products of the Histocompatibility Gene Complex in Immune Responses. D. H. Katz and B. Benacerraf, editors. Academic Press, Inc., New York. 133–146.

Strober, S. 1975. Immune function cell surface characteristics and maturation of B cell subpopulations. *Transplant. Rev.* **24**:84–102.

Tada, T., and M. Taniguchi. 1976. Characterization of the antigen-specific suppressive T cell factor with special reference to the expression of the *I* region genes. *In* The Role of Products of the Histocompatibility Gene Complex in Immune Responses. D. H. Katz and B. Benacerraf, editors. Academic Press, Inc., New York. 513–528.

Taussig, M. J., A. J. Munro, and A. L. Luzzati. 1976. *I*-region gene products in cell cooperation. *In* The Role of Products of the Histocompatibility Gene Complex in Immune Responses. D. H. Katz and B. Benacerraf, editors. Academic Press, Inc., New York. 553–563.

Unanue, E. R. 1972. The regulatory role of macrophages in antigenic stimulation. *Adv. Immunol.* **15**:95–123.

Unanue, E. R. 1976. Secretory function of mononuclear phagocytes. *Am. J. Pathol.* **83**:396–415.

Warner, N. L. 1974. Membrane immunoglobulins and antigen receptors on B and T lymphocytes. *Adv. Immunol.* **19**:67–117.

Zinkernagel, R. M. 1976. H-2 restriction of virus-specific cytotoxicity across the H-2 barrier. Separate effector T-cell specificities are associated with self-H-2 and with the tolerated allogeneic H-2 in chimeras. *J. Exp. Med.* **144**:933–945.

Zinkernagel, R. M., and P. C. Doherty. 1976. Does the apparent H-2 compatibility requirement for virus-specific T cell-mediated cytolysis reflect T cell specificity for "altered self" or physiological interaction mechanisms? *In* The Role of Products of the Histocompatibility Gene Complex in Immune Responses. D. H. Katz and B. Benacerraf, editors. Academic Press, Inc., New York. 203–212.

MITOGEN RECEPTORS AND MITOGEN RESPONSIVENESS

MICHAEL J. CRUMPTON, BRIGITTE PERLÉS, and JUDY AUGER

Lymphocytes are normally quiescent cells which on contact with specific antigen express new patterns of activity and develop new functions, including growth and proliferation, differentiation and the release of soluble factors, "mediators." Although T and B lymphocytes are induced to respond in a generally similar fashion, the functional consequences of their antigenic stimulation are profoundly different (Greaves et al., 1973). The nature and sequence of the biochemical events responsible for these changes in behavior are at present very poorly understood. The major problem concerned with elucidating the molecular events induced by antigen is the very low frequency of cells, within an unprimed lymphocyte population, that respond to a particular antigen. Thus, Nossal (this volume) has suggested that no more than 0.01% of the lymphocytes are specific for a given antigenic determinant. The difficulty of ensuring an adequate number of responding cells for biochemical studies can be circumvented by making use of various nonspecific mitogens which induce the majority of lymphocytes, irrespective of their specificity for antigen, to respond in a closely similar, if not analogous, fashion to that initiated by antigen (Möller, 1972; Ling and Kay, 1975; Wedner and Parker, 1976). This mode of lymphocyte activation (transformation) has found considerable favor as a convenient experimental model not only for investigating the biochemical mechanism of the immune response but also the biochemistry of cell growth and gene activation.

This chapter is concerned with the nature of these nonspecific mitogens, their interaction with lymphocytes, and the biochemical consequences of mitogen-lymphocyte interaction. Many of the concepts have been established by the use of lymphocytes from mouse spleen and/or human peripheral blood. These concepts are regarded as being generally applicable even though lymphocytes from individual species show marked differences in their responses to some mitogens. Mouse lymphocytes possess many advantages, especially the availability of purified T and B populations. Lymphocytes from the mesenteric lymph nodes of young pigs have also proved a particularly useful source of cells principally because of their response to a large and varied collection of mitogens.

Mitogens

MOLECULAR NATURE: In contrast to specific antigens that stimulate only a very small fraction of lymphocytes, mitogens act as polyclonal activators inducing a large proportion of the cells to grow (Möller, 1972; Ling and Kay, 1975; Rosenthal, 1975; Wedner and Parker, 1976). Because the responsiveness of individual cells is independent of their specificity for antigen, mitogens are regarded as being nonspecific. Mitogen-induced lymphocyte activation was first reported by Nowell (1960) who showed that the plant lectin *Phaseolus vulgaris* phytohemagglutinin (PHA) stimulated resting human peripheral blood lymphocytes to enlarge into blast cells that subsequently underwent mitosis. An example of PHA-induced lymphocyte transformation is shown in Fig. 1, which compares the morphology of pig mesenteric lymph node cells that had been incubated for 40 h in the absence and presence of the mitogen. The results demonstrate that PHA stimulated at least the majority of lymphocytes to approximately double in size and to show other morphological features characteristic of blast cells, such as an enlargement of the cytoplasm, an increased number of mitochondria, and a dispersion of the heterochromatin.

Various studies have revealed that PHA is just one of a large and varied collection of mitogens.

MICHAEL J. CRUMPTON and JUDY AUGER National Institute for Medical Research, London, England
BRIGITTE PERLÉS Centre d'Immunologie, Université Marseille-Luminy, Marseille, France

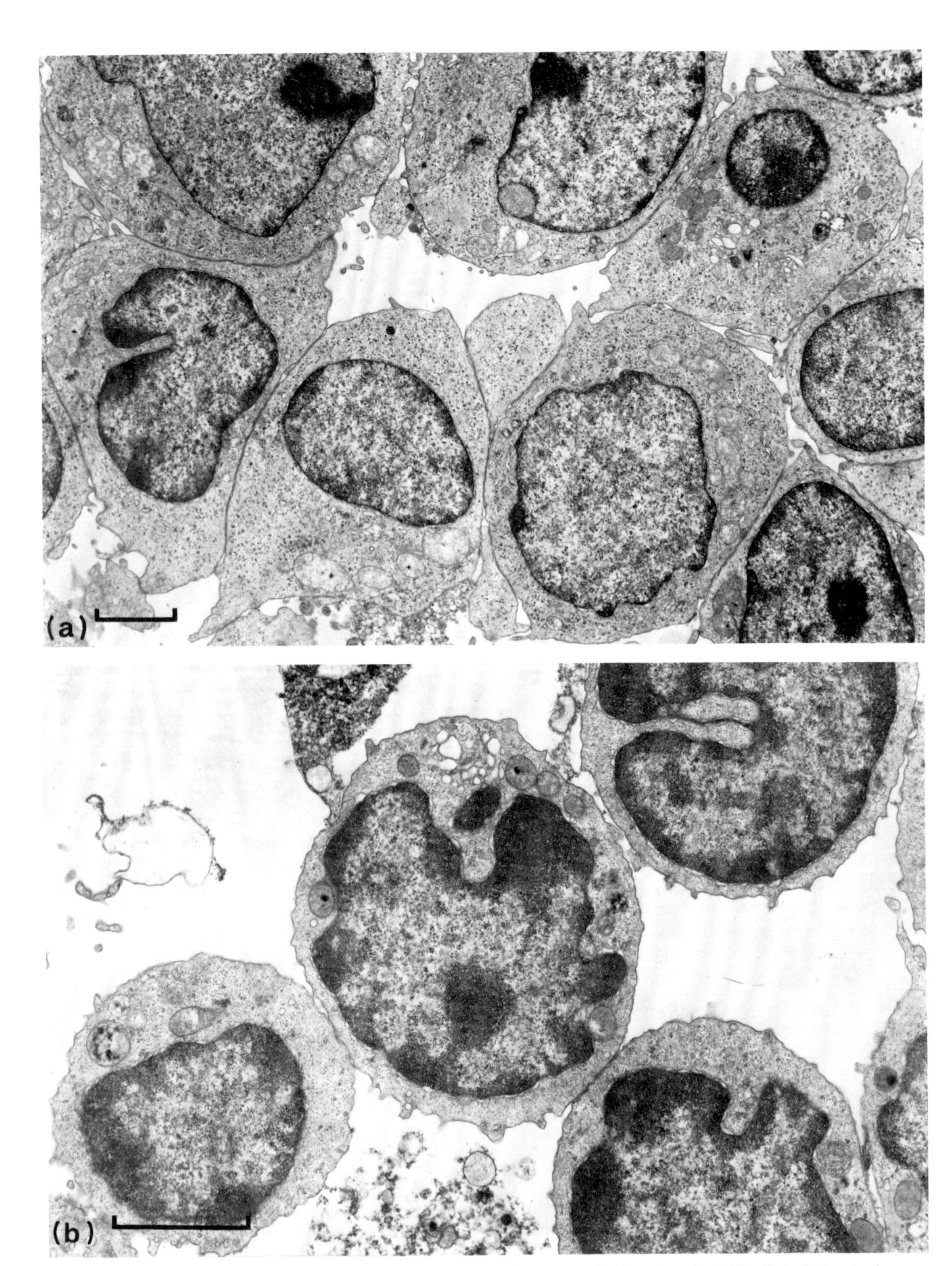

FIGURE 1 Electron micrograph of lymphocytes from pig mesenteric lymph node (10^6 cells/ml) that had been incubated for 40 h with (*a*) and without (*b*) PHA (1 μg/ml). The cells had been depleted of B lymphocytes by treatment with nylon wool and were cultured in Eagle's medium supplemented with 10% (vol/vol) fetal calf serum. PHA (leukoagglutinin) was purchased from Pharmacia Fine Chemicals, Piscataway, N. J. Magnification of PHA-treated cells (*a*) is × 7,000 compared with × 11,000 for normal cells (*b*). The bars equal 1 μm.

These include macromolecules such as lectins, bacterial products, especially Enterobacteriaceae lipopolysaccharides, dextran sulfate, proteolytic enzymes, galactose oxidase, and antilymphocytic serum, as well as small molecules such as sodium periodate, the Ca^{2+}-ionophore A 23187, heavy metal cations, especially Hg^{2+} and Zn^{2+}, and phorbol myristate acetate (Möller, 1972; Novogrodsky, 1976; Sharon, 1976; Wedner and Parker, 1976). In spite of their diverse chemical nature, all of these substances share the properties of inducing DNA synthesis and the expression of effector functions that are characteristic of the type (T or B) of lymphocyte stimulated. These properties are not, however, necessarily expressed in all animal species. Thus, antibodies against immunoglobulin stimulate transformation of rabbit and pig lymphocytes (Sell and Gell, 1965; Maino et al., 1975) but not of human and mouse cells, whereas lipopolysaccharides activate mouse spleen cells but have a minimal effect upon human and pig lymphocytes. Figure 2 compares the capacities of various mitogens to induce pig mesenteric lymph node cells to synthesize DNA. In this experiment, uptake of a pulse of radioactive thymidine into DNA after about 42 h of culture was used as a quantitative measure of the mitogenic response. The results show that increasing concentrations of each substance, above a certain critical threshold value (see also Fig. 9), stimulated increasing DNA synthesis up to a maximum level. The concentration causing optimal stimulation is a characteristic property of the mitogen and, in the case of lectins, reflects its affinity for the carbohydrate moiety of the membrane component mediating transformation (Sharon, 1976). As can be seen from Fig. 2, phorbol myristate acetate was particularly effective as a mitogen for pig lymphocytes being as active, on a molar basis, as PHA, whereas another small molecule, the Ca^{2+} ionophore A 23187 was 1,000-fold less active.

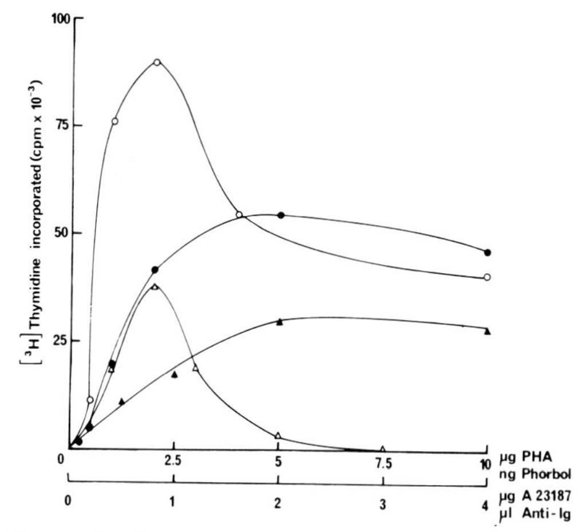

FIGURE 2 Capacities of various mitogens to stimulate pig lymphocyte transformation. Transformation was assessed in terms of the incorporation of [^{3}H]thymidine into DNA as a function of the concentration of PHA (○), phorbol myristate acetate (●), A 23187 (△), and antibodies against pig immunoglobulin (▲). Assays were performed as described by Maino et al. (1975) with 10^6 lymphocytes from pig spleen in 1 ml of medium. Cultures were incubated for about 42 h before the addition of [^{3}H]thymidine (1 μCi) and for a further 6 h before recovering the DNA. A 23187 was kindly donated by Dr. R. Hamill (Eli Lilly Research Laboratories, Indianapolis, Ind.).

SELECTIVITY FOR T OR B LYMPHOCYTES: Many mitogens show a striking selectivity for either T or B lymphocytes (Greaves and Janossy, 1972). Thus, the majority of mitogenic lectins, including PHA, concanavalin A, and *Lens culinaris* hemagglutinin, but excluding pokeweed mitogen, selectively activate T lymphocytes, whereas lipopolysaccharides, dextran sulfate, and antibodies against immunoglobulin stimulate only B cells. For instance, in Fig. 2, removal of the immunoglobulin-bearing B lymphocytes by treating the cells with nylon wool resulted in the complete abrogation of the response to antibodies against immunoglobulin with a negligible effect upon the degrees of activation by the other mitogens. One possible explanation for the failure of many mitogenic lectins to transform B lymphocytes is that they possess a very much lower affinity for the surface glycoprotein mediating transformation (by implication immunoglobulin) compared with that for the corresponding glycoprotein of T lymphocytes (see below). This suggestion is supported by the reports that B lymphocytes were stimulated by various T lymphocyte-specific mitogenic lectins (e.g., PHA and concanavalin A) when the lectins were covalently attached to a solid support such as Sepharose beads (Greaves and Janossy, 1972). Under these circumstances, the attached lectin should possess a much higher affinity for cell surface carbohydrate as a result of the increase in multivalency.

Mitogen-Lymphocyte Interaction

NATURE OF THE "RECEPTOR" MEDIATING LECTIN-INDUCED TRANSFORMATION: All mitogens probably initiate lymphocyte transformation by perturbing the cell surface membrane. Indeed, in some cases it is apparent that the effects of chemically diverse mitogens are mediated via the perturbation of a common membrane component. Thus, the enzyme galactose oxidase, the small chemical sodium periodate, and the lectin soybean agglutinin all appear to initiate transformation by interacting with the same membrane glycoprotein (Novogrodsky, 1976). Furthermore, various arguments suggest that lectin-induced transformation is mediated by the interaction of the lectins with a specific membrane glycoprotein (i.e., a unique glycoprotein) which satisfies the criteria of a "receptor" protein (Greaves, 1975). The most compelling evidence in support of the latter proposal is provided by the reports that different T lymphocyte-specific mitogenic lectins initiate transformation by interacting with the same membrane component, and by the demonstration that not all lectins that bind to the lymphocyte surface are mitogenic (Maino et al., 1975). Figure 3 shows that wheat germ and *Axinella polypoides* agglutinins failed to stimulate pig lymphocytes to synthesize DNA, whereas, under the same conditions, PHA induced considerable DNA synthesis. It is also apparent (Fig. 3) that PHA-induced DNA synthesis was not inhibited by the addition of a 10- to 200-fold molar excess of wheat germ agglutinin. The results of other studies indicate that the failure of wheat germ and *A. polypoides* agglutinins to stimulate transformation is not due to lack of binding. Thus, PHA, wheat germ, and *A. polypoides* agglutinins had similar agglutination titres for pig lymphocytes (1:32, 1:16, and 1:128 dilution of 100 μg/ml solution, respectively), and fluorescein-labeled wheat germ agglutinin induced similar degrees of staining, "patching," and "capping" to that elicited by a mitogenic lectin (*Lens culinaris* phytohemagglutinin). Also, as judged from the analysis of binding curves obtained with ^{125}I-labeled lectins, pig lymphocytes possessed similar numbers of binding sites for the different lectins. The most plausible explanation for the lack of mitogenicity of wheat germ and *A. polypoides* for pig lymphocytes is that they failed to bind to the specific glycoprotein that mediates transformation by the mitogenic lectins.

Further evidence in support of the above proposal is provided by the observation that only a small fraction (about 10%) of the total PHA-binding sites on the lymphocyte surface (about 6 × 10^5 sites/cell; Allan and Crumpton, 1973) are occupied under culture conditions, giving optimal

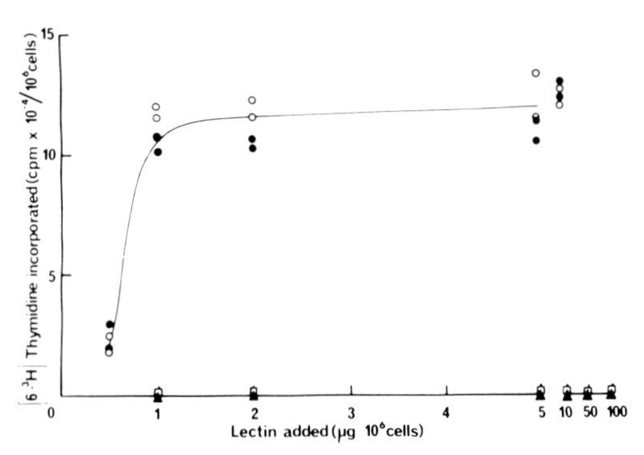

FIGURE 3 Capacities of wheat germ agglutinin (▲) and of *Axinella polypoides* agglutinin (□) to stimulate pig lymphocyte transformation. PHA (○) was included as a positive control. The effect of a constant amount of wheat germ agglutinin (20 μg/culture) on the enhanced incorporation induced by PHA (●) is also shown. Experimental details are given by Maino et al. (1975).

stimulation of DNA synthesis. This fraction apparently binds PHA with a higher affinity than, and may differ in its molecular nature from, the remaining sites (Allan and Crumpton, 1973). The relationship between the amount of PHA bound and the subsequent degree of transformation induced by various concentrations of PHA is illustrated in Fig. 4. The results indicate that maximal DNA synthesis is stimulated when 10–20% only of the potential binding sites are occupied by PHA. A conventional analysis of a Scatchard plot of the binding data provides evidence for two classes of binding sites which differ markedly in their affinities and relative distributions (Fig. 5). Of these, the small fraction with a much higher affinity (about 10% of the total sites) appears to be directly related to the transformation process (see also Allan and Crumpton, 1973). Although other interpretations of the binding data are possible if allowances are made for multivalent binding and the consequent increase in affinity (M. Flanagan, B. Perlés and M. J. Crumpton, unpublished data), the above interpretation is supported by an analysis of the molecular nature of the complexes formed on addition of different amounts of PHA. Information on the nature of the membrane-lectin complexes was obtained with radioactively labeled lectin as a marker and by solubilizing the complex in a detergent that failed to promote its dissociation. Suitable solubilizing agents are sodium deoxycholate and nonionic detergents such as Nonidet P-40 which combine extensive solubilization of cell membranes with a minimal dissociating effect

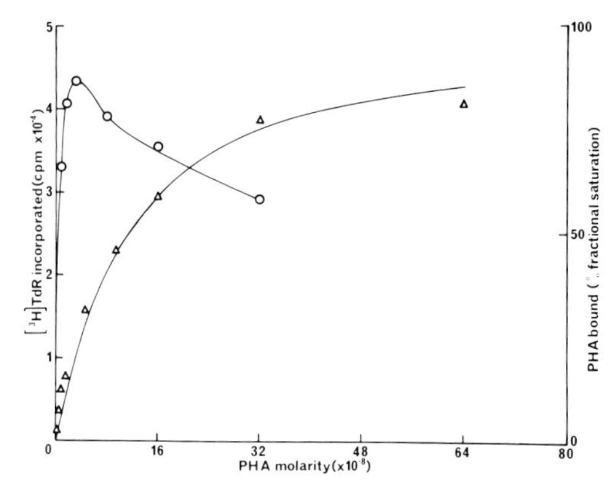

FIGURE 4 Relationship between the degree of saturation (△) and the degree of transformation (○) of pig lymphocytes induced by increasing concentrations of PHA. The cells were separated from pig mesenteric lymph nodes and were depleted of B lymphocytes by treatment with nylon wool. Amounts of PHA bound and the degree of transformation were assessed under identical culture conditions except that binding was measured after 30 min incubation at 37°C with ^{125}I-labeled PHA and transformation after incubating with PHA for 42 h. Binding of ^{125}I-labeled PHA was determined after washing the cells twice by sedimenting through a cushion of 5% bovine serum albumin and was corrected for nonspecific binding estimated in the presence of a 100-fold molar excess of unlabeled-PHA.

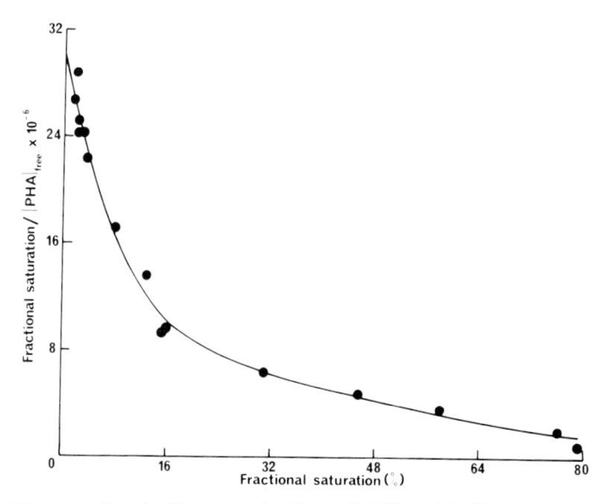

FIGURE 5 A Scatchard plot of PHA binding to pig lymphocytes. The binding data is shown in Fig. 5. A conventional analysis of the Scatchard plot reveals 8.7% of the total sites with Ka, 2.68 × 10^8 liter/mol, and Kd, 3.73 × 10^{-9} M, and 91.3% of the sites with Ka, 7.09 × 10^6 liter/mol, and Kd, 1.41 × 10^{-7} M (M. Flanagan, B. Perlés and M. J. Crumpton, unpublished results).

upon specific molecular interactions with dissociation constants of the order of 10^{-6} M or less. Figure 6 compares the molecular size, assessed by gel filtration in detergent, of the complex formed by PHA under optimal stimulating conditions with that produced with 10-fold more PHA. The complex formed under optimal conditions was eluted as a broad peak whose position corresponded to a molecular size of about 3 × 10^5. As PHA has a molecular weight of 1.2 × 10^5, the membrane component(s) has a molecular size of about 1.8 × 10^5 or, if each PHA molecule cross-links two glycoprotein molecules, 0.9 × 10^5. In contrast, lymphocytes incubated with much more PHA gave a small shoulder only in the position of the above complex. Since the amount of the complex was no greater than that formed under optimal conditions, it appears that the amount of this particular membrane component(s) was limiting and all had interacted with PHA under optimal conditions. The elution pattern also showed a major peak coincident with free PHA. This suggests that, in contrast to optimal stimulating conditions, the majority of the bound PHA had interacted with membrane components either of low molecular weight, such as glycolipids, or of relatively low affinity so that the complex dissociated on detergent solubilization.

REQUIREMENT FOR PROLONGED CONTACT: If the above arguments in support of a

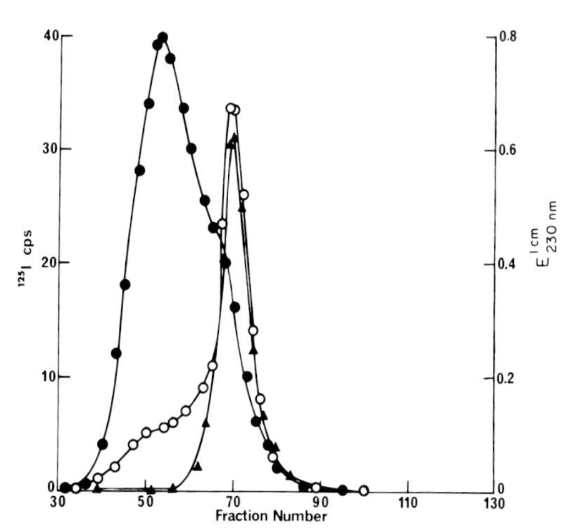

FIGURE 6 Gel filtration pattern of the complexes formed by PHA bound to the surface of pig lymphocytes after solubilization in 1% Na deoxycholate. Cells were incubated with ^{125}I-labeled PHA (100 μg) either under conditions giving optimal stimulation (●; 10^8 cells in 100 ml) or with a 10-fold excess of PHA (○; 10^7 cells in 10 ml). After 1 h at 37°C, the washed cells were suspended in 1% Na deoxycholate, and the soluble fraction was eluted from a column of Sepharose 6B in 1% Na deoxycholate. The elution position of free PHA is also shown (▲).

unique (specific) membrane glycoprotein mediating lectin-induced transformation are correct, then the initiation of lymphocyte activation corresponds to the regulation of other cells' functions by the interaction of chemical signals (ligands), such as polypeptide hormones and neurotransmitters, with their complementary cell surface receptors. There is, however, one striking difference between the two systems, namely, that whereas many chemical signals act as triggers inducing an essentially immediate response (Rodbell, 1972), prolonged contact (about 20 h) with the lectin is required before the lymphocyte is committed to grow (Lindahl-Kiessling, 1972). That is, the lectin acts as a "push." This requirement for prolonged contact is illustrated in Fig. 7 which shows the effect of removal of mitogen at various times on the subsequent capacity of the cells to synthesize DNA. The results indicate that incubation with mitogen is necessary for about 20 h before commitment to growth is established, that this requirement is independent of the chemical nature of the mitogen, and that the change from a state of no commitment to commitment is restricted to a relatively short interval of the culture period. Identical results have been observed for many mitogens and although contradictory results have been obtained

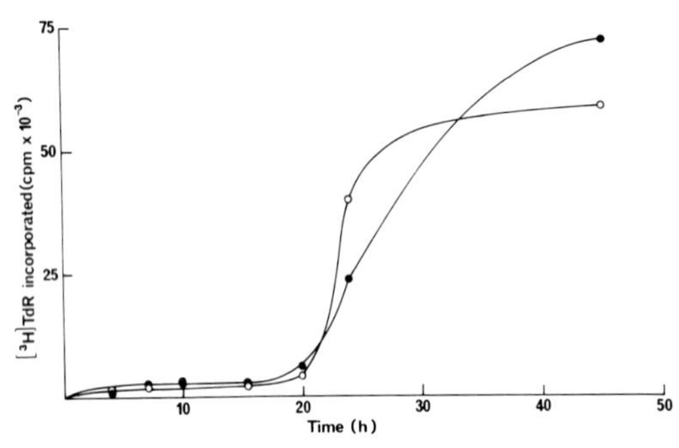

FIGURE 7 Effect of removal of mitogen ([●], soybean agglutinin [10 μg/10^6 cells in 1 ml]; ○, phorbol myristate acetate [1 ng/10^6 cells in 1 ml]) at various times on the subsequent capacity of pig lymphocytes to synthesize DNA. Soybean agglutinin is specific for *N*-acetylgalactosamine and its interaction with pig lymphocytes was reversed by adding this sugar (200 μg). Lymphocytes were separated from mitogen by centrifuging through a cushion of fetal calf serum and were resuspended in fresh medium without mitogen. DNA synthesis was determined by measuring the uptake of [^{3}H]thymidine between 42 and 48 h of culture. Soybean agglutinin was kindly donated by Professor Nathan Sharon (The Weizmann Institute, Rehovet) and represented the spontaneously aggregated species (Sharon, 1976).

for PHA (Lindahl-Kiessling, 1972) and A 23187, these most probably reflect the difficulty of removing the mitogen on account of its firm association with the cell rather than a different mechanism of action.

The significance of the apparent difference between mitogens and other chemical signals is questionable, especially since mitogens do induce various immediate biochemical responses (see below) and may be regarded as a trigger insofar as these parameters are concerned. The requirement for prolonged contact does, however, argue against a transient initial signal inducing a preprogrammed series of events which about 20 h later culminate in the commencement of DNA and protein synthesis.

Mitogen-Induced Biochemical Changes

Mitogen-lymphocyte interaction rapidly initiates (within 30 min) a series of biochemical changes that are primarily located within the surface membrane (Crumpton et al., 1976; Resch, 1976; Wedner and Parker, 1976). These changes include the enhanced selective uptake of ions and metabolites, an enhanced turnover of phosphatidylinositol, and a selective incorporation of long chain fatty acids into phospholipid. Rapid alterations in the intracellular concentrations of cyclic AMP and cyclic GMP have also been claimed (Hadden et al., 1972; Parker et al., 1974; Watson, 1975), although some confusion exists as to the exact nature of the changes. The sequence, relative importance, and control of these events have yet to be elucidated, but there is increasing evidence for an increase in intracellular Ca^{2+} concentration playing a primary role in mediating lymphocyte activation.

THE ROLE OF CA^{2+} IONS: Various workers have attempted to determine the role of Ca^{2+} in lymphocyte activation by a variety of approaches. The majority of the data is consistent with a primary role for Ca^{2+}, although the results of some of the individual studies are not in complete agreement and the significance of some of the results are questionable. Thus, the results indicate that mitogen-induced lymphocyte proliferation is dependent on the presence of greater than 10^{-5} M Ca^{2+} in the culture medium, although there is some disagreement as to the actual portion of the culture period during which extracellular Ca^{2+} is essential (Whitney and Sutherland, 1972; Diamantstein and Ulmer, 1975). Also, measurements of Ca^{2+} uptake suggest that mitogens rapidly stimulate T lymphocytes to accumulate Ca^{2+} (Whitney and Sutherland, 1973; Freedman et al., 1975). The most compelling evidence in support of calcium playing a central role is, however, provided by the observations that the divalent cation ionophore, A 23187, stimulated pig, human, and rabbit lymphocytes to transform into blast cells in an analogous manner to other mitogens (Maino et al., 1974; Luckasen et al., 1974; Greene et al., 1976; Resch, 1976).

A 23187 is a monobasic carboxylic acid, two molecules of which complex divalent cations, especially Ca^{2+}. As the complex is lipid soluble and the Ca^{2+} concentration of the medium is high compared with the cytosol, the ionophore selectively transports Ca^{2+} into cells by acting as a carrier in the plasma membrane and thereby leads to an increase in the cytoplasmic Ca^{2+} concentration and an activation of various Ca^{2+}-dependent intracellular activities. Methylation of the carboxyl group represents the minimum molecular change necessary to prevent Ca^{2+} binding and, indeed, the methyl ester failed to cause any detectable stimulation of Ca^{2+} transport in sarcoplasmic reticulum vesicles (N. M. Green, unpublished results).

Incubation of A 23187 with pig mesenteric lymph node lymphocytes caused 80–90% of the cells to show many of the morphological features characteristic of blast cells. Furthermore, it stimulated an increased incorporation of thymidine into DNA (Fig. 8) according to a similar time-course to that induced by PHA and in a manner dependent on extracellular Ca^{2+} (see Figs. 3 and 4; Maino et al., 1974). In contrast, A 23187 methyl ester failed to cause any detectable increase in thymidine incorporation (Fig. 8). Although the extent of DNA synthesis stimulated by ionophore relative to that given by PHA (Fig. 8) was lower than that predicted from the morphological results, this decrease most probably reflects the toxicity of the ionophore arising from its penetration to the mitochondria and the consequent perturbation of the cells' metabolism. This interpretation is supported by the observation that replacement of the ionophore-containing medium by fresh culture medium after about 12 h of culture resulted in a two- to threefold increase in thymidine incorporation. Rodent lymphocytes appear to be particularly sensitive to A 23187 toxicity, and this most probably accounts for the lack of reports of their transformation by ionophore. Most results are consistent with the proposal that A 23187 is mitogenic for T lymphocytes. Thus, in the case of pig lymphocytes, removal of immunoglobulin-bearing cells and macrophages by nylon wool treatment failed to abrogate or reduce the response. On the other hand, whether B lymphocytes are also stimulated to transform has yet to be proven.

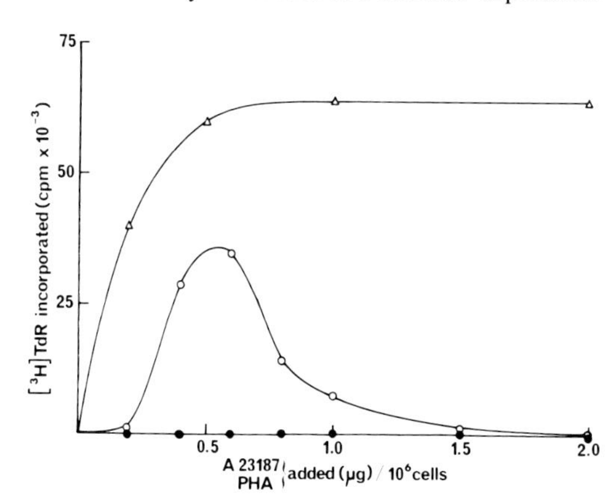

FIGURE 8 Stimulation of incorporation of radioactive thymidine by pig lymphocytes as a function of the concentration of A 23187 (○) or of A 23187 methyl ester (●). PHA (△) was included as a control. The methyl ester of A 23187 was prepared and kindly donated by Dr. N. Michael Green (National Institute for Medical Research, London).

The above results argue strongly in support of the view that the mitogenicity of A 23187 is entirely the result of an increase in intracellular Ca^{2+} concentration, especially since the failure of the methyl ester to induce a response discounts the possibility that A 23187 is acting as a membrane perturbant rather than as a Ca^{2+} ionophore. If the other mitogens also act by increasing the intracellular Ca^{2+} concentration then it should be possible to induce transformation by combining subthreshold amounts of ionophore and these mitogens. Further, the ionophore and the other mitogens should induce identical patterns of biochemical responses. Both of these predictions have been confirmed by experiment. Figure 9 shows that 0.25 μg of A 23187, which by itself had a negligible effect, markedly increased the response induced by suboptimal concentrations of PHA. Similar results have been obtained by adding a subthreshold amount of PHA to increasing concentrations of A 23187 (Maino et al., 1974). Measurement of early biochemical responses has revealed a striking similarity in phosphatidylinositol turnover, amino acid transport, and cyclic nucleotide accumulation stimulated by A 23187 and mitogenic lectins (Crumpton et al., 1976; Greene et al., 1976). This marked parallelism between the

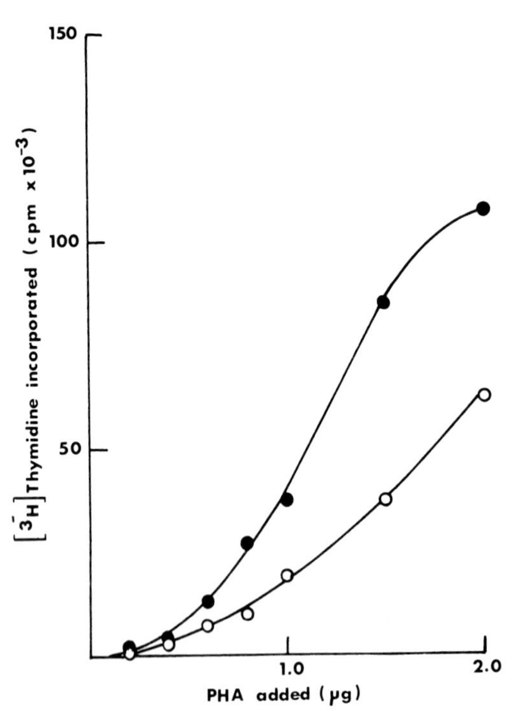

FIGURE 9 Effect of the addition of a constant, subthreshold amount (0.25 µg) of A 23187 (●) on the incorporation of radioactive thymidine stimulated by increasing concentrations of PHA, compared with that induced by PHA alone (○). The amount of A 23187 used had a negligible effect on thymidine incorporation when added alone (see Fig. 8).

biochemical responses argues in favor of the view that if the action of A 23187 is effectively restricted to an increase in intracellular Ca^{2+} concentration then the effects initiated by other mitogens can be ascribed to the same event. In this case, lymphocyte activation by lectins is the result of a direct effect of receptor-lectin interaction on the level of intracellular Ca^{2+}. A number of mechanisms are possible by which receptor-lectin interaction could mediate an increase in cytosol Ca^{2+} concentration. One plausible mechanism is based on the supposition that the specific glycoprotein mediating transformation by T lymphocyte-specific lectins spans the lipid bilayer. In this case, interaction with lectin could induce the formation of clusters that contain polar channels permitting an influx of Ca^{2+}. This mechanism presupposes that cross-linkage at the cell surface (Greaves and Janossy, 1972) and a rapid increase in Ca^{2+} uptake are essential features induced by all mitogens. These requirements are, however, not universally accepted as being essential (e.g., Diamantstein and Ulmer, 1975). Another plausible mechanism is that the rise in intracellular Ca^{2+} concentration is achieved by a redistribution from intracellular sources particularly the inner surface of the plasma membrane. This proposal is based upon a number of well-documented precedents in which modulation of cell function is controlled by a subcellular redistribution of Ca^{2+}. Although it depreciates the significance of the rapid increase in Ca^{2+} uptake which has been claimed by some workers to occur, it has the advantage of providing a rational explanation for the apparent lack of requirement for extracellular Ca^{2+} during the initial phase of the transformation process which has been documented by other workers.

Further work is essential before the validity of the above proposed mechanisms can be evaluated. In particular, more definitive measurements of the Ca^{2+} uptake and of the requirement for extracellular Ca^{2+}, as well as information on the molecular nature of the specific glycoprotein mediating lectin-induced transformation, are required.

REFERENCES

ALLAN, D., and M. J. CRUMPTON. 1973. Phytohemagglutinin-lymphocyte interaction: characterization of binding sites on pig lymphocytes for ^{125}I-labeled phytohemagglutinin. *Exp. Cell. Res.* **78:**271–278.

CRUMPTON, M. J., J. AUGER, N. M. GREEN, and V. C. MAINO. 1976. Surface membrane events following activation by lectins and calcium ionophore. *In* Mitogens in Immunobiology. J. J. Oppenheim and D. L. Rosenstreich, editors. Academic Press, Inc., New York. 85–101.

DIAMANTSTEIN, T., and A. ULMER. 1975. The control of immune response *in vitro* by Ca^{2+}. II. The Ca^{2+}-dependent period during mitogenic stimulation. *Immunology.* **28:**121–125.

FREEDMAN, M. H., M. C. RAFF, and B. GOMPERTS. 1975. Induction of increased calcium uptake in mouse T lymphocytes by concanavalin A and its modulation by cyclic nucleotides. *Nature (Lond.).* **255:**378–382.

GREAVES, M. F. 1975. Scratching the surface. *In* Immune Recognition. A. S. Rosenthal, editor. Academic Press, Inc., New York. 3–19.

GREAVES, M. F., and G. JANOSSY. 1972. Elicitation of selective T and B lymphocyte responses by cell surface binding ligands. *Transplant. Rev.* **11:**87–130.

GREAVES, M. F., J. J. OWEN, and M. C. RAFF. 1973. T and B Lymphocytes: Origins, Properties and Role in Immune Responses. Excerpta Medica Foundation, Amsterdam. 316.

GREENE, W. C., C. M. PARKER, and C. W. PARKER. 1976. Calcium and lymphocyte activation. *Cell. Immunol.* **25:**74–89.

HADDEN, J. W., E. M. HADDEN, M. K. HADDOX, and N. D. GOLDBERG. 1972. Guanosine 3′:5′-cyclic monophosphate: a possible intracellular mediator of mitogenic influences in lymphocytes. *Proc. Natl. Acad. Sci. U. S. A.* **69:**3024–3027.

LINDAHL-KIESSLING, K. 1972. Mechanism of phytohemagglutinin (PHA) action. V. PHA compared with Concanavalin A (Con A). *Exp. Cell Res.* **70:**17–26.

LING, N. R., and J. E. KAY. 1975. Lymphocyte stimulation. 2nd ed. North-Holland Publishing Co., Amsterdam. 398.

LUCKASEN, J. R., J. G. WHITE, and J. H. KERSEY. 1974. Mitogenic properties of a calcium ionophore, A 23187. *Proc. Natl. Acad. Sci. U. S. A.* **71:**5088–5090.

MAINO, V. C., N. M. GREEN, and M. J. CRUMPTON. 1974. The role of calcium ions in initiating transformation of lymphocytes. *Nature (Lond.).* **251:**324–327.

MAINO, V. C., M. J. HAYMAN, and M. J. CRUMPTON. 1975. Relationship between enhanced turnover of phosphatidylinositol and lymphocyte activation by mitogens. *Biochem. J.* **146:**247–252.

MÖLLER, G., editor. 1972. Lymphocyte activation by mitogens: models for immunocyte triggering. *Transplant. Rev.* **11:**1–267.

NOVOGRODSKY, A. 1976. A chemical approach for the study of lymphocyte activation. *In* Mitogens in Immunobiology. J. J. Oppenheim and D. L. Rosenstreich, editors. Academic Press, Inc., New York. 43–56.

NOWELL, P. C. 1960. Phytohemagglutinin: an initiator of mitosis in cultures of normal human leukocytes. *Cancer Res.* **20:**462–466.

PARKER, C. W., T. J. SULLIVAN, and H. J. WEDNER. 1974. Cyclic AMP and the immune response. *Adv. Cyclic Nucleotide Res.* **4:**1–80.

RESCH, K. 1976. Membrane associated events in lymphocyte activation. *In* Receptors and Recognition. Vol. I, series A. P. Cuatrecasas and M. F. Greaves, editors. Chapman and Hall Ltd., London, 61–117.

RODBELL, M. 1972. Cell surface receptor sites. *In* Current Topics in Biochemistry. C. B. Anfinsen, R. F. Goldberger, and A. N. Schechter, editors. Academic Press, Inc., New York. 187–218.

ROSENTHAL, A. S. editor. 1975. Immune Recognition. Academic Press, Inc., New York. 855.

SELL, S., and P. G. H. GELL. 1965. Studies on rabbit lymphocytes *in vitro*. I. Stimulation of blast transformation with an antiallotype serum. *J. Exp. Med.* **122:**423–440.

SHARON, N. 1976. Lectins as mitogens. *In* Mitogens in Immunobiology. J. J. Oppenheim and D. L. Rosenstreich, editors. Academic Press, Inc., New York. 31–41.

WATSON, J. 1975. Cyclic nucleotides as intracellular mediators of B cell activation. *Transplant. Rev.* **23:** 223–249.

WEDNER, H. J., and C. W. PARKER. 1976. Lymphocyte activation. *Prog. Allergy.* **20:**195–300.

WHITNEY, R. B., and R. M. SUTHERLAND. 1972. Requirement for calcium ions in lymphocyte transformation stimulated by phytohemagglutinin. *J. Cell. Physiol.* **80:**329–338.

WHITNEY, R. B., and R. M. SUTHERLAND. 1973. Characteristics of calcium accumulation by lymphocytes and alterations in the process induced by phytohemagglutinin. *J. Cell. Physiol.* **82:**9–19.

Cell Surface and Neoplasia

CELL SURFACE AND NEOPLASIA

MAX M. BURGER

Cell growth, control of cell movement, and control of the antigenic make-up of a neoplastic cell are different from that of a normal cell. Many basic insights into cell biology and molecular biology have come from efforts to define better these aberrations.

Making and breaking cell-cell contacts is an important aspect of cell adhesion and cell migration, and may possibly also influence growth. There is little doubt that all these phenomena depend, among other things, on the surface membrane. We have clearly been too narrow-minded in considering this membrane as a lipid bilayer with a few interspersed proteins. It may be more adequate to consider it as an entire organelle in which, besides its core or integral proteins, an outer and an inner layer have additional functional significance. The proteoglycan portion on the outside is important for direct contact with neighboring cells, and the submembranous network of filamentous proteins serves at least as linkage to the cytoskeleton. Alterations in any of these envelope layers affect the periphery as a unit and thereby, eventually, the entire cell.

Almost all of the ongoing research in the area of neoplasia and the cell surface can ultimately be reduced to either of the following two questions.

(1) Are there differences between tumor and normal cells at the morphological, supramolecular, macromolecular, or molecular level?

(2) Can the behavior of these two cells be modified by manipulating the cell surface in a reproducible and well-defined manner?

In this paper, some aspects will be raised that are not dealt with by the other members of this symposium—in particular, the role of the neglected proteoglycans, as well as a few points from our own work regarding the second question raised. Finally, the validity of the two questions, as well as some common pitfalls, will be assessed critically.

Surface Alterations

One of the serious problems in the field of experimental tumor research in general is that we still operate almost exclusively with correlations between in vitro phenomena and tumorigenicity. Very few correlations have remained without exceptions, and in the field of membrane biochemistry, certainly none could be considered clear-cut causal relationships.

Although the histologist in a tumor biopsy laboratory is well aware that there are only well-defined differences between individual groups of tumors and normal tissue, and that there are essentially no simple morphological criteria that hold for all tumor cells, such general and absolute differences are still sought by the biochemist. Many molecular differences have been observed and seemed to be amazingly common to transformed or even to neoplastic cells in general, although exceptions have been found. Thus, the carbohydrate portions of some glycolipids seem to be less elaborate in transformed cells. Some glycopeptide fractions of glycoproteins seem to be substituted with more sialic acid (see Warren, this symposium), whereas a large glycoprotein present at the periphery of untransformed cell membranes seems to be missing or reduced in transformed cells (Hynes, 1973; Vaheri and Ruoslahti, 1974). Decreased agglutinability with carbohydrate binding lectins was also first interpreted as a loss of an outermost protein component in transformed cells, because untransformed cells could be brought to agglutinate after they were treated with various types of proteases (Burger, 1971).

A recent interpretation, which offers a new outlook, considers not only compositional and structural, but also dynamic, aspects. Thus, based on the fluid mosaic model of membrane structure, an increased "fluidity" was considered for the trans-

---

MAX M. BURGER Department of Biochemistry, Biocenter, University of Basel, Basel, Switzerland

formed plasma membrane. According to the original concept, "fluidity" was based on the lipid portion of the membrane, and changes in the lipid compartment were suspected and found (Bergelson et al., 1970), particularly in cholesterol (Inbar and Shinitzky, 1974). It is, however, uncertain whether such an increased "fluidity" in the lipid portion of the membrane is typical for all tumor cells. We were unable to reduce the increased agglutinability of transformed cells after incorporation of poorly melting elaidate into the plasma membrane, and vice versa for the untransformed cell (Horwitz et al., 1974). Nor could we find any increase in the "fluidity" of the bulk lipid phase of transformed cell membranes by the use of two different physicochemical approaches (Hatten et al., submitted). Others have had similar results with transformed and untransformed fibroblasts in culture (Gaffney, 1975; Fuchs et al., 1975). More work with neoplastic, as opposed to cultured, transformed cells is required to settle this question. In the meantime the concept of an increased fluidity in transformed cells has been extended or shifted to that of an increased mobility of the intramembranous proteins and perhaps the submembranous proteins; as well. To what degree this can be deduced from a general increase in the degree of cluster-formation after lectin addition is presently questioned (Ukena et al., 1976).

It is likely that other suggestions for an explanation of the increased agglutinability of transformed and neoplastic cells will be forthcoming and will contribute additional insights into the biochemistry of the transformed cell surface. Some of the older concepts also must be reinvestigated. Thus, not only identical receptors but also receptors with heterogeneous affinities may display different availabilities to the lectins (crypticity), or alterations of the receptors may lead to small heterogeneities in K_M of receptor sites with different significances for the agglutination process (chemical alterations). A heterogeneity of receptor sites occurs almost with certainty, and has been overshadowed by the statement that the overall amount of lectin binding to the two types of cells is the same. This has not been considered sufficiently by the workers in this field, particularly not in light of the older concepts of crypticity of chemical receptor alterations (Rapin and Burger, 1974).

Much effort has been invested in the study of glycoproteins and glycolipids, whereas the glycosaminoglycans and proteoglycans have been neglected all too long. This may be because of a lack of detailed knowledge of their structure or because they are probably not covalently anchored in the plasma membrane but, to a large degree, are considered extracellular. Nevertheless, they occur on connective-tissue cells as well as on epithelial cells and belong to that class of surface molecules involved in the initial contact of two cells. They may, therefore, profoundly influence cellular behavior upon contact. For instance, if two cells migrate past each other or are sheared off from each other when leaving a primary tumor site, this proteoglycan layer may be among the determining factors that impede or promote the process.

Chiarugi (1974) has recently observed that during protease treatment, polyoma virus-transformed hamster cells released less of a sulfate and glucosamine labeled macromolecule, tentatively identified as heparansulfate. On the other hand, it seemed that the total amount that could be removed by EDTA showed no gross differences beween the two cell types. In this particular case, the alteration would lie in a greater availability of this proteoglycan fraction to proteolytic enzymes only. Roblin et al. (1975) have analyzed the cell-associated sulfate-labeled macromolecules of 3T3 mouse fibroblasts, as well as transformed and lectin revertant derivatives. The virally transformed cells contained about two- to fivefold less of a sulfate-labeled glycosaminoglycan material that was not further identified. Transformed cells, which reverted back to a lower density of growth, if selected for lectin resistance and lower agglutinability, increased their glycosaminoglycan content again by a factor of two and one half- to eightfold. This material seemed to be primarily situated at the cell surface, as about three-fourths could be removed by a relatively mild trypsin treatment.

It should be pointed out that, before these observations were made, Kraemer and Tobey (1972) reported that a loss of heparansulfate occurred at or around mitosis of Chinese hamster ovary cells (Fig. 1). In view of the fact that a similarity between the surfaces of mitotic untransformed cells and transformed interphase cells had been postulated, such comparisons might stimulate further work, although the Chinese hamster ovary cells might not be a good example for untransformed cells. Many other surface changes have been discovered in the meantime which again do not occur only in mitosis and early G_1 cells but also in interphase or all through the cell cycle of

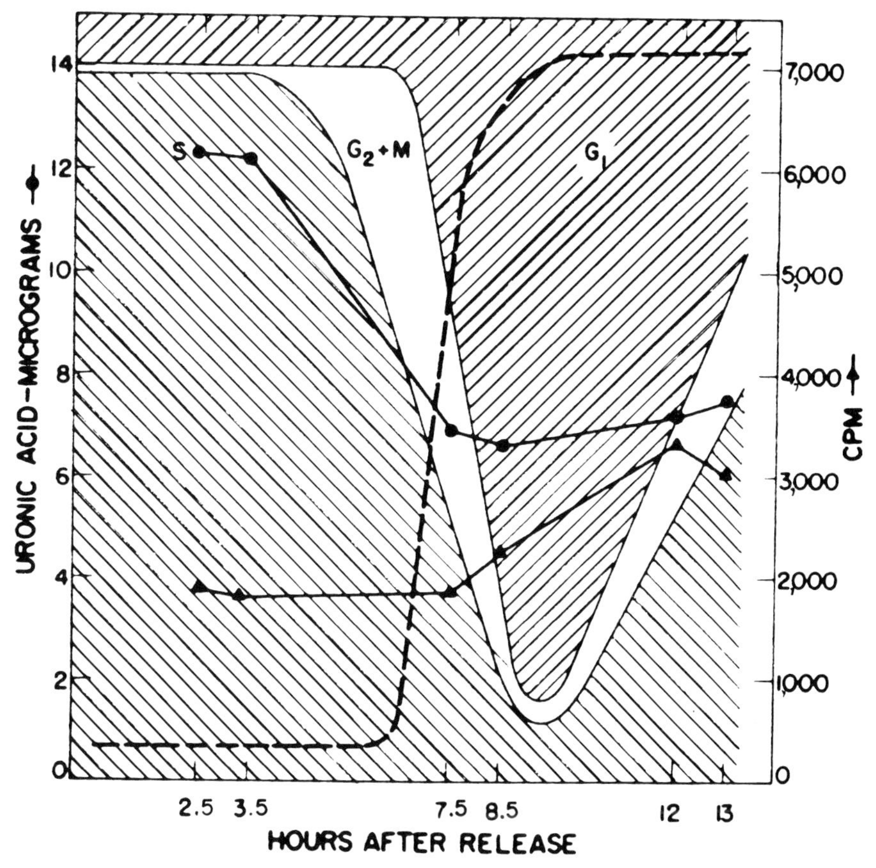

FIGURE 1 Release of a proteoglycan from the surface of mitotic cells. Shown are the increase in cell number, with a dashed line indicating mitosis to occur at 7.5 h (no scale given). The crosshatched areas give the percent cells in the three cycle phases S, G_2 + M, and G_1. Cells were released from a thymidine block at 0 h and pulsed with glucosamine-6-^{3}H for 2 h before each harvest point. Heparan-sulfate was removed from the cell surface with trypsin and isolated specifically. ● μg uronic acid; ▲ cpm ^{3}H. The mass (uronic acid) curve indicates a loss of heparan sulfate at mitosis (7.5 h). The glucosamine-6-^{3}H incorporation curve indicates an increase in biosynthesis and transport of heparan-sulfate to the cell surface at about the same time. (From Kraemer and Tobey, 1972.)

transformed cells (Mannino and Burger, 1975). We will return to this point below.

For anyone trying to establish correlative differences between neoplastic and normal cells, the two following criticisms should always be kept in mind:

Whatever criteria are used to define transformation, they will first have to stand the test of correlation with an accepted neoplastic characteristic. The best test is probably the capability of a cell to give rise to tumors if injected into a test animal. Even this criterion is not absolutely reliable, because a bona fide tumor cell may be rejected by the test animal for various reasons, or a normal, nontumorigenic cell may become tumorigenic in the test animal during the long testing period. Even human SV40 transformed cells, which had all the presently accepted neoplastic characteristics (low serum requirement, transformed morphology, and low anchorage dependence), turned out, against all predictions, to be incapable of producing tumors in athymic or nude mice (Stiles et al., 1975). We will still have a long way to go before reliable parameters will be found for the definition of a tumor cell in vitro. Exceptions to such parameters will probably always be found, provided enough types of cells and cell lines are tested. On the other hand, these exceptions do not necessarily destroy the importance of such a characteristic because artifical growth conditions may have led to the exceptional behavior of the cell in vitro or because organ or species differences may have interfered with the expression of the parameter in case, leading to the exceptional behavior of the cell.

The second criticism to keep in mind is that the best correlation cannot prove a causal relationship between the alteration found and the neoplastic behavior of the cell. Causality can be demonstrated only if one succeeds in introducing a defined chemical change into the cell that can be shown to be sufficient and necessary to lead to a neoplastic behavior of the manipulated normal cell or vice versa.

Alterations in Cellular Behavior after Surface Modification

In general untransformed cells display a more pronounced dependence on cell density for growth than do transformed cells under similar conditions (serum, pH, etc.). It is not clear whether this phenomenon is directly dependent on cell membrane contact or on the presence of a so-called unstirred boundary layer (Stoker, 1973). Be that as it may, it is generally assumed that the difference in growth control between the two types of cells is in some way due to differences in the cell surface (Holley, 1975). We searched, therefore, for means by which the surface could be modified and so lead to alterations in growth behavior and control.

Proteolytic enzymes were shown to release some untransformed 3T3 mouse cells from growth control (Burger, 1970). Inasmuch as the surface alteration was temporary, no lasting transformation could be expected and cells had to be treated again to escape control for a second time (Noonan and Burger, 1973), as seen in Fig. 2. Thus, the cells repaired their proteolytic surface alteration after 6 h, as could be monitored by a transition from the state of increased agglutinability with lectins back to a state of poor agglutinability. Whether the agglutinable state *per se* is sufficient to trigger growth of resting cells is not clear (Cunningham et al., 1974), particularly as recent ob-

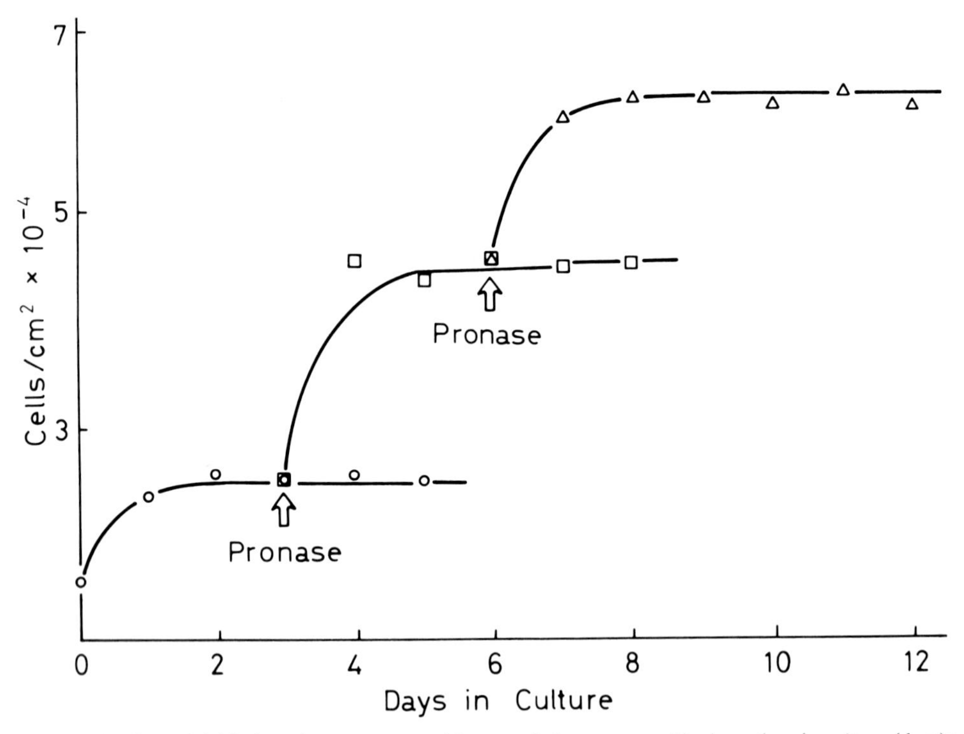

FIGURE 2 Growth initiation after treatment with proteolytic enzymes. Abscissa: time in culture (days); ordinate: cells/cm^2 × 10^{-4}. ○, 3T3 cells, control; □, 3T3 cells, first pronase pulse; △, 3T3 cells, second pronase pulse. 3T3 cells were grown to confluency with 10% calf serum and left there for at least 2 days before stimulation with 10 μg/ml pronase for the first as well as the second time. (From Noonan and Burger, 1973.)

servations point out that different cell lines require different amounts of serum to respond with growth after stimulation with protease (Noonan, 1976). It may, however, be a necessary condition. Growth stimulation of chick embryo fibroblasts with proteolytic enzymes could also be shown (Sefton and Rubin, 1970). Recently, a series of other agents, which presumably act first at the cell surface, were reported to act as triggering agents (i.e., neuraminidase and bacterial lipopolysaccharide; Vaheri et al., 1974).

In view of the decrease of some proteoglycans on transformed cell surfaces mentioned above, one would like to know the effect of the addition of this type of macromolecule to growing cultures. However, we could find only a few studies with sulfated polydextran, a polyanion that in some respects is similar to the sulfated glycosaminoglycans. Clarke and Stoker (1971) reported that dextran sulfate inhibited BHK fibroblasts, incorporating labeled thymidine into DNA. Montagnier (1971) demonstrated that BHK cells particularly were inhibited in their growth by the polyanion but apparently virally transformed cells were not. Recently, Goto et al. (1973) achieved growth inhibition with such polydextran sulfate on partially transformed 3T3, i.e., 3T6 cells in a reversible fashion. Similar well-controlled experiments unfortunately have not yet been carried out with natural glycosaminoglycans.

Reversible inhibition of growth of cultured, untransformed fibroblasts was also observed (Mannino and Burger, 1975) after the addition of a modified lectin that was converted into a nontoxic form by succinylation. This inhibition demonstrated two interesting peculiarities. First, contact between the cell surface and the carbohydrate-binding lectin had to occur only during a particular time of the cell cycle, mitosis and early G_1, i.e., the same portion of the cycle where the cells displayed an increased agglutinability and binding of the fluorescein-labeled lectin (Fig. 3). We feel that the observation by Kraemer and Tobey (1972), illustrated in Fig. 1 and indicating that proteoglycans are lost only during mitosis, is probably quite significant in this context and not simply coincidental. In a manner similar to untransformed cells, which shut down at confluency in G_1, these succinyl Con A-inhibited cells also entered G_1, but at densities lower than confluency.

As a second peculiarity, the inhibitor allowed the cells to achieve a given density that depended only on the amount of lectin and not at all on the time of incubation (Mannino and Burger, 1975). In other words, cells reached a certain preconfluent density level regardless of how densely they were seeded and only dependent on the lectin concentration. This can mean that the surface-bound lectin sensitized the cells to contact and high density, and in that sense, by replacing density as an inhibitory effect, might have acted in a way similar to the density effect. An oversimplified interpretation would be that the lectin molecule coat mimics confluency and shuts down growth of these cells at low cell densities. The question arose, of course, whether transformed cells could also be shut down temporarily in the presence of this nontoxic lectin. We found that if

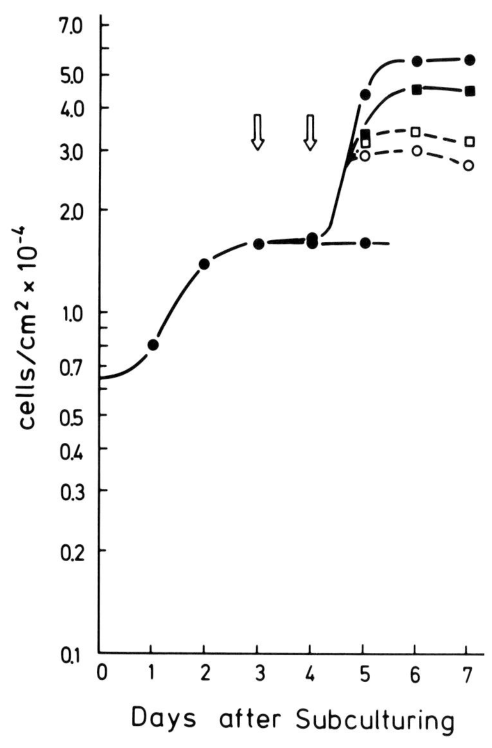

FIGURE 3 Selective sensitivity of mitotic and early G_1 cells to succinyl-Con A-induced growth inhibition. 3T3 cells were subcultured into DME + 2% calf serum. When growth had become stationary, the medium was replaced with DME + 10% calf serum ± succinyl-Con A (500 μg/ml⁻¹) and replaced 23 h later, just before mitosis. ●, control without medium change (flat) or with DME + 10% calf serum without any succinyl-Con A (full response); ■, succinyl-Con A at the first medium change, none at the second; □, succinyl-Con A at the second medium change; ○, succinyl-Con A at both medium changes. (From Mannino and Burger, 1975.)

enough lectin were added to saturate the serum glycoproteins and glucose in the medium, which neutralizes the carbohydrate binding lectin, virally transformed cells were also inhibited. Such cells were stopped in G_1, an uncommon but definitely not unprecedented finding. They could be released easily from this inhibition and were, therefore, clearly not damaged by the elevated lectin concentrations. More work will be required to follow up this observation (Mannino, Ballmer, and Burger, manuscript in preparation).

Provided the lectin acts at the cell surface, some mechanisms of actions can already now be considered. Simple explanations, such as an increase of cAMP or inhibition of single cell movement, can be ruled out thus far, at least for the untransformed cells, which up to now have been better-studied than have the transformed cells. As a working hypothesis, we would like to suggest that the lectin-treated transformed cells increase their adhesion to each other directly, via the lectin, or indirectly, by inducing an adhesive topography or the formation of adhesive material. Alternatively, the lectin may shut down growth in cells that do not meet and, in this case, adhesion would not be involved. It is tempting to speculate that the lectin molecule rendered anionic by succinylation may play a similar role as the polyanionic proteoglycans. We doubt, however, that the lectin has such an unspecific effect, in view of the fact that highly succinylated bovine serum albumin had no growth inhibitory effect whatsoever. Furthermore, one should be aware that the alterations found in proteoglycan content and release do not concern them all equally well and that even the dextran sulfate effects on growth depend on the length of the polymer. So one must consider even more subtle, and perhaps more specific, effects by such polyanions.

Another critical question, which should be raised about this work as well as about similar work that has been carried out in this field (Yamada et al., 1976), is the problem of proving beyond any doubt that the primary site of action is the cell surface. Because such work on growth control involves, by necessity, long-term incubations, the inhibitor might – and probably will – be ingested by the cell and could then act inside the cell. Preliminary studies indicate that the fluorescein-labeled lectin enters the cell, but that after reversal and inhibition of further binding by a specific sugar hapten, the labeled material remains inside even when the cells begin to grow again (Mannino and Burger, unpublished data). Nevertheless, experiments with inhibitor that are bound in an irreversible covalent form to a beaded or to a culture dish substratum should be performed; controls will have to give evidence that no inhibitor was cleaved off during the incubation. Such control experiments usually have not been conducted or, if performed, the controls were not carried out in a rigorous manner.

In this brief discussion of preliminary results on growth control after the addition of surface-binding macromolecules and particularly with the summary on proteoglycans as potentially important surface macromolecules, some long-neglected aspects were mentioned which may provide substantial help in unravelling the role of the cell surface for neoplasia as it occurs in vivo.

ACKNOWLEDGMENTS

The work from my own laboratory has been carried out over the last 10 years by many collaborators. I am grateful to them, and particularly to Dr. R. Mannino and Mr. K. Ballmer who are responsible for the results on growth inhibition by succinyl-Con A.

This work was supported by the Swiss National Foundation (grant 3-1330.73).

REFERENCES

Bergelson, L. D., E. V. Dyatlovitskaya, T. I. Torkhovskayo, I. B. Sorokina, and N. P. Gorkova. 1970. Phospholipid composition of membranes in the tumor cell. *Biochim. Biophys. Acta.* **210:**287–298.

Burger, M. M. 1970. Proteolytic enzymes initiating cell division and escape from contact inhibition of growth. *Nature (Lond.).* **227:**170–171.

Burger, M. M. 1971. Cell surfaces in neoplastic transformation. *Curr. Top. Cell. Regul.* **3:**135–193.

Chiarugi, V. P., S. Vannucchi, and P. Urbano. 1974. Exposure of trypsin-removable sulphated polyanions on the surface of normal and virally transformed $BHK^{12/C13}$ cells. *Biochim. Biophys. Acta.* **345:**283–293.

Clarke, G. D., and M. G. P. Stoker. 1971. Conditions affecting the response of cultured cells to serum. *In* Growth Control in Cell Cultures. G. E. W. Wolstenholme and J. Knight, editors. Churchill Livingstone, Edinburgh and London. p. 17–32.

Cunningham, D. D., C. R. Trash, and R. D. Glynn. 1974. Initiation of division of density-inhibited fibroblasts by glucocorticoids. *Cold Spring Harbor Conf. Cell Proliferation.* **1:**105–113.

Fuchs, P., A. Parola, P. W. Robbins, and E. R. Blout. 1975. Fluorescence polarization and viscosities of membrane lipids of 3T3 cells. *Proc. Natl. Acad. Sci. U. S. A.* **72:**3351–3354.

GAFFNEY, B. J. 1975. Fatty acid chain flexibility in the membranes of normal and transformed fibroblasts. *Proc. Natl. Acad. Sci. U. S. A.* **72:**664–668.

GOTO, M., Y. KATAOKA, T. KIMURA, K. GOTO, and H. SATO. 1973. Decrease of saturation density of cells of hamster cell lines after treatment with dextran sulfate. *Exp. Cell Res.* **82:**367–374.

HOLLEY, R. W. 1975. Control of growth of mammalian cells in culture. *Nature (Lond.).* **258:**487–490.

HORWITZ, A. F., M. E. HATTEN, and M. M. BURGER. 1974. Membrane fatty acid replacements and their effect on growth and lectin-induced agglutinability. *Proc. Natl. Acad. Sci. U. S. A.* **71:**3115–3119.

HYNES, R. O. 1973. Alteration of cell-surface proteins by viral transformation and by proteolysis. *Proc. Natl. Acad. Sci. U. S. A.* **70:**3170–3174.

INBAR, M., and M. SHINITZKY. 1974. Cholesterol as bioregulator in the development and inhibition of leukemia. *Proc. Natl. Acad. Sci. U. S. A.* **71:**4229–4231.

KRAEMER, P. M., and R. A. TOBEY. 1972. Cell-cycle dependent desquamation of heparan sulfate from the cell surface. *J. Cell Biol.* **55:**713–717.

MANNINO, R. J., and M. M. BURGER. 1975. Growth inhibition of animal cells by succinylated Concanavalin A. *Nature (Lond.).* **256:**19–22.

MONTAGNIER, L. 1971. Factors controlling the multiplication of untransformed and transformed BHK 21 cells under various environmental conditions. *In* Growth Control in Cell Cultures. G. E. W. Wolstenholme and J. Knight, editors. Churchill Livingstone, Edinburgh and London. 33–44.

NOONAN, K. D. 1976. Role of serum in protease-induced stimulation of 3T3 cell division past the monolayer stage. Nature *(Lond.).* **259:**573–576.

NOONAN, K. D., and M. M. BURGER. 1973. Induction of 3T3 cell division at the monolayer stage. *Exp. Cell Res.* **80:**405–414.

RAPIN, A. M. C., and M. M. BURGER. 1974. Tumor cell surfaces: general alterations detected by agglutinins. *Adv. Cancer Res.* **20:**1–91.

ROBLIN, R., S. O. ALBERT, N. A. GELB, and P. H. BLACK. 1975. Cell surface changes correlated with density-dependent growth inhibition: glycosaminoglycan metabolism in 3T3, SV 3T3, and Con A selected revertant cells. *Biochemistry.* **14:**347–357.

SEFTON, B. M., and H. RUBIN. 1970. Release from density dependent growth inhibition by proteolytic enzymes. *Nature (Lond.).* **227:**843–845.

STILES, CH. D., W. DESMOND, JR., G. SATO, and M. H. SAIER, JR. 1975. Failure of human cells transformed by simian virus 40 to form tumors in athymic nude mice. *Proc. Natl. Acad. Sci. U. S. A.* **72:**4971–4975.

STOKER, M. P. G. 1973. Role of diffusion boundary layer in contact inhibition of growth. *Nature (Lond.).* **246:**200–203.

UKENA, T. E., E. GOLDMAN, T. L. BENJAMIN, and M. J. KARNOVSKY. 1976. Lack of correlation between agglutinability, the surface distribution of Con A and post-confluence inhibition of cell division in ten cell lines. *Cell.* **7:**213–222.

VAHERI, A., and E. RUOSLAHTI. 1974. Disappearance of a major cell-type specific surface glycoprotein antigen (SF) after transformation of fibroblasts by Rous sarcoma virus. *Int. J. Cancer.* **13:**579–586.

VAHERI, A., E. RUOSLAHTI, and T. HOVI. 1974. Cell surface and growth control of chick embryo fibroblasts in culture. In: Control of proliferation in animal cells. B. Clarkson and R. Baserga, editors. *Cold Spring Harbor Conf. Cell Proliferation.* **1:**305–312.

YAMADA, K. M., S. S. YAMADA, and I. PASTAN. 1976. Cell surface protein partially restores morphology, adhesiveness, and contact inhibition of movement to transformed fibroblasts. *Proc. Natl. Acad. Sci. U. S. A.* **73:**1217–1221.

MODIFICATIONS IN TRANSFORMED AND MALIGNANT TUMOR CELLS

G. L. NICOLSON, G. GIOTTA, R. LOTAN, A. NERI, and G. POSTE

The cell surface is now recognized to be involved in a number of important aspects of neoplasia—uncontrolled cell growth, the invasion of normal tissues, and metastasis to secondary sites. Neoplastic cells escape many of the controls and social restraints that regulate normal cells, such as cell division, recognition, and positioning; thus they achieve varying degrees of independence from host controls which determine normal tissue interactions and organizations (Nicolson and Poste, 1976). Tumor cells are less subject to effective growth regulation by host hormones, serum factors, and other agents that act at the level of the cell surface (Holley, 1975), and this will result in progression of the primary tumor if angiogenesis occurs (Folkman, 1974). Other alterations of the cell surface are responsible for modifications in cellular recognition which contribute to invasion and also to metastasis where primary tumor cells invade surrounding normal tissues and break loose or detach from their primary locations and subsequently establish tumor foci at other sites. Metastasis is clinically the most disastrous event in the biology of neoplastic diseases, and the dissemination of tumor cells to distant parts of the host where they establish new colonies for further neoplastic growth often results in death of the host (Fidler, 1975*a*; Nicolson et al., 1976*a*).

Another important aspect in determining the outcome of neoplastic disease is the interaction of host immune systems with aberrant cells. Although it is generally thought that host immunity toward neoplasia is responsible for immune inhibition (immune surveillance), these interactions, in some cases, are stimulatory—not inhibitory (Prehn, 1976)—and may even aid in the establishment of distant secondary tumors (Fidler, 1974*a*; Fidler and Nicolson, Tumor cell and host properties affecting the implantation and survival of blood-borne metastatic variants of B16 melanoma. *Cancer Res.*, submitted for publication; Fidler et al., 1976).

Obviously the above phenomena are complex cell surface problems which are not approachable at the molecular level, unless a more thorough understanding of the cell surface properties of normal and neoplastic cells is forthcoming. Fortunately, in the last few years our knowledge concerning the structure, organization, and dynamics of cell surfaces has progressed to the point where actual biochemical differences between normal and tumor cell surfaces (Nicolson, 1976*b*; Wallach, 1975; Roblin et al., 1975) may eventually be utilized to develop new therapeutic approaches to fighting neoplastic disorders. In this brief article we will summarize some of the newest findings on cell surface architecture including transmembrane controlling mechanisms and discuss recent animal models that have been developed to learn more about the surface properties of metastatic tumors which may determine their biological behavior in vivo.

Dynamics of Cell Surface Architecture

General consensus has been achieved in recent years that the structure of biological membranes conforms to a number of basic principles (discussed by Guidotti, 1972; Singer and Nicolson, 1972; Bretscher, 1973; Edidin, 1974; Singer, 1974; Fox, 1975; Nicolson, 1976*a*; Nicolson et al., 1977). In brief, these principles are that (a) the majority of the membrane lipids are in a

G. L. NICOLSON, G. GIOTTA, and A. NERI Department of Developmental and Cell Biology, University of California, Irvine, California.

G. L. NICOLSON and R. LOTAN Department of Cancer Biology, The Salk Institute for Biological Studies, San Diego, California.

G. POSTE Department of Experimental Pathology, Roswell Park Memorial Institute, Buffalo, New York.

"fluid" bilayer state under physiological conditions, although minor classes of lipids may be immobilized in lipoprotein complexes or "solid" lipid islands; (b) the lipid bilayer is not continuous but is interrupted by numerous proteins which are inserted to varying degrees into the bilayer; (c) the lipid bilayer is asymmetric in at least certain membranes with respect to the distribution of specific phospholipids in the inner and outer halves of the bilayer as well as to the distribution of oligo- and polysaccharides on the outer surface; (d) the membrane proteins (and glycoproteins) are quite heterogeneous and can be broadly categorized into integral and peripheral (Singer and Nicolson, 1972), or intrinsic and extrinsic (Capaldi and Green, 1972). Of the two types of membrane proteins and glycoproteins, integral membrane proteins are characterized by their hydrophobic interactions with lipid hydrocarbon tails in the membrane interior. This type of interaction is driven by the favorable entropy gained through sequestering integral protein hydrophobic structures away from the aqueous phase and intercalating them to various depths into the lipid bilayer. Certain integral membrane proteins actually span the entire lipid bilayer and have regions of their structure protruding at both sides of the bilayer (Segrest et al., 1973; Morrison et al., 1974). (e) At least certain integral membrane proteins are thought to exist as oligomeric complexes (Guidotti, 1972; Singer, 1974; Nicolson, 1976*a*), and complexes between integral and peripheral membrane proteins are also possible; and (f) cell membrane components (lipids, proteins, and glycoproteins) are capable of lateral rearrangements in the membrane in response to a variety of perturbations.

The cell membrane in its most basic form can be thought of as a solution of integral membrane proteins in a fluid bilayer (Singer and Nicolson, 1972). However, localized differences in the degrees of association between proteins with proteins, lipids and lipids, and proteins with lipids may render membrane architecture less than completely nonrandom in topography. This arrangement permits rapid and reversible changes in the topography of specific components in response to both intra- and extracellular stimuli, whereas others may be restrained in their ability to move laterally in the membrane plane. In addition, most cells possess controlling mechanisms that can maintain some resemblance of topographic order even in a "fluid" membrane region by transmembrane associations to cellular cytoskeletal elements (Berlin et al., 1974; Edelman, 1976; Loor, 1977; Nicolson, 1976*a*; Nicolson and Poste, 1976; Nicolson et al., 1977). These elements appear to be microtubules, microfilaments, and intermediate filaments, and their roles in surface receptor dynamic phenomena such as receptor "capping," endocytosis, cell attachment, movement, etc., are just beginning to be understood (Unanue and Karnovsky, 1973; Berlin et al., 1974; Bretscher and Raff, 1975; Edelman, 1976; Loor, 1977; Nicolson, 1976*a*; Nicolson et al., 1977).

One of the ways in which the interplay between different classes of cytoskeletal elements and their roles in controlling cell surface receptor dynamics has been studied is in the capping of ligand-receptor complexes on lymphoid cells. The binding of multivalent ligands to the surfaces of these cells causes the ligand-receptor complexes to redistribute into clusters→patches→caps (Taylor et al., 1971). Capping of surface receptors can be inhibited to various degrees or reversed by drugs that block cellular energy systems or act on membrane-associated cytoskeletal systems such as cytochalasin B which disrupts cytoplasmic microfilaments (reviewed in Unanue and Karnovsky, 1973; Nicolson, 1976*a*; Nicolson et al., 1977). This suggests that microfilaments play an important and active role in cap formation. Binding of lectins such as concanavalin A to mouse splenic lymphocytes before addition of anti-immunoglobulin (Ig) to cap surface-Ig results in failure to cap (Loor et al., 1972). However, this lectin-mediated inhibition of capping can be overcome by drugs such as colchicine or vinblastine sulfate which impair microtubule function. This has been interpreted as indicating that microtubules "anchor" certain receptors on the cell surface and impede their lateral mobility (Yahara and Edelman, 1972; Edelman, 1974). In addition, disruption of microtubules by colchicine dramatically facilitates capping on certain cells that do not ordinarily cap well (Oliver et al., 1975). When microfilaments *and* microtubules are disrupted by combinations of cytochalasin B *plus* colchicine (De Petris, 1975; Poste et al., 1975) or by tertiary amine local anesthetics (Ryan et al., 1974; Poste et al., 1975), capping is effectively blocked and can even be reversed on precapped cells. These experiments suggest that both microfilaments and microtubules are involved in transmembrane control over the capping process with microfilaments probably providing the contractile activities and microtubules the opposing skeletal anchoring system (Poste et al., 1975; Nicolson and Poste, 1976).

The constantly evolving model of plasma membranes (our scheme, Fig. 1; Nicolson et al., 1977) is basically an elaboration of earlier models (Singer and Nicolson, 1972; Nicolson, 1976*a*) of membrane structure. The matrix of the plasma membrane is a lipid bilayer with intercalated integral proteins and glycoproteins and peripheral membrane proteins attached at both surfaces. Cytoskeletal elements (microtubules, microfilaments, intermediate filaments) are shown interacting indirectly with the plasma membrane or with each other through cross-links or bridging structures (Mooseker and Tilney, 1975). To satisfy the proposal that microfilaments and microtubules play opposing but coordinating roles in the regulation of cell surface receptor mobility and distribution, these elements are depicted in Fig. 1 as connected to one another and to similar transmembrane receptor complexes. These complexes are thought not to be large enough (<20 Å diameter) to be visualized by freeze-fracture techniques,

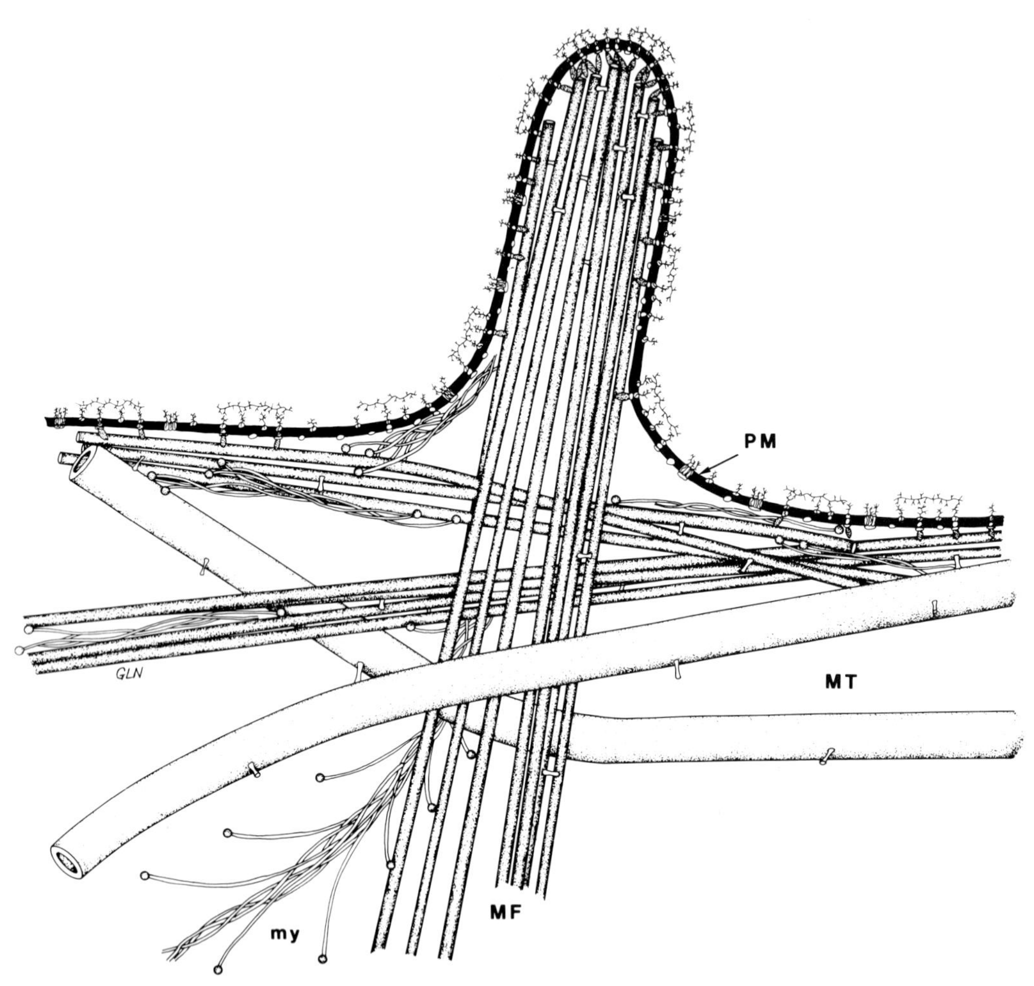

FIGURE 1 Hypothetical interactions between membrane-associated microtubule (MT) and microfilament (MF) systems involved in transmembrane control over cell surface receptor mobility and distribution. This model envisages opposite, but coordinated, roles for microfilaments (contractile) and microtubules (skeletal) and suggests that they are linked to one another or to the same plasma membrane (PM) inner surface components. This linkage may occur through myosin molecules (either in small bundles or the larger filaments [*my*]) or through cross-bridging molecules such as α-actinin. In addition, peripheral membrane components linked at the inner or outer plasma membrane surface may extend this control over specific membrane domains (from Nicolson et al., 1977).

and Edelman (1974) has additionally suggested that such structures probably exist in "free" or "attached" equilibrium states so that at any one time certain classes of receptors could be under transmembrane regulatory control.

Cell Surface Modifications after Neoplastic Transformation

An enormous catalogue of differences between normal and tumor cell surfaces has been amassed (Robbins and Nicolson, 1975; Roblin et al., 1975; Nicolson, 1976*b*; Poste and Weiss, 1976; Wallach, 1975); however, we have gained little from this exercise in the way of explaining how these surface changes arise and are maintained in tumor cell populations (Nicolson and Poste, 1976). Properties such as uncontrolled growth, tumorigenicity, invasiveness, and metastasis have yet to be explained on the basis of cell surface and cellular modifications, although we possess good evidence that the cell surface is intimately involved in all of these events. In addition, problems in obtaining uniform in vivo grown tumor cells and the unavailability of suitable normal cell controls has led to a proliferation in the use of tissue culture (particularly rodent and avian) models for neoplasia where cloned, stable, untransformed cells are transformed by oncogenic viruses, radiation, or chemical carcinogens to stable tumorigenic cell lines. This can yield large amounts of uniform cells for detailed biochemical analyses, but it is not without its own problems and artifacts attributable to in vitro culturing techniques (Nicolson and Poste, 1976; Poste and Weiss, 1976). With this reservation clearly in mind, the following alterations have been seen on a variety of neoplastic cells, but often with notable exceptions.

Of the variety of molecules present at the surfaces of untransformed and transformed cells, most attention has been focused on changes in the composition of the glycosylaminoglycans, glycolipids, glycoproteins, and proteins, enzymes, and other components (Fig. 2). Few changes in glycosylaminoglycans seem to be general or consistent, though the amounts of sulfated glycosylaminoglycans do seem to be lower on transformed cells, and hyaluronic acid and chondroitin sulfate seem

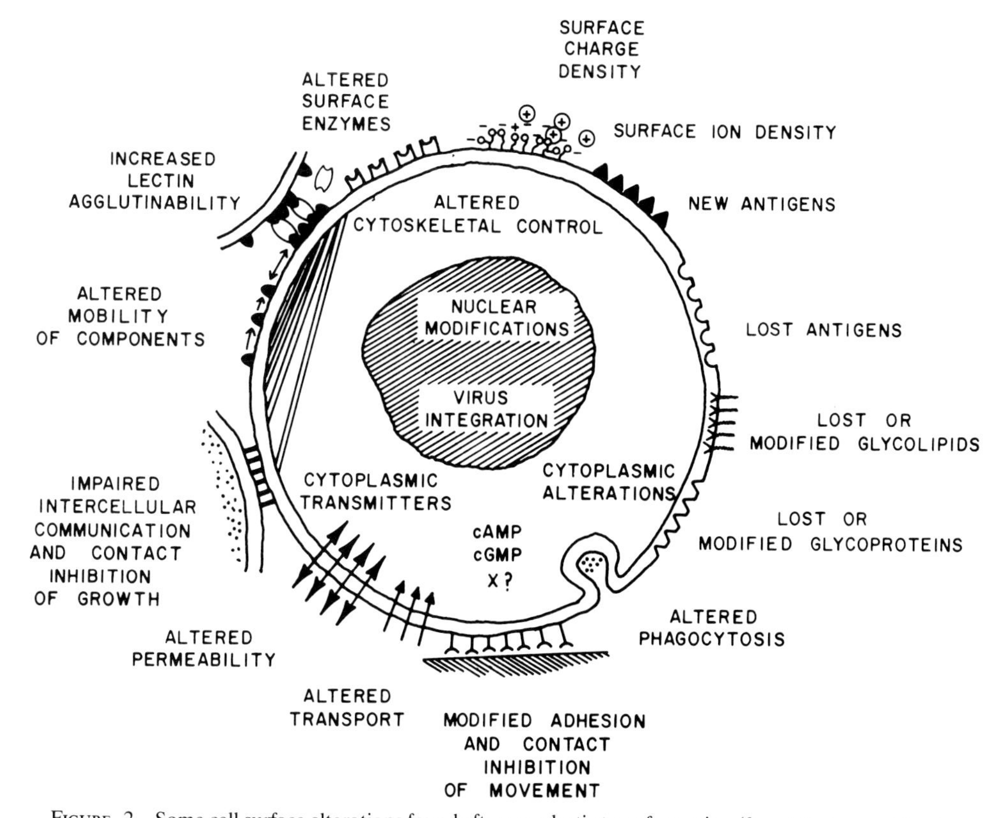

FIGURE 2 Some cell surface alterations found after neoplastic transformation (from Nicolson, 1976*b*).

to be higher in tumors compared to surrounding normal tissues (reviews in Roblin et al., 1975; Nicolson, 1976*b*).

Glycolipid changes after transformation have been documented in a wide variety of untransformed/transformed cell systems (reviews in Hakomori, 1973; Brady and Fishman, 1974). Of the many glycolipid alterations seen after transformation, the most general changes seem to be decreases in the complex glycolipids, "glycolipid simplification," resulting from deletion of the terminal saccharide residues and a loss in the ability of glycolipid synthesis to respond to cell contact, "contact-extension," (Hakomori, 1973). In addition, many of the glycolipids in transformed cell lines show increases in accessibility of their oligosaccharide residues to lectins and enzymes. These proposals have had their drawbacks, however, because some spontaneous mouse and hamster transformed cell lines do not show "glycolipid simplification" (Brady and Fishman, 1974).

Surface labeling studies utilizing lactoperoxidase-catalyzed ^{125}I-iodination and galactose oxidase [^{3}H]borohydride techniques have revealed differences between many untransformed/transformed cell pairs (reviews in Hynes, 1974, 1976; Roblin et al., 1975; Nicolson, 1976*b*). Examination of ^{125}I-tyrosine residues in cell surface proteins or [^{3}H]*D*-gal and -*D*-galNAc containing cell surface glycoproteins reveals loss of a high molecular weight (210,000–250,000) component after transformation. This component has been variously designated LETS protein (large, external, transformation-sensitive), galactoprotein a, CSP (cell surface protein), component Z, SF210, etc. It seems to be mising on a wide variety of (but not all) transformed cells (Hynes, 1974, 1976; Nicolson, 1976*b*) and could be important in cell attachment and adhesion (Hynes, 1976). This component is antigenic and appears to be localized on the cell surface in association with cellular fibrillar structures (apparently microfilaments) and surface ridges and appendages as well as on the growth substrate as a fibrous network (Wartiovaara et al., 1974). It is extremely sensitive to proteolytic enzymes, and this could lead to its loss after transformation by protease cleavage during or after its biosynthesis (Hynes, 1976; Fig. 3).

Portions or fragments of cell surface glycoproteins have been removed by proteolytic enzyme treatment and partially purified by gel filtration (Warren et al, 1973). Glycopeptides obtained from the surfaces of transformed cells chromatograph differently when compared to those from untransformed cells. Transformed cell glycopeptides usually migrate at higher apparent molecular weight, but removal of sialic acid from the glycopeptide mixture abolishes the abnormal migration, suggesting that there are at least differences at the cell surface after transformation concerning

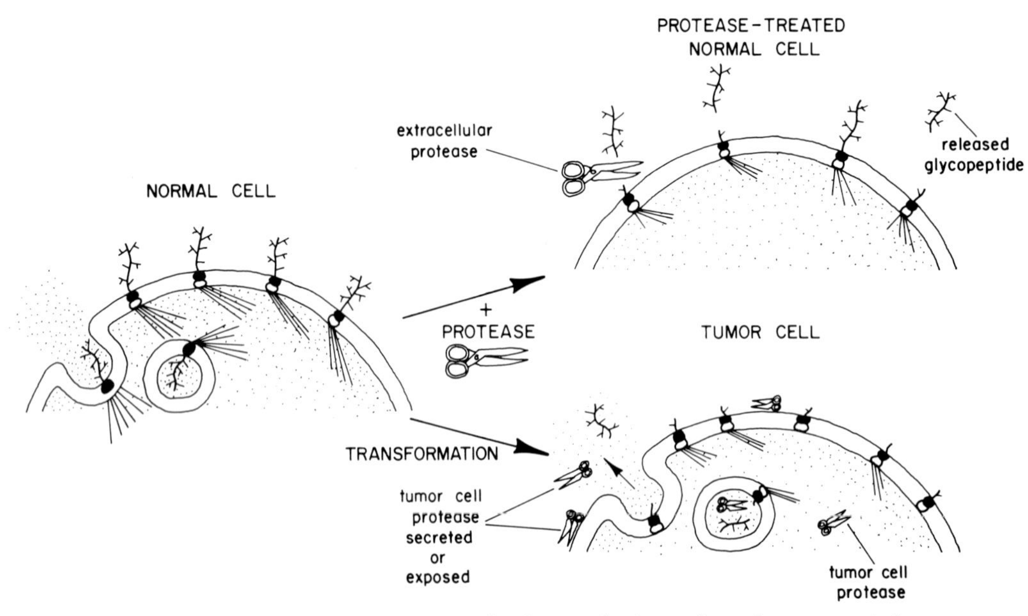

FIGURE 3 Loss of 200–250 K surface glycoprotein after neoplastic transformation or proteolytic enzyme treatment.

the amount of sialic acid in glycoproteins. More recent work has revealed additional saccharide compositional differences in the glycopeptides obtained from transformed cells (see L. Warren et al., this volume).

Several cell surface enzyme activities have been observed to change after transformation. Increases in cell surface and secreted proteases, including plasminogen activator, glycosidases, and other enzymes have been found (reviews in Hynes, 1974, 1976; Reich et al., 1975; Roblin et al., 1975; Nicolson, 1976*b*), although few of these changes seem to be universal. Several surface properties of transformed cells seem to depend on protease activities such as lectin agglutinability (reviews in Rapin and Burger, 1974; Nicolson, 1974, 1976*b*), stimulation of cell division (reviews in Hynes, 1974, 1976; Rapin and Burger, 1974; Roblin et al., 1975), mobility of surface components (reviews in Rapin and Burger, 1974; Nicolson, 1974, 1976*b*), changes in sugar and phosphate transport (Hatanaka, 1974; Pardee, 1975), and antibody reactivities (Hakomori, 1973; Nicolson, 1976*b*; and others). Poste and Weiss (1976) have proposed that increased proteolytic enzyme release by transformed cells maintains them in an activated state because of "sublethal autolysis" at the cell surface. When untransformed cells are treated with low levels of proteases, they transiently acquire many of the above properties usually associated with transformed cells (Hynes, 1974, 1976; Rapin and Burger, 1974; Poste and Weiss, 1976), and transformed cell behavior can be modified, in part, by protease inhibitors (Roblin et al., 1975; Poste and Weiss, 1976).

Proteases can also transiently mimic transformation of normal cells with respect to cellular transport (Hatanaka, 1974; Poste and Weiss, 1976). Transport of many sugars and metabolites occurs at higher rates in transformed cells (Hatanaka, 1974; Pardee, 1975). Some of these changes in transport can also be mimicked in untransformed cells by hormones, serum (Holley, 1975; Gospodarowicz and Moran, 1976), or the addition of cyclic nucleotides (Pastan and Johnson, 1974).

Virus-transformed cells generally have lower levels of cyclic AMP (Pastan and Johnson, 1974), and addition of stimulatory quantities of insulin or other growth factors to quiescent untransformed cells usually triggers cell multiplication (Holley, 1975; Gospodarowicz and Moran, 1976). Other cyclic nucleotides such as cyclic GMP have been proposed to be involved in this process in at least some cells as intracellular messengers (Goldberg et al., 1974); however, these changes do not appear to be general.

Modifications usually ocurring on transformed cell surfaces also include increased susceptibility to lectin agglutination (reviews in Nicolson, 1974, 1976*b*; Rapin and Burger, 1974), enhanced mobilities of certain cell surface receptors and disorganization of cell cytoskeletal elements (review in Nicolson, 1976*b*). These events are probably not unrelated. Transmembrane linkages of cytoskeletal elements to cell surface receptors and their control over receptor distribution and mobility may determine, in part, the cell agglutination characteristics of transformed cells (Fig. 4), although cell agglutination appears to be a complex process involving many different properties of the cell surface (reviews in Nicolson, 1974, 1976*b*; Rapin and Burger, 1974).

Cell surface antigens are, in many cases, also capable of higher rates of redistribution on transformed compared to untransformed cells (Edidin and Weiss, 1974), but cap formation is most often reduced (review in Nicolson, 1976*b*). These changes in receptor mobility could also be the result of modifications in transmembrane cytoskeletal control or sublethal autolysis by proteolytic enzymes. Transformed cells are characterized by the appearance of tumor-associated antigens (Baldwin, 1973). These antigens could be of fetal or embryonic or of viral origin, or they could be antigens entirely unique to the transformed state. The release into the surrounding media of these antigens and other cell surface components appears to be an important way in which tumor cells interfere with host immunologic responses to neoplasia (review in Alexander, 1974). The released antigens and factors can serve as "blocking factors" that neutralize immune cells involved in cell-mediated immunity, one of the most effective host defenses against certain neoplastic cells (Hellström and Hellström, 1972).

The Cell Surface and Malignancy

Of the many cell surface differences between untransformed and transformed cells briefly described above, few probably have any relevance to tumor malignancy in vivo (Nicolson, 1976*b*) which can be defined as the ability of tumor cells to metastasize to near and distant host sites. This phenomenon is usually the final, fatal course of

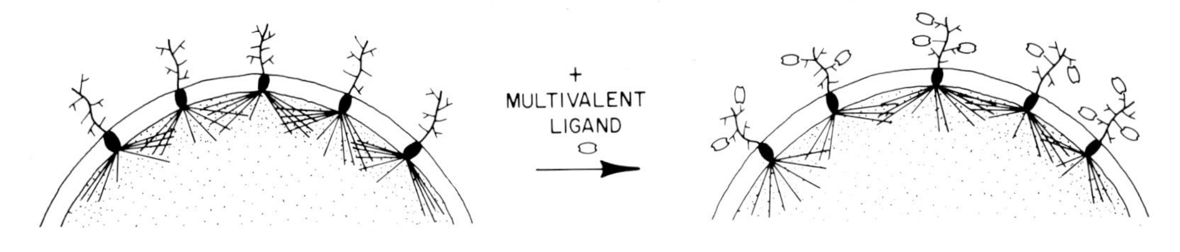

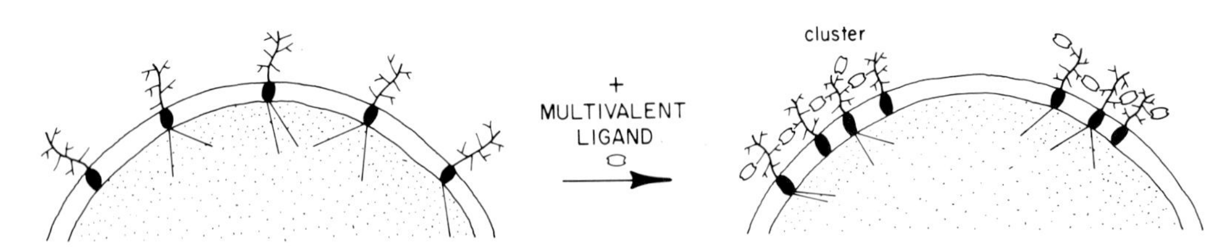

FIGURE 4 Modification of cell surface ligand-induced receptor mobility after neoplastic transformation.

clinical cancer, and this is now receiving an increasing amount of scientific attention. Tumor cells metastasize by invasion of surrounding normal host tissues, followed by entry into the circulatory system, lymphatics, or coelomic cavities where the malignant cells detach from the primary tumor and travel to distant sites and establish new secondary tumor colonies (review in Fidler, 1975*a*). Blood-borne malignant tumor spread (Fig. 5) is particularly insidious, as the major organs can become targets for tumor colonization. In blood-borne metastasis malignant cells enter the circulation, but they usually die quickly; only a small fraction survive (<1% in one study [Fidler, 1970]) and grow to eventually form gross tumor nodules. In fact, the mere presence of tumor cells in the blood is not an indication that subsequent implantation, survival, and growth will occur (Salsbury, 1975), because the transported emboli must successfully arrest in a capillary bed, invade surrounding basement membranes and tissues, establish a proper microenvironment for growth, and escape host defense mechanisms (Fidler, 1975*a*; Nicolson et al., 1976*a*). These necessary physiological properties of metastatic tumor cells, particularly loss of proper cell positioning, detachment, transport, etc., make them quite different from nonmetastasizing or benign tumor cells, and it is important to determine what tumor cell surface characteristics are essential in defining states of tumor progression and spread. Unfortunately, few experimental models exist that can be used to study metastasis.

One system that has been successfully used to study the possible tumor cell properties important in determining host tumor invasion, cell detachment, survival in the circulation, arrest, migration, vascularization, growth, and escape from host defenses is the B16 malignant melanoma variants described by Fidler (1973*a*). These variant tumor cell lines were sequentially selected in syngeneic mice by repeated intravenous injection of B16 cells removed and cultured from lung colonies. B16 lines were obtained after ten such sequential in vivo selections, and these variants show markedly enhanced abilities to form lung colonies from subcutaneous implants or after intravenous injection of a single tumor cell suspension.

The cell surface properties of these variants of low and high in vivo metastatic potential suggest that the following characteristics are important in tumor malignancy: surface enzymes, cell adhesive properties, angiogenesis, certain surface antigens, cell mechanical properties, secretion and shedding, and undoubtedly others (Nicolson et al., 1976*a*).

During primary invasion tumors must extend and break down the extracellular matrix which holds normal tissues together. In in vitro invasion assays the highly metastatic B16 variants always invade normal tissues more extensively than B16 variants of low metastatic potential (Nicolson et al., 1976*a*). Actively expanding tumors are known to contain large concentrations of degradative enzymes (Strauch, 1972), and measurement of degradative enzymes in sparse cultures of B16

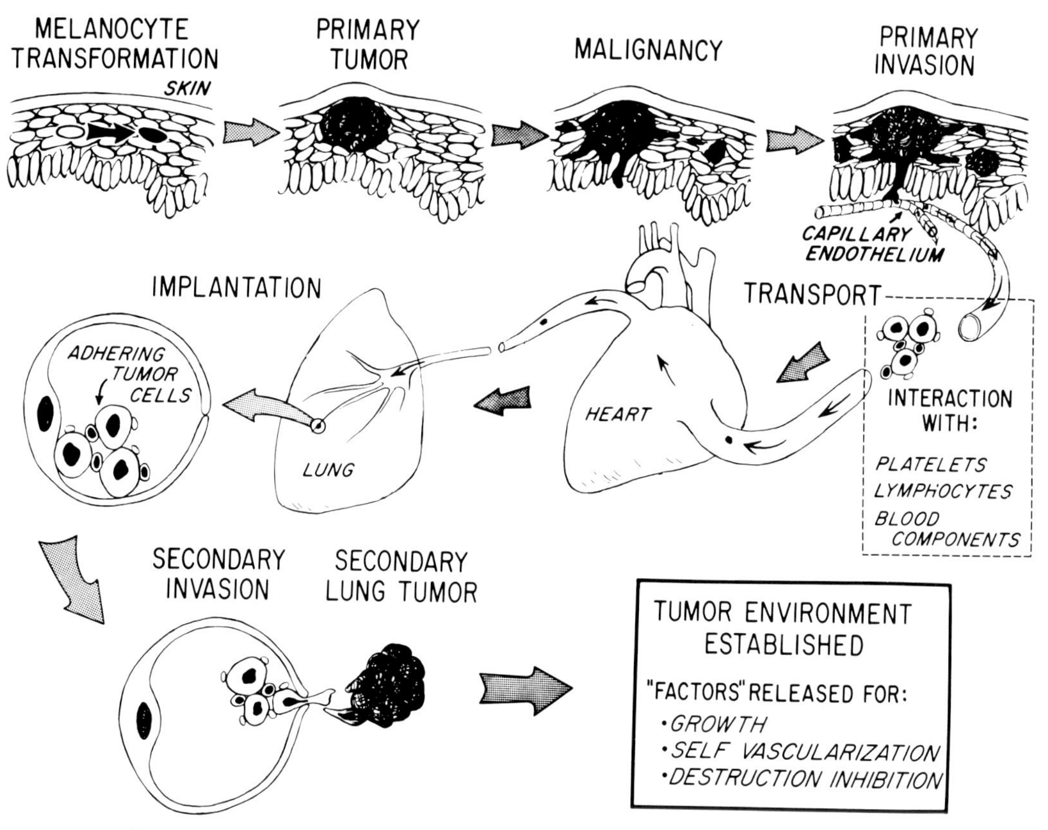

FIGURE 5 Pathogenic sequence of melanoma metastasis (from Nicolson et al., 1976*a*).

melanoma indicates that the more metastatic lines produce higher levels of some proteases and glycosidases (Bosmann et al., 1973) but produce similar amounts of others such as plasminogen activator (Nicolson et al., 1976*b*). Alternatively, the presence of dead and dying cells in tumor tissue, which release lysosomal and other enzymes, could account for the differences observed between in vivo grown tumor cells and cells grown in vitro (Poste and Weiss, 1976).

Once tumor cells invade into the lymphatics and/or circulation, they can interact with one another or with host blood cells. The B16 lines of high metastatic potential homotypically (self) aggregate at greater rates compared with variant lines of low metastatic potential (Nicolson et al., 1976*b*). Similarly, high metastatic B16 variants heterotypically aggregate at greater rates with blood lymphocytes (Fidler, 1975*b*), platelets (Gasic et al., 1973), and noncirculating organ cells (Nicolson and Winkelhake, 1975). These interactions lead to the formation of multicellular emboli, and these emboli are known to be more successful on a per cell basis in implanting and surviving to form experimental metastases (Fidler, 1973*b*).

At any time during the sequence of metastatic events the tumor cells must escape host defenses which can be immune or nonimmune in nature (Fidler, 1975*a*). In the B16 system host immunity against the primary tumor may actually aid in the formation of experimental metastases (Fidler et al., 1977; Fidler and Nicolson, submitted for publication). This appears to be primarily because of an increase in blood-borne tumor cell arrest after heterotypic tumor cell-lymphocyte aggregation. In this system, lymphocytes are not nearly as effective as macrophages in killing B16 melanoma (Fidler, 1974*b*), so the melanoma cells resistant to lymphocyte killing actually survive at higher rates.

It is impossible to discuss the many aspects of metastasis here in detail, but future work in cancer research will undoubtedly focus more on systems that have possible clinical relevance such as control of the malignant phenotype.

ACKNOWLEDGMENTS

Dr. Nicolson's studies were supported by U. S. National Cancer Institute contract CB-33879 from the Tumor

Immunology Program, U. S. Public Health Service grant CA-15122-A1, American Cancer Society grant BC-211, and U. S. National Science Foundation grant PCM76-18528. Dr. Poste was supported by U. S. Public Health Service grant CA-13393. Dr. Lotan is a Fellow of the World Health Organization. G. Giotta and A. Neri were, respectively, postdoctoral and predoctoral trainees supported by U. S. National Cancer Institute training grant CA-09054 to the University of California, Irvine.

REFERENCES

ALEXANDER, P. 1974. Escape from immune destruction by the host through shedding of surface antigens: is this a characteristic shared by malignant and embryonic cells? *Cancer Res.* **34:**2077–2082.

BALDWIN, R. W. 1973. Immunological aspects of chemical carcinogenesis. *Adv. Cancer Res.* **18:**1–76.

BERLIN, R. D., J. M. OLIVER, T. E. UKENA, and H. H. YING. 1974. Control of cell surface topography. *Nature (Lond.).* **247:**45–46.

BOSMANN, H. B., G. F. BIEBER, A. E. BROWN, K. R. CASE, D. M. GERSTEN, T. W. KIMMERER, and A. LIONE. 1973. Biochemical parameters correlated with tumour cell implantation. *Nature (Lond.).* **246:**487–489.

BRADY, R. O., and P. H. FISHMAN. 1974. Biosynthesis of glycolipids in virus-transformed cells. *Biochim. Biophys. Acta.* **355:**121–148.

BRETSCHER, M. S. 1973. Membrane structure: some general principles. Membranes are asymmetric lipid bilayers in which cytoplasmically synthesized proteins are dissolved. *Science (Wash. D. C.).* **181:**622–629.

BRETSCHER, M. S., and M. C. RAFF. 1975. Mammalian plasma membranes. *Nature (Lond.).* **258:**43–49.

CAPALDI, R. A., and D. E. GREEN. 1972. Membrane proteins and membrane structure. *FEBS (Fed. Eur. Biochem. Soc.) Lett.* **25:**205–209.

DE PETRIS, S. 1975. Concanavalin A receptors, immunoglobulins and θ antigens of the lymphocyte surface. Interactions with concanavalin A and cytoplasmic structures. *J. Cell Biol.* **65:**123–146.

EDELMAN, G. M. 1974. Surface alteration and mitogenesis in lymphocytes. *In* Control of Proliferation in Animal Cells. Vol. 1. Cold Spring Harbor Conferences on Cell Proliferation. B. Clarkson and R. Baserga, editors. Cold Spring Harbor Laboratory, New York. 357–377.

EDELMAN, G. M. 1976. Surface modulation in cell recognition and cell growth. *Science (Wash. D. C.).* **192:**218–226.

EDIDIN, M. 1974. Rotational and translational diffusion in membranes. *Annu. Rev. Biophys. Bioeng.* **3:**179–201.

EDIDIN, M., and A. WEISS. 1974. Restriction of antigen mobility in the plasma membranes of some cultured fibroblasts. *In* Control of Proliferation in Animal Cells. Vol. 1. Cold Spring Harbor Conferences on Cell Proliferation. B. Clarkson and R. Baserga, editors. Cold Spring Harbor Laboratory, New York. 213–220.

FIDLER, I. J. 1970. Metastasis: quantitative analysis of distribution and fate of tumor emboli labeled with ^{125}I-5-iodo-2′-deoxyuridine. *J. Natl. Cancer Inst.* **45:**775–782.

FIDLER, I. J. 1973*a*. Selection of successive tumor lines for metastasis. *Nat. New Biol.* **242:**148–149.

FIDLER, I. J. 1973*b*. The relationship of embolic homogeneity, number, size and viability to the incidence of experimental metastasis. *Eur. J. Cancer.* **9:**223–227.

FIDLER, I. J. 1974*a*. Immune stimulation-inhibition of experimental cancer metastasis. *Cancer Res.* **34:**491–498.

FIDLER, I. J. 1974*b*. Inhibition of pulmonary metastasis by intravenous injection of specifically activated macrophages. *Cancer Res.* **34:**1074–1078.

FIDLER, I. J. 1975*a*. Mechanisms of cancer invasion and metastasis. *In* Biology of Tumors: Surfaces, Immunology, and Comparative Pathology. Vol. 4. F. F. Becker, editor. Plenum Publishing Corp., New York. 101–131.

FIDLER, I. J. 1975*b*. Biological behavior of malignant melanoma cells correlated to their survival *in vivo*. *Cancer Res.* **35:**218–224.

FIDLER, I. J., S. CAINES, and Z. DOLAN. 1977. Survival of hematogenously disseminated allogeneic tumor cells in athymic nude mice. *Transplantation (Baltimore).* In press.

FOLKMAN, J. 1974. Tumor angiogenesis. *Adv. Cancer Res.* **19:**331–358.

FOX, C. F. 1975. Phase transitions in model systems and membranes. *In* Biochemistry of Cell Walls and Membranes. Vol. 2. MTP Int. Rev. Sci. Biochem. Series One. C. F. Fox, editor. University Park Press, Baltimore. 279–306.

GASIC, G. J., T. B. GASIC, N. GALANTI, T. JOHNSON, and S. MURPHY. 1973. Platelet-tumor cell interaction in mice. The role of platelets in the spread of malignant disease. *Int. J. Cancer.* **11:**704–718.

GOLDBERG, N. D., M. K. HADDOX, E. DUNHAM, C. LOPEZ, and J. W. HADDEN. 1974. The Yin Yang hypothesis of biological control: opposing influences of cyclic GMP and cyclic AMP in the regulation of cell proliferation and other biological processes. *In* Control of Proliferation in Animal Cells. Vol. 1. Cold Spring Harbor Conferences on Cell Proliferation. B. Clarkson and R. Baserga, editors. Cold Spring Harbor Laboratory, New York. 609–625.

GOSPODAROWICZ, D., and J. S. MORAN. 1976. Growth factors in mammalian culture. *Annu. Rev. Biochem.* **45:**531–558.

GUIDOTTI, G. 1972. Membrane proteins. *Annu. Rev. Biochem.* **41:**731–752.

HAKOMORI, S-I. 1973. Glycolipids of tumor cell membrane. *Adv. Cancer Res.* **18:**265–315.

HATANAKA, M. 1974. Transport of sugars in tumor cell membranes. *Biochim. Biophys. Acta.* **355:**77–104.

HELLSTRÖM, K. E., and I. HELLSTRÖM. 1972. Immunity to neuroblastomas and melanomas. *Annu. Rev. Med.* **23:**19–38.

Holley, R. W. 1975. Control of growth of mammalian cells in culture. *Nature (Lond.).* **258:**487–490.

Hynes, R. O. 1974. Role of surface alterations in cell transformation: the importance of proteases and surface proteins. *Cell.* **1:**147–156.

Hynes, R. O. 1976. Cell surface proteins and malignant transformation. *Biochim. Biophys. Acta.* **458:**73–107.

Loor, F. 1977. Structure and dynamics of the lymphocyte surface. *In* B and T Cells in Immune Recognition. F. Loor and G. E. Roelants, editors. John Wiley & Sons Ltd., Chichester, England. In press.

Loor, F., L. Forni, and B. Pernis. 1972. The dynamic state of the lymphocyte membrane. Factors affecting the distribution and turnover of surface immunoglobulins. *Eur. J. Immunol.* **2:**203–212.

Mooseker, M. S., and L. G. Tilney. 1975. Organization of an actin filament-membrane complex: filament polarity and membrane attachment in the microvilli of intestinal epithelial cells. *J. Cell Biol.* **67:**725–743.

Morrison, M., T. J. Mueller, and C. T. Huber. 1974. Transmembrane orientation of the glycoproteins in normal human erythrocytes. *J. Biol. Chem.* **249:**2658–2660.

Nicolson, G. L. 1974. The interactions of lectins with animal cell surfaces. *Int. Rev. Cytol.* **39:**89–190.

Nicolson, G. L. 1976*a*. Transmembrane control of the receptors on normal and tumor cells. I. Cytoplasmic influence over cell surface components. *Biochim. Biophys. Acta.* **457:**57–108.

Nicolson, G. L. 1976*b*. Transmembrane control of the receptors on normal and tumor cells. II. Surface changes associated with transformation and malignancy. *Biochim. Biophys. Acta.* **458:**1–72.

Nicolson, G. L., and G. Poste. 1976. The cancer cell: dynamic aspects and modifications in cell-surface organization. *New Engl. J. Med.* **295:**197–203; 253–258.

Nicolson, G. L., G. Poste, and T. H. Ji. 1977. The dynamics of cell membrane organization. *In* Dynamic Aspects of Cell Surface Organization. Vol. 3. Cell Surface Reviews. G. Poste and G. L. Nicolson, editors. North-Holland Publishing Co., Amsterdam. In press.

Nicolson, G. L., and J. L. Winkelhake. 1975. Organ specificity of blood-borne tumour metastasis determined by cell adhesion? *Nature (Lond.).* **255:**230–232.

Nicolson, G. L., C. R. Birdwell, K. W. Brunson, and J. C. Robbins. 1976*a*. Cellular interactions in the metastatis process. *J. Supramol. Struct.* **1**(suppl.): 237–244.

Nicolson, G. L., J. L. Winkelhake, and A. C. Nussey. 1976*b*. An approach to studying the cellular properties associated with metastasis: some *in vitro* properties of tumor variants selected *in vivo* for enhanced metastasis. *In* Fundamental Aspects of Metastasis. L. Weiss, editor. North-Holland Publishing Co., Amsterdam. 291–303.

Oliver, J. M., R. B. Zurer, and R. D. Berlin. 1975. Concanavalin A cap formation on polymorphonuclear leukocytes of normal and beige (Chediak-Higashi) mice. *Nature (Lond.).* **253:**471–473.

Pardee, A. B. 1975. The cell surface and fibroblast proliferation. Some current research trends. *Biochim. Biophys. Acta.* **417:**153–172.

Pastan, I., and G. S. Johnson. 1974. Cyclic AMP and the transformation of fibroblasts. *Adv. Cancer Res.* **19:**303–330.

Poste, G., D. Papahadjopoulos, and G. L. Nicolson. 1975. The effect of local anesthetics on transmembrane cytoskeletal control of cell surface mobility and distribution. *Proc. Natl. Acad. Sci. U. S. A.* **72:**4430–4434.

Poste, G., and L. Weiss. 1976. Some considerations on cell surface alterations in malignancy. *In* Fundamental Aspects of Metastasis. L. Weiss, editor. North-Holland Publishing Co., Amsterdam. 25–47.

Prehn, R. T. 1976. Tumor progression and homeostatis. *Adv. Cancer Res.* **23:**203–236.

Rapin, A. M. C., and M. M. Burger. 1974. Tumor cell surfaces: general alterations detected by agglutinins. *Adv. Cancer Res.* **20:**1–91.

Reich, E., D. B. Rifkin, and E. Shaw. Editors. 1975. Proteases and Biological Control. Vol. 2. Cold Spring Harbor Conferences on Cell Proliferation. Cold Spring Harbor Laboratory, New York. 1021.

Robbins, J. C., and G. L. Nicolson. 1975. Surfaces of normal and transformed cells. *In* Biology of Tumors: Surfaces, Immunology, and Comparative Pathology. Vol. 4. F. F. Becker, editor. Plenum Publishing Corp., New York. 3–54.

Roblin, R., I-N. Chou, and P. H. Black. 1975. Proteolytic enzymes, cell surface changes and viral transformation. *Adv. Cancer Res.* **22:**203–259.

Ryan, G. B., E. R. Unanue, and M. J. Karnovsky. 1974. Inhibition of surface capping of macromolecules by local anaesthetics and tranquillisers. *Nature (Lond.).* **250:**56–57.

Salsbury, A. J. 1975. The significance of the circulating cancer cells. *Cancer Treat. Rev.* **2:**55–72.

Segrest, J. P., I. Kahne, R. L. Jackson, and V. T. Marchesi. 1973. Major glycoprotein of the human erythrocyte membrane: evidence for an amphipathic molecular structure. *Arch. Biochem. Biophys.* **155:**167–183.

Singer, S. J. 1974. The molecular organization of membranes. *Annu. Rev. Biochem.* **43:**805–833.

Singer, S. J., and G. L. Nicolson. 1972. The fluid mosaic model of the structure of cell membranes. *Science (Wash. D. C.).* **175:**720–721.

Strauch, L. 1972. The role of collagenases in tumor invasion. *In* Tissue Interactions in Carcinogenesis. D. Tarin, editor. Academic Press, Inc., New York. 399–434.

Taylor, R. B., W. P. H. Duffus, M. C. Raff, and S. De Petris. 1971. Redistribution and pinocytosis of lymphocyte surface immunoglobulin molecules induced by anti-immunoglobulin antibody. *Nat. New Biol.* **233:**225–229.

UNANUE, E. R., and M. J. KARNOVSKY. 1973. Redistribution and fate of Ig complexes on surface B lymphocytes: functional implications and mechanisms. *Transplant. Rev.* **14:**184–210.

WALLACH, D. F. H. 1975. Membrane Molecular Biology of Neoplastic Cells. North-Holland Publishing Co., Amsterdam. 525.

WARREN, L., J. P. FUHRER, and C. A. BUCK. 1973. Surface glycoproteins of cells before and after transformation by oncogenic viruses. *Fed. Proc.* **32:**80–85.

WARTIOVAARA, J., E. LINDER, E. RUOSLAHTI, and A. VAHERI. 1974. Distribution of fibroblast surface antigen. Association with fibrillar structures of normal cells and loss upon viral transformation. *J. Exp. Med.* **140:**1522–1533.

YAHARA, I., and G. M. EDELMAN. 1972. Restriction of the mobility of lymphocyte immunoglobulin receptors by concanavalin A. *Proc. Natl. Acad. Sci. U. S. A.* **69:**608–612.

THE MEMBRANE GLYCOPROTEINS OF NORMAL AND MALIGNANT CELLS

L. WARREN, C. A. BUCK, G. P. TUSZYNSKI, and J. P. FUHRER

Many knowledgeable people believe that malignancy begins with one or a small number of changes (Knudson, 1973), a mutation or what might be its functional equivalent, an oncogenic viral infection (Tooze, 1973). Yet, the most striking feature of the malignant cell is that so many details of function have changed even if only in a minor, quantitative way. Differences between normal and malignant cells may be divided into two categories; first, the discrete, qualitative, primary type of change that centers in the DNA and genetic apparatus of the cell; second, the secondary and tertiary involvements that cascade to encompass most of the cell's apparatus.

It should be emphasized that, despite the extensive changes in the malignant cell, it is still alive and well. It divides too well and lacks certain subtle and specialized functions, such as interacting with neighboring cells and failing to respond to their appeals to stop moving about and dividing. The fundamental changes in malignancy seem to be small, not large, otherwise the cell could not survive the upheaval. The original change is compatible with life, as is the cascade of secondary and tertiary involvements. The malignant cell arrives at a new equilibrium, still within the confines permissive of life. Although the numerous and extensive changes within the cell considered separately may be almost trivial, the consequences for the host are disastrous.

The last few decades have witnessed many comparative structural and compositional studies of normal and malignant cells. Studies on lipids have revealed rather small and unpromising differences (Bergelson, 1972) despite the fervent hopes of those who would like to ascribe an increased "fluidity" to the surface membrane of malignant cells. Extensive work on the glycolipids has revealed widely occurring differences (Hakomori, 1973; Brady and Fishman, 1974; Sakiyama and Robbins, 1973). Generally, there is a tendency for malignant cells to leave the oligosaccharide chains of the glycolipids incomplete, resulting in a simplification of glycolipid patterns. Synthesis of some glycolipids has been found to be growth-dependent. When NIL cells become confluent and enter plateau phase they begin to synthesize certain neutral glycolipids, whereas transformed NIL cells, which continue to divide, do not (Robbins and Macpherson, 1971). The extent of exposure of glycolipids on the surface of transformed cells differs from that of control cells as detected by their susceptibility to attack by galactose oxidase (Gahmberg and Hakomori, 1974).

In the past 5 yr a number of workers have been investigating proteins of surface membranes by comparing polyacrylamide gel electrophoresis patterns of surface membranes after labeling them by iodination with lactoperoxidase, reduction with NaB^3H_4 after galactose oxidase, metabolic labeling, and other methods (Hynes, 1976). Among several differences in autoradiographic patterns that have been found, perhaps the most important is the partial or complete disappearance from the transformed cell of a large, surface glycoprotein of 210,000–250,000 mol wt. This glycoprotein is present in highest concentration during the G_1 phase of control cells and decreases during mitosis. When added back to transformed cells, it appears to restore their morphology to normal but does not alter their ability to divide or transport nutrients (Yamada et al., 1976).

We have also been looking for differences in the banding patterns of membrane proteins of normal and malignant cells. We have fractionated control hamster tissue culture cells (BHK_{21}/C_{13}) and their Rous virus-transformed counterpart, C_{13}/B_4 into

L. WARREN and CO-WORKERS The Wistar Institute, Philadelphia, Pennsylvania

various membrane components, the fractions have been electrophoresed on SDS polyacrylamide gel, and stained with Coomassie blue. In Fig. 1 are seen the results of such a procedure. Careful inspection of the patterns reveals some differences, mostly decreases or disappearances of bands in transformed cells. No claim is made that these are consistent and reproducible changes, but it is quite clear that if one looks closely enough several fine differences can be found. At the present time the critical problem is not only to find chemical, structural, and compositional differences but when found to show that they are widespread, reproducible, and can be linked to specific malignant characteristics, such as persistent cell division, characteristic morphology, decreased intercellular adhesiveness, loss of contact inhibition of motion, metabolism, and cell division, ability to grow to higher saturation densities in culture, alterations of nutrient transport, decreased dependency on serum factors, ability to grow in soft agar, etc. (Tooze, 1973; Burger, 1971; Weiss, 1967; Pitot, 1974).

Although there are several possible schemes to explain the cascade phenomenon in malignancy in which one or a small number of changes leads to many, it is not unreasonable to believe that alterations in cellular membranes could perturb numerous processes producing the pleotypic character of the malignant cell (Pitot, 1974; Wallach, 1968). My co-workers and I have been examining the membranes of normal and malignant cells to test this hypothesis. To provide a background for discussion of our latest work on this subject, it would be best to review work by me and my associates (Buck et al., 1970, 1971*a*, *b*, and 1974; Warren et al., 1972*b*, 1973) as well as by others (Meezan et al., 1969; Sakiyama and Burge, 1972; Emmelot, 1973; Beek et al., 1973, 1975; Smets et al., 1975) carried out over the past 6 yr.

We have been studying the carbohydrate components of the membrane glycoproteins of tissue

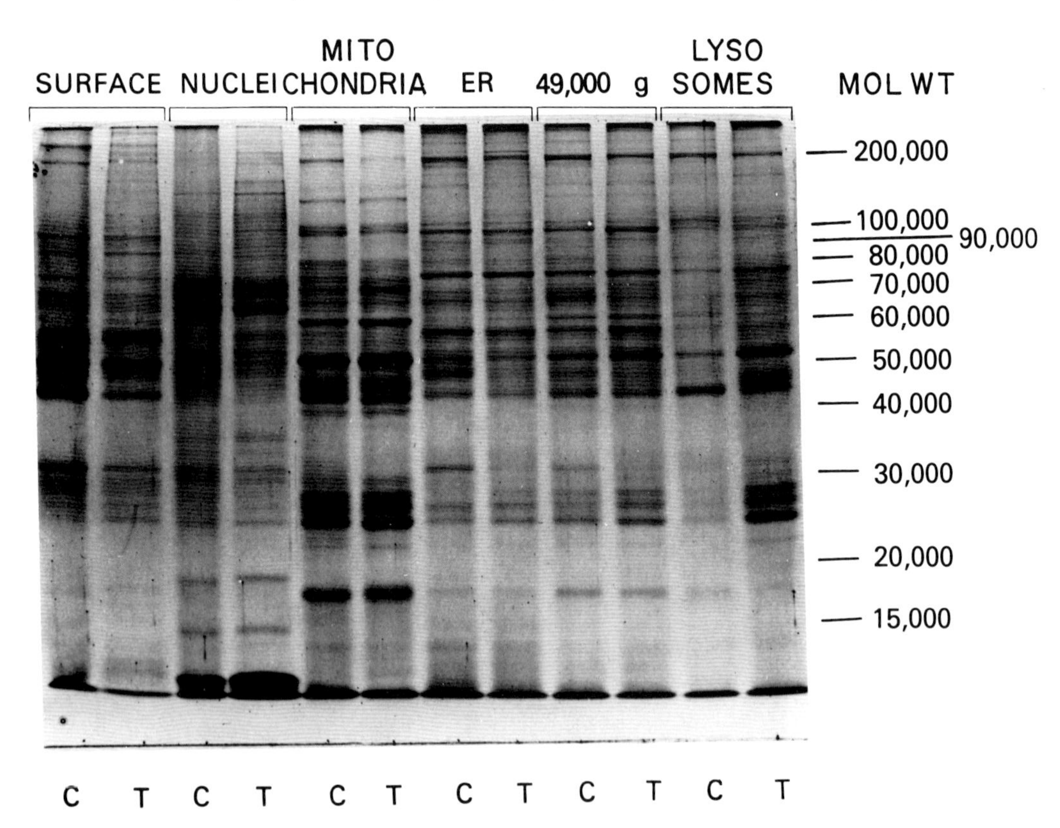

FIGURE 1 Coomassie blue staining profiles of membrane fractions from control BHK_{21}/C_{13} (*C*) and transformed cells C_{13}/B_4 (*T*) analyzed on a 10% polyacrylamide SDS slab gel. Cells were fractionated by methods previously described (Buck et al., 1974). Gel electrophoresis was performed by the method of Laemmli (1970).

culture cells. The control cell most frequently used in our studies has been the baby hamster kidney cell BHK_{21}/C_{13} which, although not really a normal cell, is virtually nontumorigenic and does not grow in soft agar medium. Our malignant cell is C_{13}/B_4 which has been transformed with Rous sarcoma virus (Bryan strain). To compare carbohydrate structures the double-label method is used. Control cells are grown in the presence of L-[^{14}C]fucose or D-[^{14}C]glucosamine, and transformed cells in the presence of L-[^{3}H]fucose or D-[^{3}H]glucosamine. Equivalent labeled fractions (e.g., surface membranes) or surface material removed by trypsin, "trypsinates," from each type of cell are exhaustively digested with pronase, and the limit digest (containing both ^{14}C and ^{3}H) is chromatographed on a column of Sephadex G50 (Buck et al., 1970, 1971*a*). The elution patterns clearly show the presence of an early-eluting peak of glycopeptides, which is considerably larger in the malignant cell than in the control (Fig. 2). We call this peak of material "group A glycopeptides" (Buck et al., 1974; Warren et al., 1972*b*).

Quantitation and Nature of Difference of A Glycopeptides in Control and Malignant Cells

The difference between control and transformed cells appears to be quantitative with re-

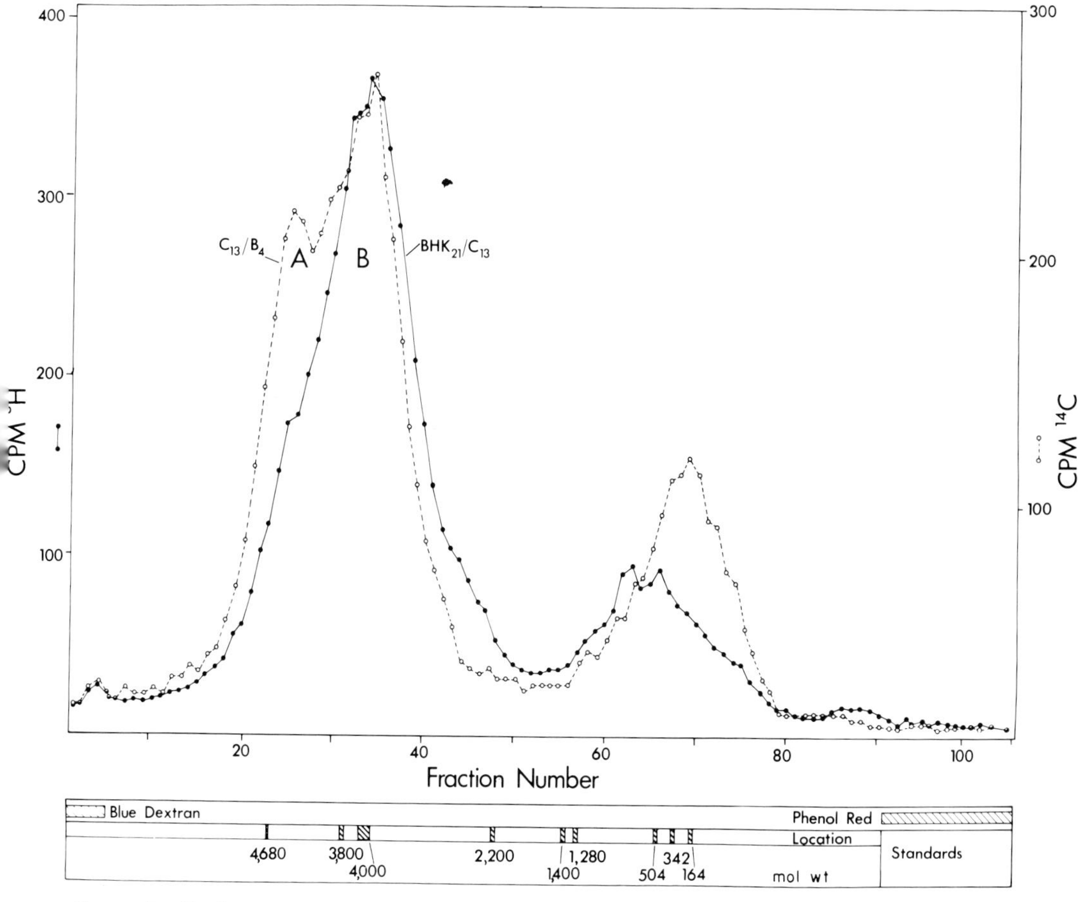

FIGURE 2 Co-chromatography on Sephadex G50 of pronase digested trypsinates from control cells, BHK_{21}/C_{13} and virus-transformed C_{13}/B_4. Control cells had been grown for 3 days in the presence of L-[^{3}H]fucose while transformed cells had been grown in the presence of L-[^{14}C]fucose. The column was 0.8 × 100 cm and was developed with buffer (0.1 M Tris acetate, pH 9.0, 0.1% SDS, 0.1% mercaptoethanol, and 0.01% EDTA). Samples of 0.7 ml were collected and counted for ^{3}H and ^{14}C. The point of elution of various carbohydrates of known molecular weight are shown below.

spect to peak A glycopeptides. Recent unpublished analytic studies have shown that there are two to four times the amount of peak A glycopeptides in transformed cells than in controls. When peak A and B type glycopeptides are further analyzed by paper electrophoresis and by chromatography on columns on DEAE Sephadex, no real qualitative differences are found between the glycopeptides of control and transformed cells despite the considerable resolving power of these methods. However, these experiments are not yet definitive. To date the differences are only quantitative.

By using radioactive fucose, balance studies have shown that the increase of peak A glycopeptides is accompanied by a corresponding decrease in the peak B glycopeptides as the total amount of glycopeptide groups derived from membrane glycoproteins of control and transformed cells is approximately the same. Although the B type glycopeptides are eluted later from the Sephadex G50 column and are smaller than the A glycopeptides, pulse-chase experiments suggest that they are not precursors of A glycopeptides. The B glycopeptides bind Concanavalin A (Con A) and behave completely differently as a group from the A glycopeptides on two-dimensional paper electrophoresis (D. Blithe and D. Wylic, unpublished data). The A glycopeptides do not bind Con A either before or after their sialic acid residues are removed.

Growth Dependence of Peak A Glycopeptide Appearance

The formation of peak A glycopeptides is clearly growth-dependent (Buck et al., 1971*b*; Muramatsu et al., 1973; Ceccarini, 1975; Ceccarini et al., 1975; Glick and Buck, 1973). They are formed in log phase of growth, in cells inhibited in M phase with vinblastine (Glick and Buck, 1973), or at the G_1/S phase in cells inhibited by thymidine (unpublished data). It is clear that for meaningful comparisons of control and transformed cells both types of cell must be labeled with isotope in full log phase of growth and with approximately the same doubling time, 21 h for BHK_{21}/C_{13} and C_{13}/B_4 cells. It is under these conditions that we have found peak A type glycopeptides to be increased two- to fourfold in transformed cells. When cells enter a plateau phase and cease to increase in number, peak A glycopeptides sharply decrease in amount (Buck et al., 1971*b*; Muramatsu et al., 1973; Ceccarini, 1975; Ceccarini et al., 1975).

Analytic Work

Peak A and B glycopeptides are complex mixtures, and a major effort has been made in this laboratory to separate and purify them before detailed chemical and structural analyses are undertaken.

Preliminary analyses show that the A glycopeptides as a whole contain about 16 sugars and the B glycopeptides about 12. These are, *N*-acetyl-D-glucosamine, D-mannose, D-galactose, L-fucose, and sialic acid. Only traces of *N*-acetyl-D-galactosamine are present. The A and B glycopeptides contain approximately three and one sialic acid residues, respectively (Warren et al., 1974*a*). The sugar-polypeptide union appears to be a β-glycosylamine linkage as it is not susceptible to cleavage with alkaline borohydride. Further, the antibiotic tunicamycin (Tkacz and Lampen, 1975) added to cell cultures (0.6 μg/ml) rapidly and *completely* inhibits the addition of sugars to proteins (unpublished data). It is interesting that whereas only 1 of 16 carbohydrate groups of glycophorin A, the major glycoprotein of the human red blood cell, is bound through a β-glycosylamine linkage (Tomita and Marchesi, 1975), our experience is that the vast bulk of the linkages in the glycoproteins of fibroblasts are of this type.

When the sialic acid of group A and B glycopeptides is removed chemically or enzymatically, they lose charge, become smaller, and elute later from columns of Sephadex G50 than do unmodified B glycopeptides. After desialylation, the A and B glycopeptides of control and transformed cells elute as a single superimposable peak (Warren et al., 1972*b*, 1973). The fact that the glycopeptides elute together may be fortuitous and clearly does not mean that they are of identical composition and structure.

Distribution of Group A-Bearing Glycoproteins within the Cell

Earlier work by Robbins and his co-workers (Meezan et al., 1969) investigating glycopeptide patterns of glycoproteins from control and transformed mouse cells with the double-label method suggested some differences. Buck et al. (1970, 1971*a*, *b*) then demonstrated reproducible differences when comparing glycopeptides from the cell surface (membranes and "trypsinates"). However, we later found that the increase in A glycopeptides is not confined to the cell surface but is also found on glycoproteins of the endoplasmic reticulum, Golgi membranes, lysosomes, inner and outer mitochondrial membranes (Buck et al., 1974), and nuclear membranes (Keshgegian and Glick, 1973; Buck et al., 1974). The change appears to be coordinate and is found in the membrane glycoproteins throughout the cell. From the elution patterns of glycopeptides derived from the *entire* cell labeled with radioactive L-fucose, it can be shown that at least 75% of all cellular membrane glycopeptides are of the A-B type.

Occurrence of Increased Peak A Glycopeptides in Malignant Cells

The enrichment of peak A glycopeptides in malignancy is widespread throughout nature (Table I). It has been found in five species of cells transformed by DNA and RNA oncogenic viruses, chemically and spontaneously transformed cells of the fibroblastic, epithelial, and lymphoid series (Beek et al., 1973), in primary cell lines and established lines, euploid and aneuploid cells, virus producers and nonproducers. Group A glycopeptides are enriched in cells transformed by temperature-sensitive virus (T5) growing at a permissive temperature where malignancy is expressed (Warren et al., 1972*a*). Further, Lai and Duesberg (1972) have shown that there is an increased amount of peak A glycopeptides in the glycoproteins of T5 Rous sarcoma virus, itself produced by transformed cells growing at a permissive temperature (36°C). They are not found when the cells are grown at the nonpermissive temperature of 41°C. Virus production is approximately equal at both temperatures (Martin, 1970).

The phenomenon occurs in solid tumors such as hepatomas (Smets et al., 1975) and mouse melanoma (Warren et al., 1975) in various sites of the body. Recently, we have found it in mammary carcinomas, a solid epithelial tumor. Dutch workers have found the enrichment in relevant cells of individuals with acute and chronic lymphatic and myelocytic leukemias and Burkitts lymphoma but not with infectious mononucleosis (Beek et al., 1975). The glycopeptides are not enriched in lymphocytes stimulated with mutagens (Beek et al., 1975).

Recently, Ceccarini compared the surface glycopeptides of the human diploid fibroblast W138 with those of its counterpart, transformed by SV40 (Ceccarini, 1975). He showed that both control and transformed cells in log phase contained type A carbohydrate in equal amounts; they disappeared in both cell types upon cessation of growth, and he concluded that type A glycopeptide formation is more closely related to the growth state than to malignancy. However, it has been recently shown that SV40-transformed W138 is not tumorigenic in nude mice (Stiles et al., 1975). It is of relevance that Glick, working with Rabinowitz and Sachs (Glick et al., 1973, 1974), has presented evidence that the extent of peak A glycopeptide production in chemically and

TABLE I
Occurrence of Glycopeptide Alterations in Malignant Cells

Species	Control cell	Malignant cell	Comments
Chicken	CEF*	CEF-SR	Virus (RNA)
	CEF-T5 (41°C)	CEF-T5 (36°)	T5 virus (RNA)
Mouse	3T3 (Balb C or Swiss)	SVT2 3T3-SV 3T3-Py 3T3-PySV	SV40, Py (DNA)
		Ka 31	Virus (RNA)
		3T3-RSV	Virus (RNA)
		3T3-f, 3T12-3	Spontaneous
	Various mouse tissues	Lymphosarcoma	Solid tumor
		Melanoma	" "
	Resting mammary gland	Mammary carcinoma	" "
	MB III (lymphoblast)	MBVIA	Lymphoid cells
Rat	Liver cells	N1S1-67 (Novikoff hepatoma)	Chemical
Hamster	BHK_{21}/C_{13}	C_{13}/SR_7, C_{13}/B_4, PyY	Virus (RNA and DNA)
		$C_{13}/SV40$, $C_{13}/SV40-M$	
		Chemically transformed lines	Chemical
Human	Lymphocytes	Acute, chronic lymphatic and myelocytic leukemias	
	PHA-stimulated lymphocytes	Burkitts lymphoma	
	Infectious mononucleosis		

* Chick embryo fibroblasts.

virally transformed cells is more closely related to in vivo tumorigenesis than to any in vitro criterion of malignancy.

Biosynthesis and Turnover of Carbohydrates of Peak A and B Glycopeptides

The rate of synthesis and degradation of peak A and B glycopeptides has been measured by observing the rate of incorporation of radioactive D-glucosamine and L-fucose into the glycopeptides and the rate of exit of isotope from prelabeled cells (unpublished data). It is evident that the rate curves are parallel. Enhancement of peak A glycopeptides in transformed cells is not the result of differential turnover. Studies of glycosidase levels, especially of neuraminidase, in control and transformed cells strongly suggest that differential levels of degradative enzymes in these cells could not account for increased peak A glycopeptides in transformed cells. Neuraminidase activity is so small it is very difficult to detect.

Glycosyl transferases have been studied. We have found a sialyl transferase that transfers *N*-acetylneuraminic acid (NAN) from its activated form, cytidine 5′-monophospho-*N*-acetylneuraminic acid (CMPNAN), to desialylated peak A and other glycopeptides, which is 3–10 times greater in activity in transformed cells than in controls (Warren et al., 1972*b*, 1973). At the same time in the same cells there are equal activities of other sialyl transferases which use desialylated mucin or fetuin as acceptors. Chick embryo fibroblasts (CEF) transformed with the temperature-sensitive mutant Rous sarcoma virus, T5 (Warren et al., 1973), manifest elevated peak A glycopeptide levels and sialyl transferase activity at a permissive temperature (36°C) where malignancy is expressed compared to control levels at 41°C (nonpermissive temperature). Peak A glycopeptide levels and sialyl transferase activity are elevated at both temperatures in CEF transformed by the wild-type virus.

We are completing work on a study which clearly shows that a requirement of the sialyl transferase elevated in malignancy is that L-fucose must be present on the peak A glycopeptide acceptor before it will accept sialic acid (NAN) from CMPNAN. It is probable that this sialyl transferase is not the only one with an L-fucose specificity.

Transformed and control cells also contain a fucosyl transferase capable of transferring L-fucose from GDP-fucose to defucosylated peak A and other glycopeptides. Activity is at least two and a half- to eightfold greater in the transformed cells (unpublished data). It would appear that the biosynthesis of the carbohydrate component of glycoproteins is complex and that alterations in the level of their biosynthesis is even more complex. Our work suggests that a whole battery of transferase activities may be altered.

Shifts in the Populations of Carbohydrate Groups on Glycoproteins

Do peak A glycopeptides increase in malignant cells because there are more glycoproteins formed that happen to be rich in peak A carbohydrate groups or do peak A carbohydrates substitute for peak B types on a specific polypeptide? Attempts to answer these questions are now being made, and this requires the isolation of homologous glycoproteins of control and transformed cells. Control and transformed cells metabolically labeled with ^{14}C and ^{3}H D-glucosamine or L-fucose have been extracted with lithium diiodosalicylate (LIS; Marchesi and Andrews, 1971). The LIS extract contains the bulk of the glycoproteins of the cell. By extracting the LIS extract with phenol, water-soluble and phenol-soluble fractions have been obtained. These consist of two distinct and separate populations of proteins and glycoproteins as shown by chromatography on SDS-hydroxyapatite columns (Warren et al., 1974*b*). The fractions have been analyzed on gels and have been resolved by preparative disk gel electrophoresis. Gel patterns as visualized by Coomassie blue staining and autoradiography by the use of materials labeled with radioactive amino acids or D-glucosamine reveal close homology of glycoproteins of control and malignant cells. Bands of similar molecular weight from control and transformed cells (BHK_{21}/C_{13} and C_{13}/B_4) labeled with ^{14}C or ^{3}H D-glucosamine have been cut out, mixed, and exhaustively digested with pronase, and the limit digest chromatographed on columns of Sephadex G50. In eight out of eight pairs of bands from different molecular weight regions of the gel of control and transformed cells, the double-label glycopeptide patterns have been shown to differ significantly (Fig. 3). In most instances there are more peak A glycopeptides in the bands from the transformed cells. Bands from the phenol-soluble fraction from control cells contain predominantly peak B glycopeptides, whereas type A, B, and C

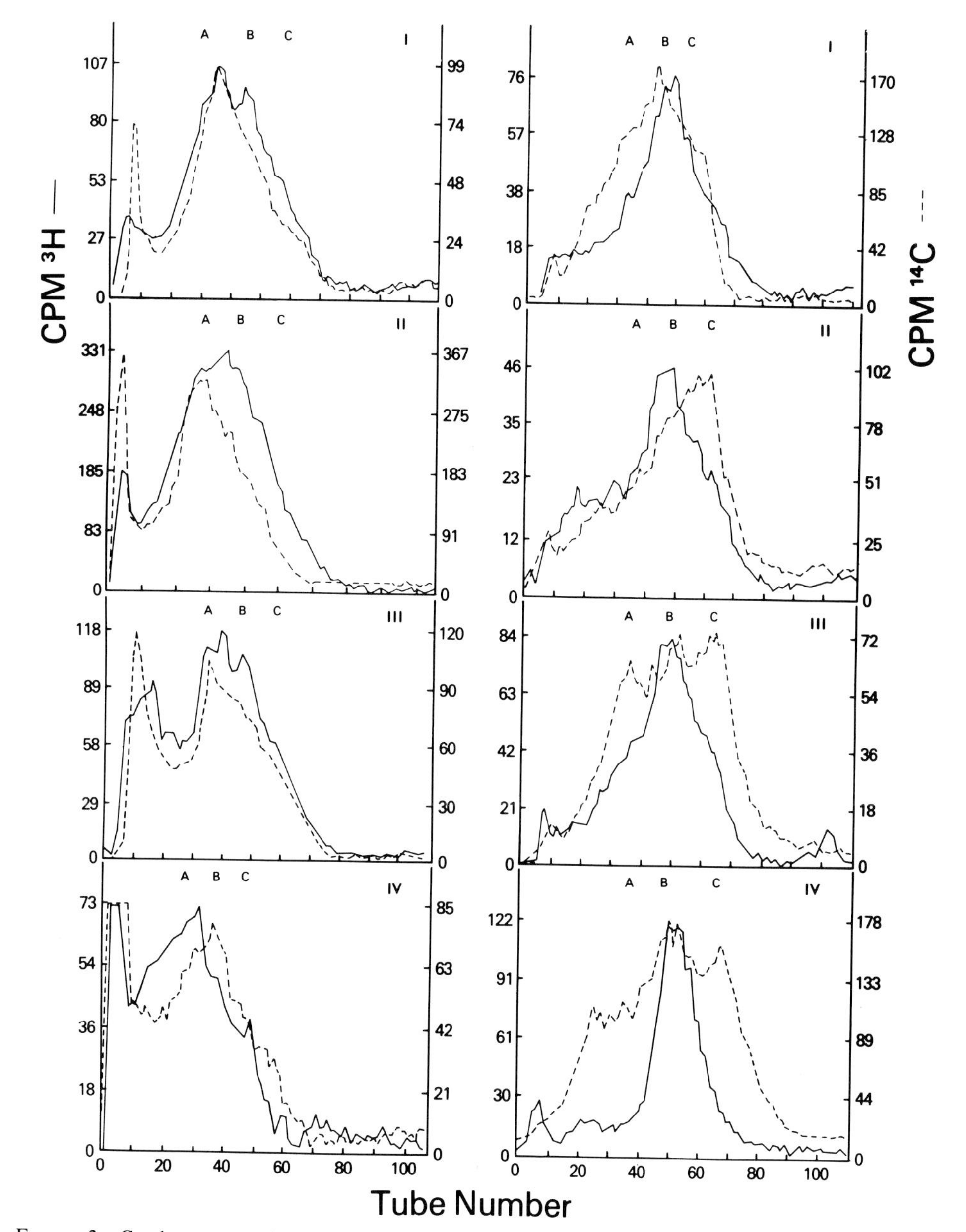

FIGURE 3 Co-chromatography on Sephadex G50 of pronase-digested glycoprotein bands cut from a preparative electrophoretic (slab) gel. Control (BHK_{21}/C_{13}; ——) and transformed (C_{13}/B_4; ---) cells had been grown in the presence of [^{3}H]- and [^{14}C]glucosamine, respectively. Cells were extracted with LIS and the extracts were electrophoresed. Bands from identical molecular weight regions were cut out, mixed, exhaustively digested with pronase, and chromatographed on a column of Sephadex G50. Bands I–IV on the left (mol wt 150,000–18,000) are from the LIS aqueous layer and bands I–IV on the right) (mol wt 150,000–18,000) are from the LIS phenol layer.

glycopeptides are derived from the bands of material from transformed cells. Type C glycopeptides are most likely core structure glycopeptides containing about six sugars, mostly D-mannose and *N*-acetyl-D-glucosamine, some D-galactose, and no sialic acid or L-fucose. These elute later, to the right of the B glycopeptides, in elution patterns of Sephadex G50 columns.

Assuming for the present that we are dealing with relatively homogeneous glycoproteins in the gel bands, these observations strongly suggest that the carbohydrate components of many (perhaps all!) membrane glycoproteins are significantly altered when the cell becomes malignant. These changes occur in every membrane system of the cell. It would appear that several of the membrane glycoproteins of control cells contain primarily B type glycopeptides. Upon transformation, larger glycopeptides (peak A) and smaller, incomplete glycopeptides (type C) are formed on given polypeptide chains. Formation of smaller incomplete sugar chains is reminiscent of Hakomori's thesis (1973) that transformed cells often fail to complete the chains of sugars of their glycolipid. The relative levels of homologous membrane glycoproteins in control and transformed cells is not known at the present time nor is it known whether the carbohydrate groups of soluble glycoproteins are similarly altered. Changes in the carbohydrates of lysosomal enzymes, some of which have been shown to be glycoproteins (Goldstone and Koenig, 1970), might alter their activities and their binding to lysosomal membranes (proteinases and proteinase activators at the cell surface?). Perhaps the altered binding of large external transformation-sensitive (LETS) protein (Hynes, 1976) to malignant cell surfaces is the result of changes in its carbohydrate components.

Conclusions

We are studying a change in the carbohydrate moiety of membrane glycoproteins of the malignant cell. The change, which appears to be a gross exaggeration of a normal state, takes place in a wide variety of tumors, in every membrane system of the cell. Our most recent work suggests that the carbohydrate units of most, if not all, membrane glycoproteins of malignant cells are altered. However, we have not looked at the carbohydrates of soluble glycoproteins where one could imagine that altered carbohydrate units would affect the binding to membranes. These extensive, structural changes could certainly be the physical basis for the cascade phenomenon and the pleotypic character of the malignant cell. However, we really do not know whether altered glycosylation patterns are related to changes observed in the malignant cell—enzyme function, immunogenicity adhesiveness, growth control, initiation of DNA synthesis, and others. Do the carbohydrate changes alter the conformation and arrangements of proteins in membranes? Do the extra sialic acid-rich peak A type of glycopeptides alter, block, or create various receptors on and in the cell?

We certainly cannot claim that carbohydrate changes are primary carcinogenic events, or even close to it. We would like to think that it is one of the important changes. Perhaps it is one of the stages in the emergence of the full-blown malignant cell. Perhaps some carcinogenic change, possibly in the membrane, causes altered glycosylation which in turn alters many other processes, some of which are integral to malignancy. These more remotely perturbed processes may reach back and affect the glycosylation mechanism causing a second shift in activity patterns. Obviously the situation is complex and it will take a long time to understand it.

REFERENCES

BEEK, W. P. VAN, L. A. SMETS, and P. EMMELOT. 1973. Increased sialic acid density in surface glycoproteins of transformed and malignant cells: a general phenomenon? *Cancer Res.* **33:**2913–2922.

BEEK, W. P. VAN, L. A. SMETS, and P. EMMELOT. 1975. Changed surface glycoprotein as a marker of malignancy in human leukemic cells. *Nature (Lond.).* **253:**457–460.

BERGELSON, L. D. 1972. Tumor lipids. *Prog. Chem. Fats Other Lipids.* **13:**1–59.

BRADY, R. O., and P. H. FISHMAN. 1974. Biosynthesis of glycolipids in virus-transformed cells. *Biochim. Biophys. Acta.* **355:**121–148.

BUCK, C. A., J. P. FUHRER, G. SOSLAU, M. M. K. NASS, and L. WARREN. 1974. Membrane glycopeptides from subcellular fractions of control and virus-transformed cells. *J. Biol. Chem.* **249:**1541–1550.

BUCK, C. A., M. C. GLICK, and L. WARREN. 1970. A comparative study of glycoproteins from the surface of control and Rous sarcoma virus-transformed hamster cells. *Biochemistry.* **9:**4567–4576.

BUCK, C. A., M. C. GLICK, and L. WARREN. 1971*a*. Glycopeptides from the surface of control and virus-transformed cells. *Science (Wash. D.C.).* **172:**169–171.

BUCK, C. A., M. C. GLICK, and L. WARREN. 1971*b*. Effect of growth on the glycoproteins from the surface of control and Rous sarcoma virus-transformed hamster cells. *Biochemistry.* **10:**2176–2180.

BURGER, M. M. 1971. Cell surfaces in neoplastic transformation. *Curr. Top. Cell Regul.* **3:**135–193.

CECCARINI, C. 1975. Appearance of smaller mannosyl-glycopeptides on the surface of a human cell transformed by simian virus 40. *Proc. Natl. Acad. Sci. U. S. A.* **72:**2687–2690.

Ceccarini, C., T. Muramatsu, J. Tsang, and P. H. Atkinson. 1975. Growth-dependent alterations in oligomannosyl cores of glycopeptides. *Proc. Natl. Acad. Sci. U. S. A.* **72:**3139–3143.

Emmelot, P. 1973. Biochemical properties of normal and neoplastic cell surfaces; a review. *Eur. J. Cancer.* **9:**319–333.

Gahmberg, C. G., and S. Hakomori. 1974. Organization of glycolipids and proteins in surface membranes: dependency of cell cycle and on transformation. *Biochem. Biophys. Res. Commun.* **59:**283–291.

Glick, M. C., and C. A. Buck. 1973. Glycoproteins from the surface of metaphase cells. *Biochemistry.* **12:**85–90.

Glick, M. C., Z. Rabinowitz, and L. Sachs. 1973. Surface membrane glycopeptides correlated with tumorigenesis. *Biochemistry.* **12:**4864–4869.

Glick, M. C., Z. Rabinowitz, and L. Sachs. 1974. Surface membrane glycopeptides which coincide with virus transformation and tumorigenesis. *J. Virol.* **13:**967–974.

Goldstone, A., and I. Koenig. 1970. Lysosomal hydrolases as glycoproteins. *Life Sci.* **9:**1341–1350.

Hakomori, S. 1973. Glycolipids of tumor cell membrane. *Adv. Cancer Res.* **18:**265–315.

Hynes, R. O. 1976. Cell surface proteins and malignant transformation. *Biochim. Biophys. Acta.* **3:**73–107.

Keshgegian, A. A., and M. C. Glick. 1973. Glycoproteins associated with nuclei of cells before and after transformation by a ribonucleic acid virus. *Biochemistry.* **12:**1221–1226.

Knudson, A. G., Jr. 1973. Mutation and human cancer. *Adv. Cancer Res.* **17:**317–352.

Laemmli, U. K. 1970. Cleavage of structural proteins during the assembly of the head of bacteriophage T4. *Nature (Lond.).* **227:**680–685.

Lai, M. M. C., and P. H. Duesberg. 1972. Differences between the envelope glycoproteins and glycopeptides of avian tumor viruses released from transformed and from nontransformed cells. *Virology.* **50:**359–372.

Marchesi, V. T., and E. P. Andrews. 1971. Glycoprotein isolation from cell membranes with lithium diiodosalicylate. *Science (Wash. D.C.).* **174:**1247–1248.

Martin, G. S. 1970. Rous sarcoma virus: a function required for the maintenance of the transformed state. *Nature (Lond.).* **227:**1021–1023.

Meezan, E., H. C. Wu, P. H. Black, and P. W. Robbins. 1969. Comparative studies on the carbohydrate-containing membrane components of normal and virus-transformed mouse fibroblasts. II. Separation of glycoproteins and glycopeptides by Sephadex chromatography. *Biochemistry.* **8:**2518–2524.

Muramatsu, T., P. H. Atkinson, S. G. Nathenson, and C. Ceccarini. 1973. Cell surface glycopeptides: growth-dependent changes in the carbohydrate-peptide linkage region. *J. Mol. Biol.* **80:**781–799.

Pitot, H. C. 1974. Neoplasia: a somatic mutation or a heritable change in cytoplasmic membranes. *J. Natl. Cancer Inst.* **53:**905–911.

Robbins, P. W., and I. Macpherson. 1971. Glycolipid synthesis in normal and transformed animal cells. *Proc. R. Soc. Lond. B. Biol. Sci.* **177:**49–58.

Sakiyama, H., and B. W. Burge. 1972. Comparative studies of the carbohydrate-containing components of 3T3 and simian virus 40 transformed 3T3 mouse fibroblasts. *Biochemistry.* **11:**1366–1377.

Sakiyama, H., and P. W. Robbins. 1973. Glycolipid synthesis and tumorigenicity of clones isolated from the NIL 2 line of hamster embryo fibroblasts. *Fed. Proc.* **32:**86–90.

Smets, L. A., W. P. van Beek, J. G. Collard, H. Temmink, B. van Gils, and P. Emmelot. 1975. Comparative evaluation of plasma membrane alterations associated with neoplastia. *In* Cellular Membranes and Tumor Cell Behaviour. Williams & Wilkins Company, Baltimore. 582.

Stiles, C. D., W. Desmond, Jr., G. Sato, and M. H. Saier, Jr. 1975. Failure of human cells transformed by simian virus 40 to form tumors in athymic nude mice. *Proc. Natl. Acad. Sci. U. S. A.* **72:**4971–4975.

Tkacz, J. S., and J. O. Lampen. 1975. Tunicamycin inhibition of polyisoprenyl *N*-acetylglucosaminyl pyrophosphate formation in calf-liver microsomes. *Biochem. Biophys. Res. Commun.* **65:**248–257.

Tomita, M., and V. T. Marchesi. 1975. Amino acid sequence and oligosaccharide attachment sites of human erythrocyte glycophorin. *Proc. Natl. Acad. Sci. U. S. A.* **72:**2964–2968.

Tooze, J. 1973. The Molecular Biology of Tumor Viruses. Cold Spring Harbor Laboratory, Cold Spring Harbor, New York. 743.

Wallach, D. F. H. 1968. Cellular membranes and tumor behaviour: a new hypothesis. *Proc. Natl. Acad. Sci. U. S. A.* **61:**868–874.

Warren, L., D. Critchley, and I. Macpherson. 1972*a*. Surface glycoproteins and glycolipids of chicken embryo cells transformed by a temperature-sensitive mutant of Rous sarcoma virus. *Nature (Lond.).* **235:**275–278.

Warren, L., J. P. Fuhrer, and C. A. Buck. 1972*b*. Surface glycoproteins of normal and transformed cells: a difference determined by sialic acid and a growth-dependent sialyl transferase. *Proc. Natl. Acad. Sci. U. S. A.* **69:**1838–1842.

Warren, L., J. P. Fuhrer, and C. A. Buck. 1973. Surface glycoproteins of cells before and after transformation by oncogenic viruses. *Fed. Proc.* **32:**80–85.

Warren, L., J. P. Fuhrer, C. A. Buck, and E. F. Walborg, Jr. 1974*a*. Membrane glycoproteins in normal and virus-transformed cells. *In* Membrane transformations in neoplasia. *Miami Winter Symp.* **8:**1–21.

Warren, L., J. P. Fuhrer, G. P. Tuszynski, and C.

A. Buck. 1974*b*. Cell-surface glycoproteins in normal and transformed cells. *Biochem. Soc. Symp.* **40:**147–157.

Warren, L., I. Zeidman, and C. A. Buck. 1975. The surface glycoproteins of a mouse melanoma growing in culture and as a solid tumor *in vivo*. *Cancer Res.* **35:**2186–2190.

Weiss, L. 1967. The Cell Periphery, Metastasis and Other Contact Phenomena. North-Holland Publishing Co., Amsterdam. 388.

Yamada, K. M., S. S. Yamada, and I. Pastan. 1976. Cell surface protein partially restores morphology, adhesiveness and contact inhibition of movement to transformed fibroblasts. *Proc. Natl. Acad. Sci. U. S. A.* **73:**1217–1222.

Cells and Hormone Action

BIOCHEMICAL BASIS OF THE MORPHOLOGIC PHENOTYPE OF TRANSFORMED CELLS

IRA PASTAN, MARK WILLINGHAM, KENNETH M. YAMADA, JACQUES POUYSSEGUR, PETER DAVIES, and IRWIN KLEIN

Cancer cells have defective growth control. However, in addition to their abnormal growth, malignant cells often have other properties that distinguish them from normal cells. These properties may enable cancer cells to spread and to invade foreign tissues. One of the principal goals of our laboratory recently has been to use tissue culture, biochemistry, and genetics to understand the biochemical basis of this abnormal behavior.

Our work in this area began when G. Johnson and I. Pastan observed that a variety of cancer cells growing in tissue culture were altered in morphology after treatment with Bt_2cAMP to a shape more similar to that of normal cells (Johnson et al., 1971). Normal fibroblastic cells propagated in tissue culture are elongated and flattened. Transformed fibroblasts are often more compact, with shortened cell processes, and in some cases are quite round (Fig. 1 A). This change in shape is usually accompanied by decreased adhesion to substratum, increased agglutinability by plant lectins, and the presence of numerous microvilli or blebs on the cell surface. These four alterations, (1) rounded shape, (2) surface microvilli, (3) high agglutinability by plant lectins, and (4) low adhesion, we shall refer to as "morphologic transformation" of fibroblasts. Our aim in this paper is to distinguish these features from loss of growth control. We further suggest that decreased cell-to-substratum adhesion is primarily responsible for the three other responses. Lowered adhesiveness may also be partially responsible for the disorganized, overlapping arrangement of cells characteristic of cultures of transformed fibroblasts.

Treatment of transformed cells with Bt_2cAMP promotes cell process extension, resulting in elongated, flattened cells that resemble normal cells (Fig. 1 B; Johnson et al., 1973; Willingham and Pastan, 1975*a*). Associated with this change in shape are increased adhesiveness to substratum (Johnson and Pastan, 1972) and decreased motility (Johnson et al., 1972). Further, the susceptibility of the transformed cells to agglutination by plant lectins such as concanavalin A is diminished (Sheppard, 1971; Willingham and Pastan, 1974), and microvilli decrease in number (Willingham and Pastan, 1975*b*). These actions of cyclic AMP are not prevented by treating the cells with actinomycin D or cycloheximide and, therefore, apparently do not require new RNA or protein synthesis (reviewed in Pastan et al., 1975). These actions of cyclic AMP may be mediated by enhanced phosphorylation of certain proteins (see below).

To establish the possible physiological relevance of these observations, we attempted to isolate mutants that were defective in synthesizing or responding to cyclic AMP. These efforts have been partially successful. More important, they have forced us to consider and begin to investigate the cellular components directly involved in cell shape and movement. These structural components include extracellular factors, such as adhesion proteins, and intracellular components, such as tubulin, actin, and myosin. Morphologic transformation probably results from alterations in those structural components, and/or changes in substances such as cyclic AMP that regulate the activity of these molecules.

External Components: Adhesion and Attachment Factors

Many of us have had the misfortune of finding our cell cultures growing poorly because we had

IRA PASTAN and CO-WORKERS Laboratory of Molecular Biology, National Cancer Institute, N.I.H., Bethesda, Maryland

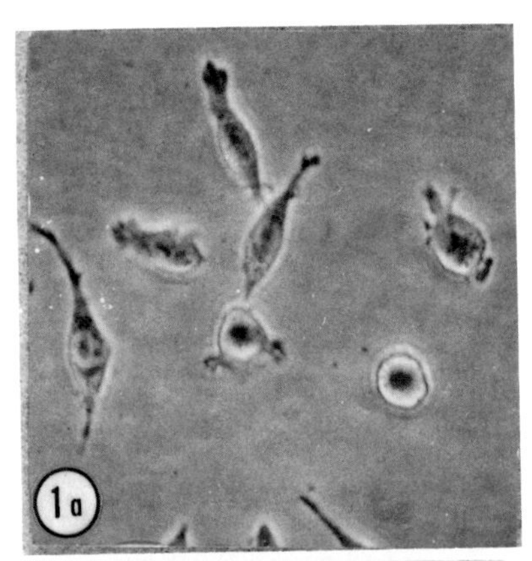

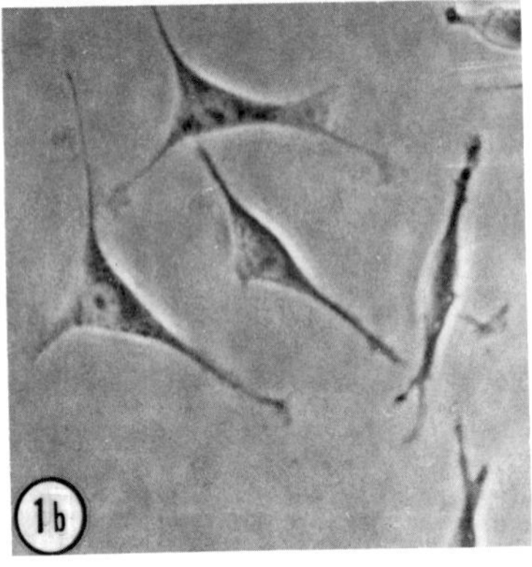

FIGURE 1 L929 cells growing on Falcon plastic. (A) Control; (B) 1 mM Bt_2cAMP for 24 h. Phase contrast × 260.

accidently used bacterial-grade Petri dishes. Fibroblastic cells adhere poorly (or not at all) to ordinary plastic dishes and appear quite round. For cells to spread out, the surface must have a negative charge (as is present on glass or "falconized" plastic), or must consist of an organic substance, such as collagen, to which the cells can adhere. Under these conditions, proteins or glycoproteins present on the cell surface interact with the substratum, apparently allowing the cells to spread out and to move. Treatment of cells with proteases cleaves such proteins, and the cells round up and detach from the substratum.

A number of workers have attempted to identify and isolate, either from cell surfaces or from serum, factors necessary for cell attachment and adhesion (Moscona, 1973). K. Yamada has identified and then purified a protein from the surface of chick embryo cells that seems to play an important role in cell adhesion. This glycoprotein is known as CSP (cell surface protein: Yamada and Weston, 1974), LETS (large external transformation-sensitive) protein (Hynes and Bye, 1974), and by other names (reviewed in Hynes, 1976; and Yamada and Pastan, 1976). Similar proteins are present in large amounts on the surfaces of early passage human, mouse, and rat fibroblasts (unpublished data). Transformed cells, such as chick embryo cells transformed by the Schmidt-Ruppin strain of Rous sarcoma virus, usually have a diminished content of CSP or LETS protein. In addition, the adhesiveness of these cells is greatly decreased. When we added CSP back to other transformed fibroblasts, it restored adhesiveness and cell shape toward normal (Yamada et al., 1976). In addition to flattening and elongating these cells, CSP treatment decreased the number of microvilli on the cell surface. In several cases, CSP also restored the ability of fibroblastic cells to align as they grow, inhibiting the marked overlapping of cells often seen after transformation. In no case has CSP inhibited the growth of transformed cells.

In comparing the amount of CSP present on different cell lines, we were surprised to find that some continuous nontransformed cell lines, such as 3T3 and NRK (normal rat kidney fibroblast), have quite low levels of CSP. In contrast, CSP constitutes 3% of the total cell protein and up to 50% of the plasma membrane-associated protein of early passage chick, mouse, and human embryo cells. Despite the low content of CSP in 3T3 cells, they are very tightly attached to the substratum. This indicates that other mechanisms of adhesion exist. With the exception of CSP, the principle surface proteins of 3T3 cells and primary mouse embryo cells are similar. Presumably, one or more of these other proteins are also involved in the adhesive process.

One clue as to the possible identity of such proteins has come from studies of mutant cells. J. Pouyssegur and his co-workers (1977) have isolated a mutant of Balb 3T3 cells (AD6) that is defective in adhesion to substratum. When the external protein components of AD6 were analyzed by surface iodination, a number of surface proteins were found to be diminished (Pouyssegur et al., 1977). Subsequently, we have found that AD6 is defective in its ability to accumulate *N*-acetylglucosamine-6-phosphate and, therefore, makes incompletely glycosylated glycoproteins (Pouyssegur and Pastan, 1977). Apparently these defective glycoproteins are not situated in the membrane in a position in which they can be iodinated from the outside. Whether their inability to be iodinated is because they do not penetrate through the membrane to the outside or is the result of other steric factors needs to be clarified. When AD6 is fed *N*-acetylglucosamine, the intermediate just distal to the site of the metabolic block, its morphology returns to normal. Further, the microvilli present on these cells decrease in number, agglutinability by concanavalin A is diminished, and adhesion to substratum increases to the level of wild-type cells. Although AD6 has the morphologic phenotype of transformed cells, it has normal growth control and is not tumorigenic (Pouyssegur et al., 1977).

Another approach to the isolation of mutants defective in cell surface glycoprotein synthesis is to use a selection procedure in which cells with normal glycoproteins are killed by various toxic lectins that bind to these glycoproteins. Gottlieb and co-workers (1975) have isolated mutants of CHO cells. These mutants have a block in glycoprotein synthesis at a much later step than does AD6; they are defective in the enzyme *N*-acetylglucosaminyltransferase. This defect also leads to a decrease in adhesion to substratum. Whether the decrease in adhesion in both types of mutants is on account of a general change in membrane structure, a general decrease in cell surface carbohydrates, or to a loss of one or more specific adhesive molecules is not yet clear. The isolation of additional adhesive-defective mutants may shed light on such questions.

Internal Components: Tubulin, Actin, and Myosin

It seems clear that possessing the external components necessary for attachment to a substratum is essential, but not by itself sufficient, for cells to spread out, assume a normal fibroblastic shape, and move about. For example, some cells will not spread normally in the presence of vinblastine (Rabinovitch and DeStefano, 1973). Conversely, colchicine treatment of fibroblastic cells results in a more rounded configuration and inhibits directional motility (Vasilieu et al., 1970).

At least two major groups of intracellular proteins participate in the maintenance of cell shape and in cell movement. One is tubulin and its associated proteins. The other group contains myosin, actin, and their associated proteins. These proteins are thought to form the cellular cytoskeleton. After transformation, fibroblastic cells reportedly contain decreased numbers of intact microtubules (Edelman and Yahara, 1976). Although the total amount of cellular actin and myosin are usually not markedly decreased after transformation, formation of microfilament bundles is often inhibited (Edelman and Yahara, 1976; Pollack et al., 1975). The decrease in bundles could be a result of decreased membrane-associated actin (Wickus et al., 1975) and myosin (Shizuta et al., 1976).

As mentioned above, treatment with plant alkaloids which disrupt microtubules can result in cell rounding. Conversely, treatment of cells with Bt_2cAMP increases the number of microtubules, which fill elongated cell processes (Porter et al., 1974; Willingham and Pastan, 1975*a*). Agents that specifically interfere with the actin-myosin system are not available. One class of compounds apparently affecting this system are the cytochalasins. These agents disrupt certain microfilament systems that may contain actin, halt cell movements, and cause bizarre alterations in cell shape (Wessells et al., 1971). Because the cytochalasins also strongly inhibit glucose transport, interpretation of experiments with these agents must be made with caution, although work with cell-free contractile systems should clarify this problem (Pollard and Weihing, 1974).

Regulators: Cyclic AMP

Effects of cyclic AMP on cell shape have already been discussed. If cyclic AMP has an important role in shape regulation, then conditions that lower cyclic AMP levels should cause flat cells to round up and manifest the other features associated with "transformed morphology": low adhesion, high agglutinability, and the presence of surface microvilli. M. Willingham and his colleagues (1973) have isolated a mutant of Swiss 3T3 cells that fulfills this prediction. The mutant cells (3T3 $cAMP^{tcs}$) have the typical flat morphology of normal 3T3 cells when maintained at a constant temperature of 39°C. However, when the temperature is lowered, the cells undergo the following sequence of changes: 0–2 min, cyclic AMP levels fall by 50% from 20 to 10 pmol/mg protein; at 2–5 min, adhesion decreases; at 5–10 min, cell processes retract, the cells round up, microvilli form, and agglutinability is increased (Willingham and Pastan, 1974). The changes in shape, adhesion, and agglutinability are all prevented by the previous addition of Bt_2cAMP.

For cyclic AMP to alter cell shape, the cells must be attached to a substratum and be actively extending and retracting their processes. This is true of both the mutant cells previously discussed and of transformed cells. From analysis of time-lapse movies of L929 cells, it appears that one principle effect of cyclic AMP is to inhibit the retraction of cell processes (unpublished data). Continued process extension with poor retraction leads to striking changes in cell shape.

The ultimate shape a cell assumes is related to the initial shape of its processes. L929 cells, for example, possess narrow processes. Extension of these processes leads to long, narrow cells. In contrast, the processes of 3T3 cell are broad and extension of these leads to large, flat cells (Willingham and Pastan, 1975*a*).

Inhibition of process retraction could be the result of a *direct* effect of cyclic AMP on the microfilamentous system (here we use the microfilamentous system as a synonym for actin, myosin, and related proteins) or on the microtubular system. An obvious way for cyclic AMP to act would be to stimulate phosphorylation of one or more of these proteins. Phosphorylation of proteins associated with both these systems has been documented in muscle cells (Rubin and Rosen, 1975) and in the nervous system (Sloboda et al., 1975), respectively.

Inhibition of process retraction could also be the result of an increase in the strength with which the processes are attached to the substratum. There is ample evidence that cyclic AMP increases the overall adhesion of cells to substratum (Johnson and Pastan, 1972). It is not clear whether cyclic AMP increases adhesion by modulating the function of internal or external components. It is also not known whether cyclic AMP is selective in its effects on adhesion points situated at different sites under the cell.

Phosphorylation

To investigate the possible mechanism by which cyclic AMP modulates the contractile and adhesive systems of fibroblastic cells, we have begun a study of the phosphoproteins of cultured cells. The approach we have used is quite simple. We grew cells for several hours in the presence of ^{32}Pi and then determined which proteins were labeled by electrophoresis in 5% polyacrylamide slab gels after denaturing the proteins with sodium dodecyl sulfate (SDS) and reducing them with DTT. When this was done, a large number of phosphorylated bands were evident.

We have also investigated phosphorylation in cell-free extracts in order to ascertain more easily which proteins might have their phosphorylation controlled by cyclic AMP. In one set of experiments, whole homogenates were incubated with [γ-^{32}P]ATP for short times in the presence and absence of cyclic AMP. Then SDS gels were performed and the gels subjected to autoradiography to locate the labeled proteins. Figure 2 shows the result of this experiment. A large number of proteins are phosphorylated.

Bands 2 and 3 have the same mobility as microtubule-associated proteins. The phosphorylation of these proteins has been studied by Sloboda and co-workers (1975) in brain and appears to be affected by cyclic AMP in vitro. Band 4 has the same mobility as a protein referred to either as actin binding protein (Hartwig and Stossel, 1975) or filamin (Shizuta et al., 1977; Wang et al., 1975). P7 has the same mobility as phosphorylase a.

In addition to these high molecular weight proteins, a number of other low molecular weight proteins are phosphorylated. However, actin, tubulin, and CSP are not phosphorylated to a significant degree.

In whole homogenates, cyclic AMP increases the rate of phosphorylation of at least four proteins, labeled P2, P3, P4, and P7. P2 and P3 have the same mobility as microtubule-associated proteins; P4 migrates with filamin. By using the method outlined by Wang et al. (1975), we have purified filamin from chicken gizzard (Shizuta et al., 1976). Chicken gizzard filamin is a substrate for cyclic AMP-dependent protein kinase (Davies et al., 1977). The physiological relevance of this phosphorylation is currently under investigation.

It is also possible to investigate phosphorylation in various subcellular fractions. Figure 2 shows the phosphorylation pattern of a partially purified plasma membrane preparation and the effect of cyclic AMP. There is obviously sufficient protein kinase present in this membrane preparation to catalyze efficient phosphorylation. Cyclic AMP enhances the phosphorylation of at least 10 bands. Whether these are truly integral plasma membrane proteins, proteins bound to integral proteins, or proteins present as a result of contamination needs to be clarified.

Summary

The "morphologic phenotype" of transformed fibroblasts includes decreased adhesion to substratum, rounded cell shape, increased microvilli or

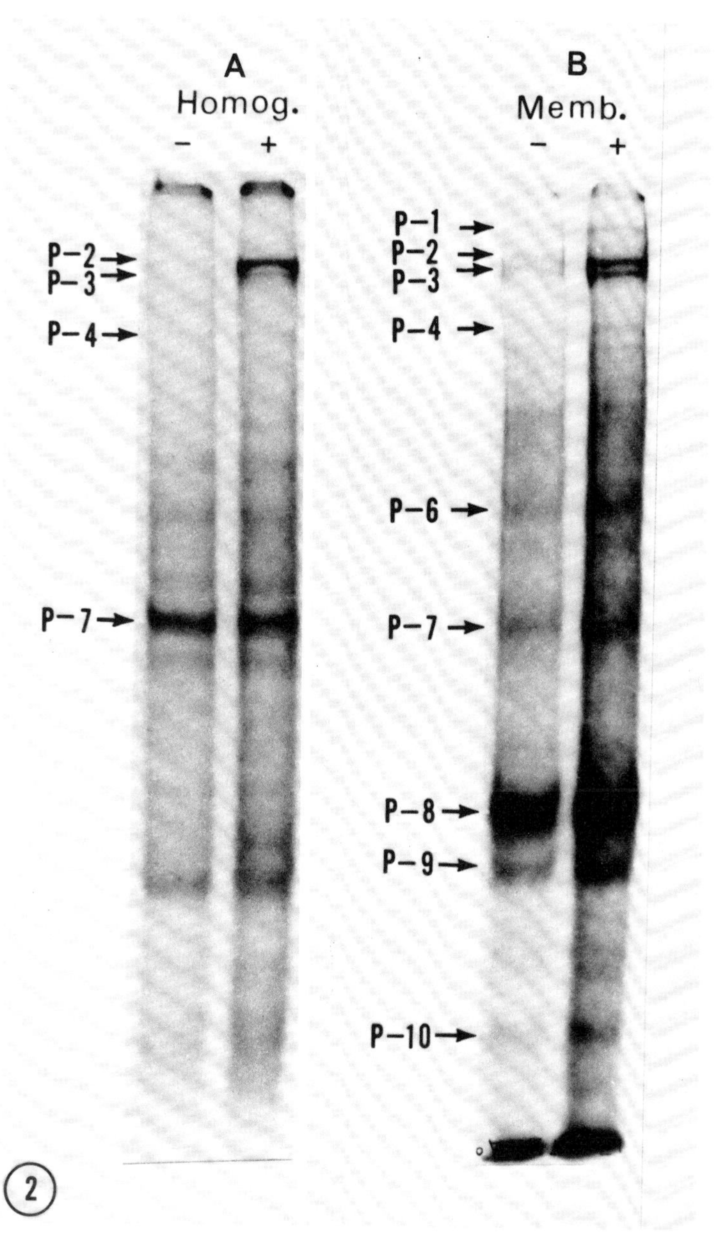

FIGURE 2 Autoradiograph of gel electrophoretogram of (A) homogenate and (B) plasma membrane preparation of normal rat kidney fibroblasts. Fractions were incubated with [γ-^{32}P]ATP in the presence (+) and absence (−) of 10^{-6} M cyclic AMP. Samples were then dissolved in SDS, reduced with DTT and electrophoresed on 5% polyacrylamide slab gels. Shown is the autoradiograph after 48 h; arrows indicate phosphoproteins whose phosphorylation is increased by cyclic AMP. (From Davies et al., 1977).

blebs, and increased agglutinability by lectins. Investigations on cells treated with cyclic AMP, on mutant cells with altered adhesion or cyclic AMP metabolism, and on a major adhesive protein (CSP) all indicate that the morphologic phenotype can be separated from growth control (Table I). We suggest that the decreased cell-to-substratum adhesiveness after transformation could account for the other three morphologic alterations as follows: (1) rounded shape would result from the

Table I

Separation of Morphologic Phenotype of Transformation from Growth Pattern

Condition or cell type	Morphologic phenotype: Adhesion	Morphology	Microvilli	ConA agglutinability	Growth
cAMP treatment	high	flat	low	low	*
CSP treatment	high	flat	low	not tested	uncontrolled
AD6	low	round	high	high	controlled
3T3 $cAMP^{tcs}$ at lowered temperature	low	round	high	high	untestable

* Growth slowed, but saturation density not affected.

inability of cells to attach firmly and spread on the substratum; (2) microvilli would form as cells become rounder to increase the total surface area of cells, and act as a reservoir for plasma membrane; (3) the increased number or size of microvilli would make the cells more agglutinable by plant lectins.

The basis of diminished adhesion varies depending on the cell type studied, and might result from the following known alterations: decreased quantities or arrangement of CSP and other adhesive proteins; altered organization of tubulin or microfilamentous systems; or decreased cyclic AMP levels resulting in altered phosphorylation of proteins that participate in the regulation of cell shape.

Elucidation of the causes of the morphologic phenotype of transformation could provide insight into the mechanisms by which cancer cells invade locally or metastasize.

REFERENCES

Davies, P., Y. Shizuta, K. Olden, M. Gallo, and I. Pastan. 1977. Phosphorylation of filamin and other proteins in cultured fibroblasts. *Biochem. Biophys. Res. Commun.* **74:**300–307.

Edelman, G. M., and I. Yahara. 1976. Temperature-sensitive changes in surface modulating assemblies of fibroblasts transformed by mutants of Rous sarcoma virus. *Proc. Natl. Acad. Sci. U. S. A.* **73:**2047–2051.

Gottlieb, C., J. Baenziger, and S. Kornfeld. 1975. Deficient uridine diphosphate-*N*-acetylglucosamine: glycoprotein *N*-acetylglucosaminyltransferase activity in a clone of Chinese hamster ovary cells with altered surface glycoproteins. *J. Biol. Chem.* **250:**3303–3309.

Hartwig, J. H., and T. P. Stossel. 1975. Isolation and properties of actin, myosin, and a new actin-binding protein in rabbit alveolar macrophages. *J. Biol. Chem.* **250:**5696–5705.

Hynes, R. O. 1976. Cell surface proteins and malignant transformation. *Biochim. Biophys. Acta.* **458:**73–107.

Hynes, R. O., and J. M. Bye. 1974. Density and cell cycle dependence of cell surface proteins in hamster fibroblasts. *Cell.* **3:**113–120.

Johnson, G. S., R. M. Friedman, and I. Pastan. 1971. Restoration of several morphological characteristics in sarcoma cells treated with adenosine 3′,5′-cyclic monophosphate and its derivatives. *Proc. Natl. Acad. Sci. U. S. A.* **68:**425–429.

Johnson, G. S., W. D. Morgan, and I. Pastan. 1972. Regulation of cell motility by cyclic AMP. *Nat. New Biol.* **235:**54–56.

Johnson, G. S., and I. Pastan. 1972. Cyclic AMP increases the adhesion of fibroblasts to substratum. *Nat. New Biol.* **236:**247–249.

Moscona, A. A. 1973. Cell Aggregation. *In* Cell Biology in Medicine. E. E. Bittar, editor. Wiley Interscience, New York. 571–591.

Pastan, I. H., G. S. Johnson, and W. B. Anderson. 1975. Role of cyclic nucleotides in growth control. *Annu. Rev. Biochem.* **44:**491–522.

Pollack, R., M. Osborn, and K. Weber. 1975. Patterns of organization of actin and myosin in normal and transformed cultured cells. *Proc. Natl. Acad. Sci. U. S. A.* **72:**994–998.

Pollard, T. D., and R. R. Weihing. 1974. Actin and myosin and cell movement. *CRC Crit. Rev. Biochem.* **2:**1–65.

Porter, K. R., T. T. Puck, A. W. Hsie, and D. Kelley. 1974. An electron microscopic study of the effects of Bt_2cAMP on CHO cells. *Cell.* **2:**145–162.

Pouyssegur, J., M. Willingham, and I. Pastan. 1977. Role of cell surface carbohydrates and proteins in cell behavior: studies on the biochemical reversion of an *N*-acetyl glucosamine deficient fibroblast mutant. *Proc. Natl. Acad. Sci. U. S. A.* **74:**243–247.

Pouyssegur, J., and I. Pastan. 1977. Mutants of mouse fibroblasts altered in the synthesis of cell surface glycoproteins: evidence for a block in the acetylation of glucosamine-6-phosphate. *J. Biol. Chem.* In press.

Rabinovitch, M., and M. J. DeStefano. 1973. Manganese stimulates adhesion and spreading of mouse sarcoma I ascites cells. *J. Cell Biol.* **59:**165–176.

Rubin, C. S., and O. M. Rosen. 1975. Protein phosphorylation. *Annu. Rev. Biochem.* **44:**831–887.

Sheppard, J. R. 1971. Restoration of contact-inhibited growth to transformed cells by dibutyryl adenosine

3′,5′-monophosphate. *Proc. Natl. Acad. Sci. U. S. A.* **68:**1316.

SHIZUTA, Y., P. DAVIES, K. OLDEN, and I. PASTAN. 1976. Diminished content of plasma membrane-associated myosin in transformed fibroblasts. *Nature (Lond.).* **261:**414–415.

SHIZUTA, Y., H. SHIZUTA, M. GALLO, P. DAVIES, I. PASTAN, and M. S. LEWIS. 1976. Purification and properties of filamin, an actin binding protein from chicken gizzard. *J. Biol. Chem.* **251:**6562–6567.

SLOBODA, R. D., S. A. RUDOLPH, J. L. ROSENBAUM, and P. GREENGARD. 1975. Cyclic AMP-dependent endogenous phosphorylation of a microtubule-associated protein. *Proc. Natl. Acad. Sci. U. S. A.* **72:**177–181.

VASILIEV, J. M., I. M. GELFAND, L. V. DOMNINA, O. Y. IVANOVA, S. G. KOMM, and L. V. OLSHEVSKAJA. 1970. Effect of colcemid on the locomotory behaviour of fibroblasts. *J. Embryol. Exp. Morphol.* **24:**625–640.

WANG, K., J. F. ASH, and S. J. SINGER. 1975. Filamin, a new high-molecular-weight protein found in smooth muscle and non-muscle cells. *Proc. Natl. Acad. Sci. U. S. A.* **72:**4483–4486.

WESSELLS, N. K., B. S. SPOONER, J. F. ASH, M. O. BRADLEY, M. A. LUDVENA, E. L. TAYLOR, J. T. WRENN, and K. M. YAMADA. 1971. Microfilaments in cellular and developmental processes. *Science (Wash. D. C.).* **171:**135–143.

WICKUS, G., E. GRUENSTEIN, P. W. ROBBINS, and A. RICH. 1975. Decrease in membrane-associated actin of fibroblasts after transformation by Rous sarcoma virus. *Proc. Natl. Acad. Sci. U. S. A.* **72:**746–749.

WILLINGHAM, M., R. CARCHMAN, and I. PASTAN. 1973. A mutant of 3T3 cells with cyclic AMP metabolism sensitive to temperature change. *Proc. Natl. Acad. Sci. U. S. A.* **70:**2906–2910.

WILLINGHAM, M. C., and I. PASTAN. 1974. Cyclic AMP mediates the concanavalin A agglutinability of mouse fibroblasts. *J. Cell Biol.* **63:**288–294.

WILLINGHAM, M., and I. PASTAN. 1975*a*. Cyclic AMP and cell morphology in cultured fibroblasts: effects on cell shape, microfilament and microtubular distribution, and orientation to substratum. *J. Cell Biol.* **67:**146–159.

WILLINGHAM, M., and I. PASTAN. 1975*b*. Cyclic AMP modulates microvillus formation and agglutinability in transformed and normal mouse fibroblasts. *Proc. Natl. Acad. Sci. U. S. A.* **72:**1263–1267.

YAMADA, K. M., and I. PASTAN. 1976. Cell surface protein and neoplastic transformation. *Trends Biochem. Sci.* **1:**222–224.

YAMADA, K. M., and J. A. WESTON. 1974. Isolation of a major cell surface glycoprotein from fibroblasts. *Proc. Natl. Acad. Sci. U. S. A.* **71:**3492–3496.

YAMADA, K. M., S. S. YAMADA, and I. PASTAN. 1976. Cell surface protein partially restores morphology, adhesiveness, and contact inhibition of movement to transformed fibroblasts. *Proc. Natl. Acad. Sci. U. S. A.* **73:**1217–1221.

SYNTHESIS AND ISOLATION OF A SPECIFIC EUKARYOTIC GENE

LARRY McREYNOLDS, JOHN J. MONAHAN, SAVIO L. C. WOO, and BERT W. O'MALLEY

The egg-white protein ovalbumin can be induced by either estrogen or progesterone in vivo. The steroids act by binding initially to high affinity, specific cytosol receptor proteins, and these complexes are then translocated to the nucleus. After the administration of the hormone, there is a rapid increase in the production of ovalbumin mRNA (Harris et al., 1975; Cox et al., 1974; Schmike et al., 1973). This and other evidence suggests that the hormone-receptor complexes act by directly enhancing the transcription of the ovalbumin gene (for a review, see O'Malley and Means, 1974). To study the detailed interactions of the ovalbumin gene with steroid receptors, chromosomal proteins, and RNA polymerase it is necessary to purify the gene.

There are at least two different approaches to obtain a purified gene; it can be either synthesized enzymatically from purified mRNA or purified from total cellular DNA.

Synthesis of the Ovalbumin Gene

Ovalbumin mRNA has been purified from the hen oviduct to greater than 95% purity. A complementary single-stranded DNA can be synthesized from the mRNA by avian myeloblastosis virus reverse transcriptase in the presence of the four nucleotide triphosphates and an oligo dT primer. After alkaline digestion to remove the RNA, the single-stranded DNA can be used as a primer for the synthesis of a double-stranded DNA. Figure 1 diagrammatically shows the steps involved in the synthesis of the double-stranded DNA and its attachment to the plasmid DNA. The single-stranded cDNA is shown schematically as having a hook at its 3′ terminus. This means that when *E. coli* DNA polymerase I is added to the molecule, it has the ability to act as its own primer for the synthesis of the second strand (illustrated with wavy lines in Fig. 1). S_1 nuclease was then used to cut the hairpin loop at the end of the molecule. The two 3′-OH groups can then act as primers for the addition of about 100 dA residues to each end of the molecule with terminal transferase. The size of the homopolymer is regulated by limiting the length of the incubation.

The plasmid used in these experiments, pMB9, was developed in Dr. H. Boyer's laboratory at the University of California Medical School, San Francisco. This plasmid confers tetracycline resistance to transformed *E. coli.* The circular plasmid DNA can be converted to the linear form by cleavage with the restriction enzyme Eco RI. Terminal transferase is used to add approximately 100 poly dT residues to the 3′-terminus of the linear plasmid DNA. The poly dT of the plasmid is then hybridized with the poly dA of the ovalbumin DNA to form a circular molecule. This procedure for joining molecules by their poly dA, poly dT "tails" was developed at Stanford University (Lobban and Kaiser, 1973; Jackson et al., 1972). Two other laboratories have used this approach to incorporate globin DNA into bacterial plasmids (Maniatis et al., 1976; Higuchi et al., 1976).

The plasmid cloning technique offers the dual advantage of purification and amplification. The cloning of individually transformed bacteria allows the separation of pieces of DNA that could not be separated by any physical technique. Bacteria can also be grown in large amounts, making it possible to obtain milligram quantities of the desired piece of the DNA (for review, see Cohen, 1975).

The bacterial cells used in the transformation with the ovalbumin containing plasmids is X 1849. This bacteria was developed in Dr. Roy Curtiss'

McREYNOLDS and CO-WORKERS Department of Cell Biology, Baylor College of Medicine, Houston, Texas

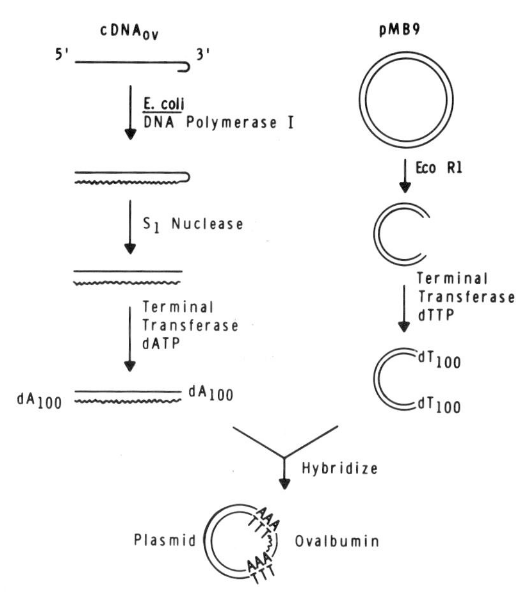

FIGURE 1 Schematic diagram of the construction of the chimeric plasmids between ovalbumin DNA and pMB9. The wavy line represents the synthesis of the DNA strand complementary to $cDNA_{ov}$.

laboratory at the University of Alabama Medical School. This strain has the limited ability to grow in the external environment. The bacteria requires diaminopimelic acid, a necessary constituent of the cell wall, for growth. As an additional precaution, the transformations were carried out in a P-3 physical containment facility.

The transformation consists of incubating the washed bacterial cells with the DNA in low ionic strength calcium salt solution at 4°C. The bacteria are then heated to 37°C for 30 min and then spread on nutrient agar containing tetracycline. The 26 colonies that grew were resistant to tetracycline, indicating that they contained the tetracycline-resistant plasmids. To test if these colonies contained ovalbumin DNA, the bacteria were transferred to a Millipore filter on top of a nutrient agar dish. The colonies were allowed to grow and were lysed with alkali, and the DNA was fixed to the filter. After deproteinization, the filters were hybridized to [^{32}P]ovalbumin RNA. The autoradiography of the filter is shown in Fig. 2. Five or six colonies were strongly positive and were grown for additional analysis. This *in situ* filter assay (Grunstein and Hogness, 1975) permits the rapid screening of large numbers of colonies.

The size of the inserted ovalbumin DNA was determined by hybridization and gel electrophoresis. The four clones studied, pOv1, pOv2, pOv3, and pOv4, were digested with the restriction enzyme Hha I. This restriction enzyme cuts the plasmid into many different fragments, but fortunately leaves the ovalbumin DNA intact. The digested fragments are separated by agarose gel electrophoresis and then stained with ethidium bromide to visualize the DNA. Standards of SV40 DNA digested with Hha I were used to calibrate the gel. The amount of flanking DNA contributed by the plasmid was subtracted from the large fragments of the chimeric plasmid. The difference is the result of the inserted DNA (Table I). To determine if the inserts are due to ovalbumin DNA, the four different plasmids were hybridized to full length [^{3}H]$cDNA_{ov}$. The values determined by gel electrophoresis agree very closely with the values obtained by hybridization. The slight difference is probably on account of the poly dA:dT terminus that joins the plasmid DNA to that of the ovalbumin DNA. As Table I shows, the size of the inserted DNA varies from 535 to 930 nucleotides. The inserted DNA is smaller than the 1,750 nucleotides of the complete transcript because of the absence of a sizing step in preparing the complete double-stranded DNA.

An interesting question that can be asked with this system is whether the ovalbumin gene can be expressed in a bacterial cell. This was tested by isolating the RNA from the minicells from clone pOv4. Minicells are small vesicles that bud off from the bacterial cell and contain only plasmid DNA not chromosomal DNA. Because the minicells have all the machinery necessary for transcription and translation, they provide a good system for studying the expression of plasmid-linked genes. The RNA from clone pOv4 was separated from the DNA by DNase treatment, then hybridized to either full-length $cDNA_{ov}$ or anti-$cDNA_{ov}$. The anti-$cDNA_{ov}$ is DNA synthesized from a $cDNA_{ov}$ template by the same technique shown in Fig. 1. A saturation hybridization analysis showed that 50% of the ovalbumin sequences in pOv4 were protected by the minicell RNA, whereas only 20% of the anticoding strand is expressed (McReynolds et al., 1977). This type of analysis should be useful in future studies designed to compare the in vivo transcription of eukaryotic versus prokaryotic genes.

Isolation of the Ovalbumin Gene by Affinity Chromatography

To isolate DNA from sequences adjacent to the transcribed portion of the ovalbumin gene, it is

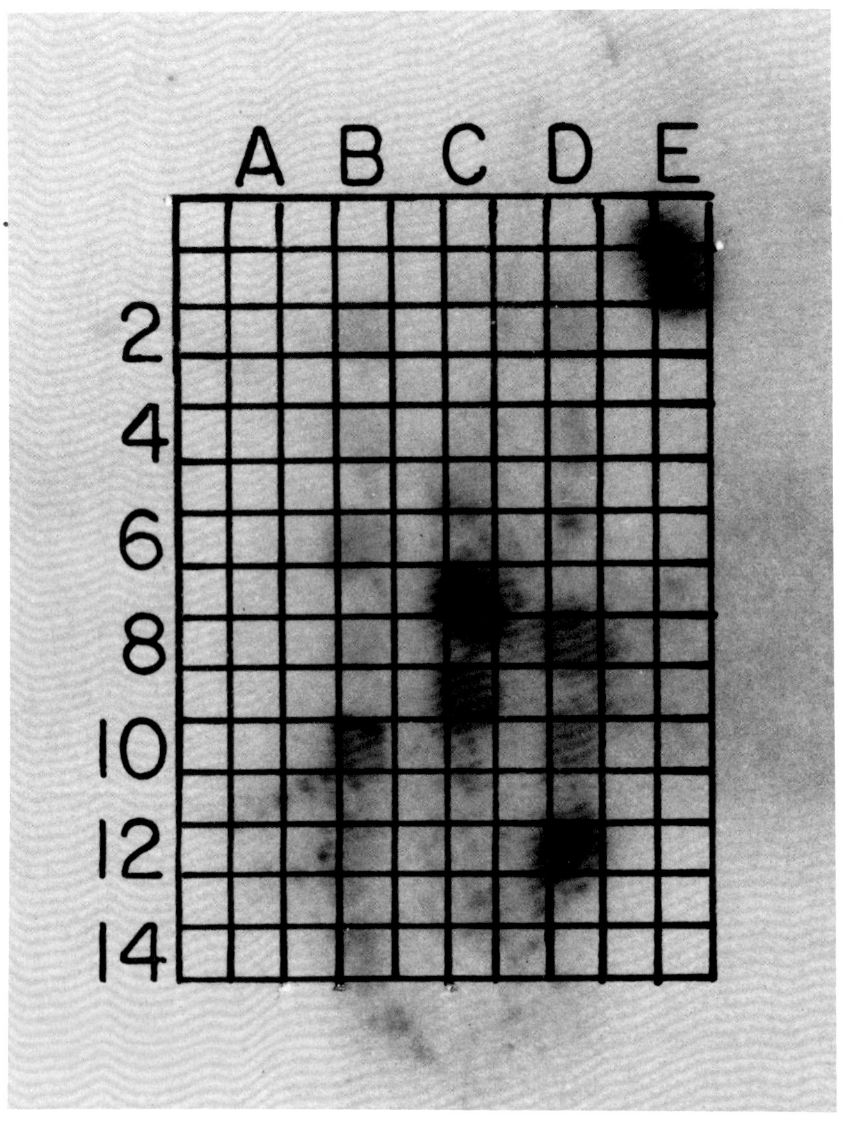

FIGURE 2 Autoradiograph of the transformed clones hybridized to [^{32}P]RNA$_{ov}$. The clones were grown on a Millipore filter containing a grid network. The lines have been redrawn over the autoradiograph.

necessary to start with total DNA. Affinity chromatography was the technique used to enrich for the ovalbumin sequences (Fig. 3). Purified ovalbumin mRNA was coupled to a phosphocellulose matrix (Shih and Martin, 1974). Sheared chick DNA was cycled through the column at 45°C in dimethylformamide buffer, followed by passage through a denaturing column at 70°C before it was returned to the affinity column. After 3 days, the DNA bound to the affinity column was eluted and assayed for the presence of ovalbumin sequences, by hybridization to ^{125}I-mRNA$_{ov}$. The unbound DNA was also hybridized to the same probe to determine if it was depleted in ovalbumin sequences. Additional purification could be obtained by rechromatography of the bound DNA on the affinity column (Woo et al., 1976). C_0t curves were used to assay the extent of purification by the column. The DNA fraction bound to the mRNA$_{ov}$-affinity column hybridized with ^{125}I-mRNA$_{ov}$ with a $C_0t_{1/2}$ value of approximately 1.25. Inasmuch as the unfractionated chick DNA gives a $C_0t_{1/2}$ value of 12,000, this means that there is a 9,600-fold purification of the coding strand of the ovalbumin gene.

To purify the anticoding strand of the ovalbumin gene, an affinity column containing $cDNA_{ov}$ was used. The sheared chick DNA which bound to the column was again passed over the column, then eluted, and the purity determined by hybridization to $[^3H]cDNA_{ov}$, a $C_0t_{1/2}$ value of about 2,300 showing that a 10,000-fold enrichment of the anticoding strand also has been effected. The

TABLE I

Size of Inserted DNA in Chimeric Plasmids

Chimeric plasmid	Hha I fragment lengths (determined from gel)	Inserted DNA	Ovalbumin DNA (determined by hybridization)	Estimated poly dA:dT per termini
pOvl	1,675 NTP	725 NTP	535 NTP	95 NTP
pOv2	1,630 NTP	680 NTP	550 NTP	65 NTP
pOv3	2,040 NTP	1,090 NTP	930 NTP	80 NTP
pOv4	1,950 NTP	1,000 NTP	790 NTP	105 NTP

The size of the inserted DNA was calculated by subtracting the 950 NTP fragment of the pMB9 from the largest band of the Hha I digest of the chimeric plasmid DNA. The amount of the ovalbumin DNA was calculated from the percentage of DNA that hybridized to full length $[^3H]cDNA$. A value of 1,750 nucleotides was used for the $cDNA_{ov}$. The length of the poly (dA:dT) added per termini was determined by subtracting the length determined by hyridization from the inserted DNA size and dividing by two.

ovalbumin gene is only present in one part per million in the genome, so that the affinity column DNA is about 1% pure. The size of the DNA from both columns was determined by alkaline sucrose sedimentation. Both the coding and the anticoding strands were 4,000–5,000 nucleotides in length. This is considerably longer than the size of the mRNA, which is approximately 1,800 nucleotides in length.

Final purification will be obtained by "tailing" the DNAs from both affinity columns with poly dA, reannealing the two complementary strands of the ovalbumin gene, annealing the hybrid with plasmid DNA containing poly dT "tails" to develop a chimera, and then selecting a transformed bacterial clone.

Summary

Two different techniques have been employed to purify and amplify the ovalbumin gene. The first technique is the enzymatic synthesis of double-stranded ovalbumin DNA from purified

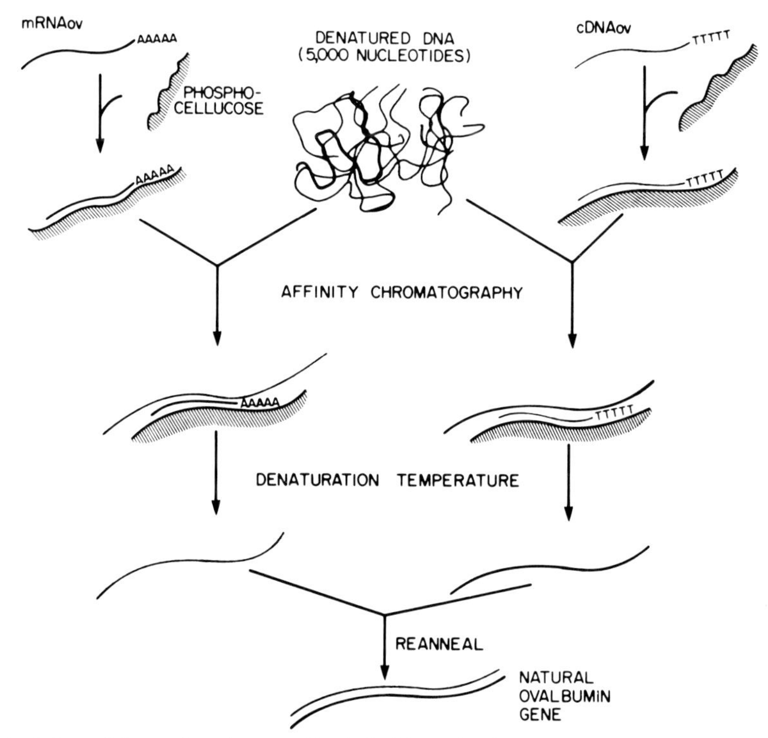

FIGURE 3 Isolation of the ovalbumin gene. The left side of the figure shows schematically the purification of the coding strand for the ovalbumin gene by the $mRNA_{ov}$ affinity column. The purified $mRNA_{ov}$ is covalently linked to the phosphocellulose column and sheared denatured DNA is recycled over the column. The DNA is then thermally eluted. The right side of the figure shows the $cDNA_{ov}$ affinity column used for the purification of the anticoding strand of the ovalbumin gene. The single-stranded DNAs eluted from both columns can then be reannealed to reform the ovalbumin gene.

mRNA. This DNA was then amplified with the aid of bacterial plasmids. The size of the inserted ovalbumin DNA was determined by electrophoresis of the restriction enzyme digest of plasmid and by hybridization. One clone studied, pOv4, was found to permit transcription of the ovalbumin sequences in the bacterial cell.

Affinity chromatography was used to purify both the coding and anticoding strands of the ovalbumin DNA about 10,000-fold. With the aid of additional plasmid purification and amplification, it should be possible to obtain milligram amounts of the ovalbumin gene. This should greatly facilitate the studying of the steps involved in steroid hormone regulation of gene transcription. The role of DNA sequences in binding steroid receptor proteins, RNA polymerase, and chromosomal proteins can be evaluated. In addition, cleavage of the DNA by restriction enzymes has the potential of allowing the isolation and sequencing of specific regions of interest in the natural gene.

REFERENCES

COHEN, S. 1975. The manipulation of genes. *Sci. Am.* **233:**24–33.

COX, R. J., M. E. HAINE, and J. S. EMTAGE. 1974. Quantitation of ovalbumin mRNA in hen and chick oviduct by hybridization to complementary DNA. *Eur. J. Biochem.* **49:**225–236.

GRUNSTEIN, M., and D. S. HOGNESS. 1975. Colony hybridization: a method for the isolation of cloned DNAs that contain a specific gene. *Proc. Natl. Acad. Sci. U. S. A.* **72:**3961–3965.

HARRIS, S. E., J. M. ROSEN, A. R. MEANS, and B. W. O'MALLEY. 1975. Use of a specific probe for ovalbumin messenger RNA to quantitate estrogen-induced gene transcripts. *Biochemistry.* **14:**2072–2081.

HIGUCHI, R., G. V. Paddocks, A. WALL, and W. SALSER. 1976. A general method for cloning eukaryotic structural gene sequences. *Proc. Natl. Acad. Sci. U. S. A.* **73:**3146–3150.

JACKSON, D., R. SYMONS, and P. BERG. 1972. Biochemical method for inserting new genetic information into DNA of Simian Virus 40: circular SV40 molecules containing lambda phage genes and the galactose operon of *Escheria coli. Proc. Natl. Acad. Sci. U. S. A.* **69:**2904–2909.

LOBBAN, R. E., and A. D. KAISER. 1973. Enzymatic end to end joining of DNA molecules. *J. Mol. Biol.* **78:**453–471.

MANIATIS, T., S. G. KEE, A. EFSTRATIADIS, and F. C. KAFATOS. 1976. Amplification and characterization of a β-globin gene synthesized *in vitro*. *Cell.* **8:**163–182.

MCREYNOLDS, L. A., J. J. MONAHAN, D. W. BENDURE, S. L. C. WOO, G. PADDOCK, W. SALSER, J. DORSON, R. MOSES, and B. W. O'MALLEY. 1977. The ovalbumin gene: insertion and transcription of ovalbumin gene sequences in chimeric bacterial plasmids. *J. Biol. Chem.* In press.

O'MALLEY, B. W., and A. R. MEANS. 1974. Female steroid hormones and target cell nuclei. *Science (Wash. D. C.).* **183:**610–620.

SCHIMKE, R. T., R. E. RHOADS, R. PALACIOS, and D. SULLIVAN. 1973. Ovalbumin mRNA complementary DNA and hormone regulation in chick oviduct. *Karolinska Symp. Res. Methods Reprod. Endocrinol.* **6:**357–379.

SHIH, T. Y., and M. A. MARTIN. 1974. Chemical linkage of nucleic acids to neutral and phosphorylated cellulose powders and isolation of specific sequences by affinity chromatography. *Biochemistry.* **13:**3411–3418.

WOO, S. L. C., R. G. SMITH, A. R. MEANS, and B. W. O'MALLEY. 1976. The ovalbumin gene: partial purification of the coding strands. *J. Biol. Chem.* **251:**3868–3874.

Organization and Assembly of Chloroplasts

GENES FOR CHLOROPLAST RIBOSOMAL RNAS AND RIBOSOMAL PROTEINS: GENE DISPERSAL IN EUKARYOTIC GENOMES

LAWRENCE BOGORAD

The objective of this symposium is to review some selected aspects of chloroplast biology. The most highly organized elements of the chloroplast are the photosynthetic membranes—the thylakoids. In her review, Dr. Anderson discusses the molecular organization of these membranes. She is concerned particularly with the identification of some of the membranes' protein components and with our current understanding of how they are organized with relation to one another in thylakoids. This aspect of chloroplast biology is extended in the discussion provided by Dr. Ohad. He, too, is interested in understanding the functions of the proteins that constitute the photosynthetic membrane, in this case, of the alga *Chlamydomonas reinhardtii,* and in how the membranes are assembled. The present paper deals primarily with chloroplast ribosomes, more specifically with the location of genes for the ribonucleic acids and proteins of these structures. Research into this problem shows clearly that the development and functioning of the chloroplast depends upon the integrated expression of genes in the nucleus and the chloroplasts.

Inheritance of Plastid Characters

Many nuclear genes affect chloroplast inheritance. For example, *ys1* (yellow stripe 1) *Zea mays* plants display alternating whitish-yellow and green stripes running the length of their leaves. *Ys1* is a nuclear gene and the character is transmitted as an autosomal recessive trait, according to the rules of Mendel. The metabolic lesion in these yellow-striped plants is in the iron uptake system of the root tips (Bell et al., 1962). The iron ion content of the cytoplasm is apparently too low for normal chloroplast maturation; *ys1* plants sprayed with iron solutions are phenocopies of wild-type.

But extranuclear genes which control chloroplast development also have been discussed for many years. In 1908, Bauer described biparental but non-Mendelian transmission of chloroplast characters in *Pelargonium*. In the same year, Correns reported that the inheritance of chloroplasts in the four-o'clock (*Mirabilis*) is strictly maternal and, thus, also non-Mendelian. Bauer and Correns suggested that some extranuclear, i.e., cytoplasmic, genes could be involved in the transmission of chloroplasts. The plastid itself is the obvious site for such genes. This, and other research on the cytoplasmic transmission of plastid characters, has been reviewed extensively by Kirk and Tilney-Basset (1967).

More is known now about the genetics of the chloroplast genome of *Chlamydomonas reinhardtii* than about any other organism; furthermore, genes in the nuclear genome have also been mapped extensively (Sager, 1972).

The two mating types of *Chlamydomonas* are designated "+" and "−." Under appropriate conditions, two gametes of opposite mating type fuse, form a zygote, undergo meiosis, and form four (or, in some strains, eight) haploid vegetative zoospores. Zoospores can be converted to gametes by appropriate nutritional manipulation. And the cycle begins again.

Chlamydomonas displays two major patterns of gene transmission. In one type, genetic markers carried in either of the two gametes are expressed in the zoospores. If the marker is carried in only one of the two gametes entering the cross, two of the four zoospores will display the character—strictly according to the rules of Mendel. Sixteen

LAWRENCE BOGORAD Biological Laboratories, Harvard University, Cambridge, Massachusetts

nuclear linkage groups have been identified. The second type is characterized by uniparental transmission. These traits are expressed in all four zoospores if introduced into the cross in the + parent, but are completely and permanently lost if introduced in the − parent. Characters transmitted in a non-Mendelian manner are carried in the plastid genome. A number of such genetic markers have been discovered and ordered on a map establishing intergenic distances on the single linkage group so far established (Sager, 1972).

Confidence in the reality of chloroplast genes came with unequivocal demonstrations of the presence of DNA. Sufficient DNA is present in the plastids of some plants to be detected by such stains as Feulgen and methyl green. Ris and Plaut (1962) used staining procedures with the green alga *Chlamydomonas moweusii*, but also demonstrated the presence of DNase-digestible 20–25 Å fibrils by electron microscopy of thin sections of this organism. DNA fibers in electron-transparent regions similar in appareance to "nucleoplasm" of bacteria have been observed in chloroplasts of a number of other plants as well (Kislev et al., 1965; Woodcock and Bogorad, 1971). Chloroplast DNA has also been isolated and shown to differ from nuclear DNA of the same plant in regard to base composition and renaturation kinetics; in some cases buoyant density differences are also striking (Chun et al., 1963; Leff et al., 1963; Woodcock and Bogorad, 1971; Sager, 1972). By the late 1960s the presence of unique DNA in chloroplasts of many species was established and accepted.

Another line of work that began to reveal the possibility of some degree of autonomy in chloroplasts was the discovery of a distinctive class of ribosomes in these organelles. Jacobson et al. (1963) observed RNase-digestible ribonucleoprotein particles in etioplasts and chloroplasts as well as the cytoplasm of *Zea mays*. The ribosomes in the plastids are smaller than those in the cytoplasm. Lyttleton (1962) found two different sedimentation classes of ribosomes in extracts of spinach leaf tissue. Only the smaller and more slowly sedimenting type was found in extracts of purified chloroplasts. Extended work along this line has established sedimentation constants of about 70S and 80S, respectively, for plastid and cytoplasmic ribosomes from many species. The major ribosomal RNAs of chloroplasts have sedimentation constants of 23S and 16S, while the RNAs of the 80S cytoplasmic ribosomes have sedimentation constants of about 25S and 18S. Finally, analyses by polyacrylamide gel electrophoresis have shown that only a few, if any, proteins are common to the two classes of ribosomes (Woodcock and Bogorad, 1971; Vasconcelos and Bogorad, 1970; Hanson et al., 1974).

By 1970 there was convincing evidence that chloroplasts contain unique DNA, ribosomes, and, indeed, the entire apparatus required for information storage, replication, and processing.

DNAs of plastids and prokaryotes are similar when viewed *in situ* with the electron microscope. Ribosomes of chloroplasts—like those of prokaryotes—tend to be smaller than cytoplasmic ribosomes of eukaryotic cells. Many antibiotics which block protein synthesis by prokaryotic ribosomes affect chloroplast ribosomes similarly. Recognition of these facts aroused increasing interest in the relatively old idea that organelles might have originated from prokaryotes which had become endosymbionts of nucleated cells (Margulis, 1967). Tied to this revival somehow was the possibility that plastids might still be genetically autonomous. Studying the interactions between the expression of chloroplast and nuclear genomes was one way to test these ideas and, simultaneously, to begin to understand better the working of eukaryotic cells.

Chloroplast DNA: Genes and Endonuclease Recognition Sites

Genes for chloroplast ribosomal RNAs have been shown to be in chloroplast DNA by molecular hybridization. Thomas and Tewari (1974) quantitated this relationship. By saturation hybridization of radioactive chloroplast rRNA with chloroplast DNA from bean, lettuce, spinach, maize, and oats, they showed that each chloroplast DNA molecule from these plants contain two copies of genes for chloroplast rRNAs. In similar types of molecular hybridization experiments with radioiodinated tRNAs prepared from *Zea mays* chloroplasts, 0.60–0.75% of maize chloroplast DNA was shown to contain sequences complementary to maize tRNAs. This corresponds to cistrons for 20–26 tRNAs. Incubating tRNAs charged with radioactive amino acids with maize chloroplast DNA under hybridizing conditions and determining which amino acids are held indirectly to the DNA show genes for tRNAs charging a total of at least 16 different amino acids to be present in chloroplasts (Haff and Bogorad, 1976).

Most, perhaps all, of the chloroplast DNA of maize, spinach, *Euglena,* and thus, probably, of many plants, is in the form of closed, supercoiled circles. Maize chloroplast DNA, for example, can be isolated as supercoiled circles with a molecular weight of 85×10^6 (Kolodner and Tewari, 1975*a*, *b*). Fragments generated from the circular chloroplast DNA of *Zea mays* by restriction endonucleases have now been ordered to provide a physical map of the location of recognition sites for these enzymes (Bedbrook and Bogorad, 1976*a*).

Terminal digestion of maize chloroplast DNA with restriction endonuclease Sal I yields 10 fragments which can be separated into 8 size classes by electrophoresis in agarose gels. When the weight of these fragments is summed, the entire DNA molecule is accounted for. Sixty percent of the mass of the maize chloroplast DNA molecule can be accounted for in the 16 pieces distinguishable on agarose gels after digestion with the restriction endonuclease Eco RI. Eighty percent is recovered after digestion with Bam I, which yields 18 fragments. The incomplete recovery probably results from formation of a number of very small fragments which are not detectable on agarose gels after electrophoresis or are otherwise lost during analysis of the digests.

The order of the Sal I fragments, together with information on the location of some DNA pieces generated by the other two endonucleases, is shown in Fig. 1. To obtain this information, maize chloroplast DNA was isolated and digested with a restriction endonuclease. Then the number and sizes of the DNA fragments produced by this enzyme were determined by electrophoresis in agarose gels. This was done separately for each of the three endonucleases used. The order of fragments produced by one enzyme was determined by finding overlapping fragments produced by another enzyme.

One of the techniques used, for example, was to determine which Eco RI restriction fragment(s) carry a Sal I recognition site. The RI DNA fragment isolated from an agarose gel was used as a template for the preparation of [^{32}P]phosphate-labeled complementary RNA for molecular hybridization tests against Sal I fragments.

The next step, in this example, is to prepare a Sal I terminal digest of whole chloroplast DNA and to separate the fragments according to size on an agarose gel by electrophoresis. Then, using the technique of Southern (1975), the DNA in the agarose gel is denatured and transferred to a sheet

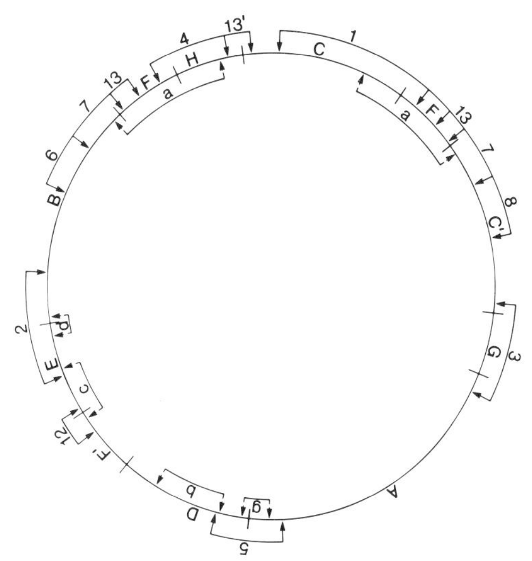

FIGURE 1 Locations of recognition sites for restriction endonucleases on *Zea mays* chloroplast DNA. Recognition sites which have been determined for Sal I, Bam I, and Eco RI are shown. Sal I fragments are designated by capital letters and recognition sites are shown in the central horizontal line by vertical lines. Bam I fragments are designated by arabic numerals on the upper horizontal line. Eco RI fragments are shown on the lower line and designated by lower case letters. Ribosomal RNA genes are contained in each of the two Eco RI fragments *a* (Bedbrook and Bogorad, 1976*a*).

of nitrocellulose filter. This is done by placing a piece of nitrocellulose filter material in contact with the gel and a piece of dry, highly absorbant filter paper on the opposite side of the filter. Then liquid is permitted to flow by capillarity through the gel, through the filter, and up into the blotting sheet. The denatured DNA is carried out of the agarose gel and sticks to the nitrocellulose filter. The filter then contains the DNA formerly present in the gel; each band, i.e., each size class, on the gel yields a band on the nitrocellulose filter strip and in the same order as it was formerly. This nitrocellulose filter is then used for hybridization with radioactive complementary RNA to determine the relationship between the particular RI fragment from which complementary RNA was made and at least two Sal fragments.

A notable feature of the maize chloroplast DNA genome is the presence of two of the same Eco RI fragments. These fragments are designated "*a*." Each of these *a* fragments accounts for about 15% of the genome. The two copies of the *a* sequence are present in inverted orientation with respect to one another, and they are separated by a nonhomologous sequence representing an additional 10% of the genome length. Each Eco RI *a* fragment contains a gene for the 23S and the 16S RNA of the chloroplast (Bedbrook and Bogorad, 1976*a*). This information was obtained by first preparing rRNA from chloroplast ribosomes. The RNA was radiolabeled in vitro by exchanging 5′ OH with [^{32}P]phosphate using radioactive ATP and the enzyme polynucleotide kinase. The Eco RI fragments obtained by terminal digestion of maize chloroplast DNA were separated by agarose gel electrophoresis and transferred to nitrocellulose filters, again by the technique of Southern (1975). The radiolabeled chloroplast rRNAs were found to hybridize to only Eco RI fragment *a*.

To extend this work, Eco RI fragment *a* was incorporated into the tetracycline resistance plasmid pMB9 and cloned in *Escherichia coli*. This permitted us to obtain large enough quantities of fragment *a* to map the position of recognition sites for restriction endonucleases Sal I, Bam I, and Hind III on it. Then, by molecular hybridization, the location of genes for 16S and 23S rRNAs were located on these fragments. The rRNA genes occupy about 50–60% of Eco RI fragment *a* (Bedbrook and Bogorad, 1976*b*). The presence of two copies of each of the rRNA genes per chloroplast DNA molecule is in line with the previous observations of Thomas and Tewari (1974) in which saturation hybridization to the total chloroplast genome was carried out.

From the experiments just described, we now know both the number of chloroplast rRNA genes in the chloroplast genome of *Zea mays* and just where they are situated. But where are genes for chloroplast ribosomal proteins?

Genes for Proteins of Chloroplast Ribosomes

Genes for chloroplast ribosomal proteins have been identified in both the nuclear and chloroplast genomes of the single-celled green alga *Chlamydomonas reinhardtii*.

At 10–20 μg/ml, erythromycin A completely inhibits the growth of *Chlamydomonas* on agar. This antibiotic blocks protein synthesis by bacterial ribosomes. It binds only to the large (52S) subunit of the chloroplast ribosome, to neither the small subunit of the chloroplast ribosome nor to either of the subunits of cytoplasmic ribosomes (Mets and Bogorad, 1971). Presumably, it inhibits the growth of the alga through its action on the chloroplast ribosomes. A single molecule of the antibiotic associates with a single 52S ribosomal subunit; the binding constant of erythromycin A with chloroplast ribosomes of wild-type *Chlamydomonas* is 3–4 $\times$ 10^6 M^{-1} (Hanson, 1976).

Nine erythromycin-resistant strains of *Chlamydomonas* were isolated by mutagenizing cells with ethyl methane sulfonate and plating them out on 5 $\times$ 10^{-4} M erythromycin in an agar medium (Mets and Bogorad, 1971). Unlike chloroplast ribosomes from wild-type cells, those isolated from the erythromycin-resistant mutants failed to bind the antibiotic in low concentrations of KCl. Thus, mutation to resistance appears to be the consequence of an alteration in the 52S ribosomal subunits of the chloroplast, rather than a loss of the ability to take up the antibiotic or the acquisition of the capacity to inactivate it. Thus, with erythromycin resistance as a genetic marker, genes which affect chloroplast ribosomal proteins or perhaps rRNAs can be tracked to the nuclear or chloroplast genome. The possible results of matings which could permit one to conclude whether a gene is in the chloroplast or in the nuclear genome have already been described.

Twenty-six different proteins have been identified in preparations from 52S subunits of *Chlamydomonas* chloroplast ribosomes by a two-dimensional polyacrylamide gel electrophoresis system (in the method of Mets and Bogorad [1974], proteins are first separated electrophoretically in the presence of urea and then separated on the basis of size in the presence of sodium dodecyl sulfate [SDS]). A map showing the position of these proteins on a two-dimensional grid is presented in Fig. 2 (Hanson et al., 1974).

Protein 4 of the large subunit (LC4) is altered in an erythromycin-resistant strain designated *ery*-U1a. This alteration is transmitted in a uniparental manner and thus appears to be the consequence of a chloroplast gene mutation. Alterations in the nuclear gene *ery*-M1 result in changes in the size or electrophoretic ability of chloroplast protein LC6. A third protein, not yet identified on the two-dimensional map, is altered in mutants of the strain *ery*-M2. Transmission of these two latter characters is in a Mendelian manner (as designated by the "M") and the genes are consequently judged to be in the nuclear genome (Mets and Bogorad, 1971, 1972; Hanson et al., 1974; Davidson et al., 1974).

Mutants of the class *ery*-M1 deserve special

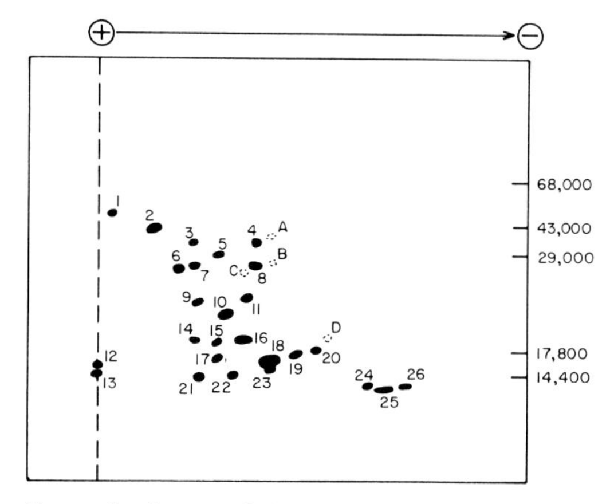

FIGURE 2 A map of the locations of proteins of the large subunit of chloroplast ribosomes of *Chlamydomonas reinhardtii* after electrophoresis in urea (left to right) and then (top to bottom) in the presence of SDS (Hanson et al., 1974). Large chloroplast ribosomal subunit protein 4 (LC4) is altered in mutant *ery*-U1a. Protein LC6 is altered in the *ery*-M1 mutants. (Units at right in mol wt.)

additional attention. We have isolated four mutants which fit into this class (*ery*-M1a, b, c, d). All four map to the same position on linkage group XI of the nuclear genome of *Chlamydomonas*. Each of the four mutants we isolated has an alteration in protein LC6. Three different types of alterations of LC6 (either in molecular weight or in electrophoretic mobility in urea) occur among the four mutants that map to the locus *ery*-M1. This provides very strong evidence that the gene at *ery*-M1 on linkage group XI is in fact the structural gene for the chloroplast ribosomal protein LC6 (Davidson et al., 1974).

This view is supported by a study of diploid strains of *C. reinhardtii* constructed and analyzed by M. R. Hanson (1976, The genetics and biochemistry of chloroplast ribosome mutants of *Chlamydomonas reinhardi*. Ph.D. thesis. Department of Biology, Harvard University, Cambridge, Mass. 219.). Using methods which had been described by others earlier, vegetative *C. reinhardtii* diploids were constructed, in this case, between wild-type cells and various *ery*-M1 mutants. The diploid +/*ery*-M1d, for example, has been shown to contain both the wild-type and *ery*-M1b forms of LC6. In each case studied, both the wild-type and the mutant form of LC6 were present in the ribosomal subunits isolated from the diploid, just as to be expected if two different structural genes for LC6 were present. The ratio of wild-type LC6 to the mutant form was generally about 3:2. The deviation from unity may result from the mutant form of the protein having a lower affinity than the normal one for its site on the ribosome.

Thus, two of these three chloroplast ribosomal proteins are the products of nuclear genes and one appears to be the product of a chloroplast gene. J. N. Davidson (1976, Genes affecting erythromycin resistance in *ery*-M1 mutants of *Chlamydomonas reinhardi*. Ph.D. thesis. Harvard University, Cambridge, Mass. 159.) has located, also on nuclear linkage group XI, a gene that increases sensitivity to erythromycin of strains carrying *ery*-M1 resistance. This and other genes which may affect chloroplast ribosomal proteins have been discussed by Bogorad et al. (1977). Information on the location of genes for chloroplast rRNAs is not available in as detailed form for *Chlamydomonas* as for maize, but Bastia et al. (1971*a, b*) have shown by hybridization against total chloroplast DNA that genes for *C. reinhardtii* chloroplast rRNAs are located in chloroplast DNA.

Thus, genes for proteins and RNAs of the 52S subunit of the chloroplast ribosome are dispersed in two genomes and, within the nuclear genome, on several linkage groups. The integrated expression of components of two genomes is required to make up the functional cell. Regulation of the production of any one of these components in either genome could affect the development and metabolism of the chloroplast or of the nuclear-cytoplasmic system either directly or indirectly. This dispersal of genes for components of a single structure, the large subunit of a chloroplast ribosome, forces us to consider again the possible modes of origin of organelles, and thus of the eukaryotic habit, as well as the evolutionary steps, which have led to the existence of the modern eukaryotic cell.

The Origin and Evolution of Eukaryotic Genomes

The dispersal of genes for components of complex organellar elements such as membranes and ribosomes or other multimeric enzymes may be a general principle of organelle biology. The case for ribosomal RNAs and some proteins of *Chlamydomonas* plastid ribosomes has been discussed.

Some proteins of photosynthetic membranes (thylakoids) of the giant single-celled alga *Acetabularia* appear to be specified by genes of the chloroplast and others by nuclear genes, judging from nuclear transfer experiments between *A. mediterranea* and *A. calyculus* (Apel and Schweiger, 1972). These two species have some thylakoid proteins which are indistinguishable by electrophoresis in phenol-acetic acid but other thylakoid proteins differ. Exchange of nuclei leads to eventual changes in some thylakoid proteins.

Gene dispersal is well-authenticated for the enzyme ribulose-1,5-diphosphate carboxylase of green plants. This enzyme has one polypeptide subunit of about 14,000 and another of about 50,000 daltons. Chan and Wildman (1972) and Kawishima and Wildman (1972) found that the small subunits of this enzyme from *Nicotiana glauca* and *N. tabaccum* differ in tyrosine content and tryptic peptides. Information for the small subunit appears to be transmitted in a Mendelian manner in crosses between these two species. There is other evidence that the large subunit of ribulose-1,5-diphosphate carboxylase is transmitted in a non-Mendelian manner in crosses between *N. gossei*, an Australian *Nicotiana*, and *N. tabaccum*. The demonstrations by Blair and Ellis (1973) that isolated pea chloroplasts can synthesize or complete the synthesis of the large subunit of ribulose-1,5-diphosphate carboxylase supports the view that the large subunit of ribulose diphosphate carboxylase is the product of a chloroplast gene.

The sites of synthesis of organelle proteins have been sought by another experimental approach. Cycloheximide inhibits protein synthesis by cytoplasmic ribosomes. Chloramphenicol prevents chloroplast (and mitochondrial) ribosomes from functioning. The dangers of concluding that a protein is made on cytoplasmic ribosomes from use of cycloheximide alone on intact organisms or that organelle ribosomes synthesize another protein because administration of chloramphenicol interferes with its production are apparent. Within the complex of unknown compartmental interactions the production of any protein may well be stopped not only by directly blocking the working of the ribosomes on which it is produced, but also by a variety of indirect events. But Ohad (1977), in another paper in this symposium, describes the sequential use of antibiotics in which normal products are finally made to demonstrate that some *Chlamydomonas* chloroplast membrane proteins are made on chloroplast ribosomes and others on cytoplasmic ribosomes. This approach also has been used very successfully for studying mitochondrial biogenesis.

To return to the ribosome case, it seems reasonable to assume that ribosomes originated only once, and that no matter how eukaryotic cells originated, the genes and their products were initially in the same compartment. It is from this base that the modern cell has evolved.

Two possible origins of eukaryotic cells have been discussed: (1) the endosymbiont hypothesis (e.g., Margulis, 1967), and (2) the cluster-clone hypothesis (Bogorad, 1975). The former posits the addition of prokaryotic organisms to an anaerobic, nucleated (?) cell incapable of oxidative respiration or photosynthesis. The latter posits the subdivision into membrane-limited compartments of genes and metabolic functions already present in an organelleless primitive cell.

Regardless of the way the eukaryotic habit was founded, we realize that it evolved to its modern forms. It is easiest to discuss gene dispersal in endosymbiont terms, although this is the least likely way for genes and their products to have been separated from the start of eukaryotism.

Two mechanisms of ribosomal gene dispersal have been suggested. The first, illustrated in Fig. 3, is by gene transfer. Thus, if an organelle or gene duplicates and the copy is incorporated into the nuclear genome, the gene would be represented twice. If the organelle or gene were to be lost, the sole gene for the organelle component would be in the nuclear genome. Transfer of a gene without prior duplication is equally likely.

Figure 4 illustrates another gene dispersal possibility, dispersal by protein and gene substitution. As illustrated, the mutation of a gene to the point where it makes a useless protein should be fatal for the organelle or the entire cell. But if the product (if any) of the useless gene can be substituted for by another protein available in the cell, the organelle or ribosome might still be able to function. If the substituted protein is a nuclear gene product, the gene for the plastid ribosomal protein would now be in the nuclear genome. Subsequent mutations and alterations of the protein could lead to its serving the organelle alone and perhaps better than at the time of its original substitution (Bogorad, 1975).

The cluster-clone hypothesis proposes that genes in the prokaryotic ancestor were associated in clusters and that each cluster was separated from the remainder of the cell by a membrane. Subsequently, each set of clustered genes replicated (i.e., was cloned) and thus gave rise to progenitors of organelles in the eukaryotic cell. If

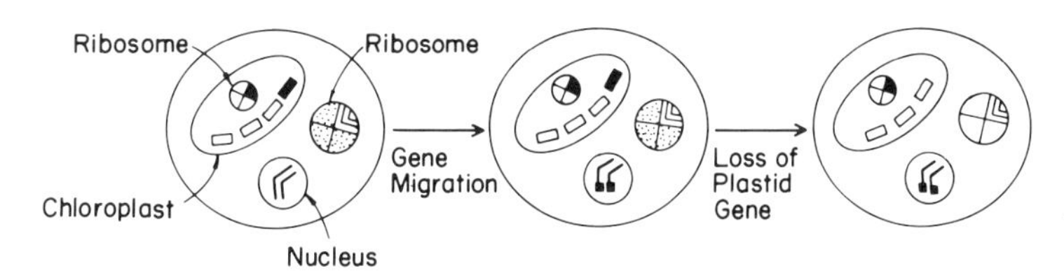

FIGURE 3 A schematic representation of gene dispersal by transfer (Bogorad, 1975).

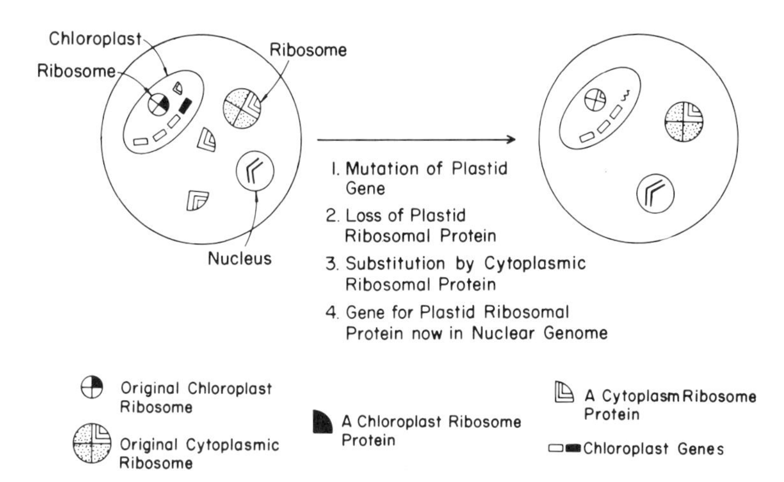

FIGURE 4 A schematic representation of gene dispersal by protein and gene substitution (Bogorad, 1975).

each gene was present only once initially, separation of the gene and its product would have occurred at the time eukaryotic cells originated. On the other hand, if each gene were present more than once at the outset, some gene reduction and sorting of the gene transfer or substitution types might have played roles in evolution.

ACKNOWLEDGMENTS

This work was made possible by grants from the National Institute of General Medical Sciences and it was also supported in part by the Maria Moors Cabot Foundation of Harvard University.

I am indebted to my colleagues who carried out the bulk of the work described from our laboratories: J. R. Bedbrook, W. D. Bell, J. N. Davidson, L. Haff, M. R. Hanson, A. B. Jacobson, N. Kislev, L. J. Mets, A. C. L. Vasconcelos, and C. L. F. Woodcock.

REFERENCES

APEL, K., and H. SCHWEIGER. 1972. Nuclear dependency of chloroplast proteins in *Acetabularia*. *Eur. J. Biochem.* **25:**229–238.

BASTIA, D., K.-S. CHIANG, and H. SWIFT. 1971*a*. Studies on the ribosomal RNA cistrons of chloroplast and nucleus in *Chlamydomonas reinhardtii*. Abstracts of Papers, 11th Annual Meeting of the American Society of Cell Biologists. 25.

BASTIA, D., K.-S. CHAING, H. SWIFT, and P. SIERSMAN. 1971*b*. Heterogeneity, complexity and repetition of the chloroplast DNA of *Chlamydomonas reinhardtii*. *Proc. Natl. Acad. Sci. U. S. A.* **68:**1157–1161.

BEDBROOK, J. R., and L. Bogorad. 1976*a*. Endonuclease recognition sites mapped on *Zea mays* chloroplast DNA. *Proc. Natl. Acad. Sci. U. S. A.* **73:**4309–4313.

BEDBROOK, J. R., and L. BOGORAD. 1976*b*. Physical and transcriptional mapping of *Zea mays* chloroplast DNA. *In* The Genetics and Biogenesis of Chloroplasts and Mitochondria. Th. Bücher, editor. Elsevier Scientific Publishing Company, Amsterdam. 369–373.

BELL, W. D., L. BOGORAD, and W. J. MCILRATH. 1962. Yellow stripe phenotype in maize. I. Effects of ys, locus on uptake and utilization of iron. *Bot. Gaz.* **124:**1–8.

BLAIR, G. E., and R. J. ELLIS. 1973. Protein synthesis in chloroplasts. I. Light-driven synthesis of the large subunit of fraction I protein by isolated pea chloroplasts. *Biochim. Biophys. Acta.* **319:**223–234.

BOGORAD, L. 1975. Evolution of organelles and eukaryotic genomes. *Science* (*Wash. D.C.*) **188:**891–898.

BOGORAD, L., J. N. DAVIDSON, and M. R. HANSON. 1977. The genetics of the chloroplast ribosome in *Chlamydomonas reinhardi*. *In* Protein Synthesis and Nucleic Acids in Plants. L. Bogorad and J. Weil, editors. Plenum Publishing Company, New York. In press.

CHAN, P., and S. G. WILDMAN. 1972. Chloroplast DNA codes for the primary structure of the large subunit of fraction I protein. *Biochim. Biophys. Acta.* **277**:677–680.

CHUN, E. H. L., M. H. VAUGHN, and A. RICH. 1963. The isolation and characterization of DNA associated with chloroplast preparations. *J. Mol. Biol.* **7**:130–141.

DAVIDSON, J. N., M. R. HANSON, and L. BOGORAD. 1974. An altered chloroplast ribosomal protein in *ery*-M1 mutants of *Chlamydomonas reinhardi. Molec. Gen. Genet.* **132**:119–129.

HAFF, L., and L. BOGORAD. 1976. Hybridization of maize chloroplast DNA with transfer ribonucleic acids. *Biochemistry.* **15**:4105–4109.

HANSON, M. R., J. N. DAVIDSON, L. J. METS, and L. BOGORAD. 1974. Characterization of chloroplast and cytoplasmic ribosomal proteins of *Chlamydomonas reinhardi* by two-dimensional gel electrophoresis. *Molec. Gen. Genet.* **132**:105–118.

JACOBSON, A. B., H. SWIFT, and L. BOGORAD. 1963. Cytochemical studies concerning the occurrence and distribution of RNA in plastids of *Zea mays. J. Cell Biol.* **17**:557–570.

KAWISHIMA, N., and S. G. WILDMAN. 1972. Studies on fraction I protein. IV. Mode of inheritance of primary structure in relation to whether chloroplast or nuclear DNA contains the code for a chloroplast protein. *Biochim. Biophys. Acta.* **262**:42–49.

KIRK, J. T. O., and R. A. E. TILNEY-BASSET. 1967. The Plastids. W. H. Freeman & Co., New York. 608.

KISLEV, N., H. SWIFT, and L. BOGORAD. 1965. DNA from chloroplasts and mitochondria of swiss chard. *J. Cell Biol.* **25**:327–333.

KOLODNER, R. D., and K. K. TEWARI. 1975*a*. The molecular size and conformation of the chloroplast DNA from higher plants. *Biochim. Biophys. Acta.* **402**:372–390.

KOLODNER, R. D., and K. K. TEWARI. 1975*b*. Denaturation mapping studies on the circular chloroplast deoxyribonucleic acid from pea leaves. *J. Biol. Chem.* **250**(7):4888–4895.

LEFF, J., M. MANDEL, H. T. EPSTEIN, and J. A. SCHIFF. 1963. DNA satellites from cells of green and aplastidic algae. *Biochem. Biophys. Res. Commun.* **13**:126–130.

LYTTLETON, J. W. 1962. Isolation of ribosomes from spinach chloroplasts. *Exp. Cell Res.* **26**:312–317.

MARGULIS, L. 1967. On the origin of mitosing cells. *J. Theor. Biol.* **14**:225–274.

METS, L. J., and L. BOGORAD. 1971. Mendelian and uniparental alterations in erythromycin binding by plastid ribosomes. *Science (Wash. D.C.)* **174**:707–709.

METS, L. J., and L. BOGORAD. 1972. Altered chloroplast ribosomal proteins associated with erythromycin-resistant mutants in two genetic systems of *Chlamydomonas reinhardi. Proc. Natl. Acad. Sci. U. S. A.* **69**:3779–3783.

METS, L. J., and L. BOGORAD. 1974. Two-dimensional polyacrylamide gel electrophoresis: an improved method for ribosomal proteins. *Anal. Biochem.* **57**:200–210.

RIS, H., and W. PLAUT. 1962. The ultrastructure of DNA-containing areas in the chloroplast *Chlamydomonas. J. Cell Biol.* **13**:383–391.

SAGER, R. 1972. Cytoplasmic Genes and Organelles. Academic Press, Inc., New York. 405.

SOUTHERN, E. M. 1975. Detection of specific sequences among DNA fragments separated by gel electrophoresis. *J. Mol. Biol.* **98**:503–517.

THOMAS, J. R., and K. K. TEWARI. 1974. Conservation of 70S ribosomal RNA genes in the chloroplast DNAs of higher plants. *Proc. Natl. Acad. Sci. U. S. A.* **71**:3147–3151.

VASCONCELOS, A. C. L., and L. BOGORAD. 1970. Proteins of cytoplasmic, chloroplast and mitochondrial ribosomes of some plants. *Biochim. Biophys. Acta.* **228**:492–502.

WOODCOCK, C. L. F., and L. BOGORAD. 1971. Nucleic acids and information processing in chloroplasts. *In* Structure and Function of Chloroplasts. M. Gibbs, editor. Springer-Verlag New York Inc., New York. 89–128.

THE MOLECULAR ORGANIZATION OF CHLOROPLAST THYLAKOIDS

JAN M. ANDERSON

An ultimate understanding of chloroplast membrane assembly and function depends on knowledge of its molecular structure. Description of the molecular organization of the inner chloroplast membranes which contain the photosynthetic apparatus is challenging because of their intricate structure and complex energy-transducing function. These inner membranes consist of flattened, saclike vesicles, termed thylakoids, which are arranged as a network of unstacked membranes (stroma thylakoids) that are connected to a series of closely contacted, stacked membranes (grana thylakoids). The outer thylakoid surface is in contact with the chloroplast matrix, the stroma, where CO_2 fixation occurs; the inner surface encloses the intrathylakoid space, which is continuous between stacked and unstacked thylakoids. In higher plant and algal chloroplasts, photosynthesis involves two separate light reactions catalyzed by light absorbed by two different pigment assemblies, each of which consists of many light-harvesting chlorophyll and carotenoid molecules and the reaction-center chlorophyll, where the primary conversion of light into chemical energy takes place. Each pigment assembly is associated with electron transport carriers to form a photosystem (Govindjee and Govindjee, 1975). Functionally, the two photosystems act in series to transfer electrons from water to $NADP^+$, photophosphorylation being coupled to this electron transport. Photosystem II (PS II) is involved in the evolution of oxygen, while photosystem I (PS I) is involved in the reduction of $NADP^+$.

Two features of the molecular organization of chloroplast thylakoids will be considered: (1) the location of the individual components within and across the membrane; and (2), the possible organization of these components into supramolecular complexes which may account for the differentiation of stacked and unstacked membranes of most higher plant chloroplasts.

JAN M. ANDERSON Commonwealth Scientific and Industrial Research Organization, Division of Plant Industry, Canberra City, A.C.T., Australia

Molecular Architecture

An imaginative conceptual basis for the understanding of the molecular organization of chloroplast thylakoids is given by the fluid lipid-protein mosaic model of Singer and Nicolson (1972). The membrane consists of a lipid bilayer continuum, in which the intrinsic proteins are firmly embedded. The hydrophobic regions of the intrinsic proteins are associated by hydrophobic interactions with the fatty acid tails of the lipid bilayer, and their hydrophilic areas interact with the polar head groups of the boundary lipids. Some intrinsic proteins may extend across the membrane and thus have hydrophilic regions at each membrane surface. Detergents, chaotropic agents, or organic solvents are required for the release of intrinsic proteins from the membrane. In contrast, the extrinsic proteins, being attached to the membrane mainly by ionic interactions with the intrinsic proteins, are more easily removed from the membrane (Singer, 1974). Most of the lipid is thought to be in the bilayer form, with a fraction being immobilized around the intrinsic proteins, the so-called boundary lipids (Lee, 1975).

The basic concepts of this molecular organization include fluidity, asymmetry, and economy. The fluidity of thylakoids allows the free lateral diffusion of some components along the membrane, which may be required for the complex function of energy transduction. Further, fluidity allows a redistribution of components during the membrane stacking process, permits the insertion of new components into the membranes during biogenesis, and allows for even distribution of membrane components during chloroplast division. In contrast to this possible movement of molecules along the membrane, there is a restriction of movement of molecules across the membrane from one half of the bilayer to the other, which permits the necessary asymmetry of components essential for thylakoid function. Finally, there is a splendid economy in placing molecules in such thin membranes, which not only greatly increases their effective concentration, but also imparts order, since random movement in three-dimensions is removed.

With chloroplast thylakoids, evidence from both low-angle X-ray diffraction (Sadler et al., 1973) and freeze-fracture studies (Arntzen and Briantais, 1975) is consistent with the presence of substantial regions of lipid bilayer. Moreover, freeze-fracture data (Ojakian and Satir, 1974) clearly show that the matrix of chloroplast thylakoids is fluid.

The lipids make up some 50% of thylakoid membrane mass. The main classes include those involved directly in photosynthesis—chlorophylls (21%), carotenoids (3%), and plastoquinones (3%); and the structural lipids of the matrix—the glycolipids, monogalactosyl diacylglycerol (27%), digalactosyl diacylglycerol (14%), and an anionic sulpholipid (4%); and lesser amounts of phospholipids (10%) (Lichtenthaler and Park, 1963). The most significant feature of the structural lipids is the high degree of unsaturation of their fatty acids, linolenic acid (18:3) being the predominant acyl group. This high degree of unsaturation of acyl groups, together with the absence of cholesterol (Lee, 1975), would allow thylakoid membranes to have very fluid domains. There is no unequivocal evidence for the localization of any of the lipid components. Anderson (1975) proposed that the neutral galactolipids would form the fluid matrix, because their acyl groups have the greatest degree of unsaturation of the various lipid classes; further, their polar head groups are uncharged and they may not be an integral part of the photosystems per se. If this were so, it would leave sulpholipid and the phospholipids to be variously immobilized as boundary lipids. They would be suited for this because they contain one saturated and one unsaturated acyl group. Moreover, their polar head groups all carry fixed charges which could be involved in interactions with the fixed charges of the hydrophilic protein domains.

Interest in the spatial arrangement of thylakoid components has been greatly stimulated by the chemiosmotic hypothesis of energy conservation (Mitchell, 1966) which requires a vectorial arrangement of photosynthetic electron transport components. Both the light-induced generation of potential across the membrane in the primary photoacts (Witt, 1971) and the uptake of protons from the stroma by thylakoids during electron transport demonstrate that these membranes are asymmetric. Strategies to explore thylakoid topology have included the use of specific group labeling reagents, hydrolytic enzymes, and the extensive use of antibody labeling techniques. From such studies, it is clear that thylakoids are indeed highly asymmetric (Trebst, 1974; Anderson, 1975), although the evidence is not as comprehensive as has been obtained for erythrocyte membranes (Bretscher and Raff, 1975).

The proteins of thylakoids include chloroplast ATPase, which is structurally and chemically similar to that of inner mitochondrial and bacterial ATPases, and consists of an inner membrane segment, CF_0, which is placed across the membrane with the photophosphorylation coupling factor, CF_1, located at the outer thylakoid surface. Accumulated evidence from electron microscopy (both freeze-etch and negative staining) and antibody labeling experiments demonstrate that CF_1 is located at the outer surface of unstacked stroma and end grana thylakoids (see Miller and Staehelin, 1976).

The other functional proteins of chloroplast thylakoids are the electron transport components of both photosystems. Some of these are accessible at or near the outer thylakoid surface, whereas some are more deeply buried within the membrane or located at or towards the inner thylakoid surface. Description of their location in thylakoids is hampered by the unavailability from thylakoids of "inside-out" vesicles, such as may be prepared from inner mitochondrial or photosynthetic bacterial membranes, and by the limited accessibility of the outer surface of the contacted thylakoids of the grana. There is some evidence for the vectorial arrangement of the electron carriers of PS I (Trebst, 1974). The adjacent electron carriers on the acceptor site of PS I, ferredoxin and ferredoxin $NADP^+$ reductase, are located toward the outer thylakoid surface; cytochrome *f* and possibly plastocyanin are located at or towards the inner thylakoid surface (Trebst, 1974). The distribution of the components of PS II is uncertain, because the evidence for their location is equivocal and the components are not fully characterized (Trebst, 1974). Recently, Renger (1976) used trypsin to explore the topology of PS II; this proteolytic enzyme does not penetrate thylakoids and attacks proteins only at the outer membrane surface. Renger showed that the primary electron acceptor X320 (Witt, 1971) is covered by a proteinaceous component susceptible to trypsin attack, but components on the oxidizing side of PS II were unaffected, suggesting they are located toward the inner thylakoid surface. Although the vectorial distribution of the electron carriers of both photosystems is not yet proved, it seems that the supramolecular complexes of each of the photosystems must extend across the membrane. Indeed, such a molecular geometry would be a necessity for the separation of charge across the membrane that occurs in the primary photoacts (Witt, 1971). Furthermore, it is likely, although unproved, that the chlorophyll-protein complexes also extend across the membrane (Anderson, 1975).

Most, if not all, of the chlorophylls are attached to specific proteins; these chlorophyll-protein complexes are the main thylakoid intrinsic proteins. Thylakoid membranes may be fragmented by nonionic detergents, sonication, or the French pressure cell into subchloroplast fragments with different photochemical properties and chemical composition (Boardman, 1970). The small fragments with a high chlorophyll *a*/chlorophyll *b* ratio are enriched in PS I, whereas the larger fragments, derived from grana thylakoids, have a low chlorophyll *a*/chlorophyll *b* ratio and are enriched in PS II. These fragments contain a large number of thylakoid polypeptides, including those of the chlorophyll-protein complexes, suggesting that each photosystem may be organized as a specific complex in the membrane. Extended detergent treatment, either by digitonin (Wessels and Borchert, 1975) or Triton X-100 (Vernon and Klein, 1975), permits the isolation of smaller particles which contain discrete chlorophyll-protein complexes: a PS I complex containing the reaction-center molecule P 700, a PS II complex containing the reaction-center molecule P 680, and a third light-harvesting complex.

A more complete solubilization of thylakoids is obtained with anionic detergents. When chloroplast thylakoids are solubilized in sodium dodecyl sulfate (SDS) without prior extraction of lipids, two main chlorophyll-protein complexes are separated by SDS polyacrylamide gel electrophoresis (PAGE) (Thornber, 1975; Brown et al., 1975). One is chlorophyll-protein complex I (CP I), which is the central complex of PS I, as it contains P 700. This complex has 1 P 700/45 chlorophyll *a* molecules, and the molecular weight of its single polypeptide is 64,000–70,000 daltons (Machold, 1975; Nelson and Bengis, 1975; Chua et al., 1975). The second, and major, complex is the light-harvesting chlorophyll-protein complex (LHCP), which represents some 40–60% of the total chlorophyll of mature thylakoids. This complex is photochemically inactive but transfers light energy to PS II and possibly PS I. LHCP contains both chlorophyll *a* and chlorophyll *b*; it is agreed that this complex contains most, if not all, of the chlorophyll *b* of mature thylakoids (Thornber, 1975). It is not established whether LHCP contains one or two polypeptides in the 25,000-dalton range (Anderson, 1975). In the case of *Acetabularia mediterranea,* a LHCP of 67,000 daltons has been shown to include two subunits of 23,000 and 21,500 daltons (Apel, 1977). The third complex, which contains the reaction center of PS II, P 680, and associated chlorophyll *a* molecules, is not detected by SDS PAGE. However, using *Chlamydomonas* mutants deficient in PS 2 reaction centers, Chua and Bennoun (1975) have circumstantial evidence that a polypeptide of 47,000 daltons is associated with this chlorophyll complex.

That the two major chlorophyll-protein complexes are indeed the main intrinsic proteins of chloroplast thylakoids was elegantly demonstrated by Machold (1975). Purified tobacco thylakoids, with a protein/chlorophyll mass ratio of 4, were extracted exhaustively with the protein peturbant, 6 M guanidine-hydrochloride, thereby removing extrinsic proteins which comprised about half of the membrane protein. The insoluble residue (protein/chlorophyll ratio of 2) still preserved its basic membrane structure; it was demonstrated by SDS PAGE that the polypeptides of both CP I and LHCP were quantitatively retained and comprised about 70% of the intrinsic polypeptides. Similarly, Apel (1977) showed that *Acetabularia mediterranea* thylakoids treated with EDTA and pronase lost some 60% of their membrane proteins. The pronase-resistant membrane retained its basic membrane structure and the chlorophyll-protein complexes were still present, confirming that they are intrinsic proteins. Enzymatic iodination of EDTA-treated thylakoids resulted in labeling of the 23,000-dalton subunit but not the 21,500-dalton subunit of the light-harvesting complex; further the chlorophyll-containing polypeptide of the PS I complex was not labeled. Consequently, two of the chlorophyll-containing polypeptides of this green alga appear to be buried in the membrane.

A question of central importance concerns the localization of the chlorophyll molecules. Fenna and Matthews (1975) have determined the X-ray structure of a water-soluble bacteriochlorophyll-protein from the green photosynthetic bacterium, *Chlorobium*. This complex consists of three identical subunits, each containing a core of seven bacteriochlorophyll molecules arranged in an irregular fashion and completely surrounded by protein. This may not be the arrangement of chlorophyll molecules in the intrinsic complexes of higher plant thylakoids. Anderson (1975) proposed that the light-harvesting chlorophylls would be located as boundary lipids of their specific intrinsic proteins (Fig. 1). In this model, the hydrophobic phytyl chains are inserted perpendicular to the membrane plane in close interaction with the hydrophobic exterior of the intrinsic proteins. The hydrophobic portion of the chlorin ring is bent over and buried in the hydrophobic interior of the protein, leaving its hydrophilic side at the membrane surface. If the concept of intrinsic proteins being surrounded by a layer of boundary lipids is correct, a substantial fraction of the thylakoid lipid must be so immobilized. One reason for proposing that chlorophyll is located as boundary lipid is that thylakoids have less structural lipid available for the bilayer than is found with many other biomembranes.

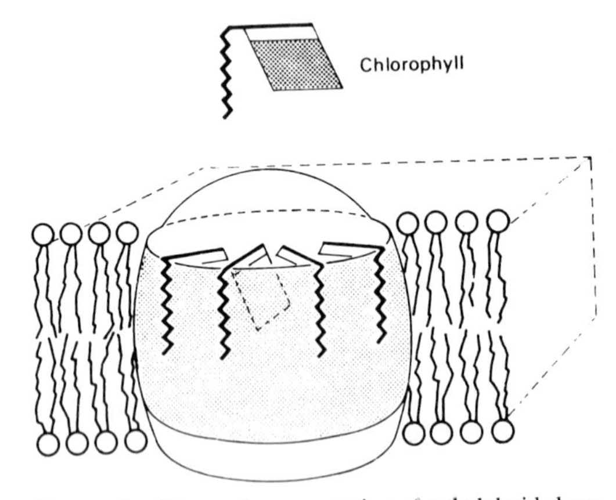

FIGURE 1 Schematic cross section of a thylakoid showing an intrinsic protein extending across the membrane. It is proposed that the light-harvesting chlorophyll molecules are part of the boundary lipids of their specific intrinsic proteins. Additional lipid molecules would be required to complete the layer of boundary lipids surrounding the intrinsic protein (Anderson, 1975).

Molecular Organization

The function of chloroplast membranes depends on protein-protein, lipid-protein, and lipid-lipid interactions. Having discussed how the individual thylakoid components may be located within the membrane, we now must consider how these components interact with each other to form supramolecular complexes. The molecular organization is complicated by the further differentiation of mature chloroplast thylakoids into grana and stroma regions which possess distinct functional and substructural properties. Further, we have to consider the interaction between PS I and PS II, interactions between the electron transport chain carriers and the more bulky chlorophyll-protein complexes within each photosystem, and the coupling of electron transport to photophosphorylation.

Much of our knowledge of the location and properties of the photosystems comes from procedures for thylakoid fragmentation that allow separation of grana and stroma thylakoids. These procedures include either mechanical methods, such as sonication or passage through the French pressure cell, or nonionic detergents, digitonin or Triton X-100 (Boardman, 1970). After thylakoid fragmentation, subchloroplast fragments enriched in PS I or PS II are isolated by differential centrifugation. In all cases, a large subchloroplast fragment is obtained which is enriched in PS II and chlorophyll *b*; this is derived from grana thylakoids (Park and Sane, 1971; Arntzen and Briantais, 1975). In contrast, the smaller vesicles contain mainly PS I and have a higher chlorophyll *a*/*b* ratio compared to chloroplasts. These small vesicles are thought to be derived from stroma thylakoids in the French pressure cell method, but are probably derived from both stroma and grana thylakoids with digitonin treatment. These results and other observations led to the proposal by Park and Sane (1971) that PS II is restricted to grana in mature chloroplasts, whereas PS I is present both in grana and stroma thylakoids. Definitive evidence for this proposal can only be obtained when the presence or absence of the photosystems can be shown in vivo in different regions of the membranes. Recent evidence shows that the PS I derived from stroma thylakoids is identical in composition to that of grana thylakoids (Wessels and Borchert, 1975; Brown et al., 1975).

This functional differentiation of the membrane system is paralleled by the gross structural differentiation. Beyond this, however, there is a difference in the distribution of the particulate subunits of chloroplast thylakoids which occur both within the membrane and at its surface. This substructure is wondrously revealed by the freeze-fracture technique, by which the frozen membrane is fractured through its hydrophobic interior, exposing complementary fracture faces. Replicas, then, show smooth areas representing the lipid domains and particles which probably represent intrinsic proteins and their associated boundary lipids. This view seems reasonable, since artifical lipid layers and vesicles have no such particles unless globular intrinsic proteins are included.

Chloroplast thylakoids possess a high density of freeze-fracture particles whose sizes, being markedly different, are referred to as "large" and "small" particles; however, there is a size distribution within these groups (Park and Sane, 1971; Arntzen and Briantais 1975). Histograms of the particle size distribution demonstrate that the particles fall into four size classes of 7.0-, 10.5-, 14.0-, and 16.0-nm diameter in *Chlamydomonas* (Ojakian and Satir, 1974). These particles are located on different fracture faces; furthermore, the appearance of the fracture faces of the stacked membrane region is unique and distinctly different from that of unstacked membrane regions. Thus large, widely spaced particles are found in the inner fracture face (EF) adjacent to the intrathylakoid space of stacked membranes only, whereas the coplanar unstacked membrane has fewer particles of a smaller size. In contrast, the outer fracture face (PF) adjacent to the stroma contains small, tightly packed particles only, in both stacked and unstacked membrane regions (Table I). The freeze-fracture data suggest that the particles are arranged asymmetrically within the thylakoid membrane, with the large particles located more toward the inner half of the bilayer and partly protruding into the intrathylakoid space, and the small particles located more toward the outside of the thylakoid membrane. Circumstantial evidence from freeze-fracture studies with other membranes indicates that particles probably must extend across the membrane in order to be visualized, and this is probably true for chloroplast thylakoids. The asymmetric distribution of the particles on the fracture faces of thylakoids may arise because the small particles have more mass in the outer half of the membrane layer and hence cleave with that fracture face, while the large particles may have more mass in the inner half of the bilayer.

The unique structural organization of thylakoids is dependent on cations. Izawa and Good (1966) first showed that spinach thylakoids, suspended in zwitterionic buffers, no longer have the characteristic areas of grana and stroma thylakoids, and the membranes become completely unstacked. This effect is reversed when monovalent or divalent cations are added back to the low-salt media and the membranes are restacked.

In an elegant study with *Chlamydomonas* membranes, Ojakian and Satir (1974) quantitatively measured the size and distribution of freeze-fracture particles. They showed that artificial unstacking of these membranes resulted in a random distribution of freeze-fracture particles along the membrane, but their total number and sizes on both fracture faces remained nearly constant (Table I). In the *ac*-31 strain of *Chlamydomonas,* the thylakoids are unstacked in vivo after isolation in Tris buffer; addition of Tris-Mg buffer will induce membrane stacking and produce a change in distribution of the particles, with no change in their sizes or number (Table I). The nearly complete conservation of both the numbers and sizes of the

TABLE I

*Particle Conservation on Fracture Faces of Chlamydomonas Thylakoids**

Strain	Fracture face†	Particles per μm^2		% membranes stacked	Particles per μm^2; unstacked regions	
		Stacked	Unstacked		Measured	Calculated§
wt	PF	3,510	3,665	57	3,440	3,565
wt	EF	1,943	606	57	1,337	1,368
ac-31	EF	2,514	888	43	1,600	1,606

* Ojakian and Satir, 1974.

† PF, outer fracture face adjacent to stroma; EF, inner fracture face.

§ Calculated particle distribution expected by a redistribution of existing components by lateral diffusion in the fluid matrix so that the particles occupy the entire membrane surface.

particles from the stacked to the unstacked configuration and vice versa indicates a rearrangement of membrane components by lateral mobility in the membrane plane. This not only demonstrates the fluid nature of thylakoids, but also points to a different distribution of particles, that is, intrinsic proteins, in grana and stroma thylakoids. As mentioned, the main intrinsic proteins are the chlorophyll-protein complexes (Machold, 1975).

Although chloroplast coupling factor, CF_1, has been detected only on unstacked membranes, it has been difficult to resolve whether it is also present on the regions of stacked membranes, but not detected there because of the close contact of the grana thylakoids. Recently, Miller and Staehelin (1976) have examined the outer surface of spinach thylakoids by antibody labeling and by the technique of freeze-etching, which exposes the outer surfaces of biological membranes by sublimitation of their frozen media. After some 30% of the large freeze-etch particles, identified as carboxydismutase, had been selectively removed, the remaining large particles were removed under conditions which resulted in the loss of ATPase activity. Addition of purified coupling factor restored the large particles and ATPase activity. Since the ATPase activity and the particle numbers were the same in control and reconstituted membranes, these particles were identified as CF_1. Spinach thylakoids, from which carboxydismutase had been previously removed, had 720 particles per μm^2 of unstacked membrane surface, which formed some 35% of the total membrane surface (Table II). Artificial unstacking of these membranes by the method of Izawa and Good (1966) led to an apparent decrease in the distribution of particles to 264 particles per μm^2 consistent with the idea that the particles originally present in the unstacked membranes had been free to move over the entire membrane surface. Restacking the membrane system by the addition of cations restored the original particle distribution. Significantly, no loss or increase in either particle size or ATPase activity is observed during these procedures (Table II). This dramatic conservation of numbers and activity of CF_1 particles during unstacking and restacking suggests that CF_1 is excluded from the stacked membrane region, where most of the photochemical activities are located. Consequently, the coupling between electron transport and photophosphorylation may be indirect.

TABLE II

*CF_1 Distribution on Spinach Thylakoids during Unstacking and Restacking**

Preparation	Particle density (per μm^2)	Percentage membrane surface in unstacked regions	Calculated particle density (per μm^2)†	ATPase-specific activity§
Stacked	710	35	251	5.33
Unstacked	261	100	261	5.78
Restacked	690	40	282	5.44

* Miller and Staehelin, 1976.
† Number of CF_1 particles per μm^2 of total membrane surface.
§ Units of enzyme required to release 1 μmol inorganic phosphate per min at 37°C.

Chloroplast thylakoids show both functional and structural differences between stacked and unstacked thylakoids, with PS II and large freeze-fracture particles being restricted to stacked membranes. The location of the chlorophyll-protein complexes has been under extensive study. Much evidence derived mainly from SDS PAGE now demonstrates that the major chlorophyll-protein complex, LHCP, is restricted mainly to stacked membrane regions (Anderson, 1975). First, fragmentation studies of chloroplast thylakoids, using either nonionic detergents or mechanical procedures, demonstrate that CP I is found in both subchloroplast fragments derived from grana and stroma thylakoids. In contrast, LHCP is restricted primarily to the stacked membrane fractions enriched in PS II and chlorophyll *b*. However, the PS I vesicles derived from the French pressure cell do contain some LHCP (Brown et al., 1975). Second, studies with mutants have been useful. These mutants generally are chlorophyll-deficient, especially in chlorophyll *b*, and possess reduced or no membrane stacking. Recent examination by SDS PAGE of mutant chloroplasts reveals that they contain little or no LHCP (Anderson and Levine, 1974). Finally, membrane biogenesis studies show that developing plastids initially have unstacked membranes which contain little or no chlorophyll *b*, and neither the polypeptides of LHCP nor LHCP itself.

Both large freeze-fracture particles and LHCP are confined mainly to stacked membranes, and over 50% of the intrinsic protein of thylakoids belongs to LHCP, so Anderson (1975) proposed that large freeze-fracture particles would include LHCP. The small, more densely packed particles of the outer fracture face found in both stacked and unstacked membranes might include CP I which represents about 20% of the thylakoid intrinsic protein.

The structural approach was used by Armond et al. (1977), who studied the development in continuous light of chloroplasts from pea seedlings which previously had been exposed only to intermittent light. Such chloroplasts (0-h plastids) are photosynthetically competent, but they have unstacked membranes and no LHCP. Continuous illumination of the pea seedlings leads to membrane synthesis and differentiation, concomitantly with synthesis of LHCP (Arntzen et al., 1977). During this membrane synthesis and differentiation, Armond et al. (1977) showed that the total number of freeze-fracture particles on both fracture faces is fairly constant; hence, incorporation of massive amounts of LHCP into the thylakoids did not result in new particles. However, there was a significant increase in the size of the particles, particularly those on the inner fracture face. The 0-h plastids had particles of 8 nm; during greening their size increased gradually, possibly in discrete jumps, so that finally a population containing four main sizes was obtained (8.0-, 10.5-, 13.2-, and 16.4-nm diameters) with the 10.5- and 16.4-nm particles being dominant in 48-h plastids. Armond et al. (1977) propose that the 8.0-nm particles represent "core complexes" of PS II, which would contain the PS II reaction center and antenna chlorophyll *a*, and that subsequently during greening, aggregate complexes of LHCP are added to these "cores" to form completed PS II units. They suggest that the small, densely packed particles of the outer fracture face may correspond to the morphological equivalent of a PS I complex (Fig. 2).

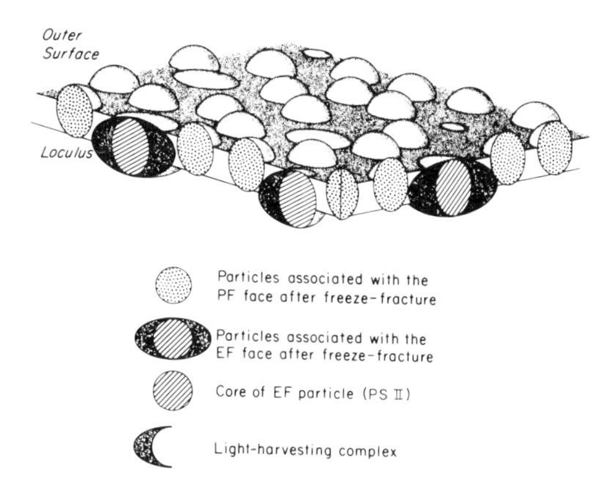

FIGURE 2 A proposed model for the localization of the particulate subunits of thylakoids obtained from a freeze-fracture study. In plastids from plants grown under intermittent illumination (0-h plastids), the membranes are unstacked and 7.5-nm particles are associated with the outer fracture face (PF), and 8.0-nm particles are associated with the inner fracture face (EF). After the pea seedlings had been greened in continuous illumination (48-h plastids), grana stacking occurs together with massive synthesis of LHCP. No new particles are formed, but particularly those on the inner fracture face increase in size. It is proposed that aggregates of LHCP are added to the 8.0-nm particles, which represent "core complexes" of PS II, to form completed PS-2 units. The PF particles may then be the morphological equivalent of PS I (Armond et al., 1977).

The complex beauty of freeze-fracture micrographs has revealed the asymmetry of distribution of particulate subunits within and along the thylakoids, thereby provoking many hypotheses concerning their chemical and functional identity. Eventually, these freeze-fracture particles must be identified either by isolation or by labeling in situ.

From evidence derived from thylakoid fragmentation studies, it has been proposed that PS II is confined principally to grana in mature thylakoids. Nevertheless, it is clear that stacked membranes are not necessarily required for PS II activity. Thus, a number of higher plant and algal mutants which have no membrane stacking have PS II activity, and this situation occurs also in developing plastids which may be photosynthetically competent while their membranes are unstacked (Anderson, 1975; Arntzen and Briantais, 1975).

What, then, is the significance of grana stacking? Environmental conditions affect grana development; growth at low light results in more membrane stacking, particularly with plants grown in densely shaded habitats. Shade-plant chloroplasts have more chlorophyll, especially chlorophyll *b*, than do sun-plant chloroplasts (Boardman et al., 1975), and they have more LHCP (Brown et al., 1975). Boardman et al. (1975) proposed that grana may simply be a means of increasing the light-harvesting capacity of membranes. Not only is light trapping possible along the membrane, but the close proximity of photosynthetic units on adjacent membranes would also allow excitation energy transfer across the membrane. Cations exert an effect on the distribution of excitation energy between the photosystems, thereby regulating quantum efficiency. Several studies suggest that LHCP may be involved in this cation-regulated control of energy distribution (see Arntzen et al., 1977). In developing chloroplasts, the onset of Mg^{++} regulation of excitation energy between the photosystems correlated with the synthesis of LHCP, and Arntzen et al. (1977) conclude that

LHCP is involved in this effect. However, the synthesis of LHCP occurs concomitantly with grana formation, and it may be the stacking of membranes, rather than LHCP itself, which is required for this cation-regulated distribution of excitation energy between the photosystems. Further study of the effect of cations in modulating the interactions and organization of the photosystems should give insight into the molecular mechanism of membrane stacking.

In summary, the combined evidence of functional, structural, and compositional differences between grana and stroma thylakoids suggests that undifferentiated membranes contain "initial" PS I and PS II and chloroplast ATPase, all of which may be free to move laterally within the membrane plane. Following syntheses of the polypeptides and chlorophylls of LHCP, part of the undifferentiated thylakoids become stacked. During this stacking process, it seems that the bulky chloroplast coupling factor is excluded from the stacked membrane region (Miller and Staehelin, 1976) and thus is left in the stroma thylakoids. Although this picture of the molecular organization of the photosystem may be satisfactory at the structural level, we are very far from a precise description of interactions of individual thylakoid components at the molecular level. Furthermore, the picture presented is static rather than dynamic. The very large, light-induced changes in ion fluxes across chloroplast membranes will have a profound influence on the molecular interactions of thylakoid components (Murakami et al., 1975).

The structural relationship between PS I and PS II and the mechanism of their interaction have yet to be elucidated. However, the photosystems may not be directly linked to one another, as the large pool of plastoquinone molecules may be mobile in the membrane and the link between the photosystems would then be a dynamic one. A complex interaction exists at two levels within each photosystem: on the one hand, there is the light-harvesting system through which excitation energy is transferred from the many light-harvesting chlorophylls to the reaction-center chlorophyll and, on the other, there is the electron transport chain which is associated also with the reaction center. The chlorophyll-protein complexes are large compared to individual components of the electron transport chain. In the central complex of PS I, each P 700 molecule is associated with some 45 chlorophyll *a* molecules and a polypeptide of 70,000 daltons, and this complex probably has both the primary donor and acceptor for PS I (Nelson and Bengis, 1975). However, there are another 150 or so chlorophyll molecules in PS I, some of which may be complexed to the 70,000-dalton polypeptide or to other, as yet unidentified, polypeptides; the remaining PS I chlorophylls probably belong to LHCP. The central complex of PS II is not characterized as yet, but it also is likely to have light-harvesting chlorophyll *a* molecules in close association with the reaction center, P 680. Both of these complexes, then, may be surrounded by aggregates of LHCP, with much more LHCP associated with the PS II complex than with the PS I complex. Interaction of LHCP aggregates with the central complexes of both photosystems may well be dynamic rather than static. It is probable that this association is influenced by the cation concentrations in local domains of the membrane. If LHCP is partly involved in membrane stacking, however, some may be anchored within the grana thylakoids.

Comparison of the electron transport chains from plants grown at different light intensities indicates that there is no universal stoichiometry between the amounts of electron carriers (e.g., cytochrome *f*) and the amounts of P 700 and Q (a marker for PS II reaction centers) (Boardman et al., 1975). Consequently, these authors suggest that individual molecules of the electron transport chain do not belong specifically to a single chain, but are part of a pool which may be common to several chains. An implication of these proposals is that some of the electron transport carriers may be mobile in the membrane. As the collision rates of membrane proteins, which are free to move laterally in the membrane, are extremely high (Lee, 1975), it is possible that interaction between the central complexes containing either P 700 or P 680 and complexes of intermediate electron carriers may also be dynamic, rather than static. Thus, there need not necessarily be a specific macromolecular complex containing all of the components of each photosystem.

The romantic era of speculative dreams of the molecular organization of membranes is almost ended, even for chloroplast thylakoids. With the investigative strategies now available, nothing but the discipline of experimentation is needed to reveal the interactions between individual thylakoid proteins and between proteins and lipids. The future challenge is to unravel the intricate mechanism of the assembly of chloroplast membranes from their diverse components and to explain how this assembly is controlled by the complex interaction of the chloroplast and nuclear genomes.

REFERENCES

ANDERSON, J. M. 1975. The molecular organization of chloroplast thylakoids. *Biochim. Biophys. Acta.* **416:**191–235.

ANDERSON, J. M., and P. R. LEVINE. 1974. The relationship between chlorophyll-protein complexes and chloroplast membrane polypeptides. *Biochim. Biophys. Acta.* **357:**118–126.

APEL, K. 1977. Chlorophyll-proteins from *Acetabularia mediterranea*. Brookhaven Symposium, Brookhaven, N. Y. In press.

ARMOND, P. A., L. A. STAEHELIN, and C. J. ARNTZEN. 1977. Spatial relationship of photosystem I, photosystem II and the light-harvesting complex in chloroplast membranes. *J. Cell Biol.* In press.

ARNTZEN, C. J., P. A. ARMOND, J. M. BRIANTAIS, J. J. BURKE, and W. P. NOVITSKY. 1977. Dynamic interactions among structural compounds of the chloroplast membrane. Brookhaven Symposium, Brookhaven, N. Y. In press.

ARNTZEN, C. J., and J. M. BRIANTAIS. 1975. Chloroplast structure and function. *In* Bioenergetics of Photosynthesis. Govindjee, editor. Academic Press, Inc., New York. 51–113.

BOARDMAN, N. K. 1970. Physical separation of the photosynthetic photochemical systems. *Annu. Rev. Plant Physiol.* **21:**115–140.

BOARDMAN, N. K., O. BJÖRKMAN, J. M. ANDERSON, D. J. GOODCHILD, and S. W. THORNE. 1975. Photosynthetic adaptation of higher plants to light intensity: relationship between chloroplast structure, composition of the photosystems and photosynthetic rates. *In* Third International Congress on Photosynthesis. M. Avron, editor. Elsevier, Amsterdam. 1809–1827.

BRETSCHER, M. S., and M. C. RAFF. 1975. Mammalian plasma membranes. *Nature (Lond.).* **258:**43–49.

BROWN, J. S., R. S. ALBERTE, and J. P. THORNBER. 1975. Comparative studies on the occurrence and spectral composition of chlorophyll-protein complexes in a wide variety of plant material. *In* Third International Congress on Photosynthesis. M. Avron, editor. Elsevier, Amsterdam. 1951–1962.

CHUA, N. H., and P. BENNOUN. 1975. Thylakoid membrane polypeptides of *Chlamydomonas reinhardtii*: wild-type and mutant strains deficient in photosystem II reaction centers. *Proc. Natl. Acad. Sci. U. S. A.* **72:**2175–2179.

CHUA, N. H., K. MATLIN, and P. BENNOUN. 1975. A chlorophyll-protein complex lacking in photosystem I mutants of *Chlamydomonas reinhardtii*. *J. Cell Biol.* **67:**361–377.

FENNA, R. E., and B. W. MATTHEWS. 1975. Chlorophyll arrangement in a bacteriochlorophyll-protein from *Chlorobium limicola*. *Nature (Lond.)* **258:**573–577.

GOVINDJEE, and R. GOVINDJEE. 1975. Introduction to photosynthesis. *In* Bioenergetics of Photosynthesis. Govindjee, editor. Academic Press, Inc., New York. 1–50.

IZAWA, S., and N. E. GOOD. 1966. Effect of salts and electron transport on the conformation of isolated chloroplasts. II. Electron microsocpy. *Plant Physiol.* **41:**544–552.

LEE, A. G. 1975. Functional properties of membranes: a physical-chemical approach. *Prog. Biophys. Mol. Biol.* **29:**3–56.

LICHTENTHALER, H. K., and R. B. PARK. 1963. Chemical composition of chloroplast lamellae from spinach. *Nature (Lond.)* **198:**1070–1072.

MACHOLD, O. 1975. On the molecular nature of chloroplast thylakoid membranes. *Biochim. Biophys. Acta.* **382:**494–505.

MILLER, K. R., and L. A. STAEHELIN. 1976. Analysis of the thylakoid outer surface. *J. Cell Biol.* **68:**30–47.

MITCHELL, P. 1966. Chemiosmotic coupling in oxidative and photosynthetic phosphorylation. *Biol. Rev. (Camb.)* **41:**445–502.

MURAKAMI, S., J. TORRES-PEREIRA, and L. PACKER. 1975. Structure of the chloroplast membrane relation to energy coupling and ion transport. *In* Bioenergetics of Photosynthesis. Govindjee, editor. Academic Press, Inc., New York. 555–618.

NELSON, N., and C. BENGIS. 1975. Reaction center P 700 from chloroplasts. *In* Third International Congress on photosynthesis. M. Avron, editor. Elsevier, Amsterdam. 609–620.

OJAKIAN, G. K., and P. SATIR. 1974. Particle movements in chloroplast membranes: quantitative measurements of membrane fluidity by the freeze-fracture technique. *Proc. Natl. Acad. Sci. U. S. A.* **71:**2052–2056.

PARK, R. B., and P. V. SANE. 1971. Distribution of function and structure in chloroplast lamellae. *Annu. Rev. Plant Physiol.* **22:**395–430.

RENGER, G. 1976. Studies on the structural and functional organization of system II of photosynthesis. The use of trypsin as a structurally selective inhibitor at the outer surface of the thylakoid membrane. *Biochim. Biophys. Acta.* **440:**287–300.

SADLER, D. M., M. LEFORT-TRAN, and M. POUPHILE. 1973. Structure of photosynthetic membranes of *Euglena* using x-ray diffraction. *Biochim. Biophys. Acta.* **298:**620–629.

SINGER, S. J. 1974. The molecular organization of membranes. *Annu. Rev. Biochem.* **43:**805–833.

SINGER, S. J., and G. L. NICOLSON. 1972. The fluid mosaic model of the structure of cell membranes. *Science (Wash. D. C.).* **175:**720–731.

THORNBER, J. P. 1975. Chlorophyll-proteins: light-harvesting and reaction center components of plants. *Annu. Rev. Plant Physiol.* **26:**127–158.

Trebst, A. 1974. Energy conservation in photosynthetic electron transport of chloroplasts. *Annu. Rev. Plant Physiol.* **25:**423–458.

Vernon, L. P., and S. M. Klein. 1975. Nature of plant chlorophylls *in vivo* and their associated proteins. *Ann. N. Y. Acad. Sci.* **244:**281–296.

Wessels, J. S. C., and M. T. Borchert. 1975. Studies on subchloroplast particles. Similarity of grana and stroma photosystem I and the protein composition of photosystem I and photosystem II particles. *In* Third International Congress on Photosynthesis. Vol. I. M. Avron, editor. Elsevier, Amsterdam. 473–484.

Witt, H. T. 1971. Coupling of quanta, electrons, fields, ions and phosphorylation in the functional membrane of photosynthesis. *Q. Rev. Biophys.* **4:**365–477.

ONTOGENY AND ASSEMBLY OF CHLOROPLAST MEMBRANE POLYPEPTIDES IN *CHLAMYDOMONAS REINHARDTII*

I. OHAD

Formation of photosynthetic membranes occurs by a process of growth of preexisting membranes. This has been demonstrated in numerous systems such as growth and multiplication of chloroplasts in dividing cells, greening of etiolated plants, or when conditional mutants of algae are transferred from restrictive to permissive conditions (Ohad, 1975). The process of membrane assembly can be resolved into several steps: synthesis of membrane components, their binding and insertion into the membrane, and their integration, together with other membrane components, into functional photosynthetic units. Although all membrane components—proteins, lipids, and pigments—should be considered, I must limit myself to the discussion of the synthesis and integration of membrane polypeptides and chlorophyll and the establishment of photosynthetic electron flow.

Ontogeny and Function of Chloroplast Membrane Peptides

Several major chloroplast membrane peptides are synthesized by 80S cytoplasmic polyribosomes. These ribosomes might be free or bound to the double, outer, chloroplast envelope. In both cases, the peptides must cross this barrier and find their way to the developing photosynthetic membrane, a process that still awaits elucidation. Moreover, nuclear messages coding amino acid sequence of membrane polypeptide might also be translated by the 70S chloroplast ribosomes. Such a case was recently demonstrated in a temperature-sensitive, nuclear mutant of *Chlamydomonas reinhardtii* (T4). In this mutant, a 44 kdalton-polypeptide coded by a nuclear gene is translated in the chloroplast and required for the formation of photosystem II (PS II; Kretzer et al., 1977).

The chloroplast membranes of *C. reinhardtii* can be resolved by the current sodium dodecyl sulfate (SDS) acrylamide electrophoretic techniques into about 20–30 distinct polypeptides, of which about 18 are present in significant amounts. Based on the results obtained in several laboratories, the possible function and corresponding transcripton and translation sites of about 12 polypeptides can be identified as follows (Table I): polypeptide(s) of chloroplast translation (63–65 kdaltons) appears to be required for the formation of photosystem I (PS I; Bar-Nun and Ohad, 1974). The fact that polypeptides in this molecular weight range are missing in nuclear mutants of *C. reinhardtii* suggests that at least one of these polypeptides might be coded by a nuclear gene (Chua et al., 1975). The formation of the chlorophyll-protein complex I (CP I) as detected by SDS gel electrophoresis requires their presence. After its dissociation, the isolated CP I complex releases a polypeptide(s) of 64 kdaltons (Chua et al., 1975), to which about 20–30 chlorophyll *a* molecules are specifically bound (Bar-Nun et al., 1977). A 49-kdalton polypeptide of chloroplastic translation is required for the development of PS I activity. Isolated membrane particles prepared by deoxycholate treatment exhibit high PS I activity, measured with methyl viologen as an electron acceptor when this polypeptide is present. Such particles isolated from membranes that lack the 49-kdalton polypeptide do not exhibit this activity (Bar-Nun and Ohad, 1974). A 47-kdalton polypeptide of chloroplastic translation and transcription is associated with the water-splitting activity of PS II (Kretzer et al., 1977). The formation of the reaction center of PS II also depends on the presence

I. OHAD Department of Biological Chemistry, The Hebrew University of Jerusalem, Jerusalem, Israel

TABLE I

Ontogeny and Function of Major Chloroplast Membrane Peptides in Chlamydomonas reinhardtii

Synthesis		Function							
Transcription	Translation	Light control	Photochemical activity	Chlorophyll binding	Stabilization of complexes	Mol wt	Gel pattern	Other nomenclatures	
Nuclear	Chloroplastic	–	PS I	20–30 chl *a*	CP I	63; 65	☆	1–3; I	
	Chloroplastic	–	PS I		CP I	49		4_1; I; II	II_1
Chloroplastic	Chloroplastic	–	PS II (H_2O)		CP I	47		4_2;	II_2
	Chloroplastic	–			CP I	45		II_b;	II_3
Nuclear	Chloroplastic	–	PS II (RC)		CP I	44			II_4
	Chloroplastic	–				40			II_5
	Cytoplasmic	±				30–32		8–10; IIa	III
Nuclear	Cytoplasmic	±				28*	★	11–12; IIb	IV
	Chloroplastic	–	PS II			26		13–14; IVa	
Nuclear	Cytoplasmic	+		8–10 chl *a*, *b*	CP II	24		15–18; IIc;	Va
Nuclear	Cytoplasmic	+			CP II	22			Vb

The data are taken from Anderson and Levine, 1974; Bar-Nun and Ohad, 1974; Brown et al., 1975; Chua and Bennoun, 1975; Chua et al., 1975; Eytan and Ohad, 1972; Hoober and Stegeman, 1973; Kan and Thornber, 1976; and Kretzer et al., 1976. Molecular weights ($\sim\pm2\%$) are estimated from electrophoretic mobility in the presence of SDS. (RC = reaction center.) The free mobility of chlorophyll-protein complexes CP I and CP II are 1.6–1.8. The respective polypeptides after dissociation from chlorophyll have free mobilities of about 1.1–1.2. This might indicate that the complexes have a much higher molecular weight than the apparent one. If one considers that CP I contains polypeptides 63–65 kdaltons and 20–30 chlorophyll *a*, the molecular weight might be 150–160 kdaltons; the apparent molecular weight is found to be about 88 kdaltons. The CP II contains polypeptide 22 and 24 kdaltons and about 10 chl (*a*, *b*) and thus might have a molecular weight of 56 kdaltons while the apparent molecular weight is 28 kdaltons.

* The 28-kdalton peptide is usually considered to bind chlorophyll *a* and *b* in CP II complex (Kan and Thornber, 1976); white star, black star, location of CP I and CP II, respectively.

of a 44-kdalton polypeptide of chloroplastic translation and nuclear transcription. In addition to the 64-kdalton polypeptide, all the other polypeptides of chloroplastic translation are detected as being present in small amounts in the isolated CP I complex. Although some of these polypeptides, such as the 47 and 44 kdaltons, are associated with PS II activity, their presence in the membrane is a prerequisite for the organization of the components that form CP I, as detected by the SDS acrylamide gel electrophoretic technique (Bar-Nun et al., 1977).

The synthesis of all the above polypeptides and their insertion into the membrane is not controlled by light, and for some it can occur independently if a second group of polypeptides of cytoplasmic transcription and translation already has been synthesized and integrated into the membrane (Ohad, 1975). This includes polypeptides of 30–32, 28, 24, and 22 kdaltons. The 24- and 22-kdalton polypeptides bind chlorophyll, and thus form the light-harvesting chlorophyll-protein complex II (CP II; Bar-Nun et al., 1977). In mutants that are deficient in chlorophyll or lack the 22 to 24-kdalton polypeptides, no CP II is detected and the thylakoid membranes are unstacked. However, PS II activity is present (Anderson and Levine, 1974). Thus, it has been inferred that the presence of the 22 to 24-kdalton polypeptides might play a role in the stacking process. Each of these two polypeptides binds about 4–6 chlorophyll molecules (both *a* and *b*). Their synthesis is strongly controlled by light of 632 nm (Ohad, 1975). When associated with chlorophyll, their migration in the electrophoretic field is retarded. The complex shows an apparent molecular weight of about 28–29 kdaltons. As a result, it comigrates with another major membrane polypeptide of 28 kdaltons, which was considered to be the chlorophyll-binding protein of CP II (Anderson and Levine, 1974; Kan and Thornber, 1976). The function of this polypeptide is not well defined. Its relative content during dilution of membranes increases after the growth of the *y-1* mutant in the dark. The polypeptide is rapidly synthesized during the greening process of this mutant. Possibly its role is to specify the recognition or binding of other membrane components and eventually to be involved in the binding site of 70S ribosomes to the thylakoid membranes during active synthesis of membrane proteins. The minor 26-kdalton polypeptide of chloroplastic translation, which can often be resolved into two distinct bands, is presumably required for the formation of an active PS II complex (Bar-Nun and Ohad, 1974). The polypeptides of 30–32 kdaltons of cytoplasmic translation have not yet been assigned definite roles, although they are present in isolated, purified particles that exhibit PS II activity. However, their presence is not required for the expression of this activity, for they can be completely digested with trypsin without loss of either PS II or PS I activity (Regitz and Ohad, 1974).

Synthesis and Assembly of Chlorophyll-Protein Complexes and Their Relation to the Formation of PS I and PS II Activities

The data presented in Table I demonstrate that certain polypeptides are required for both the formation and detection of chlorophyll-protein complexes, as well as for the development of PS I and PS II activity. Determination of the minimal requirements for establishing either chlorophyll-protein complexes or photosynthetic activity during membrane development might yield insight into the process of the synthesis and assembly of the components involved (Table II). The polypeptides and chlorophyll associated in the formation of the chlorophyll-protein complexes must be synthesized and integrated into the membrane simultaneously. The formation of a detectable CP II com-

TABLE II

*Requirement for Simultaneous De Novo Synthesis of Membrane Components Essential for Restoration of Photosynthetic Activity and Detection of Chlorophyll-Protein Complexes in Preexisting Membranes**

Membrane components		CP I	PS I	CP II	PS II
Polypeptides	63–65	+	+	–	–
(mol wt × 10^{-3}	49	+	+	–	–
kdaltons)	44–47	+	–	–	+
	28	–	–		–
	26	–	–	–	+
	24	–	–	+	–
	22	–	–	+	–
Chlorophyll	*a*	+	–	+	–
	b	–	–	+	–

* The conclusions presented in this table are based on several types of experiments: greening of *y-1* mutant, in absence or presence of chloramphenicol (CAP); repair of photosynthetic activity in *y-1* cells greened in the presence of CAP; degreening of *y-1* cells; repair of deficient membranes in T4 mutant, following transfer from restrictive to permissive conditions; CP, chlorophyll-protein complex; PS, photosystem.

plex requires only concomitant synthesis of the 22 and 24-kdalton polypeptides and chlorophyll *a* and *b*. However, the formation of the complex is not sufficient, nor is its concomitant assembly required, for the expression of PS II activity. The latter depends on the synthesis and assembly of polypeptides 44, 47, and 26 kdaltons, all of chloroplastic translation. Thus the minimal requirements for PS II activity and detection of CP II do not coincide.

A different situation is found in the development of CP I and PS I activity. The latter can be established whenever the 63 to 65-, and 49-kdalton polypeptides are synthesized and inserted into the membrane. However, this is apparently not sufficient for the formation of a CP I complex as detected by SDS gel electrophoresis. Its formation requires not only simultaneous synthesis of chlorophyll and the aforementioned polypeptides, but also the synthesis of other chloroplastic translated polypeptides, including the 47 and 44 kdaltons (Bar-Nun et al., 1977).

Although present in the membrane, chlorophyll synthesized independently of the above-mentioned polypeptides cannot be reutilized for the formation of CP I if the polypeptides are synthesized and integrated into the membrane at a later time in the absence of chlorophyll synthesis. This suggests that the polypeptides involved in the formation of the chlorophyll-protein complexes might assume more than one stable configuration within the membrane. One "irreversible" configuration could be induced when the peptides insert in the membrane and interact with hydrophobic components before the binding of chlorophyll. A second stable condition could be achieved when the polypeptides associate with the chlorophyll before or during the process of insertion into the membrane, resulting in the formation of the complex. This is supported by the following: (*a*) the aforementioned nonreutilization of chlorophyll in the membrane for binding with nonsimultaneously synthesized polypeptides; (*b*) the relatively low turnover of chlorophyll bound to the polypeptides when compared to that of the "free" chlorophyll in the membrane (Brown et al., 1975); (*c*) lack of exchange between free chlorophyll added artificially to the membranes with the chlorophyll bound to the polypeptides (Bar-Nun et al., 1977).

Synthesis and insertion of a "naked" chlorophyll-binding polypeptide into the membrane can be detected experimentally only for the 63- and 65-kdalton polypeptides of the CP I complex. These are synthesized in the chloroplast and might be directly inserted into a preformed membrane by polyribosomes bound to the membrane. It could not be demonstrated experimentally that the 22- and 24-kdalton polypeptides of cytoplasmic origin which bind chlorophyll *a* and *b* and form the CP II complex were synthesized, transported, and integrated into the membrane in substantial amounts in the absence of chlorophyll synthesis (Hoober and Stegeman, 1973). Possibly this indicates that during their formation these polypeptides associate with chlorophyll or chlorophyll precursors and that this is essential for their transport across the chloroplast outer envelope and subsequent integration into the membrane. When only chlorophyll *a* is synthesized, as is the case in mutants of higher plants or in leaves greening in flashing light, polypeptides similar to the 22 and 24 kdaltons are not present in the membrane (Anderson and Levine, 1974). Thus, it is possible that chlorophyll *b* stabilizes the structure of the polypeptides bound to both chlorophyll *a* and *b* and that this stabilization is essential for the presence of the polypeptides in the membrane. This is further supported by the fact that chlorophyll *b* is only found in the CP II complex, which accounts for all the chlorophyll *b* present in the membrane. It is also possible that mutations affecting these peptides result in the absence of chlorophyll *b*. However, peptides in this molecular weight range are found in *Euglena* chloroplast membranes which do not have detectable amounts of chlorophyll *b* and only traces of a CP II complex (Gurevitz et al., 1977).

Assembly of the Electron Transfer Chain

The photosynthetic electron-transfer chain is shown schematically in Table III as a linear sequence of segments. These can be defined by the experimental methodology employed. Measurements of photoreduction or photooxidation of natural or artificial electron acceptors and donors do not disclose electron-transfer chain components which might be present, but not interconnected, so as to allow electron flow. Moreover, polypeptides can be detected by gel electrophoresis in very small amounts. However, most of the electron carriers have not yet been identified by this technique.

A variety of methods, including measurements of fluorescence induction, flash yield, quantum

TABLE III
Sequential Assembly of Various Segments of the Electron Transfer Chain

Segment	Parameter measured	Results
1	Fl. ind. (F_v; A + DCMU); peptide 42 kdaltons	Sequential; can be inserted in preexisting membranes
2	Fl. ind. (A)	Sequential; slightly increases during greening
3	Flash yield; DPC oxidation maximum rate	Precedes H_2O splitting
4	H_2O splitting, maximum rate; flash yield; peptide 47 kdaltons	Sequential
5, 8	Cyt f oxidation/red.; $H_2O \rightarrow$ NADP	Sequential; detected after PS II, PS I activity are present
6	P_{700} photooxidation; MV reduction	Sequential; decays or rises faster than PS II activity
7	Asc. DCIP $\rightarrow$ NADP	Sequential; parallels PS I
9	Quantum yield; light saturation; Absorption at 685 nm	Sequential; develops earlier for PS II

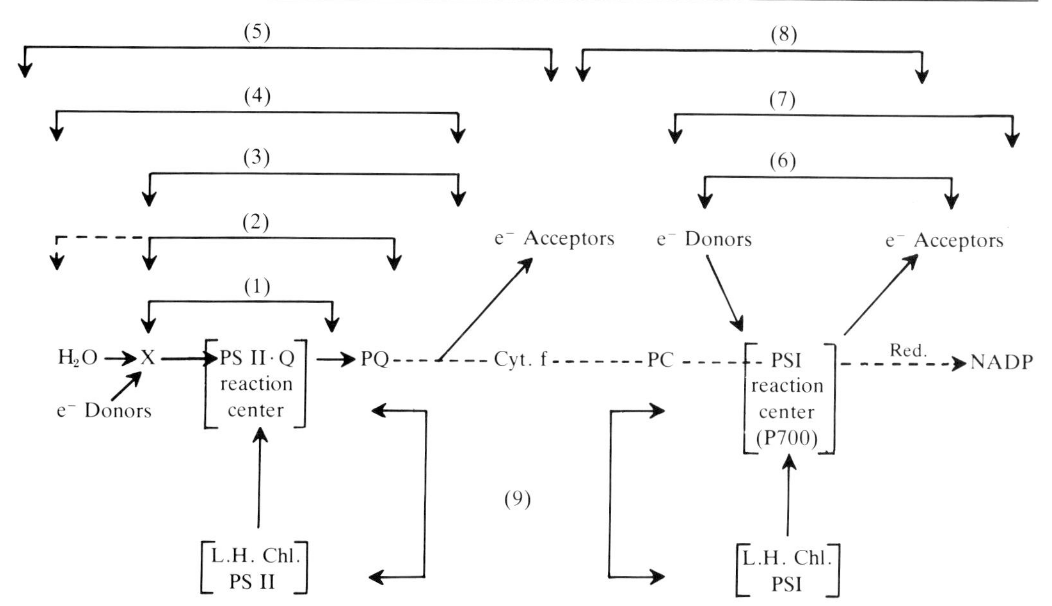

The schematic representation of the various segments of the electron transfer chain as shown here is defined by measurements of the parameters shown in the table and based on Cahen et al., 1976.

Abbreviations used in this table are: A, area under curve of fluorescence induction; Asc. DCIP, ascorbate-dichlorophenol indephenol mixture; cyt. f, cytochron f; Fl. ind., fluorescence induction; F_v, variable fluorescence; L. H., light harvesting; MV, methyl viologen; PC, plastocyanine; PQ, plastoquinone; Red., reduced.

yield, light saturation of photosynthetic activity, measurements of reduction of artificial electron donors and acceptors, and oxidation or reduction of membrane-bound carriers can be used, and the electron-transfer chain can thus be dissected as shown (Table III) into defined, discrete parts, and their interconnection evaluated. The presence, activity, and interconnection of these segments during membrane development was analyzed in detail in *C. reinhardtii y-1* mutant (Ohad, 1975; Cahen et al., 1976). During the initial phase of the greening process, detection of PS II activity precedes that of PS I. The connection between PS I reaction center and the natural donor or acceptor sites develops independently. The connection of PS II to the rest of the chain (photoreduction of cytochrome f) is achieved only after both PS I and PS II activities are significantly developed. This is especially striking, since components such as cytochrome f or plastocyanin are present in the membrane in relatively large amounts before they are photochemically active (Eytan et al., 1974). In the *y-1* mutant, the link and final proportion between light-harvesting chlorophyll and PS II reaction center is established during the first hours of the greening process. Complete development of the light-harvesting chlorophyll of PS I is achieved later during the greening process. Thus, during the early phase of the greening (1–3 h), both existing and new components are integrated in the photosynthetic-active segments, which are later interconnected and finally establish an operative, completed chain. This phase is characterized by a fast rise in specific photosynthetic activity per chlorophyll unit, reduction of the apparent size of the photosynthetic unit, and increase in the quantum yield. During the remainder of the greening process, the activity per cell continues to rise, but, on a chlorophyll basis, most parameters measured remained practically constant, indicating that during the actual growth the composition and organization of the membrane is kept practically constant. Stepwise insertion and activation of segments of the photosynthetic electron-transfer chain compatible with this scheme have been reported to occur in other algae (Senger, 1970; Dubertret and Joliot, 1974), wild-type *Chlamydomonas*, and higher plants (Thorne and Boardman, 1971).

Changes in the membrane organization during development can also be demonstrated by assessing the susceptibility of various membrane peptides and the related activities to digestion with trypsin from the outer or inner surface of the thylakoid (Regitz and Ohad, 1974). This approach yielded valuable information. Polypeptides of 63–65, 49, 30–32, 28, and 22 kdaltons exhibit transient sensitivity or resistance toward trypsin digestion at different developmental stages of the membrane. However, the activity of both PS II and PS I reaction centers is stable toward trypsin digestion during all times of the developmental process. The 63 to 65- and 28-kdalton polypeptides are sensitive to trypsin digestion in membranes from dark-grown cells in which the chlorophyll-protein complexes are not detectable. However, the polypeptides in the region 30–32 and 22–24 kdaltons are resistant. At later stages of the greening, when the relative proportion of these peptides in the membrane is changed, chlorophyll-protein complexes become detectable and photosynthetic activities are established, the 63 to 65-kdalton polypeptide become resistant while the polypeptides in the region 30–32 kdaltons become extremely sensitive. At the same time, the sensitivity of the 28-kdalton polypeptide is not significantly altered while a fragment of about 2 kdaltons is clipped off from the 22-kdalton polypeptide. However, the latter remains associated with the CP II complex (Bar-Nun et al., 1977). In membranes formed in the presence of chloramphenicol in which the CP I complex and all photosynthetic activities are absent, the 63 to 65-kdalton polypeptides remain sensitive and the 30 to 32-kdalton polypeptides remain resistant toward trypsin digestion.

Insertion and Activation of PS II Components

The *C. reinhardtii y-1* mutant, in which chlorophyll synthesis is light-dependent, does not form photosynthetic membranes when grown in the dark, and those initially present are diluted out among the daughter cells (Ohad, 1975). When such cells are exposed to the light in the presence of chloramphenicol, peptides of cytoplasmic translation are rapidly synthesized together with chlorophyll, and membranes in which a CP II complex is detectable are formed. However, because of the absence of peptide synthesis by the chloroplast 70S ribosomes, the reaction center of PS I and PS II are not formed as well as the CP I complex (Bar-Nun et al., 1977; Fig. 1 A). When the cells are washed free of the inhibitor and further incubated in the light, additional chlorophyll is synthesized together with the missing peptides of chloroplastic origin, and both PS I and PS II activities are reestablished, as is the CP I complex (Fig. 1 B). However, when the cells are incubated in the dark, although the same polypeptides are synthesized and integrated, and activities of both PS I and PS II develop, the CP I complex is not reestablished. Thus, in addition to the previous conclusions drawn from such experiments regarding the independent formation of CP I and PS I, one can conclude that when integration and activation of peptides required for PS I and PS II activity occur simultaneously, they can take place in the dark and photosynthetic electron flow is not required (Ohad, 1975). This is further supported by the fact that inhibition of electron flow during membrane development by 3(3,4 dichlorophenyl)-1,2-dimethylurea (DCMU) does not prevent formation of a complete membrane in which all activities can be detected after the inhibitor is removed.

A different situation is found in the *C. reinhardtii* T4 mutant (Kretzer et al., 1977). When these

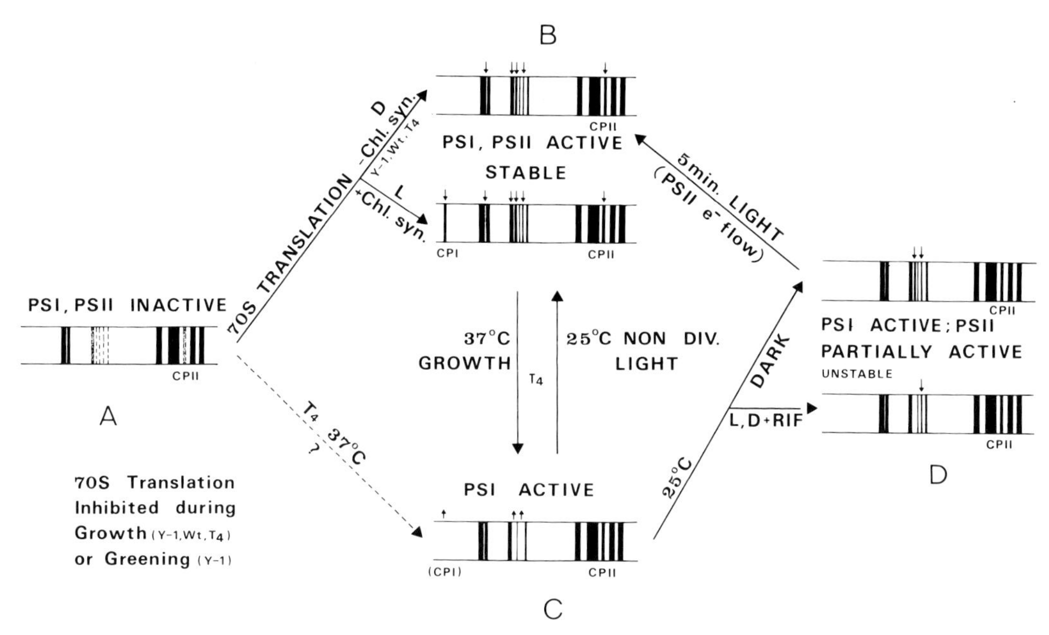

FIGURE 1 Schematic representation of the dynamics of PS II, PS I, and chlorophyll-protein complexes formation in *y-1* and T4 mutants of *C. reinhardtii*. For explanation, see text. ↓ and ↑ indicate insertion or deletion, respectively, of membrane components or complexes following previous transfer of cells from conditions A → B; A → C; C → D.

cells are grown at 37°C, PS I activity and all membrane polypeptides but two (44 and 47 kdaltons) are present, and PS II activity is absent, as is the CP I complex (Fig. 1 C). The missing peptides are synthesized and integrated into the existing membranes, and a normal PS II activity (water-splitting and 2,6 diphenylcarbaside [DPC] photooxidation) is reestablished if the cells are incubated in the light at 25°C. As expected, no detectable CP I complex is formed if no additional chlorophyll synthesis occurs. If the repair is carried out at 25°C in the dark, both polypeptides are inserted. If rifampicin (preventing chloroplast transcription) is added during the repair at 25°C, only the 44-kdalton polypeptide is inserted (Fig. 1 D). In both cases, only PS II reaction center activity can be measured with DPC as an electron donor and water-splitting activity is not detectable, indicating that the 44-kdalton polypeptide is associated with the PS II reaction center. This is further supported by measurements of fluorescence induction (Chua and Bennoun, 1975). When the 47-kdalton polypeptide is not inserted into the membranes, water-splitting activity cannot be detected in either the light or the dark. Even when this peptide is present in the membrane, it is not functional unless light-dependent electron flow through PS II is allowed for at least a short period of time. When the 44-kdalton polypeptide is inserted with or without a nonfunctional 47-kdalton polypeptide, the photooxidation of DPC is unstable. The activity, as well as the peptide(s), is lost from the membrane upon heating the cells or isolated membranes at 38°C for a short time or by treating the membrane with Tris or deoxycholate at concentrations which would not affect normally active membranes. Thus, insertion of both missing peptides does not necessarily insure their proper integration and activation.

Apparently, electron flow induces a certain rearrangement of the membrane that is required when PS II components are reinserted into a membrane that already contains an active PS I, but not when both photosystems are integrated simultaneously. The PS II activity present in T4 cells grown at 25°C is not destroyed, but is diluted among the daughter cells after growth at 37°C. In the absence of division, the activity is stable at 37°C for 48–72 h. In cells that have a significant residual activity and are incubated at 25°C in the dark, the partially repaired PS II (DPC photooxidation) is thermosensitive, whereas the initial activity (H_2O and DPC oxidation) is not. These observations demonstrate that the PS II units, when properly organized, do not dissociate and probably do not exchange components with the newly formed ones. Thus, one can consider that they exist as definite, individual entities, as also seems to be the case for the chlorophyll-protein complexes.

Stepwise addition of components required for PS II activity and dissociation between development of PS I and PS II and the synthesis of chlorophyll have also been found in *Euglena* chloroplasts (Gurevitz et al., 1977).

A similar situation might be found in the results of experiments in which etiolated higher plants are greened in a regime of intensive, extremely short light flashes (milliseconds) interspersed within longer dark periods. In such conditions, chlorophyll is synthesized and PS I develops. However, water-splitting activity is not detectable, although PS II reaction center is present. After a short, continuous illumination, water-splitting activity is restored (Remy, 1973).

Correlation between the Insertion of Membrane Peptides, Membrane Stacking, and Photosynthetic Activity

Alteration of membrane composition and activity might induce changes in the stacking pattern of the membranes. Such relationships have been described for *Chlamydomonas* mutants (Goodenough and Staehelin, 1971). An example of membrane-morphology alteration following deletion of specific polypeptides is found also in the T4 mutant. When grown at 37°C with loss of PS II activity (but displaying PS I activity and normal levels of chlorophyll) the membranes become extensively stacked (Fig. 2*a*). However, after incubation at 25°C for a few hours, during which only small amounts of the missing peptides are synthesized and inserted and only about 10% of the PS II recovered, the membrane stacks dissociate (Fig. 2*b*). Thus, large expanses of thylakoid surfaces are exposed to the chloroplast matrix. Polyribosomes, which are active in synthesizing the two missing membrane polypeptides, might bind to the membrane and cause dissociation of the stack even before the quantity of the missing peptides inserted reaches substantial amounts. Dissociation of thylakoids and increase in the surface area of thylakoids available for ribosome binding have been reported to occur at specific times during the life cycle of synchronously grown *C. reinhardtii*. This might coincide with the insertion of peptides, resulting in an increase of PS II activity.

From the data obtained from the T4 mutants, it seems that polypeptides of chloroplastic translation which are essential for PS I formation, and peptides of cytoplasmic origin, including the chlorophyll-binding peptides, can integrate in stacked membranes. However, insertion of peptides of chloroplastic translation which are required for PS II complex formation necessitates direct access of the ribosomes to the surface of the developing membrane.

It is attractive to consider as a working hypothesis that components of PS I, CP I, and CP II could diffuse in a translational movement across the lipid phase of the stacked membranes and reach membrane regions distant from the point of insertion. On the other hand, components of PSI II reaction center might be less mobile. It is possible that the water-splitting activity and PS II reaction center is connected with peptides which span the membrane. This is in agreement with previously reported data and inferred from the fact that water-splitting activity is trypsin-sensitive from either side of the membrane. The reaction center of PS II is protected against trypsin digestion at all times during the development and even in dark-repaired T4 membranes before stabilization of the complex by light. In stacked thylakoids, the diffusion of this complex in the lipid phase of the membrane might be hindered.

Regulation of Synthesis and Assembly of Membrane Components

So far, the assembly process, including synthesis and insertion of components into the membrane, might appear to be loosely regulated. Interaction among proteins, lipids, and pigments seems to be limited only by the stoichiometry of interaction and relative concentration of the components in the bilayer. However, this is not the case. Synthesis of membrane components is under strict nuclear control, which is further modulated by discrete influence exerted by the chloroplast. The control exists at least at two levels (Fig. 3). First, the cytoplasm supplies the membrane peptide that forms the membrane structure and binds chlorophyll *a* and *b* in CP II. Apparently, the supply of these peptides is rate-limiting for the synthesis of peptides translated on 70S ribosomes (Ohad, 1975). This is a specific effect on synthesis of membrane peptides. Chloroplast ribosomes that synthesize proteins unrelated to the membranes are not affected in this way (Gershoni and Ohad, manuscript in preparation). Second, the peptides

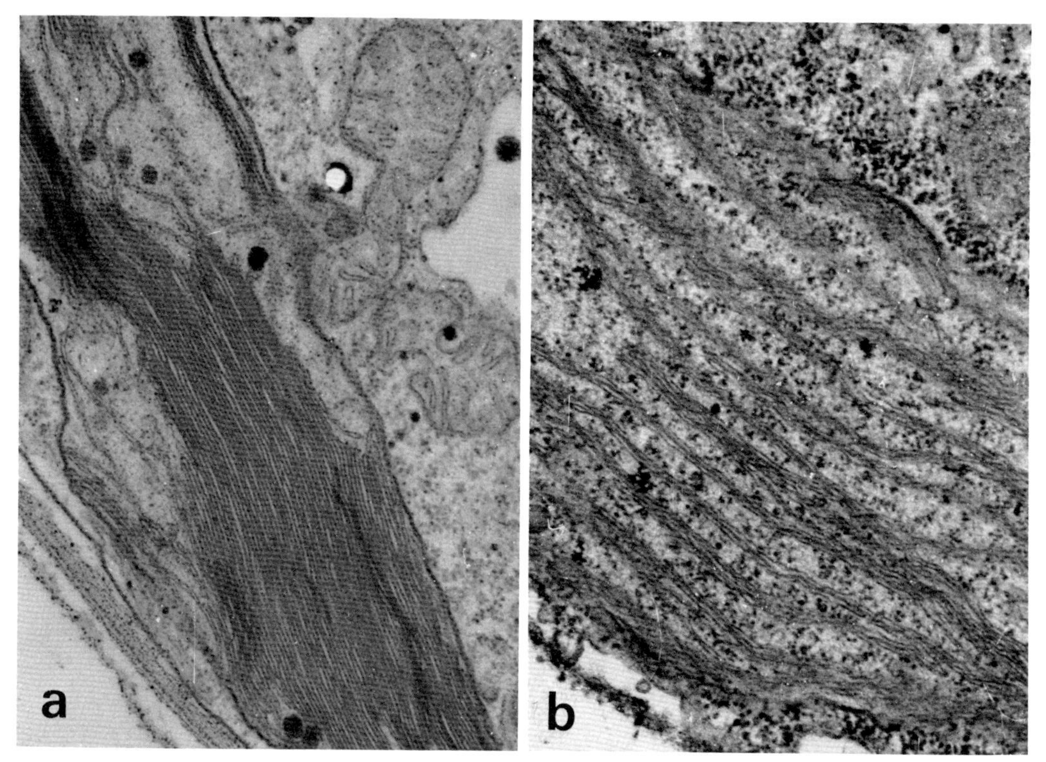

FIGURE 2 Extensive stacking of PS II-deficient thylakoid membranes in *C. reinhardtii*. T4 cells grown at 37°C for five generations (*a*). Following incubation at 25°C for 5 h, the stacked membranes dissociate, and repair of PS II activity is initiated (*b*).

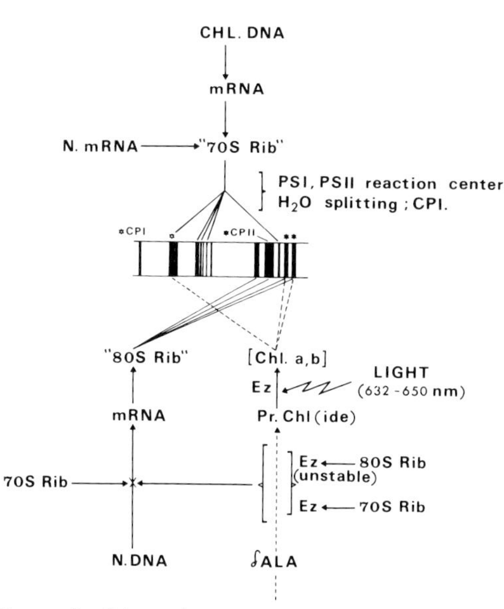

FIGURE 3 Schematic representation of the control of synthesis and assembly of chloroplast membrane polypeptides and chlorophyll required for the formation of chlorophyll-protein complexes and development of photosynthetic activity of PS I and PS II.

required for the formation of PS I and PS II, that is, performance of photochemistry by the membrane, are coded by the nucleus. Thus, the formation of the membrane proper and its activity are under extra chloroplastic control. It has been suggested that an unstable (nuclear) message(s) is needed for the synthesis of membrane peptides of cytoplasmic translation. The synthesis of this message is apparently repressed by an unidentified chlorophyll precursor and a peptide of chloroplastic translation (Hoober and Stegeman, 1973; Ohad, 1975).

The complex sequence of enzymatic reactions leading to the formation of chlorophyll includes at least two control points of interest to this discussion. One can be defined as an unstable nuclear message, coding for an unstable enzyme, the pair having a half-life of about 45 min (Eytan and Ohad, 1972). The second step can be defined as a stable chloroplast-translated enzyme(s) which can be rendered rate-limiting only after prolonged incubation of the cells in the presence of chloramphenicol or similar protein-synthesizing inhibitors (Gershoni and Ohad, manuscript in preparation).

CONCLUSION

Chloroplast membranes grow through the synthesis, insertion, and proper integration (assembly) of membrane components in a multistep process. The membrane is composed of peptides of nuclear transcription-cytoplasmic translation, nuclear transcription-chloroplast translation, and chloroplast transcription-chloroplast translation. The peptides of chloroplast translation participate in the formation of PS I and PS II active centers, the splitting enzyme, and the chlorophyll-protein binding complex, CP I. The chlorophyll-protein complex, containing both chlorophyll *a* and *b* (CP II), consists only of proteins of cytoplasmic translation. The development of PS I activity does not coincide nor require the formation of the CP I complex, as detected by SDS-gel electrophoresis.

The formation of the CP II complex is not sufficient for establishing PS II activity which can be formed subsequently. Synthesis of chlorophyll is required simultaneously with that of the peptides which bind it and form the CP complexes I and II. The formation of the former requires simultaneous synthesis and insertion into the membrane of several chloroplast-translated peptides, among them some associated with PS II activity. Simultaneous insertion of peptides required for both PS I and PS II activity can occur in the dark and does not require photosynthetic electron flow. When PS II peptides are synthesized alone and insert into a membrane which already contains a normal, active PS I, a reorganization of the membranes is required, and is induced by photosynthetic electron transfer through PS II. Organized chlorophyll-protein complexes and PS II units behave like stable individual entities and possibly do not dissociate or exchange components.

Continuous reorganization occurs during assembly and development of the membranes, as demonstrated by formation of connections among preexisting units or segments of the photosynthetic electron flow and transient sensitivity of different membrane peptides toward trypsin digestion. The synthesis of the membrane peptides by the chloroplast ribosomes depends on the presence of membrane polypeptides of cytoplasmic origin whose synthesis is, in turn, controlled by chlorophyll precursors and polypeptides of chloroplastic translation.

ACKNOWLEDGMENTS

I am grateful to Dr. Shoshana Bar-Nun and to Dr. Frank Kretzer for their criticisms, suggestions, and help in all ways during the preparation of this manuscript which is based largely on results of experiments conducted by them during the last two years. I wish also to thank Dr. D. Cahen and Dr. S. Malkin from the Weizmann Institute of Science, Rehovot, and the students and co-workers of our laboratory, Mr. M. Gurevitz, Mr. J. Gershoni, Mrs. S. Shochat, and Dr. N. Lavintman, as well as Dr. R. Schantz (presently at the University of Strasbourg, France). The work on the T4 mutant was carried out in cooperation with Dr. P. Bennoun from the Institut de Biologie Physico-Chimique (Fondation Edmond de Rothschild), Paris.

This work was supported by grants from the United States-Israel Binational Science Foundation (no. 184) and the Deutsche Forschungsgemeinschaft (no. DR 29/17).

REFERENCES

ANDERSON, J. M., and R. P. LEVINE. 1974. The relationship between chlorophyll-protein complexes and chloroplast membrane polypeptides. *Biochim. Biophys. Acta.* **357:**118–126.

BAR-NUN, S., and I. OHAD. 1974. Cytoplasmic and chloroplastic origin of chloroplast membrane proteins associated with PS II and PS I active centers in *Chlamydomonas reinhardi* y-1. *In* Proceedings of the 3rd International Congress on Photosynthesis. Vol. III. M. Avron, editor. Elsevier Scientific Publishing Co., Amsterdam. 1627–1637.

BAR-NUN, S., R. SCHANTZ, and I. OHAD. 1977. Appearance and composition of chlorophyll-protein complexes during chloroplast membrane biogenesis in *Chlamydomonas reinhardi* y-1. Biochim. Biophys. Acta In press.

BROWN, J., S. ACKER, and J. DURANTON. 1975. The difference in turnover rate between the chlorophyll *a* in the P700-chlorophyll a-protein and in the total chloroplast membrane. *Biochem. Biophys. Res. Commun.* **62:**336–341.

CAHEN, D., S. MALKIN, S. SHOCHAT, and I. OHAD. 1976. Development of photosystem II complex during greening of *Chlamydomonas reinhardi* 7-1. *Plant Physiol.* **58:**257–267.

CHUA, N. H., and P. BENNOUN. 1975. Thylakoid polypeptides of *Chlamydomonas reinhardi*: wild-type and mutant strains deficient in photosystem II reaction center. *Proc. Natl. Acad. Sci. U. S. A.* **72:**2175–2179.

CHUA, N. H., K. MATLIN, and P. BENNOUN. 1975. A chlorophyll-protein complex lacking in photosystem I mutants of *Chlamydomonas reinhardi. J. Cell Biol.* **67:**361–377.

DUBERTRET, G., and P. JOLIOT. 1974. Structure and organization of system II photosynthetic units during the greening of a dark-grown Chlorella mutant.

Biochim. Biophys. Acta. **357**:399–411.

Eytan, G., R. C. Jennings, G. Forti, and I. Ohad. 1974. Biogenesis of chloroplast membranes: changes in photosystem I activity and membrane organization during degreening and greening of a *Chlamydomonas reinhardi* mutant, y-1. *J. Biol. Chem.* **249**:738–744.

Eytan, G., and I. Ohad. 1972. Biogenesis of chloroplast membranes. Modulation of chloroplast lamellae composition and function induced by discontinuous illumination and inhibition of ribonucleic acid in protein synthesis during greening of *Chlamydomonas reinhardi* y-1 mutant cells. *J. Biol. Chem.* **247**:122–129.

Goodenough, U. W., and L. A. Staehelin. 1971. Structural differentiation of stacked and unstacked chloroplast membranes. Freeze-etch electron microscopy of wild-type and mutant strains of *Chlamydomonas*. *J. Cell Biol.* **48**:594–619.

Gurevitz, M., H. Kratz, and I. Ohad. 1977. Origin of photosystem II and chlorophyll binding peptides in *Euglena gracilis Z* chloroplasts. *Isr. J. Med. Sci.* In press.

Hoober, J. K., and W. J. Stegeman. 1973. Control of the synthesis of a major polypeptide of chloroplast membranes in *Chlamydomonas reinhardi*. *J. Cell Biol.* **56**:1–12.

Kan, K. S., and J. P. Thornber. 1976. The light-harvesting chlorophyll a/b-protein complex of *Chlamydomonas reinhardi*. *Plant Physiol.* **57**:47–52.

Kretzer, F., I. Ohad, and P. Bennoun. 1977. Ontogeny, insertion and activation of two thylakoid peptides required for photosystem II activity in the nuclear, temperature sensitive T4 mutant of *Chlamydomonas reinhardi*. *In* Symposium on Genetics and Biogenesis of Chloroplast and Mitochondria. T. Bucher, editor. Elsevier Scientific Publishing Co., Amsterdam. In press.

Ohad, I. 1975. Biogenesis of chloroplast membranes. *In* Membrane Biogenesis, Mitochondria, Chloroplast and Bacteria. A. Tzagoloff, editor. Plenum Publishing Corp., New York. 279–350.

Regitz, G., and I. Ohad. 1974. Changes in the protein organization in developing thylakoids of *Chlamydomonas reinhardi* y-1 as shown by sensitivity to trypsin. *In* Proceedings of the 3rd International Congress on Photosynthesis. Vol. III. M. Avron, editor. Elsevier Scientific Publishing Co., Amsterdam. 1615–1625.

Remy, R. 1973. Appearance and development of photosynthetic activities in wheat etioplasts greened under continuous or intermittent light–evidence for greening under intermittent light. *Photochem. Photobiol.* **18**:409–416.

Senger, H. 1970. Quantum yield and variable behavior of the two photosystems of the photosynthetic apparatus during the life cycle of *Scenedesmus obliquus* in synchronous cultures. *Planta* (*Berl.*) **92**:327–346.

Thorne, S. W., and N. K. Boardman. 1971. Formation of chlorophyll *b*, and the fluorescence properties and photochemical activities of isolated plastids from greening pea seedlings. *Plant Physiol.* **47**:252–261.

Somatic Plant Hybridization by Fusion of Protoplasts

MICROBIAL TECHNIQUES IN SOMATIC HYBRIDIZATION BY FUSION OF PROTOPLASTS

G. MELCHERS

Plans for introducing microbial techniques in plant genetics, physiology, and breeding have existed in our laboratory since the middle of the 1950s (Melchers and Engelmann, 1955; Melchers and Bergmann, 1958/59; Bergmann, 1959*a, b*, 1960; Melchers, 1960). Cloning large quantities of plant cells in agar plates, using vegetative haploid cells for mutation work, and combining mutation and selection in plant tissue and cell cultures were started at that time. However, these experiments were neglected to a great extent in the interest of our promising studies on mutation in tobacco mosaic virus (TMV), see Mundry and Gierer, 1958. The mutations were induced by nitrous acid, the mode of action of which is known, so that it was possible to correlate the mutation of the genetic material to the amino acid exchanges in the viral coat protein (Wittmann 1959; Wittmann and Wittmann-Liebold, 1963; Melchers, 1966, 1968*b*).

Thus it may be understandable why progress in cell biology was slow in our laboratory in the 1960s. We became interested in the comparison of hormone autotrophy (tumor growth) by "spontaneous habituation" and "transformation by *Agrobacterium tumefaciens*" (Sacristán, 1967; Melchers, 1971; Sacristán and Melchers, 1969). The question arose: is habituation and transformation by *A. tumefaciens* reversible? To obtain the final answer, a cloning of tumor cell material became necessary (Sacristán and Melchers, 1977).

In 1968, I visited Itaru Takebe's laboratory in Chiba, Japan, and saw the wonderful protoplast preparations of tobacco mesophyll cells (Takebe et al., 1968). I invited him, as a guest in our laboratory, to work to combine his techniques of preparation of protoplasts with our experience in cloning on agar plates and regenerating plants out of callus cultures (Melchers, 1965, 1968*a*).

In 1970, we succeeded for the first time in cultivating protoplasts to form calluses and in regenerating plants from calluses of protoplastic origin (Takebe et al., 1971; see also, Nagata and Takebe, 1971). From that moment on, it was possible to combine somatic genetics by microbial techniques with conventional sexual genetics.

Fusion of Protoplasts

At that time, some people became interested in fusing plant protoplasts (Cocking, 1960, 1972; Binding, 1966). However, for our homogeneous material of mesophyll protoplasts, the proposed fusion solutions (especially $NaNO_3$) proved to be useless (Power and Cocking, 1971; Keller and Melchers, 1973). High pH (up to 10.5) and, more importantly, high calcium concentration neutralize the normal negative charge on the surface of tobacco mesophyll protoplasts (T. Nagata, Institute of Biology, University of Tokyo, a guest in our laboratory, 1975–76, unpublished data), and lead to fusion of protoplasts with a high efficiency (Keller and Melchers, 1973). Polyethyleneglycol (PEG), introduced by Wallin and Eriksson (1973), Wallin et al. (1974), and Kao and Michayluk (1974), induces close contacts and fusion between plant protoplasts. Especially good percentages of fusions are obtained by removing PEG with a solution of high pH and high calcium concentration (Kao et al., 1974).

If you wish to hydbridize two plants, A and B, you will find in a 1:1 mixture of their protoplasts many unfused A and B protoplasts, and such fusion products as A+A,* A+B, B+B, as well as

G. MELCHERS Max-Planck-Institute für Biologie, Tübingen, Germany

* $v \times s$ means the sexual, $v + s$, the somatic hybrid, see Winkler, 1908.

more complex ones, such as A+A+A, A+A+B, and so on. In cooperation with Nagata and Hj. Eibl (MPI für Biophysikalische Chemie, Göttingen), we have tried to change the surface charge of one of the protoplasts – let us say A – not only to zero, but to positive values. Because the glass or plastic surface is normally negatively charged, it is possible to produce a monolayer of positively charged protoplasts. By placing the normally negatively charged protoplasts B in these conditions, one can produce a double layer of the two kinds of protoplasts:

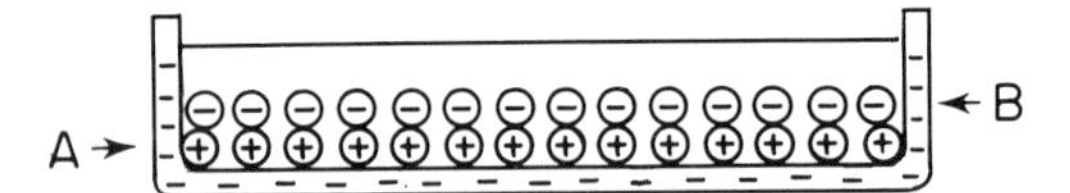

In such an arrangement, the frequency of 1:1 (A+B) fusion products should be much higher than chance events. We are developing this fusion technique, and the synthesis of positively charged nontoxic phospholipids is in progress (Eibl, unpublished data). This technique may build up a system like a secondary "artificial sexuality."

Selection of Fusion Products of Nicotiana tabacum s + v

If the percentage of (A+B) fusions among nonfused A and B, A+A, B+B, and many different combinations of large fusion products is low, we need a selection system for the (A+B) protoplasts or the callus that originated from them. A well-known microbial technique is to combine two recessive mutants, both of which have a negative influence on growth and/or development, under special external conditions or on selection media – complementing in the hybrid to normal growth and/or development

$$\frac{\mathrm{v\,S}}{\mathrm{v\,S}} \times \frac{\mathrm{V\,s}}{\mathrm{V\,s}} \rightarrow \frac{\mathrm{V\,S}}{\mathrm{v\,s}}.$$

Because genome-dependent chlorophyll defects are often found in higher plants as recessive mutants, it is probable that they complement one another to synthesize normal chlorophyll. The protoplasts, if not easily prepared from the mutants growing in soil, may be produced from heterotrophic cultures of these mutants. For our first experiments, we chose a special type of chlorophyll deficiency, namely destruction of chlorophyll by high light intensity. The varieties sl_1sl_2, "sublethal" (called by us *ss* [or s^2–] amphidiploid, s = dihaploid) and $v_1 – A_1$, "virescent" (*vv* or v^2 and v, respectively), grow very slowly in normal greenhouse light conditions (Fig. 1). The seeds used in this study were kindly supplied by D. U. Gerstel, University of North Carolina, Raleigh (Melchers and Labib, 1974). They become a more or less normal green when grown in ca. 800-lx intensity, 28°C, and high humidity. Protoplasts can be prepared from plants grown under such conditions. After fusion experiments, the protoplasts first are

FIGURE 1 Seedlings after 1.5-mo growth in a normal greenhouse in winter. From left to right: (1) "sublethal" = s; (2) (s × v) F_1; (3) (v × s) F_1; (4) "virescent" = v, demonstrating the light sensitivity of s and v, the complementation in the hybrids and the identity of (s × v) and (v × s). (No differences in the plastome.)

cultivated in low light intensity. The hybrid calluses are selected in 10,000 lx in a top layer on agar plates with reduced organic compound in the culture medium (Fig. 2). (High content of organic compounds in the culture medium and anaerobic conditions counteract the influence of light. The chlorophyll defect—especially in *v*—seems, therefore, to be induced by a photooxidation; see Melchers and Labib, 1974.)

The chromosome number of the somatic hybrids v + s

In four independent experiments, we found at least 20, but up to 43, hybrid calluses and sometimes more than one plant from one callus (Melchers and Sacristán, 1977; Figs. 3*a* and *b*). Because we started our experiments with dihaploid material produced by "anther culture" with 24 chromosomes, we obtained "normal" green tobacco plants with 48 chromosomes in most cases. Some plants have aneuploid chromosome numbers near 48, others the triploid number 72, and aneuploids in the neighborhood or tetraploids 96, and some aneuploids around them. It is possible that the triploids and tetraploids arose by fusions of more than two "haploid" protoplasts, but it is more or less impossible to prove this hypothesis (Sacristán and Melchers, 1977). Abnormal ploidy of callus cells is well known without fusion only by callus culture. A correlation between time of culture in callus conditions and level of ploidy is expressed in long-term experiments (2 yr of callus culture versus 8 yr; Melchers, 1965, 1968*a*). A correlation between time and ploidy level did not become apparent during a culture period of 13 mo (Sacristán and Melchers, 1977).

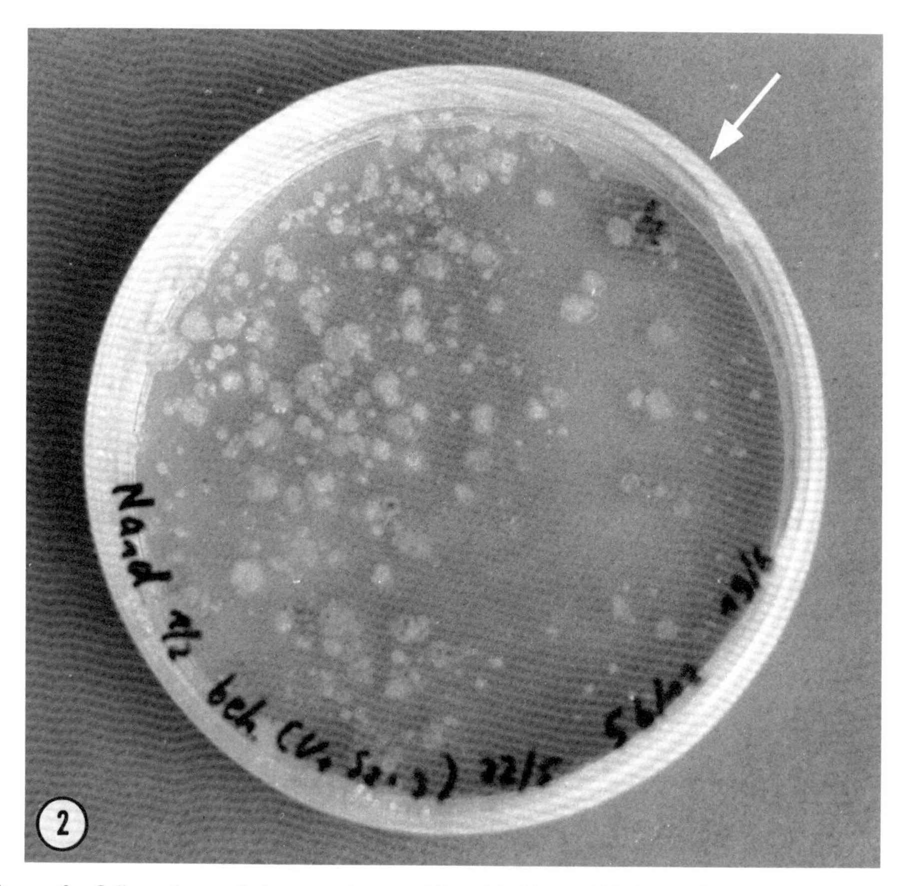

FIGURE 2 Calluses from a fusion experiment cultivated in Nagata-Takebe medium with the amount of organic constituents reduced to $^1/_5$, benzyladenine reduced to $^1/_2$, and without any auxin. Calluses indicated as v, s, v+v, s+s are indistinguishable yellowish light green. One callus, indicated by an arrow, regenerated dark green and afterwards was found to have 48 chromosomes and normal hybrid characters.

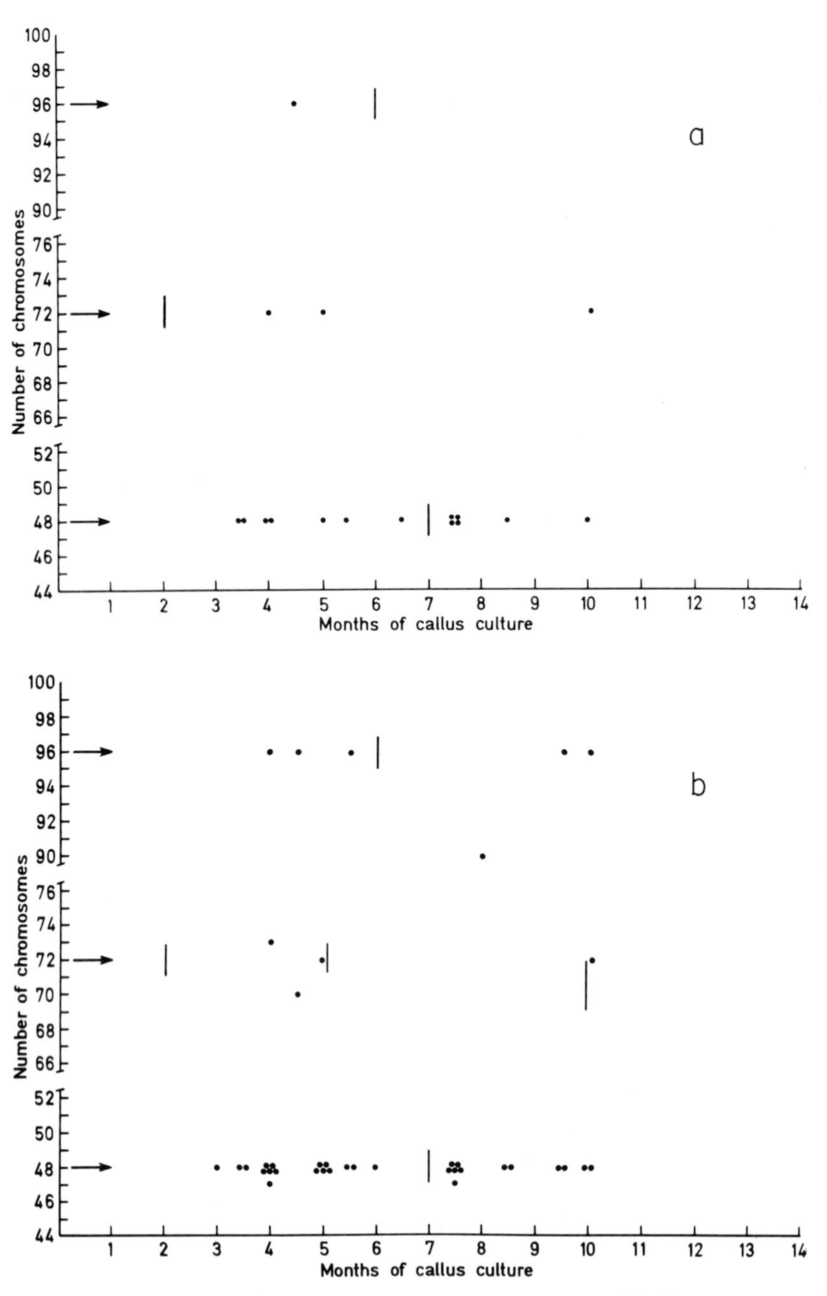

FIGURE 3 Ordinate: chromosome numbers; abscissa: time of callus culture. (*a*) Plants regenerated from calluses isolated without any possibility of mechanical division of calluses. (*b*) Plants regenerated with a low possibility of mechanical division of some calluses by the first dilution step of the culture medium.

Experiments with N. tabacum + Nicotiana silvestris

In vegetatively propagated haploid *Nicotiana silvestris* plants, P. Maliga (Biological Research Center, Szeged, Hungary), a guest in our laboratory, found a chlorophyll-deficient, light-sensitive mutant (Fig. 4); we call it *sis* (haploid with 12 chromosomes), sis^2 (diploid with 24 chromosomes). Although we were not able to cultivate calluses from sis^2 protoplasts after four independent positive fusion experiments with *s* (dihaploid tobacco

FIGURE 4 *Nicotiana silvestris*, chlorophyll-deficient mutant sis^2, left in low light intensity, ca. 800–1000 lx, in normal greenhouse condition.

with 24 chromosomes) and s^2 (amphidiploid tobacco with 48 chromosomes), we are cultivating four plants of s^2 + sis^2 and about 100 plants of s + sis^2 (Fig. 5). Many plants of s + sis^2 have 48 chromosomes, some are aneuploid, others tetraploid with abnormal morphology (Fig. 6). When we started these fusion experiments with *N. silvestris* (sis^2) and *N. tabacum* (s and s^2), we had no knowledge of the sexual genetics of the mutant *sis*. We detected the complementation of the $s-$ and $sis-$ gene only by somatic hybridization. This demonstrates clearly how useful chlorophyll-deficient mutants are for somatic hybridization.

Experiments with Petunia hybrida + N. tabacum

In 19 independent experiments with 8.10^7 protoplasts, we fused protoplasts of tobacco v (v^2), s, and s^2 with protoplasts of the light-green mutant Mu_1 (supplied by Professor J. Straub, Cologne) of *Petunia hybrida*. Unfortunately, the calluses of Mu_1 *Petunia* are green under our selective conditions, whereas s and v protoplasts are yellowish or light green, respectively. Therefore, our selection system did not work perfectly and we had to regenerate many green calluses to plants. We have found no hybrid plant as yet. Some callus lines exist with abnormal chromosome numbers that do not regenerate, or start only with abnormal regenerates. But, at the moment, it is not clear if these calluses arose from fusions between *Petunia* and tobacco or have abnormal tobacco chromosome sets. Unfortunately, the chromosomes of tobacco and *Petunia* are not distinguishable individually. Since M. Zenkteler (Institute of Biology, University of Poznan, Poland), as a guest in our laboratory, found that the sexual fusion between tobacco and *P. hybrida* leads only to a fertilized egg and then to very few cell divisions, it has been proved that incompatibility between tobacco and *P. hybrida* genomes does not take place in the process of fertilization, but during embryo formation. These findings demonstrate nothing about possible genomes mixed by some chromosomes of tobacco and others of *Petunia*. They demonstrate only that complete hybrid genomes do not cooperate to develop an embryo. It is unlikely that a complete cell hybrid between *Nicotiana glauca* and soybean (Kao, this volume) would regenerate a hybrid plant. But no one can predict at the moment if regenerates of various types of mixed genomes are possible, as, for example, a combination of mainly *Petunia* or tobacco chromosomes and only a few chromosomes of the other plant.

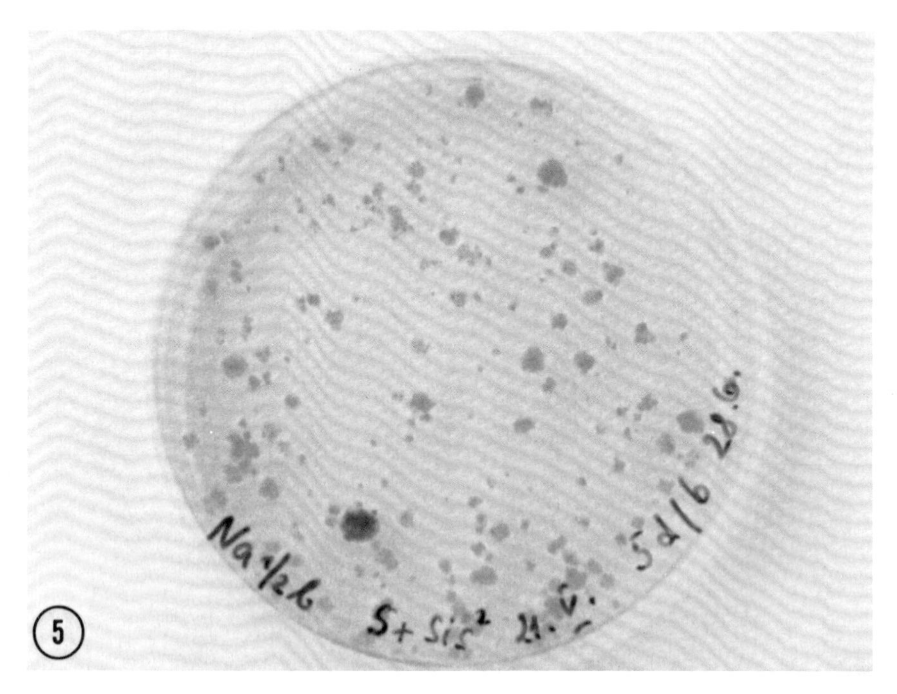

FIGURE 5 Petri dish with medium $^1/_{10}$ of organic compounds in high light intensity 8,000–10,000 lx. Calluses of *N. tabacum s* are yellow; that callus of a somatic hybrid *N. tabacum s* + *N. silvestris sis*^{2} is dark green.

FIGURE 6 (Left) *N. silvestris sis*^{2}; (right) *N. tabacum s*. (Above and below) somatic hybrid plants *N. tabacum s* + *N. silvestris sis*^{2}, both with 48 chromosomes. In the center of the picture, a somatic hybrid plant with abnormal morphology and 96 chromosomes.

Equally uncertain is the possibility of raising a plant from soybean with a few chromosomes (or parts thereof) of *N. glauca*. This field is at present highly exciting.

Discussion

Yuri Gleba (Institute of Botany, Academy of Science, Ukrainian SSR, Kiev) was to have participated in this symposium. However, illness prevented him from attending. Although experimental details are not given, he states in his abstract that the transfer of characters after somatic hybridization by fusion of protoplasts is determined by the plastome, too. That may be the case only if the relationship between the genome and plastome is not too small. Smith et al. (1976) published a report that, after fusion with PEG, somatic hybrids originated between *N. glauca* and *N. langsdorffii*. Twenty-three plants are described. None of these is identical with the amphidiploid sexual or the so-called parasexual hybrid in either morphology or chromosome number (42), see Carlson et al. (1972). H. H. Smith, in an unpublished roundtable discussion (1976) at the UNESCO-UNDP-ICRO Training Course on "Cell genetics in higher plants," July 1976, Szeged, Hungary, mentioned that only one of the somatic hybrids showed that plastome determined large subunits of the fraction-I protein properties of both parents. This plant was very weak and yellowish. All the other hybrid plants with chromosome numbers from 56 to 64, close to the triploid number 63, showed the peptide pattern of the large subunit of fraction-I protein of *N. glauca* or *N. langsdorffii*. Because the genomes of the 23 hybrid plants are not identical, as demonstrated by the variation in chromosome numbers, the plastomes may be selected in one case in the direction of *langsdorffii*, in the other in the direction of *glauca*, depending on a greater or smaller number of chromosomes of *glauca* or *langsdorffii*. This material does not confirm the finding of Kung et al. (1975) that the offspring of one specimen of the so-called parasexual hybrid (Carlson et al., 1972) shows only *N. glauca* plastome-determined large subunit of fraction-I protein. Three plants were identical with the amphidiploid sexual hybrid in morphology and chromosome number; the fraction-I protein of the offspring of only one of these plants has been analyzed. My hypothesis, that the so-called parasexual hybrid from 1972 was actually a sexual hybrid with *N. glauca* as the mother, is not disproved, and the interesting hybrids of Smith et al. (1976), induced by PEG fusions, and the report of Smith in the Szeged meeting are not confirmations for the published parasexual hybrids, said to be arisen after $NaNO_3$ "clumping." (P. Carlson, when asked at the Miles Symposium in Cambridge, Mass., June 1976, in what way "25% fusion events" [Carlson et al., 1972] were counted, answered that there were no fusions, just clumps.) If in many plants with normal amphidiploid chromosome number and identical genomes (*N. glauca* + *N. langsdorffii*), somatic hybrids would have *langsdorffii* or *glauca* fraction-I protein and no mixtures, it would be an indication—not more!—that the "parasexual hybrid" of 1972 could be a product of fusion of somatic cells.

Our material between two varieties (s^2 and v^2) of *N. tabacum*, and of the two "species" *N. tabacum* and *N. silvestris*, also shows different chromosome numbers. But, as I pointed out above, the majority in both cases has the normal chromosome numbers. It is difficult, or perhaps impossible, to prove that triploid and tetraploid chromosome numbers arise by fusions of more than two protoplasts. Abnormal chromosome numbers arise by callus culture only, so this question remains open. Our somatic hybrids are produced by fusion of protoplasts with high pH and calcium without PEG, and are selected by high light intensity, complemented by normally colored, regenerating calluses out of yellow, nonregenerating calluses, and most have normal chromosome numbers. The somatic hybrids *N. glauca* + *N. langsdorffii* (Smith et al., 1976) are produced by PEG. They are selected in two steps. First, the better-growing calluses are selected in a hormone-containing medium. Second, the calluses growing in hormone-free medium are selected. Does the selection method favor especially well-growing, nearly triploid calluses? Or is the nuclear division disturbed more by PEG + high calcium concentration than by calcium and high pH?

As Kao has shown (this volume), fusion and the division and development of fusion products may be observed directly step-by-step microscopically. We preferred the more indirect methods of microbiology. Both methods lead to good results: that of Kao, by producing cell-hybrid lines from unrelated parents, and our method, by production of hybrid plants through cell hybrids of more related parents.

Summary

(1) Fusion of protoplasts from two varieties of tobacco and from *N. tabacum* and *N. silvestris* is induced with high pH and high calcium concentration.

(2) It should be possible in the future both to reduce the surface charge of the protoplasts to zero and to change it from the normal negative values to small, positive ones.

(3) Selection of hybrids in the callus stage is performed by complementation to normal chlorophyll and resistance to high light intensity of recessive mutations to chlorophyll deficiency in high light intensity in *N. tabacum* and *N. silvestris*.

(4) After many fusion experiments with the chlorophyll-deficient varieties of tobacco and a mutant of *Petunia hybrida* – unfortunately growing as a green callus in high light intensity – a hybrid plant from calluses still could not be regenerated. Some lines of callus cultures exist with abnormal chromosome numbers and regenerate teratomalike leaves, but do not produce shoots and roots. It has not yet been established that chromosomes of tobacco and *Petunia* participate in the genomes of these calluses.

REFERENCES

BERGMANN, L. 1959*a* Über die Kultur von Zellsuspensionen von *Daucus carota*. *Naturwissenschaften*. **46:**20–21.

BERGMANN, L. 1959*b*. A new technique for isolating and cloning cells of higher plants. *Nature (Lond.)*. **184:**648–649.

BERGMAN, L. 1960. Growth and division of single cells of higher plants in vitro. *J. Gen. Physiol.* **43:**841–851.

BINDING, H. 1966. Regeneration und Verschmelzung nackter Laubmoosprotoplasten. *Z. Pflanzenphysiol.* **55:**305–321.

CARLSON, P. S., H. H. Smith, and R. DEARING. 1972. Parasexual interspecific plant hybridization. *Proc. Natl. Acad. Sci. U. S. A.* **69:**2292–2294.

COCKING, E. C. 1960. A method for the isolation of plant protoplasts and vacuoles. *Nature (Lond.)*. **187:**962–963.

COCKING, E. C. 1972. Plant cell protoplasts – Isolation and development. *Annu. Rev. Plant Physiol.* **23:**29–50.

KAO, K. N., and M. R. MICHAYLUK. 1974. A method for high frequency intergeneric fusion of plant protoplasts. *Planta (Berl.)*. **115:**355–367.

KAO, K. N., F. CONSTABEL, M. R. MICHAYLUK, and O. L. GAMBORG. 1974. Plant protoplast fusion and growth of intergeneric hybrid cells. *Planta (Berl.)*. **120:**215–227.

KELLER, W. A., and G. MELCHERS. 1973. The effect of high pH and calcium on tobacco leaf protoplast fusion. *Z. Naturforsch.* **28c:**737–741.

KUNG, S. D., J. C. GRAY, S. G. WILDMAN, and P. S. CARLSON. 1975. Polypeptide composition of fraction I protein from parasexual hybrid plants in the genus *Nicotiana*. *Science (Wash. D. C.)*. **187:**353–355.

MELCHERS, G. 1960. Haploide Blütenpflanzen als Material der Mutationszüchtung. Beispiel: Blattfarbmutanten und mutatio *wettsteinii* von *Antirrhinum majus*. *Der Züchter*. **30:**129–134.

MELCHERS, G. 1965. Einige genetische Gesichtspunkte zu sogenannten Gewebekulturen. *Ber. Dtsch. Bot. Ges.* **78:**21–29.

MELCHERS, G. 1966. Contributions of plant virus research to molecular genetics. Proceedings of the G. Mendel Memorial Symposium, Brno, 4–7 August 1965. Prague. Academic Publishing House, Czechoslovak Academy of Science. 119–136.

MELCHERS, G. 1968*a*. Genetical aspects in callus culture work. Proceedings of the International Symposium Plant Growth Substances, January 1967. S. M. Sircar, editor. Calcutta University. 89–90.

MELCHERS, G. 1968*b*. Techniques for the quantitative study of mutation in plant viruses. *Theor. Appl. Genet.* **38:**275–279.

MELCHERS, G. 1971. Transformation or habituation to autotrophy and tumor growth and recovery. *Colloq. Int. Cent. Natl. Rech. Sci.* **193:**229–234.

MELCHERS, G., and L. BERGMANN. 1958/59. Untersuchungen an Kulturen von haploiden Geweben von *Antirrhinum majus*. *Ber. Dtsch. Bot. Ges.* **71:**459–473.

MELCHERS, G., and U. ENGELMANN. 1955. Die Kultur von Pflanzengewebe in flüssigem Medium mit Dauerbelüftung. *Naturwissenschaften*. **42:**564–565.

MELCHERS, G., and G. LABIB. 1974. Somatic hybridization of plants by fusion of protoplasts. I. Selection of light resistant hybrids of "haploid" light sensitive varieties of tobacco. *Mol. Gen. Genet.* **135:**277–294.

MELCHERS, G., and M. D. SACRISTÁN. 1977. Somatic hybridization of plants by fusion of protoplasts. II. The chromosome numbers of somatic hybrid plants of 4 different fusion experiments. *In* Recueil de Travaux Dédié à la Mémoire de G. Morel. J. Gautheret, editor. Masson et Cie, Paris. In press.

MUNDRY, K. W., and A. GIERER. 1958. Die Erzeugung von Mutationen des TMV durch chemische Behandlung seiner Nukleinsäure in vitro. *Z. Vererbungsl.* **89:**614–630.

NAGATA, T., and I. TAKEBE. 1971. Plating of isolated tobacco mesophyll protoplasts on agar medium. *Planta (Berl.)*. **99:**12–20.

POWER, J. B., and E. C. COCKING. 1971. Fusion of plant protoplasts. *Sci. Prog.* **59:**181–198.

SACRISTÁN, M. D. 1967. Auxin-Autotrophie und Chromosomenzahl. *Mol. Gen. Genet.* **99:**311–321.

SACRISTÁN, M. D., and G. MELCHERS. 1969. The caryological analysis of plants regenerated from tumorous and other callus cultures of tobacco. *Mol. Gen. Genet.* **105:**317–333.

Sacristán, M. D., and G. Melchers. 1977. Regeneration of plants from habituated and *Agrobacterium*-transformed single cell clones of tobacco. *Mol. Gen. Genet.* **152,** In press.

Smith, H. H., K. N. Kao, and N. C. Combatti. 1976. Interspecific hybridization by protoplast fusion in *Nicotiana. J. Hered.* **67:**123–128.

Takebe, I., G. Labib, and G. Melchers. 1971. Regeneration of whole plants from isolated mesophyll protoplasts of tobacco. *Naturwissenschaften.* **58:**318–320.

Takebe, I., Y. Otsuki, and S. Aoki. 1968. Isolation of tobacco mesophyll cells in intact and active state. *Plant Cell Physiol.* **9:**115–124.

Wallin, A., and T. Eriksson. 1973. Protoplast cultures from cell suspensions of *Daucus carota. Physiol. Plant.* **28:**33–39.

Wallin, A., K. Glimelius, and T. Eriksson. 1974. The induction of aggregation and fusion of *Daucus carota* protoplasts by polyethyleneglycol. *Z. Pflanzenphysiol.* **74:**64–80.

Winkler, H. 1908. *Solanum tubingense*, ein echter Pfropfbastard zwischen Tomate und Nachtschatten. *Ber. Dtsch. Bot. Ges.* **26a:**595–608.

Wittmann, H. G. 1959. Vergleich der Proteine des Normalstammes und einer Nitritmutante des TMV. *Z. Vererbungsl.* **90:**463–475.

Wittmann, H. G., and B. Wittmann-Liebold. 1963. Tobacco mosaic virus mutants and the genetic coding problem. *Cold Spring Harbor Symp. Quant. Biol.* **28:**589–595.

ADVANCES IN TECHNIQUES OF PLANT PROTOPLAST FUSION AND CULTURE OF HETEROKARYOCYTES

K. N. KAO and L. R. WETTER

Hybridization of somatic cells has been suggested as a possible technique to overcome the sexual incompatibility of two unrelated plant species. Somatic hybrids might be of great value for crop improvement (Schenk and Hildebrandt, 1968). Hybridization of somatic plant cells by means of protoplast fusion requires the following steps: (*a*) removal of the cell wall to form protoplasts; (*b*) high frequency of protoplast fusion; (*c*) identification and selection of hybrids; (*d*) culture of the hybrid protoplasts; (*e*) regeneration of plants from the hybrid protoplasts.

Attempts to hybridize somatic cells were made early in this century by Winkler, Küster, and Michel (in Power et al., 1970; Melchers and Labib, 1974). Michel (1937) demonstrated the fusion of homo- and heterospecific protoplasts after their treatment with sodium nitrate solution. However, fusions were rare and the subsequent fusion products could not be cultured (Power et al., 1970).

Rapid advances in techniques of protoplast production, fusion, and culture have been made during the last few years, some of which will be reported here.

Protoplast Production

In 1960 Cocking (Cocking, 1960) demonstrated the removal of walls from plant cells by hydrolytic enzymes, thus producing large quantities of protoplasts. Since then protoplasts have been isolated from various plant organs and cultured cells of numerous species. Young, fully expanded leaves from plants grown in controlled environments, petals from buds, root and shoot tips from germinated seeds, and fast growing cells in suspension cultures usually are the best source of protoplasts.

Successful isolation of protoplasts depends not only on the source of plants but also on the composition of wall-degrading enzymes, ionic and nonionic components of the isolation medium, the pH and duration of incubation. The enzyme solution we use extensively consists of 2% Onozuka R10 cellulase (Kinki Yakult Mfg. Co. Ltd., Nishinomiya, Japan), 2% Rhozyme hemicellulase (Rohm and Haas Co. Canada Ltd., West Hill, Ontario), and 1% pectinase (Sigma Chemical Co., St. Louis, Mo. (all desalted), in a solution containing 6 mM $CaCl_2 \cdot 2H_2O$, 0.7 mM $NaH_2PO_4 \cdot H_2O$, 3 mM MES (2 [N-morpholino] ethane sulfonic acid), 700 mM glucose, the pH is adjusted to 5.7 with dilute KOH (Kao et al., 1974). Usually 1 ml of this enzyme solution is mixed with 1 ml of a cell suspension culture in a 15 × 60-mm Petri dish and incubated at 24°C for 7–8 h. Occasional gentle agitation is required. At the end of the incubation period the protoplasts are collected and washed by centrifugation.

Protoplast Culture

After the enzymes are washed away, the protoplasts can be cultured in droplets or in a thin layer of liquid medium or soft agar. The nutritional requirements of the protoplasts are very similar to cultured cells, except that a higher concentration of calcium ions and sugars as ionic and nonionic stabilizers is required. At a very low population density (1 protoplast/ml), *Vicia hajastana* protoplasts require a supplement of the mineral salt medium with organic acids, vitamins, phytohormones, and amino acids in proper concentration (Kao and Michayluk, 1975). Leaf protoplasts of pea, alfalfa, and *Vicia hajastana* can undergo cell

---

K. N. KAO and L. R. WETTER Prairie Regional Laboratory, National Research Council of Canada, Saskatoon, Saskatchewan, Canada

division much sooner in lower population densities (less than 100 protoplasts/ml) than in high population densities. The components of a medium for low population density protoplast culture are listed in Table I. The protoplasts can only be cultured in very low light intensity (50 lux) or in the dark in this medium because it becomes phytotoxic when exposed to strong light. Protoplasts from a number of different species are able to regenerate cell walls, multiply, and regenerate plants. Protoplast isolation and culture has been recently reviewed by Gamborg (1976).

TABLE I

A Medium for Culturing Protoplasts

	mg		*mg*
(A) Mineral salt			
NH_4NO_3	600	KI	0.75
KNO_3	1900	H_3BO_3	3.00
$CaCl_2 \cdot 2H_2O$	600	$MnSO_4 \cdot H_2O$	10.00
$MgSO_4 \cdot 7H_2O$	300	$ZnSO_4 \cdot 7H_2O$	2.00
KH_2PO_4	170	$Na_2MoO_4 \cdot 2H_2O$	0.25
KCl	300	$CuSO_4 \cdot 5H_2O$	0.025
Sequestrene* 330Fe	28	$CoCl_2 \cdot 6H_2O$	0.025
	mg		*mg*
(B) Sugars			
Glucose	68,400	Mannose	125
Sucrose	125	Rhamnose	125
Fructose	125	Cellobiose	125
Ribose	125	Sorbitol	125
Xylose	125	Mannitol	125
	mg		*mg*
(C) Organic acids (adjusted to pH 5.5 with NH_4OH)			
Sodium pyruvate	5	Malic acid	10
Citric acid	10	Fumaric acid	10
	mg		*mg*
(D) Vitamins			
Inositol	100	Biotin	0.005
Nicotinamide	1	Choline chloride	0.5
Pyridoxine · HCl	1	Riboflavin	0.1
Thiamine · HCl	10	Ascorbic acid	1
D-Calcium pantothenate	0.5	Vitamin A	0.005
Folic acid	0.2	Vitamin D_3	0.005
p-Aminobenzoic acid	0.01	Vitamin B_{12}	0.01
	mg	*mg*	
(E) Hormones	Soybean × barley	Soybean × pea or *N. glauca*	
2,4-D‡	1	0.2	
Zeatin	0.1	0.5	
NAA§	–	1	

(F) Vitamin-free Casamino acid‖ (*mg*) 125
(G) Coconut water (*ml*) 10
From mature fruits: heated to 60°C for 30 min and filtered
(H) Glass distilled water (*ml*) 1000
(I) pH 5.7 (NaOH) for all the media
All media were filter sterilized

* Geigy Chemical Corp., Ardsley, N. Y.
‡ dichlorophenoxyacetic acid
§ naphthaleneacetic acid
‖ Difco Laboratories, Detroit, Mich.

Protoplast Fusion

Fusion of plant protoplasts can occur spontaneously during enzymatic degradation of cell walls. This type of fusion has been attributed to the expansion of plasmodesmata between protoplasts derived from adjacent cells (Cocking, 1972). Interspecific fusion must be induced. Molecular contact between membranes is an essential prerequisite for fusion (Poste and Allison, 1973).

Earlier reports indicated that treatment of protoplasts with $NaNO_3$ (Power et al., 1970), artifical seawater (Eriksson, 1971), and deplasmolyzing osmotic shock (Keller et al., 1973) could induce fusion, but the rate of fusion by these means was very low (Potrykus, 1972; Carlson et al., 1972).

More recently, Ito (1973) claimed that high frequency protoplast fusion could be obtained when protoplasts from meiotic cells collided with each other. Kameya (1975) indicated that protoplast adhesion and fusion could be induced with dextran sulfate. Several workers (Keller and Melchers, 1973; Melchers and Labib, 1974; and Binding, 1974) successfully induced protoplast fusion in a high pH and high calcium ion solution at 37°C. High frequency protoplast fusion can also be induced by high molecular weight (1,500–6,000) polyethylene glycol (PEG; Kao and Michayluk, 1974). The PEG-induced fusion is nonspecific and has been effective for fusion of protoplasts from a large number of species, from different genera and families. Heterokaryocyte frequencies of up to 30% can occur.

The procedure of the PEG method is as follows (Kao, 1976): (*a*) suspend mixed protoplasts of two different species in a solution consisting of 0.5 M glucose, 3.5 mM $CaCl_2$, and 0.7 mM KH_2PO_4 (pH 5.5); (*b*) put a drop of the protoplast suspension on a cover slip and allow the protoplasts to settle to form a thin layer; (*c*) slowly add three drops of a PEG solution to the protoplast preparation. The PEG solution is made up by dissolving 5g of PEG 1540 in 10 ml of a solution having 10.5 mM $CaCl_2$, 0.7 mM KH_2PO_4 (pH 5.5, KOH); (*d*) incubate the protoplasts in the PEG solution at room temperature for 10–20 min; (*e*) slowly elute the PEG with a protoplast culture medium.

Higher fusion frequency was usually obtained if the PEG was eluted with a high pH and high calcium solution and then followed by washing with the protoplast culture medium (Kao et al., 1974; Constabel et al., 1976). Up to 50% heterokaryocytes of soybean-pea were found in the protoplast population after such treatment. High pH (9) and high Ca^{2+} solution alone at 24°C produced 4–5% heterokaryocytes. The pea mesophyll protoplasts were unable to tolerate pH 10.5.

The protoplasts became tightly adpressed to each other immediately after the PEG solution was introduced. Only a very low frequency of heteroplasmic fusion occurred during PEG incubation. Soon after the PEG solution was diluted with a protoplast culture medium, many heterokaryocytes were observed. Thus most of the fusion occurred during washing of the protoplasts.

Many factors can effect PEG-induced adhesion and fusion (Kao and Michayluk, 1974; Kao et al., 1974; Constabel et al., 1976; Ferenczy et al., 1975; Anné and Peberdy, 1975; Weber et al., 1976.) (*a*) No adhesion and fusion of protoplasts were observed in a solution of PEG when the mol wt was below 300. Loose adhesion at low frequency occurred in a solution of PEG having a mol wt between 380–630. Tight adhesion of protoplasts and high frequency of protoplast fusion occurred in solutions of PEG having a mol wt of over 1,000. (*b*) Protoplast adhesion and fusion can only be induced in solutions having a high PEG concentration. (*c*) PEG-induced adhesion and fusion are enhanced by enrichment of the PEG solution with Ca^{2+} (up to 10 mM) and inhibited by K^+ or Na^+ at high pH. (50 mM Na glycine buffer at pH 9 completely inhibits PEG-induced adhesion.) However, if the Ca^{2+} concentration is high enough, then adhesion and fusion can occur in a PEG solution having high Na^+ at high pH (50 mM Na glycine buffer at pH 9; see Table II). (*d*) Protoplasts from young leaves or fast growing cultures are the best material for fusion. Protoplasts from old mesophyll cells are not suitable for fusion. Old cells presumably have secondary cell walls which are not digested by the enzymes. Rapid wall regeneration on the protoplasts before they are exposed to the fusogenic agent can also reduce fusion.

Proposed Mechanism of PEG-Induced Adhesion and Fusion

PEG is very water-soluble. The general formula of PEG is $HOCH_2\text{-}(CH_2\text{-}O\text{-}CH_2)_n\text{-}CH_2OH$. The ether linkages present make the molecule slightly negative in polarity and capable of forming hydrogen bonds with water, proteins, carbohydrates, etc., which possess positively polarized groups. When the chain of the PEG molecule is long enough, it is proposed that it acts as a molecular bridge between the surfaces of adjacent proto-

TABLE II

*Effect of PEG, Ca^{2+}, Na^+, and pH on fusion of pea and soybean protoplasts**

Components in PEG solution					
$CaCl_2$	PEG 1540	Na-glycine buffer	pH	Eluting solution	Heterokaryocytes
mM	%	*mM*			%
10.5	33	–	5.5	Medium	33
10.5	33	–	5.5	High pH, Ca^{2+}‡	47
–	33	50	9.0	Medium	0
50.0	33	50	7.5	Medium	25
50.0	33	50	9.0	Medium	30

* The protoplasts were washed once in a solution consisting of 0.5 M glucose, 3.5 mM $CaCl_2$, and 0.7 mM KH_2PO_4 (pH 5.5) and resuspended in this solution shortly before PEG treatment.
‡ The solution contains 50 mM $CaCl_2$, 0.3 M glucose, and 50 mM Na-glycine buffer, pH 10.5. The final pH of the PEG and high pH and high Ca^{2+} solution mixture was about pH 9.

plasts, and adhesion occurs. PEG can bind Ca^{2+} as well as many other cations. The Ca^{2+} may form a bridge between the negatively polarized groups of protein (or phospholipids) and PEG, thus enhancing adhesion. During the washing process, those PEG molecules that bind to protoplast membranes either directly or indirectly through Ca^{2+} ions may be eluted. This could result in disturbance and redistribution of electric charges. Since the membranes of two tightly adhering protoplasts are in intimate contact over large surface areas (Kao and Michayluk, 1974; Wallin et al., 1974; Burgess and Fleming, 1974; Fowke et al., 1975), such redistribution of charges could link some of the positively charged groups of one protoplast to some of the negatively charged groups of another protoplast and vice versa, resulting in protoplast fusion. Viability of the protoplasts is essential to restore the newly fused membranes to the "normal" condition.

Cell Division and Nuclear Behavior in Heterokaryocytes and Hybrids

Mitosis in most multinucleate protoplasts of soybean (Miller et al., 1971) and brome grass (Kao et al., 1973) was synchronized. Protoplasts with one nucleus in mitosis and the other in interphase were found only occasionally. This likely resulted from fusion of protoplasts at two different phases of the cell cycle.

In combinations such as soybean with barley, corn, pea, or *Vicia*, and carrot with barley, mitosis in heterokaryocytes was observed (Kao et al., 1974; Constabel et al., 1975; Dudits et al., 1976). Generally, when protoplasts of one of the parental species were capable of undergoing cell division, the heterokaryocytes were also able to divide. Nuclear fusion usually occurred during mitosis of multinucleate protoplasts when the nuclei were in adjacent position. After a culture period of 4–5 days, over 99% of the daughter cells from heterokaryocytes contained one nucleus. Dinuclear daughter cells occurred at a very low frequency in soybean-*Nicotiana glauca* heterokaryocytes (Fig. 1). When the nuclei in a protoplast were distantly separated from each other, multiple poles were quite often observed. The multipole formations could result in formation of chimeral cell colonies (Kao et al., 1974).

Premitotic nuclear fusion has been observed in heterokaryocytes of soybean-pea (Constabel et al., 1975), barley-carrot (Dudits et al., 1976), and soybean-*N. glauca*. In synkaryons of soybean-*N. glauca*, mitosis (prophase) was observed (Fig. 2). The result indicated that hybrid cell populations could arise from heterokaryocytes with premitotically fused nuclei.

In soybean-*N. glauca* heterokaryocytes, chromatid bridges were observed in the anaphase of the first mitosis. Extremely long chromosomes (Figs. 3–5) were observed in the second as well as the subsequent mitosis. Chromosomal fragments and ring chromosomes were also observed. Mitotic activity of the hybrid cells decreased gradually when in a mixed population. Eventually only soybean and *N. glauca* cells became the dominating species. When, however, the heterokaryocytes (or hybrids) were isolated and cultivated individually, they grew well and formed small cell clusters of 100–200 cells in 2 wk. A total of 20 colonies were isolated. The first cytological examination of the isolated hybrids was made about 1 mo after fusion. Long chromosomes and chromosome fragments were found quite frequently in the cell pop-

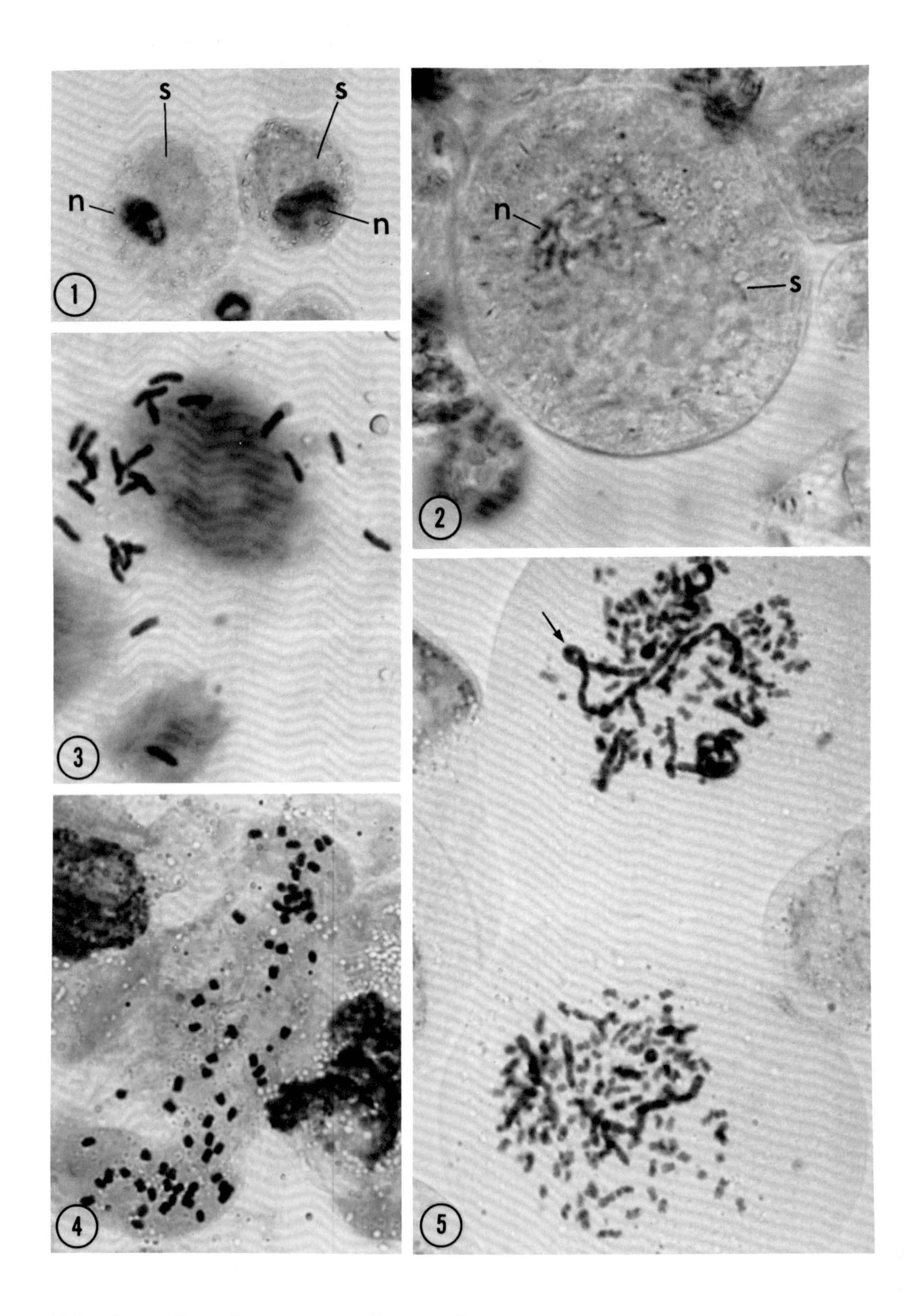
s
s
n
n
1
n
s
2
3
4
5

ulation. Chromosomal bridges were observed in anaphase. The constitution of the cell types in any of the hybrid lines became highly heterogeneous. As time went by, the number of *N. glauca* chromosomes in the hybrids was gradually reduced. Standard types of *N. glauca* chromosomes (Goodspeed, 1954) were difficult to find in the hybrid cells. Di- and triconstrictional chromosomes were found quite often (Figs. 6 and 7). After 6 mo of culturing, the hybrid cell lines (five in all) which were followed closely still retained some of the *N. glauca* chromosomes with modified structures (Figs. 8 and 9). The soybean chromosomes in the hybrid cells did not show obvious morphological changes. Occasionally, a very short chromosome was found in the cells of the 6-mo old *N. glauca* culture. No obvious change in the structure of chromosomes was observed in soybean cells during this 6-mo period.

Gene Expression in Somatic Hybrids

Carlson et al. (1972) indicated that the leaf peroxidase isozymes in the somatically produced hybrids of *N. glauca-Nicotiana langsdorffii* were identical to those of the sexually produced amphiploid. The isozyme bands of the hybrids were a composite of those found in the parental species.

Wetter and Kao (1976), employing gel electrophoresis, investigated a number of enzyme systems in the somatic hybrids of *N. glauca-N. langsdorffii* produced in their laboratory. They report only on those systems that were relatively simple. They found that a composite pattern, utilizing alcohol, glutamate, and lactate dehydrogenase, clearly demonstrated the formation of a hybrid (Fig. 10). It should be noted that in this case the hybrids are not a summation of parentally derived bands, the fast band (R_f 0.40) found in *N. langsdorffii* was not observed in the hybrid (Fig. 10). The deletion in 3 and 5 in the R_f 0.15 region could be on account of the loss of chromosomes in some of the hybrids (Smith et al., 1976). Aminopeptidase yielded a single bond in each parent with different R_fs, in the hybrid both could be detected.

In a preliminary gel electrophoresis study, we demonstrated that two enzyme systems (alcohol dehydrogenase and aspartate aminotransferase) could be employed to indicate the presence of interfamilial somatic hybrids of soybean-*N. glauca*. We could detect alcohol dehydrogenase bands derived from both parents in most of the five somatic hybrids cultured for 2–3 mo after fusion. Continued culturing showed that some of the hybrids gradually lost the band that had the same R_f (0.35) as the one found in *N. glauca*. 8 mo after the initial fusion only one hybrid of five retained the bands from both parents.

The electrophoretic pattern of aspartate aminotransferase shows that the soybean-*N. glauca* hybrid derives one of its bands from soybean (R_f 0.24), two from *N. glauca* (R_f 0.29 and 0.33) and one is common to both (R_f 0.41). The two slow bands from *N. glauca* do not appear in the hybrids (see Fig. 11). It was noted that in several hybrids the two intermediate bands (R_f 0.29 and 0.33) disappeared with time and the zymogram pattern could not be distinguished from soybean. After a 8-mo culturing period only two out of the five hybrids retained the typical banding of the hybrid.

It remains to be seen if these hybrids can regenerate plants.

CONCLUSION

Sexual incompatibility can be overcome by somatic cell hybridization, and genes from different families can be expressed in an interfamilial hybrid. We can expect that somatic cell hybridization could become a useful tool for plant breeders.

FIGURE 1 Two daughter cells, each has one nucleus from soybean (*s*) and one from *Nicotiana glauca* (*n*). × 550.

FIGURE 2 A synkaryon of soybean-*N. glauca* in mitosis. (*s*) soybean prophase chromosomes, (*n*) *N. glauca* prophase chromosomes. × 1,200.

FIGURE 3 Chromosomes of *N. glauca*. × 1,200.

FIGURE 4 Chromosomes of soybean. × 1,200.

FIGURE 5 Occurrence of extremely long chromosomes (arrow) in a cell of 2-wk old somatic hybrid of soybean-*N. glauca*. × 1,200.

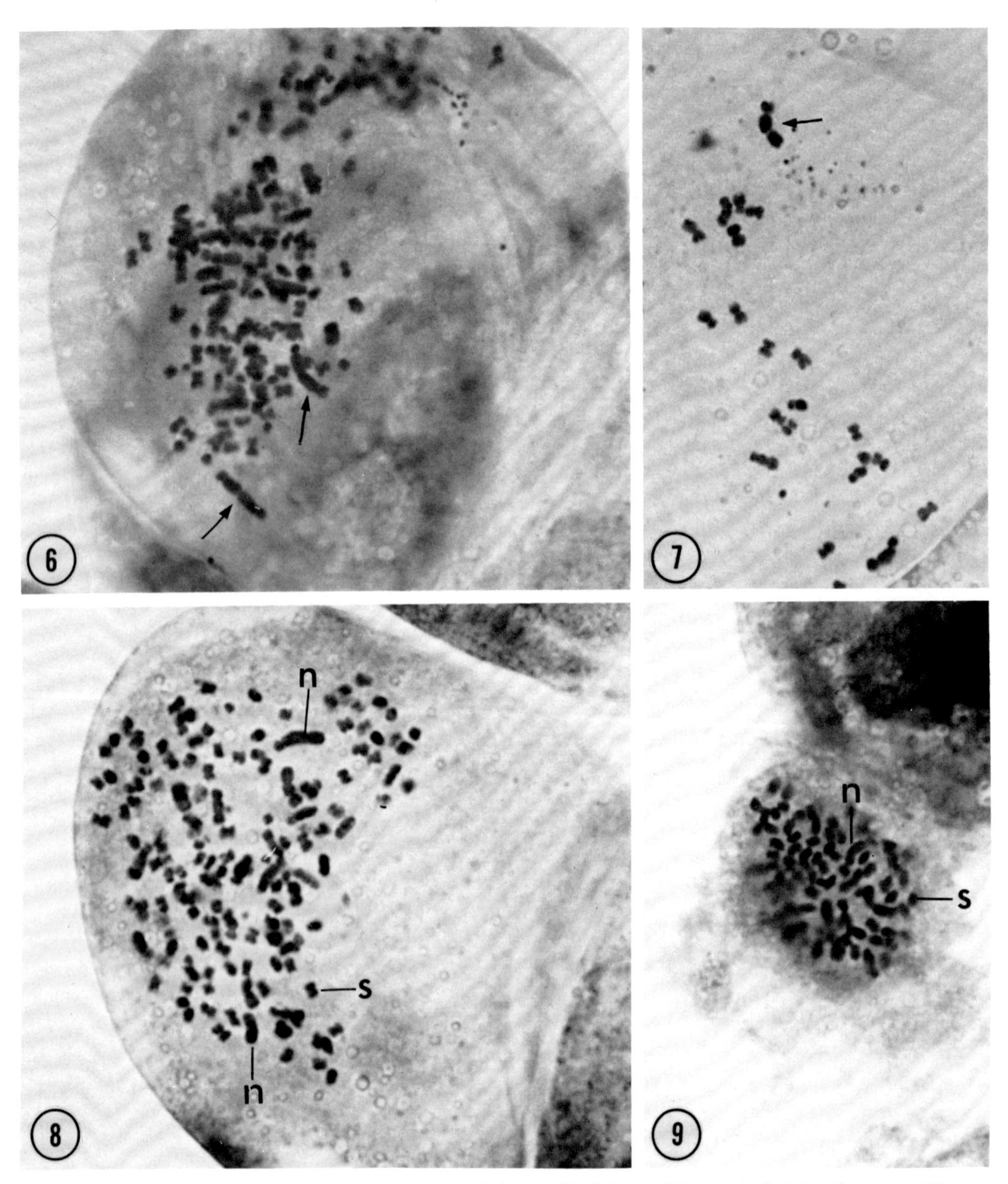

FIGURE 6 Triconstritional chromosomes (arrows) in a cell of 4-mo old somatic hybrid of soybean-*N. glauca*. × 1,200.

FIGURE 7 Diconstritional chromosomes (arrow) in a cell of 6-mo old somatic hybrid of soybean-*N. glauca*. × 1,200.

FIGURES 8 and 9 Metaphase in two cells, each from a different 6-mo old somatic hybrid cell line. The chromosomes of *N. glauca* have considerably changed in morphology. (*s*) soybean chromosome; (*n*) *N. glauca* chromosome.

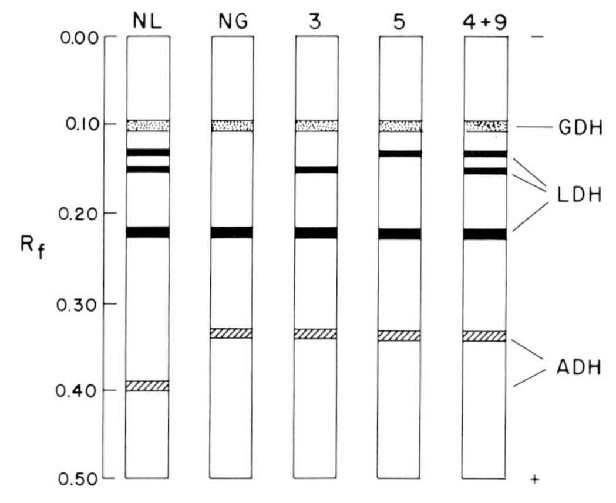

FIGURE 10 A graphic representation of dehydrogenase zymograms—glutamate (GDH), lactate (LDH), alcohol (ADH)—obtained from callus cultures of *N. langsdorffii* (NL), *N. glauca* (NG), and four somatic hybrids of *N. langsdorffii-N. glauca* (3, 4, 5, 9). The sexual amphiploid has the same zymograms as somatic hybrid line 4 and 9.

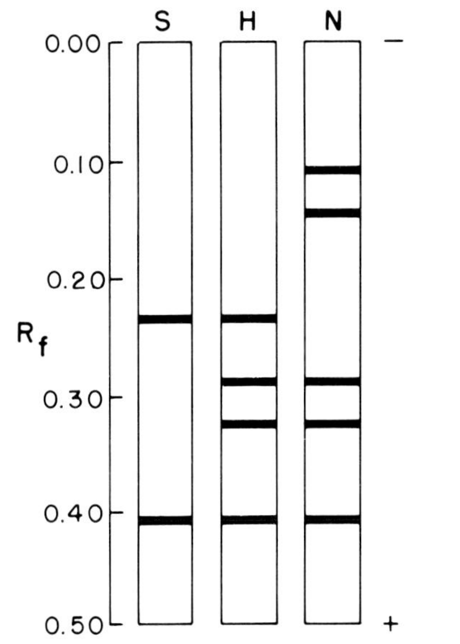

FIGURE 11 A graphic representation of aspartate aminotransferase zymograms of soybean (S), *N. glauca* (N), and of the hybrids (H).

ACKNOWLEDGMENT

This work was supported by grant no. 15772 of the National Research Council of Canada.

REFERENCES

ANNÉ, J., and J. F. PEBERDY. 1975. Conditions for induced fusion of fungal protoplasts in polyethylene glycol solution. *Arch. Microbiol.* **105:**201–205.

BINDING, H. 1974. Fusion experiments with isolated protoplasts of *Petunia hybrida* L. *Z. Pflanzenphysiol.* **72:**422–426.

BURGESS, J., and E. N. FLEMING. 1974. Ultrastructural studies of the aggregation and fusion of plant protoplasts. *Planta* (*Berl.*) **118:**183–193.

CARLSON, P. S., H. H. SMITH, and R. D. DEARING. 1972. Parasexual interspecific plant hybridization. *Proc. Natl. Acad. Sci. U. S. A.* **69:**2292–2294.

COCKING, E. C. 1960. A method for the isolation of plant protoplasts and vacuoles. *Nature* (*Lond.*). **187:**962–963.

COCKING, E. C. 1972. Plant cell protoplasts—isolation and development. *Annu. Rev. Plant Physiol.* **23:**29–50.

CONSTABEL, F., D. DUDITS, O. L. GAMBORG, and K. N. KAO. 1975. Nuclear fusion in intergeneric heterokaryons. A note. *Can. J. Bot.* **53:**2092–2095.

CONSTABEL, F., and K. N. KAO. 1974. Agglutination and fusion of plant protoplasts by polyethylene glycol. *Can. J. Bot.* **52:**1603–1606.

CONSTABEL, F., G. WEBER, W. J. KIRKPATRICK, and K. PAHL. 1976. Cell division in intergeneric protoplast fusion product. *Z. Pflanzenphysiol.* **79:**1–7.

DUDITS, D., K. N. KAO, F. CONSTABEL, and O. L. GAMBORG. 1976. Fusion of carrot and barley protoplasts and division of heterokaryocytes. *Can. J. Genet. Cytol.* **18:**263–269.

ERIKSSON, T. 1971. Isolation and fusion of plant protoplasts. *Colloq. Int. Cent. Natl. Rech. Sci.* **193:**297–302.

FERENCZY, L., F. KEVEI, and M. SZEGEDI. 1975. High-frequency fusion of fungal protoplasts. *Experientia* (*Basel*). **31:**1028–1030.

FOWKE, L. C., P. J. RENNIE, J. W. KIRKPATRICK, and F. CONSTABEL. 1975. Ultrastructural characteristics of intergeneric protoplast fusion. *Can. J. Bot.* **53:**272–278.

GAMBORG, O. L. 1976. Plant protoplast isolation, culture and fusion. *In*: Cell Genetics in Higher Plants. Proceedings of the UNDP/UNESCO/ICRO, D. Dudits, G. L. Farkas, and P. Maliga, editors. Akadémiai Kiadó (Budapest). 107–127.

GOODSPEED, T. H. 1954. The genus *Nicotiana*. *Chron. Bot.* 536.

ITO, M. 1973. Studies on the behavior of meiotic protoplasts. II. Induction of a high fusion frequency in protoplasts from liliaceous plants. *Plant Cell Physiol.* **14:**865–872.

KAMEYA, T. 1975. Induction of hybrids through somatic cell fusion with dextran sulfate and gelatin. *Jpn. J. Genet.* **50:**235–246.

KAO, K. N. 1976. A method for fusion of plant protoplasts with polyethylene glycol. In: Cell Genetics in Higher Plants. Proceedings of the UNDP/UNESCO/ICRO, D. Dudits, G. L. Farkas, and P. Maliga, editors. Akadémiai Kiadó (Budapest). 233–237.

KAO, K. N., F. CONSTABEL, M. R. MICHAYLUK, and O.

L. GAMBORG. 1974. Plant protoplast fusion and growth of intergeneric hybrid cells. *Planta (Berl.).* **120:**215–227.

KAO, K. N., O. L. GAMBORG, M. R. MICHAYLUK, W. A. KELLER, and R. A. MILLER. 1973. The effects of sugars and inorganic salts on cell regeneration and sustained division in plant protoplasts. *Colloq. Int. Cent. Natl. Rech. Sci.* **212:**207–213.

KAO, K. N., and M. R. MICHAYLUK. 1974. A method for high-frequency intergeneric fusion of plant protoplasts. *Planta (Berl.).* **115:**355–367.

KAO, K. N., and M. R. MICHAYLUK. 1975. Nutritional requirements for growth of *Vicia hajastana* cells and protoplasts at a very low population density in liquid media. *Planta (Berl.).* **126:**105–110.

KELLER, W. A., B. HARVEY, K. N. KAO, R. A. MILLER, and O. L. GAMBORG. 1973. Determination of the frequency of interspecific protoplast fusion by differential staining. *Colloq. Int. Cent. Natl. Rech. Sci.* **212:**455–463.

KELLER, W. A., and G. MELCHERS. 1973. Effect of high pH and calcium on tobacco leaf protoplast fusion. *Z. Naturforsch. Teil c Biochem.* **28c:**737–741.

MELCHERS, G., and G. LABIB. 1974. Somatic hybridization of plants by fusion of protoplasts. I. Selection of light resistant hybrids of "haploid" light sensitive varieties of tobacco. *Mol. Gen. Genet.* **135:**277–294.

MICHEL, W. 1937. Uber die experimentelle fusion pflanzlicher protoplasten. *Arch. Exp. Zellf.* **20:**230–252.

MILLER, R. A., O. L. GAMBORG, W. A. KELLER, and K. N. KAO. 1971. Fusion and division of nuclei in multinucleated soybean protoplasts. *Can. J. Genet. Cytol.* **13:**347–353.

POSTE, G., and A. C. ALLISON. 1973. Membrane fusion. *Biochim. Biophys. Acta.* **300:**421–465.

POTRYKUS, I. 1972. Intra and interspecific fusion of protoplasts from petals of *Torenia baillonii* and *Torenia fournieri. Nat. New Biol.* **231:**57–58.

POWER, J. B., S. E. CUMMINS, and E. C. COCKING. 1970. Fusion of isolated plant protoplasts. *Nature (Lond.).* **225:**1016–1018.

SCHENK, R. U., and A. C. HILDEBRANDT. 1968. Somatic hybridization: a new approach to genetic change. *Am. J. Bot.* **55:**731.

SMITH, H. H., K. N. KAO, and N. C. COMBATTI. 1976. Confirmation and extension of interspecific hybridization by somatic protoplast fusion in *Nicotiana. J. Hered.* **67:**123–128.

WALLIN, A., K. GLIMELIUS, and T. ERIKSSON. 1974. The induction of aggregation and fusion of *Daucus carota* protoplasts by polyethylene glycol. *Z. Pflanzenphysiol.* **74:**64–80.

WEBER, G., F. CONSTABEL, F. WILLIAMSON, L. FOWKE, and O. L. GAMBORG. 1976. Effect of preincubation of protoplasts on PEG – induced fusion of plant cells. *Z. Pflanzenphysiol.* **79:**459–464.

WETTER, L. R., and K. N. KAO. 1976. The use of isozymes in distinguishing the sexual and somatic hybrids in callus cultures derived from *Nicotiana. Z. Pflanzenphysiol.* **80:**455–462.

CHLOROPLAST INCORPORATION, SURVIVAL, AND REPLICATION IN FOREIGN CYTOPLASM

H. T. BONNETT and M. S. BANKS

Chloroplasts may have originated through a refinement of an endosymbiotic association, beginning with a tentative association between primitive photosynthetic and heterotrophic organisms, and evolving into the complex, interdependent relationship characteristic of nucleus and chloroplast. Investigations to support this concept of chloroplast origin by the techniques of genetics, biochemistry, and comparative physiology are limited by the range of combinations of nuclear and chloroplastic hereditary information that occurs naturally. If it becomes possible to achieve novel combinations of such information through the technique of chloroplast transplantation, these could be helpful in understanding compatibility between nucleus and chloroplast, competition in replication between genetically different chloroplasts, and developmental changes associated with the maintenance of chloroplasts in a foreign cytoplasm.

The obstacles to these experiments are substantial but the existence of endosymbiotic chloroplasts in cells of some invertebrates and the promise offered by successful chloroplast uptake encourage continuation of the study. We shall discuss conditions we have found optimal for chloroplast transplantation, constraints on the replication of chloroplasts after transplantation, and practical consequences of chloroplast transplantation.

Chloroplast Competition

Experiments involving interspecific transplantation of chloroplasts depend for their success on the ability of one or a small number of transplanted chloroplasts to compete effectively with the plastids already present in the host cell. Although there is no evidence on this constraint with chloroplast transplantation techniques, some indication may be derived from genetic experiments with plants showing biparental inheritance of plastids and from interspecific somatic hybrids of *Nicotiana*. In the genus *Oenothera,* plastids are transmitted both maternally and paternally. Schötz (1974*a*) and Epp (1973) have compared the competitive success of the various plastid types within the genus in interspecific crosses. By using normal green plants as one parent and plants with chlorophyll-defective chloroplasts as the other, Schötz distinguished three different classes of chloroplasts within the numerous species of the genus on the basis that they were either strongly, moderately, or weakly competitive. Furthermore, the mutation from green to white, which occurs within each chloroplast class, did not affect the relative competitive position of that chloroplast class with the other classes (Schötz, 1974*b*).

In the 15 interspecific somatic hybrids of *Nicotiana glauca* and *Nicotiana langsdorfii* reared by Smith et al. (1976), half of the plants contained *N. glauca* chloroplasts and the other half contained *N. langsdorfii* chloroplasts (Chen et al., 1976). Only one plant contained chloroplasts from both parents, and this was sickly and died. Thus whereas the two chloroplast types both retain the capability for replication after somatic hybridization, some form of interspecific chloroplast incompatibility seemed to prevent their simultaneous occurrence in the same plant. Chloroplast transplantation can add another dimension to somatic cell hybridization in investigations of chloroplast competition and compatibility.

Selection of Recipient Protoplast Lines

If the formation of novel plants is the goal, then protoplasts must be derived from tissues, calli, or

H. T. BONNETT and M. S. BANKS Department of Biology, University of Oregon, Eugene, Oregon cell suspensions which have the capacity for regeneration. While the best techniques of transplantation now available can induce uptake into over one-third of the protoplasts, only one to several chloroplasts are observed within a single protoplast. A selection system for those protoplasts in which chloroplast uptake is successful is advantageous, and the most promising is the use of protoplasts with chlorophyll-deficient chloroplasts. Kung et al. (1975) prepared protoplasts from white portions of a variegated *Nicotiana tabacum* and introduced normal chloroplasts isolated from *Nicotiana suaveloens*. The plant obtained possessed the fraction I protein electrophoretic mobility of *N. suaveolens*, indicating that it was a product of induced chloroplast uptake, and that the chloroplast(s) after uptake had survived and replicated.

Cell lines with defective chloroplasts are rare, and such plants generally appear by chance among young seedlings. Several workers have reported that a high proportion of variegated plants with defective plastids can be induced with *N*-nitroso-*N*-methylurea. Razoriteleva et al. (1970) showed that this compound was effective in inducing variegated seedlings after treatment of sunflower seeds. Pohlheim (1974) treated leaves of *Saintpaulia* plants with *N*-nitroso-*N*-methylurea and a high proportion of the plants regenerated were variegated. If *N*-nitroso-*N*-methylurea is generally effective in inducing plastid mutants, plant material could be selected for other experimental advantages, and lines with defective plastids could be derived subsequently.

Selection of Chloroplast Donors

For short periods after isolation, chloroplast preparations are capable of some normal functions. Even though spinach chloroplasts have undergone limited division in culture (Ridley and Leech, 1970), in general, attempts to culture angiosperm chloroplasts indicate that they are fragile and deteriorate rapidly. The use of algal chloroplasts has been more successful (Giles and Sarafis, 1972). Chloroplasts isolated from *Caulerpa* were maintained for several weeks with a gradual decline in activity; some chloroplast division occurred.

The fragility and limited in vitro survival of isolated chloroplasts emphasize the necessity of performing transplantation experiments rapidly after chloroplast isolation. We have selected plants from which chloroplasts are known to establish an endosymbiotic association with another organism in nature, and have concentrated on the cytological course of events during and after chloroplast incorporation. In the Oregon intertidal zone, the green alga *Codium fragile* is grazed by the marine opisthobranch *Placida dendritica* in a species-specific association in which *Codium* chloroplasts are incorporated into the cells of the gut diverticula of the sea slug (Trench, 1975). Because biochemical functions of the *Codium* chloroplasts within the foreign cytoplasm are maintained, these chloroplasts constitute desirable material for induced chloroplast uptake into other plant cytoplasms. In some uptake experiments, we have used chloroplasts isolated from another alga, *Vaucheria dichotoma*.

The Uptake of Codium Chloroplasts into Protoplasts of Carrot

A cell suspension of *Daucus carota* has been chosen as a recipient for *Codium* chloroplasts because, in all conditions of culture under which it has been observed, this line has neither differentiated chloroplasts nor synthesized chlorophyll. Thus, any greening or any microscopic detection of chloroplasts is circumstantial evidence for the incorporation of *Codium* chloroplasts. Definitive evidence for chloroplast uptake is based on the characteristic arrangement of chloroplast lamellae in electron micrographs of incorporated chloroplasts.

THE MECHANICS OF UPTAKE: Chloroplasts were isolated from *Codium fragile* in an isolation medium of 0.25 M sorbitol, 2% Ficoll 400, and 0.05 M phosphate, pH 7.2, at 6°C. Protoplasts were isolated from *Daucus carota* by incubating a cell suspension for 2.5 h in 0.4 M sorbitol containing 2% Driselase at 28°C in the dark. Protoplasts were resuspended in the chloroplast suspension and this mixture was combined with polyethylene glycol 4,000 (PEG). The PEG was then diluted and removed by washing with protoplast culture medium. All manipulations were carried out at 28°C. For a quantitative assessment of the treatment, the PEG-treated protoplast-chloroplast mixture was examined by light microscopy; protoplast survival and chloroplast uptake were assessed.

A number of experimental conditions have been examined in order to promote uptake frequency yet conserve protoplast viability. Previous work (Bonnett, 1976) showed that protoplast densities that favor protoplast aggregation were also conducive to chloroplast uptake. Thus, a protoplast concentration of about 10^6/ml was adopted for subsequent experiments.

Kao and Michaluk (1974) reported that the fusion of protoplasts induced by PEG occurred during slow dilution of the PEG after treatment. Two rates of PEG dilution were tested: rapid dilution over a period of a few seconds, and slow dilution over a period of 9 min. Higher uptake was consistently observed with slower dilution (Table I). Viability was not significantly different in either case.

By using slow dilution, the frequency of uptake was compared under conditions of varying chloroplast densities (Fig. 1). Uptake was proportional to chloroplast density at concentrations up to about 10^8 chloroplasts/ml, reached a peak at about 2×10^8, and then declined at yet higher chloroplast densities. Thus, a chloroplast density of 10^8/ml was selected, or a ratio of chloroplast to protoplast of about 100:1.

The optimum concentration of PEG was assessed at this chloroplast density (Fig. 2). Uptake of chloroplasts required significant concentrations

TABLE I
Chloroplast Uptake and Protoplast Survival after Rapid or Slow PEG Dilution

	Slow dilution	Rapid dilution
	%	%
Chloroplast uptake*	21	14
Protoplast viability‡	54	54

* Average of two experiments; in each experiment 400 living protoplasts were scored.
‡ Average of two experiments; in each experiment 400 protoplasts in all were scored.

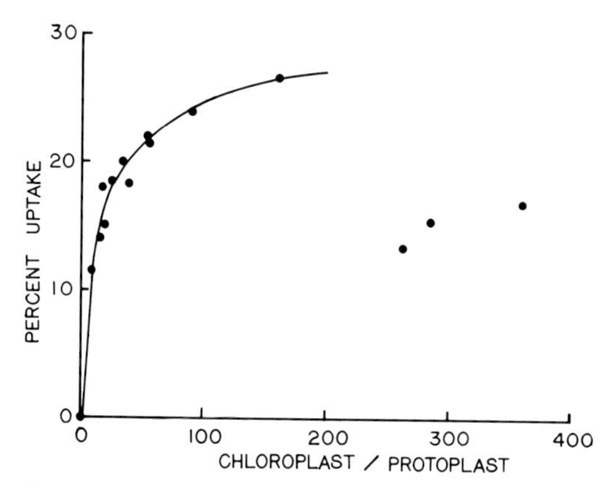

FIGURE 1 Influence of chloroplast concentration on chloroplast uptake by protoplasts treated with PEG.

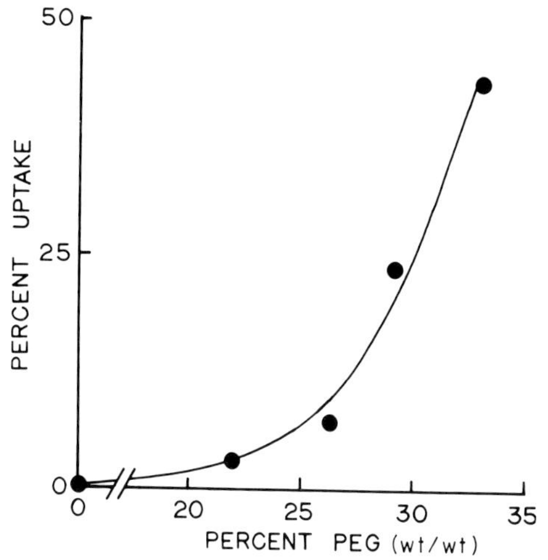

FIGURE 2 Relationship between chloroplast uptake and PEG concentration.

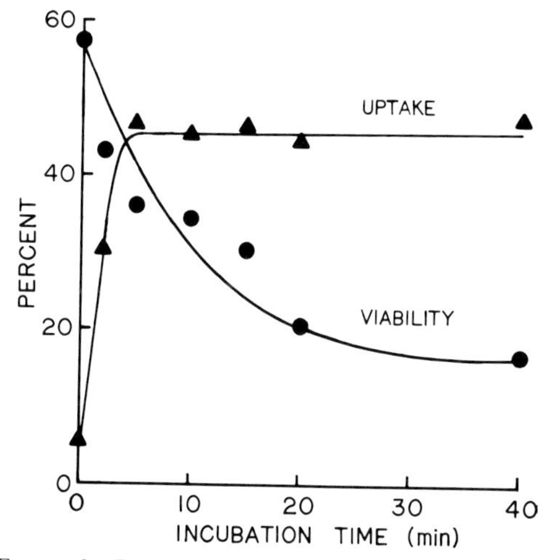

FIGURE 3 Dependence of chloroplast uptake on duration of PEG treatment.

of PEG, increasing sharply over the range of PEG concentrations tested. However, even though 35% wt/wt PEG increased uptake over lower concentrations, protoplast viability declined sharply at concentrations beyond 28% wt/wt.

Because PEG treatment of protoplasts is known to reduce viability, the time of treatment optimal for uptake was investigated. Figure 3 shows that only minimal uptake was observed at zero time of treatment (PEG added and immediately followed by stepwise slow dilution). Uptake increased with increasing PEG-treatment time to a plateau at 5–10 min in PEG. Inasmuch as protoplast viability decreased with longer exposure times, 10 min was chosen as the optimal time.

The conditions defined to promote maximum uptake are protoplast concentrations of about 1 × 10^6/ml, chloroplast densities of about 10^8/ml, and PEG treatment at a concentration of 28% wt/wt for a period of 10 min, followed by slow dilution and washing. With these conditions, we have obtained uptake frequencies as high as 45% with viability near 50%. Whereas dilution is an important treatment parameter, we believe most uptake occurs during PEG treatment, not during dilution. Chloroplasts can be observed within protoplasts while still in PEG. Moreover, Fig. 3 shows that uptake varied strikingly with time of PEG treatment under identical conditions of PEG dilution.

CULTURE: Chloroplasts have been incorporated into protoplasts of *Daucus carota* and maintained in sterile culture for several days. After 2 days, a cell wall could be distinguished around control protoplasts as well as those with incorporated chloroplasts. The chloroplasts maintained their green color and their ellipsoid shape for at least 4 days in culture, but after this time they became increasingly difficult to distinguish by light microscopy because of the increasingly refractile nature of both the cell wall and intracellular material.

CHLOROPLAST LOCALIZATION: That chloroplasts are taken up into protoplasts rather than simply closely adherent to the plasmalemma can be shown both microscopically and biochemically. The dye, nitro blue tetrazolium chloride (NBT), is reduced to an insoluble blue precipitate in the presence of photosynthetic electron transport. In populations of isolated chloroplasts, 100% of the chloroplasts turn blue after exposure to light. NBT does not penetrate the plasmalemma and so if dye is added to a PEG-treated chloroplast-protoplast mixture, chloroplasts free in suspension or merely adherent to the plasmalemma turn blue and those incorporated into protoplasts remain green. Only when the host protoplast dies and the permeability properties of the membrane are lost do such chloroplasts turn blue. Uptake frequencies assessed on the basis of dye reduction match those obtained by our standard assessment procedure, indicating that the chloroplasts are inside the protoplast.

The intracellular localization of the chloroplasts has been confirmed by electron microscope analysis. In chloroplasts of *Vaucheria,* the photosynthetic lamellae are arranged in parallel arrays which traverse the entire length of the chloroplast (Fig. 4). In contrast, the lamellae of the carrot chloroplast are arranged into discrete granal stacks and intergranal thylakoids (Fig. 5). Figure 6 shows a chloroplast isolated from *Vaucheria* which has been incorporated into a carrot protoplast. The arrangement of the photosynthetic lamellae is definitely that of the algal chloroplast. The structural preservation of the carrot cytoplasm even in the vicinity of the chloroplast indicates that the foreign body is not detrimental to the protoplast.

A Potential Role for Chloroplast Uptake in the Derivation of Cytoplasmic Male Sterile Lines

Cytoplasmic male sterility represents a failure of a plant, normally bisexual, to produce viable pollen. The inheritance of this trait follows the maternal line in those species in which the pollen contributes only the nuclear genome and is probably associated with one of the DNA-containing organelles. Even though strong correlative evidence exists that favors the mitochondria in male sterile lines of corn, in the genus *Nicotiana* the evidence favors the chloroplasts. Chen et al. (1976) have analyzed the fraction I protein of a number of male sterile tobaccos, some of which originated from interspecific crosses with *N. tabacum* as the male parent followed by repeated back-crossings with *N. tabacum* pollen. In the cases examined, the large subunit of fraction I protein, which is coded by chloroplast DNA, was distinct from that of *N. tabacum.* Figure 7 diagrammatically represents the large subunit electrophoretic mobility for male sterile *N. tabacum* cultivars derived from interspecific crosses. This figure is adapted from Chen et al. (1976), Edwardson (1970), and Berbeč (1974). All of these male sterile cultivars possess fraction I protein mobilities different from the group of species which includes *tabacum,* suggesting that the combination of chloroplasts of certain *Nicotiana* species with the genome of *N. tabacum* results in cytoplasmic male sterility. Moreover, a male sterile mutant of *N. tabacum* var. Nadwialanski Maby has been described by Berbeč (1974) and shown to have an altered electrophoretic mobility of the large subunit of the fraction I protein (Chen et al., 1976).

If the cytoplasmic genetic information leading to male sterility resides in the chloroplast genome, chloroplast transplantation experiments could provide a direct confirmation for this hypothesis. Chloroplast transplantation could then be applied as a means to induce male sterile lines in plants where there are currently none available.

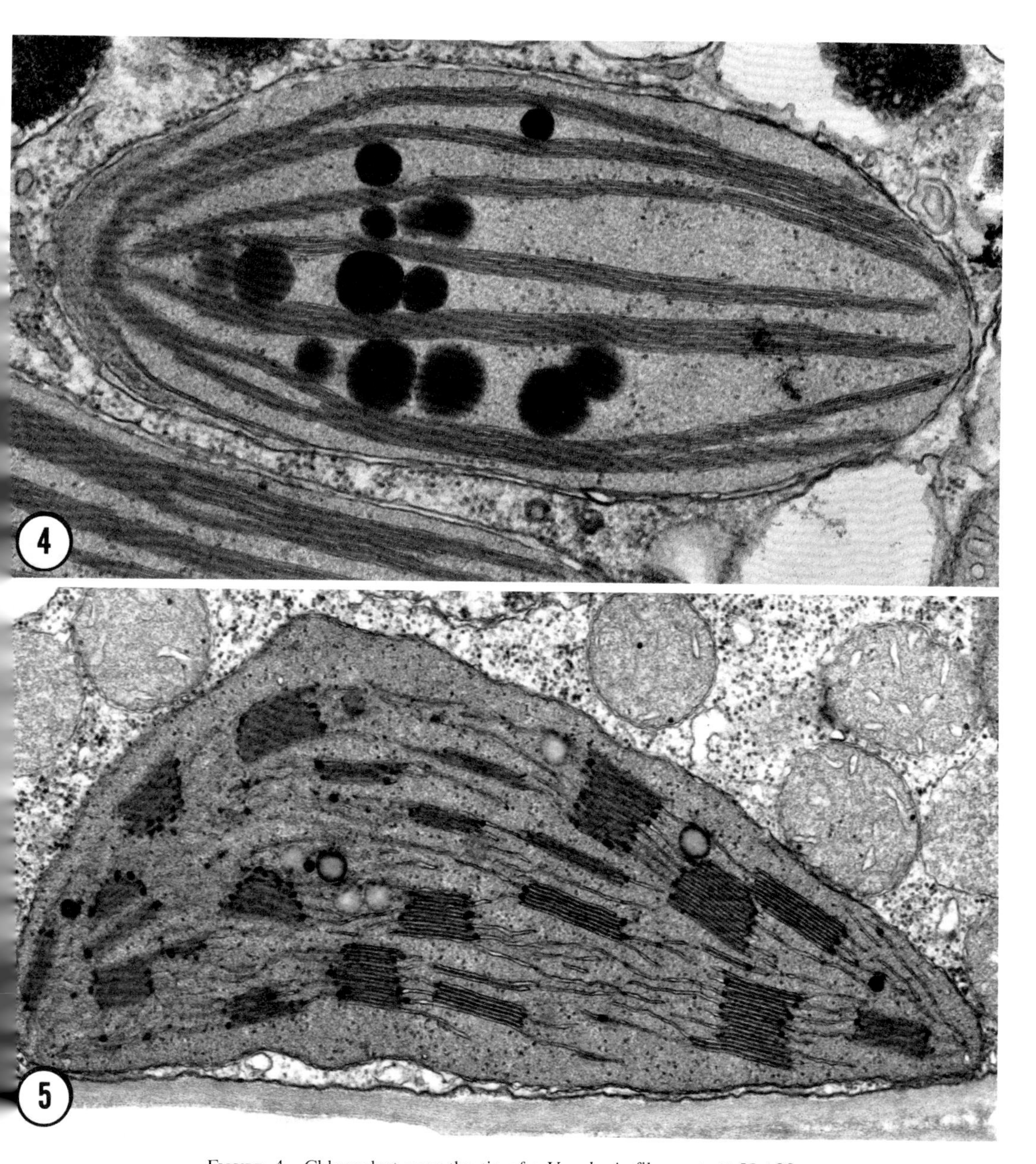

FIGURE 4 Chloroplast near the tip of a *Vaucheria* filament. × 53,100.
FIGURE 5 Chloroplast in a mesophyll cell of a carrot leaf. × 33,100.

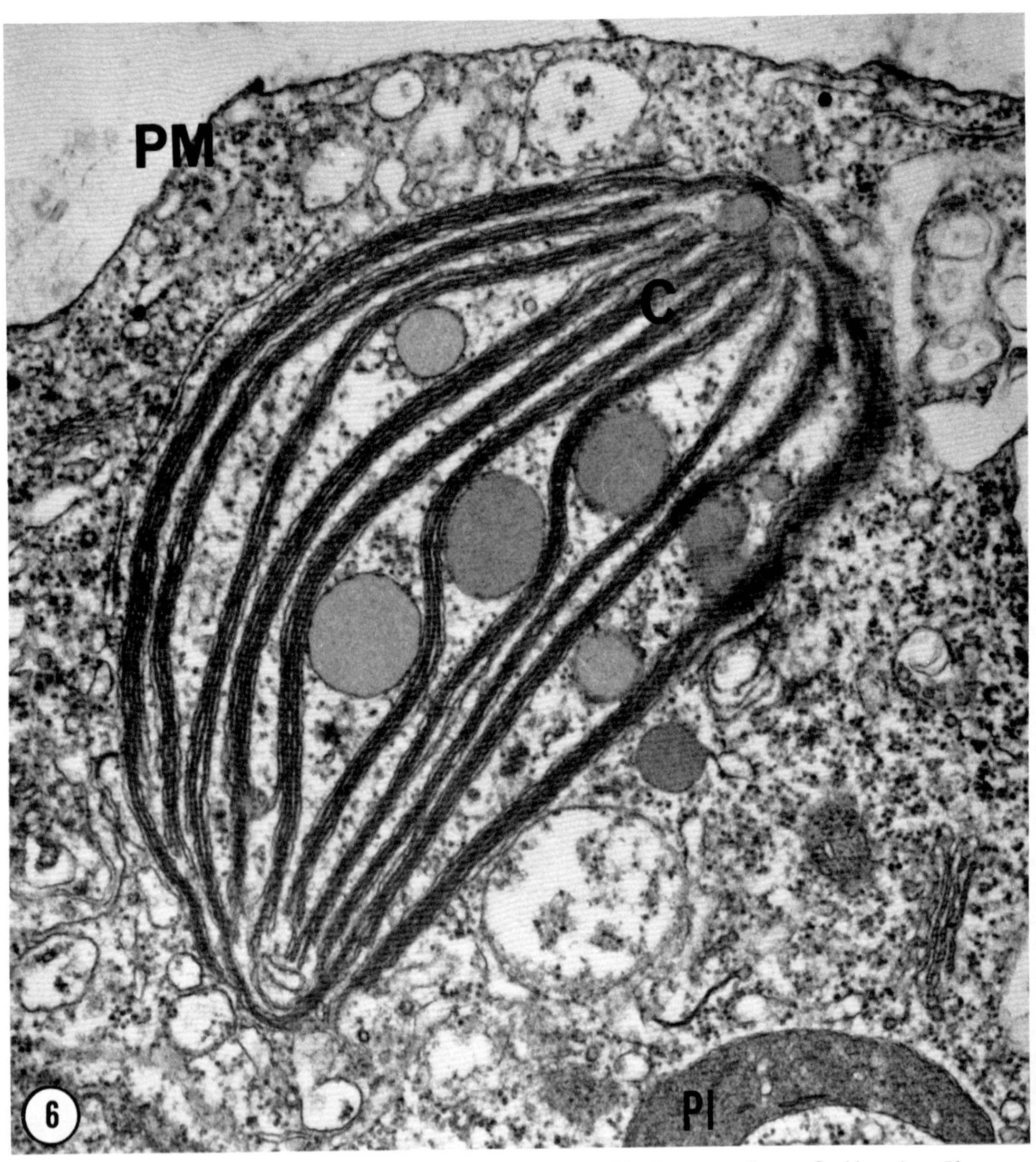

FIGURE 6 *Vaucheria* chloroplast within a carrot protoplast. *PM*, plasma membrane; *C*, chloroplast; *Pl*, carrot plastid. × 40,600.

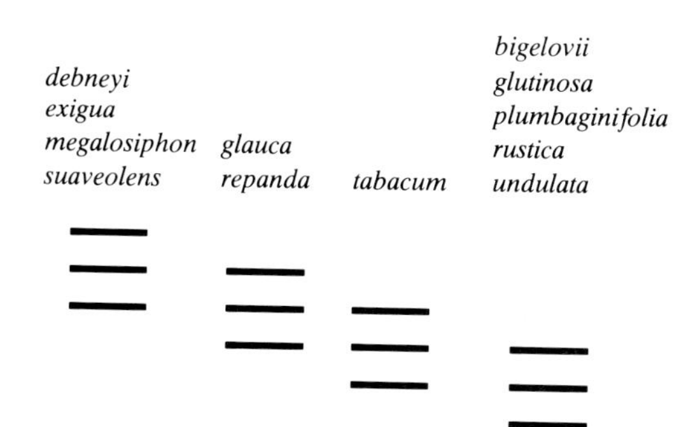

FIGURE 7 Diagrammatic representation of the position of the large subunit polypeptides of fraction I protein in male sterile cultivars of *N. tabacum* derived from interspecific crosses.

Summary

By using algal chloroplasts and carrot protoplasts, we can introduce one to several chloroplasts into a substantial proportion of the protoplasts. Preliminary culture of the resultant organelle hybrids indicates that there is no adverse effect of transplantation on either host or chloroplast. However, the long-term survival and potential for replication of chloroplasts transplanted by this method have not yet been investigated. We believe that these techniques may be useful to extend the range of nucleus-chloroplast combinations and to selectively construct combinations of organisms which will further the understanding of chloroplast evolution and interaction with the nuclear genome.

ACKNOWLEDGMENT

This research was supported by grant no. BMS75-13676 from the National Science Foundation.
Figures 4–6 are reprinted by permission of Springer-Verlag (*Planta* [Berl.] **131**:229–233).

REFERENCES

BERBEČ, J. 1974. A cytoplasmic male sterile mutation form of *Nicotiana tabacum* L. *Z. Pflanzenzuecht.* **73**:204–216.

BONNETT, H. T. 1976. On the mechanism of the uptake of *Vaucheria* chloroplasts by carrot protoplasts treated with polyethylene glycol. *Planta* (*Berl.*). **131**:229–233.

CHEN, K., S. JOHAL, and S. G. WILDMAN. 1977. Phenotypic markers for chloroplast DNA genes in higher plants and their use in biochemical genetics. *In* Nucleic Acids and Protein Synthesis in Plants. J. W. Weil and L. Bogorad, editors. Plenum Publishing Corp., Strasbourg. In press.

EDWARDSON, J. R. 1970. Cytoplasmic male sterility. *Bot. Rev.* **36**:341–420.

EPP, M. D. 1973. Nuclear gene-induced plastome mutations in *Oenothera hookeri*. I. Genetic analysis. *Genetics.* **75**:465–483.

GILES, K. L., and V. SARAFIS. 1972. Chloroplast survival and division *in vitro*. *Nat. New Biol.* **236**:56–58.

KAO, N. N., and M. R. MICHAYLUK. 1974. A method for high frequency intergeneric fusion of plant protoplasts. *Planta* (*Berl.*). **115**:355–367.

KUNG, S. D., J. C. GRAY, S. G. WILDMAN, and P. S. CARLSON. 1975. Polypeptide composition of fraction I protein from parasexual hybrid plants in the genus *Nicotiana*. *Science* (*Wash. D.C.*). **187**:353–355.

POHLHEIM, F. 1974. Nachweis von Mischzellen in variegaten Adventivsprossen von *Saintpaulia*, entstanden nach Behandlung isolierter Blätter mit N-Nitroso-N-Methylharnstoff. *Biol. Zentralbl.* **93**:141–148.

RAZORITELEVA, E. K., YU. D. BELETSKY, and YU. A. ZHDANOV. 1970. The genetical nature of mutation induced by *N*-nitroso-*N*-methylurea in sunflower. I. The variegated plants. *Genetika.* **6**(8):102–107.

RIDLEY, S. M., and R. M. LEECH. 1970. Division of chloroplasts in an artificial environment. *Nature* (*Lond.*). **227**:463–465.

SCHÖTZ, F. 1974*a*. Untersuchungen über die Plastidenkonkurrenz bei *Oenothera*. IV. Der EinfluB des Genoms auf die Durchsetzungsfähigkeit der Plastiden. *Biol. Zentralbl.* **93**:41–64.

SCHÖTZ, F. 1974*b*. Untersuchungen über die Plastidenkonkurrenz bei *Oenothera*. V. Die Stabilität der Konkurrenzfähigkeit bei Verwendung verschiedenartiger mutierter Testplastiden. *Biol. Zentralbl.* **93**:483.

SMITH, H. H., K. N. KAO, and N. C. COMBATTI. 1976. Interspecific hybridization by protoplast fusion in *Nicotiana*: confirmation and extension. *J. Hered.* **67**:123–128.

TRENCH, R. K. 1975. Of "leaves that crawl": functional chloroplasts in animal cells. *Symp. Soc. Exp. Biol.* **29**:229–265.

Biogenesis of Mitochondria

INTRODUCTORY REMARKS

PIET BORST

Twenty years ago Simpson and co-workers (Simpson and McLean, 1955; McLean et al., 1958) discovered protein synthesis in mitochondria. At first this discovery generated little enthusiasm among cell biologists. I still remember a heated discussion during the FEBS Meeting in Warsaw in 1966, only 11 years ago, in which a competent Swedish group maintained that the so-called mitochondrial protein synthesis was an artifact due to contamination of mitochondrial preparations with bacteria and that mitochondrial protein synthesis did not exist at all.

That was the last skirmish of that nature, however. In the 11 years that followed, studies of mitochondrial biogenesis became one of the fastest growing fields of molecular and cell biology. These studies have led to the four main conclusions stated in Table I and illustrated in part in Fig. 1. Three main methods have been used in these studies. First is the analysis of highly purified, isolated mitochondria: (*a*) what is there? (DNA, RNA, ribosomes, etc); (*b*) what do they make? (DNA, RNA, protein synthesis).

Second is the analysis of mitochondrial biogenesis in intact cells: (*a*) pulse-labeling experiments using specific inhibitors like D-chloramphenicol, the specific and universal inhibitor of mitochondrial protein synthesis (Fig. 1); (*b*) labeling experiments using mutants. An example is provided by the classic experiments of Luck (1965) with a choline-requiring mutant of *Neurospora* which he used to demonstrate that mitochondria multiply by growth and division of pre-existing mitochondria, rather than by *de novo* synthesis. Another example is given by the more recent experiments on mitochondrial DNA synthesis in animal cells, which use cells that lack the cell-sap thymidine kinase isoenzyme (Attardi and Attardi, 1972; Berk and Clayton, 1973). Since the mitochondrial isoenzyme is not affected in these mutants, they will incorporate added thymidine exclusively in mitochondrial DNA.

Third is the analysis of the defect in mutants defective in mitochondrial biogenesis. A detailed picture of our present knowledge of mitochondrial biogenesis is provided in a recent book (Lloyd, 1974) and in the proceedings of four symposia (Birky et al., 1975; Bandlow et al., 1976; Saccone and Kroon, 1977; Bücher et al., 1977). Here we can only illustrate some of the main research lines. In the first paper I summarize our knowledge about the structure and function of mitochondrial DNA with emphasis on primitive eukaryotes like yeast, the pet organism of most workers in the field. In the second paper by Dr. O'Brien, the emphasis will shift to animal mitochondrial DNA and to the mechanism of transcription and translation in mitochondria; and finally, in the last paper,

TABLE I

The Biosynthesis of Mitochondria: Main Characteristics

1. All functional mitochondria contain mitochondrial DNA and the enzymic machinery to replicate this DNA, to copy it into RNA, and to translate messenger RNAs into protein.
2. Mitochondrial DNA codes for mitochondrial ribosomal RNA, mitochondrial transfer RNAs, and a limited number of mitochondrial inner membrane proteins.
3. Most of the mitochondrial proteins are specified by nuclear genes, made on cell-sap ribosomes, and imported into the mitochondrion.
4. The mitochondrial genetic system and the nucleo-cell-sap genetic system (probably) have no components in common.

PIET BORST Section for Medical Enzymology and Molecular Biology, Laboratory of Biochemistry, University of Amsterdam, Amsterdam, The Netherlands

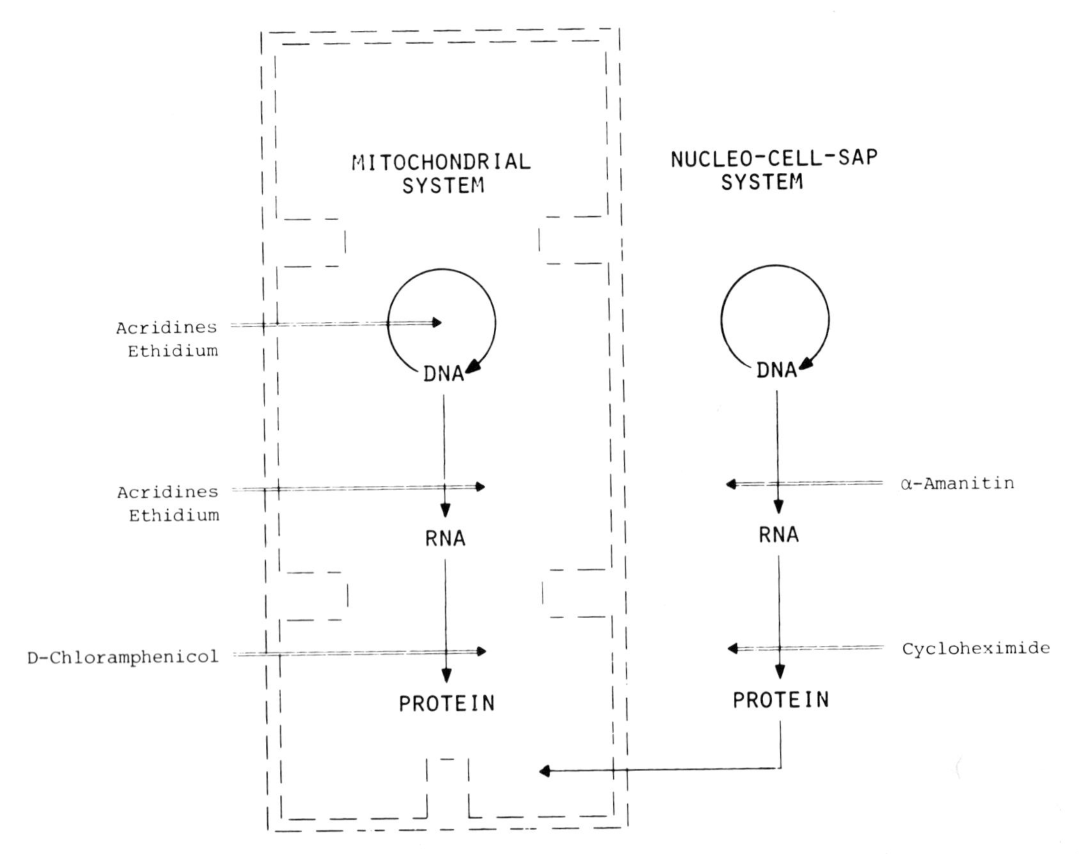

FIGURE 1 The site of action of various inhibitors on mitochondrial biogenesis.

Dr. Schatz will turn to the problem of membrane assembly and the interplay of the two genetic systems required for a balanced synthesis of mitochondria.

REFERENCES

ATTARDI, B., and G. ATTARDI. 1972. Persistence of thymidine kinase activity in mitochondria of a thymidine kinase-deficient derivative of mouse L cells. *Proc. Natl. Acad. Sci. U. S. A.* **69:**2874–2878.

BANDLOW, W., R. J. SCHWEYEN, D. Y. THOMAS, K. WOLF, and F. KAUDEWITZ. Editors. 1976. Genetics, Biogenesis, and Bioenergetics of Mitochondria. Gruyter & Co., Berlin. 418.

BERK, A. J., and D. A. CLAYTON. 1973. A genetically distinct thymidine kinase in mammalian mitochondria. *J. Biol. Chem.* **248:**2722–2729.

BIRKY, C. W., JR., P. S. PERLMAN, and T. J. BYERS. Editors. 1975. Genetics and Biogenesis of Mitochondria and Chloroplasts. Ohio State University Press, Columbus. 361.

BÜCHER, TH., W. NEUPERT, W. SEBALD, and S. WERNER. Editors. 1977. Genetics and Biogenesis of Chloroplasts and Mitochondria, North-Holland Publishing Co., Amsterdam. In press.

LLOYD, D. Editor. 1974. The Mitochondria of Microorganisms. Academic Press Inc., Ltd., London. In press.

LUCK, D. J. L. 1965. Formation of mitochondria in *Neurospora crassa.* A study based on mitochondrial density changes. *J. Cell Biol.* **24:**461–470.

MCLEAN, J. R., G. L. COHN, I. K. BRANDT, and M. V. SIMPSON. 1958. Incorporation of labeled amino acids into the protein of muscle and liver mitochondria. *J. Biol. Chem.* **233:**657–663.

SACCONE, C., and A. M. KROON. Editors. 1977. The Genetic Function of Mitochondrial DNA. North-Holland Publishing Co., Amsterdam. In press.

SIMPSON, M. V., and J. R. MCLEAN. 1955. The incorporation of labeled amino acids into the cytoplasmic particles of rat muscle. *Biochim. Biophys. Acta.* **18:**574–575.

STRUCTURE AND FUNCTION OF MITOCHONDRIAL DNA

PIET BORST

The Nature of Mitochondrial DNA

After the discovery in 1966 that the mitochondrial DNA (mtDNA) of animal tissues consists of a homogeneous population of small duplex circles (Borst, 1972) which are easy to isolate, the mtDNA from a variety of organisms has been characterized. The results of the efforts in many laboratories are summarized in Table I. As far as we now know, the molecular weight of mtDNA varies within a fairly narrow range—between the 10×10^6 in animal tissues and the 70×10^6 in higher plants. Circularity is the rule, but there are two exceptions: the mtDNAs of *Tetrahymena* and *Paramecium*.

Normal eukaryotic cells contain many mtDNA molecules, ranging from 10^2 in a yeast cell to 10^8 in a toad egg. An adult man contains even 10^{17} mtDNA molecules and one may wonder, therefore, if all these molecules are identical in nucleotide sequence. This problem has been tackled by quantitative DNA-DNA renaturation and, more recently, with restriction endonuclease digestion (Borst, 1972, and in Saccone and Kroon, 1976). The results strongly indicate that all mtDNA molecules in a single normal organism are identical and, in addition, that mtDNA contains no major gene repetitions. The potential genetic information of each mtDNA molecule is, therefore, equivalent to its (low) molecular weight. The exception is again *Tetrahymena,* because in some strains up to 30% of the total DNA may be present in duplications of unknown function (Goldbach et al. in Saccone and Kroon, 1976).

The most unusual mtDNA is found in the primitive protozoa belonging to the *Kinetoplastidae*. This mtDNA, usually referred to as kinetoplast DNA (kDNA), consists of a large network of about 10^4 topologically interlocked DNA circles, containing two components: minicircles and maxicircles. The minicircles represent >90% weight of the network and vary in size from 0.2 μm in some *Leishmania* species to 0.8 μm in *Crithidia* species. The recent discovery that these minicircles are heterogeneous in sequence has raised the possibility that they do not code for any mitochondrial RNA (mtRNA) or protein, but have a "spacer" or nongenetic function. The maxicircles are similar in size and genetic information content to the mtDNA of other protozoa (Table I) and they probably represent the "true" mtDNA of trypanosomes. This remains to be proved, however.

Within the mitochondrion, the mtDNA is present in the matrix space and probably attached to the inner mitochondrial membrane at the point where DNA replication starts. Replication of mtDNA is under nuclear control, probably involves only nuclearly coded enzymes, and the intermediates involved in replication have been characterized in detail in animal cells (Kasamatsu et al., 1974; Wolstenholme et al., 1974; Berk and Clayton, 1976), *Tetrahymena* (Arnberg et al., 1974), and *Paramecium* (Cummings et al. in Saccone and Kroon, 1976).

Genes in Yeast mtDNA Identified by DNA-RNA Hybridization

For a discussion of the function of mtDNA, I shall now focus on yeast. As an experimental organism for the study of mitochondrial biogenesis, yeast has two major advantages: first, yeast can live without functional mitochondria, allowing manipulations with mitochondrial biogenesis that are lethal in other organisms; second, the mitochondrial genetics of yeast are highly developed, allowing a combined genetic and biochemical attack on problems in mitochondrial biogenesis, not yet available in any other organism.

The first method used to identify genes in

PIET BORST Section for Medical Enzymology and Molecular Biology, Laboratory of Biochemistry, University of Amsterdam, The Netherlands

Table I
Size and Structure of mtDNAs

Species	Structure	Mol wt ($\times 10^{-6}$)
Animals (from flatworm to man)	Circular	9–12
Higher plants	Circular	70
Fungi		
Baker's yeast (*Saccharomyces*)	Circular	49
Kluyveromyces	Circular	22
Protozoa		
Acanthamoeba	Circular	27
Malarial parasite (*Plasmodium*)	Circular	18
*Paramecium**	Linear	27
Tetrahymena‡	Linear	30–36
Kinetoplastidae§	Circle network	2,000–20,000
Trypanosoma brucei	Minicircle	0.6
	Maxicircle	13
Crithidia luciliae	Minicircle	1.5
	Maxicircle	22

See Borst (1976) and Borst and Flavell (1976).
* Cummings et al. in Saccone and Kroon (1976).
‡ Goldbach et al. in Saccone and Kroon (1976).
§ See also Borst et al. in Saccone and Kroon (1976).

mtDNA is DNA-RNA hybridization. Molecular hybridization of the structural RNAs found in yeast mitochondria with mtDNA has shown that this DNA contains one gene for each of the ribosomal RNAs (rRNAs) and a minimum of 20 4S RNA genes (see Borst et al. in Bandlow et al., 1976). By using aminoacyl-transfer RNAs (tRNAs), labeled in the amino acid moiety, Martin and Rabinowitz (in Bücher et al., 1976) have already identified tRNA genes for 19 of the 20 amino acids and for formyl methionine. In at least one case, isoaccepting species are present that are specified by different genes. This brings the minimal number of tRNA genes to 21, but the search has not yet been completed and suggestive evidence for 12 additional genes is already available (Martin and Rabinowitz in Bücher et al., 1976). It is possible, therefore, that yeast mtDNA specifies all 32 tRNAs minimally required for reading 61 codons according to the "wobble" hypothesis (Crick, 1966). In animal mtDNA this may not be the case, and Dr. O'Brien will return to this point in the following paper.

Most of these RNAs have been positioned on the fragment "physical" map of yeast mtDNA, constructed with restriction endonucleases, mainly through the efforts of Sanders in our laboratory (Sanders et al. in Saccone and Kroon, 1976, and Bücher et al., 1976). The detailed map is shown in Fig. 1 and the positions of the RNAs mapped thus far are shown in the simplified map in Fig. 2. The most interesting result is that the two rRNA genes are very far (about 28,000 nucleotide pairs) apart on this mtDNA. This is most unexpected, because in all other systems studied until now, including bacterial DNA, eukaryotic nuclear DNA, and chloroplast DNA, the genes for large and small rRNA are always adjacent and transcribed into a RNA precursor containing both RNAs. The situation in yeast mtDNA may be unusual even for mtDNAs, because the rRNA genes on animal mtDNAs and *Neurospora* mtDNA are adjacent (see O'Brien, this volume).

Many of the tRNAs mapped so far are clustered in one quadrant of the map (Fig. 2). At least one tRNA gene, however, is in the left half of the map (Martin and Rabinowitz in Bücher et al., 1976).

Genes in Yeast mtDNA: Information from Mitochondrial Protein Synthesis

As a first approximation, one may expect that proteins specified by mtDNA are made on mito-

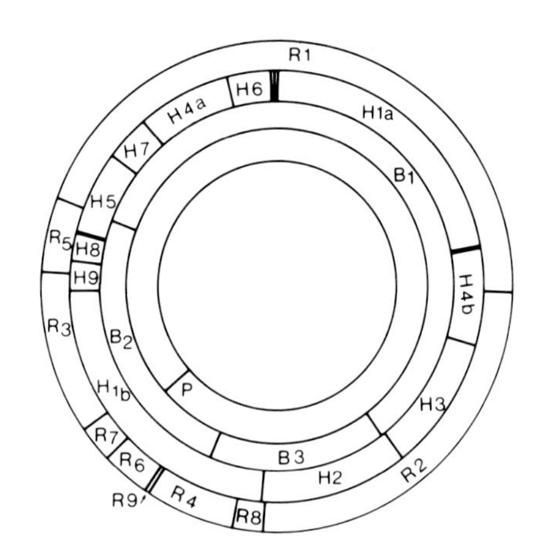

Figure 1 The physical map of mtDNA from the yeast *Saccharomyces carlsbergensis* (from Sanders et al. in Saccone and Kroon, 1976). The map shows the recognition sites on the mtDNA for endonuclease EcoR1 (giving 9 R fragments), endonucleases HindII + III (giving 15 H fragments), endonuclease BamHI (giving 3 B fragments), and endonuclease PstI (giving 1 P fragment). The total map has 28 fragments and 68,000 nucleotide pairs.

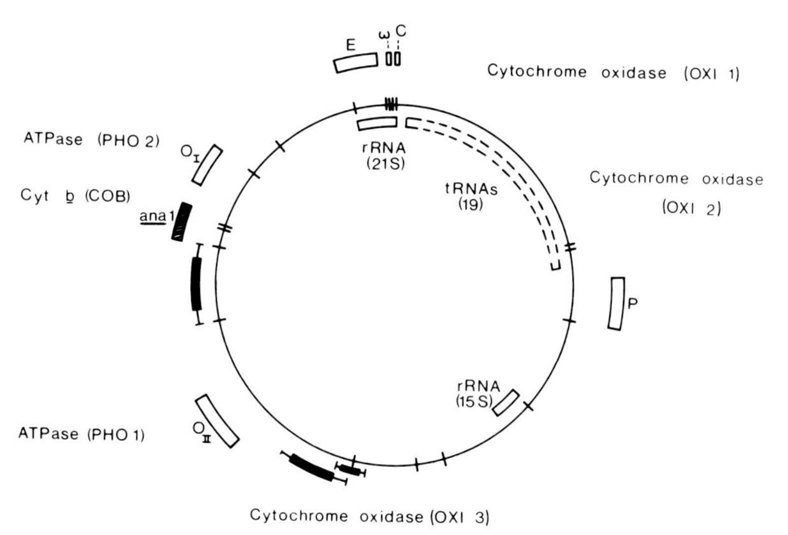

FIGURE 2 The physical and genetic map of mtDNA of *Saccharomyces cerevisiae*. The mtDNA of *S. cerevisiae* strain KL14-4A is indicated by the central solid line, with the cross lines indicating recognition sites for restriction endonucleases HindII + III. The total length is 75,000 nucleotide pairs (from Sanders et al., in Saccone and Kroon, 1976, and Bücher et al., 1976). The inner ring gives the position of the two rRNA genes (Sanders et al. in Saccone and Kroon, 1976) and the approximate position of a cluster of 19 tRNA genes (Sanders et al. in Bücher et al., 1976; Morimoto et al., in Bücher et al. 1976; Martin and Rabinowitz in Bücher et al., 1976). There is no evidence that all 19 tRNAs are adjacent and they may be interspersed with other genes (e.g., oxidase genes). The solid black boxes represent DNA sections that are absent in mtDNA of *S. carlsbergensis* and present only in part in the mtDNA of *S. cerevisiae* strain JS1-3D (Sanders et al. in Bücher et al., 1976). In the outer ring the open blocks represent genetic loci located on the map by determining the wild-type mtDNA segments present in petite mutant mtDNAs, carrying a genetic marker as described in Fig. 4 and in the text. The *ana* 1 locus (hatched block) was located in a slightly different way (Nagley et al. in Saccone and Kroon, 1976). The data for C, ω, E, O_1, and P are from Sanders et al. (in Saccone and Kroon, 1976), Heyting and Sanders (in Saccone and Kroon, 1976) and DiFranco et al. (in Saccone and Kroon, 1976); for O_{II} from Morimoto et al. (in Bücher et al., 1976); and for *ana* 1 from Nagley et al. (in Saccone and Kroon, 1976) and Linnane et al. (in Saccone and Kroon, 1976, and Bücher et al., 1976). C (RIB 1), E (RIB 2), O, P, and *ana* 1 are loci for chloramphenicol, erythromycin, oligomycin, paromomycin, and antimycin A resistance, respectively; ω is a polarity locus that affects the transmission of genetic markers in its neighborhood. In the outer ring the approximate position of genes affecting the biosynthesis of cytochrome *c* oxidase, cytochrome *b*, or the ATPase complex is indicated. These positions have either been determined by (rough) physical mapping (Nagley et al. in Saccone and Kroon, 1976; Linnane et al. in Bücher et al., 1976) or inferred by me from their relative positions on the genetic map (data from Nagley et al. in Saccone and Kroon, 1976; Linnane et al. in Bücher et al., 1976; Slonimski and Tzagoloff, 1976; Tzagoloff et al. in Saccone and Kroon, 1976, and Bücher et al., 1976). (In Nagley et al. [in Saccone and Kroon, 1976] and Linnane et al. [in Bücher et al., 1976], *ana* 1 is situated at 9 o'clock, exactly at the position of the large insertion. I assume that this insertion is absent in the mtDNA of the yeast strain studied by Linnane's group and that in our strain, shown above, the *ana* 1 locus is situated above this insertion at 9.30 o'clock.)

chondrial ribosomes. Such proteins can be identified by the judicious use of inhibitors of mitochondrial protein synthesis, as indicated in Fig. 1 of the General Introduction to this section. Proteins made on mitochondrial ribosomes continue to be made (for some time) in the presence of cycloheximide, an inhibitor of cell-sap protein synthesis, and labeling of these proteins is specifically prevented by D-chloramphenicol. The results of such experiments are summarized in Table II. The interesting point is that in each case both cell-sap and mitochondrial protein synthesis are involved in making major mitochondrial enzyme complexes. Whether the single mitochondrial ribosomal protein found to be made on mitochondrial ribosomes in yeast (Groot, 1974) and *Neurospora* (Lambowitz in Bücher et al., 1976) is essential for ribosomal function or is only a membrane protein

TABLE II

Biosynthesis of Major Mitochondrial Enzyme Complexes in Yeast

Enzyme complex	Number of subunits		
	Total	Made on cell-sap ribosomes	Made on mitochondrial ribosomes
Cytochrome oxidase	7	4	3
Cytochrome bc_1 complex	7	6	1
ATPase (oligomycin-sensitive)	9	5	4*
Large ribosomal subunit	30	30	0
Small ribosomal subunit	22	21	1

See Schatz and Mason (1974), Groot (1974), and Katan and Groot in Bücher et al. (1976).

* Essentially similar results have been obtained for *Neurospora crassa*, with the exception that only two subunits of the ATPase complex are made on mitochondrial ribosomes (Jackl and Sebald, 1975).

stuck on the ribosome during purification remains to be determined.

Obviously, products of mitochondrial protein synthesis should not be directly equated with gene products of mtDNA, because imported nuclear messenger RNAs (mRNAs) might also be translated on mitochondrial ribosomes. There is no evidence for such import, however, and the genetic evidence discussed below fully supports the notion that the structural genes for most of the proteins tabulated in Table II are in mtDNA. Nevertheless, it is important to establish this by direct experiments and two approaches are being used.

(1) mtDNA is used to program an in vitro coupled transcription-translation system, and the protein products are characterized by immunoprecipitation with specific antibodies. This approach has shown conclusively that mtDNA codes for components of cytochrome oxidase in agreement with Table II (Moorman and Grivell in Saccone and Kroon, 1976, and Bücher et al., 1976).

(2) RNA is extracted from mitochondria and the true mitochondrial mRNAs are selected by hybridizing with mtDNA. This RNA is translated in an in vitro protein-synthesizing system and the products are identified. By this same route, peptides reacting with antibodies against cytochrome oxidase have been found.

A major problem in these experiments is the hydrophobic nature of the proteins made on mitochondrial ribosomes. This makes them unusually sticky, leading to adsorption artifacts even in the most denaturing solvents. Eventually, however, this approach should allow the mapping of all proteins specified by mtDNA.

Genes on mtDNA: Information from Mutants

Three classes of mutants affect mitochondrial biogenesis:

NUCLEAR MUTANTS: These are extremely numerous. In a recent survey (Schweizer in Bücher et al., 1976) at least 38 nuclear genes (distinguishable by complementation tests) were found to be involved in cytochrome oxidase biosynthesis alone. It is clear that a further analysis of these genes will provide a powerful tool in the study of the interplay between mitochondrial and nuclear genetic systems; Dr. Schatz will return to this point.

MITOCHONDRIAL PETITE MUTANTS: These are deletion mutants in which a proportion of the DNA, varying between 50 and 100%, has been lost (Perlman, in Birky et al., 1975). Our present views on the origin of these mutants and the nature of the mtDNA sequences retained in them are summarized in Fig. 3. None of these mutants has retained a functional mitochondrial protein-synthesizing system. This is expected, because the spreading of the rRNA and tRNA genes over the

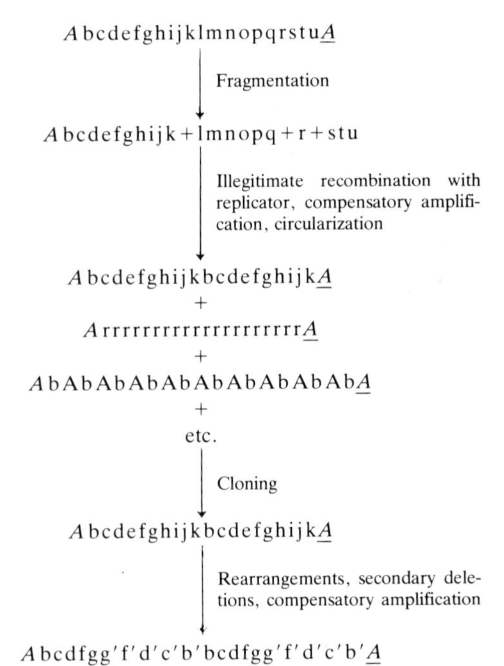

FIGURE 3 Speculative model for the origin of petite mutant mtDNA (from Borst et al. in Bücher et al., 1976). *A* indicates the start of DNA synthesis (replicator sequence); underlined terminal letters indicate circularity, i.e., in the top line *A* is joined to b and to u.

mitochondrial genome (Fig. 2) makes it impossible to remove a large part of the mtDNA without removing a gene essential for mitochondrial protein synthesis.

Mutants in which the genetic function of mtDNA is completely destroyed are very useful, because any mitochondrial protein found in such mutants must be specified by nuclear genes and made on cell-sap ribosomes. The list is quite impressive, as Table III shows. These results underline the limited contribution that the mitochondrial genetic system makes to mitochondrial biogenesis.

Petite mutants have also turned out to be very useful for mapping single-gene mutants on the physical map of mtDNA. This is discussed below.

MITOCHONDRIAL SINGLE-GENE MUTANTS: (Birky et al., 1975, and many papers in Saccone and Kroon, 1976, and Bücher et al., 1976): The most important groups are (1) mutants resistant to an antibiotic, such as chloramphenicol or erythromycin, that blocks mitochondrial protein synthesis; (2) mutants resistant to an antibiotic that blocks oxidative phosphorylation, such as oligomycin, which inhibits the ATPase complex, or antimycin A, which inhibits the respiratory chain in the region of cytochrome *b*; (3) *mit*⁻ mutants defective either in cytochrome oxidase (OXI mutants), cytochrome *b* (COB mutants), or the ATPase complex (PHO mutants).

TABLE III

Mitochondrial Components in Yeast Petite Mutants with Grossly Altered mtDNA

Absent
- A functional respiratory chain; functional cytochromes *b* and aa_3
- A functional energy transfer system
- A functional protein-synthesizing system; ribosomes

Present
- Outer membrane
- Inner membrane (altered)
- Parts of respiratory chain: cytochromes *c* and c_1, some subunits of cytochrome aa_3
- Parts of the energy transfer system: F_1-ATPase
- Permeability barrier for H^+ and K^+; translocators for adenine nucleotides, phosphate, dicarboxylic acids, and succinate
- The Krebs cycle
- Parts of the system for protein and nucleic acid synthesis: DNA and RNA polymerases, ribosomal proteins, elongation factors

Based mainly on Mahler (1973).

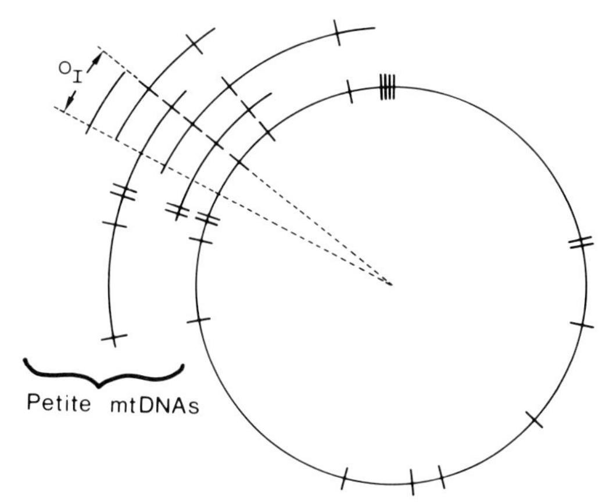

FIGURE 4 Scheme illustrating the procedure used for placing genetic markers on the physical map of mtDNA (from Borst et al. in Bandlow et al., 1976). Petite mutants are selected that can still transmit the marker for oligomycin resistance O^R, which belongs to the genetic locus O_I. The mtDNA segments still present in these petite mtDNAs are determined by physical mapping techniques. O_I lies on the segment that all mutants have in common.

Some of the mutants of groups 1 and 2 have recently been placed on the physical map with the help of cytoplasmic petite mutants, as illustrated in Fig. 4. The resulting map positions are shown in the open blocks in the outer ring of the map in Fig. 2, together with the approximate map positions of the *mit*⁻ mutants.

What do these mutations and their position on the map tell us about the genetic function of mtDNA? A speculative interpretation is presented in Table IV.

It seems likely that the mitochondrial mutations that give rise to chloramphenicol (C in Fig. 2) and erythromycin (E in Fig. 2) resistance are located in the gene for the large subunit rRNA, as I suggested in 1969 (see Borst, 1971). The mutations affect the large subunit of the mitochondrial ribosome; no alterations in the ribosomal proteins have been found in these mutants (Grivell et al., 1973), and in any case these ribosomal proteins are made on cell-sap ribosomes (and, therefore, probably specified by nuclear genes; see Table II); it is very unlikely that these mutations could affect the secondary modification of the rRNA because it is only minimally modified and contains no methylated bases (Klootwijk et al., 1975). Unfortunately, the mapping with petite mutants is not yet precise enough to determine whether the E and C loci are just inside or just outside the gene

TABLE IV
Genes on mtDNA: Facts and Speculations

Genes on mtDNA	Hybridization	Products mt protein synthesis	Genetics* (minimal number of separate genes)
Proved			
Large rRNA (21S)	1		{1 ERY resistance
Small rRNA	1		{1 CAP resistance
tRNAs	>21		
Tentative			
Subunits cytochrome oxidase		3	3 cytochrome oxidase
Subunit(s) cytochrome *b*		1	{1 cytochrome *b* {1 ANT resistance
ATPase		4 (2)	{2 ATPase genes {2 OLI resistance
Ribosomal proteins		1	1 PAR resistance ??

* See Saccone and Kroon (1976) and Bücher et al. (1976).

for rRNA. Sequence analysis of the rRNAs of resistant mutants has been started in our laboratory to decide the matter. Mutations in the primary nucleotide sequence of rRNA could be of great use in increasing our insight into the role of rRNA in ribosome function. These mutations will easily be picked up in mtDNA, which contains only a single set of rRNA genes, but not in bacterial or eukaryotic nuclear DNAs, which contain multiple sets of rRNA genes, as pointed out by Grivell et al. (1973).

The other interpretations in Table IV of possible defects in mutants are even more speculative. I give one example: the antimycin-resistant mutants affect the cytochrome bc_1 complex and map in the same region as two classes of mit^- mutants (COB mutants) that affect the biosynthesis of cytochrome *b*. The obvious interpretation is that all these mutants affect the biosynthesis of the single cytochrome *b* subunit made on mitochondrial ribosomes (Table II), that the structural gene for this subunit lies on mtDNA, and that antimycin resistance is the result of a mutation in this structural gene. It should be stressed, however, that the obvious interpretation is not always the correct one. For instance, it is also possible that the structural gene for the *b* subunit is in the nucleus and that the mRNA is imported and translated on mitochondrial ribosomes. The COB mutants could represent mutations in a gene for a regulatory protein required for cytochrome *b* synthesis, and antimycin resistance could be the result of an alteration in an enzyme that modifies the cytochrome bc_1 complex. These unlikely alternatives remain to be eliminated.

The Variability of Yeast mtDNA

When the mtDNAs from various yeast strains are compared, numerous small differences are found that can be attributed to small deletions and insertions (Bernardi et al. in Bandlow et al., 1976, Saccone and Kroon, 1976, and Bücher et al., 1976). In addition, Sanders et al. (in Bücher et al., 1976) have found some large deletions/insertions, indicated by black boxes in Fig. 2. The nature of these "extra" sequences is unknown. It seems likely that such sequence alterations are the result of faulty recombination. Notwithstanding these alterations, a careful comparison of the restriction fragment maps of the mtDNAs of three yeast strains (Sanders et al. in Saccone and Kroon, 1976) has shown that the overall gene order is the same in these mtDNAs. This suggests that the gene order has a selective advantage and that it could be present in all *Saccharomyces* strains. Indications for the conservation of gene order in mtDNA, even though extensive sequence divergence has occurred, also come from studies with animal mtDNA.

Why mtDNA?

About 16% of the nucleotide sequence of yeast mtDNA is required to code for the genes specified in Table IV. One could argue, therefore, that the majority of the proteins encoded by this mtDNA remain to be found. There is evidence, however, that animal mtDNA codes for a similar set of genes, as Dr. O'Brien will point out. Because animal mtDNA is 1/5th the size of yeast mtDNA, the genes in Table IV would take up most of its coding potential. From Table IV we should, therefore, try to understand why the mitochondrial genetic system exists at all. This question is the more pressing because the disadvantages of the system are obvious. It takes about 100 proteins to set up a genetic system with the complete machinery for DNA replication and transcription and protein synthesis. All evidence indicates that the mitochondrial and the nuclear genetic systems have no proteins in common. It takes about 50×10^6 daltons DNA to specify 100 proteins, i.e., five times the total genetic information present in animal mtDNA. Even if this rough estimate is off by a factor of two, it is clear that the genetic cost for having a separate mitochondrial genetic system is high. Moreover, this extra system is sensitive to a set of inhibitors that do not affect the nuclear system, making the cell more vulnerable. Notwithstanding these clear disadvantages, the mitochondrial genetic system has survived eukaryote evolution, and there must be a reason.

Three types of explanations have been brought forward (see Borst, 1971, 1972).

(1) mtDNA represents a useful form of gene amplification, providing multiple copies of genes for products that are required in large amounts in all cells. This idea is unattractive because it does not explain why the amplified genes have to be segregated in mitochondria and provided with a separate system for DNA replication, transcription, and translation. Moreover, it is difficult to see why only some of the subunits of, say, cytochrome *c* oxidase are coded for by mtDNA, whereas other subunits that are required in equimolar amounts are coded for by nuclear genes.

(2) The assembly of the inner membrane of double-membraned organelles—such as chloroplasts or mitochondria—requires a number of proteins that can be put in place only from the inside and cannot pass through the membrane. Hence, they must be synthesized inside, and this, in turn, requires an internal protein-synthesizing system. All known products of mitochondrial protein synthesis are rather hydrophobic and one could argue that this is the reason they must be made inside, on the site, as it were, they will occupy when finished. Before we know more about the topology of proteins in the inner mitochondrial membrane and the way this membrane is assembled, this (unattractive) possibility cannot be discounted.

(3) The existence of the mitochondrial protein-synthesizing system is a "frozen leftover." This explanation is based on the hypothesis that mitochondria arrived in the eukaryotic cell as bacterial endosymbionts (see Bogorad, this volume), that these symbionts transferred most of their genes to the nucleus, but that the process stopped before it was complete. Once the gene exchange between mitochondria and nucleus was blocked, the further reduction of the mitochondrial genetic system would require the *de novo* elaboration of "mitochondrial" functions by the nuclear system. This would be a most unlikely event, because any further transfer of functions is bound to have only a marginally beneficial effect, as long as the whole mitochondrial protein-synthetic machinery cannot be dispensed with.

Clearly, none of these explanations provides compelling arguments for the existence of a mitochondrial genetic system in nearly all eukaryotic cells. It is possible, therefore, that we are still on the wrong track and that separate organelle genes provide advantages to the cell that remain to be discovered.

Outlook

mtDNA is the smallest genome in nature that contains genes for rRNA, tRNAs, and proteins. In the coming years, the inventory of genes on mtDNA will be completed and, with the rapid improvement in DNA cloning and sequencing technology, the complete nucleotide sequence of some mtDNAs will eventually be determined. This will provide us with a complete picture of the function of mtDNA and an interesting eukaryotic model system for the study of the long-range evolution of DNA sequences involved in the regulation of DNA and RNA synthesis and in the specification of various types of RNAs. As more detailed information on the function of mtDNA is obtained, it will become possible to clarify the control of mitochondrial gene expression, to characterize the (nuclear) proteins involved, and to unravel the integration of the mitochondrial and nuclear contributions to mitochondrial biosynthesis. This will finally provide the basis for reconstructing the biogenesis of complete mitochondria in a subcellular system.

REFERENCES

Arnberg, A. C., E. F. J. Van Bruggen, R. A. Clegg, W. B. Upholt, and P. Borst. 1974. An analysis by electron microscopy of intermediates in the replication of linear *Tetrahymena* mitochondrial DNA. *Biochim. Biophys. Acta.* **361:**266–276.

BANDLOW, W., R. J. SCHWEYEN, D. Y. THOMAS, K. WOLF, and F. KAUDEWITZ, editors. 1976. Genetics, Biogenesis and Bioenergetics of Mitochondria. De Gruyter & Co., Berlin. 418.

BERK, A. J., and D. A. CLAYTON. 1976. Mechanism of mtDNA replication in mouse L-cells: topology of circular daughter molecules and dynamics of catenated oligomer formation. *J. Mol. Biol.* **100:**85–102.

BIRKY, C. W., JR., P. D. PERLMAN, and T. J. BYERS, editors. 1975. Genetics and Biogenesis of Mitochondria and Chloroplasts. The Ohio State University Press, Columbus, Ohio. 361.

BORST, P. 1971. Size, structure and information content of mitochondrial DNA. *In* Autonomy and Biogenesis of Mitochondria and Chloroplasts. N. K. Boardman, A. W. Linnane, and R. M. Smillie, editors. North-Holland Publishing Co., Amsterdam. 260–266.

BORST, P. 1972. Mitochondrial nucleic acids. *Annu. Rev. Biochem.* **41:**333–376.

BORST, P. 1976. Properties of kinetoplast DNAs. *In* Handbook of Biochemistry and Molecular Biology, 3rd edition. Vol. 2. G. D. Fasman, editor. CRC Press, Cleveland, Ohio. 375–378.

BORST, P., and R. A. FLAVELL. 1976. Properties of miochondrial DNAs. *In* Handbook of Biochemistry and Molecular Biology. 3rd edition. Vol. 2. G. D. Fasman, editor. CRC Press, Cleveland, Ohio. 363–374.

BÜCHER, TH., W. NEUPERT, W. SEBALD, and S. WERNER, editors. 1976. Genetics and Biogenesis of Chloroplasts and Mitochondria. North-Holland Publishing Co., Amsterdam. 908.

CRICK, F. H. C. 1966. Codon-anticodon pairing: the wobble hypothesis. *J. Mol. Biol.* **19:**548–555.

GRIVELL, L. A., P. NETTER, P. BORST, and P. P. SLONIMSKI. 1973. Mitochondrial antibiotic resistance in yeast: ribosomal mutants resistant to chloramphenicol, erythromycin and spiramycin. *Biochim. Biophys. Acta.* **312:**358–367.

GROOT, G. S. P. 1974. The biosynthesis of mitochondrial ribosomes in *Saccharomyces cerevisiae. In* The Biogenesis of Mitochondria. A. M. Kroon and C. Saccone, editors. Academic Press, Inc., New York. 443–452.

JACKL, G., and W. SEBALD. 1975. Identification of two products of mitochondrial protein synthesis associated with mitochondrial adenosine triphosphatase from *Neurospora crassa. Eur. J. Biochem.* **54:**97–106.

KASAMATSU, H., L. I. GROSSMAN, D. L. ROBBERSON, R. WATSON, and J. VINOGRAD. 1974. The replication and structure of mitochondrial DNA in animal cells. *Cold Spring Harbor Symp. Quant. Biol.* **38:**281–288.

KLOOTWIJK, J., L. A. GRIVELL, and I. KLEIN. 1975. Minimal post-transcriptional modification of yeast mitochondrial ribosomal RNA. *J. Mol. Biol.* **97:**337–350.

MAHLER, H. R. 1973. Biogenetic autonomy of mitochondria. *CRC Crit. Rev. Biochem.* **1:**381–460.

SACCONE, C., and A. M. KROON, editors. 1976. The Genetic Function of Mitochondrial DNA. North-Holland Publishing Co., Amsterdam. 354.

SCHATZ, G., and T. L. MASON. 1974. The biosynthesis of mitochondrial proteins. *Annu. Rev. Biochem.* **43:**51–87.

SLONIMSKI, P. P., and A. TZAGOLOFF. 1976. Localization in yeast mitochondrial DNA of mutations expressed in a deficiency of cytochrome oxidase and/or coenzyme QH_2-cytochrome *c* reductase. *Eur. J. Biochem.* **61:**27–41.

WOLSTENHOLME, D. R., K. KOIKE, and P. COCHRAN-FOUTS. 1974. Replication of mitochondrial DNA: replicative forms of molecules from rat tissues and evidence for discontinuous replication. *Cold Spring Harbor Symp. Quant. Biol.* **38:**267–280.

TRANSCRIPTION AND TRANSLATION IN MITOCHONDRIA

THOMAS W. O'BRIEN

Interest in the area of mitochondrial biogenesis grew rapidly when it was recognized that mitochondria contain DNA. Some of the questions that were raised early about the coding potential and expression of the mitochondrial genome continue to pose exciting challenges for workers in this area. The progress in our understanding of transcription and translation in mitochondria has been especially rapid during the past few years, as attested by the recent proliferation of reviews covering various aspects of the subject.

Mitochondria have been termed "semiautonomous" organelles to acknowledge their limited genetic potential and, at the same time, to indicate their dependency on cellular biosynthetic processes. What is the nature and the extent of this interdependency of the mitochondrial and nucleocytoplasmic biogenetic systems? Regardless of the organism or cell type in which they are found, and despite wide variations in their size, appearance, and enzymatic composition, all mitochondria appear to reproduce and "grow" by the same general mechanism. Mitochondrial growth, or replenishment of mitochondrial functional mass, occurs by a concerted biosynthetic process in which the mitochondrial and nucleocytoplasmic systems cooperate, temporally and spatially, to synthesize new mitochondrial components.

Mitochondria contain a complete set of macromolecules required for the transcription and translation of their genetic information, and the overall process is much the same in mitochondria of diverse origins. When examined at a finer level, however, some features of the mitochondrial biogenetic apparatus appear quite different in different organisms. In the preceding chapter, for example, we have seen that the size of mitochondrial DNA molecules tends to vary systematically along phylogenetic lines. An even more striking example is provided by mitochondrial ribosomes. Their composition and physicochemical properties vary widely among diverse organisms.

Our knowledge of mitochondrial biogenesis derives largely from studies with yeast and *Neurospora*. For the most part, these findings apply to all mitochondrial systems. In this overview of transcription and translation in mitochondria, summarizing the results from many laboratories, we will emphasize those results obtained with animal systems, especially in cases where the processes or components involved show phylogenetic variation.

Elements of Mitochondrial Transcription and Translation

The overall scheme for transcription and translation in mitochondria depicted in Fig. 1 provides a background for discussing these processes (Kroon, et al., 1972; Borst, 1972; Schatz and Mason, 1974; Mahler and Raff, 1975; Milner, 1976; O'Brien, 1976). Some of the elements required for transcription and translation (ribosomal, transfer, and messenger RNA) originate intramitochondrially, specified entirely by mitochondrial DNA, whereas others (ribosomal proteins, RNA and DNA polymerases, amino acyl-tRNA synthetases, and the protein synthesis factors) are encoded and synthesized outside the mitochondria. It should be noted that the nucleic acid molecules involved directly in mitochondrial translation are the primary gene products of mitochondrial DNA. On the other hand, essentially all of the proteins involved in the expression of mitochondrial genetic information are products of cytoplasmic protein synthesis, and all of these must necessarily be imported by mitochondria. It is remarkable that all of the molecules imported by mitochondria for use in transcribing and translating their genetic information are distinctly differ-

THOMAS W. O'BRIEN Department of Biochemistry and Molecular Biology, University of Florida, Gainesville, Florida

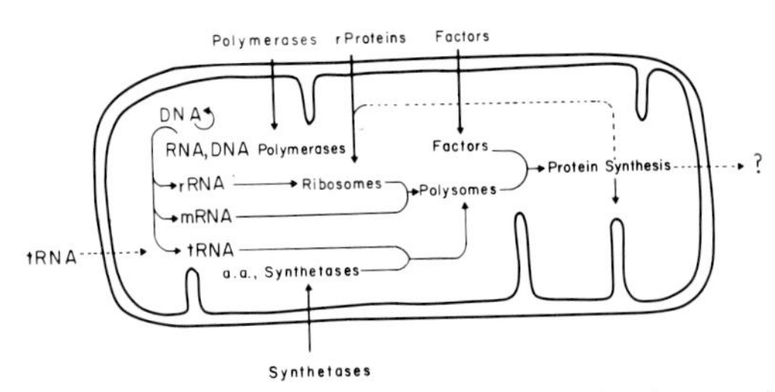

FIGURE 1 Mitochondrial enzymes, molecules, and structural components involved in mitochondrial transcription and translation. Solid arrows denote well-established interactions and pathways, less firmly documented proposals are indicated by dashed lines. *Abbreviations used are*: rRNA, ribosomal RNA; mRNA, messenger RNA; tRNA, transfer RNA; rProteins, ribosomal proteins; a.a., amino acids.

ent from the molecules that perform similar functions in the nucleocytoplasmic system. The mitochondrial and nucleocytoplasmic systems for gene expression thus appear to use no components in common. This arrangement may provide a means to separate and coordinate the genetic regulation of the two systems.

Mitochondrial DNA polymerase and mitochondrial RNA polymerase are imported to replicate and transcribe mitochondrial DNA. The DNA appears to be transcribed symmetrically, along its entire length, so that the RNA transcripts must be processed to yield the primary gene products—rRNA, tRNA, and mRNA. Further processing of the transcripts includes the addition of a short poly(A) tail to the 3′ end of mitochondrial mRNA before the mRNA participates in mitochondrial protein synthesis.

The various transfer RNAs used in mitochondrial protein synthesis are charged with amino acids through the action of a corresponding set of imported amino acyl-tRNA synthetases. In general, the mitochondrial enzymes show high specificity for mitochondrial tRNA, and low activities with cytoplasmic tRNA molecules, even though the synthetases are products of the nucleocytoplasmic system. It has been suggested that some of the mitochondrial tRNAs may be imported, as well, because presently there is no direct evidence that mitochondria (especially in higher animals) specify all of the tRNAs required in mitochondrial protein synthesis. We will return to this point later.

The initiation and elongation factors required to support protein synthesis on mitochondrial ribosomes are also imported from the cytoplasm. Despite their cytoplasmic origin, the mitochondrial factors invariably show higher activity with homologous ribosomes than when tested with cytoplasmic ribosomes. In this respect, it is noteworthy that these factors are partially interchangeable with bacterial protein-synthesis factors.

Among the components of the mitochondrial biogenetic system, the mitochondrial ribosomes provide a remarkable example of the concerted nature of mitochondrial biosynthetic processes. Ribosomal proteins that are specified by nuclear DNA and made on cytoplasmic ribosomes must assemble with rRNA transcribed from mitochondrial DNA in order to produce mitochondrial ribosomes. A similar pattern also applies to the products of mitochondrial protein synthesis. As described in detail in the following chapter, the various polypeptides synthesized on mitochondrial ribosomes combine with complementary subunits made on cytoplasmic ribosomes in order to form functional complexes that finally reside in the inner mitochondrial membrane. In addition to these membrane proteins, it has been suggested that one (or more) other mitochondrial product(s) is exported and serves in a control capacity (repressor?) to regulate the synthesis of all of the proteins mitochondria import from the nucleocytoplasmic system (Barath and Küntzel, 1972).

Coding Potential and Genetic Map of Mitochondrial DNA in Animals

From the genetic and biochemical studies reviewed in the preceding chapter, the 25 μm circles of yeast mitochondrial DNA appear to code for mitochondrial rRNA molecules, several tRNAs, and a small number of proteins. What does the DNA in animal mitochondria specify? Although the circular DNA molecules in animal mitochondria are only about 5 μm long, they appear to code for essentially the same number of products as in the yeast system.

Because of their small size, it is considerably easier to isolate intact circular molecules of DNA from animal mitochondria than from yeast or *Neurospora* mitochondria. Furthermore, it is relatively easy to separate and purify the two complementary strands of DNA, based on their buoyant density differences in CsCl. The availability of intact, single-stranded circles of DNA has simplified the construction of genetic maps for animal mitochondrial DNA with electron microscopy. In practice, the genes for rRNA or tRNA molecules are located directly by examining the single-stranded DNA circles after hybridization with the rRNA molecules or ferritin-conjugated tRNAs (Attardi et al., in Saccone and Kroon, 1976).

The genetic map for human (HeLa cell) mitochondrial DNA, derived by the above approach, is presented in Fig. 2. This DNA contains one gene each for the small and large rRNAs, arranged in tandem on the heavy (H) strand. It is interesting to note that these rRNA genes are separated and closely flanked by genes for 4S RNA. Additional genes for 4S RNA are scattered about on both of the complementary strands. So far, it has been possible to map 19 different gene loci for 4S RNA on HeLa mitochondrial DNA, 12 on the H strand, and 7 on the L strand (Attardi et al., in Saccone and Kroon, 1976).

How much of the mitochondrial DNA may be available to code for mRNA? The genes for mRNA would be restricted to gap regions between the rRNA and 4S RNA genes. As shown in Fig. 2, no more than 13 such regions exist which are large enough to accommodate an mRNA gene. The putative mRNAs from these regions, ranging in size from 165,000 (region 7) to 435,000 daltons (region 6), could code for proteins of 21,000–55,000 daltons. These regions are thus more than adequate, in both size and number, to accommodate genes for all of the known

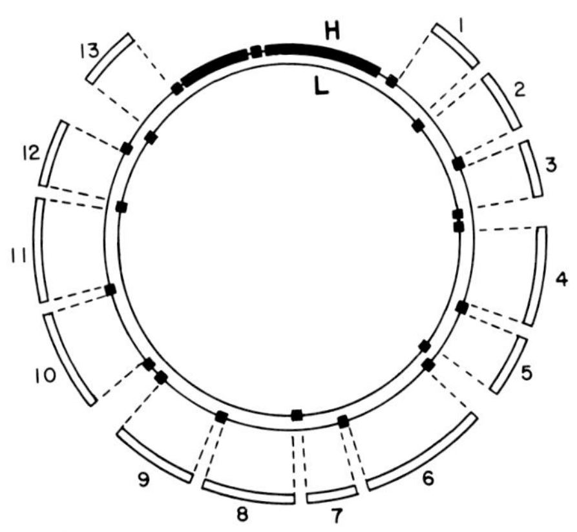

FIGURE 2 Genetic map of HeLa mitochondrial DNA. Solid bars indicate the position and the extent of genes for the rRNAs (long bars) and 4S RNAs (short bars) on both the heavy (H) and light (L) strands of the circular DNA molecule. Intergene gap regions of significant size have been extended to indicate the location and maximum span of putative mRNA genes (open bars). (Adapted from Attardi et al., in Saccone and Kroon, 1976).

products of mitochondrial protein synthesis (see Table II of Borst, preceding chapter).

The arrangement of genes in the mitochondrial DNA from other animals seems to conform to the general pattern described above for HeLa mitochondria. The rRNA cistrons have been mapped by electron microscopy on mitochondrial DNA molecules from the toad, *Xenopus,* and the insect, *Drosophila,* where they also occur in unit, tandem copies (Dawid et al., in Saccone and Kroon, 1976). The 15 different sites that have been mapped for 4S RNA genes on the H strand of *Xenopus* mitochondrial DNA are also scattered about the molecule, just as in the case of HeLa. The most striking resemblance, however, is the presence of 4S RNA genes between and flanking the rRNA genes mapped in *Xenopus,* leading to the suggestion (Dawid et al., in Saccone and Kroon, 1976) that this pattern (tandem rRNA cistrons, flanked and separated by 4S RNA genes) may be a general feature of metazoan mitochondrial DNA. This pattern differs markedly from that of the genes in yeast mitochondrial DNA; here the 4S RNA genes are clustered, and the genes for the small and large rRNAs map nearly opposite each other (preceding chapter).

The presence of a gene for 4S RNA (about 70 nucleotides) in the short space of only 120 nucleotides (in *Xenopus*) between the mitochondrial rRNA genes deserves special comment. First, it conveys an impression that maximum use is being made of a minimal genome. The small spacer regions between these genes in animal mitochondrial DNA stand in marked contrast to the vast stretches of "spacer" DNA found in the nucleus. Second, the occurrence of rRNA cistrons punctuated by tRNA genes is not unique to animal mitochondrial DNA. A gene for glutamyl tRNA was recently reported to reside between the genes for the small and large rRNAs in *Escherichia coli* (Lund et al., 1976).

Mitochondrial RNA Polymerase

Studies on mitochondrial RNA polymerase have been performed in yeast and *Neurospora,* as well as in the animal systems, *Xenopus* and the rat, and there is general agreement that the mitochondrial enzyme is clearly distinct from nuclear and bacterial RNA polymerases.

All of the well-characterized bacterial and nuclear RNA polymerases are multisubunit enzymes. The most remarkable feature described for the mitochondrial enzyme in all of these organisms is that it appears to consist of only one polypeptide chain. Its molecular weight is about 65,000 daltons in the case of yeast, *Neurospora,* and the rat, and about 45,000 daltons in *Xenopus*. In terms of antibiotic susceptibility, the mitochondrial enzymes are usually inhibited by rifampicin (or rifamycin SV), like bacterial RNA polymerase, and are insensitive to α-amanitin, an inhibitor of one of the nuclear RNA polymerases (Saccone and Quagliariello, 1975; Mahler and Raff, 1975).

Investigators should be cautioned that the (rat) mitochondrial RNA polymerase is unexpectedly sensitive to cycloheximide, an inhibitor frequently used selectively to suppress protein synthesis on cytoplasmic ribosomes (Saccone and Quagliariello, 1975). This property of the enzyme will complicate the interpretation of studies on products of mitochondrial protein synthesis when cycloheximide is used to block cytoplasmic protein synthesis for extended periods.

If indeed the "simple" mitochondrial RNA polymerase consists only of a single polypeptide chain, it is sure to be an interesting molecule. It may be similar to the multifunctional bacterial DNA polymerase I, which contains separate binding sites for the template and nucleoside triphosphates, as well as a catalytic site. On the other hand, the active enzyme may consist of more than one subunit of the same molecular weight. Indeed, there is reason to believe that the native enzyme may contain an additional subunit, possibly involved in the initiation of transcription, as rifampicin sensitivity can be eliminated during purification of the enzyme (Saccone and Quagliariello, 1975).

Symmetrical Transcription of Mitochondrial DNA

Except for a few examples in the viruses, RNA synthesis is usually initiated and terminated at specific sites so that the DNA is transcribed selectively and asymmetrically. The large RNA molecules transcribed in vivo from HeLa mitochondrial DNA during brief labeling periods can hybridize along the entire length of both the H and the L strands of the DNA (Saccone and Quagliariello, 1975). Thus, both strands of the mitochondrial DNA are transcribed completely. As a consequence, extensive processing of the transcripts is required to yield the primary gene products, and all these will be synthesized in unit amounts.

Processing of Mitochondrial RNA Transcripts

The processing of mitochondrial RNA includes polyadenylation, cleavage of large precursor molecules to yield the primary gene products, and a low level of base modification or RNA methylation (Borst, 1972; Mahler and Raff, 1975). Mitochondria contains a poly(A) polymerase, which adds tracts of poly(A) posttranscriptionally to the 3′ ends of mitochondrial RNA transcripts (Saccone and Quagliariello, 1975). By analogy with the cytoplasmic species of RNA that contain poly(A) tracts, the poly(A)-containing RNA in mitochondria is thought to be mitochondrial mRNA (see below). Polyadenylation of nascent RNA occurs rapidly, like heterogeneous nuclear RNA (HnRNA), before significant cleavage of the transcripts occurs to give the individual gene products. Some very large, polyadenylated transcripts, for example, which also contain rRNA sequences, have been detected in HeLa (Attardi et al., in Saccone and Kroon, 1976) and hamster mitochondria. A polyadenylated 20S RNA molecule has been identified as a precursor of the 17S rRNA molecule in hamster mitochondria (Cleaves et al., 1976).

Neurospora provides the best-documented example of precursor rRNA molecules in mitochondria. In this system, a 32S precursor undergoes rapid cleavage to give rise to two smaller precursor RNAs, which ultimately end up as the mature small (19S) and large (25S) rRNAs. The "poky" mutation in *Neurospora,* characterized by a deficiency of the small subunits of mitochondrial ribosomes, was thought to result from improper processing of the mitochondrial small rRNA (Saccone and Quagliariello, 1975). It now appears, however, that the molecular basis for the "poky" phenotype is a missing or altered ribosome protein (see "Products of Mitochondrial Protein Synthesis," below).

Except for the above example, none of the enzymes or the proteins involved in processing mitochondrial RNAs is known, and neither are the RNA signals to which they respond. These factors should be considered in evaluating experiments that involve the transcription and translation of mitochondrial DNA in heterologous systems (see below).

Mitochondrial Transfer RNA

It has been known for some time that mitochondrial DNA contains genes for mitochondrial tRNAs. The early saturation-hybridization experiments indicated that mitochondrial DNA might not contain enough genes for a complete set of tRNA molecules. These observations led to suggestions that some mitochondrial tRNAs may be imported, or that all 20 amino acids may not be used in mitochondrial protein synthesis (Borst, 1972). Support for the idea that mitochondria import some tRNA species comes mainly from studies in *Tetrahymena*, from the observation that the arginyl, lysyl, and valyl tRNAs extracted from these mitochondria hybridize with nuclear, instead of mitochondrial, DNA (Chiu et al., 1975).

Animal mitochondria contain genes for several tRNAs. As many as 19 genes for 4S RNA have been detected in HeLa mitochondria (Fig. 2), and *Xenopus* mitochondrial DNA contains at least 15 such genes on the H strand alone (Dawid et al., in Saccone and Kroon, 1976). However, not all of these loci on HeLa mitochondrial DNA correspond to tRNAs carrying different amino acids, as both the H and L strands code for separate isoaccepting seryl tRNAs. In fact, the HeLa tRNAs that hybridize with mitochondrial DNA can be charged (so far) with only 16 different amino acids; asparagine, glutamine, histidine, and proline fail to charge. HeLa mitochondria must contain tRNAs specific for these amino acids, though, because these amino acids are incorporated into products of mitochondrial protein synthesis (Attardi et al., in Saccone and Kroon, 1976). Whether these tRNAs are imported or whether alternative explanations will surface to account for the apparent failure of these four tRNAs to hybridize with mitochondrial DNA remains an open question.

Until very recently, it appeared that yeast mitochondria also lacked genes for a large number of the tRNAs needed for mitochondrial protein synthesis. Yet, yeast mitochondrial DNA is now known to code for tRNAs that can be charged with all amino acids (including several isoaccepting species) except threonine and glutamine (Rabinowitz et al., in Saccone and Kroon, 1976; Martin and Rabinowitz, in Bucher et al., 1976). Most interestingly, the lack of mitochondrially coded tRNAs that can be charged with glutamine may be compensated for by a novel mechanism. One of the isoaccepting tRNAs for glutamic acid responds to a glutamine codon; this observation permits the speculation that after this tRNA is charged with glutamic acid the amino acid may be enzymatically converted to glutamine (N. C. Martin, and M. Rabinowitz, unpublished data). Perhaps similar reactions in animal mitochondria circumvent the apparent absence of mitochondrially coded tRNAs that can be charged with asparagine and glutamine.

Mitochondrial Messenger RNA

It is generally agreed that animal mitochondrial DNA codes for several mRNAs, that at least eight of these molecules have poly(A) tails added post-transcriptionally, and that the mitochondrial mRNAs are translated on mitochondrial ribosomes.

Doubts raised earlier about the limited coding potential of mitochondrial DNA, and the observation that some mitochondria can translate exogenous polynucleotides in vitro, sustained the notion that mitochondria may import mRNAs of nuclear origin (discussed in Borst and Grivell, 1973; Dawid, 1972). However, with the discovery of several presumptive mitochondrial mRNA molecules that satisfy the criteria of mitochondrial origin, there are no longer compelling reasons to entertain this proposal in the absence of any conclusive supporting evidence.

What criteria can be used to characterize mitochondrial mRNA? The most convincing criteria that a messenger RNA of mitochondrial origin can satisfy are that it should hybridize with mitochondrial DNA and should direct the synthesis of a product of mitochondrial protein synthesis. Alternative criteria that may be applied to establish the mitochondrial origin of the RNA include: synthesis of the molecule in the presence of selective inhibitors of nuclear RNA synthesis (such as camptothecin), synthesis by isolated mitochondria, or inhibition of the mRNA synthesis by selective inhibitors of mitochondrial RNA synthesis (such as ethidium bromide). Lastly, if antibodies against known products of mitochondrial protein synthesis (see below) are available, they can be used to characterize the translation products of the putative mRNA in cell-free protein synthesizing systems. Unfortunately, the above criteria have not been applied systematically to characterize mitochondrial mRNA from animal cells because antibodies against (purified) products of protein synthesis in animal cells are not available.

Some of the presumptive mRNAs in animal mitochondria, partially characterized by pulse labeling in the presence of camptothecin (Attardi et al., in Saccone and Kroon, 1976) or by extraction from mitochondrial polysomes and hybridization to mitochondrial DNA (Lewis et al., 1976) may lack a poly(A) segment. However, the majority of such RNAs in animal mitochondria contain poly(A) tracts of about 55 residues at their 3′ end (Hirsch et al., 1974; Attardi et al., in Saccone and Kroon, 1976). In contrast, most of the cytoplasmic mRNAs in animal cells have longer poly(A) tails, of 100–200 residues.

Mammalian (human and hamster) mitochondria (Hirsch et al., 1974; Attardi et al., in Saccone and Kroon, 1976) and insect (*Drosophila* and mosquito) mitochondria (Hirsch et al., 1974) contain at least eight distinct, relatively stable, poly(A)-containing RNA species. High resolution analysis of the poly(A)-containing RNAs made by HeLa mitochondria in the presence of camptothecin discloses 18 different molecules, ranging in size from 93,000 to 3,400,000 daltons (Attardi et al., in Saccone and Kroon, 1976). The smallest of these hybridizes to the L strand, whereas all the others hybridize preferentially to the H strand of mitochondrial DNA. All of them may contain sequences for mitochondrial mRNA. It is interesting to note that only the eight smallest species (93,000–420,000 daltons) can be accommodated in the putative mRNA (gap) regions of the genetic map for HeLa mitochondria (Fig. 2); the larger RNAs probably correspond to incompletely processed molecules. Coincidentally, it should be noted that the known products of mitochondrial protein synthesis are also eight in number (see below).

Transcription and Translation of Mitochondrial DNA in Heterologous Systems

Studies of this sort in animals are still hampered by the lack of suitable antibodies to identify the translation products. Some workers have transcribed animal mitochondrial DNA with homologous and heterologous RNA polymerases, but the template activity of the transcripts has not been examined (Saccone and Quagliariello, 1975). In a novel attempt to characterize the gene products of animal mitochondrial DNA, restriction enzymes were used to couple mouse mitochondrial DNA with that of the *E. coli* plasmid pSC101 (Clayton, in Saccone and Kroon, 1976). In this manner, it is possible to insert an entire mitochondrial DNA molecule, as a component of the hybrid DNA, into an *E. coli* minicell, where it can be transcribed by the bacterial polymerase to give RNA molecules that can be translated directly on the minicell ribosomes. Disappointingly, the specific translation products bear no resemblance to the proteins synthesized by mouse mitochondria in vivo; they are all of low molecular weight, less than 5,000 daltons. This result is not totally unexpected, however, inasmuch as the *E. coli* RNA polymerase preferentially transcribed the L strand of animal mitochondrial DNA in minicells, as well as in vitro (Clayton, in Saccone and Kroon, 1976). One of the problems in this heterologous system, at least, is probably the failure of the bacterial polymerase to recognize the promotor sites on mitochondrial DNA, as most of the mitochondrial mRNAs appear to be transcribed from the H strand in vivo (Attardi et al., in Saccone and Kroon, 1976; Saccone and Quagliariello, 1975).

Products of Mitochondrial Protein Synthesis

Mitochondria can incorporate radioactive amino acids into protein, either in vitro or in vivo, in the presence of selective inhibitors of protein synthesis on cytoplasmic ribosomes. Most of the label is incorporated into proteins that account for 10–15% of the inner mitochondrial membrane. These translation products are unusually hydrophobic and their strong tendency to aggregate complicates their analysis. However, they can be resolved into 7–10 bands by electrophoresis on SDS-containing gels. Some of the translation products have been identified by using antibodies against purified mitochondrial components and by demonstrating specific association of the product(s) with a known, purified mitochondrial component. In the latter case, it is essential that the mitochondrial products have the opportunity to assemble with other proteins (for example, those made on cytoplasmic ribosomes during a "chase" period in the absence of inhibitors of protein synthesis) into isolatable, identifiable complexes (Wheeldon, 1973; O'Brien, 1976).

Eight of the major mitochondrial translation products have been identified in yeast (see Table II of Borst, preceding chapter). The three largest subunits of the cytochrome oxidase complex and one of the subunits of the coenzyme Q-cytochrome *c* reductase (cytochrome b complex) are made on mitochondrial ribosomes. Four other mitochondrial products constitute a specific membrane site for integration of the mitochondrial oligomycin-sensitive ATPase complex (Schatz and Mason, 1974; Tzagoloff et al., 1973). On the basis of the available evidence, it seems likely that a corresponding set of eight proteins are also made in animal mitochondria, because the translation products of animal mitochondria bear a qualitative and quantitative resemblance to those of yeast, on SDS gels. Furthermore, inhibition of mitochondrial protein synthesis leads to a deficiency in the cytochrome oxidase, ATPase, and coenzyme Q-cytochrome *c* reductase activity in animal cells (Kroon et al., 1972; Schatz and Mason, 1974; Neubert et al., 1975).

Are there other products of mitochondrial protein synthesis? The possibility was raised earlier that as many as 13 monocistronic mRNAs may be produced during the processing of mitochondrial RNA transcripts (Fig. 2). On this basis, it is unlikely that more than 5 translation products remain to be identified (at least in animal mitochondria). Possible candidates include the exported regulatory protein or proteins (Barath and Küntzel, 1972), other hydrophobic proteins in the inner membrane, and mitoribosomal proteins. Supporting evidence is available for the case of mitochondrial r-proteins (O'Brien and Matthews, 1976). Both yeast (Groot, in Kroon and Saccone, 1974) and *Neurospora* mitoribosomes appear to contain a mitochondrial translation product. This 35,000-dalton r-protein has been implicated in the assembly of the small subunit of *Neurospora* mitoribosomes (Lambowitz and Luck, 1976).

Mitochondrial Ribosomes

Mitochondrial ribosomes resemble bacterial (70S) and eukaryotic cytoplasmic ribosomes (80S) in their fundamental properties. They consist of two subunits containing RNA and proteins, and they function according to the same overall mechanism, using initiator tRNA, aminoacyl tRNAs, and soluble initiation and elongation factors to translate an mRNA molecule. In terms of their fine structure and physicochemical properties, however, they differ unexpectedly from both these kinds of ribosomes, as well as from each other.

Cytoplasmic ribosomes from organisms in the four eukaryotic taxonomic kingdoms (protists, fungi, plants, animals) all appear to be members of a single structural class having the characteristic properties shown in Table I. Similarly, prokaryotic ribosomes, including those from bacteria, mycoplasma, blue-green algae, and chloroplasts, can be adequately described by another set of values (Table I). In contrast to the relatively simple classification scheme possible for ribosomes from all other sources, mitochondrial ribosomes do not seem to fall into one or even a few structural categories. Depending upon the organism from which they are isolated, their sedimentation coefficients may range from 55S to 80S, their buoyant density from 1.40 to 1.61 g/cc, their rRNA molecules from 0.3 and 0.5 to 0.8 and 1.25 million daltons, and their guanosine and cytosine (GC) content from 19 to 43% (Table II).

Mitochondrial Ribosomal Proteins

How do the proteins of diverse mitoribosomes compare with each other, and with those of bacterial and cytoplasmic ribosomes? The technique of two-dimensional electrophoresis in polyacrylamide gels has been used to great advantage to analyze mitoribosomal proteins from a number of organisms. No obvious similarities in the two-dimensional electrophoretic patterns are observed in comparisons of mitoribosomes from distantly related organisms, such as *Neurospora* and rat. Moreover, mitochondrial r-protein patterns bear no resemblance to those of *E. coli* or to the corresponding cytoplasmic ribosomes (DeVries, in Bücher et al., 1976; O'Brien and Matthews, 1976). The mitochondrial r-protein patterns from more closely related organisms, however, such as

TABLE I

General Properties of Bacterial and Eukaryotic-Cytoplasmic Ribosomes

Ribosome class	Sedimentation coefficient	Inhibited by CAP	Inhibited by CHI	Buoyant density	Mol wt of RNA (× 10⁶ daltons)	RNA base composition
				(*g/cc*)		(*%GC*)
Eukaryotic	80S	−	4+	1.57	0.7, 1.3–1.7	45–65
Prokaryotic	70S	+	−	1.63	0.56, 1.1	45–55

Adapted from O'Brien and Matthews, 1976. *Abbreviations used are:* CAP, chloramphenicol; CHI, cycloheximide; %GC, % guanosine and cytosine.

TABLE II

Properties of Mitochondrial Ribosomes

Subclass	Kingdom	Example	Sedimentation coefficient	Buoyant density	Mol wt of RNA (× 10⁶ daltons)	RNA base composition
				(*g/cc*)		(*%GC*)
1	Animals	Vertebrates	55–60S	1.44	0.35, 0.54	40–45
2		Insects	60–71S	–	0.3, 0.5	19–32
3	Protists	*Euglena*	71S	(1.61)	0.56, 0.93	27
4		*Tetrahymena*	80S	1.46	0.47, 0.90	29
5	Fungi	*Candida*	72S	1.48	0.71, 1.21	34
6		*Neurospora*	73–80S	1.52	0.72, 1.28	38
7		*Saccharomyces*	72–80S	1.64	0.70, 1.30	30
8	Plants	Maize	77S	1.56	0.76, 1.25	–

Adapted from O'Brien and Matthews, 1976.

the mammals, cow and rat, do tend to resemble each other; still, many differences in the electrophoretic mobility of individual r-proteins are apparent (Matthews and O'Brien, unpublished data). Thus the pattern of variability, of evolutionary instability that seems to be a trademark of mitochondrial ribosomes, is apparent even in the electrophoretic mobility of mitoribosomal proteins in closely related organisms. The most remarkable example of this trend is the discovery of seven major differences between the toads *Xenopus laevis* and *Xenopus mulleri* expressed in the protein patterns from their large mitoribosomal subunits alone (Leister and Dawid, 1975).

Protein Content of Animal Mitoribosomes

Despite their unusually low sedimentation coefficient (55S), animal mitoribosomes are actually slightly larger than 70S bacterial ribosomes, both on the basis of particle molecular weight and physical dimensions (O'Brien and Matthews, 1976). It follows that a ribosome somewhat larger than bacterial ribosomes, but containing only about half as much rRNA, should have a correspondingly higher protein content. Indeed, animal mitoribosomes are exceedingly protein-rich, even compared to cytoplasmic ribosomes (Table III).

To better define the protein composition of mammalian mitoribosomes, we have correlated the function of salt-washed bovine mitoribosomes with their protein composition (O'Brien et al., in Bücher et al., 1976). At present, it appears that particles with a protein content of 67%, made up of about 90 individual proteins, represents the minimum functionally active form of the bovine mitoribosome. This unusual composition suggests

TABLE III

RNA and Protein Content of Representative Ribosomes from Bacteria (70S), Eukaryotic-Cytoplasm (80S), and Animal Mitochondria (55S)

Ribosome	Mol wt (× 10⁶ daltons)	No. of proteins	Content		Protein/RNA
			RNA	Protein	
70S	2.6	53	1.7	0.9	0.6
80S	4.5	70	2.4	2.1	0.9
55S	2.8	84–90	0.9	1.9	2.1

Adapted from O'Brien and Matthews, 1976.

that great differences will be found between the detailed structural and functional roles of the individual proteins in animal mitoribosomes and those of other kinds of ribosomes. The arrangement, or "packaging," of the components must be fundamentally different in animal mitoribosomes, which have a protein:RNA ratio of 2.1, from that found in bacterial ribosomes (0.6).

Mitochondrial Ribosomal RNA

The separate molecules of rRNA in each subunit of mitochondrial ribosomes are the best characterized primary gene products of mitochondrial DNA. Their base composition and size varies systematically, along phylogenetic lines, suggesting that they have evolved in parallel with eukaryotic cells (see below).

The rRNA molecules in animal mitoribosomes are the smallest rRNA molecules known; their aggregate molecular weight is only about half that of the smallest bacterial rRNAs (Table III). These rRNAs probably became enriched in the functionally important sequences, at the expense of "dispensable" structural segments, as they evolved to their present small size.

Certain sequences in the rRNA of bacterial ribosomes (for example, the 3′-terminal region of the small rRNA) appear to be involved in important ribosome functions, such as mRNA and initiation factor binding, subunit association, and chain termination. These vital functions are probably carried out by analogous sequences (structures) in mitochondrial rRNA. It might be expected that the rRNA, as a mitochondrial gene product, would reflect the prokaryotic nature of mitochondrial ribosomes. Surprisingly, however, the 3′-terminal sequence of the small rRNA molecule from hamster mitochondria resembles that of cytoplasmic rRNA more closely than bacterial RNA (Dubin and Shine, 1976).

Mitochondrial 5S RNA

All nonmitochondrial ribosomes contain one molecule of 5S RNA in their large subunit to serve the important function of tRNA binding. Bacterial and cytoplasmic ribosomes which lack 5S RNA are inactive in protein synthesis. Do mitochondrial ribosomes contain 5S RNA? Although mitoribosomes from higher plants (Leaver, 1975) and *Tetrahymena* appear to contain a separate 5S RNA, its presence in *Neurospora* mitoribosomes remains controversial (Agsteribbe, in Bücher et al., 1976; Michel et al., in Bücher et al., 1976). Reports from two different laboratories indicate that mammalian mitoribosomes contain a small RNA molecule (3S), which may be the functional analogue of 5S RNA in other ribosomes (O'Brien and Matthews, 1976).

All mitoribosomes probably contain the functional equivalent of 5S RNA. Definitive answers to the above question, however, will require additional testing to establish the functional or structural involvement of the putative 5S RNA analogue. The tetranucleotide probe TψCG, which competes with tRNA for the tRNA binding site on 5S RNA, might be used to establish the functional involvement of the 5S RNA analogue (O'Brien and Matthews, 1976). Alternatively, it may be possible to identify the 5S RNA analogue on the basis of sequence homologies, inasmuch as the base sequence of bacterial and eukaryotic 5S RNAs is both characteristic and highly conserved. Indeed, preliminary results from the "fingerprint" analysis of 5S RNA from wheat mitochondria show this RNA to be related more closely to prokaryotic than to eukaryotic 5S RNA (Cunningham, in Bücher et al., 1976).

Evolution of Mitochondrial Ribosomes

From a systematic comparison of the different kinds of ribosomes, it is apparent that some ribosome properties (buoyant density and sedimentation coefficient) are less conserved than others, and also that mitoribosomes show more variation in their properties than do other ribosomes. We may use these properties to define relationships between different kinds of ribosomes (O'Brien and Matthews, 1976).

Considering the more conserved properties of antibiotic susceptibility (chloramphenicol sensitivity) and exchangeability of protein-synthesis factors, the first major branch point in the evolution of ribosomes is the divergence of prokaryotic and eukaryotic-cytoplasmic ribosomes (Fig. 3). All mitochondrial ribosomes are included in the prokaryotic category, by virtue of their general susceptibility to inhibitors of bacterial protein synthesis and (partial) exchangeability of protein-synthesis factors. Properties of intermediate variability (rRNA size and GC content) segregate mitochondrial ribosomes, not only from other prokaryotic ribosomes, but also from each other, along phylogenetic lines. By comparison, only relatively minor changes in these properties are seen in the other ribosomes (Fig. 1 in O'Brien and Matthews, 1976). Further divergence of mitochondrial ribosomes within a given phylum is evident from significant changes in the structural properties of ribosomes (buoyant density and sedimentation

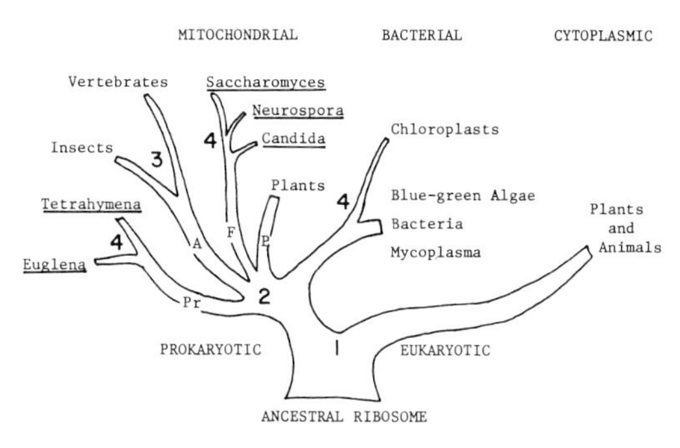

FIGURE 3 Evolutionary relationships among different kinds of ribosomes based on their functional and physicochemical properties. The numbers at branch points indicate divergence on the basis of: *1*, antibiotic susceptibility and factor exchangeability; *2*, rRNA size; *3*, rRNA GC content; and *4*, buoyant density and sedimentation coefficient. The letters indicate the eukaryotic taxonomic kingdoms: *Pr*, protists; *A*, animals; *F*, fungi; and *P*, plants.

coefficient). It should be noted that other organellar ribosomes, from chloroplasts, are more closely related to bacterial ribosomes by these criteria, sharing all properties except similar buoyant density (O'Brien and Matthews, 1976).

Concluding Remarks

Including the rRNAs, tRNAs, mRNAs, the ribosomal proteins, the synthetases, and protein-synthesis factors, well over 100 different macromolecules are involved, exclusively and directly, in mitochondrial transcription and translation. In contrast, the end product of these processes appears to be only 10 or so mitochondrial proteins. Although these few products of mitochondrial protein synthesis are vital to the integration and functioning of electron-transfer complexes in the inner mitochondrial membrane, one may understandably wonder why they could not be imported along with the hundreds of other proteins that mitochondria are known to import. Why is it necessary for the cell to maintain a separate protein-synthesizing system inside mitochondria? Perhaps the answer lies in the special topological restraints imposed on the mitochondrial products for their proper positioning within the inner mitochondrial membrane. The known products of mitochondrial protein synthesis are all extremely hydrophobic proteins. Perhaps there is no efficient mechanism in the cell for the transport of such hydrophobic proteins through the mitochondrial outer membrane, across the intermembrane space, and into the inner membrane. From thermodynamic considerations and from the vectorial aspects of electron transfer and oxidative phosphorylation within the inner membrane, it may be essential that these hydrophobic products are inserted into the inner membrane from the matrix side, as discussed in the following chapter.

REFERENCES

Barath, Z., and H. Küntzel. 1972. Cooperation of mitochondrial and nuclear genes specifying the mitochondrial genetic apparatus in *Neurospora crassa*. *Proc. Natl. Acad. Sci. U. S. A.* **69:**1371–1374.

Borst, P. 1972. Mitochondrial nucleic acids. *Annu. Rev. Biochem.* **41:**333–376.

Borst, P., and L. A. Grivell. 1973. Mitochondrial nucleic acids. *Biochimie (Paris)* **55:**801–804.

Bücher, T., W. Neupert, W. Sebald, and S. Werner. editors. 1976. The Genetics and Biogenesis of Chloroplasts and Mitochondria. North-Holland Publishing Company, Amsterdam. 895.

Chiu, N., A. Chiu, and Y. Suyama. 1975. Native and imported transfer RNA in mitochondria. *J. Mol. Biol.* **99:**37–50.

Cleaves, G. R., T. Jones, and D. T. Dubin. 1976. Properties of a discrete high molecular weight poly(A) containing mitochondrial RNA. *Arch. Biochem. Biophys.* **175:**303–311.

Dawid, I. B. 1972. Mitochondrial protein synthesis. *In* Mitochondria/Biomembranes. S. G. van den Bergh, P. Borst, L. L. M. van Deenen, J. C. Riemersma, E. C. Slater, and J. M. Tager, editors. North-Holland Publishing Company, Amsterdam. 35–51.

Dubin, D. T., and J. Shine. 1976. The 3′-terminal sequence of mitochondrial 13S ribosomal RNA. *Nucl. Acid. Res.* **3:**1225–1231.

Hirsch, M., A. Spradling, and S. Penman. 1974. The Messenger-like poly(A)-containing RNA species from the mitochondria of mammals and insects. *Cell.* **1:**31–35.

Kroon, A. M., E. Agsteribbe, and H. DeVries. 1972. Protein synthesis in mitochondria and chloroplasts. *In* The Mechanism of Protein Synthesis and Its Regulation. L. Bosch, editor. American Elsevier, New York. 551.

Lambowitz, A. M., and D. J. L. Luck. 1976. Studies on the poky mutant of *Neurospora crassa*: fingerprint analysis of mitochondrial ribosomal RNA. *J. Biol. Chem.* **251:**3081–3093.

Leaver, C. J. 1975. The biogenesis of plant mitochondria. *In* The Chemistry and Biochemistry of Plant Proteins. J. B. Harborne and C. F. van Sumere, editors. Academic Press, Inc., New York. 137–165.

Leister, D. E., and I. B. Dawid. 1975. Mitochondrial ribosomal proteins in *Xenopus laevis/X. mulleri* interspecific hybrids. *J. Mol. Biol.* **96:**119–123.

Lewis, F. S., R. J. Rutman, and N. G. Avadhani. 1976. Messenger ribonucleic acid metabolism in mammalian mitochondria: discrete poly (adenylic acid) lacking messenger ribonucleic acid species associated with mitochondrial polysomes. *Biochemistry.* **15:**3367–3372.

Lund, E., J. E. Dahlberg, L. Lindahl, S. R. Jaskunas, P. P. Dennis, and M. Nomura. 1976. Transfer RNA genes between 16S and 23S rRNA genes in rRNA transcription units of *E. coli*. *Cell.* **7:**165–177.

Mahler, H. R., and R. A. Raff. 1975. The evolutionary origin of the mitochondrion: a nonsymbiotic model. *Int. Rev. Cytol.* **43:**1–124.

Milner, J. 1976. The functional development of mammalian mitochondria. *Biol. Rev. (Camb.).* **51:**181–209.

Neubert, D., C. T. Gregg, R. Bass, and H-J. Merker. 1975. Occurrence and possible functions of mitochondrial DNA in animal development. *In* The Biochemistry of Animal Development. Vol. III. R. Weber, editor. Academic Press, Inc., New York. 387–464.

O'Brien, T. W. 1976. Mitochondrial protein synthesis.

In Protein Synthesis, Vol. II. E. H. McConkey, editor. Marcel Dekker, Inc., New York. 249–307.

O'Brien, T. W., and D. E. Matthews. 1976. Mitochondrial ribosomes. *In* Handbook of Genetics. Vol. V. R. C. King, editor. Plenum Press, New York. 535–580.

Saccone, C., and A. M. Kroon, editors. 1976. The Genetic Function of Mitochondrial DNA. North-Holland Publishing Company, Amsterdam. 354.

Saccone, C., and E. Quagliariello. 1975. Biochemical studies of mitochondrial transcription and translation. *Int. Rev. Cytol.* **43:**125–165.

Schatz, G., and T. L. Mason. 1974. The biosynthesis of mitochondrial proteins. *Annu. Rev. Biochem.* **43:**51–87.

Tzagoloff, A., M. Rubin, and M. F. Sierra. 1973. Biosynthesis of mitochondrial enzymes. *Biochim. Biophys. Acta.* **301:**71–104.

Wheeldon, L. W. 1973. Products of mitochondrial protein synthesis. *Biochimie (Paris).* **55:**805–814.

THE ASSEMBLY OF MITOCHONDRIA

J. SALTZGABER, F. CABRAL, W. BIRCHMEIER, C. KOHLER, T. FREY, and G. SCHATZ

Research on mitochondrial biogenesis is now progressing on two major fronts. On the first front, the genes localized on mitochondrial DNA are being identified and mapped by combining the following approaches: (*a*) isolation and mapping of mutants carrying specific lesions on mitochondrial DNA; (*b*) cleavage of mitochondrial DNA with restriction enzymes; (*c*) hybridization of mitochondrial RNA components to mitochondrial DNA coupled with identification of the hybridized regions in the electron microscope; (*d*) "denaturation mapping" by electron microscopy. These approaches, which have scored impressive advances, were reviewed in two recent symposia (Saccone and Kroon, 1976; Bücher et al., 1977).

On the second front, efforts are being made to understand the assembly of the mitochondrial membranes. This approach, which will be summarized here, has largely focused on a few reasonably well-defined oligomeric enzymes associated with the mitochondrial inner membrane. These enzymes are cytochrome *c* oxidase, the oligomycin-sensitive ATPase complex, and the cytochrome bc_1 complex. (Tzagoloff et al., 1973; Schatz and Mason, 1974). As mentioned in the preceding article by Borst, each of these enzyme complexes contains mitochondrially made polypeptide subunits as well as cytoplasmically made subunits. The assembly of these complexes must involve several intricate controls which are as yet almost completely unknown. Some of the most important questions are the following: (*a*) what are the chemical properties of the individual polypeptides? (*b*) how are these polypeptides positioned within the membrane? (*c*) how is the assembly of these polypeptides controlled? (*d*) how are cytoplasmically made polypeptides transported across the mitochondrial membrane? We shall deal with these questions by reviewing the asssembly of cytochrome *c* oxidase in the yeast *Saccharomyces cerevisiae*.

J. SALTZGABER and CO-WORKERS Department of Biochemistry, Biocenter, University of Basel, Basel, Switzerland

Chemical Properties of Cytochrome c Oxidase Subunits

Yeast cytochrome *c* oxidase consists of seven distinct polypeptide subunits (Fig. 1). The three large polypeptides (mol wt I, 40 K; II, 34 K; III, 23 K) are coded by mitochondrial DNA (Rabinowitz et al., 1976; Moorman and Grivell, 1976) and made on mitochondrial ribosomes (Mason and Schatz, 1973; Rubin and Tzagoloff, 1973); the four small polypeptides IV–VII (mol wt 14 K; 12 K; 12 K; 4.5 K, respectively) are coded by nuclear DNA (Ebner et al., 1973*a*) and made on cytoplasmic ribosomes (Schatz et al., 1972). When the seven subunits were purified (Poyton and Schatz, 1975*a*), the mitochondrially made subunits proved to be very hydrophobic, whereas the cytoplasmically made subunits were relatively hydrophilic. This was also reflected in the amino acid compositions of the subunits (Poyton and Schatz, 1975*a*; W. Birchmeir, A. Tsugita, and G. Schatz, manuscript in preparation). Subunits IV, VI, and VII are being sequenced in collaboration with Dr. A. Tsugita. Antibodies directed against individual subunits have so far failed to reveal any immunological cross reaction between the subunits (Poyton and Schatz, 1975*b*). Similar results have been reported for cytochrome *c* oxidase from *Neurospora crassa* (Sebald et al., 1973) and bovine heart (Briggs et al., 1975).

Three-Dimensional Arrangement of Cytochrome c Oxidase Subunits

By reacting soluble and membrane-bound cytochrome *c* oxidase from yeast and bovine heart

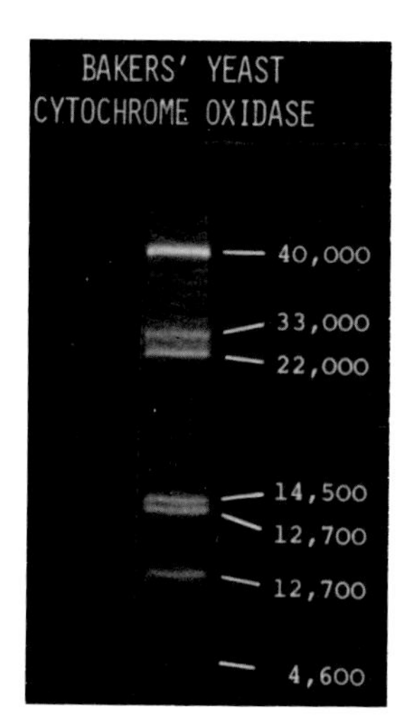

FIGURE 1 The seven polypeptide subunits of yeast cytochrome *c* oxidase. The subunits were resolved by sodium dodecyl sulfate (SDS)-polyacrylamide gel electrophoresis on a 10/20% step gel. The generally accepted molecular weights (Poyton and Schatz, 1975*a*) are indicated on the right.

with various radioactive polar "surface probes," it was found (Eytan and Schatz, 1975; Eytan et al., 1975) that subunit I and, to a lesser extent, subunit II were inaccessible to these reagents. In the membrane-bound bovine heart enzyme, subunit V was also inaccessible. By comparing the accessibility of cytochrome *c* oxidase subunits in intact bovine heart mitochondria and in inverted submitochondrial particles, cytochrome *c* oxidase was shown to span the mitochondrial inner membrane in an asymmetric manner. Subunits II and III, VI and VII are only accessible from the outer side, subunit IV is only accessible from the inner side, and subunits I and V are inaccessible from either side. (Subunits II and III of the bovine heart enzyme could not be separated by the electrophoretic methods used in these experiments.)

Subunit III of yeast cytochrome *c* oxidase can be "affinity-labeled" by activated yeast iso-1-cytochrome *c* (Birchmeier et al., 1976; Fig. 2). The cross-link is prevented by excess underivatized cytochrome *c*, is cleaved by excess mercaptoethanol, and is not merely caused by the presence of an exceptionally reactive sulfhydryl group on subunit III. (On the contrary, most of the easily accessible sulfhydryl groups are on subunit II.) Subunit III is thus at the very least close to the cytochrome *c* binding site and, by implication, on the same side of the mitochondrial inner membrane as cytochrome *c*, i.e., on the outer side. The cytochrome *c* oxidase model emerging from these studies is shown in Fig. 3.

We are currently attempting to arrive at a more refined model of the enzyme by subjecting electron micrographs of two-dimensional cytochrome *c* oxidase crystals (Vanderkooi et al., 1972) to image filtering techniques. The filtered images of the cytochrome *c* oxidase molecules obtained so far reveal distinct substructures which we are now trying to equate with some of the polypeptide subunits shown in Fig. 1. We still do not know which subunit(s) carry the heme *a* since it is detached from the protein under conditions which disaggregate the oligomeric enzyme complex.

Effect of Nuclear Genes on the Assembly of Cytochrome c Oxidase

Since the four small cytochrome *c* oxidase subunits are coded by nuclear genes (Schatz and Mason, 1974), one should be able to isolate nuclear yeast mutants specifically lacking cytochrome *c* oxidase. This is indeed true but the effect of the nuclear mutations described so far is more complex than expected. Several years ago, Ebner et al. (1973 *a, b*) isolated several cytochrome *c* oxidase-less nuclear yeast mutants and checked them for residual cytochrome *c* oxidase subunits by radioimmunochemical methods. More recently, the

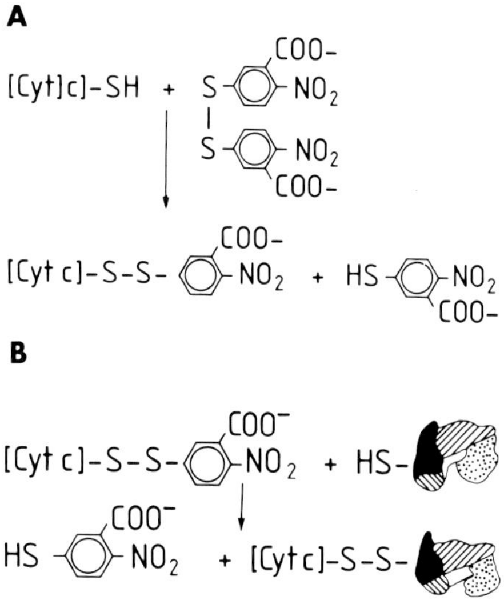

A

[Cyt]c]-SH + S-(C₆H₃)(COO⁻)-NO₂ | S-(C₆H₃)(COO⁻)-NO₂
↓
[Cyt c]-S-S-(C₆H₃)(COO⁻)-NO₂ + HS-(C₆H₃)(COO⁻)-NO₂

B

[Cyt c]-S-S-(C₆H₃)(COO⁻)-NO₂ + HS-[enzyme]
↓
HS-(C₆H₃)(COO⁻)-NO₂ + [Cyt c]-S-S-[enzyme]

FIGURE 2 Affinity-labeling of yeast cytochrome *c* oxidase with activated iso-1-cytochrome *c*. (A) Activation of iso-1-cytochrome *c*; (B) cross-linking of activated iso-1-cytochrome *c* to cytochrome *c* oxidase.

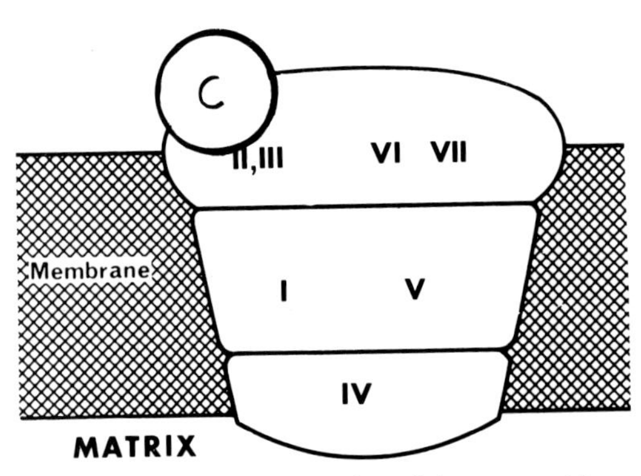

FIGURE 3 Schematic illustration of the asymmetric arrangement of the cytochrome *c*-cytochrome *c* oxidase complex across the mitochondrial inner membrane. The size and shape of the subunits (or groups of subunits) are displayed arbitrarily. The numbering of the subunits conforms to that of the yeast enzyme.

mutants were reinvestigated by three additional methods: (*a*) double-diffusion analysis, employing subunit-specific antisera; (*b*) two-dimensional electrophoresis, resolving polypeptides both by charge as well as by size (Figs. 4 and 5); (*c*) highly resolving exponential gradient gel electrophoresis in the presence of sodium dodecyl sulfate (SDS) followed by autoradiography (Douglas and Butow, 1976).

Taken together, these procedures can detect not only *assembled* cytochrome *c* oxidase subunits but also unassembled ones. They revealed that each of the two nuclear mutants investigated in detail (pet 494-1 and pet E11-1) lacked at least one of the mitochondrially made subunits while containing a complete set of the four cytoplasmically made subunits (Table I, first two horizontal columns). No alteration could be detected in any of the small

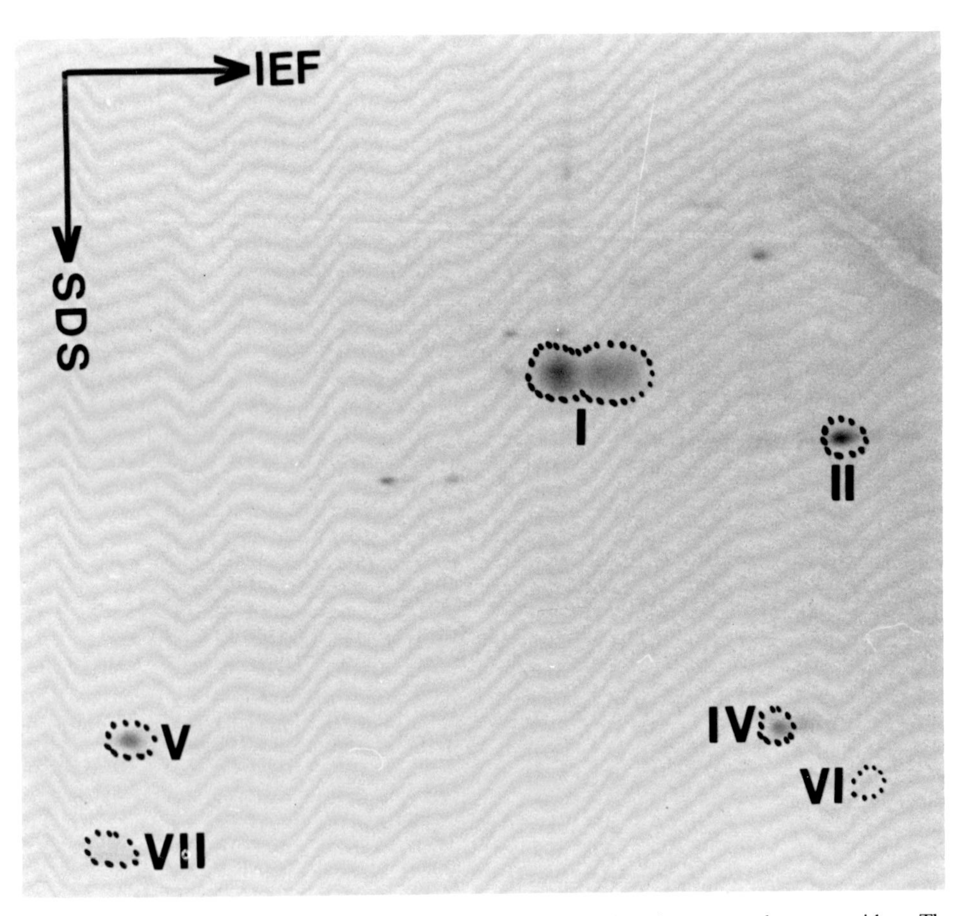

FIGURE 4 Two-dimensional analysis of radioactive immunoprecipitated yeast cytochrome *c* oxidase. The roman numerals identify the individual cytochrome *c* oxidase subunits. (Subunit III is not clearly displayed by this system.)

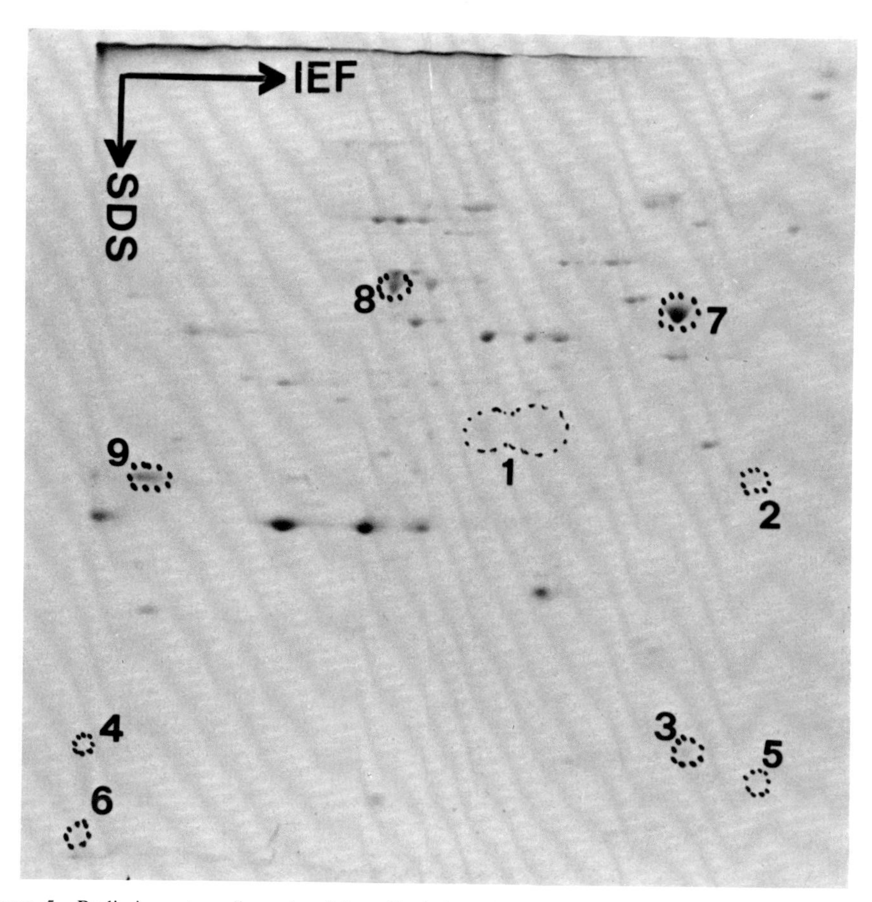

FIGURE 5 Preliminary two-dimensional "map" of the mitochondrial inner membrane. Yeast submitochondrial particles (uniformly labeled with ^{35}S) were analyzed by two-dimensional electrophoresis. (1–6) Cytochrome *c* oxidase subunits I, II, IV, V, VI, and VII, respectively; (7, 8, and 9) α, β, and γ subunits of F_1.

TABLE I
Cytochrome c Oxidase Subunits in Cytochrome c Oxidaseless Yeast Mutants

	Mutant			
Subunit	pet 494-1	pet E11-1	ρ^-	heme⁻ mutant
I	+	−	−	−
II	+	−	−	+
III	−	+	−	+
IV	+	+	+	+
V	+	+	+	−
VI	+	+	+	+
VII	+	+	+	+ (?)

subunits; however, they were only loosely bound to the mitochondrial inner membrane and were apparently not assembled with the residual mitochondrially made subunit(s). These results lead to the following conclusions: (*a*) nuclear mutations can affect accumulation of specific mitochondrially made subunits; (*b*) mitochondrially made subunits anchor the cytoplasmic subunits to the mitochondrial inner membrane (cf. also Fig. 2).

The pet 494-1 mutation can be suppressed by a nuclearly coded *amber* suppressor (Ono et al., 1975) which acts at the level of cytoplasmic translation by inserting tyrosine in place of the *amber* codon. It is thus very likely that the pet 494 locus codes for an (as yet unknown) cytoplasmic protein that is necessary for the synthesis or integration of subunit III.

Effect of Mitochondrial Genes on the Assembly of Cytochrome c Oxidase

If the mitochondrial genome is inactivated by the extrachromosomal petite mutation, the three mitochondrially made subunits are no longer made (Ebner et al., 1973*a*), and the cells lose functional cytochrome *c* oxidase (together with other enzymes containing mitochondrially synthesized polypeptides). The four cytoplasmically made subunits are still present in mitochondria (Table I) but are no longer firmly integrated into the inner membrane. Subunit VII particularly is easily lost, and this appears to account for our earlier failure to detect this subunit in petite mitochondria (Ebner et al., 1973*a*).

During the past 2 years several laboratories have isolated yeast mutants (mit^- mutants) carrying selective mutations at various loci on mitochondrial DNA (for review, see Bücher et al., 1977). The most extensive studies of these mutants were done by Tzagoloff and co-workers (1975) together with Slonimski (Slonimski and Tzagoloff, 1976). A minimum of three loci appear to specifically control cytochrome *c* oxidase formation, and it is probable, but not yet proved, that at least some of these loci are structural genes for the large cytochrome *c* oxidase subunits. In order to establish this important point, we are currently screening these mutants for an *alteration* of mitochondrially made cytochrome *c* oxidase subunits. (A *loss* would be less revealing since it could also result from indirect effects.) A closer study of these very interesting mutants and their revertants might also provide information on the possible existence of precursors to the mitochondrially made subunits and on the interaction between the various subunits during the assembly of cytochrome *c* oxidase.

Effect of Heme on the Assembly of Cytochrome c Oxidase

Small molecules such as prosthetic groups or metal cofactors can profoundly affect the conformation of oligomeric proteins. Are such interactions important for the in vivo assembly of oligomeric enzyme complexes? We have approached this question by studying cytochrome *c* oxidase assembly in a yeast mutant (Gollub et al., 1974) unable to synthesize heme in the absence of added δ-aminolevulinic acid.

When the mutant is grown in the absence of δ-aminolevulinic acid, it is respiration-deficient and completely lacks any cytochrome absorption bands. However, when δ-aminolevulinic acid is added to the growth medium, the cells are essentially normal (Fig. 6). Analysis of the heme-defi-

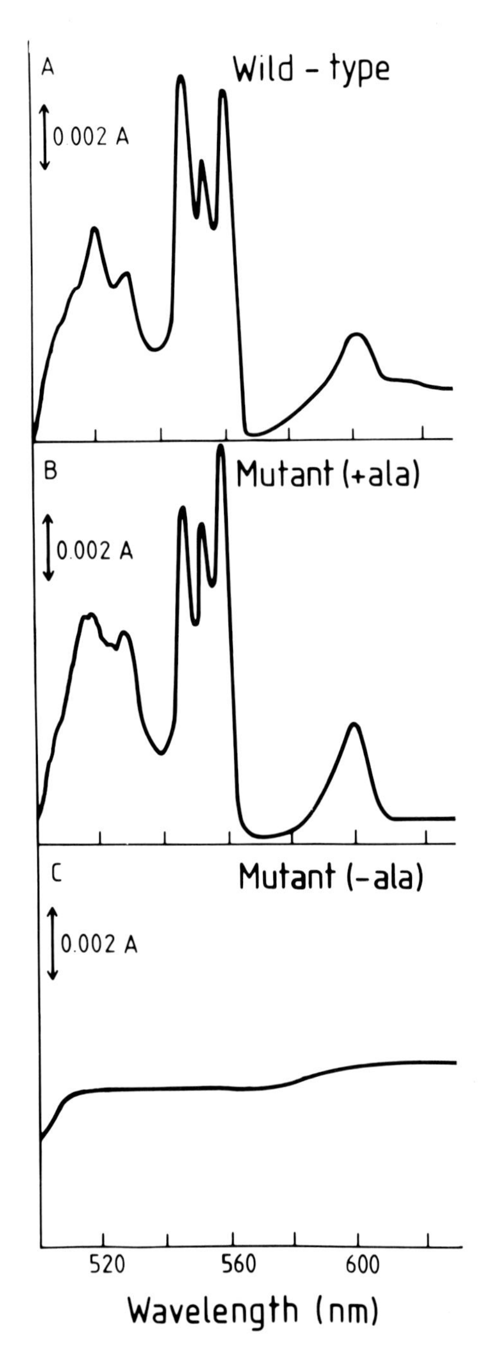

FIGURE 6 Low-temperature spectra of mitochondria from wild-type yeast (strain X-2180) and the δ-aminolevulinic acid synthetase-deficient mutant GL1-38. Reduced minus oxidized difference spectra were recorded at liquid nitrogen temperature at the following protein concentrations: wild-type, 12.0 mg/ml; mutant (+ala), 14.7. mg/ml; mutant (−ala), 14.0 mg/ml.

cient cells by the methods outlined above failed to reveal the presence of subunits I and V, whereas subunits II, III, IV, and VI were still detectable (Fig. 7 and Table I). (No clear result is as yet available on subunit VII.)

As with the nuclear mutants discussed earlier, the residual cytochrome *c* oxidase subunits are no longer firmly bound to each other. This result reemphasizes the importance of ligand interactions in the assembly of supermolecular structures.

Proteolytic Processing in the Assembly of Cytochrome c Oxidase

Poyton and Kavanagh (1975) have recently suggested that at least subunits IV and VI (and perhaps all four cytoplasmically made subunits) may be cut from a common precursor protein. They found that the synthesis of the three large cytochrome *c* oxidase subunits by isolated yeast mitochondria was stimulated by a dialyzed postribosomal supernate. The effect of the supernate could be traced to the presence of 53,000 dalton polypeptide that cross-reacted with antisubunit IV serum as well as with antisubunit VI serum. A one-dimensional tryptic fingerprint of this polypeptide appeared to contain all the peptides which were observed with a tryptic digest of a mixture of subunits IV-VII, as well as some extra peptides (Poyton, 1977). If this model of cytochrome *c* oxidase assembly can be further substantiated, it will add an important new dimension to the problem. Even more importantly, it raises intriguing new questions. Where in the cell does this processing occur, and how does the precursor protein (or its cleavage products) enter the mitochondria?

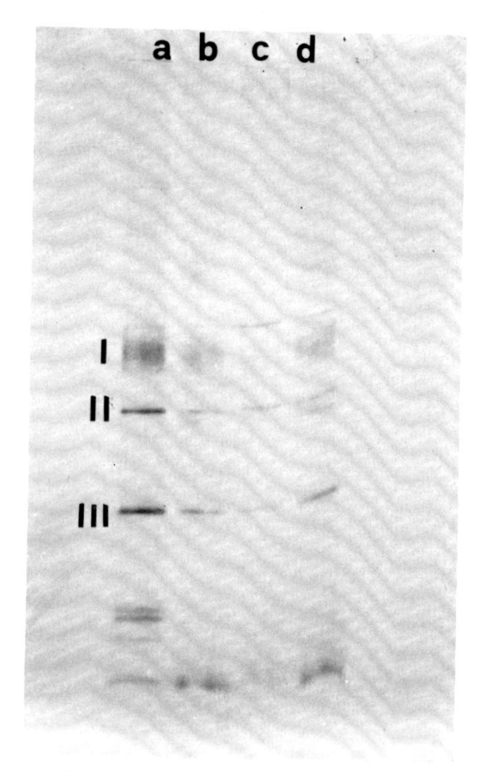

FIGURE 7 Gradient slab gel autoradiographic analysis of mitochondrial translation products. Mitochondria were isolated from cells labeled in vivo with [^{35}S]methionine in the presence of cycloheximide. (*a*) Immunoprecipitate of cytochrome oxidase; (*b*) wild-type mitochondria; (*c*) mutant (−ala) mitochondria; (*d*) mutant (+ala) mitochondria.

Transport of Cytoplasmically Synthesized Proteins into Mitochondria

Our present cytochrome *c* oxidase model (Fig. 3) places the cytoplasmically made subunit IV on the matrix side of the mitochondrial inner membrane. How does this subunit get there? This question is still unanswered. Butow and his colleagues (1975) have suggested that cytoplasmically made proteins can be channeled into mitochondria by a process akin to "vectorial translation" since yeast mitochondria contain cytoplasmic polysomes attached to those regions of their outer surface which are in close apposition to the inner membrane. The best way to clarify the role of these mitochondria-associated polysomes would be the demonstration that they synthesize the precursor to a defined mitochondrial protein which is normally found on the matrix side of the inner membrane. Such precursors (which still contain the hydrophobic "signal" sequence at the N-terminus; see Blobel and Dobberstein, 1975) have already been detected with the aid of cell-free translation systems for other proteins transported across biological membranes.

CONCLUSION

The mitochondrially made cytochrome *c* oxidase subunits have been identified as hydrophobic "cores" of a transmembranous oligomeric enzyme. They are necessary for binding the cytoplasmic "partner proteins" to the mitochondrial inner membrane, yet are also dependent on these proteins for their own accumulation. They are, as it were, part of the language in which the two genetic systems speak to each other during the assembly of a mitochondrion.

REFERENCES

BIRCHMEIER, W., C. KOHLER, and G. SCHATZ. 1976. Interaction of integral and peripheral membrane proteins: affinity labeling of yeast cytochrome oxidase by modified cytochrome *c*. *Proc. Natl. Acad. Sci. U. S. A.* In press.

BLOBEL, G., and B. DOBBERSTEIN. 1975. Transfer of proteins across membranes. II. Reconstitution of functional rough microsomes from heterologous components. *J. Cell Biol.* **67:**852–862.

BRIGGS, M., P.-F. KAMP, N. C. ROBINSON, and R. A. CAPALDI. 1975. The subunit structure of the cytochrome *c* oxidase complex. *Biochemistry.* **14:**5123–5128.

BÜCHER, T., W. NEUPERT, W. SEBALD, and S. WERNER, editors. 1977. Genetics and Biogenesis of Mitochondria and Chloroplasts. North-Holland Publishing Co., Amsterdam. In press.

BUTOW, R. A., W. F. BENNETT, D. B. FINKELSTEIN, and R. E. KELLEMS. 1975. Nuclear-cytoplasmic interactions in the biogenesis of mitochondria in yeast. *In* Membrane Biogenesis. A. Tzagoloff, editor. Plenum Publishing Corp., New York. 155–200.

DOUGLAS, M., and R. A. BUTOW. 1976. Variant forms of mitochondrial translation products in yeast: evidence for location of determinants on mitochondrial DNA. *Proc. Natl. Acad. Sci. U. S. A.* **73:**1083–1086.

EBNER, E., T. L. MASON, and G. SCHATZ. 1973*a*. Mitochondrial assembly in respiration-deficient mutants of *Saccharomyces cerevisiae*. II. Effect of nuclear and extrachromosomal mutations on the formation of cytochrome *c* oxidase. *J. Biol. Chem.* **248:**5369–5378.

EBNER, E., L. MENNUCCI, and G. SCHATZ. 1973*b*. Mitochondrial assembly in respiration-deficient mutants of *Saccharomyces cerevisiae*. I. Effect of nuclear mutations on mitochondrial protein synthesis. *J. Biol. Chem.* **248:**5360–5368.

EYTAN, G., R. C. CARROLL, G. SCHATZ, and E. RACKER. 1975. Arrangement of the subunits in solubilized and membrane-bound cytochrome *c* oxidase from bovine heart. *J. Biol. Chem.* **250:**8598–8603.

EYTAN, G., and G. SCHATZ. 1975. Cytochrome *c* oxidase from bakers' yeast. V. Arrangement of the subunits in the isolated and membrane-bound enzyme. *J. Biol. Chem.* **250:**767–774.

GOLLUB, E. G., P. TROCHA, P. K. LIU, and D. B. SPRINSON. 1974. Yeast mutants requiring ergosterol as only lipid supplement. *Biochem. Biophys. Res. Commun.* **56:**471–477.

MASON, T. L., and G. SCHATZ. 1973. Cytochrome *c* oxidase from bakers' yeast. II. Site of translation of the protein components. *J. Biol. Chem.* **248:**1355–1360.

MOORMAN, A. F. M., and L. A. GRIVELL. 1976. Coupled transcription-translation of yeast mitochondrial DNA *in vitro* as a means of gene identification and mapping. *In* The Genetic Function of Mitochondrial DNA. C. Saccone and A. M. Kroon, editors. North-Holland Publishing Co., Amsterdam. 281–289.

ONO, B.-I., G. R. FINK, and G. SCHATZ. 1975. Mitochondrial assembly in respiration-deficient mutants of *Saccharomyces cerevisiae*. IV. Effects of nuclear amber suppressors on the accumulation of a mitochondrially made subunit of cytochrome *c* oxidase. *J. Biol. Chem.* **250:**775–782.

POYTON, R. O. 1977. *In* Genetics and Biogenesis of Mitochondria and Chloroplasts. T. Bücher, W. Neupert, W. Sebald, and S. Werner, editors. North-Holland Publishing Co., Amsterdam. In press.

POYTON, R. O., and J. KAVANAGH. 1975. *In* Electron Transfer Chains and Oxidative Phosphorylation. E. Quagliariello, S. Papa, F. Palmieri, E. C. Slater, and N. Siliprandi, editors. North-Holland Publishing Co., Amsterdam. 75–80.

POYTON, R. O., and G. SCHATZ. 1975*a*. Cytochrome *c* oxidase from bakers' yeast. III. Physical characterization of isolated subunits and chemical evidence for two different classes of polypeptides. *J. Biol. Chem.* **250:**752–761.

POYTON, R. O., and G. SCHATZ. 1975*b*. Cytochrome *c* oxidase from bakers' yeast. IV. Immunological evidence for the participation of a mitochondrially synthesized subunit in enzymatic activity. *J. Biol. Chem.* **250:**762–766.

RABINOWITZ, M., S. JAKOVCIC, N. MARTIN, F. HENDLER, A. HALBREICH, A. LEWIN, and R. MORIMOTO. 1976. Transcription and organization of yeast mitochondrial DNA. *In* The Genetic Function of Mitochondrial DNA. C. Saccone and A. M. Kroon, editors. North-Holland Publishing Co., Amsterdam. 219–230.

RUBIN, M. S., and A. TZAGOLOFF. 1973. Assembly of the mitochondrial membrane system. X. Mitochondrial synthesis of three of the subunit proteins of yeast cytochrome oxidase. *J. Biol. Chem.* **248:**4275–4279.

SACCONE, C., and A. M. KROON. 1976. The Genetic Function of Mitochondrial DNA. North-Holland Publishing Co., Amsterdam, 354.

SCHATZ, G., G. S. P. GROOT, T. L. MASON, W. ROUSLIN, D. C. WHARTON, and J. SALTZGABER. 1972. Biogenesis of mitochondrial inner membranes in bakers' yeast. *Fed. Proc.* **31:**21–29.

SCHATZ, G., and T. L. MASON. 1974. The biosynthesis of mitochondrial proteins. *Annu. Rev. Biochem.* **43:**51–87.

SEBALD, W., W. MACHLEIDT, and J. OTTO. 1973. Products of mitochondrial protein synthesis in *Neurospora crassa*: determination of equimolar amounts of three products in cytochrome oxidase on the basis of amino acid analysis. *Eur. J. Biochem.* **38:**311–324.

SLONIMSKI, P. P., and A. TZAGOLOFF. 1976. Localization in yeast mitochondrial DNA of mutations expressed in a deficiency of cytochrome oxidase and/or coenzyme QH_2-cytochrome *c* reductase. *Eur. J. Biochem.* **61:**27–41.

TZAGOLOFF, A., A. AKAI, R. B. NEEDLEMAN, and G. ZULCH. 1975. Assembly of the mitochondrial membrane system: cytoplasmic mutants of *Saccharomyces cerevisiae* with lesions in enzymes of the respiratory chain and in the mitochondrial ATPase. *J. Biol. Chem.* **250:**8236–8242.

TZAGOLOFF, A., M. S. RUBIN, and M. F. SIERRA. 1973. Biosynthesis of mitochondrial enzymes. *Biochim. Biophys. Acta.* **301:**71–104.

VANDERKOOI, G., E. SENIOR, R. A. CAPALDI, and H. HAYASHI. 1972. Biological membrane structure. III. The lattice structure of membranous cytochrome oxidase. *Biochim. Biophys. Acta.* **274:**38–48.

Endoplasmic Reticulum-Golgi Apparatus and Cell Secretion (Plant Cells)

GOLGI APPARATUS AND PLASMA MEMBRANE INVOLVEMENT IN SECRETION AND CELL SURFACE DEPOSITION, WITH SPECIAL EMPHASIS ON CELLULOSE BIOGENESIS

R. MALCOLM BROWN, Jr., and J. H. MARTIN WILLISON

The cell wall is an extracellular covering of the plant cell. It functions in the protection of the protoplasm from environmental stress, and it serves as a structural skeleton. Furthermore, patterns of growth and differentiation are governed largely by the biosynthetic processes leading to the assembly of the cell wall. As a specialized type of extracellular matrix, the cell wall can be regarded as a preserved record (like geological strata) of a series of successive stages in a time-sequence of metabolic, transport, and depositional events.

The fundamental components of the cell wall include a *microfibrillar* reinforcing network which lies in a gel-like *matrix* of interlinked molecules. The matrix consists of protein, hemicelluloses, and pectins; the reinforcing rods are comprised of β 1,4-linked glucan chains crystallized into microfibrils of cellulose. Cell walls are of more than theoretical importance to cell biologists. Cellulose is the most abundant product on earth and is of great commercial value. It is a natural polymer, having features equivalent to several man-made polymers, yet its in vitro synthesis remains to be accomplished.

The purpose of this presentation is to report new observations of cell wall biogenesis, using a variety of organisms and experimental approaches. To this end, the authors are grateful to the following individuals who have contributed major facts to our group effort: Richard Santos, Dwight Romanovicz, David Montezinos, and Susette Mueller. The theme of our presentation concerns the synthesis of cellulose and the role of the endoplasmic reticulum, Golgi apparatus, and plasma membrane in its biogenesis. The noncellulosic wall components will not be discussed, and the reader is referred to a recent review (Chrispeels, 1976) for a more complete coverage of secretion and cell walls.

Clearly, an understanding of cellulose synthesis is the key to understanding cell wall biogenesis (Preston, 1974). Progress in this particular area of plant cell biology may have been slow for two reasons: (1) technological limitations; and (2) inappropriate choice of suitable experimental organisms (which has led to the failure to test hypotheses on cellulose formation critically).

The descriptions of cellulose formation that follow will be largely cytological; however, in the cases of *Pleurochrysis* (Brown and Romanovicz, 1976; Romanovicz and Brown, 1976), *Acetobacter* (Cooper and Manley, 1975), and corn (Wright and Northcote, 1976), a large compilation of supporting biochemical evidence is available.

The Golgi Apparatus and Scale Formation in Pleurochrysis Scherffelii

The system to be described first is representative of an extremely advanced and streamlined process of cell wall formation. The organism, *Pleurochrysis scherffelii,* is a representative of the haptophycean, or scale-bearing, algae (Fig. 1). The Golgi apparatus of this marine benthic alga assumes the function of scale assembly and transport (Fig. 2). Other membranes, notably the endoplasmic reticulum (ER), as well as the subcortical and plasma membranes, function in the early stages of scale product biosynthesis, exocytosis, and cell movement.

The Golgi-derived scale exhibits a characteristic

R. MALCOLM BROWN, JR. Department of Botany, University of North Carolina, Chapel Hill, North Carolina
J. H. MARTIN WILLISON Department of Biology, Dalhousie University, Halifax, Nova Scotia, Canada

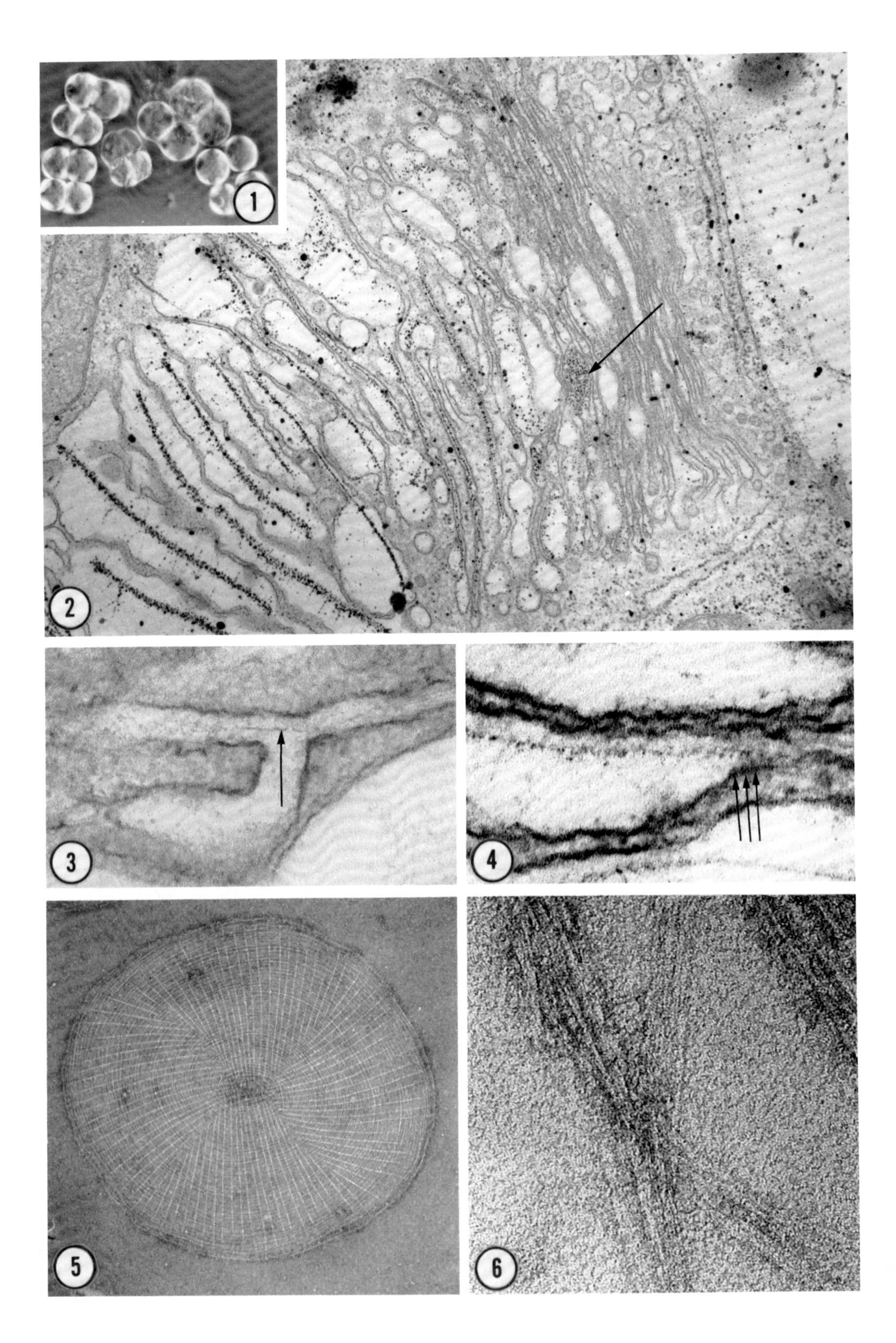
1
2
3
4
5
6

morphology (Fig. 5). It consists of a quadriradial microfibrillar network associated with a band of spiral microfibrils. In addition, amorphous coating substances serve as interscale adhesives to produce the characteristic cell wall of the vegetative cell. Scale rim modification subcomponents have been recognized with immunocytochemical approaches (A. Kavookjian and R. M. Brown, Jr., manuscript in preparation).

The scale biogenic pathway will be described briefly. For more detailed information, the reader should consult Brown and Romanovicz (1976), and Romanovicz and Brown (1976). The forming face of the Golgi apparatus is located adjacent to the amplexus, which is an extension of ER from the outer membrane of the nuclear envelope (Fig. 2). The budding face of the ER is devoid of ribosomes. Golgi membranes and early precursor pool products are derived from the amplexus ER. A topographical distribution of pre-Golgi subcomponents has been noted in association with the ER.

The earliest-detected scale product that can be recognized in thin section is the radial microfibril precursor pool (Fig. 2, arrow). This region becomes heavily labeled with reduced silver when periodic acid-silver methenamine is used. Obviously, this product must exist in polymerized form to be retained in the cisterna, but it undergoes further modification at later stages of cisternal differentiation to form the quadriradial network of microfibrils (Fig. 5). Changes leading to this state include: (*a*) crystallization of the radial precursor pool products into folded microfibrillar rods; (*b*) sulfation of the microfibrillar rods; and (*c*) unfolding into the quadriradial base plate. The quadriradial elements are noncellulosic microfibrils rich in galactose, arabinose, fucose, and some protein. Evidence supporting sulfation of newly crystallized quadriradial microfibrils is based on cytochemical localization of arylsulphatase and acid phosphatase activity (Brown and Romanovicz, 1976).

Of the known scale subcomponents, the spiral microfibrils have been investigated in greatest detail through chemical analyses, X-ray diffraction, and infrared absorption. The microfibrils were unequivocally demonstrated to contain cellulose. The microfibril composition is 37% protein (rich in serine and hydroxyproline), and 63% carbohydrate (all of which is β 1,4-linked glucan). Each microfibril is associated with a coating substance rich in galactose, glucose, mannose, and some protein. The microfibrils are organized into spiral bands, the integrity of which seems to be maintained by the spiral coating substances (Fig. 6).

FIGURES 1–6 *Pleurochrysis scherffelii.*

FIGURE 1 Vegetative cells observed with polarized light and at extinction. Note the birefringence of the cell walls. × 900.

FIGURE 2 Golgi apparatus of a vegetative cell. Periodic acid silver methenamine staining. Note the forming face (to the right) and the secreting face (to the left) of inflated cisternae containing fully assembled scales. The budding face of the amplexus ER is visible (right), and a radial precursor pool can be discerned (arrow). × 27,000.

FIGURE 3 A central feeding tubule. The secreting face projects toward the bottom of the figure. Note the radial scale subcomponent (arrow). × 106,000.

FIGURE 4 A cisterna at the stage of spiral cellulosic microfibril addition. Note the granules (arrows) associated with the distal surface of the cisternal membrane at the junction of the inflated region. Spiral microfibrils are observed in cross section (as electron-dense dots) layered upon the radial microfibrils. A fully assembled scale is shown at the bottom. × 126,000.

FIGURE 5 Bacitracin-uranyl acetate negative stain preparation of a mature vegetative cell scale. Note the distinct quadriradial microfibril network and the spiral network converging in the form of an ellipitical "eye" at the center of the scale. Note the evidence of strain or torque imposed upon the quadriradial microfibrils which are slightly bent. × 36,000.

FIGURE 6 Bacitracin-uranyl acetate negative stain preparation of 2 N trifluoroacetic-extracted scale preparation in which the spiral bands of cellulosic microfibrils have been released from the quadriradial network. × 249,000.

Until recently, very little was known of the mechanisms involved in the synthesis and assembly of the spiral cellulose microfibril in *Pleurochrysis*. The deposition of the spiral network onto the distal or secreting face of the unfolded quadriradial network had been demonstrated (Brown et al., 1973). The involvement of a distal central "feeding tubule" (Fig. 3) was suggested, but since microfibrils had never been observed in this region, the evidence that it was the actual site of cellulose synthesis was inconclusive.

Recently, Brown and Romanovicz (1976) described granules in association with the distal Golgi membranes at the locus of spiral microfibril addition (Fig. 4). However, a more attractive and realistic interpretation of the events of cellulose synthesis in the Golgi apparatus of *Pleurochrysis* can be formulated on the basis of the recent discovery in *Oocystis* of plasma membrane-bound linear complexes at termini of growing cellulosic microfibrils (Brown and Montezinos, 1976). The central "feeding tubule" seems not to be the site of cellulosic microfibril formation but, rather, could function as a directed passageway or sequestering locus for precursors of cellulose synthesis. The earlier report of polysomes associated with the central feeding tubule supports the suggestion that this region is a site of protein synthesis within cisternae of the Golgi apparatus. The proteins synthesized in this region could be either structural protein associated with the spiral microfibrils, or glycosyltransferases (or possibly both entities).

Thus, the sequence of events of cellulose synthesis within the Golgi apparatus could be viewed as follows: (1) at the appropriate stage of differentiation, glycosyltransferases might be synthesized in the central feeding tubule region, then transported and assembled in association with the distal cisternal membrane; (2) nucleotide sugars, precursors, proteins, etc., could be stored or transferred through the central feeding tubule to the appropriate catalytic sites on the Golgi membranes; (3) before the initiation of cellulose synthesis, the glycosyltransferases would aggregate to form 8–10 complexes; (4) through terminal synthesis, the β 1,4 glucan chains would be synthesized and crystallized into the cellulosic microfibril. Eight to ten microfibrils would be synthesized simultaneously. They would be immediately coated with the noncellulosic substances, including the protein subcomponents; (5) through the forces of polymerization and crystallization, the cellulose synthesizing complexes would move in the plane of the cisternal membranes. The elongating band of 8–10 microfibrils eventually would come into contact with the peripheral cisternal membrane. As the microfibrils grow, they would be constrained within the restricted space of the cisternal membranes, thus directing the terminal synthetic sites to migrate in a centripetal spiral pattern; (6) as the spiral band moves over the radial microfibrillar network, it would be firmly cemented into place at each radial-spiral junction. Continued centripetal spiral growth would become increasingly difficult as the reduced space available would force greater constraint upon the microfibrillar band. Finally, the torque generated from the last inward spiral turn would be so great that microfibrils would be fractured, thus creating the "eye" region of the scale; (7) after spiral microfibril addition is completed, the scale would become separated from the cisternal membrane.

The hypothetical series of events just described summarizes the synthesis of the cellulosic subcomponent. In vegetative cells, amorphous polysaccharide scale-coating substances rich in galactose, glucose, fucose, and some protein are added to the completed scales that are destined to become incorporated into a compact cell wall. In the zoospores, the amorphous product pathway is inhibited. Specialized exocytotic mechanisms for handling structures as large as scales serve to bring the Golgi cisternal membranes into contact with specific sites of the plasma membrane. Protoplasmic rotation insures even deposition of scales into the cell wall. These processes will not be dealt with here, and the reader is referred to Brown and Romanovicz (1976) for details.

From the above evidence with *Pleurochrysis*, it is now clear that the Golgi membranes have the capacity to polymerize and fully crystallize β 1,4 glucan chains into cellulosic microfibrils. The role of the Golgi apparatus in cellulose synthesis is not universal for plant cells. Most authors accept that the most probable site for cellulose microfibril synthesis in plant cells is the plasma membrane. This does not eliminate the probability that some of the essential functions may be carried out by the Golgi apparatus. For instance, the glycosyltransferases may be synthesized in the ER, then assembled and transferred by the Golgi apparatus to the plasma membrane, whereupon they become active in microfibril synthesis (Kiermayer and Dobberstein, 1973).

It is conceivable that, within higher plant systems, synthesis of cellulose by both Golgi apparatus and plasma membrane occurs. Recently, Wright and Northcote (1976) described cellulose in Golgi-derived, root-cap slime preparations of maize. In this example, short cellulose microfibrils are completely enclosed by hydrophilic noncellulosic polysaccharides, thereby creating a highly soluble form of microfibrils.

What is known about the mechanisms of synthesis for the extremely long, highly ordered, nonsoluble cellulosic microfibrils which comprise the cell walls of so many plant cells? As early as 1958, Roelofsen indicated that an enzyme might be located at the growing tip of a microfibril, thereby attaching glucose from its precursor to the molecule ends at the protruding tip. At that time there was no strong evidence in favor of this concept, but Preston (1964) elaborated the theme with his "ordered granule hypothesis." This hypothesis assumed that contact with a microfibril end stimulated a granule to synthesis, whereby it would grow through the granule. The hypothesis also assumed that simultaneous synthesis in two directions was possible and that the granule did not move with the growing microfibril ends.

Involvement of the Plasma Membrane in the Synthesis of Cellulose: the Case in Oocystis

The proposition of the ordered granule hypothesis (Preston, 1964) stimulated many searches for supporting examples. The studies of Robinson and Preston (1972) on the unicellular green alga, *Oocystis,* suggested that the observed "granule bands" might be microfibril synthesizers. This work was open to criticism in terms of the identification of the membrane fracture faces observed.

Recently Brown and Montezinos (1976) examined cellulose microfibril formation in *Oocystis* (Fig. 7), making use of the widely accepted idea that unit membranes split during freeze-fracturing at the central hydrophobic interface. The granule bands were found on inner membrane leaflet (P) fracture faces only (Fig. 9, 19). Fine corrugations corresponding with the granule bands are found in E fracture moieties (Fig. 10). More significantly, on E fracture faces (i.e., that half of the membrane in contact with extracellular microfibrils) there was a previously undescribed feature, in the form of linear complexes clearly associated with microfibril tips (Fig. 10). All microfibrils being synthesized were uniaxially oriented, but synthesis was bidirectional (Fig. 8). The terminal complex consisted of three rows, each of approximately 30 subunits (Fig. 10). Paired linear complexes, which presumably are inactive, were often found (Fig. 13). The micrograph indicates that, at the initiation of synthesis, the paired complexes move apart in opposite directions (Fig. 8). Treatment with cryoprotectant or glutaraldehyde resulted in the complete loss of structural integrity of the linear complexes (Montezinos and Brown, 1976).

Recent studies (Montezinos) demonstrate that the complexes are disrupted by treatment with 0.1 M EDTA (Fig. 12). If 0.1 M $MgSO_4$ is added, the complexes reassemble within 15 min (Fig. 13). However, they form only in the paired complex state and never reaggregate at the microfibril tips.

Treatment with 1% colchicine for 2 h results in the disruption of granule bands and the production of wavy microfibrils (Fig. 11). Because microtubules are not present at this stage, they can have no function in the orienting of the microfibrils. It seems probable, therefore, that the orientation of growing microfibrils is controlled, in some way, by the granule bands embedded in the plasma membrane. They might perhaps be envisioned as railroad tracks to which the linear synthesizing complex and its growing microfibril product are attached.

The Plasma Membrane and its Role in Cellulose Synthesis in Glaucocystis

In addition to *Oocystis,* several genera have highly ordered crossed-polylamellate walls (Roelofsen, 1965). The microfibrils are unusually straight, lie parallel with each other within individual layers, and have unusually large cross-sectional dimensions of about 8 × 20 nm (Figs. 14 and 15). The cell walls of these algae (notably *Valonia, Chaetomorpha, Oocystis,* and *Glaucocystis*) were extensively studied by Preston and his colleagues using X-ray diffraction and electron microscopy of dried, shadowed material (see Preston, 1974). The success of finding regular complexes at the ends of growing microfibrils in *Oocystis* (Brown and Montezinos, 1976) led us to examine another member of this group of algae. *Glaucocystis* has a disputed taxonomic position (Robinson and Preston, 1971*a*), but superficially is very similar to *Oocystis.* As in *Oocystis,* the microfibrillar component of the wall is deposited during a relatively short period of the life cycle, while the daughter cells are still enclosed within the mother-cell wall.

As in *Oocystis* (Fig. 17), microfibril ends could be found impressed into the plasma membrane

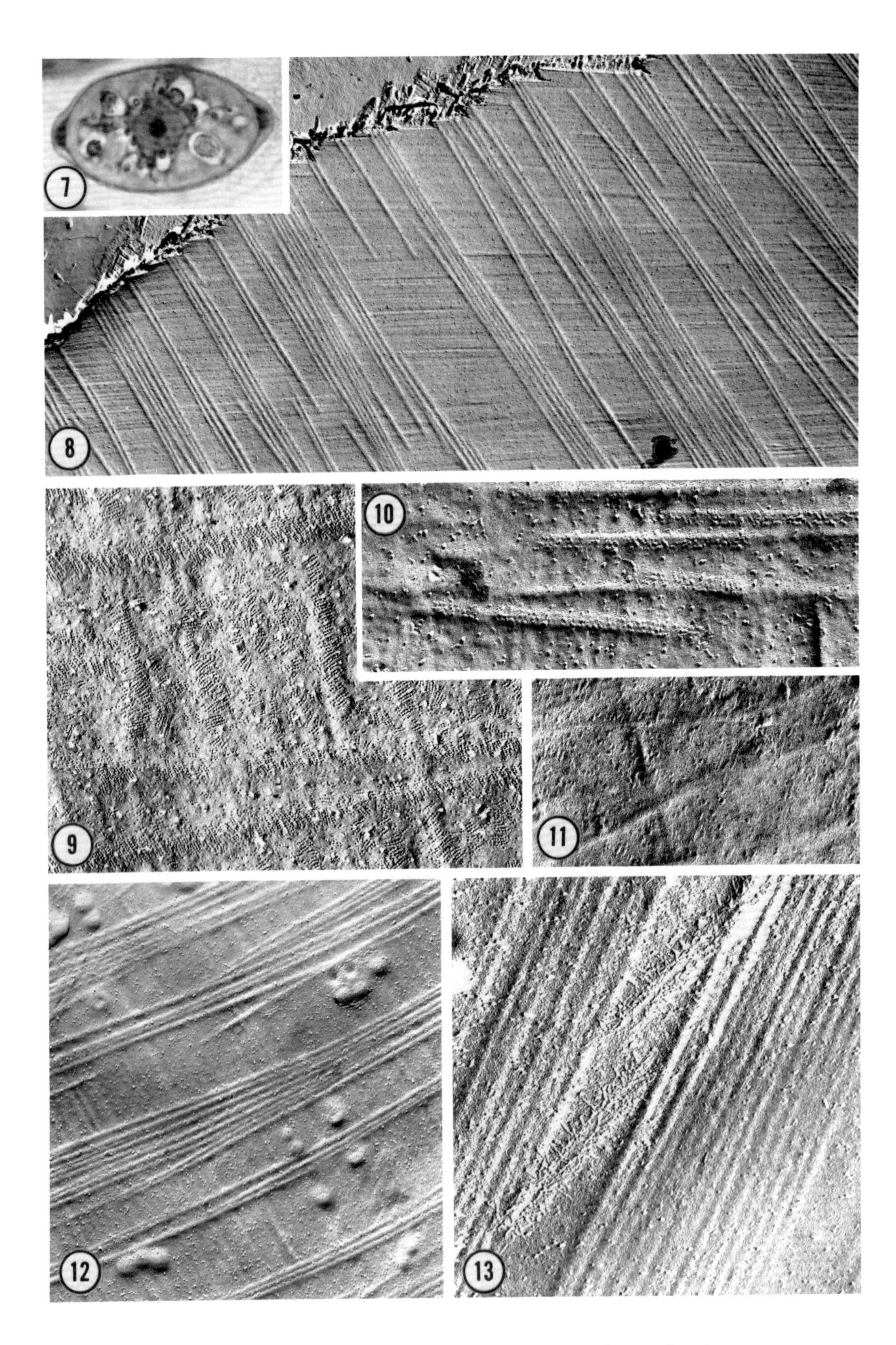
7
8
9
10
11
12
13

fracture faces (Figs. 18 and 20). Terminal complexes were present (Fig. 18), but these were less distinct and were structurally different from those of *Oocystis* (Fig. 17). The impressions of the leading ends of growing *Glaucocystis* microfibrils in E fracture faces usually consisted of a tripartite leading edge, followed by a short, indistinct region, and then a distinctly bipartite microfibril leading abruptly into the complete microfibril (Fig. 18). Microfibril ends usually were found in defined areas (Fig. 22), giving the impression of an advancing front like a wave on the sea. Lying beneath the plasma membrane of *Glaucocystis,* covering most of the surface of the cell, is a series of closely adpressed flat vesicles, which are termed "shields" (Fig. 21). Preceding the advancing front of microfibril terminal complexes, yet lying between the plasma membrane and the shields, there are fine filamentous structures which are oriented parallel with the microfibrils, but are not directly aligned with individual microfibrils (Fig. 22). "Granule bands" were not present on P fracture faces, although the impressions of microfibril ends could be discerned clearly (Fig. 20). We propose that the subplasma membrane filaments are intermediaries in determining the path taken by the microfibril-synthesizing complexes of *Glaucocystis* as they travel upon, or within, the plasma membrane. Each layer of microfibrils with the same orientation in the *Glaucocystis* wall is several microfibrils thick (Fig. 15), in contrast to that of *Oocystis*. The thickness of the layers seems to be determined by the number of microfibril-synthesizing complexes active within the depth of the progressing region of synthesis. That is, at regions distal to the leading edge of a progressing front (such as that visible in Fig. 22), synthetic complexes can be seen beneath the layer of microfibrils already formed (Willison and Brown, 1977). This situation is not found in *Oocystis*.

A Comparative Evaluation of Microfibril Orientation in Glaucocystis *and* Oocystis

The studies of Robinson and Preston (1971*b*) showed clearly that in *Glaucocystis* alternate layers of microfibrils were arranged in fast and slow helices about the cell, so that each layer of microfibrils converged upon the poles of these ellipsoidal cells. However, the poles themselves are areas of mystery. Because the microfibrils appear to be laid down by the progression of mobile complexes upon the plasma membrane surface, the disposition of microfibrils at the poles should indicate the means by which alternation of orientation be-

FIGURES 7–13 *Oocystis*.

FIGURE 7 Glycol methacrylate-embedded preparation of a mature vegetative cell which has been stained with toluidine blue. Note the prominent central nucleus and polar thickenings of the cell wall. × 2,100.

FIGURE 8 An E fracture view of the outer leaflet of the plasma membrane and impressions made by growing microfibrils of cellulose. Part of the wall microfibrils are exposed (to left). Note that the microfibrillar impressions oriented from left to right represent the previously deposited axis, those impressions running diagonally are representative of the axis under current bidirectional synthesis. × 18,000.

FIGURE 9 Fracture view of the inner leaflet of the plasma membrane demonstrating granule bands. Note that the granule band associated with the most recently synthesized axis of microfibrils cuts through the axis of the previous deposition. × 56,000.

FIGURE 10 A detailed E fracture view of the terminal synthetic complexes. Note the three rows of subunits arranged linearly. × 65,000.

FIGURE 11 An E exposure face of a colchicine-treated cell demonstrating the wavy pattern of microfibril synthesis. × 70,000.

FIGURE 12 An E fracture surface of a cell treated for 15 min with 0.1 M EDTA. Note the termini of the microfibrillar impressions lacking any associated subunits as shown in Fig. 10. × 28,000.

FIGURE 13 Same as Fig. 12, but 15 min after the addition of 0.1 M $MgSO_4$. Note the reaggregation of subunits to form the dimer complexes which are never associated with growing microfibrillar tips at this stage after treatment. × 60,000.

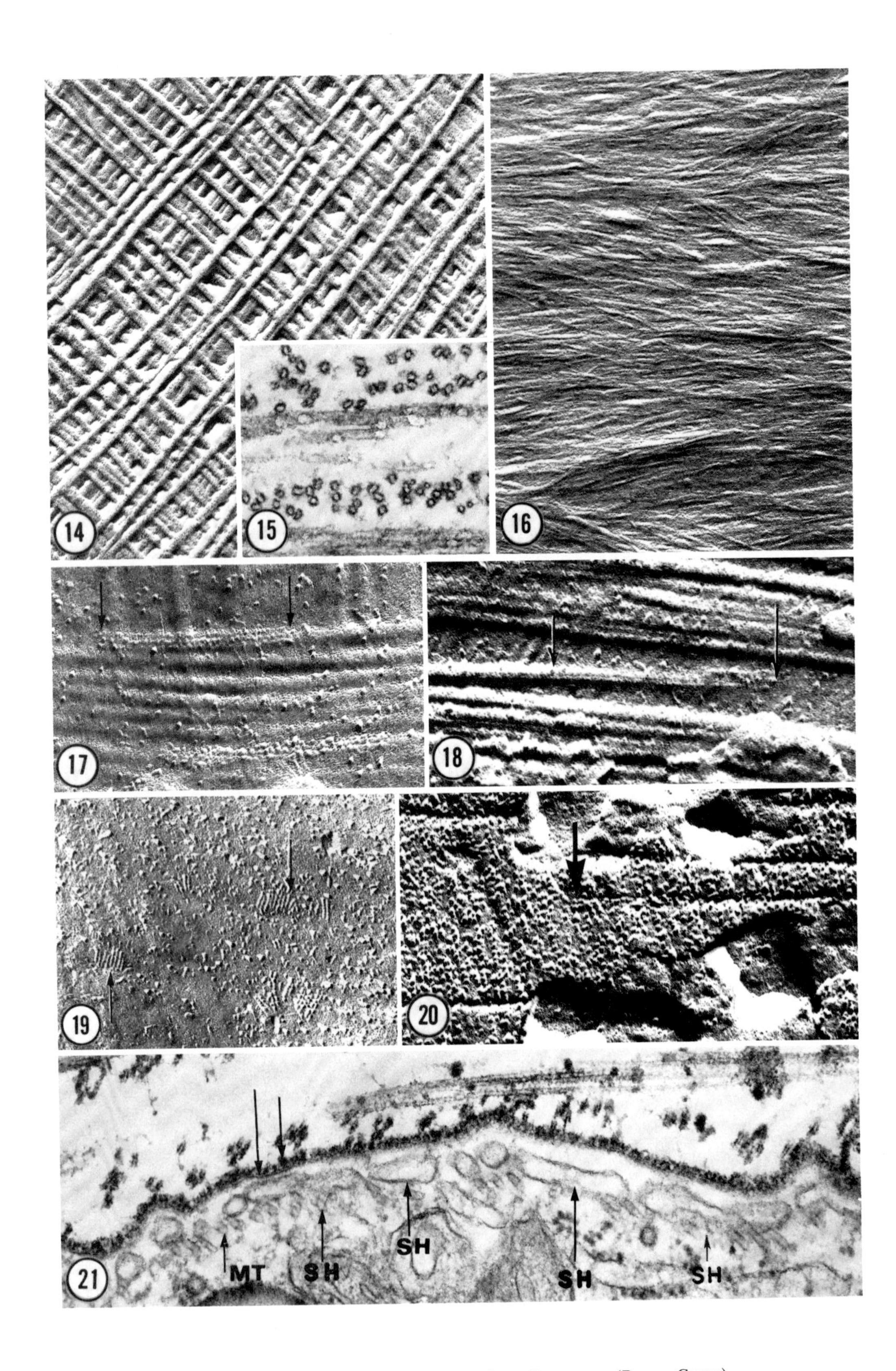
14
15
16
17
18
19
20
21
MT
SH
SH
SH
SH

tween layers occurs. Replicas of the poles of expanded mother-cell walls of *Glaucocystis* show that microfibrils make loops around the poles (Fig. 23). We have named the points about which the microfibrils loop "rotation centers." Careful analysis of replicas of mother-cell walls indicates that there are always three rotation centers at each pole, and that one-third of the microfibrils loop around each rotation center as a broad band (Figs. 23, 39, 40). This same pattern is found by freeze-fracturing the poles of unexpanded cells still enclosed within mother-cell walls (Fig. 24). There appears to be no interweaving of the microfibrils associated with one rotation center with those of another. Thus, it seems that each group of mobile complexes passes around its rotation center before those of the next group arrive at theirs. A model of the pattern of *Glaucocystis* microfibrils (Fig. 39) indicates that there is continuity of microfibrils between each successive layer of microfibrils. It seems most likely, therefore, that the plasma membrane-bound synthesizing complexes continue to follow "figure of eight" patterns from initiation to completion of the wall.

The results indicate that in *Oocystis*, by contrast with *Glaucocystis*, there is disaggregation and reaggregation of the complexes between successive wall layers. This hypothesis is supported by observation of the microfibril arrangement at the *Oocystis* poles. Replicas of fragments of *Oocystis* walls reveal that there are no rotation centers, but instead the microfibrils converge on the pole as four broad bands alternating between right-handed and left-handed helices (Fig. 25). The precise disposition of microfibril ends is obscured by matrix materials at the poles, but freeze-fracturing of the polar region during wall formation reveals microfibril impressions in the form of arrowheads (Fig. 26). David Montezinos (personal communication) interprets this to indicate that the synthesizing complex "backs up," that is, that its direction of travel, and therefore of synthesis, reverses. Arrowheaded microfibrils may be found in replicas of disrupted walls from *Oocystis* poles. The only alternative to a "backing-up" hypothesis is that separate synthesizing complexes, taking different tracks, come together and fuse at the poles. Whichever mechanism operates, it is clear that the process is finely controlled and that the traveling synthesizing complex is remarkably versatile.

The results with *Glaucocystis* and *Oocystis* may be summarized as follows: two members of the group of algae with wide microfibrils appear to have linear microfibril-synthesizing complexes terminally related to growing microfibrils. These complexes are presumed to travel upon the surface of the plasma membrane along predetermined tracks defined by structures that can be visualized. The precise structure of the terminal complexes, and of the direction-determining tracks are quite different in the two organisms. The arrangement of the microfibrils in the *Glaucocystis* walls shows that there is continuity at the poles between wall layers of different orientations, indicating that wall architecture is determined entirely by the paths taken by the motile complexes. In *Oocystis* there appears to be partial continuity between otherwise distant microfibrils within any one wall layer. However, in *Oocystis*, circumstantial and experimental evidence indicates that the

---

FIGURES 14–16 Replica of part of the mother wall of *Glaucocystis* after autospore release (Fig. 14). Note that the microfibrils are broad and straight by comparison with those of higher plants as exemplified by the secondary wall of a cotton fiber (Fig. 16). Each oriented layer of the *Glaucocystis* wall is several microfibrils in depth, and microfibrils display a rectangular cross section (Fig. 15). × 52,000 for all figures.

FIGURES 17 and 18 A comparison of the microfibril terminal plasma membrane-associated "complexes" (delimited by arrows) of *Oocystis* (Fig. 17) and *Glaucocystis* (Fig. 18) exhibited on E fracture faces of developing autospores. × 85,000 (Fig. 17) and × 100,000 (Fig. 18).

FIGURES 19 and 20 Plasma membrane P fracture faces of *Oocystis* (Fig. 19) and *Glaucocystis* (Fig. 20) compared. "Granule bands" (arrows), which may be guide tracks, are exhibited by *Oocystis*. These are lacking in *Glaucocystis* even at microfibril termini (arrow, Fig. 20). × 78,000 for both figures.

FIGURE 21 Thin section of a region of a *Glaucocystis* autospore during wall formation. Microfibrils are tightly associated with the plasma membrane (large arrows). Microtubules (*MT*) underlie a series of shields (*SH*) situated immediately beneath the plasma membrane. × 63,000.

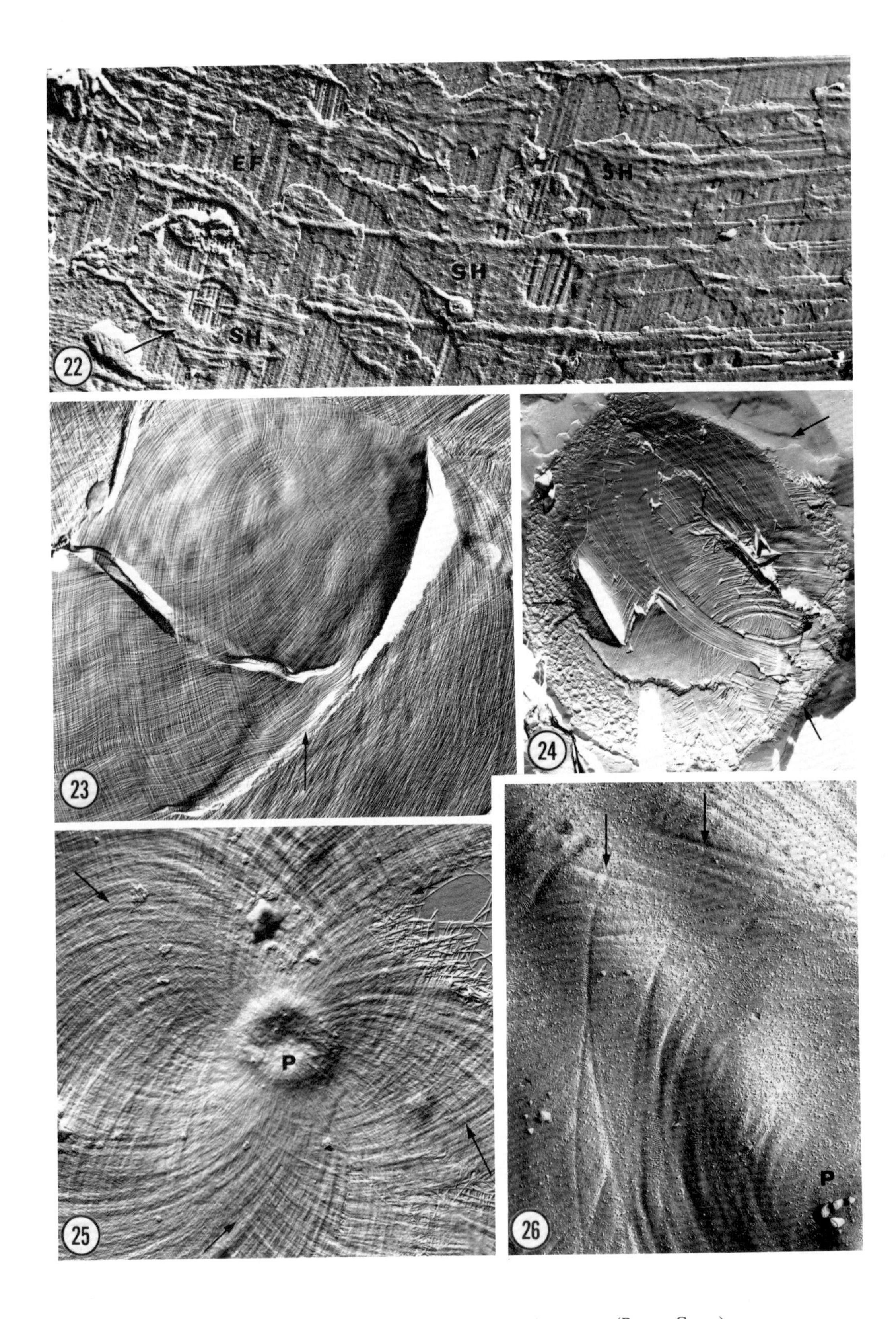
EF
SH
SH
SH
22
23
24
P
25
P
26

terminal complexes have the capacity to disaggregate and later reaggregate to resume functioning, and that this capacity allows for alternation of the orientation of wall layers. While there are striking parallels between these two algae in the basic mechanisms, the differences between them indicates the capacity for considerable diversity among organisms in the mechanism of microfibril synthesis.

Plasma Membrane Involvement in Cellulose Formation Exemplified by the Corn Root and Cotton Fiber

Most land plants retain cellulose as the principal reinforcing component of their cell walls. Unlike the cellulose of the algae described above, it is in the form of fine microfibrils, which have a poorly defined, elliptical cross section of 5–8 nm (Fig. 16). These microfibrils commonly undulate.

Several unsuccessful studies have attempted to discover morphological entities, using freeze-fracturing, which might be related to microfibril synthesis or assembly in high plants (see Willison, 1976). More recently, Mueller et al. (1976) examined untreated isolated corn-root steles. They found impressions made by distinctive globules at the ends of certain microfibrils on plasma membrane E fracture faces (Fig. 28). These were associated with the walls of an unidentified, but distinctive, cell type in regions of steles about 12 mm from the root tip. The apparent lamellar separation, and the change in orientation, of these microfibrils from those underlying them, suggest that they may represent the beginnings of secondary wall formation. This is only supposition, however, since lamellar changes in orientation are also commonplace in primary walls (Roland et al., 1975). Not uncommonly, microfibrils were partially ripped through the plasma membrane, showing clearly that the 20 nm globule was terminally associated with the microfibril (Fig. 27). The simplest interpretation is that this terminal globule is a synthesizing complex equivalent to those of the algae described above.

At the time of Mueller's discovery, we were examining cotton (*Gossypium hirsutum* L.) fibers. Cotton fibers are single cells, about 20 mm in length at maturity, which elongate from the epidermis of ovules to form an intricately intertwined mass. In order to freeze these fibers without disruption, we adapted the spray-freezing technique of Bachmann and Schmitt (1971) by freezing the whole ovule mass before transferring the fibers to butylbenzene (Willison and Brown, 1977). This technique successfully retained the turgid plasma membrane morphology of the cotton fiber (Figs. 29 and 30). On E fracture faces of fibers actively forming cell walls, 20 nm globules were found at a frequency of 5–10 μm^{-2} (Fig. 31). Unequivocal evidence that these globules were associated with microfibril ends was not found, however. The secondary walls of cotton fibers are predominantly cellulose (Fig. 16), and are readily distinguishable from primary walls, which have a low cellulose content, by the microfibril orientation (Willison and Brown, 1977). If these globules represent locomotory cellulose-synthesizing complexes, and if the system is homologous with the algal system, we can expect to find a subplasma membrane guide track. In cotton fibers, as in many other higher plants (Newcomb, 1969), the orientation

FIGURE 22 E fracture face of *Glaucocystis* autospore in the region of an advancing front of microfibril synthesizing sites. No microfibrils are present at the left-hand side of the micrograph where numerous bars (arrows) lie between plasma membrane (*EF*) and shields (*SH*). These bars are supposed guide elements. × 30,000.

FIGURES 23–26 The polar regions of *Glaucocystis* (Figs. 23 and 24) and *Oocystis* (Figs. 25 and 26). In the *Glaucocystis* wall, microfibrils encircle three symmetrically placed "rotation centers" at the poles such that there is continuity between the wall layers. Microfibrils may be traced around a rotation center in the replica of a mother wall (arrow, Fig. 23), and the three rotation centers are shown by a freeze-fractured pole (arrows, Fig. 24). The *Oocystis* pole (Figs. 25 and 26) differs from that of *Glaucocystis*. Four sets of microfibrils (arrows, Fig. 25) appear to approach the pole symmetrically but are obscured in a raised zone at the apex. Freeze-fracturing indicates that microfibrils partially encircle the *Oocystis* pole and that a reversal in the direction of synthesis may occur at the end of the tract (arrows, Fig. 26). × 5,000 (Fig. 23); × 7,000 (Fig. 24); × 9,000 (Fig. 25); × 31,000 (Fig. 26).

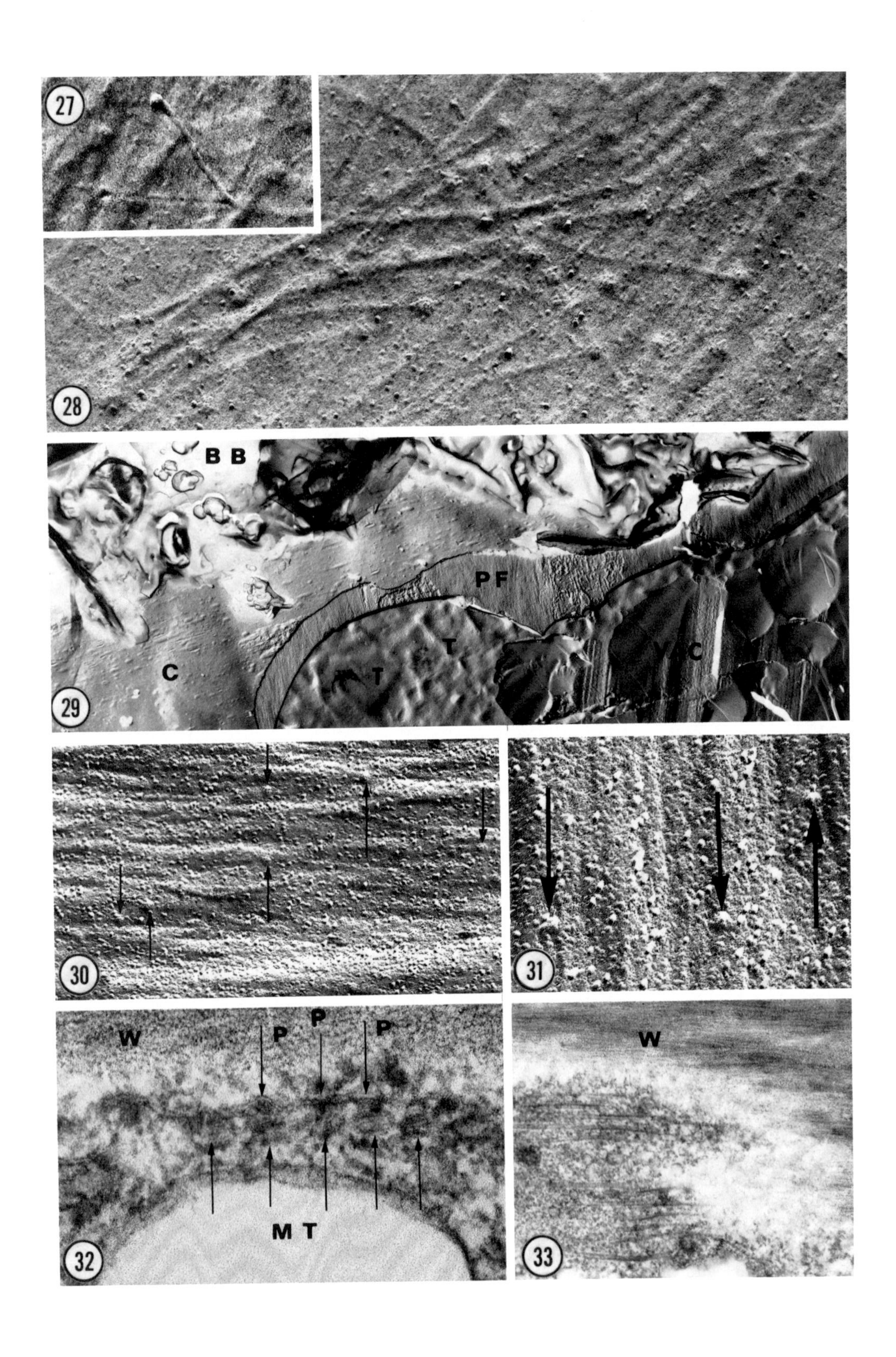
27
28
B B
PF
C
T
M T
V C
29
30
31
W
P
P
P
M T
32
W
33

of wall microfibrils and cortical cytoplasmic microtubules are similar. For this reason, Heath (1974) proposed that microtubules may represent a guide track for cellulose synthetases. This proposition has been made more likely by the finding of bridges spanning the gap between plasma membrane and cortical microtubules in *Poteriochromonas* during the formation of noncellulosic microfibrils (Schnepf et al., 1975). Similar bridges are also revealed in cotton fibers when tannic acid is used as a supplement to glutaraldehyde fixation (Figs. 32 and 33).

Our cytological findings relating to the role of plasma membrane-bound, locomotory, microfibril-synthesizing complexes in higher plants can be summarized as follows: despite earlier lack of success in finding suitable structures, recent findings suggest that a globular complex, situated at the ends of growing microfibrils, may be present in secondary walls. Evidence for plasma membrane/microtubule links supports the hypothesis of a guide-track role for microtubules in higher plants. The significance of short rows of particles on freeze-fractured plasma membranes, lying parallel with microfibrils of young primary walls, is unclear at present (Willison, 1976; Mueller et al., 1976).

Cellulose Synthesis in Acetobacter

Cellulose is not restricted to the plant kingdom, but its occurrence elsewhere in the living world is scattered. A gram-negative bacterium, *Acetobacter xylinum*, has received considerable attention because of its suitability for the study of cellulose metabolism. Several studies had been interpreted to indicate that cellulose microfibril synthesis occurs extracellularly, at a distance from the cell, mediated by a secreted enzyme and a secreted glycolipid which acts as a glucose carrier (Colvin, 1972). We reexamined *Acetobacter* because this pattern of synthesis was in marked contrast to our findings in algae and higher plants.

Bacteria were isolated from pellicles by shaking in distilled water (Brown et al., 1976). Freshly isolated bacteria were almost always free of cellulose, but within 5 min microfibril synthesis restarted. Dark-field optical microscopy, which clearly reveals microfibrillar ribbons (Figs. 34 and 35), was used to monitor in vivo microfibril growth. Microfibrils, which were always attached by one end to bacteria (Figs. 34 and 38), elongated at a steady rate of 2 μm/min. That synthesis occurred at the proximal ends of microfibrils was demonstrated by cases where the distal ends of microfibrils had become bound to the slide or cover-slip. In these cases, as elongation continued, bacteria moved away from the stationary distal end. Within 20 min, a characteristic, though flimsy, pellicle had formed (Fig. 35). In a few rare cases the sites of attachment of the cellulosic microfibrils were revealed on freshly isolated bacte-

FIGURES 27 and 28 Freeze-fractured E plasma membrane fracture faces of cells within isolated corn steles 12 mm from the root tip. Note that prominent granules appear at the ends of microfibrils (Fig. 28) and that in some cases a microfibril may be ripped through the plasma membrane so that a globule remains attached to the terminus, corresponding with the plasma membrane-impressed granule (Fig. 27). × 132,000 (Fig. 27) and × 83,000 (Fig. 28).

FIGURE 29 Part of a 10-day postanthesis cotton fiber prepared by the butylbenzene technique. Plasma membrane (*PF*) morphology is retained. *BB*, butyl benzene; *C*, cuticle; *T*, tonoplast; *Vac*, vacuole. × 7,000.

FIGURES 30 and 31 E fracture faces of plasma membranes of 21-day postanthesis cotton fibers prepared by the butylbenzene technique. Granules equivalent in size to those of Figs. 27 and 28 are commonplace (arrows). × 94,300.

FIGURE 32 Thin section of the periphery of a cotton fiber, fixed with tannic acid-glutaraldehyde. Microtubules (*MT*) underlying the plasma membrane (*P*) are linked to the plasma membrane by cross bridges. *w*, cell wall. × 119,000.

FIGURE 33 Same as Fig. 32, but a grazing section through the cell wall. Note that wall microfibrils (top and right) parallel the microtubules (left). The cross bridges are demonstrated in this projection as single electron-dense lines in association with the microtubules. × 30,000.

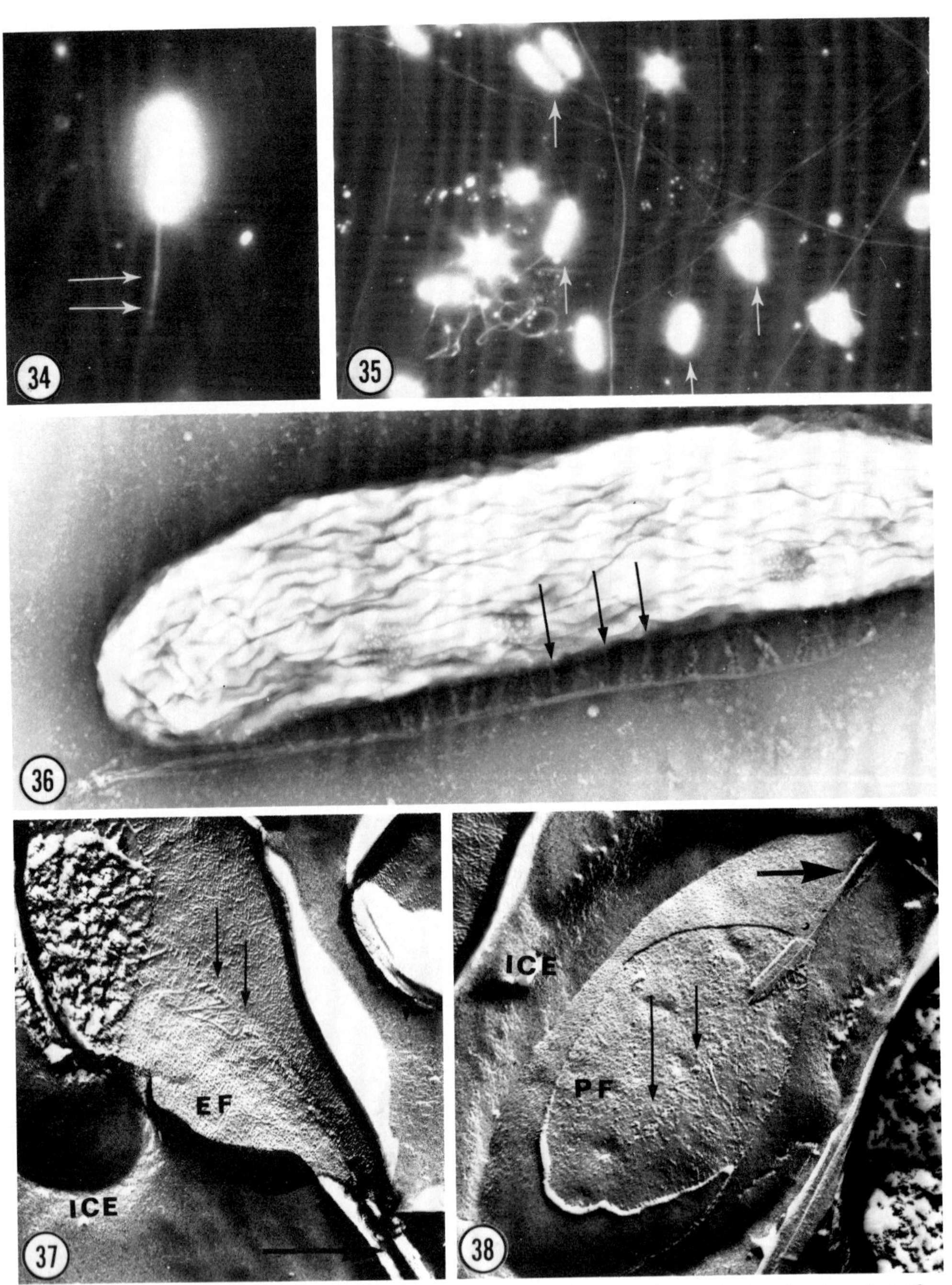

FIGURES 34 and 35 Dark-field optical micrographs of *Acetobacter xylinum* 5 min after isolation of bacteria from the pellicle (Fig. 34). A regenerated microfibrillar ribbon is visible (arrows). After 20-min (Fig. 35) growth of microfibrillar ribbons at sites in the bacterial surface produces an interweaved net. Arrows (Fig. 35) indicate bacteria at which microfibrillar ribbons clearly terminate. × 3,000 (Fig. 34) and × 1,350 (Fig. 35).

FIGURE 36 A rare case of a bacterium freshly isolated from a pellicle in which the microfibrillar ribbon has been incompletely sheared from the bacterial surface. The microfibrillar ribbon is divided into subelements (= nascent microfibrils?) which are separately related to the bacterial surface (arrows). Negatively stained micrograph. × 36,000.

FIGURES 37 and 38 Freeze-etch micrographs of *Acetobacter* 10 min after isolation from the pellicle. Concave (*EF*, Fig. 37) and convex (*PF*, Fig. 38) fracture faces of the outer envelope are illustrated. A distinctive row of particles lies along each fracture face (small arrows). In each case, ice-etching has revealed extracellular microfibrils (large arrows) which lie directly in line with the linear particle aggregate. × 55,000.

ria with negative-staining (Fig. 36). In these cases, the microfibrillar ribbon was not completely sheared from the bacterium, but about 50 subelements of the microfibrillar ribbon arose from an axial row of sites associated with the bacterial surface (Fig. 36). Subdivision of the bacterial microfibrillar ribbon is shown clearly by the distal ends of certain ribbons in which a "wavy lateral mat" is sometimes produced at the reinitiation of microfibril synthesis (Brown et al., 1976), presumably as a result of the disruption caused by shearing in isolating bacteria. About 50 microfibrils can always be delimited in these cases. These results appear to indicate that there is a row of microfibril-synthesizing sites arranged along the bacterium, and that association of these microfibrils into a ribbon occurs close to the bacterial surface.

Further details of these presumptive synthetic sites were found by freeze-etching. Distinctive axial rows of particles were present on both P and E fracture faces of the outer envelope of the bacterium (Figs. 37 and 38). These rows of particles always lay directly in line with microfibrils, where the latter had been revealed by etching (Figs. 37 and 38). Fracture occurred so rarely at the plasma membrane that the presence or absence of a comparable row of particles at this locus could not be determined (Brown et al., 1976).

In the light of these findings, the results of Cooper and Manley (1975) are of special significance. They showed that an intact outer envelope is necessary for cellulose synthesis by *Acetobacter;* that there are separate hexose phosphate pools situated on either side of the plasma membrane; and that the enzymes necessary for the interconversion of hexose 1 phosphate and uridinediphosphoglucose (UDPG) lie in the periplasmic space. We have shown that there is a row of particles in the outer membrane, which presumably corresponds with the glycosyltransferases necessary for the final transfer of glucose moieties to the growing glucan chains. We therefore propose an adaptation of Cooper and Manley's (1975) model of cellulose synthesis in *Acetobacter* (Fig. 41). In this model, the enzymes necessary for conversion of the hexose 6 phosphate substrate into the microfibril are arranged as a unitary stack (enzyme complex) across the bacterial envelope. A row of such enzyme complexes along the bacterium, acting in synchrony, leads to the production of a coordinated product—the microfibrillar ribbon. Although no direct evidence presently exists

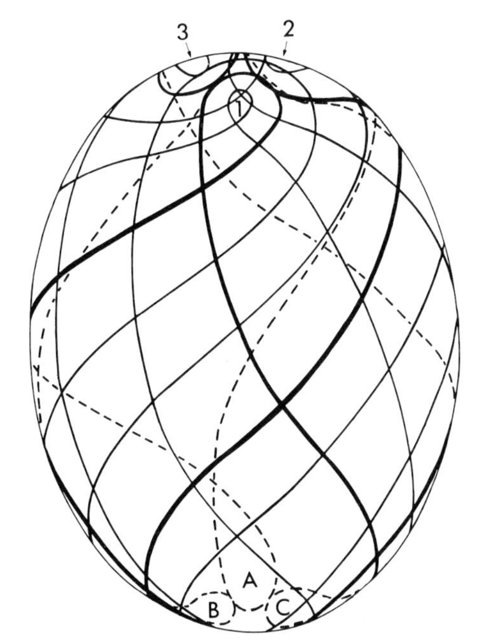

FIGURE 39 The arrangement of microfibrils in the cell wall of *Glaucocystis*. There are three microfibril "rotation centers" at each pole of the ellipsoid (labeled A, B, C, at one pole, and 1, 2, 3 at the other pole). Microfibril synthesizing centers probably follow the path A→1, B→2, C→3, A→1, etc., producing a layered wall with alternating microfibril orientations in successive wall layers.

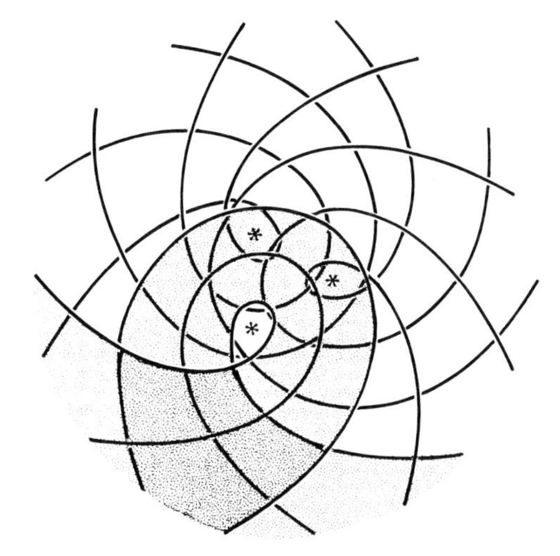

FIGURE 40 A zenithal equidistant projection of the *Glaucocystis* pole. In this "flattened out" view, the three equilaterally spaced "rotation centers" are shown. One set of microfibrils (i.e., those passing about one of the rotation centers) is shaded. This diagram should be compared with Figs. 23 and 24.

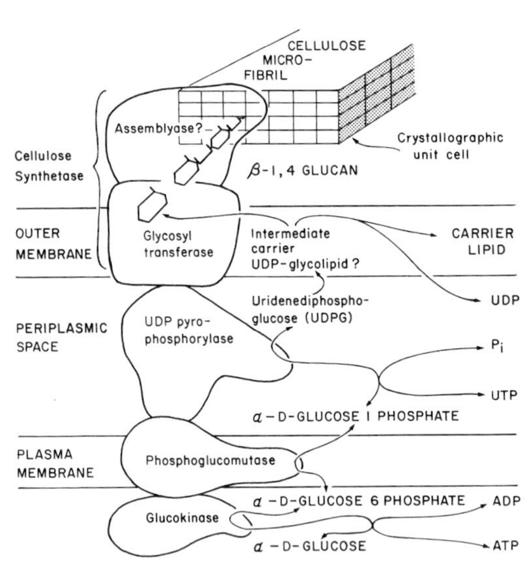

FIGURE 41 A representation of the possible location of the enzymes responsible for cellulose synthesis by *Acetobacter*. It is proposed that the enzymes may be arranged as a linked complex which spans the cell envelope, there being a row of such complexes along the bacterial axis. The "assemblyase" might correspond with the glycosyltransferase. The diagram is modified from Cooper and Manley (1975).

for the participation of a crystallization catalyst, we nevertheless propose either the necessity of such a catalyst (= assemblyase) or the duality of function of the glycosyltransferase to polymerize and crystallize the polymer.

This prokaryotic cell exhibits a major difference in cellulose synthesis from the eukaryotic systems examined. In this case, the bacterium and microfibril are the mobile entities, while the enzyme complex is stationary relative to the surface of the organism. Here, the linear element necessary for the uniform association of glucan chains, or of subelements of the completed product, is provided by the linear arrangement of the synthetic sites, rather than by the linear translocation of the synthesizing site.

CONCLUSIONS

We have at last made progress toward a better understanding of cytological aspects of cellulose synthesis. There are structures, presumably enzymes or enzyme complexes, at the growing ends of microfibrils. In eukaryotes, these are mobile, using the fluid matrix of the plasma membrane as a transport medium. They appear to be guided by linear cytoplasmic elements. The precise arrangement and nature of the synthesizing system is as variable as the crystalline fibril formed. No doubt there are more variations than those presented in this study. Once the synthetic terminal complex has started there seems no reason, at least in the case of *Glaucocystis*, for it to stop synthesizing until the microfibrillar portion of the wall is completed. Prokaryotic cellulose is somewhat different, but not as different as once believed. It, too, is bound to membranous structures (the outer envelope), and has a linear element presumably responsible in part for the linearity of the crystalline cellulose product. *Oocystis* shows us that the complexes can have the capacity to disassemble and then reassemble elsewhere to start again. Disappointingly, most of what we can guess of the synthesis of higher plant cellulose is by analogy with these less economically important organisms, but the identification of a probable synthetic structure and a fuller understanding of the essential need for a fluid matrix means that we might now be able to approach the sticky problem of in vitro synthesis of microfibrillar cellulose, nature's most abundant macromolecule.

ACKNOWLEDGMENTS

This work was supported by a grant from the National Science Foundation (GB 40937). Much of the research data presented here are based on the dissertation researches of Dwight Romanovicz, David Montezinos, and Susette Mueller. To them we are grateful for allowing presentation of their data and electron micrographs. Other thanks go to Richard Santos who played a major role in helping to coordinate our research project, to Marion Seiler for providing the drawings, and to Larry Blanton for helping copy some of the pictures. The following figures are reprinted with permission: Figs. 3–6 (*Appl. Polymer Sci. No. 28,* 1976, John Wiley & Sons, Inc., New York); Figs. 27 and 28 (*Science* (*Wash.*), 1976); and Figs. 34 and 36 (*Proc. Natl. Acad. Sci., U. S. A.,* Dec. 1976).

REFERENCES

BACHMANN, L., and W. W. SCHMITT. 1971. Improved cryofixation applicable to freeze-etching. *Proc. Natl. Acad. Sci. U. S. A.* **68:**2149–2152.

BROWN, R. M., JR., W. HERTH, W. W. FRANKE, and D. ROMANOVICZ. 1973. The role of the Golgi apparatus in the biosynthesis and secretion of cellulosic glycoprotein in *Pleurochrysis:* a model system for the synthesis of structural polysaccharides. *In* Biogenesis of Plant Cell Wall Polysaccharides. F. Loewus, editor. Academic Press, Inc., New York. 207–257.

BROWN, R. M., JR., and D. MONTEZINOS. 1976. Cellulose microfibrils: visualization of biosynthetic and orienting complexes in association with the plasma membrane. *Proc. Natl. Acad. Sci. U. S. A.* **73:**143–147.

BROWN, R. M., JR., and D. K. ROMANOVICZ. 1976. Biogenesis and structure of Golgi-derived cellulosic scales in *Pleurochrysis*. I. Role of the endomembrane system in scale assembly and exocytosis. *Appl. Polymer Symp. No.* **28:**537–585.

BROWN, R. M., JR., J. H. M. WILLISON, and C. L. RICHARDSON. 1976. Cellulose biosynthesis in *Acetobacter xylinum*. I. Visualization of the site of synthesis and direct measurement of the *in vivo* process. *Proc. Natl. Acad. Sci. U. S. A.* **73**(12)**:**4565–4569.

CHRISPEELS, M. J. 1976. Biosynthesis, intracellular transport, and secretion of extracellular macromolecules. *Annu. Rev. Plant Physiol.* **24:**19–38.

COLVIN, J. R. 1972. The structure and biosynthesis of cellulose. *C.R.C. Crit. Rev. Macromol. Sci.* **1:**47–84.

COOPER, D., and R. ST. JOHN MANLEY. 1975. Cellulose synthesis by *Acetobacter xylinum*. III. Matrix, primer, and lipid requirements and heat stability of the cellulose-forming enzymes. *Biochim. Biophys. Acta* **381:**109–119.

HEATH, I. B. 1974. A unified hypothesis for the role of membrane-bound enzyme complexes and microtubules in plant cell wall synthesis. *J. Theor. Biol.* **48:**445–449.

KIERMAYER, O., and B. DOBBERSTEIN. 1973. Membrankomplexe dictyosomaler Herkunft als "Matrizen" fur die extraplasmatische Synthese und Orientierung von Mikrofibrillen. *Protoplasma.* **77:**437–451.

MONTEZINOS, D., and R. M. BROWN, JR. 1976. Surface architecture of the plant cell: biogenesis of the cell wall, with special emphasis on the role of the plasma membrane in cellulose biosynthesis. *J. Supramol. Struct.* In press.

MUELLER, S., R. M. BROWN, JR., and T. K. SCOTT. 1976. Cellulosic microfibrils: nascent stages of synthesis in a higher plant cell. *Science (Wash. D.C.).* **194:**949–951.

NEWCOMB, E. H. 1969. Plant microtubules. *Annu. Rev. Plant Physiol.* **20:**253–288.

PRESTON, R. D. 1964. Structural and mechanical aspects of plant cell walls, with particular reference to synthesis and growth. *In* Formation of Wood in Forest Trees. M. H. Zimmerman, editor. Academic Press, Inc., New York. 169–188.

PRESTON, R. D. 1974. The Physical Biology of Plant Cell Walls. Chapman & Hall, Ltd., London. 491.

ROBINSON, D. G., and R. D. PRESTON. 1971*a*. Studies on the fine structure of *Glaucocystis nostochinearum* Itzigs. II. Membrane morphology and taxonomy. *Br. Phycol. J.* **6:**113–128.

ROBINSON, D. G., and R. D. PRESTON. 1971*b*. Studies on the fine structure of *Glaucocystis nostochinearum* Itzigs. I. Wall structure. *J. Exp. Bot.* **22:**635.

ROBINSON, D. G., and R. D. PRESTON. 1972. Plasmalemma structure in relation to microfibril biosynthesis in *Oocystis*. *Planta (Berl.).* **104:**234–246.

ROELOFSEN, P. A. 1958. Cell wall structure as related to surface growth. *Acta Bot. Neerl.* **7:**77–89.

ROELOFSEN, P. A. 1965. Ultrastructure of the wall in growing cells and its relation to the direction of growth. *Adv. Bot. Res.* **2:**69–149.

ROLAND, J. C., B. VIAN, and D. REIS. 1975. Observations with cytochemistry and ultracryotomy on the fine structure of the expanding walls in actively elongating plant cells. *J. Cell Sci.* **19:**239–259.

ROMANOVICZ, D. K., and R. M. BROWN, JR. 1976. Biogenesis and structure of Golgi-derived cellulosic scales in *Pleurochrysis*. II. Scale composition and supramolecular structure. *Appl. Polymer Symp. No.* **28:**587–610.

SCHNEPF, E., G. RODERER, and W. HERTH. 1975. The formation of the fibrils in the lorica of *Poteriochromonas stipitata*: tip growth, kinetics, site, orientation. *Planta (Berl.)* **125:**45–62.

WILLISON, J. H. M. 1976. An examination of the relationship between freeze-fractured plasmalemma and cell-wall microfibrils. *Protoplasma.* **88:**187–200.

WILLISON, J. H. M., and R. M. BROWN, JR. 1977. An examination of the developing cotton fiber: wall and plasmalemma. *Protoplasma.* In press.

WRIGHT, K., and D. H. NORTHCOTE. 1976. Identification of $\beta 1 \rightarrow 4$ glucan chains as part of a fraction of slime synthesized within the dictyosomes of maize root caps. *Protoplasma.* **88:**225–239.

THE ROLE OF THE ENDOPLASMIC RETICULUM IN THE BIOSYNTHESIS AND TRANSPORT OF MACROMOLECULES IN PLANT CELLS

MAARTEN J. CHRISPEELS

The endoplasmic reticulum, first described in plant cells in 1957 (for a review, see Buvat, 1961), is a highly differentiated membrane system consisting of an extensive network of flattened sacs or cisternae and tubules. The presence of ribosomes attached to the cytoplasmic surface of certain portions of the endoplasmic reticulum (ER) distinguishes the rough ER (RER) from the smooth ER (SER). The amount of ER in a cell varies with the type of cell, the stage of development, or the level of activity. Investigations in many different laboratories suggest that the endoplasmic reticulum plays a role in the biosynthesis of proteins, lipids, and polysaccharides, and in the formation of certain subcellular organelles, such as protein bodies, glyoxysomes, and vacuoles. The assignment of these functions to the ER is based largely on cytological studies, including autoradiography and cytochemistry at the electron microscopic level, and does not always rest on a firm analytical basis. In this review I will emphasize those functions of the ER that have been discovered by cytological studies and corroborated by analytical work involving cell fractionation and enzymology.

The Role of the Rough ER in the Biosynthesis and Transport of Extracellular Enzymes

The rough ER takes on its most characteristic form – large stacks of long cisternae studded with ribosomes – in cells actively engaged in the biosynthesis of proteins for export. This characteristic configuration is observed in both animal and plant cells. Treatments that induce the synthesis of extracellular proteins – feeding a gland of a carnivorous plant (Schnepf, 1963) or application of gibberellic acid to aleurone cells (Jones, 1969) – elicit the formation and proliferation of rough ER. However, evidence that the membrane-bound polysomes are engaged in the biosynthesis of extracellular proteins and that the ER functions in protein transport is still scant. The best-researched case involves the gibberellic acid-mediated biosynthesis and secretion of α-amylase in cereal endosperm. The aleurone cells of cereal grains synthesize and release into the starchy endosperm a number of hydrolytic enzymes when challenged with the hormone gibberellic acid (GA_3). Addition of the hormone to isolated aleurone tissue of barley greatly enhances both the synthesis and the release into the incubation medium of α-amylase and protease. The *de novo* synthesis of these enzymes and their secretion begins 8–10 h after the addition of GA_3 and is preceded and accompanied by marked ultrastructural changes: the ER proliferates and forms extensive stacks of RER, especially around the nucleus; the protein bodies swell and lose their proteinaceous matrix; and the thick cell walls of the aleurone cells start to disintegrate. The proliferation of RER is accompanied by an increased turnover of the membrane-bound phospholipids and increased activity of enzymes involved in phospholipid biosynthesis (for a review, see Varner and Ho, 1976). It appears, however, that the GA_3-induced biosynthesis and secretion of α-amylase is not accompanied by an increase in lipid biosynthesis (Firn and Kende, 1974) or dependent on ribosomal RNA synthesis (Jacobsen and Zwar, 1974). These observations suggest that the proliferation of RER may occur at the expense of preexisting lipids (membranes?) and ribosomes.

With autoradiography, Chen and Jones (1974)

MAARTEN J. CHRISPEELS Department of Biology, University of California, San Diego, La Jolla, California

showed that the RER of aleurone cells is a major site of amino acid incorporation and protein synthesis. Treatments that disrupt the normal configuration of the RER, such as actinomycin D (Virgil and Ruddat, 1973) or water stress (Armstrong and Jones, 1973), also disrupt the synthesis and/or the secretion of α-amylase, suggesting that the RER is necessary for either the synthesis or the secretion of the enzyme. Efforts to isolate α-amylase-containing ER-derived vesicles have not been entirely successful. Jones (1972) found that less than 10% of the α-amylase present in a tissue homogenate could be sedimented, and concluded that intracellular transport of the enzyme to the cell membrane is nonvesicular. Gibson and Paleg (1972, 1976), on the other hand, observed that half the intracellular α-amylase was particulate, that the enzymatic activity in the particles was latent and could be activated with Triton X-100. The particulate fraction had a density on sucrose-ficoll gradients similar to the ER and consisted of numerous vesicles ranging in size from 0.1 to 0.5 μm. These results indicate that α-amylase may be associated with the ER before it is secreted. Jones and Chen (1976) reached the same conclusion after localizing the intracellular α-amylase with fluorescent antibodies: in aleurone cells treated with GA_3 for 12 h, α-amylase was invariably located in the perinuclear region which also contained extensive stacks of RER (Fig. 1). The resolution of this technique was insufficient to conclude that the α-amylase was associated with the ER, and these preliminary results will have to be confirmed by immunocytochemistry at the ultrastructural level. It is unfortunate that all the experiments purporting to show the involvement of the ER in α-amylase secretion lack a dynamic dimension (e.g., pulse-chase). Proof that the ER is involved in α-amylase synthesis and secretion must await the demonstration that the mRNA for α-

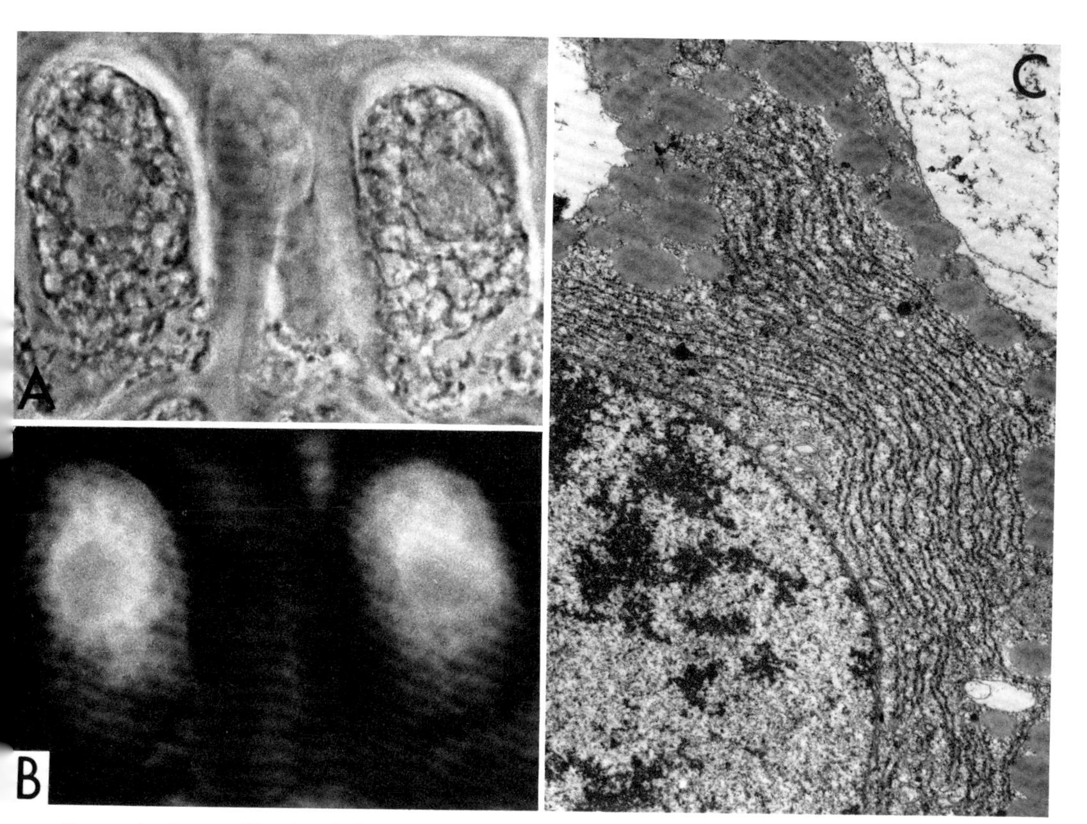

FIGURE 1 Immunohistochemical localization of α-amylase in aleurone cells. A and B are photomicrographs of sections of GA-treated aleurone tissue of barley incubated with antibodies against α-amylase and stained with rhodamine-conjugated goat-anti-rabbit IgG. (A) bright-field optics; (B) fluorescence optics. Note fluorescence in the perinuclear area. (C) an electron-micrograph of GA-treated aleurone tissue. × 13,000. Note the extensive perinuclear RER. (Jones and Chen, 1976, reprinted with permission.)

amylase is specifically associated with the RER, or that those α-amylase molecules being secreted are specifically localized in the RER cisternae.

The Role of the ER in the Biosynthesis and Transport of Polysaccharides

Plant cell walls consist of a network of cellulose microfibrils embedded in a complex matrix of pectin, hemicellulose, and glycoprotein. When tissues known to be actively engaged in cell wall biosynthesis are pulsed with [^{3}H]glucose and examined by autoradiography, silver grains are found over both the Golgi apparatus and the ER. Such experiments suggest that both organelles may function in the biosynthesis and secretion of cell wall polysaccharides. The role of the Golgi apparatus in these processes is well documented (see the article by Brown and Willison, this volume) but the role of the ER is not yet understood.

Bowles and Northcote (1972, 1974) followed the metabolic fate of [^{14}C]glucose fed to corn root tips for 45 min and determined the distribution of radioactivity in various subcellular fractions isolated by differential centrifugation and on discontinuous gradients. They observed that 90% of the radioactivity incorporated into organelle-associated macromolecules was in the microsomal fraction and only 10% in the Golgi-rich fraction. In both fractions, radioactivity was present in pectin-hemicelluloselike macromolecules, and there was no evidence that one fraction was precursor to the other. They concluded from these experiments that the ER plays a major role in the biosynthesis and transport of the pectin-hemicellulose fraction of the wall in corn root tips.

Ray et al. (1969) performed similar experiments with pea epicotyls and reached the opposite conclusion. They fractionated the subcellular organelles on isopycnic sucrose gradients and found that the metabolically active pool of cell wall precursors, and the enzyme UDP-glucose:β-1,4 glucan transferase sedimented largely with the dictyosomes and not with the ER. More recently, Ray and collaborators reexamined the possibility that the RER may be involved in cell wall polysaccharide biosynthesis and transport. When stem segments were pulsed for 10 min with [^{14}C]glucose, about half the radioactivity incorporated intracellularly was associated with the microsomal fraction. However, this radioactivity was not associated with the RER, but with a distinct class of vesicles that did not cosediment with the ER in isopycnic sucrose gradients (Fig. 2). The vesicles became labeled somewhat more slowly than did the dictyosomes. Ray and co-workers have postulated that these "microsomal" vesicles represent the dictyosome-derived secretory vesicles which transport the polysaccharides to the plasma membrane.

Additional evidence against a direct role for the ER in the biosynthesis and transport of a major portion of the cell wall polysaccharides comes from cytochemical experiments. Dictyosome cisternae and their secretory vesicles normally stain positively for carbohydrate with either the silver-hexamine or the silver-proteinate stain. Cisternae of the ER, however, do not give a positive reac-

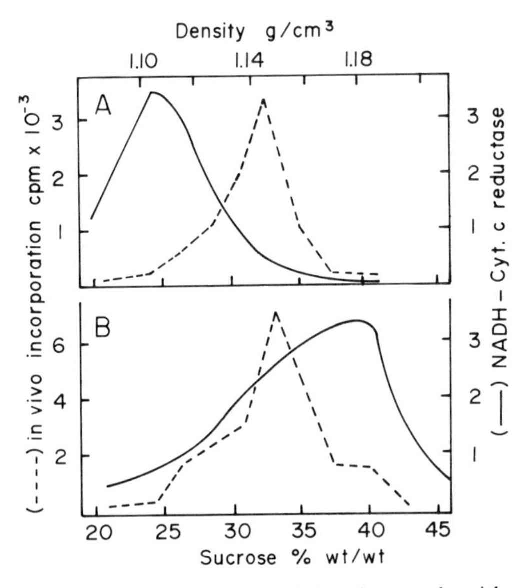

FIGURE 2 Characterization of the microsomal vesicles from pea stems which contain rapidly labeled polysaccharides. Pea stem segments were labeled with ^{14}C-glucose for 10 min, homogenized, and the microsomal fraction layered on a 20–45% (wt/wt) linear sucrose gradient. The position of the ER in these gradients—as indicated by the marker enzyme NADH-cytochrome *c* reductase, depends on the presence or absence of ribosomes. (Top) homogenization medium contains 0.1 mM $MgCl_2$, and this causes the ribosomes to dissociate; ER bands at 25% sucrose, polysaccharide-containing vesicles at 32% sucrose. (Bottom) homogenization medium contains 3 mM $MgCl_2$, and the ribosomes remain intact; ER bands at 37% sucrose and polysaccharide-containing vesicles at 32% sucrose. The experiment demonstrates that the rapidly labeled polysaccharides present in the microsomal fraction are not present in the ER. (P. M. Ray and collaborators, unpublished observations.)

tion, suggesting that they contain little polysaccharide material (for a review, see Chrispeels, 1976).

Ultrastructural studies of cells known to be engaged in the deposition of a secondary cell wall often show a characteristic distribution pattern of ER cisternae near the site of cell wall deposition. It is unlikely that the ER is directly involved in the biosynthesis of cellulose, as there is considerable circumstantial evidence which suggests that cellulose synthesis takes place at the plasma membrane (see, for example, Bowles and Northcote, 1972; and for a review, see Chrispeels, 1976). The observations are entirely consistent with an indirect role of the ER in cellulose synthesis. Such an indirect role could involve the supply of precursors, primers, enzyme complexes, or templates for the extracellular synthesis of cellulose and the assembly of cellulose microfibrils.

Role of the ER in the Biosynthesis and Transport of Lipids

Plant cells synthesize and accumulate a variety of specific lipids. For example, epidermal cells and cork cells, respectively, synthesize and secrete cutin and suberin into the cell wall. Oil glands synthesize and release terpenes and other volatile lipophilic substances, whereas the storage cells of many seeds synthesize triglycerides, which accumulate in oil droplets or oleosomes. Ultrastructural studies clearly show that the cells of glands which secrete lipophilic substances have a very extensive tubular smooth ER network (Schnepf, 1969). It is rarely possible, however, to detect the precursors of the lipophilic secretions within the ER tubules. Dumas (1973) made a study of the lipophilic stigma exudate of *Forsythia* and observed that substances which stained in the same way as the exudate were also present in the ER and the vacuole. However, the direction of transport and the site of biosynthesis could not be deduced from his experiments.

Whether the ER plays a role in the formation of oil droplets or oleosomes is not clear. The enzymes involved in triglyceride biosynthesis in developing oil-rich seeds have been characterized extensively, and 80% of the synthetic activity is found in association with the oleosome fraction. A detailed ultrastructural study on the origin of oleosomes in the developing seeds of *Crambe abyssinica* (Smith, 1974) showed that oil droplets arise from masses of electron-dense particles, rather than from ER cisternae, as suggested by Matile (1975). It is possible that these masses of dense particles contain the enzymes which synthesize triglycerides.

The Role of the ER in the Biosynthesis of Reserve Proteins and the Formation of Protein Bodies

Seed formation is accompanied by the biosynthesis and the accumulation of reserve proteins in the storage tissues of the seeds. Many of these proteins (e.g., zein in corn endosperm, vicilin and legumin in legume cotyledons) are localized in specialized organelles called protein bodies or aleurone grains. Protein bodies, which measure 2–10 μm in diameter, consist of a dense, amorphous, protein matrix surrounded by a limiting membrane. In some species, protein bodies also contain a large protein crystalloid and one or more phytin globoids. The cells of the storage tissues of developing seeds contain an extensive RER network which is involved in the biosynthesis of the storage proteins. Evidence to support this conclusion comes from cytological, autoradiographic, and biochemical studies.

The biosynthesis and deposition of reserve proteins has been studied most extensively in the protein-rich seeds of leguminous plants and in corn. In legumes, seed development consists of a cell-division phase, during which all the cells of the axis and the cotyledons are formed, followed by a cell-expansion phase, during which the proteins and other reserves are synthesized and deposited in the cotyledons (for a review on reserve protein synthesis and protein body formation, see Millerd, 1975). At first, reserve protein accumulates within the central vacuole, which simultaneously fragments, giving rise to a number of large, irregularly shaped protein bodies. Later on, smaller protein bodies arise *de novo*, possibly as distentions of the ER or of the Golgi-ER-lysosome complex (GERL). The cells of the cotyledons contain numerous stacks of RER cisternae, and the involvement of this RER in the biosynthesis of reserve proteins was first demonstrated by Bailey et al. (1970). They incubated cotyledon slices of developing *Vicia faba* beans with radioactive amino acids, and demonstrated by means of autoradiography that the RER was the major site of amino acid incorporation and that the radioactive proteins accumulated in the protein bodies. The pathway of intracellular transport from the RER to the protein bodies has not yet been elucidated. Elec-

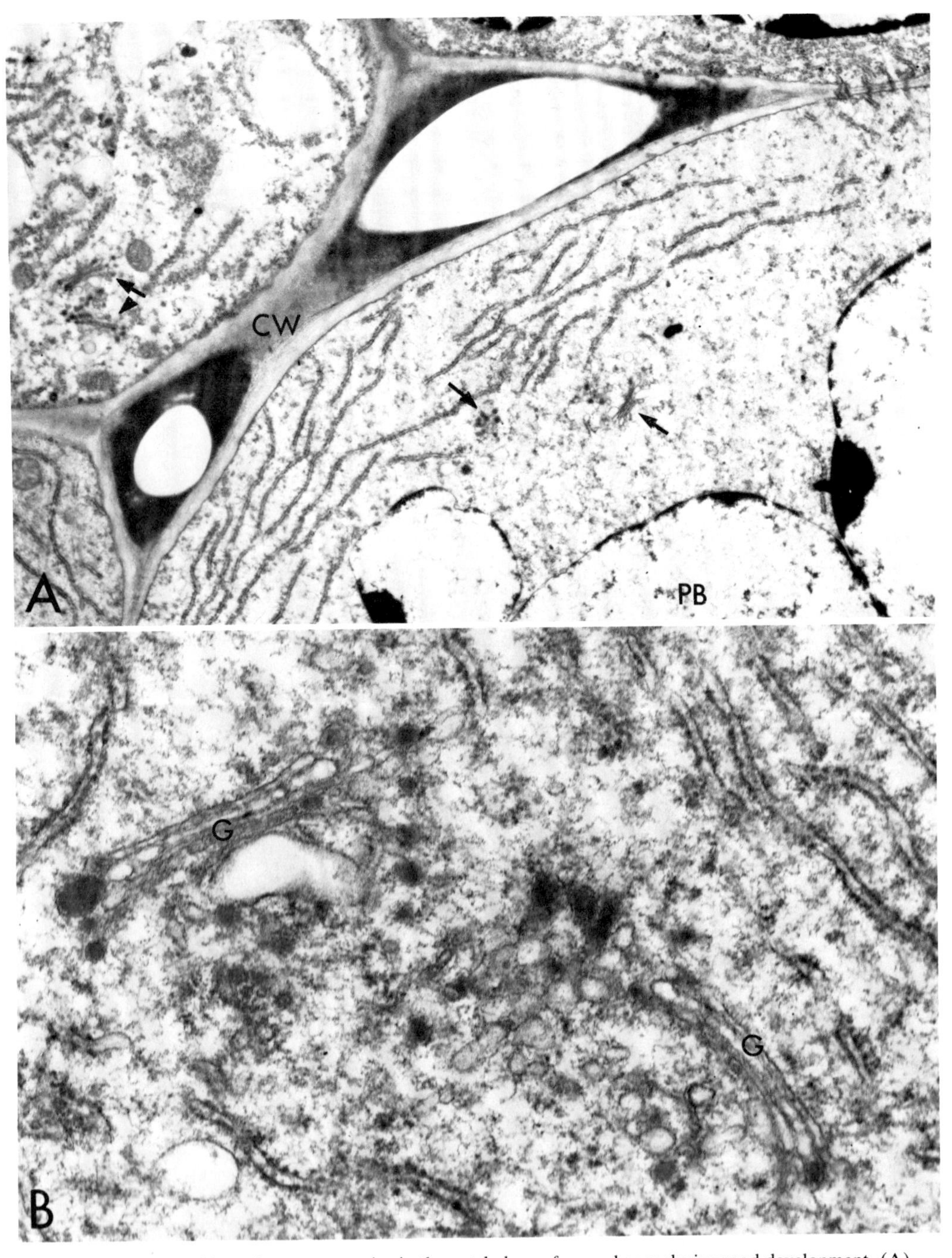

FIGURE 3 Deposition of reserve proteins in the cotyledons of mung beans during seed development. (A) low magnification view (× 9,400) showing portions of several cells separated by cell walls (*CW*). Note the presence of many RER cisternae as well as several dictyosomes (arrows). Fixation caused the reserve proteins to precipitate on the protein body (*PB*) membrane. (B) high magnification view (× 44,000) of an area of the cytoplasm showing RER cisternae and an active Golgi zone (*G*) with two dictyosomes. Golgi vesicles contain electron-dense material (reserve protein?). (Courtesy of Nick Harris, Department of Botany, University of Durham, Durham, England.)

tron-dense deposits are sometimes found within the RER cisternae, and often within the secretory vesicles of the dictyosomes (Fig. 3). The reserve proteins of the *Leguminosae* are glycoproteins which contain mannose and glucosamine. Identification of the organelles which contain the glycosyl transferases involved in the posttranslational modifications may help elucidate the transport pathway.

Larkins and Dalby (1975) demonstrated that the mRNA for zein—the major reserve protein in corn endosperm—is specifically associated with the RER, and that the RER is the major site of zein synthesis. They isolated both free- and RER-bound polysomes from the endosperm of developing corn kernels and characterized the proteins synthesized in vitro by these polysomes. The proteins were characterized on the basis of their solubility—authentic zein is soluble in 70% ethanol; on the basis of the incorporation of specific amino acids—zein contains very little lysine; and by sodium dodecyl sulfate (SDS)-acrylamide gel electrophoresis. The results (Fig. 4) suggest that zein is made only or largely by membrane-bound polysomes. Similar experiments were recently done by Burr and Burr (1976), who showed that ribosomes are attached to the outside surface of the limiting membrane of the protein bodies. Polysomes obtained from isolated protein bodies are capable of synthesizing a protein which has many of the characteristics of zein. The major difference between these two sets of experiments concerns the site of zein synthesis—the RER or the protein body membrane—and may be largely semantic. Indeed, it is far from clear when a RER cisterna becomes a protein body or when a developing protein body ceases to be a RER cisterna.

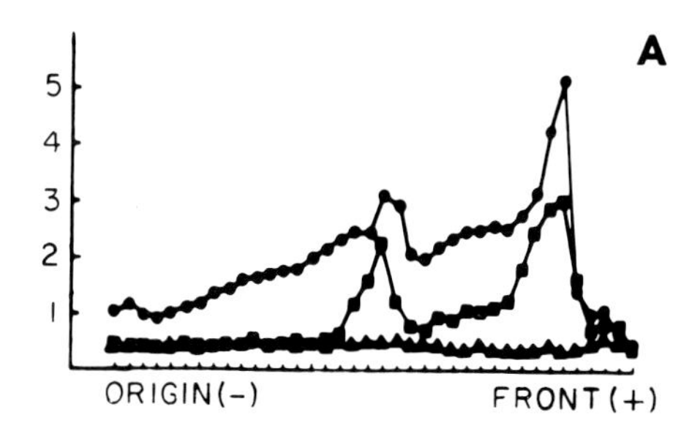

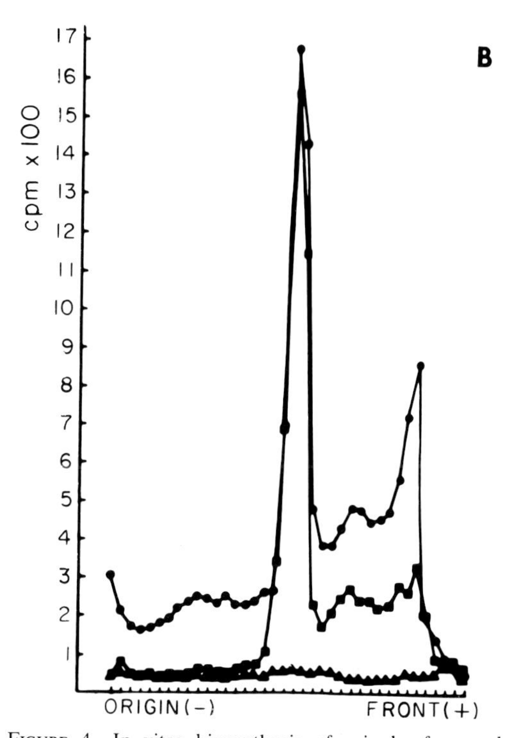

FIGURE 4 In vitro biosynthesis of zein by free and membrane-bound polyribosomes obtained from developing maize kernels. SDS-acrylamide gel electrophoresis of [^{14}C] leucine labeled proteins. (A) hot-acid insoluble (●) and hot 70% ethanol soluble (■) protein synthesized by the free polyribosomes. (B) same for polyribosomes derived from RER. The large peak in B and the small peak in A correspond to the position of authentic zein. The two polypeptides of zein (mol wt, 22,800 and 19,000) are not separated on these gels. (Larkins and Dalby, 1975, reprinted with permission.)

The Role of the ER in the Biosynthesis and Transport of Enzymes which Accumulate in Cytoplasmic Organelles

FORMATION OF GLYOXYSOMES. Seedling growth in castor beans is accompanied by the breakdown of the large amount of fat stored in the endosperm. The onset of fat breakdown is accompanied by the biogenesis of glyoxysomes, special organelles involved in triglyceride metabolism, and the *de novo* synthesis of the enzymes necessary to carry out this metabolic process. Glyoxysomes consist of an amorphous protein matrix surrounded by a limiting membrane. The phospholipid components of these membranes have been shown to be synthesized on the membranes of the ER, suggesting that the limiting membranes of the glyoxysomes may be derived from the ER (Kagawa et al., 1973). Gonzales and Beevers (1976) recently obtained evidence that the enzymes of castor bean glyoxysomes may also originate in the ER. They fractionated endosperm extracts on isopycnic sucrose gradients and observed that malate synthetase – a marker enzyme for the glyoxysomes – sedimented with the ER (density 1.12 g/cm³) as well as with the glyoxysomes (density 1.24 g/cm³). After 2 days of growth, when malate synthetase synthesis had just begun, half of the enzyme activity was in the ER fraction and half was in the glyoxysomes. This ratio shifted in favor of the glyoxysomes as growth proceeded and enzyme accumulated in the tissue. By the fourth day of growth, more than 90% of the enzyme activity was in the glyoxysomes (Fig. 5). A similar pattern was observed for citrate synthetase, except that this enzyme was also present in the mitochondria at all times. Although there are no kinetic data showing enzyme movement from one organelle to the other, the results strongly support the conclusion that glyoxysomal enzymes are synthesized on the ER.

METABOLISM OF RESERVE PROTEINS CONTAINED IN PROTEIN BODIES. The growth of a leguminous seedling is accompanied by the metabolism of the reserve proteins contained in the protein bodies of the cotyledons. Although protein bodies already contain certain proteolytic enzymes (e.g., carboxypeptidase), reserve protein metabolism depends on the appearance in the tissue of other proteases (Chrispeels and Boulter, 1975). In mung bean seedlings, growth is accompanied by the biosynthesis of an endopeptidase which accumulates in the protein bodies, where it causes the hydrolysis of the storage proteins (Chrispeels et al., 1976). Before enzyme synthesis, long RER cisternae proliferate rapidly. Whether this proliferation is the result of the synthesis of more ER during the first 3 days of growth or of a reorganization of existing membranes remains to be determined. The activity of the ER marker enzyme NADH-cytochrome *c* reductase increases more than tenfold in the cotyledons concomitant with the proliferation of the ER cisternae (Chrispeels, unpublished data). Ultrastructural investigations show that vesicles which may originate from the ER appear to merge with the limiting membrane of the protein body before reserve protein digestion. We have postulated that these vesicles could be the vehicles that carry the

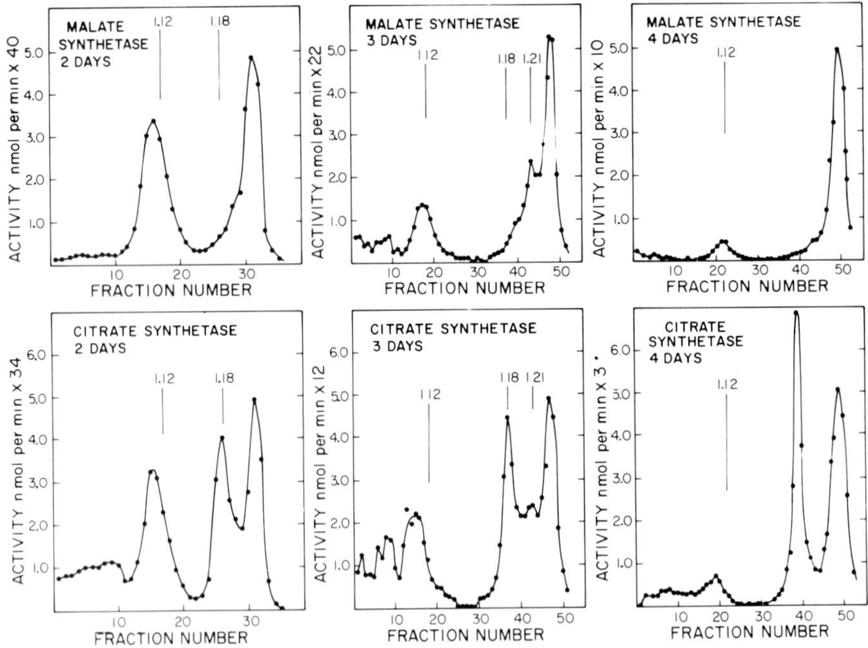

FIGURE 5 Fractionation of organelles containing malate synthetase and citrate synthetase of castor bean endosperm. Castor beans were germinated for 2, 3, and 4 days and the endosperm homogenate layered on a 20–48% (wt/wt) sucrose gradient. The ER banded at 1.12 g/cm³ (marker enzyme NADPH-cytochrome *c* reductase), the mitochondria banded at 1.19 g/cm³ (marker enzymes malate dehydrogenase and citrate synthetase), and the glyoxysomes at 1.24 g/cm³ (marker enzyme catalase), whereas the glyoxysomal ghosts had a density of 1.21 g/cm³. (Gonzalez and Beevers, 1976, reprinted with permission.)

endopeptidase from its site of synthesis—the RER—to its site of action—the protein bodies (Chrispeels et al., 1976).

FORMATION OF LYSOSOMES AND VACUOLES. Plant cells contain lysosomelike organelles consisting of a limiting membrane and an electron-translucent matrix with acid hydrolase activity. These organelles, variously called lysosomes, provacuoles, vacuoles, or spherosomes (the term "spherosome" has also been used to describe lipid droplets or oleosomes), are an integral component of the lytic compartment of the plant cell. Several investigators have suggested, on the basis of morphological evidence, that vacuoles originate as local distentions of the ER (for a recent summary of this evidence, see Matile, 1975). However, Marty (1973*a, b*) recently reexamined the ontogeny of the plant vacuole with the high voltage electron microscope and concluded that the ER is only indirectly involved in vacuole formation. He observed that GERL-derived vesicles elongate and then fuse to give rise to acid hydrolase-rich autophagic provacuoles. The provacuoles then become vacuoles. These observations suggest that the ontogeny of the vacuole may be more complex than previously postulated, and indicate once again that morphological observations alone are not sufficient to elucidate the role of the ER.

ACKNOWLEDGMENTS

Supported by a contract from the U. S. Energy Research and Development Agency (ERDA) and a grant from the National Science Foundation of the United States of America.

REFERENCES

ARMSTRONG, J. E., and R. L. JONES. 1973. Osmotic regulation of α-amylase synthesis and polyribosome formation in aleurone cells of barley. *J. Cell Biol.* **59:**444–455.

BAILEY, C. J., A. COBB, and D. BOULTER. 1970. A cotyledon slice system for the electron autoradiographic study of the synthesis and intracellular transport of the seed storage protein of *Vicia faba*. *Planta (Berl.)*. **95:**103–118.

BOWLES, D. J., and D. H. NORTHCOTE. 1972. The site of synthesis and transport of extracellular polysaccharides in the root tissues of maize. *Biochem. J.* **130:**1133–1145.

BOWLES, D. J., and D. H. NORTHCOTE. 1974. The amount and rates of export of polysaccharides found within the membrane system of maize root cells. *Biochem. J.* **142:**139–144.

BURR, B., and F. A. BURR. 1976. Zein synthesis in maize endosperm by polyribosomes attached to protein bodies. *Proc. Natl. Acad. Sci. U. S. A.* **73:**515–519.

BUVAT, R. 1961. Le réticulum endoplasmique des cellules végétales. *Ber. Dtsch. Bot. Ges.* **74:**261–267.

CHEN, R., and R. L. JONES. 1974. Studies on the release of barley aleurone cell proteins: autoradiography. *Planta (Berl.)*. **119:**207–220.

CHRISPEELS, M. J. 1976. Biosynthesis, intracellular transport, and secretion of extracellular macromolecules. *Annu. Rev. Plant Physiol.* **27:**19–38.

CHRISPEELS, M. J., B. BAUMGARTNER, and N. HARRIS. 1976. The regulation of reserve protein metabolism in the cotyledons of mung bean seedlings. *Proc. Natl. Acad. Sci. U. S. A.* **73:**3168–3172.

CHRISPEELS, M. J., and D. BOULTER. 1975. Control of storage protein metabolism in the cotyledons of germinating mung beans: role of endopeptidase. *Plant Physiol.* **55:**1031–1037.

DUMAS, C. 1973. Contribution à l'étude cyto-physiologique du stigmate VII. *Botaniste.* **56:**59–80.

FIRN, R. D., and H. KENDE. 1974. Some effects of applied gibberellic acid on the synthesis and degradation of lipids in isolated barley aleurone layers. *Plant Physiol.* **54:**911–915.

GIBSON, R. A., and L. G. PALEG. 1972. Lysosomal nature of hormonally induced enzymes in wheat aleurone cells. *Biochem. J.* **128:**367–375.

GIBSON, R. A., and L. G. PALEG. 1976. Purification of GA_3-induced lysosomes from wheat aleurone cells. *J. Cell Sci.* In press.

GONZALEZ, E., and H. BEEVERS. 1976. Role of the endoplasmic reticulum in glyoxysome formation in castor bean endosperm. *Plant Physiol.* **57:**406–409.

JACOBSEN, J. V., and J. A. ZWAR. 1974. Gibberellic acid and RNA synthesis in barley aleurone layers: metabolism of rRNA and tRNA and of RNA containing polyadenylic acid sequences. *Aust. J. Plant Physiol.* **1:**343–356.

JONES, R. L. 1969. Gibberellic acid and the fine structure of barley aleurone cells. I. Changes during the lag-phase of α-amylase synthesis. *Planta (Berl.)*. **87:**119–133.

JONES, R. L. 1972. Fractionation of the enzymes of the barley aleurone layer: evidence for a soluble mode of enzyme release. *Planta (Berl.)*. **103:**95–109.

JONES, R. L., and R. CHEN. 1976. Immunohistochemical localization of α-amylase in barley aleurone cells. *J. Cell Sci.* **20:**183–198.

KAGAWA, T., J. M. LORD, and H. BEEVERS. 1973. The origin and turnover of organelle membranes in castor bean endosperm. *Plant Physiol.* **51:**61–65.

LARKINS, B. A., and A. DALBY. 1975. *In vitro* synthesis of zein-like protein by maize polyribosomes. *Biochem. Biophys. Res. Comm.* **66:**1048–1054.

MARTY, F. 1973*a*. Mise en evidence d'un appareil provacuolaire et de son rôle dans l'autophagie cellulaire et l'origine des vacuoles. *C. R. Acad. Sci. (Paris).* **276:**1549–1552.

MARTY, F. 1973*b*. Dissemblance des faces golgiennes et activité des dictyosomes dans les cellules en cours de vacuolisation de la racine d'*Euphorbia characias* L. *C. R. Acad. Sci. (Paris).* **277:**1749–1752.

MATILE, PH. 1975. The Lytic Compartment of Plant Cells. Springer-Verlag New York Inc., New York. 1–183.

MILLERD, A. 1975. Biochemistry of legume seed proteins. *Annu. Rev. Plant Physiol.* **26:**53–72.

RAY, P. M., T. L. SHININGER, and M. M. RAY. 1969. Isolation of β-glucan synthetase particles from plant cells and identification with Golgi membranes. *Proc. Natl. Acad. Sci. U. S. A.* **64:**605–612.

SCHNEPF, E. 1963. Zur cytologie und physiologie pflanzlicher Drüsen 3. Teil. Cytologische veranderungen in den Drüsen von *Drosophyllum* während der Verdauung. *Planta (Berl.).* **59:**351–379.

SCHNEPF, E. 1969. Über den Feinbau von Öldrüsen. I. Die Drüsenhaare von *Arctium lappa. Protoplasma.* **67:**185–194.

SMITH, C. G. 1974. The ultrastructural development of spherosomes and oil bodies in the developing embryo of *Crambe abyssinica. Planta (Berl.).* **119:**125–142.

VARNER, J. E., and D. T. HO. 1976. The Role of Hormones in the Integration of Seedling Growth. Thirty-fourth Symp. Soc. Devel. Biol., Academic Press. In press.

VIGIL, E. L., and M. RUDDAT. 1973. Effect of gibberellic acid and actinomycin D on the formation and distribution of rough endoplasmic reticulum in barley aleurone cells. *Plant Physiol.* **51:**549–558.

MEMBRANE DIFFERENTIATION AND THE CONTROL OF SECRETION: A COMPARISON OF PLANT AND ANIMAL GOLGI APPARATUS

D. JAMES MORRÉ

Activities of the Golgi apparatus must be regulated to account for the many coordinated activities ascribed to this cell component during growth and secretion. Secretory vesicles are produced, loaded with secretory products, and released into the cytoplasm either as condensing vacuoles or as mature secretory vesicles. Ultimately, these vesicles migrate to and fuse with specific portions of the plasma membrane of the cell surface in a vectorial flow process. The products secreted via the contents of these vesicles are as diverse as the variety of cells that contain Golgi apparatus. A constant number of cisternae per dictyosomal stack is maintained during steady-state secretion, but the number and size of the cisternae are subject to modification as secretory activities are accelerated or diminished. Ordered progressions in membrane thickness and staining intensity are observed across the stacked cisternae, which show progressive differentiation of membranes. Connections and associations with other cell components are formed and broken. Clearly, Golgi apparatus are dynamic structures in communication with their environment and with other endomembrane components with which they function in concert. The important questions which remain concern not so much the extent to which the Golgi apparatus are regulated, but the mechanisms by which they are regulated.

Most chemical reactions of the cell, such as synthesis, catabolism, polymerization, etc., are catalyzed by specific enzymes whose origins trace ultimately to DNA. Thus, it is not surprising that present knowledge of regulation in the cell is almost entirely derived from experiments on regulation of the activity and synthesis of enzymes, especially enzymes of the cytosol. Much less is known about the regulation of membrane assembly, modification, and flux, or what factors interact to insure the coordinated operation of membrane-associated multienzyme systems (Morré, 1975).

In principle, Golgi apparatus function can be regulated at many different levels (Fig. 1). Among these are the following: 1) the genetic potential of the cell and the expression of this potential (transcriptional and translational control); 2) controlled assembly mechanisms through which the familiar form of the Golgi apparatus is achieved, maintained, and modified; and 3) modulatory controls in which availability of precursors and energy balance (input) interact with utilization pathways (output) via mechanisms mediated through as-yet-unknown detector systems and reference inputs. The latter presumably would be responsive to both homeostatic controls and environmental effectors.

In this report, information on the structure, composition, and functional properties of plant and animal Golgi apparatus will be compared with a view toward understanding what types of control might operate in the cell to insure the orderly operation of the Golgi apparatus in membrane biogenesis and differentiation and in secretion.

Structural Studies

Mollenhauer et al. (1967) first drew attention to the similarities of form among plant and animal Golgi apparatus. Both consist of dictyosome subunits (i.e., the stacks of Golgi apparatus cisternae and associated secretory vesicles; Fig. 2). The

---

D. JAMES MORRÉ Department of Medicinal Chemistry and Pharmacognosy and the Department of Biological Sciences, Purdue University, West Lafayette, Indiana

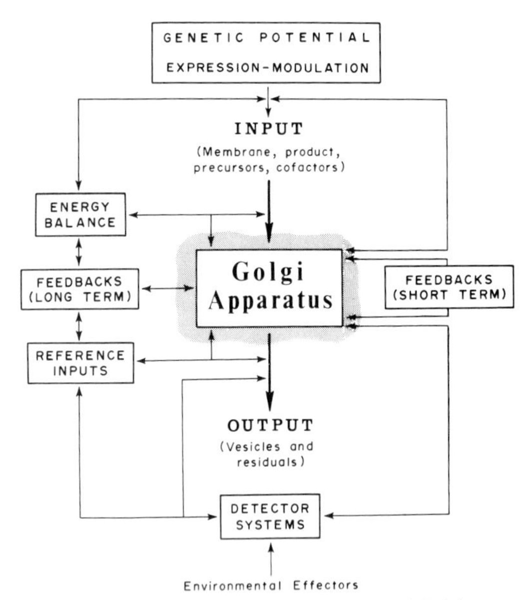

FIGURE 1 Diagrammatic representation of Golgi apparatus regulation. The shaded portion represents the specialized zone of cytoplasm ("zone of exclusion") which surround all Golgi apparatus and is considered in this report to provide the milieu for a controlled assembly mechanism through which Golgi apparatus form is achieved, maintained, and modified.

number of cisternae per stack may vary from 1 (certain fungi) to more than 20 (*Euglena*, chrysophycean algae; Brown, this volume) but is usually about 5. More important, the form of the cisternae is similar from one organism to another. Cisternae of the Golgi apparatus are flattened and consist of a central platelike region or "saccule," which is continuous with a system of tubules and secretory vesicles. The platelike regions are typically 0.5–1 μm in diameter with fenestrated margins (Fig. 2). Tubules, about 50 nm in diameter, emanate from the complex anastomotic margins and may extend several microns from the edge of the saccule. The tubules serve to interconnect adjacent dictyosomes, as attachment sites for secretory vesicles, and as regions of association with endoplasmic reticulum (Morré et al., 1971*b*). The interconnected system of plates, tubules, and vesicles allows for considerable subcompartmentation and restriction of functional activities to specific regions even within a single cisterna.

Each dictyosome is surrounded in the cytoplasm by a specialized cytoplasmic region from which glycogen, most ribosomes, and organelles such as mitochondria are excluded ("zone of exclusion," Morré et al., 1971*b*; Fig. 3*b*, *d*). Endoplasmic reticulum entering the Golgi apparatus zone of exclusion is always smooth-surfaced (lacking ribosomes), and coated membranes and vesicles of the Golgi apparatus region are restricted to this zone (Morré et al., 1971*b*; Figs. 2*a* and 3*b*). The zone of exclusion is continuous with the intercisternal region, which in plants contains rodlike elements or fibers 15 nm in diameter, the intercisternal elements (Mollenhauer and Morré, 1966; Fig. 3*a*).

Each dictyosome is a polarized structure with clearly defined faces. One, the forming face, is usually proximal to endoplasmic reticulum or nuclear envelope and is characterized by the presence of transition vesicles. The other, the maturing face, is at the opposite pole and is characterized by mature secretory vesicles. This observation has led to a dynamic concept of Golgi apparatus function, in which small transition vesicles are thought to bleb off the endoplasmic reticulum and migrate to the Golgi apparatus region where they coalesce to form new Golgi apparatus cisternae. To maintain a constant number of cisternae per stack, a compensating loss of cisternae from the opposite or mature face is accomplished by the formation and release of secretory vesicles (Beams and Kessel, 1968; Cook, 1973; Flickinger, 1975; Franke et al., 1971; Michaels and Leblond, 1976; Mollenhauer and Morré, 1966; Morré and Mollenhauer, 1974; Morré et al., 1971*b*, 1974; Novikoff et al., 1962; Schnepf, 1969). The release of mature cisternae as secretory vesicles has been observed directly with the light microscope in certain favorable unicellular organisms (Brown, 1969; Schnepf, 1969).

Dictyosome polarity is also expressed in terms of membrane differentiation (Grove et al., 1968; Morré et al., 1971*a*,*b*). This differentiation is usually observed as a progressive increase in membrane thickness (Table I) and a change in staining intensity or other properties. Golgi apparatus membranes at the forming face resemble endoplasmic reticulum, and at the maturing face resemble plasma membrane. A decrease in luminal width of the cisternae (Fig. 3*a*) is a morphological aspect of dictyosome polarity which is conspicuous only with plant cells (cf. Fig. 3*c*).

Thus, except for the presence of intercisternal elements and the marked alteration of thickness of the luminal space, plant and animal Golgi apparatus are morphologically similar (Table II) yet uniquely distinct from all other intracellular membranes. This constancy of Golgi apparatus struc-

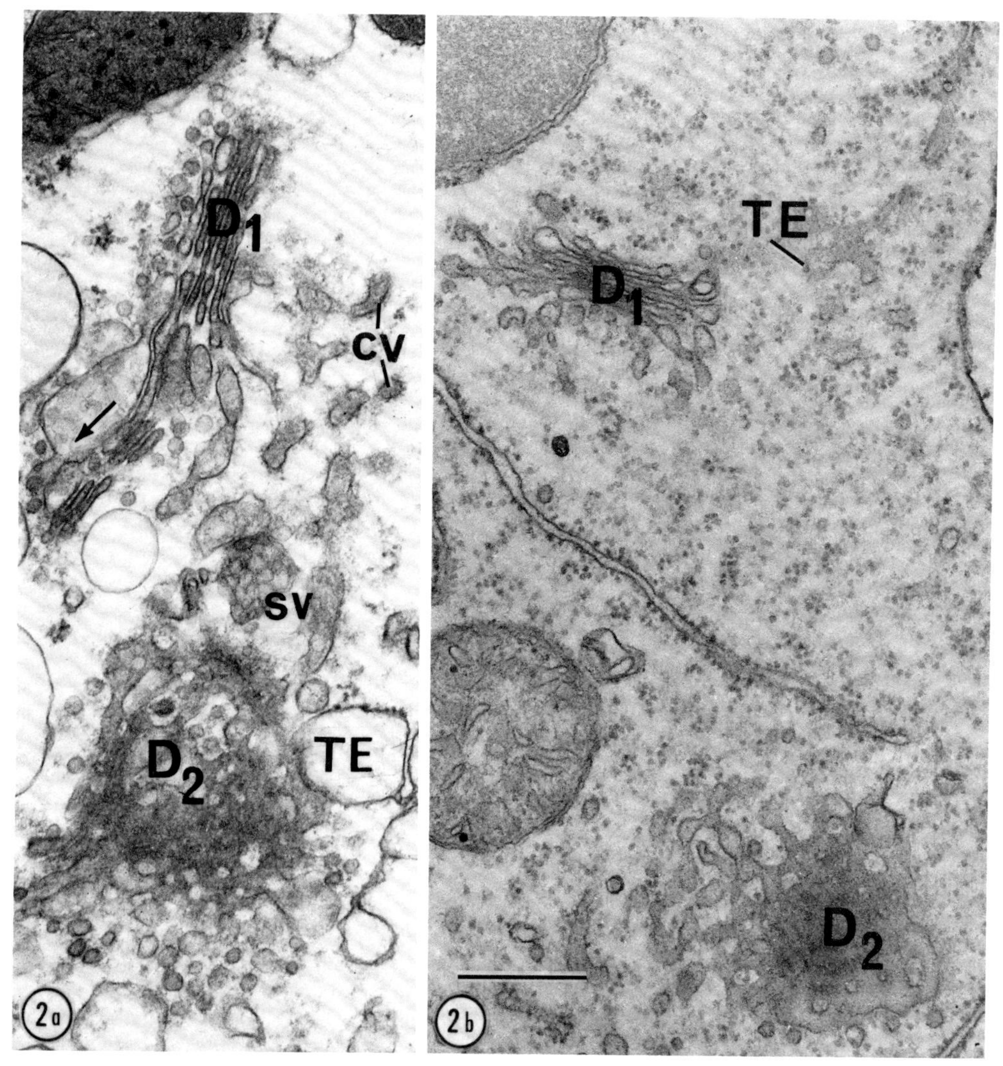

FIGURE 2 Comparison of the structure of the animal and plant Golgi apparatus. (*a*) Rat liver. (*b*) Onion (*Allium cepa*) stem. Illustrated in each electron micrograph are two cisternal stacks, the dictyosomes (*D*). D_1 is sectioned tangentially to show the familiar cross-sectioned aspects of the stacked cisternae. D_2 is a view orthogonal to D_1 and shows cisternae from the mid-region of the stack in face view. The cisternae consist of a platelike central region (the "saccule") with a fenestrated margin continuous with a system of peripheral tubules. *TE* = transition element consisting of part rough, part smooth elements of the endoplasmic reticulum adjacent to the dictyosomes; *cv* = coated vesicles; *sv* = mature secretory vesicle. The single arrow in *a* marks a connection between the boulevard périphérique (smooth endoplasmic reticulum tubules carrying lipoprotein particles) and an immature secretory vesicle. Glutaraldehyde-osmium tetroxide fixation. × 35,000. Bar = 0.5 μm.

ture contrasts markedly with the wide variety of secretory products known to be packaged for export by plant and animal Golgi apparatus. The products vary from cell to cell and within a single cell, depending on the stage of development. How is such a remarkable constancy of structure achieved within such a wide range of functional diversity among genetically different organisms? Does the underlying basis rest in the composition of the membrane? Are the chemical makeups of membranes of plant and animal Golgi apparatus, as examples of extremes, similar or different?

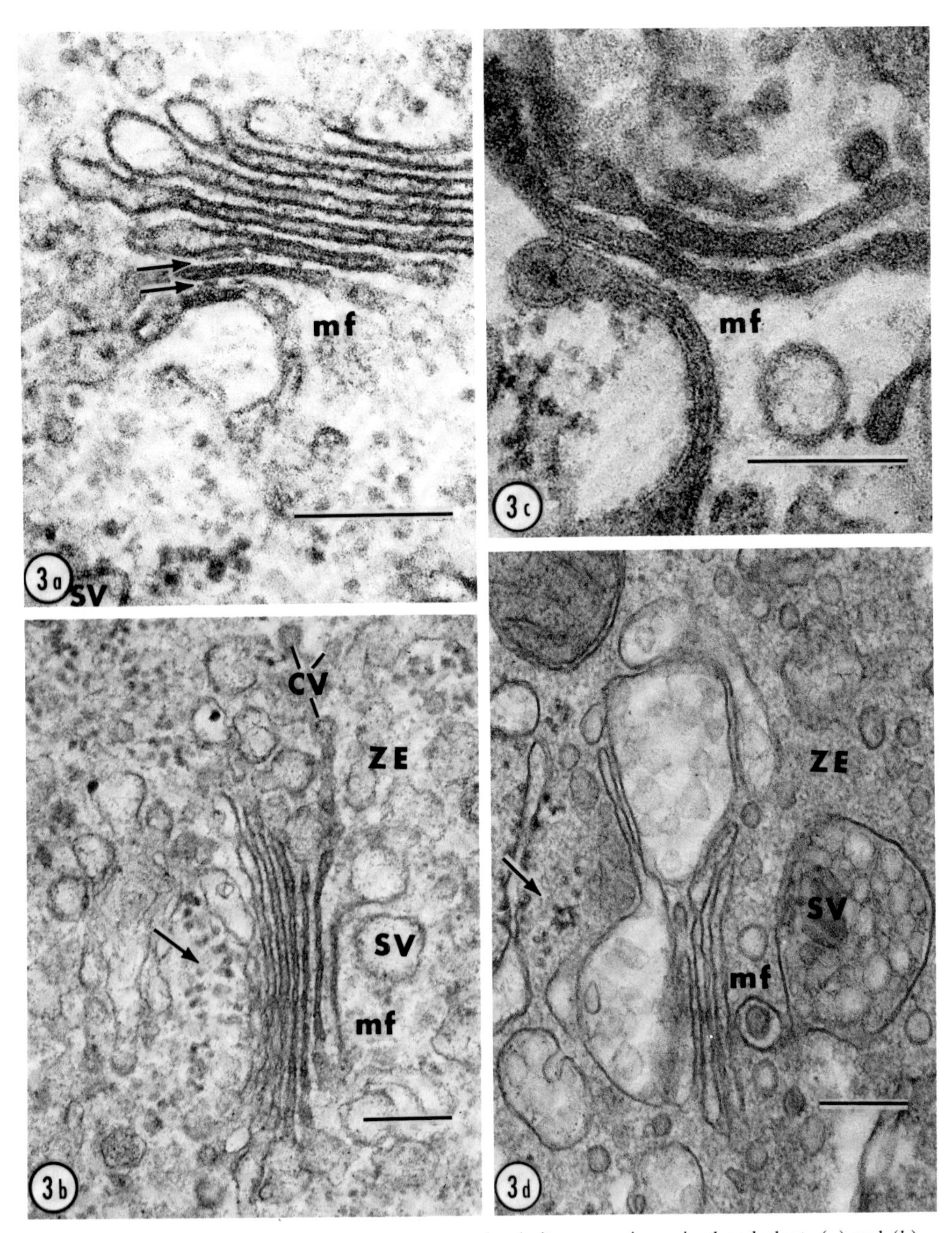

FIGURE 3 Details of Golgi apparatus structure and polarity comparing animal and plant. (*a*) and (*b*) Soybean (*Glycine max*) hypocotyl. (*c*) Rat liver. (*d*) Rat hepatoma. In *a* and *c* the dictyosomes are oriented with the maturing face (*mf*) at the bottom. Polarity is exhibited both in the thickness (Table I) and staining intensity of the membrane and in the thickness of the luminal spaces of the cisternae. The latter characteristic is most marked for plants (*a* and *b*). Intercisternal spaces are sometimes occupied by fibrous elements (intercisternal elements) with plant (*a*) but not animal (*c*) dictyosomes (arrows in *a*). Details of the zone of exclusion (*ZE*) are illustrated in *b* and *d*. Coated vesicles (*cv*) of the Golgi apparatus region are restricted to this zone. Endoplasmic reticulum membranes entering the zone are smooth. A few to several free polyribosomes occupy the zone of exclusion frequently at or near the forming face (arrows). Mature secretory vesicles (*sv*) are found near the maturing face (*mf*) of the dictyosomes. Glutaraldehyde-osmium tetroxide fixation. *a* and *c*, × 114,000. *b* and *d*, × 65,000. Bar = 0.2 μm.

TABLE I

*Membrane Differentiation in Golgi Apparatus of Animals and Plants**

Membrane type	Membrane thickness (nm): Rat liver	Rat mammary gland	Onion stem	Soybean hypocotyl
Nuclear envelope	65	60	56	56
Endoplasmic reticulum	65		53	56
Golgi apparatus:				
Cisterna 1	65	70	53	56
Cisterna 2	68		60	38
Cisterna 3	72		65	61
Cisterna 4/5	80		75	69
Secretory vesicle	83	85	88	78
Plasma membrane	85	97	93	88

* Determined from measurements of photographically enlarged electron micrographs of glutaraldehyde-osmium tetroxide fixed materials. Includes results from studies with T. W. Keenan and F. M. Twohig.

TABLE II

Summary of Structural Similarities and Differences Comparing Plant and Animal Golgi Apparatus

Characteristic	Animal	Plant
Dictyosome subunits	+	+
Stacked cisternae	+	+
Av. 5 cisternae/stack	+	+
Polarity	+	+
Membrane differentiation	+	+
Product transformations	+	+
Change in luminal thickness	±	+
Flattened cisternae with fenestrated margins continuous with tubules	+	+
Central plate (0.5–1 μm diameter)	+	+
Peripheral tubules (ca. 50 nm diameter)	+	+
Secretory vesicles	+	+
Transition vesicles	+	+
"Boulevard périphérique"	+	+
Coated vesicles and membranes	+	+
Intercisternal region; "zone of exclusion"	+	+
Golgi apparatus-associated polysomes	+	+
Intercisternal elements	–	+

Biochemical Studies

Information on the biochemical nature of endomembranes has come from studies in which highly purified fractions of Golgi apparatus from rat liver or mammary gland have been compared with endoplasmic reticulum and plasma membrane fractions isolated in parallel (e.g., Keenan and Morré, 1970). Although fraction purity is not as good with plant sources, comparisons are of interest in view of the close morphological similarities of plant and animal dictyosomes.

PROTEIN COMPOSITION: Fig. 4 compares densitometer scans of proteins of stripped Golgi apparatus- and plasma membrane-rich fractions from onion (*Allium cepa*) stem, soybean (*Glycine max*) hypocotyl, and rat liver. While there are certain obvious similarities between Golgi apparatus and plasma membrane fractions from each source, there is no obvious commonness between Golgi apparatus or plasma membrane fractions from one species to another. There appear to be at least five bands present in both Golgi apparatus and plasma membrane fractions of onion and soybean (arrows, Fig. 4), but even these few bands do not appear to be shared with rat liver. Thus, the composition of Golgi apparatus appears to be characterized by marked differences in proteins among different species.

Although enzymatic information on plant Golgi apparatus is rather sparse, certain phospholipid biosynthesizing activities appear to be common to both (e.g., P-choline cytidyl transferase; Morré, 1975) as do inosine diphosphatase and thiamine pyrophosphatase activities (Dauwalder et al., 1972). However, the latter two enzymes have different properties comparing plant and animal

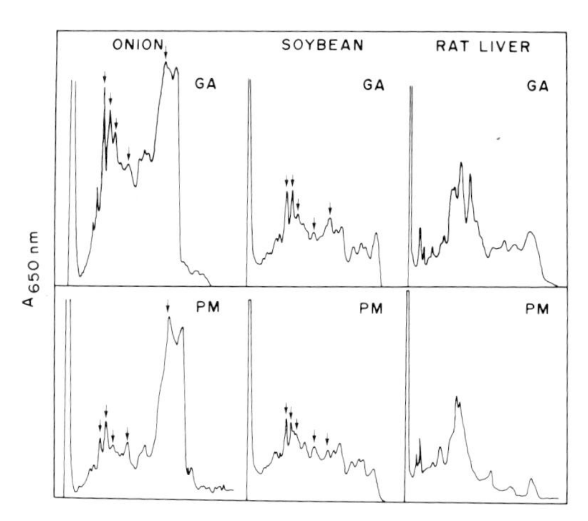

FIGURE 4 Densitometer tracings of sodium dodecyl sulfate (SDS) polyacrylamide gel electrophoretograms of purified and stripped membranes comparing Golgi apparatus (*GA*) and plasma membrane (*PM*) fractions of onion (*Allium cepa*) stem, soybean (*Glycine max*) hypocotyl, and rat liver. Arrows indicate bands common to both onion and soybean fractions. Gels were stained with Coomassie blue and scanned at 650 nm. Unpublished data of Wayne N. Yunghans.

cells and clearly represent different proteins. Both plant and animal Golgi apparatus seem to contain glycosyltransferases, but for the most part with different donor and acceptor specificities (Schachter, 1974). A comparison of plasma membranes of rat liver and soybean hypocotyls (Table III), where information is more complete, suggests that enzymatic differences may be the rule rather than the exception.

PHOSPHOLIPID COMPOSITION: Plant and animal Golgi apparatus have three major phospholipids in common and show a phospholipid composition intermediate between that of the endoplasmic reticulum and the plasma membrane (Keenan and Morré, 1970; Fig. 5). Yet plant Golgi apparatus (and other plant membranes) lack sphingomyelin, one of the major phospholipids of mammalian Golgi apparatus and plasma membranes (see also Fleischer et al., 1974). Phosphatidylserine, if present at all in plant fractions, is a minor constituent. In contrast to rat liver, the plant Golgi apparatus is comprised to the extent of 30–40% by phosphatidic acid, phosphatidylglycerol, and a few constituents that remain unidentified. Thus, within a species, phospholipid compositions of Golgi apparatus membranes are intermediate between those of endoplasmic reticulum and plasma membrane in support of the concept of membrane differentiation, but, in comparing plant and animal Golgi apparatus, major chemical differences are evident.

SUGARS, CARBOHYDRATES, AND HETEROGLYCANS: Plant membranes, including Golgi apparatus, appear to lack sialic acid, whereas this sugar accounts for about 20% of the total carbohydrate of rat liver Golgi apparatus and plasma membranes (Franke and Kartenbeck, 1976; Morré et al., 1974). Instead, uronic acids (largely glucuronate) are present. The sugars of plant membranes consist mainly of hexoses, whereas animal membranes are built around sialic acid and hexosamines with hexoses in lesser amounts. Pentoses (e.g., xylose and arabinose) are also found in plant Golgi apparatus, but are absent from Golgi apparatus of rat liver. Although superficially it may appear that plant and animal Golgi apparatus share common carbohydrate constituents (both contain glucosamine, galactose, glucose, mannose, and fucose, as examples), it is expected that the linkages will be very different. Although plants contain glycolipids, glycosphingolipids, a major glycolipid fraction of mammalian Golgi apparatus, are absent from plants.

SUMMARY OF BIOCHEMICAL STUDIES: Plant and animal Golgi apparatus are morphologically similar but compositionally different. Because no Golgi apparatus protein has been sequenced, it is possible that common sequences in proteins of differing electrophoretic mobility or some other as-yet-unknown factors might account for structural uniformity within the concept of a self-assembly mechanism of membrane biogenesis. However, as an alternative explanation, perhaps plant and animal cells share a common mechanism by which Golgi apparatus form is regulated in a manner at least partly independent of the composition of the membranes. Obviously, certain proteins and most phospholipids of the Golgi apparatus are shared with endoplasmic reticulum and plasma membranes. Yet all three cell components are morphologically distinct, and only the Golgi apparatus exists in the form of stacks of platelike cisternae with attached tubules and secretory vesicles.

TABLE III

Comparison of Enzymatic and Other Biochemical "Markers" of Plasma Membranes of Plant Stems and Rat Liver

Rat liver	Soybean and onion stem	Both
Na^+, K^+, Mg^{++}-ATPase	K^+, Mg^{++}-ATPase	Mg^{++}-ATPase
Adenylate cyclase	Glucan synthetase	Phosphatidylcholine
5′-Nucleotidase	PACP*-staining	Phosphatidylethanolamine
Sphingomyelin	Hexose-rich	Phosphatidylinositol
Phosphatidylserine	Auxin receptors	Cholesterol
Sialic acid	NPA‡ receptors	Glucosamine
Gangliosides		
Glucagon receptors		
Insulin receptors		

* Periodate-phosphotungstate-chromate.

‡ N-1-naphthylphthalamic acid.

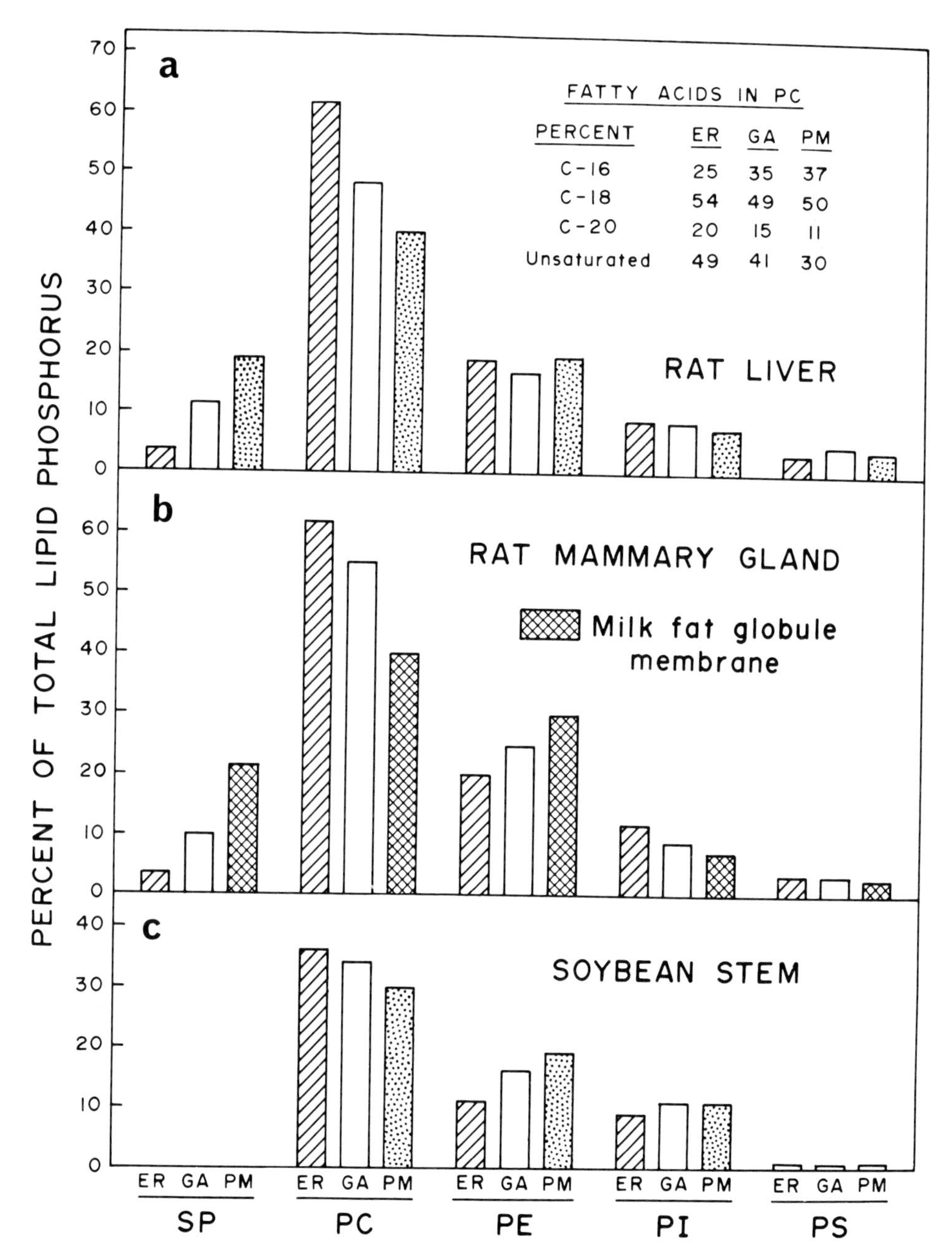

FIGURE 5 Phospholipid composition of animal and plant endomembranes. (*a*) Rat liver. *Inset* gives characteristics of the fatty acid composition of phosphatidylcholine (*PC*) (Keenan and Morré, 1970). (*b*) Rat mammary gland. Golgi apparatus (*GA*) and endoplasmic reticulum (*ER*) are compared with membranes of milk fat globules, a plasma membrane derivative, rather than with plasma membrane (*PM*); Keenan et al., 1974. (*c*) Elongating hypocotyls of etiolated soybeans (*Glycine max*). Data for endoplasmic reticulum (*ER*) based on total rough microsomes. (Unpublished data of W. J. Hurkman, Purdue University.) *SP* = sphingomyelin; *PC* = phosphatidylcholine; *PE* = phosphatidylethanolamine; *PI* = phosphatidylinositol; *PS* = phosphatidylserine.

A HYPOTHESIS: How can such a widely varying chemical composition of Golgi apparatus be rationalized in the context of a constant pattern of Golgi apparatus architecture among different species, while at the same time within a species, the composition of the discrete stacks of Golgi apparatus cisternae is largely intermediate between that of endoplasmic reticulum lamellae and sheets of plasma membrane?

The required controlling assembly mechanism may be provided by the specialized zone of cytoplasm or zone of exclusion that surrounds all Golgi apparatus. The existence of a "Golgi field" was postulated in 1958 by Whaley; Sjöstrand spoke of a Golgi ground substance (for literature, see Whaley, 1975; Morré et al., 1971*b*). Most recently, Giulian and Diacumakos (1976) have reported that the cytoplasm surrounding the Golgi apparatus has electrical properties different from the bulk cytoplasm, and Shivers and Locke (1976) present evidence from orthopod Golgi apparatus for "rings of beads" in this region which stain with bismuth salts at presumptive sites of formation of transition vesicles. The chemical properties of the Golgi apparatus zone of exclusion are unknown, as only recently have attempts begun to isolate the material. A group of fibrous proteins may be a major constituent, but doubtless its composition will be quite complex.

Additionally, Golgi apparatus zones of exclusion contain few to several polyribosomes, usually (but not necessarily) oriented at the forming face of individual dictyosomes (Fig. 3*b* and *d*). They are not attached to membranes per se but, rather, represent a class of free polyribosomes. Golgi apparatus-associated polyribosomes isolated with Golgi apparatus of rat liver support the synthesis of proteins in vitro, and the products of synthesis have electrophoretic mobilities similar to those of certain proteins absent from endoplasmic reticulum but present in plasma membranes (Elder and Morré, 1976). At least zones of exclusion and their associated polyribosomes add the new dimension of protein synthesis to the overall capacity for Golgi apparatus to modify and transform membranes.

Regulation of Golgi Apparatus Function

Before discussing other types of controls that might operate to control Golgi apparatus function, it is important to review the overall secretory process. The major Golgi apparatus output is a secretory vesicle normally destined to fuse with the plasma membrane. The membrane of the vesicle is incorporated at least transiently into the plasma membrane. In certain tip-growing and dividing cells it is a major source of new plasma membrane (Morré and VanDerWoude, 1974; Whaley, 1975). Thus, a major Golgi apparatus function is in the biosynthesis, modification, transformation, and transport of membranes.

Secondly, the vesicles have contents that may range from simple fluids, such as water or dilute salt solutions, to various macromolecular complexes or even preformed wall units, such as are encountered in chrysophycean algae (Brown, this volume). A vast diversity of materials are packaged by plant and animal Golgi apparatus into secretory vesicles for transport to the cell surface and discharged into the extraprotoplastic environment. Virtually every major class of macromolecule is secreted. Each cell type seems to produce a slightly different type of vesicle, condensing vacuole, or secretory granule with content in keeping with cellular function (Mollenhauer and Morré, 1966; Beams and Kessel, 1968; Palade, 1975; Whaley, 1975). Thus, a corresponding diversity of control mechanisms which may influence secretion is also expected.

Finally, it is necessary to emphasize that production of secretory vesicles is a dynamic process involving the bulk movement or flow of membranes (Franke et al., 1971). Additionally, flow mechanisms must be selective, so that although some components are transferred others are not (Morré et al., 1971a; Franke and Kartenbeck, 1976). How chemical energy is transduced into a selective and vectorial movement of membranes or membrane constituents remains a mystery. Clearly, the question of regulation of membrane flow must await elucidation of the mechanism. Golgi apparatus contain a complement of electron transport chain components (flavoproteins, cytochromes, quinones), and their involvement as part of the energy-transducing mechanism in membrane flow is being investigated (F. L. Crane and D. J. Morré, manuscript in preparation).

INDUCTION-REPRESSION: As with microsomal electron-transport chains, the Golgi apparatus system shows inducible components in response to the drug phenobarbital. This is most evident in terms of the NADPH mixed-function oxidase activity, which may be concentrated in the boulevard périphérique of connecting smooth endoplasmic reticulum but is nonetheless a part of the total Golgi apparatus complex in hepatocytes (Morré et al., 1974).

FEEDBACK REGULATION: Rat liver Golgi apparatus contain the complete biosynthetic pathway for synthesis of the major mono- and disialogangliosides from dihexosylceramide (see Richardson et al., 1975, for literature; Morré, in press). Recent findings suggest that this pathway may be subject to a complex set of regulatory restraints, with the net effects of both positive and negative feedback control. Doubtless other examples of this type of regulation will be forthcoming as additional anabolic and catabolic pathways are ascribed to Golgi apparatus. The numerous glycoprotein glycosyltransferase activities of Golgi apparatus membranes have been reviewed (Schachter, 1974).

Pollen tubes and fungal hyphae reveal another type of regulation. When the accumulation of secretory vesicles in the apical zone is monitored in living cells with the light microscope, a relationship is seen between the size of the apical zone of vesicles and the growth rate. The relationship observed suggests that vesicle production, vesicle coalescence at the growing tip, and growth rate are closely coupled. Growth rate is necessarily tied to the rate of surface increase, so a rapid incorporation of vesicles accompanies rapid growth and vesicles do not accumulate extensively. With declining growth rate, the increase in extent of the apical zone may reflect a decreased frequency of vesicle fusion. Yet, it is not known whether frequency of vesicle incorporation and/or production control growth rate or whether growth rate influences the rates of vesicle production and/or incorporation. In spite of this uncertainty, environmental factors that alter growth rates are expressed ultimately in an altered rate of vesicle production. Communication between cytoplasm and cell surface is achieved rapidly in tip-growing cells and indicates some form of feedback regulation which modulates vesicle production to keep pace with the requirements for surface growth.

INHIBITOR STUDIES: A large number of studies have accumulated over the years in which Golgi apparatus are reported to respond in one way or another to various metabolic inhibitors, drugs, effectors, or environmental stimuli. Examples are listed in Table IV. With few exceptions, Golgi apparatus persist although frequently in modified forms. What is perhaps most impressive is the resistance of the Golgi apparatus to morphological alteration. When secretion is inhibited, for example, the extent of the cisternae may increase and one or more additional cisternae per stack may be added. Similarly, when secretion is stimulated, the Golgi apparatus may become slightly smaller with, on the average, one less cisterna per stack. In either situation, the Golgi apparatus quickly adapts to the new steady state. Beyond suggesting a requirement of Golgi apparatus functioning for energy (see Jamieson and Palade, 1968) and an ultimate dependency on RNA and protein synthesis (see Flickinger, 1969), inhibitor studies have added little additional information to our knowledge of Golgi apparatus regulation. Most appear to affect either input (most metabolic inhibitors) or output (e.g., colchicine, Vinca alkaloids, cytochalasins; see Mollenhauer and Morré, 1976). None is known to exert its primary action directly on the Golgi apparatus, although Treolar et al. (1974) reported direct inhibition of a Golgi apparatus galactosyltransferase by puromycin.

TABLE IV

Examples of Inhibitors, Promoters, and Modulators of Golgi Apparatus Structure and/or Function

Protein synthesis inhibitors (cycloheximide, puromycin)
RNA synthesis inhibitors (actinomycin D, enucleation)
Respiratory inhibitors (cyanide, DNP)
Antimicrotubular agents (colchicine, Vinca alkaloids)
Antimicrofilament agents (cytochalasins)
Antimetabolites (ethionine, fluorophenylalanine)
Sugars and glycosides (mannose, ouabain)
Ions and metals (Ca^{++}, Pb)
Hormones; miscellaneous drugs (estrogens, antipyrine)
Environmental factors (temperature, water stress)
Pathogens (viruses, fungi)

CONCLUDING COMMENTS

Much remains to be learned about regulation of Golgi apparatus and Golgi apparatus activities. Only a few examples at each of the various levels show clearly that controls operate to affect the synthesis and differentiation of membranes and to regulate secretion. Because of the overall complexity and integrated nature of the secretory process, it may be difficult to determine precisely at what level(s) each component of a functional Golgi apparatus is regulated and to what extent apparent control influences are direct or indirect in their action. Yet the Golgi apparatus affords an opportunity to study regulation of a dynamic intracellular membrane system in which a common structural plan is assembled and maintained. Despite a perplexing diversity of membrane composition, it allows, at the same time, for a degree of functional heterogeneity necessary to account for the many secretory activities in which Golgi apparatus participate.

REFERENCES

Beams, H. W., and R. G. Kessel. 1968. The Golgi apparatus: structure and function. *Int. Rev. Cytol.* **23**:209–276.

Brown, R. M. 1969. Observations on the relationship of the Golgi apparatus to wall formation in the marine Chrysophycean alga, *Pleurochrysis scherffelii* Pringsheim. *J. Cell Biol.* **41**:109–123.

Cook, G. M. W. 1973. The Golgi apparatus: form and function. *In* Lysosomes in Biology and Pathology. J. T. Dingle, editor. North-Holland Publishing Co., Amsterdam. 237–277.

Dauwalder, M., W. G. Whaley, and J. E. Kephart. 1972. Functional aspects of the Golgi apparatus. *Sub-Cell. Biochem.* **1**:225–276.

Elder, J. H., and D. J. Morré. 1976. Synthesis *in vitro* of intrinsic membrane proteins by free, membrane-bound, and Golgi apparatus-associated polyribosomes from rat liver. *J. Biol. Chem.* **251**:5054–5068.

Fleischer, B., F. Zambrano, and S. Fleischer. 1974. Biochemical characterization of the Golgi complex of mammalian cells. *J. Supramol. Struct.* **2**:737–750.

Flickinger, C. J. 1969. The development of Golgi complexes and their dependence upon the nucleus in Amebae. *J. Cell Biol.* **43**:250–262.

Flickinger, C. J. 1975. The relation between the Golgi apparatus, cell surface, and cytoplasmic vesicles in Amoeba studied by electron microscope radioautography. *Exp. Cell Res.* **96**:189–201.

Franke, W. W., and J. Kartenbeck. 1976. Some principles of membrane differentiation. *In* Progress in Differentiation Research. N. Müller-Berat, editor. American Elsevier Publishing Co., New York. 213–243.

Franke, W. W., D. J. Morré, B. Deumling, R. D. Cheetham, J. Kartenbeck, E. D. Jarasch, and H. W. Zentgraf. 1971. Synthesis and turnover of membrane protein in rat liver: an examination of the membrane flow hypothesis. *Z. Naturforsch.* **26b**:1031–1039.

Giulian, D., and E. G. Diacumakos. 1976. The study of intracellular compartments by micropipette techniques. *J. Cell Biol.* **70**:332*a* (Abstr.).

Grove, S. N., C. E. Bracker, and D. J. Morré. 1968. Cytomembrane differentiation in the endoplasmic reticulum-Golgi apparatus-vesicle complex. *Science (Wash. D. C.).* **161**:171–173.

Jamieson, J. D., and G. E. Palade. 1968. Intracellular transport of secretory proteins in the pancreatic exocrine cell. IV. Metabolic requirements. *J. Cell Biol.* **39**:589–603.

Keenan, T. W., and D. J. Morré. 1970. Phospholipid class and fatty acid composition of Golgi apparatus isolated from rat liver and comparison with other cell fractions. *Biochemistry.* **9**:19–25.

Keenan, T. W., D. J. Morré, and C. M. Huang. 1974. Membranes of the mammary gland. *In* Lactation: A Comprehensive Treatise. Vol. II. B. L. Larson and V. R. Smith, editors. Academic Press, Inc., New York. 191–233.

Michaels, J. E., and C. P. Leblond. 1976. Transport of glycoprotein from Golgi apparatus to cell surface by means of "carrier" vesicles, as shown by radioautography of mouse colonic epithelium after injection of ^{3}H-fucose. *J. Microscopie Biol. Cell.* **25**:243–248

Mollenhauer, H. H., and D. J. Morré. 1966. Golgi apparatus and plant secretion. *Annu. Rev. Plant Physiol.* **17**:27–46.

Mollenhauer, H. H., and D. J. Morré. 1976. Cytochalasin B, but not colchicine, inhibits migration of secretory vesicles in root tips of maize. *Protoplasma.* **87**:39–48.

Mollenhauer, H. H., D. J. Morré, and L. Bergmann. 1967. Homology of form in plant and animal Golgi apparatus. *Anat. Rec.* **158**:313–317.

Morré, D. J. 1975. Membrane biogenesis. *Annu. Rev. Plant Physiol.* **26**:441–481.

Morré, D. J. 1977. The Golgi apparatus and membrane biogenesis. *In* Cell Surface Reviews. G. Poste and G. L. Nicholson, editors. Elsevier, North Holland, Amsterdam. In press.

Morré, D. J., W. W. Franke, B. Deumling, S. E. Nyquist, and L. Ovtracht. 1971*a*. Golgi apparatus function in membrane flow and differentiation: origin of plasma membranes from endoplasmic reticulum. *Biomembranes.* **2**:95–104.

Morré, D. J., T. W. Keenan, and C. M. Huang. 1974. Membrane flow and differentiation: origin of Golgi apparatus membranes from endoplasmic reticulum. *In* Advances in Cytopharmacology. Vol. II. B. Ceccarelli, F. Clementi, and J. Meldolesi, editors. Raven Press, New York. 107–125.

Morré, D. J., H. H. Mollenhauer, and C. E. Bracker. 1971*b*. The origin and continuity of Golgi apparatus. *In* Results and Problems in Cell Differentiation. II. Origin and Continuity of Cell Organelles. T. Reinert and H. Ursprung, editors. Springer-Verlag, Berlin. 82–126.

Morré, D. J., and H. H. Mollenhauer. 1974. The endomembrane concept: a functional integration of endoplasmic reticulum and Golgi apparatus. *In* Dynamic Aspects of Plant Ultrastructure. A. W. Robards, editor. McGraw-Hill Book Co., New York. 84–137.

Morré, D. J., and W. J. VanDerWoude. 1974. Origin and growth of cell surface components. *In* Macromolecules Regulating Growth and Development. E. D:

Hay, T. J. King, and J. Papacanstantinou, editors. Academic Press, Inc., New York. 81–111.

NOVIKOFF, A. B., E. ESSNER, S. GOLDFISCHER, and M. HEUS. 1962. Nucleoside-diphosphatase activities of cytomembranes. *In* The Interpretation of Ultrastructure. R. J. C. Harris, editor. Academic Press, Inc., New York. 149–192.

PALADE, G. E. 1975. Intracellular aspects of the process of protein secretion. *Science (Wash. D. C.).* **189:**347–358.

RICHARDSON, C. L., S. R. BAKER, D. J. MORRÉ, and T. W. KEENAN. 1975. Glycosphingolipid synthesis and tumorigenesis: a role for the Golgi apparatus in the origin of specific receptor molecules of the mammalian cell surface. *Biochim. Biophys. Acta.* **417:**175–186.

SCHACHTER, H. 1974. The subcellular sites of glycosylation. *Biochem. Soc. Symp.* **40:**57–71.

SCHNEPF, E. 1969. Sekretion und Exkretion bei Pflanzen. *Protoplasmatologia Handbuch der Protoplasmaforschung.* **8:**1–181.

SHIVERS, R., and M. LOCKE. 1976. Golgi complex-endoplasmic reticulum transition region has rings of beads in freeze-fracture too. *J. Cell Biol.* **70:**339*a* (Abstr.).

TRELOAR, M., J. M. STURGESS, and M. A. MOSCARELLO. 1974. An effect of puromycin on galactosyltransferase of Golgi-rich fractions from rat liver. *J. Biol. Chem.* **249:**6628–6632.

WHALEY, W. G. 1975. The Golgi Apparatus. Cell Biology Monographs. Springer-Verlag, Berlin. **2:**1–190.

Endoplasmic Reticulum-Golgi Apparatus and Cell Secretion (Animal Cells)

INTRODUCTORY REMARKS

GEORGE E. PALADE

Available information on the morphology of the endoplasmic reticulum and Golgi complex is the result of extensive studies on a wide variety of animal and plant cells which have established that these subcellular structures are ubiquitous, and presumably obligatory components in the organization of all eukaryotes. By contrast, most of the information concerning their function has been obtained by studying cells specifically involved in the secretion of proteins in mammals. In such cells, the endoplasmic reticulum (ER)-Golgi system is definitely involved in the production and subsequent processing of proteins for secretion or export to the extracellular medium. Some years ago, a generalization of this function to all eukaryotes seemed hardly possible, because it implied that all of them are protein-secreting cells. In the meantime, it was realized that all plant cells secrete, indeed, the polysaccharides and proteins of their cell walls (as discussed in some detail in the companion symposium), and that secretory activities are widespread, if not ubiquitous, among animal eukaryotes. Then it was established that the production of lysosomal hydrolases involves the same ER-Golgi system. Inasmuch as the lysosomes are ubiquitous in their distribution among animal and plant cells, the initial difficulties in generalizing the secretory function seemed to be resolved. Moreover, it was realized that secretory activities appear already in prokaryotic cells in connection with the production of their cell walls, the segregation of paraplastic proteins, and the secretion of a variety of hydrolases. The original mechanisms for transferring large molecules across membranes were undoubtedly developed by these "primitive" cells.

It should be pointed out, however, that the ER-Golgi system of eukaryotic cells most likely has additional functions. In a number of cell types, for instance, it is convincingly established that most of the enzymic equipment required for the synthesis of membrane lipids (phospholipids and cholesterol) is located in the membranes of the ER. Moreover, available evidence suggests that the ER-Golgi system is involved in the production of proteins and glycoproteins for membranes.

Hence, in focusing on the secretory process, this symposium deals with that function of the ER-Golgi system which happens to be the one best understood at present, without implying in any way that secretion of macromolecules is the only activity of the system.

GEORGE E. PALADE Section of Cell Biology, Yale University School of Medicine, New Haven, Connecticut

PRODUCTION OF SECRETORY PROTEINS IN ANIMAL CELLS

JAMES D. JAMIESON and GEORGE E. PALADE

The aim of this chapter is to review briefly the current state of knowledge concerning the mechanisms whereby animal cells synthesize and process proteins for export. The review sets the stage for the subsequent papers in this symposium which fill out the details of several of the steps outlined here.

Much of what we know concerning the process that cells have evolved for the production of secretory proteins (abbreviated for convenience to the "secretory process") comes from studies on the mammalian pancreatic exocrine cell and on other cells of similar structure specialized to produce exportable proteins in large amounts and at high rates. Although our discussion will be focused on the structure-function correlates relating to the secretory process in a specific cell type, i.e., the pancreatic exocrine cell of the guinea pig, it appears that many of the basic findings to be described can be generalized to most, if not all, eukaryotic cells, as we will point out later on. A recent review (Palade, 1975) summarizes studies on the secretory process in the exocrine pancreas, and should be consulted for details.

We will assume, for the sake of brevity, that the reader is familiar with the fine structural details of the pancreatic exocrine cell, which have been the object of extensive studies published elsewhere (Jamieson, 1972) and will refer to a diagrammatic representation of the cell shown in Fig. 1.

Two features of the intracellular organization of this exocrine cell are immediately apparent. First, it is evident that the cell is divided into two main compartments: (1) the cytosol containing soluble enzymes and macromolecular assemblies concerned with the intermediary metabolism, the maintenance of the intracellular environment, and most of the biosynthetic activities of the cell, and (2) a series of compartments separated from the cytosol by rough- and smooth-surfaced membranes. These compartments comprise, respectively, the cisternae of the rough-surfaced endoplasmic reticulum (RER) and the cisternae, vesicles, and vacuoles of the Golgi complex, including its derived secretory granules. Table I provides quantitative morphometric data on the relative volumes of these intracellular compartments and surface areas of their membranes (Bolender, 1974; Amsterdam and Jamieson, 1974). The data emphasize the structural amplification of the RER which reflects the specialized function of the cell, i.e., the production of exportable proteins in large amounts and at high rates.

Second, in glandular epithelial cells, the membrane-enclosed compartments which belong to the secretory pathway (RER-Golgi complex-secretory granules) are polarized, i.e., they are disposed in a constant sequence from the base to the apex of the cell. This sequence reflects functional interactions between successive compartments during the processing of secretory proteins, which in turn parallels the temporal sequence of posttranslational operations and chemical modifications undergone by the secretory proteins while in transit along the pathway. The remainder of the chapter will focus on these interactions.

An Analysis of the Secretory Process

For convenience, and as indicated in the figure, the secretory process can be divided into six steps which have been studied individually, but which should be considered as distinct sets of operations in a continuous process.

STEPS 1 AND 2. SYNTHESIS AND SEGREGATION OF EXPORTABLE PROTEINS IN THE

JAMES D. JAMIESON and GEORGE E. PALADE Section of Cell Biology, Yale University School of Medicine, New Haven, Connecticut

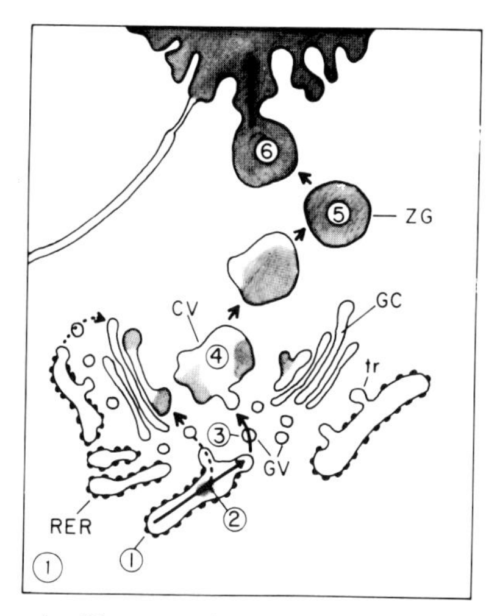

FIGURE 1 Diagrammatic representation of typical glandular epithelial (exocrine) cell indicating intracellular compartments involved in the processing of exportable proteins. *RER*, rough-surfaced endoplasmic reticulum; *tr*, transitional elements of the RER; *GV*, Golgi transporting vesicles; *GC*, Golgi cisternae; *CV*, condensing vacuoles; *ZG*, zymogen granules. The numbers indicate steps in the secretory process discussed in the text. The solid line shows the pathway of intracellular transport determined for the resting guinea pig pancreatic exocrine cell; the dotted line indicates the more typical transport pathway observed in other cell types, including the pancreatic exocrine cell of other species and the stimulated guinea pig pancreatic exocrine cell.

TABLE I

Relative Cytoplasmic Volumes and Membrane Surface Areas of Secretory Compartments in Resting Guinea Pig Pancreatic Exocrine Cells

Compartment	Relative cytoplasmic* volume	Membrane surface area
	%	μm^2/cell
RER	~20	~8,000
Golgi complex	~8	~1,300
Condensing vacuoles	~2	~150
Secretory granules	~20	~900
Apical plasmalemma		~30
Basolateral plasmalemma		~600

Data compiled from Bolender, 1974, and Amsterdam and Jamieson, 1974.

* Percent relative to cytoplasmic volume exclusive of the nucleus.

CISTERNAL SPACE OF THE RER: Based on the early studies of Siekevitz and Palade (1960) and extended by Redman et al. (1966) and Redman (1969), it is now clear that exportable proteins are synthesized on attached polysomes. This step actually takes place in the cytosol environment in which are located (or available) the necessary substrates and the soluble macromolecules involved in protein synthesis, as well as all the other factors required for this operation. During the elongation of each nascent polypeptide chain, its carboxyl terminus segment is protected against attack by exogenously added proteolytic enzymes, presumably because of its location in a channel or groove (inaccessible to proteases) within the large ribosomal subunit (Blobel and Sabatini, 1970). Then, at some stage in elongation, after it begins to emerge from the large subunit, the chain is vectorially directed through the RER membrane toward the cisternal space. While in transit, and with elongation continuing, longer and longer amino terminal segments of the polypeptide chain acquire protection against proteolysis because they have already reached the cisternal space (Sabatini and Blobel, 1970). Upon termination, the whole chain is released and thereby segregated into the cisternal space. Segregation is, therefore, the final outcome of a vectorial transport that occurs concomitantly with elongation. At present, the most reliable assay for this step is partial or full protection against proteolysis.

Recent studies (Blobel and Dobberstein, 1975*a, b*) based on a hypothesis put forth originally by Blobel and Sabatini (1971), and now termed the "signal hypothesis," have shown that attachment to the membrane of the RER of polysomes programmed with mRNAs for exportable proteins apparently involves a signal sequence on the leading end of the amino terminus of the growing peptide. The sequence is efficiently removed by proteolysis during the vectorial transport of the nascent chain. This key aspect in the life history of an exportable protein is dealt with in detail by Günter Blobel in the following chapter, in which it is to be noted that a common, hydrophobic signal sequence has been detected on the amino terminus of a series of exportable proteins synthesized in vitro by polysomes programmed with pancreatic mRNAs (Devillers-Thiery et al., 1975). Additional factors involved in ribosome attachment to the RER membrane are now under active investigation. Two polypeptides have been identified in the RER membrane (not in that of the smooth ER) in hepatocytes, and shown to separate with detached polysomes upon detergent solubilization of the RER (microsome) membrane

(Kreibich et al., 1975). Moreover, evidence which suggests that the mRNAs for exportable proteins become themselves attached to the RER membrane (and thereby facilitate polysome attachment) has been published (Lande et al., 1975; Adesnik et al., 1976).

In summary, segregation of exportable proteins into the cisternal space of the RER depends on: (1) the nature of the mRNAs which program polysomes for synthesis of exportable proteins (these mRNAs include a translated signal sequence [Blobel and Dobberstein, 1975*a, b*]), and (2) the presence of specific recognition sites on the membrane of the RER for the signal sequence (Blobel and Dobberstein, 1975*b*), for the large ribosomal subunit (Kreibich et al., 1975), and, at least in some cases, for the 3′ end of the mRNA (Lande et al., 1975; Adesnik et al., 1976). The complex interactions involved in the segregation of secretory proteins begin to provide an answer to a general problem of considerable importance in cell biology, namely, the regulation of the intracellular traffic of proteins from the cytosol, the common compartment in which they are synthesized, to a variety of compartments or structural entities (membranes, for instance) to which they are destined. In the case of secretory proteins, traffic regulation appears to be based on the ability of a specific sequence ("the signal") originally built in the primary structure of each protein to recognize (and be recognized by) a specific site ("the address") in the RER membrane.

STEP 3. INTRACELLULAR TRANSPORT: After segregation in the cisternal spaces of the RER, secretory proteins move, most likely by diffusion, to the transitional elements of the RER. These elements are located at the boundary between the RER and the Golgi complex and are characterized by their special morphology: they have attached polysomes on most of their surface except for the areas facing the cis side of the Golgi complex; these are smooth and provided with vesicular buds or protrusions similar in size to peripheral Golgi vesicles. By a combination of cell fractionation and electron microscope autoradiography (Jamieson and Palade, 1967*a, b*), we have obtained evidence which has led us to hypothesize that small vesicles, each with a small enclosed sample of RER content, bud off from the transitional elements of the RER, likely by membrane fusion and fission, become transporting Golgi vesicles (usually found in large numbers at the periphery of the Golgi complex), and ferry their contents to the next way-station on the secretory pathway, viz., the condensing vacuoles of the Golgi complex. The mechanism of delivery of the contents of transporting vesicles to the cavities of condensing vacuoles is unknown, although morphologic evidence suggests that it is the reverse of the formation of transporting vesicles from the transitional elements of the RER. Autoradiographic evidence from a variety of cell types indicates that the pathway from the RER to the Golgi complex is generally the same in all cells producing secretory proteins (e.g., parotid acinar cells [Castle et al., 1972], thyroid follicular cells [Bjorkum et al., 1974], odontoblasts [Weinstock and Leblond, 1974]), but characteristic variations from one cell type to another are encountered and will be discussed later on.

Because transport from the RER to the Golgi complex is against an apparent concentration gradient, we next examined this link of the pathway for possible requirements for metabolic energy and protein synthesis, because it might be assumed that sustained or continuous synthesis is needed to maintain the flow of products along this segment of the secretory pathway. By using cycloheximide to terminate abruptly further input of secretory proteins into the cisternae of the RER, we observed that intracellular transport over this link of the pathway and, indeed, over the entire pathway up to and including exocytosis (step 6), was not dependent on ongoing protein synthesis (Jamieson and Palade, 1968*a;* Jamieson and Palade, 1971*a*). These results suggested that pari passu synthesis of enzymes or other proteins was not required to effect transport and, more importantly, that synthesis of proteins for the membrane containers, in which the exportable proteins are transported and packaged, was also not tightly coupled to the synthesis of secretory proteins. The dissociation of synthesis from transport of secretory proteins obtained by using protein synthesis inhibitors opened up the possibility of independently examining the energy requirements for step 3. To summarize, we (Jamieson and Palade, 1968*b*) found that inhibition of ATP production by a variety of means led to the abrupt arrest of intracellular transport. The most proximal site for energy requirement was localized at the level of the transitional elements of the RER, but the nature of the energy-requiring reactions remains unknown. These reactions may be connected with membrane fusion and fission taking place as transitional elements generate transporting vesicles, or as these vesicles fuse with the condensing vacuoles, or they may be involved in the propulsion of transporting vesicles to the condensing vacuoles of the Golgi complex in preparation for their unloading at that point. In any event, the cell appears to be provided with the equivalent of a lock or valve at the level of the transitional elements that is energized by ATP, and the opening of which permits one intracellular compartment (the RER cisternae) to communicate with the next compartment in the series (the elements of the Golgi complex) that forms the secretory pathway.

STEP 4. CONCENTRATION OF SECRETORY PROTEINS: The transport of secretory proteins to condensing vacuoles occurs concomitantly with the latter's progressive filling and concentration of content. In the process, these vacuoles are converted into mature secretion granules (also known as zymogen granules in the case of the pancreatic exocrine cell). Within the time limits investigated (up to 1 h), this conversion is independent of protein synthesis and does not require energy (Jamieson and Palade, 1971*a*). The latter finding ruled out the involvement of membrane-associated ion pumps in the concentration of the vacuole content, but provided no further insight into the mechanisms operating at this step. Earlier autoradiographic studies had shown that [^{35}S]sulfate is incorporated in vivo into Golgi elements and zymogen granules in the pancreatic exocrine cell of the mouse (Berg and Young, 1971). Similar findings have been recorded in the guinea pig (Reggio and Palade, 1976), and more recent investigations suggest that concentration is the result of the interaction of the predominantly basic secretory proteins (in the guinea pig pancreas) with a large polyanion (Tartakoff et al., 1974), namely a sulfated peptidoglycan synthesized in the Golgi complex and present in condensing vacuoles (Reggio and Palade, 1976). As presently envisaged, this interaction leads to the formation of osmotically inactive aggregates, resulting in the passive flow of water from the forming secretory granule content to the relatively hyperosmotic cytosol.

The general functional implication of this hypothesized concentration mechanism is that the cell possesses a "sink" for incoming proteins at the level of the Golgi complex, which explains the lack of requirement for energy (except for that initially needed for the synthesis of the sulfated polyanion) and the apparent irreversibility of this step (Jamieson and Palade, 1971*a*); it also renders unlikely the assumption that transport from the RER to the condensing vacuoles is effected against an actual (as opposed to an apparent) concentration gradient. The hypothesis evidently has some attractive features, but remains to be confirmed by further work in this case and in the case of other secretory products in other cell types. Suggestive evidence in support of the hypothesis has been obtained recently in pituitary mammotrophs (Giannattasio and Zanini, 1976). Moreover, other types of ionic interactions, paracrystal formation (Blundell et al., 1971), and binding of small peptides to carriers (Kirshner, 1974; Breslow, 1974) described in other secretion granules are expected to lead in principle to the same end result.

STEP 5. INTRACELLULAR STORAGE: After their formation, secretory granules reside in the cytoplasm for varying times until the appropriate stimulus for release impinges on the cell. Temporary storage of secretory products before discharge likely represents a means whereby a cell can cope with short-term demands for secretory product output at a rate in excess of that which can be maintained by synthesis alone. This is the case with the majority of endocrine cells that produce a peptide hormone or a small amine complexed to a protein matrix. It is also the case for exocrine cells, whose function involves cycles of discharge and reaccumulation of secretory proteins. In all these cases, a stimulus (hormone or neurotransmitter) causes large and sudden increases in the rate of discharge. Concentration at the preceding step can be viewed as a space-saving solution required for extensive intracellular accumulation of secretory products. Yet, this rationalization is not without exceptions, because there is concentration and a limited amount of intracellular storage of secretory products (albumin, lipoprotein) in the mammalian hepatocyte (Redman et al., 1975), although, as far as we know, their discharge is not hormonally controlled.

STEP 6. EXOCYTOSIS: The objective and culmination of the previous steps in the secretory process is the exocytosis of the secretion granule content.

Morphologically, exocytosis implies the movement of the secretory granule from its site of storage in the apical region of the cell to the immediate vicinity of the luminal membrane, the lateral (side-to-side) fusion of the secretion granule membrane to the luminal plasmalemma, and the fission of the fused membranes leading to the formation of an opening or orifice in the area of fusion through which secretory discharge takes place (Palade, 1959, 1975). These operations remove all membrane barriers between the secretion granule content and the lumen and, at the same time, establish bilayer continuity between the two interacting membranes, thus enabling the cell to discharge macromolecular products without disrupting the selective permeability of its membranes (Fig. 2).

Biochemically, exocytosis depends on the intracellular elevation of Ca^{++} derived from intra- or extracellular sources (Schramm and Selinger, 1977), requires continuous ATP production (Jamieson and Palade, 1971*a*), and is initiated by an appropriate neural or hormonal stimulus interacting with its corresponding receptor in the cell membrane. The sequence of events that leads from stimulus to discharge is termed stimulation-secretion coupling (Douglas, 1968); it involves, as a second intracellular messenger, Ca^{++} and/or a cyclic nucleotide generated as a result of the hormone-receptor interaction (Rasmussen et al., 1975; Schramm and Selinger, 1977). The type of nucleotide appears to differ among cells (Schramm and Selinger, 1977), and its effects on, or interaction with, Ca^{++} fluxes is still a matter of debate (Rasmussen et al., 1975). In any case, it is clear that exocytosis is energy-dependent, and this dependency defines a second energy-requiring gate in the secretory pathway which, when open, puts in communication an intracellular compartment (secretion granule or secretion vacuole) with the extracellular space (glandular lumen for exocrine cells, interstitia for endocrine cells). Precisely where and how energy is consumed in exocytosis is unknown, but movement of secretory granules, membrane fusion and fission, and generation of cyclic nucleotides are all likely candidates for energy-dependent processes or reactions involved in the final discharge of secretory products.

Posttranslational Modifications

Besides being segregated, transported, and concentrated as described in the preceding sections of this chapter, the secretory polypeptide chains are the object of varied and often extensive chemical modifications which occur either concurrently with translation and segregation or after segregation while the chains are in transit through specific compartments of the secretory pathway.

The first set of modifications occurs at the level of the RER and begins with the proteolytic cleavage of the signal sequence from the growing polypeptide chain (Blobel and Dobberstein, 1975*a*,

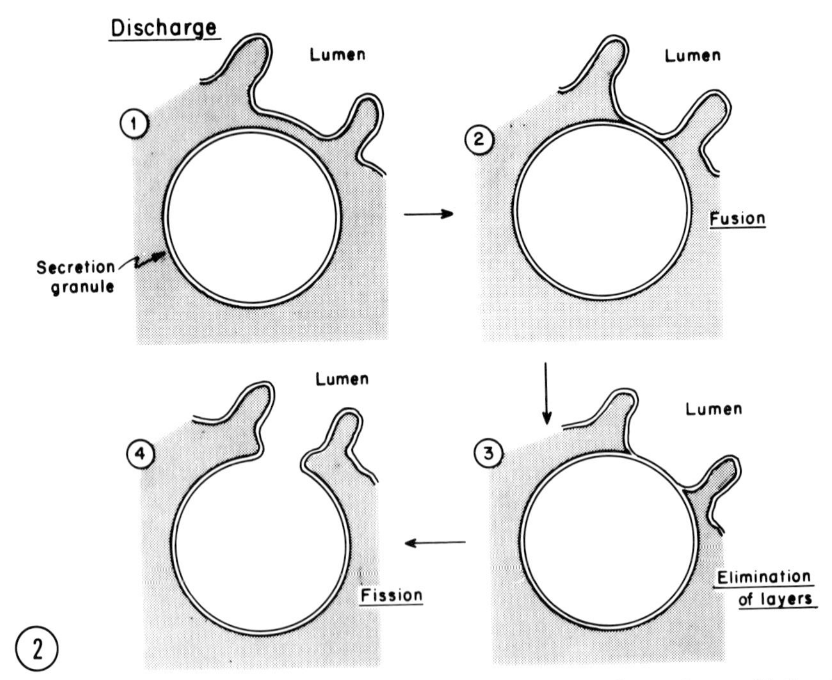

FIGURE 2 Schematic representation of interactions of the secretory granule membrane with the plasmalemma that occurs during exocytosis.

b). This modification, which is discussed in detail in Günter Blobel's chapter, is presumably catalyzed by an endopeptidase associated with the membrane of the RER. In addition, as the growing peptide enters the RER cisternal space, intramolecular disulfide bonds are formed (Anfinsen, 1973), converting the linear peptide to a globular molecule of rather large diameter (≳20Å). This conversion presumably renders irreversible the segregation of the secretory polypeptide in the cisternal space of the RER, because the limiting membrane of this compartment is freely permeable to molecules smaller than ~20Å in diameter (Palade, 1975). With the formation of disulfide bonds, which do not necessarily involve cysteine residues in their order of emergence from the RER membrane (e.g., Steiner et al., 1972; Anfinsen, 1973), the secretory protein begins to acquire its tertiary structure, which will be perfected by subsequent modifications.

Removal of the signal peptide and formation of disulfide bridges can be considered as general modifications that probably affect all secretory proteins. In addition, specific modifications are incurred by special classes of secretory proteins. Addition of proximal or core sugars such as *N*-acetylglucosamine and mannose to those proteins which will become secretory glycoproteins also occurs within the RER (Wuhr et al., 1969; Zagury et al., 1970) and appears to involve dolicholphosphates as carriers of the corresponding core sugar residues (Lucas et al., 1975). Hydroxylation of lysyl and prolyl residues of collagens also takes place in the same compartment in fibroblasts (Olsen et al., 1975) and probably in other cell types producing the same or similar proteins. Finally, quaternary structure begins to be acquired in the cisternal space of the RER in a process which appears to be coordinated with the chemical modifications of individual polypeptide chains (Byers et al., 1975; Vassalli et al., 1971) mentioned above.

Further modifications of the secretory products take place during their transport through the Golgi complex. Foremost among them is the completion of the oligosaccharide side chains of glycoproteins and peptidoglycans by the sequential action of membrane-associated glycosyltransferases, which include galactosyl-, fucosyl-, *N*-acetylglucosaminyl-, and sialyltransferases (e.g., Schachter et al., 1970; Munro et al., 1975). Galactosyltransferase, for instance, has become a useful and generally accepted "marker" for Golgi fractions (Fleischer et al., 1969; Bergeron et al., 1973) and Golgi membranes (e.g., Bergeron et al., 1973). The function of the Golgi complex in terminal glycosylation of glycoproteins and glycolipids will be dealt with in detail by Charles Leblond and Gary Bennett (this volume).

Even though these Golgi functions are well documented and generally recognized, it does not necessarily follow that all secretory proteins that pass through the Golgi complex are glycosylated. Notable exceptions are the pancreatic secretory proteins trypsinogen and chymotrypsinogen, which have been located in Golgi cisternae by immunocytochemical tests at the electron microscope level (Kraehenbuhl et al., 1977). Evidently, passage through the Golgi complex is mandatory regardless of the nature of the protein that is processed.

As mentioned previously, in many cells (Lane et al., 1964; Young, 1973), including those of the exocrine pancreas (Berg and Young, 1971; Reggio and Palade, 1976), inorganic sulfate is incorporated into macromolecular components in the Golgi complex, and in the few cases so far studied it appears to be added to certain sugars of the oligosaccharide side chains of peptidoglycans. The possible function of these polyanions in the concentration of secretory proteins in condensing vacuoles has been discussed.

Finally, further proteolysis of the transported products appears to take place in the Golgi complex, the classic example being the cleavage out of the connecting peptide of proinsulin to form two-chain active insulin (Steiner et al., 1972).

Variations in the Function of the Golgi Complex

A basic, and as yet little-understood, function of the Golgi complex pertains to its ability to sort out and package secretory proteins along separate pathways, depending upon their ultimate destination: export to some extracellular compartment or discharge within a distinct, membrane-bound intracellular compartment as typified by the lysosomal system.

The formation of lysosomes is a ubiquitous function of all eukaryotic cells and appears to involve the same basic sequence of synthesis, segregation, transport, and packaging operations as discussed for the exocrine pancreatic cell (de Duve and Wattiaux, 1966; Bainton et al., 1976). In "professional" phagocytes, i.e., polymorphonuclear leukocytes, these primary lysosomes take the form of intracellular granules produced in the Golgi complex and maintained in storage in the cytoplasm until used (Bainton et al., 1976), but in many other cells the concentration step (and perhaps the intracellular storage step) is omitted: morphologically, primary lysosomes appear as small vesicles with or without a coat and with a content of low density, so that their identification depends primarily on cytochemical tests (Nichols et al., 1971; Friend and Farquhar, 1967).

The transport route already discussed was established by work carried out on the pancreatic exocrine cell of the guinea pig. The solid routing line in Fig. 1, which is based on this work, indicates that secretory proteins are transferred from the RER to Golgi condensing vacuoles. The evidence concerning this pathway, which is peculiar for the "resting" pancreatic exocrine cell of the guinea pig, has been taken by some to demonstrate that the Golgi cisternae of this cell type are generally bypassed in the transport operation. This interpretation is not correct, because we regard the condensing vacuoles as part of the Golgi complex and consider them as dilated Golgi cisternae occupying an extreme trans position in Golgi cisternal stacks. Moreover, when this cell is stimulated, the accumulation and concentration of secretory proteins takes place in the dilated rims of most (if not all) stacked Golgi cisternae (Jamieson and Palade, 1971*b*), as it does, in fact, in all other cells, as indicated by the dotted routing line in Fig. 1. In the latter, the involvement of Golgi cisternae in the concentration step is clearly established by morphological (Bainton et al., 1976), cytochemical (Herzog and Miller, 1970), and autoradiographic (Leblond and Bennett, this volume) observations.

In the case of cells in which morphologically recognizable secretory granules are absent (such as plasma cells and fibroblasts), post-Golgi events in the secretory process cannot be followed easily, and the means of transport and discharge of secretory products remain a matter of debate (Ross and Benditt, 1965; Olsen and Prokop, 1974). Recently, however, dilated Golgi vacuoles containing secretory products have been identified in related cell types (Weinstock and Leblond, 1974) or in certain experimental situations (Olsen and Prokop, 1974), a finding which suggests that under normal conditions transport to the cell surface and discharge may be effected by small vesicles analogous to those implicated in step 3, i.e., transport from RER to condensing vacuoles.

On the trans side of the Golgi complex, Novikoff et al. (1971) have described a special compartment which is assumed to connect Golgi elements to the RER and to be involved in lysosome formation, hence the acronym GERL proposed to designate it. This compartment has histochemically demonstrable acid phosphatase activity and, on this account as well as on account of its topography and morphology, it has been identified in a variety of cell types (Bentfield and Bainton, 1975; Nichols, 1975; Hand, 1973). Recently, Novikoff (1976) has postulated that GERL and Golgi-condensing vacuoles are the same structure and that secretion granules, in general, are formed in this compartment. This interpretation raises a number of questions concerning the nature of the local phosphatasic activity, the ability of glandular cells to separate lysosomal from secretory proteins efficiently, and the possible functional meaning of the imperfect separation of the two types of proteins (assuming that this is actually the case).

Conclusions

From the studies outlined above we can conclude the following: (1) After segregation in the cisternal space of the RER, secretory proteins are transported through the cell to the extracellular space within the confines of a series of connectable membrane-bounded compartments; they cross a membrane directly only once at the time of synthesis and segregation.

(2) As a corollary to (1), we and others (e.g., Redman et al., 1975; Bainton and Farquhar, 1970; Bainton et al., 1976; Herzog and Miller, 1970) have found no evidence for transport of secretory proteins in soluble form through the cytosol. These conclusions have been challenged by Rothman (1975), but the evidence he marshals in favor of transport through the cytosol and against segregation and exocytosis is questionable because it is seriously affected by cell fractionation artifacts and is not adequately supported by a microscope survey of experimental specimens (cf. Palade, this volume).

(3) Transport along the secretory pathway is: (a) vectorial, i.e., it proceeds regularly in the direction of secretion granules (apically) in exocrine cells; (b) irreversible, i.e., there is no evidence of back flow; and (c) functionally discontinuous, i.e., the secretory products are moved from one compartment to the next in series by discrete membrane containers (vesicles or vacuoles) and the pathway behaves as if provided with locks whose

opening requires energy and involves membrane fusion-fission.

(4) Transport involves highly specific membrane – membrane interactions. For instance, in exocrine cells secretory granules always recognize and fuse with the apical plasmalemma, although proximity to the lateral plasmalemma is frequent, and transporting vesicles, derived from transitional RER elements appear to deliver their contents into Golgi condensing vacuoles and cisternae and not elsewhere.

(5) Transport is associated with translocations of membranes between compartments. This occurs at the RER-Golgi step of intracellular transport, where it is mediated by Golgi vesicles, but is particularly obvious upon secretory granule discharge. During exocytosis, large amounts of secretory granule membranes are inserted into the apical plasmalemma (Amsterdam et al., 1969), yet the cell is able to remove this excess luminal membrane and, at the same time, maintain its intracellular membrane pool.

Perspectives

In the past, our studies have been focused on the route and timetable of transport and processing of secretory proteins. The results of these studies have enabled us to identify the compartments involved in the secretory process, and to define, to some extent, their interactions. For the future, it will be important to examine the membranes of the compartments and of the containers themselves in order to determine, at the molecular level, how this precise and controlled set of transport operations is effected. It is particularly important to understand how the cell is able to maintain both balanced membrane distribution and specificity of individual compartmental membranes in the face of extensive membrane translocations, which, in the framework of our current ideas on membrane organization (Singer and Nicolson, 1972), are expected to lead to intermixing and eventual equilibration of molecular components among interacting membranes. The evidence so far available suggests that translocated membranes are specifically removed from their receiving compartments, but gives no clue as to the mechanisms involved in this apparently nonrandom removal (Palade, 1975; Bergeron et al., 1973).

REFERENCES

Adesnik, M., M. Lovele, T. Martin, and D. D. Sabatini. 1976. Retention of mRNA on the endoplasmic reticulum membranes after in vivo disassembly of polysomes by an inhibitor of initiation. *J. Cell Biol.* **71**:307–313.

Amsterdam, A., and J. D. Jamieson. 1974. Studies on dispersed pancreatic exocrine cells. I. Dissociation technique and morphologic characteristics of separated cells. *J. Cell Biol.* **63**:1037–1056.

Amsterdam, A., I. Ohad, and M. Schramm. 1969. Dynamic changes in the ultrastructure of the acinar cell of the rat parotid gland during the secretory cycle. *J. Cell Biol.* **41**:753–773.

Anfinsen, C. B. 1973. Principles that govern folding of protein chains. *Science (Wash. D. C.).* **181**:223–230.

Bainton, D. F., and M. G. Farquhar. 1970. Segregation and packaging of granule enzymes in eosinophilic leucocytes. *J. Cell Biol.* **45**:54–73.

Bainton, D. F., B. A. Nichols, and M. G. Farquhar. 1976. Primary lysosomes of blood leukocytes. *In* Lysosomes in Biology and Pathology. Vol. 5. J. T. Dingle and R. T. Dean editors. North-Holland Publishing Co., Amsterdam. 3–32.

Bentfield, M. E., and D. F. Bainton. 1975. Cytochemical localization of lysosomal enzymes in rat megakaryocytes and platelets. *J. Clin. Invest.* **56**:1635–1649.

Berg, N. B., and R. W. Young. 1971. Sulfate metabolism in pancreatic acinar cells. *J. Cell Biol.* **50**:469–483.

Bergeron, J. J. M., J. H. Ehrenreich, P. Siekevitz, and G. E. Palade. 1973. Golgi fractions prepared from rat liver homogenates. II. Biochemical characterization. *J. Cell Biol.* **59**:73–88.

Bjorkum, U., R. Ekholm, L. G. Elmquist, L. E. Erickson, A. Melander, and S. Smeds. 1974. Induced unidirectional transport of protein into the thyroid follicular lumen. *Endocrinology.* **95**:1506–1517.

Blobel, G., and B. Dobberstein. 1975*a*. Transfer of proteins across membranes. I. Presence of proteolytically processed and unprocessed nascent immunoglobulin light chains on membrane-bound ribosomes of murine myeloma. *J. Cell Biol.* **67**:835–851.

Blobel, G., and B. Dobberstein. 1975*b*. Transfer of proteins across membranes. II. Reconstitution of functional rough microsomes from heterologous components. *J. Cell Biol.* **67**:852–862.

Blobel, G., and D. D. Sabatini. 1970. Controlled proteolysis of nascent peptides in rat liver cell fractions. I. Location of polypeptides within ribosomes. *J. Cell Biol.* **45**:130–145.

Blobel, G., and D. D. Sabatini. 1971. Ribosome membrane interaction in eukaryotic cells. *In* Biomembranes. Vol. II. L. A. Mason, editor. Plenum Publishing Co., New York. 193–195.

Blundell, T. L., J. F. Cutfield, I. M. Cutfield, E. J. Dodson, G. G. Dodson, D. C. Hodgkin, D. A. Mercola, and M. Vijayan. 1971. Atomic positions in rhombohedral 2-zinc insulin crystals. *Nature (Lond.).* **231**:506–511.

Bolender, R. P. 1974. Stereological analysis of the

guinea pig pancreas. I. Analytical model and quantitative description of nonstimulated pancreatic exocrine cells. *J. Cell Biol.* **61:**269–287.

Breslow, E. 1974. The neurophysins. *Adv. Enzymol. Relat. Areas Mol. Biol.* **40:**271–333.

Byers, P. H., E. M. Click, E. Harper, and P. Bornstein. 1975. Interchain disulfide bonds in procollagen are located in a large non-triple-helical COOH-terminal domain. *Proc. Natl. Acad. Sci. U. S. A.* **72:**3009–3013.

Castle, J. D., J. D. Jamieson, and G. E. Palade. 1972. Radioautographic analysis of the secretory process in the parotid acinar cell of the rabbit. *J. Cell Biol.* **53:**290–311.

de Duve, C., and R. Wattiaux. 1966. Functions of lysosomes. *Annu. Rev. Physiol.* **28:**435–492.

Devillers-Thiery, A., T. Kindt, G. Scheele, and G. Blobel. 1975. Homology in aminoterminal sequence of precursors to pancreatic secretory proteins. *Proc. Natl. Acad. Sci. U. S. A.* **70:**1554–1558.

Douglas, W. W. 1968. Stimulus-secretion coupling: the concept and clues from chromaffin and other cells. *Br. J. Pharmacol.* **34:**451–474.

Fleischer, B., S. Fleischer, and H. Ozawa. 1969. Isolation and characterization of Golgi membranes from bovine liver. *J. Cell Biol.* **43:**59–79.

Friend, D. S., and M. G. Farquhar. 1967. Function of coated vesicles during protein absorption in the rat vas deferens. *J. Cell Biol.* **35:**357–376.

Giannattasio, G., and A. Zanini. 1976. Presence of sulfated proteoglycans in prolactin secretory granules isolated from the rat pituitary gland. *Biochim. Biophys. Acta.* **439:**349–357.

Hand A. R. 1973. Secretory granules, membranes and lysosomes. *In* Symposium on the Mechanism of Exocrine Secretion. S. S. Han, L. Sreebny, and R. Suddick, editors. University of Michigan Press, Ann Arbor. 129–151.

Herzog, V., and F. Miller. 1970. Die Lokalisation endogener Peroxydase in der Glandula Parotis der Ratte. *Z. Zellforsch. Mikrosk. Anat.* **107:**403–420.

Jamieson, J. D. 1972. Transport and discharge of exportable proteins in pancreatic exocrine cells: *in vitro* studies. *Curr. Top. Membranes Transp.* **3:**273–338.

Jamieson, J. D., and G. E. Palade. 1967*a*. Intracellular transport of secretory proteins in the pancreatic exocrine cell. I. Role of the peripheral elements of the Golgi complex. *J. Cell Biol.* **34:**577–596.

Jamieson, J. D., and G. E. Palade. 1967*b*. Intracellular transport of secretory proteins in the pancreatic exocrine cell. II. Transport to condensing vacuoles and zymogen granules. *J. Cell Biol.* **34:**597–615.

Jamieson, J. D., and G. E. Palade. 1968*a*. Intracellular transport of secretory proteins in the pancreatic exocrine cell. III. Dissociation of intracellular transport from protein synthesis. *J. Cell Biol.* **39:**580–588.

Jamieson, J. D., and G. E. Palade. 1968*b*. Intracellular transport of secretory proteins in the pancreatic exocrine cell. IV. Metabolic requirements. *J. Cell Biol.* **39:**589–603.

Jamieson, J. D., and G. E. Palade. 1971*a*. Condensing vacuole conversion and zymogen granule discharge in pancreatic exocrine cells: metabolic studies. *J Cell Biol.* **48:**503–522.

Jamieson, J. D., and G. E. Palade. 1971*b*. Synthesis, intracellular transport, and discharge of secretory proteins in stimulated pancreatic exocrine cells. *J. Cell Biol.* **50:**135–158.

Kirshner, N. 1974. Molecular organization of the chromaffin vesicles of the adrenal medulla. *Adv. Cytopharmacol.* **2:**265–272.

Kraehenbuhl, J-P., L. Racine, and J. D. Jamieson. 1977. Immunocytochemical localization of secretory proteins in bovine pancreatic exocrine cells. *J. Cell Biol.* **72:**406–423.

Kreibich, G., B. Ulrich, and D. D. Sabatini. 1975. Polypeptide compositional differences between rough and smooth microsomal membranes. *J. Cell Biol.* **67:**225a (Abstr.).

Lande, M. A., M. Adesnik, M. Sumida, Y. Tashiro, and D. D. Sabatini. 1975. Direct association of messenger RNA with microsomal membranes in human diploid fibroblasts. *J. Cell Biol.* **65:**513–528.

Lane, N., L. Caro, L. R. Otero-Vilardebó, and G. C. Godman. 1964. On the site of sulfation in colonic goblet cells. *J. Cell Biol.* **21:**339–351.

Lucas, J. J., C. J. Wachter, and W. J. Lennarz. 1975. The participation of lipid-linked oligosaccharide in synthesis of membrane glycoproteins. *J. Biol. Chem.* **250:**1992–2002.

Munro, J. R., S. Narasimhan, S. Wetmore, J. R. Riordan, and H. Schachter. 1975. Intracellular localization of GDP-L-fucose: glycoprotein and CMP-sialic acid: apolipoprotein glycosyl transferase in rat and pork livers. *Arch. Biochem. Biophys.* **169:**269–277.

Nichols, B. A. 1975. Phagocytosis and degradation of surfactant by alveolar macrophages. *J. Cell Biol.* **67:**307a(Abstr.).

Nichols, B. A., D. F. Bainton, and M. G. Farquhar. 1971. Differentiation of monocytes. Origin, nature, and fate of their azurophil granules. *J. Cell Biol.* **50:**498–515.

Novikoff, A. B. 1976. The endoplasmic reticulum: a cytochemist's view (a review). *Proc. Natl. Acad. Sci. U. S. A.* **73:**2781–2787.

Novikoff, P. M., A. B. Novikoff, N. Quintana, and J-J. Hauw. 1971. Golgi apparatus, GERL, and lysosomes of neurons in rat dorsal root ganglia, studied by thick section and thin section cytochemistry. *J. Cell Biol.* **50:**859–886.

Olsen, B. R., R. A. Berg, Y. Kishida, and D. J. Prokop. 1975. Further characterization of embryonic tendon fibroblasts and the use of immunoferritin techniques to study collagen biosynthesis. *J. Cell Biol.* **64:**340–355.

OLSEN, B. R., and D. J. PROKOP. 1974. Ferritin-conjugated antibodies used for labeling of organelles involved in the cellular synthesis and transport of procollagen. *Proc. Natl. Acad. Sci. U. S. A.* **71:**2033–2037.

PALADE, G. E. 1959. Functional changes in the structure of cell components. *In* Subcellular Particles. T. Hayashi, editor. The Ronald Press Company, New York. 64–83.

PALADE, G. E. 1975. Intracellular aspects of the process of protein secretion. *Science (Wash. D. C.).* **189:**347–358.

RASMUSSEN, H., P. JENSEN, W. LAKE, N. FRIEDMANN, and D. B. P. GOODMAN. 1975. Cyclic nucleotides and cellular calcium metabolism. *Adv. Cyclic Nucleotide Res.* **5:**375–394.

REDMAN, C. M. 1969. Biosynthesis of serum proteins and ferritin by free and attached ribosomes of rat liver. *J. Biol. Chem.* **244:**4308–4315.

REDMAN, C. M., D. BANERJEE, K. HOWELL, and G. E. PALADE. 1975. Colchicine inhibition of plasma protein release from rat hepatocytes. *J. Cell Biol.* **66:**42–59.

REDMAN, C. M., P. SIEKEVITZ, and G. E. PALADE. 1966. Synthesis and transfer of amylase in pigeon pancreatic microsomes. *J. Biol. Chem.* **241:**1150–1158.

REGGIO, H., and G. E. PALADE. 1976. Sulfated compounds in the secretion and zymogen granule content of the guinea pig pancreas. *J. Cell Biol.* **70:**360a (Abstr.).

ROSS, R., and E. P. BENDITT. 1965. Wound healing and collagen formation. V. Quantitative electron microscope radioautographic observations of proline-H^3 utilization by fibroblasts. *J. Cell Biol.* **27:**83–106.

ROTHMAN, S. S. 1975. Protein transport by the pancreas. *Science (Wash. D. C.).* **190:**745–753.

SABATINI, D. D., and G. BLOBEL. 1970. Controlled proteolysis of nascent polypeptides in rat liver cell fractions. II. Localization of the polypeptides in rough microsomes. *J. Cell Biol.* **45:**146–157.

SCHACHTER, H., I. JABBAL, R. L. HUDGIN, L. PINTERIC, E. J. MCGUIRE, and S. ROSEMAN. 1970. Intracellular localization of liver sugar nucleotide glycoprotein glycosyltransferases in a Golgi-rich fraction. *J. Biol. Chem.* **245:**1090–1100.

SCHRAMM, M., and Z. SELINGER. 1977. Neurotransmitters, receptors, second messengers, and responses in parotid gland and pancreas. *J. Cyclic Nucleotide Res.* In press.

SIEKEVITZ, P., and G. E. PALADE. 1960. A cytochemical study on the pancreas of the guinea pig. V. *In vivo* incorporation of leucine-1-^{14}C into the chymotrypsinogen of various cell fractions. *J. Biophys. Biochem. Cytol.* **7:**619–630.

SINGER, S. J., and G. L. NICOLSON. 1972. The fluid mosaic model of the structure of cell membranes. *Science (Wash. D. C.).* **175:**720–731.

STEINER, D. F., W. KEMMLER, J. L. CLARK, P. E. OYER, and A. H. RUBENSTEIN. 1972. The biosynthesis of insulin. *Handb. Physiol. Endocrinology.* **1:**175–198.

TARTAKOFF, A. M., L. J. GREENE, and G. E. PALADE. 1974. Studies on the guinea pig pancreas: fractionation and partial characterization of exocrine proteins. *J. Biol. Chem.* **249:**7420–7431.

VASSALLI, P., B. LISOWSKA-BERNSTEIN, and M. E. LAMM. 1971. Cell-free synthesis of rat immunoglobulin. IV. Analysis of the cell-free made chains and their mode of assembly. *J. Mol. Biol.* **56:**1–19.

WEINSTOCK, M., and C. P. LEBLOND. 1974. Synthesis, migration, and release of precursor collagen by odontoblasts as visualized by radioautography after [^3H] proline administration. *J. Cell Biol.* **60:**92–127.

WUHR, P., A. HERSCOVICS, and C. P. LEBLOND. 1969. Radioautographic visualization of the incorporation of galactose-^3H and mannose-^3H by rat thyroid in vitro in relation to the stages of thyroglobulin synthesis. *J. Cell Biol.* **43:**289–311.

YOUNG, R. W. 1973. The role of the Golgi complex in sulfate metabolism. *J. Cell Biol.* **57:**175–189.

ZAGURY, D., J. W. UHR, J. D. JAMIESON, and G. E. PALADE. 1970. Immunoglobulin synthesis and secretion. II. Radioautographic studies of intracellular transport and sites of addition of the carbohydrate moieties. *J. Cell Biol.* **46:**52–63.

SYNTHESIS AND SEGREGATION OF SECRETORY PROTEINS: THE SIGNAL HYPOTHESIS

GÜNTER BLOBEL

The synthesis, passage through various cellular compartments (referred to as the secretory pathway [Palade, 1975]), and finally exocytosis of secretory proteins has been reviewed in the preceding paper (Jamieson and Palade). This paper deals with the early events along this pathway and focuses primarily on two questions: first, how are secretory proteins transferred across the endoplasmic reticulum, given the fact that they are assembled by ribosomes in the cytosol, and second, what mechanism discriminates between proteins, which are transferred to the cisternal space, and other proteins, such as cytosol proteins, which are excluded?

Functional Significance of the Ribosome-Membrane Junction

A first clue to the mechanism of transfer of secretory proteins across the endoplasmic reticulum (ER) membrane came from the discovery in the mid 1950s that ribosomes occur either free in the cytosol or bound to the ER membrane and that membrane-bound ribosomes are a characteristic feature of cells that actively secrete proteins (Palade, 1955, 1958). Only a decade later, after suitable methods had been developed for cell fractionation as well as for cell-free protein synthesis, it was possible to design in vitro experiments to probe the functional significance of the ribosome-membrane junction. Thus Redman et al. (1966) showed that the bulk of amylase synthesized in vitro by rough microsomes isolated from pigeon pancreas was not released into the incubation medium (the equivalent of the cytosol), but instead remained associated with the rough microsomes, from which it could be "solubilized" after treatment with detergent. These data suggested that the completion of nascent amylase chains in vitro was coupled to their segregation across the microsomal membrane into the cavities (content) of microsomal vesicles. However, it remained to be determined whether this apparent segregation resulted from a transfer of the entire chain across the microsomal membrane or merely from adsorption to the outer aspect of the membrane, and whether transfer had occurred during or after completion of the nascent chain. Progress toward answering these questions came from experiments with proteolytic enzymes as probes. It was found that nascent chains still associated with free ribosomes were degraded by proteolytic enzymes except for a fragment comprising 30–40 amino acid residues of the carboxyl terminal portion of the nascent chain (Malkin and Rich, 1967; Blobel and Sabatini, 1970). These results led us to propose (see Fig. 1, upper panel) that the nascent chain grows in a groove or tunnel located within the large ribosomal subunit (Blobel and Sabatini, 1970). Proteolysis of rough microsomes, on the other hand, left most of the nascent chains protected in an acid-insoluble form, although the ribosomes were detached from the microsomal membrane. It was found that the detached ribosomes again contained a fragment of the nascent chain, comprising 40 amino acids, whereas the remainder of the chain down to the amino terminal was presumably protected within the intravesicular space to which proteolytic enzymes had no access (Sabatini and Blobel, 1970).

These data led us to propose a model (Sabatini and Blobel, 1970) in which the putative tunnel in the large ribosomal subunit is linked directly to a tunnel in the membrane (see Fig. 1, lower panel). The model assigned an important role to the ribosome-membrane junction, namely, to provide the topological conditions which, concomitantly with translation, will result in a unidirectional, i.e.,

GÜNTER BLOBEL The Rockefeller University, New York.

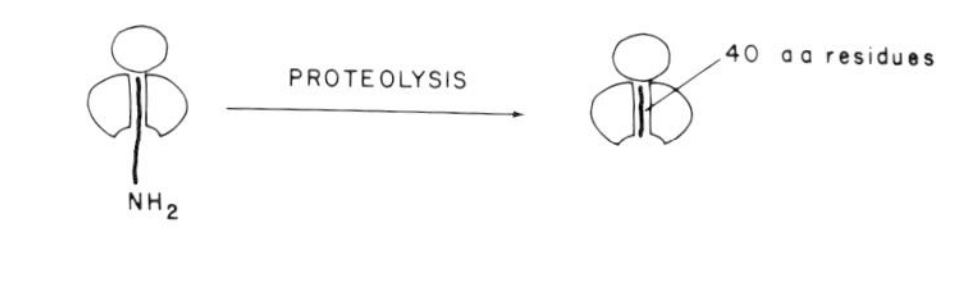

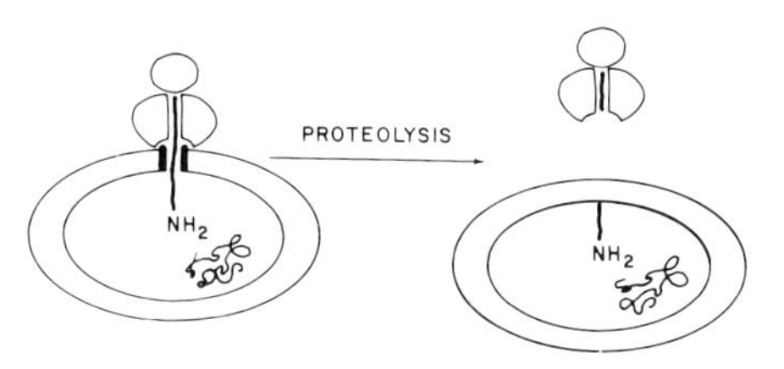

FIGURE 1 Postulated location of nascent polypeptide chains within free ribosomes and rough microsomes, based on experiments with proteolytic enzymes.

irreversible, transfer of the nascent chain into the intracisternal space, without compromising the intrinsic impermeability of the membrane. Recently, we accumulated further evidence in support of this model. Light chains of immunoglobulin, completed in vitro by rough microsomes from murine myeloma, were subjected to posttranslational proteolysis and subsequently to polyacrylamide gel electrophoresis (PAGE) in sodium dodecyl sulfate (SDS). They were shown to be protected, i.e., not reduced in size (Blobel and Dobberstein, 1975*a*). Furthermore, we demonstrated (Blobel and Dobberstein, 1975*b*) that chain segregation in vitro is a cotranslational, not a posttranslational event (see below).

On the Nature of the Ribosome-Membrane Junction

Detailed information on the nature of the ligands in the ribosome as well as in the membrane is still lacking. The results obtained with hydrolytic enzymes as probes suggest that ribosomal RNA (Blobel and Potter, 1967) does not play a direct role in ribosome attachment but that the linkage is mediated by ribosomal and membrane proteins (Sabatini and Blobel, 1970). From disassembly studies (Adelman et al., 1973) using increasing concentrations of monovalent ions in the presence of magnesium, it could be concluded that the interaction is electrostatic in nature. However, with rat liver rough microsomes, it was found that at high salt concentrations more than half of the ribosomes remained bound to the membrane. Release of these ribosomes from the membrane required discharge of the nascent chain by puromycin, in addition to high salt concentrations. There was no ribosome release if the puromycin discharge of the nascent chain was performed at low salt concentrations. Thus a role of the nascent chain in ribosome-membrane interaction is indicated by these experiments although this was revealed only at high and unphysiological salt concentrations. It is conceivable that the formation of the native tertiary structure of the protein is initiated in the luminal space by those amino terminal portions of the nascent chain which are already transferred. Folding of this segregated portion of the nascent chain could explain its role in anchoring the ribosome to the membrane under conditions of high salt treatment. On the basis of these disassembly studies, it was possible to define two classes among membrane-bound ribosomes: one group released by high salt alone and the other group requiring high salt as well as puromycin for release.

Ribosome-Membrane Junction and mRNA: The Signal Hypothesis

After it became evident that the ribosome-membrane junction may function in a translation-coupled transfer of nascent polypeptide chains across the microsomal membrane and in this capacity constitute an important element in the secretory process, the intriguing question arose as to the mechanism underlying the selective translation of mRNAs for secretory proteins on membrane-bound ribosomes. In an attempt to provide a working hypothesis, we proposed (Blobel and Sabatini, 1971) that mRNAs for secretory proteins contain information that is translated into a unique amino terminal sequence of the nascent chain that in turn triggers ribosome attachment to the membrane. Recently (Blobel and Dobberstein, 1975*a*), we have incorporated this postulate into a more comprehensive scheme which we call the "signal hypothesis" (see Fig. 2). We postulated that all mRNAs for secretory proteins contain a sequence of codons—referred to as "signal codons"—localized on the 3′ side of the AUG initiation codon. Initiation of protein synthesis and translation of the signal codons, as well as the following codons, takes place on free ribosomes. Only after the so-called signal sequence of the nascent chain has emerged from within the large ribosomal subunit is the ribosome attached to the membrane. The signal sequence is thought to interact with the membrane and to cause the association of several ribosome receptor proteins which form a tunnel through the membrane. In their

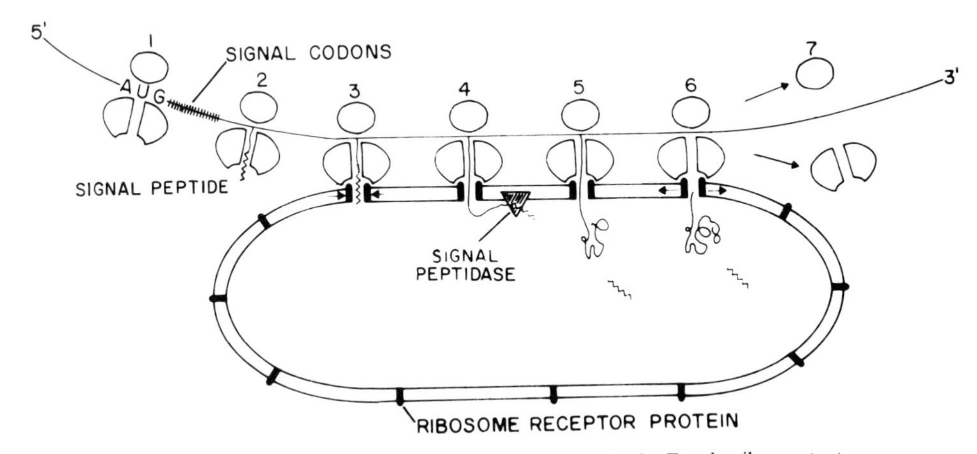

FIGURE 2 Schematic illustration of the signal hypothesis. For details, see text.

associated form these proteins interact with multiple sites on the large ribosomal subunit linking up the tunnel in the large ribosomal subunit with the newly formed tunnel in the membrane. This coordinated binding of the ribosome to the associated ribosome receptor proteins would "cross-link" the latter and thus maintain the tunnel even after the signal sequence has passed into the cisternal space. The association of ribosome receptor proteins induced by the signal peptide effectively segregates the latter from the lipid bilayer in preparation for its own crossing as well as the passage of the contiguous remainder of the nascent chain. Soon after the signal peptide region has been transferred into the luminal space, an enzyme—referred to as "signal peptidase"—will cleave it from the uncompleted remainder of the nascent chain. The "processed" nascent chain continues to grow and after chain termination and release will be entirely segregated within the cisternal space. It is proposed that the ribosome is then detached from the membrane. Displacement is probably mediated by a factor which was recently isolated and designated "detachment factor" (Blobel, 1976). Detachment of the ribosome results in dissociation of the ribosome receptor proteins, their lateral diffusion in the plane of the membrane, and elimination of the tunnel.

Thus, tunnel formation in the membrane, coordinated binding of the ribosome to the membrane, and removal of the signal sequence from the nascent, uncompleted chain by signal peptidase are envisioned as cotranslational events. The posttranslational removal of the ribosome and elimination of the tunnel in the membrane restore the original condition and, therefore, complete the cycle.

Evidence for the Signal Hypothesis

Data in support of predictions made on the basis of the signal hypothesis have so far come essentially from three different approaches. In the first approach, mRNAs for a variety of secretory proteins have been translated in vitro in the absence of microsomal membrane, i.e., in the absence of signal peptidase (see Fig. 2). Unlike their "authentic" secreted counterparts, these in vitro-synthesized proteins could be expected to retain their signal sequence. Unequivocal evidence for this was provided by radiosequence analysis of these proteins with consecutive Edman degradation. It was found that the amino terminal sequence of the authentic protein or proprotein was reached anywhere after 16–23 steps. Detailed analysis of this sort has so far been performed on several light chains of mouse immunoglobulin (Burstein et al., 1976), bovine parathyroid hormone (Kemper et al., 1976), dog pancreas trypsinogen (Devillers-Thiery et al., 1975), and rat insulin (Chan et al., 1976). Furthermore, a considerable number of other secretory proteins synthesized by translation of mRNA in vitro in the absence of membranes have been shown to be synthesized as larger precursors. Although sequence data have not yet been reported, it is likely that these in vitro-synthesized polypeptides again differ from their authentic counterparts by retaining their amino terminal signal sequence. However, it is conceivable that not all secretory proteins are synthesized in vitro as larger precursors. Some secretory proteins could have evolved which retained their signal peptide region (e.g., as a result of an alteration of the site for signal peptidase) for some additional, posttranslational function. On the other hand, sequence alterations rendering the signal peptide region resistant to signal peptidase may play a role in pathology. Retention of the signal peptide may cause partial or total loss of activity of the secreted proteins.

To distinguish the in vitro-synthesized precursors from the authentic secreted proteins and their proforms (such as proinsulin, proalbumin, proparathyroid hormone, etc.), the prefix "pre" has been adopted (Kemper et al., 1974). Thus "pre" refers to a primary polypeptide which includes the short-lived signal sequence cleaved by the microsomal signal peptidase and "pro" refers to the first proteolytic product which—as a result of subsequent proteolytic events involving intra- or extracellular proteases—will yield the final secretory protein.

In view of the signal hypothesis, the synthesis of presecretory proteins in vitro is due to the absence of membranes (containing signal peptidase) in the protein-synthesizing system. In vivo, in the presence of ER, the signal sequence will be removed from the nascent chain and, therefore, is not present in the completed chain. One cannot exclude, however, the possibility that completed preproteins are synthesized also in vivo. The signal sequence could escape processing by signal peptidase, but this is an unlikely event in view of the observed tight coupling of chain segregation and processing (see below). Alternatively, because of a limited number of ribosome receptor proteins, translation could continue on a free ribosome without attachment to the membrane. Thus "nontopological" translation would result in the synthesis of presecretory proteins which would be neither segregated nor processed. Considerations of this sort make it evident that in the normal cell there must be a regulated relationship between the amount of translated mRNAs for secretory protein and the amount of ribosome receptor proteins in the ER membrane. Imbalance in this ratio, i.e., reduction in ribosome receptor proteins resulting in nontopological translation, would abort secretion.

The second approach which has generated a wealth of data in favor of the signal hypothesis is based on in vitro reconstitution of functional rough microsomes from heterologous components (Blobel and Dobberstein, 1975*b*). For these experiments, free ribosomes as well as rough microsomes and soluble factors from a variety of sources have been utilized to test whether the results obtained with reassembled rough microsomes follow the predictions of the signal hypothesis. It was found that native small ribosomal subunits from rabbit reticulocytes (containing initiation factors) and large ribosomal subunits derived from free polysomes of rabbit reticulocytes by the puromycin-KCl procedure can function with ribosome-stripped microsomal vesicles derived from dog pancreas rough microsomes in a protein-synthesizing system in vitro (by using pH 5 enzymes from Krebs ascites cells), resulting in translation of mRNA for the light chain of immunoglobulin (isolated from murine myeloma) and the segregation of the translation product in a proteolysis-resistant space, presumably the intravesicular space of the stripped dog pancreas microsomes. No such segregation took place for the translation products of rabbit globin mRNA. These in vitro experiments established clearly that the information for segregation is encoded in the mRNA and not in any other component (such as factors or ribosomes) of the reconstituted system. Furthermore, it was found that the segregated translation product of the light chain mRNA was already proteolytically processed and reduced to the same size as the authentic light chain of immunoglobulin. These results were remarkable in that they demonstrated a widespread equivalence of sites. Thus, the signal sequence of a heterologous nascent presecretory protein was recognized by the membrane of dog microsomes, resulting in the attachment of heterologous ribosomes and transfer of the nascent chain into the intravesicular space. Moreover, the heterologous signal sequence of the nascent light chain was recognized by the dog microsomal signal peptidase and apparently removed in a correct manner.

Recently, we have extended our investigations concerning the equivalence of sites. We have shown (Dobberstein and Blobel, 1977) that wheat germ ribosomes translating mRNA for the light chain of murine immunoglobulin can attach to dog pancreas microsomal membranes, again resulting in segregation as well as processing of the translation product. Studies have been extended also to mRNAs from other mammalian secretory cells. We found (Lingappa, Devillers-Thiery, and Blobel, manuscript in preparation) that there is segregation and processing of the translation products of the mRNAs for bovine pituitary prolactin and growth hormone when dog pancreas microsomal membranes were present in the wheat germ system, whereas preprolactin and pregrowth hormone were synthesized in the absence of the dog microsomal membrane. Amino terminal sequence analysis of the in vitro-synthesized and processed prolactin revealed the same amino terminal sequence as authentic bovine prolactin. These experiments established that the dog pancreas microsomal signal peptidase removes the signal sequence of a heterologous nascent presecretory protein correctly, thereby generating the amino terminal sequence of the authentic secreted protein. The demonstration (see above) that plant ribosomes (wheat germ) can establish a functional junction with animal membranes (dog pancreas) indicates that the corresponding sites have been highly conserved in evolution. To determine whether there is conservation also of the signal sequence, we investigated the synthesis of fish insulin (Shields and Blobel, manuscript in preparation). It was shown that fish preproinsulin was synthesized in the wheat germ system in the absence of dog pancreas microsomal membranes, whereas in the presence of the latter, segregation and processing of preproinsulin to proinsulin took place. These data indicate that the signal sequence of fish may share common structural features with that of mammals so as to secure ribosome attachment to the mammalian microsomal membrane and recognition by mammalian signal peptidase.

From a comparison of the data so far available (see above), it is not yet possible to decide what particular features of the signal sequence trigger the postulated aggregation of ribosome receptor proteins and which provide the site for signal peptidase. From the heterologous reconstitution experiments described above, however, it is clear that signal sequences of a large variety of presecretory proteins may have features in common which were highly conserved through evolution. In this respect it will be interesting to investigate whether the signal mechanism has evolved in bacteria, in order to achieve exocytosis of secretory proteins, with the bacterial plasma membrane being the equivalent and the precursor of the ER, and whether the putative signal sequence of prokaryotes resembles that of eukaryotes.

The reconstitution experiments provided other circumstantial evidence supporting the signal hypothesis. It was shown (Blobel and Dobberstein, 1975*b*) that segregation and processing are strictly cotranslational events. Posttranslational incubation with dog pancreas microsomal membranes did not result in segregation of presecretory proteins, nor was there any removal of the signal sequence.

A third approach for testing predictions made on the basis of the signal hypothesis was to investigate the nature of the nascent chains in rough microsomes. For these experiments, we used murine myeloma rough microsomes which synthesize predominantly one secretory protein, the light chain of immunoglobulin. Our strategy was to operate under conditions which preclude initiation in vitro; to complete (in the presence of labeled amino acids) only those native chains already initiated in vivo; and to detect the completed chains by SDS-PAGE and autoradiography, by virtue of the labeled amino acid residues they had acquired during their in vitro completion. We also utilized "detached" polysomes prepared from rough microsomes after detergent solubilization of the microsomal membrane. Detached polysomes, unlike rough microsomes, presumably have lost the signal peptidase during membrane solubilization. We found (Blobel and Dobberstein, 1975*a*) that chain completion in rough microsomes only yielded processed chains, whereas detached polysomes also yielded some unprocessed precursor. Thus, in rough microsomes there are clearly some nascent chains which still contain the signal sequence. Excision of the sequence did not occur upon completion of these chains on detached polysomes, presumably because the endogenous signal peptidase had been removed or inactivated by the detergent treatment. Analysis of the completed chains and their characterization as processed or precursor forms during the course of in vitro chain completion revealed that ribosomes localized near the 3′ side of mRNA had already lost their signal sequence in vivo, whereas those near the 5′ side still retained it. These results demonstrated that in vivo the signal peptide is removed during translation, that is, before the chain is completed, but it remains to be determined when precisely during translation this event occurs. In deviation from the proposed scheme, it is, for instance, conceivable that the aggregated ribosome receptor proteins themselves contribute the signal peptidase activity so that ribosome binding and removal of the signal sequence would be synchronous events. The fact that chain completion by detached polysomes yielded some precursor chains of the same size as those synthesized by translation of mRNA in a heterologous, membrane-free system in vitro provided the definitive evidence that the latter are not an artifact resulting, for example, from faulty in vitro initiation.

We also used aurintricarboxylic acid in concentrations shown to interfere only with the attachment of ribosomes to the membrane (Borgese et al., 1974), and with chain initiation but not with chain elongation (Lodish et al., 1971). In the presence of this compound, chain completion by rough microsomes in vitro also yielded some precursor, presumably because ribosomes near the 5′ side of mRNA were prevented from attaching to the microsome membrane during chain completion, depriving their nascent chain from segregation as well as processing (Blobel and Dobberstein, 1975*a*). These results provide support for the proposal that chain initiation takes place on free ribosomes and that chain elongation must proceed to a certain extent before the translating ribosome can be attached to the membrane.

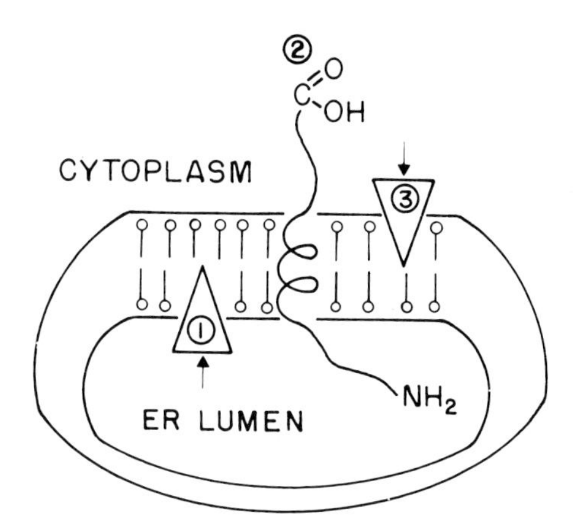

FIGURE 3 Scheme for three principal modes in the asymmetric distribution of membrane proteins in the lipid bilayer of the ER. For details, see text.

Outlook

The evidence accumulated so far has established the signal hypothesis as a useful working hypothesis. It should be emphasized, however, that some of the proposed details may have to be modified as more data become available. The aspect least documented by available experimental evidence is the proposed mechanism(s) for signal peptide induction of ribosome attachment and tunnel formation in the membrane; it should be clear that a few other alternatives are conceivable.

The signal hypothesis provides a formula by which specific proteins, such as secretory proteins, are transferred across the microsomal membrane. Transfer through a membrane from their cytoplasmic site of synthesis to their final destination is required also for a number of other proteins, some of which do not qualify as secretory in nature, e.g., lysosomal, peroxisomal, certain mitochondrial and chloroplast proteins, and even for certain membrane proteins. With respect to the latter, it is conceivable that the asymmetric distribution of membrane proteins in the lipid bilayer is to a large extent generated biosynthetically. In principle, one could conceive three different modes of deposition of proteins in the lipid bilayer (Fig. 3). Proteins localized on the cytoplasmic aspect of the membrane may be synthesized on free ribosomes. Subsequent folding could generate a hydrophobic domain which causes the insertion of the protein into the cytoplasmic aspect of the membrane. Proteins localized on the luminal aspect of the ER membrane (Fig. 3, case 1) or its derivatives may be synthesized essentially like secretory proteins by the signal mechanism, and subsequently inserted from the luminal aspect. Proteins that span the membrane, with their amino and carboxyl terminals separated by the lipid bilayer (Fig. 3, case 2), may also utilize the signal mechanism. However, in this case a portion of the nascent chain may become locked in the membrane during translation, as indicated in Fig. 4. Fusion of an ER vesicle, or derivative thereof, with the plasma membrane would then result in a reversal of polarity (see Fig. 5), i.e., the amino terminal would face the extracellular space, whereas the carboxyl terminal would point to the intracellular space.

It remains to be investigated whether the signal mechanism is utilized also for the synthesis of the other above-mentioned classes of proteins. If lysosomal and peroxisomal proteins were to be synthesized and segregated by the rough ER using the signal mechanism, the problem arises of how these proteins are subsequently sorted out from secretory proteins with which they presumably coexist in the intracisternal space of the ER.

For some of the mitochondrial and chloroplast proteins synthesized in the cytoplasm but localized in the innermost compartment of these semiautonomous cell organs, the problem arises of how these proteins are transferred across two membranes. In yeast, cytoplasmic ribosomes bound to the outer mitochondrial membrane (Kellems et al., 1975) have been demonstrated at areas where both outer and inner membranes are joined. It is conceivable, therefore, that ribosome binding to these joined membranes is triggered by the signal mechanism and results in the cotranslational transfer of the nascent protein into the innermost compartment of mitochondria. Junctions between cytoplasmic ribosomes and fused portions of the chloroplast envelope, on the other hand, have not been demonstrated. A formula other than the

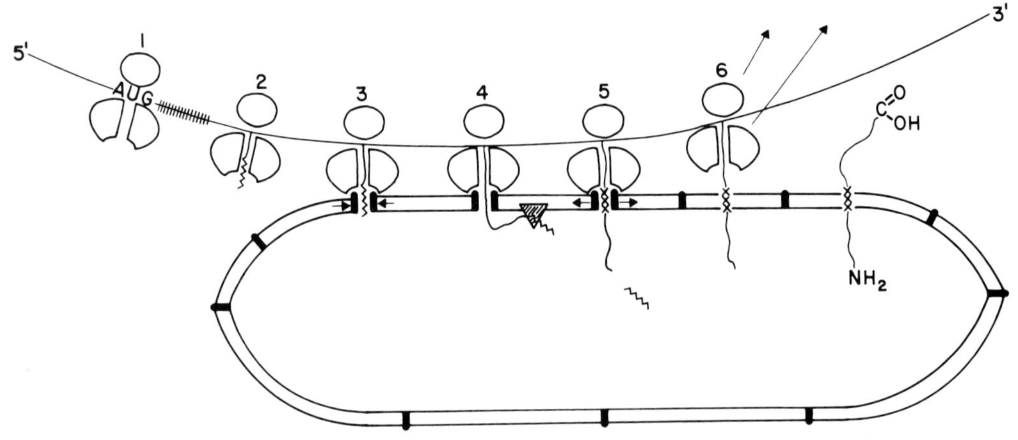

FIGURE 4 Scheme for a cotranslational deposition of a transmembrane protein into the ER membrane (see also Fig. 3). For details, see text.

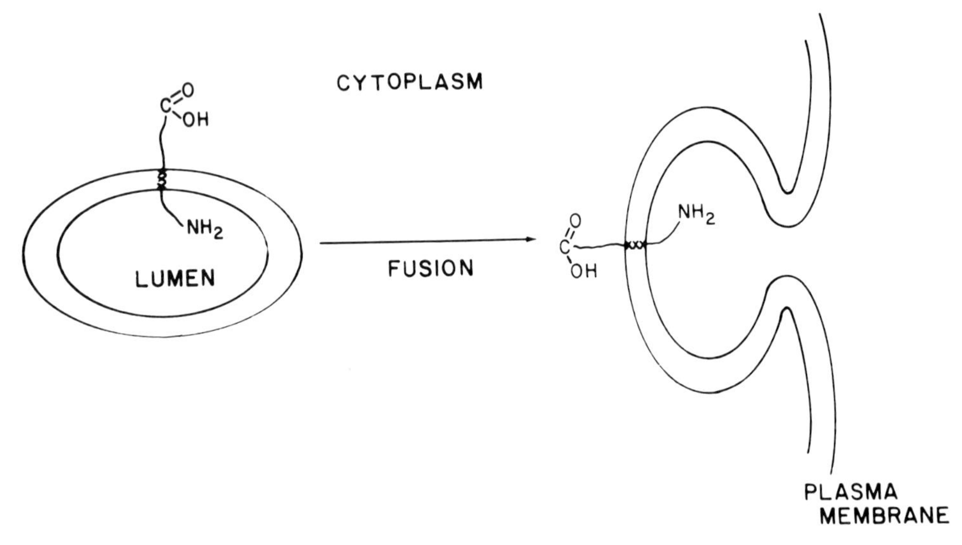

FIGURE 5 Localization of amino and carboxyl terminals of a transmembrane protein synthesized by rough ER (see Fig. 4) and transported by vesicle fusion to the plasma membrane.

signal mechanism, therefore, may have evolved to deal with the problem of transferring proteins synthesized in the cytoplasm across two membranes into the innermost compartment of mitochondria and chloroplasts. For the synthesis of proteins within semiautonomous cell organs, the signal mechanism may be utilized to achieve cotranslational insertion into membranes, because junctions between chloroplast ribosomes and thylakoid membranes (Chua et al., 1973) and mitochondrial ribosomes and inner mitochondrial membranes (Kuriyama and Luck, 1973) have been demonstrated.

The emergence in evolution of the lipid bilayer as the basic structural element of biological membranes provided an effective means to compartmentalize large as well as small molecules. The intrinsic impermeability of the lipid bilayer, however, also imposed severe restrictions. These were apparently overcome by the emergence of membrane proteins, able to communicate with the outside (receptor proteins) and able to facilitate transport (transport proteins). The evolution of the complex signal mechanism and other putative formulas (see above) for the unidirectional and selective transfer of proteins across membranes represents a fundamental solution to an important biological problem.

The synthesis of macromolecules via larger precursors is not a new concept. Most RNA molecules are known to be synthesized as precursors although it is not possible yet to assign functions to these precursor sequences. The synthesis of the precursors for proteins described here clearly points to a new concept, namely, that precursor sequences may have topological functions important for the intracellular or extracellular distribution of proteins.

ACKNOWLEDGMENT

I thank Dr. G. E. Palade for his many helpful comments on this manuscript.

REFERENCES

ADELMAN, M. R., D. D. SABATINI, and G. BLOBEL. 1973. Ribosome-membrane interaction: nondestructive disassembly of rat liver rough microsomes into ribosomal and membranous components. *J. Cell Biol.* **56:**206–229.

BLOBEL, G. 1976. Extraction from free ribosomes of a factor mediating ribosome detachment from rough microsomes. *Biochem. Biophys. Res. Commun.* **68:** 1–7.

BLOBEL, G., and B. DOBBERSTEIN. 1975*a*. Transfer of proteins across membranes. I. Presence of proteolytically processed and unprocessed nascent immunoglobulin light chains on membrane-bound ribosomes of murine myeloma. *J. Cell Biol.* **67:**835–851.

BLOBEL, G., and B. DOBBERSTEIN. 1975*b*. Transfer of proteins across membranes. II. Reconstitution of functional rough microsomes from heterologous components. *J. Cell Biol.* **67:**852–862.

BLOBEL, G., and V. R. POTTER. 1967. Studies on free and membrane-bound ribosomes in rat liver. II. Interaction of ribosomes with membranes. *J. Mol. Biol.* **26:**293–301.

BLOBEL, G., and D. D. SABATINI. 1970. Controlled proteolysis of nascent polypeptides in rat liver cell fractions. I. Location of the polypeptides within ribosomes. *J. Cell Biol.* **45:**130–145.

BLOBEL, G., and D. D. SABATINI. 1971. Ribosome-membrane interaction in eukaryotic cells. *In* Biomembranes. L. A. Manson, editor. Plenum Publishing Corp., New York. **2:**193–195.

BORGESE, N., W. MOK, G. KREIBICH, and D. D. SABATINI. 1974. Ribosomal-membrane interaction: *in vitro* binding of ribosomes to microsomal membranes. *J. Mol. Biol.* **88:**559–580.

BURSTEIN, Y., F. KANTOR, and I. SCHECHTER. 1976. Partial amino-acid sequence of the precursor of an immunoglobulin light chain containing NH_2-terminal pyroglutamic acid. *Proc. Natl. Acad. Sci. U. S. A.* **73:**2604–2608.

CHAN, S. J., P. KLEIN, and D. F. STEINER. 1976. Cell-free synthesis of rat preproinsulins: characterization and partial amino acid sequence determination. *Proc. Natl. Acad. Sci. U. S. A.* **73:**1964–1968.

CHUA, N. H., G. BLOBEL, P. SIEKEVITZ, and G. E. PALADE. 1973. Attachment of chloroplast polysomes to thylakoid membranes in *Chlamydomonas reinhardtii. Proc. Natl. Acad. Sci. U. S. A.* **70:**1554–1558.

DEVILLERS-THIERY, A., T. KINDT, G. SCHEELE, and G. BLOBEL. 1975. Homology in amino-terminal sequence of precursors to pancreatic secretory proteins. *Proc. Natl. Acad. Sci. U. S. A.* **72:**5016–5020.

DOBBERSTEIN, B., and G. BLOBEL. 1977. Functional interaction of plant ribosomes with animal microsomal membranes. *Biochem. Biophys. Res. Commun.* In press.

KELLEMS, R. E., V. F. ALLISON, and R. A. BUTOW. 1975. Cytoplasmic type 80S ribosomes associated with yeast mitochondria. IV. Attachment of ribosomes to the outer membrane of isolated mitochondria. *J. Cell Biol.* **65:**1–14.

KEMPER, B., J. F. HABENER, M. D. ERNST, J. T. POTTS, JR., and A. RICH. 1976. Pre-proparathyroid hormone: Analysis of radioactive tryptic peptides and amino acid sequence. *Biochemistry.* **15:**15–19.

KEMPER, B., J. F. HABENER, R. C. MULLIGAN, J. T. POTTS, JR., and A. RICH. 1974. Pre-proparathyroid hormone: a direct translation product of parathyroid messenger RNA. *Proc. Natl. Acad. Sci. U. S. A.* **31:**3731–3735.

KURIYAMA, Y., and D. LUCK. 1973. Membrane-associated ribosomes in mitochondria of *Neurospora crassa. J. Cell Biol.* **59:**776–784.

LODISH, H. F., D. HOUSMAN, and M. JACOBSEN. 1971. Initiation of hemoglobin synthesis: specific inhibition by antibiotics and bacteriophage ribonucleic acid. *Biochemistry.* **10:**2348–2356.

MALKIN, L. I., and A. RICH. 1967. Partial resistance of nascent polypeptide chains to proteolytic digestion due to ribosomal shielding. *J. Mol. Biol.* **26:**329–346.

PALADE, G. E. 1955. A small particulate component of the cytoplasm. *J. Biophys. Biochem. Cytol.* **1:**59–68.

PALADE, G. E. 1958. Microsomes and ribonucleoprotein particles. *In* Microsomal Particles and Protein Synthesis. First Symposium of the Biophysical Society. Richard B. Roberts, editor. Pergamon Press, Inc., New York. 36–61.

PALADE, G. E. 1975. Intracellular aspects of the process of protein synthesis. *Science (Wash., D. C.)* **189:**347–358.

REDMAN, C. M., P. SIEKEVITZ, and G. E. PALADE. 1966. Synthesis and transfer of amylase in pigeon pancreatic microsomes. *J. Biol. Chem.* **241:**1150–1158.

SABATINI, D., and G. BLOBEL. 1970. Controlled proteolysis of nascent polypeptides in rat liver cell fractions. II. Location of the polypeptides in rough microsomes. *J. Cell Biol.* **45:**146–157.

ROLE OF THE GOLGI APPARATUS IN TERMINAL GLYCOSYLATION

C. P. LEBLOND and G. BENNETT

The presence of secretory granules within and next to the Golgi apparatus has long been interpreted as indicating the involvement of this organelle in processes of secretion (for review, see Whaley, 1975). The existence of carbohydrate in a wide variety of secretory products was revealed by staining with the periodic acid-Schiff technique (Leblond, 1950). The Golgi apparatus also reacted positively with this technique, suggesting that it played a role in the production of the carbohydrate material. When this technique was adapted to electron microscopy by substituting silver methenamine for the Schiff reagent (van Heyningen, 1965), the saccules of the Golgi apparatus, as well as secretory products, lysosomes, and the plasma membrane were positively stained in more than 20 cell types; it was proposed that the Golgi apparatus played a major role in the addition of carbohydrate residues to glycoproteins destined for migration to secretory products, lysosomes, or the plasma membrane (Rambourg et al., 1969).

Structure of the Golgi Apparatus

This organelle consists of several stacks of flattened saccules. The stacks are continuous with one another, as first suggested by observation in ameloblasts (Kallenbach et al., 1963) and mucous cells (Neutra and Leblond, 1969), and later demonstrated in a variety of cells by Rambourg et al. (1974). Each Golgi stack is formed of several superimposed saccules, as illustrated in cross section in Fig. 1. It is usually possible to distinguish in the stack a convex surface variously referred to as the forming, cis, or entry face, and a concave surface referred to as the mature, trans, or exit face. The lowest of the four saccules of the stack (forming face) appears to be periodically interrupted on account of its fenestrated nature, whereas the saccule above it possesses a more expanded lumen than do the others. Associated with the saccule on the forming face are small transfer vesicles (arrow), which are believed to bud off nearby cisternae of rough endoplasmic reticulum. Associated with the mature face of the Golgi stack is an elongated saccular structure exhibiting regional enlargements of its lumen that contain flocculent material. This saccular structure was originally described as GERL by Novikoff, who showed that, unlike the regular Golgi saccules, it was rich in acid phosphatase (see Novikoff et al., 1971). Novikoff, as well as Hand (1971), showed that the enlargements of the GERL saccule become secretory granules; they may accordingly be referred to as presecretory granules, but they are commonly called condensing vacuoles.

The carbohydrates of Golgi apparatus and other organelles will be examined in a thyroid follicular cell stained by the periodic acid acid-chromic acid silver methenamine technique (hereafter referred to as PA-silver) and viewed at rather low magnification in Fig. 2. The unstained forming face of a Golgi stack (Gf) is separated by faint lines, corresponding to stained saccules, from the mature face (Gm), where dark dots represent the stained condensing vacuoles. In the neighborhood of the Golgi apparatus and also in the apical cytoplasm are scattered dark dots which represent the secretory granules (usually called apical vesicles in thyroid cells; AV). The lumen of the thyroid follicle is filled by homogeneously stained secretory material, the colloid. Finally, two other cell components are stained: the contents of the lysosomal bodies scattered throughout the cytoplasm and the plasma membrane.

Similar results were obtained with the columnar cells of the intestinal epithelium. Examination of

C. P. LEBLOND and G. BENNETT Department of Anatomy, McGill University, Montreal, Canada

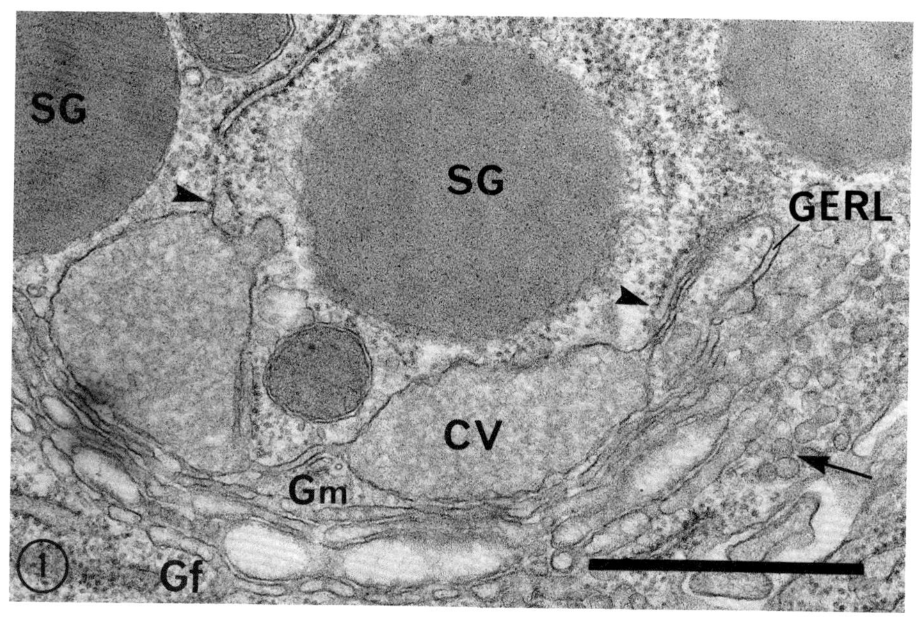

FIGURE 1 Golgi apparatus in a secretory cell of the rat parotid gland (courtesy of Arthur R. Hand). The forming face (*Gf*) is at the base of the picture. At the arrow the transfer vesicles are seen. The first of the four saccules shows periodic interruptions; the second is somewhat distended, whereas the third and fourth are flattened. Then at the mature face (*Gm*), the GERL saccule shows distensions, the condensing vacuoles (*CV*). Toward the two ends of the GERL saccule, cisternae of rough endoplasmic reticulum are in close proximity (arrowheads), at the right with the saccule itself and at the left with a condensing vacuole. Above, several secretory granules (*SG*) may be seen. 1 mm = 1,000. × 37,500.

the Golgi apparatus (Fig. 3) at a higher magnification than of the thyroid cell shows the individual saccules of the stack as solid profiles, indicating that the luminal contents are stained as well, perhaps, as the membranous wall of the saccules. A gradient in staining may also be seen with the saccule at the forming face of the stack, staining only very lightly, whereas those near the mature face are heavily stained. This staining gradient suggests that increasing amounts of carbohydrate material are added to substances as they pass from the forming to the mature face of the Golgi apparatus. The condensing vacuoles have a stained wall; similar smooth-surfaced vesicles are scattered throughout the cytoplasm and concentrated in the apical region as in the thyroid cell. Lysosomes and the plasma membrane are also stained (Fig. 3). The staining is prominent at the apical surface of the plasma membrane (Fig. 4). Not only does a dense line of stain outline the surface of microvilli, but stained filaments project outward from it. The stained external surface of the plasma membrane and the filaments make up a carbohydrate-rich layer, which has been referred to either as "cell coat" or "glycocalyx" (Rambourg and Leblond, 1967). In the cytoplasm below the apical microvillar surface, the smooth-surfaced vesicles show a positively stained membranous wall and also exhibit stained filaments projecting inward from the membranous wall. One of the vesicles at the left of Fig. 4 possesses a long neck in continuity with the apical plasma membrane. It can be visualized that a fusion process may be occurring here, in which the filaments lining the vesicle will become those of the cell coat covering the microvilli.

The nature of the carbohydrate material stained in the above micrographs may be inferred from its staining properties and from the steps involved in the preparation of the tissue. The PA-silver technique stains the 1, 2 glycol groups of glycoproteins and glycolipids as well as glycogen. The staining reactions were unaffected by salivary amylase treatment, indicating that glycogen was not involved. The use of such nonpolar solvents as acetone or propylene oxide in the tissue preparation steps would be expected to dissolve out most, although perhaps not all, of the tissue glycolipids.

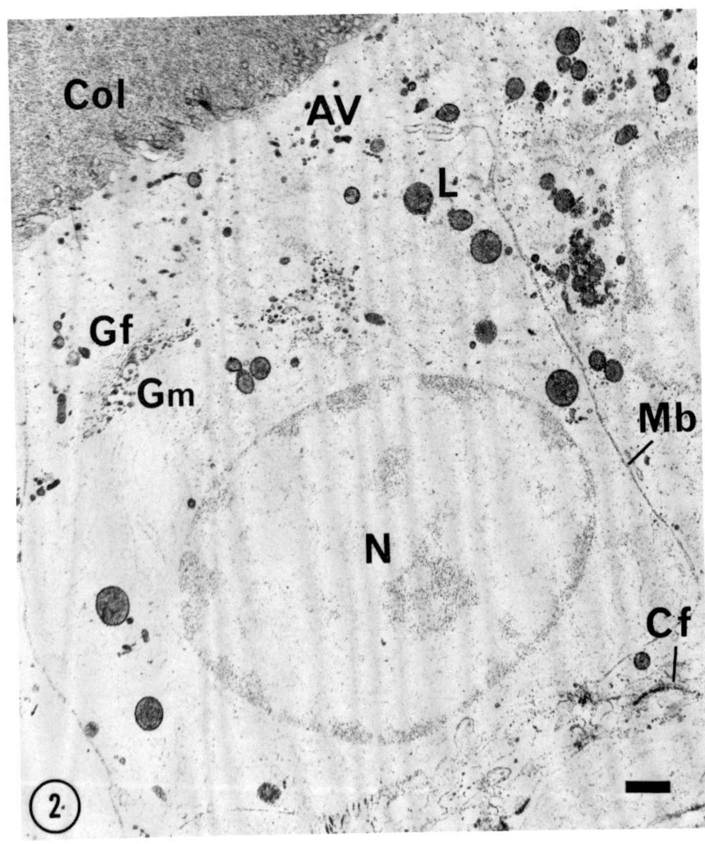

FIGURE 2 Follicular cell in rat thyroid gland, stained by the PA-silver technique (courtesy of W. Hernandez). From top left down, a series of stained structures are seen: the luminal colloid (*Col*), the apical plasma membrane with microvilli, the dotlike apical vesicles (*AV*) in the apical cytoplasm, the Golgi apparatus with the forming face (*Gf*) and the mature face (*Gm*; the latter includes dotlike condensing vacuoles), the relatively large lysosomal bodies (*L*) scattered throughout the cytoplasm, the lateral plasma membrane (*Mb*), and, outside the cell, groups of collagen fibers (*Cf*). × 6,000.

Thus, the remaining stained material could be expected to be composed mainly, if not exclusively, of glycoproteins.

Structure and Formation of Glycoprotein Side Chains

Glycoproteins comprise a group of widely diversified compounds, but all consist of a polypeptide backbone linked covalently to one or more oligosaccharide side chains. Figure 5 illustrates some typical side chains. Example A shows a side chain from human IgM; this is similar to the short side chains of thyroglobulin to be examined below. Note that all terminal sugars are mannose residues. Example B depicts the larger side chains of thyroglobulin, in which one may distinguish a core rich in mannose; the terminal sugars consist of galactose, sialic acid, and, attached to the core, fucose residues. Example C represents a glycoprotein from submaxillary mucus with several terminal residues, that is, *N*-acetylgalactosamine, sialic acid, and fucose.

To examine the role of the Golgi apparatus in the formation of glycoproteins, we have injected labeled sugars into young rats and studied the initial intracellular site of their incorporation into glycoproteins by autoradiography of various tissues. It was known from biochemical evidence that the synthesis of carbohydrate side chains begins soon after or even before completion of synthesis of the polypeptide backbone. The individual monosaccharide residues are then added one by one to form the growing carbohydrate side chain. To be added, each monosaccharide must first be phosphorylated and then complexed with a specific nucleoside. Then the nucleoside donates its sugar residue to the end of the growing carbohydrate side chain (or a branch thereof) in a reaction

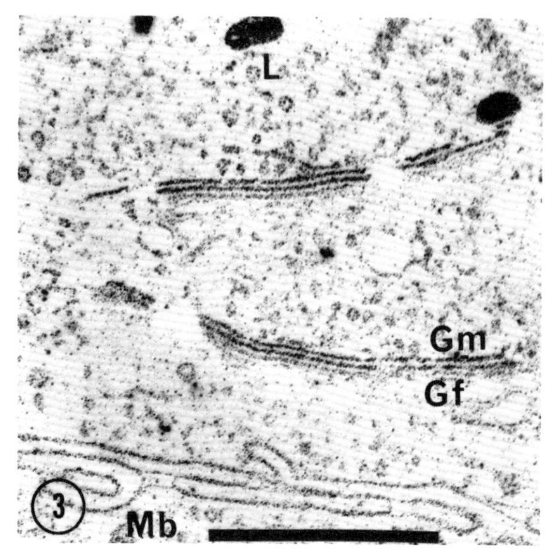

FIGURE 3 Columnar cell from rat small intestine, stained by the PA-silver technique (courtesy of W. Hernandez). At the base of the picture, the lateral plasma membranes (*Mb*) of adjacent cells are stained. Above, two Golgi stacks show a staining gradient from the forming face (*Gf*) to the mature face (*Gm*) where vesicles with a stained wall are interpreted as condensing vacuoles. The densely stained bodies at top of the picture are lysosomes (*L*). × 30,000.

catalyzed by a glycosyltransferase, which is highly specific for both the monosaccharide to be added and the acceptor monosaccharide residue of the incomplete glycoprotein. In the case of the core *N*-acetylglucosamine and mannose residues of asparagine-linked side chains (examples A and B in Fig. 5), a dolicholphosphate lipid intermediate is involved in the addition reaction.

Synthesis of Secretory Glycoproteins with Thyroglobulin as a Model

Previous autoradiographic studies utilizing labeled amino acids had shown that, in thyroid follicular cells, protein—presumed to be mainly thyroglobulin—was synthesized on the ribosomes of the rough endoplasmic reticulum. This newly synthesized protein then migrated with time through the lumen of the cisternae of the rough endoplasmic reticulum to the Golgi apparatus and from there, via condensing vacuoles and secretory granules (called apical vesicles), to the lumen of the thyroid follicle (Nadler et al., 1964; Bjorkman et al., 1974).

When [^{3}H]mannose was injected into young rats or incubated with thyroid tissue from similar animals, the initial uptake of the label into glycoproteins was found to occur exclusively in the rough endoplasmic reticulum (Fig. 6; Whur et al., 1969). The uptake of label was completely inhibited by puromycin, indicating that the sugar

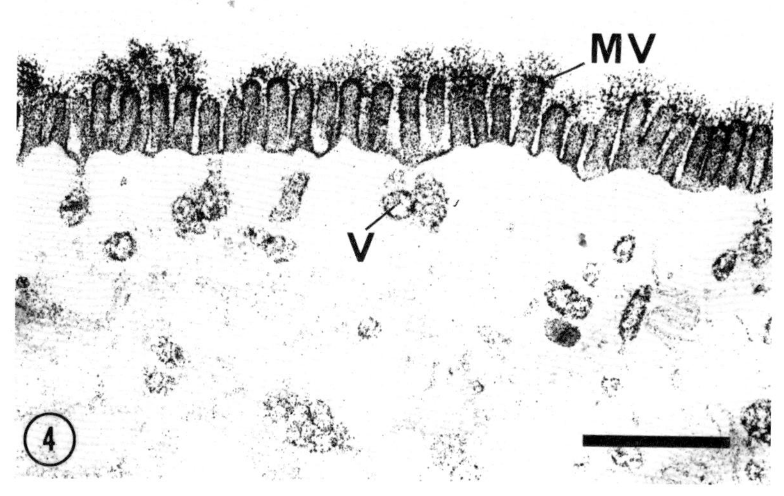

FIGURE 4 Columnar cell of rat ascending colon stained by the PA-silver technique (courtesy of John Michaels). The microvilli (*MV*) show staining of the plasma membrane as well as of filaments projecting outwards. Below, vesicles (*V*) show staining of the wall as well as of their content, a fibrillar material. One vesicle (left) exhibits a long neck which is in continuity with the apical plasma membrane. (Notice that the vesicles are larger in the ascending colon than in small intestine.) × 20,000.

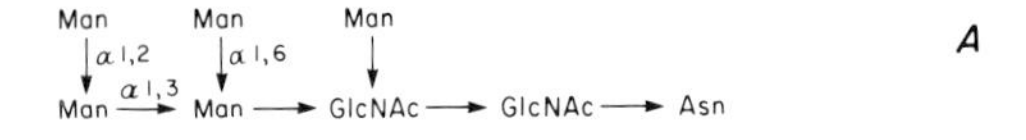

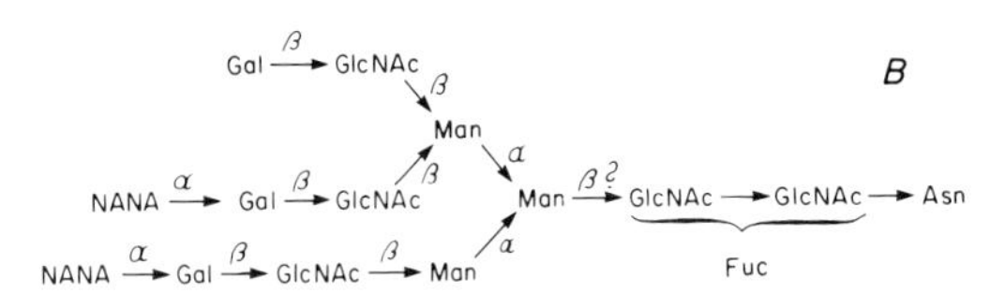

C

GalNAc α→ Gal β→ GalNAc → Ser (Thr)
α Fu
α NANA

FIGURE 5 Typical side chains of glycoproteins: (A) human IgM; (B) large side chain of thyroglobulin; (C) side chain of submaxillary mucus. Man, mannose; GlcNAc, *N*-acetylglucosamine; Asn, asparagine; Gal, galactose; NANA, sialic acid; Fuc, fucose; GalNAc, *N*-acetylgalactosamine; Ser, serine; Thr, threonine.

was being added to growing glycoproteins very soon after the completion of their polypeptide chains. Inasmuch as some of the short side chains of thyroglobulin (similar to example A in Fig. 5) end in mannose residues, terminal glycosylation of these side chains is probably occurring in the rough endoplasmic reticulum. In the case of the more complex side chains of thyroglobulin (example B in Fig. 5), only the core region would be completed in the rough endoplasmic reticulum.

When, on the other hand, labeled sugars in the terminal sequences of the more complex side chains were injected into young rats, the label was initially taken up in the Golgi apparatus. For example, at 5 min after [^{3}H]fucose injection (Haddad et al., 1971), the silver grains occurring over thyroid follicular cells were found to be almost exclusively localized over the Golgi saccules (Fig. 7). No label was observed over the colloid secretion product in the follicle lumen, over the secretion granules, or even over the condensing vacuoles seen within the GERL structure. This evidence indicated that fucose residues were being added to the more complex side chains of thyroglobulin as the molecules passed through the saccules of the Golgi apparatus. Because fucose always occurs in a terminal position at the end of a side chain or one of its side branches, the completion of synthesis of these side chains or branches, i.e., terminal glycosylation, would take place in the Golgi saccules. Only after the incorporation of [^{3}H]fucose do the glycoproteins migrate to the GERL structure to be packaged into secretion granules.

The question as to whether terminal glycosylation of all carbohydrate side chains occurred before the glycoproteins left the Golgi saccules could not be answered from the above evidence, however, because, as seen in Fig. 5, many side chains or their branches end not in fucose but in galactose or sialic acid (NANA). [^{3}H]Galactose was shown by Whur et al. (1969) to be also mainly incorporated into glycoproteins within Golgi saccules. Investigation of the site of incorporation of sialic acid was more difficult, for administered exogenous sialic acid does not enter cells effectively; after intracerebral injection into mice, less than 1% of the administered dose was incorporated (DeVries and Barondes, 1971), and recent in vitro experiments have similarly detected only a small incorporation of exogenous-labeled sialic acid into glycoproteins and glycolipids of cultured fibroblasts (Hirschberg and Goodman, 1976). However, biochemical studies have shown that *N*-

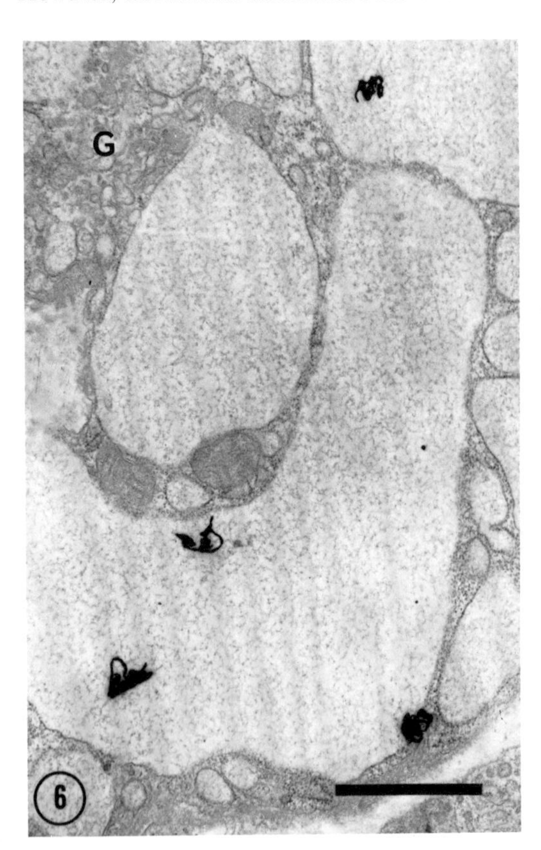

FIGURE 6 Autoradiograph of a portion of a follicular cell from a thyroid lobe incubated with [^{3}H]mannose for 15 min. Silver grains occur over rough endoplasmic reticulum cisternae; none occur over the Golgi apparatus (*G*). × 20,500.

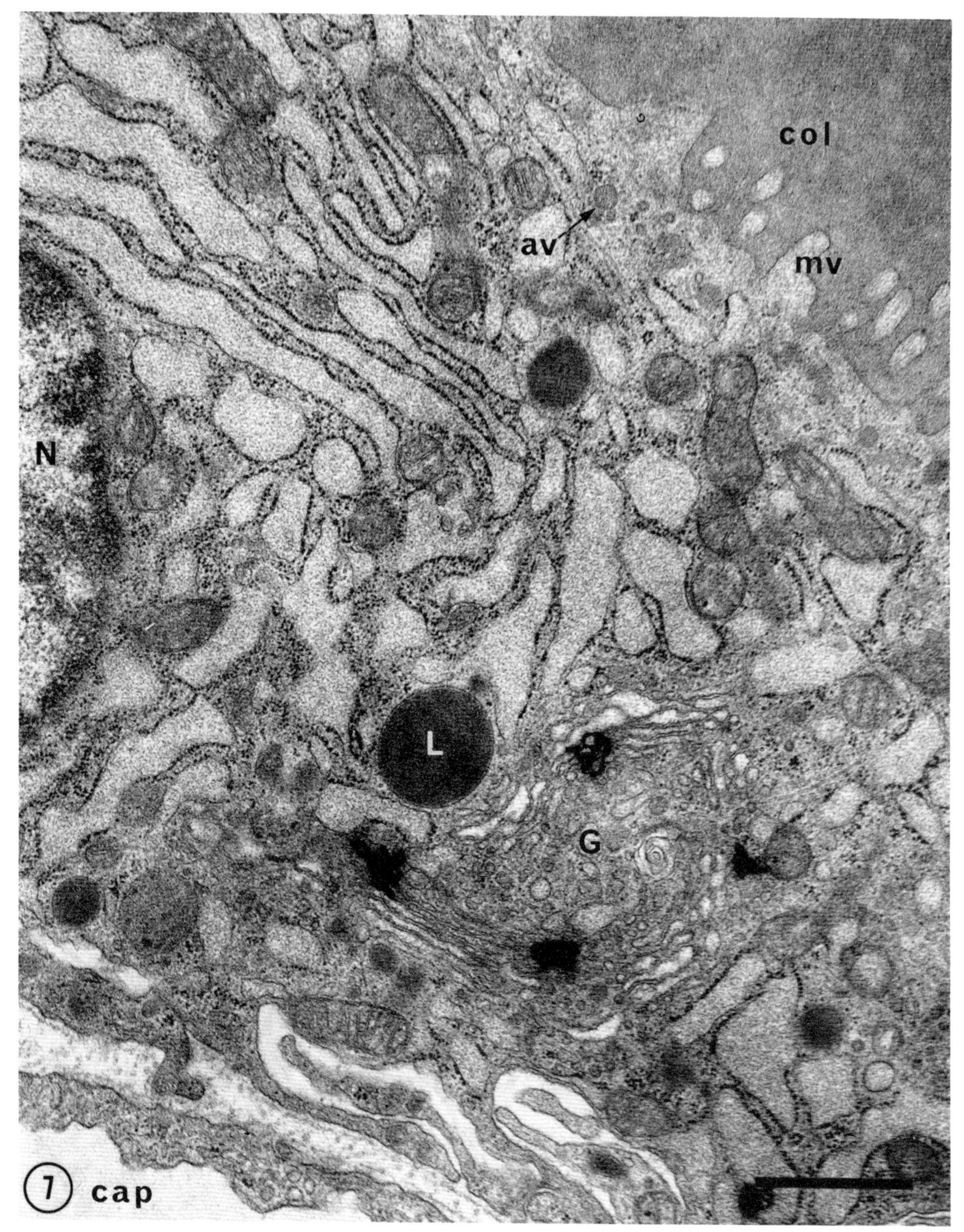

FIGURE 7 Autoradiograph of a thyroid follicular cell from a young rat sacrificed 5 min after [^{3}H]fucose injection. The Golgi apparatus in lower center shows four silver grains overlying the saccules. The central region (labeled *G*) containing condensing vacuoles is not labeled at this early time. *N*, nucleus; *L*, lysosome; *av*, apical vesicle; *mv*, microvillus; *col*, colloid; *cap*, capillary. × 23,000.

acetylmannosamine can serve as a fairly specific precursor for the sialic acid residues of glycoproteins. This sugar is an intermediate compound produced in the normal intracellular biosynthesis of sialic acid. When [^{3}H]*N*-acetylmannosamine was administered in vitro to thyroid hemilobes (Monaco and Robbins, 1973) and the glycoproteins were analyzed, virtually all of the label resided in sialic acid residues.

We have injected [^{3}H]*N*-acetylmannosamine

into young rats intravenously. In light microscope autoradiographs of thyroid follicular cells of animals sacrificed 10 min after injection, the reaction was relatively weak and its visualization required long autoradiographic exposure times; it could be seen, however, that the label was localized in the supranuclear region of the cytoplasm, a region normally occupied by the Golgi apparatus (in collaboration with F. Kan and D. O'Shaughnessy).

In conclusion, the fucose, galactose, and sialic acid residues of the large chains of thyroglobulin all appear to be added within the Golgi apparatus. These results are in accord with the localization of transferases determined by Schacter and his colleagues (see Schacter, 1974), showing that galactosyl, fucosyl, and sialyltransferases are concentrated in Golgi-enriched cell fractions of hepatocytes. Inasmuch as sialic acid and fucose, along with some galactose, account for the majority of the terminal sugar residues of the side chains of thyroglobulins and many other glycoproteins, the addition of these sugar residues to glycoproteins in the Golgi apparatus confirms that the intracellular synthesis of glycoproteins is completed at this site.

Close examination of the distribution of silver grains over the Golgi stack at early time intervals after [^{3}H]fucose injection (Fig. 7) showed that the grains were not preferentially localized over the forming or mature face of the stack, but rather were randomly distributed, suggesting that material is added throughout the stack. The fact that the saccules near the forming face are only very lightly stained with PA-silver, however, indicates that relatively poorly glycosylated substances from the rough endoplasmic reticulum enter at this face. As they pass to the mature face they would acquire near-terminal and terminal residues, thus accounting for the observed staining gradient and the resultant heavy staining of saccules near the mature face.

With time, thyroglobulin, whether labeled by [^{3}H]mannose in the rough endoplasmic reticulum or by [^{3}H]fucose, galactose, or sialic acid in the Golgi apparatus, is transported by apical vesicles to the apex of follicular cells, where it is secreted into the colloid-containing lumen (as shown in the case of fucose by Haddad et al., 1971).

Synthesis of Plasma Membrane and Lysosomal Glycoproteins with the Columnar Cells of the Intestinal Epithelium as a Model

As in the case of thyroid follicular cells, studies with labeled amino acids in these columnar cells had shown that the synthesis of the polypeptide portion of glycoproteins occurred in the rough endoplasmic reticulum. After injection of young rats with labeled mannose, a weak reaction in light microscope autoradiographs was initially scattered over most of the cell cytoplasm, suggesting an uptake in the rough endoplasmic reticulum. By 4 h, some reaction was seen over microvilli.

After injection of [^{3}H]galactose or [^{3}H]fucose, on the other hand, a strong localized reaction appeared at early time intervals over the saccules of the Golgi apparatus (Fig. 8). A similar localization was recently seen after [^{3}H]*N*-acetylmannosa-

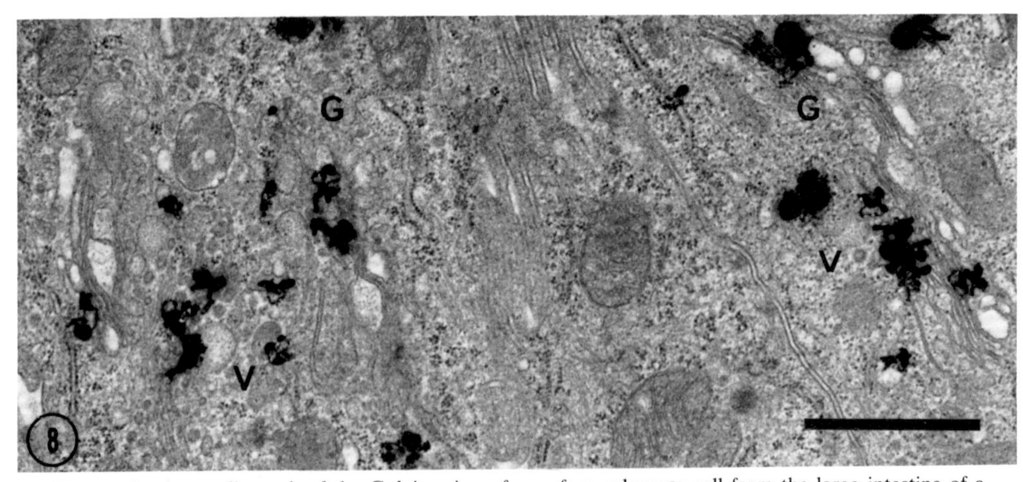

FIGURE 8 Autoradiograph of the Golgi region of a surface columnar cell from the large intestine of a young rat sacrificed 20 min after [^{3}H]fucose injection (from Michaels and Leblond, 1976). Most of the silver grains are localized over the Golgi apparatus (*G*). Some occur over vesicles (*V*) along the mature face. × 24,000.

mine injection (Fig. 9). With all of these sugars, no significant number of silver grains was observed at the cell surface at early time intervals after injection. It thus appears that in the case of membrane glycoprotein synthesis, as in the case of secretory glycoprotein synthesis, terminal glycosylation occurred before the glycoprotein had left the Golgi apparatus.

With time after injection, the labeled glycoproteins were seen to migrate to the apical and lateral plasma membrane as well as to lysosomes (Fig. 10). An analysis of grain counts carried out after different time intervals in columnar cells of the large intestine (Fig. 11) indicated that the labeled glycoproteins destined for the apical plasma membrane migrated from the Golgi apparatus to vesicles of the type shown in Fig. 3, and thence to the apical cell surface. These vesicles thus appear to carry membrane glycoproteins to the cell surface.

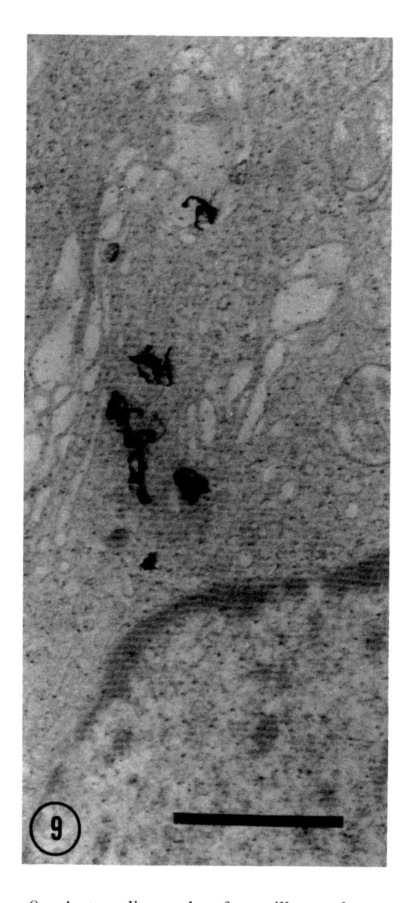

FIGURE 9 Autoradiograph of a villus columnar cell from the small intestine of a young rat sacrificed 10 min after an injection of [^{3}H]*N*-acetylmannosamine. The Golgi apparatus is labeled. × 23,400.

Although the precise mechanism of this transport is not known, it is clear that the vesicles are derived from the mature face of the Golgi apparatus. Labeled glycoproteins in the membranous wall of the Golgi saccule would thus become incorporated into the membranous wall of the vesicle. After passage to the apex of the cell, the vesicle would fuse with the plasma membrane and the labeled glycoproteins would thus become incorporated into the plasma membrane itself (Michaels and Leblond, 1976). It is, therefore, likely that there is a continuous flow of membranes from Golgi apparatus to plasma membrane (with the mode of disposal of the excess membrane not yet clarified). However, as it is known that glycoproteins may diffuse laterally within the plane of membranes, it is also possible that no net conversion of membrane from Golgi to vesicle wall to plasma membrane occurs, but that the glycoproteins migrate from the membrane of one compartment to that of the next during periods of continuity.

It is clear that the fucose label in some cells appears with time in all of the three types of glycoproteins: secretory, plasma membrane, and lysosomal. Thus, in liver cells, the [^{3}H]fucose label originally taken up into the Golgi was later observed at the cell surface, over lysosomes, and in blood plasma, where it presumably represents plasma proteins elaborated by these cells. It appears, therefore, that the Golgi apparatus synthesizes the three types of glycoprotein simultaneously.

In conclusion, the three main types of glycoproteins, secretory, plasma membrane, and lysosomal, are elaborated in a similar pattern (Fig. 12). The polypeptide backbone is synthesized on the ribosomes, and, at least in the case of asparagine-linked side chains, the addition of carbohydrate begins immediately. Short side chains ending in mannose are completed in the cisternae of the rough endoplasmic reticulum, whereas the near-terminal and terminal sugars of large side chains are added in the saccules of the Golgi apparatus.

The gradient of periodic acid-silver methenamine staining from forming to maturing faces indicates that poorly glycosylated proteins enter the forming face and migrate toward the mature face while accumulating periodic acid reactive sugar residues. They then pass into condensing vauoles, which, in at least some cell types, appear to form from a specialized GERL structure; these condensing vacuoles are then released as secretory granules.

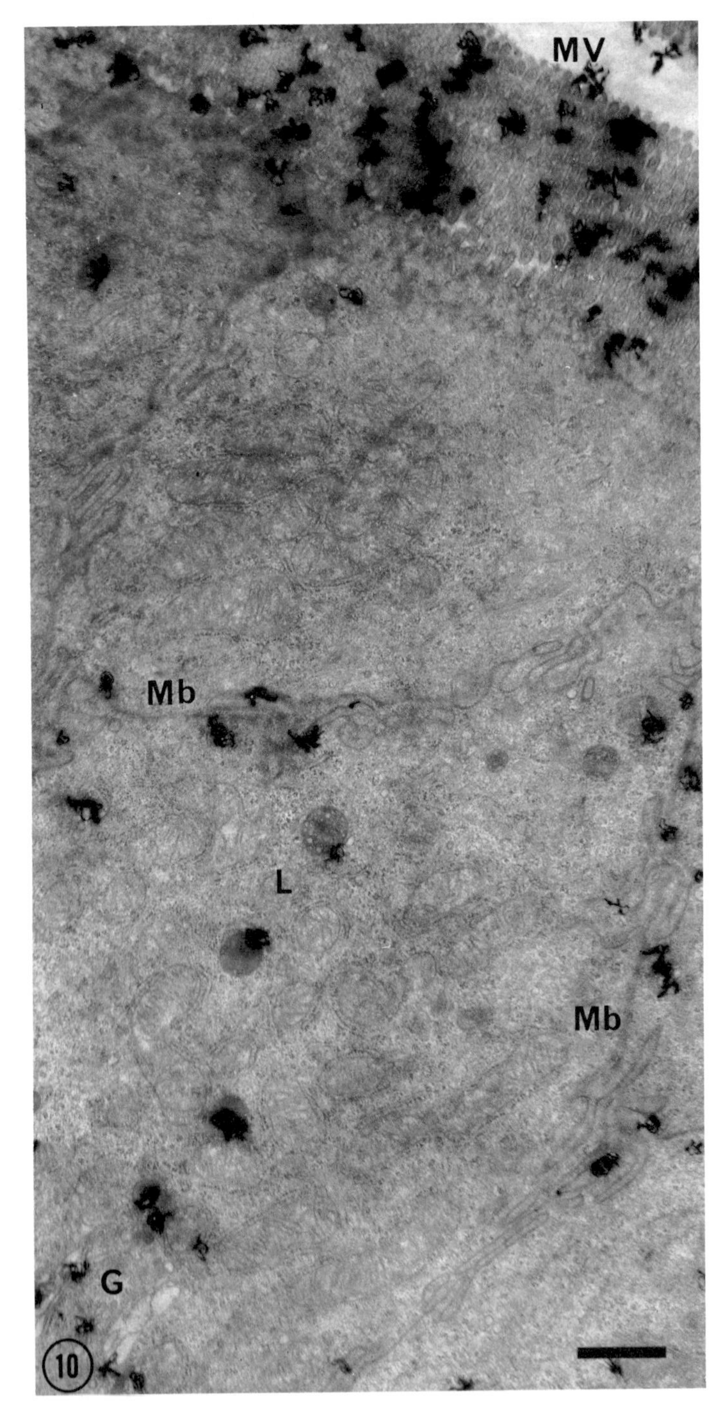

FIGURE 10 Autoradiograph of villus columnar cells from the small intestine of a young rat sacrificed 4 h after [^{3}H]fucose injection. In this oblique section, a few grains remain over the Golgi apparatus (*G*) but most grains are localized over lateral plasma membrane (*Mb*), lysosomes (*L*), and the apical microvillus membrane (*MV*). × 12,000.

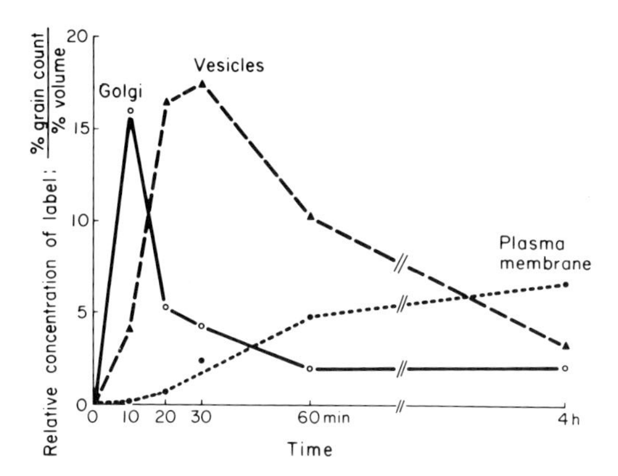

FIGURE 11 Changes in the relative concentration of [³H]fucose with time after injection in the epithelial cells of the mouse ascending colon. The initial peak of radioactivity is in the Golgi apparatus; it is followed by a peak in the smooth vesicles. Finally by 4 h, the peak is in the plasma membrane.

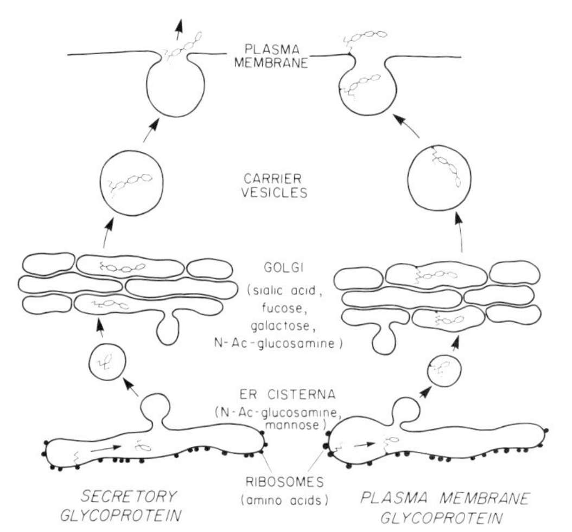

FIGURE 12 Diagram (modified from Schacter, 1974) comparing current hypotheses on the elaboration of secretory glycoproteins (left) and plasma membrane glycoproteins (right). Starting from the base, it is seen that both arise on ribosomes. The secretory glycoprotein is then released free to the lumen of ER cisternae (although there is evidence that some secretory glycoproteins may remain membrane-bound until their transfer to the Golgi apparatus; Redman and Cherian, 1972). The plasma membrane glycoprotein remains inserted into the wall of ER cisternae, in which it may move by lateral diffusion without becoming detached of the membrane. In the case of glycoproteins whose side chains contain a core of *N*-acetylmannosamine and mannose residues linked to asparagine (example B in Fig. 5), the core sugars are added in the rough endoplasmic reticulum, as symbolized by a hexagon. In the case of side chains linked to serine or threonine by *N*-acetylgalactosamine, the site of addition of the proximal sugar residue has not been as well established. Both secretory and membrane glycoproteins are then transferred to the Golgi apparatus by transfer vesicles; there, near-terminal and terminal sugar residues are inserted into the free secretory glycoprotein and the attached plasma membrane glycoprotein. The completed glycoproteins are introduced into carrier vesicles within the GERL saccules and thus transported to the surface. In the case of secretory material, the vesicle filled with the free glycoprotein is known as a secretory granule and releases its content to the outside by exocytosis. Similarly, lysosomal glycoproteins are probably carried from Golgi saccules in vesicles and, after fusion of the vesicular wall with the lysosomal membrane, released into the lysosome (not illustrated). In the case of plasma membrane glycoprotein, the vesicle wall fuses with the plasma membrane and the glycoprotein becomes incorporated into the plasma membrane either by conversion of the vesicle wall into plasma membrane or by lateral diffusion.

ACKNOWLEDGMENT

The work reported by the authors and their collaborators was done with the support of grants from the Medical Research Council of Canada.

REFERENCES

BJORKMAN, U., R. EKHOLM, L.-G. ELMQUIST, L. E. ERICKSON, A. MELANDER, and S. SMEDS. 1974. Induced unidirectional transport of protein into the thyroid follicular lumen. *Endocrinology*. **95:**1506–1517.

DE VRIES, G., and S. BARONDES. 1971. Incorporation of [¹⁴C]*N*-acetyl neuraminic acid into brain glycoproteins and gangliosides in vivo. *J. Neurochem.* **18:**101–105.

HADDAD, A., M. D. SMITH, A. HERSCOVICS, N. J. NADLER, and C. P. LEBLOND. 1971. Radioautographic study of *in vivo* incorporation of fucose-³H into thyroglobulin by rat thyroid follicular cells. *J. Cell Biol.* **49:**856–882.

HAND, A. R. 1971. Morphology and cytochemistry of the Golgi apparatus of rat salivary gland acinar cells. *Am. J. Anat.* **130:**141–157.

HIRSCHBERG, C., and S. GOODMAN. 1976. Free sialic acid uptake by fibroblasts and subsequent incorporation into glycoproteins and glycolipids. *J. Cell. Biol.* **70:**149*a* (Abstr.).

KALLENBACH, E., E. SANDBORN, and H. WARSHAWSKY. 1963. The Golgi apparatus of the ameloblast of the rat at the stage of enamel matrix formation. *J. Cell Biol.* **16:**629–632.

LEBLOND, C. P. 1950. Distribution of periodic acid-reactive carbohydrates in the adult rat. *Am. J. Anat.* **86:**1–49.

MICHAELS, J., and C. P. LEBLOND. 1976. Transport of glycoprotein from Golgi apparatus to cell surface by means of "carrier vesicles," as shown by radioautography of mouse colonic epithelium after injection of ^{3}H-fucose. *J. Microscop. Biol. Cell.* **25**:243–248.

Monaco, F., and J. Robbins. 1973. Incorporation of *N*-acetylmannosamine and *N*-acetylglucosamine into thyroglobulin in rat thyroid *in vitro*. *J. Biol. Chem.* **248**:2072–2077.

Nadler, N. J., B. A. Young, C. P. Leblond, and B. Mitmaker. 1964. Elaboration of thyroglobulin in the thyroid follicle. *Endocrinology.* **74**:333–354.

Neutra, M., and C. P. Leblond. 1969. The Golgi apparatus. *Sci. Am.* **220**:100–107.

Novikoff, P., A. Novikoff, N. Quintana, and J. Hauro. 1971. Golgi apparatus, GERL, and lysosomes of neurons in rat dorsal root ganglia, studied by thick section and thin section cytochemistry. *J. Cell Biol.* **50**:859–886.

Rambourg, A., Y. Clermont, and A. Marraud. 1974. Three-dimensional structure of the osmium impregnated Golgi-apparatus as seen in the high voltage electron microscope. *Am. J. Anat.* **140**:27–46.

Rambourg, A., W. Hernandez, and C. P. Leblond. 1969. Detection of periodic acid-reactive carbohydrate in Golgi saccules. *J. Cell Biol.* **40**:395–414.

Rambourg, A., and C. P. Leblond. 1967. Electron microscope observations on the carbohydrate-rich cell coat present at the surface of cells in the rat. *J. Cell Biol.* **32**:27–53.

Redman, C. M., and M. G. Cherian. 1972. The secretory pathways of rat serum glycoproteins and albumin. *J. Cell Biol.* **52**:231–245.

Schacter, H. 1974. The subcellular sites of glycosylation. *Biochem. Soc. Symp.* **40**:57–71.

van Heyningen, H. 1965. Correlated light and electron microscope observations on glycoprotein-containing globules in the follicular cells of the thyroid gland of the rat. *J. Histochem. Cytochem.* **13**:286–296.

Whaley, G. W. 1975. The Golgi apparatus. *In* Cell Biology Monographs. Vol. II. Springer-Verlag, New York. 190.

Whur, P., A. Herscovics, and C. P. Leblond. 1969. Radioautographic visualization of the incorporation of galactose-^{3}H and mannose-^{3}H by rat thyroids in vitro in relation to the stages of thyroglobulin synthesis. *J. Cell Biol.* **43**:289–311.

CONCLUDING REMARKS

GEORGE E. PALADE

The task assigned at this congress to the chairman of each symposium is to assess critically the evidence available in the corresponding field, and to point out the avenues along which future research promises to provide a better understanding of important but still unresolved issues. The main interest in our case centers on the process by which proteins and other macromolecules are produced, processed, and finally discharged by eukaryotic cells.

The first point to be made in relation to current and future work in this field (or any other) is that no data can be better than the preparations and procedures used to generate them. Research on macromolecular secretion in animal eukaryotic cells has been—and still is—heavily dependent on a few techniques, primarily cell-fractionation procedures and autoradiography, as clearly illustrated by the symposium's papers.

In principle, cell fractionation has the signal advantage of making possible the analysis of the secretory process at the level of individual subcellular components. Moreover, in a second step, via appropriate biochemical assays and separation procedures, the inquiry can be focused on the molecules and macromolecules present within every type of subcellular component, thereby allowing an analysis of the secretory process at the molecular level. In practice, however, the data used for the analysis of the process are collected on cell fractions, i.e., on preparations known to be heterogeneous in most cases. Attempts to eliminate or control heterogeneity are being made, and the results of new or modified cell-fractionation schemes are being monitored by morphological survey (electron microscopy) and biochemical assays (for "marker components"), but—with the exception of a few preparations that approach homogeneity—cell fractions currently in use are still quite heterogeneous mixtures. Some new problems are emerging because many of the traditional marker enzymes now appear to be associated with more than one subcellular component: this is particularly true, for instance, for the plasmalemma and the Golgi complex. But new monitoring approaches are also being developed: cytochemical (Widnell, 1972; Farquhar et al., 1974; Cheng and Farquhar, 1976) and autoradiographic (Reggio and Palade, 1976) procedures are applied to cell fractions to differentiate between activity contributed by particulate contaminants and activity associated with the indigenous particulates of the fraction under investigation. Of particular interest are cell-fractionation procedures (Leighton et al., 1968; Thines-Sempoux et al., 1969; Ehrenreich et al., 1973) that rely on the selective experimental modification of the density of certain subcellular components for improving either resolution or yield. It appears, therefore, that progress is being made—albeit slowly—on these techniques. At least the problems connected with the particulate contamination of cell fractions are generally recognized and reasonably well understood in the field.

Less generally recognized, although potentially more important and difficult, are problems caused by molecular contamination of particulates collected in various cell fractions. It is widely assumed that subcellular components survive, with little or no damage, the homogenization of cells or tissues; that leakage of soluble components from ruptured compartments is limited; and, hence, that relocation of such components by adsorption on other subcellular structures can be ignored. In fact, all these assumptions are wrong. Although there are no published systematic studies of this problem, there are already strong indications that leakage and relocation are quantitatively important artifacts (Tartakoff et al., 1975) of current cell-fractionation procedures which differ in extent from protein to protein and from one subcellular

GEORGE E. PALADE Section of Cell Biology, Yale University School of Medicine, New Haven, Connecticut

component to another. The problem of molecular contamination is compounded when the content of various membrane-bounded particles (or vesicles) is extracted in order to isolate the corresponding membranes. Current extraction procedures are far from fully efficient (Kreibich and Sabatini, 1974) and, as a result, the isolated membranes, potentially contaminated on their cytoplasmic side by proteins adsorbed from the final supernate, are contaminated on their cisternal (or inner) side by proteins absorbed or incompletely extracted from the content (Castle et al., 1975). Traditional procedures used in monitoring particulate contamination are, naturally, of no help in coping with these problems. New detection procedures must be devised to detect and assess molecular contamination of cell fractions.

Finally, another widely used but unwarranted assumption is that subcellular vesicular components—presumably isolated undamaged—remain intact past the end of experiments in which they are subjected to a wide variety of treatments. The structural and functional integrity of these components is rarely examined at the end of the experiments, although it is implied in interpreting the results.

At present, in investigating either old (e.g., distribution of secretory products) or new (functional interactions among the membranes of the secretory pathway) research problems, we are often trying to operate at levels of precision hardly compatible with the cell-fractionation procedures in current use. These procedures must be improved, the main goals being to increase the degree of homogeneity of the fractions, to reduce damage during homogenization, and to reduce relocation artifacts, or at least to quantitate them for adequate correction; in other words, to minimize and control particulate and molecular contamination of the cell fractions of interest. In any case, it should be evident that the assumption that cell fractions are reasonably free of particulate and molecular contaminants and, hence, can be equated to specific cell components as they exist *in situ,* is not justified and should not be accepted without adequate proof.

The still-lingering controversy (Rothman, 1975) about alternative pathways for the intracellular transport of secretory proteins (through the cytosol rather than cisternal compartments) and about other means for their discharge (by specific transport of individual macromolecules across the plasmalemma, rather than by exocytosis) is a good example of the types of problems generated when current cell-fractionation procedures are applied and their results accepted, without being critically assessed and checked by other independent methods of investigation.

The other technique of major importance to the field is autoradiography. In this case, the danger of artificial relocation of macromolecular secretory products is considerably smaller than in cell fractionation, but the information obtained is affected by limitations in resolution, specificity, and sensitivity. The resolution of the current procedures is not high enough for many subcellular components of interest; the technique locates classes of compounds, rather than specific molecules, and is highly dependent on differences in biosynthetic rates. The results are clear-cut when secretory proteins are synthesized at a considerably higher rate than are proteins for intracellular use; when the difference in rates is small, extensive statistical analysis of the data becomes necessary, and an unambiguous interpretation of the results is often impossible.

During the last few years, a considerable amount of work has been done to define the resolution of the technique and to put the distribution of autoradiographic grains on a reliable quantitative basis for radioactive isotopes and radioactivity detectors (photographic emulsions) in current use (Salpeter et al., 1969, 1977), but much more than this is needed for further progress. Improvements in resolution and sensitivity of radioactivity detectors would be highly desirable; in principle, such improvements are possible.

Moving now from methodology to new findings, it is clear that the signal hypothesis and the evidence already brought forward in its support (Blobel, this volume) are major developments in the field. They promise to provide an answer to many questions which have been afloat for a number of years. These concern the traffic of mRNAs to the two main classes of cytoplasmic polysomes and the means of producing, from common precursor-subunits, either free or attached polysomes. As always, a problem solved generates a few new problems in need of solutions. In this case, the immediate problems concern, on the one hand, the exact interactions among the signal sequence, the large ribosomal subunit, and the endoplasmic reticulum (ER) membrane that lead to attachment, and, on the other hand, the ultimate fate of the large number of sequences clipped by the endopeptidase of the ER membrane from the initial translation product.

Of course, the generality of the mechanisms of

attachment and the extent of conservation of the signal sequence through evolution remain to be explored further and established beyond the limits of the attractive sketch we have at present. But the entire development is appealing, because it reveals another aspect of the innate logic of the cellular systems and illustrates the relatively simple means by which basic controls are achieved.

Moreover, it opens new vistas: researchers are already looking for other signals specific for other "addresses"; it strengthens considerably the concept of segregation by vectorial transport; and it adds another example of large-size precursors for macromolecular products. In this case (as in the case of rRNAs, for instance), the removable segment of the chain plays the role of recognition site, tool, or countertool for the agonists involved in the sequence of processing operations that will eventually yield a final product. In fact, the same "principle" appears after discharge in the conversion of zymogens to active enzymes.

This development has been preceded by a relatively long period of accumulation of observations and data on the attached polysome-ER membrane system, but the final demonstration of a signal sequence came from work carried out in vitro by using mRNAs of known proteins and reconstituted systems of increasing degrees of complexity (from polysomes to rough microsomes). In this respect, the work on the signal sequence may set an example for the investigation of other problems connected with the secretory process.

By comparison with the synthesis and segregation steps, the progress made during the last few years in the analysis of the rest of the secretory process is modest. We have identified the main operations, and have succeeded in localizing to specific compartments a number of posttranslational modifications to which the secretory proteins are subjected. These modifications involve enzymic reactions which are either reasonably well understood or are accessible to investigation in terms of available biochemical information. They are, however, sequential in nature and, hence, coordinated with the transport of secretory proteins from one compartment to the next in series along the secretory pathway. Transport operations are dominated by specific membrane interactions which lead to membrane fusion-fission and eventually to continuity established between the interacting compartments. Somewhere in this chain of reactions energy is required, but the biochemistry of the whole process of specific membrane fusion-fission is essentially unknown and remains to be elucidated with the help of information obtained on model systems, i.e., artificial phospholipid vesicles (Papahadjopoulos et al., 1974), and on isolated secretion granules whose fusion can be induced in vitro under apparently simple conditions (Gratzl and Dahl, 1976).

Another problem to be settled concerns the exact pathway followed by secretory proteins in the Golgi complex and the means involved in their movement from a cisterna to its neighbor in the trans direction, asuming that the entire stack of cisternae is traversed by the pathway (cf. Leblond and Bennett, this volume). Finally, a traffic regulator must exist in the Golgi complex to channel usual secretory proteins to secretion granules and lysosomal enzymes to primary lysosomes. For a while it seemed plausible that the traffic control for lysosomal hydrolases involved the structure designated GERL (Novikoff et al., 1971), but more recent work by Novikoff (1976) equates GERL with Golgi condensing vacuoles. This view implies that the traffic control is either imperfect or is located somewhere else in the Golgi complex.

It should be clear that much work remains to be done on the secretory process, and it is reasonable to assume that the next phase in the analysis of this process will deal extensively with membrane interactions along the secretory pathway.

REFERENCES

Castle, J. D., J. D. Jamieson, and G. E. Palade. 1975. Secretion granules of the rabbit parotid gland: isolation, subfractionation, and characterization of the membrane and content subfractions. *J. Cell Biol.* **64:**182–210.

Cheng, H., and M. G. Farquhar. 1976. Presence of adenylate cyclase activity in Golgi and other fractions from rat liver. II. Cytochemical localization within Golgi and ER membranes. *J. Cell Biol.* **70:**671–684.

Ehrenreich, J. H., J. J. M. Bergeron, P. Siekevitz, and G. E. Palade. 1973. Golgi fractions prepared from rat liver homogenates. I. Isolation procedure and morphological characterization. *J. Cell Biol.* **59:**45–72.

Farquhar, M. G., J. J. M. Bergeron, and G. E. Palade. 1974. Cytochemistry of Golgi fractions prepared from rat liver. *J. Cell Biol.* **60:**8–25.

Gratzl, M., and G. Dahl. 1976. Ca^{2+}-induced fusion of Golgi-derived secretory vesicles isolated from rat liver. *FEBS* (*Fed. Eur. Biochem. Soc.*) *Lett.* **62:**142–145.

Kreibich, G., and D. D. Sabatini. 1974. Selective release of content from microsomal vesicles without membrane disassembly. II. Electrophoretic and immunological characterization of microsomal subfractions. *J. Cell Biol.* **61:**789–807.

LEIGHTON, F., B. POOLE, H. BEAUFAY, P. BAUDHUIN, J. W. COFFEY, S. FOWLER, and D. DE DUVE. 1968. The large-scale separation of peroxisomes and lysosomes from the livers of rats injected with Triton WR-1339. *J. Cell Biol.* **37:**482–513.

NOVIKOFF, A. B. 1976. The endoplasmic reticulum: a cytochemist's view (a review). *Proc. Natl. Acad. Sci. U. S. A.* **73:**2781–2787.

NOVIKOFF, P. M., A. B. NOVIKOFF, N. QUINTANA, and J. J. HAUW. 1971. Golgi apparatus, GERL, and lysosomes of neurons in rat dorsal root ganglia studied by thick section and thin section cytochemistry. *J. Cell Biol.* **50:**859–886.

PAPAHADJOPOULOS, D., G. POSTE, B. E. SCHAEFFER, and W. J. VAIL. 1974. Membrane fusion and molecular segregation in phospholipid vesicles. *Biochim. Biophys. Acta.* **352:**10–28.

REGGIO, H., and G. E. PALADE. 1976. Sulfated compounds in the secretion and zymogen granule content of the guinea pig pancreas. *J. Cell Biol.* **70:**360a (Abstr.).

ROTHMAN, S. S. 1975. Protein transport by the pancreas. *Science (Wash. D.C.).* **190:**747–753.

SALPETER, M. M., L. BACHMAN, and E. E. SALPETER. 1969. Resolution in electron microscope radioautography. *J. Cell Biol.* **41:**1–20.

SALPETER, M. M., H. C. FERTUCK, and E. E. SALPETER. 1977. Resolution in electron microscope autoradiography. III. Iodine-125 and the effect of heavy metal staining. *J. Cell Biol.* **72:**161–173.

TARTAKOFF, A. M., J. D. JAMIESON, G. A. SCHEELE, and G. E. PALADE. 1975. Studies on the pancreas of the guinea pig: parallel processing and discharge of exocrine proteins. *J. Biol. Chem.* **250:**2671–2677.

THINES-SEMPOUX, D., A. AMAR-COSTESEC, H. BEAUFAY, and J. BERTHET. 1969. The association of cholesterol, 5′-nucleotidase, and alkaline phosphodiesterase I with a distinct group of microsomal particles. *J. Cell Biol.* **43:**189–192.

WIDNELL, C. C. 1972. Cytochemical localization of 5′-nucleotidase in subcellular fractions isolated from rat liver. I. The origin of 5′-nucleotidase activity in microsomes. *J. Cell Biol.* **52:**542–558.

Microtubules and Flagella

MICROTUBULES AND FLAGELLA

J. B. OLMSTED

In the second edition of his classic book on the biology of cells, Wilson (1900) reviewed the observations of many of the nineteenth-century cytologists on the nature of cellular "ground plasm," and, in synthesizing their theories, postulated that there might be "a common element in the origin and function of the mitotic fibrillae, the centrosome and midbody, and the contractile substances of cilia, flagella and muscle fibers" (p. 323). The definition of the structural and biochemical nature of this "common element" has undergone considerable evolution and refinement in the intervening decades, and at these meetings has served as the basis for this symposium, for one on the molecular basis of motility, and for more than 200 contributed papers. This symposium focuses on microtubules, and although the cellular phenomena with which these structures are now known to be associated had been described for many years, it was only with the development of modern ultrastructural and biochemical techniques that their role in flagellar motility and in the subcellular organization of cytoplasm was more specifically defined. This paper briefly outlines the historical background which served as the conceptual basis for the current studies on flagellar motility and microtubule assembly. More detailed accounts of many aspects of microtubule structure and function are available in a number of recent books and reviews (Olmsted and Borisy, 1973; Hepler and Palevitz, 1974; Roberts, 1974; Wilson and Bryan, 1974; Borgers and de Brabander, 1975; Inoué and Stephens, 1975; Soifer, 1975; and Snyder and McIntosh, 1976).

Observations on the fibrillar nature of flagellar substructure were first made in the early twentieth century with silver staining to reveal 7–11 filaments in the tails of macerated sperm. However, a clearer definition was not possible until the mid-1940's, when whole mount preparations for electron microscopy and chromium-shadowing demonstrated the uniform diameters of the fibers in *Paramecium* cilia (Jakus and Hall, 1946). Subsequently, in a survey of a number of motile plant sperm, Manton and Clarke (1952) proposed that the basic substructure was composed of nine doublet and two singlet fibers. In the same decade, Hoffman-Berling (1955) used the techniques employed for studying muscle contraction to examine motility in other cellular systems. His work on glycerinated models of sperm, as well as the later work of Brokaw (1961), suggested that flagellar motility required ATP and served as a foundation for the more detailed analysis of mechanochemical transduction in flagella to be described subsequently.

During the same era of the 1950s, the nature of Wilson's "common element" in the cytoplasm was more difficult to elucidate, primarily because of the much greater lability of structures such as the mitotic spindle as compared to cilia and flagella. Although by the mid-1950s it had been suggested that a tubular element was a common feature of the mitotic spindle (Porter, 1954; de Harven and Bernhard, 1956) and axonal cytoplasm (Palay, 1956), the quality of fixation made further definitions of structure difficult. In addition, earlier attempts by Mazia and Dan (1952) to isolate and biochemically characterize the spindle fibers had been hampered by failures to stabilize the structure adequately. However, despite these problems encountered in studying the structure and chemistry of the cytoplasmic fibers, a number of observations were made by Inoué in the early 1950s which proved to be seminal in defining the association properties of the spindle fibers. Also, because these studies were carried out on living cells, the arguments prevalent since Wilson's time about the fixation-induced, artifactual nature of the spindle fibers could be dispelled. By the use of polarization optics, Inoué (1953) suggested that the bire-

J. B. OLMSTED Department of Biology, University of Rochester, Rochester, New York.

fringence seen in the mitotic apparatus was derived from the ordered arrangement of fibrous elements, and that shifts in birefringence during the course of mitosis were indicative of the labile nature of these structures. In addition, Inoué (1952) observed that the mitotic inhibitor colchicine caused the dissolution of the spindle birefringence, a reaction that could be reversed by washing out the drug; similar dissolution of the spindle was also seen with shifts to lower temperature. These observations led to the suggestion that the fibrous birefringent elements of the mitotic spindle were in equilibrium with a monomeric unit, and that this equilibrium could be shifted towards monomer by treatment with colchicine or low temperature. Although the chemical nature of the fibers was still undefined, these concepts laid the foundation for much of the subsequent work done on elucidating the function and assembly of microtubules in cells.

In the 1960s, much of the information on the chemistry and structure of both flagellar and cytoplasmic microtubules was obtained. In addition, numerous studies on the mechanism by which these organelles were involved in ciliary and flagellar motility and in cytoplasmic functions were initiated. As before, the studies on cilia and flagella slightly antedated those on cytoplasmic microtubules because of the greater ease of isolation and fixation. Thin sections of flagella confirmed Manton and Clarke's (1952) original prediction that the axonemal core was composed of nine doublet and two singlet fibers; the complete A-tubule of each doublet and the central pair of tubules had outer diameters of 240 Å and a clear lumen of 150 Å (Afzelius, 1959). Negatively stained preparations revealed that the tubule consisted of rows of protofilaments running parallel to the long axis with each row having a beaded substructure comprised of 40 Å globular units. In addition, structures were observed attached to the A-tubule of the doublets and interconnecting outer doublets to the region of the central pair; as described subsequently, these structures were found to have a role in motility.

The first detailed chemical analysis of the subunit in the flagellar tubules was carried out by Gibbons (1963, 1965) in the early 1960s. By the use of selective solubilization techniques to preferentially fractionate different parts of *Tetrahymena* cilia, the axonemal core was separated from membrane and matrix components and then further solubilized to remove the prominent projections on the A-tubule. The isolated arms from the A-tubule had ATPase activity, could be restored to the stripped axoneme, and were termed "dynein" in reference to their postulated force-generating function in ciliary motion. The major protein in the microtubular axoneme was defined as having a denatured sedimentation coefficient of 3–4S, and in subsequent work (Renaud et al., 1968; Mohri, 1968; Stephens, 1968) was found to have a native sedimentation coefficient of 6S, a denatured molecular weight of 60,000 and to bind 1–2 mol of guanine nucleotide. This protein was termed "tubulin" by Mohri (1968), the name now generally applied to the subunit of microtubules.

Given the highly ordered nature of the fibers in flagella, the mechanism by which these structures were involved in motility was also under investigation. Afzelius (1959) had originally suggested that, in analogy to muscle contraction, motility might be mediated by the relative sliding of the outer fibers. In detailed ultrastructural analyses of gill cilia in various stages of bend formation, Satir (1968) found that the relative displacement of doublets on opposite sides of an individual cilium was directly proportional to the degree of curvature. These observations, which resulted in the formulation of the "sliding filament" theory of flagellar motility, are discussed in greater detail in the following paper.

Work on cytoplasmic microtubules also began to expand rapidly during the last decade. The general preservation of microtubules in the cytoplasm was facilitated by the introduction of glutaraldehyde as a fixative (Sabatini et al., 1963), and in the same year, the term "microtubule" was coined to describe these straight cylinders, 240 Å in diameter with a clear lumen of 150 Å (Ledbetter and Porter, 1963; Slautterback, 1963). These tubular elements were noted not only to be similar to the flagellar fibers but also to be prevalent in the majority of cell types examined. The widespread distribution of these organelles prompted investigations to define their role in a variety of cellular processes. The best characterized cytoplasmic component of which microtubules constituted the major framework was the mitotic spindle; the mechanisms by which these structures are involved in chromosome movement is still the subject of current experimentation. However, since ultrastructural analyses now confirmed that Inoué's observations of spindle birefringence dissolution in the presence of cold or colchicine probably corresponded to the destruction of the microtubules, these agents were used extensively in defining other tubule-related functions. Because microtubules often parallel the axes of asymmetric cellular processes, a cytoskeletal role was postulated; clear documentation of this hypothesis was provided in a series of papers by Tilney and Porter (reviewed in Tilney, 1971) on the effects of various microtubule-disruptive agents on the maintenance of axopodial processes in *Actinosphaerium*. Studies with a number of other cell systems corroborated the role of microtubules in the development and maintenance of cell anisometry. In addition, as discussed more fully in several of the reviews and books cited earlier, disruption of microtubules by various agents was found to interfere with processes of intracellular transport (pigment granule migration, axonal flow, secretion), and recent studies suggest cytoplasmic microtubules may also have a role in surface membrane topography.

In view of these observations on multiplicity of distribution and function, Porter had already suggested in a review written in 1966 that the cellular localization of microtubules would have to be rigidly controlled both temporally and spatially. The concept that the formation of arrays of microtubules might shift temporally with various cellular activities had its antecedent in the proposal by Boveri that the fibrillae were not "permanent structures but formations which come and go with different phases of protoplasmic activity" (Wilson, 1924, p. 723). In addition, it had been postulated that the "archoplasm might pre-exist as a specific homogenous substance which, though not ordinarily visible, may become so by taking on the form of granules or fibrillae that crystallize" (Wilson, 1924, p. 723). Although too simplistic to explain the multitude of proteins now known to comprise the cytoplasm, these ideas were important in suggesting that an equilibrium existed between subunits and formed polymers which might be temporally shifted as a function of cell activity. Such a "dynamic equilibrium" was specifically postulated in Inoué's model for chromosome movement during mitosis (Inoué and Sato, 1967), and elucidation of the factors controlling this equilibrium is the focus of current research on the assembly of microtubules in cells. In addition to temporal regulation, Porter (1966) also observed that the formation of microtubules might be dictated spatially by sites that initiated and/or organized microtubule assembly. For example, centrioles were obvious candidates for the polar organization of mitotic spindle microtubules and, using flagellar regeneration systems (Rosenbaum and Child, 1967; Tamm, 1968), basal bodies could be confirmed as the sites for flagellar tubule growth. However, microtubules also appeared to arise from more amorphous, nonmicrotubular structures, such as chromosomal kinetochores, primitive spindle plaques (see Kubai, 1975, for review), and less localized centrosomal regions (for example, see Tilney and Goddard, 1970). The identification and molecular characterization of these sites is also of considerable current interest in understanding how microtubule assembly is regulated.

In addition to being widely defined morphologically, the chemistry of cytoplasmic microtubules was also elucidated during the 1960s, although by the use of methods less direct than those for flagellar characterization. There had been several attempts to isolate the mitotic apparatus and cytoplasmic microtubules in various stabilizing media, but no fraction was sufficiently pure to identify the microtubule subunit unambiguously. However, on the basis of kinetic studies on the uptake and release of colchicine from cultured cells, Taylor (1965) proposed that there was a cellular site to which colchicine bound; coupled with Inoué's observations on the effect of colchicine in destruction of the mitotic spindle, it was postulated that the site might constitute the subunit of the spindle fibers. Procedures were subsequently devised to enrich for the colchicine-binding fraction (Borisy and Taylor, 1967; Shelanski and Taylor, 1968; Weisenberg et al., 1968), which was characterized as a protein with a native sedimentation coefficient of 6S and a molecular weight of 110,000 daltons. In addition, the native molecule bound 1 mol of colchicine and 2 mol of guanine nucleotide, one exchangeable and the other tightly bound. The correlation of maximum colchicine binding activity to tissues with high mitotic indices or large amounts of microtubules, as well as the similarity of chemical characteristics to those determined for the tubule protein from flagella, led to the conclusion that the colchicine binding moiety was the subunit of microtubules. Tubulin has now been characterized from a variety of sources and found to be composed of two distinct polypeptide chains, termed α- and β-tubulin, each of 55,000 mol wt. Although the existence of the native molecule of 110,000 daltons as a heterodimer or a homodimer is still under investigation, partial sequence analyses of the α and β peptides from various species indicate that tubulin has been highly conserved during evolution (Luduena and Woodward, 1975). In addition, as reviewed by Wilson and Bryan (1974), the tubulin molecule has multiple ligand binding sites, and can also be phosphorylated (Eipper, 1974); the basic subunit therefore shows considerable molecular complexity.

With the basic biochemical and structural characterization of microtubules having been elucidated during the last decade, recent work has concentrated on more detailed analyses of the involvement of these structures in cellular processes. The development of model systems in which the role of microtubules and associated proteins in the generation of flagellar motion could be more readily studied has allowed substantiation and extension of the theories proposed earlier. In addition, the discovery by Weisenberg (1972) of conditions under which microtubule protein would assemble in vitro has made possible more detailed analyses of the mechanisms that might govern the formation of microtubules in cells, and the questions of equilibrium, nucleation, and ligand-mediated assembly have started to be studied at the molecular level.

REFERENCES

AFZELIUS, B. 1959. Electron microscopy of the sperm tail: results obtained with a new fixative. *J. Biophys. Biochem. Cytol.* **5**:269–278.

BORGERS, M., and M. DE BRABANDER, editors. 1975. Microtubules and Microtubule Inhibitors. American Elsevier Publishing Co. Inc., New York. 533.

BORISY, G. G., and E. W. TAYLOR. 1967. The mechanism of action of colchicine: binding of colchicine-^{3}H to cellular protein. *J. Cell Biol.* **34**:525–533.

BROKAW, C. 1961. Movement and nucleoside polyphosphatase activity of isolated flagella from *Polytoma uvella*. *Exp. Cell Res.* **22**:151–162.

DE HARVEN, E., and W. BERNHARD. 1956. Etude au microscope électronique de l'ultrastructure du centriole chez les vértebrés. Z. Zellforsch. **45**:378–398.

EIPPER, B. A. 1974. Rat brain tubulin and protein kinase activity. *J. Biol. Chem.* **249**:1398–1406.

GIBBONS, I. R. 1963. Studies on the protein components of cilia from *Tetrahymena pyriformis*. *Proc. Natl. Acad. Sci. U. S. A.* **50**:1002–1010.

GIBBONS, I. R. 1965. Chemical dissection of cilia. *Arch. Biol.* **76**:317–352.

HEPLER, P., and B. PALEVITZ. 1974. Microtubules and microfilaments. *Annu. Rev. Plant Physiol.* **25**:309–362.

HOFFMAN-BERLING, H. 1955. Geisselmodelle und adenosine triphosphat. *Biochim. Biophys. Acta.* **16**:146–154.

INOUÉ, S. 1952. The effect of colchicine on the microscopic and submicroscopic structure of the mitotic spindle. *Exp. Cell Res.* **2**(suppl.):305–318.

INOUÉ, S. 1953. Polarization optical studies of the mitotic spindle. I. The demonstration of spindle fibers in living cells. *Chromosoma (Berl.).* **5**:487–500.

INOUÉ, S., and H. SATO. 1967. Cell motility by labile association of molecules: the nature of mitotic spindle fibers and their role in chromosome movement. *J. Gen. Physiol.* **50**(suppl.):259–288.

INOUÉ, S., and R. STEPHENS, editors. 1975. Molecules and Cell Movement. Raven Press, New York. 450.

JAKUS, M. A., and C. E. HALL. 1946. Electron microscopic observations of the trichocysts and cilia in *Paramecium*. *Biol. Bull. (Woods Hole).* **91**:141–144.

KUBAI, D. 1975. The evolution of the mitotic spindle. *Int. Rev. Cytol.* **43**:167–227.

LEDBETTER, M., and K. PORTER. 1963. A "microtubule" in plant cell fine structure. *J. Cell Biol.* **19**:239–250.

LUDUENA, R. F., and D. O. WOODWARD. 1975. α and β tubulin: separation and partial sequence analysis. *Ann. N. Y. Acad. Sci.* **253**:272–283.

MANTON, I., and B. CLARKE. 1952. An electron microscopic study of the spermatozoid of *Sphagnum*. *J. Exp. Bot.* **3**:265–289.

MAZIA, D., and K. DAN. 1952. The isolation and biochemical characterization of the mitotic apparatus in dividing cells. *Proc. Natl. Acad. Sci. U. S. A.* **38**:827–838.

MOHRI, H. 1968. Amino acid composition of "tubulin" constituting microtubules of sperm flagella. *Nature (Lond.)* **217**:1053–1054.

OLMSTED, J. B., and G. G. BORISY. 1973. Microtubules. *Annu. Rev. Biochem.* **42**:507–540.

PALAY, S. 1956. Synapses in the central nervous system. J. Biophys. Biochem. Cytol. **2**(suppl.):193–201.

PORTER, K. R. 1954. Changes in cell fine structure accompanying mitosis. *In* Fine Structure of Cells. Wiley Interscience, New York. p. 236.

PORTER, K. R. 1966. Cytoplasmic microtubules and their functions. *In* Principles of Biomolecular Organization. G. E. W. Wolstenholme and M. O'Connor, editors. Little Brown & Co., Inc., Boston, Mass. 308–345.

RENAUD, F. L., A. J. ROWE, and I. R. GIBBONS. 1968. Some properties of the protein forming the outer fibers of cilia. *J. Cell Biol.* **36**:79–90.

ROBERTS, K. 1974. Cytoplasmic microtubules and their functions. *Prog. Biophys. Mol. Biol.* **28**:371–420.

ROSENBAUM, J. L., and F. CHILD. 1967. Flagellar regeneration in protozoan flagellates. *J. Cell Biol.* **34**:345–364.

SABATINI, D., K. BENSCH, and R. BARRNETT. 1963. Cytochemistry and electron microscopy: the preservation of cellular ultrastructure and enzymatic activity by aldehyde fixation. *J. Cell Biol.* **17**:19–58.

Satir, P. 1968. Studies on cilia. III. Further studies on the cilium tip and a "sliding filament" model of ciliary motility. *J. Cell Biol.* **39:**77–94.

Shelanski, M. L., and E. W. Taylor. 1968. Properties of the protein subunit of central pair and outer doublet microtubules of sea urchin flagella. *J. Cell Biol.* **38:**304–315.

Slautterback, D. 1963. Cytoplasmic microtubules. I. *Hydra. J. Cell Biol.* **18:**367–388.

Snyder, J., and J. R. McIntosh. 1976. Biochemistry and physiology of microtubules. *Annu. Rev. Biochem.* **45:**669–720.

Soifer, D., editor. 1975. The biology of cytoplasmic microtubules. *Ann. N. Y. Acad. Sci.* **253:**1–848.

Stephens, R. E. 1968. On the structural proteins of flagellar outer fibers. *J. Mol. Biol.* **32:**277–283.

Tamm, S. 1968. Flagellar development in the protozoan *Peranema trichophorum. J. Exp. Zool.* **164:**163–186.

Taylor, E. W. 1965. The mechanism of colchicine inhibition of mitosis. *J. Cell Biol.* **25:**145–160.

Tilney, L. G. 1971. Origin and continuity of microtubules. *In* Origin and Continuity of Cell Organelles. J. Reinert and H. Ursprung, editors. Springer-Verlag New York Inc., N. Y. 222–260.

Tilney, L. G., and J. Goddard. 1970. Nucleating sites for the assembly of cytoplasmic microtubules in the ectodermal cells of blastulae of *Arabacia punctulata. J. Cell Biol.* **46:**564–575.

Weisenberg, R. C. 1972. Microtubule formation in solutions containing low calcium concentrations. *Science (Wash. D. C.).* **177:**1104–1105.

Weisenberg, R. C., G. G. Borisy, and E. W. Taylor. 1968. The colchicine-binding protein of mammalian brain and its relation to microtubules. *Biochemistry.* **7:**4466–4479.

Wilson, E. B. 1900. The Cell in Development and Inheritance. The Macmillan Company, New York. 2nd edition. 483.

Wilson, E. B. 1924. The Cell in Development and Heredity, The Macmillan Company, New York. 3rd edition. 1232.

Wilson, L., and J. Bryan. 1974. Biochemical and pharmacological properties of microtubules. *Adv. Cell Mol. Biol.* **3:**21–72.

STRUCTURE AND FUNCTION OF FLAGELLAR MICROTUBULES

I. R. GIBBONS

Cilia and eukaryotic flagella are specialized microtubular organelles that project from the surface of cells and undergo regular bending movements. Inasmuch as they can be isolated in reasonable quantity by procedures which permit retention of their ability for movement, they constitute highly favorable material for the study of microtubular function.

Numerous electron microscope studies have established that cilia and flagella from nearly all eukaryotic organisms possess the same, highly uniform, substructure of a cylinder of nine doublet tubules surrounding two singlet central tubules. This 9+2 tubular framework, together with its complex assortment of connecting structures, is known as the axoneme. In the intact cell, the flagellar axoneme is enclosed within the cell membrane and is supplied with ATP by diffusion from mitochondria at its base.

A typical and much studied example of flagellar movement is that of the sea urchin sperm. Sea urchin sperm are readily available in quantities suitable for biochemical study, and they have a bending pattern that is relatively easy to record and analyze. Their flagella propagate regular, nearly symmetrical, planar waves along their length. Such 2-dimensional symmetrical bending waves appear to be an efficient means of propulsion in sperm, but many other types of bending patterns are found in flagella and cilia. Cilia typically have bending patterns that are asymmetrical and 3-dimensional as an adaptation for movement of fluid over a surface covered by a field of cilia. In many cases a cilium appears to have two distinct phases in its beat cycle. During the effective stroke, the cilium appears rigid and straight while bending near its base through an angle of about 2 radians. During the recovery stroke, the cilium returns to its original position usually by propagating a 3-dimensional wave of bending along its length. That the same 9+2 structure is capable of producing either a ciliary or a flagellar type of beating is well illustrated by *Chlamydomonas* where the *flagella,* as they are called, beat with a ciliary-type movement during normal forward swimming, and change to an undulatory movement like that of sperm flagella when the organism reverses direction and swims backward (Ringo, 1967). These observations indicate that the apparently uniform 9+2 structure of the flagellum or cilium is capable of considerably modifying its bending pattern to adapt to the needs of a particular biological situation, both between species and even at different times within the same species. One is tempted to wonder why evolution has not produced variations of the 9+2 pattern that are adapted specifically to flagellar or ciliary-type movement. Calculations indicate that the efficiency of mechanochemical conversion in sperm flagella is of the order of 30%, as compared to efficiencies of 60% or more in striated muscle working under optimal conditions (Brokaw and Gibbons, 1975). At the moment, we can only speculate that the overall economy to the cell of using a single basic structure for several purposes may outweigh the increased mechanochemical efficiency that might be achieved from more specialized organelles. Returning to the structure of the flagellar axoneme, Fig. 1 shows a cross-sectional diagram of the various accessory structures that are associated with the 9+2 tubular framework: the two arms on one side of each doublet, the radial spokes that link the A-tubule of each doublet to the central sheath that envelops the two central tubules, and the circumferential nexin links

I. R. GIBBONS Pacific Biomedical Research Center, University of Hawaii, Honolulu, Hawaii.

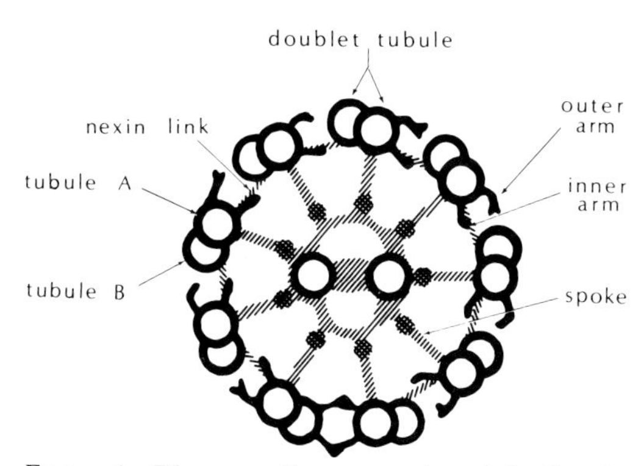

FIGURE 1 Diagrammatic cross section of the flagellar axoneme (from Brokaw and Gibbons, 1975).

that run around the inside of the axonemal cylinder linking adjacent doublet tubules.

Since the axonemal structure is complex, it is not surprising that it is composed of numerous protein components. A sodium dodecyl sulfate (SDS)-polyacrylamide gel pattern of isolated axonemes shows about 20 distinct bands (Fig. 2), and even more could probably be detected on a 2-dimensional gel. Of these bands, two are considerably more intense than the rest. One is the 55,000 mol wt band corresponding to tubulin, the principal structural protein of the tubules, and the other is a band with a very high molecular weight of 300,000–400,000. When this high molecular weight band is examined at higher resolution, it can be seen that it is not a single band, but rather that it is, at least in sea urchin sperm flagella, a cluster of four closely spaced bands (Fig. 3) (Gibbons et al., 1976). Recent work involving the differential extraction from the axoneme of the proteins corresponding to these high molecular weight bands has made it possible to identify two of them (*A* and *D*) as deriving from isoenzymic forms of the axonemal ATPase protein dynein, while the other two (*B* and *C*) are derived from proteins that appear to have no ATPase activity under the conditions tested (Ogawa and Gibbons, 1976). It seems possible that some of the high molecular weight proteins from flagellar axonemes may be related to the high molecular weight proteins associated with cytoplasmic microtubules.

Of all the axonemal proteins, only the two present in largest quantity, tubulin and dynein, have been studied in any detail so far. The native molecule, which has a molecular weight of about 110,000 daltons, is a heterodimer composed of an α and a β subunit each of 55,000 mol wt. The packing arrangement of subunits in the structure of flagellar doublets has been studied by Amos and Klug (1974) who have shown that the packing is different in the two tubules of the doublet, the A-tubule having a half-staggered array of structural dimers, whereas in the B-tubule the structural dimers are lined up obliquely at a shallow angle. The structural dimers visible in electron micrographs may well be identical to the chemical $\alpha\beta$ tubulin dimers, although this is not to be taken for granted. One puzzling factor has been the reports from two laboratories that under special conditions the α-tubulin can be preferentially extracted from the tubules into solution, leaving a tubule remnant of about 3 of the original 24 protofilaments which appears to consist predominantly of β-tubulin (Witman et al., 1972; Linck, 1976). If this is not some kind of artifact, it is not obvious how it will be reconciled with the $\alpha\beta$ dimer structural model.

Regardless of how this point turns out, the structural model of the tubule has a basic 80 Å periodicity. When one examines the reported periodicities of the various accessory structures, the arms, radial spokes, central sheath, and nexin links (Warner and Satir, 1974), they do not at first sight appear to be co-periodic with the 80 Å period of the tubule. However, if one adjusts the values derived from thin sections upward by 11%, then the values all fall almost exactly as multiples of the basic 80 Å period (Table I). The 80 Å value for the period is likely to be the valid one because it derives from X-ray crystallography of unfixed hydrated tubules (Cohen et al., 1971). The need for an 11% adjustment factor to the thin-section values could well be accounted for by section compression or by shrinkage during fixation, dehydration, and embedding.

Although all the accessory structures appear to bind at multiples of the basic 80 Å period, the tubulin lattice does not provide unique binding sites with the appropriate spacings for their attachment. Therefore, it seems not unlikely that the binding of the various accessory structures is effected by means of some of the minor protein components that are associated with tubulin in the tubule. It may be noted that the accessory structures mostly attach to the A-tubule of the doublet, and Linck (1976) has recently reported that all 10 of the minor protein components associated with the doublet are localized in the A-tubules.

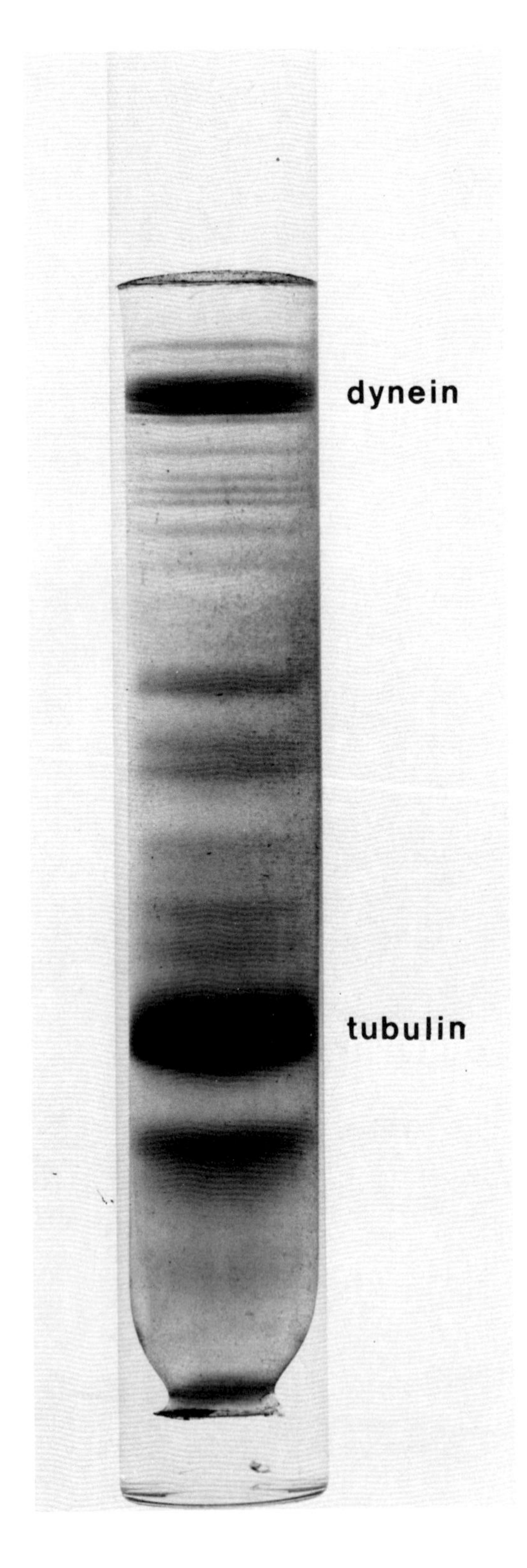

FIGURE 2 SDS-polyacrylamide electrophoresis gel pattern of intact axonemes isolated from sea urchin sperm. Axonemes were heated to 100°C in 1% SDS, 1% mercaptoethanol, and then a sample containing 50 μg protein was applied to the top of the gel column. 4% gel. Running buffer: 0.1% SDS, 0.1% mercaptoethanol, 50 mM phosphate buffer, pH 7.0. Running time 2 h at 5 ma/tube. Stained with brilliant Coomassie Blue, and destained by diffusion.

The axonemal ATPase protein dynein plays a vital role in flagellar function because it is involved in the transduction of the chemical energy provided by ATP hydrolysis into the mechanical energy of bending. In sea urchin sperm flagella, dynein exists in two isoenzymic forms, dynein 1 and dynein 2. Dynein 1, the major form, which accounts for about 80% of the total axonemal ATPase activity, gives rise to the A-electrophoretic band on SDS-polyacrylamide gels. Selective extraction and reconstitution experiments combined with electron microscopy have demonstrated that the dynein 1 is localized in the arms on the doublet tubules of the axoneme (Kincaid et al., 1973; Gibbons and Gibbons, 1976). When extracted from the axoneme with 0.6 M NaCl under appropriate conditions, dynein 1 is obtained in a form that has a low specific ATPase activity of about 0.4 μmol P_i/min × mg. The ATPase activity of this form can be activated about 10-fold by a variety of treatments, including heating at 42°C for 10 min, or incubating with 0.1% Triton X-100 for 10 min at room temperature (Gibbons et al., 1976). The form of dynein 1 with a low ATPase activity will be referred to as latent activity dynein 1 (LAD-1). As extracted, LAD-1 has a sedimentation coefficient of 19S and a Stokes radius of 17 nm which correspond to a molecular weight of about 1,500,000. When LAD-1 is dialyzed against a low ionic strength buffer containing EDTA, it is converted to another form sedimenting at 11S, with about half the original molecular weight, and it has its ATPase activity partially activated. When examined on SDS-polyacrylamide electrophoresis gels, LAD-1 shows the high molecular weight A-band mentioned previously and in addition minor components with apparent molecular weights of 130,000, 93,000, 74,000, and 55,000. The 130,000, 93,000, and 74,000 mol wt components copurify with the LAD-1 on sucrose density centrifugation and chromatography on Sepharose but their significance is not yet known. LAD-1 appears to retain many of the properties of native dynein 1, and its ATPase activity is increased 6- to 10-fold when it is rebound to NaCl-extracted axonemes. Dynein 1 from sea urchin sperm appears to have many properties in common with the 30S form of dynein from *Tetrahymena* cilia (Blum and Hayes, 1974).

The other isoenzymic form, dynein 2, accounts for about 15% of the axonemal ATPase activity in sea urchin sperm. It has been discovered only recently, and relatively little information about its properties and function are yet available. Dynein 2 is isolated by dialysis of salt-extracted axonemes against a low ionic strength buffer, followed by purification by chromatography on Sepharose 4B and hydroxylapatite (Ogawa and Gibbons, 1976). It has a molecular weight of about 700,000, and gives rise to the high molecular weight D-electrophoretic band. Its enzymic properties are similar, although not identical, to those of dynein 1; but no form of dynein 2 having latent ATPase activity has yet been identified. Dynein 2 from sea urchin sperm appears to have some properties in common with the 14S form of dynein from *Tetrahymena* cilia.

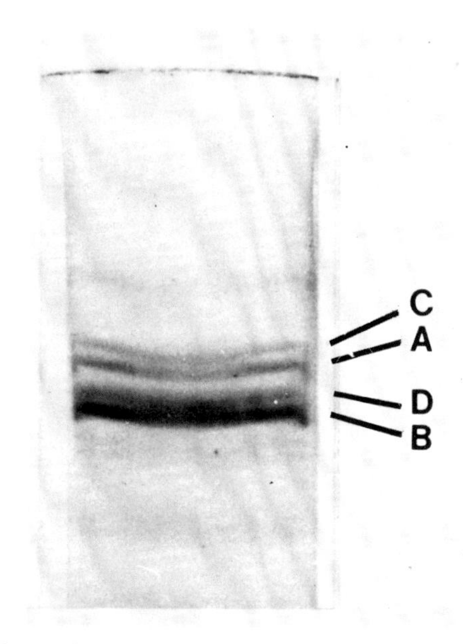

FIGURE 3 SDS-polyacrylamide electrophoresis gel pattern of axonemes extracted with 0.5 M NaCl to remove the bulk of the dynein 1 (*A* band) and render the *C* and *D* bands more visible. Bands *A*, *B*, *C*, and *D* are indicated. Preparation and electrophoresis were the same as in Fig. 2, except that running time was 3 h. Only top region of gel is shown.

Let us consider the mechanism of movement of cilia and flagella. It is by now generally accepted that these organelles move by a mechanism involving relative sliding movements between the tubules of the axoneme. Such a sliding tubule mechanism was first proposed by Afzelius (1959) who, when he discovered the arms on the doublets, postulated that they might function to produce sliding in a manner analogous to the cross bridges on the thick filaments in striated muscle. This hypothesis has been amply confirmed by subsequent evidence. The studies of ciliary tips by Satir (1968) showed that the lengths of ciliary tubules do not change during bending—a finding that has since been confirmed in more detail by Warner and Satir (1974) who used the irregular spacings of the radial spokes as a vernier to measure longitudinal displacements between doublet tubules. More direct evidence of sliding was obtained in my laboratory by Keith Summers, who showed directly that ATP induces active sliding between tubules by using dark-field light microscopy to visualize the process of ATP-induced disintegration of axonemes that had been briefly digested with trypsin to remove connecting structures that

TABLE I

Longitudinal Periodicities in Axonemes

	Thin section	X-ray diffraction	Optical diffraction	Corrected value*
	Å	Å	Å	Å
Tubule	–	40(80)	80(160, 480)	80
Dynein arms	215			240
Radial spokes	860			955
	(290/222/360)			(320/245/400)
Control sheath	142			160
Nexin link	866			960
	Warner and Satir, 1974	Cohen et al., 1971	Amos and Klug, 1974	

* Thin-section values increased by 11% and rounded to nearest 5 Å.

normally maintain the structural integrity of the axoneme (Summers and Gibbons, 1971). These observations have combined to place the sliding tubule hypothesis for flagellar bending, in which bending is generated as a result of sliding between tubules by an ATP-driven mechanochemical cycle involving the dynein arms, on nearly as substantial a basis as the sliding filament theory of muscle contraction.

When performing experiments on the mechanism of flagellar movement, it is frequently helpful to first remove the permeability barrier presented by the flagellar membrane in order that the axoneme may be directly accessible to chemical manipulation. The currently preferred technique is to treat with a nonionic detergent, such as Triton X-100, which completely solubilizes the flagellar membrane while leaving the motile apparatus apparently undamaged. Sea urchin sperm that have been demembranated with Triton X-100 in this way can be reactivated by placing them in a suitable solution containing ATP, and their movements closely resemble those of live sperm (Gibbons and Gibbons, 1972). Similar demembranated preparations have been described for cilia and flagella of a variety of other organisms, including bull sperm (Lindemann and Gibbons, 1975), *Chlamydomonas* (Allen and Borisy, 1974), *Crithidium* (Holwill and McGregor, 1977), and *Paramecium* (Naitoh and Kaneko, 1972).

There are several important approaches that can be made by using demembranated flagella. Their frequency can be controlled by changing the concentration of ATP, which enables one to examine the effects of changing frequency on the other wave parameters. Such experiments have shown that the waveform is little affected by change in frequency, with the bend angles and wavelength changing by only 10–20% for a 10-fold change in beat frequency (Gibbons and Gibbons, 1972; Brokaw, 1975*a*).

If the ATP concentration is abruptly lowered to below 10^{-7} M, then the swimming sperm become stiff, and set into stationary rigor waves which closely resemble the waveforms occurring at different phases of the bending cycle in beating flagella (Gibbons and Gibbons, 1974). The form of these rigor waves is believed to be maintained by the dynein arms, which, in the absence of ATP, form fixed cross bridges between adjacent doublet tubules, and so force the axoneme to remain in approximately the same waveform as existed at the time of removal of ATP. Improved techniques for specimen preparation have made it possible to visualize these dynein cross bridges by electron microscopy (Gibbons, 1975).

Another useful approach has been to examine the effect of partially removing the dynein arms from axonemes of the sea urchin *Colobocentrotus atratus* by brief extraction with 0.5 M KCl. Such extraction removes preferentially the outer of two dynein arms on the doublet tubules, and upon subsequent reactivation with ATP it is found that the best frequency is reduced proportionately to the fraction of arms removed, whereas the waveform appears largely unchanged (Gibbons and Gibbons, 1973). Addition of LAD-1 to these KCl-extracted sperm causes their frequency to increase back to nearly that of control demembranated sperm which have not been extracted with KCl (Gibbons and Gibbons, 1976). This property of functional recombination is characteristic of LAD-1 and it is lost if the dynein 1 is subjected to treatments such as heating to 42°C or exposure to Triton X-100, which activate its ATPase activity.

Unlike striated muscle, cilia and flagella do not require trace amounts of Ca^{2+} for activation, and they beat perfectly well in the presence of the Ca^{2+}-chelating agent EGTA (Gibbons and Gibbons, 1972). In many cases, however, Ca^{2+} does have a regulating effect upon the waveform of flagella and cilia (Fig. 4). As shown in Table II, the effects of Ca^{2+} are highly varied in cilia and flagella of different organisms. For instance, sea urchin sperm flagella have a symmetrical beat at low Ca^{2+} and an asymmetrical beat at high Ca^{2+} (Brokaw et al., 1974), whereas in *Chlamydomonas* flagella, it is the other way around (Hyams and Borisy, 1975). This variability indicates that the Ca^{2+} regulatory system, unlike the axonemal structure visible by electron microscopy, has been highly plastic in evolution in adapting to particular biological situations. In spite of the variability, certain regularities may be noted. Firstly, the changeover from the low to the high Ca^{2+} form always appears to occur at a Ca^{2+} concentration of the order of 10^{-6} M. Secondly, it is always the low Ca^{2+} beating pattern that is the normal one for the organism, while the high Ca^{2+} pattern is the transitory modified form exhibited in response to an appropriate stimulus such as in the avoidance response, or during photo- or chemotaxis.

In order for the sliding movements between tubules to produce regular bending waves, it is necessary for the sliding to be properly localized and coordinated. Relatively little is yet known regarding the properties of this system of coordination. Since the plane of bending in cilia and flagella is perpendicular to the plane containing the two central tubules, it has been postulated that the central tubules, together with the central sheath and the radial links that join them to the outer doublets, are involved in the coordination process (Summers and Gibbons, 1971). As demonstrated by Warner and Satir (1974), the radial spokes undergo longitudinal displacement along the central sheath during bending, and they may, therefore, be involved in providing a resistance that localizes the sliding of the outer doublets and converts it into bending.

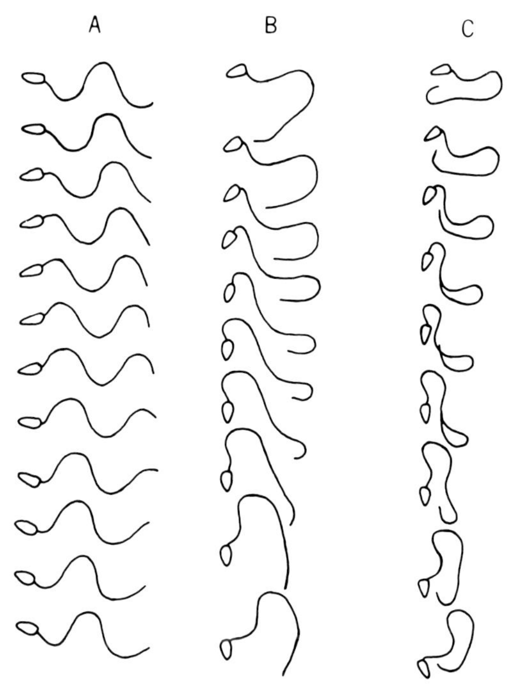

FIGURE 4 Tracings of consecutive frames of movie films showing demembranated sperm of the sea urchin *Tripneustes gratilla*. The sequences show the effect of variation in the Ca^{2+} concentration in the Triton-extraction solution. Sequence *A*, extracted in 0.04% Triton X-100, 0.15 M KCl, 2 mM $MgCl_2$, 0.5 mM EDTA, 4 mM $CaCl_2$, 1 mM dithiothreitol, 10 mM Tris-HCl buffer, pH 8.1. Sequences *B* and *C*, same except that the 4 mM $CaCl_2$ was replaced by 2 mM EDTA. In all cases sperm were reactivated in 0.5 M KCl, 2 mM $MgSO_4$, 0.5 mM EDTA, 1 mM dithiothreitol, 10 mM Tris-HCl buffer, pH 8.1, and 1 mM ATP. Sequences *B* and *C* represent variations in the amount of asymmetry of different sperm in the same preparation. Framing rates, 300/s (*A*) and 250/s (*B* and *C*).

In the absence of direct evidence concerning the mechanisms that regulate sliding, it is necessary to consider indirect evidence. One such approach is to examine the effects of various perturbing agents on the parameters of flagellar oscillatory bending (Gibbons, 1974). When one examines the effects of a variety of such perturbing agents, one is struck by the fact that they tend to fall into three groups. Those in the first group affect principally the beat frequency and have only a small effect on the flagellar waveform; these agents include ATP concentration, temperature, and changes in the number of dynein arms present. Those in the second group affect principally the flagellar waveform and have only a small effect on the beat frequency; these include attachment of the sperm head to the slide, Ca^{2+} concentration, and flagellar length. The third group is represented by the viscosity of the medium which has only a relatively small effect upon both the beat frequency and the waveform. In sea urchin sperm, for example, a twofold increase in the viscosity of the medium causes only about a 15% change in beat frequency and bend angle (Brokaw, 1975*a*). Considering that a twofold increase in viscosity corresponds to a 100% increase in the forces of viscous drag on the flagellum, the relative insensitivity of the frequency and the waveform to this increase in viscous drag is quite striking.

The existence of one group of perturbing agents that primarily affects the beat frequency, and another group that primarily affects the waveform, suggests that the mechanisms regulating the fre-

TABLE II

Regulation of Cilia and Flagella by Ca^{2+}

	$Ca^{2+} < 10^{-6}$ M	Ca $> 10^{-6}$ M	Reference
Sea urchin sperm flagella	Symmetric waves	Asymmetric waves	Brokaw et al., 1974
Chlamydomonas flagella	Ciliary-type movement	Flagella-type movement	Hyams and Borisy, 1975
Crithidium flagellum	Wave propagation tip → base	Wave propagation base → tip	Holwill and McGregor, 1976
Paramecium cilia	Forward swimming	Backward swimming	Naitoh and Kaneko, 1972

quency and the waveform function separately, with only a small degree of interaction (Gibbons, 1974).

In addition to these perturbations of the steady-state beating, we can also look at the effects of transient perturbations. One such transient involves the waves produced as a flagellum is starting or stopping its motion. Sperm of the sea urchin *Tripneustes* that are swimming in seawater containing 0.2 mM EDTA alternate between periods of quiescence and periods of normal movement. During a quiescent period, the sperm flagellum is completely stationary, with one sharp bend near the head and a second gentle bend near the tip of the tail. During a motile period, the sperm flagellum moves with normal waveform and beat frequency. High-speed movies made by Dr. Barbara Gibbons show that at the end of a quiescent period, the first sign of movement is an increase in the curvature of the bend near the sperm head (Fig. 5). This sharp principal bend then begins to propagate distally and a shallow reverse bend is formed close to the sperm head. Both the principal and reverse bends decrease in angle as they propagate and die away without passing along the full length of the flagellum. The starting flagellum usually propagates about three asymmetrical waves of this type which propagate progressively further along the flagellum before it reaches its steady state of symmetrical waves propagating along the full length. A somewhat analogous situation is observed in reverse as a sperm stops moving at the beginning of a quiescent period. Similar observations have been reported by Dr. Stuart Goldstein (1976) who started quiescent sperm moving by changing the local pH of the seawater with a micropipet. A striking feature of these observations is the asymmetry of bending during the transient state, which resembles that produced in the steady state in the presence of high Ca^{2+}. Whether the asymmetry of the transient state results from a pulse of Ca^{2+} ions leaking into the cell, or whether it is an intrinsic property of the transient state is not yet known, but in any case the properties of these asymmetrical waves suggest there is a fundemental difference between the principal and reverse bends both as regards their formation and their propagation.

A different approach to elucidating control mechanisms is the study of mutant flagella. Mutants of *Chlamydomonas* with paralyzed flagella were first reported by Ralph Lewin over 20 yr ago

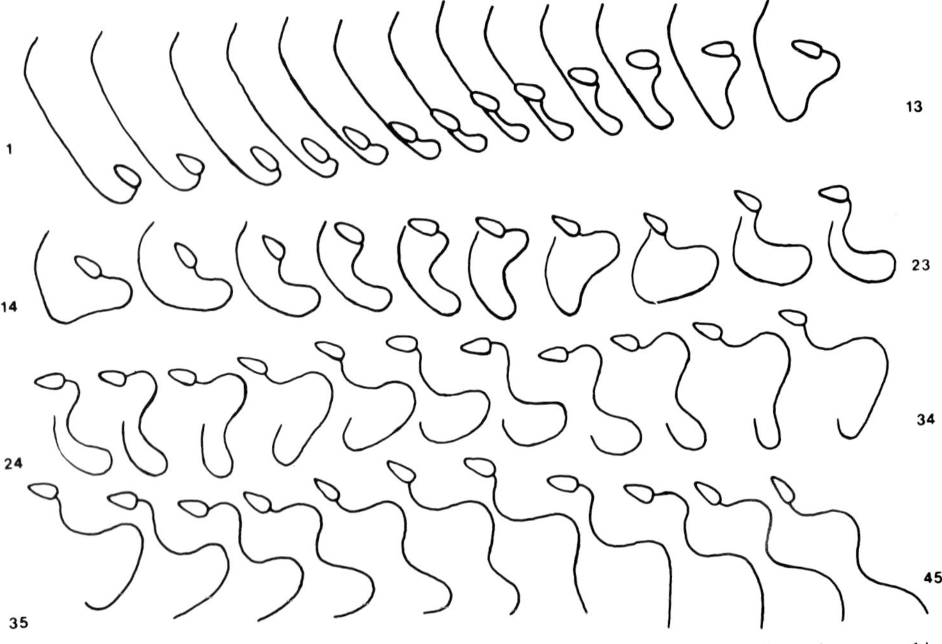

FIGURE 5 Tracings of consecutive frames of a movie film showing a live sperm from the sea urchin *Tripneustes gratilla* in seawater containing 0.2 mM EDTA and adjusted to pH 8.3. At the beginning of the series, the sperm is lying quiescent in a partially bent shape, and at the end of the series it is swimming normally. Framing rate, 275/s. Temperature = 23°C.

(Lewin, 1954), but it is only recently that they have become the subject of detailed biochemical and physiological investigation. A considerable variety of mutants, including some that are temperature-sensitive, have been discovered, among them being ones with structurally deficient axonemes, such as one lacking the two central tubules (Randall et al., 1964), and another lacking the radial spokes (Witman et al., 1975). In view of all the work being done with *Chlamydomonas,* it seems ironic that the first mutant involving the dynein arms should have been discovered in humans. Pedersen and Rebbe (1975) in Copenhagen and Afzelius (1976) in Stockholm have described patients with nonmotile sperm whose doublets are completely lacking their dynein arms. These patients also suffer from a variety of other symptoms known as Kartagener's syndrome, involving chronic bronchitis and sinusitis, which apparently results from the nonmotility of the cilia of the respiratory tract. Although studies of mutant flagella are still in the preliminary stages, they have the promise of making major contributions to our knowledge of the mechanisms involved with control of motility as well as those involved with flagellar morphogenesis.

Another approach to the study of the control of flagellar bending has been the development of quantitative theoretical models. In these models, the movement is determined by the balance between the active bending moment, and the bending moments resulting from the viscous resistance of the surrounding fluid medium and the elastic resistance of the axoneme. In sliding tubule models, the active bending moment (M_a) is produced by summation of the local shear moment (m), resulting from activity of cross bridges distributed along the length of the axoneme: $m = -dMa/ds$. Two main classes of model have been considered for flagellar movement. In the first, the metachonous activation needed for production of flagellar bending waves is provided by a local feedback control of active shear moment by the curvature of the axoneme (Brokaw, 1972). In the second, kinetic properties of the cross bridges cause the active shear moment to depend on the velocity of sliding in a way which can establish metachronous activation (Brokaw, 1975*b*, 1976). In the second type of model, as in one suggested earlier by Rikmenspoel and Rudd (1973), oscillation is a local property of the flagellar axoneme, and a feed-back control by curvature is not required for oscillation or coordinated wave propagation. Computer simulations by Brokaw (1972, 1975*b*) have shown that both types of model are capable of generating wave movements generally similar to those of real flagella.

Even the best of the current models is still a long way from providing a completely satisfactory explanation of the mechanisms that control bending in cilia and flagella. They have particular difficulty in explaining the observed independence between the waveform and the best frequency. There is a real need for more experimental data about the basic properties of the mechanisms involved. At the level of the dynein cross bridges, we need to know whether their action is polarized so that force is always exerted in the same direction between a particular pair of tubules, as the subunit structure of the tubule and the analogy to cross-bridge behavior in muscle would suggest. We also need to know whether cross-bridge action is controlled in a nonlinear, on-off manner, with the shear moment being affected only by the shear velocity, or whether there is a modulated control by a macroscopic variable such as curvature. The roles of dynein 2 and of the longitudinal displacements of the radial spokes relative to the central sheath remain to be clarified. There are obvious dangers in proceeding too far with theoretical modeling until more basic experimental evidence is available.

More detailed examination of hydromechanical influences, the effects of perturbing agents on beating, and the properties of theoretical models will certainly be valuable, but it seems likely that adequate understanding of the control mechanisms will require independent information about the nature and properties of the components present in the feed-back loop. This information is perhaps most likely to be obtained by further study of either modified flagella, such as the trypsin-treated axonemes, or paralyzed mutant flagella, in which the feed-back loop has been opened to facilitate the investigation of its individual components.

ACKNOWLEDGMENTS

I am grateful to Dr. C. J. Brokaw for helpful discussions. This work has been supported in part by NIH grants HD 06565, HD 09707, and HD 10002.

REFERENCES

AFZELIUS, B. A. 1959. Electron microscopy of the sperm tail: results obtained with a new fixative. *J. Biophys. Biochem. Cytol.* **5**:269–278.

Afzelius, B. A. 1976. A human syndrome caused by immotile cilia. *Science* (*Wash. D. C.*). **193**:317–319.

Allen, C., and G. B. Borisy. 1974. Flagellar motility in *Chlamydomonas:* reactivation and sliding *in vitro*. *J. Cell Biol.* **63**:5*a* (Abstr.).

Amos, L. A., and A. Klug. 1974. Arrangement of subunits in flagellar microtubules. *J. Cell Sci.* **14**:523–549.

Blum, J. J., and A. Hayes. 1974. Effect of *N*-ethylmaleimide and of heat treatment on the binding of dynein to ethylenediaminetetraacetic acid extracted axonemes. *Biochemistry*. **13**:4290–4298.

Brokaw, C. J. 1972. Computer simulation of flagellar movement. I. Demonstration of stable bend propagation and bend initiation by the sliding filament model. *Biophys. J.* **12**:564–586.

Brokaw, C. J., 1975*a*. Effects of viscosity and ATP concentration on the movement of reactivated sea urchin sperm flagella. *J. Exp. Biol.* **62**:701–719.

Brokaw, C. J. 1975*b*. Molecular mechanism for oscillation in flagellar and muscle. *Proc. Natl. Acad. Sci. U. S. A.* **72**:3102–3106.

Brokaw, C. J. 1976. Computer simulation of flagellar movement. IV. Properties of an oscillatory two-state cross bridge model. *Biophys. J.* **16**:1029–1042.

Brokaw, C. J., and I. R. Gibbons. 1975. Mechanisms of movement in flagella and cilia. *In* Swimming and Flying in Nature. Vol. 1. T. Y.-T. Wu, C. J. Brokaw, and C. Brennan, editors. Plenum Publishing Co., New York. 89–125.

Brokaw, C. J., and R. Josslin. 1973. Maintenance of constant wave parameters by sperm flagella at reduced frequencies of beat. *J. Exp. Biol.* **59**:617–628.

Brokaw, C.J., R. Josslin, and L. Bobrow. 1974. Calcium ion regulation of flagellar beat symmetry in reactivated sea urchin spermatozoa. *Biochem. Biophys. Res. Commun.* **58**:795–800.

Cohen, C., S. C. Harrison, and R. E. Stephens. 1971. X-ray diffraction from microtubules. *J. Mol. Biol.* **59**:375–380.

Gibbons, I. R. 1974. Mechanisms of flagellar motility. *In* The Functional Anatomy of the Spermatozoon. B. A. Afzelius, editor. Pergamon Press Ltd., Oxford. 127–140.

Gibbons, I. R. 1975. The molecular basis of flagellar motility in sea urchin spermatozoa. *In* Molecules and Cell Movement. S. Inoue and R. E. Stephens, editors. Raven Press, New York. 207–231.

Gibbons, I. R., E. Fronk, B. H. Gibbons, and K. Ogawa. 1976. Multiple forms of dynein in sea urchin sperm flagella. *In* Cell Motility. R. Goldman, T. Pollard, and J. Rosenbaum, editors. Cold Spring Harbor Laboratory, Cold Spring Harbor, New York. 915–932.

Gibbons, B. H., and I. R. Gibbons. 1972. Flagellar movement and adenosine triphosphatase activity in sea urchin sperm extracted with Triton X-100. *J. Cell Biol.* **54**:75–97.

Gibbons, B. H., and I. R. Gibbons. 1973. The effect of partial extraction of dynein arms on the movement of reactivated sea-urchin sperm. *J. Cell. Sci.* **13**:337–357.

Gibbons, B. H., and I. R. Gibbons. 1974. Properties of flagellar "rigor waves" produced by abrupt removal of adenosine triphosphate from actively swimming sea urchin sperm. *J. Cell Biol.* **63**:970–985.

Gibbons, B. H., and I. R. Gibbons. 1976. Functional recombination of dynein I with demembranated sea urchin sperm partially extracted with KCl. *Biochem. Biophys. Res. Comm.* **73**:1–6.

Goldstein, S. F. 1976. Bend initiation in quiescent sperm flagella. *J. Cell Biol.* **70**:71*a* (Abstr.).

Holwill, M. E. J., and J. L. McGregor. 1977. Effects of calcium on flagellar movement in the trypanosome *Crithidia oncopelti*. *J. Exp. Biol.* In press.

Hyams, J. S., and G. G. Borisy. 1975. The dependence of the waveform and direction of beat of *Chlamydomonas* flagella upon calcium ions. *J. Cell Biol.* **67**:186*a* (Abstr.).

Kincaid, H. L., B. H. Gibbons, and I. R. Gibbons. 1973. The salt-extractable fraction of dynein from sea urchin sperm flagella: an analysis by gel electrophoresis and by adenosine triphosphatase activity. *J. Supramol. Struct.* **1**:461–470.

Lewin, R. A. 1954. Mutants of *Chlamydomonas moewusii* with impaired motility. *J. Gen. Microbiol.* **11**:358–363.

Linck, R. W. 1976. Flagellar doublet microtubules: fractionation of minor components and α-tubulin from specific regions of the A-tubule. *J. Cell Sci.* **20**:405–439.

Lindemann, C. B., and I. R. Gibbons. 1975. Adenosine triphosphate-induced motility and sliding of filaments in mammalian sperm extracted with Triton X-100. *J. Cell Biol.* **65**:147–162.

Naitoh, Y., and H. Kaneko. 1972. Reactivated triton-extracted models of *Paramecium:* modification of ciliary movement by calcium ions. *Science* (*Wash. D. C.*). **176**:523–524.

Ogawa, K., and I. R. Gibbons. 1976. Dynein 2: a new adenosine triphosphatase from sea urchin sperm flagella. *J. Biol. Chem.* **251**:5793–5801.

Pedersen, H., and H. Rebbe. 1975. Absence of arms in the axoneme of immobile human spermatozoa. *Biol. Reprod.* **12**:541–544.

Randall, J., J. R. Warr, J. M. Hopkins, and A. McVittie. 1964. A single-gene mutation of *Chlamydomonas reinhardii* affected motility: a genetic and electron microscope study. *Nature* (*Lond.*). **203**:912–914.

Rikmenspoel, R., and W. G. Rudd. 1973. The contractile mechanism in cilia. *Biophys. J.* **13**:955–993.

Ringo, D. L. 1967. Flagellar motion and fine structure of the flagellar apparatus in *Chlamydomonas*. *J. Cell Biol.* **33**:543–571.

Satir, P. 1968. Studies on cilia. III. Further studies on

the cilium tip and a "sliding filament model" of ciliary motility. *J. Cell Biol.* **39:**77–94.

SUMMERS, K. E., and I. R. GIBBONS. 1971. Adenosine triphosphate-induced sliding of tubules in trypsin-treated flagella of sea-urchin sperm. *Proc. Natl. Acad. Sci. U. S. A.* **68:**3092–3096.

WARNER, F. D., and P. SATIR. 1974. The structural basis of ciliary bend formation: radial spoke positional changes accompanying microtubule sliding. *J. Cell Biol.* **63:**35–63.

WITMAN, G. B., K. CARLSON, and J. L. ROSENBAUM. 1972. *Chlamydomonas* flagella. II. The distribution of tubulins 1 and 2 in the outer doublet microtubules. *J. Cell Biol.* **54:**540–555.

WITMAN, G., R. FAY, J. PLUMMER, and G. BORISY. 1975. Studies on *Chlamydomonas* mutants lacking the radial spokes and the central sheath and tubules. *J. Cell Biol.* **67:**458*a* (Abstr.).

Molecular Basis of Motility

INTRODUCTORY REMARKS

N. KAMIYA

Various types of movements exhibited by eukaryotic cells may be classified into two major groups with respect to the proteins involved, i.e., the actin-myosin system and the tubulin-dynein system. To the first group belong ameboid movement, cytoplasmic streaming, movement of tissue cells, cytokinesis, etc. On the other hand, ciliary and flagellar movements, migration of pigment granules in some chromatophores, bending of axostyles, etc., belong to the second group. Whether anaphase chromosome movements also utilize an actomyosin system is a question under investigation.

The scope of this symposium is restricted to the molecular mechanisms of actomyosin-based cell motility. Although forms of actomyosin-based biological movements appear diverse, their energy converting mechanisms are related. The question to be asked is to what extent the motile and regulatory mechanisms they use are analogous to those in muscular systems, and in what respects they are unique.

A quarter of a century has elapsed since A. Loewy (1952) showed that actomyosinlike proteins are present in myxomycete plasmodia, a nonmuscular motile system. During this period, especially in the last decade, research on actomyosin-based cell motility has made great strides (for key references, see Pollard and Weihing, 1974; Hepler and Palevitz, 1974; Inoué and Stephens, 1975; Weihing, 1976) in isolating and characterizing the main proteins related to movement of many kinds, in discovering regulatory proteins, in developing new methods for identifying these proteins, *in situ*—e.g., through heavy meromyosin arrowhead formation (Ishikawa et al., 1969) and through fluorescent antibody staining—in preserving or reactivating movement in model systems, etc. Characterizations of contractile proteins in a variety of nonmuscular cells, for instance, myxomycete plasmodia (Hatano, 1973), *Acanthamoeba,* blood cells, cellular slime mold, etc., are thoroughly summarized by Pollard and Weihing in their 1974 review. Actin now seems to be a ubiquitous and common protein not only in muscle but also in nonmuscle cells in animals, plants, fungi, and protists. Our knowledge is also growing about cytoplasmic myosin and cofactors, their biochemical characterization, mode of action, and localization in cells.

Dr. Taylor, Dr. Pollard, and Dr. Tilney will review the "State of the Art" in some important aspects of cell motility. It is encouraging that we are now in a position to present some convincing evidence regarding molecular aspects of several different types of cell motility. One of the intriguing manifestations of nonmuscular cell motility which will not be discussed in this symposium is the phenomenon of cytoplasmic streaming in plant cells. Thus I feel it might be pertinent to give brief consideration to this topic.

Recent work on cytoplasmic streaming has been concentrated on two major materials. One is the classic characean cell in which Corti (1774) first discovered cytoplasmic streaming two centuries ago. The other material is the plasmodium of acellular slime molds, especially that of *Physarum polycephalum*. Here, however, I should like to consider the streaming primarily in characean cells.

The endless rotation of the endoplasm in giant cells of *Nitella* or *Chara* is not only simple in pattern but beautiful. The flow continues steadily day and night throughout the whole life-span of a cell. It is easy to understand why this endless rotation has attracted the keen interest of workers in the past. Although the dynamics of streaming on the cellular level have now been pretty well analyzed (Kamiya and Kuroda, 1973), biochemical approaches, especially isolation and characterization of proteins important for streaming, have

N. KAMIYA Department of Biology, Faculty of Science, Osaka University, Japan

not progressed very far.

As to the site of the motive force, it was concluded by Kamiya and Kuroda (1956) that the force causing streaming must be generated in the form of active shearing at the boundary between the stationary ectoplasmic layer and the outermost edge of the flowing endoplasm. Our conclusion was arrived at primarily through analysis of velocity distributions in the cytoplasm. No special structure was found at that time in the region where the motive force was supposed to be produced. But this conclusion led to further experiments to determine the magnitude of the active shearing force produced there. It was determined to be as small as 1–2 dyn/cm^2 (Kamiya and Kuroda, 1958, 1973; Tazawa, 1968). It was in 1966 that fibrillar structures were discovered at the very site where the motive force was predicted to be generated. The discovery was made almost simultaneously by light microscopy (Kamitsubo, 1966, 1972*a*) and by electron microscopy (Nagai and Rebhun, 1966).

Fig. 1 shows a photomicrograph taken by Kamitsubo with differential interference optics. The fibrils are attached to the inner surface of chloroplast files and run parallel to the direction of streaming. Each of the fibrils is composed of 50-100 microfilaments, 5-6 nm in diameter (Nagai and Rebhun, 1966). These microfilaments were identified as F-actin through formation of arrowhead structures with heavy meromyosin (Palevitz et al., 1974; Palevitz and Hepler, 1975) or subfragment 1 (Williamson, 1974). Further, it was shown by Kersey et al. (1976) that all these arrowheads point upstream.

An interesting approach to studying functional aspects of the subcortical fibrils was the application of either microbeam irradiation or centrifugation as a means of locally destroying or disintegrating the fibrils. It was shown by Kamitsubo (1972*b*) that the endoplasmic streaming is either stopped or rendered passive just at the site where the filaments disappear. The streaming restarts only after the filaments have regenerated. The new streaming always occurs alongside the newly regenerated filaments and not elsewhere.

The technique of vacuolar perfusion of characean cells developed in our laboratory by Tazawa (1964) is also an extremely useful means of gaining insight into the functional aspects of cytoplasmic streaming. Recently, experiments were done by Williamson (1975), and independently by Tazawa et al. (1976), to remove the vacuolar membrane by perfusing the cell with media containing EGTA. Because there is no longer any vacuolar membrane in this case, chemicals in the perfusing solutions are believed to be able to diffuse readily into the cytoplasm. These experiments show that streaming can continue without any vacuolar membrane. Williamson reported instructive observations on the morphology of subcortical fibrils and motile behavior of endoplasmic organelles along them in relation to the concentrations of several ions, nucleotides, cytochalasin B, etc. Mo-

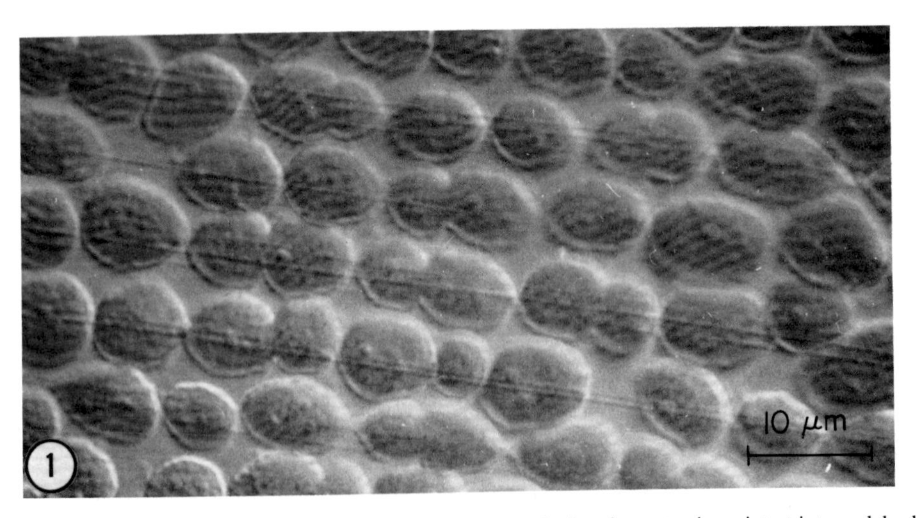

FIGURE 1 Cytoplasmic fibrils on the files of chloroplasts attached to the cortex in an intact internodal cell of *Nitella axillaris*. These fibrils are in direct contact with the endoplasm (Kamitsubo, 1972*b*).

tility requires ATP and Mg^{2+} at millimolar levels, whereas Ca^{2+} concentrations in excess of 10^{-7} M are inhibitory to streaming.

What is still unknown is the cellular localization of myosin and the mode of myosin action. To shed some light on this problem, Chen and Kamiya (1975) tried to treat the streaming endoplasm and the stationary cortex of the living internodal cell of *Nitella* separately with the SH-reagent, *N*-ethylmaleimide (NEM). A method combining the double-chamber technique with centrifugation made such differential treatment possible.

When an internodal cell is centrifuged gently, the endoplasm collects at one end of the cell while the cortex, including chloroplasts and subcortical fibrils, remains *in situ*. Under these conditions, we treat either the centrifugal half or centripetal half of the cell with the appropriate reagent. It is known that NEM inhibits F-actin-activated ATPase of myosin, forming a covalent bond with SH-1 of myosin (Shibata-Sekiya and Tonomura, 1975), but this reagent has little effect on polymerization or depolymerization of actin. After washing free of NEM, the cell is centrifuged in the opposite direction to bring the untreated endoplasm into contact with the treated cortex, or the treated endoplasm into contact with untreated cortex.

When only the cell cortex is treated with NEM and the endoplasm is left untreated, streaming occurs normally (Fig. 2). When only endoplasm is treated with the same reagent, and the ectoplasm is left untreated, no streaming takes place (Fig. 3). If we use cytochalasin B as a reagent, the situation is exactly the opposite. Cytoplasmic streaming is stopped only when the cortex is treated with cytochalasin B (Nagai and Kamiya, unpublished data).

The implication of these results is that the component, whose function is readily abolished by NEM, must be present in the endoplasm, not in the cortex. At present we tentatively suppose that this component may be a myosinlike protein which is as yet not identified biochemically. A possible lowering in ATP level as a result of NEM action may be an additional cause for the cessation of streaming. The component whose function is impaired by cytochalasin B, supposedly actin, must reside in the cortex, not in the endoplasm.

Thus, expressed in molecular terms, cytoplasmic streaming in the characean cell is supposed to be caused by the sliding of the endoplasmic myosin molecules alongside the stationary F-actin filaments lying on the cortex. This is our tentative conclusion. This is, however, not the only current theory of cytoplasmic streaming in characean cells. (Recently, Nina Allen [1974] observed bending

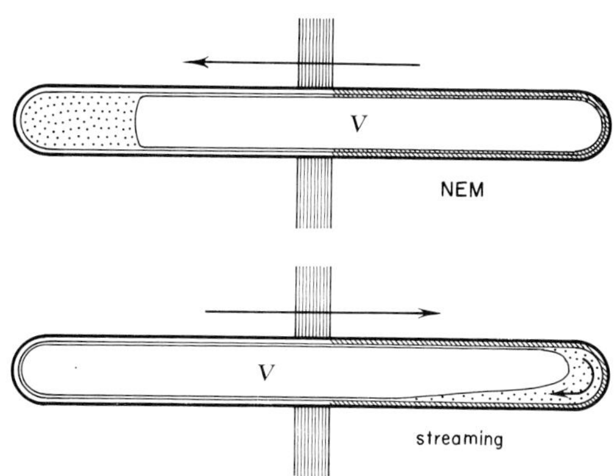

FIGURE 2 Differential treatment of the cell with NEM. (Top) The internodal cell in the double-chamber is centrifuged toward the left to evacuate the endoplasm from the centripetal side. The centripetal half was treated with NEM while the entire cell was chilled to prevent back streaming of the dislocated endoplasm. (Bottom) The same cell was centrifuged in the opposite direction to bring the untreated endoplasm into contact with NEM-treated cortex. The arrows indicate the direction of centrifugation. The treated portion of the cell is shown by the hatched lines. Note the occurrence of the streaming (Chen and Kamiya, 1975).

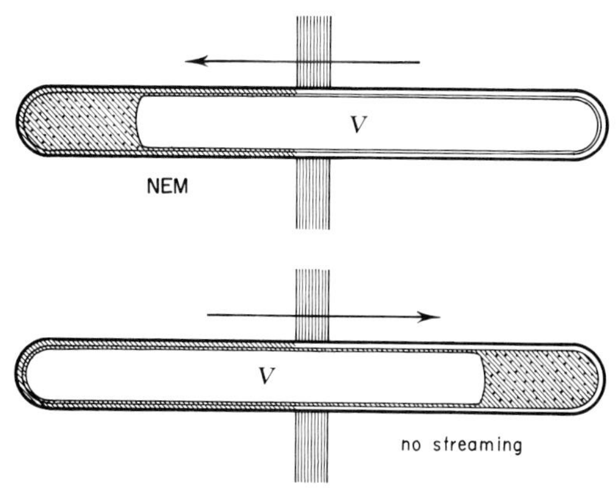

FIGURE 3 Differential treatment of the cell with NEM under arrangements reciprocal to the procedures shown in Fig. 2. (Top) The centrifugal half of the cell where the endoplasm accumulated was treated with NEM while the entire cell was chilled, as in the procedure demonstrated in Fig. 2. (Bottom) The treated endoplasm was moved by reverse centrifugation to the other end of the cell where the cortical layer remained untreated. The arrows show the direction of centrifugation. The treated area is hatched. No streaming was observed (Chen and Kamiya, 1975).

waves of filaments in the endoplasmic layer. She believes that they may be a contributing cause for producing the force driving the endoplasm. This idea is novel and interesting, but must await further evidence.)

For an approach to elucidating the molecular basis of cell motility, the use of models of various motile systems has become more important recently. I should like to report an experiment in which Dr. Kuroda and I have been engaged recently in collaboration with members of Dr. Tonomura's laboratory (Kuroda and Kamiya, 1975).

As an object of observation, we selected spinning of chloroplasts instead of streaming of the endoplasm, because we know that these two phenomena are caused by the same active shearing mechanism (cf. Kamiya, 1959). As has been observed by many authors in the past, chloroplasts in isolated cytoplasmic droplets obtained from characean internodal cells rotate or move around independently (Jarosch, 1956; Kamiya and Kuroda, 1957; Fig. 4*a*). The first step of the experiment was to search for a condition which might allow chloroplasts to rotate after the membrane of the droplet was removed, so that the cytoplasmic bulk was exposed to the external medium. Our preliminary experiments showed that the following low Ca^{2+} solution met this requirement:

KNO_3	80 mM	ATP	1 mM
NaCl	2 mM	DTT	2 mM
$Mg(NO_3)_2$	1 mM	Sorbitol	160 mM
$Ca(NO_3)_2$	1 mM	Ficoll	3%
EGTA	30 mM	Pipes buffer (pH 7.0)	5 mM

The following results were shown by means of a motion picture film. On removal of the surface membrane of a cytoplasmic droplet with the tip of a fine glass needle, it can no longer form a new membrane in this medium. The cytoplasm gradually disperses into the surrounding solution, and the earlier clear demarcation of the droplet disappears. A point of special emphasis here is that chloroplasts continue to rotate almost as actively as they did before the surface membrane was removed (Fig. 4*b*). When a chloroplast comes out spontaneously from the cytoplasm-dense region to the cytoplasm-sparse area beyond the original boundary site, the rotation becomes slow and sporadic. One reason for this may be the contact of the chloroplast with the glass surface. Sometimes it takes up to 2 h for the chloroplasts in the cytoplasm-sparse area to come to a complete standstill.

The rotation of chloroplasts that have come out from the core area stops completely within 15 min on addition of 1 mM NEM to the above solution. After chloroplasts have ceased to rotate, the NEM-containing solution is partially replaced with a large excess of 12 mM dithiothreitol (DTT) so that free NEM is made ineffective. Chloroplast rotation does not recover after this procedure (Fig. 4*c*) as expected inasmuch as NEM would be expected to form a covalent bond with the SH-1 group in a putative *Nitella* myosin molecule and inhibit any actomyosin-type ATPase activity irreversibly as in muscle (Shibata-Sekiya and Tonomura, 1975).

The next step of our experiment was to observe whether rabbit heavy meromyosin (HMM) added then to the solution could restore the rotation in the presence of Mg-ATP. As a matter of fact, HMM did restart rotation of the chloroplasts, even though the rate of rotation was extremely slow; it took 1 min or longer for a chloroplast to complete one revolution (Fig. 4*d*). The rotation of chloroplasts which we could observe under these conditions lasted only 10 min or so. It may be that we can improve the conditions for reactivation of chloroplast rotation with rabbit muscle HMM with more systematic investigation of the medium and of experimental procedures.

We are still not quite certain about the molecular mechanism of this movement, but the experimental fact that the rotation of chloroplasts that has been once stopped with NEM in the presence of Mg-ATP can be reactivated with muscle HMM may imply that chloroplast rotation in vitro, and hence cytoplasmic streaming in vivo, may result from the interaction between F-actin attached to the gel phase (chloroplast) and myosinlike protein in the sol phase (endoplasm) in the presence of Mg-ATP. It also shows that the putative *Nitella* myosin, which has not yet been isolated or characterized, can be functionally replaced, to some extent at least, with muscle HMM.

It is now established that the subcortical filaments are composed of bundles of actin filaments all with the same polarity over their entire length. As is well known, the streaming occurs steadily in the form of an endless belt only in one direction alongside these filaments. There is no chance for the endoplasm to move backward; the track is strictly "one-way."

It would be of great interest to know what kind of aggregation pattern the presumptive *Nitella* myosin has and what kind of interaction occurs

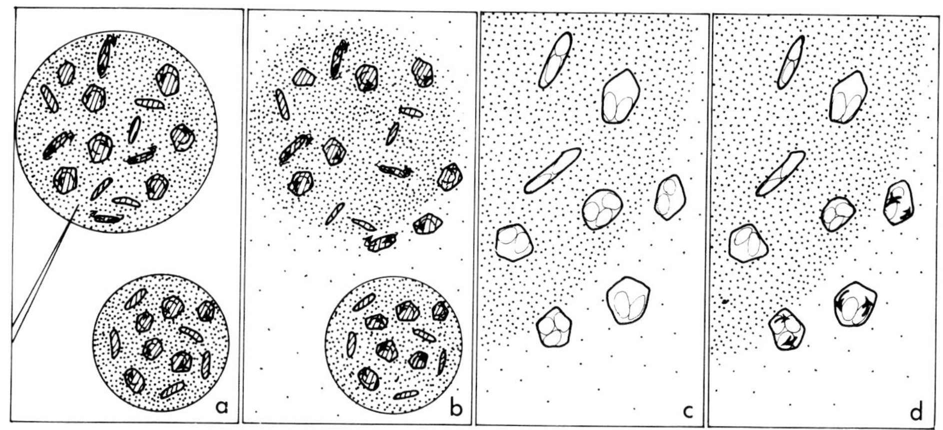

FIGURE 4 Semidiagrammatic sketches of motile behavior of extracellular chloroplasts. (*a*) Squeezed out cytoplasmic droplets with surface membrane. (*b*) Removal of the surface membrane with a glass microneedle in a Ca^{2+}-deficient solution containing Mg-ATP. Chloroplasts come out to the cytoplasm-sparse area spontaneously while rotating. (*c*) Cessation of rotation on application of NEM. (*d*) Restart of slow rotation on application of heavy meromyosin (HMM) after NEM was washed off with DTT from the whole cytoplasmic bulk (Kuroda and Kamiya, 1975).

with the subcortical fibrils to produce the continuously active shear force responsible for this endless rotation. It is possible that *Nitella* myosin links up with endoplasmic organelles as suggested by Bradley (1973) and Williamson (1975), or might form some special and as yet unseen type of thick filaments for this situation.

Taking into consideration the facts so far known about the streaming in Charophyte cells, it seems to me that the most plausible and reasonable image of the streaming in these cells is that it is caused by active shearing between F-actin in the stationary subcortical filaments and myosin in the moving endoplasm. The problem left to be solved is the molecular mechanism of the active shearing and its control. It is not known how far the active shear mechanism is applicable to other systems exhibiting cytoplasmic streaming, for example myxomycete plasmodium, ameba, etc.

The molecular basis of movement may be common to many different kinds of cells from muscle to ameba or plant cells. How the varieties of visual motile phenomena are related to events on the molecular level is a cardinal problem to be dealt with in cell motility.

REFERENCES

ALLEN, N. S. 1974. Endoplasmic filaments generate the motive force for rotational streaming in *Nitella*. *J. Cell Biol.* **63:**270–287.

BRADLEY, M. O. 1973. Microfilaments and cytoplasmic streaming: inhibition of streaming with cytochalasin. *J. Cell Sci.* **12:**327–343.

CHEN, J. C. W., and N. KAMIYA. 1975. Localization of myosin in the internodal cell of *Nitella* as suggested by differential treatment with *N*-ethylmaleimide. *Cell Struct. Funct.* **1:**1–9.

CORTI, B. 1774. Osservationi microscopiche sulla tremella e sulla circulazione del fluido in una pianta aquajuola. Lucca.

HATANO, S. 1973. Contractile proteins from the myxomycete plasmodium. *Adv. Biophys.* **6:**143–176.

HEPLER, P. K., and B. A. PALEVITZ. 1974. Microtubules and microfilaments. *Annu. Rev. Plant Physiol.* **25:**309–362.

INOUÉ, S., and R. E. STEPHENS, editors. 1975. Molecules and Cell Movement. Raven Press, New York.

ISHIKAWA, H., R. BISCHOFF, and H. HOLTZER. 1969. Formation of arrowhead complexes with heavy meromyosin in a variety of cell types. *J. Cell Biol.* **43:**312–328.

JAROSCH, R. 1956. Plasmaströmung und Chloroplastenrotation bei Characeen. *Phyton* (*Argentina*) **6:**87–107.

KAMITSUBO, E. 1966. Motile protoplasmic fibrils in cells of Characeae. II. Linear fibrillar structure and its bearing on protoplasmic streaming. *Proc. Jpn. Acad.* **42:**640–643.

KAMITSUBO, E. 1972*a*. Motile protoplasmic fibrils in cells of the Characeae. *Protoplasma.* **74:**53–70.

KAMITSUBO, E. 1972*b*. Destruction and restoration of the protoplasmic fibrillar structure responsible for streaming in the *Nitella* cell. (Japanese with English abstract) *In:* Symposium for Cell Biology (Japan Soc. Cell Biol.) **23:**123–130.

KAMIYA, N. 1959. Protoplasmic streaming. *Protoplas-*

matologia. VIII, **3a:**1–199.

Kamiya, N., and K. Kuroda. 1956. Velocity distribution of the protoplasmic streaming in *Nitella* cells. *Bot. Mag.* (*Tokyo*). **69:**544–554.

Kamiya, N., and K. Kuroda. 1957. Cell operation in *Nitella*. II. Behavior of isolated endoplasm. *Proc. Jpn. Acad.* **33:**201–205.

Kamiya, N., and K. Kuroda. 1958. Measurement of the motive force of the protoplasmic rotation in *Nitella*. *Protoplasma*. **50:**144–148.

Kamiya, N., and K. Kuroda. 1973. Dynamics of cytoplasmic streaming in a plant cell. *Biorheology*. **10:**179–187.

Kersey, Y. M., P. K. Hepler, B. A. Palevitz, and N. K. Wessels. 1976. Polarity of actin filaments in Characean algae. *Proc. Natl. Acad. Sci. U. S. A.* **73:**165–167.

Kuroda, K., and N. Kamiya. 1975. Active movement of *Nitella* chloroplasts *in vitro*. *Proc. Jpn. Acad.* **51:**774–777.

Loewy, A. G. 1952. An actomyosin-like substance from the plasmodium of a myxomycete. *J. Cell. Comp. Physiol.* **40:**127–256.

Nagai, R., and L. I. Rebhun. 1966. Cytoplasmic microfilaments in streaming *Nitella* cells. *J. Ultrastruct. Res.* **14:**571–589.

Palevitz, B. A., J. F. Ash, and P. K. Hepler. 1974. Actin in the green alga, *Nitella*. *Proc. Natl. Acad. Sci. U. S. A.* **71:**363–366.

Palevitz, B. A., and P. K. Hepler. 1975. Identification of actin *in situ* at the ectoplasm-endoplasm interface of *Nitella:* microfilament-chloroplast association. *J. Cell Biol.* **65:**29–38.

Pollard, T. D., and R. R. Weihing. 1974. Actin and myosin and cell movement. *CRC Crit. Rev. Biochem.* **2:**1–65.

Shibata-Sekiya, K., and Y. Tonomura. 1975. Desensitization of substrate inhibition of acto-H-meromyosin ATPase by treatment of H-meromyosin with *p*-chloromercuribenzoate: relation between extent of desensitization and amount of bound *p*-chloromercuribenzoate. *J. Biochem.* (*Tokyo*). **77:**543–557.

Tazawa, M. 1964. Studies on *Nitella* having artificial cell sap. I. Replacement of the cell sap with artificial solutions. *Plant Cell Physiol.* **5:**33–43.

Tazawa, M. 1968. Motive force of the cytoplasmic streaming in *Nitella*. *Protoplasma*. **65:**207–222.

Tazawa, M., M. Kikuyama, and T. Shimmen. 1976. Electric characteristics and cytoplasmic streaming of Characeae cells lacking tonoplast. *Cell Struct. Funct.* **1:**165–176.

Weihing, R. R. 1976. Occurrence of microfilaments in non-muscle cells and tissues (pp. 341–346). Physical and chemical properties of microfilaments in non-muscle cells and tissues (pp. 346–352). Biochemistry of microfilaments in cells and tissues (pp. 352–356). *In* Biological Handbooks, Vol. 1, Cell Biology. P. L. Altman and D. D. Katz, editors. Fed. Amer. Soc. Exp. Biol.: Bethesda, Md.

Williamson, R. E. 1974. Actin in the alga, *Chara corallina*. *Nature* (*Lond.*). **248:**801–802.

Williamson, R. E. 1975. Cytoplasmic streaming in *Chara:* a cell model activated by ATP and inhibited by cytochalasin B. *J. Cell Sci.* **17:**655–668.

DYNAMICS OF CYTOPLASMIC STRUCTURE AND CONTRACTILITY

D. LANSING TAYLOR

Various cytoskeletal and potentially motile structures may exist in nonmuscle cells. Figure 1 is a diagram of an idealized cell containing free F-actin filaments, myosin, filament bundles containing actin, actin in reticular networks, as well as microtubules and ca. 100 Å filaments of unknown chemical identity. Some cells have all of these components, while some have mainly, or exclusively, one system. However, actin, myosin, and associated proteins can perform both cytoskeletal and motile functions even in the absence of microtubules and ca. 100 Å filaments.

There are two major types of cellular movements based on actin, myosin, and associated proteins. First, there are movements involving rapid and extensive changes in cytoplasmic consistency which are involved directly in locomotion. Examples of this form of movement include the cytoplasmic streaming observed in the giant free living amebas, *Chaos carolinensis* and *Amoeba proteus* (Allen, 1972), as well as the acellular slime mold *Physarum polycephalum* (Komnick et al., 1973). Second, there are movements involving more localized motile events. These movements include cytokinesis (Schroeder, 1975), as well as the extension and retraction of filopodia in cells such as most tissue cells in culture (Wessells *et al.*, 1973) and the ameboid stage of the cellular slime mold *Dictyostelium discoideum* (Shaffer, 1964).

The giant amebas have been studied in detail for well over 100 yr. Dujardin, followed by Mast, Pantin, Harvey, Allen, Wohlfarth-Bottermann, Kamiya, and others (Allen, 1972; Komnick et al., 1973) have emphasized the role of changes in cytoplasmic consistency and have related these changes to cell movement. Movement of the giant amebas occurs when the less structured endoplasm streams through the more structured ectoplasmic tube. The endoplasm everts to form the ectoplasmic tube at the tips of advancing pseudopods, while the ectoplasm primarily in the uroid is recruited to form the streaming endoplasm. Therefore, the streaming cycle is related directly to a cycle of consistency changes in the cytoplasm.

In contrast, the ameboid cells of *D. discoideum* are capable of extending and retracting long slender pseudopodia (filopodia) as part of their motile behavior. The extension and retraction of filopodia appear to involve localized regions of the cell at any one time. Therefore, if changes in cytoplasmic consistency occur, they must be highly localized.

Cytoplasmic consistency and contractility in nonmuscle cells have been investigated in vitro with various types of cellular extracts. The earliest such studies were based on an experiment first performed on vertebrate striated muscle (Szent-Györgyi, 1947). Szent-Györgyi had demonstrated that a high ionic strength extract of muscle containing actin and myosin would "superprecipitate" or contract when the ionic strength was lowered in the presence of ATP. Similar results were obtained from high ionic strength extracts of nonmuscle cells including *Physarum* (Loewy, 1952) and blood platelets (Bettex-Galland and Lüscher, 1961). These early experiments with extracts prepared from different cell types have been discussed in detail by Wolpert et al. (1964). Such model systems suggested that "actin- and myosin-like" proteins might be present in nonmuscle cells.

Later studies on cell extracts were aimed at identifying the structures and ionic regulation responsible for cytoplasmic contractions and streaming. These models were prepared from both single cells and mass cultures of the giant amebas *C. carolinensis* and *A. proteus* (Table I). These studies demonstrated the presence of actin filaments in

D. LANSING TAYLOR The Biological Laboratories, Harvard University, Cambridge, Massachusetts

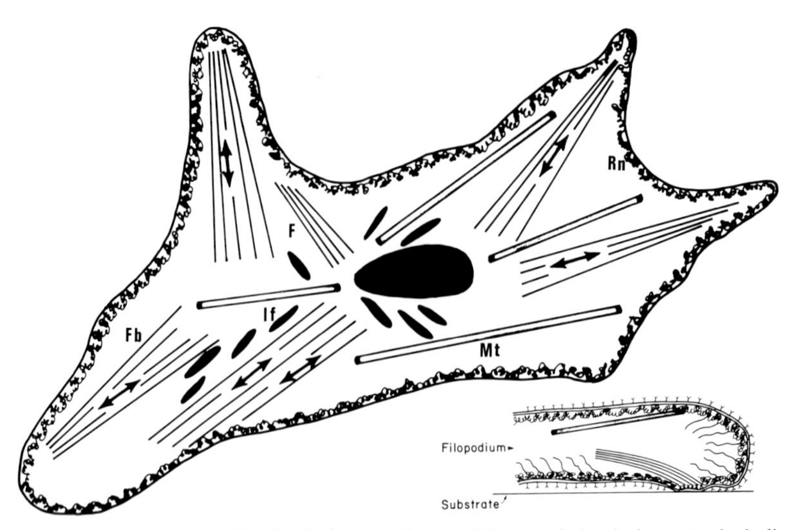

FIGURE 1 A diagram of an idealized cell showing the possible cytoskeletal elements, including free F-actin filaments (*F*), filament bundles containing actin (*Fb*), reticular networks containing actin (*Rn*), microtubules (*Mt*), and intermediate filaments (*If*) which are ca. 100 Å diameter. Some of the actin-containing supramolecular structures are associated with the membrane. (*Inset*) The cell surface (*Y*) is in close proximity to the substrate at several points on the cell during movement. Furthermore, initiation of dynamic events probably involves the relay of information from the environment to the cell surface, through the plasmalemma, to the dynamic cytoskeleton.

TABLE I
Early Cell Extracts

Cell type	Extraction solution
Chaos carolinensis	
Allen et al., 1960	Single cell fractured under oil
Taylor et al., 1973	Single cell ruptured in 5.0 mM pipes, 5.0 mM EGTA, 30 mM KCl
Amoeba proteus	
Gicquand and Couillard, 1970	Single cell squash
Thompson and Wolpert, 1963	2,000 *g* supernate of cells homogenized in D-H_2O
Pollard and Ito, 1970	1,000 *g* supernate of cells homogenized in 10 mM Tris-maleate

the motile models (Pollard and Ito, 1970) as well as thick filaments similar morphologically to myosin aggregates (Taylor et al., 1973). ATP was implicated in both the nonmotile cytoplasmic structure and contractions (Thompson and Wolpert, 1963; Pollard and Ito, 1970). Furthermore, calcium was shown to regulate contractility and streaming in single cell extracts that maintained the in vivo geometry of cytoplasmic streaming (Taylor et al., 1973). This later study was the first to demonstrate a calcium regulated contractile process as the basis of ameboid movement.

More recently, bulk extracts have been prepared from many types of cells to determine the molecular basis of the changes in cytoplasmic consistency and contractile events. These extracts are capable of forming a nonmotile cytoskeletal structure (gel) which can be induced to contract (Fig. 2). The solutions used for extractions have varied (Table II), as have the experimental results (Table III). The extracts from *A. proteus* (Taylor *et al.*, 1976*b*) and *D. discoideum* (Taylor *et al.*, 1976*a*; Condeelis and Taylor, 1976) formed a nonmotile gel at pH 7.0, I = 0.05 in the presence of Mg-ATP and a submicromolar concentration of free calcium ions. Contraction of the gel was elicited by the addition of a contraction solution containing a micromolar concentration of free calcium ions. However, the addition of micromolar calcium at 0°C inhibited the formation of the gel (Taylor *et al.*, 1976*a*; Condeelis and Taylor, 1976, The contractile basis of ameboid movement. V. The sol-gel-contraction cycle of extracts from *D. discoideum*, submitted for publication) during warming to room temperature. Furthermore, a myosinless extract would gel under the conditions described above but would solate upon the addition of the contraction solution containing ca. 10^{-6} M free calcium ions. Thus it appears that changes in the concentration of free calcium ions around ca. 10^{-} M can regulate both cytoplasmic consistency and contractility in vitro.

The major proteins involved in gelation and contraction in the various extracts are shown i

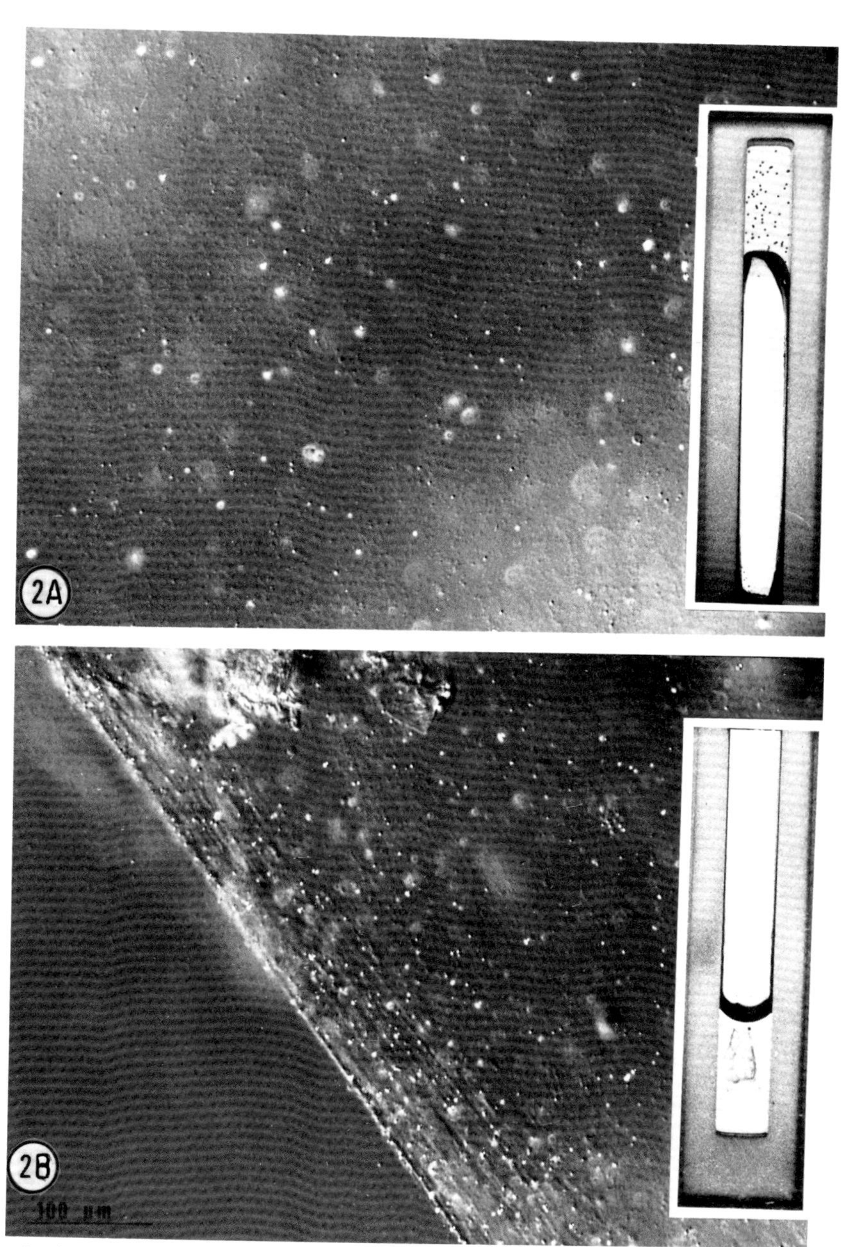

FIGURE 2 (A) Extracts from *A. proteus* and *D. discoideum* form nonmotile gels at pH 7.0, I = ca. 0.05, in the presence of submicromolar free calcium ion concentrations and Mg-ATP. × 185. *Inset* shows gelled extract in inverted curvette. (B) Upon the addition of a contraction solution containing ca. 10^{-6} M free calcium and Mg-ATP, the gel contracts. × 185. *Inset* shows contracted pellet in a curvette.

Table IV. It is apparent that whereas actin and myosin are the major components in the contracted pellets, several proteins other than myosin appear to "bind" actin in the nonmotile complex (gel). Much progress has been made, but more work is required to demonstrate both the minimal molecular components and the regulation of the reconstructed complex. The aim of these investigations has been to prepare a functional motile model from purified proteins.

Cytoplasmic Consistency In Vitro and In Vivo

The viscoelasticity of cytoplasm has been demonstrated by inducing strain birefringence in the cytoplasm of intact amebas (Francis and Allen, 1971). By the use of a similar strain birefringence assay, a gradient of increasing endoplasmic viscoelasticity from the tail to the tips of extending pseudopods has been demonstrated in intact *C. carolinensis* (Taylor, 1976, 1977). The same assay has shown similar changes in consistency during the gelation of extracts (Condeelis and Taylor, 1976, submitted). These results indicated that the changes in the consistency of extracts could reflect the changes occurring during normal cell movements.

Various experimental agents have been used in order to modulate the cytoplasmic consistency, especially of the cortex, in living cells (see Allen, 1961, for review). Table V summarizes some of the effects of these agents on intact cells. The most consistent effect was a loss of evidence of gel structure in the cortex during the application of the agent followed by contraction during the recovery. Essentially the same results were obtained on gelated extracts (Table VI). An initial solation was followed by contraction. It was apparent from these studies that (1) contractions followed the breakdown of the rigid gel, and (2) calcium ions

TABLE II
Recent Cell Extracts

Cell type	Extraction solution
Sea urchin eggs (Kane, 1975)	0.9 M glycerol, 0.1 M pipes buffer, 5.0 mM EGTA
Amoeba proteus (Taylor et al., 1976*b*)	5.0 mM pipes buffer, 5.0 mM EGTA, 1.0 mM $CaCl_2$, 30 mM KCl, 1.0 mM $MgCl_2$
Acanthamoeba (Pollard, 1976*a*)	0.34 M sucrose, 10 mM imidazole chloride, 1.0 mM ATP, 1.0 mM EGTA, 1.0 mM DTT
Rabbit pulmonary macrophages (Stossel and Hartwig, 1976)	0.34 M sucrose, 10 mM DTT, 5.0 mM ATP, 1.0 mM EDTA, 20 mM Tris-maleate
Dictyostelium discoideum	
Taylor *et al.*, 1976*a*	5.0 mM pipes, 5.0 mM EGTA, 1.0 mM $CaCl_2$, 30 mM KCl, 1.0 mM MgATP, 1.0 mM DTT, trasylol
Condeelis and Taylor, 1976, submitted	5.0 mM pipes, 5.0 mM EGTA, 1.0 mM DTT, trasylol

TABLE III
Summary of Recent Motile Extracts

Gelation	Contraction
ATP and EGTA required in most extracts	ATP and $MgCl_2$ induced contractions *Acanthamoeba* (Pollard, 1976*a*) Sea urchin eggs (Kane, 1975) Rabbit pulmonary macrophages (Stossel and Hartwig, 1976) An absolute requirement of $>10^{-6}$ M Ca^{++} in addition to ATP and $MgCl_2$ was identified *C. carolinensis* (Taylor et al., 1973) *A. proteus* (Taylor et al., 1976*b*) *D. discoideum* (Taylor et al., 1976*a*; Condeelis and Taylor, 1976, submitted)
Gel highly birefringent Sea urchin eggs (Kane, 1975) *Acanthamoeba* (Pollard, 1976*a*)	—
Gel weakly birefringent; almost isotropic *A. proteus* (Taylor et al., 1976*b*) *D. discoideum* (Taylor et al., 1976*a*; Condeelis and Taylor, 1976, submitted)	Large increase in birefringence during contraction *A. proteus* (Taylor et al., 1976*b*) *D. discoideum* (Taylor et al., 1976*a*; Condeelis and Taylor, 1976, submitted)
F-actin has been identified in the gelled extract *Acanthamoeba* (Pollard, 1976*a*) Sea urchin eggs (Kane, 1975) Rabbit pulmonary macrophages (Stossel and Hartwig, 1976)	F-actin has been identified in the contracted extract *Acanthamoeba* (Pollard, 1976*a*) Sea urchin eggs (Kane, 1975) Rabbit pulmonary macrophages (Stossel and Hartwig, 1976)
Polymeric network containing actin identified in gelled extract *A. proteus* (Taylor et al., 1976*b*) *D. discoideum* (Taylor et al., 1976*a*; Condeelis and Taylor, 1976, submitted)	Increase in the relative number of free F-actin filaments during contraction *A. proteus* (Taylor et al., 1976*b*) *D. discoideum* (Taylor et al., 1976*a*; Condeelis and Taylor, 1976, submitted)

TABLE IV
Reconstitution of Gelation and Contraction

Cell type	Major proteins in gel	Major proteins concentrated during contraction	Minimal proteins identified in gel
	mol wt/1,000	*mol wt*	
Sea urchin eggs (Kane, 1975)	Actin, 220, 58	–	Actin, 220, 58
Acanthamoeba (Pollard, 1976*a*)	Actin, 160 (myosin), 280, 93, 68, 50, 35	Actin, 160 (myosin), 280, 95	Actin
Rabbit pulmonary macrophages (Stossel and Hartwig, 1976)	Actin, 200 (myosin), 280, 90 +	Actin, myosin, 280, 155, 90	Actin, 280
D. discoideum (Taylor et al., 1976 a)	Actin, 230 (myosin), 280, 240, 220, 95, 75, 55	Actin, myosin, 32, 95, 20	Actin, myosin, 280, 240, 220, 95, 90, 75, 55, 35, 20
(Condeelis and Taylor, 1976, submitted)	Actin, myosin, 250, 160, 145, 95, 75, 55	Actin, myosin, 95, 35, 20	Actin, 95, 90, 75, 35, 20

TABLE V
Experimental Manipulation of the Cell Cortex in Intact Cells

Experimental parameter and cell type	Effect of application	Effect upon recovery
Pressure 22°C *A. proteus* (Landau et al., 1954)	4,000–5,000 psi induced solation	Ectoplasmic region contracted ca. 15 s. after decompression
Low temperature (10°C)-pressure *A. proteus* (Landau et al., 1954)	3,000 psi induced solation	Ectoplasmic region contracted after decompression
Mechanical stimulation *A. proteus* (Angerer, 1936)	Agitation of cells by shaking caused a decrease in ectoplasmic viscosity	Ectoplasmic region contracted during the recovery from agitation
Microinjection of 1/10 cell volume with ca. 10^{-6} M Ca^{++} and Mg-ATP *C. carolinensis* (Taylor, 1976)	Local contraction of ectoplasm	Normal streaming
Microinjection of 1/10 cell volume with <ca. 10^{-6} M Ca^{++} and Mg-ATP *C. carolinensis* (Taylor, 1976)	Local loss of separate endoplasmic and ectoplasmic structure	Normal streaming
Increase extracellular pH from 7.0 to 7.8 *A. proteus* (Mast and Prosser, 1932)	Increased rate of movement	Normal rate of movement
Cytochalasin	Variable	Normal cell movement

and pH were the most "physiological" parameters examined that affected both cytoplasmic consistency and contractility. The free calcium ion concentration modulated around the micromolar level at pH 7.0 could control both cytoplasmic consistency and contractility (Taylor et al., 1976*a*, 1976*b*; Condeelis and Taylor, 1976, submitted).

Furthermore, varying the pH of the cytoplasmic extracts around pH 7.0 when the calcium ion concentration was maintained just below 1 μmol also controlled both cytoplasmic consistency and contractility. Gelation and contraction were inhibited below pH 7.0, while gelation was maximized at pH 7.0. Furthermore, contractions occurred above pH 7.0 after the breakdown of the gel (Taylor, 1977; Taylor et al., 1976*a*; Condeelis and Taylor, 1976, submitted). Therefore, both the cytoplasmic pH and free calcium ion concentration can be involved in regulating cellular structure and contraction.

Changes in the free calcium ion concentration have been detected in intact cells, as well as in cell extracts. When the bioluminescent protein aequorin, which luminesces upon binding free calcium ions, was microinjected into *C. carolinensis*, the monopodial cells exhibited luminescence while moving in an electric field (Taylor et al., 1975*a*). However, the monopodial cells lost obvious ectoplasmic tubes after more than a few minutes in the electric field. It is not presently known what other ionic conditions were altered when the cells were placed in the electric field. One major possibility is the involvement of protons.

Extracts from *D. discoideum* gelled without contracting when warmed to room temperature in the presence of aequorin. The luminescence was

TABLE VI

Experimental Manipulation of the Extract Structure

Experimental parameter and type of extract	Effect of application	Effect upon recovery
Pressure 22°C		
D. discoideum (Condeelis and Taylor, 1976, submitted)	2,000 psi induced solation of previously gelled extract	Contraction 2 min after decompression
Low temperature (0–4°C)		
Macrophages (Stossel and Hartwig, 1976)	Solation	Gelation
Acanthamoeba (Pollard, 1976*a*)	Solation	Gelation
Sea urchin eggs (Kane, 1975)	No effect	No effect
D. discoideum (Taylor et al., 1976*a*; Condeelis and Taylor, 1976, submitted)	Solation followed by contraction	–
Mechanical stimulation		
Acanthamoeba (Pollard, 1976*a*)	Gel fractures	–
Macrophages (Stossel and Hartwig, 1976)	Gel fractures	–
D. discoideum (Condeelis and Taylor, 1976, submitted)	Contraction after agitation of gel	–
Addition of contraction solution (ca. 10^{-6} M Ca^{++}, Mg-ATP)		
D. discoideum (Taylor et al., 1976*a*; Condeelis and Taylor, 1976, submitted)	Contraction of gel	–
Addition of 7.5 mM EGTA		
D. discoideum (Condeelis and Taylor, 1976, submitted)	Solation of gel within 10 min	–
Increase pH from 7.0 to 7.5		
D. discoideum (Taylor et al., 1976*a*; Condeelis and Taylor, 1976, submitted)	Contraction upon warming	–
Cytochalasin		
Macrophages (Hartwig and Stossel, 1976)	>1 μM inhibits gelation	–
Hela cell (Weihing, 1976)	>0.25 μM inhibits gelation	–
D. discoideum (Condeelis and Taylor, 1976, submitted)	>1.0 μM induces solation of gel followed by very slow contraction >5 h	–
Acanthamoeba (Pollard, 1976*b*)	>1.0 μM inhibits gelation	–

very low as a result of the calcium buffering of the extraction solution. However, the addition of the contraction solution (ca. 1.0 × 10^{-6} M free calcium ions) induced a dramatic rise in the luminescence during the contraction of the gel (Fig. 3; Condeelis and Taylor, 1976, submitted). However, pH-induced contractions did not elicit increases in luminescence (Fig. 3), demonstrating that endogenous calcium was not being released.

The effects of the free calcium ion concentration at pH 7.0 on cytoplasm have been demonstrated directly on intact cells through microinjection experiments (Taylor, 1976, 1977). The normal differentiation between the more structured ectoplasmic tube and the streaming endoplasm, as well as movement, was lost when *C. carolinensis* was microinjected with ¹/₁₀ the cell volume with a relaxation solution (submicromolar calcium and Mg-ATP, pH 7.0; Taylor, 1976, 1977). In contrast, the cytoplasm contracted into a central mass when *C. carolinensis* was microinjected with the same volume of the contraction solution (ca. micromolar calcium and Mg-ATP). Both of these experimental states were reversible, indicating that the intact cells regulated the internal free calcium ion concentration. Furthermore, changes in pH produced the same results.

Is the Cytoskeleton a Dynamic Structure?

The ultrastructural appearance of actin and associated proteins depends in part on the ionic environment (Taylor et al., 1976*a*, 1976*b*; Condeelis and Taylor, 1976, submitted). For example, Hatano demonstrated that actin from *Physarum* would form a polymer other than F-actin in the presence of magnesium ions and another protein (Hatano, 1972). In addition, partially purified actin from *D. discoideum* formed various supramolecular structures under different ionic conditions (Spudich and Cooke, 1975). The extracts from

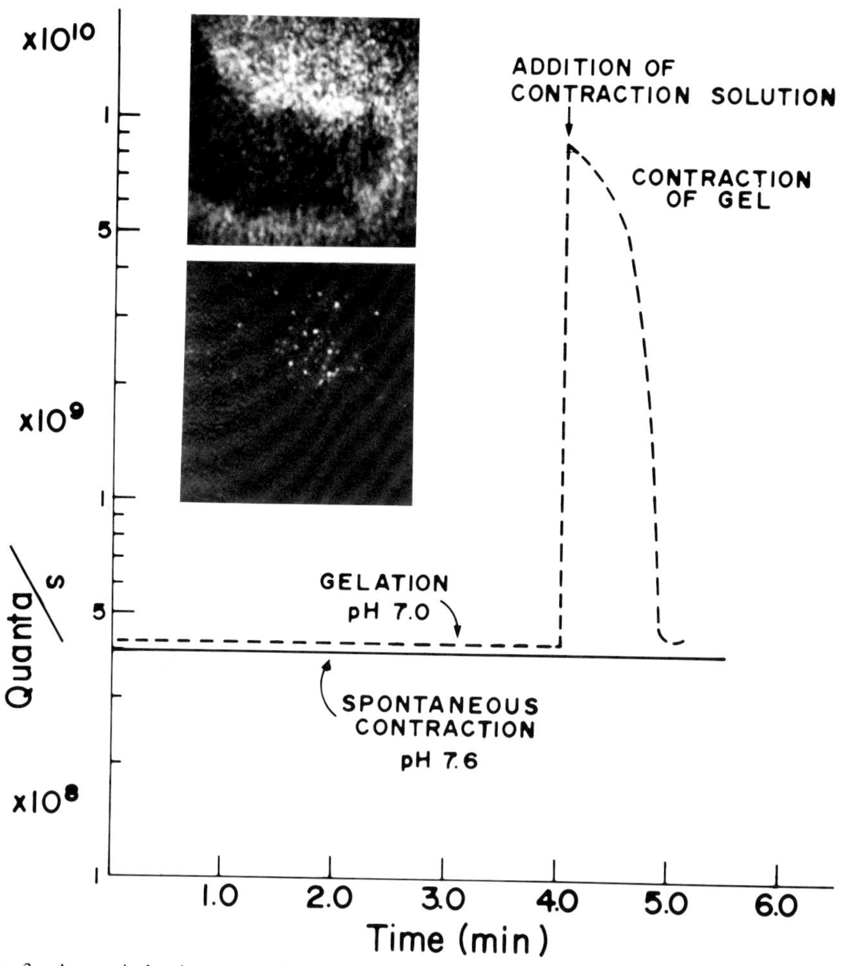

FIGURE 3 Aequorin luminescence of a gelled extract from *D. discoideum* during contraction. Contractions induced by raising the pH did not cause luminescence. The upper *inset* shows a single specimen of *C. carolinensis* after microinjection with aequorin viewed in an image intensifier. The lower *inset* demonstrates the luminescence induced by orientation in an electric field.

both *A. proteus* and *D. discoideum* exhibited the same structures under the same ionic conditions. The gelled extracts did not contain readily identifiable free F-actin filaments. Instead, latticeworks containing actin were demonstrated. However, free F-actin filaments were identified after the addition of ca. 10^{-6} M calcium ions to the gelled extracts at pH 7.0, or by raising the pH from 7.0 to ca. 7.4 (Fig. 4). These results suggested that a calcium and/or pH regulated "transformation" of actin-containing structures occurred as part of the transition from the cytoskeletal to the motile state.

The Cytoskeleton and Calcium Regulation

It has been demonstrated that the free calcium ion concentration, along with a closely regulated pH, controls the formation and the breakdown of the nonmotile cytoskeleton, as well as contractions (Taylor et al., 1976*a*, 1976*b*; Condeelis and Taylor, 1976, submitted). Therefore, the simplest possible regulation of movement would be the calcium and/or pH regulated gelation-contraction cycle. At just submicromolar free calcium ion concentrations and/or at pH 7.0, actin could form a latticework (gel) with the actin-binding protein(s) which would physically inhibit interaction with myosin. Raising the free calcium ion concentration to about the micromolar level and/or raising the pH above 7.0 would dissociate the actin from the actin-binding protein(s), permitting actin to form free F-actin filaments that could interact with

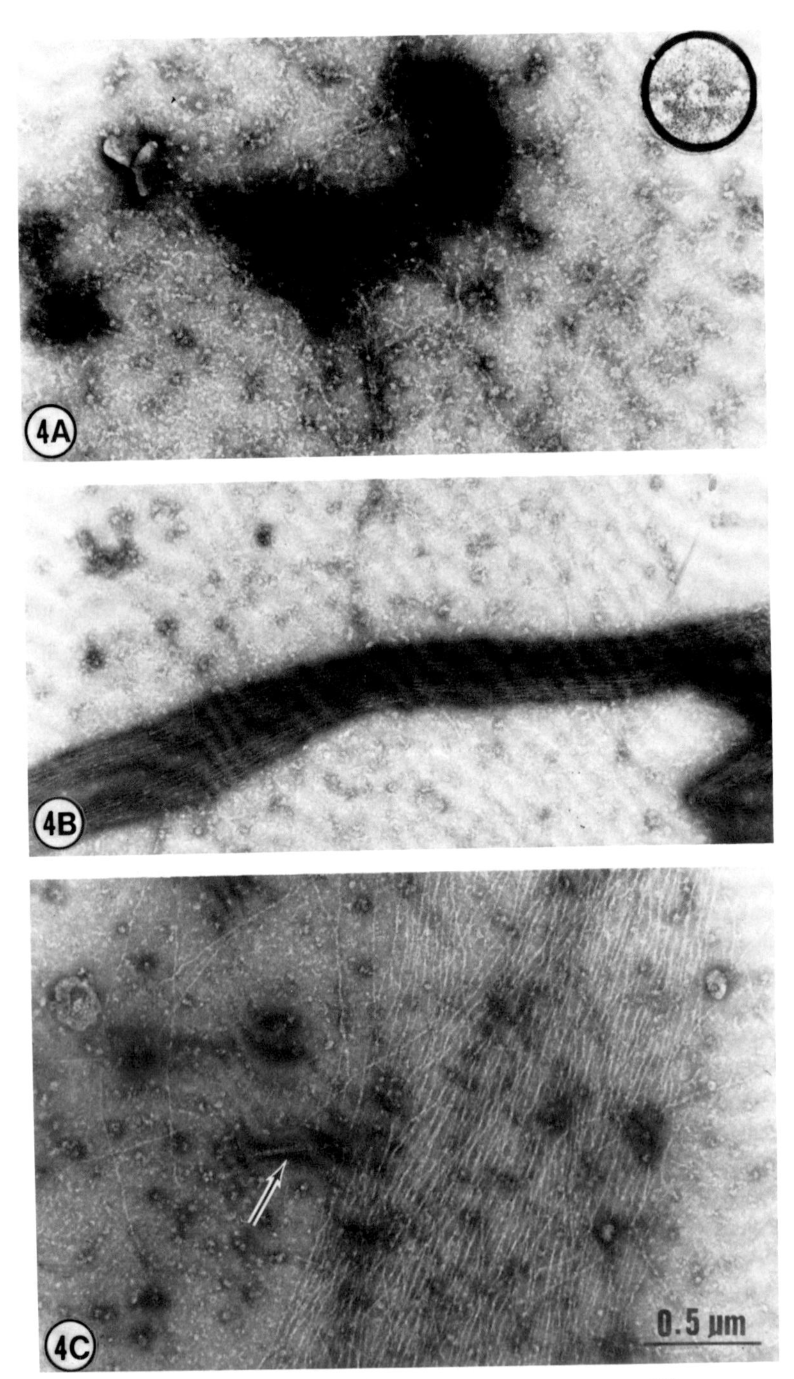

FIGURE 4 Actin in extracts from *A. proteus* and *D. discoideum* can assume different supramolecular structures depending on the ionic environment. (A) Reticular networks are the most prevalent structures observed in the presence of 1.0 mM Mg-ATP and just submicromolar free calcium ion concentrations. (Conditions for gelation.) *Inset* shows a doughnut-shaped structure in close association with a filament. (B) Raising the $MgCl_2$ concentration to ca. 3–5 mM induces the formation of filament bundles. (C) Free F-actin filaments are formed by the addition of ca. 10^{-6} M free calcium ions to the preparation in (A). × 45,000.

myosin. Finer regulation could also be accomplished by dual mechanisms involving either actin- or myosin-associated control proteins (see discussion by T. D. Pollard, this volume). However, the simplest mechanism could be the complete control.

Membranes and Cell Movement

In order for cellular movements to occur, the contractile cytoskeleton must be attached at least transiently to the membrane. Actin has been shown to be "associated" with membranes from many different cell types (see Tilney, this volume). Table VII is a representative list of the cells in which actin has been shown in association with the plasmalemma. Many motile events occur or are initiated at a cytoplasm-membrane interface, and it appears that actin in association with other proteins forms different supramolecular structures during a motile event. For example, cytokinesis (Schroeder, 1975) involves the "transformation" of the cell cortex from a rigid structure not containing many free F-actin filaments to an actively contracting ring containing F-actin filaments. Furthermore, the activation of blood platelets (Behnke et al., 1971) involves the rapid formation of F-actin filaments that are associated with the membrane (Taylor et al., 1975*b*). Calcium ion and/or pH transients could also regulate the cytoskeletal structure and contractions in these and other cellular movements (Taylor et al., 1976*b*).

Some Future Challenges

Determining the molecular basis of the formation of the actin-based cytoskeleton and contraction in vitro should aid in defining the dynamics and regulation of these events in vivo. The minimal components that form the nonmotile gel and contracting complex must be isolated and reconstructed. Furthermore, the physiological control of the solation-gelation-contraction cycle of these proteins must be demonstrated experimentally.

A complete understanding of the ionic requirements for solation, gelation, and contraction should suggest local ionic conditions in living cells. Micro environments can be sought in vivo by the use of specific indicators of calcium, ATP, pH, and other physiological parameters. Furthermore, the localization, interaction, and supramolecular structure of actin, myosin, and their associated proteins, as well as microtubules and 100 Å filaments, could be identified in vivo by incorporating properly labeled proteins into living cells and examining them with quantitative optical microscopy (Taylor, 1976; Taylor and Zeh, 1977).

TABLE VII
Association of Actin with Membrane

Cell type and membrane	Method of identification
Plasmalemma	
Red blood cells (Tilney and Detmers, 1975)	HMM label
Mammalian intestinal brush border (Mooseker and Tilney, 1975)	HMM label
Acanthamoeba castelanii (Pollard and Korn, 1973)	HMM label
Platelets (Taylor et al., 1975*b*)	Biochemical
C. carolinensis (Comly, 1973; Taylor et al., 1976*b*)	HMM label EM
D. discoideum (Spudich, 1974)	Biochemical
Mouse 3T3 cells (Gruenstein et al., 1975)	Biochemical
Cultured nerve cells (Wessells et al., 1973)	HMM label
Guinea pig peritoneal macrophages (Allison et al., 1971)	EM

The mechanochemistry of cell movement appears to be regulated ultimately at the cytoplasm-plasmalemma-cell surface complex. The cell surface receives "signals" from the environment that indicate directions of food, light, moisture, presence of specific ligands, etc. In order for the cells to react, these "signals" must be transmitted directly or indirectly through the plasmalemma to the contractile machinery, part of which appears to be associated with the cytoplasmic side of the plasmalemma (Fig. 1, *inset*). Therefore, the complete understanding of cellular controls of movement will require an integral knowledge of the cell surface, the membrane, and the contractile and cytoskeletal proteins.

Summary

Dynamic cell structures and movements can be reconstructed and regulated by calcium ions, ATP, and pH in vitro, and the results can be correlated with cytoplasmic dynamics in vivo. Actin in association with other proteins can assume different supramolecular structures under physiological conditions involving gelation, contraction, and solation. The tension generated by cytoplasmic contractions in intact cells is transmitted to the substrate through the association of actin with the membrane.

REFERENCES

ALLEN, R. D. 1961. Amoeboid movement. *In* The Cell. J. Brachet and A. E. Mirsky, editors. Academic Press, Inc., New York. 135–216.

ALLEN, R. D. 1972. Biophysical aspects of pseudopodium formation and retraction. *In* The Biology of Amoeba. K. Jeon, editor. Academic Press, Inc., New York. 201–247.

ALLEN, R. D., J. W. COOLEDGE, and P. J. HALL. 1960. Streaming in cytoplasm dissociated from the giant amoeba *Chaos chaos*. *Nature (Lond.)*. **187:**896.

ALLISON, A. C., P. DAVIES, and S. DEPETRIS. 1971. Role of contractile microfilaments in movement and endocytosis. *Nat. New Biol.* **232:**153–155.

ANGERER, C. A. 1936. The effects of mechanical agitation on the relative viscosity of amoeba protoplasm. *J. Cell Comp. Physiol.* **8:**327–345.

BEHNKE, O., B. I. KRISTENSEN, and L. E. NIELSEN. 1971. Electron microscopical observations of actinoid and myosinoid filaments in blood platelets. *J. Ultrastruct. Res.* **37:**351-363.

BETTEX-GALLAND, M., and E. F. LÜSCHER. 1961. Thrombostenin: a contractile protein from thrombocytes: its extraction from human blood platelets and some of its physical properties. *Biochim. Biophys. Acta.* **49:**536–551.

COMLY, L. 1973. Microfilaments in *Chaos carolinensis*. *J. Cell Biol.* **58:**230–237.

FRANCIS, D. W., and R. D. ALLEN. 1971. Induced birefringence as evidence of endoplasmic viscoelasticity in *Chaos carolinensis*. *J. Mechanochem. Cell Motility.* **1:**1–10.

GICQUAND, C. R., and P. COUILLARD. 1970. Preservation des mouvements dans le cytoplasme demembrane d'*Amoeba proteus*. *Cytobiologie.* **1:**460–467.

GRUENSTEIN, E., A. RICH, and R. R. WEIHING. 1975. Actin associated with membranes from 3T3 mouse fibroblast and HeLa cells. *J. Cell Biol.* **64:**223–234.

HARTWIG, J. H., and T. P. Stossel. 1976. Interactions of actin, myosin, and an actin-binding protein of rabbit pulmonary microphages. III. Effects of cytochalasin B. *J. Cell Biol.* **71:**295–302.

HATANO, S. 1972. Conformational changes of plasmodium actin polymers formed in the presence of Mg^{++}. *J. Mechanochem. Cell Motility.* **1:**75–80.

KANE, R. E. 1975. Preparation and purification of polymerized actin from sea urchin egg extracts. *J. Cell Biol.* **66:**305–315.

KOMNICK, H., W. STOCKEM, and K. E. WOHLFARTH-BOTTERMANN. 1973. Cell motility: mechanisms in protoplasmic streaming and amoeboid movement. *Int. Rev. Cytol.* **34:**169–249.

LANDAU, J. V., A. M. ZIMMERMAN, and D. A. MARSLAND. 1954. Temperature-pressure experiments on *Amoeba proteus*: plasmagel structure in relation to form and movement. *J. Cell. Comp. Physiol.* **44:**211–232.

LOEWY, A. G. 1952. An actomyosin-like substance from the plasmodium of a myxomycete. *J. Cell. Comp. Physiol.* **40:**127–135.

MAST, S. O., and C. L. PROSSER. 1932. Effect of temperature, salts, and hydrogen-ion concentration on rupture of the plasmagel sheet, rate of locomotion, and gel/sol ratio in *A. proteus*. *J. Cell. Comp. Physiol.* **1:**333–354.

MOOSEKER, M. S., and L. G. TILNEY. 1975. Organization of an actin filament-membrane complex: filament polarity and membrane attachment in the microvilli of intestinal epithelial cells. *J. Cell Biol.* **67:**725–743.

POLLARD, T. D. 1976*a*. The role of actin in the temperature-dependent gelation and contraction of extracts of Acanthamoeba. *J. Cell Biol.* **68:**579–601.

POLLARD, T. D. 1976*b*. Cytoskeletal functions of cytoplasmic contractile proteins. *J. Supramol. Struct.* In press.

POLLARD, T. D., and S. ITO. 1970. Cytoplasmic filaments of *Amoeba proteus*. I. The role of filaments in consistency changes and movement. *J. Cell Biol.* **46:**267–319.

POLLARD, T. D., and E. D. KORN. 1973. Electron microscopic identification of actin associated with isolated amoeba plasma membranes. *J. Biol. Chem.* **218:**448–450.

SCHROEDER, T. 1975. Dynamics of the contractile ring. *In* Molecules and Cell Movement. Raven Press, New York. 305–334.

SHAFFER, B. M. 1964. Intracellular movement and locomotion of cellular slime-mold amoebae. *In* Primitive Motile Systems in Cell Biology. R. D. Allen and N. Kamiya, editors. Academic Press, Inc., New York. 387–403.

SPUDICH, J. A. 1974. Biochemical and structural studies of actomyosin-like proteins from non-muscle cells. *J. Biol. Chem.* **249:**6013–6020.

SPUDICH, J. A., and R. COOKE. 1975. Supramolecular forms of actin from amoebae of *Dictyostelium discoideum*. *J. Biol. Chem.* **250:**7485–7491.

STOSSEL, T. P., and J. H. HARTWIG. 1976. Interactions of actin, myosin, and a new actin-binding protein of rabbit pulmonary macrophages. II. Role in cytoplasmic movement and phagocytosis. *J. Cell Biol.* **68:**602–619.

SZENT-GYÖRGYI, A. 1947. Chemistry of Muscle Contraction. Academic Press, Inc., New York. 150.

TAYLOR, D. G., V. M. WILLIAMS, and N. CRAWFORD. 1975*b*. Platelet membrane actin: solubility and binding studies with ^{125}I-labelled actin. *Biochem. Soc. Trans.* **4:**156–160.

TAYLOR, D. L. 1976. Motile models of amoeboid movement. *In* Cell Motility. R. Goldman, T. Pollard, and J. Rosenbaum, editors. Cold Spring Harbor Laboratory, New York. 797–821.

TAYLOR, D. L. 1977. The contractile basis of amoeboic movement. IV. The viscoelasticity and contractility o amoeba cytoplasm. *Exp. Cell Res.* In press.

Taylor, D. L., J. S. Condeelis, P. L. Moore, and R. D. Allen. 1973. The contractile basis of amoeboid movement. I. The chemical control of motility in isolated cytoplasm. *J. Cell Biol.* **59:**378–394.

Taylor, D. L., J. S. Condeelis, and J. A. Rhodes. 1976*a*. The contractile basis of amoeboid movement. III. Structure and dynamics of motile extracts and membrane fragments from *D. discoideum* and *A. proteus*. *J. Supramol. Struct.* In press.

Taylor, D. L., G. T. Reynolds, and R. D. Allen. 1975*a*. Aequorin luminescence in *Chaos carolinensis*. *Biol. Bull.* (*Woods Hole*). **149:**448.

Taylor, D. L., J. A. Rhodes, and S. A. Hammond. 1976*b*. The contractile basis of amoeboid movement. II. Structure and contractility of motile extracts and plasmalemma-ectoplasm ghosts. *J. Cell Biol.* **70:**123–143.

Taylor, D. L., and R. Zeh. 1977. Methods for the measurement of polarization optical properties. I. Birefringence. *J. Microsc.* (*Oxford*). In press.

Thompson, C. M., and L. Wolpert. 1963. The isolation of motile cytoplasm from *Amoeba proteus*. *Exp. Cell Res.* **32:**156–160.

Tilney, L. G., and P. Detmers. 1975. Actin in erythrocyte ghosts and its association with spectrin: evidence for a nonfilamentous form of these two molecules *in situ*. *J. Cell Biol.* **66:**508–517.

Weihing, R. R. 1976. Cytochalasin B inhibits actin-related gelation of HeLa cell extracts. *J. Cell Biol.* **71:**303–306.

Wessells, N. K., B. S. Spooner, and M. A. Luduena. 1973. Locomotion in tissue cells. *CIBA Found. Symp.* **14:**53–82.

Wolpert, L., C. M. Thompson, and C. H. O'Neill. 1964. Studies on the isolated membrane and cytoplasm of *Amoeba proteus* in relation to amoeboid movement. *In* Primitive Motile Systems in Cell Biology. R. D. Allen and N. Kamiya, editors. Academic Press, Inc., New York. 43–65.

CYTOPLASMIC CONTRACTILE PROTEINS

THOMAS D. POLLARD

The discovery of the contractile proteins, actin and myosin, in nonmuscle cells has revolutionized thinking about cell motility by suggesting for the first time plausible molecular mechanisms for some cellular movements such as cytokinesis. At the present time, investigators are trying to complete the catalogue of protein components involved with motility and are just beginning to characterize them in some detail. At the same time, important work has begun on localizing the cytoplasmic contractile proteins within cells undergoing various movements. Together with experiments on model systems (see Taylor, this volume), these approaches provide strong evidence for the participation of actin and myosin in a variety of important cellular movements.

In this brief summary of recent work on cytoplasmic contractile proteins, I will enumerate the proteins that have been identified as components of cytoplasmic contractile systems and review the evidence that these proteins power cellular movements. In addition, I will discuss some issues that remain unresolved and are under active investigation. Readers in search of more detailed information might consult the review by Pollard and Weihing (1974) or the new three-volume *Cell Motility*, edited by Goldman, Pollard, and Rosenbaum (1976).

Components of Cytoplasmic Contractile Systems

Twelve proteins have been identified as components of cytoplasmic contractile systems by isolation and characterization (Table I). Only 10 yr ago, Hatano and Oosawa (1966) first purified a cytoplasmic contractile protein, *Physarum* actin. The catalogue of cytoplasmic contractile proteins has grown rapidly in recent years, but there is little doubt that it is incomplete. Cells with these various proteins come from all parts of the eukaryotic phylogenetic tree including plants, protista, and animals, but no single cell has all of these proteins.

The force-generating proteins are actin and myosin. They are found in all eukaryotic cells; myosin in low concentration and actin in high concentration. The high concentration of actin is somewhat of a puzzle. One explanation is that it is involved with more than cell motility, perhaps serving a cytoskeletal role by virtue of its ability to control cytoplasmic consistency (see Pollard, 1977, and Taylor, this volume, for more information on this point).

The proteins in the second group are thought to participate in the regulation of actin-myosin interaction. Because these regulatory mechanisms are far from understood, they are considered below as an unsolved problem.

The proteins in the third group seem to have structural roles, although none has been characterized well enough to rule out other functions, The "actin-binding" proteins come in several sizes. They can cross-link actin filaments, which leads to the formation of gels and/or stable filament bundles (see both Taylor and Tilney, this volume). The role of the erythrocyte actin-binding-protein, spectrin, is controversial (see Marchesi et al., 1976), but it may link actin to the red cell membrane. Alpha-actinin is found in the Z line of muscle, and in nonmuscle cells it may participate in actin filament-membrane interactions (see Tilney, this volume). All of these proteins, except beta-actin, can stabilize actin filaments; beta-actinin can cause their fragmentation (Hatano and Owaribe, 1976).

Inasmuch as the detailed understanding of the molecular basis of cellular movements depends on the thorough characterization of all the involved proteins, it is regretable that not a single protein listed in Table I has been well characterized. Consequently, there is justification for examining criti-

THOMAS D. POLLARD Anatomy Department, Harvard Medical School, Boston, Massachusetts

TABLE I

*Cytoplasmic Contractile Proteins—1976**

Force generation	
	Actin
	Myosin
Control	
	Tropomyosin
	Troponin-C
	Slime mold regulatory factor
	Cofactor
	Light chain kinase
Structure	
	Spectrin
	280,000-dalton actin-binding protein
	55,000-dalton actin-binding protein
	Alpha-actinin
	Beta-actinin

* Pollard, 1976.

cally in the following section the evidence that these proteins are responsible for biological movements. The accumulated evidence strongly supports this view, but it cannot be regarded as conclusive.

Evidence that Actomyosin Powers Cell Movements

Four general types of results argue that cellular motile force is generated by cytoplasmic actin, myosin, and associated proteins: (1) comparison of the properties of muscle and cytoplasmic contractile proteins; (2) hybridization of muscle and cytoplasmic contractile proteins; (3) test tube contractile models of purified proteins and crude cytoplasmic extracts; (4) localization of the proteins at sites of motile force generation. While none of these approaches can prove rigorously that cytoplasmic actin and myosin cause cell movements, collectively they are quite persuasive.

COMPARISONS

Because cytoplasmic actin and myosin are similar to their muscle counterparts which generate the force for muscle contraction, it can be argued that the cytoplasmic actin and myosin must also be capable of generating motile force. The data supporting this case is very strong.

Taking actin first, we find that all actins are remarkably similar in their primary structure. For example, Lu and Elzinga (1976) have completed about half of the *Acanthamoeba* actin amino acid sequence, and find that 94% of the residues are identical to those in rabbit-muscle actin. All of the differences are conservative and isopolar. Only one of these differences requires more than a single base change in the gene. Clearly, very strong evolutionary constraints have limited the changes in both of these actin genes since the time that amebas and rabbits had a common ancestor. Less extensive sequence data show that vertebrates have three closely related actin genes. Isoelectric focusing experiments from several laboratories suggest that individual species may have two or more types of actin (Whalen et al., 1976).

Given these similarities in primary structure, it is not surprising that all actins form identical 6 nm-wide filaments consisting of a double helical array of globular actin molecules (Fig. 1). All actin filaments bind myosin in the same way, forming the familiar arrowhead-shaped structure (Fig. 1).

There is widespread suspicion that cytoplasmic actins do not polymerize well, but in the case of the actins from *Dictyostelium* (Spudich and Cooke, 1975) and *Acanthamoeba* (Gordon et al., 1976), the polymerization is quite similar to that of muscle actin. For example, *Dictyostelium* and rabbit-muscle actin polymerize to the same extent at several concentrations, both having the same critical concentration below which no polymerization occurs. The critical concentration is a measure of the equilibrium constant for the elongation reaction, which, therefore, is the same in the two cases.

Compared with actin, there is much diversity among myosins. Even among the muscles of a single vertebrate species, up to four myosin isozymes can be distinguished by sequencing and serological analysis (see Mannherz and Goody, 1976, for more details). In addition, nonmuscle cells contain different myosin isozymes. For example, Burridge and Bray (1975) showed by peptide

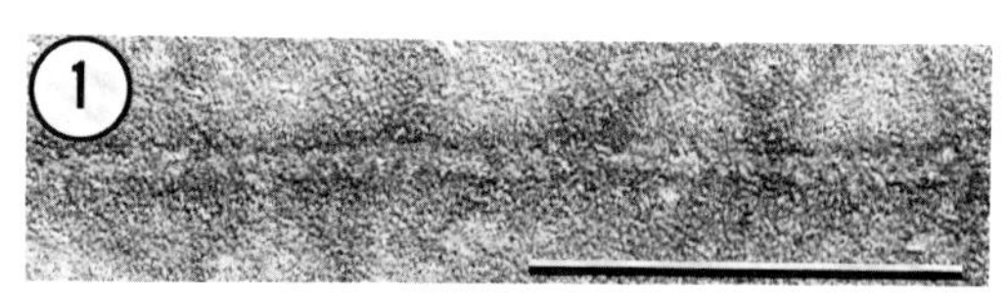

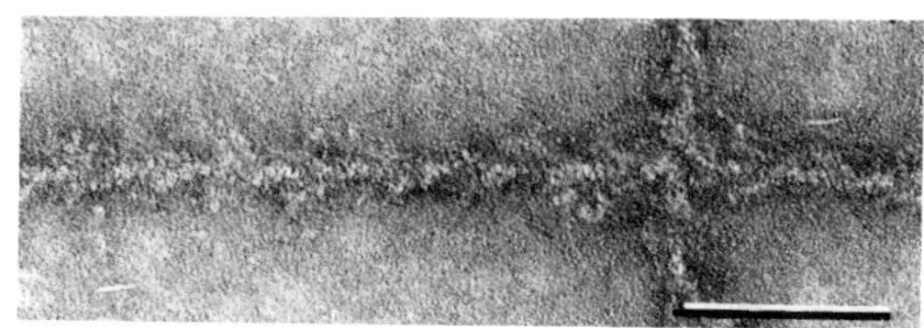

FIGURE 1 Electron micrographs of negatively stained ameba actin filaments, before (above) and after (below) reaction with muscle-heavy meromyosin. Bars are 1 μm.

mapping that chicken tissues contain two different cytoplasmic myosins, both of which are different from the myosin isozymes found in chicken muscles.

In surveying the cytoplasmic myosins found in different species, we find three general classes (Table II). In metazoan species, the cytoplasmic myosin isozymes are conventional. They have a native molecular weight of about 470,000 daltons and are composed of two heavy chains with molecular weights of 200,000 daltons. In slime molds and giant amebas, the myosins have larger heavy chains of about 230,000 daltons. The myosin in *Acanthamoeba* is apparently smaller, consisting of a single 160,000-dalton heavy chain. All of the myosins have low molecular weight light chains.

In spite of major differences in their physical properties, all myosins have two common features which are vital for force generation in muscle and which presumably are involved with force production in nonmuscle cells, as well. All myosins bind reversibly to actin filaments. All myosins catalyze the hydrolysis of ATP in a reaction which requires the presence of actin filaments. Circumstantial evidence suggests that this is the force-generating reaction (see Mannherz and Goody, 1976).

Except for the globular *Acanthamoeba* myosin, all of the myosins have two globular heads and a long tail (Fig. 2). By isolating the separate parts of platelet myosin after proteolytic cleavage, it was shown that the heads have the actin-binding sites and the ATPase sites, as in muscle myosin (Adelstein et al., 1971). The tail of the myosin molecule is necessary for filament formation. Consequently, *Acanthamoeba* myosin does not form filaments.

Under conditions similar to those which probably exist inside cells, platelet and other vertebrate cytoplasmic myosins aggregate to form short bipolar filaments that are composed of about 30 myosin molecules (Fig. 2; Niederman and Pollard, 1975). These filaments are considerably shorter and thinner than the myosin-thick filaments found in striated muscle. The bipolar geometry of these filaments is thought to be important for their role in motility. Because the myosin molecules face in opposite directions at the two ends, the projecting heads can cross-link oppositely polarized actin filaments (Fig. 3). Although different in gross appearance, this platelet actin-myosin complex is geometrically identical to a muscle sarcomere, so that the structure is compatible with a sliding filament contractile mechanism.

The drawing at the bottom of Fig. 3 shows the amount of actin filament for each myosin filament in muscle and platelets, as if both were arranged in a sarcomere and all of the actin and myosin were assembled into filaments. Of course, such a precise arrangement has not been found in platelets. Nevertheless, the drawing makes two points: the muscle sarcomere should develop more force and the contractile unit in platelets should be capable of more extensive shortening. This is true because the size of the myosin filament determines the maximum force per contractile unit, and the length of the actin filaments limits the extent of shortening. The power output per gram is also higher in muscle because the myosin concentration is about 50 times higher than in nonmuscle cells.

To summarize the first argument, the structure of cytoplasmic actin and myosin molecules and the filaments that they form, together with evidence for actomyosin enzyme activity, all argue, by analogy with muscle, that these proteins should be able to generate cellular contractions, probably by a sliding filament mechanism.

TABLE II

Cytoplasmic Myosins

Class	Actin binding	Mg-ATPase (−)Actin(+)		Composition	Mol wt
Orthodox myosin	+	−	+	2 × 200,000	460,000
Vertebrate cells				2 × 19,000	
Invertebrate cells				2 × 16,000	
Macromyosin	+	−	+	2 × 230,000	460,000
Myxomycetes				? × light chains	
Carnivorous amebas					
Minimyosin	+	−	+	1 × 140,000	180,000
Acanthamoeba				1 × 16,000	
				1 × 14,000	

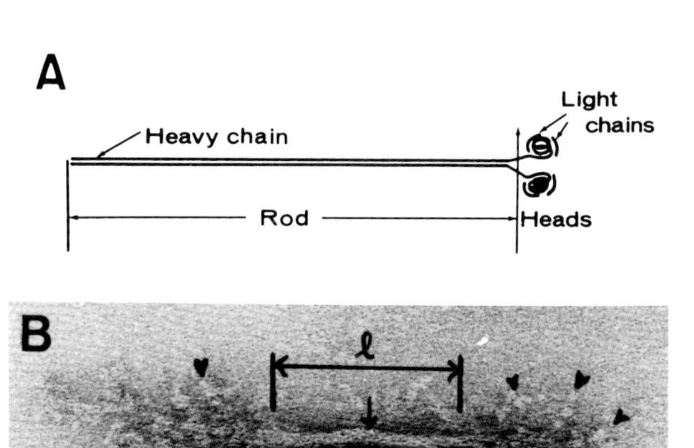

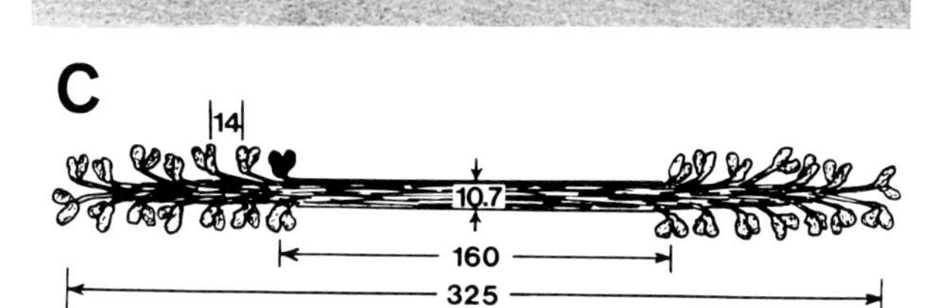

FIGURE 2 (A) Model of platelet myosin. The vertical arrow separating the rod and head regions indicates the place where the molecule can be split during storage of the platelets in the blood bank. (B) An electron micrograph of a filament formed from human platelet myosin. The length of the bare zone (l) is 0.16 μm. The width (D) is 10.5 nm. (C) A model for the platelet myosin filament (From Niederman and Pollard, 1975).

HYBRIDIZATION

These experiments show that cytoplasmic contractile proteins will form active hybrids when mixed with muscle contractile proteins, so one can argue that the cytoplasmic contractile proteins must be able to interact with each other to power cellular movements.

There is evidence that actin in both slime mold and ameba can copolymerize with muscle actin and that platelet myosin can copolymerize with muscle myosin. More importantly, it is clear that muscle myosin binds to cytoplasmic actin filaments and that cytoplasmic myosins bind to muscle actin filaments. Thus, the actin-actin, myosin-myosin, and actin-myosin binding sites are all sufficiently similar in the various proteins for hybridization. The fidelity of these hybridizations is less well studied, but one example shows that the fit may be imperfect. Gordon et al. (1976) compared the ability of muscle actin and *Acanthamoeba* actin to activate muscle myosin ATPase. Both gave the same V_{max} at infinite actin, but attaining half the maximum rate required three times as much ameba actin. This difference in K_a shows that the efficiency of interaction in the hybrid is lower and that the fit at the active sites must be imperfect.

MOTILE MODELS

Mixtures of isolated contractile proteins with ATP can produce several types of gross streaming and contraction, so it is reasoned that they must be capable of doing something similar inside living cells. Related experiments with cell extracts containing crude mixtures of the contractile proteins make the same point (see Taylor, this volume, for review).

An early example of this approach with isolated contractile proteins comes from the pioneering work of Bettex-Galland and Luscher (1961). They formed threads of platelet actomyosin and observed their contraction in the presence of ATP.

Recently, Stossel and Hartwig (1976) have done similar, but more sophisticated, experiments with purified macrophage contractile proteins. The mixture of actin, actin-binding protein, myosin, and cofactor with ATP forms a gel which subsequently contracts. The actin and actin-binding protein are required for gelation, the myosin is

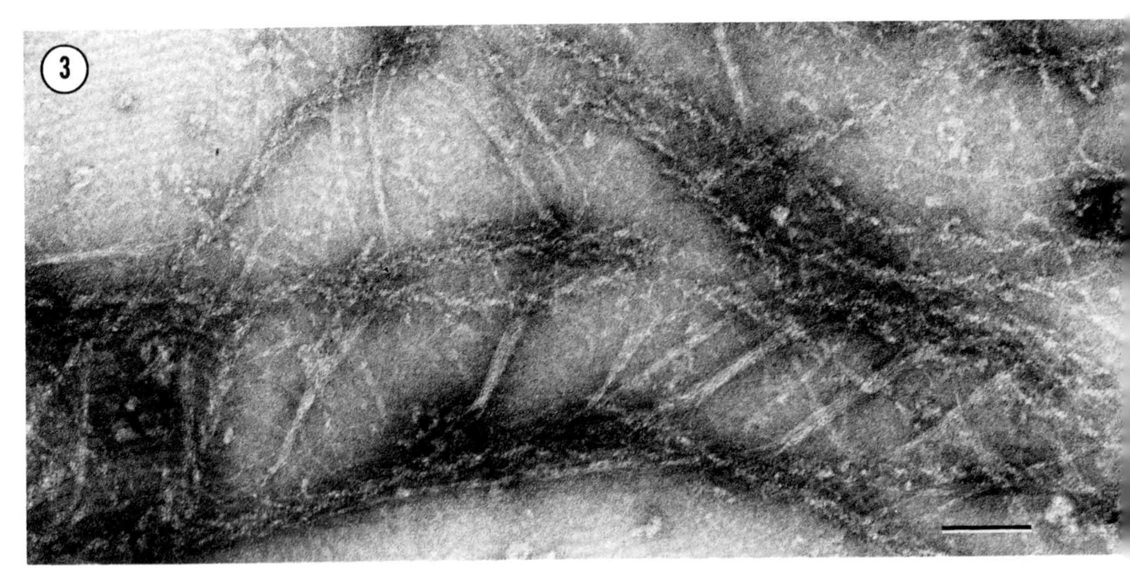

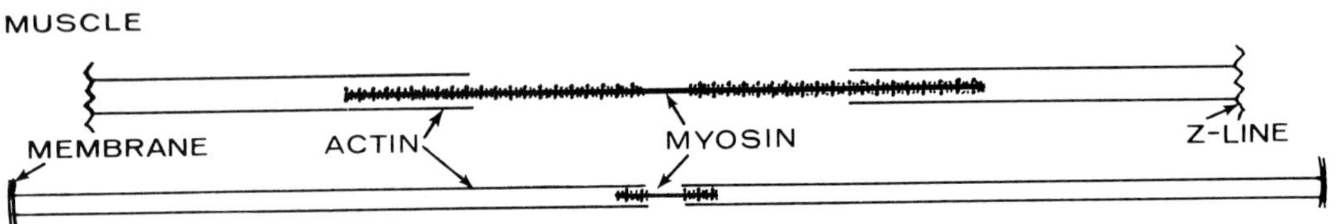

FIGURE 3 An electron micrograph of negatively stained platelet actin and myosin filaments showing how the bipolar myosin filaments can cross-link actin filaments. Bar is 1 μm. Below are drawings comparing the actin:myosin filament ratio in skeletal muscle and in platelets. It is assumed that all of the actin and myosin molecules are polymerized into filaments.

required for contraction, and the cofactor accelerates the rate of contraction. Although the relation of gross contraction in these in vitro systems to the refined movements of living cells is not established, it has been important to show that these proteins can interact in the test tube to produce some sort of movement.

LOCALIZATION WORK

A fourth approach to show that contractile proteins cause a certain movement is to demonstrate that they are present at the site of force development. This is, of course, a necessary, but insufficient, condition to prove that the contractile proteins produce the force for the movement.

At the present time, there is only one good example of the coexistence of actin and myosin in a cellular contractile structure, the cytokinetic contractile ring. This ring forms around the equator of dividing cells in late anaphase and pinches the cell in two by contracting like a purse string. Schroeder showed that the contractile ring is composed of 6-nm wide filaments, later identified as actin by Perry et al. (1971) and Schroeder (1973) by their ability to bind muscle-heavy meromyosin. Myosin has not been seen in the contractile ring by electron microscopy, although dividing Hela cells stained with rhodamine-antimyosin antibody have fluorescence concentrated in the cleavage furrow, in the region of the contractile ring (Fig. 4 and Fujiwara and Pollard, 1976). Presumably the myosin is concentrated in the contractile ring itself, along with the actin filaments.

In addition to this example, a large body of electron microscope data shows that similar bundles of actin are found at other sites of cell contraction, such as the apex or base of cells in folding embryonic epithelia. Myosin localization work must still be done in these cases.

The distribution of the contractile proteins is dynamic, changing with the activity of the cells. For example, in Hela cells with a motile morphology, staining with fluorescent antimyosin is found throughout the cytoplasm (Fig. 5). The same is true of other highly motile cells, such as leukocytes. At other times, Hela cells spread out and develop stress fibers. Then most of the antimyosin staining is found in the stress fibers (Fig. 5). Lazarides (1975) has made similar observations regarding actin and tropomyosin through the use of fluorescent antibodies. In spread cells, the actin and tropomyosin antibody staining is concentrated in stress fibers, whereas in a cell with a more active morphology, the antiactin staining, at least, is found throughout the cytoplasm. This suggests, that during locomotion, all regions of the cytoplasm are potentially contractile and may cooperate in powering the movement.

Although much more detailed correlations of contractile protein localization with movements are needed, the observations on the contractile ring are an encouraging beginning. The actin and myosin are concentrated together in the right place at just the right time to cause this vital cell movement.

Unsolved Problems

In spite of this spectacular progress in defining the molecular basis of cell movements, major problems remain unsolved. In this section I will

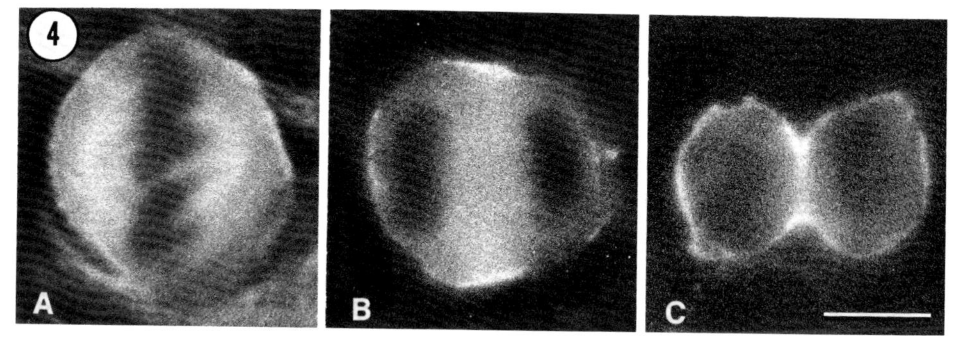

FIGURE 4 Fluorescence micrographs of dividing Hela cells stained with rhodamine-antiplatelet myosin. (A) Metaphase showing concentration of staining in the mitotic spindle between the kinetochores and the poles. (B and C) Telophase showing the concentration of staining in the cleavage furrow. Bar is 10 μm (Data of Keigi Fujiwara).

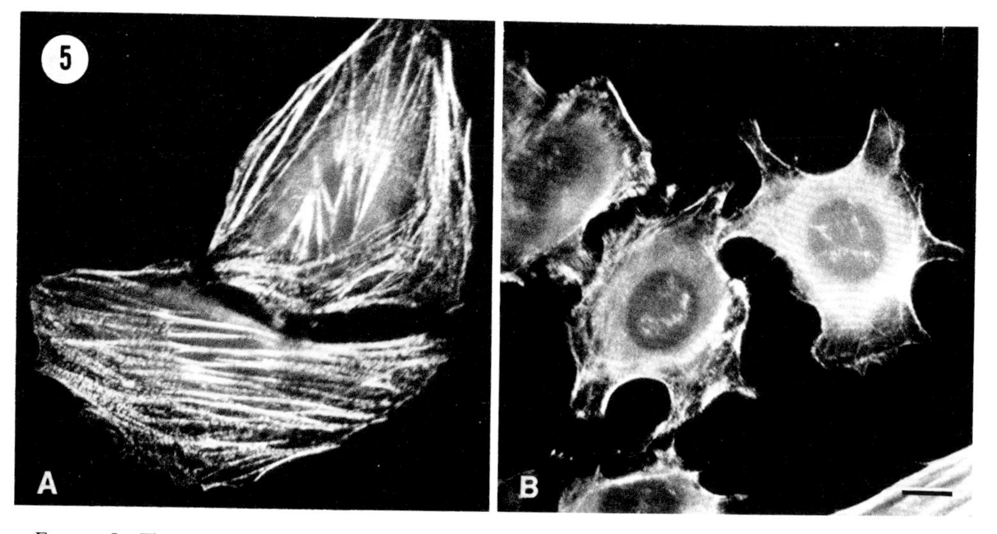

FIGURE 5 Fluorescence micrographs of interphase Hela cells stained with rhodamine-antiplatelet myosin. (A) Cells with stress fibers and a sessile morphology. (B) Cells with more generalized cytoplasmic staining and a motile morphology. Bar is 10 μm (Data of Keigi Fujiwara).

consider three. Others are covered in the accompanying papers from this symposium.

Regulatory Mechanisms

The mechanisms controlling the activity and assembly of the contractile proteins have been difficult to decipher. As described by Taylor (this volume), Ca^{++} is involved but neither the mechanism of its action on the contractile proteins, nor its mechanism of entry into the cytoplasm is known. There is, however, good evidence for avid Ca^{++}-sequestering vesicles inside cells, comparable to the muscle sarcoplasmic reticulum (Robblee et al., 1973). These vesicles probably keep the cytoplasmic Ca^{++} concentration very low.

Both negative and positive regulators of actomyosin are known in cells (Table III). As far as this catalogue has gone, the two negative regulators are Ca^{++} modulated, whereas the two positive regulators are Ca^{++} independent.

TROPOMYOSIN-TROPONIN: Beginning with the pioneering work of Cohen and Cohen (1972), there is good evidence for tropomyosin in vertebrate nonmuscle cells. Tropomyosin has not been purified from protozoa or plants. Although one-seventh shorter than muscle tropomyosin, the cytoplasmic tropomyosins have properties similar to their muscle counterparts. They have a very high alpha-helix content and can bind muscle troponin to form an active hybrid capable of regulating the muscle actin-myosin ATPase.

Recently troponin-C has been isolated from brain (Fine et al., 1975) and adrenal gland (Kuo and Coffee, 1976). Like muscle troponin-C, it binds Ca^{++} and interacts with other troponin components. Quite surprisingly, cytoplasmic troponin-C is the Ca^{++}-modulated regulator of cyclic-AMP phosphodiesterase (Stevens et al., 1976; Watterson et al., 1976). Its primary structure closely resembles muscle troponin-C.

In addition to this work with purified components, Puszkin and Kochwa (1975) obtained an unfractionated mixture of actin, tropomyosin, and troponin components from brain. This mixed fraction behaved like muscle tropomyosin-troponin in regulating actin-myosin interaction.

In spite of these similarities to muscle tropomyosin-troponin, there is some question about whether cytoplasmic tropomyosin-troponin constitutes, by itself, an adequate regulatory system. In muscle, tropomyosin-troponin controls contraction by blocking actin-myosin interaction when the Ca^{++} concentration is low. To be effective there must be enough tropomyosin-troponin to tie up all of the actin. Any actin without it is unregulated. This is the problem in nonmuscle cells. It is not yet clear that they have enough tropomyosin or troponin to regulate their vast supply of actin. This suspicion seems to be confirmed by two types of localization studies reported at the International Congress on Cell Biology. At least some parts of vertebrate cells have actin without tropomyosin (Sanger and Sanger, 1976; Lazarides, 1976).

SLIME MOLD CONTROL PROTEINS: Different molecules may control contraction in the slime molds. Nachmias (1975) and Kato and Tonomura (1975) isolated regulatory protein fractions from the acellular slime mold, *Physarum,* and Mockrin and Spudich (1976) reported related work with the cellular slime mold *Dictyostelium.* All of these preparations are heterogeneous, but all are capable of inhibiting actin-myosin ATPase when the Ca^{++} concentration is low, and allowing full activity when the Ca^{++} concentration is high. Inasmuch as the active molecules have not been purified, it is not known whether these regulatory proteins are similar to each other or related to muscle tropomyosin-troponin. The *Physarum* proteins, but not the *Dictyostelium* proteins, can regulate muscle myosin.

COFACTOR: We discovered cofactor in *Acanthamoeba* (Pollard and Korn, 1973), and it has also been found in rabbit macrophages by Stossel and Hartwig (1975). In both cases, the myosin ATPase is activated little by actin, unless the cofactor is present. The mechanism of this stimulation is unknown, largely because little of this cofactor protein has ever been purified. Because the availability of the cofactor can turn the actomyosin on or off, it seems likely to be a component of a regulatory system. By itself, cofactor is not modulated by calcium.

LIGHT CHAIN KINASE: A second way to activate these sluggish cytoplasmic actomyosin ATPases is to phosphorylate one of the myosin light chains.

Table III
Control Proteins

	Distribution
Inhibitory, calcium modulated	
Tropomyosin-troponin	Vertebrates
Slime mold factor	Slime molds
Activating, calcium independent	
Cofactor	Ameba, vertebrate
Light chain kinase	Vertebrates

Adelstein and his colleagues (Adelstein and Conti, 1975; Daniel and Adelstein, 1976) isolated a specific light chain kinase from platelets and showed that the phosphorylated platelet myosin has a higher actin-activated ATPase activity than the dephosphorylated form. In contrast to some other kinases, the light chain kinase is not modulated by either Ca^{++} or cyclic AMP.

SUMMARY: None of the available biochemical experiments on these four types of regulatory mechanism can begin to explain how cells control the activity of their contractile proteins, much less how they control their movements. Extensive additional work is needed in this area. The diversity already found in these regulatory systems shows that the control of movement in nonmuscle cells is more complex than the simple on-off switch found in muscle. This is, of course, appropriate because of the simplicity of the one-dimensional shortening motion of muscle and the complexity of motion in nonmuscle cells, where spatial, as well as temporal, control must be exerted. In addition, there must be mechanisms controlling the assembly and disassembly of the contractile filaments, which are still incompletely understood (see both Taylor and Tilney, this volume).

Form of Cytoplasmic Actin and Myosin

There is biochemical evidence that part of the actin is depolymerized inside cells, but some is present as straight actin filaments which are seen by electron microscopy. In addition, some of the actin may be present in a particulate storage form found first by Tilney (this volume) in sperm. Finally, there are microfilament networks composed of actin, which look little like native actin filaments. These networks could be composed of actin filaments fragmented as a result of their association with the protein beta-actinin or some other filament destabilizing factor. On the other hand, they could be a fixation artifact, as Maupin-Szamier and I have shown in model studies with pure actin. The viscosity of solutions of both native and glutaraldehyde cross-linked actin filaments is rapidly lost when exposed to osmium tetroxide under the conditions routinely used for electron microscopy. The filaments are broken into progressively smaller fragments over the period of about 30 min. At the same time, osmium cleaves the actin molecule into peptides by an incompletely understood oxidative mechanism. Actin-tropomyosin filaments are more resistant to fragmentation, accounting for the presence of straight thin filaments in muscle and in the stress fibers of cultured cells. In other cases where actin-containing filaments are well preserved after fixation, they may be stabilized by myosin or actin-binding proteins (see Tilney, this volume). Until this fixation problem is solved, electron micrographs of actin filaments will need to be interpreted cautiously. Moreover, new experiments are needed to determine how much of a cell's actin is polymerized during various activities.

Myosin forms filaments under the conditions one expects to find inside cells, and yet no myosin filaments are seen in electron micrographs of most vertebrate cells. New work is necessary here, as well, to determine whether this is also a result of a fixation artifact, or whether there are cellular factors blocking myosin filament assembly, or whether this is simply a statistical problem arising because there are only a few myosin filaments and it is difficult to distinguish them from the more numerous actin filaments (Niederman and Pollard, 1975).

Participation of Contractile Proteins in Microtubule-Dependent Movements

A new area of active investigation is the possible relationship of actomyosin to such microtubule-dependent movements as mitosis, axoplasmic flow, and saltatory movements. Because there is little strong evidence for an energy-transducing enzyme, like dynein, associated directly with cytoplasmic microtubules, it is natural to consider the possibility that these microtubule-dependent movements are powered by actomyosin muscles associated with the microtubule cytoskeleton. There is no biochemical evidence to support this idea, but the discovery of actin and myosin in the mitotic spindle suggests that the possibility is worth taking seriously.

Mitotic spindle actin was first identified by electron microscope observations of heavy meromyosin-binding filaments made by Forer and Behnke (1972), Gawadi (1974), and Hinkley and Telser (1974). Subsequently, Sanger (1975) stained spindles in glycerinated cells with fluorescent heavy meromyosin, and recently Cande et al. (1975) showed that actin antibodies stain spindles. The actin seems to be concentrated in the region between the kinetochores and the poles. There is also a single report (Fig. 4 and Fujiwara and Pollard, 1976) that antibodies to myosin also stain the spindle of Hela cells. Like the actin, the antimyosin staining is concentrated between the kinetochores and the poles.

The perpetrators of these tantalizing observations are aware that the experiments are plagued by the possibility of artifact and open to a number of interpretations. At the very least, this work has provided the impetus for a careful search for actomyosin-microtubule cooperation. If found, this could provide a simple explanation for the variety of cellular movements that require intact microtubules.

Concluding Remarks

The preceding section deals with only three of the frontiers in this burgeoning field. Others include the participation of the contractile proteins in cell structure, in the modulation of membrane activities, and in other nonmotility functions. The involvement of troponin-C in cyclic-AMP metabolism is but one recent example. It would appear that the field, rather than narrowing down as it approaches the solution of a specific problem (like the focused work leading to the discovery of the genetic code), is expanding into many areas of cellular function. If the presence of the contractile proteins in the nucleus can be confirmed to be authentic, these proteins could well be involved with all aspects of cellular metabolism.

For those of us interested in cell motility, the major challenge remains the elucidation of the mechanism of cell motion. I feel that the submolecular details of actin-myosin interaction should be left to those working with striated muscle. It is up to us cell biologists to collect the additional biochemical, morphological, and physiological evidence that will show exactly how the activity of the contractile proteins is expressed as cell movement.

ACKNOWLEDGMENTS

I am grateful for the helpful comments of Keigi Fujiwara and Mark Mooseker.

The preparation of this review was supported by Research Career Development Award GM-70755 from the National Institutes of Health.

REFERENCES

ADELSTEIN, R. S., and M. A. CONTI. 1975. Phosphorylation of platelet myosin increases actin-activated myosin ATPase activity. *Nature* (*Lond.*). **256:**597–598.

ADELSTEIN, R. S., T. D. POLLARD, and W. M. KUEHL. 1971. Isolation and characterization of myosin and two myosin fragments from human blood platelets. *Proc. Natl. Acad. Sci. U. S. A.* **68:**2703–2707.

BETTEX-GALLAND, M., and E. F. LUSCHER. 1961. Thrombosthenin – a contractile protein from thrombocytes: its extraction from human blood platelets and some of its physical properties. *Biochim. Biophys. Acta.* **49:**536–547.

BURRIDGE, K., and D. BRAY. 1975. Purification and structural analysis of myosins from brain and other non-muscle tissues. *J. Mol. Biol.* **99:**1–14.

CANDE, W. Z., E. LAZARIDES, and J. R. MCINTOSH. 1975. Visualization of actin in functional lysed mitotic cell preparations. *J. Cell Biol.* **67:**54*a* (abstr.).

COHEN, I., and C. COHEN. 1972. A tropomyosin-like protein from platelets. *J. Mol. Biol.* **68:**383–387.

DANIEL, J. L., and R. S. ADELSTEIN. 1976. Isolation and properties of platelet myosin light chain kinase. *Biochemistry.* **15:**2370–2377.

FINE, R., W. LEHMAN, J. HEAD, and A. BLITZ. 1975. Troponin-C in brain. *Nature* (*Lond.*). **258:**260–262.

FORER, A., and O. BEHNKE. 1972. An actin-like component in spermatocytes of a crane fly (*Nephrotoma suturalis Loew*). I. The spindle. *Chromosoma* (*Berl.*). **39:**145–173.

FUJIWARA, K., and T. D. POLLARD. 1976. Fluorescent antibody localization of myosin in the cytoplasm, cleavage furrow and mitotic spindle of human cells. *J. Cell Biol.* **71:**848–875.

GAWADI, N. 1974. Characterization and distribution of microfilaments in dividing locust testis cells. *Cytobios.* **10:**17–35.

GOLDMAN R., T. D. POLLARD, and J. ROSENBAUM, editors. 1976. Cell Motility. Cold Spring Harbor Laboratory, Cold Spring Harbor, N. Y. 1500.

GORDON, D. J., E. EISENBERG, and E. D. KORN. 1976. Characterization of cytoplasmic actin isolated from *Acanthamoeba castellanii* by a new method. *J. Biol. Chem.* **251:**4778–4786.

HATANO, S., and F. OOSAWA. 1966. Isolation and characterization of plasmodium actin. *Biochim. Biophys. Acta.* **127:**488–498.

HATANO, S., and K. OWARIBE. 1976. Actin and actinin from myxomycete plasmodia. *In* Cell Motility, R. D. Goldman, T. D. Pollard, and J. Rosenbaum, editors. Cold Spring Harbor Laboratory, Cold Spring Harbor, N. Y. 499–512.

HINKLEY, R., and A. TELSER. 1974. Heavy meromyosin binding filaments in the mitotic apparatus of mammalian cells. *Exp. Cell Res.* **86:**161–165.

KATO, T., and Y. TONOMURA. 1975. Ca^{++} sensitivity of actomyosin ATPase purified from *Physarum polycephalum*. *J. Biochem.* **77:**1127–1134.

KUO, I. C. Y., and C. J. COFFEE. 1976. Purification and characterization of a troponin-C like protein from bovine adrenal medulla. *J. Biol. Chem.* **251:**1603–1609.

LAZARIDES, E. 1975. Tropomyosin antibody: the specific localization of tropomyosin in nonmuscle cells. *J. Cell Biol.* **65:**549–561.

LAZARIDES, E. 1976. Two general classes of cytoplasmic

actin filaments in tissue culture cells: the role of tropomyosin. *J. Cell Biol.* **70:**359*a* (abstr.).

LU, R., and M. ELZINGA. Comparison of amino acid sequences of actins from bovine brain and muscles. *In* Cell Motility. R. Goldman, T. Pollard, and J. Rosenbaum, editors. Cold Spring Harbor Laboratory, Cold Spring Harbor, N. Y. 487–492.

MANNHERZ, H. G., and R. S. GOODY. 1976. Proteins of contractile systems. *Annu. Rev. Biochem.* **45:**427–466.

MARCHESI, V. T., H. FURTHMAYR, and M. TOMITA. 1976. The red cell membrane. *Annu. Rev. Biochem.* **45:**667–698.

MOCKRIN, S. C., and J. A. SPUDICH. 1976. Calcium control of actin-activated myosin ATPase from *Dictyostelium discoideum. Proc. Natl. Acad. Sci. U. S. A.* **73:**2321–2325.

NACHMIAS, V. T. 1975. Calcium sensitivity of hybrid complexes of muscle myosin and *Physarum* proteins. *Biochim. Biophys. Acta.* **400:**208–221.

NIEDERMAN, R., and T. D. POLLARD. 1975. Human platelet myosin. II. In vitro assembly and structure of myosin filaments. *J. Cell Biol.* **67:**72–92.

PERRY, M. M., H. A. JOHN, and N. S. T. THOMAS. 1971. Actin-like filaments in the clevage furrow of newt eggs. *Exp. Cell Res.* **65:**249–252.

POLLARD, T. D. 1976. Contractile proteins in non-muscle cells. Handbook of Biochemistry and Molecular Biology – Proteins. Vol. II. G. D. Fasman, editor. CRC Press, Cleveland, Ohio. 307–324.

POLLARD, T. D. 1977. Cytoskeletal functions of cytoplasmic contractile proteins. *J. Supramol. Struct.* In press.

POLLARD, T. D., and E. D. KORN. 1973. *Acanthamoeba* myosin. II. Interaction with actin and with a new cofactor protein required for actin activation of Mg^{++} ATPase activity. *J. Biol. Chem.* **248:**4691–4697.

POLLARD, T. D., and R. R. WEIHING. 1974. Actin and myosin and cell motility. *CRC Crit. Rev. Biochem.* **2:**1–64.

PUSZKIN, S., and S. KOCHWA. 1975. Regulation of neurotransmitter release by a complex of actin with relaxing protein isolated from rat brain synaptosomes. *J. Biol. Chem.* **249:**7711–7714.

ROBBLEE, L. S., D. SHEPRO, and F. A. BELAMARICH. 1973. Calcium uptake and associated ATPase activity of isolated platelet membranes. *J. Gen. Physiol.* **61:**462–481.

SANGER, J. W. 1975. Presence of actin during chromosomal movement. *Proc. Natl. Acad. Sci. U. S. A.* **72:**2451–2455.

SANGER, J. W., and J. M. SANGER. 1976. Two microfilament systems in rat kangaroo cells. *J. Cell Biol.* **70:**277*a* (abstr.).

SCHROEDER, T. E. 1973. Actin in dividing cells: contractile ring filaments bind heavy meromyosin. *Proc. Natl. Acad. Sci. U. S. A.* **70:**1688.

SPUDICH, J. A., and R. COOKE. 1975. Supramolecular forms of actin from amoebae of *Dictyostelium discoideum. J. Biol. Chem.* **250:**7485–7491.

STEVENS, F. C., M. WALSH, H. C. HO, T. S. TEO, and J. H. WANG. 1976. Comparison of calcium binding proteins: bovine heart and brain protein activators of cyclic nucleotide phosphodiesterase and rabbit skeletal muscle troponin-C. *J. Biol. Chem.* **251:**4495–4500.

STOSSEL, T. P., and J. HARTWIG. 1975. Interactions between actin, myosin and an actin-binding protein from rabbit alveolar macrophages: alveolar macrophage myosin Mg^{++} ATPase requires a cofactor for activation by actin. *J. Biol. Chem.* **250:**5706–5712.

STOSSEL, T. P., and J. HARTWIG. 1976. Interactions of actin, myosin and a new actin-binding protein of rabbit pulmonary macrophages. II. Role in cytoplasmic movement and phagocytosis. *J. Cell Biol.* **68:**602–619.

WATTERSON, D. M., W. G. HARRELSON, P. M. KELLER, F. SHARIEF, and T. C. VANAMAN. 1976. Structural similarities between Ca^{++} dependent regulatory proteins of 3′-5′-cyclic nucleotide phosphodiesterase and actomyosin ATPase. *J. Biol. Chem.* **251:**4501–4513.

WHALEN, R. G., G. S. BUTLER-BROWNE, and F. GROS. 1976. Protein synthesis and actin heterogeneity in calf muscle cells in culture. *Proc. Natl. Acad. Sci. U. S. A.* **73:**2018–2122.

ACTIN: ITS ASSOCIATION WITH MEMBRANES AND THE REGULATION OF ITS POLYMERIZATION

LEWIS G. TILNEY

There are two features of actin-associated motility in nonmuscle cells with which I would like to concern myself in this brief review. One is how actin (both polymeric and monomeric) may be associated with membranes and/or membrane precursors. The other is how actin may be stored in cells in a nonfilamentous form and what the necessary events might be in vivo for the organization of an actin backbone on which or by which force may be produced. The other members of the symposium will concern themselves with different features of nonmuscle motility, such as the interaction of actin with myosin, the regulation of motility, the possible role of actin and myosin in the movement of chromosomes, movement of amebas, etc. I have decided for three reasons to concentrate on these two topics rather than to write a general review. The first two involve limitations as to the length of this review and the number of references allowed. The third reason is the fact that at the present moment these two features of actin are, to me, the most exciting areas. I will draw from my own work only because I know it best. I am hoping that it will provide *an* example, not necessarily *the* example, to illustrate a particular phenomenon.

In nonmuscle cells the motile event, whether it is cytokinesis, phagocytosis, ruffling of the cell surface, pseudopod or filopod formation, or cell elongation as during the formation of the acrosomal process, clearly requires a tight coupling of the contractile apparatus with the cell surface. What one needs to know, then, is how this coupling is brought about and what controls the appropriate assembly of actin filaments relative to the cell surface, filament polarity, distribution of filaments, etc. I will begin with a description of how actin filaments may be connected to the plasma membrane and later ask questions about how other states of actin are attached to membranes. The experimental system with which I will illustrate these connections is the long fingerlike projections or microvilli which extend from intestinal epithelial cells and, in fact, from many, if not most, animal cells.

Within each microvillus is a bundle of actin filaments (Ishikawa et al., 1969; Tilney and Mooseker, 1971). One end of each filament is attached to the limiting membrane at the tip of the microvillus where the filaments are embedded in an electron-dense material (Fig. 1). This material stains with an antibody prepared against the main component of the "Z" line of skeletal muscle, α actinin (Schollmeyer et al., 1974). The actin filaments are also attached to the membrane along their lengths by threadlike connectors (Mukherjee and Staehelin, 1971; Mooseker and Tilney, 1976; Fig. 1). The chemical nature of these connectors remains unestablished, but it may also be in part α actinin.

The next questions to which we should address ourselves are: exactly how are the actin filaments or substances attached to the actin filaments, such as α actinin or the hairlike connectors, attached to the plasma membrane? What other substances are involved and do these substances connect to the lipid bilayer by being attached to peripheral membrane proteins or to integral membrane proteins, or to both? Does the end of the actin filament which attaches to the membrane always have the same polarity? So far there is no information available with which to answer the first question. In regard to the second, we (Tilney and Mooseker, 1976) have recently examined freeze-fractures of regions where actin filaments are attached to membranes in an attempt to see if membrane particles, which indicate the presence of integral

LEWIS G. TILNEY Department of Biology, University of Pennsylvania, Philadelphia, Pennsylvania

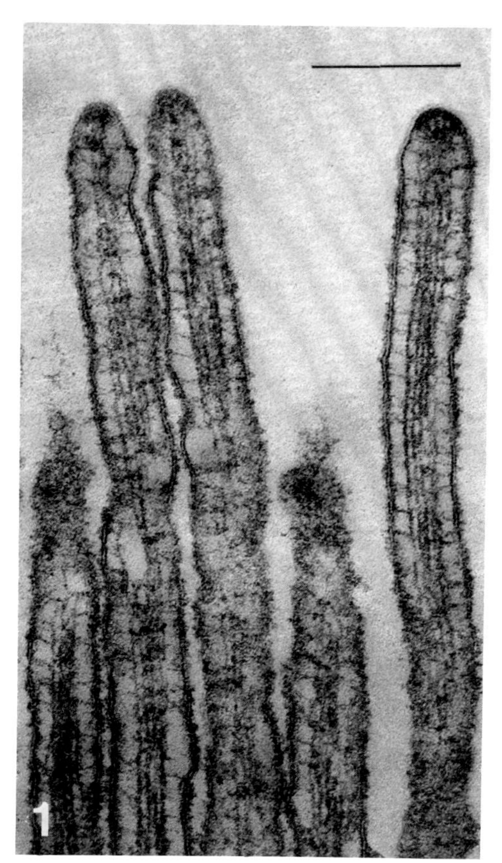

FIGURE 1 Thin section through a microvillus from an isolated brush border of a chicken intestine. The brush border was incubated in 15 mM Mg^{++}. Note that the filaments not only are connected to the membrane at the microvillus tip but are also attached to the membrane along their lengths by cross bridges. Where they are attached to the membrane at the tips, they are embedded in a dense material that stains with an antibody prepared against α actinin. (From Mooseker and Tilney, 1976.) Bar, 0.2 μm. $\times$ 100,000.

membrane proteins, are attached to actin filaments. A careful choice of the biological system is essential for these studies to enable one to identify the point at which actin filaments attach to the membrane in a replica. We can determine by the membrane topography the point of insertion of the actin filaments from replicas of microvilli and of two species of sperm. In the three systems studied there were either no particles at the point of attachment of the actin filaments to the membrane or fewer particles in this region than in nearby regions. Furthermore, we know by the periodicity of the connectors to the membrane in the microvilli (Fig. 1) that these connectors cannot all be attached to membrane particles, and, in fact, most would not be connected. However, these observations really tell us only that the actin filaments and/or their associated proteins are either connected to peripheral proteins that do not span the lipid bilayer or are connected either directly or indirectly to integral membrane proteins that are too small to be resolved by the technique of freeze-fracturing. Clearly, peripheral proteins are involved, as for example the α actinin at the tips of microvilli, but exactly how these molecules function is not obvious. We do know, however, that they must bind to the lipid bilayer because the membrane at the tip resists detergent solubilization, a characteristic of protein-lipid interactions (see Helenius and Simon, 1975).

Actin filaments also must be attached to membranes with the requisite polarity to perform useful work. Huxley (1963) was the first to demonstrate that actin filaments are polarized. He examined the polarity of the actin filaments in skeletal muscle and was able to show, with heavy meromyosin or subfragment 1 of myosin decoration, that all these filaments have the same polarity with respect to the Z line. Two nonmuscle systems have been studied in which the polarity of the actin filaments relative to a membrane can be defined: microvilli (Mooseker and Tilney, 1976), and *Mytilus* sperm (Tilney, 1975). (In some cases, such as the polarity of actin filaments relative to isolated secretory granules, unidirectional polarity does not seem to be present [Burridge, 1976], but if the temperature is controlled, indeed unidirectional polarity has been observed [personal communication]. It is not clear at this time if all the filaments are really bound to the granules or if they are just associated with the granules.) In the case of microvilli and *Mytilus* sperm the polarity of the filaments is such that the membrane could be said to replace the Z line of muscle (Fig. 2). Equally interesting is that in both cases, *Mytilus* sperm and intestinal microvilli, examination of stages in the formation of the actin bundles, i.e., during spermiogenesis (Longo and Dornfeld, 1967) or microvillar growth (Tilney and Cardell, 1970), reveals that the actin filaments begin to polymerize from the membrane and gradually elongate (Fig. 3). Thus selected areas of the membrane, presumably by their association with proteins such as α actinin, appear to nucleate the assembly of actin filaments from the membrane (Tilney and Cardell, 1970). These observations on the growth and polarity of the actin filaments in these two cases suggest that not only does nucleation of actin polymerization occur but it may also dictate the polarity of actin filaments relative to the membrane. The nonrandomness of the actin filaments and the polarity would, therefore, enable the cell to produce asymmetric movements. From these results we know that actin filaments are bound to membranes in two ways, i.e., from one end and laterally, and that, at least in microvilli and sperm, polarity and nonrandomness can be determined by spatially controlling the sites of nucleation. Clearly much more work must be carried out to see whether or not this pattern is universal.

I will now discuss what might control the polymerization of actin in nonmuscle cells. Purified actin, albeit from skeletal muscle cells or from nonmuscle cells, will polymerize essentially to completion with physiological concentrations of salt (i.e., 50–100 mM KCl and/or 0.1–10 mM $MgCl_2$). Yet, in contrast to skeletal muscle where the actin filaments are stable, in most nonmuscle cells the actin filaments are transitory, appearing when needed, disappearing at a later developmental stage. Frequently these stages are separated by only a minute or two. For example, during cytokinesis actin

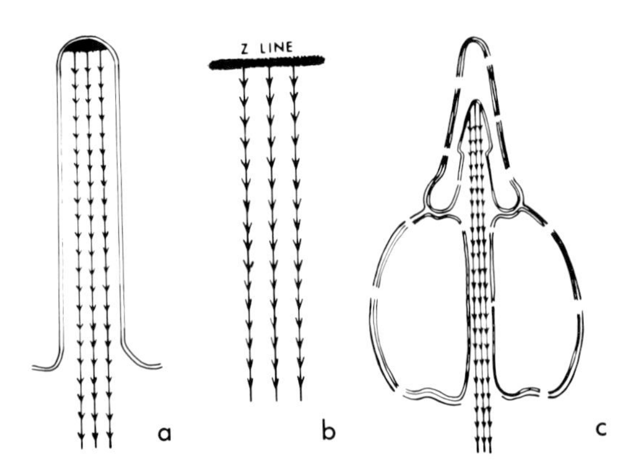

FIGURE 2 Drawing illustrating the polarity of actin filaments in microvilli (*a*), from the Z line of muscle (*b*), and in *Mytilus* sperm (*c*).

filaments appear in the cleavage furrow, yet before and after cytokinesis these filaments are not observed. Thus the actin in nonmuscle cells must be capable of rapid assembly and disassembly. We also know that actin comprises 10–15% of the total protein of many cell types such as the ameba, blood platelets, nerve cells, and leukocytes. Yet there are not enough filaments in these cell types to account for the amount of actin present. It is possible that there are few filaments in thin sections because of inadequacies of current methods of fixation (see Tilney, 1976*a*; Szamier et al., 1975). Even so, we know that when cells are broken, a large percentage of the actin is not sedimentable by conditions which should sediment F-actin (see Tilney, 1975, for references). Yet this same actin can be polymerized into filaments (Hinssen, 1972) if purified further. These facts, coupled with the changes in actin filament distribution, lead us to suspect that the cell has some mechanism for controlling the assembly and disassembly of its actin; in fact, what controls the state of the actin, the monomer, polymer, or a storage state of some kind, must be one of the most important mechanisms for the regulation of motility in nonmuscle cells, particularly since the myosin concentration is so low (see Pollard's discussion in this symposium). Although Bray and Thomas (1976) have suggested that brain actin does not polymerize as well as muscle actin, because the critical concentration for polymerization is higher, the sequence data available suggests that the changes observed are very conservative (Lu and Elzinga, 1976) and thus not likely to change the kinetics of polymerization. Also in vitro polymerization studies of others (Spudich, 1974; Kyrka, 1976) show values for polymerization not appreciably different from muscle. Thus from the available data it appears that some other components are important in regulating polymerization of actin rather than the actin alone.

I would like to begin a discussion of this regula-

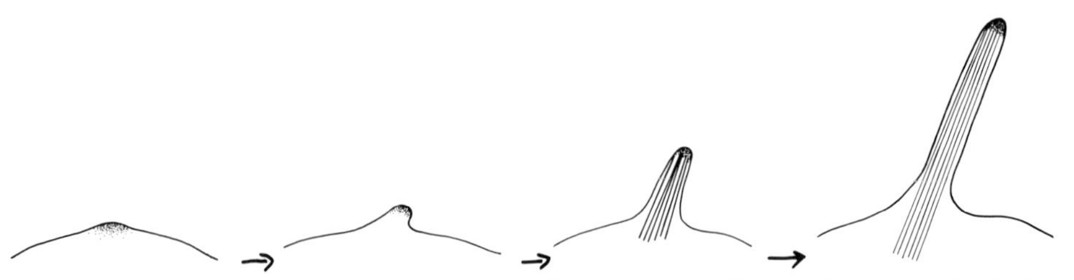

FIGURE 3 Drawing illustrating the growth of microvilli along the apical border of an epithelial cell of an intestine. The data for this drawing can be found in Tilney and Cardell (1970).

tion with a description of some experiments on the acrosomal reaction of echinoderm sperm which indicate that the actin is inhibited from polymerizing by its association with other proteins which sequester it in a storage state. When echinoderm sperm such as those from the starfish (*Marthasterias, Asterias*) or the sea cucumber (*Thyone*) come in contact with the extracellular material surrounding the egg of the same species, an extraordinary reaction ensues. Within 10 s a process (which, in *Thyone,* is 90 μm in length) is generated from the anterior end of the sperm. With this process the sperm pierces the jelly of the egg, the membrane at its tip then being able to fuse with the plasma membrane limiting the egg. We (Tilney et al., 1973) demonstrated that the newly formed process is filled with actin filaments (Fig. 4), yet before induction no actin filaments could be found in the sperm. Because membranes have little mechanical rigidity, we concluded that the change in the state of the actin, from the monomeric (or short oligomeric state) to the filamentous state, is responsible for this rapid cell extension. We also provided biochemical evidence to strengthen the fine structural observations of the absence of actin filaments before induction. We found that when sperm were extracted with Triton X 100 at pH 8.0 (the pH of seawater), most of the actin was released; furthermore, this actin was not sedimentable (85% of it) under conditions which should sediment F-actin (80,000*g* for 2 h). Inasmuch as this same detergent solution did not affect the integrity of F-actin isolated and polymerized from *Thyone* muscle, nor did it inhibit muscle actin polymerization, we concluded that the actin in the sperm did not exist as filaments, but rather in the monomeric or short oligomeric state.

Let us examine the unreacted sperm more carefully. The nucleus is spherical in form except at the anterior end where it is indented (Fig. 5). Lying within this indentation is a vacuole, the acrosomal vacuole, and between the vacuole and the nucleus is some amorphous material. When the acrosomal reaction is induced, the contents of this vacuole become externalized; thus this vacuole might be

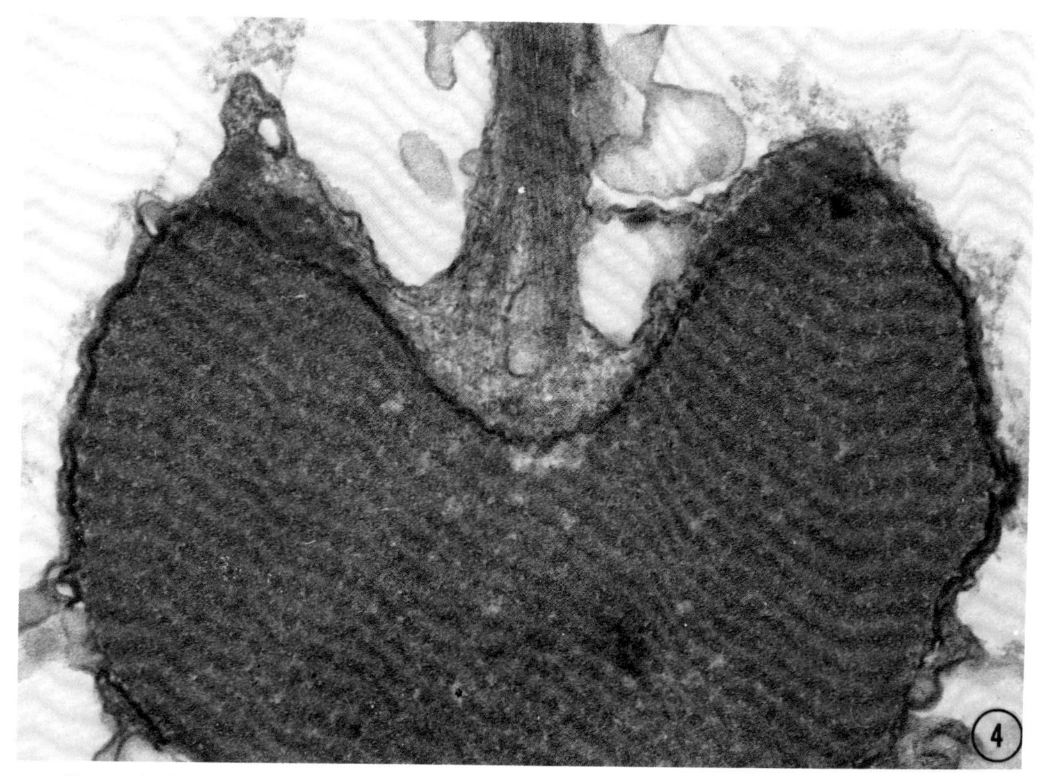

FIGURE 4 Thin section of the sperm of *Asterias amurensis* which has undergone the acrosomal reaction. The basal portion of the acrosomal process is included in this micrograph. Note that it is filled with actin filaments. (From Tilney et al., 1973.) × 82,000.

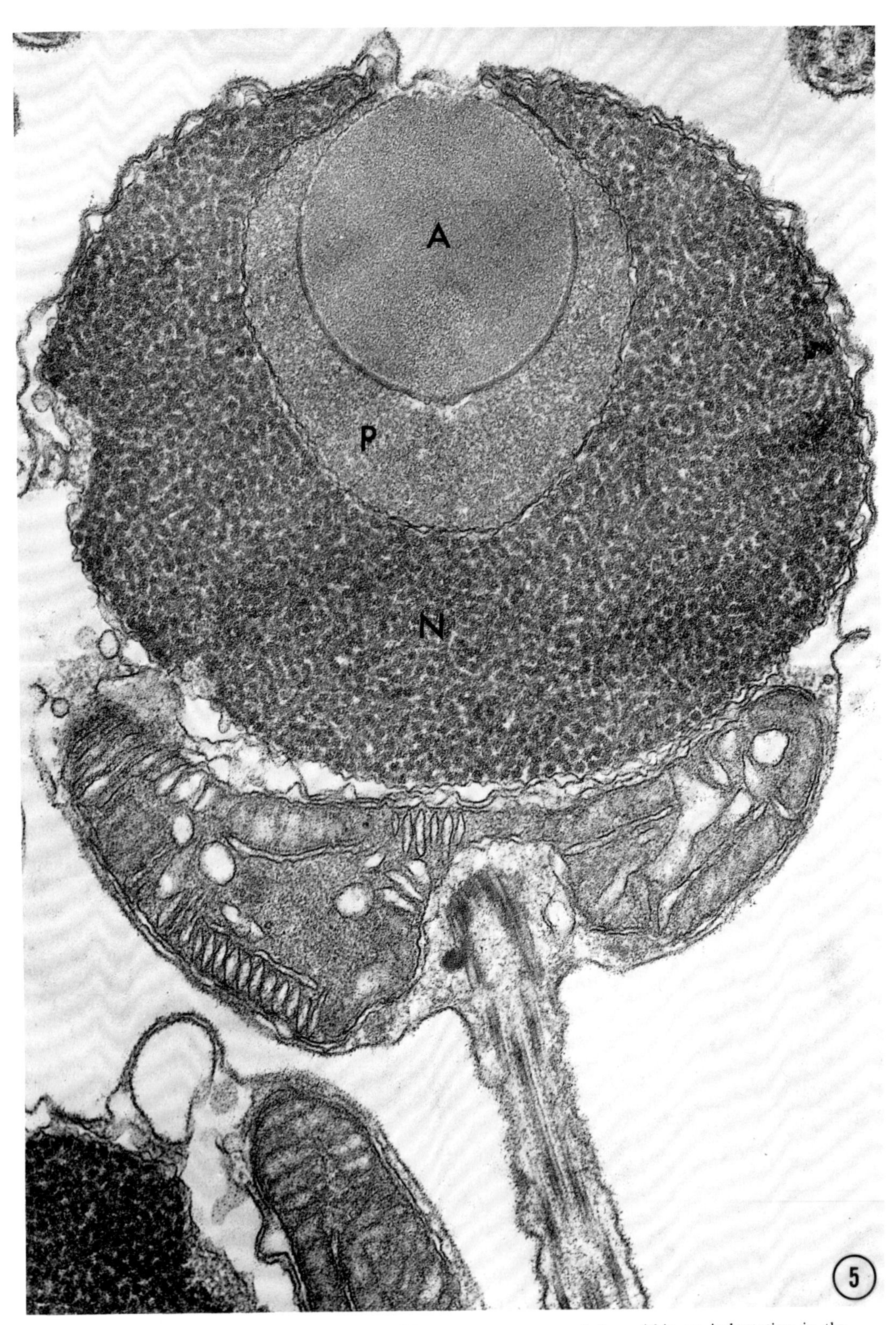

FIGURE 5 Thin section through a mature *Thyone* spermatozoon. Lying within an indentation in the nucleus (*N*) is the spherical acrosomal vacuole (*A*) and beneath and lateral to it is the profilamentous actin (*P*). Note that filaments are not present in the profilamentous actin. (From Tilney, 1976*a*.) × 92,000.

better thought of as a secretion vacuole; in fact it is thought to contain enzymes helpful in digesting the jelly surrounding the egg. The interesting area for us is the region (periacrosomal) containing the amorphous material, for it is this material that ultimately becomes the contents (the filaments) of the acrosomal process. Since the amorphous material becomes polymerized into filaments, I will refer to this material as profilamentous actin or profil. actin.

When sperm are extracted with the detergent, Triton X 100, at low pH (6.4) and low ionic strength (10 mM phosphate) buffer, the membranes solubilize but the profil. actin does not disperse. Instead it remains as a compact cup readily identifiable in the light microscope because of its phase density (Fig. 6). These cups can, in fact, be isolated by differential centrifugation after Triton extraction by treating the demembranated sperm with DNase. Solubilization of the cups can be achieved merely by raising the pH to pH 8.0; changing the type of buffer (from phosphate to Tris) also encourages solubilization. By sodium dodecyl sulfate (SDS) gel electrophoresis the cups appear to be composed of actin and two high molecular weight components of 250,000 and 230,000 daltons (Fig. 7). There is also considerable material running with the dye front. I thought this material might be contaminating protamines as the quantity varied from preparation to preparation; because DNA is the major contaminant, this is reasonable. However, this material may, in fact, be protamines plus a low molecular weight component of considerable importance. My conception here is influenced by a recent paper by Carlson et al. (1976) which demonstrates that actin purified from the spleen will crystallize if a low molecular weight component is present. This component apparently is important in maintaining the actin in an unpolymerized state. A similar situation may exist in sperm.

FIGURE 7 5% SDS gel. *Thyone* sperm were first demembranated with 1% Triton X 100 in 10 mM phosphate buffer at pH 6.4, washed twice in buffer, and incubated in 30 mM Tris HCl at pH 8.0 containing 3 mM $MgCl_2$ and 0.1 mM EDTA. The last incubation solubilized most of the profil. actin. The unsolubilized material was removed by centrifugation and the supernate run on this gel. Four major bands are seen here. The upper two have molecular weights of 250,000 and 230,000. The arrow indicates material running with the dye front. (From Tilney, 1976*b*.)

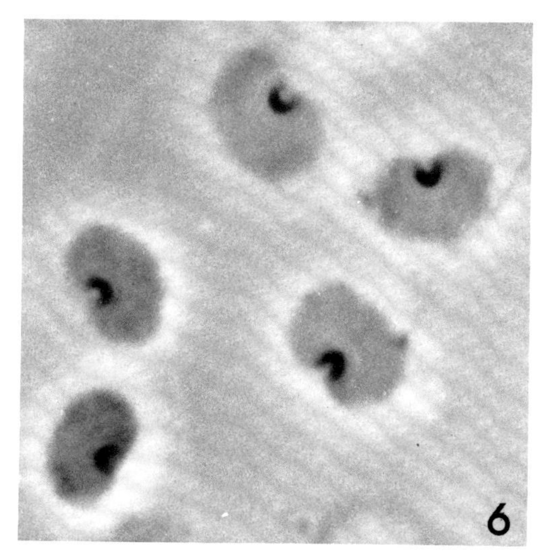

FIGURE 6 Phase contrast micrograph of *Thyone* sperm that have been demembranated with 1% Triton X 100 in 10 mM phosphate buffer at pH 6.4. Attached to the chromatin is a phase-dense cup. This cup corresponds to the profil. actin seen in electron micrographs. (From Tilney, 1976*b*.) × 3,000.

When the cups are treated with trypsin, they rapidly disappear. SDS gels of the liberated proteins demonstrate that the high molecular weight bands are lost, yet the actin remains largely unaffected (actin is notoriously trypsin-insensitive and, in fact, actin filaments were often isolated from other components in muscle by enzyme digestion). Inasmuch as the cups disappeared, we concluded that the high (and possibly low) molecular weight components must be important in holding the actin in this insoluble complex and also in keeping the actin from polymerizing. Thus, all the evidence taken together indicates that in sperm there is a storage form of actin, profil. actin, which is composed of actin, two high molecular weight proteins, and perhaps a low molecular weight component. The high molecular weight components appear to be involved in sequestering the actin. If the pH is raised, these high molecular weight components separate from the actin which must now be in the form of monomeric or short oligomeric actin based upon sedimentation studies. This actin can then be induced to polymerize. Even so, the actin in solution will not polymerize to any marked extent (Tilney et al., 1973). What inhibits the polymerization of the actin once it is in the monomeric state remains unclear, but certainly it is one of the most important areas in which future work must be carried out. Nevertheless we can draw the following simplified scheme:

$$\text{Profil. actin} \rightarrow \text{G actin} \rightarrow \text{F actin}$$
$$+$$
$$\text{other components.}$$

In the sperm one control must be at the level of sequestering the actin in an insoluble state, yet another control must exist as well because once released, the actin polymerizes poorly. Perhaps it is the small molecular weight protein recently reported by Carlson et al. (1976).

The question immediately becomes: what are the necessary requirements for the polymerization of the actin from the isolated cups? Unfortunately, I have only a partial answer here as the actin, as already mentioned, is difficult to polymerize in vitro. I do know that some polymerization from the isolated cups can be achieved with ATP and salt. Furthermore, Mg^{++} ions tend to compact the profil. actin (it forms a series of small dense particles), and high pH and ATP tend to solubilize it. Obviously separation of the actin from its high molecular weight components must occur before polymerization proceeds.

The Colwins in 1956 discovered that the addition of ammonium hydroxide to seawater to pH 9.2 would induce the acrosomal reaction. NaOH at the same pH is ineffective. Presumably NH_4OH works because the ammonium (NH_4^+) ion is in equilibrium with ammonia gas (NH_3) to which the cell membrane is permeable. Once inside the cell the NH_3 could ionize and in so doing would raise the intracellular pH. These observations are interesting in view of the fact that solubilization of the cup in vitro occurs by raising the pH.

J. C. Dan in 1954 was the first to demonstrate that the acrosomal reaction could not be induced in calcium free seawater. Eggs also cannot be fertilized without Ca^{++}. Ca^{++} is not a requisite for polymerization of purified actin; instead any of a variety of monovalent or divalent ions will work. We are left with the conclusion that Ca^{++} and/or an increase in pH may be important in the release of the actin from the high molecular weight components and/or other binding components so that the actin can polymerize. Similar conclusions have been postulated recently by Taylor et al. (1976) for filament formation in amebas.

In other cell types as well, actin appears to be associated with either one or a pair of high molecular-weight components, and in this way it is stored until the time is ready for actin filament formation. For example, in erythrocyte ghosts actin and two high molecular weight components (spectrin) are present. Not only will the actin and spectrin associate in vitro, but also methods that permit release of spectrin from the purified ghost release actin as well (Tilney and Detmers, 1975). Some evidence exists that the actin in erythrocyte ghosts, as in the case of sperm, is not in the form of filaments, but rather in the monomeric or short oligomeric state forms a net by its association with spectrin.

In macrophages, Stossel and Hartwig (1975) were the first to demonstrate the presence of a protein ("the actin binding protein") that binds actin and has a high molecular weight (280,000 daltons). A similar high molecular weight protein has been described in other cells as well, i.e., amebas, sea urchin eggs, gizzards, and many others. It is not clear if this protein acts like the pair of high molecular weight proteins in sperm and erythrocytes by binding to monomeric or short oligomeric actin. The name given to this protein by Stossel and Hartwig (1975) is an inappropriate one because many proteins bind actin, i.e., spectrin and sperm proteins, troponin, tropomyosin, myosin, α actinin, the small molecular weight component in the spleen, and a 55,000 molecular weight protein. It is interesting that the actin binding protein may also be found attached to the cell surface (Boxer et al., 1976).

The transformation of a storage form of actin, monomeric actin coupled to other components, to actin filaments is presumably under ionic control. To investigate what ions are important in vivo Christian Sardet and I attempted to induce the acrosomal reaction by the addition to a sperm suspension of compounds that allowed ions to flux into the cell. The "magic bullets" we used were ionophores which have been defined as antibiotics which bind and transport cations across water inaccessible regions (Case et al., 1974). We used two ionophores, A23187 and X537A. A23187 transports mainly Ca^{++}, Mg^{++}, and H^+ (Pfeiffer and Lardy, 1976), trying to bring the inside and outside of the cell to equilibrium for these ions. X537A transports monovalent ions as well as Mg^{++}, Ca^{++}, and H^+ (Case et al., 1974). Although the purpose of these experiments was to demonstrate what ions are necessary for the polymerization of actin, they really have just emphasized the complexity of the acrosomal reaction and by so doing have enabled us to define what specific questions should be pursued. In fact, I now realize that there are a number of parameters associated with actin polymerization in vivo which must be considered. I will discuss only the situation with sperm, but many of these observations have parallels in other systems, as the regulation of ions probably triggers actin polymerization in many other systems.

If A23187 or X537A is applied to sperm at the pH of seawater (7.8–8.0), the acrosome reaction is induced in 100% of *Pisaster* (starfish) or *Thyone* sperm. The actin under these conditions polymerizes into filaments. However, in all cases the acrosomal reaction was abnormal. In general, the reaction consisted of a mushroomlike growth extending from the anterior end of the sperm (Fig. 8). Sometimes a short process 2–3 μm in length could be seen; a few produced a process somewhat longer. However, if the pH of the seawater was maintained at pH 6.5 (the presumed intracellular pH), none of the sperm reacted.

From the published literature (see Case et al., 1974; Célis et al., 1974; Pfeiffer and Lardy, 1976), we know that the ionophores act as carriers exchanging one ion for another. The resting potential of the cell is maintained so that if one divalent cation is carried in (i.e., Ca^{++}), two monovalent cations are carried out (i.e., $2H^+$). Because the concentration of calcium in seawater is much higher than calcium in the cell and because the concentration of H^+ in the cell is much higher (pH 6.5) than in the seawater (pH 8.0), it stands to reason that Ca^{++} will be carried in and $2H^+$ will be carried out. In fact one can measure the efflux of H^+ when the sperm is placed into the ionophores. It is not a trivial efflux because within 2 min a fairly dense suspension of sperm

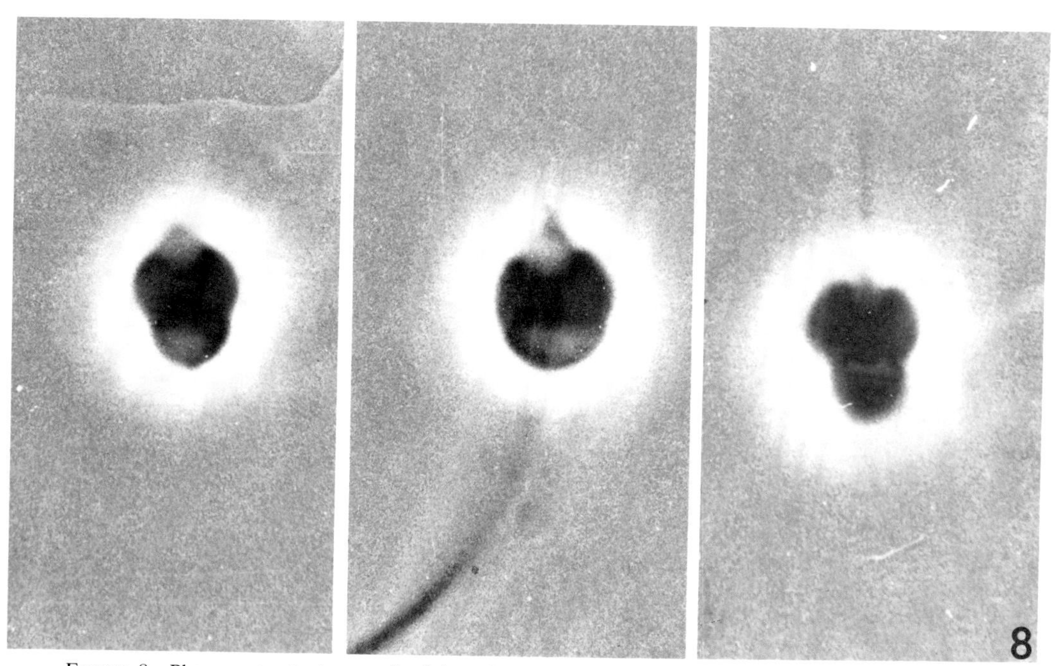

FIGURE 8 Phase contrast micrograph of three *Marthasterias* sperm treated with the ionophore, X537A. The most common situation is seen in the sperm on the left which has a mushroomlike growth. × 5,000.

treated with the ionophores will change the pH of the seawater from pH 8.0 to pH 6.8. Thus either the influx of Ca^{++} or other ions (remember the ionophores are not Ca^{++} ionophores but will transport other ions as well; see Pfeiffer and Lardy 1976; Case et al., 1974) or the efflux of H^+, and thus a rise in internal pH, or both are important in the acrosomal reaction. Inasmuch as H^+ is carried as efficiently as Ca^{++} (actually more efficiently as $2H^+$ and carried for every Ca^{++}) from experiments published to date, it is unclear if Ca^{++} is important or if lack of H^+ (pH) is important, or both, for the polymerization of actin or in fact if other ions such as Na^+ could be used. By regulating the species of ions outside the sperm and by monitoring what ions are carried out of the sperm, it is possible to dissect out what ions do what.

Examination of sperm that had "mushroomed" when treated with X537A or reacted with A23187 at pH 8.0 revealed that the profil. actin had been transformed into filaments 50 Å in diameter. In most cases the filaments were randomly oriented with respect to each other. Often the acrosomal vacuole was intact, having failed or partially failed to externalize its contents (Fig. 9). Examination of images such as these impressed upon us that the acrosomal reaction was a complex one and one in which at least six separate events must take place. These events are listed below: some must take place sequentially, others clearly occur simultaneously.

(1) Fusion of the acrosomal vacuole membrane with the cell surface.

(2) Increase in cell volume.

(3) Polymerization of the actin.

(4) Formation of membrane to limit the acrosomal process.

(5) Alignment of the actin filaments.

(6) Change in the shape of the nucleus.

It should be remembered that the polymerization of the actin cannot take place before the release of the actin from its binding proteins. It is interesting to recall that in vitro a change in pH (6.5 → 8.0) solubilizes the cup so that the actin could polymerize. Polymerization must be followed by lateral association of the actin filaments. Both these events may require different ions or lack of ions.

In preparations of sperm which were suspended in isotonic NaCl or KCl (with EGTA present) we found, using ionophores, that a rise in internal pH is essential for actin polymerization. The species of cation entering the cell is not important,

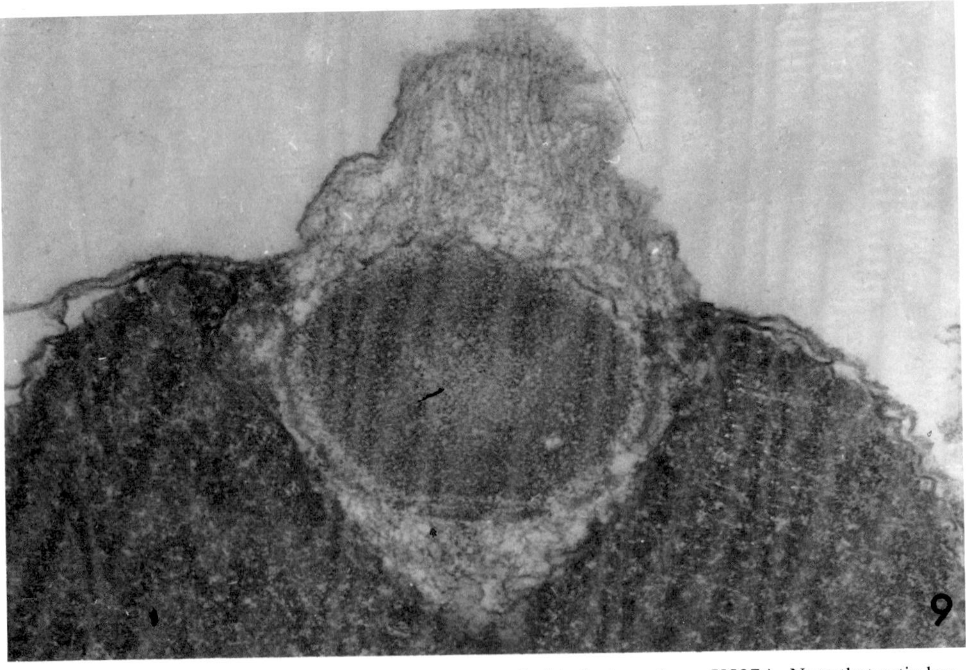

FIGURE 9 Thin section of a *Marthasterias* sperm treated with the ionophore, X537A. Note that actin has polymerized into filaments, but the acrosomal vacuole has not fused with the cell surface. × 67,000.

albeit Na^+, K^+, Mg^+, or Ca^{++}. Even though Ca^{++} is not required for actin polymerization, fusion of the acrosomal vacuole with the cell surface (step 1) requires Ca^{++}. We still have no information about what ions may be important for step 2 (volume increase) or step 5 (alignment of the filaments so that a thin process will form, not a mushroomlike growth). It seems reasonable to suspect that each of the six events listed above may require different quantities and/or different combinations of ions.

In other cell types as well many of the above events must take place upon the induction of polymerization. For example, we know that actin polymerization in many systems, e.g., microvilli, pseudopods, and the cleavage furrow, is coupled to parallel alignment of actin filaments, presumably by being cross-linked. It is also true that in many nonmuscle systems an increase in cell surface accompanies filament formation and alignment as, for example, in cytokinesis, in phagocytosis, or in pseudopod formation. In these systems as well ion fluxes are probably critical.

For the remainder of this article I will present evidence demonstrating that profil. actin in sperm is membrane-bound, as is the "net" of actin and spectrin in erythrocytes. I assume that in many other cells as well the cell cortex which contains actin, perhaps also in a nonfilamentous form, is associated with the membrane. Finally, I would like to end on a very speculative note, suggesting that "membrane precursors" may be associated with profil. actin.

What is, in fact, the data suggesting that profil. actin is membrane-bound? In studying how the actin and its associated proteins are sequestered in the anterior end of the sperm during spermiogenesis, I found that at no stage in the accumulation of the actin and its associated proteins was the actin sequestered in a membrane-bound compartment such as a vacuole or vesicle. Instead, the actin and its associated proteins must be transported from the basal end of the cell to the anterior end through the cytoplasm. I discovered that the membranes in the periacrosomal region (the region occupied by the profil. actin), specifically a portion of the nuclear envelope and the basal half of the acrosomal vacuole membrane, became specialized morphologically in advance of the accumulation of the profil. actin in this region. The basal half of the acrosomal vacuole membrane is specialized in having an electron-dense material associated with its internal surface. The nuclear membrane is also specialized in the periacrosomal region. Instead of the usual separation of 150–200 Å between the outer and inner nuclear envelopes in this region, this space is eliminated and, in fact, the outer leaflets of the adjacent unit membranes fuse in an analogous fashion to a tight junction. I suggested that actin and its associated proteins might accumulate in this region by associating with these specialized membranes and with themselves. Diffusion then would be sufficient to move these substances to this region. In support of this hypothesis is the fact that when sperm are treated with detergents, glycols, and hypotonic media (Fig. 10), agents which solubilize or lift away the plasma membrane, the profil. actin remains attached to these specialized membranes. In fact, these regions where the profil. actin is attached are remarkably resistant to detergent treatment. Thus the basal half of the acrosomal vacuole membrane (the specialized half) and the nuclear envelope in

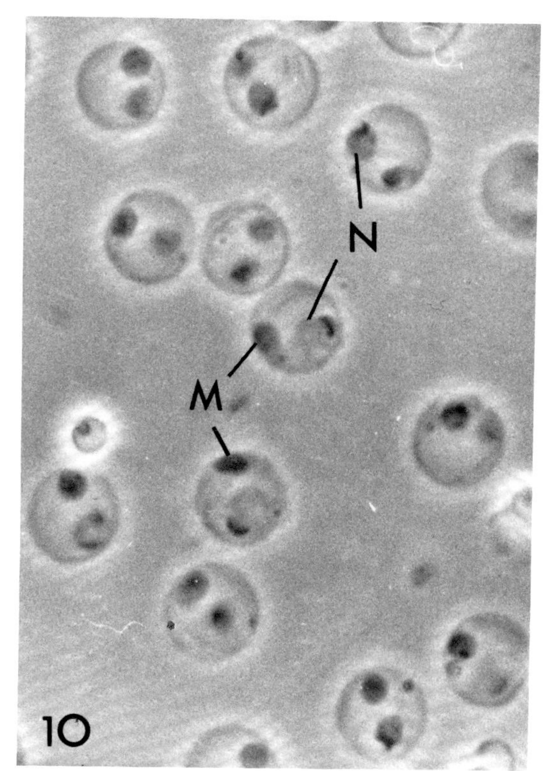

FIGURE 10 Phase contrast micrograph of *Thyone* sperm whose seawater had been diluted with distilled water. The flagellar axoneme retracted. Within the membrane are two bodies; the nucleus (*N*), with its associated phase-dense cap of profil. actin, and the mitochondrion (*M*). (From Tilney, 1976*a*.) × 3,300.

the periacrosomal region do not solubilize readily (Tilney, 1976*a*). From our studies, therefore, we know that actin in a nonfilamentous form can be associated with membranes. When these specialized membranes were examined in freeze-fracture preparations, the particle distribution did not differ from that on the unspecialized regions of the same membrane, indicating that the profil. actin is not associated with a particular type of particle.

From the experiments with the ionophores and other data, we know that the acrosomal reaction must include a large increase in cell surface. We calculated that for *Marthasterias* sperm the membrane limiting the acrosomal process is 5.2 μm^2 (18 μm in length; 0.1 μm in width) and the plasma membrane that limits the cell body exclusive of the flagellum is 12.5 μm^2. (Because an acrosomal process of normal length will form in the absence of a flagellum [see Tilney et al., 1973], the membrane limiting the flagellar axoneme must not be used for the increase in cell surface necessary for this process.) For *Thyone* sperm the surface area of the acrosomal process is 22 μm^2 (90 μm in length; 0.075 μm in diameter) and the surface area of the cell body is 16 μm^2. Thus during the acrosomal reaction one needs to increase the surface area from 40 to 150% depending upon the species.

The impressive feature of echinoderm sperm is that there are no vesicles or vacuoles in the sperm other than the acrosomal vacuole (the surface area of the acrosomal vacuole in both species is 1.1 μm^2 and, in fact, will contribute little new surface even if all of it is used, which turns out not to occur) which could contribute to this increase in membrane surface. Furthermore, in thin sections, by scanning microscopy, and in freeze-fracture preparations, it is clear that there are no surface projections such as microvilli that could add membrane to the acrosomal process; instead, the plasma membrane follows the contour of the sperm rather precisely. Furthermore, the membrane limiting the acrosomal process resembles a typical "unit membrane" both in freeze-fractures (it cleaves normally like other membranes and has a few particles on its P face, Fig. 11), and in thin sections where one gets the three-layered struc-

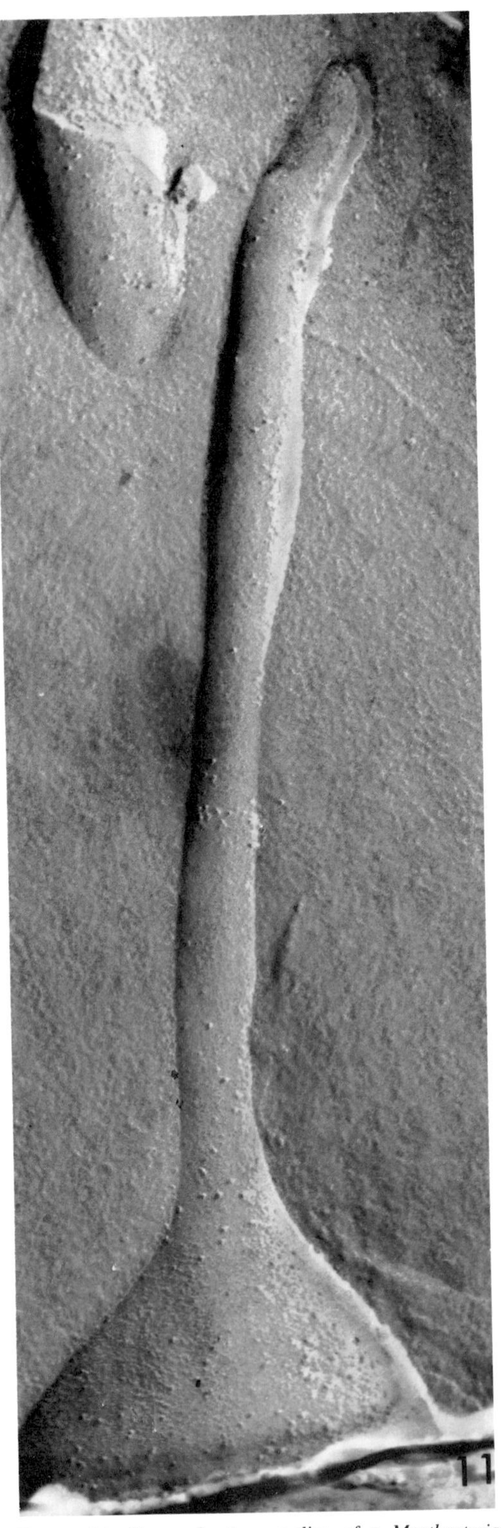

FIGURE 11 Freeze-fracture replica of a *Marthasterias* sperm that had been induced to undergo the acrosomal reaction. Note that there are particles along the process. From Sardet and Tilney, 1977, *Cell Biology International Reports*. In press. Reprinted by permission. × 70,000.

ture. Inasmuch as there are no vesicles or surface projections which could account for the increase in surface, we suspect that the new membrane is generated from precursors stored with the profil. actin. This view concurs with the speculation of the Colwins (1963, 1975). Stretching of the existing membrane is unlikely on account of the amount of increase in cell surface (up to 150%). Biosynthesis is also unlikely on account of the rapidity of the reaction (10 s). Because there is little unoccupied space in the sperm, each cell body containing a mitochondrion, a nucleus, an acrosomal vacuole, and some periacrosomal material (the region where the profil. actin is stored), we suggested that membrane precursors are stored with the profil. actin. To test our idea we first stained sperm with Sudan Black B, a stain used to demonstrate lipids by light microscopy. Both the profil. actin and the mitochondrion stained. We then applied it to sperm compounds such as didansyl cysteine and *N* phenyl-1-napthylamine (NPN) which fluoresce when they enter a hydrophobic environment. Two regions of the sperm fluoresce; the mitochondrion and the cup of profil. actin (Fig. 12). The nucleus,

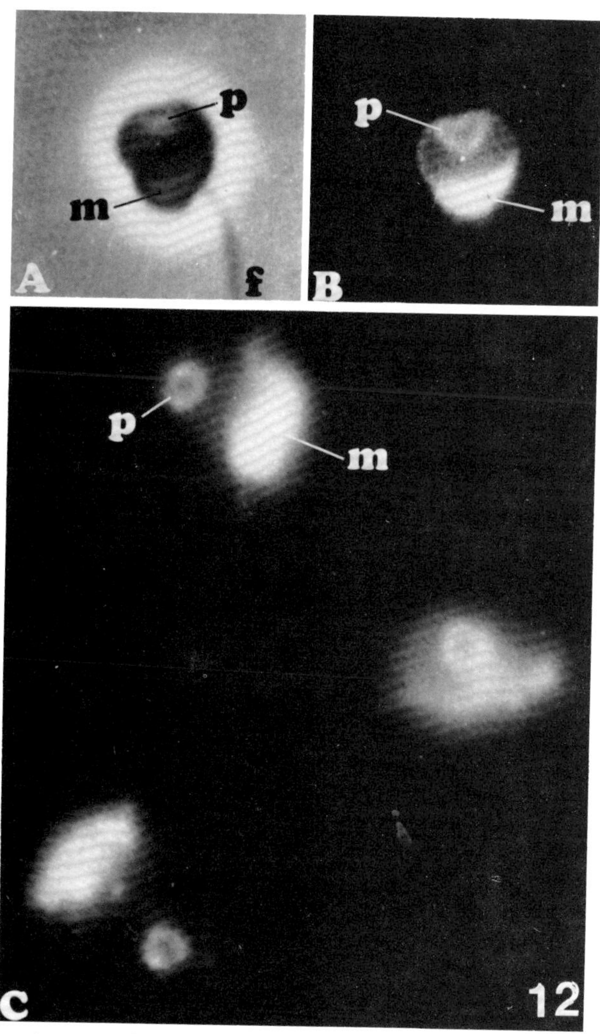

FIGURE 12 *Marthasterias* sperm observed by phase contrast microscopy (*A*) and fluorescence microscopy (*B* and *C*). The pelleted sperm were suspended in seawater containing 100 μM didansyl cysteine. The periacrosomal material (*p*), the mitochondrion (*m*), and the flagellum (*f*) are indicated. A and B are photographs of the same sperm which was fixed in 1% glutaraldehyde in seawater. C is a photograph of unfixed sperm; the periacrosomal material in these flattened cells appears as a fluorescent donut. (From Sardet and Tilney, 1977, *Cell Biology International Reports*. In press. Reprinted by permission.) × 5,000.

the flagellar axoneme, and the acrosomal vacuole do not fluoresce. We then placed sperm in hypotonic media (Fig. 10). Under these conditions the axoneme retracts; this allows the plasma membrane to lift away from the nucleus with its attached profil. actin cup. (No swelling occurs with deflagellated sperm; thus the increase in surface must be due to the membrane formerly surrounding the axonemes.) The bloated sperm were then homogenized which liberated the nuclei with the attached profil. actin cups. When NPN or didansyl cysteine was added, the cup fluoresced vividly (Fig. 13). The only membrane, then, in this preparation is the nuclear envelope which surrounds the margin of the whole nucleus. From these experiments it appears that the profil. actin would provide the proper hydrophobic environment for membrane precursor lipids. Assuming that there are precursor lipid molecules in this material, we calculate that these precursors should represent one-third of the weight of the periacrosomal material. These calculations were made from estimates of the amount of actin in the process, the amount of high molecular weight proteins seen on SDS gels, and the amount of surface area that a phospholipid occupies in a membrane. It is interesting that lipoproteins with comparable lipid:protein ratios have been described in human blood serum and in egg yolks, materials that appear amorphous, not as little vesicles, in electron micrographs. Furthermore, the very low density lipoproteins in serum appear as little dense granules, as does the isolated profil. actin when divalent salts are present.

The exciting aspect of an association of actin and its associated proteins with membrane precursors is that in many other nonmuscle cells the motile event not only involves the formation of filaments but also necessitates an increase in cell surface. Examples include cytokinesis, phagocytosis, pseudopod formation, ruffling of the cell surface, and clot retraction. What more likely place is there to store membrane precursor molecules than with a storage form of actin? Admittedly, in some systems the increase in membrane can, and probably is, associated with the fusion of vesicles with the plasma membrane. In sperm this mechanism cannot be used as there simply are not any vesicles or surface infoldings. Indeed if such a storage form of membranes associated with actin can be shown to occur in sperm, other systems as well may employ it.

CONCLUSIONS

I have concentrated on two features of actin in nonmuscle cells. One, how and in what state is actin associated with (bound to) membranes or possible membrane precursor molecules, and two, how may actin be maintained in a stored state and what are the known conditions which allow it to polymerize and perform useful work. I have suggested that filaments of actin are bound to membranes both at their ends and along their lengths. By inducing the actin to polymerize (nucleation) from special sites on membranes, the actin filaments are not only nonrandomly distributed, but their polarity can be precisely determined. Nonfilamentous actin or actin associated with several

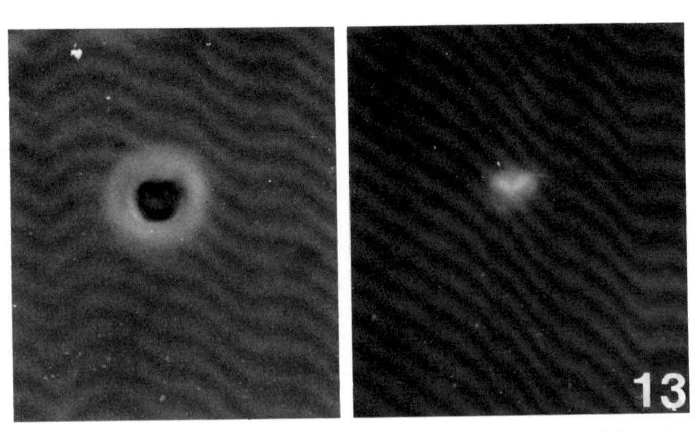

FIGURE 13 Phase contrast and corresponding fluorescent image of sperm nuclei incubated in didansyl cysteine (100 μM). Sperm were subjected to an osmotic shock which causes the flagellar axoneme to retract. This allows a swelling to occur as a result of the increase in surface provided by the flagellar membrane. The bloated sperm is then homogenized, the nuclei float free with the periacrosomal material attached. × 2,200.

other proteins, at least in sperm and erythrocytes, are also bound to membranes and may bind membrane precursors to them. In all cases the actin, albeit in the filamentous state or in the nonfilamentous state, is not associated with particles in the plane of the bilayer. The membrane-associated actin does, however, appear to stabilize the membranes to which it is attached so that the membrane lipids resist solubilization with detergents. Finally, I have described a storage form of actin or profil. actin and have suggested that in order to polymerize the actin it must first be liberated from its proteins into the G state. In vivo and in vitro experiments show that a rise in internal pH is essential for the transformation of profil. actin into filaments. Calcium is not required for this transformation but is a requisite for exocytosis. For the formation of a long process such as the acrosomal process or for the performance of unidirectional force, a series of steps are required which involve not only the polymerization of actin (once it has been released from its bound proteins), but an increase in volume by the entry of water, an increase in membrane surface, a cross-linking of filaments, and an initial secretory event. All these events require the proper concentration and sequence of entry of ions which may also include a change in local intracellular pH.

ACKNOWLEDGMENTS

I want to thank Doris Bush, my colleague and friend for fantastic assistance over the past few years. Sadly, Mrs. Bush died tragically this past summer. With her help and enthusiasm, any project we set out to do was exciting and fun.

This work was supported by grant GB22863 from the National Science Foundation.

REFERENCES

Boxer, L. A., S. Richardson, and A. Floyd. 1976. Identification of actin binding protein in membrane of polymorphonuclear leucocytes. *Nature (Lond.).* **263:**249–251.

Bray, D., and C. Thomas. 1976. Unpolymerized actin in tissue cells. *Cold Spring Harbor Conf. Cell Profileration.* **3:**461–473.

Burridge, K. 1976. Multiple forms of non-muscle myosin. *Cold Spring Harbor Conf. Cell Profileration.* **3:**739–747.

Carlson, L., L. E. Nystrom, W. Lindberg, K. Kannan, H. Cid-Dresdner, S. Lovgren, and H. Jornvall. 1976. Crystallization of a non-muscle actin. *J. Mol. Biol.* **105:**353–366.

Case, G. D., J. M. Vanderkooi, and A. Scarpa. 1974. Physical properties of biological membranes determined by the fluorescence of the calcium ionophore A23187. *Arch. Biochem. Biophys.* **162:**174–185.

Célis, H., S. Estrada-O., and M. Montal. 1974. Model translocators for divalent and monovalent ion transport in phospholipid membranes. I. The ion permeability induced in lipid bilayers by the antibiotic X 537A. *J. Membr. Biol.* **18:**187–199.

Colwin, L. H., and A. L. Colwin. 1956. The acrosome filament and sperm entry in *Thyone briareus* (Holothuria) and *Asterias. Biol. Bull. (Woods Hole).* **110:**243–257.

Colwin, A. L., and L. H. Colwin. 1963. Role of the gamete membranes in fertilization in *Saccoglossus kowalevskii* (Enteropneusta). I. The acrosomal region and its change in early stages of fertilization. *J. Cell Biol.* **19:**477–500.

Colwin, L. H., A. L. Colwin, and R. G. Summers. 1975. The acrosomal region and the beginning of fertilization in the holothurian, *Thyone briareus. In* Functional Anatomy of the Spermatozoon. B. Afzelius, editor. Pergamon Press, Oxford. 27–38.

Dan, J. C. 1954. Studies on the acrosome. III. Effect of calcium deficiency. *Biol. Bull. (Woods Hole).* **107:**335–349.

Helenius, A., and K. Simon. 1975. Solubilization of membranes by detergents. *Biochim. Biophys. Acta.* **415:**29–79.

Hinssen, H. 1972. Actin in isoliertem Grundplasma von *Physarum polycephalum. Cytobiologie.* **5:**146–164.

Huxley, H. E. 1963. Electron microscope studies on the structure of natural and synthetic protein filaments from striated muscle. *J. Mol. Biol.* **7:**281–308.

Ishikawa, H., R. Bischoff, and H. Holtzer. 1969. The formation of arrowhead complexes with heavy meromyosin in a variety of cell types. *J. Cell Biol.* **43:**312–315.

Kyrka, R. 1976. The gelation of a soluble protein extract of embryonic chick brain involving supramolecular associations of actin filaments and the isolation of brain actins. Ph.D. Thesis. University of Pennsylvania.

Longo, F. J., and E. J. Dornfeld. 1967. The fine structure of spermatid differentiation in the mussel, *Mytilis edulis. J. Ultrastruct. Res.* **20:**462–480.

Lu, R., and M. Elzinga. 1976. Comparison of amino acid sequences of actins from bovine brain and muscles. *Cold Spring Harbor Conf. Cell Profileration.* **3:**487–492.

Mooseker, M. S., and L. G. Tilney. 1976. Organization of an actin filament-membrane complex: filament polarity and membrane attachment in the microvilli of intestinal epithelial cells. *J. Cell Biol.* **67:**724–743.

Mukherjee, T. M., and L. A. Staehelin. 1971. The fine-structural organization of the brush border of intestinal epithelial cells. *J. Cell Sci.* **8:**573–600.

Pfeiffer, D. R., and H. A. Lardy. 1976. Ionophore A23187: the effect of H^+ concentration on complex formation with divalent and monovalent cations and the demonstrtion of K^+ transport in mitochondria mediated by A23187. *Biochemistry*. **15:**935–943.

SCHOLLMEYER, J. V., D. E. GOLL, L. G. TILNEY, M. MOOSEKER, R. ROBSON, and M. STROMER. 1974. Localization of α actinin in non-muscle material. *J. Cell Biol.* **63:**304*a* (Abstr.).

SPUDICH, J. A. 1974. Biochemical and structural studies of actomyosin-like proteins from non-muscle cells. II. Purification, properties and membrane association of actin from amoebae of *Dictyostelium discoideum*. *J. Biol. Chem.* **249:**6013–6020.

STOSSEL, T. P., and J. H. HARTWIG. 1975. Interactions between actin, myosin and an actin binding protein from rabbit alveolar macrophages. *J. Biol. Chem.* **260:**5706–5712.

SZAMIER, P., T. D. POLLARD, and K. FUJIWARA. 1975. Tropomyosin prevents the destruction of actin filaments by osmium. *J. Cell Biol.* **67:**424*a* (Abstr.).

TAYLOR, D. L., J. S. CONDEELIS, and J. A. RHODES. 1976. The contractile basis of amoeboid movement. III. Structure and dynamics of motile extracts and membrane fragments from *Dictyostelium discoideum* and *Amoeba proteus*. *J. Supramol. Struct.* In press.

TILNEY, L. G. 1975. The role of actin in non-muscle cell motility. *In* Molecules and Cell Movement. S. Inoue and R. E. Stephens, editors. Raven Press, New York. 339–388.

TILNEY, L. G. 1976*a*. The polymerization of actin. II. How non-filamentous actin becomes nonrandomly distributed in sperm: evidence for the association of this actin with membranes. *J. Cell Biol.* **689:**51–79.

TILNEY, L. G. 1976*b*. The polymerization of actin. III. Aggregates of nonfilamentous actin and its associated proteins: a storage form of actin. *J. Cell Biol.* **69:**73–89.

TILNEY, L. G., and R. R. CARDELL, JR. 1970. Factors controlling the reassembly of the microvillus border of the small intestine of the salamander. *J. Cell Biol.* **47:**408–422.

TILNEY, L. G., and P. DETMERS. 1975. Actin in red blood cell ghosts and its association with spectrin: evidence for a nonfilamentous form of these two molecules *in situ*. *J. Cell Biol.* **66:**508–520.

TILNEY, L. G., S. HATANO, H. ISHIKAWA, and M. MOOSEKER. 1973. The polymerization of actin: its role in the generation of the acrosomal process of certain echinoderm sperm. *J. Cell Biol.* **59:**109–126.

TILNEY, L. G., and M. MOOSEKER. 1971. Actin in the brush border of epithelial cells of the chicken intestine. *Proc. Natl. Acad. Sci. U. S. A.* **68:**2611–2615.

TILNEY, L. G., and M. S. MOOSEKER. 1976. Actin filament-membrane attachment: are particles involved? *J. Cell Biol.* **71:**402–416.

CONCLUDING REMARKS

ROBERT D. ALLEN

In his introduction to this symposium, Professor Kamiya indicated that I would probably have more to say in my concluding remarks about recent work on rotational streaming in characean cells. I will concentrate on some investigations that have been carried out recently in our laboratory and the ideas they have suggested.

E. Kamitsubo, who discovered the subcortical fibrils at the suspected site of the postulated active shearing process in Professor Kamiya's laboratory, later worked in our laboratory to devise a simple, elegant method for making "windows" in the chloroplast layer (i.e., depleting the cortex of chloroplasts) by exposing a portion of the cell to a microbeam of intense light (Kamitsubo, 1966, 1972*a*). Through these windows, Kamitsubo (1972*b*) obtained an uninterrupted view of the surface layers of the cell, and was able to document on cine-films the details of the saltation of particles along the subcortical fibrils.

By the use of similar *Nitella* internodal cells with windows, N. S. Allen (1973, 1974) used an improved, phase-randomized, laser illumination (Hard et al., 1976) and a selected high extinction differential interference contrast (DIC) system to obtain an unparalleled view of serial optical sections of the endoplasm. In a published film, N. S. Allen (1973) documented the existence of previously undetected endoplasmic filaments that exhibited sinusoidal undulations in a plane perpendicular to the optical axis of the microscope when focused on the upper surface of the cell. Although the demonstration of the filaments themselves required sophisticated optical technique, their undulation could be visualized in stroboscopic light by the behavior of particles transported along the filaments at the velocity that characterizes both the wave propagation and particle transport. N. S. Allen (1974) calculated from hydrodynamic theory, which had originally been developed for the analysis of the swimming sperm, that the force produced by active undulations of filaments with the characteristics measured would be more than sufficient to account for the measured motive force of streaming.

The endoplasmic filaments in many cases could be seen to originate as branches of subcortical fibrils, and when streaming was inhibited temporarily by stimulation, it could be seen that the filaments were distributed more or less evenly throughout the endoplasm. Counts of filaments in various layers of the endoplasm led to the estimate that over 50 m of filaments are present in an internodal cell 2 cm in length. Each of these filaments throughout the endoplasm could not only undulate but also transport particles in contact with them at velocities up to the streaming velocity. Thus the properties described for the subcortical fibrils (Kamitsubo, 1972*b*) were shared by the endoplasmic filaments.

Inasmuch as the filaments were so numerous and so easy to visualize under appropriate conditions, it has seemed curious that ultrastructural studies have failed to confirm their existence. Although the osmium tetroxide used in some of the early transmission electron microscope studies might have been expected, in retrospect (Szamier et al., 1975), to have destroyed certain systems of unstabilized actinlike filaments, even the more recent results of Kersey and Wessells (1976) with a scanning electron microscope failed to show more than the subcortical fibrils.

In an effort to substantiate the light-microscopical results and confirm the reality of endoplasmic filaments, N. S. Allen et al. (1976) have recently prepared *Nitella* internodal cells in a variety of ways for examination with the scanning electron

ROBERT D. ALLEN Department of Biological Sciences, Dartmouth College, Hanover, New Hampshire

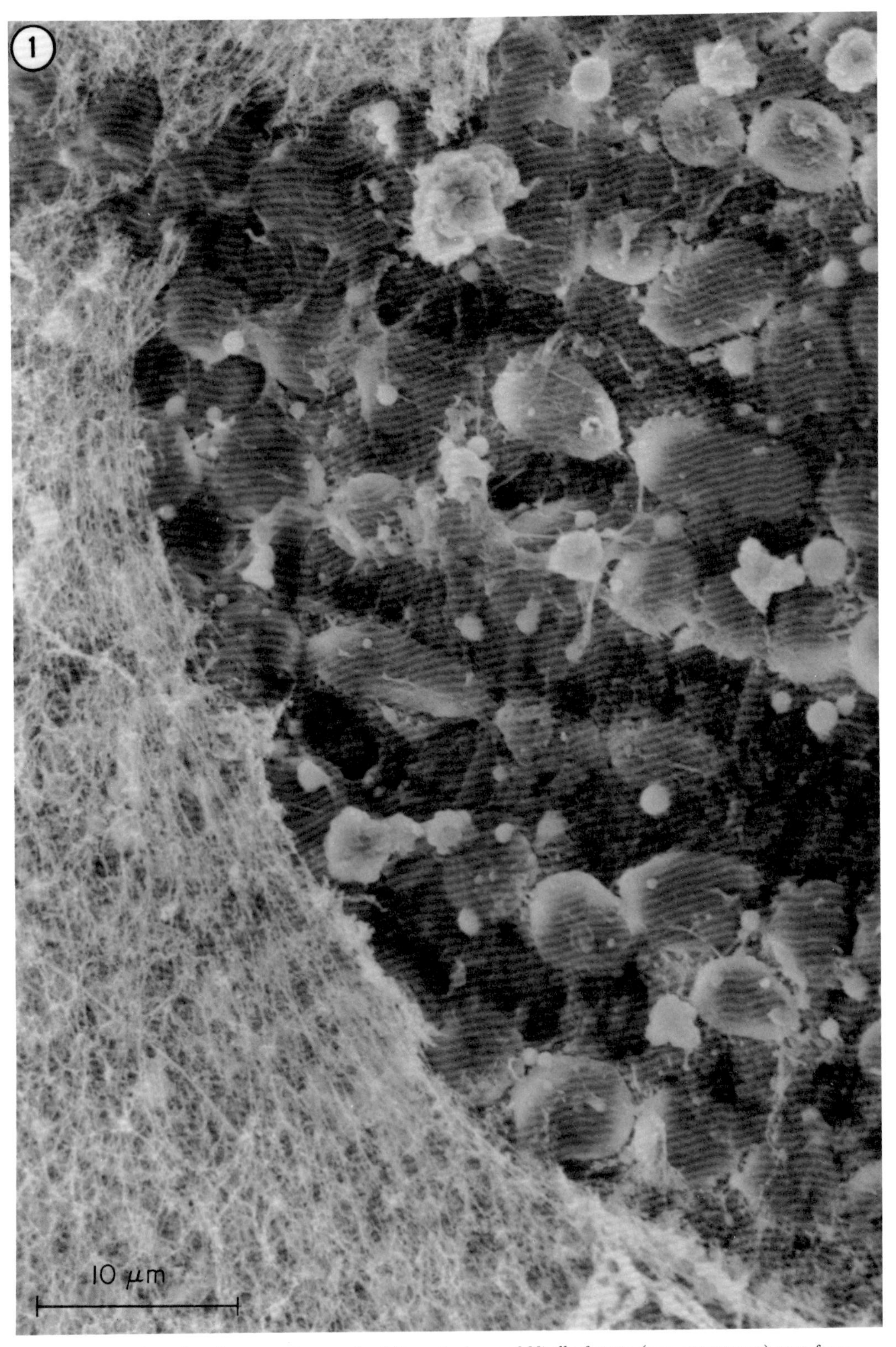

FIGURE 1 Scanning electron micrograph of the cytoplasm of *Nitella furcata* (var. *megacarpa*) seen from the vacuole side. Masses of endoplasmic filaments have been "blown aside" revealing the chloroplast rows below. × 3,200.

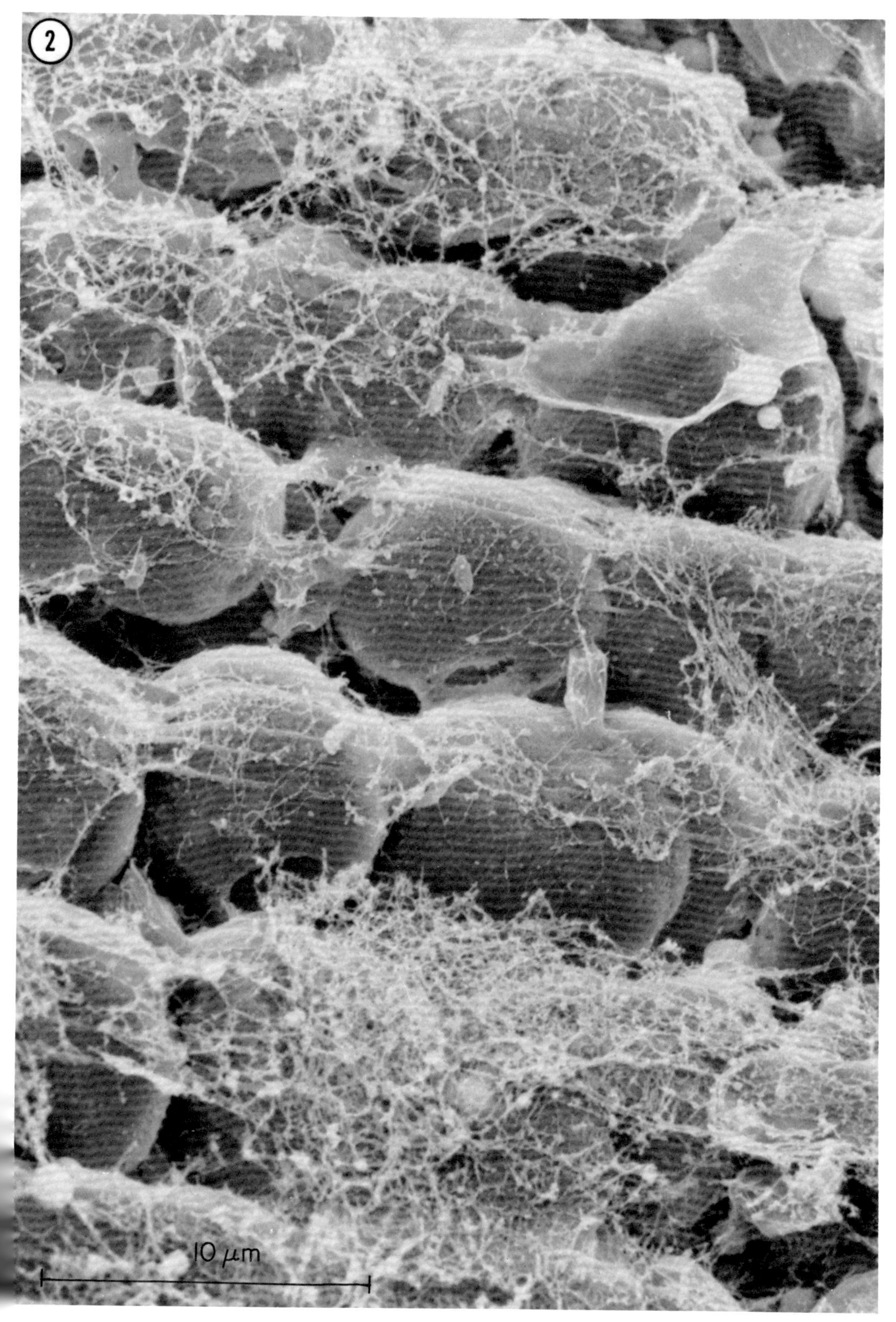

FIGURE 2 Chloroplast rows connected by subcortical fibrils with many thinner endoplasmic filaments forming a loose network. × 5,400.

microscope. The results of this study have amply confirmed the existence of endoplasmic filaments.

Figure 1 shows a view of the cytoplasm from the vacuolar side. The tonoplast, often present in patches in these preparations, is missing, in the region shown, and a dense mat of endoplasmic filaments has been locally "blown aside" to reveal the chloroplast layer to which the subcortical fibrils and their branches are still attached. Many filaments thinner than the subcortical fibrils are visible throughout the endoplasm.

Figure 2 is a view at higher magnification of several chloroplast rows with their subcortical fibrils connecting chloroplasts in each row. These, in turn, connect with many smaller filaments from 80 to over 400 nm in diameter. Even allowing for the 40–50 nm gold-palladium coating sputtered onto these preparations, the small diameters of the filaments adequately account for the difficulty of documenting their presence in the light microscope. They are easily destroyed by preparation for electron microscopy unless extraordinary caution is used at every stage of the process.

Although the results so far confirm the light microscopical observations, the filaments have not yet been characterized either ultrastructurally or biochemically; such work is in progress.

To speculate a little, one might hazard a guess that the endoplasmic filaments, because they are apparently branches of the subcortical fibrils, are also bundles of microfilaments composed, at least in part, of actin. If so, it would be expected that actin would emerge as one of the main cytoplasmic proteins of these algal cells. This would not be surprising, as the cytoplasm of many protists and tissue cells contains bundles of actin in locations corresponding to "tracks" where saltatory motions of particles are observed. The observation that endoplasmic filaments of *Nitella* undulate has not yet been extended to other cells. It is possible either that undulation is unique to filaments of *Nitella* and other characean cells, in which the streaming velocity is considerably greater than in other green plant cells, or that undulation may occur widely in still thinner microfilament bundles that are below the threshold of detection with present techniques.

In conclusion, Allen's (1974) results show clearly that particles are transported (i.e., saltate) at up to streaming velocities while in contact with endoplasmic filaments as well as subcortical fibrils. This observation would appear to suggest at least a modification to the active shearing theory (Kamiya and Kuroda, 1956) in which the postulated motive force would be delivered to particles not only at the corticoendoplasmic interface but thoughout the endoplasm as well.

Allen's (1974) data also support an alternative and/or additional hypothesis that active bending waves along endoplasmic filaments produce a force more than adequate to transport the fluid portion of the cytoplasm around the cell.

Because both the particles and the fluid around them are transported, it is important to determine the respective contributions of these plausible mechanisms (active shearing and bending wave propagation) to the phenomenon of rotational streaming. It is quite possible that both are manifestations of the same underlying process, which we seek to understand as the molecular mechanism of saltation and cytoplasmic streaming.

REFERENCES

Allen, N. S. 1973. Endoplasmic filaments in *Nitella translucens*. Film available at cost from Calvin Communications, Inc., 1105 Truman Road, Kansas City, Mo. 64106.

Allen, N. S. 1974. Endoplasmic filaments generate the motive force for rotational streaming in *Nitella*. *J. Cell Biol.* **63:**270–287.

Allen, N. S., R. D. Allen, and T. E. Reinhart. 1976. Confirmed: the existence of abundant endoplasmic filaments in *Nitella*. *Biol. Bull.* (*Woods Hole*). **154:**398.

Hard, R. Zeh, and R. D. Allen. 1976. Phase-randomized laser illumination for microscopy. *J. Cell Sci.* **23:**1–9.

Kamitsubo, E. 1966. Motile protoplasmic fibrils in cells of characeae. II. Linear fibrillar structure and its bearing on protoplasmic streaming. *Proc. Jpn. Acad.* **42:**640–643.

Kamitsubo, E. 1972*a*. A "window technique" for the detailed observation of Characean cytoplasmic streaming. *Exp. Cell Res.* **74:**613–616.

Kamitsubo, E. 1972*b*. Motile protoplasmic fibrils in cells of the Characeae. *Protoplasma*. **74:**53–70.

Kamiya, N., and K. Kuroda. 1956. Velocity distribution of the protoplasmic streaming in *Nitella* cells. *Bot. Mag.* (*Tokyo*). **69:**544–554.

Kersey, Y. M., and N. K. Wessells. 1976. Localization of actin filaments in internodal cells of characean algae: a scanning and transmission electron microscope study. *J. Cell Biol.* **68:**264–275.

Szamier, P., T. Pollard, and K. Fujiwara. 1975. Tropomyosin prevents the destruction of actin filaments by osmium. *J. Cell Biol.* **67:**424*a* (abstr.).

Eukaryotic Cell Cycle

NUCLEAR CONTROL OF CELL PROLIFERATION

RENATO BASERGA, PEN-MING L. MING, YOSHIHIRO TSUTSUI, SANDRA WHELLY, HELENA CHANG, MARA ROSSINI, and CHENG-HSIUNG HUANG

The control of cell proliferation in mammalian cells is exerted at several levels: the environment (with growth factors, hormones, and inhibitory factors), the cell membrane, the cytoplasm, and finally the nucleus, where the genome receives the extra- and intracellular signals and responds with the appropriate messages for the regulation of cell growth. This paper will deal with the role of the nucleus in the control of cell proliferation and, more specifically, will cover two selected areas of interest: 1) the changes occurring in the nucleolus of resting cells (G_0 cells) stimulated to proliferate, and 2) a first crude attempt to a chromosomal mapping of the cell cycle.

The Role of the Nucleolus

When G_0 cells are stimulated to proliferate, a number of nuclear changes indicative of increased transcriptional activity are known to occur in the early prereplicative phase, that is, several hours before the onset of DNA synthesis. These changes, which have recently been reviewed (Baserga and Nicolini, 1976) and are summarized in Table I, are detectable in isolated chromatin, isolated nuclei, nuclear monolayers, and intact cells. Some of these changes (increased RNA synthesis, increased binding of intercalating dyes, increased molar ellipticity in circular dichroism [CD] spectra, and decreased thermal stability of chromatin) can be considered the expression of increased transcriptional activity (for a discussion, see review by Baserga and Nicolini, 1976). These findings have been repeatedly confirmed, but for several years they have constituted a source of perplexity for the student of cell proliferation. For instance, RNA synthesis in isolated nuclei often increases more than 100% within a few hours after G_0 cells are stimulated to proliferate. Yet, it is hardly believable that the number of genes active in transcription may actually double in proliferating cells (RNA synthesis in isolated nuclei measures essentially the number of growing RNA chains and is not influenced by *rate* of transcription). Indeed, in those few instances in which the number and the diversity of cytoplasmic mRNA were investigated, the differences between growing and resting cells were found to be consistently small (Grady and Campbell, 1975; Getz et al., 1976). If the number of genes activated by the proliferating stimulus is small, it is difficult to explain how we can detect a 100% increase in RNA synthesis (the magnitude of increase is roughly the same for other events indicative of increased transcriptional activity). A way out of this puzzle is offered by the recent demonstration of a markedly increased template activity in nucleoli isolated from cells stimulated to proliferate. The nucleolus contains 3–5% of the total DNA of a mammalian nucleus (Busch and Smetana, 1970; Schmid and Sekeris, 1975), but the only active genes that have been localized in the nucleolus are those for ribosomal RNA, more specifically those for the precursor of 28S and 18S rRNA. Although it has been known for a long time that rRNA synthesis is increased in proliferating cells (Tsukada and Lieberman, 1964; Zardi and Baserga, 1974; Epifanova et al., 1975; Rovera et al., 1975), it is only with the measurement of template activity in isolated nucleoli that it has been possible to accurately determine the magnitude of the increase. The data are summarized in Table II. Although rRNA genes constitute only 0.04% of the total genome, they account for 30–40% of the total nuclear RNA synthesis (Marzluff and Huang, 1975). It is easy to see that a three- to fourfold increase in rRNA synthesis can account

RENATO BASERGA and CO-WORKERS Department of Pathology and Fels Research Institute, Temple University Health Services Center, Philadelphia, Pennsylvania

TABLE I

Changes Occurring in the Nucleus (or Chromatin) of Resting Cells Stimulated to Proliferate

Increased RNA synthesis	Nuclei	Chiu and Baserga, 1975
	Nuclear monolayers	Mauck and Green, 1973
	Chromatin	Barker and Warren, 1966
Increased binding of intercalating dyes (acridine orange, ethidium bromide)	Intact cells	Rigler and Killander, 1969
	Nuclei	Chiu and Baserga, 1975
	Chromatin	Nicolini et al., 1975
Increased molar ellipticity in the 250–300 nm region of CD spectra	Nuclei	Chiu and Baserga, 1975
	Chromatin	Nicolini et al., 1975
Decreased thermal stability of chromatin	Intact cells	Rigler et al., 1969
Increased synthesis of nonhistone chromosomal proteins	Nuclei	Rovera and Baserga, 1971
Increased accumulation of nonhistone chromosomal proteins	Intact cells	Auer and Zetterberg, 1972
	Nuclei	Whelly and Baserga, unpublished data
Accumulation of phosphoproteins	Nuclei	Johnson et al., 1974
Increased phosphokinase activity	Chromatin	Ishida and Ahmed, 1974

All these changes have been described in the early prereplicative phase several hours before the onset of DNA synthesis. The second column gives the methodology used, but it should be understood that all changes are nuclear, even in those described in intact cells.

TABLE II

Increased Template Activity of Nucleoli Isolated from Resting Cells Stimulated to Proliferate

Tissue	Cpm/mg DNA		References
	Nuclei	Nucleoli	
Rat uterus, resting	68	536	Nicolette and Babler, 1974
Estrogen stimulated	127	1,050	
Rat liver, resting	20,200	200,000	Schmid and Sekeris, 1975
Regenerating	36,800	650,000	
TS AF8 hamster cells, resting	98,000	100,000	Rossini and Baserga, unpublished data
Serum stimulated	220,000	430,000	

Conditions for the isolation of nuclei and nucleoli and for the template assay varied in the different references. The times after stimulation were 2 h for the uterus, 12 h for the regenerating liver, and 18 h for the serum-stimulated cells.

for almost all of a twofold increase in total nuclear RNA synthesis. At any rate, it explains why an increase in RNA synthesis is detectable in nuclei of stimulated cells. The increased activity of rRNA genes does not even involve new gene activation, since the synthesis of rRNA can be increased by simply increasing the number of RNA polymerase molecules (RNA polymerase I) per rRNA (Alberghina et al., 1975).

Although the molar ellipticity in CD spectra of the isolated nucleolus also increases in AF8 cells stimulated to proliferate (Baserga et al., 1976) the magnitude of the increase is not sufficient to account completely for the increase in molar ellipticity of nuclei (Huang and Baserga, unpublished data) or chromatin. Yet, it is clear that some of the quantitative changes in transcriptional activity described in G_0 cells stimulated to proliferate can be attributed to the nucleolus. How the nucleolus is turned on and whether its activation is regulated by extranucleolar genes remain to be elucidated.

An Attempt to a Chromosomal Mapping of the Cell Cycle

This brings us to the second point of this paper, i.e., a first timid attempt to assign certain steps of the cell cycle to specific human chromosomes. The use of somatic cell hybridization and chromosome analysis for the localization of human genes on specific chromosomes need not be reviewed here (for a discussion, see Ruddle, 1974). In terms of chromosomes and cell proliferation, the first such study of interest to us is the report by Croce and Koprowski (1974) that human chromosome 7 from SV-40-transformed Lesch-Nyhan fibroblasts (LNSV cells) confers to mouse peritoneal macrophages the ability to grow in nonpermissive conditions. Although human chromosome C7 from LNSV cells contains the SV-40 genome (hardly a component of the normal genome), the important points here are two, i.e., 1) human C7 chromosome (from LNSV cells) contains information that allows mouse peritoneal macrophages to transit from a resting to a growing state; and 2) under these experimental conditions mouse peritoneal macrophages are arrested at a point located 20 h before the onset of DNA synthesis (Virolainen and Defendi, 1967), which is given as G_0 in Fig. 1 (it is really immaterial whether we call it G_0 or G_1, the crucial point is the 20 h).

In another experiment, Ming et al. (1976) found that human chromosome A3 (from LNSV cells) conferred to temperature-sensitive (TS) AF8 hamster fibroblasts the ability to grow at the nonpermissive temperature. TS AF8 cells are a temperature-sensitive cell line derived from BHK cells that grow normally at 34°C but arrest in G_1 at 39.5°C (Burstin et al., 1974). The mid-G_1 block at the nonpermissive temperature has been localized at a point 6–8 h before the onset of DNA synthesis (Kane et al., 1976), clearly different from the block point of mouse peritoneal macrophages (Fig. 1). The hybrid clones from TS AF8 × LNSV are T-antigen negative, but react with anti-human chromatin antibodies (Tsutsui and Baserga, unpublished data). Thus, they contain human information that is presumably different from the information contained in the C7 chromosome and that apparently controls a different step in the cell cycle.

These two are the only direct experiments reported thus far, but with a little bit of good will and imagination, we can gather from the literature at least two other cell-cycle steps for which a chromosome assignment can tentatively be made (Fig. 1). The relevant information for human

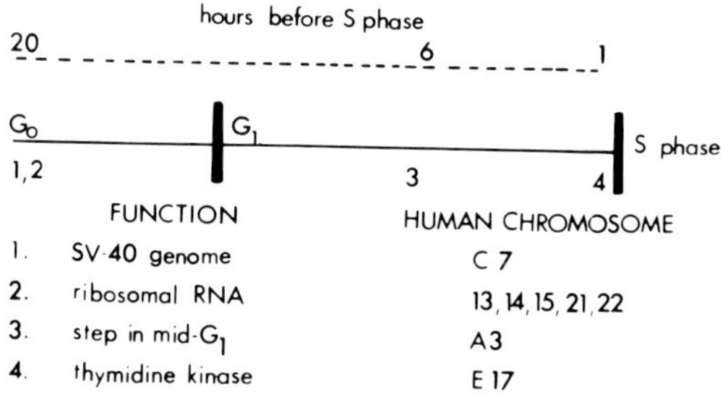

FIGURE 1 Cell cycle steps for which a chromosome assignment can tentatively be made. Explanation in the text.

TABLE III

Assignment of a Late G_1 Step in the Cell Cycle to Human Chromosome E17

	References
TK* activity makes its appearance at the G_1/S boundary	Brent et al., 1965
TK genes are also activated immediately before the onset of DNA synthesis	Littlefield, 1966
TK genes have been localized on the long arm of human chromosome E17	Boone et al., 1972

* Cytosol thymidine kinase.

chromosome E17 is summarized in Table III. The cytosol thymidine kinase (TK) step can be reasonably located at the very end of G_1. Since rRNA synthesis is increased within minutes after serum stimulation (Mauck and Green, 1973), and since the genes for 45S rRNA have been localized on human chromosomes 13, 14, 15, 21, and 22 (Henderson et al., 1972; Evans et al., 1974), we can make in Fig. 1 another assignment for a very early step in the cell cycle.

The outline of Fig. 1 is meant rather as an invitation to pursue the topic in depth than as a statement of facts. It simply indicates that a genetic analysis of the cell cycle is now feasible in mammalian cells with its important implication in terms of the possibilities for a biochemical dissection of the various steps, in much the same way as microbial genetics have opened the way to molecular biology. The availability of a number of cell cycle-specific conditional mutants is, of course, the key to our chromosomal mapping of the cell cycle.

Even at this crude stage, though, it is apparent that different steps of the cell cycle have different degrees of "restriction," to use the terminology introduced by Pardee (1974). Thus, the steps controlled by C7 and A3 are obligatory, and the cell-cycle traverse is blocked if those funcitons are altered or inhibited. Conversely, inhibition of rRNA synthesis does not prevent a cell from entering the S phase (Studzinski and Ellem, 1966; Ringertz and Bolund, 1974), and TK activity is not necessary for the growth of a cell. In this respect, the experiments of Kit and co-workers (1974; Dubbs and Kit, 1976) are illuminating. When TK^- cells ($LMTK^-$ mouse cells, deficient in TK) are fused with chick erythrocytes, both chick and mouse nuclei in heterokaryons are capable of entering the S phase (ordinarily, chick erythrocytes do not synthesize DNA and LTK^- cells cannot enter S in HAT medium). However, when K12 hamster cells (a TS mutant of Chinese hamster cells that is blocked in mid-G_1 at the nonpermissive temperature) are fused with chick erythrocytes at the temperature nonpermissive for K12 cells, neither nuclei become capable of DNA synthesis. These results indicate that a function present in late G_1 (LTK^- cells) can rescue chick erythrocytes from their G_0 state. The same function is absent in cells blocked in mid-G_1 (K12 hamster cells), and chick erythrocytes cannot be stimulated to synthesize DNA at the temperature nonpermissive for K12 cells. This mid-G_1 function, in other words, is more "restrictive" than is the absence of TK activity.

In conclusion, we have attempted to define, both from a biochemical and a genetic point of view, some steps in the cycle of mammalian cells. Although there are many gaps in our knowledge, the fragmentary evidence is sufficient to encourage us in our studies toward a complete molecular and genetic analysis of the cell cycle. This, in turn, ought to throw some light on our understanding of the biochemical control of cell proliferation, in both normal and abnormal conditions.

ACKNOWLEDGMENT

This work was supported by U. S. Public Health Service Research grants CA-12923 and CA-08373 from the National Cancer Institute, GM-22359 from the National Institute of General Medical Science, and HD-06323 from the National Institute of Child Health and Human Development. Dr. Huang is a recipient of a Leukemia Society Special Fellowship.

REFERENCES

ALBERGHINA, F. A. M., E. STURANI, and J. R. GOHLKE. 1975. Levels and rates of synthesis of ribosomal ribonucleic acid, transfer ribonucleic acid and protein in *Neurospora crassa* in different steady states of growth. *J. Biol. Chem.* **250:**4381–4388.

AUER, G., and A. ZETTERBERG. 1972. The role of nuclear proteins in RNA synthesis. *Exp. Cell Res.* **75:**245–253.

BARKER, K. L., and J. C. WARREN. 1966. Template capacity of uterine chromatin: control by estradiol. *Proc. Natl. Acad. Sci. U. S. A.* **56:**1298–1302.

BASERGA, R., C. H. HUANG, M. ROSSINI, H. CHANG, and P. M. L. MING. 1976. The role of nuclei and nucleoli in the control of cell proliferation. *Cancer Res.* 36:4297–4300.

BASERGA, R., and C. NICOLINI. 1976. Chromatin structure and function in proliferating cells. *Biochim. Biophys. Acta.* **458:**109–134.

BOONE, C., T. R. CHEN, and F. H. RUDDLE. 1972. Assignment of three human genes to chromosomes (LDH-A to 11, TK to 17, and IDH to 20) and evidence for translocation between human and mouse chromosomes in somatic cell hybrids. *Proc. Natl. Acad. Sci. U. S. A.* **69:**510–514.

BRENT, T. P., J. A. V. BUTLER, and A. R. CRATHORN. 1965. Variations in phosphokinase activities during the cell cycle in synchronous populations of HeLa cells. *Nature (Lond.).* **207:**176–177.

BURSTIN, S. J., H. K. MEISS, and C. BASILICO. 1974. A temperature-sensitive cell cycle mutant of the BHK cell line. *J. Cell. Physiol.* **84:**397–408.

BUSCH, H., and K. SMETANA. 1970. The Nucleolus. Academic Press, Inc., New York.

CHIU, N., and R. BASERGA. 1975. Changes in template activity and structure of nuclei from WI-38 cells in the pre-replicative phase. *Biochemistry.* **14:**3126–3132.

CROCE, C. M., and H. KOPROWSKI. 1974. Somatic cell hybrids between mouse peritoneal macrophages and SV-40 transformed human cells. *J. Exp. Med.* **140:**1221–1229.

DUBBS, D. R., and S. KIT. 1976. Reactivation of chick erythrocyte nuclei in heterokaryons with temperature-sensitive Chinese hamster cells. *Somatic Cell Genet.* **2:**11–19.

EPIFANOVA, O. I., M. K. ABULADZE, and A. I. ZOSIMOVSKA. 1975. Effects of low concentrations of actinomycin D on the initiation of DNA synthesis in rapidly proliferating and stimulated cell cultures. *Exp. Cell Res.* **92:**25–30.

EVANS, H. J., R. A. BUCKLAND, and M. L. PARDUE. 1974. Location of the genes coding for 18S and 28S ribosomal RNA in the human genome. *Chromosoma (Berl.).* **48:**405–426.

GETZ, M. J., P. K. ELDER, E. W. BENZ, JR., R. E. STEPHENS, and H. L. MOSES. 1976. Effect of cell proliferation on levels and diversity of poly (A)-containing mRNA. *Cell.* **7:**255–265.

GRADY, L. J., and W. P. CAMPBELL. 1975. Non-repetitive DNA transcripts in nuclei and polysomes of polyoma-transformed and non-transformed mouse cells. *Nature (Lond.).* **254:**356–358.

HENDERSON, A. S., D. WARBURTON, and K. C. ATWOOD. 1972. Location of ribosomal DNA in the human chromosome complement. *Proc. Natl. Acad. Sci. U. S. A.* **69:**3394–3398.

ISHIDA, H., and K. AHMED. 1974. Studies on chromatin-associated protein phosphokinase of submandibular gland from isoproterenol-treated rats. *Exp. Cell Res.* **84:**127–136.

JOHNSON, E. M., J. KARN, and V. G. ALLFREY. 1974. Early nuclear events in the induction of lymphocyte proliferation by mitogens. *J. Biol. Chem.* **249:**4990–4999.

KANE, A., C. BASILICO, and R. BASERGA. 1976. Transcriptional activity and chromatin structural changes in a temperature-sensitive mutant of BHK cells blocked in early G_1. *Exp. Cell Res.* **99**:165–173.

KIT, S., W. C. LEUNG, G. JORGENSEN, D. TRKULA, and D. R. DUBBS. 1974. Acquisition of chick cytosol thymidine kinase activity by thymidine kinase-deficient mouse fibroblast cells after fusion with chick erythrocytes. *J. Cell Biol.* **63**:505–514.

LITTLEFIELD, J. W. 1966. The periodic synthesis of thymidine kinase in mouse fibroblasts. *Biochim. Biophys. Acta.* **114**:398–403.

MARZLUFF, W. F., JR., and R. C. C. HUANG. 1975. Chromatin directed transcription of 5S and tRNA genes. *Proc. Natl. Acad. Sci. U. S. A.* **72**:1082–1086.

MAUCK, J. C., and H. GREEN. 1973. Regulation of RNA synthesis in fibroblasts during transition from resting to growing state. *Proc. Natl. Acad. Sci. U. S. A.* **70**:2819–2822.

MING, P. M. L., H. L. CHANG, and R. BASERGA. 1976. Release by human chromosome 3 of the G_1 block in hybrids between tsAF8 hamster and human cells. *Proc. Natl. Acad. Sci. U. S. A.* **73**:2052–2055.

NICOLETTE, J. A., and M. BABLER. 1974. The selective inhibitory effect of NH_4Cl on estrogen-stimulated rat uterine in vitro RNA synthesis. *Arch. Biochem. Biophys.* **163**:656–665.

NICOLINI, C., S. NG, and R. BASERGA. 1975. Effect of chromosomal proteins extractable with low concentrations of NaCl on chromatin structure of resting and proliferating cells. *Proc. Natl. Acad. Sci. U. S. A.* **72**:2361–2365.

PARDEE, A. B. 1974. A restriction point for control of normal animal cell proliferation. *Proc. Natl. Acad. Sci. U. S. A.* **71**:1286–1290.

RIGLER, R., and D. KILLANDER. 1969. Activation of deoxyribonucleoprotein in human leucocytes stimulated by phytohemagglutinin. *Exp. Cell Res.* **54**:171–180.

RIGLER, R., D. KILLANDER, L. BOLUND, and N. R. RINGERTZ. 1969. Cytochemical characterization of deoxyribonucleoprotein in individual nuclei. *Exp. Cell Res.* **55**:215–224.

RINGERTZ, N. R., and L. BOLUND. 1974. Reactivation of chick erythrocyte nuclei by somatic cell hybridization. *Int. Rev. Exp. Pathol.* **13**:83–116.

ROVERA, G., and R. BASERGA. 1971. Early changes in the synthesis of acidic nuclear proteins in human diploid fibroblasts stimulated to synthesize DNA by changing the medium. *J. Cell Physiol.* **77**:201–212.

ROVERA, G., S. MEHTA, and G. MAUL. 1975. Ghost monolayers in the study of the modulation of transcription in cultures of CV_1 fibroblasts. *Exp. Cell Res.* **89**:295–305.

RUDDLE, F. H. 1974. Human genetic linkage and gene mapping by somatic cell genetics. *In* Somatic Cell Hybridization. R. L. Davidson and F. de la Cruz, editors. Raven Press, New York. 1–12.

SCHMID, W., and C. E. SEKERIS. 1975. Nucleolar RNA synthesis in the liver of partially hepatectomized and cortisol-treated rats. *Biochim. Biophys. Acta.* **402**:244–252.

STUDZINSKI, G. P., and K. A. O. ELLEM. 1966. Relationship between RNA synthesis, cell division, and morphology of mammalian cells. *J. Cell Biol.* **29**:411–421.

TSUKADA, K., and I. LIEBERMAN. 1964. Metabolism of nucleolar ribonucleic acid after partial hepatectomy. *J. Biol. Chem.* **239**:1564–1568.

VIROLAINEN, M., and V. DEFENDI. 1967. Dependence of macrophage growth in vitro upon interaction with other cell types. *Wistar Inst. Symp. Monogr.* **7**:67–83.

ZARDI, L., and R. BASERGA. 1974. Ribosomal RNA synthesis in WI-38 cells stimulated to proliferate. *Exp. Mol. Pathol.* **20**:69–77.

STUDIES ON *KAR1*, A GENE REQUIRED FOR NUCLEAR FUSION IN YEAST

GERALD R. FINK and JAIME CONDE

When vegetative haploid cells of yeast are mixed together under conditions that allow cell-to-cell contact, efficient fusion occurs between pairs of cells of opposite mating type. In standard laboratory strains of *Saccharomyces cerevisiae,* nuclear fusion follows immediately after cell fusion, with no intervening cell or nuclear division. Once the nuclei have fused, the zygote buds off diploid cells by successive mitotic divisions (Fig. 1). Nuclear fusion normally occurs at a high incidence, although the spontaneous failure of nuclear fusion has been reported (Fowell, 1951; Wright and Lederberg, 1957; Zakharov et al., 1969).

We have described a mutation (*kar1-1*) that causes a defect in nuclear fusion during conjugation (Conde and Fink, 1976). The failure of nuclear fusion leads to the formation of several novel cell types as exconjugants. Zygotes from a *kar1-1* x wild-type cross form diploids at a low frequency. The majority of the exconjugants are "heteroplasmons," strains which contain the cytoplasmic components of both parents and the nuclear genotype of only one parent. In addition, heterokaryons containing the unfused parental nuclei segregate off from the zygote and continue to propagate under appropriate conditions.

The production of heteroplasmons and heterokaryons by strains carrying *kar1-1* permits new approaches to problems of cell biology in yeast. In this report, we show that the *kar1⁻* defect is corrected only when the *KAR1⁺* gene product is in the same nucleus. The dominance of *kar1⁻* in crosses where the other nucleus is *KAR1⁺* has allowed us to investigate the transmission of two well known non-Mendelian traits, killer and [PSI⁺]. The question we have asked is: do these non-Mendelian traits show cytoplasmic inheritance?

GERALD R. FINK Department of Genetics, Development, and Physiology, Cornell University, Ithaca, New York

JAIME CONDE Departamento de Genetica, Universidad de Sevilla, Sevilla, Spain

MATERIALS AND METHODS

Strains and Media

The yeast strains used and their genotypic designations have been described in a previous publication (Conde and Fink, 1976). The designation *kar1-1* is for the mutation which prevents nuclear fusion, and *KAR1⁺* is the normal wild-type allele. [RHO⁺] stands for the presence of the cytoplasmic determinant for respiratory ability. The [RHO⁻] strains lack mitochondrial function and are unable to grow on such nonfermentable carbon sources as glycerol and ethanol. [PSI⁺] stands for the presence of the non-Mendelian determinant which enhances nonsense suppression (Cox, 1965). The [PSI⁺] phenotype can be seen when a strain has an inefficient ochre nonsense suppressor such as *SUQ5* and several ochre alleles. Because of the inefficiency of the suppressor in [PSI⁻] backgrounds, the ochre suppressor fails to suppress the ochre alleles, and the strain manifests the auxotrophic requirements caused by the nonsense mutations. The introduction of the [PSI⁺] determinant into this strain increases the efficiency of the suppressor and the strain no longer shows the auxotrophic requirements. [KIL-K] stands for the non-Mendelian determinant that confers the ability of a yeast strain to kill a sensitive strain. Strains harboring the [KIL-K] determinant form a clear zone of killing on a lawn of a sensitive [KIL-o] indicator strain (Fink and Styles, 1972).

The composition of complete medium (YEPD) and synthetic medium (Difco YNB-amino acids, Difco Laboratories, Detroit, Mich.) have been described (Fink, 1970).

Cytology

The techniques for sampling, fixing, and staining the cells were described in Conde and Fink (1976).

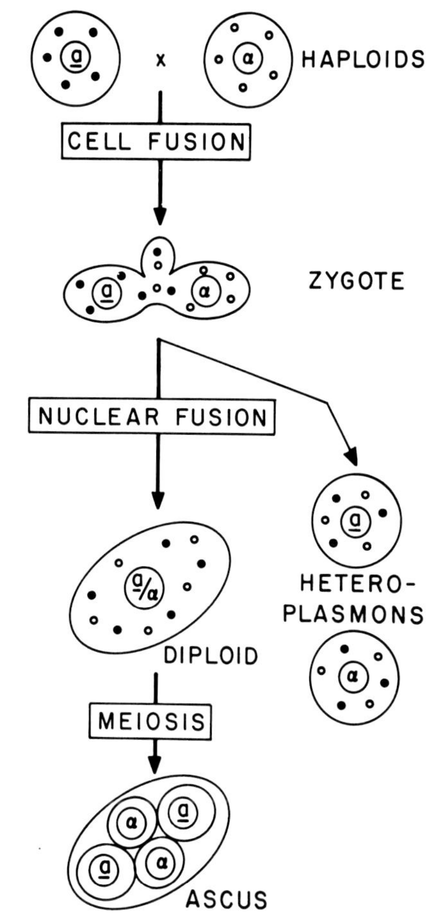

FIGURE 1 The sexual cycle in *Saccharomyces cerevisiae*. a and α are the two alleles of the mating type locus. The large circles represent cells and the small ones which encircle the mating type genotypes represent nuclei. The closed and open circles represent mitochondria from one or the other of the two haploid parents. The arrow going from the zygote to the heteroplasmons is thinner as an indication of the low frequency of this event in wild type.

RESULTS

$KAR1^+$ *is a Nuclear-Limited Trait*

In crosses of haploid strains a *kar1-1* x α $KAR1^+$ the nuclei fail to fuse. This aberration yields heteroplasmons and heterokaryons, but very few diploids (Figs. 2 and 3). This result means that when $KAR1^+$ and $kar1^-$ are in separate nuclei, the $KAR1^+$ gene fails to supply the function missing in the $kar1^-$ nucleus. Diploids monosomic for chromosome III (2n-1) and of the following composition, a $KAR1^+/kar1^-$ and α $KAR1^+/kar1^-$, were constructed. These diploids were mated, and the resulting zygotes were micromanipulated. All of the 22 zygotic clones were homogeneous nonmaters, that is to say, tetraploids resulting from the fusion of the two nuclei. A comparison of this result with the outcome of a $KAR1^+$ x $kar1^-$ cross shown in Tables I and II emphasizes the dramatic effect of the *kar1-1* mutation. These experiments show that *kar1-1* is recessive to $KAR1^+$ when both are in the same nucleus, and dominant to $KAR1^+$ when the alleles are in separate nuclei.

Killer is Transmitted Through the Cytoplasm

Crosses were made between a number of killer

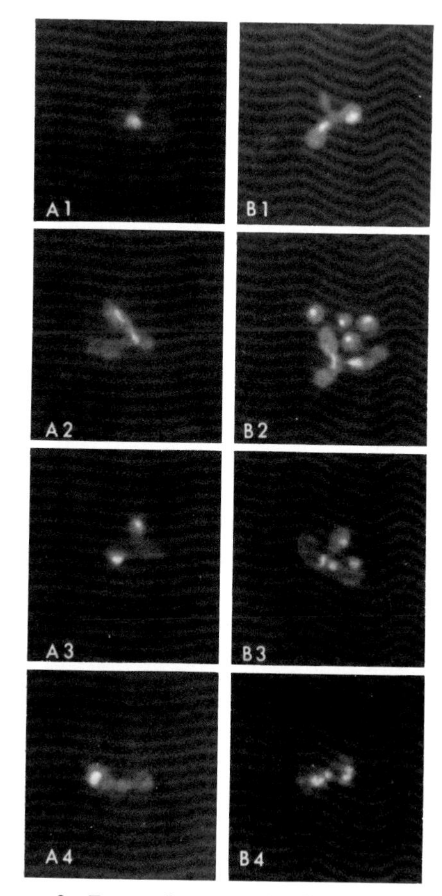

FIGURE 2 Zygotes from KAR^+ x KAR^+ (A1 to A4) and *kar1-1* x $KAR1^+$ (B1 to B4) crosses. KAR^+ x KAR^+ and *kar1-1* x $KAR1^+$ mating mixtures were incubated at 30°C for different periods of time, and stained with mithramycin according to the procedure described in Slater (1976). A1–A3 and B1–B3 were sampled after 5 h of mating; A4 and B4, after 15 h. × 1,600.

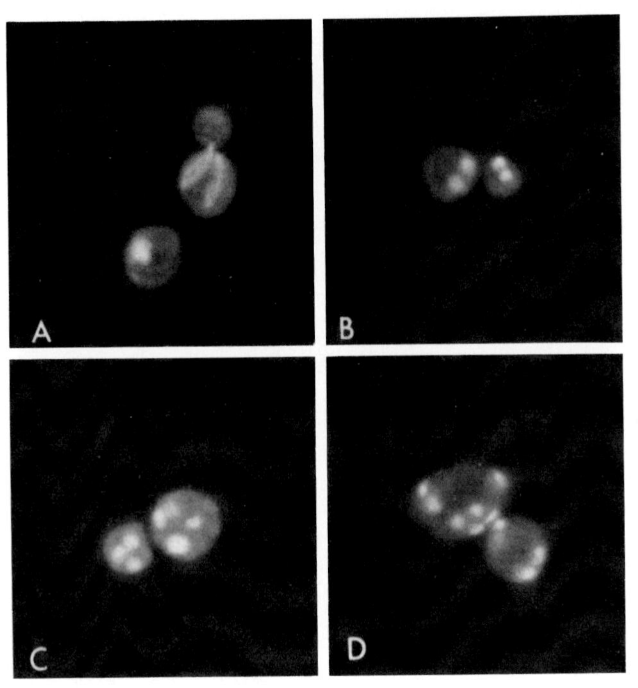

FIGURE 3 *kar1-1* x *kar1-1* heterokaryons stained with mithramycin. In A, two nuclei are dividing synchronously. × 1,600.

TABLE I

Results of the Cross X5P29 α kar1-1 his4-15 ade2-1 can1^R [KIL-K] x 5082-2A a̲ CAN1^S [KIL-o]

Pure heteroplasmons		Pure synkaryons*	Mixed clones			
X5P29	*5082-2A*	*a/α*	X5P29 + *5082-2A*	X5P29 + *synkaryons*	5082-2A + *synkaryons*	5082-2A + X5P29 + *synkaryons*
10	19	0	2	2	3	2

Zygotes from the mating mixture were manipulated and the resulting zygotic clones analyzed for nuclear and cytoplasmic markers as described in the text. We were unable to classify two of the clones.

* Cells with both nuclear contributions; heterokaryons and diploids would be classified as synkaryons.

KAR1$^+$ strains and sensitive *kar1*$^-$ strains. A typical cross was:

X5P29 a̲ *kar1-1 his4-15* ade2-1 can1^R [KIL-K] × 5082-2A a̲ CANI^S [KIL-o].

(CAN1^R, canavanine resistant; CAN^S, canavanine sensitive.)

Fifty zygotes were micromanipulated and 40 of these formed viable zygotic colonies. The composition of each of these clones was analyzed for its nuclear and cytoplasmic genotype (Table I). The killer phenotype of the 24 5082-2A heteroplasmons (5 of them came from mixed zygotic clones) was determined. Of these, 22 were killers, 1 was an unstable killer which segregated many nonkillers during vegetative growth, and 1 was a nonkiller. All of the X5P29 heteroplasmons and all synkaryons were good killers.

Killer strains of yeast contain two double-stranded RNAs, a high molecular-weight species called "L," and a lower molecular-weight species called "M." The experiment on killer could be phrased in the following way: in KAR$^+$ x kar$^-$

TABLE II

A Cross of X25P5a a kar1-1 ade2-1 leu1 thr1 can^R[RHO$^+$][PSI$^+$] X F401 α kar1$^+$ ade2-1 leu1-12 lys1-1 trp5-48 can1-100 SUQ5 [RHO$^-$][PSI$^-$]

Zygotic clone	X25P5a heteroplasmons	F401 heteroplasmons				Synkaryons
		RHO$^+$ PSI$^+$	*RHO$^+$ PSI$^-$*	*RHO$^-$ PSI$^+$*	*RHO$^-$ PSI$^-$*	
1	0*	2	0	0	30	0
2	0*	0	0	0	32	0
3	32	0	0	0	0	0
4	16	15	0	0	0	1
5	3	29	0	0	0	0
6	0*	0*	0	0	32	0
7	0*	2	0	0	30	0
8	0	32	0	0	0	0
9	0	32	0	0	0	0
10	0	32	0	0	0	0
11	0	32	0	0	0	0
12	0	0	0	0	32	0
13	0	0	0	0	32	0
14	0	1	0	0	31	0
15	0*	1	0	0	31	0
16	18	13	0	0	0	1
17	0	31	0	1	0	0
18	0	31	0	1	0	0
19	22	6	0	0	0	4
20	0	31	0	0	1	0
21	0	32	0	0	0	0

The zygotes from this cross were micromanipulated. After they grew into colonies, the zygotic clones were resuspended and plated. Thirty-two colonies from each clone were picked and analyzed. In addition the entire clone was analyzed to determine if any cells of a particular genotype were present.

* Indicates that cells of that genotype were present in the clone even though they were not present among the 32 isolated colonies. The two [RHO$^-$] [PSI$^+$] strains could have resulted from the spontaneous loss of the [RHO$^+$] factor.

crosses, are both double-stranded RNAs characteristic of killer strains transmitted cytoplasmically? Our previous examination of killer transmission suffered from the problem that the nonkiller strains used in the experiment still contain the L double-stranded RNA (Sweeney et al., 1976). Recently, we have obtained nonkiller strains from Dr. Michael Vodkin (University of South Carolina) that have no detectible L or M double-stranded RNA. Crosses of KAR$^+$ killer strains to these strains were used to study the transmission of the double-stranded RNAs. A strain with double-stranded RNA was crossed to one lacking double-stranded RNA:

JM9 a *ade2-1 lys1-1 his4-385 met8-1 CAN1^S* SUP4-3 [KIL-o] [RHO$^-$] (no double-stranded RNA);

X5P29 a *kar1-1 his4-15 ade2-1 can1^R* [KIL-k] [RHO$^+$] (L, and M double-stranded RNA).

Zygotes from this cross were micromanipulated and the composition of the resulting zygotic clones analyzed. Of the 17 zygotic clones analyzed, 6 were JM9 heteroplasmons, 2 were X5P29 heteroplasmons, and 9 were a mixture of JM9 and X5P29 heteroplasmons. The killer trait was transmitted at high frequency. All but 1 of the JM9 heteroplasmons were killers. The nonkiller JM9 heteroplasmon was also [RHO$^-$], and therefore received neither cytoplasmic element. The double-stranded RNA composition of several of the JM9 heteroplasmons which segregated from the zygotes was analyzed by agarose gel electrophoresis. As can be seen in Table III the analysis of the double-stranded RNA in crosses of killer x nonkiller (no double-stranded RNA) shows that both the L and the M double-stranded RNAs are transmitted to the nonkiller strains through the cytoplasm. The single nonkiller heteroplasmon is interesting because it emerged from the zygote without either of the double-stranded RNAs.

Table III
Double-Stranded RNA Composition of Heteroplasmons

Strains	Killer phenotype	Double-stranded RNA L	M
Parents			
JM9	nonkiller	–	–
X5P29	killer	+	+
Heteroplasmons (JM9 × X5P29)			
1. JM9	killer	+	+
2. JM9	killer	+	+
3. JM9	killer	+	+
4. JM9	killer	+	+
5. JM9	killer	+	+
6. X5P29	killer	+	+
7. X5P29	killer	+	+
8. X5P29	killer	+	+

The double-stranded RNA composition of heteroplasmons arising from crosses of JM9 (sensitive strains without L or M double-stranded RNA) x X5P29 (a killer strain with both L and M). Agarose gel electrophoresis was carried out as described by Sweeney, Tate, and Fink (1976). L stands for the large (2.4×10^6 daltons) and M for the small (1.1×10^6) double-stranded RNA.

PSI Factor is Transmitted Through the Cytoplasm

The transmission of PSI factor, another non-Mendelian trait, was also analyzed by *KAR*$^+$ x *kar*$^-$ crosses. One of the crosses and the results of its analysis are shown in Table II. Transmission of [PSI$^+$] could be followed because all of the markers harbored by F401, with the exception of *leu1-12*, are ochre mutations suppressible by *SUQ5* in the presence of [PSI$^+$]. Twenty-six zygotes were micromanipulated from the cross, 21 of which gave viable zygotic clones. Each zygotic clone was resuspended in water and plated on complete medium. From each clone, 32 isolated colonies were picked and their genotypes determined. In 17 of the 21 clones [PSI$^+$] was transmitted cytoplasmically to F401 heteroplasmons. None of the zygotes gave rise to [RHO$^+$] [PSI$^-$] heteroplasmons. These data suggest that [PSI$^+$] is a cytoplasmic element.

Not all of the segregants from the zygotes received the [PSI$^+$] element. Nine of the clones showed some [PSI$^-$] heteroplasmons. Six of the clones containing F401 [PSI$^-$] heteroplasmons also contained F401 [PSI$^+$] [RHO$^+$] heteroplasmons, indicating that cell fusion did occur. Segregation of [PSI$^-$] diploid clones is rare in KAR$^+$ x KAR$^+$ crosses (Young and Cox, 1972; Conde and Fink, unpublished). For this reason, the appearance of [PSI$^-$] heteroplasmons is likely to result from events specific to *KAR1*$^+$ x *kar1*$^-$ crosses.

DISCUSSION

The dominance of *kar1-1* in diploids has allowed us to show that both the killer trait and [PSI$^+$] are cytoplasmic genetic systems. Both of these non-Mendelian traits are transmitted to diploids at high frequencies in *KAR*$^+$ x *KAR*$^+$ crosses. In certain KAR1$^+$ x *kar1-1* crosses, [PSI$^-$] and [KIL-o] segregants appear. The appearance of cells without these cytoplasmic elements suggests that *KAR1*$^+$, *kar1-1* zygotes may bud off cells with haploid nuclei before cytoplasmic mixing. If these cytoplasmic elements were in a compartment, then haploid buds emerging from the zygote might not receive the element. The mitochondrial [RHO$^+$] element was used as a general cytoplasmic marker. If [PSI$^+$] and [KIL-K] were independent, compartmentalized cytoplasmic replicons, then some of the [RHO$^+$] heteroplasmons might segregate without receiving the [PSI$^+$] or [KIL-K] elements. None of the heteroplasmons was [RHO$^+$] [KIL-o] or [RHO$^+$] [PSI$^-$]. The absence of these heteroplasmons suggests that the [PSI$^-$] and [KIL-o] cells are early segregants which emerge before complete cytoplasmic mixing. We have noticed that the frequency with which the [PSI$^-$] and [KIL-o] segregants occur is quite strain-dependent.

The defect in *kar1-1* nuclei appears to be nuclear-limited. It is possible to imagine a number of different mechanisms that would result in the inability of *KAR1*$^+$ from one nucleus to correct the defect in the *kar1-1* nucleus. One explanation suggests that the *KAR1*$^+$ protein is synthesized in a compartment associated with the nucleus from which its message originated. This compartment model would require that the *KAR1*$^+$ gene product remain associated with only that nucleus and not be freely diffusible. An alternative model is the "catena" model. This model says that, once the chain of developmental events leading to a functional nucleus is over, the presence of the *KAR1*$^+$ gene product is unable to rectify the *kar1-1* defect. It is easy to imagine that some structure associated with the nuclear membrane is defective in the fusion process. Much of our knowledge concerning the nuclear structures associated with mating comes from the work of Byers and Goetsch (1975). From serial sections of mating yeast cells, they have reconstructed a model of the nuclear events that occur before and during conjugation. In the presence of mating hormones, yeast cells arrest at G1 of the cell cycle. The nuclei of both a and α cells arrested in G1 contain satellite-bearing single plaques. Cytoplasmic microtubules can be seen attached to the surface of the half-bridge, a structure adjacent to the plaque. Byers suggests that these cytoplasmic microtubules have some role in the interactions between nuclei because they interconnect the respective plaques of the two nuclei during certain stages of conjugation. The KAR1$^+$ gene could be involved in some aspect of synthesis, assembly, or attachment of these cytoplasmic microtubules. Since most of the visible structures involved in nuclear fusion, including the extranuclear microtubules, are present before cellular fusion, the *KAR1*$^+$ gene product may be built into a structure before the two nuclei are present in the same cytoplasm. The catena model is not sufficient to explain all the features of *KAR1*$^+$ x *kar1-1* crosses. For example, *KAR*$^+$, *kar1-1* zygotes continue to bud haploid heteroplasmons long after the first zygotic bud. Therefore, some additional restriction is necessary to account for the fact that the failure to fuse is not a transient phenomenon. Perhaps the *KAR1*$^+$ gene functions before cell fusion but is inactive after this event.

ACKNOWLEDGMENT

This work was supported by National Science Foundation grant PCM 76-11667.

REFERENCES

BYERS, B., and L. GOETSCH. 1975. Behavior of spindles and spindle plaques in the cell cycle and conjugation of *S. cerevisiae*. *J. Bacteriol.* **124:**511–523.

CONDE, J., and G. R. FINK. 1976. A mutant of *Saccharomyces cerevisiae* defective for nuclear fusion. *Proc. Natl. Acad. Sci. U. S. A.* **73:**3651–3655.

COX, B. S. 1965. PSI, a cytoplasmic suppressor of super-suppressor in yeast. *Heredity.* **20:**505–521.

FINK, G. R. 1970. Biochemical genetics of yeast. *Methods Enzymol.* **17A:**59–78.

FINK, G. R., and C. STYLES. 1972. Curing of a killer factor in *Saccharomyces cerevisiae*. *Proc. Natl. Acad. Sci. U. S.A.* **69:**2846–2849.

FOWELL, R. F. 1951. Hybridization of yeast by Lindegren's technique. *J. Inst. Brew.* **180:**180.

SLATER, M. L. 1976. A rapid staining method for *Saccharomyces cerevisiae*. *J. Bacteriol.* **126:**1336–1341.

SWEENEY, T. K., A. TATE, and G. R. FINK. 1976. A study of the transmission and structure of double stranded RNAs associated with the killer phenomenon in *Saccharomyces cerevisiae*. *Genetics.* **84:**27–42.

WRIGHT, R. E., and J. LEDERBERG. 1957. Extranuclear inheritance in yeast heterokaryons. *Proc. Natl. Acad. Sci. U. S. A.* **43:**919–923.

YOUNG, C. S. H., and B. S. COX. 1972. Extrachromosomal elements in a super suppression system in yeast. II. Relations with other extrachromosomal elements. *Heredity.* **28:**189–199.

ZAKHAROV, J. A., L. V. YURCHENKO, and B. F. YAROVOT. 1969. Cytoduction: the autonomous transfer of cytoplasmic hereditary factors during pairing of yeast cells. *Genetika.* **5:**136–141.

HISTONE PHOSPHORYLATION RELATED TO CHROMATIN STRUCTURE

L. R. GURLEY, R. A. WALTERS, C. E. HILDEBRAND, P. G. HOHMANN, S. S. BARHAM, L. L. DEAVEN, and R. A. TOBEY

It has been proposed that the control of DNA activities (such as gene expression, gene replication, and gene segregation) may be mediated by changes in chromatin structure caused by modifications of the chromatin proteins (see review by Elgin and Weintraub, 1975). If this proposal is true, one might expect to find correlations between histone modifications and variations in chromatin structure. Using cultured mammalian cells, we have searched for such correlations and have found that, in general, growth-related histone phosphorylation is associated with chromatin condensation. This report summarizes the experimental results that have led us to this conclusion.

MATERIALS AND METHODS

Cell Cultures and Synchronization

Chinese hamster cells (line CHO) were cultured in either suspension or monolayer using F-10 medium supplemented by 10% calf and 5% fetal calf sera, as previously described (Tobey et al., 1966). Cells in suspension culture were synchronized in early G_1-arrest by isoleucine deprivation (Tobey and Ley, 1971). Resumption of synchronous cell-cycle traverse was accomplished by addition of isoleucine.

Resynchronization of cells near the G_1/S boundary was accomplished by treating G_1 synchronized cells with hydroxyurea for 10 h while they were traversing G_1 (Tobey and Crissman, 1972). Resumption of synchronous cell-cycle traverse was then accomplished by resuspending the cells in fresh growth medium without hydroxyurea.

To obtain synchronized cells in mitosis, either exponential monolayer cultures or synchronized suspension cultures were treated with Colcemid (Ciba Pharmaceutical Company, Summit, N. J.; Gurley et al., 1975).

L. R. GURLEY and CO-WORKERS Cellular and Molecular Biology Group, Los Alamos Scientific Laboratory, University of California, Los Alamos, New Mexico

Synchronized metaphase cells were then obtained by the mitotic selection of cells from monolayer cultures, as previously described (Tobey et al., 1967*a*).

Peromyscus cells were obtained from ear clippings of two closely related strains of mice: *Peromyscus crinitus* and *Peromyscus eremicus* (Pathak et al., 1973). These two cell lines, which differ greatly in their heterochromatin content, were grown exponentially in monolayer culture using F-10 medium supplemented with 20% fetal calf serum (Gurley et al., Heterochromatin and histone phosphorylation, submitted for publication).

Measurement of Cell Growth and Cell-Cycle Parameters

The progress of cell division in a suspension culture was monitored by measuring the cell concentration with an electronic particle counter (Tobey et al., 1967*b*). The number of cells in S phase was measured by autoradiography in cells labeled with [^{3}H]thymidine, as previously described (Tobey and Ley, 1970). The cell-cycle distribution of cells in a given population was determined by flow microfluorometry (Crissman and Tobey, 1974).

Growth kinetics of *Peromyscus* cells were determined from a series of small replicate monolayer cultures; at varying times during the experiment, monodispersed suspensions were prepared from one of these cultures (Deaven and Petersen, 1974), and the concentration of cells was determined with an electronic particle counter. To determine the relative G_1 DNA content of *P. crinitus* and *P. eremicus* cells, monolayer cultures were accumulated in G_1 by growth to confluency. Monocellular suspensions of these cells were mixed in equal proportions, and their relative DNA contents were then determined by flow microfluorometry (Crissman and Tobey, 1974).

Cytology and Electron Microscopy

Constitutive heterochromatin in chromosomes and interphase nuclei was visualized in cells using the C-band Giemsa method (Deaven and Petersen, 1974) which specifically stains constitutive heterochromatin dark compared to the remaining chromatin (Hsu, 1973).

Cells for electron microscopy were fixed in phosphate-buffered glutaraldehyde and postfixed in osmium tetroxide (Millonig, 1962). The cells were then prestained with aqueous uranyl acetate, dehydrated, infiltrated, and embedded in epoxy resin (Spurr, 1969). Thin sections were cut and poststained with lead citrate and examined with a Phillips EM 200 transmission electron microscope.

Histone Phosphorylation, Isolation, Purification, and Fractionation

Cellular histones were labeled by exposing cultures to 50 μCi [^{3}H]lysine per liter for three generations before and during cell synchronization. During exponential growth or at various times in the cell cycle, the cultures were pulse-labeled with $^{32}PO_4$ for 1 or 2 h (as indicated) to label their phosphoproteins. Histones were isolated as previously described (Gurley et al., 1974*a*, 1975), then fractionated and purified on a preparative polyacrylamide gel electrophoresis column (Gurley and Walters, 1971). Histone phosphorylation rate was measured from the incorporation of $^{32}PO_4$ into the [^{3}H]lysine-labeled histone fractions eluted from the electrophoresis column as illustrated in Fig. 1.

The number of phosphates incorporated into histone H1 during the cell cycle was determined by analytical electrophoresis on 25-cm urea/acetic acid polyacrylamide gels (Hohmann et al., 1976). The regional location of the phosphates within the H1 molecule and the amino acids phosphorylated was determined as previously described by Hohmann et al. (1975, 1976).

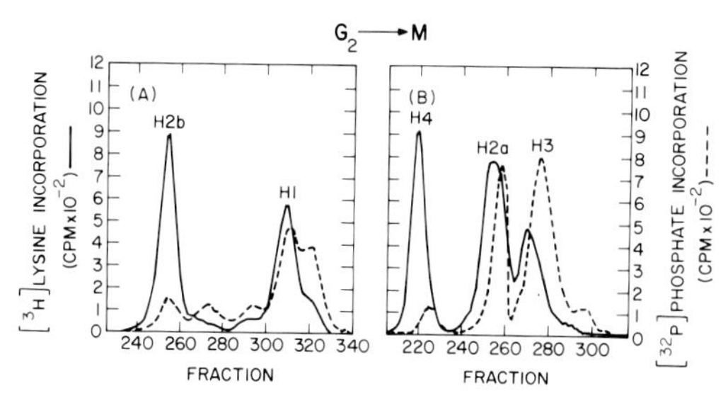

FIGURE 1 Preparative electrophoresis of phosphorylated histones isolated from CHO cells. Exponential monolayer cultures labeled for 52 h with [^{3}H]lysine were treated with Colcemid and $^{32}PO_4$ for 2 h, as previously described (Gurley et al., 1974*a*). Metaphase cells were then removed from the monolayer by mitotic selection. This collection contained only cells labeled with $^{32}PO_4$ during the transition from interphase (G_2) to mitosis (M). Electrophoretic migration proceeds from right to left. (A) Electrophoresis of histones H1 and H2b. (B) Electrophoresis of histones H2a, H3, and H4. Individual histone fractions are indicated by the [^{3}H]lysine incorporation (—). Phosphorylation of each histone is indicated by the incorporation of $^{32}PO_4$ (-).

Heparin Probe of Chromatin Structure

The resistance of chromatin structure to the heparin-mediated release of DNA from the deoxyribonucleoprotein complex in nuclei was measured as previously described by Hildebrand and Tobey (1975). Briefly, nuclei isolated from [^{14}C]thymidine labeled synchronized cells were treated with various concentrations of heparin at 2°C which removed histones from the chromatin and released DNA from the nuclei (Hildebrand and Tobey, 1975). The [^{14}C]DNA remaining in the supernatant solution after centrifugation constituted the released DNA. The concentration of heparin required to release 50% of the dissociatable DNA from the nucleus was measured at different stages of the cell cycle. Cell-cycle changes in the resistance of chromatin structure to heparin-mediated release of DNA were presented as [heparin]/[heparin]$_0$, which is the ratio of heparin concentration at a given time in the cell cycle, [heparin] to heparin concentration at the beginning of the cell cycle in early G_1, [heparin]$_0$.

RESULTS

Histone Phosphorylation in the Cell Cycle

The histone phosphorylation events associated with cell proliferation were identified in synchronized cultures of line CHO Chinese hamster cells by preparative electrophoresis of $^{32}PO_4$-labeled histones in the manner illustrated in Fig. 1. Phosphorylation events were found to occur on three different types of histones: fractions H1, H2a, and H3. (We use the new international nomenclature for histone fractions proposed at the CIBA Foundation Symposium in London [Bradbury, 1975]).

By this method, the cell-cycle dependency of histone phosphorylation was determined by measuring histone phosphorylation in each phase of the cell cycle. The results of these studies are summarized in Fig. 2. The phosphorylation of histone H2a was found to be constitutive with respect to the cell cycle, occurring in G_1-arrested CHO cells (Gurley et al., 1973*a*), as well as in all phases of the cell cycle of proliferating cells (Gurley et al., 1973*a*, *b*, 1974*a*, *b*, 1975). Thus, studies on synchronized CHO cells have not been very successful in providing insight into the function of this histone phosphorylation.

In contrast, the phosphorylation of histone H1 was found to be cell-cycle-dependent and very complex, consisting of three different phosphorylation events, each of which began in a different phase of the cell cycle (Fig. 2). When cells were arrested in early G_1 by isoleucine deprivation, no

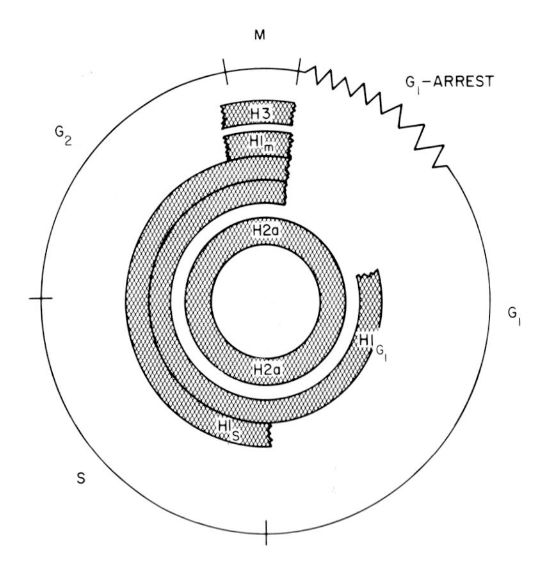

FIGURE 2 Summary of histone phosphorylation in proliferating and growth-arrested CHO cells. The periods of the cell cycle in which histones H1, H2a, and H3 are phosphorylated are indicated by the shaded bands inside the cell-cycle diagram. $H1_{G_1}$, $H1_S$, and $H1_M$ designate different specific histone H1 phosphorylation events which are initiated in the G_1, S, and M phases of the cell cycle. G_1-Arrested cells are those whose cell-cycle traverse has been arrested in early G_1 by isoleucine deprivation.

H1 phosphorylation was observed (Gurley et al., 1973*a*). However, when these cells were allowed to traverse their cell cycle, H1 phosphorylation was initiated in G_1 far ahead of DNA replication (as demonstrated in Fig. 5). A second H1 phosphorylation event was initiated during S phase (Fig. 2). This $H1_S$ phosphorylation was initially distinguished from the $H1_{G_1}$ phosphorylation by ion exchange chromatography (Gurley et al., 1975). A third H1 phosphorylation event was observed to occur only during mitosis (Fig. 2). This $H1_M$ phosphorylation is a superphosphorylation event involving the addition of many phosphate groups (Gurley et al., 1975). During mitosis, $H1_M$ phosphorylation was accompanied by the phosphorylation of histone H3 (Figs. 1 and 2). Mitosis is the only time in the cell cycle during which any significant H3 phosphorylation has been observed in CHO cells (Gurley et al., 1973*a*, *b*, 1974*a*, *b*, 1975).

Cell-Cycle-Specific H1 Phosphorylation Sites

Structural studies were performed on histone H1 to determine the differences between $H1_{G_1}$, $H1_S$, and $H1_M$ phosphorylations. Analytical 25-cm gel electrophoresis of H1 isolated from synchronized CHO cultures (Hohmann et al., 1976) demonstrated that as cells traversed from early G_1 to S, a portion of the two unphosphorylated parent H1 molecules (I and II) received a single phosphate (Fig. 3). During S phase, more of the parent H1 molecules were phosphorylated with one phosphate, and a small fraction received two

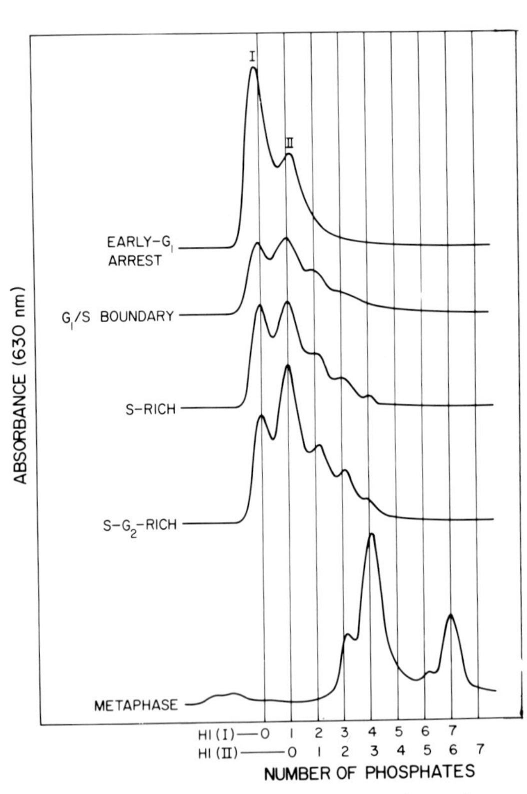

FIGURE 3 Analytical polyacrylamide gel electrophoresis of histone H1 isolated from synchronized CHO cells. The various H1 bands were detected by stained gel absorbance at 630 nm. Alignment of the various bands was accomplished with a mobility marker (not shown). Two unphosphorylated parental H1 subfractions (I and II) are present in CHO cells arrested in G_1. The mobility of these unphosphorylated H1 subfractions is expected to be reduced by 1% per phosphate added, as shown on the abscissa. Since unphosphorylated H1(I) and H1(II) are displaced from one another by 1%, each H1 subfraction must have its own abscissa as shown.

or three phosphates (Fig. 3). These conclusions are in agreement with previous studies using ion exchange chromatography of H1 from synchronized cells (Gurley et al., 1975). In contrast to interphase cells, mitotic cells contained no unphosphorylated, monophosphorylated, or diphosphorylated H1. Most of the subfraction I of H1 was tetraphosphorylated, and most of the subfraction II of H1 was hexaphosphorylated in mitosis (Fig. 3).

The regional location of phosphates within the H1 molecule was determined in each phase of the cell cycle by bisecting the H1 with N-bromosuccinimide (NBS) (Hohmann et al., 1976). This reagent cleaves the H1 molecule at its only tyrosine residue, producing two parts: an NH_2-terminal portion containing 73 residues and a COOH-terminal portion about twice as large (Bustin and Cole, 1969; Rall and Cole, 1970). After separating these two parts by exclusion chromatography, both portions were subjected to tryptic digestion and high-voltage paper electrophoresis to resolve phosphopeptides which were assayed for phosphoserine and phosphothreonine (Hohmann et al., 1975, 1976). A summary of these H1 analyses is presented in Fig. 4, and the following conclusions were drawn from these data. (1) During mid-G_1, $H1_{G_1}$ phosphorylation was initiated in the COOH-terminal part of the H1 molecule on serine. Cumulative phosphorylation of this site persisted throughout interphase and on into mitosis. (2) In S phase, $H1_S$ phosphorylation was initiated to a limited extent at two additional serine sites in the COOH-terminal portion of the H1 molecule. (3) In mitosis, $H1_M$ superphosphorylation occurred and consisted of the phosphorylation of interphase serine sites in the COOH-terminal part of the molecule, plus the phosphorylation of a new threonine site in the COOH-terminal part. In addition, both a threonine and a serine site were phosphorylated in the NH_2-terminal part of the molecule as well.

Thus, it is concluded that the $H1_{G_1}$, $H1_S$, and $H1_M$ phosphorylation events are distinctly different from one another and must be suspected of having different functions.

Correlation of $H1_{G_1}$ Phosphorylation with Presumptive Interphase Chromatin Condensation

When heparin is added to nuclei, this polyanion specifically interacts with histones in chromatin, causing DNA to be released (Hildebrand et al., 1975, 1976). When such measurements were made on nuclei isolated from synchronized CHO cells, it was found that there was an increased resistance of chromatin to heparin-mediated release of DNA as cells traversed interphase from G_1 to mitosis (Fig. 5). These measurements were interpreted as indicating that a progressive inaccessibility of histones to heparin occurs as cells traverse their cell cycle. The simplest model to explain these observations is that the chromatin structure becomes more compact at some submicroscopic level of organization as cells traverse interphase.

This presumptive interphase chromatin condensation was observed to occur simultaneously with the increase in interphase $H1_{G_1}$ phosphorylation which was initiated in the G_1 phase (Fig. 5). The close correspondence in cell-cycle timing of these two measurements suggests that $H1_{G_1}$ phosphorylation may be related to the interphase chromatin

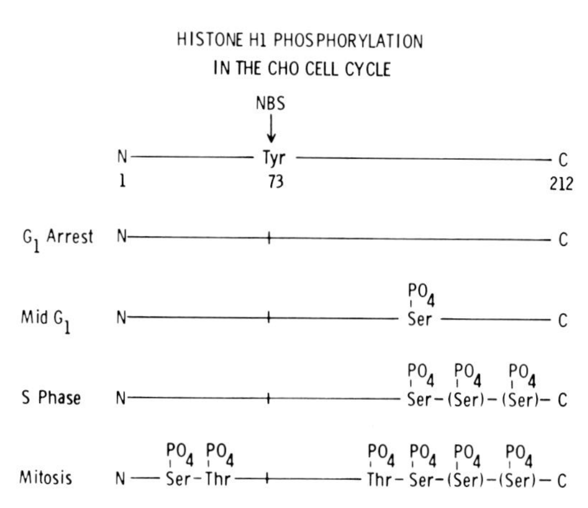

FIGURE 4 Diagrammatic representation of phosphorylation sites within the H1 molecule and the amino acids phosphorylated during different phases of the cell cycle. Positions of the sites are known only with respect to the N-bromosuccinimide (NBS) cleavage point; therefore, no significance should be attached to the order of sites as illustrated. Histone H1 isolated from $^{32}PO_4$-labeled synchronized CHO cells was purified by BioRex-70 ion exchange chromatography. The H1 was then bisected at its sole tyrosine residue with NBS and the two parts separated by Sephadex G-100 chromatography. Tryptic peptides were prepared and phosphopeptides fractionated by high-voltage paper electrophoresis. After acid hydrolysis of the phosphopeptides, serine and threonine phosphates were separated by high-voltage paper electrophoresis. Incorporation of $^{32}PO_4$ into serine and threonine was determined by autoradiography.

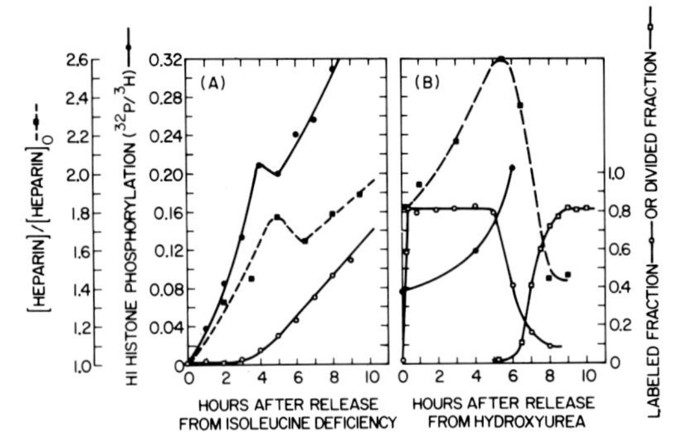

FIGURE 5 $H1_{G_1}$ phosphorylation occurring simultaneously with presumptive submicroscopic interphase chromatin condensation detected by heparin titration. An increase in $H1_{G_1}$ phosphorylation is observed as cells traverse interphase (●——●). This is accompanied by an increased resistance of chromatin to its disintegration by heparin (■——■) as described in Methods. (A) CHO cells synchronized in early G_1 by isoleucine deficiency were released to traverse the G_1 and S phases. Entry into S phase is designated by the increase in [^{3}H]thymidine labeling of cells (○——○). (B) CHO cells synchronized near the G_1/S boundary with hydroxyurea were released to traverse the S and G_2 phases. Entry into S phase is designated by the increase in [^{3}H]thymidine labeling (○——○). Cell division at mitosis is designated by the increase in divided fraction (□——□).

structural changes detected with the heparin probe.

Correlation of H2a Phosphorylation with Heterochromatin Structure

Cytochemical measurements of ear cells from two closely related strains of mice, *Peromyscus crinitus* and *Peromyscus eremicus*, have shown that these two cell lines have the same euchromatin content but grossly different heterochromatin contents (Pathak et al., 1973). We have compared histone phosphorylation in these two cell lines to determine whether any correlation exists between the condensed heterochromatin state and histone modification (Gurley et al., 1976).

Staining the chromosomes of these two cell lines by the C-band Giemsa method demonstrated that *P. crinitus* and *P. eremicus* both contain 48 chromosomes but that *P. crinitus* (Fig. 6 A) lacks the extra short chromosomal arms present in the chromosomes of *P. eremicus* (Fig. 6 B). The dark staining character of these extra chromosomal arms indicated that they are composed of constitutive heterochromatin (Hsu, 1973). Flow microfluorometric measurements of these two cell lines indicated that *P. eremicus* contains 36.4% more DNA than *P. crinitus* (Fig. 7*a*). This is in excellent agreement with the chromosome arm length measurements of Pathak et al. (1973) and indicates that essentially all the extra DNA of *P. eremicus* exists as highly condensed constitutive heterochromatin in the short chromosome arms.

Examination of these cells in interphase by C-band staining (Fig. 6 C, D) and by electron microscopy (Fig. 6 E, F) demonstrated that this highly condensed heterochromatin structure persists during interphase. Interphase *P. crinitus* cells contain small clumps of heterochromatin dispersed throughout the nucleus, while interphase *P. eremicus* cells contain large, dense clumps of heterochromatin.

To determine whether histone phosphorylation is associated with the degree of interphase chromatin condensation, we compared the incorporation of $^{32}PO_4$ into histones of these two cell lines growing under identical culture conditions (Fig. 7*b–e*). The H1 phosphorylation rates of *P. eremicus* and *P. crinitus* were identical, both having a ^{32}P/^{3}H ratio of 1.20 (Fig. 7*b*, *d*). However, the phosphorylation rate of H2a was much greater in *P. eremicus* (^{32}P/^{3}H ratio of 0.78) than in *P. crini-*

FIGURE 6 Cytology and electron microscopy of *Peromyscus crinitus* and *P. eremicus* cells. (A and B) C-band Giemsa-stained chromosomes. Both cell lines have 48 chromosomes with essentially the same amount of light staining (C-band negative) chromosomal material containing euchromatin, but *P. eremicus* cells have grossly more dark staining (C-band positive) constitutive heterochromatin existing as extra short chromosome arms (B) which are absent in the chromosomes of *P. crinitus* (A). (C and D) C-band Giemsa-stained interphase nuclei. The dark staining constitutive heterochromatin of both cell types persists during interphase, there being an excessive amount of this heterochromatin in *P. eremicus* nuclei (D) compared to *P. crinitus* nuclei (C). (E and F) Electron microscopy of thin sections of interphase nuclei. *P. crinitus* nuclei (E) contain small dispersed particles of heterochromatin, while *P. eremicus* nuclei (F) contain very large, dense clumps of heterochromatin. Heterochromatin is designated by the arrows and the nucleolus by *N*.

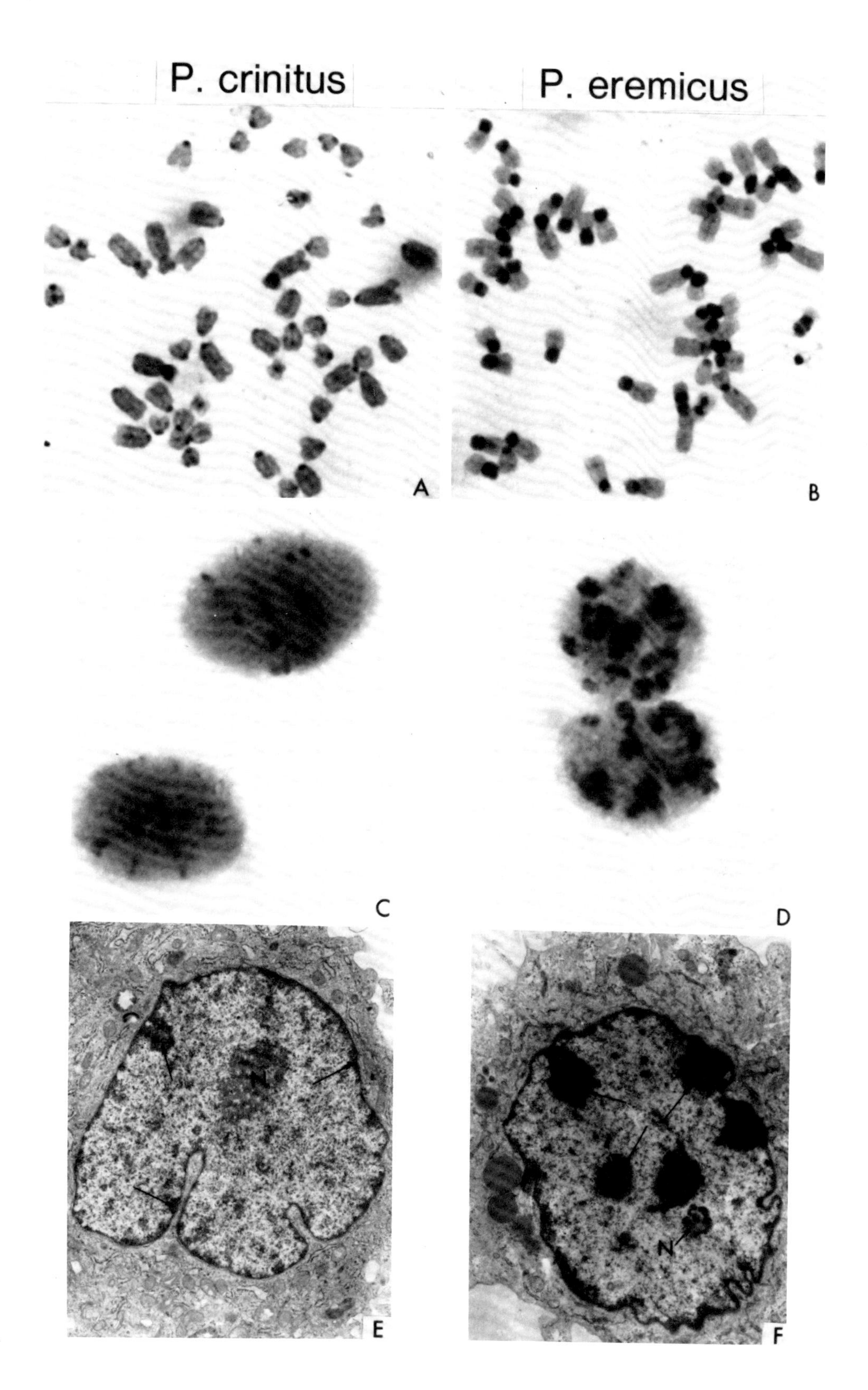
P. crinitus
P. eremicus
A
B
C
D
E
F
N

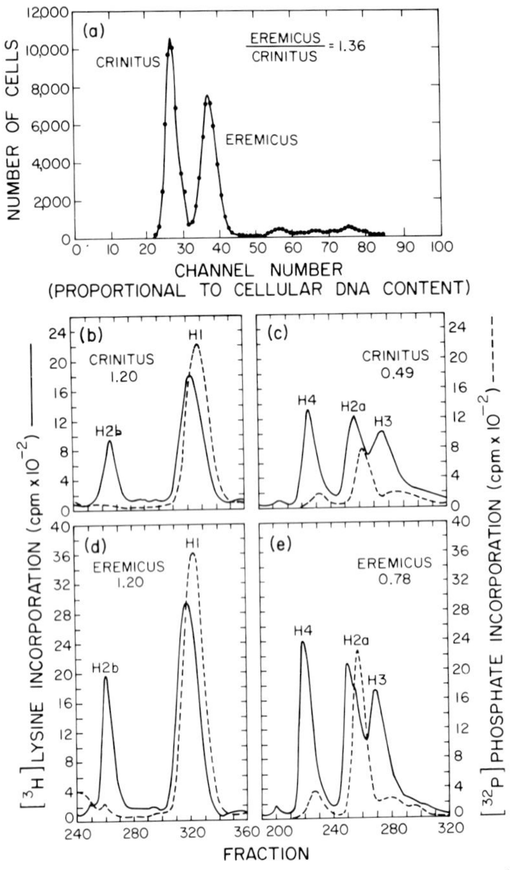

FIGURE 7 Relative DNA analysis of G_1 *P. crinitus* and *P. eremicus* cells by flow microfluorometry (*a*). Analysis of the difference between the peak fluorescence of these two cell types indicates that *P. eremicus* contains 36.4% more DNA than does *P. crinitus*. Preparative electrophoresis profiles of histone phosphorylation in *Peromyscus* cells (*b–e*). Monolayer cultures of *P. crinitus* (*b* and *c*) and *P. eremicus* (*d* and *e*) were both labeled with [^{3}H]lysine for exactly 2.3 generations. During exponential growth, cells were labeled with $^{32}PO_4$ for 2 h just before histone preparation and electrophoresis. Individual histone fractions, H1 and H2b (*b* and *d*) and H2a, H3, and H4 (*c* and *e*), are indicated by the [^{3}H]lysine incorporation (—). Phosphorylation of each fraction is indicated by the $^{32}PO_4$ incorporation (- - -). Electrophoretic migration proceeds from right to left.

tus (^{32}P/^{3}H ratio of 0.49). An average of three experiments showed the H2a phosphorylation rate to be 68% greater in the highly heterochromatic chromatin of *P. eremicus* than in the lesser heterochromatic chromatin of *P. crinitus*. These results suggest that H2a phosphorylation may be involved in some way with the condensed heterochromatin structure of interphase cells.

Correlation of $H1_M$ and H3 Phosphorylations with Mitotic Chromatin Condensation

When histone phosphorylation was measured in synchronized CHO cells, it was observed that significant histone H3 and $H1_M$ phosphorylations occurred only during mitosis (Figs. 1–4). When cells labeled with $^{32}PO_4$ during mitosis were removed from the $^{32}PO_4$-containing medium and resuspended in unlabeled growth medium, the $H1_M$ and H3 ^{32}P-phosphates were rapidly lost from these histones as the cells entered G_1 interphase (Gurley et al., 1974*a*). The correlation of these phosphorylations and dephosphorylations with chromatin condensation and decondensation observed during mitosis (Fig. 8) suggests that $H1_M$ and H3 phosphorylations are in some way associated with these mitotic chromatin structural changes.

Electron microscope examination of CHO cells undergoing mitosis revealed that chromatin was gathered into dense chromatin patches before the nuclear envelope disintegrated and chromosomal bodies formed (Fig. 8 A, B). After metaphase (Fig. 8 C), the reverse of this process appeared to occur (Fig. 8 D–F). The chromatin remained in a condensed state after the highly condensed metaphase chromosomes had disorganized and the nuclear envelope had reformed. Whether mitotic histone phosphorylations are associated with aggregation and disaggregation of chromatin at the G_2/M and M/G_1 boundaries, as shown in Fig. 8 (B, D, E), or with the final organization into metaphase chromosomes, as shown in Fig. 8 *C*, is unclear at this time.

DISCUSSION

Eucaryotic cells possess an enormous amount of DNA which, if extended, is meters in length (DuPraw, 1970). This DNA is somehow compacted into the cell nucleus which has dimensions on the order of microns. This feat is apparently accomplished by complexing DNA with histone proteins, which facilitates folding of the DNA into the substance we know as chromatin. This chromatin structure is exceedingly complex, having many orders of organization ranging from molecular and submicroscopic configurations to structures visible in the microscope. For example, the complexing of histones with DNA to form nucleosomes (Olins and Olins, 1974) introduces a tertiary structure to DNA (Bahr, 1975; Bram, 1975), and higher order folding (coiling) of the

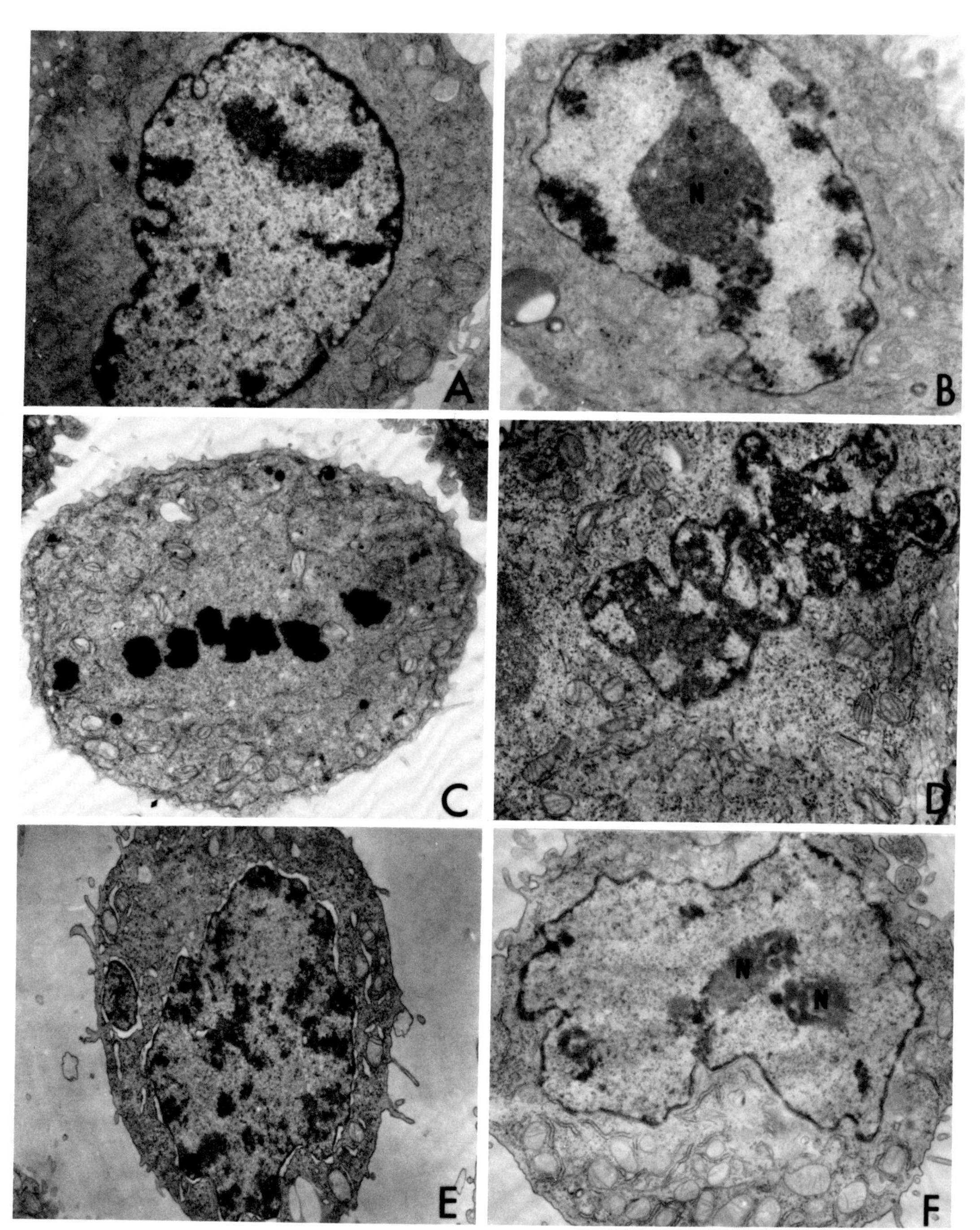

FIGURE 8 Electron microscopy of CHO chromatin condensation and decondensation during mitosis. Cells were synchronized by mitotic selection without use of mitotic arresting drugs such as Colcemid or Colchicine. (A) Interphase cell with nucleus containing dispersed euchromatin and small clumps of heterochromatin mostly associated with the nuclear envelope. The nucleolus is designated *N*. (B) Early prophase cell with chromatin condensing into aggregates near the nuclear envelope which is still intact. The disintegrating nucleolus is designated *N*. (C) Metaphase cell. Chromatin organized into highly condensed chromosomes is lined up in an equatorial plane. (D) Late telophase cell. The nuclear membrane has reformed around chromosomes which have disorganized into condensed aggregates of chromatin. (E) M/G_1 transition cell. Dilated nuclear membrane surrounds still partially condensed aggregates of chromatin. (F) G_1 cell 1 h after metaphase. Mitotic chromatin has completely dispersed into interphase chromatin, and the nucleoli (*N*) have reformed.

nucleosome particles introduces a quaternary structure. These molecular structures may be further condensed to varying degrees in the interphase nucleus to form chromatin aggregates we know as euchromatin and heterochromatin (Frenster, 1974). Finally, during mitosis, these chromatin structures are folded and compacted into the highly condensed structures we observe in the microscope as chromosomes.

Theoretically, one might expect the addition of negatively charged phosphates to positively charged histones to weaken the DNA-histone interaction in chromatin, resulting in an expansion of the chromatin structure. Therefore, it was rather unexpected that, during the course of our studies on histone metabolism in cultured mammalian cells, we repeatedly observed that histone phosphorylation was associated with some form of condensing, or condensed, chromatin structure. Nevertheless, repetition of this observation in a wide variety of experiments (outlined above) has led us to consider the possibility that growth-related histone phosphorylation, in general, is associated with the condensation of chromatin structure at a variety of levels of chromatin organization. Specifically, our observations suggest that (*a*) interphase histone $H1_{G1}$ phosphorylation is associated with a presumptive chromatin condensation at a submicroscopic or molecular level which can be detected with the heparin probe of chromatin structure; (*b*) histone H2a phosphorylation is associated with chromatin organization at the level of condensation into heterochromatin; (*c*) mitotic histone $H1_M$ and H3 phosphorylations are associated with the condensation of diffuse interphase chromatin into the dense chromatin which is organized into chromosomes.

It should be understood that our observations and the conclusions drawn from them are based on data correlating histone phosphorylation with cell biology phenomena that do not distinguish between cause and effect. Thus, the nature of the relationship between histone phosphorylation and the chromatin structure with which it is identified remains to be elucidated. Most importantly, it must be determined whether histone phosphorylation causes chromatin structural changes or whether histones are phosphorylated as a result of chromatin structural changes which expose histone phosphorylation sites to histone kinase activity. In either case, however, this protein modification may be involved in the control of DNA activities. If histone phosphorylation is found to cause chromatin structural changes, DNA activities may be controlled by mechanisms that modulate kinase and phosphatase activities. If histones are phosphorylated as a result of chromatin structural changes, DNA activities may be controlled through changes in histone-nonhistone chromosomal protein interactions induced by the phosphorylation.

ACKNOWLEDGMENT

This work was performed under the auspices of the U. S. Energy Research and Development Administration.

REFERENCES

BAHR, G. F. 1975. The fibrous structure of human chromosomes in relation to rearrangements and aberrations: a theoretical consideration. *Fed. Proc.* **34:**2209.

BRADBURY, E. M. 1975. Foreword: Histone nomenclature. *In* The Structure and Function of Chromatin. D. W. Fitzsimons and G. E. W. Wolstenholme, editors. CIBA Foundation Symposium. Elsevier-Excerpta Medica, North Holland, Amsterdam. **28:**1.

BRAM, S. 1975. A double coil chromatin sub-unit model. *Biochimie (Paris).* **57:**1301.

BUSTIN, M., and R. D. COLE. 1969. Bisection of a lysine-rich histone by N-bromosuccinimide. *J. Biol. Chem.* **244:**5291.

CRISSMAN, H. A., and R. A. TOBEY. 1974. Cell cycle analysis in 20 minutes. *Science (Wash. D. C.).* **184:**1297.

DEAVEN, L. L., and D. F. PETERSEN. 1974. Measurements of mammalian cellular DNA and its localization in chromosomes. *Methods Cell Biol.* **8:**1979.

DUPRAW, E. J. 1970. Chromosome chemistry and DNA replication in macroorganisms. *In* DNA and Chromosomes. E. J. DuPraw, editor. Holt, Rinehart & Winston, Inc., New York. 118.

ELGIN, S. C. R., and H. WEINTRAUB. 1975. Chromosomal proteins and chromatin structure. *Annu. Rev. Biochem.* **44:**725.

FRENSTER, J. H. 1974. Ultrastructure and function of heterochromatin and euchromatin. *In* The Cell Nucleus. Vol. 1. H. Busch, editor. Academic Press, Inc., New York. 565.

GURLEY, L. R., and R. A. WALTERS. 1971. Response of histone turnover and phosphorylation to X-irradiation. *Biochemistry.* **10:**1588.

GURLEY, L. R., R. A. WALTERS, and R. A. TOBEY. 1973*a*. The metabolism of histone fractions. VI. Differences in the phosphorylation of histone fractions during the cell cycle. *Arch. Biochem. Biophys.* **154:**212.

GURLEY, L. R., R. A. WALTERS, and R. A. TOBEY. 1973*b*. Histone phosphorylation in late interphase and mitosis. *Biochem. Biophys. Res. Commun.* **50**:744.

Gurley, L. R., R. A. Walters, and R. A. Tobey. 1974*a*. Cell cycle-specific changes in histone phosphorylation associated with cell proliferation and chromosome condensation. *J. Cell Biol.* **60**:356.

Gurley, L. R., R. A. Walters, and R. A. Tobey. 1974*b*. The metabolism of histone fractions: the phosphorylation and synthesis of histones in late G_1-arrest. *Arch. Biochem. Biophys.* **164**:469.

Gurley, L. R., R. A. Walters, and R. A. Tobey. 1975. Sequential phosphorylation of histone subfractions in the Chinese hamster cell cycle. *J. Biol. Chem.* **250**:3936.

Hildebrand, C. E., L. R. Gurley, R. A. Tobey, and R. A. Walters. 1975. Cell-cycle-specific changes in organization of chromatin detected by polyanion-mediated chromatin decondensation. *Fed. Proc.* **34**:581.

Hildebrand, C. E., and R. A. Tobey. 1975. Cell-cycle-specific changes in chromatin organization. *Biochem. Biophys. Res. Commun.* **63**:134.

Hildebrand, C. E., R. A. Walters, R. A. Tobey, and L. R. Gurley. 1976. Changes in chromatin organization during the cell cycle. *Biophys. J.* **16**:226*a*.

Hohmann, P., R. A. Tobey, and L. R. Gurley. 1975. Cell-cycle-dependent phosphorylation of serine and threonine in Chinese hamster cell F1 histones. *Biochem. Biophys. Res. Commun.* **63**:126.

Hohmann, P., R. A. Tobey, and L. R. Gurley. 1976. Phosphorylation of distinct regions of F1 histone: relationship to the cell cycle. *J. Biol. Chem.* **251**:3685.

Hsu, T. C. 1973. Longitudinal differentiation of chromosomes. *Annu. Rev. Genet.* **7**:153.

Millonig, G. 1962. Further observations on a phosphate buffer for osmium solutions in fixation. *In* Electron Microscopy. Vol. 2. Sect. P-8. S. S. Breese, editor. Academic Press, Inc., New York.

Olins, A. L., and D. E. Olins. 1974. Spheroid chromatin units (ν bodies). *Science (Wash. D. C.).* **183**:330.

Pathak, S., T. C. Hsu, and F. E. Arrighi. 1973. Chromosomes of *Peromyscus* (rodentia, cricetidae). IV. The role of heterochromatin in karyotypic evolution. *Cytogenet. Cell Genet.* **12**:315.

Rall, S. C., and R. D. Cole. 1970. Quantitative cleavage of a protein with N-bromosuccinimide. *J. Am. Chem. Soc.* **92**:1800.

Spurr, A. R. 1969. A low-viscosity epoxy resin embedding medium for electron microscopy. *J. Ultrastruct. Res.* **26**:31.

Tobey, R. A., E. C. Anderson, and D. F. Petersen. 1967*a*. Properties of mitotic cells prepared by mechanically shaking monolayer cultures of Chinese hamster cells. *J. Cell. Physiol.* **70**:63.

Tobey, R. A., E. C. Anderson, and D. F. Petersen. 1967*b*. The effect of thymidine on the duration of G_1 in Chinese hamster cells. *J. Cell Biol.* **35**:53.

Tobey, R. A., and H. A. Crissman. 1972. Preparation of large quantities of synchronized mammalian cells in late G_1 in the pre-DNA replicative phase of the cell cycle. *Exp. Cell Res.* **75**:460.

Tobey, R. A., and K. D. Ley. 1970. Regulation of initiation of DNA synthesis in Chinese hamster cells. I. Production of stable, reversible G_1-arrested populations in suspension culture. *J. Cell Biol.* **46**:151.

Tobey, R. A., and K. D. Ley. 1971. Isoleucine-mediated regulation of genome replication in various mammalian cell lines. *Cancer Res.* **31**:46.

Tobey, R. A., D. F. Petersen, E. C. Anderson, and T. T. Puck. 1966. Life cycle analysis of mammalian cells. III. The inhibition of division in Chinese hamster cells by puromycin and actinomycin. *Biophys. J.* **6**:567.

MOLECULAR EVENTS IN THE REPLICATION OF DNA AND CHROMATIN

GERALD C. MUELLER, KAZUTO KAJIWARA, ATSUSHI ICHIKAWA, and STEPHEN PLANCK

The molecular mechanisms that operate in the replication of DNA and chromatin clearly play a fundamental role in the organized growth of multicellular eukaryotes. Whereas the enzymatic mechanisms effecting the replication of DNA provide a new and exact copy of genetic material for each daughter cell, the mechanisms attending the assembly of newly replicated DNA and proteins in the formation of chromatin appear to play an important role in specifying which genes shall reside in an inducible state, and thus specify cell phenotype. In the living cell, these processes are tightly coupled, and attempts to disengage them usually lead to cell disaster. Operating normally, and in balance with the cell's nutrition and other environmental factors, however, they specify the nature of cell-to-cell interactions, determine the responsiveness to mitogenic stimuli, and provide for the cell functions that characterize organized tissue growth.

Impressed by the integration of these processes, our laboratory has engaged in a study of DNA and chromatin replication in the nuclear setting. These studies have as their goal the identification of the enzymatic compounds making up the DNA replicase system, a description of the role that incoming proteins play in chromatin assembly, and the characterization of aspects of DNA chain growth that might affect the open or closed state of DNA in newly replicated chromatin with respect to its potential for transcription. It is proposed that knowledge of these mechanisms will provide the ground rules for remedially shaping cell growth and developing better therapy for cancer and other growth dyscrasias.

GERALD C. MUELLER and KAZUTO KAJIWARA McArdle Laboratory for Cancer Research, the University of Wisconsin, Madison, Wisconsin

ATSUSHI ICHIKAWA Department of Biochemistry, University of Kyoto, Kyoto, Japan

STEPHEN PLANCK Laboratory of Biochemistry, National Cancer Institute, Bethesda, Maryland

A Nuclear System for Studying DNA Replication

HeLa cells growing logarithmically in suspension cultures provided an easily controlled system for staging nuclei in different parts of the cell cycle. In the present studies, the cells have usually been synchronized for entry into S phase by growth for 16 h in the presence of amethopterin (Mueller et al., 1962); reversal of the thymidineless state by the addition of exogenous thymidine (dTR) permits the synchronized cells to engage in DNA replication. At the indicated times, the cells are swelled in hypotonic buffer and lysed with a Dounce homogenizer to liberate nuclei; the latter are isolated by sedimentation and washed delicately to remove cytoplasmic debris. The cytoplasmic fraction is sedimented for 60 min at 105,000 *g* to remove particulates and yield the soluble cytoplasmic protein fraction (CF; Friedman and Mueller, 1968; Hershey et al., 1973*a*).

Previous publications from this laboratory have described the optimum conditions for studying DNA replication in nuclei of S-phase synchronized cells (Hershey et al., 1973*a*, *b*). Supplemented with optimal levels of the four deoxynucleoside triphosphates, ATP, Mg^{++}, NaCl, and the cytoplasmic soluble proteins (CF), the nuclei replicate up to 5% of their DNA at initial rates which approximate that of the living cells. The system has an absolute requirement for ATP and is highly sensitive to the ionic strength of the assay medium (i.e., 90–100 mM NaCl is optimum).

DNA replication in well-washed nuclei is 90% dependent on the presence of the soluble proteins in the CF; however, the activity of CF from cells in the S phase and G_1 intervals of the cell cycle are similar. DNA replication in the complete system continues from sites that were active in the living cells and is absolutely dependent on the S-phase state of the nuclei. Lysates of S-phase nuclei and CF from S-phase cells have failed to initiate DNA replication in G_1-state nuclei.

The Triggering of Nuclear Replication

Studies of the cell cycle support the view that the triggering of G_1 cells for nuclear replication is regulated by events occurring in the cytoplasm. In growth-restricted cells, the cytoplasmic membranes appear to contain receptors for specific mitogens which regulate the specific triggering of certain cell phenotypes. In all cases, the cells respond to mitogenic stimuli with the synthesis of new RNAs and proteins; some of the latter migrate to the nucleus and lead in a few hours to the activation of genes for DNA replication. Whether these genes for DNA replication are under some common operon control is not known; however, this appears likely, because the inductive mechanism seems to operate in a similar manner in different cell types despite vast differences in the primary recognition of mitogenic stimuli.

HeLa, and other continuously growing cell types, appear to be deficient in a mechanism that can hold them up in G_1; however, the progression of the cells through the G_1 interval depends in a similar manner on the sequential synthesis of specific RNAs and proteins. Triggering of nuclear replication is accompanied by the synthesis of enzymes for deoxyribotide metabolism. By using HeLa cells made permeable to nucleotides (Seki et al., 1975) and the isolated nuclear system (Seki and Mueller, 1975), it has been demonstrated that RNA and protein synthesis are also required both to establish the DNA replication complexes in nuclei and to maintain them during the S period. The data support the concept that a short-lived protein which is continuously synthesized in S-phase cells is required to stabilize the DNA replicase complexes. Whether the requirement for RNA synthesis relates to the availability of the messenger RNA for this protein or concerns another step in the DNA replication process (i.e., initiation of DNA synthesis at new replicons) is not known.

Dissociation and Reconstitution of the DNA Replicase System of Nuclei

Although studies in living cells point out the general requirements for RNA and protein synthesis in the assembly and operation of the DNA replicase system, the identification of the actual components and their functions requires methods for dissociating the nuclear system reversibly. In a wide range of experiments, it was found that disruption of nuclei by sonication, freeze-thawing, nonionic detergent treatment, or organic solvent extraction yielded inactive systems. However, extraction with low concentrations of NaCl at 0°C yielded an inactive nuclear residue that could be reactivated by recombination with the salt extract followed by dialysis to reduce the salt concentration or by recombination with separately dialyzed extract (Seki and Mueller, 1976). In a series of experiments, 0.3 M NaCl proved to be the optimum salt concentration for solubilizing certain components of the DNA replicase system, still leaving the nuclear residue, containing the chromosomal DNA template, in a reactivatable state; exposure of the nuclei even momentarily to higher concentrations of NaCl resulted in irreversible loss of activity. Because a higher salt concentration also disrupts the nucleosome structures in chromatin, it is concluded that the nucleosomic character of chromatin may be necessary for the replication of chromosomal DNA. The nuclear residues had to be from S-phase nuclei in order to be reactivated by the salt extracts. Similarly, active extracts could be obtained only from nuclei that had been triggered for S phase, although the actual establishment of DNA-replicase sites was not required in order to yield active salt extracts. In fact, the activity of the salt extracts was proportional to the number of nuclei that had undergone the $G_1 \rightarrow S$ transition event of the cell cycle. In contrast, the activity of the nuclear residues was proportional to the rate of DNA replication in the cells (i.e., number of active replication sites) at the time of nuclei isolation. Salt-extracted nuclear residues from S-phase cells, treated in vivo with cycloheximide to reduce the DNA replicase activity of the starting nuclei, exhibited poor responsiveness to otherwise active salt extracts. From these observations, it was concluded that the unstable component of the DNA replicase system or the product of its action, described above, resides in the nuclear residue after salt extraction, and that the synthesis of this component may normally limit the progress of DNA replication in living cells. The data also suggest that the DNA replicase components of the salt extract are produced in response to the triggering of the cells for nuclear replication during the $G_1 \rightarrow S$ transition and, as such, these components are likely to be under the control of a replication operon which, in turn, responds to an inducer of cytoplasmic origin.

Identification of the Active Components in the Nuclear Salt Extracts

In the course of these studies, it was observed that 0.02 M ATP facilitated the dissociation of the DNA replicase system in lymphocyte nuclei (Thompson and Mueller, 1975). The presence of this agent during the extraction of HeLa nuclei with 0.3 M NaCl also yielded more active extracts, although the residues were less active in reconstitution studies. Accordingly, many of the fractionation studies were carried out on the ATP-NaCl extracts. Early chromatographic experiments with DEAE cellulose revealed that the salt extracts contained DNA polymerase β (classification according to Weissbach et al., 1975), DNA polymerase α, a protein activator(s) that stimulated the activity of DNA polymerase α on an activated calf-thymus DNA template, and a dialyzable component eluting with 0.02 M NaCl from the DEAE that also stimulated the partially purified DNA polymerase α from the column. The identity of this small molecular-weight component is still unknown.

The DNA polymerase α, eluting from the DEAE column with 0.25 M $KHPO_4$, pH 7.5, was further purified by chromatography on phosphocellulose, where it eluted with 0.18 M $KHPO_4$ at pH 7.5. Purification was attended by a progressive loss of DNA polymerase activity, which could be restored by the addition of the concentrated activator fractions eluting from DEAE with 0.02 M-0.1 M $KHPO_4$, pH 7.5. Properties of the activator protein(s) are given below.

Sedimentation of the column-purified nuclear DNA polymerase α in 5–20% sucrose gradients containing 0.5 M NaCl, 20 mM Tris-HCl, pH 7.5, 1 mM EDTA, and 5 mM β-mercaptoethanol revealed a major peak of activity at 7.5S (150,000–160,000 mol wt) and a shoulder at 6.5S; this was distinctly different for the DNA polymerase α of the cytoplasmic-soluble protein fraction (CF) which sedimented sharply at 5.8S (96,000–100,000 mol wt). However, both types of DNA polymerase α exhibited a preference for activated calf-thymus DNA as a template, and were inhibited strongly by N-ethylmaleimide. The amount of the nuclear enzyme closely paralleled the DNA replicase activity of the nuclei before extraction and therefore was strictly S-phase related; on the other hand, the DNA polymerase α of the CF was relatively constant throughout the cell cycle. The nuclear DNA polymerase α of the 0.3 M NaCl extracts, in contrast to the CF enzymes, was also stimulated 50–100% by 5 mM ATP; however, UTP, GTP, or CTP were also effective. This effect may have been caused by the presence of an inhibitor or another factor, because this response was lost with purification of the polymerase.

Taken together, these observations point to a difference between the DNA polymerase α of nuclei and that of the CF; this conclusion is further supported by the high affinity of the nuclear enzyme for the activator protein described below. Thus, although both enzymes may be localized initially in the nucleus, as suggested by studies in which the nuclei were isolated in anhydrous glycerol (Spaeren et al., 1975), they nonetheless appear to be two distinct forms of DNA polymerase α. The possibility that the DNA polymerase α of the nuclear salt extract may be a metabolic derivative of the CF enzyme, which can function amid the nucleosomic organization of chromosomal DNA, projects some interesting speculations on the distinctions between DNA repair and DNA replication. Extending this line of thinking, the modification of this enzyme might then be one of the critical events in the initiation of DNA synthesis in nuclei.

Properties of the Activator Protein(s) of the Nuclear Extracts

As stated above, purification of the nuclear DNA polymerase α by chromatography on DEAE and phosphocellulose columns results in a progressive loss of activity, which can be recovered and even improved if the enzyme is supplemented with the 0.02 M-0.1 M fractions from the DEAE column. The component(s) that stimulate DNA polymerase α have been further purified by chromatography on BioRex 70 and denatured DNA cellulose columns. In each case three activity peaks were eluted. It was soon observed that rechromatography of each peak again gave rise to all three peaks and that this also was true when transferring from BioRex 70 to denatured DNA cellulose columns. In addition, filtration of the purified activators through G-100 Sephadex yielded three peaks of activity corresponding to 75,000, 43,000, and 25,000 mol wt dimensions. However, electrophoresis of the 75,000 and 43,000 mol wt fractions in SDS gels revealed the presence of a single protein band at a position corresponding to a molecular weight of 38,000. Too little of the 25,000 mol wt fraction was obtained for electrophoretic analysis; however, a 20,000 mol wt component was observed on electrophoresis by the partially purified activator fractions from the DNA cellulose columns.

Three striking observations were made in these studies: (1) the purified activator protein adsorbed tenaciously to single-stranded DNA and was not retained by native DNA; (2) this protein formed physical complexes with the nuclear DNA polymerase α, which increased the sedimentation rate of the latter in sucrose density gradients; (3) the nuclear DNA polymerase α as extracted was already in a physical association with this activator protein, the properties of which are similar to a DNA-binding protein with an unwinding activity. In addition, this DNA-binding protein from HeLa resides in the nuclei, whereas a protein with similar properties from thymocytes (Herrick et al., 1976) is located in the cytoplasm.

Despite these preliminary successes in identifying S-phase-related components in the nuclear extracts, attempts to reconstitute DNA replicase activity in salt-extracted nuclear residues with the purified DNA polymerase α of nuclei and the activator protein were variably successful; the data point to the existence of additional components in the salt extract that are either dispersed by the fractionation procedures or are variably retained by the nuclear residues. The identity and resolution of these components are the subject of current study. The possibility that the additional components are multiple and play structural roles in the replication of specific chromosomal DNA complexes is a serious consideration.

Role of the CF in DNA Replication in Isolated Nuclei

In the absence of the soluble proteins of the cytoplasm, DNA replication in isolated nuclei fails rapidly; however, with high levels of CF, DNA replication continues for more than 60 min and replicates as much as 5% of the HeLa genome in vitro (Hershey et al., 1973*a*). Fractionation of the active principles in CF by chromatography on DEAE and phosphocellulose columns revealed that the activity was distributed among numerous fractions which, on further fractionation, appeared to subdivide. Combination studies gave evidence of cooperativity among the fractions, as well as a differential ability to stimulate nuclei in different sections of the S interval.

This multicomponent nature of the CF suggested the possibility that the different proteins may be involved in the structural character of the nucleochromatin and, as such, may play a role in the formation of new chromatin. To test this concept, labeled CF proteins were prepared from cells grown in [^{3}H]leucine. Nuclei from unlabeled S-phase cells were then incubated with the usual [^{3}H]leucine-labeled CF under the DNA replication assay conditions. A temperature-dependent uptake of the labeled proteins was observed; however, only a small fraction of the uptake was dependent on the continuation of DNA replication during the incubation interval. Sheared chromatin, which was prepared and fixed with formalin according to Jackson and Chalkley (1974), was sedimented in CsCl-guanidinum chloride density gradients. In this procedure, a significant fraction of the labeled CF proteins which was taken up by the nuclei was found to be present in the chromatin band; the majority of the labeled proteins sedimented nonrandomly to the light side of the chromatin band. A similar experiment with labeled histones revealed that the latter distributed in a constant ratio of the DNA across the entire DNA-chromatin band.

Electrophoretic analysis of the nuclear proteins revealed that the uptake of labeled CF proteins was nonrandom. S-phase nuclei—in addition to binding much more of the labeled CF proteins—also bound a very different spectrum of proteins. Furthermore, certain of the proteins were subject to dilution by unlabeled CF, suggesting limited binding sites for such entities.

These studies suggest that the associations among chromosomal proteins are in a dynamic state of revision under the conditions that are optimal for DNA replication, and that the CF proteins, in contrast to histones, enter primarily into associations with proteins of existing chromatin complexes. The observations that CF proteins are essential for DNA chain growth (see next section) raises the possibility that the incoming proteins play a major role in exposing or presenting the chromosomal DNA for the replication process. On the other hand, our observation that CF proteins also protect newly replicated DNA from digestion by neurospora DNAse supports the concept that the in vitro DNA replication system in isolated nuclei is also engaged in the assembly of new chromatin, which depends on incoming proteins from the cytoplasm. This possibility is being studied further by use of two new procedures for releasing and studying the protein associations with newly replicated DNA. A challenging aspect of these studies is that it may be possible to inquire by direct experimentation into the manner in which chromosomal proteins determine how specific genes are expressed during differentiation.

DNA Chain Growth in Isolated Nuclei

Inherent in these studies of DNA replication in isolated nuclei is the need to demonstrate that the replication process is representative of that occurring in nuclei of living cells. Huberman and Riggs (1968) have shown, by autoradiographic studies and sedimentation analysis, that DNA replication in living eukaryotic cells is a bidirectional process within a single replicon, involving the direct elongation along one template strand with a discontinuous repair-back type of replication along the opposite template strand. The latter process appears to involve the synthesis of short DNA segments, Okazaki fragments, which may grow intermediately, but are ultimately ligated together to form the mature opposite strand. To test whether this situation operated during DNA replication in isolated nuclei, advantage was taken of a new procedure developed in our laboratory for studying DNA chain growth. Basically, the procedure involves reversing the thymidineless state with bromodeoxyuridine (BUdR) to label all active replicons with this thymidine analogue. The nuclei are then isolated and caused to synthesize DNA with [^{3}H]dTTP in the complete in vitro system. At the end of the incubation the nuclei are sedimented from the reaction mixture, lysed with alkali on top of an alkaline sucrose gradient, and the size of the newly replicated single-stranded DNA determined by velocity sedimentation. Under the usual conditions of nuclear DNA replication, this analysis has revealed that approximately one-half of the newly replicated DNA sedimented has a peak at 10S (1,500–3,000 nucleotides); the other half sediments as a spectrum of sizes ranging from 20–80S. When the nuclei are irradiated with UV_{313} light to photolyze the BUdR leaders, which were introduced into each active replicon in the living cells, the newly replicated DNA segments of the 20–80S DNA chains are released and shift completely to the 16–20S (5,000–8,500 nucleotides) region of the gradient.

This shift in the DNA sedimentation pattern for the DNA segments which were synthesized in vitro provides absolute proof that one-half of the newly replicated DNA extended from the BUdR leaders introduced in active replicons in living cells. Inasmuch as 60-pulse-labeling studies also showed the presence of typical Okazaki fragments (100–200 nucleotides) that could be chased into both the 10S and the 20–80S regions of the gradient, it appears most likely that the 10S peak of DNA represents Okazaki fragments that have grown dramatically, but are not attached to their origins of replication. The observation that higher levels of CF resulted in a greater fraction of the newly replicated DNA sedimenting in the 20–80S regions of the gradients is in accord with the concept that one role of CF is to protect newly replicated DNA from chain scission. This is in addition to its role of promoting the DNA replication process positively.

Thus, product analysis provides evidence that the isolated nuclei continue DNA replication at sites that were active in the living cells by processes which appear similar to those operating in vivo. In addition, the presence of a new intermediate in the DNA replication process appears to accumulate in the nuclear system as a result of the limitations of some process that is critical to the maintenance of the ligated state of certain segments in DNA replication. Resolution of these molecular events and processes is a high-priority concern in our current research.

Closing Comments

The progress in these studies of DNA and chromatin replication in isolated nuclei of HeLa cells synchronized for S phase shows that it is possible to dissociate and reconstitute the DNA replicase system of eukaryotic cells. A few of the components have been identified, as have the properties of the DNA chain growth within active replicons. Preliminary studies on the uptake of labeled CF protein suggest that it may be possible to study chromatin replication or assembly in the in vitro systems, as well. Because the incoming proteins play a critical role in the progress of DNA replication, it is tempting to speculate that this coupling amid the discontinuous events of DNA replication may play a critical role in the formulation of chromatin character—a process that may be fundamental in the molecular biology of differentation.

REFERENCES

FRIEDMAN, D. L., and G. C. MUELLER. 1968. A nuclear system of DNA replication from synchronized HeLa cells. *Biochim. Biophys. Acta.* **161**:454–468.

HERRICK, G., H. DELIUS, and B. ALBERTS. 1976. Single-stranded DNA structure and DNA polymerase activity in the presence of nucleic acid helix-unwinding proteins from calf thymus. *J. Biol. Chem.* **251**:2142–2146.

HERSHEY, H. V., J. F. STIEBER, and G. C. MUELLER. 1973*a*. DNA synthesis in isolated HeLa nuclei: a system for continuation of replication *in vivo*. *Eur. J. Biochem.* **34**:383–394.

HERSHEY, H. V., J. F. STIEBER, and G. C. MUELLER. 1973*b*. Effect of inhibiting the cellular synthesis of RNA, DNA and protein on DNA replicative activity of isolated S-phase nuclei. *Biochim. Biophys. Acta.* **312**:509–517.

HUBERMAN, J. A., and A. D. RIGGS. 1968. On the mechanism of DNA replication in mammalian chromosomes. *J. Mol. Biol.* **32**:327–341.

JACKSON, V., and R. CHALKLEY. 1974. Separation of newly synthesized nucleohistone by equilibrium centrifugation in cesium chloride. *Biochemistry.* **13**:3952–3956.

MUELLER, G. C., K. KAJIWARA, E. STUBBLEFIELD, and R. R. RUECKERT. 1962. Molecular events in the reproduction of animal cells. *Cancer Res.* **22**:1084–1090.

SEKI, S., M. LEMAHIEU, and G. C. MUELLER. 1975. A permeable cell system for studying DNA replication in synchronized HeLa cells. *Biochim. Biophys. Acta.* **378**:333–343.

SEKI, S., and G. C. MUELLER. 1975. A requirement for RNA, protein, and DNA synthesis in the establishment of DNA replicase activity in synchronized HeLa cells. *Biochim. Biophys. Acta.* **378**:354–362.

SEKI, S., and G. C. MUELLER. 1976. Dissociation and reconstitution of the DNA replicase system of HeLa cell nuclei. *Biochim. Biophys. Acta.* **435**:236–250.

SPAEREN, U., K. SCHROEDER, C. SUDBERY, E. BJORKLID, and H. PRYDZ. 1975. DNA synthesis in HeLa cell nuclei isolated in a non-aqueous medium. *Biochim. Biophys. Acta.* **395**:413–421.

THOMPSON, L. R., and G. C. MUELLER. 1975. DNA replication in nuclei isolated from bovine lymphocytes. *Biochim. Biophys Acta.* **378**:344–353.

WEISSBACH, A., D. BALTIMORE, F. BOLLUM, R. GALLO, and K. KORN. 1975. Nomenclature of eukaryotic DNA polymerases. *Eur. J. Biochem.* **59**:1–2.

Cytoplasmic Control of Nuclear Expression

NUCLEOCYTOPLASMIC INTERACTIONS IN AMPHIBIAN OOCYTES

J. B. GURDON, E. M. DE ROBERTIS, G. A. PARTINGTON, J. E. MERTZ, and R. A. LASKEY

For many years, amphibian eggs have been extensively used by cell biologists for experimental work concerned with nucleocytoplasmic interactions. Their large size and tolerance to microsurgery make them particularly suitable for nuclear transplantation experiments. In more recent years, it has become clear that oocytes have many advantages over eggs for such experiments. In this article we summarize work from this laboratory in which oocytes have been used for the transplantation of nuclei or for the microinjection of purified macromolecules.

The term oocyte is used to refer to the growing egg cells contained in the ovary of a female. An oocyte has a single, very large nucleus, called a germinal vesicle (GV), which contains highly extended "lampbrush" chromosomes. These are highly active in RNA synthesis but are inactive in DNA synthesis during the growth of the oocyte, a process which, in most species, lasts for several months. In amphibians, and in most other animals, oocytes are intimately associated with hundreds or thousands of follicle cells. Oocytes accumulate in the ovary in a fully grown state until stimulated by pituitary gonadotropic hormones to undergo "maturation," which involves two meiotic divisions of chromosomes, the rupture of the GV, and the release of the oocyte from the ovary. The oocyte passes down the oviduct, where it is surrounded by jelly, and when released by the mother it is called an egg and is now, but not before, capable of being fertilized.

The potential value of *Xenopus* oocytes for experimental work was demonstrated some years ago; it was shown that embryonic as well as adult brain nuclei remain transcriptionally active for a few days after injection into oocytes (Gurdon, 1968). A few years later, the rather surprising result was obtained that purified mRNA molecules injected into *Xenopus* oocytes are efficiently translated (Gurdon et al., 1971). Subsequent work (reviewed by Gurdon, 1974) has shown that nearly all kinds of mRNA are translated in oocytes into the proteins for which they code, a process that can continue efficiently for a few weeks. It has been clear to us for some years that the usefulness of oocytes for the analysis of nucleocytoplasmic interactions and of gene control in general would be enormously extended if it were possible to recognize (1) the activity of *individual* genes in nuclei that have been transplanted to oocytes, and (2) the activity of genes introduced into oocytes as purified molecules of DNA, rather than as components of whole nuclei. During the past year, substantial progress toward both these objectives has been achieved. This is partly due to refinement in our methods of injecting and handling oocytes. But it is also dependent on the use of methods that provide great precision in recognizing proteins and RNAs, and particularly in distinguishing gene products synthesized endogenously by oocyte genes from those synthesized by injected nuclei or DNA molecules.

The Fate and Activity of Nuclei Transplanted to Oocytes

The fate and morphological appearance of nuclei transplanted to oocytes have been followed autoradiographically by the use of nuclei whose DNA has been prelabeled with [^{3}H]thymidine. The results, which have been summarized by Gurdon et al. (1976*a*) and described in detail by Gurdon (1976), are affected by the location within an oocyte at which nuclei are deposited. By careful positioning of the injection pipette, nuclei

J. B. GURDON and CO-WORKERS Medical Research Council Laboratory of Molecular Biology, Hills Road, Cambridge, England

may be inserted into the cytoplasm, the GV, or into the region of the dispersed contents of the GV. In all cases, nuclei swell progressively during the first few days after injection, so that HeLa nuclei with an initial diameter of 12 μm are often of 30-μm diameter after 3 days. However, nuclei deposited in the dispersed contents of a GV may undergo immense enlargement up to 70 μm in diameter, which represents a volume increase of 200 times.

Several changes are associated with nuclear enlargement. These include chromatin dispersal, nucleolar size change, and exchange of proteins between nucleus and cytoplasm. The dispersal of chromatin is seen in nuclei with prelabeled DNA. Even the most highly swollen nuclei have DNA dispersed throughout their volume. The fate of nucleoli in injected nuclei depends on the relatedness of the nuclear and cytoplasmic species. HeLa or other kinds of mammalian nuclei lose their nucleoli when transplanted to frog oocytes, whereas *Xenopus* nuclei in *Xenopus* oocytes show a substantial enlargement of nucleoli. Unlike most other nuclear characteristics, nucleolar development appears to be controlled by species—or at least order-specific factors. The exchange of proteins between injected nuclei and oocyte cytoplasm has been demonstrated by autoradiography of sectioned oocytes. The uptake of cytoplasmic proteins by nuclei is easily detected if oocytes are incubated in labeled amino acids for several hours before nuclear injection, or if ^{125}I-labeled histones are injected into oocytes previously injected with nuclei (see Gurdon et al., 1976*b*, for details). The passage of proteins from injected nuclei to oocyte cytoplasm is seen by prelabeling the proteins of donor nuclei and following by autoradiography the loss of labeled proteins from injected nuclei. Because a substantial loss of proteins takes place when they are prelabeled with [^{3}H]lysine and [^{3}H]arginine or with [^{3}H]tryptophan, it appears that basic proteins, as well as nonhistone proteins, are lost as injected nuclei are incubated in oocyte cytoplasm.

Much the most important activity of transplanted nuclei, for the present discussion, is their continuing synthesis of RNA. If oocytes containing transplanted nuclei are cultured for up to 4 wk, and then given a labeled RNA precursor for a few hours, their injected nuclei can be seen, by autoradiography, to be synthesizing RNA. Furthermore, injected nuclei appear to increase their rate of RNA synthesis nearly in proportion to their increase in volume (Gurdon et al., 1976*b*). Although it is hard to determine the absolute rate of RNA synthesis, it would seem that injected nuclei undergo a great activation of RNA synthesis above the level of cultured cell nuclei.

Gene Expression by Nuclei Transplanted to Oocytes

The very active synthesis of RNA by nuclei injected into oocytes does not necessarily show that this is meaningful RNA capable of coding for normal gene products. To demonstrate gene expression by transplanted nuclei, we have analyzed proteins synthesized by oocytes containing nuclei, using high-resolution, two-dimensional (2-D) electrophoresis according to O'Farrell (1975), coupled with fluorography under the conditions specified by Laskey and Mills (1975). The initial results of this work were summarized by Gurdon et al. (1976*a*); these and other experiments are described in detail by De Robertis et al. (1977*a*). After a fluorographic exposure time of up to 2 wk, we can see 200–300 clearly distinct spots in 2-D analyses of oocytes incubated for a few hours in labeled amino acids. If the proteins synthesized by HeLa cells are analyzed by the same technique, most of these proteins can be distinguished from those of *Xenopus* oocytes. When oocytes injected with HeLa nuclei are analyzed, the synthesis of three HeLa proteins can be detected a few days after the injection.

There are four independent reasons for believing that the HeLa proteins synthesized in frog oocytes depend upon new gene transcription of the HeLa nuclei in oocytes, and the synthesis of these proteins cannot be attributed to the translation of mRNA or nuclear RNA carried over with the injected nuclei. It therefore appears that oocyte cytoplasm contains components that permit the continued transcription of injected HeLa nuclei, and that this effect is selective, causing the expression of a particular minority of HeLa genes.

The experiments so far described have not told us whether oocyte cytoplasm is selective to the extent of being able to turn on oocyte-active genes from a previously inactive state, or only that it can maintain the transcription of appropriate genes which have already been activated. To answer this question, we have transplanted nuclei from cultured *Xenopus* kidney cells into oocytes; a preliminary account of these experiments has been published by De Robertis et al. (1977*a*). Twelve major proteins are synthesized by uninjected oocytes, but not by cultured kidney cells, and six major proteins by cultured cells, but not by oocytes. The experiment consisted of asking whether kidney-cell nuclei are induced to synthesize oocyte-active genes after injection into oocytes. In order to detect the induced transcription of oocyte-active genes, we have transplanted *Xenopus* kidney cell nuclei to oocytes of *Pleurodeles,* a different amphibian, many of whose oocyte-synthesized proteins do not coincide in 2-D gels with the oocyte-synthesized proteins of *Xenopus*. In the three successful experiments so far carried out, several new spots appeared in *Pleurodeles* oocytes 3–7 days after nuclear injection. Three of these spots corresponded, on 2-D gels, to *Xenopus* oocyte-synthesized proteins. There are several reasons (De Robertis et al., 1977*b*) for believing that the synthesis of these new proteins depends upon transcription.

Therefore, the present state of these experiments encourages the view that oocyte cytoplasm not only maintains the expression of oocyte-active genes, but also is able to induce the new activity of oocyte-active genes, even though these were previously inactive in the cells from which donor nuclei were taken.

The Transcription of DNA Injected into Oocytes

The eventual aim of our experiments is to understand in molecular terms the mechanism of selective gene expression by nuclei injected into oocytes. This would be greatly simplified if we could obtain selective transcription of purified molecules of DNA injected into oocytes instead of whole nuclei. It was shown some years ago that the induction of DNA synthesis in nuclei injected into unfertilized eggs could also be seen if purified DNA were injected instead of nuclei (Gurdon et al., 1969; Laskey and Gurdon, 1973). Colman (1975) was able to find transcripts from artificial DNA templates injected into oocytes or eggs. Transcription from natural DNA templates has been described by Gurdon and Brown (1977) for ribosomal and 5S DNAs of *Xenopus müIleri* injected into fertilized eggs of *Xenopus laevis*. Some indication of the selectivity of transcription of injected DNAs comes from the observation that mouse satellite DNA is not transcribed under the same conditions as those which apply to ribosomal and 5S DNAs (Table I). A detailed investigation of the transcription of purified DNAs in oocytes has been carried out by Mertz and Gurdon (1977). It is found that many different kinds of DNA are transcribed in oocytes as long as they are injected into the GV (Table II). Selectivity of transcription is suggested by the observation that the naturally transcribed (minus) strand of ϕX174 replicative form DNA is made in oocytes in very much larger quantities than the naturally nontranscribed (plus) strand. We do not yet know whether transcription is initiated in oocytes at natural promoter sites and terminated correctly. Nevertheless, the results so far obtained encourage us to hope that correct transcription may be obtained from injected DNAs.

CONCLUSIONS AND PROSPECTS

We have emphasized the use of oocytes because they provide a convenient experimental system in which selective gene expression can be obtained from injected nuclei. We have preliminary evidence to suggest that selectivity of gene expression

TABLE I

Transcription of Purified Natural DNA Molecules Injected into Fertilized Eggs of Xenopus laevis

Type of DNA injected	Stage of labeling transcripts	Hybridization of [^{3}H]RNA to DNA on filters: Type of DNA on filter	[^{3}H] RNA hybridized
			cpm
*X. müller*i ribosomal DNA	cleavage	*X. müller*i ribosomal DNA	100
No DNA injected	"	" "	5
*X. müller*i 5S DNA	"	*X. müller*i 5S DNA	74
No DNA injected	"	" "	4
Mouse satellite DNA	"	Mouse satellite DNA	3
No DNA injected	"	" " "	4

For details, see Gurdon and Brown, 1977.

TABLE II

Transcription of Purified DNAs Injected into Oocytes of Xenopus laevis

		Hybridization of [³H]RNA to DNA on filters	
Type of DNA injected	Site of injection	Type of DNA on filter	% of total [³H]RNA hybridized
SV40 form I*	GV	SV40	11.2
" " "	Cytoplasm	"	0.01
ColE1‡	GV	ColE1	21.6
Cloned ColE1-*Drosophila* histone gene recombinant plasmid	GV	ColE1	7.9
		ColE1-*Drosophila* histone genes	15.0
ϕX174 RFI§	GV	ϕX174 RFI	10.9
" "	GV	ϕX174 "+" strand	0.6
ϕ80*plac*¶	GV	ϕ80*plac*	2.3
Adenovirus 5**	GV	Adenovirus 5	4.7

For details, see Mertz and Gurdon, 1977.
* Simian virus 40 DNA in a closed circular supercoiled configuration.
‡ From the bacterial plasmid ColE1.
§ Bacteriophage ϕX174 DNA in a two-stranded replicative form configuration.
¶ From the bacteriophage ϕ80 which contains the lactose operon of E. coli inserted into it in reverse polarity.
** DNA from whole genomes of adenovirus serotype 5.

by injected nuclei operates not only to turn off unwanted genes, but also to turn on previously inactive genes. If this result can be substantiated, we would have the basis of an experimental system by which we could hope to identify the oocyte components responsible for this selective gene control. The first step in this direction has been to try to obtain selective transcription from injected purified DNAs. If we find correct transcription from many different kinds of DNA, we could use injected oocytes to identify promoter and other important regions of purified DNA sequences that contain known genes. As a longer-term objective, we hope to pursue this work in two directions. One is to test the function of nonDNA components of chromosomes, the effect of different configurations of DNA, etc., on selective transcription. The other is to use purified segments of DNA, injected into oocytes and subsequently re-isolated with associated molecules, to "fish out" the components of oocytes that determine selective transcription.

REFERENCES

COLMAN, A. 1975. Transcription of DNAs of known sequence after injection into the eggs and oocytes of *Xenopus laevis*. *Eur. J. Biochem.* **57:**85–96.

DE ROBERTIS, E. M., J. B. GURDON, G. A. PARTINGTON, J. E. MERTZ, and R. A. LASKEY. 1977*a*. Injected amphibian oocytes: a living test tube for the study of eukaryotic gene transcription? *Biochem. Soc. Symp.* In press.

DE ROBERTIS, E. M., G. A. PARTINGTON, R. LONGTHORNE, and J. B. GURDON. 1977*b*. Somatic nuclei in amphibian oocytes: Evidence for selective gene expression. *J. Embryol. Exp. Morphol.* In press.

GURDON, J. B. 1968. Changes in somatic cell nuclei inserted into growing and maturing amphibian oocytes. *J. Embryol. Exp. Morphol.* **20:**401–414.

GURDON, J. B. 1974. The Control of Gene Expression in Animal Development. Harvard University Press, Cambridge, Mass. 160.

GURDON, J. B. 1976. Injected nuclei in frog oocytes: fate, enlargement and chromatin dispersal. *J. Embryol. Exp. Morphol.* **36:**523–540.

GURDON, J. B., M. L. BIRNSTIEL, and V. A. SPEIGHT. 1969. The replication of purified DNA introduced into living egg cytoplasm. *Biochim. Biophys. Acta.* **174:**614–628.

GURDON, J. B., and D. D. BROWN. 1976. Toward an *in vivo* analysis of gene control and function. Symposium on Molecular Biology of the Genetic Apparatus. P. T'so, editor. North-Holland Publishing Co., Amsterdam. **2:**111–123.

GURDON, J. B., E. M. DE ROBERTIS, and G. A. PARTINGTON. 1976*a*. Injected nuclei in frog oocytes provide a living cell system for the study of transcriptional control. *Nature (Lond.)*. **260:**116–120.

Gurdon, J. B., C. D. Lane, H. R. Woodland, and G. Marbaix. 1971. The use of frog eggs and oocytes for the study of messenger RNA and its translation in living cells. *Nature (Lond.).* **233:**177–182.

Gurdon, J. B., G. A. Partington, and E. M. De Robertis. 1976*b*. Injected nuclei in frog oocytes: RNA synthesis and protein exchange. *J. Embryol. Exp. Morphol.* **36:**541–553.

Laskey, R. A., and J. B. Gurdon. 1973. Induction of polyoma DNA synthesis by injection into frog-egg cytoplasm. *Eur. J. Biochem.* **37:**467–471.

Laskey, R. A., and A. D. Mills. 1975. Quantitative film detection of ^{3}H and ^{14}C in polyacrylamide gels by fluorography. *Eur. J. Biochem.* **56:**335–341.

Mertz, J. E., and J. B. Gurdon. 1977. Purified DNAs are transcribed after injection into oocytes. *Proc. Natl. Acad. Sci. U.S.A.* In press.

O'Farrell, P. H. 1975. High resolution two-dimensional electrophoresis of proteins. *J. Biol. Chem.* **250:**4007–4021.

RECONSTRUCTION OF VIABLE CELLS FROM CELL FRAGMENTS

N. R. RINGERTZ, N. BOLS, T. EGE, A. KANE, U. KRONDAHL, S. LINDER, and K. SHELTON

One of the fundamental problems of cell biology is to understand the interaction between the nucleus and the cytoplasm. A new experimental approach has been developed to study this problem. Through the use of enucleation techniques, it is possible to divide cells into nuclear and cytoplasmic cell fragments (Prescott et al., 1972). Thus it is possible to obtain large numbers of *minicells* (karyoplasts), each consisting of an intact nucleus and a thin rim of cytoplasm (Ege et al., 1973, 1974*a*), and *anucleate cells* (cytoplasms or cytoplasts), containing most of the cytoplasm of the intact cell (Prescott et al., 1972; Ege et al., 1974*a*; Shay et al., 1974). Both types of cell fragments are metabolically stable for limited periods, varying between a few hours and a few days, but ultimately they die. Because they are surrounded by a cell membrane carrying receptors for Sendai virus they may, however, be recombined into viable cells by Sendai virus-induced fusion. Enucleation or disruption of "micronucleated" cells, cells in which the genome has been fragmented into many small micronuclei by prolonged exposure to antimitotic agents such as colchicine, results in *microcells*. These cell fragments are subdiploid and may contain as little as a single chromosome. Fusion of microcells from normal cells with intact mutant cells may offer a means of transferring complementing chromosomes into genetically defective cells.

The aim of this presentation is to review briefly the methods used in generating cell fragments and in identifying reconstituted cells formed by virus-induced fusion of such fragments. (For more detailed information and specific references, see Ringertz and Savage, 1976.)

N. R. RINGERTZ and CO-WORKERS Institute for Medical Cell Research and Genetics, Medical Nobel Institute, Karolinska Institutet, Stockholm, Sweden

MATERIALS AND METHODS

Cells

Cell fragments have been prepared from a variety of human, rat, mouse, hamster, frog, and chick cells. Some of the cell types have been chosen because they show interesting phenotypes, while others have been selected because of technical advantages in reconstruction experiments. Several of the cell types used are mutant cells lacking hypoxanthine-guanine phosphoribosyl-transferase activity (HGPRT). Because of this defect, these cells are unable to grow on HAT selective medium and are unable to incorporate exogenous [^{3}H]hypoxanthine into nucleotides. [HAT selective medium contains hypoxanthine, aminopterin, and thymidin. It can be used to isolate hybrid cells arising from the fusion of cells deficient in hypoxanthine guanine phosphoribosyl transferase (HGPRT$^-$) with cells deficient in thymidine kinase (TK$^-$)]. In most of the work described in this report, we used L6 rat myoblasts and mutant sublines of mouse L cells ("fibroblasts"). A9 is a subline of mouse L929 cells that originally was described as fibroblastic. The A9 cells, however, have not been tested for collagen and hyaluronic acid syntheses, two markers characteristic of the differentiated fibroblast phenotype.

Preparation of Cell Fragments

Several methods of enucleating large numbers of cells efficiently have now been described (see reviews in Poste, 1973; Ringertz and Savage, 1976). All of these methods employ centrifugation in the presence of cytochalasin B. Enucleation is usually performed with cells attached to glass or plastic surfaces, but enucleation of cells in suspension has now been described. We most commonly use disks of 25-mm diameter and centrifuge, with the disks inverted, in phosphate buffered saline (PBS) containing 10 μg/ml cytochalasin B and 10% calf serum. Different types of cells vary in the conditions required for optimal enucleation. Centrifugal forces of between 3,000–48,000 *g* for periods of 20–60 min have been used. In favorable circumstances >99% enucleation is achieved. Cell fragments obtained by this method have been examined for dye exclusion (trypan blue viability test), dry mass (microinterferometry), DNA and protein content (microspectrophotometry), nucleic acid and protein syntheses (autoradiography), and ultrastructure (electron microscopy).

The anucleate cells remain attached to the culture surface. Although they appear grossly abnormal immediately after centrifugation, they recover within 1–2 h and resemble the parent cell morphologically. They are motile and initially can synthesize protein at rates similar to the intact parent cells. However, these activities steadily diminish, and most cells round up and die within 72 h.

The nucleated fragments of the cells (minicells) are recovered from the pellet in the centrifuge tubes. As measured by microinterferometry, minicells from rat L6 myoblasts contain less than 10% of the original cytoplasm of the parent cell. Eighty percent exclude trypan blue and incorporate [^{3}H]uridine. Some also incorporate [^{3}H]thymidine. Minicells do not, however, attach to surfaces as readily as do intact parent cells. Minicell preparations, therefore, may be purified from contaminating intact cells by allowing them to attach and then recovering the suspended or loosely attached minicells. Contrary to the findings of some other workers (Lucas et al., 1976), we have not found that minicells prepared under these conditions regenerate their lost cytoplasm. Our studies, which concentrated on minicells prepared from rat L6 myoblasts, show instead that, under normal culture conditions employed in our experiments, minicells lyse and die within 48 h.

Microcells are obtained by centrifuging micronucleated cells in the presence of cytochalasin. Evidence recently reviewed elsewhere (Ege et al., 1976) indicates that micronucleated cells arise from abnormal mitoses induced by microtubular poisons (colchicine, Colcemid, vinblastines, etc.) or cold shock. Under these conditions, nuclear membranes reassemble around individual or small groups of chromosomes to give rise to micronuclei. Enucleation of micronucleated cells results in subdiploid microcells containing one or several micronuclei surrounded by a rim of cytoplasm and a plasma membrane. As with minicells, most of the microcells are metabolically intact in the sense that they persist in culture for a short time and exclude trypan blue.

RESULTS

Three different types of reconstructed cells have been prepared, using Sendai virus to induce fusion of cell fragments. (1) Reconstituted cells: anucleate cells + minicells; anucleate cells + nucleated erythrocyte ghosts; (2) cytoplasmic hybrids ("cybrids"): anucleate cells + intact cells; (3) microcell heterokaryons: microcells + intact cells. One of the chief problems in these experiments lies in recognizing and distinguishing the intact parental cells, the reconstituted cells, and cytoplasmic hybrids (cybrids). Four types of markers have been used to identify the origin of individual cells and small colonies of cells (15–100 cells) arising from reconstructed cells: (*a*) morphological markers (distinct nuclear shape), (*b*) artificially produced markers (ingestion of polystyrene beads of different size classes), (*c*) DNA content, and (*d*) functional markers. The functional markers (mutant cells defective in specific enzyme functions) are useful not only for identification purposes, but also to obtain the progeny of the desired reconstructed cells by growth on selective media.

Identification of Single Reconstituted Cells Containing L6 Nuclei

L6 minicells have been fused with cytoplasms prepared from three different cell lines (Ege et al., 1974*b*; Ege and Ringertz, 1975; Krondahl et al., 1977): a mutant subline of L6 cells (L6 HGPRT$^-$) and two mutant sublines of mouse L cells (A9 HGPRT$^-$ and B82 TK$^-$). The technique used in identifying reconstituted cells and cybrids after fusion of minicells prepared from normal L6 cells (nuclear donor) with cytoplasms from L6 HGPRT$^-$ cells (cytoplasmic donor) is illustrated in Fig. 1. The cytoplasm of the nuclear donor cell is labeled with 150–300 small polystyrene beads with a uniform diameter of 0.4 μm. Since this cell is normal with respect to the gene specifying HGPRT, it is able to use exogenous [^{3}H]hypoxanthine for nucleotide and nucleic acid synthesis. The cytoplasmic donor, on the other hand, is a mutant cell (HGPRT$^-$) with no detectable HGPRT activity and, therefore, is unable to use exogenous [^{3}H]hypoxanthine for nucleic acid synthesis. Before enucleation, the cytoplasmic donor cells are labeled with 30–70 large (1.0 μm) polystyrene beads. Bead-labeled fragments are fused using UV-inactivated Sendai virus. After 2 days of incubation, most of the unfused minicells and cytoplasms have lysed, and the remaining cells are exposed to a pulse of [^{3}H]hypoxanthine. Reconstituted cells and cybrids can then be identified and distinguished from contaminating intact nuclear and cytoplasmic donor cells (Fig. 1) by autoradiography and microscopic identification of the beads in the cytoplasm. Reconstituted cells incorporate [^{3}H]hypoxanthine and contain large beads

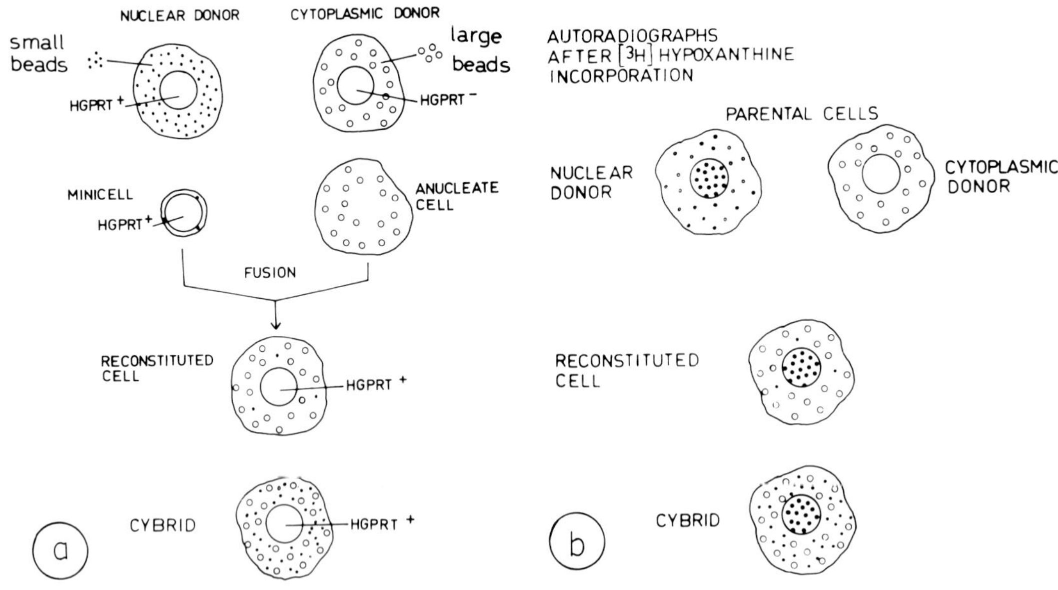

FIGURE 1 (*a*) Schematic summary of markers used in the identification of single reconstituted cells. (*b*) Appearance of parental and reconstructed cells in autoradiograms (Ringertz and Savage, 1976).

in the cytoplasm; cybrids are radioactively labeled and contain both large and small beads.

By these techniques, we established that reconstituted cells that are metabolically active for at least 2 days can be formed (Ege and Ringertz, 1975). In addition, with the use of quantitative cytochemical methods (Ege et al., 1974*a*) and labeling with small plastic beads, we demonstrated that the nuclear donor cells contributed very little cytoplasm. A few cells were observed in varying stages of mitosis, but it was not established whether these cells would multiply and form colonies. This has, however, been shown for another type of reconstitution, which is discussed in the following section.

Identification of Colonies Arising from Reconstituted Cells

The same techniques that were used for the identification of single reconstituted cells could, in principle, also be used for the identification of cell clones of 50–100 cells arising from reconstituted cells. The main difference is that, instead of examining single cells for [^{3}H]hypoxanthine incorporation and beads, whole colonies of cells would be examined. However, the presence of a large number of beads may be inhibitory for cell division. For this reason, a slightly different technique was used in identifying colonies arising from reconstituted cells. This technique is illustrated in Fig. 2.

In these experiments, rat L6 myoblasts were used to prepare minicells, which were then fused with cytoplasms from mouse A9 (HGPRT$^-$) fibroblasts. The nuclear donor cells were labeled in the cytoplasm with approximately 200 large (1.0 μm) polystyrene beads and carried a nuclear gene marker in the form of HGPRT$^+$. Minicells prepared from the nuclear donor contained <20 beads (mean: 6 beads/minicell) and had lost approximately 90% of the cytoplasmic dry mass (determined by microinterferometry). The minicell preparations were contaminated by approximately 7% intact L6 cells. When these preparations were cultured, the intact cells gave rise to cell colonies containing >100 beads/colony. Therefore, under the conditions used, the minicells did not regenerate cytoplasm and were unable to form cell colonies.

After fusion with unlabeled anucleate A9 cells (enucleation efficiency >95%), the reconstituted cells were allowed to recover for 24 h on normal culture medium. At this time, the medium was changed to selective HAT medium on which intact A9 cytoplasmic donor cells are unable to grow and form colonies. After 5 days, the cultures were exposed to a pulse of [^{3}H]hypoxanthine and fixed for autoradiography and Feulgen microspectrophotometry (DNA content/cell). The numbers of beads/colony and cells/colony were determined by phase contrast microscopy.

The possible types of colonies that could arise

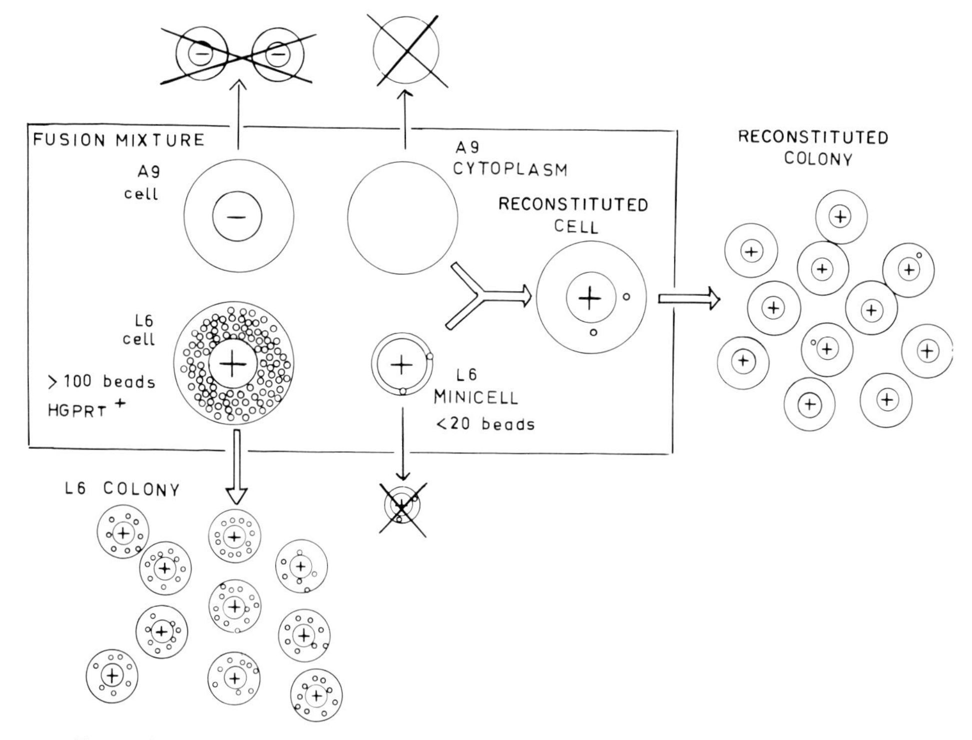

FIGURE 2 Identification of colonies of reconstituted cell colonies (for explanation, see text).

under these conditions were (*a*) reconstituted cells, (*b*) L6 cells (from intact cells contaminating the minicell preparation), and (*c*) hybrid cells arising from rare fusions of L6 minicells with contaminating intact A9 cells. Only the first two types of colonies were, in fact, observed. The results can be summarized as follows: in the absence of the fusing agent, inactivated Sendai virus, only one type of colony was observed. Each colony contained a total of >80 beads and had incorporated [^{3}H]hypoxanthine. Therefore, these colonies arose from intact L6 cells contaminating the minicell preparations. In the presence of virus, a new type of colony was found in addition to the L6 colonies. The new colonies consisted of 30–50 cells which had incorporated [^{3}H]hypoxanthine and had the same DNA content/cell as L6 nuclei. The colonies contained <20 beads/colony (as compared to >80 beads/colony for L6 colonies). This new type of colony that arose only in the presence of Sendai virus must therefore have originated from reconstituted cells formed by the fusion of L6 minicells with A9 cytoplasms. Thus, viable cells capable of multiplication have been reconstructed from two nonviable cell fragments derived from two different species and two different cell types.

Microcell Heterokaryons

Microcell heterokaryons can be clearly identified when microcells prepared from [^{3}H]thymidine cells are fused with unlabeled cells using UV-inactivated Sendai virus. Little is yet known about the properties of such heterokaryons.

DISCUSSION

Reconstructed cells may provide new insight into the interactions between the nucleus and cytoplasm. The cell reconstitution technique has the advantage over conventional cell fusion experiments with intact cells that the interaction between a single nucleus and a foreign cytoplasm can be studied. The subsequent analysis is, therefore, less complex than the usual situation in heterokaryons, in which two nuclei and two cytoplasms are mixed. Further, heterokaryons do not always give rise to hybrid cells, and when they do, the investigator has little control over the parental cytoplasmic and nuclear contributions to the hybrids. Obviously, accurate identification of reconstructed cells is essential to the realization of these benefits. This is a difficult problem technically. In this presentation, we have described how several types of reconstructed cells and cell colonies arising from reconstructed cells can be prepared and, further, how they can be identified satisfactorily.

Specific areas of nucleocytoplasmic interactions in which reconstitution techniques should prove useful include regulation of gene expression, the stability of the differentiated state, and the dependence of mitochondria and other cytoplasmic organelles on nuclear genes. Microcell hybrids, while only tentatively identified, could be useful in chromosome mapping, in gene complementation analysis designed to distinguish structural and regulatory mutations, and in the analysis of integrating sites for tumor viruses.

ACKNOWLEDGMENTS

This work was supported by grants from the Swedish Cancer Society.

REFERENCES

EGE, T., H. HAMBERG, U. KRONDAHL, J. ERICSSON, and N. R. RINGERTZ. 1974*a*. Characterization of minicells (nuclei) obtained by cytochalasin enucleation. *Exp. Cell Res.* **87**:365–377.

EGE, T., U. KRONDAHL, and N. R. RINGERTZ. 1974*b*. Introduction of nuclei and micronuclei into cells and enucleated cytoplasms by Sendai virus induced fusion. *Exp. Cell Res.* **88**:428–432.

EGE, T., and N. R. RINGERTZ. 1975. Viability of cells reconstituted by virus induced fusion of minicells with anucleate cells. *Exp. Cell Res.* **94**:469–473.

EGE, T., N. R. RINGERTZ, H. HAMBERG, and E. SIDEBOTTOM. 1976. Preparation of microcells. *Methods Cell Biol.* **15**:339–357.

EGE, T., J. ZEUTHEN, and N. R. RINGERTZ. 1973. Cell fusion with enucleated cytoplasms. Nobel Symposium 23 on Chromosome Identification. T. Caspersson and L. Zech, editors). Academic Press Inc., New York. 189–194.

KRONDAHL, U., N. BOLS, T. EGE, S. LINDER, and N. R. RINGERTZ. 1977. Cells reconstituted from cell fragments of two different species multiply and form colonies. *Proc. Natl. Acad. Sci. U.S.A.* In press.

LUCAS, J. J., E. SZEKELY, and J. R. KATES. 1976. The regeneration and division of mouse L-cell karyoplasts. *Cell.* **7**:115–122.

POSTE, G. 1973. Anucleate mammalian cells: applications in cell biology and virology. *Methods Cell Biol.* **8**:211–250.

PRESCOTT, D. M., D. MYERSON, and J. WALLACE. 1972. Enucleation of mammalian cells with cytochalasin B. *Exp. Cell Res.* **71**:480–485.

RINGERTZ, N. R., and R. E. SAVAGE. 1976. Cell Hybrids. Academic Press, Inc., New York.

SHAY, J. W., K. R. PORTER, and D. M. PRESCOTT. 1974. The surface morphology and fine structure of CHO (Chinese hamster ovary) cells following enucleation. *Proc. Natl. Acad. Sci. U. S. A.* **71**:3059–3063.

GENETIC AND EPIGENETIC STUDIES WITH SOMATIC CELL HYBRIDS

F. H. RUDDLE

Somatic cell hybrids can be used for both genetic analysis and developmental studies. I shall report on both, but emphasize the latter.

In *interspecific* cell hybrids one chromosome set is usually lost. One can use such a system to map genes by correlating the presence or loss of phenotypes with particular chromosomes. In man, more than 100 genes have been mapped in this way (Ruddle and Creagan, 1975), and we believe in excess of 1,000 will be mapped within the next decade. The system can also be extended to other mammalian species.

At the outset, I would like to describe a new procedure which we call microcell-mediated chromosome transfer which we believe will serve as a useful technique in both the mapping of genes, and testing the epigenetic effects of chromosomes. The technique was originally developed by Ringertz and Ege, and we are indebted to them for their help in getting us started in its use. As shown in Fig. 1, one prepares microcells from mouse $HPRT^+$ donor cells by sequential Colcemid and cytochalasin treatments. The microcells containing a partial genome of one to several chromosomes are fused to the $HPRT^-$ Chinese hamster recipient cells with Sendai virus as the fusogen. One can detect the delivery of the micronuclei readily, and the frequency of transfer is relatively high ($\sim 10^{-2}$). Within several generations, the donor chromosomes can be detected in the nuclei as a consequence of the special staining properties of the mouse chromosomes. Hoechst dye 33258 specifically fluoresces with the constitutive (satellite) DNA of the mouse chromosomes so that by counting chromocenters one can estimate the number of chromosomes delivered in the independently selected clones in HAT medium (Moser et al., 1975). We believe our work represents the first successful isolation of microcell hybrids (Fournier and Ruddle, 1977). The advantages of the procedure are that one transfers only a small subset of donor chromosomes into a recipient cell. Thus, the direction of chromosome segregation is determined by the choice of the donor. Since only one or several chromosomes are transferred, gene mapping is greatly simplified. Already several genes have been mapped by this procedure; and in addition, microcell hybrids have been produced which express gene products from three different species—man, mouse, and the Chinese hamster (trihybrids). Another possible advantage relates to the assignment of facultative genes. It can be assumed that the insertion of a small fraction of the donor genome will less likely upset the epigenetic program of a recipient cell. If this is correct, we expect to see donor facultative genes related to the epigenotype of the recipient activated. Precedent exists for the activation of genes in standard hybridization studies (Darlington et al., 1974). The possibility of gene activation is currently being tested by the use of the microcell technology.

Turning now to a developmental system of analysis, I shall discuss two fusion experiments in which all parents are clearly defined in terms of their epigenetic status. In the first experiment, the parents are the Friend erythroleukemia line (745) and Hepatoma la (Conscience et al., 1977; Fig. 2). Both cell populations are derived from inbred mice: Friend (DBA/2) and Hepa-la (L57 L/J). Thus the cross is *intraspecific*, but because of inbred strain differences, constitutive isozyme marker differentials exist for the enzymes MOD and PGM-1, and these thereby serve to verify hybrid cell formation. The chromosome constitutions of the two lines differ substantially: the Friend cells have a near diploid count (39) and the Hepa have a near triploid count (59).

F. H. RUDDLE Department of Biology, Yale University, New Haven, Connecticut

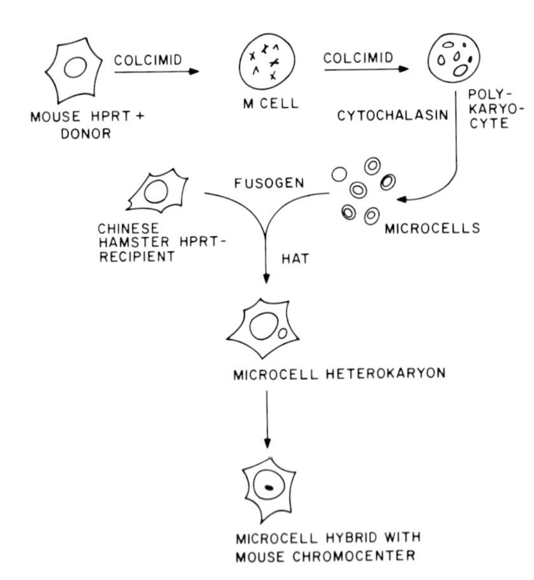

FIGURE 1 Microcell chromosome transfer system.

The differentiated phenotypes of the two lines are distinctive. The Friend cells, among other traits, produce hemoglobin, show iron uptake, and have elevated specific activities for acetylcholinesterase and carbonic anhydrase. As shown originally by Charlotte Friend, the expression of these phenotypes can be elevated by treatment with 1.5–2.0% dimethyl sulfoxide. This type of treatment thereby allows one to distinguish between constitutive and induced levels of expression. The ensemble of traits is consistent with the view that the uninduced cell corresponds to a proerythroblast.

The hepatoma parent (Hepa-la), among other traits, produces certain serum proteins such as albumin, transferrin, and α-fetoprotein. The production of albumin can be quantitated by various immunochemical procedures (Bernhard et al., 1973).

Eighteen independent hybrid clones (FHP) have been analyzed. All are true hybrids on the basis of chromosome constitution (~92) and isozyme expression. Practically all the input chromosomes are retained, and it would seem unlikely that the physical loss of genes could explain phenotype extinction.

The results clearly show that erythroid functions are extinguished and that hepatoma phenotypes continue to be expressed at levels comparable with the parent Hepa-la – especially in the case of albumin and transferrin. However, a number of the clones did not produce α-fetoprotein. The erythroid phenotypes appeared to be completely extinguished. mRNA globin assays were particularly informative in this connection, since the cDNA hybridization assay is sensitive at a level of about 10 molecules of globin mRNA per cell.

These data allow us to draw the following conclusions: (1) The extinction of the erythroid traits is very nearly, if not totally, complete. (2) Extinction is not a trivial, nonspecific event, since the hepatoma traits continue to be expressed. (3) The mRNA data suggest that extinction is not mediated at a translational level – but most probably involves a nuclear process. (4) The results provide additional evidence that genome dosage importantly influences extinction. In this instance, the triploid parent (Hepa) is dominant over the diploid (Friend cell) parent. These kinds of results suggest that whatever molecules mediate control, they are not present in very great excess over that concentration required to exert their influence (5). The data are also consistent with the view that developmental controls are designed in such a way so as to exclude the coexpression of two different epigenetic programs within a single cell.

The second fusion experiment, as in the previous case, makes use of mouse parental cell populations (Miller and Ruddle, 1976; Fig. 3). These are (1) an embryonal carcinoma cell (PCC4aza1; kindly supplied by François Jacob, Institut Pasteur), and (2) a normal thymocyte population ob-

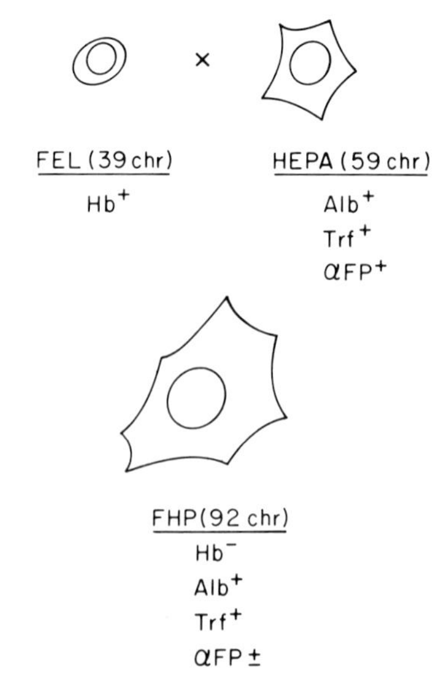

FIGURE 2 Erythroleukemia cell and hepatoma cell hybrid system.

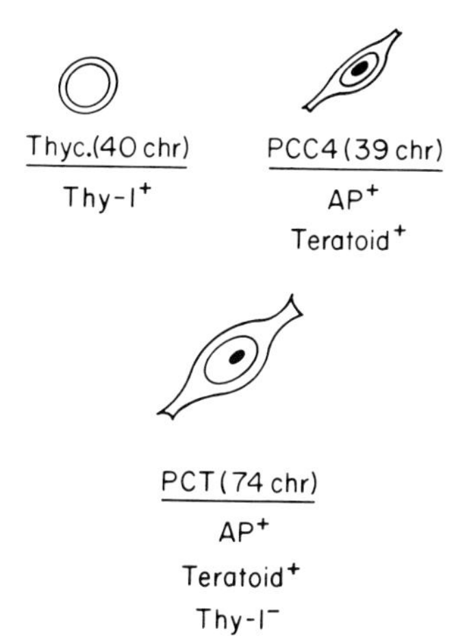

FIGURE 3 Thymocyte and teratocarcinoma cell hybrid system.

tained directly from the thymus. The thymocytes have a balanced diploid chromosome constitution, whereas PCC4azal has a very nearly balanced diploid constitution with 39 chromosomes.

PCC4aza1 is a pluripotent cell derived from a mouse teratocarcinoma. It possesses a number of unique properties, the most interesting of which is its pluripotency. When injected subcutaneously, the embryonal cells give rise to a broad spectrum of tissue types including skeletal muscle, cartilage, and nerve cells. The cells are also characterized by high alkaline phosphatase activity, unique surface properties (F9 antigen), and resistance to replication by the murine parvovirus MVM (minute virus of mouse). The morphology of these embryonal cells is also distinctive in that such cells possess a large nucleolus and limited cytoplasm. Thymocytes lack these properties, and, in addition, they possess a different characteristic morphology and express the thymocyte-specific Thy-1 surface antigen. Two separate hybridization experiments gave a total of four hybrid cell populations which could be verified as hybrid both by isozyme and chromosome analysis.

When we designed this experiment, we expected the more developmentally plastic embryonal cell to conform to the more stably differentiated thymocyte. We were surprised by the opposite result which is supported by the following data. The alkaline phosphatase results show that the hybrids (PCT) possess enzyme activity comparable to that found in the PCC4 cells. The morphology of the PCT hybrids also resembled that of embryonal cells: large nucleolus and scant cytoplasm depleted in organelles. Cytotoxicity tests showed that the Thy-1 antigen was absent from the hybrid cells. The replication of MVM virus was tested immunochemically, the presence of virus antigens being detected within nuclei by immunofluorescence (Miller et al., 1977).

The most convincing evidence that PCT hybrids conform to the epigenotype of the embryonal carcinoma cell parent comes from studies on their pluripotency. The results show that when PCT hybrids are transplanted into nude mice—or histocompatible 129 × c34 F_1 hybrids—a broad spectrum of tissue types are generated, including neuronal, muscle, glandular, and cartilage tissues.

Other experiments have shown that hybrids between PCC4aza1 and Friend erythroleukemia cells are also embryonal in nature (Miller and Ruddle, 1977).

The data show that the pluripotent cell, far from being unconditionally susceptible to whatever signals differentiated cells employ to maintain their stable phenotype, may itself be able to "reset" the genome of the differentiated cell. In this way, our experiments may bear a correspondence to the nuclear transplantation experiments carried out by Gurdon (1974). These results also suggest the technical feasibility of introducing small pieces of a foreign genome into embryonal cells in normal embryos (Mintz and Illmensee, 1975). Possibly the transfer of such genome fragments could provide information on specific genetic factors which influence the developmental process in particular ways.

ACKNOWLEDGMENTS

A number of collaborators have contributed to these studies. I would especially like to acknowledge: at Yale, Dr. Keith Fournier, Dr. Jean-Francois Conscience, and Dr. Richard Miller; at the Cornell Medical Center, New York, Dr. Gretchen Darlington; at Albert Einstein College of Medicine, Bronx, N. Y., Dr. Arthur Skoultchi.

This work is supported by grant GM09966 and National Cancer Institute contract no. N01CP55673.

REFERENCES

BERNHARD, H. P., G. J. DARLINGTON, and F. H. RUDDLE. 1973. Expression of liver phenotypes in cultured mouse hepatoma cells: synthesis and secretion of serum albumin. *Dev. Biol.* **35:**83–96.

CONSCIENCE, J.-F., F. H. RUDDLE, A. SKOULTCHI, and G. J. DARLINGTON. 1977. Suppression of co-expression of different sets of epigenetic traits in somatic cell hybrids between the Friend erythroleukemia cells and mouse hepatoma. *Som. Cell Genet.* In press.

DARLINGTON, G. J., H. P. BERNHARD, and F. H. RUDDLE. 1974. Human serum albumin phenotype activation in mouse hepatoma-human leucocyte cell hybrids. *Science (Wash. D. C.).* **185:**859–862.

FOURNIER, R. E., and F. H. RUDDLE. 1977. Microcell mediated transfer of murine chromosomes into mouse, Chinese hamster and human somatic cells. *Proc. Natl. Acad. Sci. U. S. A.* In press.

GURDON, J. B. 1974. The Control of Gene Expression in Animal Development. Harvard University Press, Cambridge, Mass. 166.

MILLER, R. A., and F. H. RUDDLE. 1976. Pluripotent teratocarcinoma-thymus somatic cell hybrids. *Cell.* **9:**45–55.

MILLER, R. A., and F. H. RUDDLE. 1977. Teratocarcinoma × Friend erythroleukemia cell hybrids resemble their pluripotent embryonal carcinoma parent. *Dev. Biol.* In press.

MILLER, R. A., D. C. WARD, and F. H. RUDDLE. 1977. Embryonal carcinoma cells (and their somatic cell hybrids) are resistant to infection by murine parvovirus MVM, which does infect other teratocarcinoma-derived cell lines. *J. Cell. Physiol.* In press.

MINTZ, B., and K. ILLMENSEE. 1975. Normal genetically mosaic mice produced from malignant teratocarcinoma cells. *Proc. Natl. Acad. Sci. U.S.A.* **72:**3585–3589.

MOSER, F. G., B. P. DORMAN, and F. H. RUDDLE. 1975. Mouse-human heterokaryon analysis with a 33258 Hoechst-giemsa technique. *J. Cell Biol.* **66:**676–680.

RUDDLE, F. H., and R. P. CREAGAN. 1975. Parasexual approaches to the genetics of man. *Annu. Rev. Genet.* **9:**407–497.

CYTOPLASMIC CONTROL OF NUCLEAR DNA REPLICATION IN *XENOPUS LAEVIS*

ROBERT M. BENBOW, HANS JOENJE, SONIA H. WHITE, CAROL B. BREAUX, MARC R. KRAUSS, CHRISTOPHER C. FORD, and RONALD A. LASKEY

Nuclear DNA replication during early development of the frog, *Xenopus laevis,* is controlled by components found in the cytoplasm of eggs before fertilization (reviewed by Gurdon and Woodland, 1968). These cytoplasmic components can induce *normal* DNA replication in nuclei microinjected into unfertilized eggs – as was convincingly proved by the elegant nuclear transplantation experiments of Gurdon (1960). Merriam (1969) has subsequently shown by autoradiography that some of these components are proteins which enter into the microinjected nuclei before the induction of nuclear DNA replication. However, identification of the *specific* cytoplasmic components involved in the initiation and control of DNA replication has proved quite difficult (reviewed by Edenberg and Huberman, 1975).

About 2 years ago we developed the first efficient and reproducible assay for the cell-free initiation of DNA replication in isolated nuclei (Benbow and Ford, 1975). By the use of this assay, we identified a protein (or proteins) of high molecular weight which appeared to initiate nuclear DNA replication in vitro. This protein, which we called "I-factor," was found at very high levels in rapidly dividing cells, but only at very low levels in nonproliferating cells (Benbow and Ford, 1975). Recently, Jazwinski, Wang, and Edelman (1976) have confirmed our results using cytoplasm prepared from mouse and avian cultured cell lines to apparently initiate DNA replication in isolated *X. laevis* nuclei.

During our initial attempts to purify and characterize I-factor and other cytoplasmic molecules controlling nuclear DNA replication, we encountered two serious problems: firstly, the isolated nuclei contained varying levels of endogenous DNA replication enzymes which interfered with reconstitution experiments; secondly, it was difficult to prove rigorously that *de novo* initiation at the correct origin for normal DNA replication occurred in the nuclei during incubation. To circumvent these problems we have extended our original assay (Benbow and Ford, 1975) to the replication of purified DNA molecules in cell-free cytoplasm prepared from unfertilized *Xenopus laevis* eggs (Benbow and Laskey, 1977*a*, 1977*b*; Benbow et al., 1977, Reconstitution of polyoma DNA replication using components from unfertilized eggs of *Xenopus laevis,* manuscript in preparation).

A Cell-Free DNA Replication System

The basic assay utilizes purified, closed, circular DNA molecules as templates, components purified from the cytoplasm of oocytes or unfertilized eggs of *X. laevis* as the replication machinery, and the appearance of DNA replication intermediates monitored by electron microscopy as the criterion by which the occurrence of DNA replication is defined.

Two small, well-defined closed circular DNA molecules were selected as templates: polyoma viral DNA (contour length 1.6 μm) and *Escherichia coli* plasmids containing *X. laevis* ribosomal DNA (contour lengths 4.6 and 5.1 μm). Polyoma viral DNA was chosen because its mechanism of

ROBERT M. BENBOW AND CO-WORKERS Department of Biology, Johns Hopkins University, Baltimore, Maryland
CHRISTOPHER C. FORD Department of Biological Sciences, University of Sussex, Brighton, Falmer, Sussex, England
RONALD A. LASKEY Medical Research Council Laboratory of Molecular Biology, Cambridge, England

DNA replication in vivo has been extensively studied (reviewed by Kornberg, 1974), and because the origin of normal polyoma DNA replication has been determined by restriction enzyme mapping (Crawford et al., 1973). *X. laevis* ribosomal plasmid DNA was selected because it is larger than polyoma DNA (making electron microscope autoradiography experiments possible), and it has a high content of *X. laevis* DNA (making it homologous with the cytoplasmic components). The input template DNA molecules were isolated as supercoils by a procedure which involves banding in ethidium bromide-CsCl density gradients, followed by denaturation, renaturation, and chromatography on benzoylated, napthoylated DEAE cellulose (Levine et al., 1970). These preparative procedures make it highly unlikely that proteins are introduced into the assay along with the input template DNA molecules, or that previously initiated DNA replication intermediates are present at high levels in the input DNA (Benbow and Laskey, 1977*b*).

Cytoplasm from unfertilized eggs of *X. laevis* was chosen as the source of I-factor (Benbow and Ford, 1975) and of DNA polymerase X-II (Benbow et al., 1975). Cytoplasm from whole ovaries was used as the source of DNA polymerases X-I, X-III* (Benbow et al., 1975), and the other DNA replication components described below. The most obvious advantages of *X. laevis* eggs and oocytes as a source of the DNA replication machinery include the availability of kilogram quantities at reasonable cost, the extraordinarily high levels of DNA polymerases and other DNA replication components in the cytoplasm (Benbow et al., 1975), the negligible level of nuclear DNA and endogenous nuclear components in the egg cytoplasm (Dawid, 1965), and the inability of cytoplasm from oocytes to support DNA replication, either in vivo (Gurdon, 1974) or in vitro (Benbow and Ford, 1975). An additional potentially valuable advantage of *X. laevis*, which we have not yet exploited, is the possibility of testing putative control factors in vivo by microinjection into intact oocytes or eggs (Gurdon, 1974).

Electron microscopy was used to monitor the time-course of appearance of DNA replication intermediates during the assay (Benbow and Ford, 1975; Benbow and Laskey, 1977*a, b;* Benbow et al., 1977, in preparation). Closed circular DNA replication intermediates, which we call "Cairns structures" (Cairns, 1963), have an unambiguous appearance in the electron microscope (Figs. 1 and 2) which can be clearly distinguished from the input template DNA molecules and from other structures generated during incubation with the egg cytoplasm. To guard against artifacts and distortions which might be introduced during the extraction of the DNA molecules and their preparation for electron microscopy, we split each sample into several identical aliquots to which we add an internal standard DNA molecule at various times during the procedure. This internal standard makes use of the fact that small closed circular DNA molecules of similar but distinguishable size behave identically during spreading and extraction (Raleigh and Davis, 1976). By adding the internal standard DNA molecules either before or after extraction, we were able to monitor the percentage recovery for each molecular species; for example, we were able to quantitate the percentage of Cairns structures generated during the experiment by following the recovery of the known percentage of Cairns structures in the internal standard. Moreover, we were able to establish the *absolute* concentration of each species, and to detect distortions in the relative proportions of each molecular species introduced by the extraction or spreading procedures. We have shown (Benbow et al., 1977, in preparation) that the observation of three or more Cairns structures per 1,000 molecules examined is significantly above ($P = 0.001$) either the input level or the number generated with cytoplasm prepared from large oocytes of *X. laevis*.

Fractionation and Reconstitution of Xenopus laevis Egg Cytoplasm

X. laevis eggs (20–200 g) were dejellied, homogenized in low ionic strength buffer, centrifuged at low (2,500 *g*) and high (25,000 *g*) speeds to remove yolk and mitochondria, and were applied to a 5 × 30-cm DEAE cellulose column as described by Benbow et al. (1975). This fractionation procedure was selected because it resolves three distinct DNA polymerase activities (Grippo and Lo Scavo, 1972; Ford et al., 1975), and also removes potentially troublesome nucleic acids. The column was washed with low ionic strength buffer, eluted with a 0.1–0.4 M Tris gradient, and washed with 1.0 M Tris buffer. Small fractions were collected and assayed for DNA polymerase activity (Fig. 3); these small fractions were pooled into nine large fractions (Table I), using the arbitrary definition that DNA polymerases X-II, X-I, and X-III were entirely contained within fractions

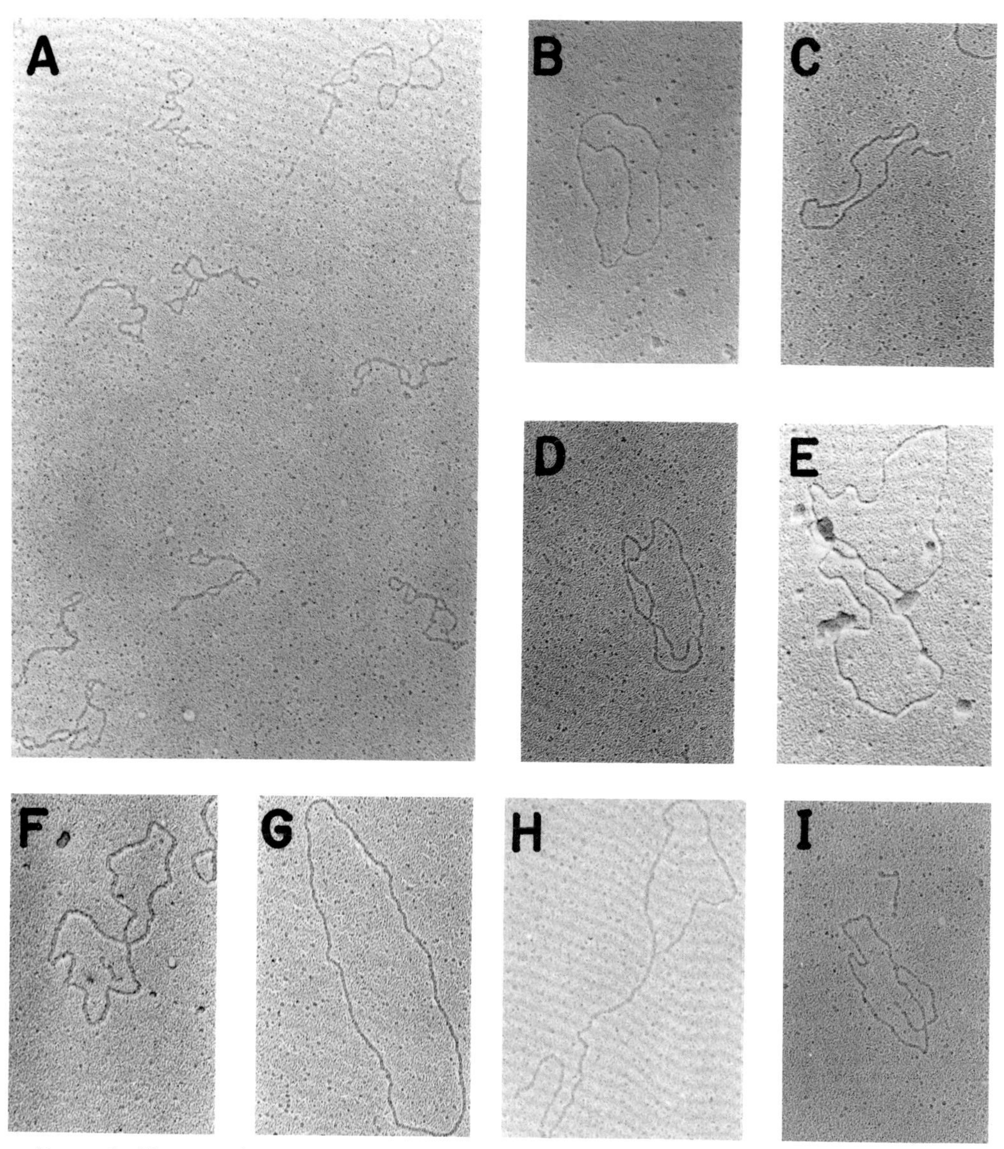

FIGURE 1 Electron microscopy of polyoma DNA molecules. (A) A field showing input polyoma DNA molecules; 10 supercoiled molecules are visible. (B) An early Cairns structure. Structures more completely replicated than this one would be classified as late Cairns structures. (C) A very early Cairns structure. Structures less completely replicated than this were not classified as Cairns structures in order to minimize the inclusion of artifactual Cairns structures in our data. (D) A late Cairns structure. (E) A late Cairns structure. Structures more completely replicated than this were not classified as Cairns structures because they could also be figure-8 or catenated structures. (F) A figure-8 structure. Note the "stiff" joint. This structure presumably is generated by genetic recombination (Benbow et al., 1975). (G) A circular dimer. These structures were *not* classified as DNA replication intermediates even though they have been identified as products of aberrant DNA replication (Benbow et al., 1972). (H) A circular monomer with a unit-length tail. These structures may arise by breakage of a fully replicated molecule or may be formed by displacement synthesis. They were *not* counted as DNA replication intermediates. (I) A circular monomer with a (probable) shorter-than-unit-length tail. This structure could also be a branch-migrated Cairns structure; such structures were *not* counted as DNA replication intermediates. Eco R1 endonuclease cleavage was carried out on the preparations containing Cairns structures that were used to generate the data in Table II. Between 40 and 60% of the Cairns structures apparently began DNA synthesis at a site consistent with the normal origin for polyoma DNA replication (Benbow et al., Reconstitution of polyoma DNA replication using components from unfertilized eggs of *Xenopus laevis*, manuscript in preparation).

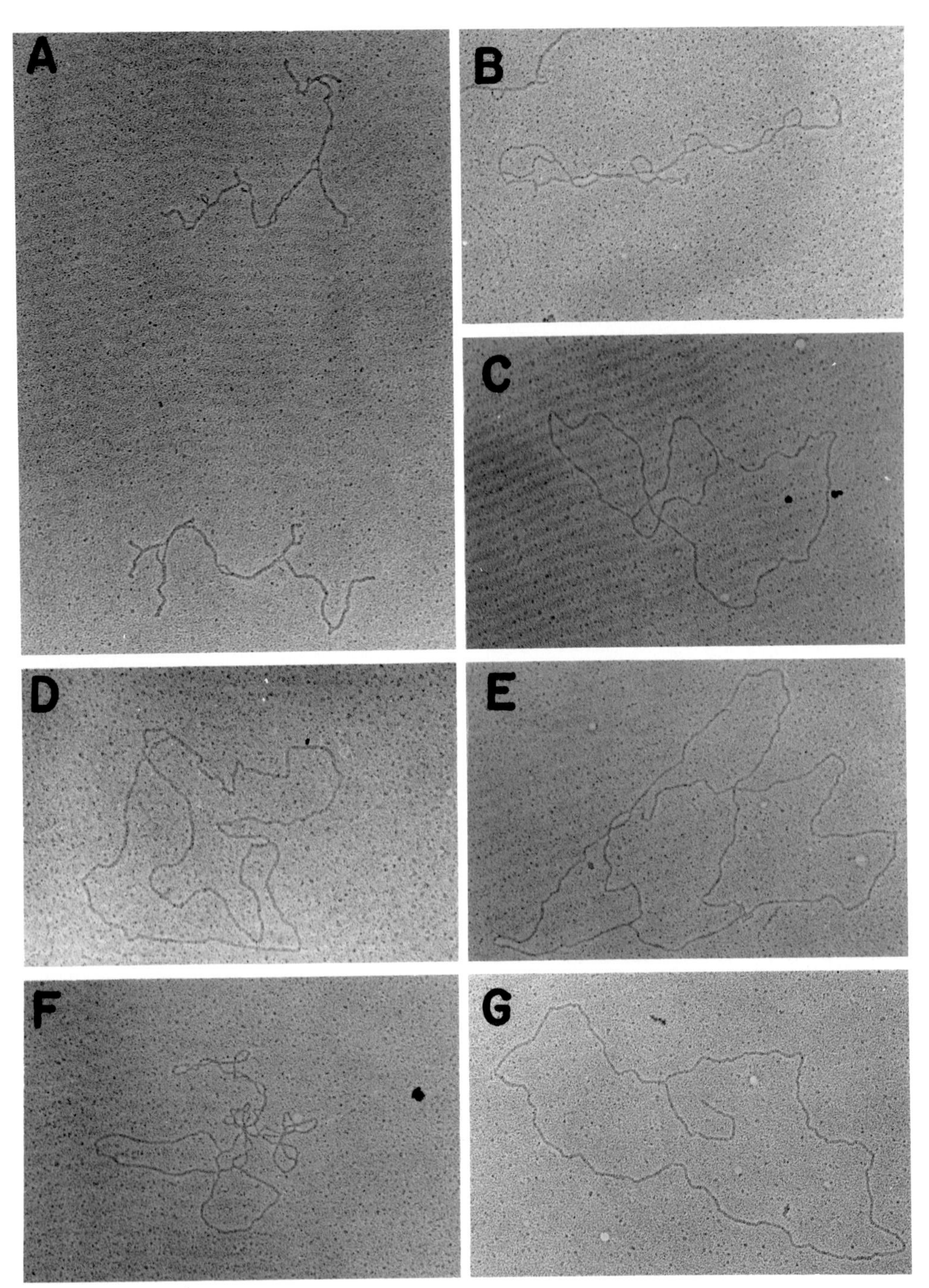

FIGURE 2 Electron microscopy of *Xenopus laevis* ribosomal DNA plasmids. These plasmids were the generous gift of Ronald Reeder of the Carnegie Institution of Embryology, Baltimore, Maryland. Two plasmids were used in these experiments: Xlr 101 and Xlr 14 which contained 65% and 56% *Xenopus laevis* ribosomal DNA, respectively. (A) Two supercoiled Xlr plasmids (the input DNA). (B) A very early Cairns structure; note that the unreplicated region remains supercoiled. (C) An early Cairns structure. (D) A late Cairns structure. (E) A very late Cairns structure. (F) An early Cairns structure in which the unreplicated region has remained supercoiled (see Kornberg, 1974). (G) A circular monomer with a tail.

IV, VI, and VIII, respectively. Each of the nine fractions was then concentrated 20-fold by pressure dialysis.

Purified closed circular DNA at 80 μg/ml was incubated with each fraction singly and in various combinations for 4 h at 30°C. Assays also contained 50 μM dNTPs, 250 μM NTPs, 10 mM $MgCl_2$, 200 μg/ml bovine serum albumin, 2 mM 2-mercaptoethanol, and an additional 2 mM ATP. The DNA was extracted and examined by electron microscopy by the aqueous Kleinschmidt procedure of Davis et al. (1971). Cairns structures were quantitated by the internal standard described above.

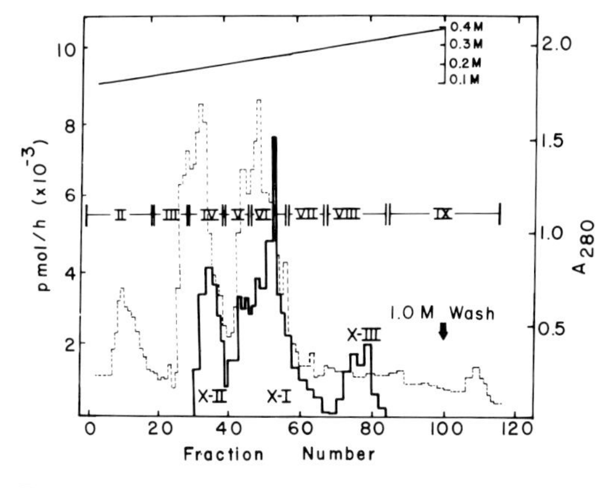

FIGURE 3 Fractionation of *Xenopus laevis* egg cytoplasm by DEAE cellulose column chromatography. An extract of 21 g of eggs was fractionated on a 5 × 30 column as described by Benbow et al. (1975). The absorbance at 280 nm and the DNA polymerase activity in each fraction was measured (Benbow et al., 1975). Other cytoplasmic components (see Table II), including unwinding protein, relaxation protein, I-factor, RNAase H, RNA polymerases I and II, nucleases, DNA ligase, and figure-8 forming activity, were assayed by published procedures (Ford et al., 1975; Benbow and Ford, 1975; Benbow and Laskey, 1977*b*; Beebee and Butterworth, 1974; Benbow et al., 1977).

The number of early and late Cairns structures observed per 1,000 molecules examined are presented in Table II for various combinations of fractions I through IX. No single fraction was able to catalyze the formation of Cairns structures, and III plus IV was the only pairwise combination to do so (Table II). Most of the observed Cairns structures were early Cairns as defined in Fig. 1. For the formation of high levels of late Cairns structures, fractions II, VI, and IX were required, and fraction I was helpful if not essential (Table II). It should be pointed out that the percentage of Cairns structures formed by various fractions varied substantially from preparation to preparation, but the same fractions were active in all preparations. Our most successful single experiment with fractions I through IV plus VI and IX generated 267 early and late Cairns structures in 1,000 examined molecules; however, in other experiments the percentage observed was 10-fold lower. As our working hypothesis, therefore, we assumed the reconstitution scheme outlined in Fig. 4. Since all attempts to optimize this crude reconstituted

TABLE I

Fractionation of Cytoplasm from Unfertilized Eggs of Xenopus laevis by DEAE Cellulose Column Chromatography

Fraction	Fractions pooled	Total volume pooled	Protein	Identifiable components
		ml	*mg/ml*	
I	wash-on	430	~0.40	Relaxation protein; unwinding protein
II	1–19	360	0.03	Relaxation protein; RNA polymerase I; DNA ligase (?)
III	20–29	196	0.35	I-factor; "restriction-type" enzyme; RNA polymerase I and II; unwinding protein
IV	30–39	196	0.60	DNA polymerase X-II; DNA polymerase X-III*, RNAase H; RNA polymerase II
V	40–46	145	0.33	Endonuclease; inhibitor (?)
VI	47–57	236	0.18	DNA polymerase X-I
VII	58–67	180	0.10	DNA ligase; figure 8-forming enzyme
VIII	67–84	280	0.11	DNA polymerase X-III
IX	85–end	315	~0.065	?

Fractions from the DEAE cellulose column in Fig. 3 were pooled as indicated; the protein concentrations were determined by the Lowry method on aliquots precipitated with TCA as described by Benbow, et al. (1975). (The protein concentrations in fraction I and IX are only approximate since these fractions contain substances which interfere with Lowry estimations.)

TABLE II

Formation of Early and Late Cairns Structures with Pooled DEAE Cellulose Column Fractions Prepared from Unfertilized Eggs of Xenopus laevis

Fraction(s)	Early Cairns*	Late Cairns*
I	0	0
II	0	0
III	0	0
IV	3	0
V	0	0
VI	0	0
VII	0	0
VIII	0	0
IX	0	0
III + IV	94	7
III + IV + I	87	7
III + IV + II	68	13
III + IV + V	0	0
III + IV + VI	83	17
III + IV + VII	68	9
III + IV + IX	57	11
III + IV + II + VI + IX	7	37
III + IV + II + VI + IX + I	14	112
III + IV + II + VI + IX + I + 0.1 μg/ml α amanitin	9	17

At least 16 other classes of molecules are also generated by combination of the various fractions, including catenated and figure-8 molecules, linear molecules, tailed molecules, etc. (Benbow et al., 1977). This table presents only the data for Cairns structures because these are the only unambiguous and unequivocal DNA replication intermediates (Benbow and Laskey, 1977*a*, *b*).
* Number observed per 1,000 molecules examined.

system failed to give significantly better results, we have proceeded directly to the identification and characterization of the active components of each fraction.

A Provisional Pathway for DNA Replication in Xenopus laevis

Our attempts to identify the active cytoplasmic components in the initiation, chain elongation, and termination steps of DNA replication are summarized in Fig. 5. It must be emphasized that this pathway is provisional. Only in the case of DNA polymerase X-I has an electrophoretically homogeneous enzyme been obtained for which a direct requirement has been demonstrated. And even in this case we have not rigorously proved that other DNA polymerases are unable to substitute.

Our overall strategy has been to purify each component, except for DNA polymerase X-II and I-factor, from whole ovaries rather than from egg cytoplasm. In this way we hope to identify additional control factors which are specifically found in eggs, and which are required for DNA replication. The properties of each of the partially purified components tentatively implicated in DNA replication are described below. (The fraction or fractions from which the component can be isolated are indicated in parentheses.)

Initiation of DNA Replication in Xenopus laevis

I-factor (III) was required for the cell-free generation of Cairns structures (Table II). The most purified preparation of I-factor (about 160-fold purified) contained no DNA polymerase activity, but did exhibit an activity which generated linear molecules from the input supercoiled DNA (Ford et al., 1975). The activity was trypsin-sensitive, and heat-sensitive; I-factor eluted from Sephadex G-200 with a Stokes radius of about 39 Å (Benbow and Ford, 1975). The molecular mechanism by which I-factor catalyzes the initiation of DNA replication is not known at present although the properties of our most purified preparation are consistent with the palindrome-specific endonuclease proposed by Benbow and Ford (1975). Further purification cannot be achieved until the other components necessary for the visualization of Cairns structures in the electron microscope

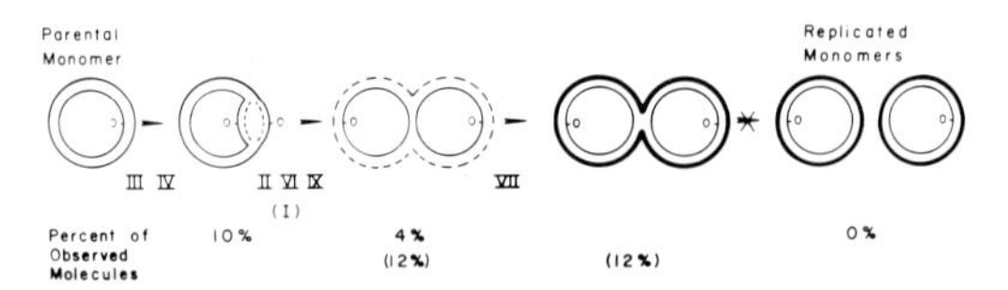

FIGURE 4 Reconstitution of polyoma DNA replication from DEAE cellulose column fractions prepared from unfertilized eggs of *Xenopus laevis*. This pathway represents the simplest interpretation of the reconstitution experiments described in Table II. The percent of all observed molecules converted to either early Cairns structures (fractions III plus IV) or to early plus late Cairns structures (fraction III plus IV plus II, VI, IX, and I) is designated beneath the diagram of the observed structure. The ligation step was deduced from alkaline sucrose velocity sedimentation patterns together with the percent of early plus late Cairns structures observed. The failure to segregate replicated monomers was also deduced from alkaline sucrose velocity sedimentation patterns (Benbow and Laskey, 1977*b*; Benbow et al., 1977).

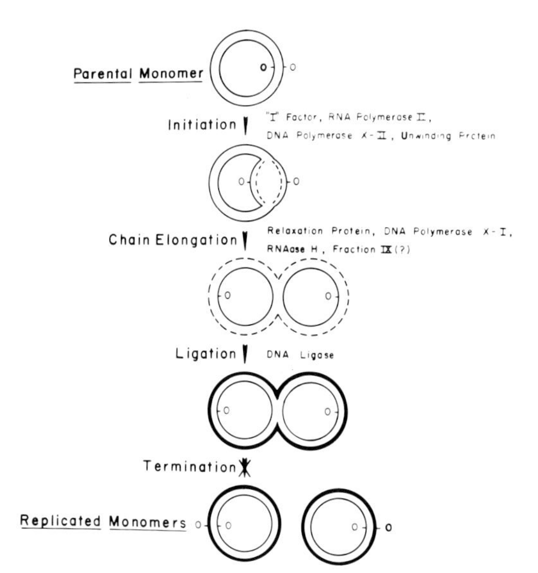

FIGURE 5 A provisional pathway for cell-free DNA replication with cytoplasmic components purified from unfertilized *Xenopus laevis* eggs. This pathway represents our tentative identification of the active components in fractions I, II, III, IV, VI, VII, and IX. This pathway should be considered as an organizational aid to our analysis and not as an established mechanism for DNA replication. Fraction VI (see Fig. 4) has been entirely replaced by electrophoretically homogeneous DNA polymerase X-I, which was then shown to be required for the formation of late Cairns structures. All of the other putative DNA replication components are assigned solely on the basis that partially purified components were able to replace one of the DEAE cellulose fractions in Fig. 4. For example, 800-fold purified DNA polymerase X-II (which still contained a trace level of DNA polymerase X-III*) was able to replace fraction IV completely; however, the activity was not electrophoretically homogeneous and may have contained additional active factors.

(presumably DNA polymerase X-II, RNA polymerase II, and unwinding protein) are available as electrophoretically homogeneous preparations.

DNA polymerase X-II (IV) was also required for the formation of visible Cairns structures (Table II). DNA polymerase X-II has been purified over 800-fold from unfertilized eggs of *X. laevis* (Benbow et al., manuscript in preparation). DNA polymerase X-II had an $s_{20,w}$ of 5.6 and a Stokes radius on Sephadex G-200 of about 60 Å. Assuming that the Siegel and Monty (1966) relationship applies, DNA polymerase X-II is an asymmetric molecule with a native molecular weight of about 160,000. Higher molecular weight aggregates were observed at low ionic strength. DNA polymerase X-II was sensitive to sulfhydryl group inhibitors such as *N*-ethylmaleimide and *p*-chloromercuribenzoate, was sensitive to monovalent cation concentrations above 50 mM, and was unable to efficiently copy the RNA strand of synthetic RNA-DNA hybrids. Purified DNA polymerase X-II did not exhibit any nuclease activity. If we assume that the α-β-γ-mt classification for vertebrate DNA polymerases (Weissbach, 1975) applies to *X. laevis,* X-II would tentatively be designated as an α polymerase.

DNA polymerase X-II has been reported either to be absent (Grippo and LoScavo, 1972) or to be present at low levels (Benbow et al., 1975) in extracts of oocytes of *X. laevis,* whereas it was found at high levels in extracts of eggs. Recently, however, we have been able to purify large quantities of DNA polymerase X-II from ovaries of *X. laevis* (Joenje and Benbow, unpublished observations). We have since discovered that DNA polymerase X-II activity was masked on the previously described DEAE cellulose columns by an inhibitor of DNA polymerase activity which elutes at nearly the same ionic strength as DNA polymerase X-II. The discovery of one or more inhibitors of DNA polymerase activity in oocytes of *X. laevis* raises the intriguing possibility of negative control of DNA replication during oogenesis.

*DNA polymerase X-III** (IV) is a low molecular-weight DNA polymerase activity which nearly coelutes on DEAE cellulose with DNA polymerase X-II. DNA polymerase X-III* was apparently not required for the formation of visible Cairns structures, although we have not rigorously ruled out its involvement in DNA replication (Fig. 5). DNA polymerase X-III* has been purified over 200,000-fold from oocytes of *X. laevis* (Joenje and Benbow, manuscript in preparation). The activity had an $s_{20,w}$ of about 3.5S, a Stokes radius of about 30 Å, and an apparent molecular weight of 45,500 calculated from these parameters. The activity was insensitive to N-ethylmaleimide, functioned optimally at high (>100 mM) concentrations of monovalent cations, and very efficiently copied the RNA strand of synthetic RNA-DNA hybrids. On the basis of these properties DNA polymerase X-III* was tentatively designated as a β polymerase.

DNA polymerase X-III (no star) was identified by Benbow et al. (1975) as a low molecular weight enzyme which was moderately resistant to monovalent cation concentrations. However, they reported that it eluted from DEAE cellulose at

0.26–0.29 M homogenization buffer (i.e., in fraction VIII) rather than with DNA polymerase X-II (i.e., in fraction IV, from which we purified this low molecular weight enzyme). Since in other eukaryotic organisms only one low molecular weight DNA polymerase has been found, we suggest that these data can be reconciled by assuming that the activity reported by Benbow et al. (1975) was DNA polymerase X-III* bound to DNA or chromatin. All of the known properties of DNA polymerase X-III are compatible with the properties of the highly purified DNA polymerase X-III*; however, because we have not yet proved the identity of DNA polymerase X-III as defined by Benbow et al. (1975) with the low molecular weight enzyme found in fraction IV, we have designated the purified enzyme as DNA polymerase X-III*.

RNA polymerase II (III) was probably required for the cell-free formation of Cairns structures: this conclusion is based on the α-amanitin sensitivity (to 0.1 μg/ml) of early Cairns structure formation (Table II). RNA polymerase II has been purified by a modification of published procedures (Roeder, 1974; Beebee and Butterworth, 1974), but we have not shown a direct requirement because I-factor has not been freed of contaminating RNA polymerase activities.

Unwinding protein (III, possibly elsewhere) has been partially purified by the affinity chromatography procedure of Herrick and Alberts (1975) on single-stranded DNA cellulose. A direct requirement for DNA replication has not yet been demonstrated, but is postulated because it is present in a required fraction and it has a proven role in prokaryotic DNA replication (reviewed by Kornberg, 1974).

The Chain Elongation Step of DNA Replication in Xenopus laevis

Of the enzymes postulated to be involved in chain elongation only DNA polymerase X-I has been purified to electrophoretic homogeneity.

DNA polymerase X-I was required for the cell-free formation of late Cairns structures (Table II) but not early Cairns structures. DNA polymerase X-I has been purified 1,100-fold to electrophoretic homogeneity from oocytes of *X. laevis* (Benbow et al., 1977, in preparation). DNA polymerase X-I had an $s_{20,w}$ of 6.2 and a Stokes radius on Sephadex G-200 of 72 Å. Assuming the Siegel and Monty relationship (1966) applies, DNA polymerase X-I is a highly asymmetric molecule with a native molecular weight of about 200,000. On SDS polyacrylamide gels, subunits of about 55,000 and 110,000 were observed. DNA polymerase X-I was sensitive to sulfhydryl group inhibitors, to monovalent cation concentrations above 100 mM, and was able to efficiently copy the RNA strand of synthetic RNA-DNA hybrids. RNA initiators were at least as efficient as DNA initiators for DNA polymerase X-I. An exonuclease activity copurified with DNA polymerase X-I.

The most interesting observation about DNA polymerase X-I was the sigmoidal enzyme kinetics exhibited by the purified enzyme as a function of enzyme concentration (Fig. 6). In view of the apparent subunit structure, it is tempting to postulate cooperativity as the cause of the sigmoidal enzyme kinetics; however, we are unable at present to rigorously rule out the presence of trace amounts of a DNA polymerase stimulating protein in our most purified preparations.

DNA polymerase X-I exhibited some of the properties of the γ polymerases, and was clearly not an α or β polymerase; however, the γ polymerases are so ill-defined that we believe it would be premature to classify DNA polymerase X-I as a γ polymerase.

The other cytoplasmic components implicated in *X. laevis* DNA replication are less well characterized at present.

Relaxation protein (I, II) has been purified over 500-fold by monitoring the relaxation of supercoiled DNA on agarose-acrylamide gels by the procedure of Keller (1975). *RNAase H* (III, IV) has been purified over 10,000-fold by monitoring the release of ^{3}H-labeled nucleotides from synthetic RNA-DNA hybrids. *Fraction IX* (IX) contains an unidentified component which is moderately heat-resistant, but which appears to be sensitive to trypsin.

Other unidentified components are undoubtedly required for cell-free DNA replication, but their identification and characterization require that the known and putative components first be purified to homogeneity.

Cytoplasmic Localization of DNA Replication Components

The localization of these cytoplasmic DNA replication components in large oocytes of *X. laevis* has been investigated by manual dissection of the germinal vesicle followed by the appropriate assays. Over 90% of measurable DNA po-

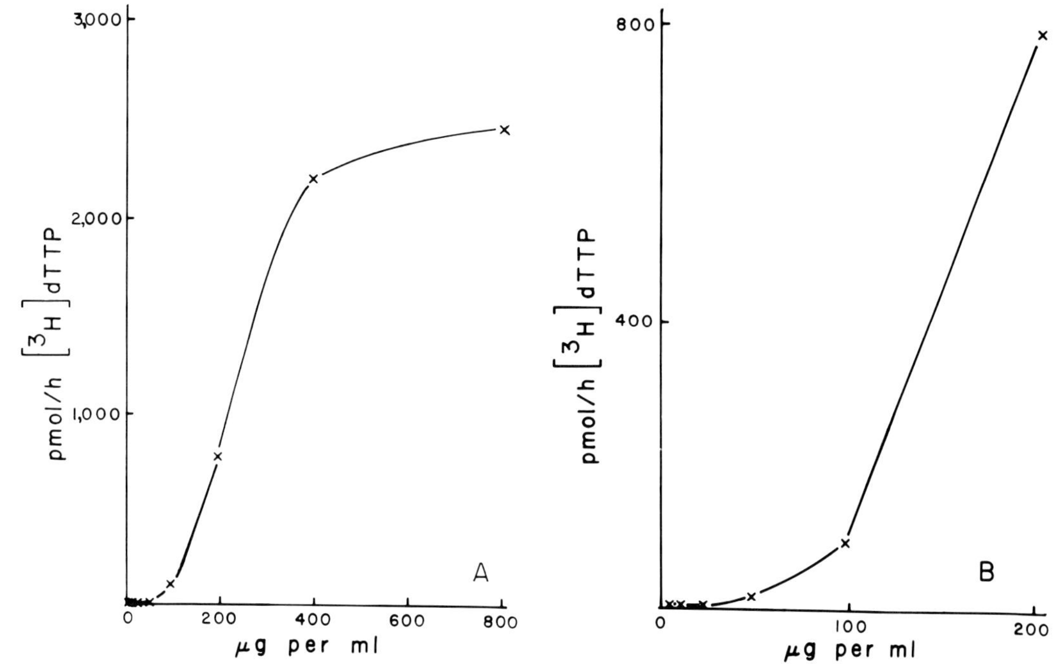

FIGURE 6 Incorporation of [^{3}H]dTTP by *Xenopus laevis* DNA polymerase X-I as a function of enzyme concentration: sigmoidal kinetics. (A) Poly dA-oligo dT (100:1) was utilized as template in this particular experiment, although the effect is observed with a variety of templates, including activated calf thymus DNA. The concentration of DNA polymerase X-I should be considered approximate in view of the difficulties in quantitation of single protein species at low concentrations. (B) The lower concentration range in Fig. 6B was measured in more detail. Note the apparent existence of a "critical concentration" below which DNA polymerase X-I appears to be inactive. Experimental details and control experiments showing that the sigmoidal kinetics do not result from stabilization of DNA polymerase X-I activity at higher enzyme concentrations are given by Benbow et al., DNA polymerase X-I from large oocytes of the frog, *Xenopus laevis*, submitted for publication.

lymerase X-I and X-II activities were localized in the germinal vesicle, whereas only about 20% of the DNA polymerase X-III* activity was in the nucleus relative to the cytoplasm. Over 90% of the low level of I-factor and almost all of the RNAase H activity were also found in the germinal vesicle. Relaxation protein was found at high levels in the nucleoli of the germinal vesicle (Hipskind and Reeder, personal communication).

Mechanisms for the Cytoplasmic Control of Nuclear DNA Replication in Xenopus laevis

At least three potential cytoplasmic control mechanisms were suggested by the experiments described above.

(1) The appearance of I-factor in an active form is an obvious candidate for the primary mechanism controlling the shift from no nuclear DNA replication in large oocytes to rapid DNA replication in eggs and embryos. Whether I-factor is synthesized *de novo*, is activated, or is released from repression is not known at present.

(2) The presence of inhibitors of DNA polymerases in extracts of oocytes suggests the possibility of the negative control of DNA replication during oogenesis. The existence of these inhibitors was indirectly shown by our failure to detect high levels of DNA polymerase X-II in extracts of oocytes (Benbow et al., 1975), even though DNA polymerase X-II can be purified from these same oocytes. We have subsequently shown with purified enzymes that an inhibitor to DNA polymerase X-III* also exists (Joenje and Benbow, unpublished observations), and we are actively attempting to identify the number and nature of these inhibitors.

(3) The sigmoidal kinetics exhibited by DNA polymerase X-I as an apparent function of en-

zyme concentration suggest that fairly small changes in the local concentration of DNA polymerase X-I (or any protein causing the "activation" of DNA polymerase X-I) would cause qualitative (on-off) changes in DNA replication. In this context the suggestion by Fansler and Loeb (1972) of nuclear-cytoplasmic shuttling of DNA polymerase becomes a very attractive candidate for a potential control mechanism for DNA replication during early embryogenesis.

Although other cytoplasmic control mechanisms can be derived from the experiments described above, it is highly probable that positive identification of the specific cytoplasmic components involved in the initiation of DNA replication will emerge as we purify to homogeneity each of the components involved in cell-free DNA replication.

ACKNOWLEDGMENTS

We would like to thank Anthony Mills and Vera Dunne for expert technical assistance, and John B. Gurdon, Ronald Lennox, Eric Nelson, and Nancy Wang for criticism of the manuscript.

This work was initiated while RMB was a fellow of the Helen Hay Whitney Foundation at the Medical Research Council Laboratory of Molecular Biology, Hills Road, Cambridge, England. Subsequent research was supported by the Medical Research Council, the Cancer Research Campaign, and National Institutes of Health grant GM/CA 22610-01. MRK was supported by National Institutes of Health Training grant GM-57. This is publication #890 of the Biology Department of Johns Hopkins University.

REFERENCES

Beebee, T. J. C., and P. H. W. Butterworth. 1974. Purification of DNA dependent RNA polymerases A and B from ovaries of *Xenopus laevis*. *Eur. J. Biochem.* **44:**115–122.

Benbow, R. M., M. Eisenberg, and R. L. Sinsheimer. 1972. Multiple length DNA molecules of bacteriophage ϕX174. *Nat. New Biol.* **237:**141–144.

Benbow, R. M., and C. C. Ford. 1975. Cytoplasmic control of nuclear DNA synthesis during early development of *Xenopus laevis*: a cell free assay. *Proc. Natl. Acad. Sci. U.S.A.* **72:**2437–2441.

Benbow, R. M., and R. A. Laskey. 1977*a*. Replication of polyoma DNA microinjected into *Xenopus laevis* eggs. *J. Mol. Biol.* In press.

Benbow, R. M., and R. A. Laskey. 1977*b*. Replication of polyoma DNA incubated in *Xenopus laevis* egg cytoplasm. *J. Mol. Biol.* In press.

Benbow, R. M., R. Q. W. Pestell, and C. C. Ford. 1975. Appearance of DNA polymerase activities during early development of *Xenopus laevis*. *Dev. Biol.* **43:**159–174.

Benbow, R. M., A. J. Zuccarelli, and R. L. Sinsheimer. 1975. Recombinant DNA molecules of bacteriophage ϕX174. *Proc. Natl. Acad. Sci. U. S. A.* **72:**235–239.

Cairns, J. 1963. The bacterial chromosome and its manner of replicating as seen by autoradiography. *J. Mol. Biol.* **6:**208–224.

Crawford, L. V., C. Syrett, and A. Wilde. 1973. The replication of polyoma DNA. *J. Gen. Virol.* **21:**515–521.

Davis, R. W., M. Simon, and N. Davidson. 1971. Electron microscope heteroduplex methods for mapping regions of base sequence homology in nucleic acids. *Methods Enzymol.* **210:**413–428.

Dawid, I. B. 1965. Deoxyribonucleic acid in amphibian eggs. *J. Mol. Biol.* **12:**581–599.

Edenberg, H. J., and J. A. Huberman. 1975. Eukaryotic chromosome replication. *Annu. Rev. Genet.* **12:**245–284.

Fansler, B., and L. A. Loeb. 1972. Sea urchin DNA polymerase. IV. Reversible association of DNA polymerase with nuclei during the cell cycle. *Exp. Cell Res.* **75:**433–444.

Ford, C. C., R. Q. W. Pestell, and R. M. Benbow. 1975. Template preferences of DNA polymerase and nuclease activities appearing during early development of *Xenopus laevis*. *Dev. Biol.* **43:**175–188.

Grippo, P., and A. Lo Scavo. 1972. DNA polymerase activity during maturation in *Xenopus laevis* oocytes. *Biochem. Biophys. Res. Commun.* **48:**280–285.

Gurdon, J. B. 1960. The developmental capacity of nuclei taken from differentiated endoderm cells of *Xenopus laevis*. *J. Embryol. Exp. Morphol.* **8:**505–526.

Gurdon, J. B. 1974. The Control of Gene Expression in Animal Development. The Clarendon Press, Oxford. 160.

Gurdon, J. B., and H. R. Woodland. 1968. The cytoplasmic control of nuclear activity in animal development. *Biol. Rev.* (*Camb.*). **43:**233–267.

Herrick, G., and B. Alberts. 1975. Purification and physical characterization of nucleic acid helix-unwinding proteins from calf thymus. *J. Biol. Chem.* **25:**2124–2132.

Jazwinski, S. M., J. L. Wang, and G. M. Edelman. 1976. Initiation of replication in chromosomal DNA induced by extracts from proliferative cells. *Proc. Natl. Acad. Sci. U. S. A.* **73:**2231–2235.

Keller, W. 1975. Characterization of purified DNA-relaxing enzyme from human tissue culture cells. *Proc. Natl. Acad. Sci. U.S.A.* **72:**2550–2554.

Kornberg, A. 1974. DNA Synthesis. W. H. Freeman & Company, San Francisco, Calif. 399.

Levine, A. J., H. S. Kang, and F. E. Billheimer.

1970. DNA replication in SV-40 infected cells. I. Analysis of replicating SV-40 DNA. *J. Mol. Biol.* **50:**549–568.

MERRIAM, R. W. 1969. Movement of cytoplasmic proteins into nuclei induced to enlarge and initiate DNA or RNA synthesis. *J. Cell Sci.* **5:**333–349.

RALEIGH, E. A., and R. W. DAVIS. 1976. Determination of DNA concentration by electron microscopy. *Anal. Biochem.* **72:**460–467.

ROEDER, R. G. 1974. Multiple forms of deoxyribonucleic acid-dependent ribonucleic acid polymerase in *Xenopus laevis*. *J. Biol. Chem.* **249:**241–248.

SIEGEL, L. M., and K. J. MONTY. 1966. Determination of molecular weights and frictional ratios of proteins in impure systems by use of gel filtration and density gradient centrifugation. Application to crude preparations of sulfite and hydroxylamine reductases. *Biochim. Biophys. Acta.* **112:**346–362.

WEISSBACH, A. 1975. Vertebrate DNA polymerases. *Cell.* **5:**101–108.

Chromatin Structure and Function

PROTEIN-DNA INTERACTIONS IN CHROMATIN

GARY FELSENFELD, RAFAEL D. CAMERINI-OTERO, and
BARBARA SOLLNER-WEBB

The study of chromatin structure has been stimulated greatly by the discovery of its fundamental subunit, the nucleosome or ν-body (Hewish and Burgoyne, 1973; Olins and Olins, 1974; Woodcock, 1973; Sahasrabuddhe and Van Holde, 1974; Kornberg, 1974). The nucleosome is a particle containing approximately two molecules each of the slightly lysine-rich histones H2A and H2B, two molecules each of the arginine-rich histones H3 and H4, and one molecule of the lysine-rich histone H1. There is 1 nucleosome for approximately every 200 base pairs of DNA. The more or less regular distribution of these particles along the chromatin fiber gives rise to its familiar "beaded string" appearance in the electron microscope (Olins and Olins, 1974; Woodcock, 1973).

The most direct chemical evidence for the existence of nucleosomes comes from studies of the action of various nucleases on chromatin. Hewish and Burgoyne (1973) first showed that digestion of nuclei by an endogenous nuclease present in rat liver nuclei results in the appearance of DNA molecules of discrete sizes, each an integral multiple of a fundamental unit about 200 base pairs long. Similar results are obtained when nuclei are digested with staphylococcal nuclease (Noll, 1974*a*).

A number of detailed studies of the action of staphylococcal nuclease on nuclei have shown that nucleosomes are composed of a relatively nuclease-accessible region and a relatively nuclease-resistant 140 base pair core (Sollner-Webb and Felsenfeld, 1975; Axel, 1975; Shaw et al., 1976). The enzyme first cleaves DNA at sites between the cores, giving rise to nucleoprotein particles that contain several nucleosomes. The DNA isolated from such oligomeric structures is roughly a multiple of 190 base pairs in length (Fig. 1). As digestion proceeds, each oligomer appears to lose about 40 to 50 base pairs; monomer is reduced from 190 to 140 base pairs (Fig. 2). The 140 base pair "core" is relatively resistant to further staphylococcal nuclease digestion. Thus, in reticulocytes and erythrocytes, the DNA between the nucleosome cores, which is kinetically more accessible to nuclease, is about 50 base pairs long. Although some variation has been reported in the length of the DNA between nucleosome cores, the 140 base-pair core size seems to be invariant in nuclei from a variety of species and tissue sources. It has recently been shown by a number of workers that the core particle lacks H1 (Shaw et al., 1976; Varshavsky et al., 1976; Sollner-Webb, Organization of histones in nuclei and chromatin. 1976 Ph.D. thesis, Stanford University, Stanford, Calif. 161 pp.).

What is the arrangement of DNA and histones within the core particle? That question remains to be answered. Although crystals of nucleosome monomers have been obtained, it will probably be several years before high-resolution X-ray diffraction methods are capable of providing detailed information about the location of the individual histone molecules.

In our own laboratory, we have been interested primarily in the effects of the histones on DNA structure, and in determining the role of the individual histone species in organizing the DNA of the nucleosome. The tools for these studies have been several enzymatic probes. We have described above the way in which staphylococcal nuclease preferentially cleaves the DNA between nucleosomes. The reaction does not cease at this point, however. The enzyme next attacks sites within the nucleosome. The products of this attack are not random in size, but consist of a series of discrete fragments (Axel et al., 1974). During the early stages of attack on chromatin, the double-

GARY FELSENFELD and CO-WORKERS Laboratory of Molecular Biology, National Institute of Arthritis, Metabolism, and Digestive Diseases, Bethesda, Maryland

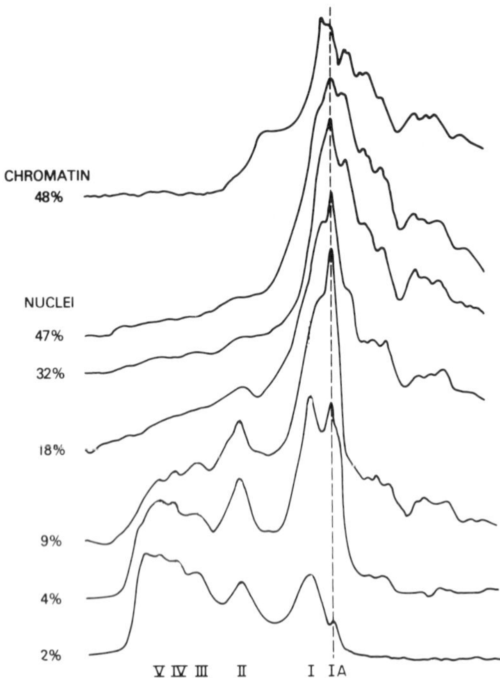

FIGURE 1 Kinetics of nuclear digestion. Duck reticulocyte nuclei were digested with staphylococcal nuclease in 1 mM Tris-HCl (pH 8), 0.1 mM $CaCl_2$ to various extents. The amount of acid-soluble material is shown on the left. Isolated DNA was run on a 4% acrylamide slab gel. Densitometer scans of a photographic negative are shown; migration is from left to right. Monomer through pentamer positions are shown, as well as the 140 base pair core DNA, labeled IA. On the top is a limit digest of isolated chromatin. From Sollner-Webb and Felsenfeld, 1975.

stranded fragments generated by digestion range in size from 160 to 40 base pairs, and they occur regularly at 10 base pair intervals (Camerini-Otero et al., 1976). As digestion proceeds, each fragment decreases by two base pairs in length, and then three of the fragments are reduced by another two base pairs (Fig. 3). In the case of staphylococcal nuclease, the reaction stops when about half the DNA is digested.

Other nucleases also reveal a regular pattern of DNA protection. Pancreatic DNase (DNase I), spleen acid DNase (DNase II), and a Ca^{++}-Mg^{++}-dependent nuclease isolated from rat liver nuclei all make single-strand cleavages in the DNA of nucleosomes at intervals of 10 nucleotides (Noll, 1974*b*; Sollner-Webb et al., 1976; Simpson and Whitlock, 1976). The pattern of these cleavages can be seen in gel electropherograms of double-stranded material, but is most clearly evident when the DNA is denatured and the single-stranded product examined (Noll, 1974*b*). Typical single-strand patterns are shown in Fig. 4.

What is the origin of this regular array of DNase I fragments? It has been suggested (Crick and Klug, 1975) that the DNA, wrapped around the histones, may be regularly kinked at 10 or 20 base pair intervals, and that the kinks may be more susceptible to nuclease action. The prediction is that cuts on the two DNA strands should be out of register by 0 or multiples of 10 bases. Another suggestion (Noll, 1974*b*) is that the DNA, assumed to be approximately in the B form, is wrapped around the histones and protected from attack except where it projects above the surface of the protein. This might reasonably be expected to happen once in every 10 nucleotides on *each* strand. If the latter model is correct, the cuts made by pancreatic DNase on one of the strands should be out of register with those made on the other

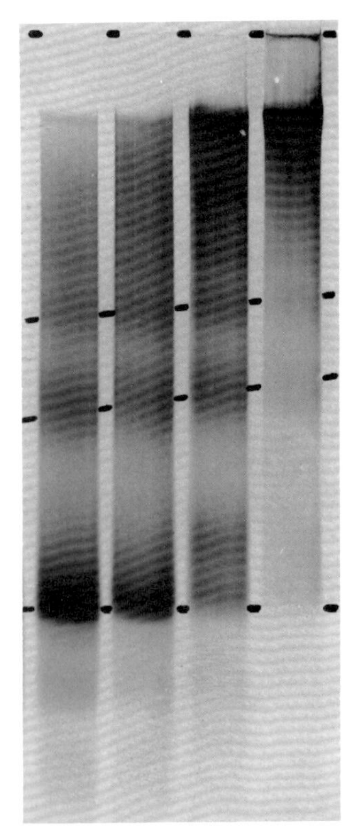

FIGURE 2 Decrease of oligomer sizes. Nuclei were digested with staphylococcal nuclease and the isolated DNA was run on a 4% slab gel. From right to left are 2, 4, 9, and 17% acid-soluble digests. Lines indicate the decrease in dimer and trimer sizes. From Sollner-Webb and Felsenfeld, 1975.

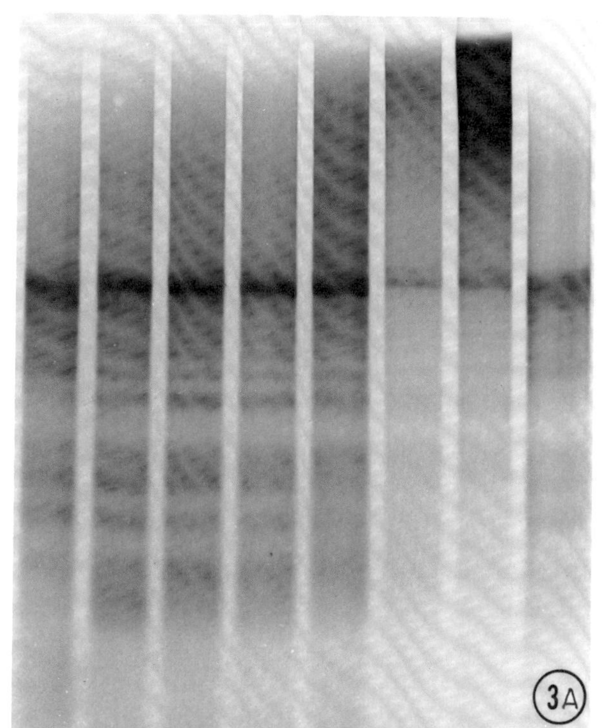

FIGURE 3A Kinetics of chromatin digestion. The DNA extracted from partial staphylococcal nuclease digests of chromatin was electrophoresed on a 6% acrylamide slab gel. At the far right is a limit digest of chromatin (50% acid soluble). From right to left the other seven samples are 8, 16, 21, 30, 31, 35, and 43% acid-soluble digests of the same chromatin preparation. From Camerini-Otero et al., 1976.

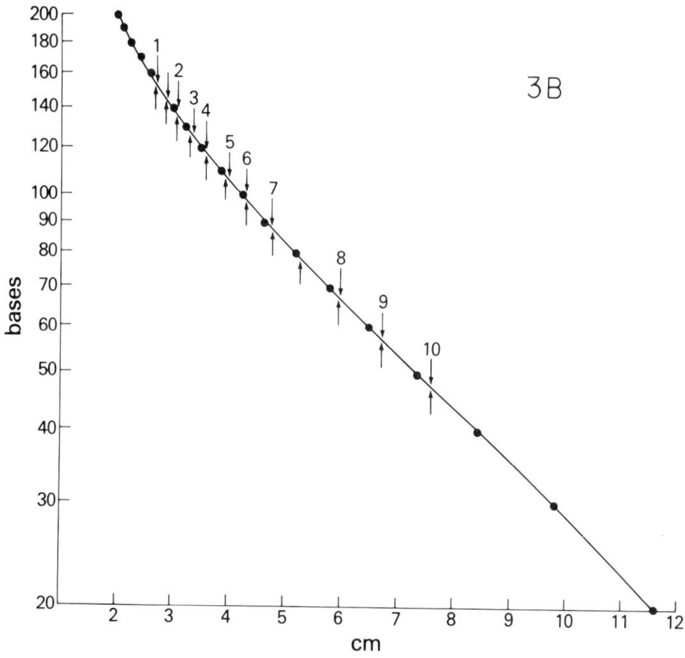

FIGURE 3B Sizings of chromatin digests. The DNA of partial and limit staphylococcal nuclease digests of chromatin was denatured and electrophoresed on 10% acrylamide slab gels with denatured DNA from a DNase I digest of nuclei. The $10 \cdot n$ base sizes of the DNase I digests (●——●) is plotted against electrophoretic migration. The arrows from below indicate the positions of the denatured DNA from a 35% acid-soluble staphylococcal nuclease digest ($10 \cdot n - 2$ bases). The arrows from above indicate the positions of the denatured DNA of a staphylococcal nuclease limit digest. From Camerini-Otero et al. 1976.

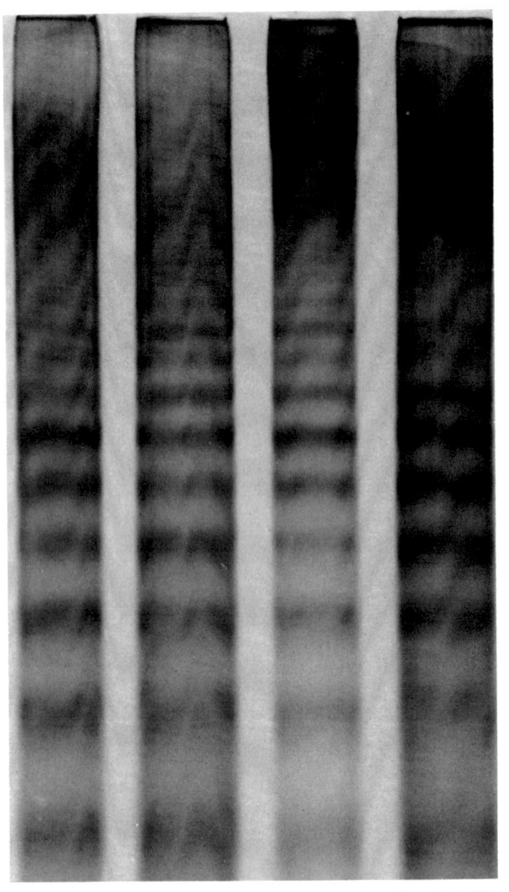

FIGURE 4 DNase I and DNase II digests of nuclei. The DNA of partial DNase I and DNase II digests of nuclei was denatured and run on 10% acrylamide gels. From left to right the samples are: DNase I, 23% acid soluble; DNase II, 60% acid soluble (digested in the presence of 0.5 mM EDTA); DNase I, 14% acid soluble; DNase II, 48% acid soluble (digested in the presence of 0.5 mM $MgCl_2$). From Sollner-Webb et al., 1976.

strand, since the two strands of B form DNA will rise to the surface successively at intervals of 6 base pairs across the major groove and 4 base pairs across the minor groove. (We will call this the "accessibility model.")

We have known for some time that partial DNase I digests contain staggered cuts. If the DNA product of such a digestion is electrophoresed without denaturation, a pattern of bands is obtained. If this material is treated with enzymes that specifically hydrolyze single-stranded regions of DNA, both the single-strand and double-strand gel patterns are altered in a way consistent with the presence of single-stranded "tails" at the end of double-stranded segments in the DNase I digest (Sollner-Webb, Organization of histones in nuclei and chromatin. 1976 Ph.D thesis, Stanford University, Stanford, Calif.).

What is the length of these tails? Is there a regular cleavage pattern and, if so, what is the distance between cuts on the two DNA strands? These questions can be answered by the use of DNA polymerases that are capable of repairing such single-stranded regions by adding complementary bases onto a recessed 3′OH end. Individual native DNA fragments were isolated from a partial DNase I digest of duck erythrocyte nuclei, and the partially double-stranded DNA was treated with *E. coli* DNA polymerase II in the presence of α-^{32}P-labeled nucleoside triphosphates (Sollner-Webb and Felsenfeld, 1977). The product was denatured, electrophoresed, and autoradiographed. We find strongly radioactive bands (Fig. 5) at 10 nucleotide intervals, centered on sizes $(10 \cdot n) + 8$, where n is an integer. Thus, the overlaps accessible to the polymerase involve a single-strand region 8 nucleotides long with a recessed 3′ end of the chain, and by inference there should be a stagger of 2 and 12 nucleotides at a recessed 5′ end (Fig. 6).

These results are consistent with any model in which the binding site of DNase I occurs at regular, 10-nucleotide intervals along each chain, provided that the site of cleavage is offset in one direction from the binding site by the appropriate number of bases. For example, if the accessibility model described above is correct, we suppose that binding would occur at the accessible site, and cleavage would occur one nucleotide in the 5′ direction from each binding site. The effect would be to convert the 6 nucleotide separations between strands across the major groove into an 8 nucleotide separation between cleavage points. Similarly, if the kinky helix model is correct, we suppose that the binding sites on both strands coincide at 10 base pair intervals but that cutting occurs one nucleotide in the 3′ direction, reducing the stagger to eight nucleotides. It is evident that experiments of this kind cannot distinguish between the structural models unless a great deal more is known about the mechanism of nuclease action.

The second problem in which we have been interested is the role played by various histones in generating the regular pattern of DNA digest frag-

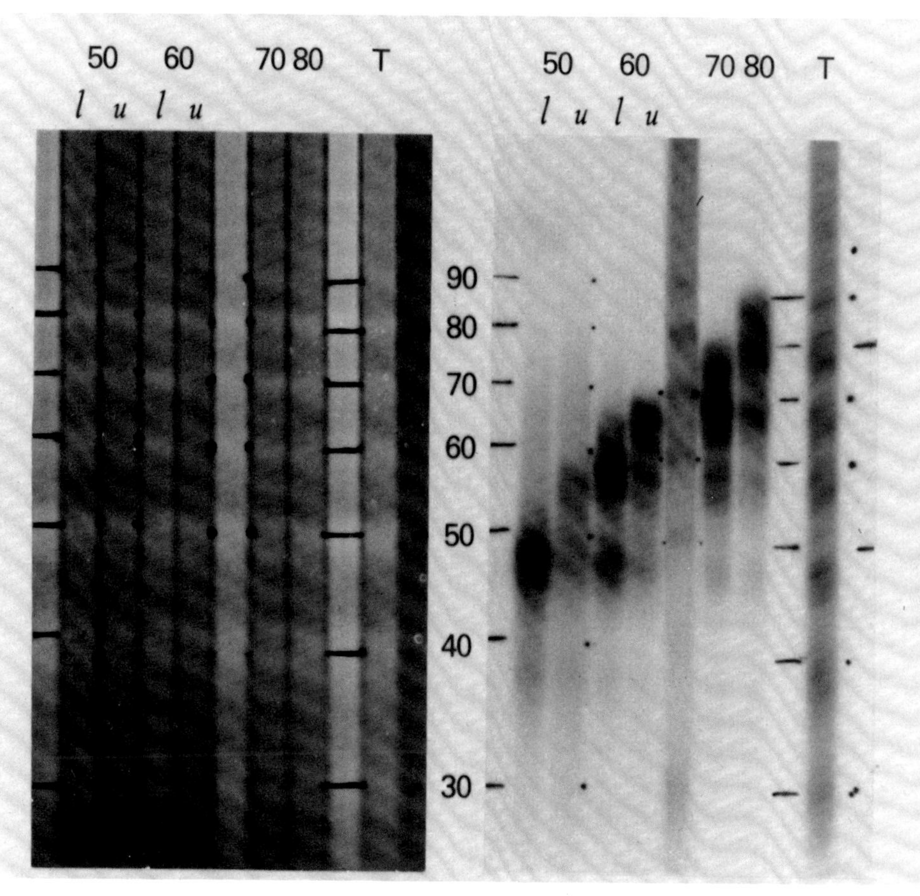

FIGURE 5 DNA polymerase II determination of the stagger of DNase I cuts. Native (nondenatured) DNase I digest DNA fragments were isolated corresponding to the lower and upper halves of the doublets which migrate as approximately 50 (50 *l* and 50 *u*) and 60 (60 *l* and 60 *u*) base pairs, as well as the bands which migrate approximately as 70 and 80 base pairs. These fragments, and total DNase I digest native DNA (*T*), were treated with *E. coli* DNA polymerase II in the presence of radioactive deoxynucleoside triphosphates. The DNA was denatured and electrophoresed with carrier total DNase I digest DNA. On the left the gel is stained for total (carrier) DNA; on the right is an autoradiograph of the same gel. By aligning densitometer scans of the two, filled in regions are determined to be eight nucleotides in length. DNA polymerase II has been shown to fill in to the end of a template, without detectable 3′ → 5′ or 5′ → 3′ exonuclease action. From Sollner-Webb and Felsenfeld, 1977.

ments. Reconstitutes of DNA with a mixture of the four histones H2A, H2B, H3, and H4, when digested with staphylococcal nuclease or pancreatic DNase, give rise to DNA fragment patterns quite similar to those described for chromatin above. To determine which histone species are essential to this pattern formation, we have formed complexes of DNA and almost all possible combinations of histones, and digested them with nucleases. We find that when single histones or most combinations of histones are reconstituted onto DNA, no discrete fragments are protected from digestion. Thus, the association of histones and DNA per se is not sufficient to give rise to this specific protection. The generation of discrete staphylococcal nuclease digestion fragments in high yields depends upon the presence of both H3 and H4, the arginine-rich histones. Furthermore, they alone with DNA are able to form a nucleoprotein structure that protects discrete fragments of DNA. A summary of the staphylococcal nuclease digest patterns of reconstitutes is shown in

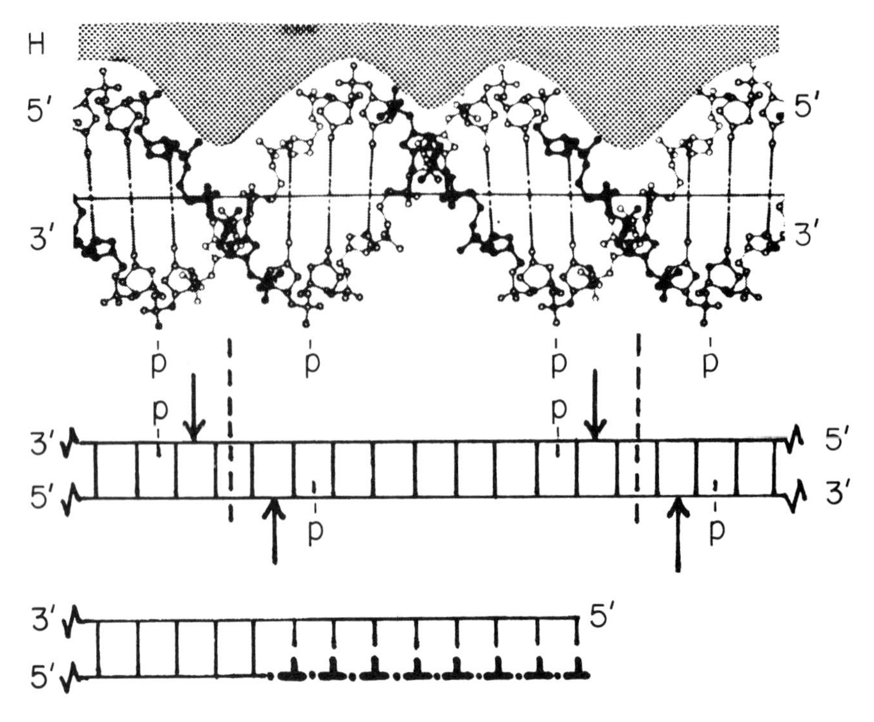

FIGURE 6 Schematic DNase I cutting sites. (Top) The DNA of the nucleosome may be modeled as lying on a histone surface (*H*, dotted region). In the accessibility model, DNA is cut only at the phosphodiester bonds most exposed to the surroundings. (Middle) Cutting at these peaks (*p*) would create a stagger of 6 and 4 bases, in the orientation that polymerase could fill in 6 nucleotides. If kinks (¦) occur every 10 base pairs, cutting there would create a stagger of 0 or multiples of 10 bases, and polymerase could fill in 10 bases. In fact, DNase I cuts may be envisioned as occurring at the arrows, creating a stagger of 8 and 2 bases. (Bottom) A DNA fragment cut to produce a stagger of 8 (with a recessed 3′OH end), which may be filled in by DNA polymerase (. ⊥ .), is shown. From Sollner-Webb and Felsenfeld, 1977.

Fig. 7 (Camerini-Otero et al., 1976). Similar results are obtained using DNase I or DNase II (Sollner-Webb et al., 1976); the presence of both H3 and H4 is again necessary and sufficient to generate discrete digest fragments $10 \cdot n$ nucleotides long.

We have examined the kinetic constants for the digestion of chromatin and partial reconstitutes by staphylococcal nuclease. The apparent V_{max} and apparent K_m of the reaction with reconstitutes which do not generate discrete DNA bands upon digestion (Camerini-Otero et al., 1976) are, respectively, 6 and 4.5 times that for chromatin. All reconstitutes that generate discrete DNA digest bands in high yields (those that contain both arginine-rich histones) have the same kinetic constants as does chromatin (Sollner-Webb et al., 1976). Furthermore, we find that the resistance to trypsin (Weintraub and Van Lente, 1974) and chymotrypsin attack of histones bound to DNA depends upon which histones are present (Sollner-Webb et al., 1976). Individual histones or most combinations of histones free in solution or bound to DNA do not generate discrete, large, protein fragments on trypsin digestion. We find once again that, when bound to DNA, the arginine-rich histones, H3 and H4, are both required and sufficient for the formation of trypsin-resistant structures (Sollner-Webb et al., 1976). A summary of our results with the different enzymatic probes is shown in Table I.

We have also examined the digestion kinetics of DNA/H3/H4 reconstitutes, and find that at early times of digestion with either staphylococcal nuclease or DNase I, fragments as large as 130 nucleotide pairs and 180 nucleotides, respectively are produced. The results indicate that the arginine-rich histone pair, H3 and H4, can organize

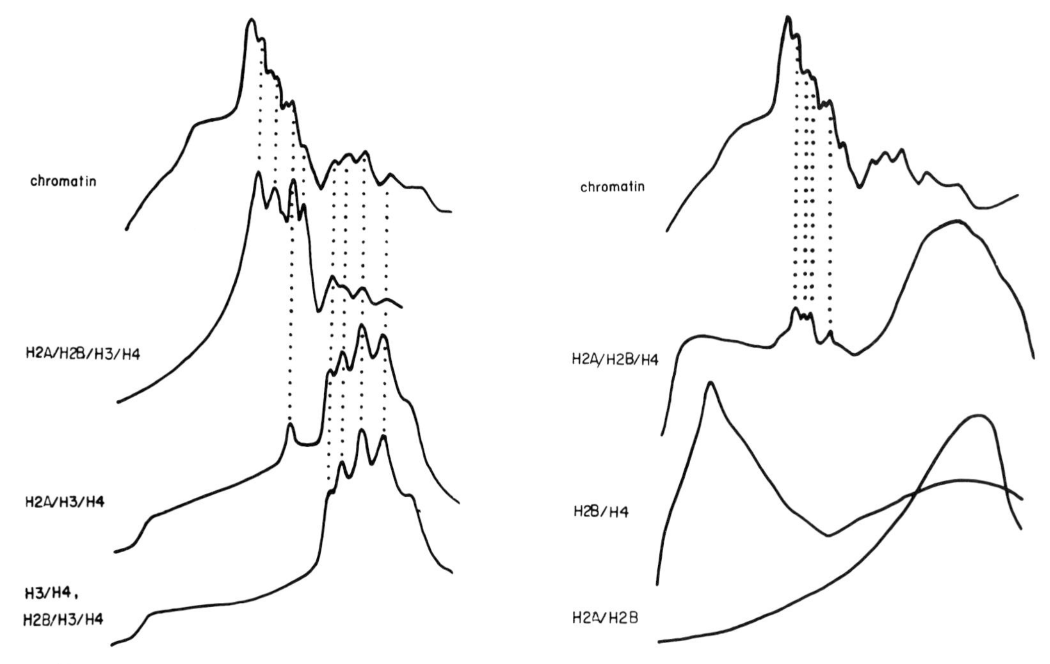

FIGURE 7 Staphylococcal nuclease digests of chromatin and partial reconstitutes. Chromatin and partial reconstitutes, formed from DNA and the histone species indicated at the left, were digested to the limit with staphylococcal nuclease and the isolated DNA electrophoresed on 6% acrylamide gels. H3 and H4 are both seen to be necessary for high yield discrete fragment protection. Other combinations either produce discrete fragments only at very low yield (H2A/H2B/H4) or not at all (all other combinations lacking H3 or H4). From Camerini-Otero et al., 1976.

TABLE I
Summary of Enzymatic Digestions of Partial Reconstitutes

	DNAase I		Trypsin		Staphylococcal nuclease			
Reconstitute	Regular bands	Relative intensities	Stable bands	Band numbers	Relative V_{max}	Apparent K_m	Limit digest bands	Band number
H2A/H2B/H3/H4/ H1/H5	+	+	+	2,3,4,5	1	1	+	1–11
H2A/H2B/H3/H4	+	+	+	2,3,4,5	1	0.9	+	2,4,6,8–11
H2A/H3/H4	+	0	+	4,5	1.5	ND*	+	6,8–11
H2B/H3/H4	+	0	+	4,5	1.5	ND*	+	8–11
H3/H4	+	0	+	4,5	1	0.8	+	8–11
H2A/H2B/H4	0	–	0		5	4.4	(+)	(2–4,6)
H2A/H2B/H3	0	–	0		5.6	4.9	0	
H2A/H2B	0	–	0		6	ND*	0	
H2A/H4	0	–	0		6	ND*	0	
H2B/H4	0	–	0		6	ND*	0	
H3	0	–	0		6	ND*	0	
H4	0	–	0		6	ND*	0	
(Naked DNA)	0	–	–		6	4.4	0	
(H3/H4/DNA mix)	0	–	0		5.6	4.7	0	

Summary of enzymatic digestions of partial reconstitutes. Reconstituted chromatin and partial chromatins were digested with DNase I, trypsin, and staphylococcal nuclease. Protein-free (naked) DNA and DNA mixed with H3/H4 directly in low ionic strength without a stepwise reconstitution (H3/H4/DNA mix) were also digested. + and 0 indicate the presence or absence of stable digestion product bands. (+) indicates low yield band formation. Relative intensities and band numbers indicate the similarity of that digest to the product from whole chromatin. The kinetic constants of staphylococcal nuclease digestion are reported relative to chromatin ≡ 1. From Sollner-Webb et al., 1976; and Camerini-Otero et al., 1976.

* Not determined.

albeit weakly, DNA stretches as large as the 140 base pair nucleosome core.

Recently, we have obtained further evidence of this central role by studying the effects of H3/H4 on the supercoiling of closed circular DNA. It has been shown (Germond et al., 1975) that the four histones H2A, H2B, H3, and H4, when reconstituted together onto closed circular DNA, deform the DNA in a manner topologically equivalent to the induction of about one superhelical turn per nucleosome. We confirm this result, and furthermore, in our hands, it is possible to induce about one superhelical turn for each nucleosome equivalent of the four histones added (1 g of protein per gram of DNA). We have now used similar methods to examine the effect of the individual histones and histone pairs on supercoiling (Camerini-Otero and Felsenfeld, manuscript in preparation). Histones are reconstituted onto closed circular DNA that has previously been relaxed by treatment with a eukaryotic nicking-closing enzyme. The complex is again treated with nicking-closing enzyme, deproteinized, and the DNA electrophoresed in a gel system that separates circular species according to their topological winding number (Keller and Wendel, 1974).

We find that no single histone can induce supercoils, but that mixtures containing both H3 and H4 together do supercoil DNA. On a weight basis, the H3/H4 pair is as effective as total histone: 1 g of H3/H4 complex per gram of DNA produces about the same number of superhelical turns as 1 g of H2A/H2B/H3/H4. A nucleosome equivalent (0.5 g/gram DNA) of H3/H4 is roughly half as effective as a nucleosome equivalent (1 g/gram DNA) of all four histones.

Our results all support the idea that the H3/H4 pair is able to combine with DNA to give a complex with many of the properties of the nucleosome with respect to a variety of enzymatic probes. The structural and energetic details of this interaction remain to be determined.

REFERENCES

Axel, R. 1975. Cleavage of DNA and chromatin with staphylococcal nuclease. *Biochemistry.* **14:**2921–2925.

Axel, R., W. Melchior, B. Sollner-Webb, and G. Felsenfeld. 1974. Specific sites of interaction between histones and DNA in chromatin. *Proc. Natl. Acad. Sci. U. S. A.* **71:**4101–4105.

Camerini-Otero, R. D., B. Sollner-Webb, and G. Felsenfeld. 1976. The organization of histones and DNA in chromatin: evidence for an arginine-rich histone kernel. *Cell.* **8:**333–347.

Crick, F. H. C., and A. Klug. 1975. Kinky helix. *Nature (Lond.).* **255:**530–533.

Germond, J. E., B. Hirt, P. Oudet, M. Gross-Bellard, and P. Chambon. 1975. Folding of the DNA double helix in chromatin-like structures from Simian Virus 40. *Proc. Natl. Acad. Sci. U. S. A.* **72:**1843–1847.

Hewish, D., and L. Burgoyne. 1973. Chromatin substructure: the digestion of chromatin DNA at regularly spaced sites by a nuclear deoxyribonuclease. *Biochem. Biophys. Res. Commun.* **52:**504–510.

Keller, W., and I. Wendel. 1974. Stepwise relaxation of supercoiled SV40 DNA. *Cold Spring Harbor Symp. Quant. Biol.* **39:**199–208.

Kornberg, R. 1974. Chromatin structure: repeating unit of histones and DNA. *Science (Wash. D. C.).* **184:**868–871.

Noll, M. 1974*a*. Subunit structure of chromatin. *Nature (Lond.).* **251:**249–251.

Noll, M. (1974*b*). Internal structure of the chromatin subunit. *Nucl. Acids Res.* **1:**1573–1578.

Olins, A., and D. Olins. 1974. Spheroid chromatin units (ν bodies). *Science (Wash. D. C.).* **184:**330–332.

Sahasrabuddhe, C. G., and K. E. Van Holde. 1974. The effect of trypsin on nuclease-resistant chromatin fragments. *J. Biol. Chem.* **249:**152–156.

Shaw, B. R., T. M. Herman, R. T. Kovacic, G. S. Beaudreau, and K. E. Van Holde. 1976. Analysis of subunit organization in chicken erythrocyte chromatin. *Proc. Natl. Acad. Sci. U. S. A.* **73:**505–509.

Simpson, R. T., and J. P. Whitlock. 1976. Chemical evidence that chromatin DNA exists as 160 base beads interspersed with 40 base pair bridges. *Nucl. Acids Res.* **3:**117–127.

Sollner-Webb, B., R. D. Camerini-Otero, and G. Felsenfeld. 1976. Chromatin structure as probed by nucleases and proteases: evidence for the central role of histones H3 and H4. *Cell.* **9:**179–193.

Sollner-Webb, B., and G. Felsenfeld. 1975. A comparison of the digestion of nuclei and chromatin by staphylococcal nuclease. *Biochemistry.* **14:**2915–2920.

Sollner-Webb, B., and G. Felsenfeld. 1977. Pancreatic DNase cleavage sites in nuclei. *Cell.* In press.

Varshavsky, A. J., V. V. Bakayev, and G. P. Georgiev. 1976. Heterogeneity of chromatin subunits *in vitro* and location of histone H1. *Nucl. Acids. Res.* **3:**477–492.

Weintraub, H., and F. Van Lente. 1974. Dissection of chromosome structure with trypsin and nucleases. *Proc. Natl. Acad. Sci. U. S. A.* **71:**4249–4253.

Woodcock, C. 1973. Ultrastructure of inactive chromatin. *J. Cell Biol.* **59:**368a.

HISTONE-HISTONE NEIGHBOR RELATIONSHIPS WITHIN THE NUCLEOSOME

HAROLD G. MARTINSON and BRIAN J. MCCARTHY

The fundamental chromatin fiber of higher eukaryotes is now generally believed to be composed of discrete histone-DNA subunits (Elgin and Weintraub, 1975; Van Holde and Isenberg, 1975; Kornberg, 1974; Olins and Olins, 1974). Each subunit is thought to possess a core comprised primarily of the C-terminal portions of two each of the histones H2A, H2B, H3, and H4. A length of DNA, comprising about 180 base pairs, is complexed with 25 or so residues of each histone N-terminus and collapsed about this core. Association of this fundamental subunit with H1 and an unknown amount of nonhistone protein completes the structure (Elgin and Weintraub, 1975; Van Holde and Isenberg, 1975).

The arrangement of histones in chromatin has been studied widely, using cross-linking reagents capable of spanning the short distance between neighboring lysine residues in the chromatin (Olins and Wright, 1973; Ilyin et al., 1974; Hyde and Walker, 1975; Chalkley, 1975; Chalkley and Hunter, 1975; Van Lente et al., 1975; Thomas and Kornberg, 1975). Such data yield information on histone proximity presumably over the surface of the subunit, and all of the histones have been shown to become involved in such cross-links. For example, use of formaldehyde, the shortest of these "spanner" cross-linking agents, has shown that H2B is very close to both H2A and H4 in chromatin (Van Lente et al., 1975).

Also required in the study of chromatin structure is information about the internal structure of the subunit and the actual histone-histone interactions involved. A thorough study of the interactions of histones in solution has revealed that the three most significant interactions are those between the pairs H2A-H2B, H2B-H4, and H4-H3 (D'Anna and Isenberg, 1974; Table I).

We have recently shown that highly selective protein cross-linking can be induced within the chromatin subunits of whole cells by zero-length cross-linking agents, which convert noncovalent interactions between two histones into covalent bonds without the interposition of bridges. The two agents, tetranitromethane (TNM) and UV light are able to penetrate hydrophobic clusters and activate tyrosine. TNM induced an H2B-H4 cross-link and UV light cross-linked H2B to H2A (Martinson and McCarthy, 1975; Martinson et al., 1976). The third strong interaction predicted by D'Anna and Isenberg (1974), H3-H4, has been demonstrated by cross-linking with another zero-length reagent, carbodiimide (Bonner and Pollard, 1975).

This report summarizes our studies of histone-histone interactions in cells, nuclei, chromatin, and reconstituted complexes, using the above cross-linking approach. In addition, we present evidence that H2A and H4 interact with different portions of the H2B molecule and that H2B is complexed simultaneously to both H2A and H4 in vivo.

Cross-Linking of H2B to H4 Induced by TNM

The reaction of TNM with HeLa or mouse L-cell chromatin results in the rapid formation of a cross-linked histone product which appears as a new species in polyacrylamide gel electrophoretograms of the acid-extracted chromosomal proteins. As the reaction proceeds, this new species, together with the histone monomers, disappears as it becomes incorporated into cross-linked aggregates of very high molecular weight (Martinson and McCarthy, 1975).

In order to determine the in vivo significance of

HAROLD G. MARTINSON Department of Chemistry, University of California, Los Angeles, Los Angeles, California

BRIAN J. MCCARTHY Department of Biochemistry and Biophysics, University of California, San Francisco, San Francisco, California

TABLE I
*Strong Interactions between Histones**

	Zero-length cross-linking agent
H2A†-H2B	UV light§
H2B-H4	Tetranitromethane‖
H3-H4	Carbodiimide¶

* D'Anna and Isenberg, 1974.
† Nomenclature formerly used: H2A = F2a2, IIb1, or LAK; H2B = F2b, IIb2, or KAS; H3 = F3, III, or ARE; H4 = F2al, IV, or GRK.
§ Martinson et al., 1976.
‖ TNM; Martinson and McCarthy, 1975.
¶ Bonner and Pollard, 1975.

the cross-linked product produced in isolated chromatin, living HeLa cells growing in culture were treated with TNM. An identical cross-linked histone product readily forms in whole cells upon treatment with the reagent. The amount of TNM added to the culture medium was minimized to preclude chemical perturbations of the system although the cells were doubtless rapidly killed. At 50 ppm by volume (0.4 mM) the TNM was limiting, and the increase in TNM product between 25 min and 2 h was not very great. When TNM at 100 ppm was used, more TNM product, estimated at about 3% of total histone, was produced. Cross-linking for 24 h in the presence of excess TNM resulted in the disappearance of most of the monomeric histones together with the specific cross-linked product, as was the case for chromatin.

Having obtained the same TNM product in whole cells as in isolated chromatin, we next asked whether reconstituted nucleohistone might yield similar results. We found not only that reconstituted nucleohistone yields a TNM product but that sophisticated reconstitution procedures are unnecessary. Rapid mixing of acid-extracted histones with DNA produced a DNA-histone complex which, when treated with TNM in the standard manner, yielded the cross-linked product. Moreover, commercial histone and DNA preparations were equally effective. The simplicity of the reconstitution procedure encouraged us to investigate the properties of the system further with pure preparations of the individual histones obtained by gel exclusion chromatography of total calf thymus histones on columns of Bio-Gel P-30.

The separate histones were mixed with each other in various combinations and then DNA was added, followed by TNM. The cross-linked product was produced when the five histones were combined with DNA in this fashion. Moreover, both H1 and H3 could be excluded from the reconstitution mix without affecting the yield of cross-linked product, but H2A, H2B, and H4 were essential. In the absence of DNA, a high background was observed on the gels but no specific TNM product was evident. Thus, these results show that any or all of histones H2A, H2B, and H4 were contained in the TNM product and that, furthermore, in our buffer system, the presence of DNA was required in order for the correct histone configurations to be assumed for specific cross-linking by TNM.

In order to determine which histones actually constitute the TNM product, radioactive H2A, H2B, or H4 were individually incorporated into separate reconstitution mixtures for cross-linking. The results, summarized in Table II, show that when either radioactive H4 or radioactive H2B is present the TNM product is radioactive, but that when radioactive H2A is used, the TNM product does not incorporate significant radioactivity. Nevertheless, if the H2A is omitted from the reaction mixture, no TNM product forms. Thus, H2A facilitates the formation of the TNM complex but does not itself become incorporated. This shows that the TNM product contains only H2B and H4.

Cross-Linking of H2B to H2A Induced by UV Light

Irradiation of whole L-cells with UV light at wavelengths greater than 260 nm results in rapid formation of a cross-linked histone product (Martinson et al., 1976). This new series appears in the

TABLE II
Production of TNM Complex from Nucleohistone Prepared with Separated Ratioactive Histones

Radioactive histone in reconstituted nucleohistone	TNM region	[^{3}H]Histone in TNM region
	cpm	*pmol*
H4	326	11.7
H2A	52	1.4
H2B	824	11.4

Reconstitution of DNA with separated histones was carried out with each of three mixtures containing a different radioactive histone species as indicated. After treating the mixtures with TNM, the histones were extracted and electrophoresed on a thin polyacrylamide slab. The stained bands corresponding to the TNM product were excised and the amount of radioactivity was determined.

dimer region of acid-urea polyacrylamide gel electrophoretograms of the acid-extracted nuclear proteins (Fig. 1). A dimer molecular weight is also indicated by its rate of migration in SDS gels and its elution position after chromatography on Sephadex G200 (not shown). With increased irradiation of the cells, the dimer photoproduct is produced in high yield, whereas histones H2A and H2B preferentially and quantitatively disappear. This suggests that the photoproduct is an H2A-H2B heterodimer, an assignment to be substantiated below. Additional uncharacterized photoproducts, presumably trimers, appear after long periods of irradiation. With the exception of "trimer" production, the irradiation of chromatin at low ionic strength yields nearly identical results.

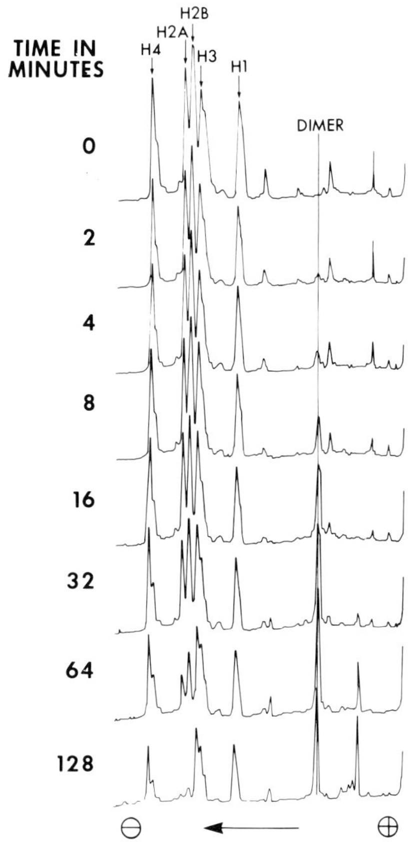

FIGURE 1 Electrophoresis of histones isolated from cells irradiated at 280 nm. About 2–3 × 10^7 cells were spun down, washed in 10 ml phosphate buffered saline (PBS), pelleted again, resuspended with 0.5 ml PBS and transferred to a quartz tube. A stream of humid nitrogen was passed over the cell slurry for 5 min. The introduction of air was prevented for the remainder of the experiment. 15 ml of PBS, which had been bubbled with humid nitrogen for 0.5 h, was then added with mixing and a 0 time sample was withdrawn. At 0 time, the remaining cells were exposed to the 450 W Hanovia UV source through a Corex water jacket maintained at room temperature. Uniform irradiation was ensured by continuous stirring of the sample. At the indicated times, 1.8 ml-aliquots were withdrawn and left at room temperature until the last time-point. The cells from each time-point were then collected for extraction of histones.

The H2A-H2B dimer can be produced by irradiation of chromatin at wavelengths less than 260 nm, as well (i.e., at 254 nm). In this case, however, the yield is much lower and all the histones, including the dimer, soon begin to disappear. Presumably the major chromophore excited at wavelengths >260 nm is tyrosine, whereas at 254 nm, it is a base in DNA. In the former case, proteins become cross-linked to each other, whereas in the latter case, DNA-protein adduct formation may dominate.

All irradiations were performed anaerobically. Control experiments showed that recoveries of all histone species were higher when the cells or chromatin were photolysed under an atmosphere of nitrogen, rather than air. The effect was particularly striking for the dimer, whose relative yield was increased by 50% and whose absolute yield was more than doubled by the exclusion of oxygen. Presumably oxidative side reactions occur which lead to general polymerization and loss of all histones and which, in addition, compete specifically with the H2A-H2B dimer formation. Fig. 2 presents a quantitative summary of dimer production as a function of elapsed irradiation time. The data for the time-points were obtained from polyacrylamide gels by excising the bands, eluting the stain, and taking the absorbance at 590 nm. The amount of stain eluted from each histone band is expressed as a percent of the total. No attempt has been made to correct for the small differences in stain/protein ratio that arise from variations in band intensity or protein stain affinity.

Panels A and B of Fig. 2 show that, with the exception of trimer production, the major features of photolysis are the same, whether intact cells at physiological ionic strength or isolated chromatin at low ionic strength are irradiated. In whole cells, the decline in rate of dimer production, accompanied by the appearance of increasing quantities of the product in the trimer region of the acrylamide gel, suggests a precursor-product relationship. Whether the production of this presumptive tri-

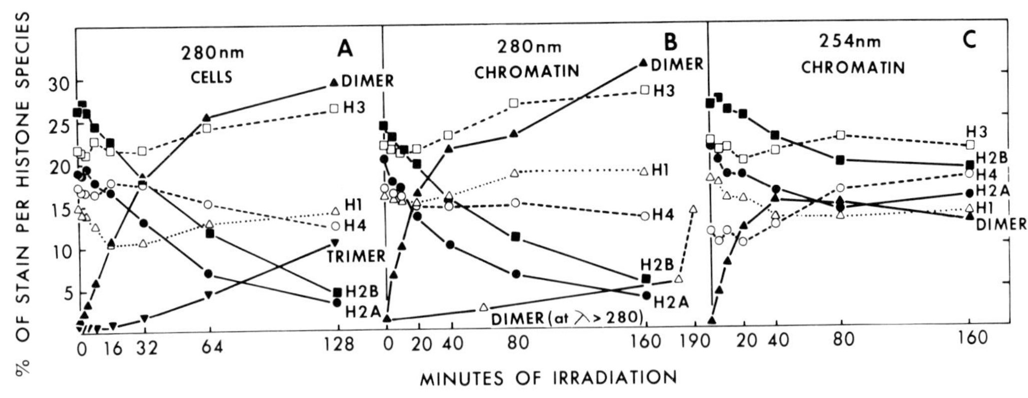

FIGURE 2 Yield of dimer with time of irradiation. Panels A and C are a quantitative summary of the data in Fig. 1 and a parallel experiment done with chromatin irradiated at 254 nm. Panel B summarizes an experiment in which chromatin was irradiated as described for whole cells in the legend to Fig. 1. The amount of stain in each band of the polyacrylamide gel was quantitated by incubating the excised band in 0.5 ml of 1% SDS and determining the absorbance of the eluted stain at 590 nm. The amount of stain in a band is expressed as a percent of the total for all bands at each time-point. In panel B, additional data are shown to demonstrate the rate of dimer production in the presence (lower dimer curve, ——) and subsequent absence, (lower dimer curve, – – –) of a Kimax filter which is essentially opaque below 280 nm.

mer is a reflection of the condensed state of chromatin in vivo has not yet been ascertained.

The solid portion of the bottom line in panel B of Fig. 2 demonstrates the consequence of excluding light of 280 nm or less by use of a Kimax glass filter. The dashed portion of the line shows the result during the first 10 min after removal of the filter. Clearly, filtering out the light between 260 and 280 nm severely depresses dimer production. Since the absorption maximum for histones lies between 270 and 280 nm and a significant emission band of the light source occurs at 280 nm, we infer that absorption of the 280 nm emission into the tyrosine chromophore is principally responsible for the histone-histone cross-links. Furthermore, since irradiation at 254 nm is near the DNA absorption maximum and is known to induce DNA-protein cross-links (Strniste and Smith, 1974; Anderson et al., 1975), we attribute the general loss of histones at this lower wavelength to DNA-histone adduct formation resulting from UV excitation of the DNA bases. That irradiation at 254 nm results in loss of each of the histones with very similar kinetics would be expected for a wavelength absorbed primarily into the DNA to which each of the histones is bound.

The yield of H2A-H2B dimer can be estimated from panels A and B of Fig. 2 to be about 80% based on recoverable histone (H2A+H2B decreases from 45 to 8% of the total, whereas dimer increases to 30%). If the presumptive trimer of panel A is a further photoproduct of the dimer, then 100% of the lost H2A and H2B can be accounted for by these two products. However, it should be recognized that the recovery of histones after photolysis even at 280 nm is incomplete, as would be expected if the action spectrum for histone loss were equivalent to the absorption spectrum of DNA. Thus, the yield estimate rests on the assumption that loss is equivalent for all of the histones, monomers, and oligomers alike.

In order to identify the dimer unambiguously as H2A-H2B, some simple DNA-histone reconstitution experiments of the type previously reported (Table II) were performed. Purified L-cell histones were mixed with DNA at low ionic strength and then irradiated in the usual way. Reisolation and electrophoresis of the histones on polyacrylamide gels showed that even in this crude system, good yields of a product with the same mobility as "native" dimer were produced. Furthermore, as shown in Fig. 3, only H2A and H2B were required for dimer formation (Martinson et al., 1976), whereas the absence of either resulted in no dimer. While this further substantiated the H2A-H2B assignment, the possibility still remained that the presence of one of the pair (H2A or H2B) was required to facilitate the production of homodimer from the other. This possibility was examined by the use of radiolabeled H2A or H2B in the reconstitution mixtures. As shown in Table III, equal quantities of both H2A and H2B are

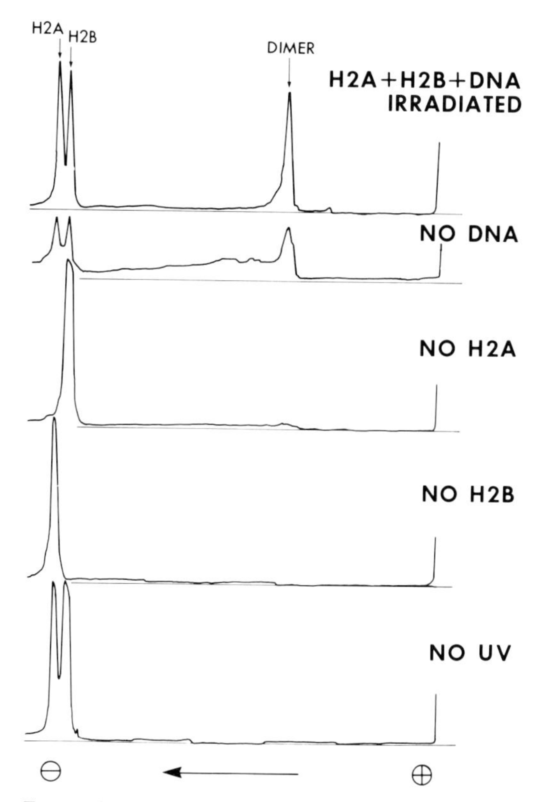

FIGURE 3 Electrophoresis of histones isolated from reconstituted nucleohistone irradiated at 280 nm. For the first panel, purified histone fractions 2A and 2B were mixed first with each other and then directly with DNA in 1 mM EDTA, pH 8, as previously described (Martinson and McCarthy, 1975). ½ ml of such a mixture, containing 100 μg of each of the two histones, was purged with N_2 for 90 min and then irradiated at 280 nm for 20 min in a CMS tube. Subsequent panels in the figure represent parallel samples in which DNA, one of the histones or the irradiation itself was omitted. The volume as well as the ratio of DNA to total histone in the various samples was kept constant.

incorporated into the dimer. The cross-linked product is thus almost certainly an H2A-H2B heterodimer.

The presence of DNA is not an absolute requirement for dimer production (Fig. 3). Nevertheless, if DNA is not present during irradiation, the yield of dimer is low and the background of heterogeneous products in the gel is high. Thus, H2A and H2B associate with each other in the absence of DNA but, under these conditions, apparently only a small proportion of the associated pairs have a conformation resembling that in the native chromsome. Thus, the DNA binding event rectifies the conformations of the various associated pairs of H2A and H2B, improving the yield of dimer and eliminating the nonspecific background. Indeed, as for the H2B-H4 interaction, preliminary association of the histones with themselves in solution must precede their deposition on DNA in order for their appropriate native conformations to be adopted.

Characterization of H2B-H4 and H2B-H2A Dimers

To examine the sites at which histone H2B interacts with H4 and with H2A, the two dimers were purified and subjected to cyanogen bromide (CNBr) peptide mapping. The H2B-H4 and H2B-H2A dimers were prepared by treating L-cells with TNM or UV light and purified by successive fractionation on Sephadex G200 and Bio-Gel P-30 (Martinson and McCarthy, 1976).

Pure preparations of H2A-H2B and H2B-H4 dimers were cleaved at their methionines using CNBr. The resulting CNBr peptides were then displayed on polyacrylamide slabs by means of gel electrophoresis, and the electrophoretic patterns obtained were interpreted by comparison with reference patterns obtained from CNBr digests of the purified monomers.

Fig. 4 demonstrates that in the TNM dimer the H2B and H4 are cross-linked to each other in their C terminal regions. This figure shows the electrophoretic profile of a preparation of CNBr-treated TNM dimer. Each peak in the pattern is assigned to one or more of the digestion products, each of which is illustrated diagrammatically above the appropriate peak. The H2B and H4 polypeptide chains are drawn to scale, together with the sites at

TABLE III

Dimer Production from Nucleohistone in which Either H2A or H2B was Radiolabeled

Radioactive histone(s)	Specific activity	Dimer region	A_{640} of dimer extract $\times 10^2$		cpm/A_{640} $\times 10^{-1}$
	cpm/pmol	*cpm*			
H2A	45	60	5.7		106
H2B	90	140	8.3		168
				sum	274
H2A + H2B	45 & 90	226	8.7		260

Tritium-labeled histones were those previously described (Martinson and McCarthy, 1975). After irradiation and polyacrylamide gel electrophoresis, the stained dimer bands were excised. The stain was extracted into 0.5 ml 20% TCA-25% isopropanol and quantitated by taking the absorbance at 640 nm. The gel slices were subsequently removed and counted.

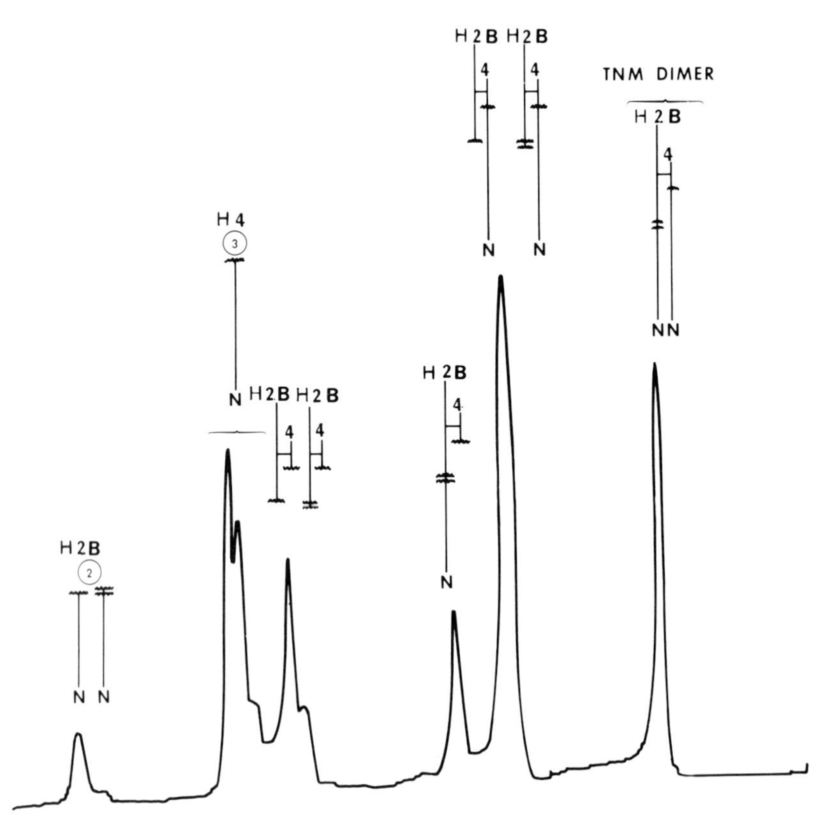

FIGURE 4 Polyacrylamide gel electrophoresis of CNBr-treated TNM dimer. Slab gels were run in acetic acid and urea essentially as described (Martinson and McCarthy, 1975).

which they can be cleaved by CNBr. The N-termini of the polypeptides are indicated. These peak assignments will be documented in a subsequent publication. Fig. 4 shows that two peaks can be identified as the CNBr-cleaved N-termini of H2B and H4. No peak corresponds to the C-terminus of either histone when the pattern of Fig. 4 is compared with the monomer cleavage patterns. Thus, H2B and H4 are cross-linked in their C-terminal portions. For the experiment illustrated, digestion was carried to only a limited extent. The resulting reaction mixture thus contained, in addition to the normal fully cleaved CNBr peptides, partially cleaved as well as uncleaved dimer material. As shown in the figure, all possible partial cleavage products can also be identified.

Inspection of a similar electrophoretic profile of CNBr-treated UV dimer (H2A-H2B) revealed that, in this case, it is the N-terminal region of H2B which is involved in the cross-linking event (Martinson and McCarthy, 1976). Thus, a region in the C-terminal half of H2B interacts with H2A.

The involvement of separate regions of H2B in the cross-links to H2A and H4 suggests that H2B interacts with the two other histones through separate domains. Further evidence that the binding sites for H2A and H4 do not overlap was obtained by treating cells sequentially with UV and then TNM (Fig. 5). After UV treatment, the typical pattern arises in which a prominent dimer (H2A-H2B) and a trimer are seen in the gel electrophoretic profile. Subsequent treatment with TNM generates the usual H2B-H4 dimer together with a new band (marked by an arrow) in the trimer region of the gel, but distinguishable from the UV trimer. We presume this new product to be an H2A-H2B-H4 trimer resulting from linkage of the H2A-H2B dimer to H4 at a separate H2B binding site. This is the most reasonable, and certainly the simplest, interpretation of these data, since treatment with TNM normally produces only a single identifiable product, namely, the H2B-H4 dimer. Attempts are currently under way to identify this trimer positively.

CONCLUSION

The two cross-linking reactions we have described are extraordinarily specific. In each case, only one

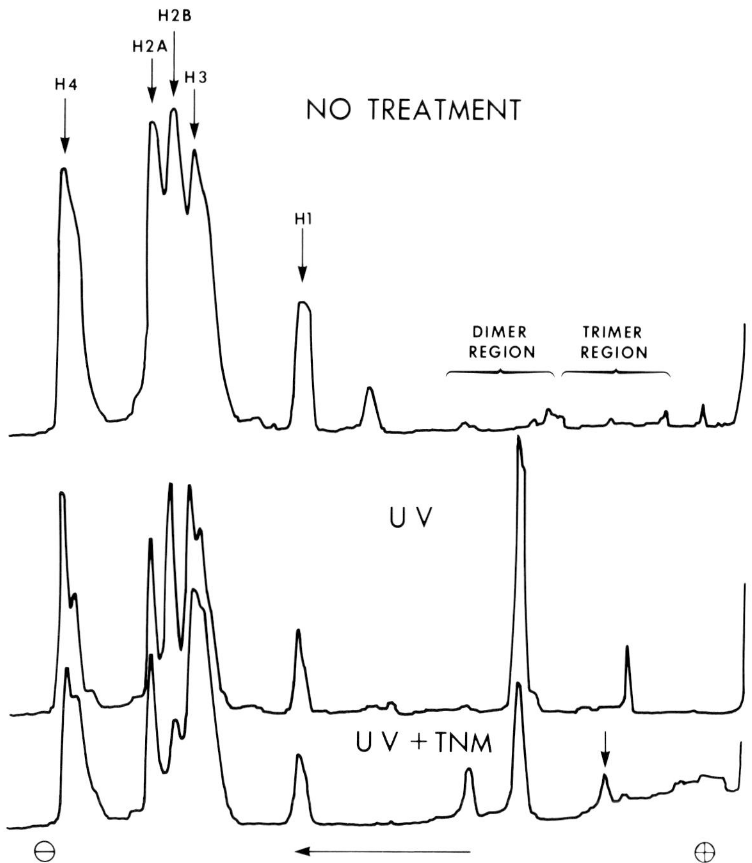

FIGURE 5 Production of putative H2A-H2B-H4 trimer (indicated by arrow) by sequential treatment with UV and TNM.

pair of histones is cross-linked. This specificity is to be expected of cross-linking reagents of zero length, which act at hydrophobic regions of the histones. Furthermore, these agents are sensitive probes of the conformations of the interacting histones. For example, the TNM-induced H2B-H4 dimer arose in reconstituted chromatin only if H2A were also present. Since H2A and H2B are known to interact specifically in solution (D'Anna and Isenberg, 1974), it was proposed that this interaction altered the structure of H2B, and hence of the H2B-H4 part of the complex, in such a way as to render H2B and H4 susceptible to cross-linking by the TNM when deposited on DNA. The results obtained with UV cross-linking are consistent with this interpretation and suggest an assembly hierarchy. Thus, the association of H2A with H2B can proceed with fidelity in the absence of other histones and is, moreover, apparently a prerequisite for the correct association of the H2B with H4. H3 would presumably be the fourth member of the series.

Weintraub et al. (1975) have recently reported the association of H2A, H2B, H3, and H4 to yield the expected tetramer when they are mixed under conditions of high salt and pH (1–3 M NaCl, pH 7–9). The histones in this tetramer exhibited properties indistinguishable from the histones of chromatin by several tests, but underwent drastic conformational changes when the salt concentration was lowered by dialysis. These changes were prevented when DNA was present during dialysis. These results confirm not only the role of DNA in maintaining histone nativity but also the existence of the H2A-H2B-H3-H4 tetramer predicted by the above assembly hierarchy and originally postulated on the basis of histone pair associations at low ionic strength (D'Anna and Isenberg, 1974).

We do not believe that the less stringent requirements for proper H2A-H2B reconstitution compared to that for H2B-H4 reflects a correspondingly lower degree of specificity for the UV reaction relative to that of TNM. The primary site of action for both cross-linking agents is presumably a tyrosine residue, yet each agent gives rise to a different cross-linked histone pair. Furthermore, CNBr peptide mapping shows that the cross-linkages in H2A-H2B and H2B-H4 involve different sites on the H2B (Martinson and McCarthy, 1976). Finally, both cross-linking reactions, even when conducted on whole cells, yield only a single dimer as their major oligomeric product. This contrasts with reagents of low specificity such as imidoesters or aldehydes, which yield mixtures of dimers and oligomers (Olins and Wright, 1973; Ilyin et al., 1974; Hyde and Walker, 1975; Chalkley, 1975; Chalkley and Hunter, 1975; Van Lente et al., 1975; Thomas and Kornberg, 1975). It should be emphasized that the cross-linking agents we have used are sensitive specifically to conformation at the histone-histone binding sites. The data, therefore, do not imply that our direct-mix reconstitution procedure yields nucleohistone which is native in all aspects of conformation.

It is noteworthy that the N-terminal portion of H4 cleaved from the TNM dimer yields, upon acid-urea electrophoresis, the usual triplet band configuration characteristic of a mixture of H4 molecules acetylated to various extents. A similar heterogeneity is revealed in the electrophoretic profile of CNBr-treated UV dimer (Martinson and McCarthy, 1976), in which case at least two of the polymorphic forms of H2A are linked to H2B. In light of the high specificity of the cross-linking agents and particularly in view of the great conformational selectivity of cross-linking exhibited by TNM, this shows that the H2B-H4 and the H2B-H2A binding interactions are fundamental structural features of chromatin. Furthermore, together with the findings that H2B-H2A cross-linking is nearly quantitative and that an H2A-H2B-H4 trimer can be produced in whole cells (Fig. 5), these results imply that the H2A-H2B-H4 association is a general feature of chromatin structure. This presumably reflects the existence in chromatin of the tetrameric H2A-H2B-H4-H3 cluster originally predicted by D'Anna and Isenberg (1974) and later demonstrated in solution by Weintraub et al. (1975).

REFERENCES

Anderson, E., Y. Nakashima, and W. Konigsberg. 1975. Photoinduced crosslinkage of gene-5 protein and bacteriophage fd DNA. *Nucl. Acids Res.* **2:**361–371.

Bonner, W. M., and H. B. Pollard. 1975. The presence of F3-F2a1 dimers and F1 oligomers in chromatin. *Biochem. Biophys. Res. Commun.* **64:**282–288.

Chalkley, R. 1975. Histone propinquity using imidoesters. *Biochem. Biophys. Res. Commun.* **64:**587–594.

Chalkley, R., and C. Hunter. 1975. Histone-histone propinquity by aldehyde fixation of chromatin. *Proc. Natl. Acad. Sci. U. S. A.* **72:**1304–1308.

D'Anna, J. A., Jr., and I. Isenberg. 1974. A histone cross-complexing pattern. *Biochemistry.* **13:**4992–4997.

Elgin, S. C. R., and H. Weintraub. 1975. Chromosomal proteins and chromatin structure. *Annu. Rev. Biochem.* **44:**725–774.

Hyde, J. E., and I. O. Walker. 1975. Covalent cross-linking of histones in chromatin. *FEBS (Fed. Eur. Biochem. Soc.) Lett.* **50:**150–154.

Ilyin, Y. V., A. A. Bayev, Jr., A. L. Zhuze, and A. J. Varshavsky. 1974. Histone-histone proximity in chromatin as seen by imidoester cross-linking. *Mol. Biol. Rep.* **1:**343–348.

Kornberg, R. D. 1974. Chromatin structure: a repeating unit of histones and DNA. *Science (Wash. D.C.).* **184:**868–871.

Martinson, H. G., and B. J. McCarthy. 1975. Histone-histone associations within chromatin: crosslinking studies using tetranitromethane. *Biochemistry.* **14:**1073–1078.

Martinson, H. G., and B. J. McCarthy. 1976. Histone-histone interactions within chromatin. Preliminary characterization of presumptive H2B-H2A and H2B-H4 binding sites. *Biochemistry.* **15:**4126–4131.

Martinson, H. G., M. B. Shetlar, and B. J. McCarthy. 1976. Histone-histone interactions within chromatin. Crosslinking studies using ultraviolet light. *Biochemistry.* **15:**2002–2007.

Olins, A. L., and D. E. Olins. 1974. Spheroid chromatin units (ν bodies). *Science (Wash. D.C.).* **183:**330–332.

Olins, D. E., and E. B. Wright. 1973. Glutaraldehyde fixation of isolated eucaryotic nuclei. Evidence for histone-histone proximity. *J. Cell Biol.* **59:**304–317.

Strniste, G. F., and D. A. Smith. 1974. Induction of stable linkage between the DNA dependent RNA polymerase and $d(A\text{-}T)_n \cdot (A\text{-}T)_n$ by ultraviolet light. *Biochemistry.* **13:**485–493.

Thomas, J. O., and R. D. Kornberg. 1975. An octomer of histones in chromatin and free in solution. *Proc. Natl. Acad. Sci. U. S. A.* **72:**2626–2630.

Van Holde, K. E., and I. Isenberg. 1975. Histone interactions and chromatin structure. Accounts Chem. Res. **8:**327–335.

Van Lente, F., J. F. Jackson, and H. Weintraub. 1975. Identification of specific crosslinked histones after treatment of chromatin with formaldehyde. *Cell.* **5:**45–50.

Weintraub, H., K. Palter, and F. Van Lente. 1975. Histones H2a, H2b, H3 and H4 form a tetrameric complex in solutions of high salt. *Cell.* **6:**85–110.

THE CELL NUCLEUS: DIGITAL OR ANALOGUE COMPUTER

JOHN PAUL

Although the underlying mathematical logic in different kinds of computers may be the same, the working characteristics are dictated by the technology used in their construction. We are most accustomed to electronic computers, but the first real computer, Burridge's engine, was mechanical; hydraulic computers have been designed and, in principle, there is no reason why one should not construct a chemical computer. Indeed, the proposition I wish to discuss is that, in the course of evolution, the nucleus of the eukaryotic cell has emerged as a chemical computer. It has a core store of information (the DNA in the chromosomes), a machinery for selecting and processing that information in response to specific input signals, and, finally, an output in the form of coded messages (messenger RNA molecules) which specify the proteins that characterize the cell. Mechanical computers are characteristically large and very slow in action. Electronic computers are relatively small and very fast. The chemical computer of the nucleus is exceedingly compact, but fairly slow. One reason for considering this proposition is that, if cell nuclei and the computers with which we are familiar have common mathematical logic, this may suggest ways in which to tackle the problems of unraveling their circuitry. It is readily appreciated that it is in this circuitry that the keys to normal cellular function, particularly differentiation, and abnormal cellular function, as in cancer cells, lie.

Electronic computers, broadly speaking, are of two major types, digital and analogue. A digital computer is essentially a network of switches; the magnitude of the electrical signal is, in principle, unimportant and carries no information. An analogue computer contains electronic components such as variable resistors and capacitors, and information is conveyed by the magnitude of the signal, the output pattern being related to the input by a mathematical formula. Geneticists have frequently thought in terms of digital computers, that is, in terms of signals which permit or prevent the transcription of specific genes, which, in turn, influence other genes. Biochemists, on the other hand, have frequently thought in terms of analogue situations. The Michaelis/Menten relationship between substrate concentrations and enzyme velocity and similar relationships between the concentration of repressors and enzyme velocity are examples of typical analogue phenomena. In this paper, I wish to address myself to experimental findings which seem to lead to paradoxical conclusions concerning the role of the nucleus as either a digital or analogue computer.

First, let us consider the methods to be used in trying to understand the "hardware" of a computer of unknown type. We can employ three main approaches. The first is to identify the signal used in the computer and design a device that will enable it to be traced. This could be said to be the tool used by the molecular biologist when he employs nucleic acid hybridization to follow the fate of nucleotide sequences. The second approach is to isolate components of the machinery, study how these work, and attempt to reassemble the machine. This could be said to be the approach of the biochemist. Finally, one can try to modify the machine by damaging or modifying components and observing the effects on its performance. This is, of course, the approach of the geneticist. In this communication, I will be concerned mainly with molecular biological and biochemical approaches.

The examples to be discussed relate to erythropoiesis. During the development of erythrocytes from proerythroblasts, the cytoplasm first becomes strongly basophilic as a result of the accumulation of ribosomes in basophilic erythroblasts.

JOHN PAUL Beatson Institute for Cancer Research, Wolfson Laboratory for Molecular Pathology, Glasgow, Scotland

Hemoglobin begins to accumulate in polychromatic erythroblasts. The ribosomes begin to diminish in reticulocytes and eventually disappear completely. Associated with these cytoplasmic changes, the nucleus becomes progressively more condensed until, in mammals, it is almost completely inactive in the orthochromatic erythroblast and is then extruded to form a reticulocyte. We are concerned with knowing what changes occur in gene expression during this process. Are there extensive changes in the numbers of genes expressed or are they confined to a few genes? Is control exercised in the course of transcription, during processing, or in later steps? Do these controls involve switching on and off of gene complexes or alterations in reaction rates?

First, we can ask direct questions about globin messenger RNA itself. Is the accumulation of hemoglobin the result of a change in the concentration of globin messenger RNA? Theoretically, this could be determined directly by titrating messenger RNA from different maturation stages with complementary DNA made by copying globin messenger RNA with reverse transcriptase. In normal erythroid tissue, there is difficulty in doing these titrations directly because it is exceedingly difficult to obtain sufficiently large numbers of pure fractions at each stage of development. Consequently, we devised a method of *in situ* hybridization that depends on detecting messenger RNA in cells fixed to a slide by hybridizing with cDNA and subsequently detecting RNA/DNA hybrid molecules by autoradiography. Experiments of this kind (Harrison et al., 1974) show rather conclusively that nonerythroid cells and early proerythroblasts have little, if any, globin messenger RNA but that late in the proerythroblast stage, and perhaps in some cases at the transition from proerythroblast to basophilic erythroblast, a massive accumulation of globin messenger RNA occurs.

Since it is difficult to obtain fractions of erythroid cells from bone marrow, great interest has been directed recently to the Friend cell culture system. Friend cells are erythroleukemic cells which originated in mice after inoculation with the Friend virus. In a normal culture, they proliferate rapidly and can be grown in bulk. They can also be cultured as colonies from single cells, making it possible to undertake genetic studies. When cultured continuously, they resemble proerythroblasts quite closely, and there is good evidence to suggest that they represent either transformed early proerythroblasts or transformed erythropoietin-sensitive cells. Differentiation beyond this stage seems to be blocked, but treatment with any one of a number of inducers, notably dimethyl sulfoxide (DMSO) permits them to mature to a stage resembling orthochromatic erythroblasts. Simultaneously, they accumulate hemoglobin and globin messenger RNA (Gilmour et al., 1974). In considering the possible analogy with a computer, we would like to know whether the alteration in globin messenger RNA level is accompanied by the appearance of other messenger RNAs and the disappearance of yet others, or whether there are quantitative changes only in RNA populations.

This kind of question can be tackled by analyzing the complexity of the messenger RNA populations and the abundance of different messenger RNA classes. By "complexity" is meant the number of different gene transcripts, and this can be measured in several ways. One method is by hybridizing a great excess of RNA to a very small amount of highly labeled unique DNA. In these experiments, it is common to use an RNA excess of several million times. When experiments of this kind are carried out with nuclear and cytoplasmic RNA from growing Friend cells, it is found that a great many more kinds of RNA sequences occur in the nucleus than in the cytoplasm; in growing Friend cells, only about 20% of the different kinds of sequences represented in the nucleus are present in the cytoplasm. This provides convincing evidence for extensive processing of RNA. It may be noted that those sequences in the nucleus that contain a polyadenylic acid sequence are almost as complex as those that do not (Birnie et al., 1974; Getz et al., 1975).

Another way to investigate this question is to isolate all the polyadenylated RNA sequences from either the nucleus or the cytoplasm and to transcribe these with reverse transcriptase to give a cDNA copy of all the RNA sequences. By measuring the rate at which RNA/DNA hybrids form between this cDNA and the RNA, it is possible to obtain an estimate of the abundance (relative concentration) of different sequences. Those sequences present in the highest abundance form hybrids more rapidly than sequences present in low abundance. We undertook experiments with RNA from both uninduced and induced Friend cells. When cDNA made from uninduced Friend cells was hybridized to RNA from both induced and uninduced cells, we found that the pattern of hybridization in both the homologous and heterologous reaction were very similar (A. Minty, unpublished results). In particular, it is clear that all the sequences in the cDNA from uninduced cells are present in both induced and uninduced ones. When the converse experiment was performed with cDNA made from a population of induced cells, we again found that all sequences were present in both induced and uninduced cells, but in the induced cells, there was a population of sequences in much higher abundance than in the uninduced cells (A. Minty, unpublished results). Quantitatively, these high-abundance sequences could be accounted for entirely by the increased concentration of globin messenger RNA which had been measured directly. The only other major difference that could be demonstrated between the RNA of cells in the two different states was in the total concentration, the total amount of RNA in the induced cells being a quarter to a half of that in uninduced cells. Apparently, the differences between induced and uninduced cells can be accounted for almost entirely by the increase in globin messenger and possibly a few other RNAs.

The Friend cell is perhaps a somewhat artificial system, and it is necessary to ask whether these phenomena occur generally. When cDNA made from Friend polysomal RNA is hybridized with RNA from the polysomes in other tissues, from normal liver tissue or from a fibroblastic line of cells, for example, we obtain evidence that almost exactly the same population of messengers is present in all cells. For example, in experiments designed to investigate this by comparing directly RNA from mouse and liver cells, the same kind of finding is obtained, indicating that the major differences in RNA populations between the two cell types are not in the appearance or disappearance of specific RNA sequences but in changes in their abundance (Young et al., 1976). Similar results have been obtained in other laboratories and this general conclusion seems to be reasonably established.

This finding conflicts with some popular ideas, for it is generally postulated that transcriptional control is important. Indeed, there is good evidence that in cells of this type, transcriptional controls do operate. Perhaps the most convincing are studies on transcription of RNA from isolated chromatin. Some of these experiments were conducted many years ago and indicated tissue differences in RNA transcribed from repetitive DNA in chromatin from different tissues (Paul and Gilmour, 1968). These old findings have now been extensively substantiated by experiments in which the transcript from chromatin has been measured by using cDNA to a known messenger. Most of these experiments have been conducted with globin cDNA and it has been shown by ourselves and by other groups (Gilmour and Paul, 1973; Axel et al., 1973; Wilson et al., 1975) that the chromatin from erythroid tissue gives rise to a transcript containing globin messenger RNA, whereas in the transcript from chromatin from nonerythroid tissue, globin messenger RNA is not measurable. The factors involved in this control have not yet been clearly identified, but the evidence suggests that, in part, these reside in the structure of the chromatin itself. This interpretation has perhaps been strengthened by recent observations which demonstrate that the active genes in chromatin are specifically sensitive to digestion with DNAse I (Weintraub and Groudine, 1976).

Hence, we are left with the paradox that, whereas studies on RNA populations in the nucleus and cytoplasm strongly suggest some kind of posttranscriptional modulation akin to an analogue computer, studies with chromatin strongly suggest that there is "on and off" transcriptional regulation. However, there is a possibility that this latter control may be confined to certain specialized genes such as those for globin and ovalbumin.

The question, therefore, remains open concerning the type of computer which the cell nucleus may represent. At the transcriptional level, we apparently discern a digital logic whereas processing appears to be essentially an analogue function.

REFERENCES

Axel, R. H., H. Cedar, and G. Felsenfeld. 1973. Synthesis of ribonucleic acid from duck reticulocyte chromatin *in vitro*. *Proc. Natl. Acad. Sci. U. S. A.* **70:**2029–2032.

Birnie, G. D., E. MacPhail, B. D. Young, M. J. Getz, and J. Paul. 1974. The diversity of the messenger RNA population in growing Friend cells. *Cell Differ.* **3:**221–232.

Getz, M. J., G. D. Birnie, B. D. Young, E. MacPhail, and J. Paul. 1975. A kinetic estimation of base sequence complexity of nuclear poly(A)-containing RNA in mouse Friend cells. *Cell.* **4:**121–129.

Gilmour, R. S., P. R. Harrison, J. D. Windass, N. A. Affara, and J. Paul. 1974. Globin messenger RNA synthesis and processing during haemoglobin induction in Friend cells. I. Evidence for transcriptional control in clone M2. *Cell Differ.* **3:**9–22.

Gilmour, R. S., and J. Paul. 1973. Tissue-specific

transcription of the globin gene in isolated chromatin. *Proc. Natl. Acad. Sci. U. S. A.* **70:**3440–3442.

HARRISON, P. R., D. CONKIE, N. AFFARA, and J. PAUL. 1974. In situ localization of globin messenger RNA formation. *J. Cell Biol.* **63:**402–413.

PAUL, J., and R. S. GILMOUR. 1968. Organ-specific restriction of transcription in mammalian chromatin. *J. Mol. Biol.* **34:**305–316.

WEINTRAUB, H., and M. GROUDINE. 1976. Chromosomal subunits in active genes have an altered conformation. *Science (Wash. D.C.).* **193:**848–856.

WILSON, G. N., A. W. STEGGLES, and A. W. NIENHUIS. 1975. Strand-selective transcription of globin genes in rabbit erythroid cells and chromatin. *Proc. Natl. Acad. Sci. U. S. A.* **72:**4835–4839.

YOUNG, B. D., G. D. BIRNIE, and J. PAUL. 1976. Complexity and specificity of polysomal polyA+ RNA in mouse tissue. *Biochemistry.* **15:**2823–2829.

Functional Organization of Chromosomes

INTRODUCTORY REMARKS

HEWSON SWIFT

Although our attention has been focused recently on our nation's bicentennial, and the ghost of Ben Franklin has walked among us (looking strange and yet vaguely familiar), one should not begin a symposium on chromosome structure without pointing out that, for the chromosome, it is an important anniversary too. The year 1876 was a particularly productive one in the history of cell biology. In that year Oscar Hertwig (1876) first saw the sperm and egg pronuclei fuse in the sea urchin egg. A few months earlier, Strasburger published the first edition of his monograph on "Zellbildung und Zelltheilung," in which he clearly identified the spindle and metaphase plate in spruce embryos, and Balbiani described chromosomes as "batonnets étroits" that formed from the substance of the nucleus before cell division in grasshopper ovarioles. There are a few recognizable drawings of chromosomes in the earlier literature, but in each case they are either passed over without comment or wrongly interpreted. Hofmeister (1848) clearly figured a metaphase plate in a dividing stamen hair cell of *Tradescantia,* but described it as a collection of mucilaginous granules, formed from the remnant of a dissolving nucleus. Balbiani in 1861 published detailed drawings of mitotic figures in micronuclei of *Paramecium,* but thought the chromosomes were bundles of sperm lined up within a minute "testicule" of the ciliate, which he considered as a multicellular organism. Not until 100 yr ago were chromosomes first correctly seen to derive from the nucleus at prophase, to be aligned on the mitotic spindle, and to be divided into two daughter groups at anaphase.

In the few years after 1876, progress in the study of chromosomes was immense. Although fixatives (acetic acid) and chromosome stains (carmine) had been employed by Balbiani in 1861, both were greatly improved by Flemming (1879) in his detailed and influential studies on the epidermal cells in the tail fin and gill filaments of salamander larvae, to the extent that "Flemming's fixative" (formalin, dichromate, acetic acid) is still in use today. Because portions of the nucleus showed strong staining with basic dyes, Flemming coined the word "chromatin" in 1879. A few years before Meischer (1871) had isolated an acidic phosphorus-containing substance from pus cells, unwittingly contributed by soldiers wounded in the Franco-Prussian War. Flemming thus suggested that the affinity of chromatin for basic dyes was a result of their content of "nuclein": "possibly chromatin is identical with nuclein, but if not . . . one carries the other." Waldeyer (1888) later used the word "chromosome" to emphasize the continuity between chromatin and the "coiled threads" described by Flemming. Both Strasburger in *Tradescantia* stamen hairs, and Flemming in *Salamandra* epidermis, also made careful sequential observations on living cells in mitosis. Flemming clearly saw the longitudinal split in metaphase chromosomes, and corrected the mistake made earlier by Balbiani (1876), who thought that chromosomes broke transversely in anaphase. He thought, however, that all prophase chromosomes were joined in a continuous thread, an error later corrected by Rabl (1885) in a beautifully illustrated paper.

The giant polytene chromosomes of dipteran larval salivary glands were described for *Chironomus* by Balbiani in 1881 (Fig. 1), and shortly thereafter also by Carnoy (1884). Although Balbiani thought the bands could be resolved into gyres, he correctly figured the relation between chromosome and nucleolus and described nucleolar fusion. He also clearly figured the giant puffed regions on chromosome 4, which he called "renflement discoïde," stating that they were apparently composed of a substance different from the chromosome. In the cytology of polytene chromo-

HEWSON SWIFT Department of Biology, University of Chicago, Chicago, Illinois

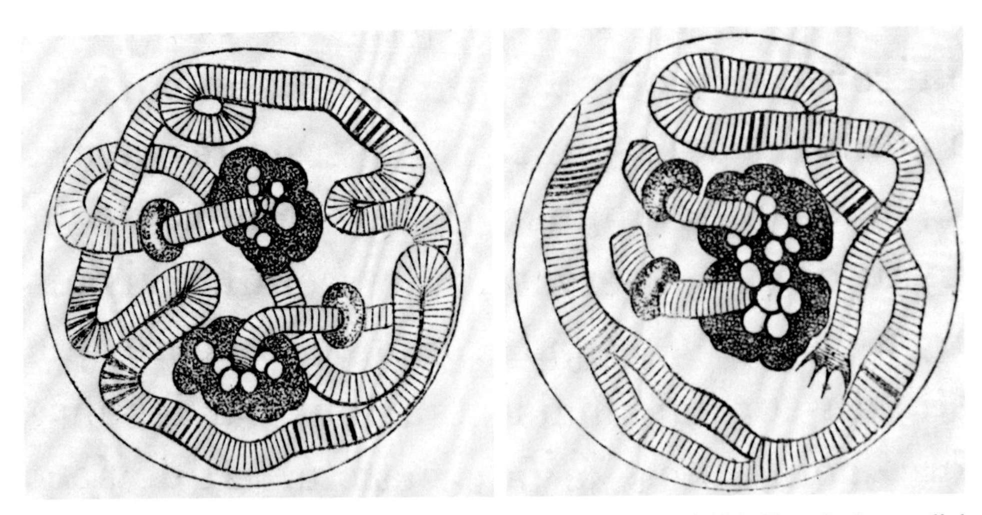

FIGURE 1 Salivary gland nuclei of *Chironomus* as drawn by Balbiani (1881). Note the large puffed regions (Balbiani rings) indicated between the site of the nucleolus and the end of the chromosome, as well as the region of chromosome asynapsis indicated in the right-hand figure.

somes, few papers were published between the initial descriptions almost a century ago and the explosive interest kindled by their rediscovery by Kostoff (1930), Painter (1933), King and Beams (1934), Heitz (1934), and Bauer (1935), and the reinterpretation of their banded structure in genetic terms. But one feels that Balbiani and Carnoy would doubtless have recognized the work of these investigators, had they lived to see it, as a logical extension of their own. Now, almost another half century later, we are entering a new dimension in the consideration of chromosome structure, with major emphasis on problems of molecular orientation and composition..

A primary problem in the current analysis of chromosome structure is the localization within chromosomes of specific DNA sequences. As is now well known, DNA can be divided into four different classes by the kinetics of renaturation, after the fragmentation of molecules and separation of the two strands by heating. (1) Some base sequences renature immediately, regardless of their concentration, as they fold back on themselves to form loops or hairpin-shaped molecules. This zero-time, or foldback, DNA comprises roughly 3–4% of the *Drosophila* genome (Schmid et al., 1975) and 10% in *Xenopus* (Perlman et al., 1976). (2) Highly repetitive or simple sequence DNAs are often simple polymers of a small number of base pairs repeated in tandem thousands or millions of times. As expected from their abundance, they show exceptionally rapid but concentration-dependent renaturation. Their average base composition is often markedly different from the main component, and thus they frequently form separate bands or satellites when DNA is centrifuged to buoyancy in cesium chloride. (3) Moderately repetitive DNAs are a heterogeneous group of sequences, longer than simple sequence DNA. They occur in families containing hundreds or thousands of copies, and thus renature at intermediate rates. Some moderately repetitive DNAs are of known function (coding for histones, ribosomal and transfer RNAs), but most are unidentified. (4) Unique DNAs renature at the slow rate characteristic of their occurrence in single copies per genome. Most structural genes are in this class, together with whatever individual adjacent spacer regions they may possess.

The genome as we presently glimpse it is a curious mixture of constancy and plasticity. Inasmuch as organic evolution is based on the behavior of DNA, an understanding of the interaction between the genome and the selective pressures that have shaped it is essential to our reading of the patterns of biological diversity. Some DNA sequences are doubtless as ancient as the earliest eukaryotes. For example, the genes for proteins, such as histones and cytochromes, appear to have been invented once in evolution, and have ever since been under strong positive selection. Others, such as the satellite DNAs in *Drosophila,* show a remarkable variability among related species, indicating marked alterations associated with speciation in comparatively recent times. The kinds of evolutionary restraints that influence these sequences are totally unknown. Even within the developmental stages of a single organism, certain portions of the genome appear to be differentially replicated, such as the well-known cases of ribosomal RNA amplification in amphibian oocytes and magnification in bobbed *Drosophila* (Tartof, 1975). A possibly related phenomenon is the differential replication of satellite DNAs in different tissues of *Drosophila virilis* (Endow and Gall, 1975). A fascinating finding concerning foldback DNAs is the indication that they may possess variable locations within the genome (Perlman et al., 1976). If chromosome structure is to be interpreted in molecular terms, we must make sense out of the mixture of extreme stability, short-term lability, and shifting developmental patterns that characterize the genome.

In our continued probing of genetic mechanisms, polytene chromosomes are clearly of great importance. Not only is the genome laid out with a detail unmatched in other nuclei, but each locus is laterally amplified 1,000–8,000 times. This provides a tremendous increase in sensitivity for studies involving cytological hybridization, and has facilitated the localization of 5S and histone loci (Wimber and Steffensen, 1970; Birnstiel et al., 1974), as well as the indication that the giant 75S RNA of *Chironomus* salivary gland cell cytoplasm is, in all probability, the synthetic product of the site of Balbiani's "renflement" (Lambert, 1974). When random fragments of the *Drosophila* genome have been isolated into bacterial plasmids, and then cloned in plasmid-infected bacterial cultures, cytological hybridization on polytene chromosomes is also important in the localization of the chromosomal site from which the DNA fragments are derived (Glover et al., 1975). Thus the site of origin of several unique sequences, inserted and cloned in plasmid DNA, has already been identified. Of great interest has been the discovery that certain sequences occur in a large number of discrete places within the karyotype. Such sequences thus possess a character expected of elements in a system for gene control, where a single factor may be involved in the simultaneous initiation or repression of transcription of a family of noncontiguous loci. Because cloned DNA is also amenable to base sequencing, clearly we are launched upon a base-by-base analysis of specific gene sequences, whose precise position within the genome can be identified.

As reviewed by Laird (1973), *Drosophila* genomes are characterized by a significant portion of simple sequence or satellite DNA. In some cases (Gall, 1974), these are simple seven-nucleotide sequences, repeated millions of times within the genome, tending to be species-specific, and clustered at specific sites, particularly the centromere. Reasons for the presence of these highly repetitive DNAs of *Drosophila* and other eukaryote genomes are, in all respects, enigmatic and obscure. Why an organism such as *Drosophila virilis* should possess 40% of its total DNA in sequences apparently untranscribed and clustered into great blocks of heterochromatin is one of the chief current problems of nuclear cell biology. Clearly, the determination of the precise localization of these components within the genome, their detailed base sequences, and their comparative anatomy among related species, are of importance to the untangling of this puzzling problem. It is obviously impossible, with our present understanding of DNA replication and function, to understand the mechanisms whereby such simple repetitive sequences have arisen—or why, indeed, they exist at all—as an important fraction of numerous genomes. The research of Peacock, as reviewed in the following paper, has been directed to these problems.

The probable relation between bands of polytene chromosomes and the chromomere structure demonstrated by many chromosomes, particularly during stages of the meiotic prophase, is a frequently discussed subject. The chromomeric or beadlike structure of prophase chromosomes was described only a few months after Balbiani's discovery of polytene chromosomes. Pfitzner (1882) attributed his discovery, that chromosomes of salamander larvae were composed of a row of "Körnchen," to the excellent quality of an oil immersion lens, newly purchased for 200 marks. As can be seen in Fig. 2, Pfitzner indicated that all the granules were of approximately the same size, although he considered that chromosome replication took place by the simultaneous division of each granule. The following year Van Beneden (1883), in studying the early cleavage of the parasitic worm *Ascaris,* pointed out not only the variation in size between different chromomeres, but also the identical patterns displayed by the two daughter chromosomes: "There is not the least peculiarity of one that is not found exactly duplicated in its fellow." This was one of the most striking early demonstrations of the intricate individuality that is a characteristic of chromosome structure. With the structure of nucleosomes, dis-

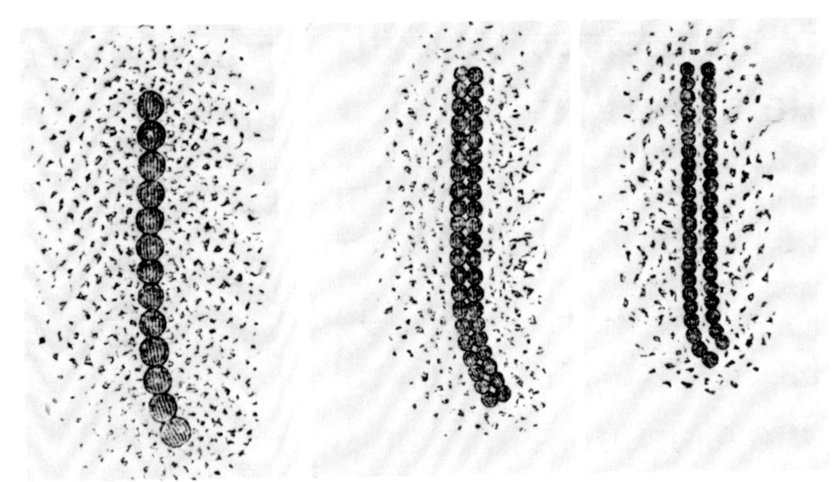

FIGURE 2 Chromomeres from epidermal cells of the salamander (*Salamandra*), as drawn by Pfitzner (1882). Stages in late prophase, evidencing chromatid separation, are shown.

cussed elsewhere in this conference, we are again confronted with a row of "Körnchen," albeit smaller by a factor of between 10 and 100 times. In modern times, the electron microscope, after many disappointing earlier years, is at last beginning to add significantly to our knowledge of the arrangement of DNA and protein molecules in the chromosome, as attested by several contributions to our symposium. For recent discussions on the electron microscopy of chromosomes, see Ris and Kubai (1970) and Olins et al. (1975). Ris, like Pfitzner, has also been directing a newly purchased optical system to the problem of chromosome structure. The million-volt electron microscope possesses a resolving power roughly 1,000 times greater than Pfitzner's oil immersion lens (and a price tag enlarged by a factor of 10,000). With its ability to penetrate relatively thick specimens at high resolution, this instrument has already provided provocative images that suggest a promising future.

The study of chromosome structure, as in most other areas of cell biology, demands a close coordination of multiple techniques. Among those useful in the past and with promise for the future are the tools of the morphologist (light and electron microscopes of diverse kinds) in parallel with manipulations of the molecular biologist, and at least for the present, particularly involving renaturation kinetics, restriction enzyme cleavage, gel separation of specific sequences, the amplification of specific nucleotide sequences in the laboratory by plasmid cloning, and the detailed analysis, piece by piece and base by base, of the vast lengths of DNA in the eukaryote genome. Chromosome structure is clearly one of the most challenging and complex problems in all of biology. Where else do tightly integrated molecular complexes occur as enormous as a chromosome, where the absence or substitution of one seemingly insignificant component can alter the growth patterns of an organism, or for that matter make the difference between viability and lethality? And because the chromosomes provide continuity from generation to generation, within their chemical structure lie clues to the complex patterns of evolution.

The symposium also contained papers by Gerald M. Rubin (Stanford University), who spoke on "The arrangement of DNA sequences within the genome of *Drosophila melanogaster*," and Hans Ris (University of Wisconsin), who spoke on "Levels of chromosome organization." The material presented in these talks will be published elsewhere.

REFERENCES

BALBIANI, E. G. 1861. Récherches sur les phénomènes sexuels des infusoires. *Jour. de la Physiol. de l'homme et des Animaux.* **4:**102–130, 431–448, 465–520.

BALBIANI, E. G. 1876. Sur les phénomènes de la division du noyau callulaire. *C. R. Acad. Sci.* (Paris). **83:**831–834.

BALBIANI, E. G. 1881. Sur la structure du noyau des cellules salivaires chez les larves de *Chironomus. Zool. Anz.* **4:**637–641, 662–666.

BAUER, H. 1935. Der Aufbau der Chromosomen aus den Speicheldrüsen von *Chironomus thummi* Keifer. *Z. Zellforsch. Mikrosk. Anat.* **23:**280–313.

BIRNSTIEL, M. L., E. S. WEINBERG, and M. L. PARDUE.

1974. Evolution of 9S mRNA sequences. *In* Molecular Cytogenetics. B. Hamkalo and J. Papaconstantinou, editors. Plenum Press, New York. 75–93.

Carnoy, J. B. 1884. "La Biologie Cellulaire." Fasc. 1. J. Van In et cie, Lierre.

Endow, S. A., and J. G. Gall. 1975. Differential replication of satellite DNA in polyploid tissues of *Drosophila virilis*. *Chromosoma (Berl.)*. **50:**175–192.

Flemming, W. 1879. Beiträge zur Kentniss de Zelle und inre Lebenserscheinungen. *Arch. Mikro. Anat.* **16:**302–406.

Gall, J. G. 1974. Repetitive DNA in *Drosophila*. *In* Molecular Cytogenetics. B. Hamkalo and J. Papaconstatinou, editors. Plenum Press, New York. 59–74.

Glover, D. M., R. L. White, D. J. Finnegan, and D. S. Hogness. 1975. Characterization of six cloned DNAs from *Drosophila melanogaster* including one that contains the genes for rRNA. *Cell.* **5:**149–157.

Heitz, E. 1934. Uber α- und β-heterochromatin sowie Konstanz und Bau der chromomeren bei *Drosophila*. *Biol. Zentralbl.* **54:**588–609.

Hertwig, O. 1876. Beitrage zur Kenntniss der Bildung. Befuchtung und Theilung des thierschen Eies. *Gegenbaur's Morphol. Jahrb.* **1:**347–432.

Hofmeister, W. 1848. Ueber die Entwicklung des Pollens. *Bt. Zeit.* **6:**425–434, 649–658, 670–674.

King, R. L., and H. W. Beams. 1934. Somatic synapsis in *Chironomus,* with special reference to the individuality of the chromosomes. *J. Morphol.* **56:**577–592.

Kostoff, D. 1930. Discoid structure of the spireme, and irregular cell divisions in *Drosophila melanogaster*. *J. Hered.* **21:**323–324.

Laird, C. D. 1973. DNA of *Drosophila* chromosomes. *Annu. Rev. Genet.* **7:**177–204.

Lambert, B. 1974. Repeated nucleotide sequences in a single puff of *Chironomus tentans* polytene chromosomes. *Cold Spring Harbor Symp. Quant. Biol.* **38:**637–644.

Miescher, F. 1871. Ueber die chemische Zusammensetzung der Eiterzellen. Hoppe-Seyler med. chem. Untersuch. **4:**441–460.

Olins, A. L., R. D. Carlson, and D. E. Olins. 1975. Visualization of chromatin substructure: nu bodies. *J. Cell Biol.* **64:**528–537.

Painter, T. S. 1933. A new method for the study of chromosome rearrangements and the plotting of chromosome maps. *Science (Wash. D.C.)*. **78:**585–586.

Perlman, S., C. Phillips, and J. O. Bishop. 1976. A study of foldback DNA. *Cell.* **8:**33–42.

Pfitzner, W. 1882. Über den feineren Bau der bei Zelltheilung auftretenden fadenförmigen Differenzirungen des Zellkernes. *Morphol. Jahrb.* **7:**289–311.

Rabl, C. 1885. Uber Zelltheilung. *Morph. Jahrb.* **10:**214–330.

Ris, H., and D. F. Kubai. 1970. Chromosome structure. *Annu. Rev. Genet.* **4:**263–294.

Schmid, C. W., J. E. Manning, and N. Davidson. 1975. Inverted repeat sequences in the *Drosophila* genome. *Cell.* **5:**159–172.

Strasburger, E. A. 1875. "Zellbildung und Zelltheil," Jena. G. Fischer.

Tartof, K. D. 1975. Redundant genes. *Annu. Rev. Genet.* **9:**355–385.

Van Beneden, E. 1883. Recherches sur la maturation de l'oeuf et la fécondation. *Arch. Biol.* **4:**265–640.

Waldeyer, W. 1888. Ueber Karyokinese und ihre Beziehungen zu den Befruchtungsvorgängen. *Arch. Mikro. Anat.* **32:**1–122.

Wimber, D., and D. Steffensen. 1970. Localization of 5S RNA genes on *Drosophila* chromosomes by RNA-DNA hybridization. *Science (Wash. D.C.)* **170:**639–641.

HIGHLY REPEATED DNA SEQUENCES: CHROMOSOMAL LOCALIZATION AND EVOLUTIONARY CONSERVATISM

W. J. PEACOCK, R. APPELS, P. DUNSMUIR, A. R. LOHE, and W. L. GERLACH

Concepts of genome organization in higher organisms changed radically eight years ago when Britten and Kohne (1968) showed that, in all the plants and animals they examined, a large proportion of the genome consisted of DNA nucleotide sequences repeated thousands and even millions of times. This observation conflicted with the expectation, on genetic grounds, that the whole of the genome would be divided into genes or into gene sequences which appeared only once per haploid complement. A great deal of research effort has been centered on defining the principles of organization and function of these highly repeated DNA sequences.

The first experiments localizing highly repeated sequences to chromosomes, *in situ*, showed them to be in heterochromatin near the centromeres (Pardue and Gall, 1970; Jones, 1970). This observation led to the suggestion that they were of fundamental importance to centromere action (Walker, 1971), but when, on the basis of buoyant density studies, it appeared that these DNA sequences were rapidly evolving (Hennig and Walker, 1970), a general fundamental function for them was questioned. In this paper we review what is known about the organization of highly repeated sequence DNA in *Drosophila melanogaster*, draw conclusions about possible functions, and consider whether these conclusions may apply to higher organisms in general.

W. J. PEACOCK and CO-WORKERS Division of Plant Industry, Commonwealth Scientific and Industrial Research Organization and Research School of Biological Sciences, Australian National University, Canberra, A. C. T., Australia

The Highly Repeated DNA of Drosophila melanogaster

Analysis of nuclear DNA from embryonic *D. melanogaster* on a CsCl density gradient shows two satellite fractions (Fig. 1*a*) at densities 1.672 g/cc and 1.688 g/cc (Peacock et al., 1973). When analyzed in the presence of the antibiotic actinomycin D (Fig. 1*b*), this same DNA shows additional satellites in CsCl density gradients (Fig. 1*c*). All six satellite species can be isolated with the subsequent use of another DNA-binding antibiotic, netropsin sulphate (Zimmer, 1975), and have the following buoyant densities in neutral CsCl density gradients – 1.705 g/cc, 1.697 g/cc, 1.690 g/cc, 1.688 g/cc, 1.686 g/cc, and 1.672 g/cc. All of these satellite species, except 1.690 g/cc, have been purified to homogeneity as judged by buoyant-density analysis and have been characterized in some detail. Physical characterization of the satellites has shown each to be homogeneous and to be composed of highly repeated simple nucleotide sequences. The basic sequences of the satellites are sufficiently different from each other not to interact in mixed reannealing experiments. These properties are illustrated in Fig. 2*a*, which shows a buoyant-density analysis of a mixture of three satellites before and after a denaturation/renaturation cycle for both sheared and unsheared DNA. These results apply to all of the satellites other than the 1.697 satellite which is sensitive to shearing (Fig. 2*b*).

The data for the 1.697 satellite indicate a sequence complexity within this satellite similar to that described by Hearst et al., (1972) for what they considered to be the entire complement of highly repeated DNA of *Drosophila*. The com-

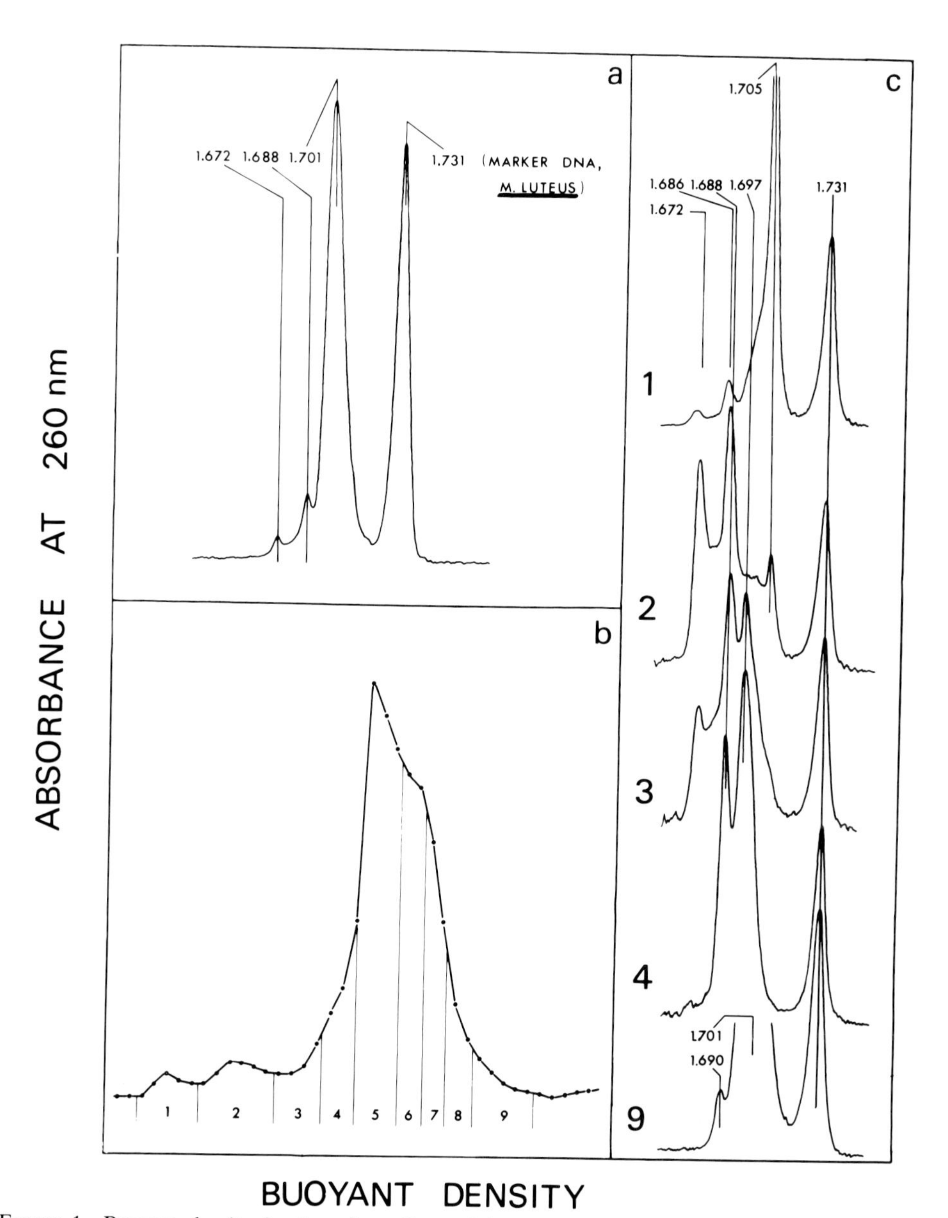

FIGURE 1 Buoyant density fractions from *Drosophila melanogaster* nuclear DNA. *D. melanogaster* nuclear DNA was isolated as described by Brutlag et al. (1976). Addition of actinomycin D to obtain further fractionation was also carried out as described by Brutlag et al. (1976). (*a*) A standard preparation of *D. melanogaster* nuclear DNA in a CsCl buoyant density gradient analyzed in an analytical ultracentrifuge. Bottom of gradient is on right-hand side of panel. (*b*) A preparative gradient with a mixture of actinomycin D and DNA centrifuged in a CsCl buoyant density gradient. Fractions were collected and pooled as indicated by numbers 1-9. Bottom of gradient is on left-hand side of panel. (*c*) Analysis of fractions 1, 2, 3, 4, and 9, obtained from the gradient shown in Fig. 1*b*, in CsCl buoyant density gradients in an analytical ultracentrifuge after the removal of actinomycin D (as described by Brutlag et al., 1977).

plexity is related in part to the presence of the ribosomal cistrons in that satellite, and we currently are investigating this further.

Primary sequence data for three of the six satellites confirm that each has a very short basic repeating unit (Table I). The 1.688 satellite has a repeating unit of 350 base pairs and so far has been analyzed only by restriction enzymes. The

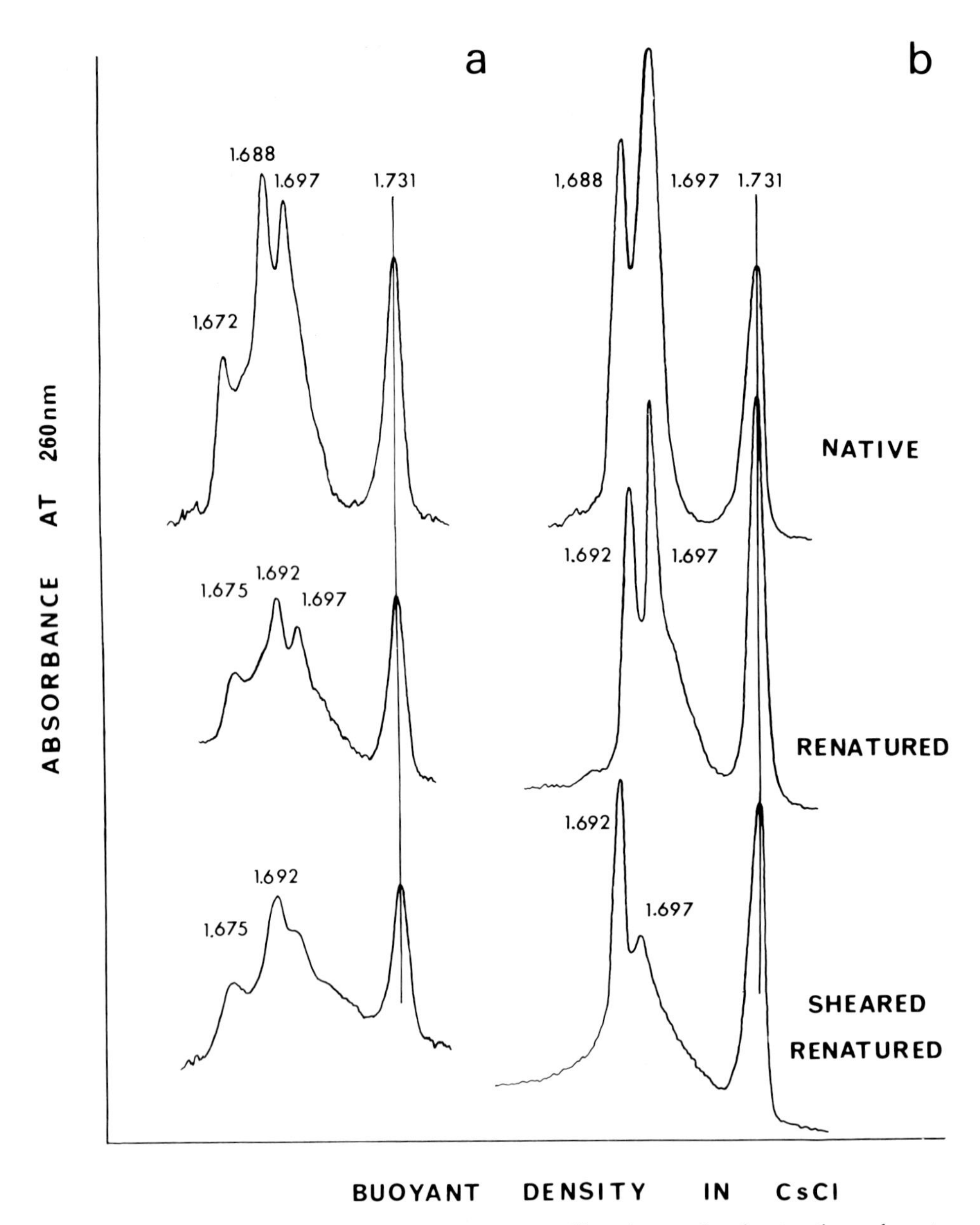

FIGURE 2 Buoyant density analysis of *D. melanogaster* satellite mixtures after denaturation and renaturation. All DNA fractions shown in Fig. 1*c* were analyzed for their buoyant density behavior when denatured and then allowed to renature to a C_0t of approximately 0.5 in 0.18 M NaCl. The effect of shearing the DNA fractions was assessed by sonicating DNA samples (3 × 30-s bursts, at 0–4°C, in a Branson-sonifier B-12) before denaturation and renaturation. Results are shown for fractions 3 (Fig. 2*a*) and 4 (Fig. 2*b*). Fraction 3 shows two satellites unaffected by the shearing, whereas fraction 4 shows 1.697 to be shear-sensitive.

sequencing data have shown that the basic repeats of the 1.672, 1.686, and 1.705 satellites are held constant in the genome throughout their many repeats, but we have detected the presence of closely related sequence isomers, as shown in Table I (Brutlag and Peacock, 1975). The two isomers identified in the 1.672 satellite are present in approximately equal proportions in the genome, whereas the basic repeat of the 1.686 satellite comprises at least 90% of this satellite, with the

TABLE I

Nucleotide Sequence Data for Satellite DNAs in Drosophila

Species	Satellite	Sequence of one strand	Estimated fraction of satellite DNA
			%
*Drosophila melanogaster**	1.672	-ATAAT-	~50
		-ATATAAT-	~50
	1.686	-AATAACATAG-	90
		$(A_7T_1C_1G_1)$	5
		$(A_5T_3C_1G_1)$	5
	1.705	-AGAAG-	>75
		-AGAGAAGAAG-	<25
Drosophila virilis†	I	-ACAAACT-	>90
	II	-ATAAACT-	>90
	III	-ACAAATT-	>93

* Data from Brutlag and Peacock (1975); Endow et al. (1975); Birnboim and Sederoff (1975).

† Data from Gall and Atherton (1974).

other isomers each accounting for approximately 5%. In the 1.705 satellite Endow et al. (1975) have reported the -AGAAG- sequence to be the predominant form. This is supported by the polypyrimidine tract analysis of Birnboim and Sederoff (1975). All of these sequences and their isomers are closely related to each other and conform to the general sequence pattern of $(AAN)_m$ $(AN)_n$. The three highly repeated DNA species isolated from the *Drosophila virilis* genome by Gall and Atherton (1974) also show close sequence relatedness (Table I).

The highly repeated species in *D. melanogaster* account for 20–22% of the genome, with the four major satellite species (1.705, 1.688, 1.686, and 1.672) amounting to approximately 16% of the total genome. Detailed analyses of the recoveries of satellites from DNA of varying molecular weights have been carried out (Goldring et al., 1975; Brutlag et al., 1976). These experiments have shown for four of the six satellites—1.672, 1.686, 1.688, and 1.705—that their recoveries are not dependent upon the initial molecular weight of the DNA, in the range of 4.5–45 kilobases (kb), indicating an arrangement of repeat sequences in blocks of at least 800 kb (Brutlag et al., 1976). Evidence that these long, uninterrupted blocks of each basic repeat are contiguous and covalently linked in the genome has been obtained in detail for the 1.705 and 1.672 satellites. Isolation of highly repeated DNA from unfractionated DNA by the use of the hydroxylapatite technique (Britten et al., 1974) under conditions which should prevent the recovery of the 1.672 sequences as a renatured species (renaturation conducted above the melting point [T_m] of this species), still results in the recovery of 1.672 sequences which band at a density of 1.705 g/cc in CsCl gradients (Fig. 3). The DNA used in this experiment was 1.5 kb long, indicating that the 1.705 and 1.672 repeat sequences must occur on the same 1.5 kb fragment of DNA. Such *junction molecules* between the different satellite species have now been detected by a number of different experimental procedures, and lead to the expectation that the long arrays of repeating sequences should often be adjacent in the genome.

Chromosomal Locations of Highly Repeated Sequence DNA

The technique of *in situ* hybridization, which showed satellite DNA to be localized to pericentromeric heterochromatic regions of mouse chromosomes, has been used in *D. melanogaster* to provide a detailed map of the heterochromatic regions by using radioactive probes to each of the satellites discussed in the preceding section (except for the 1.690 satellite). Experiments establishing the validity of the *in situ* hybridization technique have been carried out; the satellites of *D. melanogaster* are particularly well suited to such a check because each satellite has a different and well-defined melting temperature. The T_m characteristics are reflected in complementary RNA (cRNA)-DNA hybrids (where the RNA is an *Escherichia coli* RNA polymerase copy of the satellites) and can be determined by using a nitrocellulose filter technique (Brutlag et al., 1977). The hybrid formed *in situ*, by the use of a complementary cRNA probe, can have its T_m estimated by quantitative autoradiography. This type of experiment has been carried out for two satellites, 1.705 and 1.672, and shows that the melting temperatures of *in situ* hybrids fall within 1°C of the hybrids formed on nitrocellulose filters (Fig. 4). Thus, the *in situ* technique allows a quantitative analysis of the locations of satellite sequences within chromosomes via the formation of accurate RNA-DNA hybrids. Figure 5 shows the *in situ* hybridization pattern for the 1.705 sequences.

A summary of the distribution of *D. melanogaster* satellites (Fig. 6) shows: (1) Each satellite is present in the heterochromatin of most chromosomes. (2) The heterochromatin of each chromo-

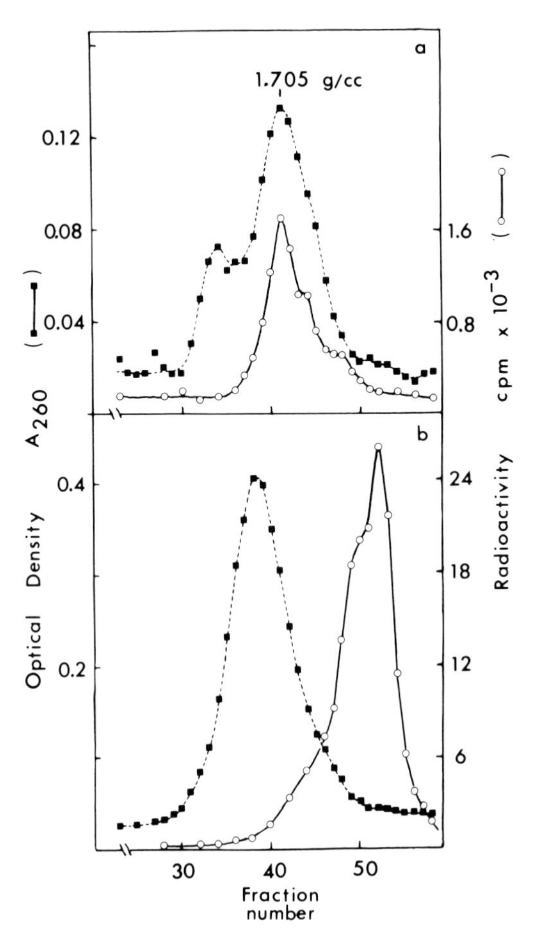

FIGURE 3 Covalent linkage of the 1.705 and 1.672 satellites. Unfractionated *D. melanogaster* DNA (average length 1.5 kb) was denatured in 0.12 M sodium phosphate buffer, pH 6.8, and renatured to a C_0t of 0.02. Single-strand and double-strand fractions were then prepared by the hydroxylapatite technique of Britten et al. (1974), and centrifuged to equilibrium in CsCl buoyant density gradients. Fractions from such gradients were monitored for DNA (by absorbance at 260 nm) and for 1.672 sequences by loading the DNA in 50 μl aliquots from appropriate fractions onto nitrocellulose filters and hybridizing to such filters a cRNA probe synthesized from 1.672 satellite using *Escherichia coli* RNA polymerase (core). The procedures for loading DNA onto filters have been described by Birnstiel et al. (1972). The hybridization was carried out in 3 × standard saline citrate (SSC)/50% formamide at 25°C. The T_ms of the hybrids formed in various parts of the gradient were checked to ensure 1.672 sequences were being assayed. (*a*) Double-strand fraction recovered at 78°C, with 1.672 sequences banding at 1.705 g/cc. (*b*) Single-strand fraction showing the main peak of 1.672 sequences at ~1.675 g/cc.

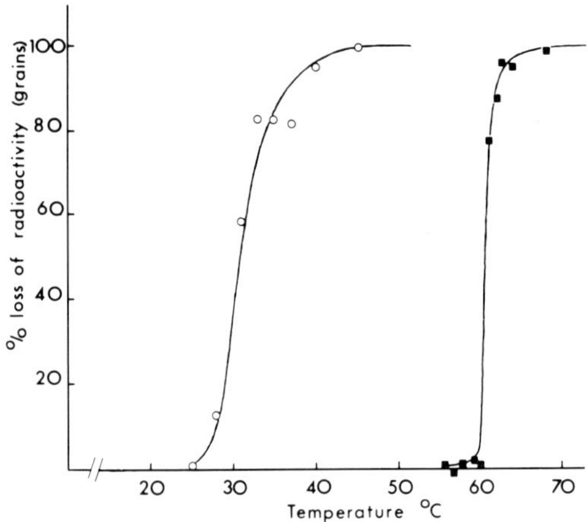

FIGURE 4 Melting points of hybrids formed in chromosomes *in situ*. *In situ* hybridization of radioactive cRNAs to salivary gland nuclei (polytene chromosomes) and larval brain nuclei (mitotic chromosomes) was carried out as described in the legend to Fig. 5. To obtain the melting points, duplicate slides were incubated for 5 or 10 min in 3 × SSC/50% formamide at the appropriate temperatures, washed in 2 × SSC and then covered with Ilford K2 emulsion (see legend to Fig. 5). Grain counts for each temperature point were estimated by scoring at least 30 nuclei per slide and the percentage loss of radioactivity computed by taking the difference between the first temperature point and subsequent ones. The melting curve shown for 1.672 (○–○) was obtained from salivary gland nuclei—the same T_m was obtained from larval brain nuclei. The melting curve for 1.705 (■–■) was obtained from larval brain nuclei with the T_m from salivary gland nuclei consistently lower by 1–2°C. The latter result is discussed in detail elsewhere (Peacock, Appels, and Steffensen, manuscript in preparation).

some is distinguished in two ways: (a) the quantity of a particular satellite present differs greatly between chromosomes, (b) the nearest neighbor relationship of satellites differ between chromosomes. (3) The same satellites (1.686, 1.688, and 1.697) are localized near the nucleolar organizer regions of both the X and Y chromosomes. (4) The euchromatic sites for the satellites, detectable only in the polytene chromosomes from salivary glands, are at 21 D 1,2 for 1.705, 1.686, and 1.672, and the tip of the X chromosome for 1.688. However, the quantities of satellites located at these regions are only minor compared to the heterochromatic sites.

The map of chromosome 3 heterochromatin appears somewhat incomplete at present, and it may be that the 1.690 satellite or the minor satel-

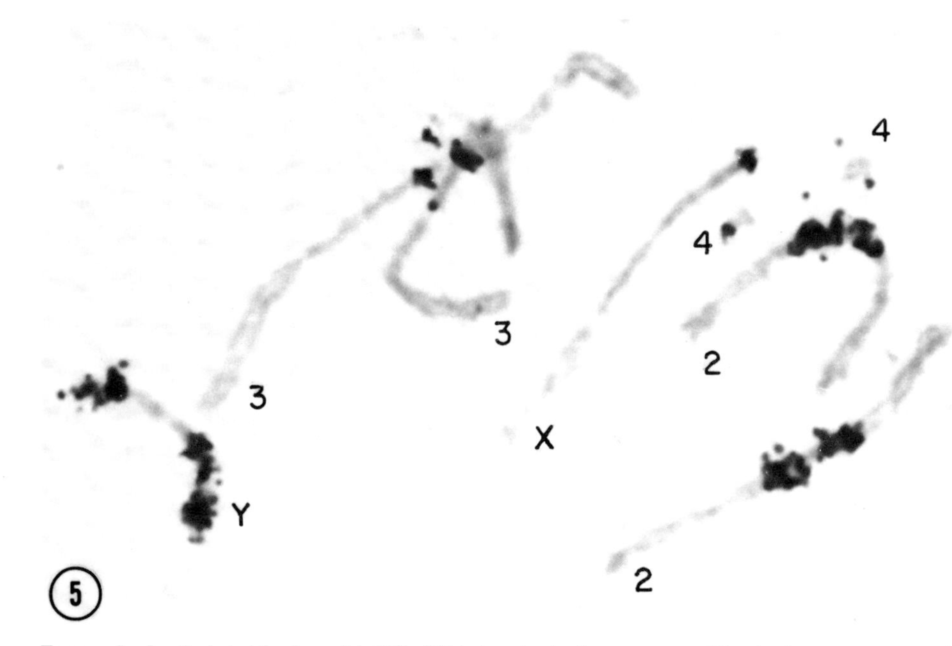

FIGURE 5 *In situ* hybridization of 1.705 cRNA to mitotic chromosomes. Mitotic chromosome squashes were obtained from the brain cells of 3rd instar larvae. *In situ* hybridization of a cRNA probe transcribed from 1.705 DNA was carried out as described by Peacock et al. (manuscript in preparation).

lites that are evident in netropsin sulphate/CsCl gradients but have not yet been purified are localized on this chromosome.

The existence of a wide range of chromosome rearrangements with at least one breakpoint in heterochromatin will enable a set of cytogenetic coordinates to be applied to the satellite map. For example, the position of the left boundary of the 1.705 sequences on the X chromosome has been analyzed with a series of inversions. The sequence boundary is proximal to the sc^4, sc^{L8}, sc^8, and sc^{S1} breakpoints, whereas the sc^{V2} breakpoint lies within the 1.705 block.

The question of whether the different sites for a given satellite are identical has been examined for the 1.672 species. Sequencing data (Table I) has shown that this satellite consists of approximately equal quantities of a pentamer-repeating unit species and a heptamer-repeating unit species. A recent experiment in this laboratory has shown that *E. coli* RNA polymerase holoenzyme preferentially copies the pentamer when presented with a double-stranded 1.672 template. In contrast, *E. coli* RNA polymerase core enzyme copies both isomers equally. We have thus been able to analyze the distribution of one of the isomers of 1.672 and have shown that this repeating sequence is localized on all the sites indicated in Fig. 6 *except* those on chromosome 2. The X, Y, 3, and 4 chromosomes therefore contain identical sites of the pentamer 1.672 species.

The core enzymes result of equal copying of both isomers was reproducibly obtained with the enzyme prepared as described by Burgess (1969). Subsequent preparations of *E. coli* RNA polymerase core enzyme following the procedure of Burgess and Jendrisak (1975) has given *in situ* hybridization patterns indistinguishable from holoenzyme cRNA patterns (Peacock, Appels, and Steffensen, manuscript in preparation). The basis for the difference between core enzyme preparations has not been determined.

The melting temperatures of the *in situ* hybrids of the other satellites appear to be the same in all sites for each particular satellite, suggesting that identical sites for these satellites exist on different chromosomes. This raises the question of the mechanism by which sequence homogeneity of these satellites is maintained. Carroll and Brown (1976) have suggested that interchromosomal exchange events may be operative in the case of the numerous sites of the 5S RNA gene sequences in *Xenopus laevis*. They find this mechanism attractive because of the observed juxtaposition of the

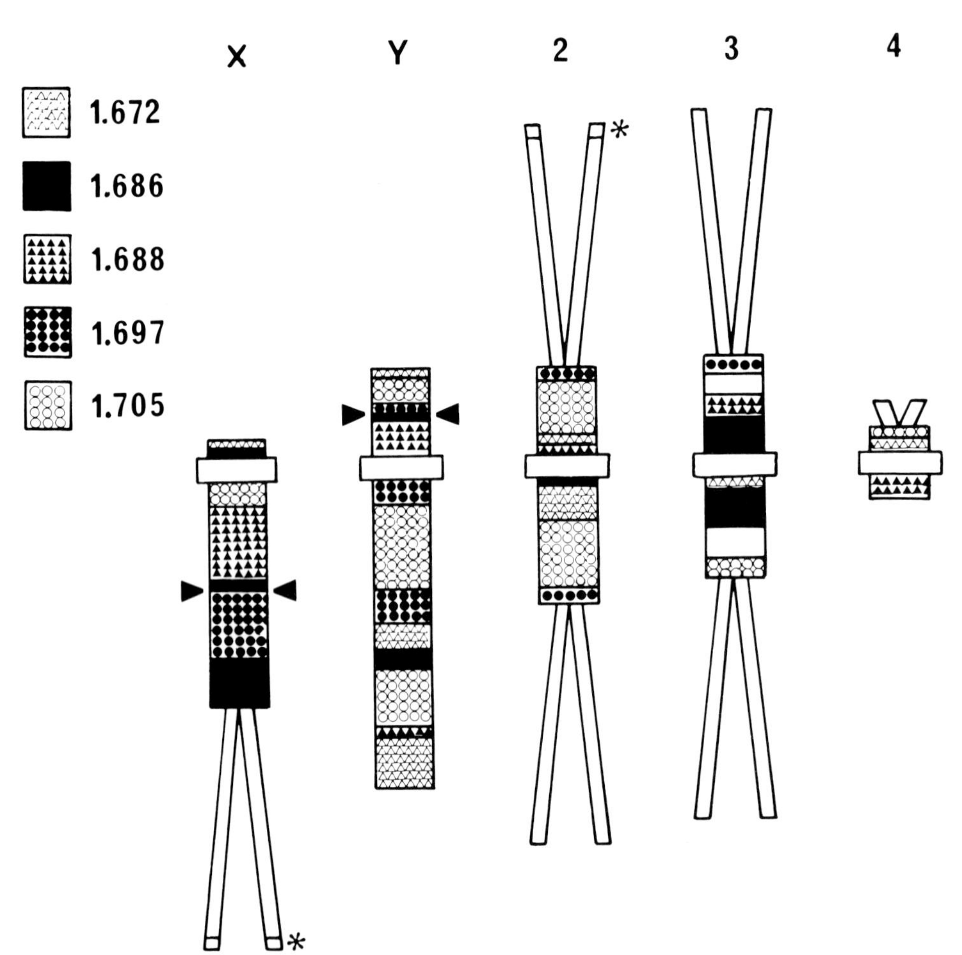

FIGURE 6 Map of *D. melanogaster* satellies with mitotic chromosomes. The *in situ* hybridization patterns for five of the six *D. melanogaster* satellites are shown in a qualitative manner. The relative positions of blocks of satellites within heterochromatin is approximate because it relies only on microscope observations of satellites hybridized in separate experiments. Accurate mapping of the X chromosome is currently being carried out (as indicated in the text) with a series of genetically altered X chromosomes. The asterisk on the X chromosome indicates a site for 1.688 (1A1,2) which is detectable only in polytene chromosomes. The asterisk on chromosome 2 denotes band 21D1,2 where sequences closely related to 1.672, 1.686, and 1.705 have been detected in polytene chromosomes.

5S DNA sites at a certain stage of meiosis and because their analysis showed that adjacent 5S repeats have slightly different spacers. Interchromosomal exchange would appear not to be the case for the ribosomal genes in *D. melanogaster*, because even though Maden and Tartof (1974) have shown that the 18S and 28S ribosomal RNA genes are the same on the X and Y chromosomes, the two chromosomes do present some differences in the spacer regions (Tartof and Dawid, 1976). Unequal intrachromosomal exchange between sister chromatids may be a homeostatic process in this case (cf. Smith, 1973, 1976), and could account for the homogeneity of a particular block of repeats on any one chromosome. The problem of interchromosomal homogeneity still remains for the 18S and 28S coding sequences as it does for the satellite sequences. It may be possible to approach the problem in *Drosophila* by deliberately introducing sequence heterogeneity in a given chromosomal location (by rearrangements) and following the fate of the heterogeneity through subsequent generations.

The Evolutionary Stability of Highly Repeated Sequences in Drosophila

Before considering possible functions for highly repeated sequence DNA, we first should evaluate the data which would implicate the existence of a function or functions. Smith (1976) has argued that unequal sister chromatid exchange, together with rare interchromosomal exchange, could insure homogeneity of repeated sequences without the need to assume a positive selection pressure and, by implication, a cellular function. However, his argument requires a certain segment length to be maintained, and this assumption in itself implies a functional significance for the DNA segment. The case for selection operating on highly repeated sequence DNA is strengthened by a comparison of the satellite complements of two sibling species of *Drosophila*, *D. simulans* and *D. melanogaster*. The buoyant densities of the *D. melanogaster* and *D. simulans* satellites in actinomycin D/CsCl and netropsin sulphate/CsCl are shown in Fig. 7. Three of the eight *D. simulans* satellites (1.707 contains two distinct satellites, a and b) are identical to the *D. melanogaster* 1.672, 1.697, and 1.705 satellites, with the identity extending to chromosomal locations. Figure 8 shows a T_m analysis which indicates that the 1.672 and 1.705 sequences in *D. simulans* and *D. melanogaster* are indistinguishable. Both the sibling species have a 1.686 satellite, but these do not contain the same sequence as judged by T_m analyses. Nucleotide sequence information is not yet available on any *D. simulans* satellite (other than 1.672), but it will be of interest to determine if they follow the general sequence rule found to be operative in *D. melanogaster*. It is known that the 1.695 *D. simulans* satellite sequence does occur in *D. melanogaster* at a much reduced level and that the amount of 1.705 sequence in *D. simulans* is only small (approximately 1/10 the amount present in *D. melanogaster*). The general picture that is emerging is one of modulation of sequence representation with concomitant conservation of sequence and chromosomal location. A measure of sequence conservation over the 10^6 yr since the two sibling species appeared as distinct taxa can be obtained from a comparison of the less highly repeated and unique DNA sequences. Laird and McCarthy (1968) showed that 20% of the DNA of *D. simulans* and *D. melanogaster* did not cross-hybridize, whereas 80% of the sequences did cross-hybridize but with base mismatching leading to a drop of 3°C in the T_m, relative to the homologous hybrids. This minimal estimate of 3% nucleotide divergence between the two species is in accord with estimates from mutation rates. In contrast, we have shown, as discussed above, that *D. melanogaster* and *D. simulans* contain *identical* highly repeated sequences as judged by T_m analyses, and conclude that there are processes which have prevented highly repeated sequence DNA from diverging as much as the bulk of the DNA. This strongly implies a function for the highly repeated sequences. A similar conclusion was

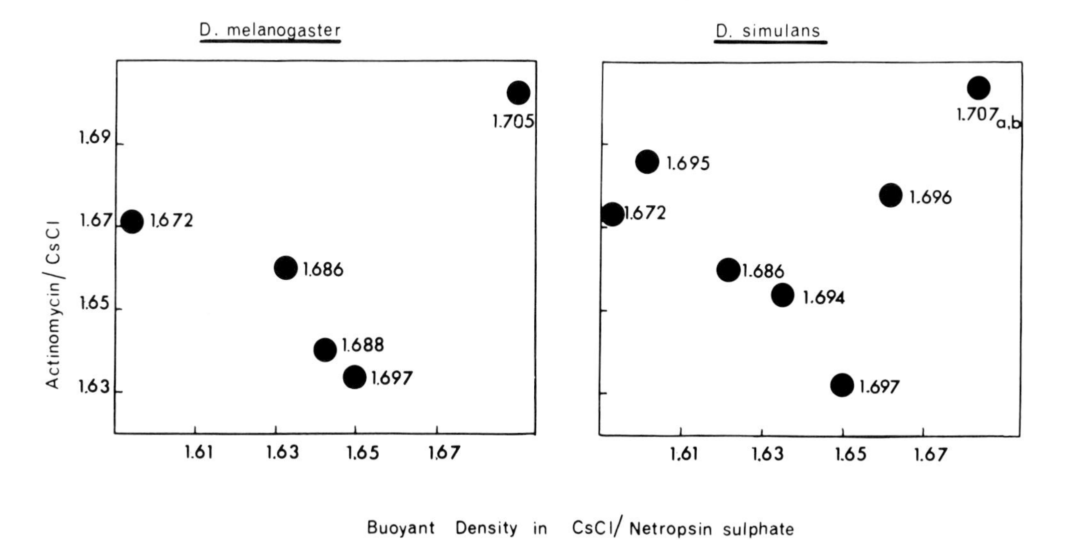

FIGURE 7 Comparison, of *D. melanogaster* and *D. simulans* satellites. The satellites of *D. melanogaster* and *D. simulans* are compared in a two-dimensional plot by using their buoyant densities in actinomycin D/CsCl and netropsin sulphate/CsCl. Because these antibiotics are base-specific in their interaction with DNA, the plot approximates a direct nucleotide sequence comparison.

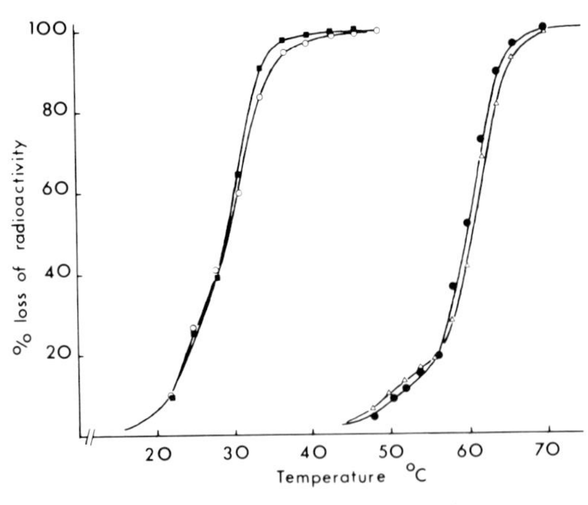

FIGURE 8 Comparison of 1.672 and 1.705 sequences in *D. simulans* and *D. melanogaster* DNA. *D. simulans* DNA (closed symbols) and *D. melanogaster* DNA (open symbols) were loaded on separate nitrocellulose filters (see legend to Fig. 3 for details of loading) and hybridized with *D. simulans* 1.672 [^{3}H]cRNA (■–■, ○–○) and *D. melanogaster* 1.705 [^{3}H]cRNA (●–●, △–△) at 25°C and 50°C, respectively. After a 3-h incubation, the filters were treated with pancreatic and T1 ribonucleases and washed in 2 × SSC. The melting profile was determined by serial transfers of filters to aliquots of 3 × SSC/50% formamide at desired temperatures with monitoring of released radioactivity.

reached by Smith (1973) on the basis of sequence studies on satellite I from *Drosophila virilis* and *Drosophila americana*.

Functions of the DNA Satellites in Drosophila

On the basis of the chromosome-mapping data, several possibilities arise for the functions of highly repeated sequence DNA. First, the fact that each satellite occurs on most chromosomes could indicate that the satellites are involved in recognition processes which lead to association of all chromosomes at certain stages of development. Two examples are the chromocentral formation in polytene nuclei of salivary glands (and other tissues) and at the commencement of meiotic pairing in female oocytes, where a fusion of centromeric regions appears to occur (Dävring and Sunner, 1976).

The second point to arise from the chromosome-mapping data is the chromosomal specificity of the distribution and arrangement of satellites. This specificity again suggests a function in recognition processes. It seems reasonable to suggest that when homologous chromosomes are required to be in a certain relation to one another either at meiosis, mitosis, or in interphase, satellite sequences may play a critical role. An instance of particular interest in relation to juxtapositioning of chromosomes at interphase is the observation that the nucleolar organizers of both the X and Y chromosomes are flanked by the 1.686 and 1.688 sequences. These sequences are not dispersed among ribosomal genes, as judged from density-shift experiments using ribosomal RNA to change the density of complementary DNA after forming DNA-RNA hybrids. The 1.697 sequences, on the other hand, are dispersed among 18S and 28S ribosomal genes because these do show shifts in density in such experiments. It is possible, therefore, that these three highly repeated sequences position the X and Y nucleolar organizers properly, with respect to each other, to allow for correct function (Steffensen et al., 1974).

The mechanism by which the postulated recognition processes operate could well be through the interaction of chromosomal proteins unique to these regions. Highly repeated sequence DNA has physical properties which are critically dependent on the sequence of the basic repeat (Wells and Wartell, 1974), and so could generate unique DNA-protein complexes as suggested by Sutton (1972), because chromosomal proteins do show sequence specificity in interacting with DNA (e.g., the HI histone; Sponar and Sormova, 1972; Renz and Day, 1976).

The Generality of a Chromosomal Specific Distribution of Satellite DNA

The rules of highly repeated sequence DNA distribution in *Drosophila* – namely, that they occur on more than one chromosome, show chromosomal specificity in distribution and arrangement, and are found in large blocks – have been examined in our laboratory in two plants, *Secale cereale*, rye (2n = 14) and *Triticum aestivum*, hexaploid wheat (2n = 42, genome constitution AABBDD), as well as in the marsupial species, *Macropus rufogriseus*, the red-necked wallaby (2n = 16).

The distribution of highly repeated sequences within rye chromosomes parallels the *D. melanogaster* situation rather closely. At present, two sets of highly repeated sequences have been isolated, a C_0t 0.02 DNA fraction isolated as a double-stranded fraction on hydroxylapatite after denaturation and renaturation of unfractionated DNA

(ScA), and long pyrimidine tracts isolated as described by Birnboim et al. (1975) from unfractionated DNA (ScB). Both ScA (approximately 10% of the genome) and ScB (approximately 0.1% of the genome) have been characterized (manuscripts in preparation), and although detailed sequence data are not yet available, a number of lines of evidence indicate that they are composed of simple repeating sequences. The observation of defined sites in *in situ* hybridization studies suggests that the highly repeated sequences are localized in large blocks. Both ScA and ScB have specific chromosomal distributions (Fig. 9). All chromosomes carry one or both of ScA and ScB, but show a distinctive "fingerprint" by virtue of quantitative and site differences between chromosomes. ScA shows telomeric locations on all chromosomes but varies markedly between chromosomes with regard to quantity. In contrast, ScB is located at well-defined sites on only three chromosomes, these sites accounting for 20–25% of the total ScB sequences. The ScA pattern is likely to

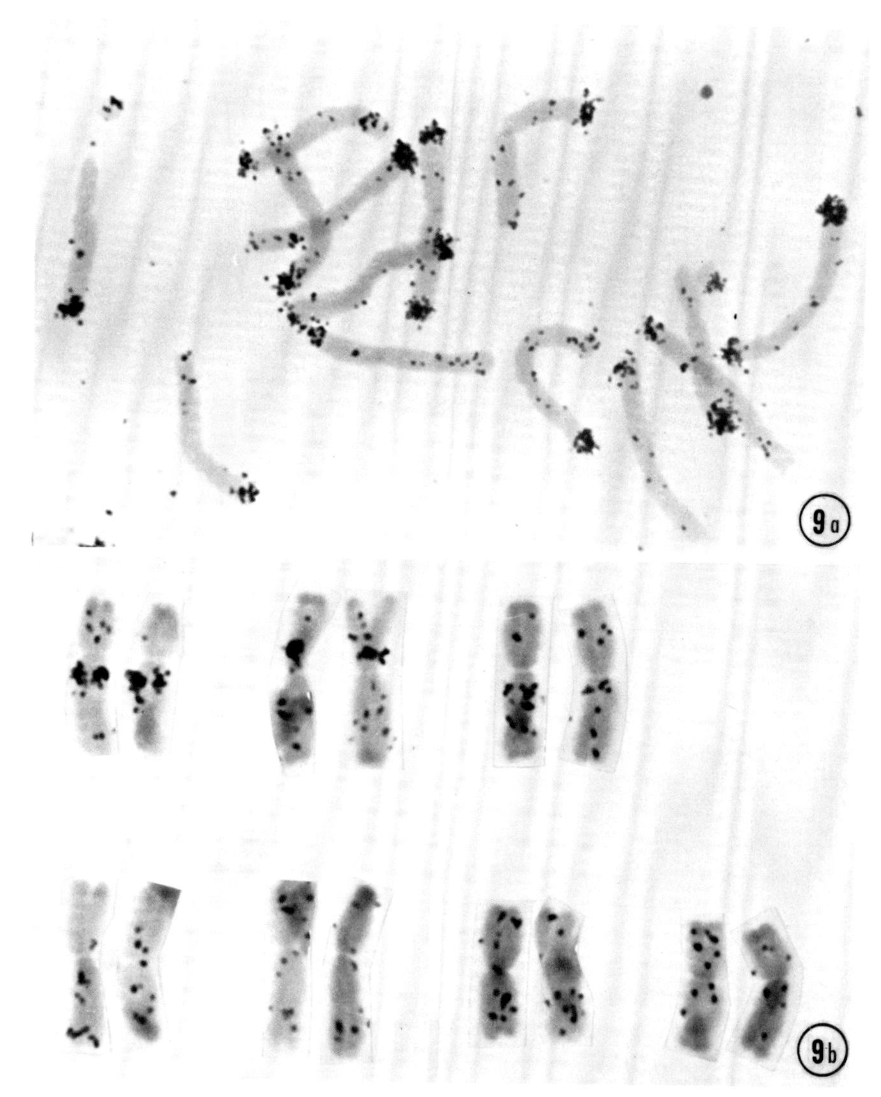

FIGURE 9 *In situ* hybridization of rye satellite probes. Hybridization of [^{3}H]cRNA synthesized from ScA and ScB satellite sequences to rye chromosomes was carried out in 3 × SSC/50% formamide at 40°C. (*a*) Hybridization of ScA [^{3}H]cRNA. (*b*) Hybridization of ScB [^{3}H]cRNA.

be more complex, as it is known to contain several buoyant density species of DNA; chromosome-specific patterns may well be found among the more defined satellites recovered from Ag^+/Cs_2SO_4 density gradients (work in progress).

A satellite prepared by fractionating DNA from hexaploid wheat in Ag^+/Cs_2SO_4 density gradients contains sequences very similar to ScB (i.e., the polypyrimidine tracts from rye) and, when hybridized to wheat chromosomes, shows specificity for the seven B-genome chromosomes, with each chromosome having a distinctive segmental pattern of distribution of this particular sequence. Two A-genome chromosomes also have a distinctive arrangement of this satellite, whereas the remaining A- and D-genome chromosomes contain very little of the sequence.

Two satellites have been isolated from the DNA of the red-necked wallaby (Dunsmuir, 1976). Although neither of these is as simple in sequence structure as the *D. melanogaster* satellites, *in situ* hybridization yields data that can be interpreted to show chromosomal specificity for the distribution of these DNA species. Figure 10 presents the quantitative distribution of each satellite on the different chromosomes of the complement. The sex chromosomes are particularly well characterized by the distribution of the satellites and most autosomes are separable.

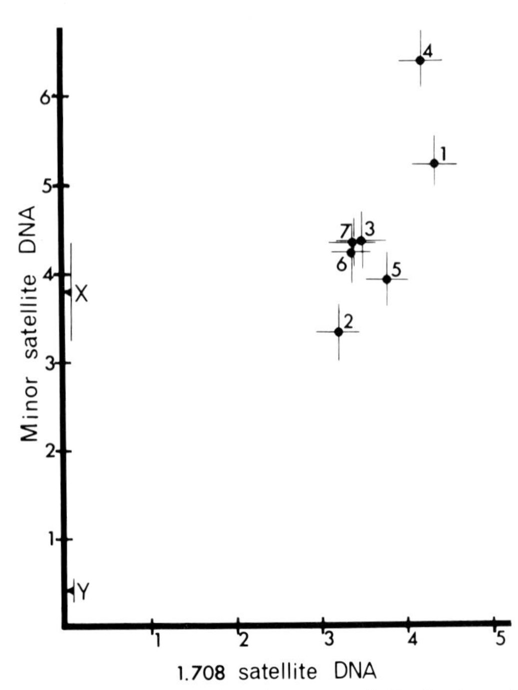

FIGURE 10 Quantitative chromosomal distribution of two satellite DNAs in the red-necked wallaby. The grid shows the relative amounts of the 1.708 and minor satellite sequences at the centromere of each chromosome (morphologically distinct) in the *Macropus rufogriseus* complement. Standard errors for the grain counts are shown by the vertical and horizontal bars.

The examples discussed so far conform to the general rules found in *D. melanogaster*, viz., that any one highly repeated sequence is distributed on most chromosomes of the complement, but that chromosomes are distinguishable by differences in quantities of highly repeated sequences present and by the arrangement of such sequences. Apparent exceptions to these rules occur in organisms in which there appears to be a single satellite distributed on all or most chromosomes. An example would be mouse, where the AT-rich satellite occurs on all chromosomes other than the Y (Pardue and Gall, 1970); thus it would seem unlikely that the chromosomes are distinguishable on the basis of the highly repeated sequences they contain. However, sufficiently detailed *in situ* hybridization analyses have not yet been carried out in the mouse system to assess whether this satellite is differentially distributed quantitatively among chromosomes. Furthermore although several related tracts have been identified (Biro et al., 1975), it is possible that the mouse satellite is a mixture of sequence isomers, which, if separated, would yield chromosome-specific patterns of distribution (analogous to the 1.672 isomers of *D. melanogaster* described earlier), as discussed by Brutlag and Peacock (1975). Evidence for such a situation is found in the red-necked wallaby, where restriction enzyme sites are distributed within the major satellite in such a way as to suggest that a *series* of different molecule populations are present. This particular satellite is cut by the Bam restriction enzyme to yield a basic 1,800 base pair unit. As in mouse and other systems, there is a distribution of monomer, dimer, etc. segments consistent with a random mutational decay of a preexisting, regular arrangement of the 1,800 base pair unit (cf. Southern, 1975). Inasmuch as the Bam restriction enzyme leaves none of the satellite undigested, the inference would have to be that the above distribution of Bam sites occurs at each autosomal centromeric region. However, when this same satellite is examined with the HindIII restriction enzyme, another 1,800 base pair unit is defined within one fraction that comprises only 20% of the satellite, whereas the other 80% of the satellite is not affected.

Similarly, the EcoR1 restriction enzyme again defines an 1,800 base pair unit, but here only 5% of the satellite is affected. The HindIII sites are not distributed randomly with respect to the populations of segments defined by digestion with a further enzyme HaeIII; preliminary data indicate a similar nonrandomness with respect to the EcoR1 sites. Inasmuch as partial digests show the HindIII sites to be grouped in long segments, the restriction data indicate that at least four different tandem arrays of the 1,800 base pair unit exist.

We suggest that the different populations of molecules within the red-necked wallaby satellite may prove to be chromosome-specific. An analogous situation regarding the distribution of restriction enzyme sites has now been shown to be true for the mouse satellite (Zachau, personal communication), so that a similar prediction may be made for a chromosome-specific distribution of satellite in mouse chromosomes.

It will be important to determine whether the segmental sequence populations as defined by restriction analysis do yield chromosomal patterns comparable to the patterns of the *Drosophila* satellites in which the differentiation of segments, on a nucleotide sequence basis, is, of course, more pronounced. The less-pronounced sequence divergence of the proposed segments in satellites such as those of the mouse and red-necked wallaby, offer difficulties in establishing chromosomal patterns by hybridization. However, we suggest that the cereals will be favorable for such an analysis because a complete set of addition lines exists in which each rye chromosome (singly) has been transferred to an otherwise wheat genome. The restriction enzyme analysis will, in these plants, be able to be conducted separately for each of the seven chromosomes.

Concluding Remarks

D. melanogaster satellites show characteristics of organization within chromosomes that appear to provide a broad set of rules followed in other organisms. This, together with the apparent evolutionary stability of satellite sequences, suggests that they may serve specific functions. The modulation of the levels of a given satellite sequence in phylogenetically related species, as first pointed out for kangaroo rat and guinea pig by Salser et al. (1976) and further discussed here for *D. melanogaster* and *D. simulans,* suggests a level of control over these sequences about which nothing is known at present. A related problem is the maintenance of identical sites of highly repeated sequences on different chromosomes. Genetic manipulation of chromosomes may prove of value in approaching the problem in *D. melanogaster*. For example, in the sc^4sc^8 chromosome (Lindsley and Grell, 1968), where much of the X heterochromatin is removed, we have noted 1.705 sequences at the euchromatic/heterochromatic junction where these have not been detected in wild-type chromosomes. If this proves to be a repeatable response in newly synthesized sc^4sc^8 chromosomes, it would suggest that a particular cellular pressure has caused a rapid increase of 1.705 sequences from a preexisting low level at this site. Analysis of this phenomenon may thus lead to evidence suggesting a specific functional requirement for a sequence at a certain position in the genome. The cereals (wheat and rye) also provide a system that allows genetic manipulation of chromosomes. We have found, for example, two diploid species of wheat each containing a specific satellite on only one pair of homologues; in one species, the location is near the centromere, whereas in the other it is telomeric. Inasmuch as it is possible to hybridize these species, a situation can be generated that allows analysis of possible associations between these unique satellite regions. In general, the ability to produce interspecies hybrid plants provides a number of test situations relating to the modulation of levels of satellite sequences at defined locations on chromosomes.

REFERENCES

BIRNBOIM, H. C., and R. SEDEROFF. 1975. Polypyrimidine segments in *Drosophila melanogaster* DNA. I. Detection of a cryptic satellite containing polypyrimidine/polypurine DNA. *Cell.* **5**:173–181.

BIRNBOIM, H. C., N. A. STRAUS, and R. SEDEROFF. 1975. Characterization of polypyrimidines in *Drosophila* and L-cell DNA. *Biochemistry.* **14**:1643–1647.

BIRNSTIEL, M. L., B. H. SELLS, and I. F. PURDOM. 1972. Kinetic complexity of RNA molecules. *J. Mol. Biol.* **63**:21–39.

BIRO, P. A., A. CARR-BROWN, E. M. Southern, and P. M. B. WALKER. 1975. Partial sequence analysis of mouse satellite DNA: evidence for short range periodicities. *J. Mol. Biol.* **94**:71–86.

BRITTEN, R. J., D. E. GRAHAM, and B. R. NEUFELD. 1974. Analysis of repeating DNA sequences by reassociation. *Methods Enzymol.* **29**:363–418.

BRITTEN, R. J., and D. E. KOHNE. 1968. Repeated sequences in DNA. *Science (Wash. D. C.).* **161**:529–540.

Brutlag, D. L., R. Appels, E. Dennis, and W. J. Peacock. 1977. Highly repeated DNA in *Drosophila melanogaster*. *J. Mol. Biol.* In press.

Brutlag, D. L., and W. J. Peacock. 1975. Sequences of highly repeated DNA in *Drosophila melanogaster*. *In* The Eukaryote Chromosome. W. J. Peacock and R. D. Brock, editors. Australian National University Press, Canberra. 35–46.

Burgess, R. R. 1969. A new method for the large-scale purification of *Escherichia coli* deoxyribonucleic acid-dependent ribonucleic acid polymerase. *J. Biol. Chem.* **244:**6160–6167.

Burgess, R. R., and J. J. Jendrisak. 1975. A procedure for the rapid, large-scale purification of *Escherichia coli* DNA-dependent RNA polymerase involving polymin P precipitation and DNA-cellulose chromatography. *Biochemistry*. **14:**4634–4638.

Carroll, D., and D. Brown. 1976. Adjacent repeating units of *Xenopus laevis* 5S RNA can be heterogeneous in length. *Cell*. **7:**477–486.

Dävring, L., and M. Sunner. 1976. Early prophase in female meiosis of *Drosophila melanogaster*. *Hereditas*. **82:**129–131.

Dunsmuir, P. 1976. Satellite DNA in the kangaroo *Macropus rufogriseus*. *Chromosoma (Berl.)*. **56:**111–125.

Endow, S. A., M. L. Polan, and J. G. Gall. 1975. Satellite DNA sequences of *Drosophila melanogaster*. *J. Mol. Biol.* **96:**665–692.

Gall, J. G., and D. D. Atherton. 1974. Satellite DNA sequences in *Drosophila virilis*. *J. Mol. Biol.* **85:**633–664.

Goldring, E. S., D. L. Brutlag, and W. J. Peacock. 1975. Arrangement of the highly repeated DNA of *Drosophila melanogaster*. *In* The Eukaryote Chromosome. W. J. Peacock and R. D. Brock, editors. Australian National University Press, Canberra. 47–60.

Hearst, J. E., M. Botchan, and R. Kram. 1972. Arrangement of the highly reiterated DNA sequences in the Centric Heterochromatin of *Drosophila melanogaster:* evidence for interspersed spacer DNA. *J. Mol. Biol.* **64:**103–117.

Hennig, W., and P. M. B. Walker. 1970. Variations in the DNA from two rodent families (Cricetidae and Muridae). *Nature (Lond.)*. **225:**915–919.

Jones, K. W. 1970. Chromosomal and nuclear location of mouse satellite DNA in individual cells. *Nature (Lond.)*. **225:**913–915.

Laird, C. D., and B. J. McCarthy. 1968. Magnitude of interspecific nucleotide sequence variability in *Drosophila*. *Genetics*. **60:**303–322.

Lindsley, D. L., and E. H. Grell. 1968. Genetic variations of *Drosophila melanogaster*. Carnegie Institution of Washington. no. 627. 1–471.

Maden, B. E. H., and K. P. Tartof. 1974. Nature of the ribosomal RNA transcribed from the X and Y chromosomes of *Drosophila melanogaster*. *J. Mol. Biol.* **90:**51–64.

Pardue, M. L., and J. G. Gall. 1970. Chromosomal localization of mouse satellite DNA. *Science (Wash. D. C.)*. **168:**1356–1358.

Peacock, W. J., D. L. Brutlag, E. Goldring, R. Appels, C. W. Hinton, and D. L. Lindsley. 1973. The organization of highly repeated DNA sequences in *Drosophila melanogaster* chromosomes. *Cold Spring Harbor Symp. Quant. Biol.* **38:**405–416.

Renz, M., and L. A. Day. 1976. Transition from non-cooperative to cooperative binding of histone H1 to DNA. *Biochemistry*. **15:**3220–3228.

Salser, W., S. Bowen, D. Browne, F. Adli, N. Federoff, K. Fry, H. Heindell, G. Paddock, R. Poon, B. Wallace, and P. Whitcome. 1976. Investigation of the organization of mammalian chromosomes at the DNA sequence level. *Fed. Proc.* **35:**23–35.

Smith, G. P. 1973. Unequal crossover and the evolution of multigene families. *Cold Spring Harbor Symp. Quant. Biol.* **38:**507–514.

Smith, G. P. 1976. Evolution of repeated DNA sequence by unequal crossover. *Science (Wash. D. C.)*. **191:**528–535.

Southern, E. M. 1975. Long range periodicities in mouse satellite DNA. *J. Mol. Biol.* **94:**51–91.

Sponar, J., and Z. Sormova. 1972. Complexes of histone Fl with DNA in 0.15 M NaCl: selectivity of interaction with respect to DNA composition. *Eur. J. Biochem.* **29:**99–103.

Steffensen, D. M., P. Duffey, and W. Prensky. 1974. Localisation of 5S ribosomal RNA genes on human chromosome I. *Nature (Lond.)*. **252:**741–743.

Sutton, W. D. 1972. Chromatin packing, repeated DNA sequences and gene control. *Nat. New Biol.* **237:**70–71.

Tartof, K. D., and I. B. Dawid. 1976. Similarities and differences in the structure of X and Y chromosome rRNA genes of *Drosophila*. *Nature (Lond.)*. **263:**27–30.

Walker, P. M. B. 1971. Repetitive DNA in higher organisms. *In* Progress in Biophysics and Molecular Biology. Vol. 23. J. A. V. Butler and D. Noble, editors. Pergamon Press, New York. 145–190.

Wells, R. D., and R. M. Wartell. 1974. The influence of DNA sequence on DNA properties. *In* MTP International Review of Science, Biochemistry. Series 1. Vol. 6. K. Burton, editor. Butterworth & Co. Ltd., London. 41–64.

Zimmer, C. 1975. Effects of the antibiotics netropsin and distamycin A on the structure and function of nucleic acids. *In* Nucleic Acid Research and Molecular Biology. Vol. 15. W. E. Cohn, editor. Academic Press, Inc., New York. 285–318.

Molecular Cytogenetics of Eukaryotes

DROSOPHILA SALIVARY GLAND POLYTENE CHROMOSOMES STUDIED BY *IN SITU* HYBRIDIZATION

M. L. PARDUE, J. J. BONNER, J. A. LENGYEL, and A. C. SPRADLING

Polytene chromosomes have contributed much to our understanding of the chromosomes of higher organisms. The giant size and extensive linear differentiation of these chromosomes permit cytogenetic studies not possible in other systems. The polytene structure is the result of multiple rounds of replication without chromatid separation. These chromosomes appear to be banded, probbly because the chromatids are so precisely aligned that localized foldings in the individual nucleoprotein strands join to form bands. The pattern of this banding on the chromosomes is the same in each polytene tissue (Beermann, 1952, 1962) and is probably the reflection of a fundamental structure of the chromatid which may exist even in nonpolytene nuclei. The exact relation between this fundamental structure and units of genetic function is still unclear (Beermann, 1972); however, the banding patterns produced on the polytene chromosomes allow the genome to be mapped cytologically with a resolution that has not been possible in higher organisms that lack such chromosomes.

Polytene nuclei are also remarkable in that they are essentially interphase nuclei. The chromosomes do not decondense during stages of DNA transcription and replication. Thus it is possible to visualize genetic activity and to relate it to specific chromosomal sites. One type of evidence for such genetic activity is puffing, a localized alteration of the structure of the polytene chromosome. Puffs behave in ways expected of specific gene transcription. Individual puffs appear and disappear at predictable times in the life cycle of the insect (Ashburner, 1972). Some puffs are tissue-specific; others are common to all polytene tissues (Berendes, 1967). Autoradiography of polytene nuclei after [^{3}H]uridine incorporation shows that puffs are sites of active uridine incorporation; however, many nonpuffed sites are also labeled with uridine (Pelling, 1959, 1964). In several cases the correlation between the presence of a specific polytene puff and the occurrence of a particular cytoplasmic protein suggests a causal relationship (Beermann, 1961; Baudisch and Panitz, 1968; Grossbach, 1974; Korge, 1975; Tissieres et al., 1974). Although there is still much to be learned about the molecular basis of the puffing, these remarkable regions have contributed many insights into genetic activity during development.

The high local concentrations of particular DNA sequences caused by the alignment of many identical chromatids make polytene chromosomes ideal material for *in situ* hybridization experiments. Because of lateral redundancy of the multiple chromatids, it is possible to detect hybridization to unique, as well as repetitive, DNA sequences under the conditions of our experiments (Fig. 1). The linear differentiation of the chromosome banding pattern allows one to resolve a large number of RNA species on the basis of the pattern of *in situ* hybridization. By doing *in situ* hybridization experiments with [^{3}H]RNA populations from a variety of cell types, it is possible to make comparisons between polytene and nonpolytene cells.

A number of workers have studied the RNA associated with polytene chromosomes in the living animal by a technique which we call *transcription autoradiography*. For a transcription autoradiograph, polytene cells are incubated with [^{3}H]uridine, and cytological preparations are made directly. Presumably the RNA is detected

M. L. PARDUE and J. J. BONNER Department of Biology, Massachusetts Institute of Technology, Cambridge, Massachusetts

J. A. LENGYEL Department of Biology, University of California, Los Angeles, Los Angeles, California

A. C. SPRADLING Department of Zoology, Indiana University, Bloomington, Indiana

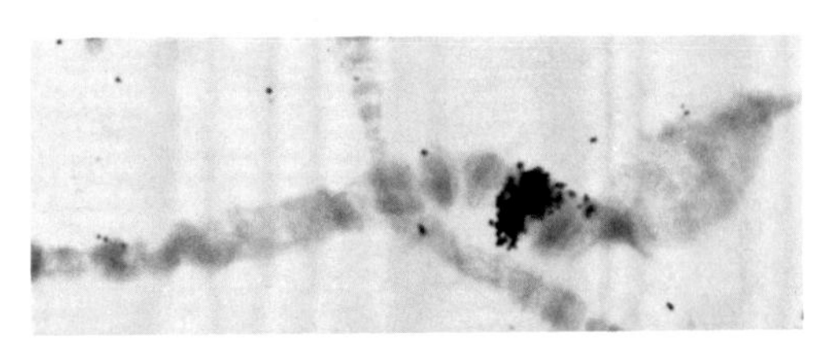

FIGURE 1 Autoradiograph of a polytene chromosome showing *in situ* hybridization to DNA of low repetition frequency. The [^{3}H]cRNA used was transcribed in vitro from a segment of *Drosophila melanogaster* DNA isolated and amplified by insertion into the *Escherichia coli* plasmid pSC101 (Wensink et al., 1974). The hybrid plasmid so produced, pDm2, maps to polytene region 84D and reassociation kinetic studies show a repetition frequency of 3.5 for this segment of *D. melanogaster* DNA (Wensink et al., 1974). The *in situ* hybridization was carried out in a buffer of 0.3 M NaCl, 0.01 Tris; pH 6.8 at 65°C for 10 h. The same conditions have been used for the other *in situ* hybridization experiments described in this paper. Giemsa stain. Autoradiographic exposure 38 days. × 980.

primarily at its site of transcription, although transport cannot be ruled out, particularly in long incubations.

It is of interest to compare transcription autoradiographs with *in situ* hybridization done with nuclear or cytoplasmic [^{3}H]RNA synthesized under the same incubation conditions. The two techniques reflect different aspects of the RNA populations they detect. Labeling of polytene bands by [^{3}H]RNA in a transcription autoradiograph is determined in part by the specific radioactivity of the RNA, the rate of transcription at a locus, the amount of DNA transcribed at that locus, and the length of time the transcript is associated with the chromosome. On the other hand, the level of *in situ* hybridization detected at a locus is determined by the amount of DNA complementary to the RNA probe, the abundance of the complementary RNA in the hybridizing solution, and the specific radioactivity of that RNA. The specific radioactivity of any species of RNA in the hybridizing solution depends not only on the level of [^{3}H]uridine incorporation, but also on the existence and turnover rate of any nonradioactive molecules of that RNA present in the population before the period of [^{3}H]uridine incorporation. Finally, it is possible that an RNA species could hybridize to several chromosomal regions, though it was transcribed from only one of them. Comparison of results obtained from these two techniques (transcription autoradiography and *in situ* hybridization) provides another approach to understanding the structure and function of polytene chromosomes, and possibly, through the polytenes, other chromosomes as well.

Response to Heat Shock: The Experimental Induction of A Small Set of Genes

When *Drosophila melanogaster* larvae or their salivary glands are shifted from the normal growth temperature of 25° to 37°C, some nine new puffs are induced on the polytene chromosomes within a few minutes. This is apparently a response to the metabolic disturbance due to the temperature shift since the same puffs can be induced by interfering with energy metabolism in other ways (Ritossa, 1962, 1963, 1964; van Breugel, 1966; Ashburner, 1970; Leenders and Berendes, 1972). Correlated with the induction of the new puffs in the polytene chromosomes is the synthesis of approximately seven new proteins in the cytoplasm of the salivary gland cells (Tissieres et al., 1974; Lewis et al., 1975). New RNAs which hybridize within the regions of heat-induced puffs can be prepared from the salivary gland cells (Bonner and Pardue, 1976*a*).

The response of the salivary gland cells to heat shock (a shift from 25° to 37°C) is apparently a general one shared by all the *Drosophila* tissues examined. Although the puffing response cannot be studied in diploid tissues, imaginal disks, adult tissues, and cultured cell lines resemble salivary glands in the heat-induced changes (at 37°C) in the synthesis of both RNA (Fig. 2; Spradling et al., 1975; McKenzie et al., 1975; Bonner and Pardue, 1976*a*) and protein (Tissieres et al., 1974; McKenzie et al., 1975; Lewis et al., 1975). RNAs which hybridize *in situ* with the six largest heat-shock polytene puffs have been found associated with polysomes after heat shock, demonstrating directly the linkage between chromosomal regions in which puffing can be induced and protein synthesis (McKenzie et al., 1975; Spradling et al., 1977). It is interesting that the level of hybridization of cytoplasmic RNA from diploid tissues at the various heat-shock loci is approximately proportional to the size of the heat shock-induced puff at each of these loci on polytene chromosomes.

As in the polytene tissues, the primary effect of the heat shock in diploid tissues is likely to be at the level of transcription. When cells from an

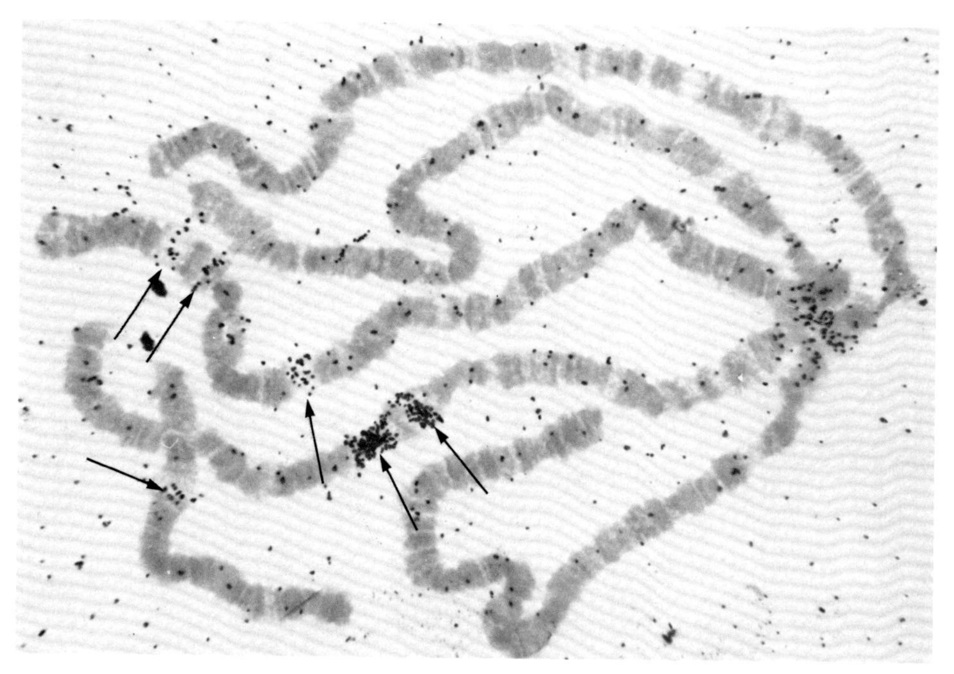

FIGURE 2 *In situ* hybridization of poly(A)$^+$ cytoplasmic RNA from heat-shocked *Drosophila* cultured cells. Cells were shifted from 25 to 37°C. Five min later [^{3}H]uridine was added and the incubation continued for 1 h. The RNA hybridizes *in situ* to six of the nine puffs induced in polytene chromosomes by the temperature shift, 63BC, 64F, 67B, 87A, 87C, and 95D (arrows). Hybridization to the puff at 93D is seen only with poly(A)$^-$ cytoplasmic RNA. The two smallest puffs, 33B and 70A, bind only a small amount of [^{3}H]RNA in these experiments. 1.5 × 10^4 cpm/slide. Giemsa stain. Exposure 172 days. × 650.

established *Drosophila* cell line are incubated for 20 min with [^{3}H]uridine at 25°C, the ^{3}H-nuclear RNA hybridizes to many sites distributed over the polytene chromosomes. However, ^{3}H-nuclear RNA from cells incubated with [^{3}H]uridine for 20 min at 37°C gives a different pattern of *in situ* hybridization (Fig. 3), even when the labeling period begins as early as 2 min after the cells are shifted to the higher temperature. The ^{3}H-nuclear RNA made at 37°C hybridizes heavily to the same polytene bands that bind the cytoplasmic RNA produced under conditions of heat shock (Lengyel and Pardue, 1975). In addition, poly(A)$^+$ ^{3}H-nuclear RNA prepared from heat-shocked cells shows significant hybridization to a number of other bands which have not yet been seen to hybridize with poly(A)$^+$cytoplasmic RNA preparations from the same cells (Lengyel and Pardue, manuscript in preparation). It is possible that some of the RNAs hybridizing to this second set of bands may never leave the nucleus. However, competition experiments with unlabeled cytoplasmic RNA will be necessary to study this point since the apparent absence of these [^{3}H]RNAs in cytoplasmic fractions might also be the result of dilution of the radioactive RNA with preexisting unlabeled RNA.

The apparent change in RNA transcription in both diploid and polytene tissues, although dramatic, is not complete. Several types of RNA which were transcribed at 25°C continue to be produced at 37°C. These include the precursor to 19S and 26S ribosomal RNA (Lengyel and Pardue, 1975); 5S RNA, and histone mRNAs (Spradling et al., 1975). Although heat shock does not stop the transcription of the large ribosomal RNA precursor, it apparently does stop further processing and transport to the cytoplasm (Lengyel and Pardue, 1975). The 5S RNA made at 37°C is larger than normal *Drosophila* 5S by some 15 nucleotides at the 3′ end (Rubin and Hogness, 1975). It is possible that this, too, is the result of an effect on processing although 5S RNA is not known to have such a precursor.

The nine puffs induced in the polytene chromosomes of *D. melanogaster* by heat shock appar-

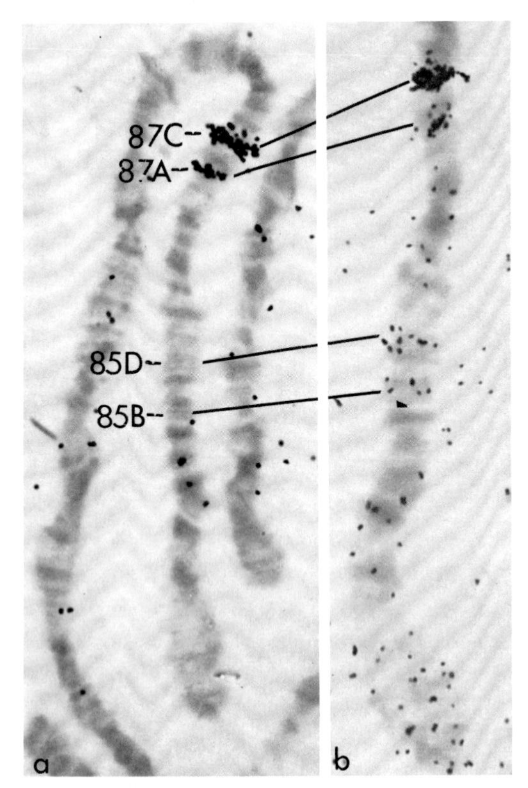

FIGURE 3 *In situ* hybridization of poly(A)$^+$ nuclear and cytoplasmic heat shock RNA. Cultured *D. melanogaster* cells were shifted to 37°C and labeled with [^{3}H]uridine for 20 min. Giemsa stain. Exposure 46 days. × 1,000. (*a*) Hybridization by poly(A)$^+$ cytoplasmic RNA. Regions 87A and 87C are heavily labeled. (*b*) Hybridization by poly(A)$^+$ nuclear RNA larger than 20S on a sucrose gradient. Regions 87A and 87C are the most heavily labeled. Significant hybridization is also seen over 85B and 85D. 85B and 85D are labeled in experiments with cytoplasmic poly(A)$^+$ RNA made during longer heat shocks.

ently represent a coordinated set of genes. Ashburner (1970) reported that the set of puffs could be induced in the several strains of *D. melanogaster* he studied as well as in the related *D. simulans*. The puffing pattern was unaffected by the developmental stage of the larvae. In addition, the nine puffs can be induced by agents interfering with respiratory metabolism (Ritossa, 1962, 1963; Ashburner, 1970; Leenders and Berendes, 1972) or by recovery from anaerobiosis (Ritossa, 1964; van Breugel, 1966). Despite the evidence that these nine loci are coordinately controlled, the expression of some members of the set can be differentially modulated. This is clearly illustrated by the locus in region 93D. When the heat shock is given under certain culture conditions, region 93D is induced well beyond the level normally seen at 37°C (Bonner and Pardue, 1976*a*). The increase in induction at 93D can be detected in RNA extracted from either diploid or polytene tissues. The same culture conditions cause a large increase in the size of the puff produced in region 93D on polytene chromosomes, implying that the effect on 93D occurs at the chromosomal level.

Leenders and Berendes (1972) have reported that individual members of the set of loci induced by heat shock in *D. hydei* show differential sensitivity to certain inhibitors of respiratory activity. One of these puffs is specifically induced by vitamin B_6 or its derivatives (Leenders et al., 1973). In the experiments on *D. hydei,* the response of the loci was measured by the degree of puffing of polytene chromosomes. Measurement of polytene puff size in *D. melanogaster* after heat shocks at five different temperatures from 29 to 39°C shows variations in the responses of individual heat-shock puffs (Ashburner, 1970). In cultured *D. melanogaster* cells the relative amounts of cytoplasmic RNA species show specific variations when the temperature of the heat shock is increased by one degree intervals from 34 to 38°C (Spradling et al., 1977), suggesting that differential temperature sensitivity also exists among the heat-shock loci in diploid cells. Thus it appears that the expression of these nine loci is not strictly coordinately controlled but may be modulated in a number of different ways.

In summary, studies of the heat-shock response in polytene tissues correlate well with studies of the response in diploid tissues, both with respect to the basic response and to modulation by specific agents. The results show that the heat-shock response is an ideal system in which to study RNA metabolism, beginning at the induction of a set of genes (puffing), followed by the rapid appearance in the nuclear RNA of previously undetected RNA species, which are complementary to DNA sequences within the induced puffs. These RNAs may be found in the cytoplasm, on polysomes, at the same time that synthesis of a new set of proteins is detected.

Response to Ecdysone: Hormonally Regulated Genes

In *Drosophila* many developmental processes are closely regulated by hormones. Increases in the level of the steroid hormone ecdysone are thought to trigger each molt. Ecdysone concentration affects most, if not all, of the larval tissues; the specificity of the response is dependent on the target tissue and not on the hormonal stimulus.

One of the earliest detectable effects of ecdysone is a change in the pattern of puffing of the polytene chromosomes in salivary glands (Berendes, 1967; Ashburner, 1972). Mid-third instar larvae show the beginning of the prepupal puffing sequence within 15 min after being injected with ecdysone. Ecdysone apparently acts as a trigger; it initiates a sequence of changes in the patterns of puffing of the polytene chromosomes but its presence is not required for the completion of the later parts of the sequence (Ashburner et al., 1974). Berendes (1967) studied the early stages of the prepupal puffing sequence in several polytene tissues, salivary gland, malphigian tubules, and midintestine cells. He found the ecdysone-induced puffs were the same in all tissues, suggesting that the early puffs, at least, represent a generalized response to the hormone.

Ecdysone also initiates changes in diploid tissues. It induces metamorphosis in imaginal disks both in vivo and in vitro (Oberlander, 1972). Imaginal disk cells do not have polytene nuclei, but it is possible to study ecdysone-induced changes in their RNA populations by *in situ* hybridization of [^{3}H]RNA from those cells (Bonner and Pardue, 1975*b*). Imaginal disks cultured in minimal Robb's medium can be induced to undergo complete evagination if β-ecdysone is added to the medium (Fristrom et al., 1973). We have used such cultures to obtain [^{3}H]RNA at various points in the evagination sequence.

In situ hybridization experiments with cytoplasmic RNA labeled in wing disks during the first few hours of ecdysone stimulation show the induction of a polyadenylated RNA species which hybridizes to polytene band 67B11 (Bonner and Pardue, 1967*b*). Our early experiments also detected six other loci, 12E, 31A, 53C, 63B, 66B, and 92B, which showed a low level of hybridization with cytoplasmic [^{3}H]RNA from ecdysone-stimulated wing disks. Later experiments with higher levels of hybridization show that some RNA binding to these six regions is also produced in wing disk cells that have not been exposed to ecdysone. However, ecdysone treatment specifically increases the level of [^{3}H]RNA binding to 12E, 63B, 66B, and 92B as compared to the levels of [^{3}H]RNA binding to control regions not affected by ecdysone.

The *in situ* experiments were done with RNA labeled during the first 4 h of ecdysone induction of the wing disks. This is the same time period during which the early puffs are active in polytene tissues, the period for which Berendes (1967) reported that the three polytene tissues he studied had identical puffing patterns. Despite the generality of the early response to ecdysone in different polytene tissues, the early response to ecdysone in wing disks results in a different set of changes in the cytoplasmic RNA of these diploid cells. Cytoplasmic [^{3}H]RNA from ecdysone-stimulated wing disk cells does not hybridize with polytene regions which puff during the early puff stages (e.g., 74EF, 75B). Nor has cytoplasmic [^{3}H]RNA from wing disk cells, synthesized either before or after ecdysone stimulation, been detected binding to the regions of the two major polytene puffs (25AC, 68C) which regress when ecdysone is added. The band that hybridizes ecdysone-stimulated RNA from the disk cells (67B11) does not puff during the early response to ecdysone stimulation, and also does not hybridize salivary gland RNA. Thus the initial response to ecdysone detected in the imaginal disk appears to be quite different from the response seen in the polytene tissues.

Because of the complex nature of RNA processing, it is quite probable that the genetic activity visualized directly on polytene chromosomes as either puffs or [^{3}H]uridine incorporation does not give an exact representation of the RNA population in the cytoplasm. For that reason we have examined the pattern of *in situ* hybridization of [^{3}H]RNA from the cytoplasm of salivary glands incubated with [^{3}H]uridine in the presence or in the absence of β-ecdysone (Bonner and Pardue, manuscript in preparation). Grain counts were made over region 68C, which regresses when ecdysone is added, and 74EF and 75B, which are induced as an early response to ecdysone, as well as over several control sites which show no effect of ecdysone. The results show that, for these RNAs at least, the induction of a puff does result in the presence of [^{3}H]RNA in the cytoplasm which can be detected by *in situ* hybridization. [^{3}H]RNA hybridizing to 68C is found in the cytoplasm of glands labeled without ecdysone. After ecdysone stimulation, the puff at 68C regresses and newly synthesized cytoplasmic RNA no longer hybridizes to this region. Cytoplasmic [^{3}H]RNA hybridizing to 74EF and 75B is seen only after the puffs have been induced by ecdysone.

All of the polytene puffs studied thus far do produce RNAs which appear in the cytoplasm. The developmental puffs 68C, 74EF, and 75B, the major heat-shock puffs discussed earlier, and *Balbiani* Ring 2 in *Chironomus tentans* (Lambert, 1974), have all been shown to bind ^{3}H-labeled cytoplasmic RNA synthesized while the puffs were present.

The Relation Between Polytene Puffs and Cytoplasmic RNA Sequences

Under the conditions of heat shock, the number of large cytoplasmic RNAs produced is sufficiently limited so these RNAs can be separated into discrete groups by polyacrylamide gel electrophoresis. This makes it possible to investigate the arrangement of sequences complementary to the individual cytoplasmic RNAs among the heat-shock puff sites. In a coordinated set of genes such as the heat-shock loci, it is of particular interest to determine whether the chromosomal arrangement of individual coding sequences gives insight into the integration of the response.

When the large RNA species found in the cytoplasm of *D. melanogaster* cells grown at 37°C are separated by electrophoresis on polyacrylamide gels, they appear to fall into three classes (Spradling et al., 1977). One class of 12 bands on the acrylamide gel contains RNA species that we have identified as mitochondrial messenger RNAs. Similar sets of RNAs can be produced by the methods used to prepare mitochondrial messenger RNAs in other systems (Hirsch and Penman, 1973). These RNAs have also been studied by hybridization. The mitochondrial messenger RNAs bind specifically to purified *Drosophila* mitochondrial DNA but have shown no detectable hybridization *in situ* to polytene chromosomes. Transcription of mitochondrial RNA is much less affected by a 37°C heat shock than is much of the nuclear transcription. Thus mitochondrial mRNAs make up a significant part of the cytoplasmic high molecular weight RNA produced when cells are shifted to 37°C.

Another group of five bands seen when cytoplasmic RNA from heat-shocked *Drosophila* cells is run on polyacrylamide gels appears to contain mRNAs coding for histones. Most of the RNA in this set of bands lacks poly(A) sequences, as judged by lack of binding to oligo(dT) columns. The two bands of this group which we have studied by *in situ* hybridization bind specifically to the region of the polytene chromosomes that has been identified as the histone locus, region 39DE (Pardue et al., 1972). These seven bands move on the polyacrylamide gel at almost the same rate as histone messenger RNAs from sea urchins. Although polytene region 39DE does not puff in response to heat shock, [^{3}H]uridine incorporation can be seen at this region in transcription autoradiographs of heat-shocked salivary gland cells. Cytoplasmic RNA from these cells hybridizes *in situ* to region 39DE (Fig. 4).

The other six bands seen on a polyacrylamide gel of cytoplasmic RNA from heat-shocked *Drosophila* cells comprise the majority of the polyadenylated RNA produced at 37°C. These sequences are almost entirely polysome associated and are in the size range which might be expected of RNAs coding for the new proteins produced at 37°C. This set of RNA bands hybridizes *in situ* with the heat-shock puff regions on polytene chromosomes.

Because the RNA that hybridizes to the heat-shock puffs can be separated electrophoretically into six fractions, it is possible to investigate the relationship between puffs and individual cytoplasmic mRNAs. There are several possible kinds of relationships: (1) a 1:1 association between a puff and a cytoplasmic RNA species; (2) two or

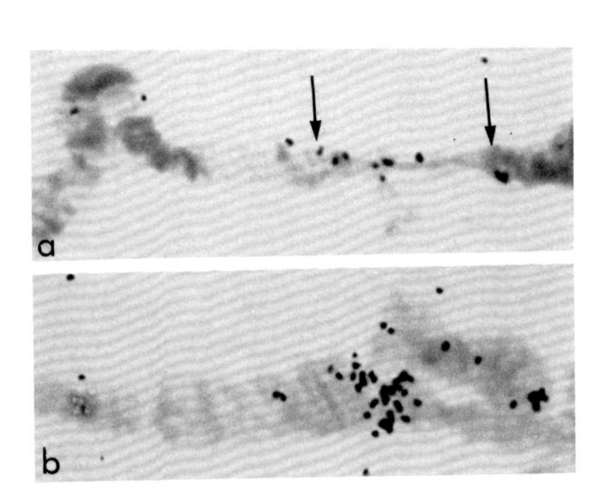

FIGURE 4 (*a*) Transcription autoradiograph showing [^{3}H]uridine incorporation across a stretched region 39 DE (between the arrows). This is the region that has been identified as the histone locus (Pardue et al., 1972). The salivary glands were heat shocked in vitro at 37°C. 22 μCi/ml of [^{3}H]uridine were added before the last 10 min of the 35-min heat shock. Glands were then fixed and prepared for autoradiography. Giemsa stain. Exposure 7 days. × 1,350. (*b*) *In situ* hybridization of [^{3}H]RNA prepared from salivary glands incubated with [^{3}H]uridine at 37°C. There is heavy hybridization over region 39DE. Exposure 26 wk. × 1,350.

TABLE I
In Situ Hybridization of Purified RNA Fractions to Polytene Chromosomes

Gel fraction	Polytene region							
Experiment A	63C	67B	87A	87C	95D	56F	39E	Ch.*
–	0	0	0.76	4.8	0	0	0	0
A1	7.3	0	3.1	7.4	0	0	0	0.98
A2	3.3	0	31	57	2.5	0	·0	15
A3	0.36	0	7.3	12	4.7	0	0	4.5
–	0	0	8.3	15	2.5	0	0	4.1
A5, A6	0	10	5.8	15	0.51	0	0	7.4
–	0	3.6	2.5	5.1	0	0	0	4.7
Experiment B								
–	1.1	0	5.7	10	1.3	0	0	7.2
A4	2.2	3.3	4.4	19	1.8	0	0	18
A5, A6	0	7.0	3.1	7.7	0	0	0	2.6

Each of the two blocks of entries in the table represents the results of *in situ* hybridization of the fractions, eluted from a single lane of a polyacrylamide gel of RNA labeled in heat-shocked cells. The gel fractions are listed in the order in which they ran on the gel. Fractions that did not show the presence of any prominent bands of RNA are denoted by a dash. The results are expressed as the mean number of grains (based on an analysis of at least 20 fields) at each site, normalized to a 60-day exposure period. (From Spradling, et al., 1977. Reprinted with permission.)
Experiment A: 20S and 13S RNA labeled at 37°C was purified by two cycles of sucrose gradient centrifugation followed by electrophoresis on 7 M urea gels. The indicated fractions were eluted and hybridized. Experiment B: RNA labeled after a 30-min heat shock at 37°C, conditions which result in a significant level of A4 labeling, was eluted after electrophoresis as above. Little A4 RNA is detected after the longer heat shock used in experiment A.
* Ch. = chromocenter.

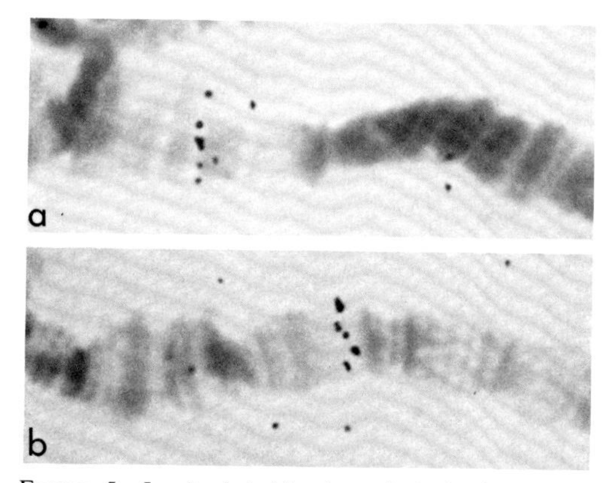

FIGURE 5 *In situ* hybridization of single fractions of poly(A)$^+$ cytoplasmic RNA from heat-shocked cultured cells. RNA from cells labeled at 37°C was run on 3% polyacrylamide gels and single bands were eluted for hybridization (Spradling et al., 1977). (*a*) Gel band A3 hybridized to region 95D. 3.5 × 10^2 cpm/slide. Exposure 52 days. × 1,400. (*b*) Gel band A5/A6 hybridized to region 67B. 1.7 × 10^3 cpm/slide. Exposure 52 days. × 1,400.

more puffs coding for the same RNA; (3) one puff coding for more than one RNA.

We have eluted individual RNA bands from a polyacrylamide gel and hybridized them *in situ* to polytene chromosomes (Spradling et al., 1977). The results of the *in situ* hybridization experiments are given in Table I and Figure 5. In Table I the results have been arranged to show the order in which the RNA fractions ran on the polyacrylamide gel as well as the polytene bands to which each hybridized. Although each fraction hybridizes, to some extent, with more than one polytene band, much of the hybridization appears to represent the spread of RNA molecules in the polyacrylamide gel. For each polytene band, hybridization is maximum with a single fraction of RNA and falls off rapidly with RNA fractions taken increasingly further away on the gel. (The one exception to this is region 87C, which shows relative peaks of hybridization with two RNA fractions well separated on the gel.) This intriguing result, obtained with RNA transcribed during the first 1/2 h at 37°C but not detectable after longer heat shocks, is discussed below.

If we consider only those gel fractions that show peaks in the level of hybridization with a polytene band, it appears that the heat-shock loci give examples of all three of the possible relationships between puff and cytoplasmic RNA mentioned above. Three loci show a 1:1 relationship with specific RNA fractions: region 63BC hybridizes with RNA A1, region 95D hybridizes with A3, and region 67B hybridizes with the combined A5/

A6 fraction. In an experiment in which A5 and A6 RNAs were separated more completely, only the A5 fraction hybridized with 67B.

Regions 87A and 87C appear to share a sequence. Both show maximum hybridization with RNA fraction A2. It is possible that gel fraction A2 is a mixture of two or more RNAs. However, all fractions of the polyacrylamide gel show some hybridization to regions 87A and 87C. Although the level of the hybridization decreases as the RNA fraction is taken from parts of the gel more distant from A2, the ratio of the hybridization of 87A to that at 87C remains 1:2. This failure to obtain any indication that the hybridization of 87A and 87C might be separable strengthens the possibility that regions 87A and 87C are binding the same RNA.

There is some evidence that region 87C may code for a second RNA in addition to the one apparently shared with 87A. One RNA band is detected on polyacrylamide gels only after heat shocks at 35°C or short exposures to 37°C. When this A4 RNA fraction was used for *in situ* hybridization, no new polytene bands were labeled. Instead the ratio of hybridization at region 87C to that at 87A increased from 2:1 to 4:1, suggesting that the A4 RNA fraction binds sequences at 87C which are not shared by 87A. Because of the unexpected spread of other sequences complementary to 87A and 87C throughout the gel, our conclusions about both of these regions remain tentative.

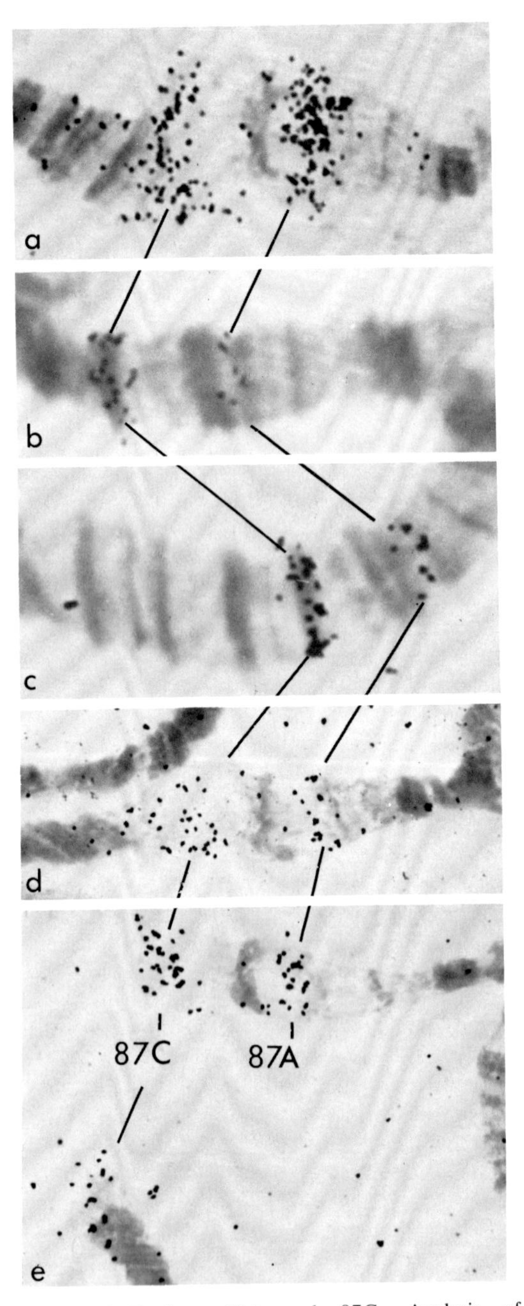

FIGURE 6 Regions 87A and 87C. Analysis of [^{3}H]RNA, made during 37°C heat shock, by transcription autoradiography (*a*) and *in situ* hybridization (*b–e*). (*a*) Salivary glands were labeled in vitro with 40 μCi/ml [^{3}H]uridine at 37°C for 10 min after a 10-min preincubation at 37°C. They were then fixed and prepared for autoradiography. In both regions 87A and 87C the [^{3}H]RNA is concentrated over one end of the puff. The region near the heavy bands of 87B between the two puffs shows no label. Giemsa stain. Exposure 5 days. × 1,500. (*b*) *In situ* hybridization of poly(A)$^+$ RNA prepared from heat-shocked imaginal disks. Hybridization in region 87C is directly over the heavy band 87C1. Hybridization in region 87A is scattered over the region of fine bands between 87A4 and 87B1. 6 × 10^3 cpm/slide. Exposure 14 days. × 1,950. (*c*) *In situ* hybridization with polyacrylamide gel band A4 prepared from heat-shocked cultured cells. The A4 RNA hybridized to 87A and 87C in a ratio of 1:4 instead of the 1:2 ratio seen with other isolated RNA fractions in the same experiment. The distribution of the *in situ* hybrid is undetectably different from that seen with unfractionated poly(A)$^+$ RNA in (*b*). 4.2 × 10^3 cpm/slide. Exposure 52 days. × 1,950. (*d* and *e*) *In situ* hybridization of poly(A)$^+$ cytoplasmic RNA from heat-shocked cultured cells to heat shock-induced puffs on polytene chromosome 3. The distribution of the hybrid closely parallels the distribution of label in transcription autoradiograph of heat-shocked polytene nuclei (*a*). In *e* the chromosome has broken in the 87C region. 1.5 × 10^4 cpm/slide. Exposure 37 days. × 850.

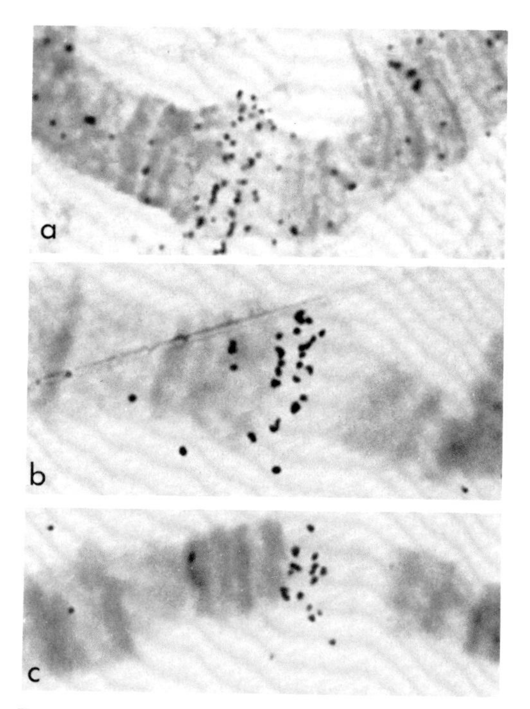

FIGURE 7 Region 93D. Analysis of [^{3}H]RNA, made during 37°C heat shock, by transcription autoradiography (*a*) and *in situ* hybridization (*b* and *c*). (*a*) Salivary glands were incubated as described for Fig. 6*a*, fixed and prepared for autoradiography. The heat shock-induced puff at 93D shows two bands of [^{3}H]RNA across the center of the puff. Giemsa stain. Exposure 5 days. × 1,500. (*b* and *c*) *In situ* hybridization of poly(A)$^-$ RNA, prepared from heat-shocked imaginal disks, to polytene chromosomes. Both the puffed and the nonpuffed chromosomes show two bands of *in situ* hybridization, corresponding to the bands in *a*. Giemsa stain. 8 × 10^3 cpm/slide. Exposure 28 days. × 1,700.

Localization of Coding Sequences within Puffs

Chromosomal puffs originate from single or double bands although a number of surrounding bands later become involved as the puffs enlarge (Beermann, 1952). We have used *in situ* hybridization to confirm the early cytological observations on puff structure and to unambiguously identify the regions complementary to cytoplasmic RNA. The set of puffs we have studied are the major heat-shock puffs 87A, 87C, 93D, 95E, 63BC, and 67B (Figs. 6, 7 and 8).

The sequences giving rise to the puffs were determined by hybridizing ^{3}H-labeled cytoplasmic RNA from heat-shocked cultured cells to nonpuffed polytene chromosomes. The results of these experiments are given in Table II. The band designations given refer to the revised map of Bridges (1941).

One unexpected finding of these experiments was that the relation between the polytene band that hybridized with cytoplasmic RNA and the bands that were involved in the puffing varied from puff to puff. In some cases, such as the puff at 95D, the sequences that hybridized the [^{3}H]RNA were toward the center of the group of bands that eventually made up the puff. In other cases, such as the puff at 87C, the sequences that hybridized [^{3}H]RNA were very close to one end

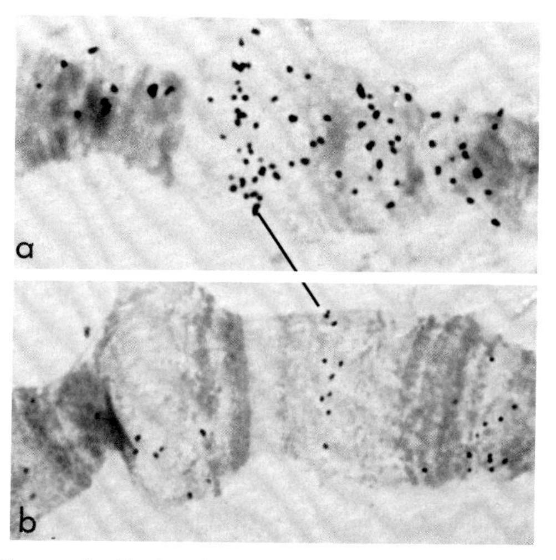

FIGURE 8 Region 63BC. Analysis of [^{3}H]RNA, made during 37°C heat shock, by transcription autoradiography (*a*) and *in situ* hybridization (*b*). (*a*) Salivary glands were incubated as described for Fig. 6*a*, fixed and prepared for autoradiography. There is a major site of labeling in the center of puff 63BC. Other label is detected over 63D and 63E. Giemsa stain. Exposure 5 days. × 1,450. (*b*) *In situ* hybridization of [^{3}H]RNA from heat-shocked imaginal disks. The principal region of hybridization corresponds to the band of label on the puff in *a*. Giemsa stain. 1.2 × 10^4 cpm/slide. Exposure 12 wk. × 1,200.

TABLE II

Puff	Sites of hybridization of cytoplasmic RNA
63 BC	One (possibly two) fine bands between 63B12 and 63C1
67B	67B1
87A	One (possibly two) fine bands between 87A5 and 87A10
87C	87C1
93D	93D3 and 93D5
95D	95D10, 11, or 95E1

of the set of bands that eventually became involved in the puff. In puff 87C the hybridizing sequences are so asymmetrically placed that they are actually in a different lettered division of the Bridges map from the major part of the puff. The puff originating from band 87C was originally identified as 87B because it includes most of the 87B region. When the region puffs, the band of origin becomes very faint if it is visible at all. The area covered by sequences binding cytoplasmic RNA becomes much larger. However, when the chromosome preparation is well stretched, it can be seen that the hybridization does not cover the entire area of the puff in any of the regions studied. The distribution of [^{3}H]RNA over the puff appears much the same whether the preparation was made by *in situ* hybridization with cytoplasmic RNA, by *in situ* hybridization with heterogeneous nuclear RNA, or by transcription autoradiography of [^{3}H]uridine incorporation directly on the chromosomes. Longer autoradiographic exposures of either *in situ* hybridization or transcription experiments show broader distributions of [^{3}H]RNA, suggesting a gradient of compaction of the transcribed sequences spreading from the band of origin into the puff.

Our observations on the major heat-shock puffs are consistent with the accepted model of a puff arising by localized uncoiling of sequences. However, they give no explanation for the apparently nontranscribed sequences included in the puff. Nor do they explain why the puff may develop symmetrically around the transcribed region in some cases and quite asymmetrically in others.

SUMMARY

In situ hybridization of [^{3}H]RNA to polytene chromosomes makes it possible to resolve many components in complex populations of RNA. With this technique we have begun to compare RNAs of several types of diploid *Drosophila* cells with those of salivary gland cells in an effort to extend the insights into eukaryotic chromosome structure and function which have been obtained from other cytogenetic studies on polytene chromosomes.

ACKNOWLEDGMENTS

This work has been supported by grants from the National Institutes of Health and the National Science Foundation.

REFERENCES

Ashburner, M. 1970. Patterns of puffing activity in the salivary gland chromosomes of *Drosophila*. V. Responses to environmental treatments. *Chromosoma* (*Berl.*). **31**:356–376.

Ashburner, M. 1972. Puffing patterns in *Drosophila melanogaster* and related species. *In* Developmental Studies on Giant Chromosomes. Results and Problems in Cellular Differentiation. Vol. 4. W. Beermann, editor. Springer-Verlag, Berlin. 101–151.

Ashburner, M., C. Chihara, P. Metzer, and G. Richards. 1974. Temporal control of puffing activity in polytene chromosomes. *Cold Spring Harbor Symp. Quant. Biol.* **38**:655–662.

Baudisch, W., and R. Panitz. 1968. Kontrolle eines biochemischen Merkmals in den Speicheldrüsen von Acricopus lucidus durch einen Balbiani-Ring. *Exp. Cell Res.* **49**:470–476.

Beermann, W. 1952. Chromosomen Konstanz und spezifische Modifikationen der Chromosomenstruktur in der Entwicklung und Organdifferenzierung von Chironomus tentans. *Chromosoma* (*Berl.*). **5**:139–198.

Beermann, W. 1961. Ein Balbianiring als Locus einer Speicheldrüsen mutation. *Chromosoma* (*Berl.*). **12**:1–25.

Beermann, W. 1962. Riesenchromosomen. *Protoplasmatologia.* Handbuch der Protoplasmaforschung. Band IV d. Springer (Wien).

Beermann, W. 1972. Chromomeres and genes. *In* Developmental Studies on Giant Chromosomes. Results and Problems in Cellular Differentiation. Vol. 4. W. Beermann, editor. Springer-Verlag, Berlin. 1–33.

Berendes, H. D. 1967. The hormone ecdysone as effector of specific changes in the pattern of gene activities of *Drosophila hydei. Chromosoma* (*Berl.*). **22**:274–293.

Bonner, J. J., and M. L. Pardue. 1976*a*. The effect of heat shock on RNA synthesis in *Drosophila* tissues. *Cell.* **8**:43–50.

Bonner, J. J., and M. L. Pardue. 1976*b*. Ecdysone-stimulated RNA synthesis in imaginal discs of *Drosophila melanogaster*: assay by in situ hybridization. *Chromosoma* (*Berl.*). **58**:87–99.

Breugel, F. M. A. van. 1966. Puff induction in larval salivary gland chromosomes of *Drosophila hydei* Sturtevant. *Genetica* (*The Hague*). **37**:17–28.

Bridges, P. N. 1941. A revision of the salivary gland 3R-chromosome map. *J. Hered.* **32**:299–300.

Fristrom, J. W., W. R. Logan, and C. Murphy. 1973. The synthetic and minimal culture requirements for evagination of imaginal discs of *D. melanogaster* in vitro. *Dev. Biol.* **33**:441–456.

Grossbach, U. 1974. Chromosome puffs and gene expression in polytene cells. *Cold Spring Harbor Symp. Quant. Biol.* **38**:619–627.

Hirsch, M., and S. Penman. 1973. Mitochondrial polyadenylic acid-containing RNA: localization and characterization. *J. Mol. Biol.* **80**:379–391.

Korge, G. 1975. Chromosome puff activity and protein synthesis in larval salivary glands of *D. melanogaster. Proc. Natl. Acad. Sci. U. S. A.* **72**:4550–4554.

Lambert, B. 1974. Repeated nucleotide sequences in a single puff of *Chironomus tentans* polytene chromosomes. *Cold Spring Harbor Symp. Quant. Biol.* **38:**637–644.

Leenders, H. J., and H. D. Berendes. 1972. The effect of changes in the respiratory metabolism upon genome activity in *Drosophila*. I. The induction of gene activity. *Chromosoma (Berl.).* **37:**433–444.

Leenders, H. J., J. Derksen, P. M. J. M. Maas, and H. J. Berendes. 1973. Selective induction of a giant puff in *Drosophila hydei* by vitamin B_6 and derivatives. *Chromosoma (Berl.).* **41:**447–460.

Lengyel, J. A., and M. L. Pardue. 1975. Analysis of hnRNA made during heat shock in *Drosophila melanogaster* cultured cells. *J. Cell Biol.* **67:**240*a* (Abstr.).

Lewis, M., P. J. Helmsing, and M. Ashburner. 1975. Parallel changes in puffing activity and patterns of protein synthesis in salivary glands of *Drosophila*. *Proc. Natl. Acad. Sci. U. S. A.* **73:**3604–3608.

McKenzie, S. L., S. Henikoff, and M. Meselson. 1975. Localization of RNA from heat-induced polysomes at puff sites in *Drosophila melanogaster*. *Proc. Natl. Acad. Sci. U. S. A.* **72:**1117–1121.

Oberlander, H. 1972. The hormonal control of development in imaginal disks. *In* The Biology of Imaginal Disks. Results and Problems in Cell Differentiation. Vol. 5. H. Ursprung and R. Nöthiger, editors. Springer-Verlag, Berlin. 155–172.

Pardue, M. L., E. Weinberg, L. H. Kedes, and M. L. Birnstiel. 1972. Localization of sequences coding for histone messenger RNA in the chromosomes of *Drosophila melanogaster*. *J. Cell Biol.* **55:**199*a* (Abstr.).

Pelling, C. 1959. Chromosomal synthesis of ribonucleic acid shown by incorporation of uridine labeled with tritium. *Nature (Lond.).* **184:**655–656.

Pelling, C. 1964. Ribonukleinsaüre-synthese der Riesenchromosomen. Autoradiographische Untersuchungen an Chironomus tentans. *Chromosoma (Berl.).* **15:**71–122.

Ritossa, F. 1962. A new puffing pattern induced by temperature shock and DNP in *Drosophila*. *Experentia (Basel).* **18:**571–573.

Ritossa, F. 1963. New puffs induced by temperature shock, DNP and salicilate in salivary chromosomes of *D. melanogaster*. *Drosophila Information Service* **37:**122–123.

Ritossa, F. 1964. Experimental activation of specific loci in polytene chromosomes of *Drosophila*. *Exp. Cell Res.* **35:**601–607.

Rubin, G. M., and D. S. Hogness. 1975. Effect of heat shock on the synthesis of low molecular weight RNAs in *Drosophila*: accumulation of a novel form of 5S RNA. *Cell.* **6:**207–213.

Spradling, A., M. L. Pardue, and S. Penman. 1977. mRNA in heat shocked *Drosophila* cells. *J. Mol. Biol.* In press.

Spradling, A., S. Penman, and M. L. Pardue. 1975. Analysis of *Drosophila* mRNA by in situ hybridization: sequences transcribed in normal and heat shocked cultured cells. *Cell.* **4:**395–404.

Tissieres, A., H. K. Mitchell, and U. M. Tracy. 1974. Protein synthesis in salivary glands: relation to chromosome puffs. *J. Mol. Biol.* **84:**389–398.

Wensink, P. C., D. J. Finnegan, J. E. Donelson, and D. S. Hogness. 1974. A system for mapping DNA sequences in the chromosomes of *Drosophila melanogaster*. *Cell.* **3:**315–325.

IN VITRO AND IN VIVO ANALYSES OF CHROMOSOME STRUCTURE, REPLICATION, AND REPAIR USING BrdU-33258 HOECHST TECHNIQUES

SAMUEL A. LATT, JAMES W. ALLEN, and GAIL STETTEN

Bromodeoxyuridine (BrdU)-dye techniques for detecting DNA synthesis have led to an increased understanding of chromosome structure and replication, and have provided a convenient means of assessing certain forms of chromosome damage and repair. Early experiments utilized the ability of 5-BrdU, incorporated biosynthetically into chromosomes in place of thymidine (dT), to quench the fluorescence of DNA-binding dyes such as 33258 Hoechst (Latt, 1973) or acridine orange (Kato, 1974). Subsequent methods, using differential staining with Giemsa, were based on the associated structural lability of BrdU-substituted chromatin with (Perry and Wolff, 1974) or without (Korenberg and Freedlender, 1974) enhanced photosensitization by previously bound fluorochrome. More recently, immunochemical detection of BrdU incorporation has been achieved (Gratzner et al., 1976). These techniques can serve as convenient, high-resolution alternatives to autoradiography for detection of DNA in fixed cytological preparations. Suppression of 33258 Hoechst fluorescence can also be utilized to detect BrdU incorporation into unfixed chromatin and intact cells (see Latt, in press, for a review). Initially applied to in vitro systems, BrdU techniques have now been adapted to in vivo studies with intact animals (Bloom and Hsu, 1975; Vogel and Bauknecht, 1976; Allen and Latt, 1976). The resulting cytological information about chromosome structure and function has been of both basic and practical interest, and it has suggested related directions for biochemical research.

SAMUEL A. LATT and CO-WORKERS Clinical Genetics Division, Mental Retardation Program, Children's Hospital Medical Center and the Department of Pediatrics, Harvard Medical School, Boston, Massachusetts

The bis-benzimidazole dye 33258 Hoechst was selected during a search for DNA-binding fluorochromes, whose emission intensity was suppressed by BrdU substitution (Latt, 1973). The rationale was based on the potential ability of a heavy, polarizable atom such as bromine to quench the fluorescence of a nearby bound dye. A variety of spectroscopic studies have been used to characterize this quenching and have also been employed to investigate other aspects of the interaction between 33258 Hoechst and DNA.

Fluorescence, absorption, and circular dichroism measurements indicate the existence of at least two modes of 33258 Hoechst-nucleic acid interactions. At high ionic strength (e.g., 0.4 M NaCl, pH 7), dye binding is highly adenine-thymine (A-T) specific, saturating at 1 dye/3 or 4 A-T base pairs, and is associated with efficient fluorescence. Substitution of BrdU for deoxythymidine (dT) increases dye-binding affinity but reduces fluorescence efficiency by about a factor of 10. At this higher ionic strength, the predominant fluorescence decay time in the dye-poly (deoxyadenine[dA]-BrdU) complex is approximately one-fifth of that of the dye complexed with poly (dA-dT). Such a reduction in fluorescence lifetime, indicating an interaction at the first excited singlet state of the dye, suggests that heavy atom-induced intersystem crossing might contribute to the observed fluorescence quenching. Reduction in 33258 Hoechst fluorescence intensity and lifetime has also been detected in complexes of the dye with poly(dA)-poly(BrdU), and comparable effects occur in complexes involving polymers composed of dA and 5-iodo-deoxyuridine (IdU). BrdU-dependent fluorescence quenching has also been detected with a number of Hoechst dyes structurally related to 33258.

At lower ionic strength (e.g., 0.01 M NaCl, pH 7), dye binding retains some A-T specificity, although fluorescence quenching as a result of BrdU incorporation is reduced. Under these conditions, extensive dye binding (up to about 1 dye/3 phosphates) can occur even to guanine-cytosine (G-C) rich DNA, although such binding exhibits little fluorescence. Thus far, there have been no data reported indicating that 33258 Hoechst binding to DNA involves intercalation.

Chromosomal proteins occlude approximately half of the DNA from 33258 Hoechst. The fluorescence of 33258 Hoechst is quenched when the dye is bound to natural DNA or to unfixed chromatin totally substituted with BrdU, although, relative to unsubstituted chromatin, there is no significant difference in dye occlusion by proteins in this chromatin (Latt and Wohlleb, 1975). Chromosomal proteins were initially presumed to underlie the ability of modified Giemsa techniques to detect BrdU incorporation (Perry and Wolff, 1974). However, trypsin pretreatment of chromosomes does not abolish differential Giemsa staining of substituted chromatids (Goto et al., 1975), as would be expected were BrdU substitution to increase the binding of proteins which, in turn, excluded the dye. Moreover, recent studies have shown that BrdU-substituted chromatin can be fragmented by exposure to light (Taichman and Freedlender, 1976). Were DNA breakage to be enhanced by BrdU substitution in chromosomes treated with the modified Giemsa methods, the buffer incubation might then serve to wash away fragmented chromatin. The use of 33258 Hoechst in these methods probably serves to enhance the photosensitized degradation of substituted chromatin.

Quenching of 33258 Hoechst fluorescence as a result of BrdU incorporation into DNA has served as an alternative to autoradiography for determining the fraction of cells responding to a stimulus for DNA synthesis. More recently, reduction in 33258 fluorescence has been used to differentiate unfixed cells substituted with BrdU from unsubstituted cells. In the course of these experiments, it was noticed that addition of 33258 Hoechst markedly increased the photosensitivity of cells which had incorporated BrdU. This has subsequently formed the basis of a system for selectively killing proliferating cells which have incorporated BrdU (Stetten et al., 1976).

Fluorometric detection of BrdU incorporation into unfixed cells should permit the analysis of DNA synthesis in these cells and ultimately their sorting according to replication kinetics. Such experiments have been initiated in collaboration with Dr. Joe W. Gray of the Lawrence Livermore Laboratory, Livermore, Calif. In addition, changes in 33258 Hoechst fluorescence as a result of selective incorporation of BrdU or dT into late-replicating chromosomes, such as the X, may afford analogous, DNA synthesis-based fluorescent activated chromosome sorting. Biochemical characterization of isolated, late-replicating X chromosomes might, in turn, provide insight about X inactivation.

Three types of cytological applications of BrdU-dye techniques have thus far been carried out with metaphase chromosomes (Latt, 1976). Administration of BrdU for a fraction of one DNA synthesis period, the remainder of which is carried out in the presence of dT, differentiates regions according to the timing of DNA synthesis (Latt, 1973, 1975). Incorporation of BrdU for one complete replication cycle permits identification of chromosomal regions that contain species of satellite DNA with an asymmetric distribution of thymine between complementary polynucleotide chains (Lin et al., 1974; Latt et al., 1974). If a replication in BrdU is followed by a cycle either in the presence or absence of this analogue (Latt, 1973, 1974; Kato, 1974; Korenberg and Freedlender, 1974; Perry and Wolff, 1974), sister chromatids can be differentiated. This provides an optical demonstration of the semiconservative distribution of newly synthesized DNA between sister chromatids, a finding originally reported by J. Herbert Taylor and co-workers (Taylor et al., 1957), using autoradiography. Sister chromatid exchanges, detected in BrdU-substituted chromosomes, can serve as a sensitive index of chromosome damage and repair (Latt, 1974; Perry and Evans, 1975) and permit differentiation between diseases characterized by chromosome fragility (Chaganti et al., 1974; Latt et al., 1975).

If cells incorporate BrdU for part of one replication cycle, which is completed in the presence of dT but not BrdU, and metaphase chromosome preparations are stained with 33258 Hoechst, late-replicating regions exhibit bright fluorescence against a BrdU-suppressed background (Latt, 1975). A converse pattern is obtained if the sequence of BrdU and dT incorporation is reversed. As illustrated for the human D group chromosomes (Fig. 1), late-replicating regions generally correspond to a subset of the structural bands as

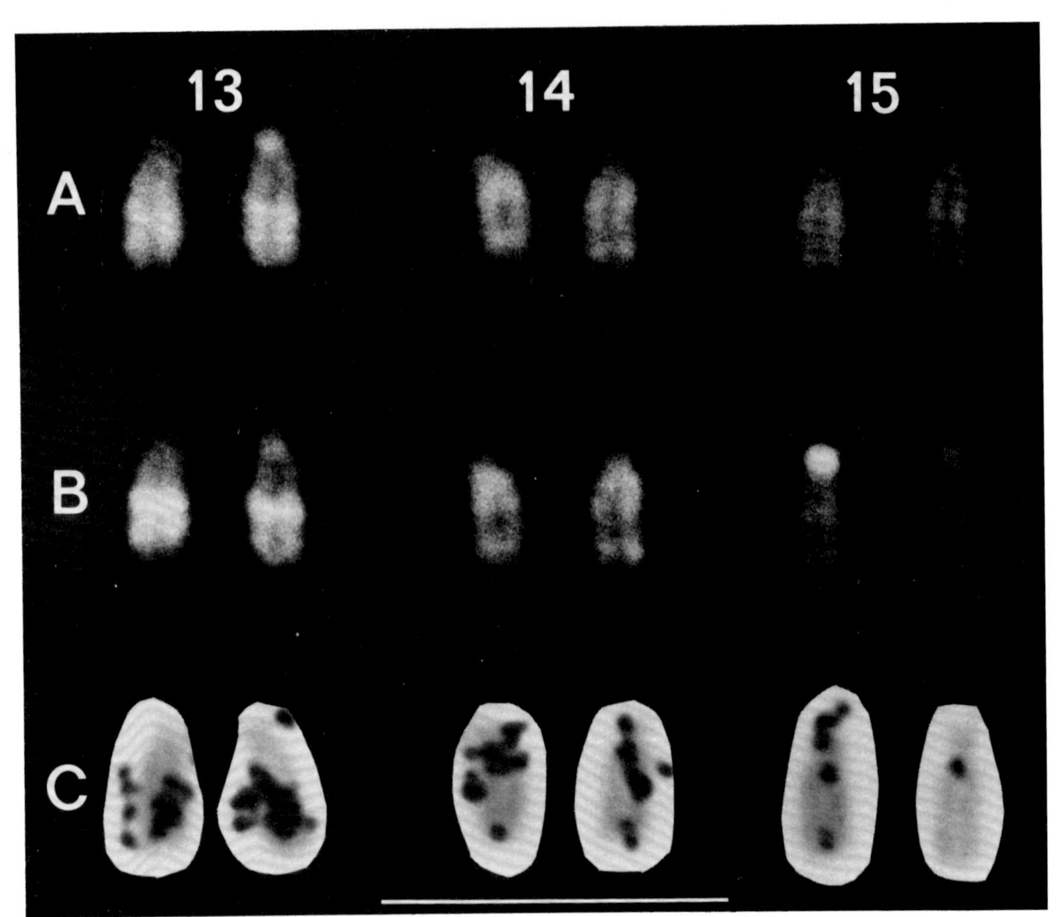

FIGURE 1 Patterns of late replication in human D group chromosomes. These chromosomes are from a 46, XY human lymphocyte grown in medium containing BrdU for the start of one replication cycle, which was completed in medium containing tritiated dT. The chromosomes were stained with quinacrine (Q-band) to reveal standard banding patterns (A), destained, and restained with 33258 Hoechst (B), which exhibits bright fluorescence in late replicating regions. Incorporation of tritiated dT into these chromosomes was subsequently demonstrated by autoradiography (C). Bar equals 10μm. (Latt, 1975.)

defined by quinacrine (Q-band) staining. Regions highlighted by 33258 Hoechst fluorescence (or related Giemsa staining) correlate well with those identified as late replicating by autoradiography on the same chromosomes. However, optical techniques provide much higher resolution, giving the further impression that replication within bands is highly coordinated. That is, bands appear to be units of chromosome replication, as well as structure. Marked fluctuations at the level of bands occur about the predominant sequence of late replication. These are most frequently evidenced as homologue asynchrony (e.g., chromosome 15 in Fig. 1). In addition, variations have been detected in the distribution of bands last to complete replication in a given chromosome from different lymphocytes of the same individual (Latt, 1975). The biochemical basis for the control of DNA replication timing within and between structural bands remains to be determined.

The most dramatic interhomologue difference in replication kinetics occurs in the female X chromosome pair. In the late-replicating, predominantly inactive X, both the distribution and extent of late-replicating regions differ from that of the other X, which exhibits a pattern indistinguishable from the single male X (Fig. 2; Willard and Latt, 1976). Studies in human lymphocytes showed that BrdU-Hoechst methods allow identification of a late-replicating X in a much larger proportion of cells than was possible with autoradiography. Moreover, significant fluctuations around the pre-

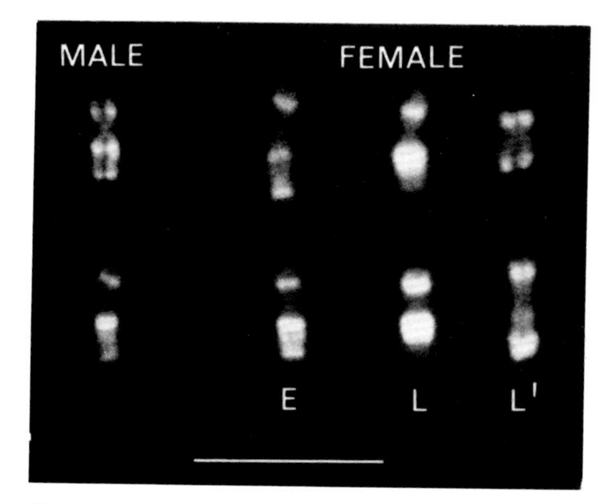

FIGURE 2. Late replication patterns in human lymphocyte X chromosomes. Culture conditions were as described for Fig. 1; chromosomes were stained with 33258 Hoechst. The distribution of brightly fluorescing late replicating regions in the single male X resembles that of the early replicating female X (*E*). A different fluorescent pattern reflecting late replication is found in the other X in the majority of female lymphocytes (*L*), while a small fraction of lymphocytes exhibit an alternative late replication pattern (L^1). Bar equals 10 μm.

dominant late-replication pattern were observed, consistent with the idea that the control of replication in X chromosomes is multifocal.

Perhaps the most curious observation thus far about X chromosome replication with BrdU-Hoechst techniques is the identification of an alternative late-replication pattern in 3–15% of lymphocytes from all females examined. This pattern highlights the terminus of the long arm of the late X, rather than the proximal region (Fig. 2). The other X invariably exhibits a typical, early replicating distribution. Further proof that this pattern is associated with late replication derives from studies with supernumerary or structurally abnormal X chromosomes (Latt et al., 1976). In 47 XXX or 48 XXXX cells, as many as 2 or 3 X chromosomes with alternative replication patterns can be found, although, in some cells, both types of replication patterns are seen. Isochromosomes for the long arm of the X can exhibit the alternative pattern, and in such cases, this usually occurs in both chromosome arms. We have recently obtained preliminary evidence that, in human fibroblasts, the predominant late X replication pattern resembles the alternative pattern found in lymphocytes. It would be interesting were variations in late X replication patterns to reflect tissue-specific differences in X chromosome expression.

If mouse cells are grown for one cycle in medium containing BrdU, lateral asymmetry in 33258 Hoechst fluorescence is observed in centromeres of all but a few chromosomes (Fig. 3). This fluorescence asymmetry corresponds in location to mouse satellite DNA, which is known to contain a thymine-rich chain. The contralateral distribution of these thymine-rich chains in metacentric mouse chromosomes has been interpreted in terms of a correspondence between DNA and chromatid polarity, together with the preservation of this polarity, perhaps via DNA continuity, through the centromere (Lin et al., 1974). Fluorescence asymmetry has also been observed in the human Y and in the secondary constriction of 16 (Latt et al., 1974). In this latter study, the contralateral orientation of thymine-rich regions in a dicentric Y was similarly interpreted in terms of chromatid polarity and DNA continuity.

Subsequent Giemsa methods have confirmed the fluorescence results and have identified additional regions of asymmetry in human chromosomes, e.g., the centromere of 15 and the secondary constriction of 1 (Angell and Jacobs, 1975; Galloway and Evans, 1975). In the chromosome 1 secondary constriction of some individ-

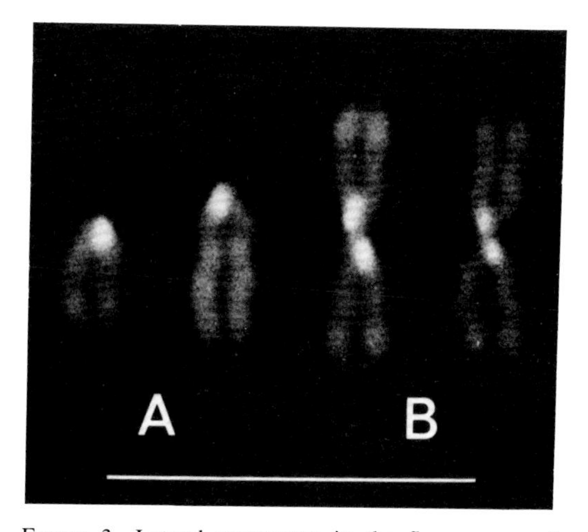

FIGURE 3 Lateral asymmetry in the fluorescence of mouse chromosome centromeric regions after staining with 33258 Hoechst. These chromosomes are from a mouse renal adenocarcinoma cell that had undergone one cycle of BrdU incorporation. Bright centromeric fluorescence in only one chromatid of each acrocentric chromsome (A) reflects the location of the thymine-rich chain of mouse satellite DNA (Lin et al., 1974). In the metacentric chromosomes (B), two contralaterally arranged, brighlty fluorescing regions are apparent. Bar equals 10 μm.

uals, compound alternating lateral asymmetry has been detected (Angell and Jacobs, 1975). It has been suggested that factors other than BrdU incorporation might contribute to the observed asymmetry (Galloway and Evans, 1975). For example, in at least one region (the secondary constriction of human chromosome 16), the satellite species (human satellite II) which is thought to be present possesses only slight (4%) thymine asymmetry. Both base sequence and overall base composition may contribute to the observed contrast.

Practical interest in BrdU-dye methods has centered thus far on sister chromatid exchanges (SCE). These exchanges can be detected in metaphase chromosomes in which sister chromatids have been differentiated. This was originally accomplished by autoradiography (Taylor et al., 1957), and more recently has been effected by BrdU-dependent alterations in dye fluorescence or Giemsa staining. In BrdU-dye methods, cells incorporate BrdU for one replication cycle and then undergo a second cycle of replication in which the presence of BrdU is optional. Sister chromatid exchanges are apparent as sharp reciprocal alterations in fluorescence of Giemsa staining along these chromosomes (Fig. 4). Multiple, closely spaced exchanges can easily be resolved.

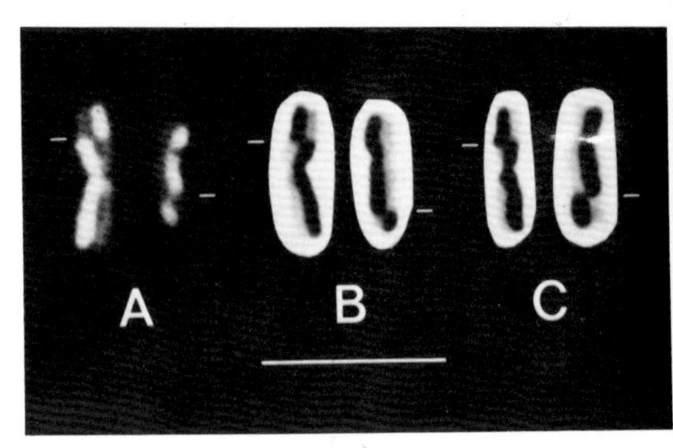

FIGURE 4 SCES in chromosomes from human lymphocytes which had replicated twice in medium containing BrdU. Chromosomes in A were stained with 33258 Hoechst and examined by fluorescence microscopy. Those in B were previously photographed to record fluorescence, as in A, and then restained with Giemsa. Chromosomes in C were exposed to visible light while mounted in a solution containing 33258 Hoechst and then stained with Giemsa. SCEs are indicated by horizontal lines. Bar equals 10 μm.

BrdU-33258 Hoechst and related methodologies have been used to show that low doses of clastogens, in particular alkylating agents, e.g., mitomycin C or psoralen-plus light, can induce large numbers of SCEs at concentrations well below those causing significant numbers of chromosome breaks (Latt, 1974; Perry and Evans, 1975; Latt and Juergens, 1976; Latt et al., in press). SCE analysis was thus proposed as a highly sensitive means of detecting chromosome damage and repair. BrdU itself was observed to induce exchanges (Latt, 1974; Kato, 1974; Perry and Wolff, 1974), although the drug-induced increments above this baseline are usually pronounced. Perry and Evans (1975) surveyed a number of exchange-inducing agents, many of which are known mutagens and/or carcinogens. On the basis of these observations, it was suggested by Perry and Evans that SCE analysis could be used as an assay for mutagenesis and carcinogenesis.

The use of SCEs to survey for potential genetic damage would benefit from a better understanding both of the molecular events involved in SCE formation and of the biological importance of SCE formation in the repair of DNA. It has been observed that psoralen-plus light treatment is very effective in inducing SCEs (Latt and Juergens, 1976; Latt et al., in press. Fig. 5). Recent studies in our laboratory with Chinese hamster ovary (CHO) cells indicate that extensive SCE induction can occur with very little cell killing. Cell death corresponds to the appearance of chromosome breaks, which occur at relatively high doses of light. We are currently using this system to investigate the relative effectiveness of DNA alkylation in inducing SCEs and chromosome aberrations at different times during the cell cycle, and parallel mutagenesis studies are planned. Biochemical characterization of SCE formation (Moore and Holliday, 1976), as well as a more precise analysis of the fidelity of DNA repair in damaged regions, might provide further information about the relationship of SCE formation to mutagenesis.

A number of inherited diseases characterized by chromosome fragility and a predisposition for the development of neoplasia have been differentiated by analysis of SCE formation. For example, Chaganti et al. (1974) detected an elevated SCE frequency in BrdU-cultivated cells from patients with Bloom's syndrome. In contrast, normal SCE frequencies have been detected in Fanconi's anemia, ataxia telangiectasia, and xeroderma pigmentosum. However, lymphocytes from Fanconi's anemia patients do not respond to mitomycin C treatment with a normal increase in SCE formation (Latt et al., 1975). The reduced stimulation by mitomycin C of SCE formation in Fanconi's

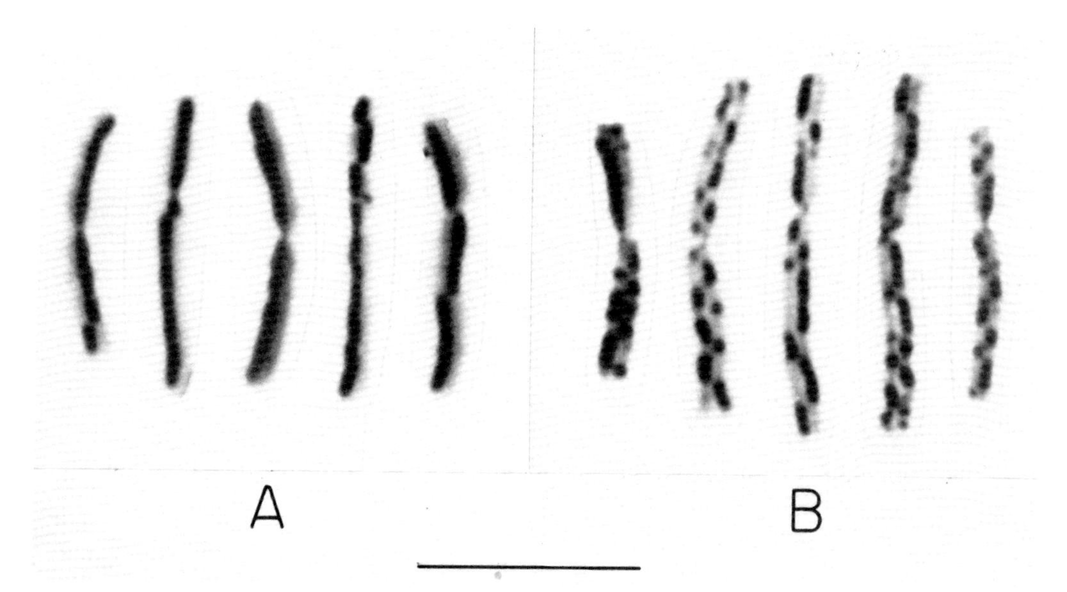

FIGURE 5 Extensive SCE induction by psoralen plus light treatment in CHO cells. Chromosomes were stained essentially as described by Perry and Wolff (1974). Chromosomes in A represent the control. Those in B are from cells exposed to 10^{-5} M 8-methoxypsoralen and approximately 10^4 ERG/mm^2 near UV light (principally at 365 mm) before addition of BrdU. Bar equals 10μm.

anemia is associated with an increased number of chromatid breaks. In the initial report of these observations, the induction of SCEs in fibroblasts from different individuals (obtained from cell repositories) was noted to be subnormal, though less so than in lymphocytes. We have recently found that SCE induction by mitomycin C is significantly reduced in dermal fibroblasts from the same individuals with Fanconi's anemia whose lymphocytes had previously been examined. Approximately half of the chromatid breaks induced in Fanconi's anemia lymphocytes by mitomycin C occurred at sites of incomplete SCE formation (Fig. 6). This observation is compatible with the hypothesis that the break increment and exchange deficit are causally related. We interpreted these results to suggest that Fanconi's anemia cells are defective in a form of DNA repair.

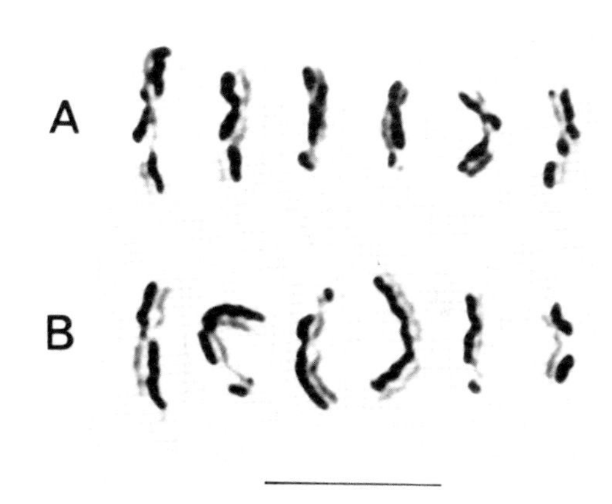

FIGURE 6 Aberrations in chromosomes from Fanconi's anemia lymphocytes treated with mitomycin C. Chromosomes in the top row (A) exhibit chromatid breaks at sites of incomplete SCE, while those in the bottom row (B) exhibit breaks away from such sites. Bar equals 10 μm.

The scope of BrdU-dye studies of DNA replication in animal cells has been extended from in vitro to in vivo systems. Bloom and Hsu (1975) described the formation of SCEs in ovo in chick embryos. More recent reports (Vogel and Bauknecht, 1976; Allen and Latt, 1976) have described the induction by alkylating agents of SCE in marrow or spermatogonia of mice that had received repeated doses of BrdU (Fig. 7). This system includes host-mediated activation properties as evidenced by the effectiveness of cyclophosphamide in vivo, but not in vitro. It also permits a comparison of the susceptibilities of multiple animal cell types to a given drug. The obvious importance of spermatogonial damage to germ cell formation makes this approach unique for studying potential mutagens.

The in vivo approach has also been used to

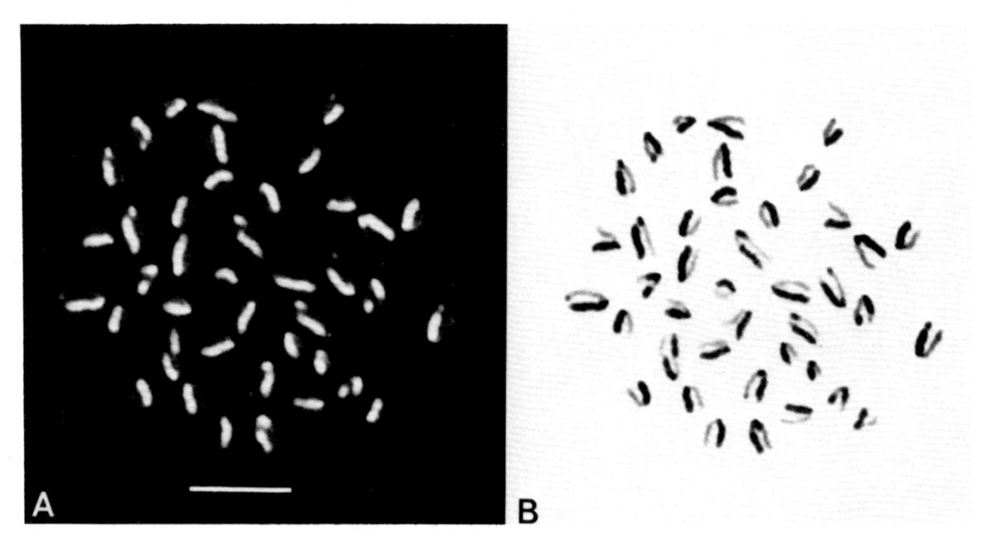

FIGURE 7 Induction by mitomycin C of SCE formation in vivo in mouse bone marrow. A mouse was sacrificed 19 h after 9 hourly injections of BrdU followed by a single injection of 0.3 mg/kg mitomycin C. The results of both fluorescence (A) and fluorescence followed by Giemsa staining (B) are shown. Bone marrow cells from animals untreated with mitomycin C exhibited an average of less than 4 SCE (Allen and Latt, 1976). Bar equals 10 μm.

examine and compare DNA replication kinetics in bone marrow and spermatogonia. In addition, sister chromatid differentiation in late-replicating chromosomes has been observed in cells at the first meiotic metaphase, and the relative orientation of the mouse X and Y at this stage determined (Allen and Latt, 1976). Future studies will attempt to develop this system to the point at which SCE formation in somatic cells can be compared with meiotic interchange.

ACKNOWLEDGMENTS

The technical assistance of Ms. Lois Juergens, Mr. Will Rogers, and Ms. Susan Brefach is greatly appreciated. 33258 Hoechst and 8-methoxypsoralen were generous gifts of Dr. H. Loewe, Hoechst, Frankfurt, Germany, and the Paul Elder Co., Bryan, Ohio, respectively.

This work was supported by research grants from the National Institutes of Health (GM 21121), National Foundation March of Dimes (1-353), and American Cancer Society (VC-144A). Dr. Latt is the recipient of a Research Career Development Award (GM 00122), and Dr. Allen is supported by funds from a postdoctoral training grant (GM 00156).

REFERENCES

ALLEN, J. W., and S. A. LATT. 1976. Analysis of sister chromatid exchange formation in vivo in mouse spermatogonia as a new test system for environmental mutagens. *Nature* (*Lond.*). **260:**449–451.

ALLEN, J. W., and S. A. LATT. 1976. In vivo BrdU-33258 Hoechst analysis of DNA replication kinetics and sister chromatid exchange formation in mouse somatic and meiotic cells. *Chromosoma* (*Berlin*). **58:**325–340.

ANGELL, R. R., and P. A. JACOBS. 1975. Lateral asymmetry in human constitutive heterochromatin. *Chromosoma* (*Berlin*). **51:**301–310.

BLOOM, S. E., and T. C. HSU. 1975. Differential fluorescence of sister chromatids in chicken embryos exposed to 5-bromodeoxyuridine. *Chromosoma* (*Berlin*). **51:**261–267.

CHAGANTI, R. S. K., S. SCHONBERG, and J. GERMAN. 1974. A many-fold increase in sister chromatid exchanges in Bloom's syndrome lymphocytes. *Proc. Natl. Acad. Sci. U. S. A.* **71:**4508–4512.

GALLOWAY, S. M., and H. J. EVANS. 1975. Asymmetrical C-bands and satellite DNA in man. *Exp. Cell Res.* **54:**454–459.

GOTO, K., T. AKEMATSU, H. SHIMAZU, and T. SUGIYAMA. 1975. Simple differential Giemsa staining of sister chromatids after treatment with photosensitive dyes and exposure to light and the mechanism of staining. *Chromosoma* (*Berlin*). **53:**223–230.

GRATZNER, H. G., A. POLLACK, D. J. INGRAM, and R. C. LIEF. 1976. Detection of deoxyribonucleic acid replication in single cells and chromosomes by immunologic techniques. *J. Histochem. Cytochem.* **24:**34–39.

KATO, H. 1974. Spontaneous sister chromatid exchanges detected by a BUDR-labelling method. *Nature* (*Lond.*). **251:**70–72.

KORENBERG, J., and E. FREEDLENDER. 1974. Giemsa technique for detection of sister chromatid exchanges. *Chromosoma (Berlin).* **48:**355–360.

LATT, S. A. 1973. Microfluorometric detection of DNA replication in human metaphase chromosomes. *Proc. Natl. Acad. Sci. U. S. A.* **70:**3395–3399.

LATT, S. A. 1974. Sister chromatid exchanges, indices of human chromosome damage and repair: detection by fluorescence and induction by mitomycin C. *Proc. Natl. Acad. Sci. U. S. A.* **71:**3162–3166.

LATT, S. A. 1975. Fluorescence analysis of late DNA replication in human metaphase chromosomes. *Somat. Cell Genet.* **1:**293–321.

LATT, S. A. 1976. Optical studies of metaphase chromosome organization. *Annu. Rev. Biophys. Bioeng.* **5:**1–37.

LATT, S. A. 1977 Fluorescent probes of DNA microstructure and synthesis. *In* Flow Cytometry. M. Melamed, P. Mullaney, and M. Mendelsohn, editors. John Wiley & Sons, Inc., New York. In press.

LATT, S. A., J. W. ALLEN, W. E. ROGERS, and L. A. JUERGENS. *In vitro* and *in vivo* analysis of sister chromatid exchange formation. *Mutat. Res.* In press.

LATT, S. A., R. L. DAVIDSON, M. S. LIN, and P. S. GERALD. 1974. Lateral asymmetry in the fluorescence of human Y chromosomes stained with 33258 Hoechst. *Exp. Cell Res.* **87:**425–429.

LATT, S. A., and L. A. JUERGENS. 1976. Determinants of sister chromatid exchange frequencies in human chromosomes. *In* Population Cytogenetics: Studies In Humans. E. Hook and I. Porter, editors. Academic Press, Inc. New York. 217–236.

LATT, S. A., G. STETTEN, L. A. JUERGENS, G. R. BUCHANAN, and P. S. GERALD. 1975. The induction by alkylating agents of sister chromatid exchanges and chromatid breaks in Fanconi's anemia. *Proc. Natl. Acad. Sci. U. S. A.* **72:**4066–4070.

LATT, S. A., H. F. WILLARD, and P. S. GERALD. 1976. BrdU-33258 Hoechst analysis of DNA replication in human lymphocytes with supernumerary or structurally abnormal X chromosomes. *Chromosoma (Berlin).* **57:**135–153.

LATT, S. A., and J. C. WOHLLEB. 1975. Optical studies of the interaction of 33258 Hoechst with DNA, chromatin, and metaphase chromosomes. *Chromosoma (Berlin).* **52:**297–316.

LIN, M. S., S. A. LATT, and R. L. DAVIDSON. 1974. Microfluorometric detection of asymmetry in the centrometric region of mouse chromosomes. *Exp. Cell Res.* **86:**392–395.

MOORE, P. D., and R. HOLLIDAY. 1976. Evidence for the formation of hybrid DNA during mitotic recombination in Chinese hamster cells. *Cell.* **8:**573–579.

PERRY, P., and H. J. EVANS. 1975. Cytological detection of mutagen-carcinogen exposure by sister chromatid exchange. *Nature (Lond.).* **258:**121–124.

PERRY, P., and S. WOLFF. 1974. New Giemsa method for the differential staining of sister chromatids. *Nature (Lond.)* **251:**256–258.

STETTEN, G., S. A. LATT, and R. L. DAVIDSON. 1976. 33258 Hoechst enhancement of the photosensitivity of bromodeoxyuridine-substituted cells. *Somat. Cell Genet.* **2:**285–290.

TAICHMAN, L., and E. FREEDLENDER. 1976. Separation of chromatin containing bromodeoxyuridine in one or both strands of the DNA. *Biochemistry.* **15:**447–451.

TAYLOR, J. H., P. S. WOODS, and W. L. HUGHES. 1957. The organization and duplication of chromosomes as revealed by autoradiographic studies using tritium-labeled thymidine. *Proc. Natl. Acad. Sci. U. S. A.* **43:**122–128.

WILLARD, H. F., and S. A. LATT. 1976. Analysis of deoxyribonucleic acid replication in human X chromosomes by fluorescence microscopy. *Am. J. Hum. Genet.* **28:**213–227.

VOGEL, W., and T. BAUKNECHT. 1976. Differential chromatid staining by *in vivo* treatment as a mutagenicity test system. *Nature (Lond.).* **260:**448–449.

Viral Gene Function in Cell Transformation

THE GENETIC STRUCTURE OF TUMOR VIRUSES

DANIEL NATHANS

To introduce this symposium on viral gene function in cell transformation, I would like to make a few remarks about viruses that transform cells to tumorigenicity and the structure of their genomes. Nearly all types of animal viruses with duplex DNA genomes intracellularly are potentially tumorigenic, regardless of the extracellular form of viral nucleic acid. These include almost every type of DNA virus and that class of RNA viruses with reverse transcriptase. The two essential features of transforming viruses are a specific gene or genes necessary for transformation and an ability to recombine with cellular DNA leading to integration and, therefore, stable inheritance of transforming gene(s).

The best studied examples of DNA and RNA tumor viruses are simian virus 40 (SV40) and Rous or avian sarcoma virus (ASV), respectively (for recent reviews, see Kelly and Nathans, 1976, and Bishop and Varmus, 1975). SV40 is one of the small papovaviruses, made up of a simple icosahedral capsid and a genome that is about 3 million daltons, i.e., about 5,000 nucleotide pairs, present in the virus as a nucleohistone or minichromosome. Aside from cellular histones, the virus particle has three proteins: the major capsid protein, called VP1, and two minor related proteins, VP2 and VP3. In contrast to this, the more complex and larger ASV particle has a core of ribonucleoprotein surrounded by capsid proteins and a lipoprotein envelope. Within the core is the RNA genome, consisting of two identical 35S single-stranded linear RNA molecules, each of about 3 million daltons or around 10,000 nucleotides, i.e., roughly twice the genetic content of SV40. Also within the core are small primer RNA molecules, the RNA to DNA polymerase that can produce a duplex DNA copy of the 35S RNA, and four or five different proteins that contain group-specific antigens and are probably all derived from a large common precursor. Within the envelope are two virus-specified glycoproteins.

How many genes do SV40 and ASV have, and which of them are involved in transformation? Figure 1 shows a map of the SV40 genome, constructed by restriction enzyme analysis and by mapping mRNAs and mutants with defined physiological defects (see review by Kelly and Nathans, 1976). Three genes have been identified, two of which (D and B/C genes) code for virion proteins (VP1, 2, and 3) and one (A gene) codes for the A protein or T antigen. As discussed by Tegtmeyer (this volume), temperature-sensitive and deletion mutants of the A gene are defective in viral DNA replication and in transformation, and the A gene segment shown in Fig. 1, together with small contiguous segments at each end, has been shown to be sufficient for cell transformation. Moreover, active A protein is needed to maintain the transformation phenotype, as inferred from the growth properties of tsA mutant-transformed cells. Therefore, the A gene appears to be the SV40 transforming gene.

A map of the genome of nondefective ASV is shown in Fig. 2, deduced from physical mapping of recombinants (Duesberg et al., 1976). So far, four genes have been identified within the viral 35S RNA molecule, one of which, (*gag* for group antigens), specifies the precursor to the core proteins; another, (*env*), the envelope glycoproteins and a third, (*pol*), the virion DNA polymerase. The fourth gene (*src*), whose protein product has not yet been identified, is present in transforming ASV, but is missing in nontransforming, viable mutants, thus defining the *src* gene (see Varmus et al., this volume). Furthermore, temperature-sensitive *src* mutants have been isolated in which transformation, but not virus production, is tem-

DANIEL NATHANS Department of Microbiology, Johns Hopkins University School of Medicine, Baltimore, Maryland

perature-dependent. As in SV40, then, a specific transforming gene is present, the product of which appears to be continuously needed for transformation. Unlike SV40, however, the transforming gene of ASV is not needed for virus production, and may have been acquired from cellular DNA by recombination (see Varmus et al., this volume).

What emerges from this brief analysis of SV40 and ASV genes (and also by similar analysis of the more complex DNA genome of adenovirus; see Sambrook et al., this volume) is summarized in Fig. 3. Tumor viruses contain vegetative genes needed for virus replication, and among these

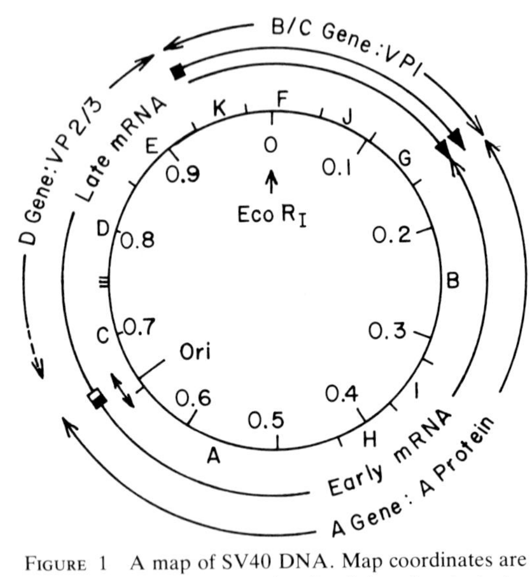

FIGURE 1 A map of SV40 DNA. Map coordinates are given inside the circle as fractional lengths clockwise from the Eco RI cleavage site. *F*, *J*, *G* etc., are fragments produced by Hind II + III restriction endonucleases. *Ori* is the origin of DNA replication, which is bidirectional. For other symbols, see the text (adapted from Kelly and Nathans, 1976).

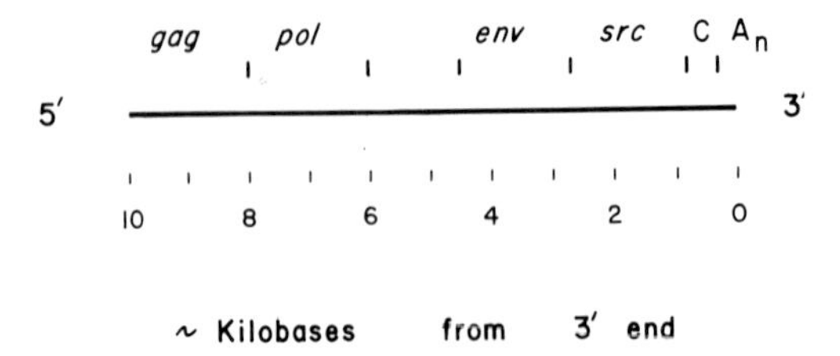

FIGURE 2 A map of ASV RNA. A_n is polyadenylate at the 3′ end of the RNA; *C* is a constant sequence shared by different strains of ASV. Other symbols are described in the text (adapted from Duesberg et al., 1976).

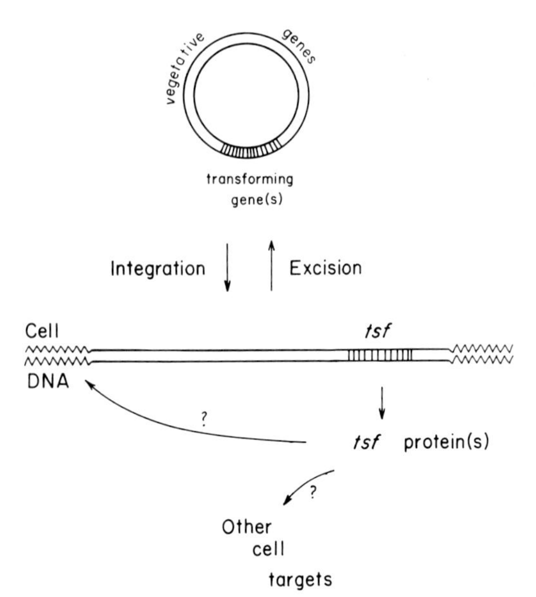

FIGURE 3 A general picture of integration and expression of transforming (*tsf*) genes of SV40 and ASV. Circle at the top is viral or proviral DNA.

genes, or in addition to them, is one (or perhaps more than one in some cases) transforming gene. In all cases, cellular inheritance of the transforming gene is assured by its integration into cellular DNA, with or without other, vegetative genes. Integrative recombination with cellular DNA, therefore, appears essential for stable transformation, but it must occur in such a way that the transforming gene(s) are transcribed into active messenger RNA for transforming protein or proteins (see Sambrook et al., this volume). How does integration occur? Does it require viral gene products or only cellular enzymes? How is the expression of integrated transforming genes regulated? How do transforming proteins lead to transformation? Do they act directly on cellular DNA to de-repress or repress a series of genes related to growth control? Do they initiate cellular DNA replication at unusual sites? Do they interact with other cellular components? What are the biochemical pathways to transformation? The title of this symposium notwithstanding, these are questions largely for the future.

REFERENCES

BISHOP, J. M., and H. E. VARMUS. 1975. The molecular biology of RNA tumor viruses. *In* Cancer: A Comprehensive Treatise. Vol. 2. I. F. Becker, editor.

Plenum Press, New York. 3–48.
DUESBERG, P. H., L. H. WANG, P. MELLOW, W. S. MASON, and P. K. VOGT. 1976. Towards a complete genetic map of Rous sarcoma virus. *In* Animal Virology, ICN-UCLA Symposium on Molecular and Cellular Biology. D. Baltimore, A. S. Huang, and C. F. Fox, editors. Academic Press, Inc., New York. 107–125.
KELLY, T. J., JR., and D. NATHANS. 1976. The genome of Simian Virus 40. *Adv. Virus Res.* **21:**86–173.

SV40 GENE A FUNCTION IN THE VIRAL GROWTH CYCLE AND TRANSFORMATION

PETER TEGTMEYER

Infection by SV40 may lead to a complete viral growth cycle and death of the permissive host cell or to a limited expression of the viral genome and transformation of the restrictive host cell. The covalent integration of viral DNA into the DNA of transformed cells without interruption of the early region of SV40 DNA insures the stable expression of early but not late viral genes (Botchan et al., 1976). This fact strongly suggests that one or more early genes are required to maintain the new growth characteristics of the altered cells. A complete understanding of the early molecular events in productive infection might suggest the proper approaches to identify the transforming events which are more difficult to study at present.

The Viral Growth Cycle

In productive infection, leading to the synthesis of infectious progeny, the total expression of viral function occurs in distinct stages regulated at the level of transcription (Khoury et al., 1972*a, b*; Sambrook et al., 1973). Early in infection, before the onset of DNA replication, stable RNA is transcribed from 45 to 50% of one strand of viral DNA. After the onset of viral DNA synthesis, stable late RNA is also transcribed from 50 to 55% of the opposite strand of viral DNA. The early genome codes for at least one protein, known as tumor (T) antigen or the A protein, that is also consistently present in SV40-transformed cells (Del Villano and Defendi, 1973; Prives et al., 1975; Roberts et al., 1975; Tegtmeyer et al., 1975; Ahmad-Zadeh et al., 1976). The late genome codes for structural proteins of the virus particle not found in transformed cells. Although synthesis of the A protein and viral DNA continue throughout productive infection, their rates of synthesis reach a peak long before cell death (Fig. 1). In contrast, the rate of synthesis of structural proteins remains at a high level for at least 72 h. Thus the viral growth cycle consists of a series of distinct events which are logical in sequence and limited in extent. In short, the cycle appears to be regulated in a coordinated way.

PETER TEGTMEYER Department of Microbiology, State University of New York, Stony Brook, New York

Autoregulation of Early Gene Expression

Studies of protein synthesis in cells productively infected by wild-type (WT) virus or temperature-sensitive (ts) mutants of SV40 show that the early A protein regulates its own synthesis (Fig. 2). After a 1-h exposure to [^{35}S]methionine at the restrictive temperature, cells infected by tsA mutants contain four- to eightfold more radiolabeled A protein than do cells infected by WT virus. The overproduction of the A protein, when the A gene is not functional, can be demonstrated in unpurified cell extracts or by immunoprecipitation of the A protein from these extracts using serum from hamsters bearing SV40-induced tumors (Fig. 2). Furthermore, pulse-chase studies (Tegtmeyer et al., 1975) show that the mutant A protein is more rapidly turned over than is the wild-type A protein. Taken together, these findings indicate that functional A protein regulates its own synthesis directly or indirectly. Recent studies by Reed et al. (1976) suggest that the autoregulation operates at the level of transcription. Both the rate of synthesis and the intracellular accumulation of early viral RNA are higher in cells infected by tsA mutants than in cells infected by wild-type virus at the restrictive temperature. Why control of early gene expression occurs is unknown. However, this type of regulation may favor an efficient balance be-

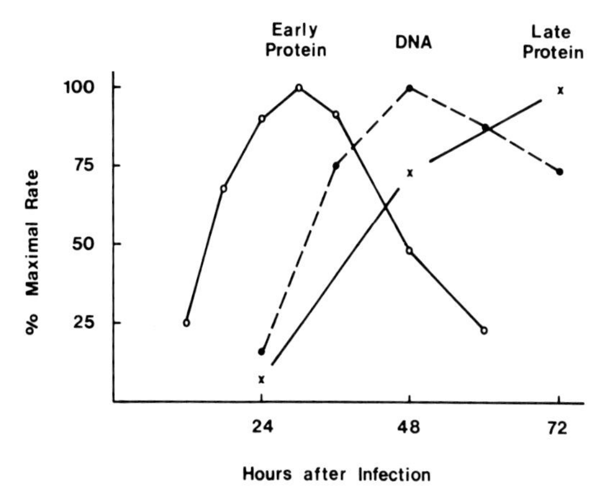

FIGURE 1 The viral growth cycle. Rates of synthesis of early protein, viral DNA and late protein were determined by radiolabeling infected cultures with [^{35}S]methionine or [^{3}H]thymidine for 1 h at various intervals after infection. After extraction, the protein or DNA was purified by electrophoresis and quantitated by liquid scintillation counting.

tween early and late functions in the synthesis of viral components.

Regulation of DNA Replication

The replication cycle of SV40 DNA has been well characterized and is summarized in Fig. 3 A. Initiation of synthesis of the closed circular DNA starts at a specific site located at 0.67 map units on a physical map of the genome (Nathans and Danna, 1972). Propagation proceeds bidirectionally and symmetrically (Fareed et al., 1972). After segregation of the relaxed daughter molecules, the DNA is restored to a superhelical configuration (Fareed et al., 1973). Thus, each round of replication can be divided into at least four distinct stages: initiation, propagation, segregation, and completion. The duration of the entire process lasts 10–15 min for the average molecule. Most of this time is required for propagation.

Continuous infection with tsA mutants at the restrictive temperature completely blocks viral DNA replication. The specific stage at which DNA replication is defective was localized by temperature shift experiments in which the A function was blocked after the onset of DNA replication (Tegtmeyer, 1972; Chou et al., 1974; Fig. 3 B). Replicating DNA in cells infected by WT or mutant virus was labeled with radioactive precursors at the permissive temperature immediately before a shift to the restrictive temperature. The fate of labeled replicative intermediates (RI) was followed after the shift to the restrictive temperature. Mutant replicating DNA was converted into mature form I DNA at the same rate as WT replicating DNA, thereby excluding a block in propagation, segregation, or completion of the cycle. In contrast, the radiolabeling of mutant and WT DNA immediately after the temperature shift showed that no new rounds of viral DNA replication were initiated in mutant infection. Thus the A function appears to be required exclusively for the initiation of viral DNA replication. In view of the

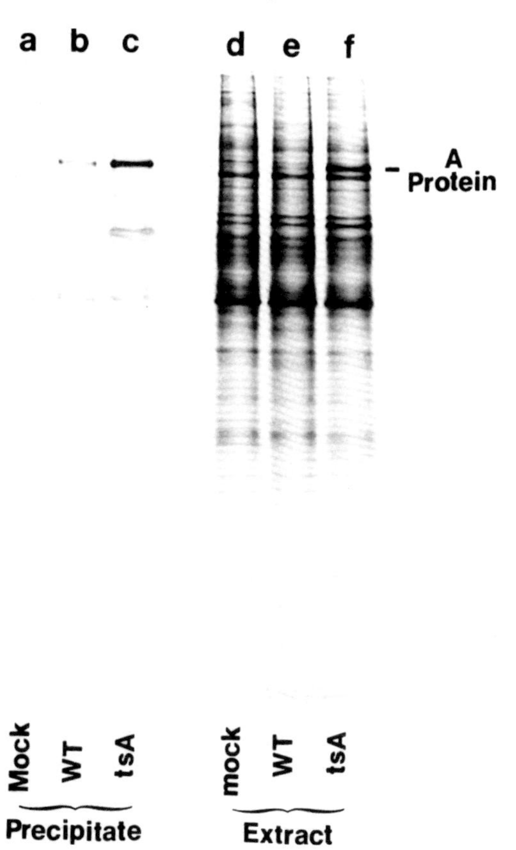

FIGURE 2 Overproduction of the A protein in infection by tsA mutants. Proteins in control or infected cells were labeled for 1 h with [^{35}S]methionine. The radioactivity incorporated into the A protein was determined by gel electrophoresis of cell extracts with or without purification by immunoprecipitation with serum prepared against SV40-induced tumors. The sample order is: (*a*) immunoprecipitated proteins from control cells, (*b*) immunoprecipitated proteins from cells infected by WT virus, (*c*) immunoprecipitated proteins from cells infected by a tsA mutant, (*d*) extract from control cells, (*e*) extract from cells infected by WT virus, and (*f*) extract from cells infected by a tsA mutant.

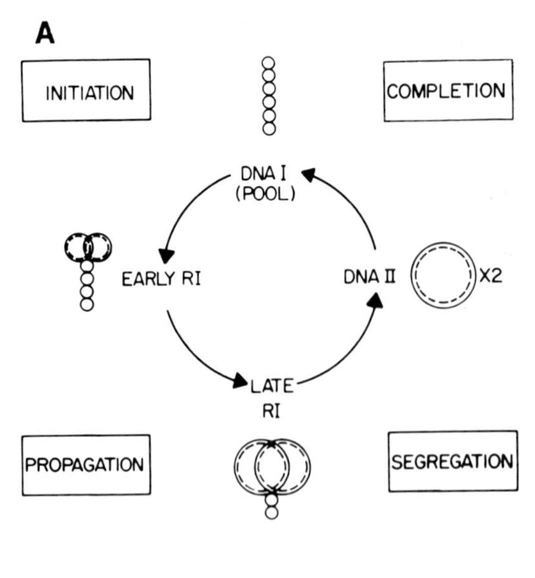

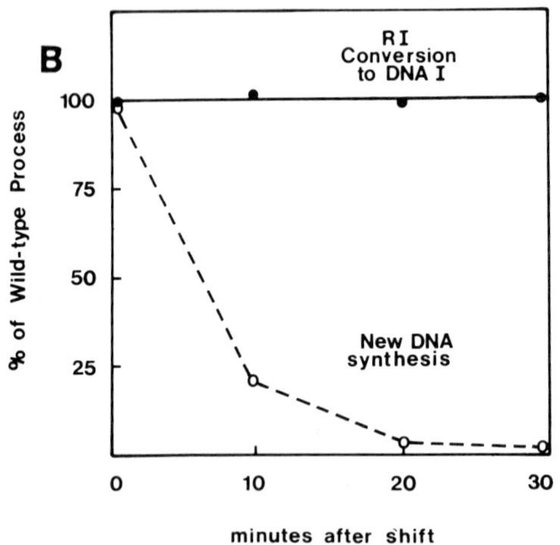

FIGURE 3 (A) Diagram summarizing the steps in the cycle of SV40 DNA replication at which the cycle could be blocked by a shift of cells infected by tsA mutants to the restrictive temperature. (B) Summary of data showing that the replication cycle is blocked exclusively at the initiation step. The conversion of replicative intermediates (*RI*) to mature DNA I after a temperature block was determined by pulse labeling infected cells with [^{3}H]thymidine for 10 min, shifting the cells to the restrictive temperature, and chasing for increasing periods of time. The labeled products were extracted and analyzed by gel electrophoresis. The new synthesis of viral DNA after a temperature block was determined by pulse labeling infected cells for 10-min periods after a shift of the cells from the permissive to the restrictive temperature. The residual incorporation of isotope into mutant DNA during the first 10 min after the shift is accounted for by rounds of DNA replication which had been initiated immediately before the temperature block.

limited coding capacity of the virus, the host presumably supplies functions required after the initiation step. How the A protein acts is not known in detail, but binding of the protein to viral DNA at a specific site appears to be a necessary step.

Regulation of Late Gene Expression

In cells infected with tsA mutants at the restrictive temperature, late viral RNA and proteins are not synthesized. When infected cells are incubated at the permissive temperature, viral DNA synthesis takes place and late viral RNA is transcribed. If the cells are then shifted to the restrictive temperature, viral DNA replication stops rapidly but synthesis of late mRNA continues (Cowan et al., 1973; Reed et al., 1976). Thus, the A protein is required to initiate but not to maintain the late phase of viral transcription. The mechanism whereby transient action of the A protein is sufficient to establish steady transcription of late viral genes is unknown. It seems most likely that cells infected by tsA mutants at permissive temperatures accumulate templates for late transcription that remain active after a shift to restrictive conditions. The nature of these viral transcription complexes remains unclear, although evidence suggests that late viral RNA is transcribed, at least in part, from free nonintegrated viral DNA (Gariglio and Mousset, 1975).

Transformation

During continuous incubation at the restrictive temperature, the tsA mutants of SV40 are uniformly defective in establishing the transformation of cells of several species, including rat, mouse, hamster, rabbit, and man. When cells are first transformed at the permissive temperature and then shifted to the restrictive temperature, the new growth properties of most, but not all, of the transformed cell lines are temperature-sensitive (Brugge and Butel, 1975; Kimura and Itagaki, 1975; Martin and Chou, 1975; Osborn and Weber, 1975; Tegtmeyer, 1975; W. W. Brockman, personal communication). The data showing the requirement for the A function of SV40 for maintaining two of the growth properties of transformed cells are summarized in Table I. Most cell lines transformed by tsA mutants at the permissive temperature lose their capacity to grow to high saturation densities and to form colonies in dilute suspensions after a shift to the restrictive temperature. Thus, the transformed cell seems dependent on the A gene product to maintain a full-blown transformed phenotype, but not all cell lines are dependent for reasons that are not entirely clear.

Discussion

In productive infection, the A protein initiates viral DNA replication, regulates its own synthesis, apparently by inhibiting early transcription, and initiates late viral transcription. The requirement for A function in DNA replication and early transcription appears to be continuous and direct, because both processes are rapidly affected by a shift from the permissive to the restrictive temperature in infection by ts mutants. The requirement of A function for late transcription is transient because a temperature shift has no effect after initiation.

As shown in Fig. 4, viral DNA replication originates at a specific site and then proceeds bidirectionally and symmetrically. Although the locations of promoters for early and late transcription have not yet been identified, the 5′ ends of cytoplasmic early and late viral RNA correspond to a site on the DNA at or near the origin of DNA replication (Khoury et al., 1975). Furthermore, the A protein of SV40 binds preferentially to SV40 DNA in vitro at the same area of the genome (Carroll et al., 1974; Reed et al., 1975; Jessel et al., 1976). The localization of the A protein binding site to a region of viral DNA that is functionally important for both DNA replication and transcription supports the idea that the A protein may regulate these processes in a direct and coordinated way. The activity of the A protein after binding to viral DNA is not known.

The requirement of the A function for the initiation and possibly the maintenance of the transformed state further raises the question of the interaction of the A protein with cellular DNA resulting in altered patterns of cellular replication or transcription. Clearly, other interactions of the A protein with transformed cells cannot be excluded.

TABLE I

Properties of Cells Transformed by SV40 at the Permissive Temperature and Shifted to the Restrictive Temperature

	Number of temperature-sensitive cell lines	
Transformed cells	Saturation density	Colony formation
WT virus	0/12	0/26
A mutants	11/11	24/34

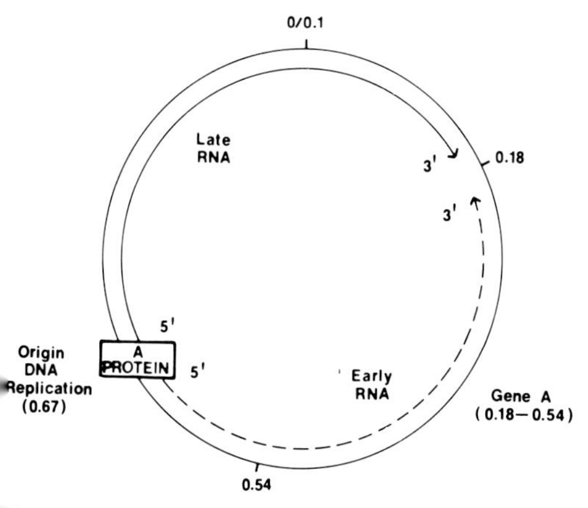

FIGURE 4 Diagram showing the origin of DNA replication, the site on DNA corresponding to the 5′ ends of early and late stable viral RNA, and the preferred binding site of the A protein at 0.67 map units from the Eco RI endonuclease cutting site.

ACKNOWLEDGMENTS

This investigation was supported by grant PRA-113 from the American Cancer Society, grant 1256 from the Damon Runyon Fund, and Public Health Service grant CA16497 from the National Cancer Institute.

REFERENCES

AHMAD-ZADEH, C., B. ALLET, J. GREENBLATT, and R. WEIL. 1976. Two forms of simian-virus-40-specific T-antigen in abortive and lytic infection. *Proc. Natl. Acad. Sci. U. S. A.* **73:**1097–1101.

BOTCHAN, M., W. TOPP, and J. SAMBROOK. 1976. The arrangement of simian virus 40 sequences in the DNA of transformed cells. *Cell.* **9:**269–287.

BRUGGE, J. S., and J. S. BUTEL. 1975. Involvement of the simian virus 40 gene A function in the maintenance of transformation. *J. Virol.* **15:**619–635.

CARROLL, R. B., L. HAGER, and R. DULBECCO. 1974. SV40 T antigen binds to DNA. *Proc. Natl. Acad. Sci. U. S. A.* **71:**3754–3757.

CHOU, J. Y., J. AVILA, and R. G. MARTIN. 1974. Viral DNA synthesis in cells infected by temperature-sensitive mutants of simian virus 40. *J. Virol.* **14:**116–124.

COWAN, K., P. TEGTMEYER, and D. D. ANTHONY. 1973. Relationship of replication and transcription of simian virus 40 DNA. *Proc. Natl. Acad. Sci. U. S. A.* **70:**1927–1930.

DEL VILLANO, B. C., and V. DEFENDI. 1973. Characterization of SV40 T antigen. *Virology.* **51:**34–46.

FAREED, G. C., C. F. GARON, and N. P. SALZMAN. 1972. Origin and direction of simian virus 40 deoxyribonucleic acid replication. *J. Virol.* **10:**484–491.

FAREED, G. C., C. F. GARON, and N. P. SALZMAN. 1973. Characterization of simian virus 40 DNA component II during viral DNA replication. *J. Mol. Biol.* **74:**95–111.

GARIGLIO, P., and S. MOUSSET. 1975. Isolation and partial characterization of a nuclear RNA polymerase-SV40 DNA complex. *FEBS (Fed. Eur. Biochem. Soc.) Lett.* **56:**149–155.

JESSEL, D., T. LANDAU, J. HUDSON, T. LALOR, D. TENEN, and D. M. LIVINGSTON. 1976. Identification of regions of the SV40 genome which contain preferred SV40 T antigen binding sites. *Cell.* **8:**535–545.

KHOURY, G., J. C. BYRNE, and M. A. MARTIN. 1972*a*. Patterns of simian virus 40 DNA transcription after acute infection of permissive and nonpermissive cells. *Proc. Natl. Acad. Sci. U. S. A.* **69:**1925–1928.

KHOURY, G., J. C. BYRNE, K. K. TAKEMOTO, and M. A. MARTIN. 1972*b*. Patterns of simian virus 40 DNA transcription. II. In transformed cells. *J. Virol.* **11:**54–60.

KHOURY, G., P. HOWLEY, D. NATHANS, and M. MARTIN. 1975. Post-transcriptional selection of simian virus 40-specific RNA. *J. Virol.* **15:**433–437.

KIMURA, G., and A. ITAGAKI. 1975. Initiation and maintenance of cell transformation by simian virus 40: a viral genetic property. *Proc. Natl. Acad. Sci. U. S. A.* **72:**673–677.

MARTIN, R. G., and J. Y. CHOU. 1975. Simian virus 40 functions required for the establishment and maintenance of malignant transformation. *J. Virol.* **15:**599–612.

NATHANS, D., and K. J. DANNA. 1972. Specific origin in SV40 DNA replication. *Nat. New Biol.* **236:**200–202.

OSBORN, M., and K. WEBER. 1975. Simian virus 40 gene A function and maintenance of transformation. *J. Virol.* **15:**636–644.

PRIVES, C., H. AVIV, E. GILBOA, E. WINOCOUR, and M. REVEL. 1975. The cell-free translation of early and late classes of SV40 messenger RNA. *Colloq. Inst. Natl. Santé Rech. Med.* **47:**305–312.

REED, S. I., J. FERGUSON, R. W. DAVIS, and G. R. STARK. 1975. T antigen binds to simian virus 40 DNA at the origin of replication. *Proc. Natl. Acad. Sci. U. S. A.* 1605–1609.

REED, S. I., G. R. STARK, and J. C. ALWINE. 1976. Autoregulation of simian virus 40 gene A by T antigen. *Proc. Natl. Acad. Sci. U. S. A.* **73:**3083–3087.

ROBERTS, B. E., M. GORECKI, R. C. MULLIGAN, K. J. DANNA, S. ROZENBLATT, and A. RICH. 1975. Simian virus 40 DNA directs the synthesis of authentic viral polypeptides in a linked transcription-translation cell-free system. *Proc. Natl. Acad. Sci. U. S. A.* **72:**1922–1926.

SAMBROOK, J., B. SUGDEN, W. KELLER, and P. A. SHARP. 1973. Transcription of simian virus 40. III. Mapping of early and late species of RNA. *Proc. Natl. Acad. Sci. U. S. A.* **70:**3711–3715.

TEGTMEYER, P. 1972. Simian virus 40 deoxyribonucleic acid synthesis: the viral replicon. *J. Virol.* **10:**591–598.

TEGTMEYER, P. 1975. Function of simian virus 40 gene A in transforming infection. *J. Virol.* **15:**613–618.

TEGTMEYER, P., M. SCHWARTZ, J. K. COLLINS, and K. RUNDELL. 1975. Regulation of tumor antigen synthesis by simian virus 40 gene A. *J. Virol.* **16:**168–178.

THE ARRANGEMENT OF VIRAL DNA SEQUENCES IN THE GENOMES OF CELLS TRANSFORMED BY SV40 OR ADENOVIRUS 2

JOE SAMBROOK, DENISE GALLOWAY, WILLIAM TOPP, and MICHAEL BOTCHAN

During the process of transformation by adenoviruses and SV40, rodent cells acquire new genetic information: viral DNA sequences recombine with those of the host and are then passed on to the cell's descendants like any other part of the cellular genome (Sambrook et al., 1968). The integrated sequences are transcribed into messenger RNA which is, in turn, translated. It is the presence of the resulting protein(s) that causes cells to display the panoply of characteristic phenotypic changes associated with the transformed state (for review, see Levine, 1976).

Rodent cells transformed by SV40 carry no trace of infectious virus. However, many lines of evidence indicate that they almost invariably contain the entire sequences of the viral DNA: if SV40-transformed cells are fused with uninfected permissive cells, infectious virus usually becomes detectable a day or so later (Watkins and Dulbecco, 1967; Koprowski et al., 1967); furthermore, virus replication has been shown to occur in permissive cells that have been exposed to high molecular weight DNA that has been extracted from transformed cells (Boyd and Butel, 1972). The presence of the total sequences of the viral genome is rather puzzling, for there are ample data available to show that only a subset of SV40 genes is required for transformation. First, the only viral transcripts detectable in many lines of transformed cells are derived entirely from the early region of the viral genome (Ozanne et al., 1973; Khoury et al., 1976). Second, cells transformed by temperature-sensitive mutants of the early gene of SV40 (ts A mutants) exhibit a partially temperature-sensitive phenotype (Brugge and Butel, 1975; Kimura and Hagaki, 1975; Tegtmeyer, 1975; Martin and Chou, 1975; Osborn and Weber, 1975). Third, transformation can be obtained by infecting cells with subgenomic fragments that span the entire early gene carrying transforming activity (Graham et al., 1974); however, no fragment that contains only a segment of the gene is able to transform cells. Thus it seems that expression of the single early gene of SV40 is necessary and sufficient for the establishment and maintenance of the transformed state.

By contrast to those transformed by SV40, rodent cells transformed by adenovirus 2 do not contain a complete set of viral genes. Experiments in which DNA extracted from transformed cells is hybridized to specific fragments of the viral genome show that different lines of rat cells transformed by adenovirus 2 carry different sets of viral DNA, with no cell line containing the entire genome (Sambrook et al., 1974). All cell lines, however, carry the segment of viral DNA that stretches from the left-hand end to a point about 14% along the viral genome. In many cases, these are the only viral sequences present. They must, therefore, code for any viral functions that may be required for the maintenance of the transformed state. It is hardly surprising that the techniques which are so successful with SV40 fail to elicit the production of infectious virus in cells transformed by adenovirus 2 (Dunn et al., 1973).

There is no fixed quantity of viral DNA in cells transformed by adenovirus 2 or SV40. In fact, the amount of viral DNA differs widely from cell line to cell line: some SV40 transformants contain as

JOE SAMBROOK and CO-WORKERS Cold Spring Harbor Laboratory, Cold Spring Harbor, New York little as one or two copies of viral DNA per diploid quantity of cell DNA (Gelb et al., 1971), others of indistinguishable phenotype carry as much as 8–10 copies (Ozanne et al., 1973). A similar situation obtains with cell lines transformed by adenovirus 2 (Sambrook et al., 1974). Furthermore there is no requirement that all segments of viral DNA carried by transformed cells be present in equimolar quantities. At least one line of SV40-transformed mouse cells contains six times as many copies of early viral gene sequences as late sequences (Botchan et al., 1974), and even more variation has been noted in the case of rodent cells transformed by adenovirus 2 (Sambrook et al., 1974).

In summary, integration is, in one sense, the central phenomenon of transformation of nonpermissive cells by adenovirus 2 and SV40: it provides simultaneously for the stable inheritance of viral genes and for their expression. Yet it is also a complicated matter, and our appreciation of the molecular events involved remains largely conjectural. There are two technical reasons for our ignorance. On the one hand, the detection of small quantities of viral DNA in transformed cells demands the application of stringent and sensitive hybridization techniques; on the other, the size of the mammalian genome is so vast and its organization so complex that few reliable methods have been available for its fractionation. Recently, both of these problems have been at least partially overcome. High molecular weight cellular DNA can be cleaved at specific sites with restriction endonucleases and the resulting fragments fractionated by electrophoresis through agarose gels denatured *in situ* and transferred to nitrocellulose sheets (Southern, 1975); highly purified viral DNA can be labeled in vitro with ^{32}P to specific activities greater than 10^8 cpm/μg (Rigby et al., 1977). The presence of as little as 10^{-13} g DNA of sequences among the different-sized fragments of transformed cell DNA can then be detected by hybridization and autoradiography of the nitrocellulose sheet. In this paper we apply these techniques to elucidate the arrangement of viral DNA sequences in rat cells transformed by adenovirus 2 or SV40.

The Arrangement of SV40 Sequences in the DNA of Transformed Cells

We have analyzed an isogenic set of 11 lines of Fisher rat cells, independently transformed either by particles of SV40 or by purified viral DNA. All are free of infectious virus and contain intranuclear T antigen and have been assayed by indirect immunofluorescence. Many of their biological properties have been described in detail elsewhere (Risser and Pollack, 1974). High molecular weight DNAs were purified from the cells and hydrolyzed to completion by various restricting endonucleases. The resulting fragments were separated by electrophoresis through agarose gels directly transferred to sheets of nitrocellulose and hybridized to ^{32}P-labeled SV40 DNA as described in the legend to Fig. 1. The results obtained with three enzymes, Bal I, Hha I, and Hpa I are shown in Figs. 1, 2, and 3.

Endonuclease Bal I does not cleave SV40 DNA at all (R. J. Roberts and B. S. Zain, unpublished results). Therefore, the total number of radioactive bands detected among the set of fragments of transformed cell DNA provides an estimate of the number of separate sites at which SV40 is integrated. Each radioactive band contains a sequence of SV40 DNA that, in the cell genome, was separated from other viral insertions by a tract of DNA which included at least one Bal I cleavage site. By this test, 7 of the 11 lines of SV40-transformed rat cells (clones 14C, 5, 13C,

FIGURE 1 Detection of fragments of DNA that contain SV40 sequences after hydrolysis of SV40-transformed rat cell DNAs with endonuclease Bal I. Separate preparations of DNA were used in the experiments shown in Figs. 1*a* and *b*. The products of digestion were fractionated by electrophoresis through 0.7% agarose gels for 15 h at a potential of 1.5 V/cm, transferred to nitrocellulose sheets and hybridized to ^{32}P-labeled SV40 DNA as described in detail elsewhere (Botchan et al., 1976). The nitrocellulose sheet was placed in contact with Kodak No-Screen Film for a period of 6 days. In this experiment and in those shown in Figs. 5 and 6, the slot labeled "control" contained 5 μg of DNA extracted from Fisher rat primary rat embryo cells. The slot labeled "SV40-reconstruction" contained 2 μg of adenovirus 2 DNA mixed with 10^{-6} μg of SV40 component II DNA. The obvious imperfections of the experiment stem from the difficulties imposed by the use of $>10^8$ cpm of ^{32}P-labeled SV40 DNA in the hybridization mixture and the problems attendant upon washing large sheets of nitrocellulose.

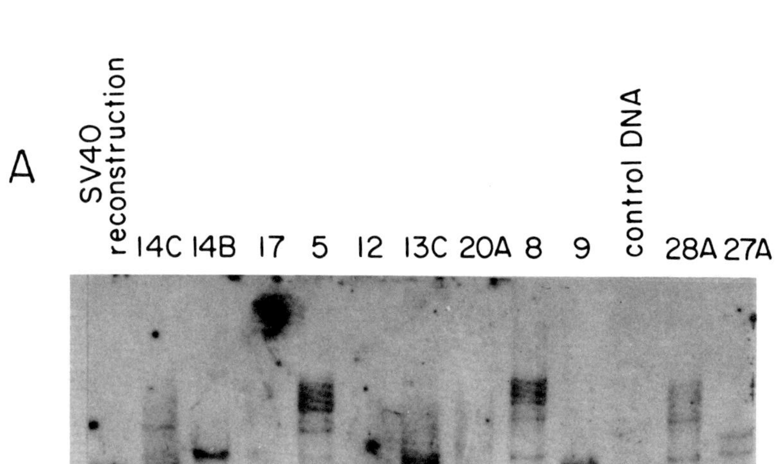
A
SV40 reconstruction
14C
14B
17
5
12
13C
20A
8
9
control DNA
28A
27A

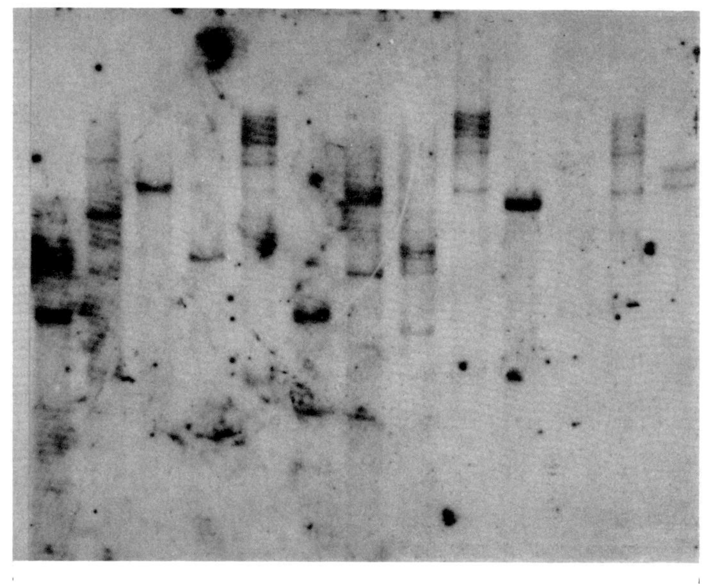

B

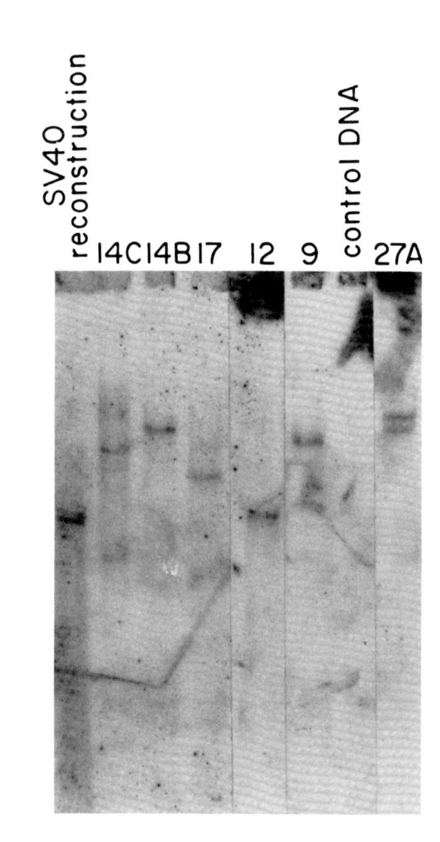
SV40 reconstruction
14C
14B
17
12
9
control DNA
27A

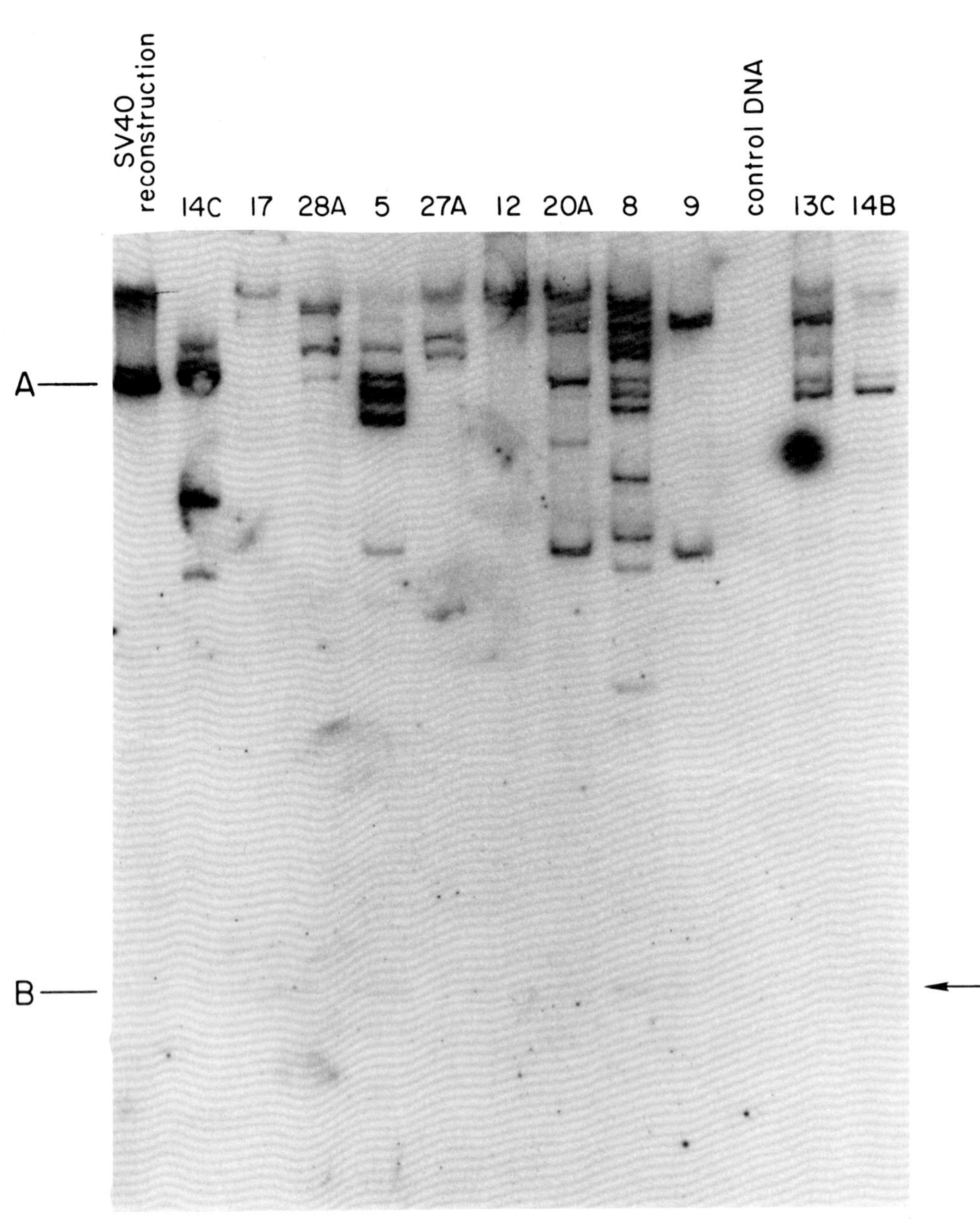

FIGURE 2 Detection of fragments of DNA that contain SV40 sequences after hydrolysis of SV40-transformed rat cell DNAs by endonuclease Hha I. The products of digestion were fractionated by electrophoresis through 1.4% agarose gels for 15 h at a potential of 1.5 V/cm, and analyzed for SV40 DNA sequences as described in detail elsewhere (Botchan et al., 1976). The autoradiograph shown was allowed to develop for a period of 20.5 days.

20A, 8, 28A, and 27A) contained more than one (and in one case at least six) separate insertions of viral DNA (see Fig. 1). The molecular weights of the bands, determined by comparing their electrophoretic mobilities with those of a reference set of fragments of adenovirus 2 DNA, differ from cell line to cell line and range from a minimum of 2×10^6 (equivalent in length to about 60% of the

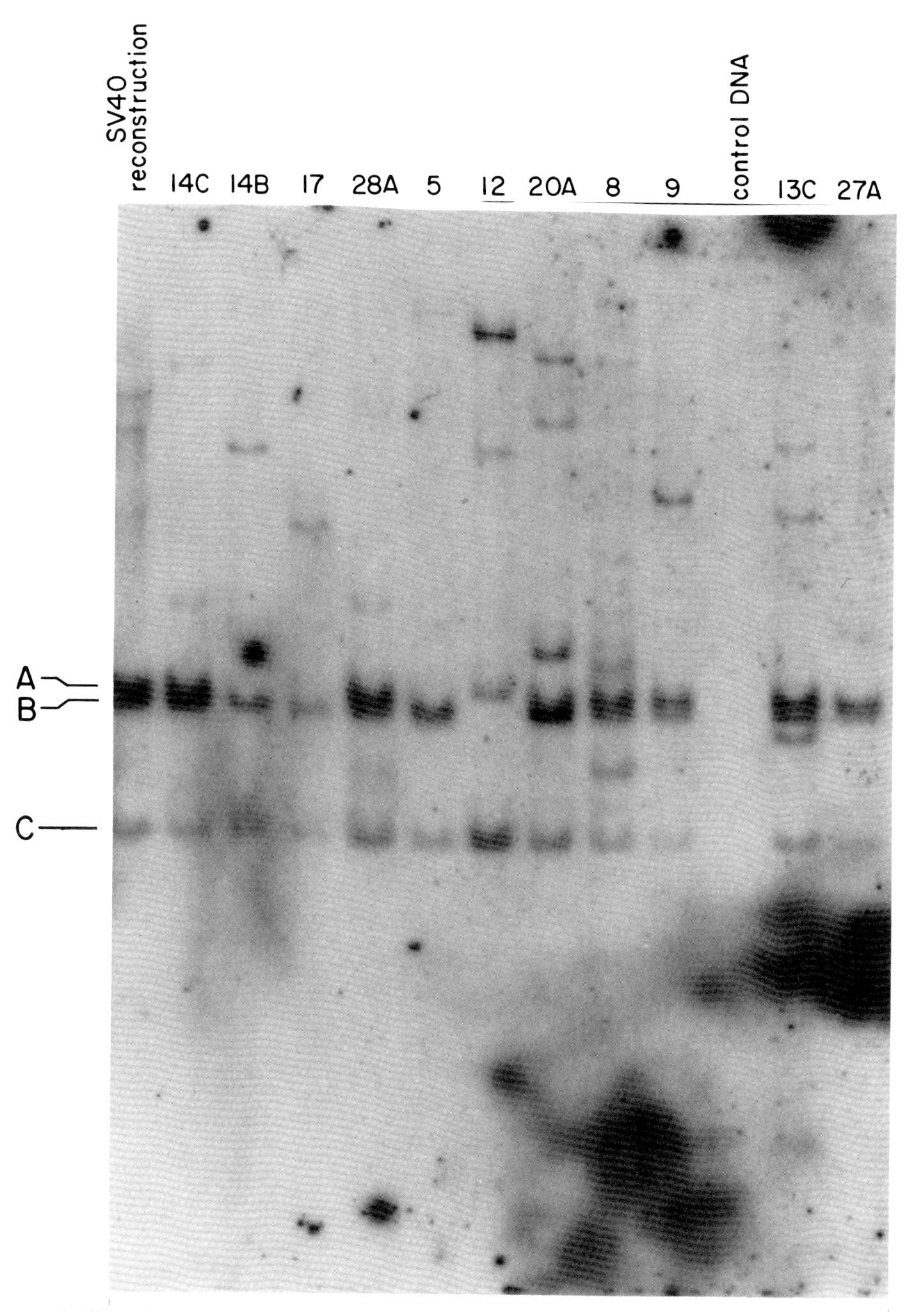

FIGURE 3 Detection of fragments of DNA that contain SV40 sequences after hydrolysis of SV40-transformed rat cell DNA by endonuclease Hpa I. The products of digestion were analyzed as described in the legend to Fig. 2. The autoradiograph shown was developed for 3.6 days.

SV40 genome) to a maximum several times the size of the viral DNA. The DNAs of the remaining 4 lines (clones 14B, 17, 12, and 9) of transformed cells yield, after cleavage with Bal I, only one detectable band that hybridizes with SV40 DNA. Thus the genomes of cells of these lines appear to contain a single insertion of viral DNA. In no case, however, are the SV40 sequences present in DNA fragments of the same size. These data provide strong evidence that SV40 DNA can integrate at many sites into the rat cell genome and that integrated viral sequences are carried at different sites in different cell lines.

As expected, there was no hybridization detectable in the channel of the gel that contained DNA extracted from untransformed cells – in this case primary cultures of rat embryo cells. In previously published work (Botchan and McKenna, 1973), and in additional experiments not shown here, we have used mouse, human, and hamster cells as sources of control DNA: in no case was there detectable hybridization to ^{32}P-labeled SV40 DNA. A reconstruction experiment in which 10^{-6} μg of relaxed circular SV40 DNA was analyzed in the presence of 2 μg of unlabeled carrier adenovirus 2 DNA is shown in the channel labeled "SV40 reconstruction." After hybridization, a single band of radioactivity was detected at the position expected for component II SV40 DNA.

Figures 1*a* and 1*b* show analyses of preparations of DNA that were extracted from SV40-transformed rat cells after growth for different periods of time in our laboratory. At least three separate preparations of DNA of each cell line were made during the course of a year, and identical results were obtained with each preparation. We conclude that the chromosomal location(s) of SV40 DNA within any one cell line is stable.

Endonuclease Hha I cleaves SV40 DNA at two sites (Fig. 4) closely spaced within the late region of the viral genome to produce two fragments (Hha A and Hha B) that are 92% and 8%, respectively, of the viral DNA in length (K. N. Subramanian, S. M. Weissman, B. S. Zain, and R. J. Roberts, unpublished results). When the enzyme digests an integrated copy of the complete SV40 genome, three DNA fragments will be produced that contain viral sequences. Whether all of these are detectable in our assay will depend upon the distance between the point at which the SV40 genome is broken during integration and the nearest recognition site for endonuclease Hha I within the viral DNA sequences. If the point of insertion is within the sequences of Hha B, two of the three fragments produced would contain extremely small quantities of viral DNA and might easily escape notice. If the point of breakage is within the sequences of Hha A but close to one or the other of the cleavage sites, then two of the three fragments should contain sufficient SV40 sequences to permit their detection. Finally, if the

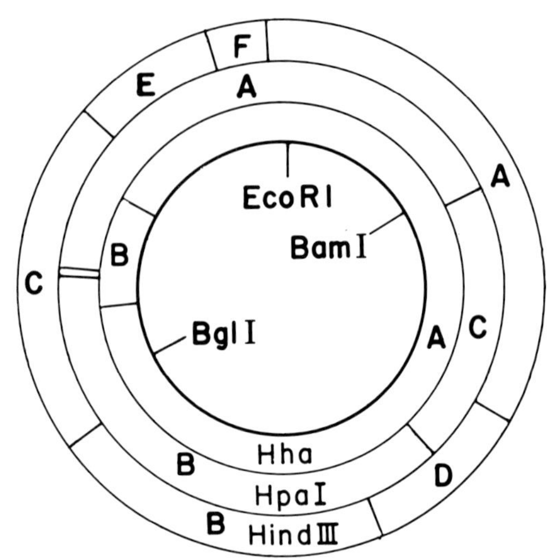

FIGURE 4 A map of the positions at which SV40 DNA is cleaved by several restriction enzymes. The data were taken from the following sources: Eco RI, Morrow and Berg (1972), Mulder and Delius (1972); Hpa I, Sharp et al. (1973); Hind III, Danna et al. (1973); Bam HI, Mathews and Sambrook (unpublished results); Bgl I and II, Zain and Roberts (unpublished results); Hha I, Subramanian, Zain, Roberts, and Weissman (unpublished results).

point of insertion lies some distance from either of the Hha sites, then all three fragments will be found. Although the situation will be complicated if partial or multiple copies of SV40 DNA are inserted into the cell's genome individually or in tandem, it nevertheless remains true that preparations of transformed cell DNA which contain single insertions of simple viral sequences should, after cleavage with endonuclease Hha I, yield between one and three fragments that hybridize with ^{32}P-labeled SV40 DNA; moreover, at least one of these fragments should comigrate through agarose gels with one of the two fragments obtained by digesting circular viral DNA with the enzyme.

When DNA isolated from SV40-transformed rat cells was cleaved with endonuclease Hha I and the location of fragments that contain SV40 DNA was determined by hybridization, the results shown in Fig. 2 were obtained. SV40 DNA sequences are seen in several major locations. In addition, minor bands (containing in total less than 5% of the hybridized radioactivity) were consistently found. We do not know the origin of these bands. They could result from microheterogeneity in and around the major set of integrated viral DNA sequences, from some looseness in the specificity of the restriction endonucleases, from methylation or other modification of cellular DNA, or from the insertion of a tiny part (<5%) of the SV40 genome.

The overall impression is again one of complexity—many of the clones of transformed cells yield large numbers of fragments that hybridize with SV40 DNA—and variety—there are few SV40-containing fragments that are common to more than one cell line. However, the patterns of hybridization are highly consistent with those obtained after hydrolysis of the same set of cell DNAs with endonuclease Bal I. Thus the genome of every cell line that, on the basis of analysis with endonuclease Bal I, was thought to contain multiple insertions of SV40 DNA displays more than three bands of hybridization after cleavage with endonuclease Hha I. On the other hand, none of the four cell lines (clones 9, 12, 14B, and 17) that appear to contain a single insertion yields more than three bands that hybridize to ^{32}P-labeled SV40 DNA. From two of these four cell lines (clones 9 and 17), a band was detected whose electrophoretic mobility through an agarose gel corresponded to that of the small (8%) fragments obtained by cleavage of circular SV40 DNA with endonuclease Hha I. In the photograph (Fig. 2), the band is not visible in the digests of DNA of clones 12 and 14B. However, a longer autoradiographic exposure (not shown) of the nitrocellulose sheet revealed the presence of the small 8% fragment in digests of the DNAs of these cell lines. The molecular weights of the other SV40-containing bands were different in DNAs of the four cell lines: none of the bands comigrated through agarose gels with the Hha A (92%) fragment of SV40 DNA. These results show that the arrangement of integrated viral sequences varies from cell line to cell line.

Again no hybridization occurred between untransformed cell DNA and ^{32}P-labeled SV40. The slot marked "reconstruction" contains 10^{-6} μg of a mixture of relaxed circular, unit length linear, and Hha fragment A SV40 DNA.

Endonuclease Hpa I cleaves circular SV40 DNA at map positions 17, 36, and 76 (where there are two very closely spaced sites) to yield three large DNA fragments (A, B, and C) whose molecular weights are 1.5×10^6, 1.4×10^6, and 0.6×10^6 daltons, respectively (Sharp et al., 1973; see Fig. 4). By comparing the electrophoretic mobilities of these fragments with those of fragments in digests of transformed cell DNA which hybridize with ^{32}P-labeled SV40 DNA, it is possible to identify the segments of the viral genome that are present in the cells.

Six of the cell lines (clones 14C, 28A, 20A, 8, 13C, and 27A) that contain complex sets of viral sequences yield three DNA fragments that comigrate with fragments A, B, and C as well as an assortment of additional species that presumably contain both SV40 and host cell sequences (see Fig. 3). These data provide further evidence that viral DNA can be inserted into the genome of rat cells at several locations. However, they give no clue concerning the arrangement of viral sequences at these sites, and we do not know whether the three fragments (A, B, and C) of SV40 DNA present in the digest are derived from different insertions or whether they all come from a single insertion of tandemly duplicated viral DNA.

Three of the remaining cell lines yield only two DNA species that correspond in mobility to fragments obtained by hydrolyzing SV40 DNA with endonuclease Hpa I. Digests of DNA extracted from clones 14B and 17 contain fragments B and C; fragment A is missing. DNA extracted from clone 12 yields fragments A and C; fragment B is missing. All three cell lines yield additional fragments, either one (clone 14B and 17) or two (clone 12), that hybridize with ^{32}P-labeled SV40 DNA and which are assumed to span the junction(s) between viral and host sequences. Presumably, this junction lies within the sequence of the fragment of viral DNA that is lacking from the cells. The data provide further evidence that these three lines of transformed cells carry a single insertion of viral DNA; and they suggest that the junction between host and viral DNA occurs at different locations in the viral genome.

The genome of the final cell line (clone 9) also contains a single insertion of SV40 (see Fig. 1). Digestion with endonuclease Hpa I (Fig. 3) yields (1) a DNA fragment of 8.6×10^6 daltons that probably contains both cellular and viral sequences, and (2) a set of fragments indistinguishable in their electrophoretic mobility from those derived from circular SV40 DNA. Therefore, we conclude that clone 9 cells contain a tandem repetition of viral DNA—a result that is entirely consistent with data obtained from restriction analysis with other endonucleases.

DNA extracted from the four lines of rat cells that contain a single insertion of SV40 DNA have been analyzed for viral sequences after digestion with four additional enzymes. Three of these, Eco RI, Bam HI, and Bgl I cleave SV40 DNA once (at positions 0, 14, and 67, respectively; see Figs. 4 and 5). In digests of transformed cell DNA of clones 9, 12, 14B, and 17, no more than two fragments were detected that hybridized with SV40 DNA – a result that can be explained only by the insertion of the viral genome at a single site.

Finally, endonuclease Hind III cleaves circular SV40 DNA at six sites to yield six fragments whose locations on the viral genome are shown in Fig. 4 and whose approximate molecular weights are 1.20×10^6 (fragment A); 0.76×10^6 (B); 0.70×10^6 (C); 0.38×10^6 (D); 0.28×10^6 (E); 0.22×10^6 (F); (Danna et al., 1973). Reconstruction experiments in which 10^{-6} μg of SV40 DNA was digested and analyzed in the presence of 2 μg of adenovirus 2 DNA showed that the hybridization technique was sensitive enough to detect the presence in diploid transformed cells of one copy of all except the smallest of these fragments. The location of SV40 sequences among the fragments produced by digesting the DNAs of clones 9, 12, 14B, and 17 with endonuclease Hind III are shown in Fig. 5. From all these data we can deduce the arrangement of viral DNA in the genomes of four lines of transformed cells (see Fig. 6). In constructing these maps, we have assumed that coidentity has been established when fragments of transformed cell DNA that hybridize with SV40 DNA are shown to migrate through agarose gels at the same rate as authentic fragments of viral DNA, obtained by digestion with the same restriction enzymes. For any individual fragment, this assumption is clearly fragile. But it is strengthened considerably by the results obtained with several enzymes that yield overlapping sets of fragments and is justified by the fact that we are able to deduce coherent maps which contain cleavage sites of several different restriction enzymes.

We draw the following conclusions from these experiments: (1) Many lines of SV40-transformed rat cells contain multiple insertions of SV40 DNA. By definition each insertion is separated from its neighbor by a tract of cellular DNA which contains at least one site that is cleaved by endonucleases which do not attack SV40 DNA. From the type of analysis used we cannot estimate the lengths of these tracts with accuracy. However, we know that they must be long, because the sum of the molecular weights of the DNA fragments which contain SV40 sequences is large. Very probably, then, these insertions result from integration events which occur independently of one another at distant sites in the cellular genome. There is no indication that the viral DNA has integrated at identical sites in different cell lines.

(2) Several lines of cells contain a single insertion of SV40 DNA sequences. In three cases (SVR12, SVR17, and SVR14B), these consist of a single copy of viral DNA that is unit length or close to it, with junctions between cell and viral sequences mapping at different places in the "late" region of the viral genome. The fourth cell line (SVR9) contains about 1.5 copies of viral DNA. One end of the viral array maps near the site at which DNA synthesis originates during the lytic infection; the other is located in the "early" region of the SV40 genome. This partial duplication of viral sequences provides an intact copy of the early gene, whose continued expression is necessary for maintenance of the transformed state. Thus, in each of four independently transformed cell lines, junctions between viral and cellular sequences map at different places on the viral genome: in one case even the two ends of a single viral insertion are not identical. From these results we conclude that there is no single position on the viral genome that serves as an obligatory site of connection of SV40 DNA to the cellular chromosome.

(3) There is within the rat genome no single site at which SV40 DNA resides. Not only do several cell lines contain more than one independent insertion of SV40 DNA but also those cell lines that contain single insertions of viral DNA carry them at different locations in their genomes. Thus, when the DNAs of these four cell lines are digested with endonuclease Bal I, SV40 sequences are found in fragments which differ in size from cell line to cell line. The molecular weights of these fragments which are composed partly of host and partly of viral sequences can be calculated from the mobility of the fragments through agarose gels. It is a simple matter to subtract the contribution of the viral DNA and show that the approximate molecular weights of the flanking host sequences are different for all four of the cell lines: we conclude that SV40 DNA sequences are present at different sites in the genomes of different cell lines (see Fig. 6). A similar conclusion can be drawn from arguments based on the molecular weights of DNA fragments that appear to consist

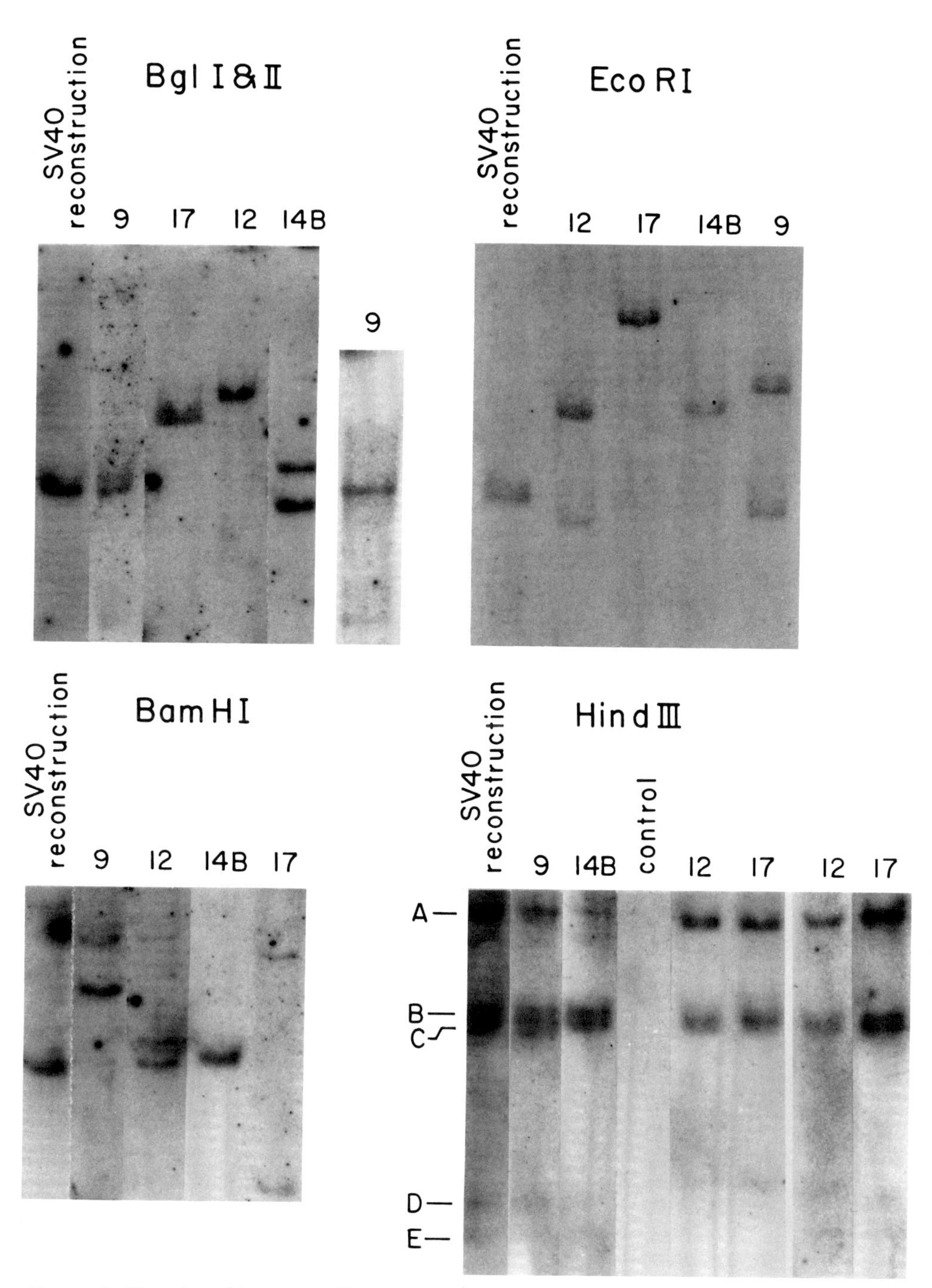

FIGURE 5 Detection of fragments of DNA that contain SV40 sequences after hydrolysis of four lines of SV40-transformed rat cells (clones 9, 12, 14B, and 17) by endonucleases Eco RI, Bam HI, Bgl I and II, and Hind III. The products of digestion were analyzed as described in the legend to Fig. 1. To illustrate points made in the text, two separate analyses are shown of the Bgl I and II digests of clone 9 DNA, and two different exposures are shown of the autoradiograph of the analysis by Hind III of clone 12 and clone 17 DNAs.

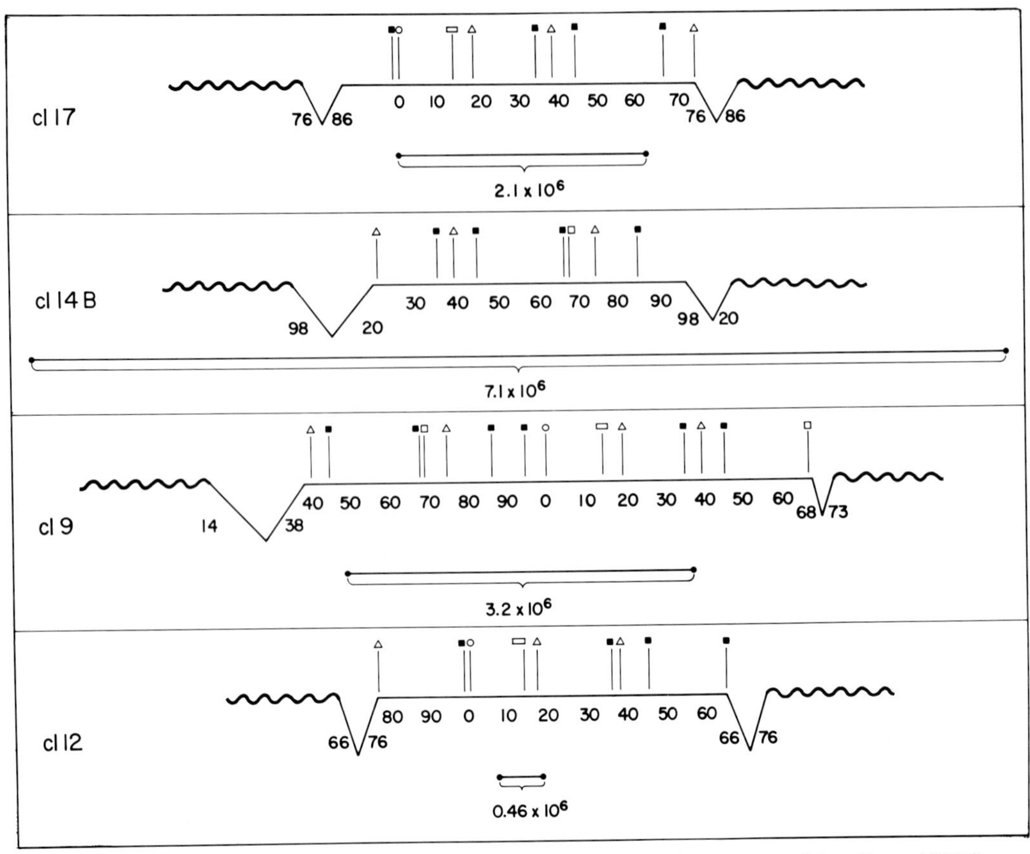

FIGURE 6 Restriction maps of viral DNA sequences integrated into the genomes of four lines of SV40-transformed rat cells. The maps were constructed as described in detail in the text. The wavy lines represent flanking cellular DNA sequences; the solid lines represent integrated SV40 DNA. The *Vs* show the regions on the SV40 map which contain the junctions between cellular and viral DNA sequences. The sites of cleavage within the integrated viral sequences were identified from the data shown in Figs. 2, 3, and 5. The symbols represent the various restriction enzymes used: Hind III (■); Hpa I (△); Eco RI (○); Bgl I (□); Bam HI (▭). The lines marked ●——● represent the length of the Bal I segment of cellular DNA into which the SV40 DNA is inserted. The molecular weights of these segments, calculated as described in the text, are given in daltons.

of host and viral sequences in preparations of cell DNA that have been treated with endonucleases Eco RI, Bam HI, Bgl I and II, and Hpa I.

The Arrangement of Adenovirus 2 DNA Sequences in the Genome of Transformed Cells

The three adenovirus 2 transformants examined are part of the extensive collection that Gallimore (1973) isolated after infection of Sprague-Dawley or Hooded Lister rat cells with virus particles. All are virus-free and contain intranuclear adenovirus T antigen. Detailed descriptions of their biological properties may be found in Gallimore (1973, 1974) and of their content of viral DNA in Sambrook et al. (1974).

Cells of the F18 and 2T4 lines contain, respectively, 2.9 and 6.9 copies of the left-hand 14% of the adenovirus genome; cells of the F4 line contain about 15 copies of the segment of the viral genome that extends from the left-hand terminus to about position 60, as well as a small stretch of sequences derived from the extreme right-hand end of the viral DNA (see Table 1). The location of adenovirus 2 sequences among the products of digestion of transformed cell DNA by restriction nuclease Hind III is shown in Fig. 7. To help

TABLE I

Viral DNA Sequences in Rat Cells Transformed by Adenovirus 2

Hpa I

4.0 24.2 26.5 58.8 86.4 98.6

Cell Line	E	C	F	A	B	D	G
F17	3.5	3.5 0.0	0.0	0.0	0.0	0.0	0.0
F18	2.9	2.9 0.0	0.0	0.0	0.0	0.0	0.0
2T4	6.9	6.9 0.0	0.0	0.0	0.0	0.0	0.0
F4	ND	15.1	ND	16.5	5.1	2.2	20.4

Eco RI

59.7 71.9 76.5 83.9 89.8

Cell Line	A	B	F	D	E	C
F17	<∿14%> 0.0	0.0	0.0	0.0	0.0	0.0
F18	<∿14%> 0.0	0.0	0.0	0.0	0.0	0.0
2T4	<∿14%> 0.0	0.0	0.0	0.0	0.0	0.0
F4	16.3	12.3	0.0	0.0	0.0	3.0

simplify the analysis, we have used as ^{32}P-labeled probe, the nick-translated DNA of Eco RI fragment A of adenovirus 2, which maps between positions 0 and 58 on the adenovirus 2 genome.

Adenovirus 2 DNA is cleaved at 11 sites by restriction endonuclease Hind III. Seven of the resulting 12 fragments map within Eco RI fragment A: these are shown in the slot labeled "reconstruction" on Fig. 7. When the products of digestion of F17 DNA by endonuclease Hind III are analyzed for their content of adenovirus DNA sequences, two bands are detected (Fig. 7). Neither of these comigrates with any of the authentic viral fragments obtained by cleavage of purified adenovirus 2 DNA with Hind III. Because the segment of viral DNA present in F17 cells contains only one cleavage site for endonuclease Hind III (at position 7.5), the results shown in Fig. 7 strongly indicate that there is a single site at which viral sequences are integrated.

The DNA of 2T4 cells yields, after cleavage with endonuclease Hind III, four bands which hybridize with ^{32}P-labeled Eco RI fragment A of adenovirus 2. Once again none of these comigrates with authentic Hind III fragments of adenovirus 2 DNA. Neither do they comigrate with fragments of F17 DNA that contain viral sequences. We conclude that adenovirus DNA sequences are integrated at two separate sites in the chromosome of 2T4 cells, and that neither of these is identical to the site of viral integration in F17 cells.

The DNA of F4 cells, after cleavage with Hind III, yields many fragments that contain adenovirus 2 DNA sequences. A comparison with the reconstruction experiments shows that bands are present in the digest of F4 DNA which comigrate with Hind III fragments C, B, I, J, and D of adenoviral DNA. Bands corresponding to Hind III fragments G and A are missing; instead, two novel bands are detected that do not comigrate with any known Hind III fragment of adenovirus 2 DNA. We

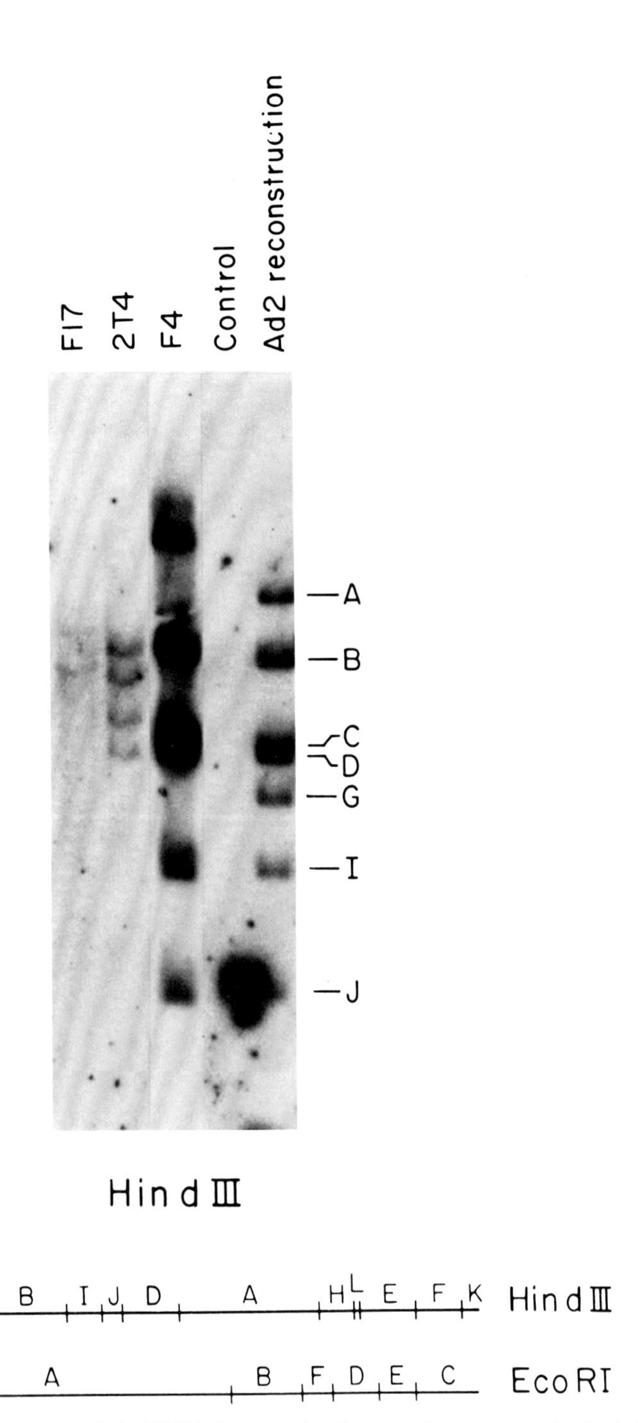

FIGURE 7 Detection of fragments of viral DNA that contain adenovirus 2 sequences after hydrolysis of transformed cell DNAs with endonuclease Hind III. Cell DNAs were prepared as described by Botchan and McKenna (1973) and analyzed as described in the legend to Fig. 1, except that the probe DNA was Eco RI fragment A prepared from adenovirus 2 DNA and nick-translated in vitro as described by Galloway and Sambrook (submitted for publication). The slot labeled "control" contained 5 μg of DNA extracted from Fisher rat primary embryo cells. The slot labeled "adenovirus 2 reconstruction" contained 5 μg of control DNA mixed with 10^{-6} μg of adenovirus 2 DNA.

conclude that the junctions between viral and cellular DNA sequences lie within Hind III fragments A and G, and that the novel bands detected in the digest of F4 DNA contain both viral and cellular sequences.

We have analyzed the DNAs of lines F17, 2T4, and F4 after cleavage with several other restriction endonucleases (Bal I, Eco RI, and Bam HI) whose sites of cleavage within purified adenovirus 2 DNA are known (Galloway and Sambrook, submitted for publication). The results obtained are entirely consistent with those shown here with endonuclease Hind III. They provide independent proof of our conclusions, but do not extend them in any significant way.

From these studies we conclude that (1) there is no single obligatory site for integration, because the viral sequences map in different locations in the genomes of different lines of transformed cells; (2) there are no strict limitations on the viral sequences that are retained because different lines of transformed cells contain different sets of viral DNA; (3) each cell line contains only a small number of insertions of viral DNA. Even in the case of F4 cells, which carry a large number of copies of viral sequences as measured by renaturation kinetics, the distribution of viral DNA among the various-sized restriction enzyme fragments is very simple. The best explanation of this paradox is that the portion of the cellular chromosome which contains viral DNA has been amplified differentially to produce a tandem array of identical repeating units consisting of both viral and cellular DNA sequences. However, we cannot at present rule out less exciting alternatives.

SUMMARY

Integration of adenovirus 2 and SV40 into the genome of mammalian cells is quite different from that of bacteriophages like λ into their prokaryotic hosts. It now seems quite likely that the cellular integration sites for these eukaryote viral DNAs occur at many places in the host genome and that the attachment site on the viral chromosome is nonspecific. Unless there is something special about their treatment of adenovirus and SV40 DNA, the possibility now becomes strong that mammalian cells may have the ability to integrate any piece of foreign DNA in a nonspecific fashion. In some cases this event would be lethal; in others, presumably the majority, no alteration in cellular phenotype would occur; in a small minority the integration event may become manifest as a consequence of a change in cellular behavior resulting from expression of the newly installed gene(s). Adenovirus and SV40 DNA fall into the last class: attached to the cellular chromosome perhaps by accident, they code for proteins that cause a vast and coordinate alteration in many aspects of the cellular phenotype. Thus what superficially may appear to be a masterpiece of genetic design may turn out to be an empirical policy of muddling through.

ACKNOWLEDGMENTS

We thank Ronni Greene for her kind gifts of restriction endonucleases.

This work was supported by a grant (CA 13106) from the National Cancer Institute.

REFERENCES

ALLET, B., and A. BUKHARI. 1975. Analysis of bacteriophage Mu and λ-Mu hybrid DNAs by specific endonucleases. *J. Mol. Biol.* **92:**529–540.

BOTCHAN, M., and G. MCKENNA. 1973. Cleavage of integrated SV40 by RI restriction endonuclease. *Cold Spring Harbor Symp. Quant. Biol.* **38:**391–395.

BOTCHAN, M., B. OZANNE, W. SUGDEN, P. A. SHARP, and J. SAMBROOK. 1974. Viral DNA in transformed cells. III. The amounts of different regions of the SV40 genome present in a line of transformed mouse cells. *Proc. Natl. Acad. Sci. U. S. A.* **71:**4183–4187.

BOTCHAN, M., W. TOPP, and J. SAMBROOK. 1976. The arrangement of Simian Virus 40 sequences in the DNA of transformed cells. *Cell.* **9:**269–287.

BOYD, V. A. L., and J. BUTEL. 1972. Demonstration of infectious deoxyribonucleic acid in transformed cells. *J. Virol.* **10:**399–409.

BRUGGE, J. S., and J. S. BUTEL. 1975. Role of Simian Virus 40 gene A: function in maintenance of transformation. *J. Virol.* **15:**619–635.

DANNA, K., G. J. SACK, and D. NATHANS. 1973. Studies of Simian Virus 40 DNA. VII. A cleavage map of the SV40 genome. *J. Mol. Biol.* **78:**363–376.

DUNN, A. R., P. H. GALLIMORE, K. W. JONES, and J. K. MCDOUGALL. 1973. *In situ* hybridization of adenovirus RNA and DNA. II. Detection of adenovirus-specific DNA in transformed and tumor cells. *Int. J. Cancer.* **11:**628–636.

GALLIMORE, P. H. 1973. Studies with adenovirus type 2: cellular transformation and oncogenicity. Ph.D. thesis, University of Birmingham, England.

GALLIMORE, P. H. 1974. Interactions of adenovirus type 2 with rat embryo cells: permissiveness, transformation and *in vitro* characteristics of adenovirus transformed rat embryo cells. *J. Gen. Virol.* **25:**263–273.

GELB, L. D., D. KOHNE, and M. A. MARTIN. 1971. Quantitation of Simian Virus 40 sequences in African green monkey, mouse and virus-transformed cell genomes. *J. Mol. Biol.* **57:**129–145.

Graham, F. L., P. J. Abrahams, C. Mulder, H. L. Heijneker, S. O. Warnaar, F. A. J. de Vries, W. Fiers, and A. J. van der Eb. 1974. Studies on *in vitro* transformation by DNA and DNA fragments of human adenoviruses and Simian Virus 40. *Cold Spring Harbor Symp. Quant. Biol.* **39:**637–650.

Khoury, G., B. J. Carter, F.-J. Ferdinand, P. M. Howley, M. Brown, and M. A. Martin. 1976. Genome localization of Simian Virus 40 RNA species. *J. Virol.* **17:**832–840.

Kimura, G., and A. Hagaki. 1975. Initiation and maintenance of cell transformation by Simian Virus 40: a viral genetic property. *Proc. Natl. Acad. Sci. U. S. A.* **72:**673–677.

Koprowski, H., F. C. Jensen, and Z. Steplewski. 1967. Activation of production of infectious tumor virus SV40 in hetero-karyon cultures. *Proc. Natl. Acad. Sci. U. S. A.* **58:**127–133.

Levine, A. J. 1976. SV40 and adenovirus early functions involved in DNA replication and transformation. *Biochim. Biophys. Acta.* **458:**213–241.

Martin, R. G., and J. Y. Chou. 1975. Simian Virus 40 functions required for the establishment and maintenance of malignant transformation. *J. Virol.* **15:**599–612.

Morrow, J., and P. Berg. 1972. Cleavage of Simian Virus 40 DNA at a unique site by a bacterial restriction enzyme. *Proc. Natl. Acad. Sci. U. S. A.* **69:**3365–3369.

Mulder, C., and H. Delius. 1972. Specificity of the break produced by restriction endonuclease RI in Simian Virus 40 DNA as revealed by partial denaturation mapping. *Proc. Natl. Acad. Sci. U. S. A.* **69:**3215–3219.

Osborn, M., and K. Weber. 1975. Simian Virus 40 gene A function and maintenance of transformation. *J. Virol.* **15:**636–644.

Ozanne, B., P. A. Sharp, and J. Sambrook. 1973. Transcription of Simian Virus 40. II. Hybridization of RNA extracted from different lines of transformed cells to the separated strands of Simian Virus 40 DNA. *J. Virol.* **12:**90–98.

Rigby, P., J. Rhodes, M. Dieckman, and P. Berg. 1977. In press.

Risser, R., and R. Pollack. 1974. A nonselective analysis of SV40 transformation of mouse 3T3 cells. *Virology.* **59:**477–489.

Sambrook, J., M. Botchan, P. Gallimore, B. Ozanne, U. Pettersson, J. Williams, and P. A. Sharp. 1974. Viral DNA sequences in transformed cells. *Cold Spring Harbor Symp. Quant. Biol.* **39:**615–632.

Sambrook, J., H. Westphal, P. R. Srinivasan, and R. Dulbecco. 1968. The integrated state of SV40 DNA in transformed cells. *Proc. Natl. Acad. Sci. U. S. A.* **60:**1288–1295.

Sharp, P. A., W. Sugden, and J. Sambrook. 1973. Detection of two restriction endonuclease activities in *Hemophilus parainfluenzae* using analytical agarose ethidium bromide electrophoresis. *Biochemistry.* **12:**3055–3063.

Southern, E. M. 1975. Detection of specific sequences among DNA fragments separated by gel electrophoresis. *J. Mol. Biol.* **98:**503–517.

Tegtmeyer, P. 1975. Function of Simian Virus 40 gene A in transforming infection. *J. Virol.* **15:**613–618.

Watkins, J. F., and R. Dulbecco. 1967. Production of SV40 virus in heterokaryons of transformed and susceptible cells. *Proc. Natl. Acad. Sci. U. S. A.* **58:**1396–1403.

THE FUNCTION AND ORIGIN OF THE TRANSFORMING GENE OF AVIAN SARCOMA VIRUS

HAROLD E. VARMUS, DEBORAH H. SPECTOR, DOMINIQUE STEHELIN, CHUN-TSAN DENG, THOMAS PADGETT, ELTON STUBBLEFIELD, and J. MICHAEL BISHOP

RNA tumor viruses are enveloped viruses widely distributed in nature, with a polyploid, single-stranded RNA genome which is replicated via a DNA intermediate. They are useful reagents for studies of oncogenesis, in view of the rapidity and efficiency with which they produce tumors in animals and transformation of cultured cells (Tooze, 1973; Gross, 1970; Bishop and Varmus, 1975). Transformation of fibroblasts by avian sarcoma virus (ASV) is probably directed by a single gene (*src*) which has been mapped near the 3′ terminus of ASV RNA (Duesberg et al., 1976). Although the function of the *src* gene has been extensively studied by the use of conditional and deletion mutants of ASV (Wyke et al., 1975; Vogt, 1977), the product of the gene, presumably a protein, has not yet been identified. We have prepared DNA complementary to a portion of the genome which includes some or all of the *src* gene (Stehelin et al., 1976*a*). This DNA ($cDNA_{sarc}$) serves as a molecular hybridization reagent with which to study the organization of the viral genome, the distribution and expression of the viral *src* gene in infected cells, the derivation of the gene from related ("sarc") sequences present in normal cells, the evolution of those sequences during speciation, and their expression in uninfected cells (Stehelin et al., 1976*b*). In this brief report, we summarize our recent progress with this work, providing references to more detailed accounts. In addition, more complete reviews of this work have recently appeared (Varmus et al., 1976; Bishop et al., 1976, 1977).

HAROLD E. VARMUS and CO-WORKERS Department of Microbiology, University of California, San Francisco, California

DOMINIQUE STEHELIN Institute National de la Santé et de la Recherche Médical, Paris, France

ELTON STUBBLEFIELD Department of Biology, University of Texas Cancer Center, Houston, Texas

Preparation and Characterization of $cDNA_{sarc}$

To prepare $cDNA_{sarc}$, we have exploited the availability of transformation-defective deletion mutants (tdASV) which have lost ca. 15% (1,600 nucleotides) of the ASV genome and are unable to transform fibroblasts or induce sarcomas in animals (Vogt, 1971; Duesberg and Vogt, 1970, 1973). Inasmuch as all conditional mutations in the *src* gene have been mapped in the deleted region (Bernstein et al., 1976), it is likely that the deletion is coextensive with most or all of the *src* gene. $cDNA_{sarc}$ was selected from the short, single-stranded DNA products of in vitro transcription of ASV RNA by exhaustive hybridization with RNA from tdASV (Stehelin et al., 1976*a*). $cDNA_{sarc}$ has a complexity of about 1,500 nucleotides and does not anneal with tdASV RNA, but it hybridizes completely with RNA from all tested strains of ASV. In addition, $cDNA_{sarc}$ does not hybridize to the genomes of avian leukosis viruses, including the endogenous chicken virus, RAV-0, several mammalian RNA tumor viruses, or representatives of the major classes of DNA tumor viruses (Stehelin et al., 1976*a*; Varmus et al., 1976). Thus, nucleotide sequences homologous to $cDNA_{sarc}$ are likely to be implicated in transformation by all strains of ASV, but unrelated genes are presumably responsible for tumors or transformation induced by the other oncogenic viruses we have tested.

Regulation of Expression of the Viral src *Gene in ASV-Infected Cells*

Infection of permissive (avian) or nonpermissive (mammalian) cells with ASV is accompanied by the synthesis of viral DNA and its integration into the host-cell genome (Varmus et al., 1975). The viral *src* gene is transcribed into DNA along with other viral genes, and its expression is then presumably controlled at the transcriptional level by host-cell RNA polymerase II; additional controls may be imposed during processing of primary RNA transcripts, translation, or cleavage of the products of translation. In ASV-infected avian cells (which are permissive for virus replication), large amounts of RNA transcripts containing the *src* gene are readily detected (Stehelin et al., 1977, Bishop et al., 1976); a large portion of these transcripts are 35S RNA molecules destined to become subunits of the viral genome, but 35S, plus 28S and 21S species of RNA are likely to serve as messenger RNA in these cells. Although $cDNA_{sarc}$ anneals to 35S, 28S, and 21S RNA in such cells, it is not yet known which of these species serves as messenger in the translation of the *src* gene product (S. Weiss et al., unpublished results).

Study of the expression of the *src* gene in ASV-infected mammalian cells has proved more tractable because analyses are not confused by large amounts of RNA for viral progeny and because interesting phenotypic variants of infected cells have been described. Thus, after infection by ASV, mammalian cells may be transformed (a relatively rare event), or they may show no alteration in growth properties despite acquisition of a DNA copy of the viral genome (Boettiger, 1974*b*; Varmus et al., 1973, 1974). In addition, ASV-transformed cells may revert to a normal phenotype (Macpherson, 1965).

We have asked whether reversion of ASV-transformed baby hamster kidney (BHK) cells to normal behavior is related to alterations in expression of the *src* gene. The reverted cells, like the transformed cells, contain one to two copies of ASV DNA integrated into the host genome (Deng et al., 1974); the entire viral genome in a genetically unaltered form is likely to be present in the DNA of reverted cells, because fully infectious ASV can be rescued upon fusion of the revertants with permissive cells (Boettiger, 1974*a*). On the other hand, the revertants contain from 2- to 80-fold less viral RNA than do the transformed parental cells (Deng et al., 1974). To determine whether this relatively modest reduction in viral RNA could be responsible for the alterations in phenotype, we compared viral RNA metabolism in transformed and reverted cells (Deng et al., 1977). In both cell types, most or all of the viral RNA was polyadenylated: about 75% was present in the cytoplasm, 24S and 35S RNA were present in the nucleus, but only 24S species were detectable in the cytoplasm; and similar proportions of the cytoplasmic RNA were associated with polyribosomes. More specifically, $cDNA_{sarc}$ hybridized to 24S RNA from the cytoplasm of both cell types and to polyribosomal RNA in proportion to the total amount of viral RNA per cell. We have, therefore, reached the provisional conclusion that the phenotype was determined by these modest quantitative differences in expression of the *src* gene (Deng et al., 1977).

Nucleotide Sequences Related to the src *Gene are Present in the DNA of Normal Chickens*

RNA tumor viruses are widely disseminated among vertebrates, and many viruses are transmitted genetically as well as horizontally (Todaro and Huebner, 1972). This conclusion is now supported in many species by the identification of viral DNA in normal cellular DNA by molecular hybridization and by the induction and spontaneous release of virus from apparently normal cells (Tooze, 1973). Although ASV is a horizontally transmitted virus, it is closely related to a nononcogenic virus, RAV-0, whose genome is encoded in the DNA of normal chicken cell (Vogt and Friis, 1971; Neiman, 1973). It has, therefore, been suggested that ASV arose from genetic components of normal cells (Temin, 1976; Martin and Weiss, 1973; Bishop et al., 1977; Varmus et al., 1976). Because RAV-0 does not transform cells (Motta et al., 1975) and because its genome does not contain the nucleotide sequences corresponding to $cDNA_{sarc}$, we asked whether $cDNA_{sarc}$ was capable of detecting some other component of the chicken genome from which the transforming gene of ASV might have arisen. We found that $cDNA_{sarc}$ can anneal extensively (50–60%) and with the kinetics of single copy DNA to DNA extracted from normal chicken embryos (Stehelin et al., 1976*b*). Interestingly and importantly, the thermal stability of duplexes formed between $cDNA_{sarc}$ and normal chicken-cell DNA was 3–

4°C below the stability of homologous duplexes formed between $cDNA_{sarc}$ and DNA from an ASV-infected rat cell containing 20 copies of the viral genome introduced as a consequence of infection. This result implies that the sequence in normal chicken cells, which is related to the transforming gene of ASV, is significantly (2–3%) different from the viral *src* gene; thus, if the viral *src* gene in the ASV genome is derived from the cellular "sarc" sequence, the process which generated ASV must have produced alterations in base sequence of the "sarc" sequence (cf. Varmus et al., 1976, for further discussion and references).

"Sarc" DNA and RAV-0 DNA are Located on Different Chromosomes

We have determined that the "sarc" sequence and the RAV-0 provirus are unlinked in the chicken genome by annealing $cDNA_{sarc}$ and virus-specific cDNA (capable of detecting RAV-0, but not sarc, DNA) to DNA extracted from chicken chromosomes fractionated in sucrose gradients (Padgett et al., submitted manuscript; Stubblefield et al., 1977). The chicken karyotype consists of 14 pairs of variably sized macrochromosomes and 30–40 pairs of indistinguishable microchromosomes. We found that the RAV-0 provirus is located in the largest group of macrochromosomes, whereas the sarc sequence is present in the microchromosomal fraction, indicating that RAV-0 DNA and the sarc DNA are unlinked in the chicken genome. This finding implies that the two sets of sequences may have evolved independently (see below), may be independently regulated, and would require special mechanisms for recombinational events which may have led to the creation of ASV.

Conservation of the Sarc Sequence in Evolution of Vertebrates

The finding of DNA related to the *src* gene of ASV in normal chicken cells does not, of itself, suggest a function for such DNA; because the product of the viral *src* gene is as yet unidentified, attempts to determine the function of the cellular sarc sequence have thus far been indirect. To date, we have asked two questions of these sequences: (1) to what extent are they conserved during evolution and (2) to what extent are they expressed as RNA in various types of cells?

By annealing $cDNA_{sarc}$ to DNA from a wide variety of organisms and determining the thermal stability of duplexes formed, we have established that sequences related to the viral transforming gene are present in vertebrates from fish to primates (including man) and that the divergence of those sequences is approximately proportional to the evolutionary distance of each species from chicken (Stehelin et al., 1976*b*; Spector et al., manuscript in preparation). $cDNA_{sarc}$ does not, however, anneal to *E. coli* or *Drosophila* DNA. These results indicate that the ancestral sarc sequence was present in the earliest vertebrates and that the sequence has been more highly conserved than the bulk of single copy DNA during evolution. In contrast, most or all of the provirus of RAV-0 is undetectable in birds other than chickens (Varmus et al., 1974; Neiman, 1973, Tereba et al., 1975), indicating that RAV-0 either has evolved very rapidly or was introduced into the germ line of chickens after the major speciation events.

The Ubiquitous Expression of Sarc Sequences in Avian RNA

Conservation of sarc sequences during evolution suggests that the sequence might be important to the survival of organisms. To explore this idea further, we asked whether the sequence is transcribed into RNA and, if so, whether control of the level of sarc RNA can be correlated with some aspect of cell growth, differentiation, or transformation by nonviral reagents. Thus far, RNAs extracted from a large variety of avian tissues and cells have been tested for their capacity to hybridize with $cDNA_{sarc}$; in every case, we have observed hybridization with kinetics consistent with the presence of about one to eight sarc RNA molecules per cell (Spector et al., manuscript in preparation). The tests have been performed with chicken, duck, and quail embryo fibroblasts at several levels of passage and at several times after plating; with quail embryo fibroblasts arrested in the G_0 stage of the growth cycle for over 1 wk; with chicken embryos from the 3rd to the 16th day after fertilization; with chicken embryos that did and did not contain proteins antigenically related to the structural proteins of ASV (presumptive evidence for partial expression of the RAV-0 provirus); with organs from adult chickens; and with tumors (and cultured cells derived from them) induced in chickens and quail with chemical carcinogens (methylcholanthrene and dimethylbenzanthracene). This extensive survey shows that the sarc DNA is rarely, if ever, silent, because it is ubiquitously copied into RNA, but the survey provides no clear indication of the function the sarc sequence might serve. The ubiquity of its expression, like its conservation during evolution, might argue that the sarc sequence is important for survival, but does not suggest a specific mechanism by which it might function.

Metabolism of Sarc RNA is Similar in Normal and Chemically Transformed Cells

Because the viral *src* gene is clearly responsible for the initiation and maintenance of viral transformation, we have been particularly interested in the question of whether the cellular sarc sequence might mediate transformation by nonviral agents. However, as noted in the previous section, we have not observed significant and reproducible differences in the concentrations of sarc-containing RNA in normal embryo fibroblasts and in cells derived from chemically induced neoplasms. Inasmuch as our studies with revertants of ASV-transformed hamster cells suggested that subtle differences in the expression of the viral *src* gene might have major effects upon cell behavior (see above), we have carefully compared the distribution and size of sarc-containing RNA in normal Japanese quail embryo fibroblasts and in cells derived from a fibrosarcoma induced in quail with methylcholanthrene (Spector et al., manuscript in preparation). In both cell types, under similar growth conditions, virtually identical concentrations of sarc RNA were found in nuclear, cytoplasmic, and polyribosomal fractions. In addition, most of the sarc RNA was polyadenylated and associated with polyribosomes in a form releasable with EDTA (indicating it was mRNA). RNA from both cell types sedimented as 30–31S RNA in rate-zonal gradients under denaturing conditions. Although this study did not reveal differences in the metabolism of sarc RNA that would implicate the sarc sequence in chemical carcinogenesis, it did reveal that the sarc sequence (with a complexity of about 450,000 daltons) is part of a larger transcriptional unit, as it is present in an RNA molecule of about 2×10^6 daltons. Moreover, the location of this RNA in polyribosomes and its linkage to poly(A) suggest that it acts as mRNA. Of course, we do not know whether the sarc sequence itself is translated or the nature of the protein(s) it might encode.

Summary

We have taken advantage of deletion mutants in the transforming gene of ASV to study the function and origin of that gene in the absence of information about its product. A nucleic acid hybridization reagent specific for most or all of the *src* gene ($cDNA_{sarc}$) has been used to demonstrate modulation of *src* gene expression in phenotypically varied ASV-infected hamster cells. In addition, annealing of $cDNA_{sarc}$ to nucleic acids from normal (uninfected) cells has established that DNA related (but not identical) to the *src* gene of ASV is present in the genomes of all vertebrates. This sarc sequence is highly conserved in vertebrate evolution and ubiquitously expressed in avian cells and tissues; however, its function and the rationale for its conservation remain unknown.

ACKNOWLEDGMENTS

This work was supported by grants from the U.S. Public Health Service (CA 12705, CA 19287, and IT32 CA 09043), the American Cancer Society (VC-70), contract no. NO1 CP 33293 within the Virus Cancer Program of the National Cancer Institute, National Institutes of Health, Public Health Service. H.E.V. is a recipient of a Research Career Development Award (CA 70193) from the National Cancer Institute. This report is an abridged version of a paper that will appear in Volume 4 of the Cold Spring Harbor Conferences on Cell Proliferation.

REFERENCES

BERNSTEIN, A., R. MACCORMICK, and G. S. MARTIN. 1976. Transformation-defective mutants of avian sarcoma viruses: the genetic relationship between conditional and non-conditional mutants. *Virology*. **70:** 206–209.

BISHOP, J. M., C. T. DENG, B. W. J. MAHY, N. QUINTRELL, E. STAVNEZER, and H. E. VARMUS. 1976. Synthesis of viral RNA in cells infected by avian sarcoma viruses. *In* Animal Virology. ICN-UCLA Symposium on Molecular and Cellular Biology. Vol. 4. D. Baltimore, A. S. Huang, and C. F. Fox, editors. Academic Press, Inc., New York. 1–20.

BISHOP, J. M., D. STEHELIN, J. TAL, D. FUJITA, D. ROULLAND-DUSSOIX, T. PADGETT, and H. E. VARMUS. 1977. The transforming gene of avian sarcoma virus. *In* Symposium on the Molecular Biology of the Mammalian Genetic Apparatus. California Institute of Technology. 8–10 December 1975. In press.

BISHOP, J. M., and H. E. VARMUS. 1975. The molecular biology of RNA tumor viruses. *In* Cancer: A Comprehensive Treatise. Vol. 2. I. F. Becker, editor. Plenum Publishing Corp., New York. 3–48.

BOETTIGER, D. 1974*a*. Reversion and induction of Rous sarcoma virus expression in virus-transformed baby hamster kidney cells. *Virology*. **62:**522–529.

BOETTIGER, D. 1974*b*. Virogenic nontransformed cells isolated following infection of normal rat kidney cells with B77 strain Rous sarcoma Virus. *Cell*. **3:**71–75.

DENG, C. T., D. BOETTIGER, I. MACPHERSON, and H. E. VARMUS. 1974. The persistence and expression of virus-specific DNA in revertants of Rous sarcoma virus-transformed BHK-21 cells. *Virology*. **62:**512–521.

DENG, C. T., D. STEHELIN, J. M. BISHOP, and H. E. VARMUS. 1977. Characteristics of virus-specific RNA in avian sarcoma virus-transformed BHK-21 cells and revertants. *Virology*. **76:**313–325.

DUESBERG, P. H., and P. K. VOGT. 1970. Differences between the ribonucleic acids of transforming and nontransforming avian tumor viruses. *Proc. Natl. Acad. Sci. U. S. A.* **67:**1673–1680.

DUESBERG, P. H., and P. K. VOGT. 1973. RNA species obtained from clonal lines of avian sarcoma and avian leukosis virus. *Virology*. **54:**207–219.

DUESBERG, P. H., L.-H. WANG, P. MELLON, W. S. MASON, and P. K. VOGT. 1976. Towards a complete genetic map of Rous sarcoma virus. *In* Animal Virology. ICN-UCLA Symposium on Molecular and Cellular Biology. Vol. 4. D. Baltimore, A. S. Huang, and C. F. Fox, editors. Academic Press, Inc., New York. 107–125.

GROSS, L. 1970. *Oncogenic Viruses*. 2nd edition. Pergamon Press, Elmsford, N.Y. 991.

MACPHERSON, I. 1965. Reversion in hamster cells transformed by Rous sarcoma virus. *Science (Wash. D.C.)*. **148:**1731–1733.

MARTIN, G. S., and R. WEISS. 1973. Genetics and evaluation of RNA tumor viruses. *Proc. Can. Cancer Res. Conf.* **11:**10–30.

MOTTA, J. V., L. B. CRITTENDEN, H. G. PURCHASE, H. A. STONE, W. OKAZAKI, and L. WITTER. 1975. Low oncogenic potential of avian endogenous RNA tumor virus infection or expression. *J. Natl. Cancer Inst.* **55:**685–689.

NEIMAN, P. E. 1973. Measurement of endogenous leukosis virus nucleotide sequences in the DNA of normal avian embryos by RNA-DNA hybridization. *Virology*. **53:**196–204.

STEHELIN, D., D. FUJITA, T. PADGETT, H. E. VARMUS, and M. J. BISHOP. 1977. Detection and enumeration of transformation defective strains of avian sarcoma virus with molecular hybridization. *Virology*. In press.

STEHELIN, D., R. V. GUNTAKA, H. E. VARMUS, and J. M. BISHOP. 1976*a*. Purification of DNA complementary to nucleotide sequences required for neoplastic transformation of fibroblasts by avian sarcoma viruses. *J. Mol. Biol.* **101:**349–365.

STEHELIN, D., H. E. VARMUS, J. M. BISHOP, and P. K. VOGT. 1976*b*. DNA related to the transforming gene(s) of avian sarcoma viruses is present in normal avian cells. *Nature (Lond.)*. **260:**170–173.

STUBBLEFIELD, E., S. LINE, F. FRANOLICH, and L. Y. LEE. 1977. Analytical techniques for isolated metaphase chromosome fractions. *Methods in Cell Biol.* In press.

TEMIN, H. 1976. The DNA provirus hypothesis. *Science (Wash. D.C.)*. **192:**1075–1080.

TEREBA, A., L. SKOOG, and P. K. VOGT. 1975. RNA tumor virus specific sequences in nuclear DNA of several avian species. *Virology*. **65:**524–534.

TODARO, G. J., and R. J. HUEBNER. 1972. The viral oncogene hypothesis: new evidence. *Proc. Natl. Acad. Sci. U. S. A.* **69:**1009–1015.

TOOZE, J., editor. 1973. The Molecular Biology of Tumor Viruses. Cold Spring Harbor Laboratory, Cold Spring Harbor, N.Y. 743.

VARMUS, H. E., R. V. GUNTAKA, C. T. DENG, and J. M. BISHOP. 1975. Synthesis, structure, and function of avian sarcoma virus-specific DNA in permissive and non-permissive cells. *Cold Spring Harbor Symp. Quant. Biol.* **39:**987–996.

VARMUS, H. E., S. HEASLEY, and J. M. BISHOP. 1974. Use of DNA-DNA annealing to detect new virus-specific DNA sequences in chicken embryo fibroblasts after infection by avian sarcoma virus. *J. Virol.* **14:**895–903.

VARMUS, H. E., D. STEHELIN, D. SPECTOR, J. TAL, D. FUJITA, T. PADGETT, D. ROULLAND-DUSSOIX, H.-J. KUNG, and J. M. BISHOP. 1976. Distribution and function of defined regions of avian tumor virus genomes in viruses and uninfected cells. ICN-UCLA Symposium on Molecular and Cellular Biology. Vol. 4. D. Baltimore, A. S. Huang, and C. F. Fox, editors. Academic Press, Inc., New York. 339–358.

VARMUS, H. E., P. K. VOGT, and J. M. BISHOP. 1973. Integration of Rous sarcoma virus-specific DNA following infection of permissive and non-permissive hosts (RNA/tumor viruses/reassociation kinetics). *Proc. Natl. Acad. Sci. U. S. A.* **70:**3067–3071.

VOGT, P. K. 1971. Spontaneous segregation of nontransforming viruses from cloned sarcoma viruses. *Virology*. **46:**939–946.

VOGT, P. K. 1977. The genetics of RNA tumor viruses. *In* Comprehensive Virology. H. Fraenket-Conrat and R. R. Wagner, editors. Plenum Publishing Corp., New York. In press.

VOGT, P. K., and R. R. FRIIS. 1971. An avian leukosis virus related to RSV(0): properties and evidence for helper activity. *Virology*. **43:**223–234.

WYKE, J. A., J. G. BELL, and J. A. BEAMAND. 1975. Genetic recombination among temperature-sensitive mutants of Rous sarcoma virus. *Cold Spring Harbor Symp. Quant. Biol.* **39:**897–907.

Morphogenesis of Gametes

CYTOLOGICAL DIFFERENTIATION OF THE FEMALE GAMETE

EVERETT ANDERSON, DAVID F. ALBERTINI, and RICHARD F. WILKINSON

Whereas the transformation of a mammalian spermatid into an architecturally complicated motile spermatozoon is accompanied by a dramatic cellular metamorphosis, the development of the mammalian egg is not distinguished by such a spectacular cellular conversion (Fawcett et al., 1971). Initially, it is an irregularly shaped cell whose final profile is spherical and whose plasmalemma (oolemma) is organized into many specialized units known as microvilli (Fig. 1 and *inset*).

The eggs of some invertebrates and most lower vertebrates possess a visible axial differentiation, which is established primarily by nonorganelle components of the cytoplasm known as yolk. The pole of large yolk accumulations is called the vegetative pole; its antipole, the animal pole (Anderson, 1974). As one studies mammalian oogenesis, one notices also a change in the polarity of the oocyte that is established by the ooplasmic localization of either one or several organelles. As shown in Fig. 1, the irregularly shaped presumptive mammalian oocyte possesses organelles that are randomly dispersed within the ooplasmic matrix. As differentiation proceeds, many of the organelles aggregate in a juxtanuclear position around two centrioles, thereby making the oocyte somewhat bilaterally symmetrical. During further differentiation, the Golgi complex increases in number; some saccules become associated with cisternae of the rough endoplasmic reticulum (RER) and eventually become situated in the peripheral ooplasm; others not associated with RER are randomly scattered in the ooplasmic matrix. When the latter events are accomplished, the egg appears radially symmetrical. The morphogenic movements of the organelles of the differentiating oocyte, coupled with specific nuclear happenings, render it different from its presumptive cell type and distinct from other cell types. The uniqueness of the primordial germ cell is characterized by its ability to undergo meiosis.

In the study of mammalian oogenesis, it is only natural to investigate not only the oocyte, but also the associated granulosa cells with which it shares common territory. This, then, becomes a study of the Graafian follicle. This paper presents a correlative microscope analysis of the developing Graafian follicle (exclusive of the theca externa and interna) in a number of mammalian species (mouse, rat, rabbit).

EVERETT ANDERSON Department of Anatomy and Laboratory of Human Reproduction and Reproductive Biology, Harvard Medical School, Boston, Massachusetts

DAVID F. ALBERTINI Department of Physiology, School of Medicine, University of Connecticut Health Center, Farmington, Connecticut

RICHARD F. WILKINSON Department of Biological Sciences, University of Southern California, Los Angeles, California

MATERIALS AND METHODS

The mammalian ovaries used in this study were processed for light, transmission, scanning electron microscopy, and freeze-fracture observations according to the methodologies outlined in Albertini and Anderson (1974) and Anderson et al. (1976).

Observations and Discussion

During this investigation we used the ovaries of animals of varying ages; however, it should be pointed out that follicle initiation and development is a continuum regardless of age, stage of the cycle, or whether or not the animal is pregnant (Peters et al., 1975). Superimposed on this continuum are the surges of follicle stimulating hormone (FSH) and luteinizing hormone (LH) that lead to

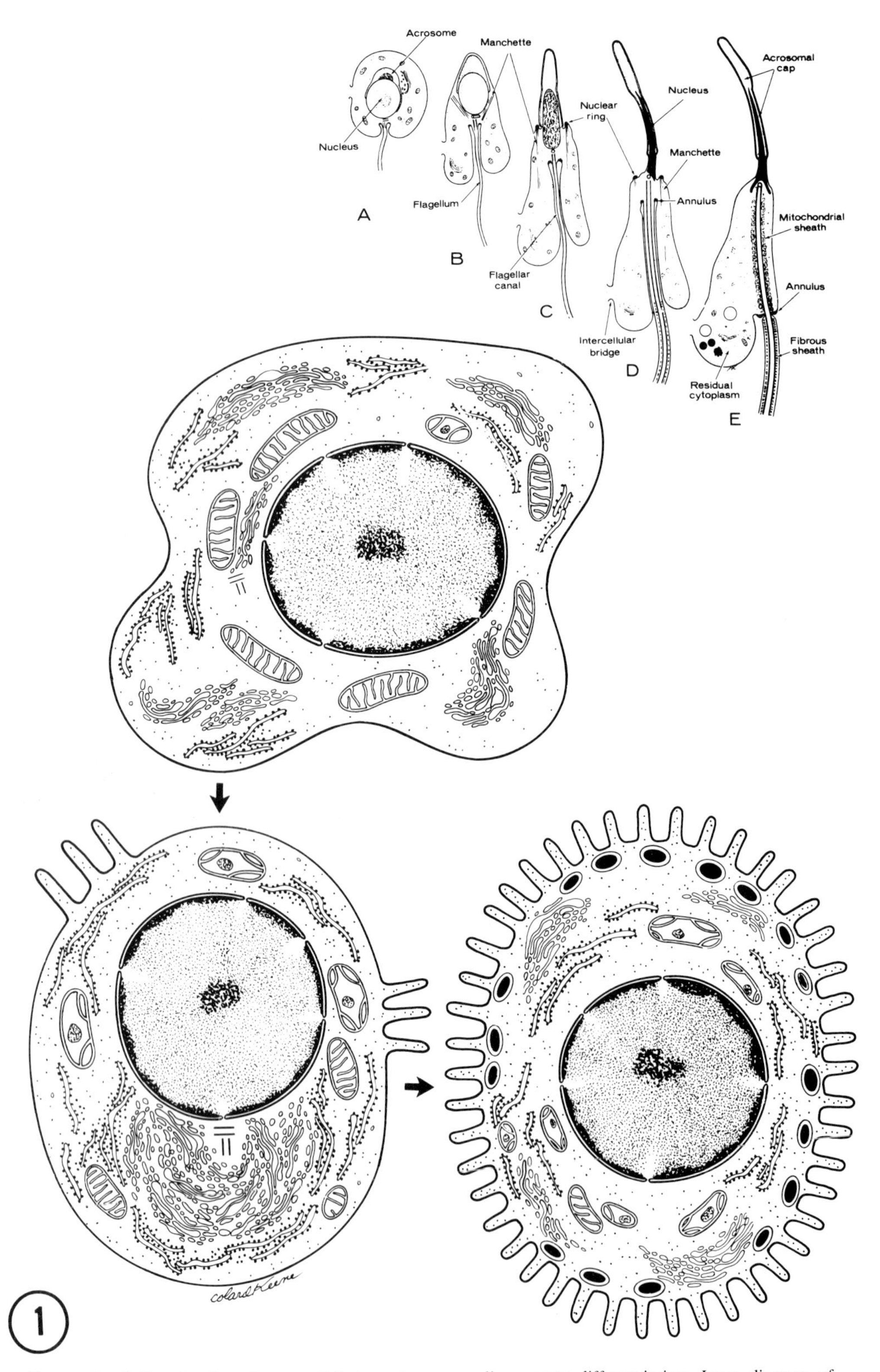

FIGURE 1 A line drawing of sequential stages in mammalian oocyte differentiation. *Inset,* diagram of successive stages in guinea pig spermatid differentiation (from D. W. Fawcett et al., 1971).

the cyclic event of ovulation (Richards and Midgley, 1976).

Primordial germ cells of mammals arise outside of the body and subsequently migrate to the gonadal analage, where they increase their numbers by mitosis (Baker and Sum, 1977). During fetal life, or immediately thereafter, mitosis ceases and oogonia enter the prophase of meiosis, a time when both X-chromosomes become functional (Gartler et al., 1975). The oogonium may now be referred to as an oocyte. Upon initiation of meiosis, the oocyte is incompletely encompassed by a layer of granulosa cells, which are thought to be derived from the rete ovari (Byskov and Moore, 1973). Once meiosis has been initiated, the process proceeds to the prolonged diplotene stage of meiosis, a stage characterized by an intense accumulation of macromolecular substances essential for future embryogenesis (Davidson, 1968). It is during the diplotene stage that organelles commence their reorganization alluded to earlier.

As is the case in other cell types, the oocyte is also distinguished by an oolemma of diverse functions that delineates an ooplasmic matrix richly compartmentalized by a system of multifunctional membranes. The introduction of the freeze-fracture technique now makes it possible to cleave a membrane through its hydrophobic interior, thereby allowing for visualization of intramembranous structural differentiation. Some investigators believe that the number of intramembranous particles may be related to the metabolic activity of the membrane and that these particles may represent protein constituents of the membrane (Branton, 1966; De Camilli et al., 1976). If this be the case, the density of particles may reflect metabolic activity of membranes during oogenesis or any other differentiative process. We have found differential particle distribution of certain membranes of the oocyte during its differentiation, which we shall explore presently.

Nuclear Envelope

If one fractures the nuclear envelope of an oocyte that is surrounded by a single layer of granulosa cells, it is found to be composed of few pores and the E face is densely covered with particles; the P face being particle-poor (Fig. 2). As the oocyte continues its differentiation, there is an increase in the number of pores and particles. The number of nuclear pores may be indicative of the "nuclear pore flow rate" (NPFR). Franke (1970) defines NPFR to mean the total mass or number of molecules of a certain substance transferred through an average pore per minute. Although a positive correlation between the number of nuclear pores and the NPFR has been shown for the nucleus of HeLa cells, *Tetrahymena pyriformis,* and the oocytes of *Xenopus laevis,* nothing is known for the nucleus of developing mammalian oocytes.

Mitochondria

Even though it is true that the differentiating mammalian oocyte contains all of the regularly occurring organelles, a portion of the mitochondrial population appears to change its internal structure as oogonia become oocytes. Most of the mitochondria of oogonia contain many transversely oriented cristae, whereas those of oocytes contain a small number of variously oriented cristae, presumably reflecting a stage of low metabolic activity. Within the mitochondria of rodent oocytes is a vacuole containing a filamentous network (Fig. 9). These filaments are interpreted to be mitochondrial DNA (Rabinowitz and Swift, 1970). Freeze-fracture replicas reveal the vacuole to possess a relatively smooth surface (Fig. 3).

Golgi Complex and Endoplasmic Reticulum

As indicated previously, the Golgi complex in young oocytes is rather large and is located in a paranuclear position (Fig. 4). Associated with the periphery of the Golgi are cisternae of the endoplasmic reticulum of the rough variety; in the center of the complex are two centrioles. These are present up to the pachytene stage of meiosis, after which time they disappear (Szollosi et al., 1972). Coincident with the disappearance of the centriole there is an increase in the volume of the oocyte accompanied by an increase in the number of Golgi complexes and cisternae of the endoplasmic reticulum. A freeze-fracture replica of a paranuclear Golgi complex of the mouse oocyte is shown in Fig. 5. Exposed in this micrograph are the E and P faces of the fenestrated lamella, or plate, of the organelle. The E face has more particles than the P face. This micrograph also serves to illustrate the associated tubular endoplasmic reticulum depicting the particle-rich E face and the relatively smooth P face. In the young oocytes of rats, the paranuclear Golgi is composed of cisternae and a network of interconnecting tubules.

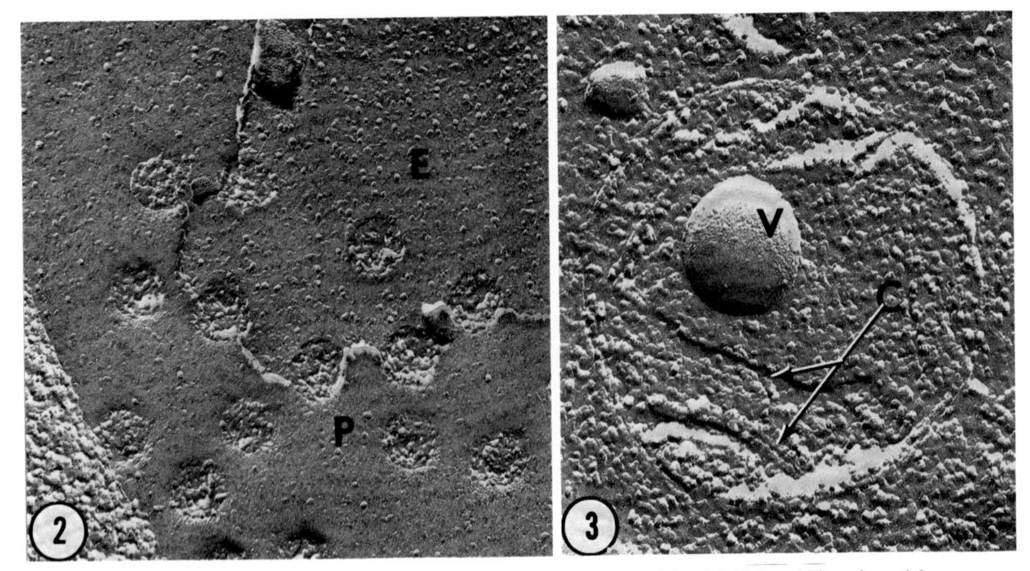

FIGURE 2 A small portion of the nuclear envelope showing the particle-rich E face (*E*) and particle-poor P face (*P*). (Mouse). × 90,000.

FIGURE 3 A mitochondrion from a mouse oocyte showing the relatively smooth surface of the vacuole. *C*, cristae. × 210,000.

Some of the Golgi complexes are involved in the synthesis and concentration of precursors utilized in the construction of cortical granules (Fig. 6). In freeze-fracture replicas, some of the Golgi saccules appear to contain presumptive cortical granules (Fig. 7). Note the relative scarcity of particles on the membrane of the cortical granule. During the formation of cortical granules, there is a close association of the endoplasmic reticulum with the organelle. The cisternae of the endoplasmic reticulum facing the forming face of the Golgi show evagination; the intervening vacuoles are thought to be derived from the endoplasmic reticulum. Experimental evidence has cradled the belief that the protein portion of secretory bodies, like cortical granules, is synthesized by the endoplasmic reticulum and subsequently packaged by the Golgi, an organelle known for its role in the glycosylation of proteins and lipids (Sturgess et al., 1974).

Oolemma

During differentiation, the oolemma becomes morphologically specialized by the production of microvilli. Figure 8 reveals a large region through the E face of the oolemma of a young oocyte and a small colony of microvilli. Also shown on the oolemma are clusters and linearly arranged particles on the P face. These particle aggregates are gap junctions between granulosa cells and the oocyte (Anderson and Albertini, 1976).

Granulosa Cells

When the oocyte of the mouse and the rat is surrounded by a single layer of granulosa cells, the granulosa cells are coupled to each other by gap junctions (Fig. 9). In the rabbit, gap junctions between granulosa cells appear to be formed coincident with the formation of the antrum. Here it is interesting to note that, after antrum formation, specific receptors for LH are integrated into the granulosa cell membrane which are thought to mediate the gonadotropin-directed transformation of granulosa cells into lutein cells (Channing and Kammerman, 1974). During further differentiation of the follicle, gap junctions between granulosa cells increase in size and number subsequent to antrum formation (Fig. 11). If one prepares the ovary by hardening in a 1.0% solution of thiosemicarbazide and subsequently dissects the follicle and prepares it for scanning electron microscopy, one notices ovoid regions on granulosa cells that are isomorphic with gap junctions (Fig. 10).

Conclusions

This study has shown that during oogenesis

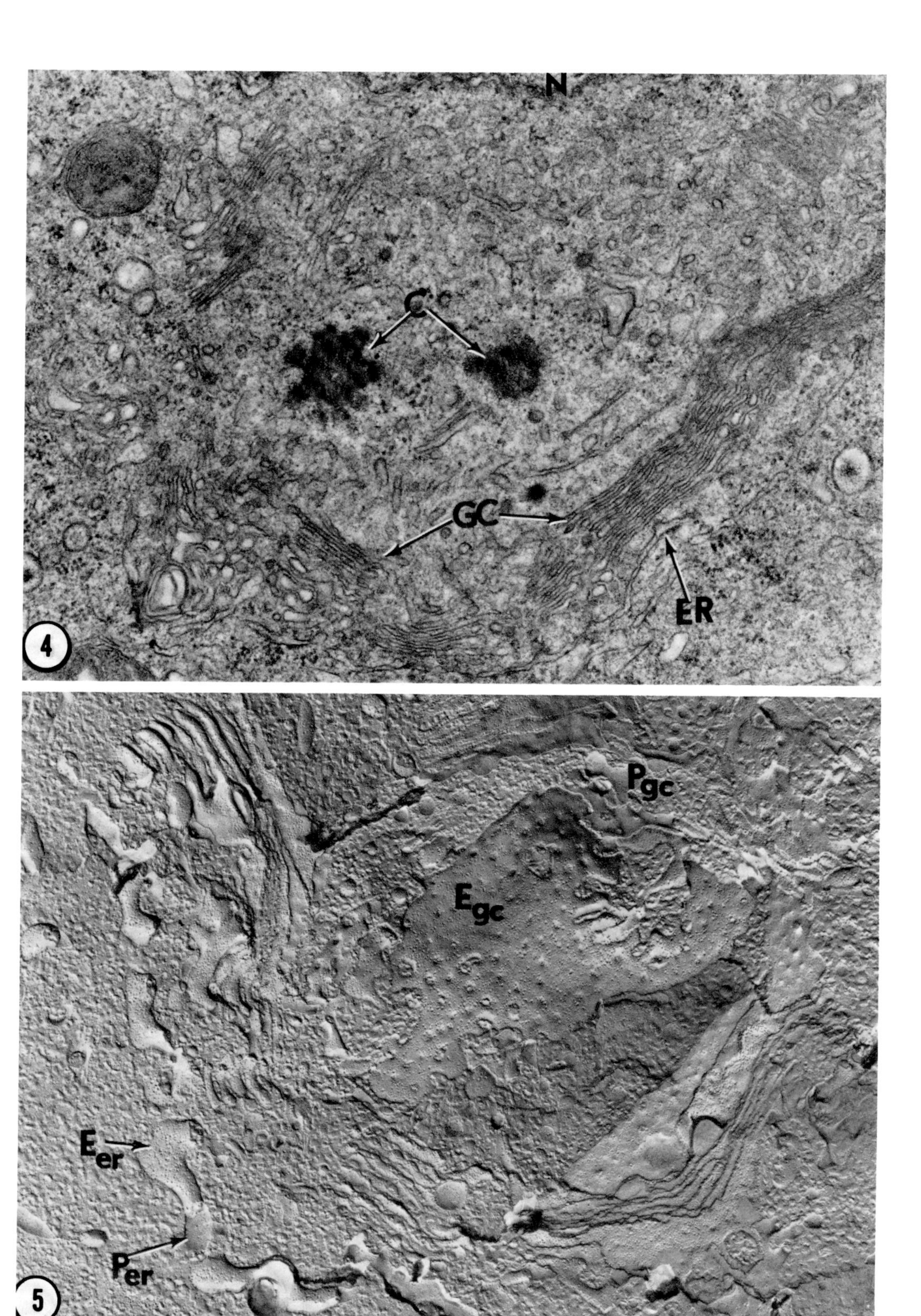

FIGURE 4 A region through the paranuclear Golgi complex of a young mouse oocyte. *N*, nuclear envelope; *C*, centrioles; *GC*, Golgi complex; *ER*, endoplasmic reticulum. × 30,000.

FIGURE 5 Freeze-fracture replica through the paranuclear Golgi complex of a young mouse oocyte. P_{gc}, P face of a Golgi lamella; E_{gc}, E face of a Golgi lamella; E_{er}, E face of the endoplasmic reticulum; P_{er}, P face of the endoplasmic reticulum. × 40,000.

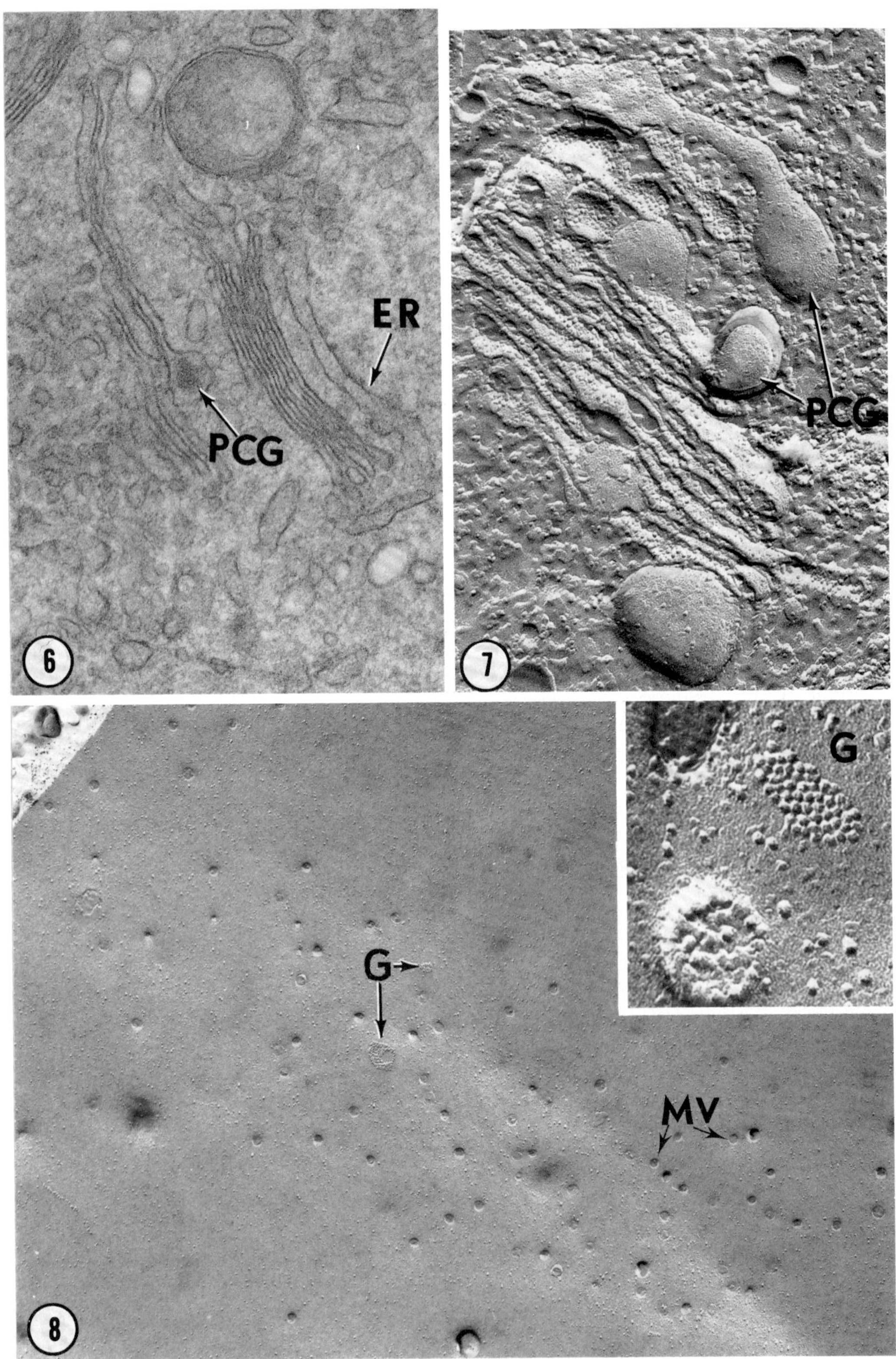

FIGURE 6 An electron micrograph showing one saccule of a Golgi complex with some electron opaque material (presumptive cortical granule, *PCG*). *ER*, endoplasmic reticulum. (Mouse). × 40,000.

FIGURE 7 Freeze-fracture replica through a small Golgi complex. Note the relatively few particles on the membrane of the presumptive cortical granules, *PCG*. (Mouse). × 60,000.

FIGURE 8 A relatively large region through the E face of the oolemma of a young oocyte. *G*, gap junctions (also see *inset*); *MV*, microvilli. × 36,000; *inset* × 165,000.

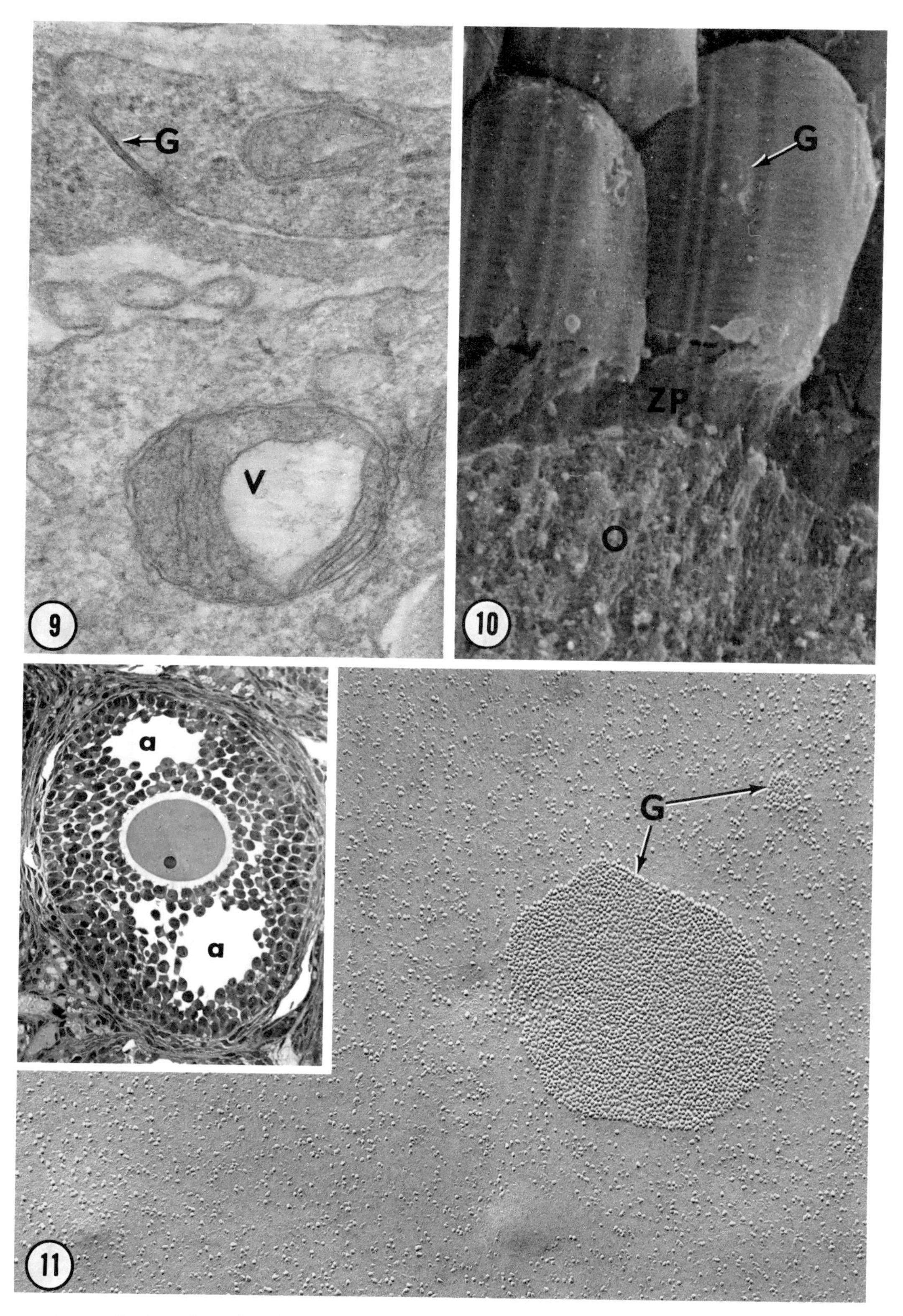

FIGURE 9 A small portion of an oocyte (mouse) with a single layer of encompassing follicle-granulosa cells. *G*, gap junction; *V*, vacuole of an oocyte mitochondrion. × 40,000.

FIGURE 10 A scanning electron micrograph of a small portion of a rabbit follicle. *G*, gap junction; *ZP*, zona pellucida; *O*, oocyte. × 30,000.

FIGURE 11 Freeze-fracture replica of a granulosa cell from an antral follicle (*a*, see *inset*) showing gap junctions (*G*) in different stages of development. (Mouse). × 60,000; *inset* × 300.

there is a redistribution of organelles coincident with the disappearance of the centrioles. Freeze-fracture replicas of organelles during oocyte differentiation reveal differential particle distribution that may be related to organelle function during this differentiative process.

The developing oocyte is coupled to the surrounding granulosa cells by gap junctions. Moreover, granulosa cells are also conjoined by gap junctions. The morphological and physiological evidence of Epstein et al. (1976) and of Potter, Fawcett, and Woodrum (unpublished observations) permits one to conclude that the oocyte and granulosa cells are in communication with each other via gap junctions. This membrane specialization may be involved in regulating the maturation of the oocyte, as well as the ultimate transformation of the follicular epithelium into the corpus luteum.

ACKNOWLEDGMENTS

This investigation was supported by grants (HD-06822 and Center Grant HD-06645) from the National Institutes of Health, U.S. Public Health Service.

REFERENCES

ALBERTINI, D. F., and E. ANDERSON. 1974. The appearance and structure of the intercellular connections during the ontogeny of the rabbit ovarian follicle with particular reference to gap junctions. *J. Cell Biol.* **63:**234–250.

ANDERSON, E. 1974. Comparative aspects of the ultrastructure of the female gamete. *Int. Rev. Cytol.* **4**(suppl.):1–70.

ANDERSON, E., and D. F. ALBERTINI. 1976. Oocyte follicle cell gap junctions in the mammalian ovary. *J. Cell Biol.* **70**(pt. 2):89a(Abstr.).

ANDERSON, E., R. WILKINSON, and G. LEE. 1976. SEM observations on differentiating ovarian follicles in mammals. *Anat. Rec.* **184:**344.

BAKER, T. G., and O. WAI SUM. 1977. Development of the ovary and oogenesis. *In* Clinics in Obstetrics and Gynaecology. Vol. III. No. 1. H. C. MacNaughton and J. D. Govan, editors. W. B. Saunders and Company, Philadelphia. In press.

BRANTON, D. 1966. Fracture faces of frozen membranes. *Proc. Natl. Acad. Sci. U. S. A.* **55:**1048–1056.

BYSKOV, A. G. S., and S. LINTERN MOORE. 1973. Follicle formation in the immature mouse ovary: the role of the rete ovarii. *J. Anat.* **116:**207.

CHANNING, C. P., and S. KAMMERMAN. 1974. Binding of gonadotropins to ovarian cells. *Biol. Reprod.* **10:**179.

DAVIDSON, E. H. 1968. Gene Activity in Early Development. Academic Press, Inc., New York. 375.

DE CAMILLI, P., D. PELUCHETTI, and J. MELDOLESI. 1976. Dynamic changes of the luminal plasmalemma in stimulated parotid acinar cells: a freeze-fracture study. *J. Cell Biol.* **70:**59–74.

EPSTEIN, M. L., W. H. BEERS, and N. B. CEILULA. 1976. Cell communication between the rat cumulus oophorus and the oocyte. *J. Cell Biol.* **70:**302a(Abstr.).

FAWCETT, D. W., W. ANDERSON, and D. M. PHILLIPS. 1971. Morphogenetic factors influencing the shape of the sperm head. *Dev. Biol.* **26:**220–251.

FRANKE, W. W. 1970. Nuclear pore flow rate. *Naturwissenschaften.* **57:**44–45.

GARTLER, S. M., R. ANDINA, and N. GANT. 1975. Ontogeny of X-chromosome inactivation in the female germ line. *Exp. Cell Res.* **91:**454–457.

PETERS, H., A. G. BYSKOV, R. HIMELSTEIN-BRAW, and M. FABER. 1975. Follicular growth: the basic event in the mouse and human ovary. *Reprod. Fertil.* **45:**559–566.

RABINOWITZ, M., and H. SWIFT. 1970. Mitochondrial nucleic acids and their relation to the biogenesis of mitochondria. *Physiol. Rev.* **50:**376–427.

RICHARDS, J. S., and A. R. MIDGLEY. 1976. Protein hormone action: a key to understanding ovarian follicular and luteal development. *Biol. Reprod.* **14:**82–94.

STURGESS, J. M., M. A. MOSCARELLO, and W. J. VAIL. 1974. The fine structure of membranes of the Golgi complex. 8th International Congress of Electron Microscopy. Vol. II. J. V. Sanders and D. J. Goodchild, editors. Australian Academy of Science, Canberra, A.C.T., Australia. 194–195.

SZOLLOSI, D., P. CALARCO, and R. P. DONAHUE. 1972. Absence of centrioles in the first and second meiotic spindles of mouse oocytes. *J. Cell Sci.* **11:**521–541.

SPERMATOGENESIS IN *MARSILEA:* AN EXAMPLE OF MALE GAMETE DEVELOPMENT IN PLANTS

PETER K. HEPLER and DIANA GOLD MYLES

With the exception of the higher flowering plants, flagellated, motile male gametes are found widely throughout the plant kingdom. These cells are especially interesting to us because, like their counterparts in the animal kingdom, they are specialized for the function of delivering the male genome to the egg during fertilization. It is not surprising, therefore, that the plant male gametes possess many similarities to animal gametes. Both usually contain an elongated or asymmetrically shaped nucleus with highly condensed chromatin, an organized mitochondrion often tightly coupled with the motile apparatus, and, finally, a motile apparatus itself, composed of one to many flagella.

It is the purpose of this chapter to look at the plant male gamete and to focus specifically on those from the water fern, *Marsilea*. Emphasis will be placed on examining the origin of the motile apparatus and on describing the events of morphogenesis involved in shaping the cell, especially its nucleus and mitochondrion.

The Overall Developmental Plan in Marsilea

Marsilea is a heterosporous fern. It contains numerous microspores and megaspores within a sporocarp. Upon hydration, the sporocarp releases both types of spores, and development is initiated: the microspores give rise to sperm, while the megaspore forms a mature egg. Each microspore begins as a single cell and during development it divides nine times to produce 32 spermatids and 7 sterile cells (Hepler, 1976; Sharp, 1914). During the last division, a unique spindle pole body, the blepharoplast, arises *de novo* and participates in spindle formation (Hepler, 1976). This results in an astral spindle, as compared to the typical anastral spindle organization found in the earlier divisions.

After the cell division phase, each spermatid then undergoes a differentiation phase, wherein it acquires the highly specialized and asymmetric shape of the sperm (Myles and Hepler, 1977). The mature sperm is a pear-shaped cell with a spiral coil of integrated organelles in its anterior end (Myles and Bell, 1975; Fig. 1). Between 100 and 150 flagella are distributed along the length of this coil, which makes about 10 turns around the edge of the cell. The coil is constructed as a left-handed spiral (Fig. 17) and, during swimming, the sperm moves with a helical twist in a left-handed direction. Its swimming rate is around 150 μm/s (Bilderback et al., 1974).

Spermatogenesis in microspores of *Marsilea* shows many properties that make it a propitious system for experimental studies of development. The cell division and differentiation processes are synchronous and temperature-dependent. For example, at 20°C the entire development requires 11 h. The nine cell divisions occur in intervals of 30 min each, thus occupying the first 4.5 h; differentiation takes the remaining 6.5 h (Hepler, 1976). Development occurs without the complications of growth and furthermore can take place in distilled water. Finally, it is important to note that of the total 39 cells produced in a microspore, 32 become motile gametes; thus, the bulk of the cellular processes are directed specifically to the task of spermatogenesis.

Formation of the Basal Bodies and Motile Apparatus

In one important aspect, developing plant sperm generally differ markedly from animal

PETER K. HEPLER and DIANA GOLD MYLES Department of Biological Sciences, Stanford University, Stanford, California

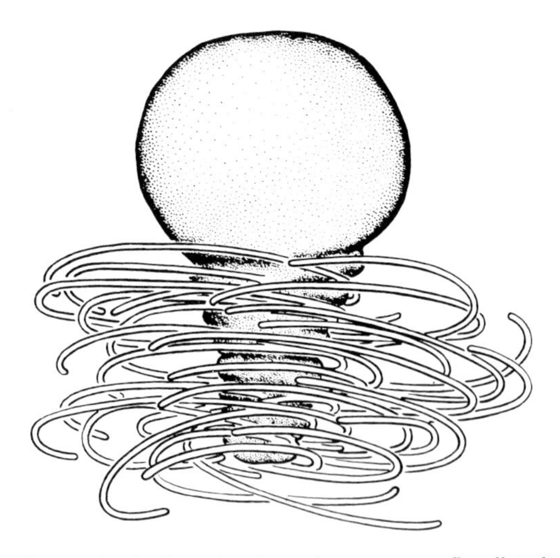

FIGURE 1 A line drawing of a mature, flagellated sperm (drawing by Teppy Williams, from Myles and Hepler, 1977).

sperm. Plant cells, including those from the archegoniates (ferns, mosses, etc.), usually do not contain basal bodies. Thus, the basal bodies, needed for the formation of the flagella, arise *de novo*. The process of formation has been examined in some detail in *Marsilea* (Hepler, 1976), where it is known that basal bodies are derived from a spherical structure called a blepharoplast. This organelle (Fig. 6) arises late in the division phase of *Marsilea* in cells possessing no structures that resemble it in either form or position. The sequential stages in blepharoplast formation have been reconstructed from electron micrographs, and appear to proceed as follows: as the cell enters telophase of the next-to-last division, an aggregation of flocculent material occurs in a small indentation in the distal surface of the telophase nucleus (Fig. 2). Two densely staining plaques (~45 nm apart) arise within the flocculent material, and these become the focal points for the construction of two blepharoplasts (Fig. 3). Each plaque is itself a triple-layered structure. The two dense layers of a single plaque each measure 20 nm and are separated by a light zone of 10 nm (Fig. 3). The distal surface of each plaque progressively acquires more material and increases in thickness and electron density. Shortly thereafter, two hemispherical blepharoplasts appear (Fig. 4). These enlarge, become rounded (Fig. 5), and finally separate from each other, moving to opposing poles of the spindle apparatus during prophase of the last (ninth) division.

The mature blepharoplast is an electron-dense, spherical structure (0.5–1 μm in diameter) interpenetrated by numerous lightly stained channels (Fig. 6). Close examination of these channels and their surrounding dense material reveals that they have the same dimensions as a procentriole, and it has also been shown (Hepler, 1976; Mizukami and Gall, 1966) that the central, light-staining channel may contain a rod and radiating spoke material which further give it a structural similarity to the hub and central spoke structure of a developing procentriole. Mizukami and Gall (1966) have calculated that a single blepharoplast is sufficient to produce the 100–150 basal bodies needed for development of the flagellar apparatus.

FIGURES 2–6 Sequential stages of *de novo* blepharoplast formation.

FIGURE 2 The blepharoplast arises in a small indentation on the distal surface of nucleus during telophase of the next-to-last division. It first appears as a sphere of flocculent material (from Hepler, 1976). × 37,000.

FIGURE 3 Two parallel plaques emerge from within the flocculent material. Each plaque is triple layered and is separated from its partner by 45 nm. The distal layer of each plaque becomes more densely stained during subsequent stages of development (from Hepler, 1976). × 37,000.

FIGURE 4 Continued condensation of material onto the distal surfaces of each plaque produces a pair of young hemispherical blepharoplasts (from Hepler, 1976). × 37,000.

FIGURE 5 The two enlarging blepharoplasts begin to separate from each other as the cell enters prophase of the last division. Some residual material still connects the pair (from Hepler, 1976). × 37,000.

FIGURE 6 The blepharoplast is the focal point for spindle microtubules during prophase of the final division. A mature blepharoplast is 0.5–1.0 μm in diameter and is interpenetrated by numerous lightly stained channels. × 30,000.

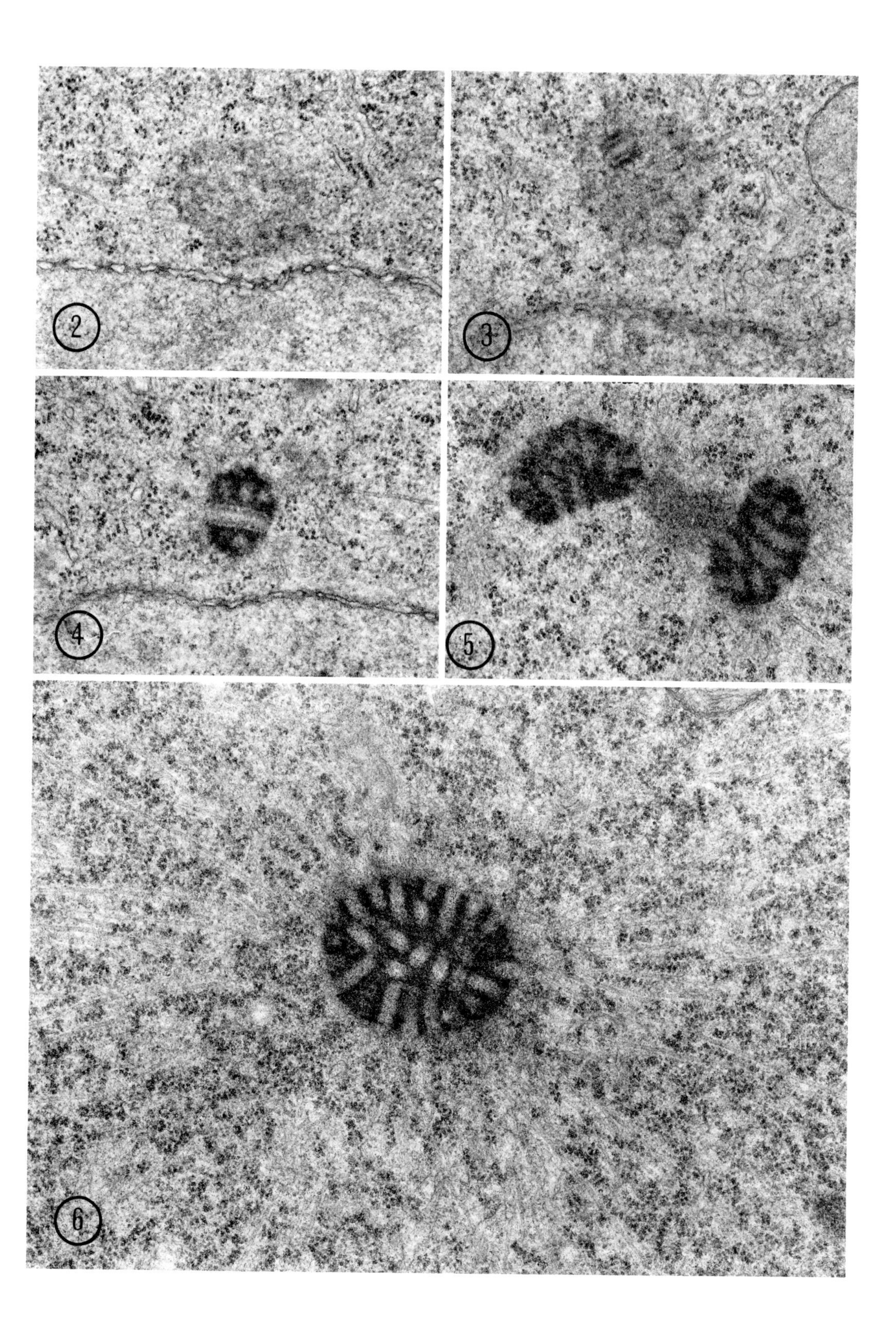
2
3
4
5
6

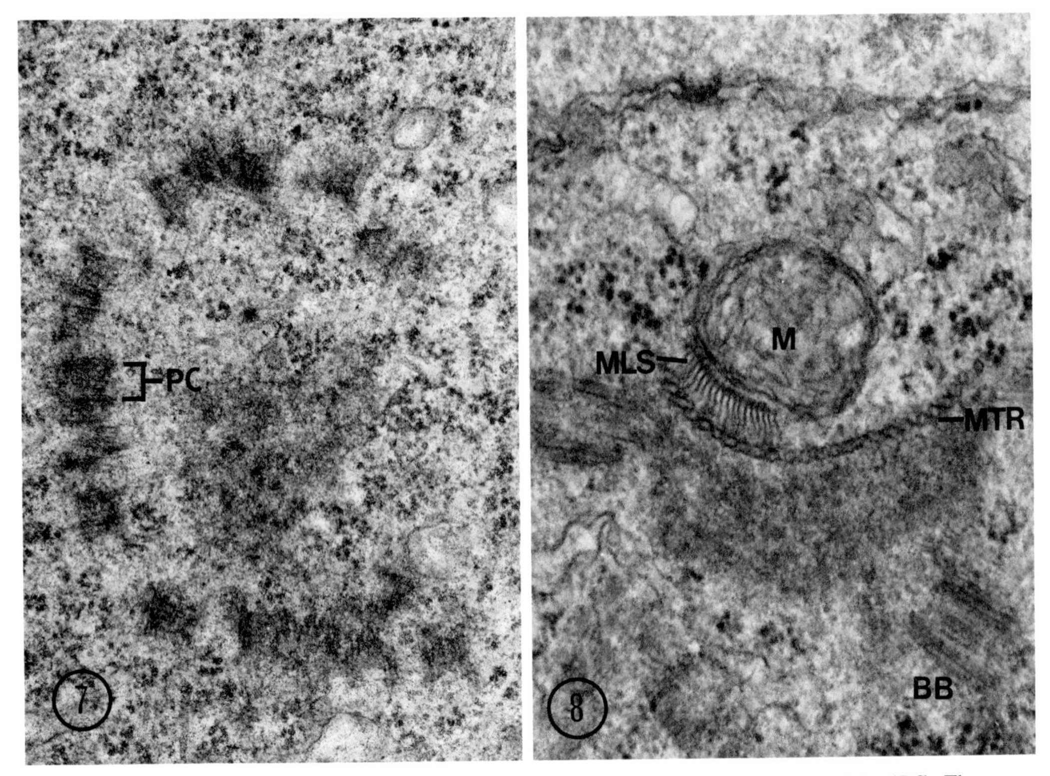

FIGURE 7 The spherical blepharoplast enlarges and transforms into numerous procentrioles (*PC*). These will subsequently mature into basal bodies with the typical nine triplet tubule structure (from Hepler, 1976). × 42,000.

FIGURE 8 The multilayered structure (*MLS*) arises at telophase of the last division. It is composed of a plaque which overlies a mitochondrion (*M*), numerous thin partitions, and a ribbon of closely aligned microtubules (*MTR*). Flocculent material occurs on the distal surface of the microtubule ribbon and may contribute to the continued growth and maturation of the basal bodies (*BB*). × 63,000.

During metaphase and anaphase of the final division, the blepharoplast begins its transformation into basal bodies. First, it becomes more lightly stained and the channel structure previously evident becomes obscured. At the surface of the blepharoplast, in the material that had stained densely, singlet tubules appear and the beginnings of a discernible procentriole structure emerge. By telophase of the last division, the blepharoplast becomes markedly swollen and young basal bodies

---

FIGURE 9 A low magnification micrograph reveals the rounded nucleus of the spermatid cell after the last cell division phase but before differentiation. The mitochondrion (*M*), multilayered structure (*MLS*), and basal body (*BB*) complex occur close to a shallow depression of the nucleus. × 19,000.

FIGURE 10 The microtubule ribbon (*MTR*) engages the nuclear envelope and nuclear shaping begins. The mitochondrion (*M*) and multilayered structure (*MLS*) form the tip of the advancing anterior spiral coil. × 45,000.

FIGURE 11 A micrograph from a transverse section through a gyre of the nucleus during an early stage of spermiogenesis. The nucleus has moved close to the cell surface. The microtubule ribbon overlies the nuclear envelope. Distal to the microtubule ribbon and emerging from the cell surface at angles to each other are two young flagella. × 31,000.

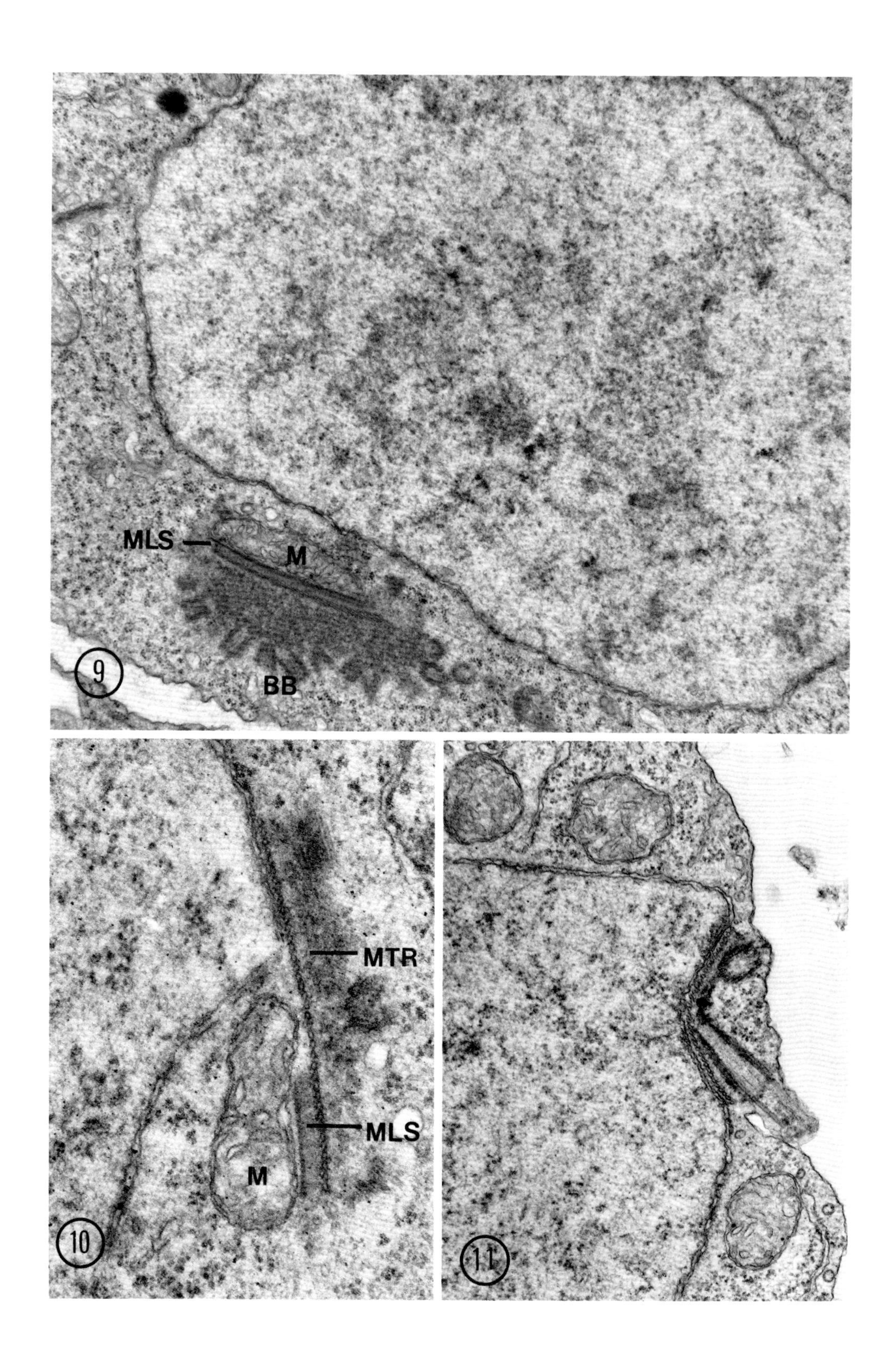
MLS
M
BB
9
MTR
MLS
M
10
11

are now clearly visible (Fig. 7). Thus, by the end of the last division, near the distal surface of the telophase nucleus, there exists a sphere of basal bodies.

Shaping of the Spermatid into a Motile Gamete

The cell division phase produces 32 spermatids, but these show almost none of the structural specializations that characterize the mature sperm. The cell shape at this time is defined by the packing of 32 cells into tetrad clusters within a spherical compartment. In the next 6 h the spermatid cell develops into an asymmetric and highly differentiated, swimming spermatozoid (Myles and Hepler, 1977).

At the beginning of sperm differentiation, the nucleus is approximately spherical with a slight indentation on its outward facing surface (Fig. 9). The developing blepharoplast situated in this indentation has now given rise to 100–150 basal bodies. Associated with the breaking-up blepharoplast is an entirely new organelle, the multilayered structure (MLS; Figs. 8–10). It arises adjacent to a mitochondrion and is composed of distinct layers, including a dense plaque near the surface of the mitochondrion, a layer of thin partitions, and finally a layer of microtubules (Fig. 8).

The complex of MLS, associated mitochondrion, and basal bodies becomes associated with a lateral edge of the nucleus (Fig. 10). The microtubule ribbon of the MLS extends posteriorly and engages the nucleus on the surface towards the outside of the cell. While the nucleus elongates along the inner surface of the microtubule ribbon, the basal bodies are being spread out along the outer surface of the ribbon between it and the plasmalemma of the cell (Figs. 10 and 11). Next, the entire organelle complex (MLS, microtubule ribbon, mitochondrion, nucleus, and basal bodies) moves toward the surface, so that the basal bodies are positioned immediately beneath the plasmalemma (Figs. 11 and 12). Thus, the MLS along with the microtubule ribbon seems to integrate the different cellular components of the future anterior spiral coil. The MLS, as the tip-most structure, may control both the initiation and subsequent growth of the coil.

Development proceeds with the continued spiral growth of the anterior coil. In the middle stages of development, the multilayered structure occupies about one gyre, and the nucleus extends through the posterior two to three gyres (Fig. 14). At the point where the nuclear envelope is closely appressed to the microtubule ribbon, one begins to see a small amount of condensation of chromatin material (Figs. 13 and 14); however, the bulk of chromatin condensation will not occur until much later in development. As the nucleus is being shaped, mitochondria appear to aggregate around it (Fig. 13). These mitochondria will probably fuse with one another and to the MLS-associated mitochondrion to produce the single, long mitochondrion which will extend the entire length (10 gyres) of the anterior spiral coil. During the middle stage of development, the flagellar band first appears (Figs. 13 and 14), possibly from material originally derived from the amorphous substance of the blepharoplast. Initially, the basal

FIGURE 12 A scanning electron microscope view of the cell surface at a stage in development comparable to that shown in Fig. 11. The flagella are disposed in a double row. However, at the tip of the advancing coil and extending backward for 2 μm, flagella grow out only on the outer edge of the curving spiral (arrow). × 10,000.

FIGURE 13 Cross section through the organelle coil showing the positional relationship of nucleus (*N*), mitochondrion (*M*), microtubule ribbon (*MTR*), and the beginnings of the flagellar band (*FB*). Chromatin condensation occurs along the nuclear envelope especially where the latter is associated with the microtubule ribbon (arrow). × 39,000.

FIGURE 14 A low magnification view shows five gyres of the developing anterior coil. The nucleus occupies the posterior three gyres whereas the mitochondrion is seen in all five. Dense material, which has aggregated between the microtubule ribbon and the plasma membrane, forms the flagellar band (*FB*). Chromatin condensation (arrow) is most conspicuous along the surface of the nucleus (*N*) associated with the microtubule ribbon (*MTR*) but nodules are seen elsewhere. A plastid (*P*) with prominent starch grains is present. × 18,000.

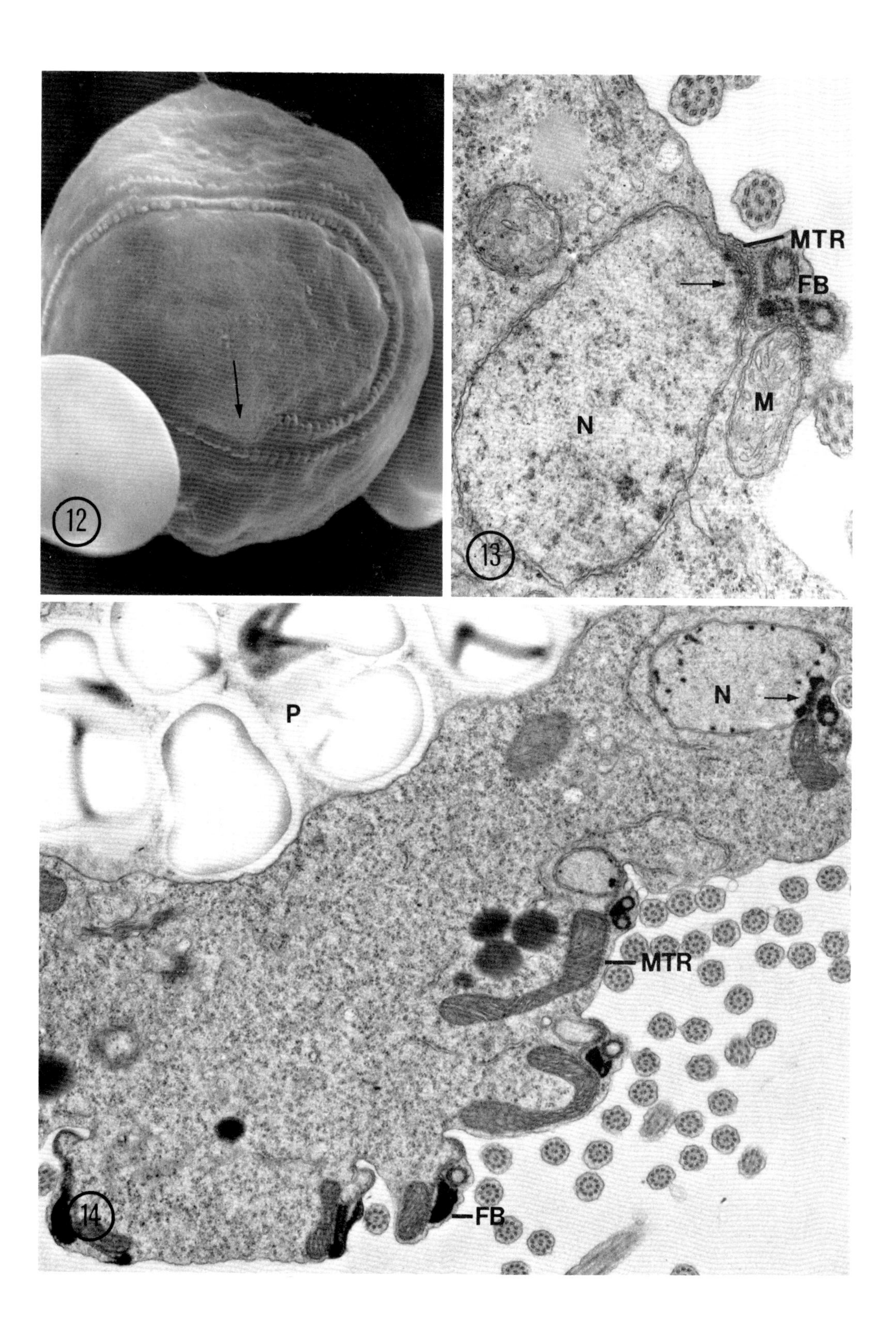
12
13
MTR
FB
M
N
14
P
N
MTR
FB

bodies are distributed at an angle to one another (Figs. 11 and 12). They are found along the entire length of the coil in two rows except in the anteriormost portion, where they occur only on the outside edge of the curving spiral (Fig. 12). This asymmetric displacement of the basal bodies at the tip appears to correspond specifically to the region of the multilayered structure, possibly indicating an underlying structural arrangement or association which is primarily responsible for introducing the curve into the growing microtubule ribbon.

In the final stages of cell shaping, the nucleus extends through the posterior four to five gyres of the anterior coil. Concomitant with the completion of shaping, the bulk of chromatin finishes condensing (Fig. 15). Condensation occurs predominately near that region of the nuclear envelope most closely associated with the microtubular ribbon. The mitochondrion continues its growth beyond the elongated nucleus until it is extended for a full 10 gyres. The basal bodies, initially distributed in V-shaped pairs, now come to lie parallel to one another, embedded in the condensed material of the flagellar band.

Accompanying the late stages of cell development is a process of cellular inversion. A thin bridge of cytoplasm grows around and completely envelopes the anterior coil of the spermatid (Figs. 15 and 16). The flagellar and anterior coil are thus confined briefly to an extracellular, but internal, space. The narrow bridge of cytoplasm expands in volume until approximately one-half of the original spermatid cytoplasm is included in it (Fig. 15). When the sperm are released, the excess cytoplasm is pinched off and shed from the sperm.

Control of Shaping

The complex series of cytomorphogenetic events, taking place during spermiogenesis in *Marsilea,* appears to be integrated and controlled by the multilayered structure and its growing ribbon of microtubules. The microtubule ribbon arises at the beginning of development and is associated with both the nucleus and the mitochondrion in a way that strongly implicates it in the shaping process. It may be the cellular component that actually generates the spiral axis along which subsequent development occurs.

We know from our studies in progress with microtubule inhibitors, including colchicine, that when we block normal formation of the microtubule ribbon, the anterior coil does not form. When colchicine is given at later stages in development, after some shaping has taken place, the microtubule ribbon is stopped from further growth and the shape of the organelle coil becomes abnormal, forming branches or loops instead of a regular spiral. Chromatin condensation, which may be instrumental in shape-generation in other kinds of sperm (Fawcett et al., 1971; Myles and Hepler, 1977, for review), appears not to be functional in shape-generation in *Marsilea*, inasmuch as it occurs to a large extent after shaping has taken place. In fact, the microtubule ribbon may even be involved in the condensation of the chromatin itself, because we have observed that chromatin condenses largely along the region of the nuclear envelope associated with the microtubule ribbon (Myles and Hepler, 1977). A close structural relationship between microtubules and condensing chromatin has been reported in other species, as well (Myles and Hepler, 1977, for review).

We think that a portion of the control of nuclear shaping may reside within the nuclear envelope itself. It seems possible that a force-generating system—for example, actin filaments or an actomyosin system—may be associated with the nuclear envelope. Microtubules, to the outside, could provide guidance for this membrane-localized, force-generating system and thus give directionality to the shaping process, while chromatin, on the inside, might interact with these hypothetical fibrous proteins and use them as foci for condensation. These ideas differ in an important way from those currently under discussion, by suggesting that shaping during spermiogenesis is driven neither by the microtubules nor by the condensing chromatin, but by an intermediary membrane-associated component.

Our results to date have not allowed us to dissect the active and passive processes of shaping, but they are generally consistent with the idea that the nuclear envelope plays an important role, possibly in providing forces for shaping, and that the microtubules impart directionality and thus coordinate the events.

Conclusions

Motile male gametes among plants and animals show diversity in outward morphology, however, closer examination reveals a great similarity in design and structure. The gamete has been stripped of excess baggage and at the same time has been equipped with a motile system, the flagella; a centrally organized energy producer, the mitochondrion; and a highly condensed rod of

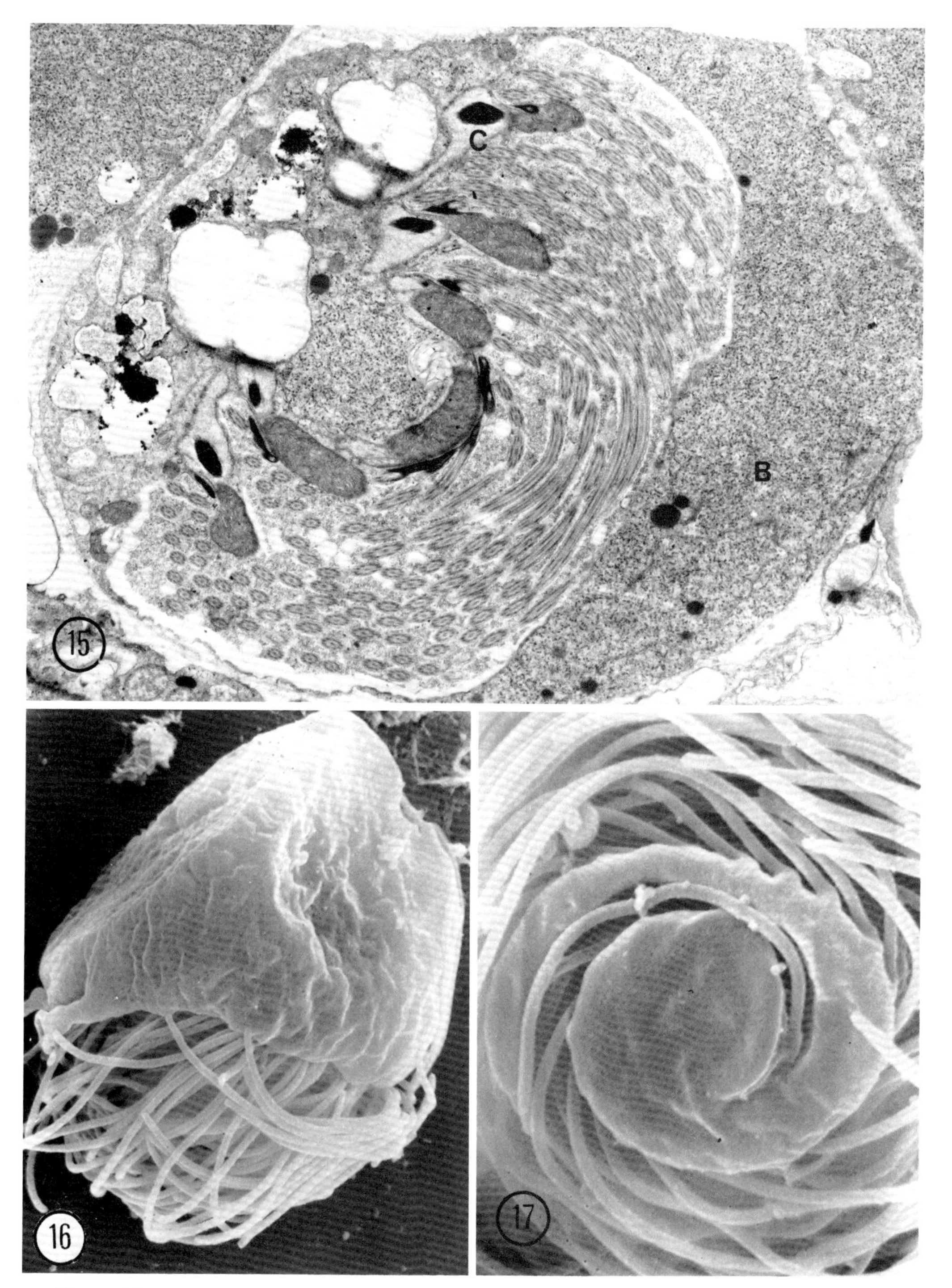

FIGURE 15 In late development the differentiating spermatid forms a bridge of cytoplasm (*B*) which encloses the anterior coil and flagella in an extracellular but internal space. In this micrograph the anterior coil is viewed obliquely revealing only the posterior three to four coils. Chromatin (*C*) is completely condensed. × 11,000.

FIGURE 16 A scanning electron micrograph depicts a stage in late development in which the cytoplasmic bridge has partially grown over the anterior coil. Great numbers of flagella obscure the coil morphology. × 10,000.

FIGURE 17 The anterior coil observed in face view by scanning electron microscopy reveals the left-handed nature of the spiral. × 17,000.

chromatin. The function of delivering the genetic material to the egg cell appears to demand these unique structural features found so commonly in plants and animals.

Development of the plant male gamete also involves structures and processes common to animal gametes. Microtubules, for example, appear in *Marsilea,* as they do in many other organisms, as one of the prominent structural components involved in the generation and maintenance of cell shape (Hepler and Palevitz, 1974; McIntosh and Porter, 1967; Porter, 1966). The microtubule ribbon in *Marsilea* appears to coordinate nuclear shaping, mitochondrial fusion and elongation, and distribution and displacement of the basal bodies. Microtubules may also participate in controlling chromatin condensation.

The complex multilayered structure, typical of all plant spermatozoids so far studied (Bell, 1974; Carothers, 1975; Duckett, 1975; Lal and Bell, 1975; Norstog, 1975), as well as some algal zoospores (Pickett-Heaps, 1975), is not found in animals and thus stands out as a unique and fascinating organelle. It appears to contain the microtubule organizing center (MTOC), because microtubules first appear as an integral part of it and extend posteriorly from it during development.

The multiflagellate condition may also be unique to plants, and even here it is restricted to ferns, fern allies, cycads, and ginkgo. Although this produces some very dramatic cells (for example, *Zamia* sperm have 10–12,000 flagella [Norstog, 1975]), there would seem to be very little difference conceptually between sperm with one or two flagella and those with many. With regard to the motile apparatus, the plants provide the best example of the formation *de novo* of basal bodies (Hepler, 1976; Paolillo, 1975; Sharp, 1914). The recent electron-microscope studies (Hepler, 1976) confirm the extensive older literature on sperm formation in plants, showing that basal bodies arise in cells that contain no apparent preexisting structure or template. The structural steps in *de novo* formation have been established for *Marsilea*, but the control of that dramatic process continues to elude us.

Finally, we emphasize the efficacy of plant gametes for cytological, pharmacological, and biochemical studies of the complex process of cytomorphogenesis. The ease with which the developing *Marsilea* microspore can be modulated and controlled gives us a strong lean on this system and provides us with many different ways of attacking the problem of cytomorphogenesis experimentally. These organisms may help unravel some of the major problems still outstanding in spermatogenesis and cell shaping.

ACKNOWLEDGMENTS

We thank Mr. Neal Burstein for helping us prepare the scanning electron micrographs. We also thank Ms. Deborah Stairs for technical assistance.

This work was supported by a National Institutes of Health postdoctoral fellowship, no. 5 F22 HD00636-03 to Diana Myles, National Science Foundation grants BMS 74-15245 and PCM 74-15245-AO2 to Peter Hepler.

REFERENCES

Bell, P. R. 1974. The origin of the multilayered structure in the spermatozoid of *Pteridium aquilinum*. *Cytobiologie*. **8:**203–212.

Bilderback, D. E., T. L. Jahn, and J. R. Fonseca. 1974. The locomotor behavior of *Lygodium* and *Marsilea* sperm. *Am. J. Bot.* **61:**888–890.

Carothers, Z. B. 1975. Comparative studies on spermatogenesis in bryophytes. *Biol. J. Linn. Soc.* **6**(suppl. 1)**:**71–84.

Duckett, J. G. 1975. Spermatogenesis in pteridophytes. *Biol. J. Linn. Soc.* **6**(suppl. 1)**:**97–127.

Fawcett, D. W., W. A. Anderson, and D. M. Phillips. 1971. Morphogenetic factors influencing the shape of the sperm head. *Dev. Biol.* **26:**220–251.

Hepler, P. K. 1976. The blepharoplast of *Marsilea:* its *de novo* formation and spindle association. *J. Cell Sci.* **21:**361–390.

Hepler, P. K., and B. A. Palevitz. 1974. Microtubules and microfilaments. *Annu. Rev. Plant. Physiol.* **25:**309–362.

Lal, M., and P. R. Bell. 1975. Spermatogenesis in mosses. *Biol. J. Linn. Soc.* **6**(suppl. 1)**:**85–95.

McIntosh, J. R., and K. R. Porter. 1967. Microtubules in the spermatids of domestic fowl. *J. Cell Biol.* **35:**153–173.

Mizukami, I., and J. Gall. 1966. Centriole replication. II. Sperm formation in the fern, *Marsilea,* and the cycad, *Zamia*. *J. Cell Biol.* **29:**97–111.

Myles, D. G., and P. R. Bell. 1975. An ultrastructure study of the spermatozoid of the fern, *Marsilea vestita*. *J. Cell Sci.* **17:**633–645.

Myles, D. G., and P. K. Hepler. 1977. Spermiogenesis in the fern *Marsilea:* microtubules, nuclear shaping and cytomorphogenesis. *J. Cell Sci.* In press.

Norstog, K. 1975. The motility of cycad spermatozoids in relation to structure and function. *Biol. J. Linn. Soc.* **6**(suppl. 1)**:**135–142.

Paolillo, D. J. 1975. Motile male gametes in plants. *In* Dynamic Aspects of Plant Ultrastructure. A. W. Ro-

bards, ed. McGraw-Hill Co. Ltd. (U.K.) 504–531.
PICKETT-HEAPS, J. D. 1975. Structural and phylogenetic aspects of microtubular systems in gametes and zoospores of certain green algae. *Biol. J. Linn. Soc.* **6**(suppl. 1)**:**37–44.
PORTER, K. R. 1966. Cytoplasmic microtubules and their functions. *In* Principles of Biomolecular Organization. *CIBA Found. Symp.* 308–334.
SHARP, L. W. 1914. Spermatogenesis in *Marsilea*. *Bot. Gaz.* **58:**419–431.

UNUSUAL FEATURES OF INSECT SPERMATOGENESIS

BACCIO BACCETTI

The insect sperm cell has typical features that make it readily recognizable among the other sperm models. These characteristics have been acquired through a series of evolutionary steps which are still represented in some of the lower apterygote orders, and are evident in almost all the most important groups of Pterygota. Generally speaking, in other arthropodan classes, the sperm evolved quickly from the aquatic, primitive form toward immotility, either assuming an "encysted" form, as in Arachnida, or directly losing the axoneme and organelles, as in Crustacea or Myriapoda (Baccetti, 1970, 1977). In Insecta, on the other hand, it typically develops an extremely elongated flagellum. In fact, among insects, the encysted form is found only in Collembola (Dallai, 1970), a primitive, isolated line; absence of the flagellum is also a rare phenomenon, occurring only at the end of a few highly evolved and predominantly motile lines, where it usually represents the final step of evolution. It is only known in the highest proturans, in the highest homopterans, in the highest termites, and in the psychodid flies.

The long, filiform, vigorously motile insect spermatozoon has developed distinctive characteristics that assure the movement of these extremely elongated cells in a viscous medium. These types of spermatozoa acquire new organelles, not commonly found in the other phyla. On the other hand, the aflagellate, immotile insect spermatozoon simply reduces its organelles and sometimes consists of only a nucleus and a chondriome. In only one case did motility evolve from the immotile ancestors, but an axoneme did not appear again; instead, other systems are employed.

Sperm morphogenesis clearly reflects the peculiar structural features of the mature spermatozoon. It is unusually complicated in the motile models. In the aflagellate spermatozoa, it is regressive, sometimes conserving traces of organelles that are not present in the mature cell. Let me give some examples.

The most important features of a typical insect spermatozoon are an extremely thin, elongated form with little cytoplasm; a very long axoneme endowed with nine accessory tubules and flanked by two mitochondrial derivatives containing crystalline material; and two crystalline accessory bodies. The crystalline intramitochondrial material is not confined to insect sperm, for it is also seen in Gastropoda (André, 1962), and accessory bodies have been demonstrated in two other groups, the Chaetognatha (Van Deurs, 1972), and the Elasmobranchia (Stanley, 1971); nevertheless, the morphogenesis and fine structure of these organelles are well known only in insects.

The first consequence of the elongation and thinning is a remodeling of the plasma membrane. In sperm of several species of insects, this remodeling is performed in the spermatid, which, in its late stages, contains the fully developed, mature spermatozoon delimited by an almost complete new membrane (Baccetti et al., 1973*a*; Baccetti, 1975). It is interesting to see how this new intracellular plasma membrane develops. At first, large, flat cisternae arise from the Golgi complex (Fig. 1), and, like the Golgi complex, these are rich in thiamine pyrophosphatase (TPPase) activity (Fig. 3). This system of membranes becomes disposed around the organelles of both the tail and the nucleus, and demarcates the limits of the future mature spermatozoon from the remaining portion of the spermatid cytoplasm (Figs. 2–4), which may include most of the transient spermatid organelles. These cisternae fuse to form a cylindrical double wall that finally splits (Fig. 2), so that the inner surface of the cisterna becomes the outer surface of the new sperm membrane, and the residual cytoplasm is enclosed between the spermatid plasma membrane and the

BACCIO BACCETTI Institute of Zoology, University of Siena, Siena, Italy

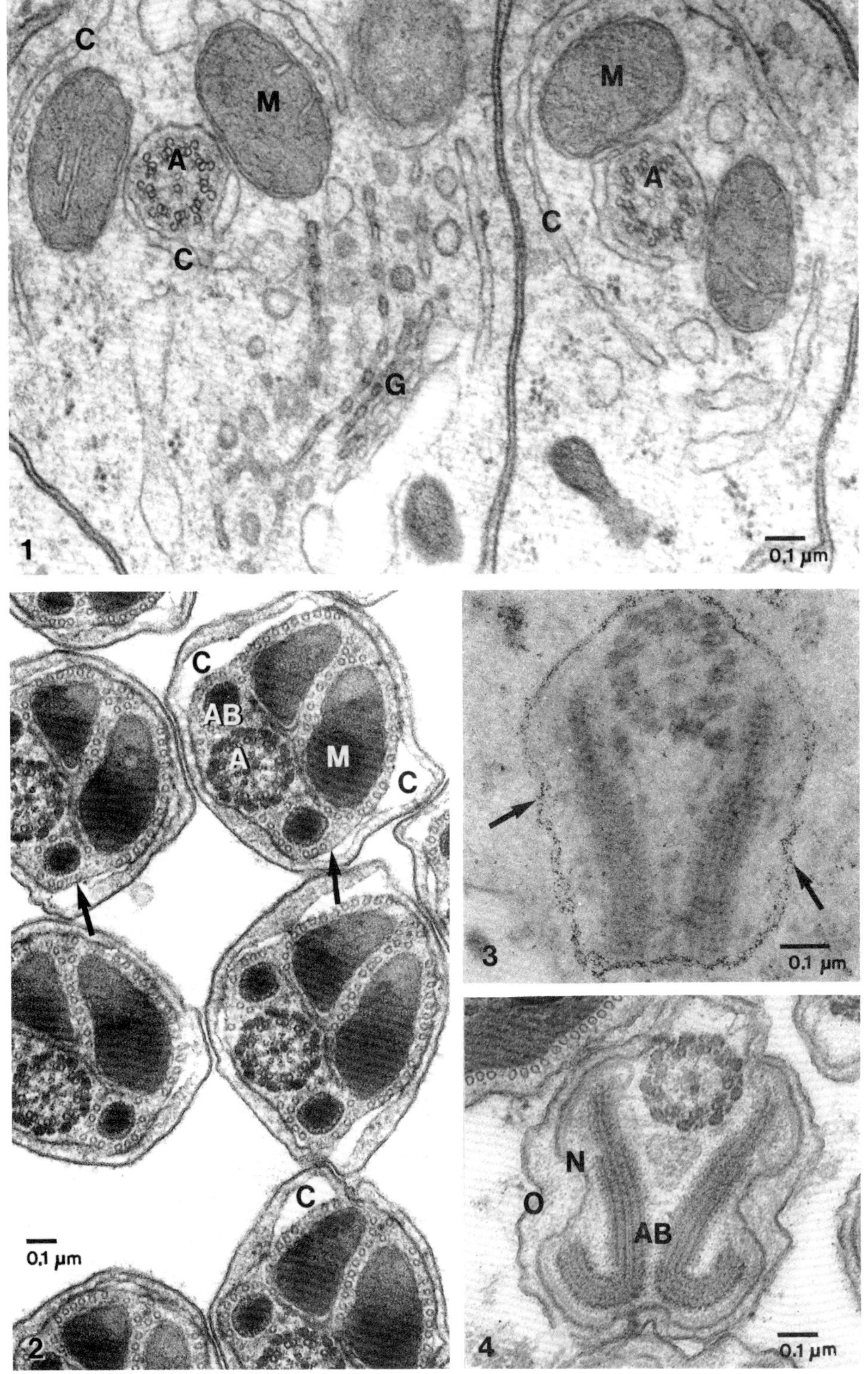

FIGURE 1 Young spermatids of *Tenebrio molitor*. Mitochondrial derivatives (*M*) and axoneme (*A*) are surrounded by a system of cisternae (*C*) originated by the Golgi complex (*G*). × 60,000.

FIGURE 2 Late spermatids of *Tenebrio*. Axoneme (*A*), mitochondrial derivatives (*M*), and accessory bodies (*AB*) are completely surrounded by a Golgi-derived cisterna (*C*), split at many points and originating a new plasma membrane (arrows). × 45,000.

FIGURE 3 TPPase activity (arrows) in the Golgi-derived cisterna surrounding the sperm organelles in a spermatid of *Bacillus rossius*. × 75,000.

FIGURE 4. Late spermatid of *Bacillus rossius*. The tail organelles, among them accessory bodies (*AB*), are evident, surrounded by the old (*O*) and the new (*N*) plasma membrane. × 60,000.

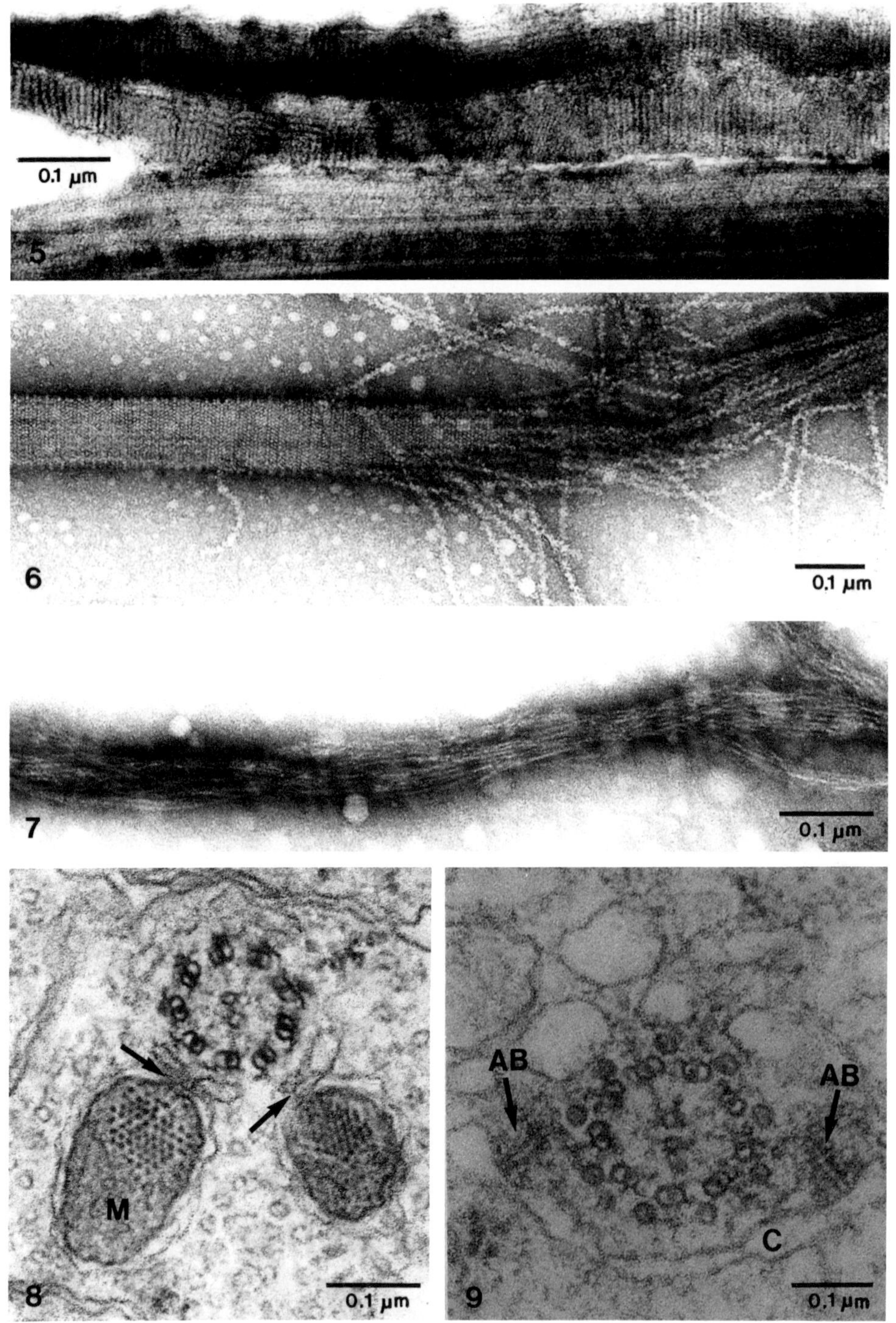

FIGURE 5 Fragmented mitochondrial derivative of *Notonecta glauca,* negatively stained by PTA (from Baccetti et al., 1977). × 135,000.

FIGURE 6 The same, after papain digestion. Coils of filaments are evident. PTA negative staining (from Baccetti et al., 1977). × 100,000.

FIGURE 7 The same, after EDTA treatment. Individual filaments are resolved. PTA negative staining (from Baccetti et al., 1977). × 135,000.

FIGURE 8 Spermatid of *Musca domestica*. Crystalline material is precipitated in the mitochondrial derivatives (*M*), starting from the points of fusion between mitochondrial membrane and Golgi-derived cisternae (arrows). × 135,000.

FIGURE 9 Young spermatid of *Bacillus rossius*. Accessory bodies (*AB*) are formed by Golgi-derived cisternae (*C*), (from Baccetti et al., 1973*b*). × 120,000.

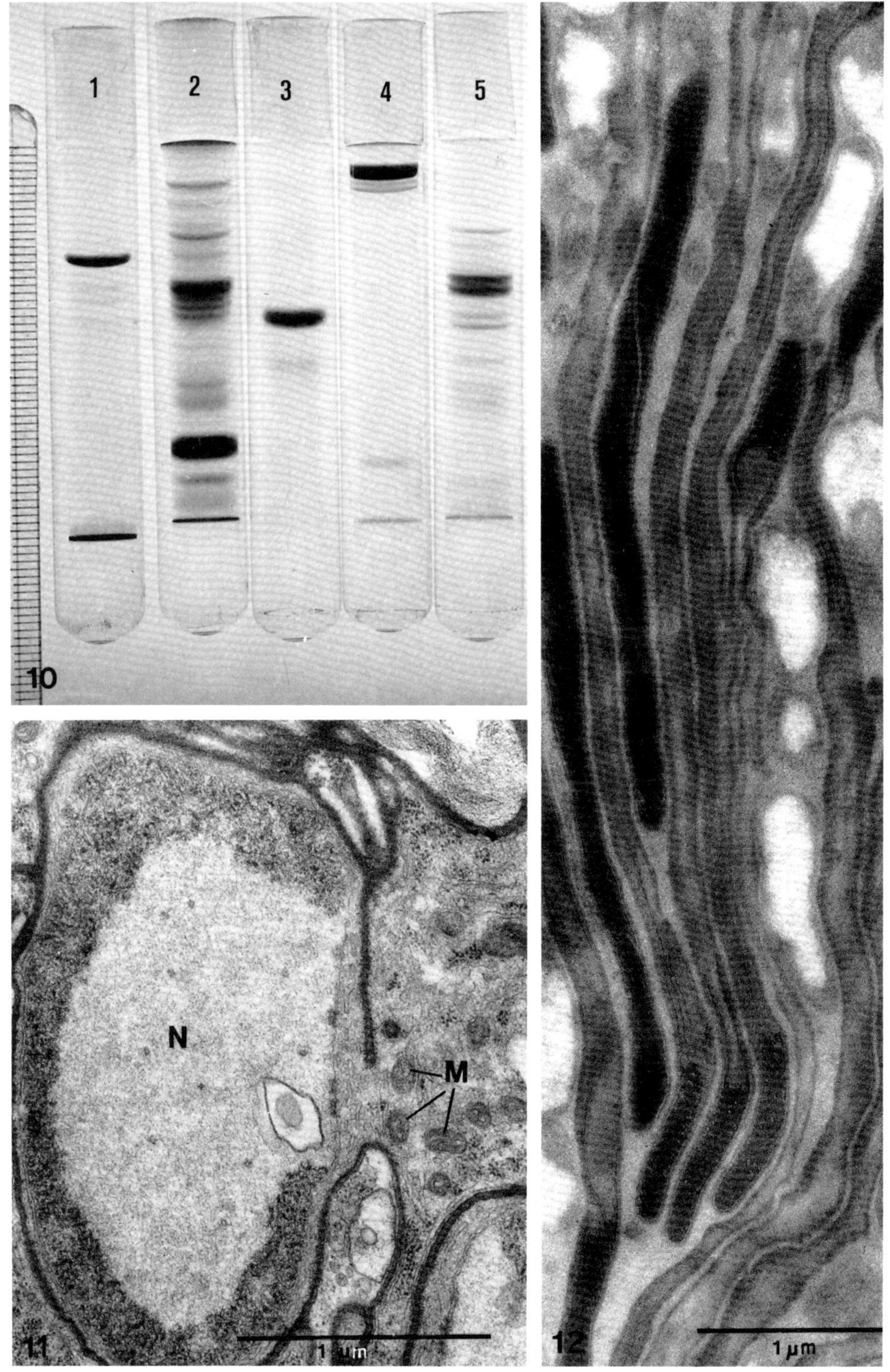

FIGURE 10 Spermatozoon of *Notonecta glauca* after SDS extraction. SDS polyacrylamide gel electrophoresis. In (5), the two bands, 52,000 and 56,000 mol wt, of crystallomitin are evident. In the other columns, cytochrome C + seralbumin (1), sea urchin sperm flagellum + papain (2), ovalbumin (3), and myosin (4) are used as standards (from Baccetti et al., 1977).

FIGURE 11 Young spermatid of *Eosentomon transitorium*. The nucleus (*N*) occupies most of the cell volume; the other organelles are only the mitochondria (*M*) located at the periphery of the cell (courtesy of R. Dallai). × 27,000.

FIGURE 12 Mature spermatozoa of *Eosentomon*. The flattened shape is evident (from Baccetti et al., 1973*d*). × 37,500.

outer membrane of the split cisterna. Later, this sheath of the spermatid cytoplasm is eliminated, and the spermatozoon emerges, enveloped now by what was formerly the inner lamina of the fused cisternae. Eventually, new glycocalyx formations develop in association with this Golgi-derived cisterna. The most conspicuous surface differentiations of insect sperm, the appendages of lepidopteran sperm (Phillips, 1971), have an origin accounted for by this process of development.

According to Friedländer (1977), the microtubular manchette located at the periphery of the persisting central portion of the spermatid remains adherent to the new plasma membrane. It becomes extracellular during exfoliation of the residual cytoplasm and is transformed into the crystalline surface decoration of the new sperm plasma membrane.

The Golgi complex also seems to be involved in the crystallization of insect mitochondria. The complicated process of rearrangement and fusion among spermatid mitochondria, from which originate one or two large, longitudinal mitochondrial derivatives, has been studied by André (1962) and Pratt (1968). In most insect sperm, these mitochondrial derivatives are filled with a crystalline protein, with a longitudinal periodicity of 450 Å, divided into two subperiods of 225 Å (Fig. 5). This protein is assembled during the late spermatid stage. It has been demonstrated (Baccetti et al., 1973*a*; Baccetti, 1975) that the crystalline material appears at a time when two flattened Golgi-derived cisternae come into contact with the wall of the recently reorganized mitochondrial derivatives, where the space within the two membranes is narrower (Fig. 8). At this point, osmiophilic filaments appear within the mitochondrial matrix, and these increase in number during the final stages of sperm maturation. The nature of the crystals has recently been elucidated (Baccetti et al., 1977). The crystals are devoid of any enzymatic activity; they are made up of a peculiar, filamentous protein named by us *crystallomitin*. It contains two polypeptide chains, 52,000 and 55,000 daltons, respectively, with high proline and cysteine content. The chains are evident in SDS polyacrylamide gel electrophoresis after the removal of tubulin (Fig. 10). After EDTA treatment, the substructure of the crystals is resolved into 2-nm globules aligned into filaments 2 nm thick (Fig. 7). After proteolysis, the filaments appear as coiled coils (Fig. 6) with a 225 Å pitch, which accounts for the longitudinal periodicity of the crystals.

Another important morphogenetic event in insect spermatids is the assembly of the two accessory bodies. They also are longitudinal, crystalline structures, flanking the axoneme for almost their whole length, as do the mitochondrial derivatives. In some instances, such as in *Tenebrio* (Fig. 2), they are thin (Baccetti et al., 1973*a*), but in *Bacillus* (Fig. 4), where the mitochondrial derivatives are absent, these structures become enormous and occupy most of the tail volume (Baccetti et al., 1973*b*). A general characteristic of the accessory bodies is an intense ATPase activity (Baccetti et al., 1973*a*, *b*). Whereas the crystallomitin of mitochondrial derivatives plays only an elastic, passive role in relation to the flagellar beat, or is merely a site of storage of material needed in the first stages of embryogenesis, the accessory bodies probably play an active role in tail movement,

FIGURE 13 Young spermatid of *Reticulitermes lucifugus*. Centriole (*C*) and two mitochondria (*M*) are located near the nucleus (*N*). Lysosomes (*L*) are frequent in the cytoplasm, where a rich system of cisternae develops from the Golgi complex (*G*). × 48,000.

FIGURE 14 A later stage of *Reticulitermes lucifugus* spermatid. The sperm organelles are surrounded by a new plasma membrane (*PM*). × 61,500.

FIGURE 15 Mature spermatozoon of *Trialeyrodes vaporariorum*. Scanning electron microscope. × 15,000.

FIGURE 16 Cross section of mature spermatozoon of *Trialeyrodes vaporariorum*. Most of the volume is occupied by the nucleus (*N*). × 75,000.

FIGURE 17 *Trialeyrodes* spermatid. Cross section of the transient axoneme (*A*). × 50,000.

FIGURE 18 *Trialeyrodes* spermatid. Longitudinal section of the transient axoneme (*A*) located near the nucleus (*N*). × 75,000.

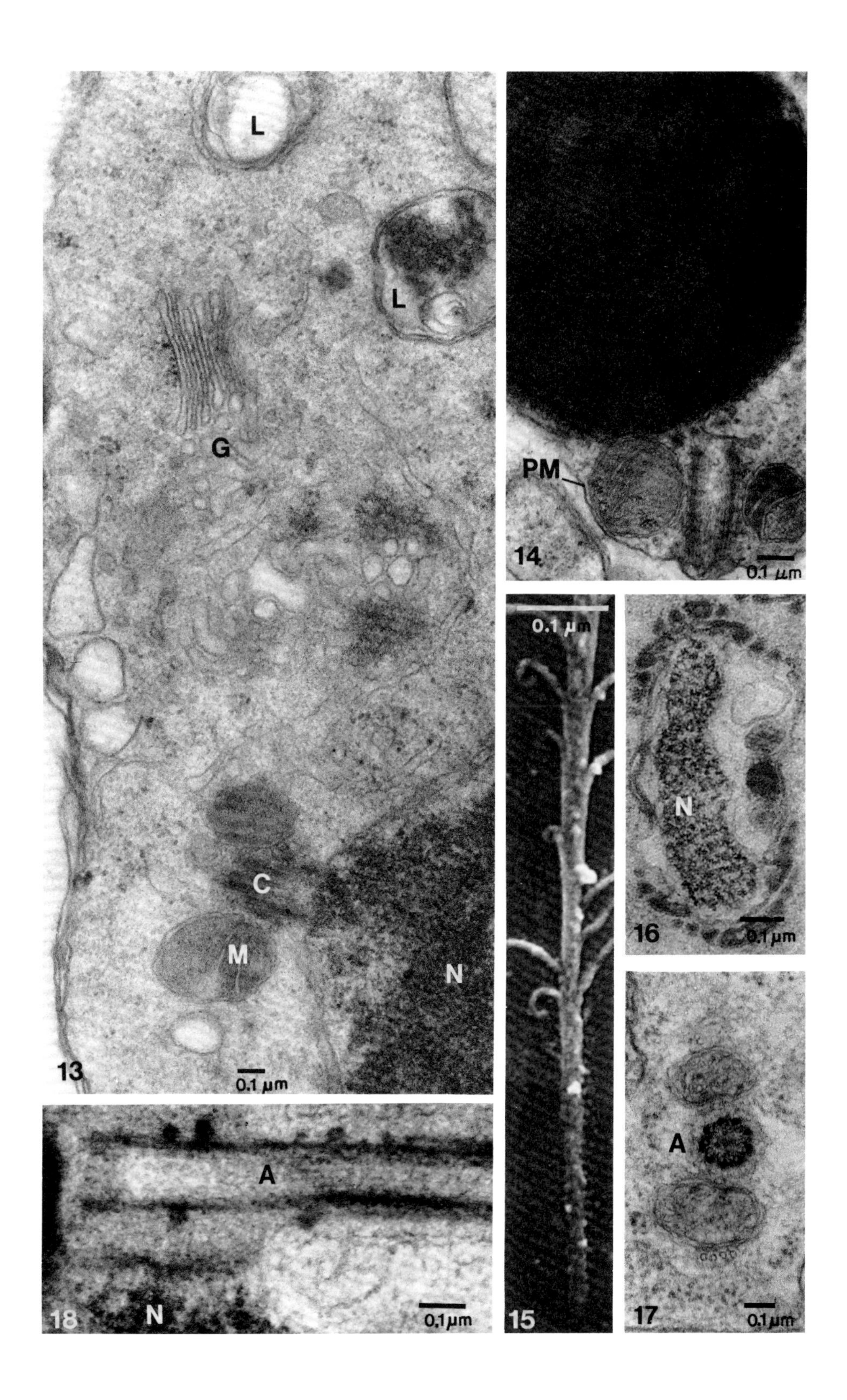
L
L
G
C
M
N
13
0.1 μm
PM
14
0.1 μm
0.1 μm
15
N
16
0.1 μm
A
17
0.1μm
A
N
18
0.1μm

giving the tail waves specific geometric characteristics (Baccetti et al., 1973*a*, *b*; Baccetti, 1972; Baccetti and Afzelius, 1976). They also are formed in the spermatid stage by cisternae arising from the Golgi complex, and show TPPase activity; they start as fingerlike evaginations and progressively acquire added complexity by the juxtaposition of several membrane layers and the assembly of transverse septa in their interior (Fig. 9). Their protein composition is under investigation in our laboratory: it is filamentous, SDS-resistant, contains several electrophoretic bands between 40,000 and 200,000 mol wt, and is rich in sulfur. In the SDS extract, the *Bacillus* spermatozoon does not show any band comigrating with actin.

The three examples described (rapid membrane substitution, crystallization of mitochondrial derivatives, and assembly of two accessory crystalline bodies) represent morphogenetic events in normal spermatogenesis peculiar to insects, and all involve participation of the Golgi complex. Other unusual features occur when classic organelles are absent from the mature sperm. Two such situations can be found. In one, the organelles are absent from the outset. In the other, they are present in the young spermatid, and disappear later. The difference may reflect the antiquity of the evolutionary process. The first case is exemplified by the proturan *Eosentomon*, or by the psychodid dipterans. The *Eosentomon* spermatozoon consists of merely a disk-shaped cell (Fig. 11), including only a flattened nucleus and scattered mitochondria (Baccetti et al., 1973*d*). The mature psychodid dipteran sperm is fusiform, and includes an elongated nucleus, an acrosomal complex, and a mitochondrial derivative (Baccetti et al., 1973*c*). In both instances, spermatogenesis directly prepares the mature sperm; namely, in *Eosentomon*, the spermatid has only a few mitochondria that remain unchanged, and a nucleus that progressively flattens (Fig. 12); in psychodids we also found an unusual elaboration of the acrosomal complex, and the fusion of mitochondria into a single derivative. At any rate, in both the psychodids and the *Eosentomon*, the centriole and axoneme are consistently lacking, and no transient organelles are found.

The second case, in which organelles normally present in conventional motile sperm are initially present but disappear later in the spermatid, is the more common. In the termite *Reticulitermes*, which has only two normal mitochondria in the mature sperm (Fig. 14), the other spermatid mitochondria (Fig. 13) are eliminated by lysosomal digestion (Baccetti et al., 1974); in the phasmid *Bacillus*, which completely lacks mitochondria, the spermatid chondriome is isolated and eliminated with the remnants of cytoplasm (Baccetti et al., 1973*b*). Perhaps the most peculiar occurrence is the absence of the axoneme from the mature sperm in aleyrodids. These peculiar hemipterans, which are closely related to the scale insects, have an elongated, aflagellate spermatozoon (Figs. 15 and 16), but in the spermatid stage a short axoneme and accessory tubules (Fig. 17) are formed near the nucleus (Fig. 18) and are then eliminated at maturity (Baccetti, unpublished data).

Finally, let me add a few words on the secondary acquisition of motility. We mentioned above that at the ends of lines that evolved towards the absence of flagella, motility appears again in a few instances, evidently as a result of new exigencies of evolution, as, for example, in ticks or in the Ostracoda. In insects, the classic case is that of coccids, reviewed by Robison (1970). The axoneme, lost in the aleyrodids, which can be considered ancestors of coccids, does not evolve anew. Instead, a new motile system appears, made up of a palisade of microtubules activated to slide along one another by the adenosine triphosphatase activity in the interstices between them (Moses, 1966). The origin of the new motile device is clear in the termite *Calotermes* (Baccetti et al., 1974), which also develops a peripheral palisade of microtubules in its aflagellate spermatozoon. The system has a persistent manchette, a structure common to almost all the elongating spermatids; this structure usually disappears at maturity. Intertubular ATPase activity has been detected in *Calotermes*, although movement has not been demonstrated. It would be important to know whether the ATPase belongs to the class of dyneins. At this time, the secondary acquisition of motility in a formerly simplified spermatozoon seems to depend mainly on the retention of juvenile characteristics.

ACKNOWLEDGMENT

Research performed under C.N.R. "Biology of Reproduction" project.

REFERENCES

André, J. 1962. Contribution à la connaissance du chondriome: étude de ses modifications ultrastructurales pendant la spermatogénèse. *J. Ultrastruct. Res.* **3**(suppl.)**:**1–185.

BACCETTI, B. 1970. The spermatozoon of Arthropoda. IX. The sperm cell as an index of arthropod phylogenesis. *In* Comparative Spermatology. B. Baccetti, editor. Accademia dei Lincei, Roma. Academic Press, Inc., N.Y. 169–182.

BACCETTI, B. 1972. Insect sperm cells. *Adv. Insect Physiol.* **9:**315–397.

BACCETTI, B. 1975. The role of the Golgi complex during spermiogenesis. *Curr. Top. Dev. Biol.* **10:**103–122.

BACCETTI, B. 1977. Ultrastructure of sperm and its bearing on Arthropod phylogeny. *XV Int. Congr. Entomol. Proc.* (In press).

BACCETTI, B., and B. A. AFZELIUS. 1976. The biology of the sperm cell. *Monogr. Dev. Biol.* **10:**1–254.

BACCETTI, B., A. G. BURRINI, R. DALLAI, F. GIUSTI, M. MAZZINI, T. REINERI, R. ROSATI, and G. SELMI. 1973*a*. Structure and function in the spermatozoon of *Tenebrio molitor*: the spermatozoon of Arthropoda. XX. *J. Mechanochem. Cell Motility.* **2:**149–161.

BACCETTI, B., A. G. BURRINI, R. DALLAI, V. PALLINI, P. PERITI, F. PIANTELLI, F. ROSATI, and G. SELMI. 1973*b*. Structure and function in the spermatozoon of Arthropoda. XIX. *J. Ultrastruct. Res.* **44**(suppl. 12)**:**1–73.

BACCETTI, B., R. DALLAI, and A. G. BURRINI. 1973*c*. The spermatozoon of Arthropoda. XVIII. The non-motile bifurcated sperm of Psychodidae flies. *J. Cell Sci.* **12:**287–311.

BACCETTI, B., R. DALLAI, and B. FRATELLO. 1973*d*. The spermatozoon of Arthropoda. XXII. The 12+0, 14+0 or aflagellate sperm of Protura. *J. Cell Sci.* **13:**321–335.

BACCETTI, B., R. DALLAI, V. PALLINI, F. ROSATI, and B. AFZELIUS. 1977. The protein of insect sperm mitochondria: crystallomitin. *J. Cell Biol.* (In press).

BACCETTI, B., R. DALLAI, F. ROSATI, F. GIUSTI, F. BERNINI, and G. SELMI. 1974. The spermatozoa of Arthropoda. XXVI. The spermatozoon of Isoptera, Embioptera and Dermaptera. *J. Microsc.* (*Paris*). **21:**159–172.

DALLAI, R. 1970. The spermatozoon of Arthropoda. XI. Further observations on Collembola. *In* Comparative Spermatology. B. Baccetti, editor. Accademia dei Lincei, Roma. Academic Press, Inc., N.Y. 275–279.

FRIEDLÄNDER, M. 1976. The role of transient perinuclear microtubules during spermiogenesis of the warehouse moth *Ephestia cautella*. *J. Submicz. Cytol.* **8:**319–326.

MOSES, M. J. 1966. Cytoplasmic and intranuclear microtubules in relation to development, chromosome morphology and motility in an aflagellate spermatozoon. *Science* (*Wash. D.C.*). **154:**424.

PHILLIPS, D. M. 1971. Morphogenesis of the lacinate appendages of lepidopteran spermatozoa. *J. Ultrastruct. Res.* **34:**567–585.

PRATT, S. 1968. An electron microscope study of nebenkern formation and differentiation in spermatids of *Murgantia histrionica* (Hemiptera, Pentatomidae). *J. Morphol.* **126:**31–66.

ROBISON, W. G. 1970. Unusual arrangement of microtubules in relation to mechanisms of sperm movement. *In* Comparative Spermatology, B. Baccetti, editor. Accademia dei Lincei, Roma. Academic Press, Inc., N.Y. 311–320.

STANLEY, H. P. 1971. Fine structure of spermiogenesis in the elasmobranch fish *Squalus suckleyi*. II. Late stages of differentiation and structure of the mature spermatozoon. *J. Ultrastruct. Res.* **36:**103–118.

VAN DEURS, B. 1972. On the ultrastructure of the mature spermatozoon of Chaetognath, *Spadella cephaloptera*. *Acta. Zool.* (*Stockh.*). **53:**93–104.

UNSOLVED PROBLEMS IN MORPHOGENESIS OF THE MAMMALIAN SPERMATOZOON

DON W. FAWCETT

Other participants in this symposium have reviewed recent progress in our understanding of oogenesis and the morphogenesis of the male gamete in plants and in insects. I shall concentrate on some of the intriguing unsolved problems in the development and functional morphology of the mammalian spermatozoon.

Transient Organelles in the Morphogenesis of Spermatozoa

A neglected aspect of spermatogenesis is concerned with those transient organelles that appear at certain stages of germ cell differentiation, carry out their mission, and then disappear, leaving behind no residue or derivative in the mature spermatozoon. These include the microcisternae of the spermatocytes, the chromatoid body, the manchette, and the centriolar adjunct of spermatids.

Parallel arrays of two to six very small flattened vesicles or *microcisternae* (Figs. 1 and 2) are found in the peripheral cytoplasm of primary spermatocytes but in no other cells of the seminiferous epithelium. These were first described by Nicander and Ploen (1969) and have been observed by most investigators of the ultrastructure of the testis. They have been misinterpreted by some as small dictyosomes associated with fragmentation and reconstitution of the Golgi complex in the spermatocyte divisions. However, they are present in early primary spermatocytes with an intact Golgi, and they increase in number through the zygotene and pachytene stages and then disperse in the latest stages of prophase. They are most abundant at the periphery of the cell. The Golgi complex has not been clearly implicated in their formation. In freeze-fracture replicas, they have much the same configuration as in thin sections and they constitute a useful feature for identification of spermatocyte cytoplasm in such preparations. In contrast to the Golgi, their membranes have very few intramembrane particles. Their origin, function, and exclusive occurrence in spermatocytes remain unexplained.

The cylindrical array of microtubules that form the *manchette* of spermatids was one of the earliest examples of cytoplasmic microtubules to be described (Burgos and Fawcett, 1955). This is attributable to the fact that they are more stable than most microtubules and are preserved by osmium-containing primary fixatives, whereas visualization of other microtubules, including those of the neighboring Sertoli cells, had to await the introduction of aldehyde fixatives. The manchette has been more extensively studied than the other transient organelles and has been assigned a function in the morphogenesis of the spermatid. It appears at the onset of spermatid elongation and is believed to provide a track directing the flow of cytoplasmic organelles and inclusions from the nuclear pole of the cell to the caudal pole (Fig. 3). When spermatid elongation is completed, the manchette dissociates to permit aggregation of mitochondria around the base of the flagellum. Although participation of this microtubule complex in cell shape change is widely accepted, the role of the nuclear ring of the plasmalemma in induction of polymerization and orientation of the microtubules remains to be clarified, and the alleged involvement of the manchette in nuclear shaping remains controversial (Fawcett et al., 1971).

One of the most enigmatic and possibly one of the most important features of the germ cell linc is a dense fibrogranular component that first appears in the interstices of the mitochondrial clusters of

DON W. FAWCETT Department of Anatomy and Laboratory of Human Reproduction and Reproductive Biology, Harvard Medical School, Boston, Massachusetts

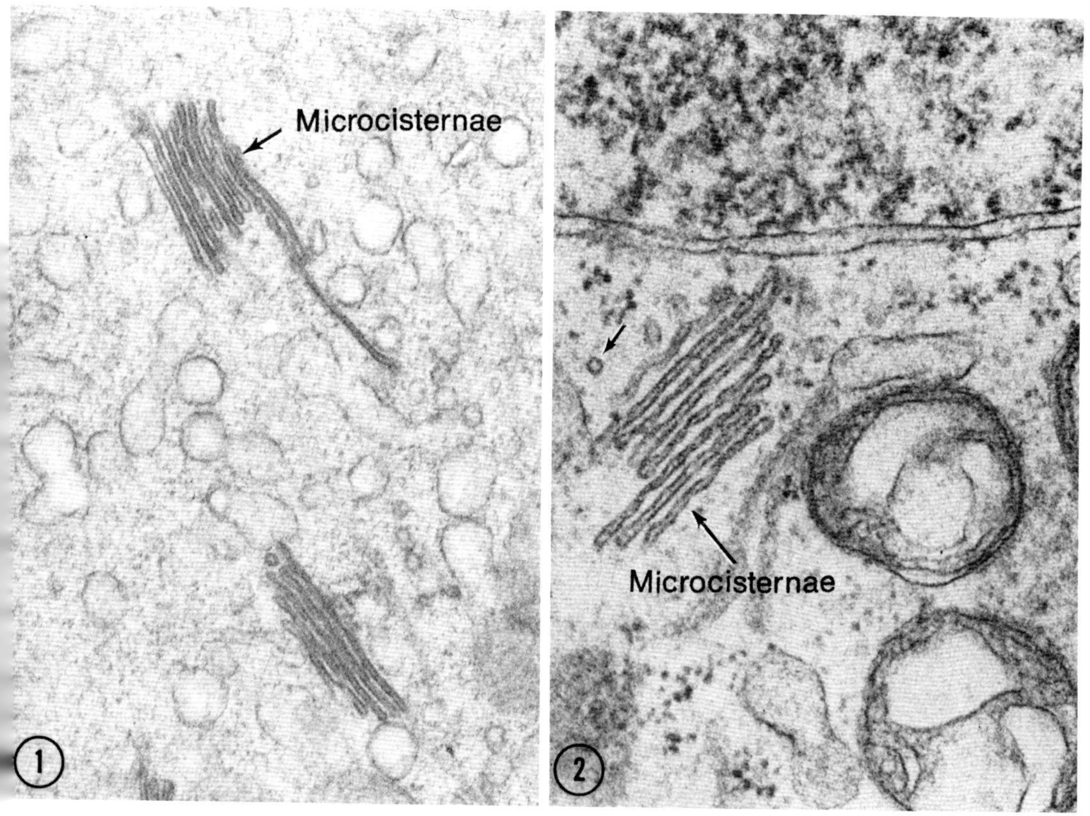

FIGURE 1 Two stacks of microcisternae in the peripheral cytoplasm of a spermatocyte from ram testis. Their membranes often stain more intensely than those of the endoplasmic reticulum.

FIGURE 2 Another example of spermatocyte microcisternae. Their small size can be judged by comparison with the profiles of the nuclear envelope above, a microtubule (small arrow) and cross sections of mitochondria.

spermatocytes (Figs. 4 and 5). This material is thought by some investigators to arise in the nucleus and to pass through nuclear pores into the cytoplasm where it becomes associated with the mitochondria (Comings and Okada, 1972). Its nuclear origin is still a subject of controversy but there is general agreement that in late spermatocytes the mitochondrial clusters dissociate and the dense material aggregates into the sizable *chromatoid bodies* found in the spermatids (Fig. 6). The chromatoid body establishes a transient close relationship to pores in the nuclear envelope, suggesting some kind of interaction or exchange of material with the nucleoplasm (Fawcett, 1971). During the acrosome phase of spermiogenesis, the chromatoid body gradually migrates to the caudal pole of the spermatid nucleus and forms a loose ring around the base of the flagellum in close association with the developing annulus (Fig. 3). Then, as the mitochondrial sheath is assembled, the annulus migrates caudad and the associated residue of the chromatoid body gradually disperses and ultimately disappears. No explanation has been forthcoming for the intriguing sequential relationships of this dense material which migrates from the nucleus to the mitochondrial aggregates, thence to the spermatid nucleus in the form of the chromatoid body, and finally to the base of the flagellum.

The chromatoid substance of spermatocytes and spermatids has its counterpart in perinuclear and intermitochondrial dense bodies of mammalian oocytes. Thus, this dense material seems to be characteristic of both the male and the female germ line. On the basis of histochemical studies, it appears to consist of basic proteins and it may contain some ribonucleic acid (Söderström and

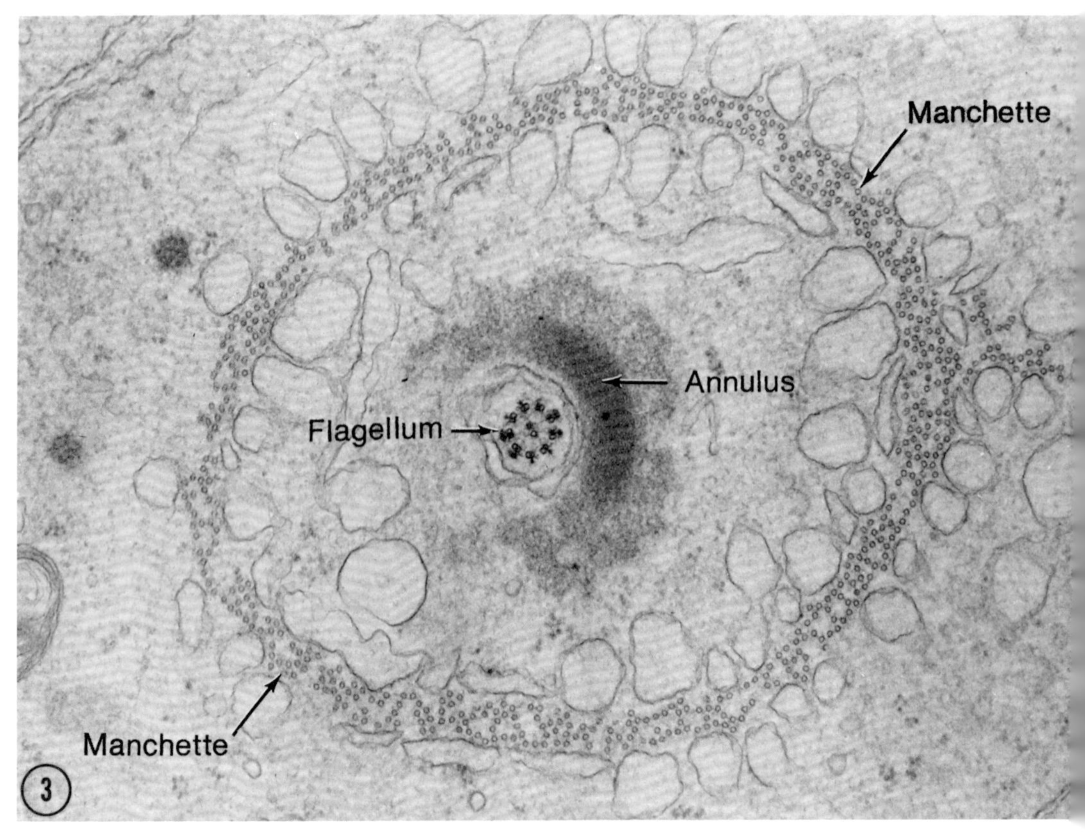

FIGURE 3 Cross section of the spermatid manchette, a cylindrical array of microtubules. Notice the close association of cytoplasmic vesicles with the inside and outside of the manchette. These are presumed to be in transit along the microtubules. In the center of the figure is the flagellum and a portion of the annulus which was slightly oblique to the plane of section. The fine fibrillar material around the dense annulus is a residue of the chromatoid body (micrograph by David Phillips).

Parvinen, 1976). Beyond this crude approximation, the nature and significance of this dense fibrogranular material are unknown. A high priority should be placed upon its isolation and characterization inasmuch as it may prove to be a unique cytoplasmic component essential for germ cell differentiation in both sexes (Beams and Kessel, 1974).

An especially puzzling transient organelle is the *centriolar adjunct*. Soon after the initiation of axoneme formation in the early spermatid, fine filamentous material begins to gather around the distal end of the juxtanuclear centriole and begins to polymerize into a cylindrical structure that resembles the centriole proper in many respects (Fig. 7). There are significant differences however. It possesses a highly ordered lining layer that gives it a slightly narrower lumen than the centriole, and subunits b and c of the triplets in its wall are often incomplete (Fawcett, 1971). Centriole replication takes place at right angles to the cylinder, never at its end. In the formation of the centriolar adjunct, the proximal centriole serves as the initiation site and template just as does the distal centriole in formation of the axoneme, but the product of its nucleation is not a typical centriole. It disappears completely later in spermatid differentiation (Fig. 8). What function it serves during its brief existence remains a complete mystery.

Problems of Interpretation in the Structure of the Mature Spermatozoon

It is widely accepted that the sperm acrosome is, in effect, a highly specialized secretory granule. It arises in the Golgi complex like other cell products that are synthesized for export. In the acrosome reaction, it releases its contents by fusion of its

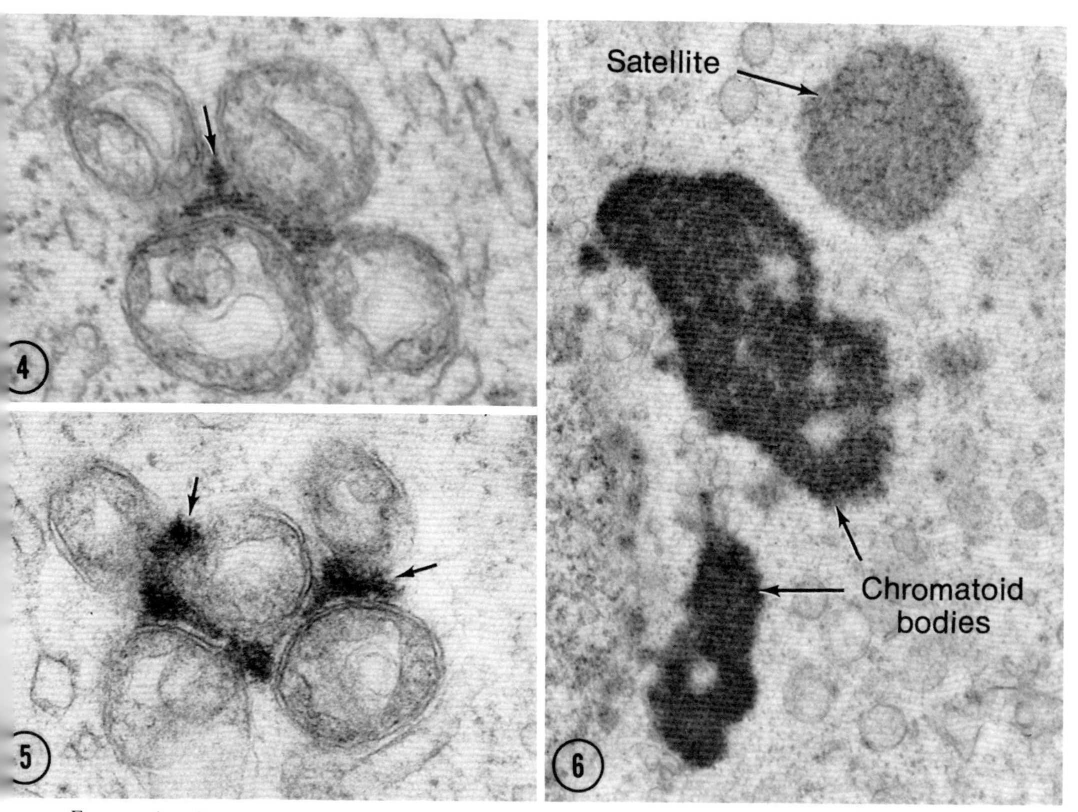

FIGURES 4 and 5 Mitochondrial aggregates from chinchilla spermatocytes showing dense fibrogranular material in the interstices (arrows).

FIGURE 6 Chromatoid bodies and a chromatoid satellite in a guinea pig spermatid. The chromatoid bodies are irregular in shape and often send extensions to nuclear pores. The satellite is spherical, less dense, and often has a reticular fine structure.

limiting membrane with the overlying plasmalemma – a process analogous to exocytosis during secretion by a glandular cell. In its supravital staining with acridine orange and its content of hyaluronidase, and the proteolytic enzyme acrosin and other hydrolases, the acrosome also shares a number of properties of the lysosomes in somatic cells. In fertilization, this sperm organelle appears to play an indispensable role in facilitating dispersion of the cumulus and penetration of the zona pellucida of the recently ovulated egg. These aspects of acrosomal function have been clearly established by ultrastructural and experimental studies, but we still do not know what triggers the acrosome reaction. Also unexplained are the remarkable differences from species to species in the volume and shape of the acrosome. The broad range in volume can be illustrated by comparing the sperm head of a primate such as man or rhesus monkey (Fig. 9 B) where there is no appreciable thickening of the apical portion of the acrosome with the sperm of a musk shrew (Fig. 9 A), an insectivore in which the apical portion of the acrosome is nearly four times the length of the sperm nucleus (Green and Dryden, 1976). If the content of hydrolytic enzymes is directly related to acrosomal volume, then one might logically expect some corresponding species differences in the barriers around the female gamete. Few comparative studies are available, but there appears to be no obvious difference in the size of the cumulus or thickness of the zona pellucida in these species that would account for the remarkable difference in size of the acrosomes on their spermatozoa.

The significance of the extraordinarily complex shape of the acrosome in some species is even more puzzling. Shape would seem to be irrelevant to function in an organelle whose primary mission

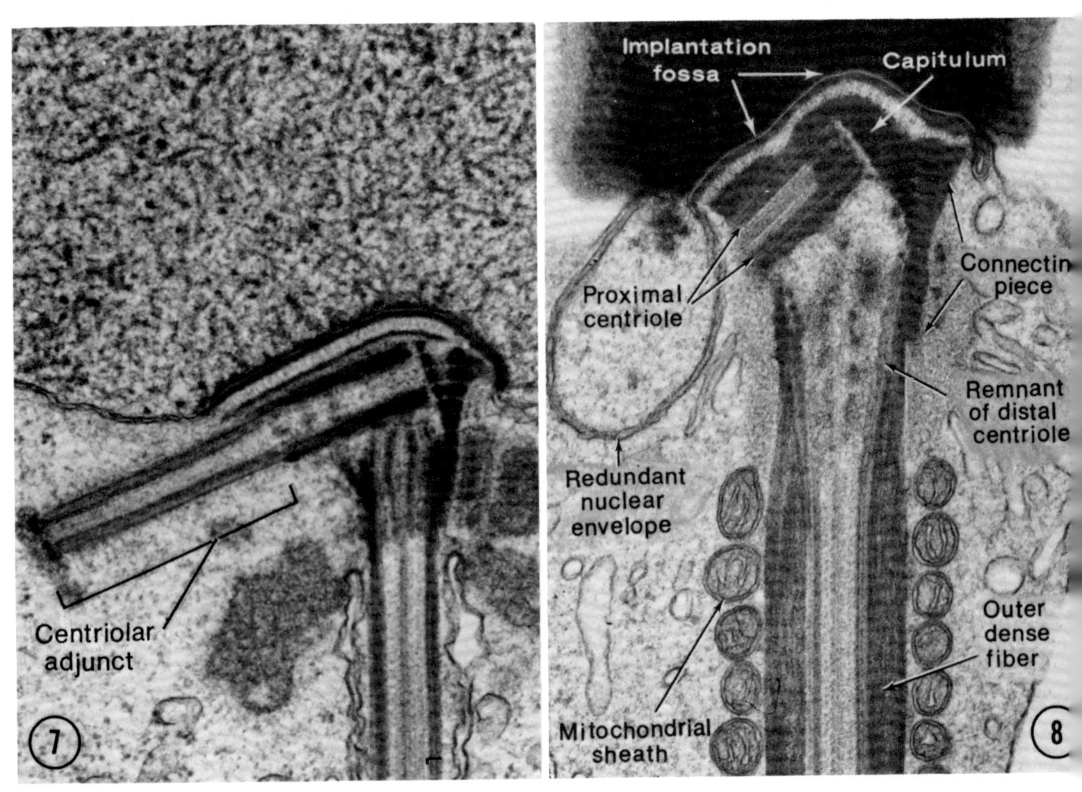

FIGURE 7 Electron micrograph of base of the flagellum in a spermatid in an intermediate stage of nuclear condensation. The proximal centriole has given rise to a long centriolar adjunct at its free end. Uncompacted fibrillar material at its end indicates that it is still elongating (from Fawcett, 1971).

FIGURE 8 A late spermatid with a condensed nucleus. The centriolar adjunct has disappeared but the proximal centriole persists in a niche in the capitulum of the connecting piece (from W. Bloom and D. W. Fawcett, Textbook of Histology. W. B. Saunders Co., Philadelphia. 10th edition. 1975).

is to self-destruct during the final approach to the egg. If it be accepted that a specific shape is of no advantage in relation to the secretory function of the acrosome, then its shape should at least conform to sound principles of streamlining so that it would not be an impediment to progress of the spermatozoon toward the egg.

In the guinea pig and the ground squirrel, *Citellus,* the sperm head is inclined at an angle to the axis of the tail and there is a further angulation of the apical segment of the acrosome with respect to the long axis of the sperm nucleus (Fig. 10). In the case of the ground squirrel, there is a deep, keyholelike recess on the concave side of the acrosome. The outline of this pit is shown in both sagittal and transverse sections in Fig. 10. It is also evident from thin sections that the entire head has a compound transverse curvature, concave centrally and convex bilaterally. The hydrodynamic effects of this elaborate shape are difficult to assess. Angulation of the head and the curvature of its margins may conceivably deflect the sperm from the surface of mucosal folds in the oviduct and minimize chances of an impasse that might result from edge-on engagement of the mucosa by a straight sperm head coinciding with the axis of progression. No suggestion can be offered as to the significance of the deep pit on the concave surface of the acrosome of *Citellus* spermatozoa.

The morphogenesis of these complex acrosomes presents intriguing problems. Evidence has been marshaled elsewhere for the view that the shaping of the sperm nucleus is the result largely of internal forces associated with condensation of the chromatin in a highly ordered lamellar packing (Fawcett et al., 1971). This interpretation cannot easily be applied to the acrosome which has no ordered internal structure the assembly of which

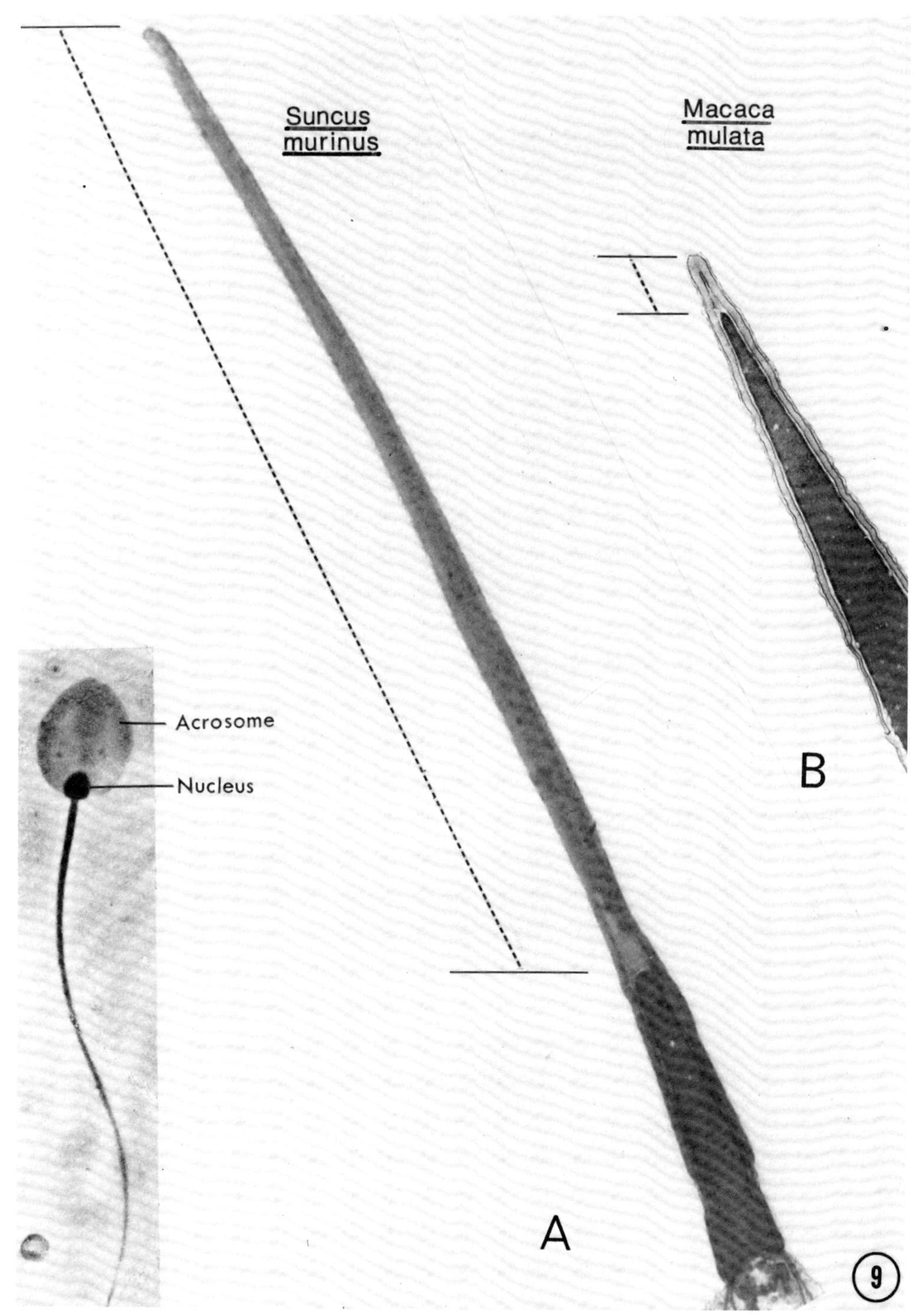

FIGURE 9 Electron micrographs illustrating some extremes of the variation in size of the sperm acrosome. (A) In the Asiatic musk shrew, *Suncus,* the apical segment of the acrosome is several times the area of the nucleus (see *inset*) and nearly four times as long in sagittal section. (B) In *Macaca*, the acrosome extends only a very short distance beyond the anterior margin of the nucleus. (Blocks of *Suncus* sperm made available by courtesy of J. A. Green and G. L. Dryden, University of Missouri, Columbia, Mo.).

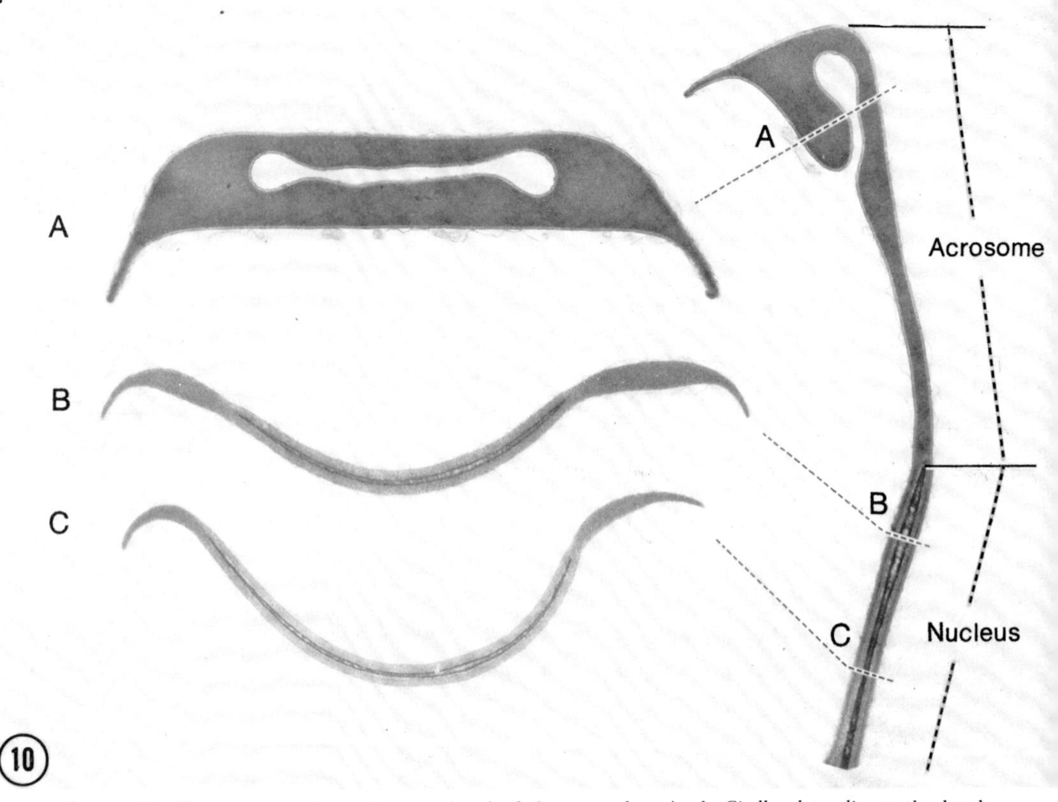

FIGURE 10 Transverse sections of sperm head of the ground squirrel, *Citellus lateralis*, at the levels indicated on the sagittal section at the right. The biological significance of this complex acrosomal and nuclear shape is obscure.

could serve as a basis for intrinsic generation of form. Moreover, in the case of chinchilla and guinea-pig spermatozoa, the acrosome acquires its definitive shape during epididymal transit (Fawcett and Hollenberg, 1963). In the epididymal duct, the sperm are suspended in fluid epididymal plasma, and therefore no external morphogenetic forces can be invoked to explain their acquisition of specific form. The mechanism by which they do acquire their shape thus remains unexplained.

The interspecific differences in the length of the sperm middle piece are also difficult to explain. Spermatozoa of marine invertebrates which are released into seawater for external fertilization, have a simple 9 + 2 flagellum and a midpiece consisting of a single ring of 2 to 4 mitochondria. With the evolution of internal fertilization, an outer row of dense fibers was acquired, resulting in the familiar 9 + 9 + 2 pattern characteristic of sperm flagella of higher forms. Correlated with the appearance of the outer dense fibers, there was a greatly increased development of the mitochondrial sheath of the midpiece. The very close relationship of a long mitochondrial sheath with the outer dense fibers led to the interpretation of the latter as accessory motor elements. It was assumed that a longer mitochondrial sheath was required to generate the necessary energy in the form of ATP for their contraction (Fawcett, 1970). This interpretation is no longer tenable. The recent isolation and chemical characterization of the outer dense fibers of mammalian sperm (Baccetti et al., 1976; Olson and Sammons, The substructure and composition of rat sperm outer dense fibers, in preparation) has provided no evidence for their contractility. They seem instead to be merely resilient stiffening structures. The only component of the sperm flagellum known to be capable of generating movement is the axoneme and inasmuch as this exhibits no significant structural difference from species to species, we are left with no satisfactory explanation for the remarkable variations in the length of the energy-generating mitochondrial sheath (Fig. 11). In man

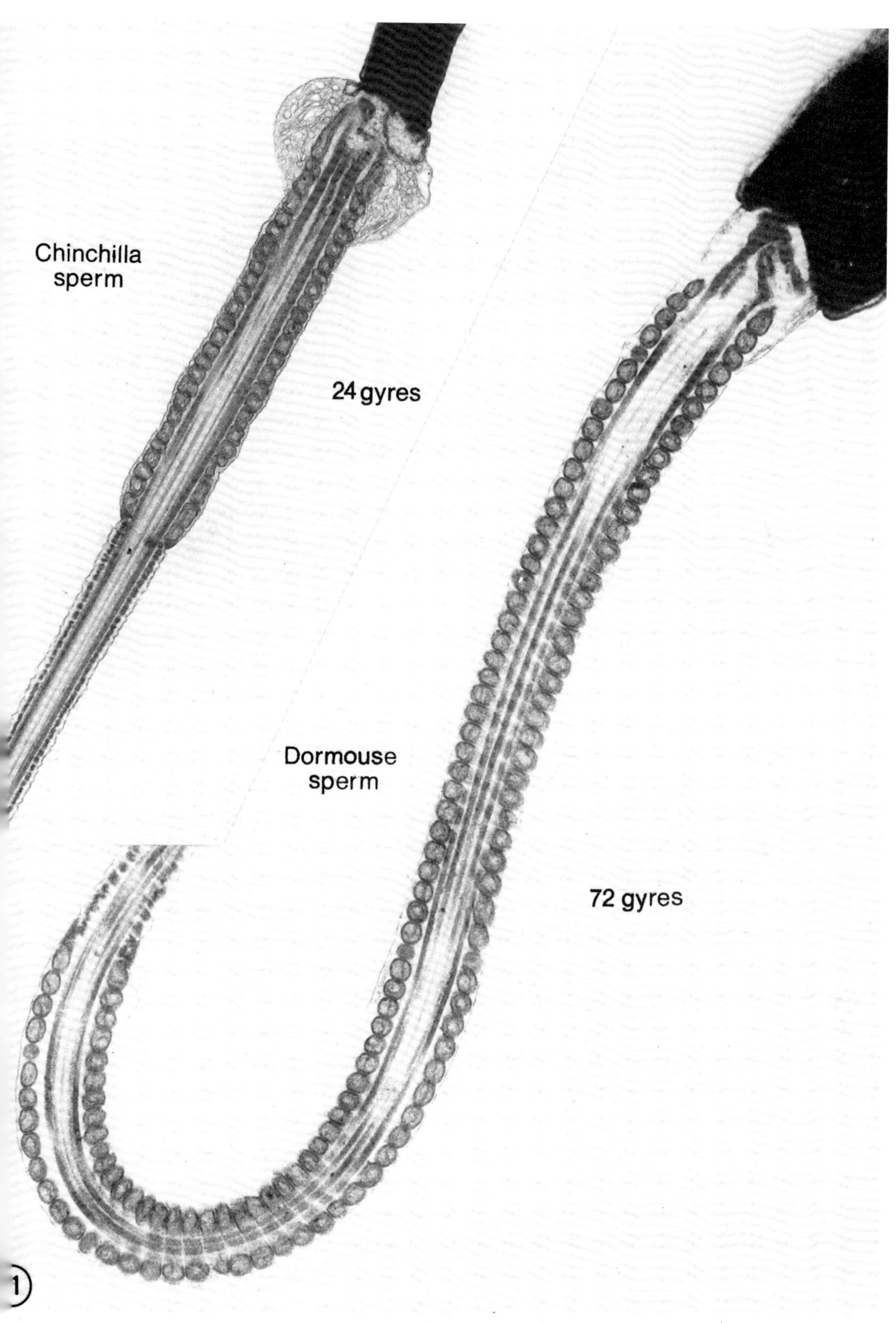

FIGURE 11 Electron micrographs comparing the length of the midpiece in chinchilla sperm with about 24 turns of the mitochondrial sheath, and the dormouse with about 72. In other rodent species, there may be up to 300 gyres in the mitochondrial helix.

and bull, there are about 12 gyres in the mitochondrial helix; 24 in chinchilla; about 75 in dormouse and suni antelope; 87 in Russian hamster; 115 in the brown bat; and about 350 in the rat (Fawcett, 1970). There appears therefore to be no correlation between the degree of development of the energy-generating apparatus of the spermatozoon and tail length, diameter of outer fibers, or the size of the animal. Although transport to the site of fertilization is not primarily dependent upon sperm motility in mammals, motility is required for penetration of the cumulus and zona pellucida in fertilization. In the absence of corresponding differences in the female gamete and its envelopes, the striking species differences in midpiece length still defy explanation.

There are several aspects of the structure of the mammalian sperm tail that are difficult to bring into accord with current theories of flagellar motion. Compelling evidence has now been adduced for a sliding-tubule mechanism for generation of the bending movements of cilia and flagella – a mechanism comparable to the sliding-filament model of muscle contraction (Satir, 1968; Summers and Gibbons, 1973; Lindemann and Gibbons, 1975). When such a mechanism is considered in relation to the attachments of the structural elements that are presumed to slide in sperm tails, some troublesome mechanical problems immediately become apparent. In those species that have thick sperm tails, it is hard to believe that sliding of the doublets in the slender axoneme could overcome the resistance to bending imposed by the thick mitochondrial sheath and the very large outer fibers that are often placed some distance from the axis of bending. The problem is compounded by the fact that the outer fibers of the sperm tail are fixed to the head anteriorly via the connecting piece and also to the correspondingly numbered doublets at their caudal end (Fawcett, 1965; Fawcett and Phillips, 1970). Moreover, the fibrous sheath of the principal piece is attached at its proximal end to outer fibers 3 and 8 and after termination of these fibers, the dorsal and ventral columns of the fibrous sheath appear to be attached to doublets 3 and 8 of the axoneme throughout the greater part of the length of the tail. Experiments involving addition of ATP to demembranated and trypsin-treated sperm tails leave little doubt that a sliding mechanism is involved (Lindemann and Gibbons, 1975), but it is also clear that much more research remains to be done before we will have a satisfactory explanation of sperm motility that will take into account the attachment of the doublets to the massive and relatively stiff accessory structural components of the tail. Why the stiff outer fibers and the resistant fibrous sheath have evolved is also obscure.

The development of methods for revealing the fine structure and chemical properties of sperm membranes has extended the scope of morphogenetic studies and has drawn attention to the spermatozoon as a very favorable subject for investigation by membrane biologists (Friend and Fawcett, 1974). Traditionally, much of our knowledge of membranes has been based upon the study of erythrocytes because they are available in bulk as separate units in suspension. Erythrocytes are, however, merely the residues of erythroblasts, devoid of organelles, and reduced to little more than a membrane-limited bag of hemoglobin. Erythrocyte ghosts are therefore a kind of "minimal" membrane preparation. In contrast, spermatozoa, also available as free cells, are highly differentiated, and endowed with motility which serves as a useful index of viability and functional integrity. Interest of membrane biologists in the spermatozoon is heightened by the fact that its surface is regionally specialized for different functions. The membrane overlying the anterior part of the head is sensitive to stimuli inducing the acrosome reaction and it responds by fusion with the membrane of the underlying acrosomal cap. The postacrosomal surface membrane is unique in being the site of recognition and initial fusion with the oolemma. The membrane enclosing the midpiece is presumably involved in access of substrate to the mitochondrial enzymes that generate the energy for sperm locomotion. And, the membrane of the principal piece may possibly play some ancillary role in propagation of the waves of bending along the tail.

Studies of spermatozoa by freeze-fracturing have revealed within the plane of the membrane in each of these regions characteristic differences in organizational patterns (Friend and Fawcett, 1974; Koehler and Gaddum-Rosse, 1975). Some of these intramembrane patterns are common to the majority of mammalian spermatozoa but others appear to be confined to particular species. The membrane of the principal piece generally has a heterogeneous population of randomly distributed intramembrane particles but in all of the species examined to date, there is a row of particles running longitudinally within the membrane over outer fiber 1 (Figs. 12 and 13). In the guinea pig, this zipperlike differentiation is a double row of particles, but in other species it varies from a single row of particles to one three or four particles in width. The functional significance of this stable differentiation of the membrane is still obscure. If it is merely involved in attachment of the membrane to the underlying fibrous sheath, the constancy of its location with respect to fiber 1 is puzzling.

In the midpiece of the guinea pig spermatozoon (but not in the rat), particles of uniform size form linear aggregates like strings of beads oriented circumferentially or obliquely in relation to the long axis of the tail. These are more concentrated over the gyres of the mitochondria than elsewhere (Fig. 14). Maintenance of this unusual pattern of intramembrane particles appears to depend upon proximity of the plasma membrane to the underlying mitochondria for where the membrane diverges from the mitochondrial sheath to enclose the cytoplasmic droplet, the linear strands of particles dissociate into individual particles (Fig. 14). The functional significance of this pattern of particle aggregation and its occurrence in some species and not in others are intriguing unsolved problems in the membrane biology of the spermatozoon.

In thin cross sections through the sperm midpiece of the opossum, *Didelphys*, the plasma membrane has a very regular scalloped contour and a delicate layer of filamentous material on its inner aspect. This layer varies slightly in thickness, in register with the ridges and grooves of the plasma membrane (Fig. 15). In freeze-fractured preparations of these spermatozoa, the midpiece membrane shows a very regular pattern of closely packed particles arranged in longitudinal rows, three to four particles wide, separated

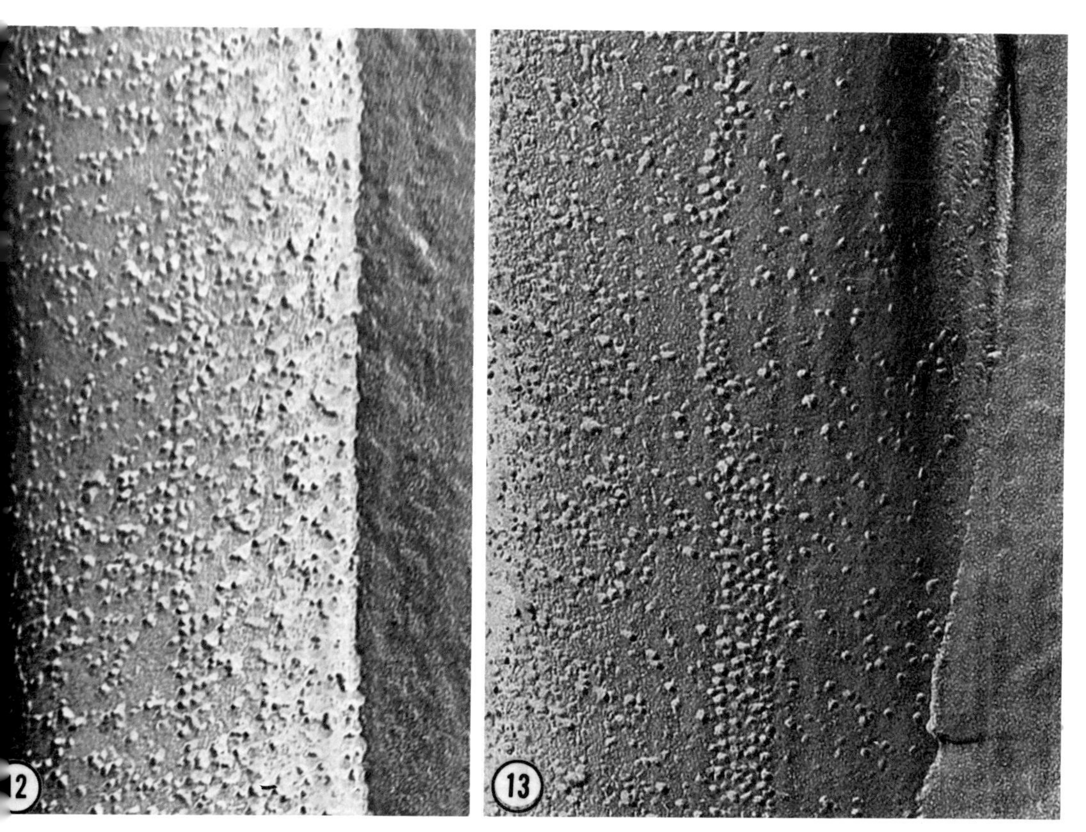

FIGURE 12 Freeze-cleave preparation of the principal piece of a guinea pig spermatozoon showing the longitudinal double row of intramembrane particles.

FIGURE 13 A corresponding view of the principal piece of opossum sperm showing some variation in the number of rows of particles (micrograph by G. Olson).

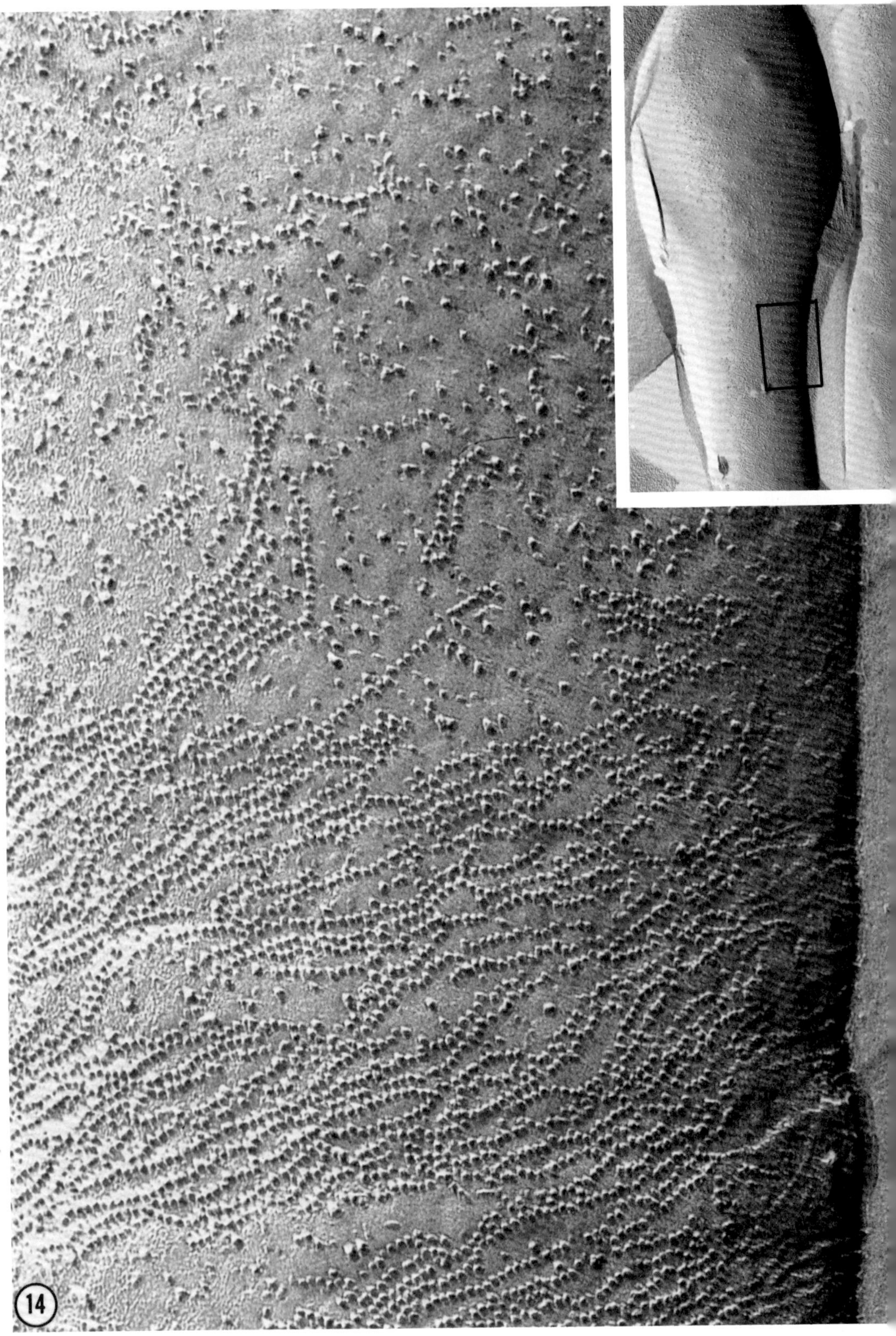

FIGURE 14 Freeze-fractured preparation of the midpiece of a guinea pig spermatozoon in the region where the membrane diverges from the underlying mitochondria to enclose a fusiform cytoplasmic droplet (see *inset*). In the lower half of the figure, the linear aggregates of particles are concentrated over the mitochondria. In the region of the droplet, rows become disoriented and dissociate into individual particles.

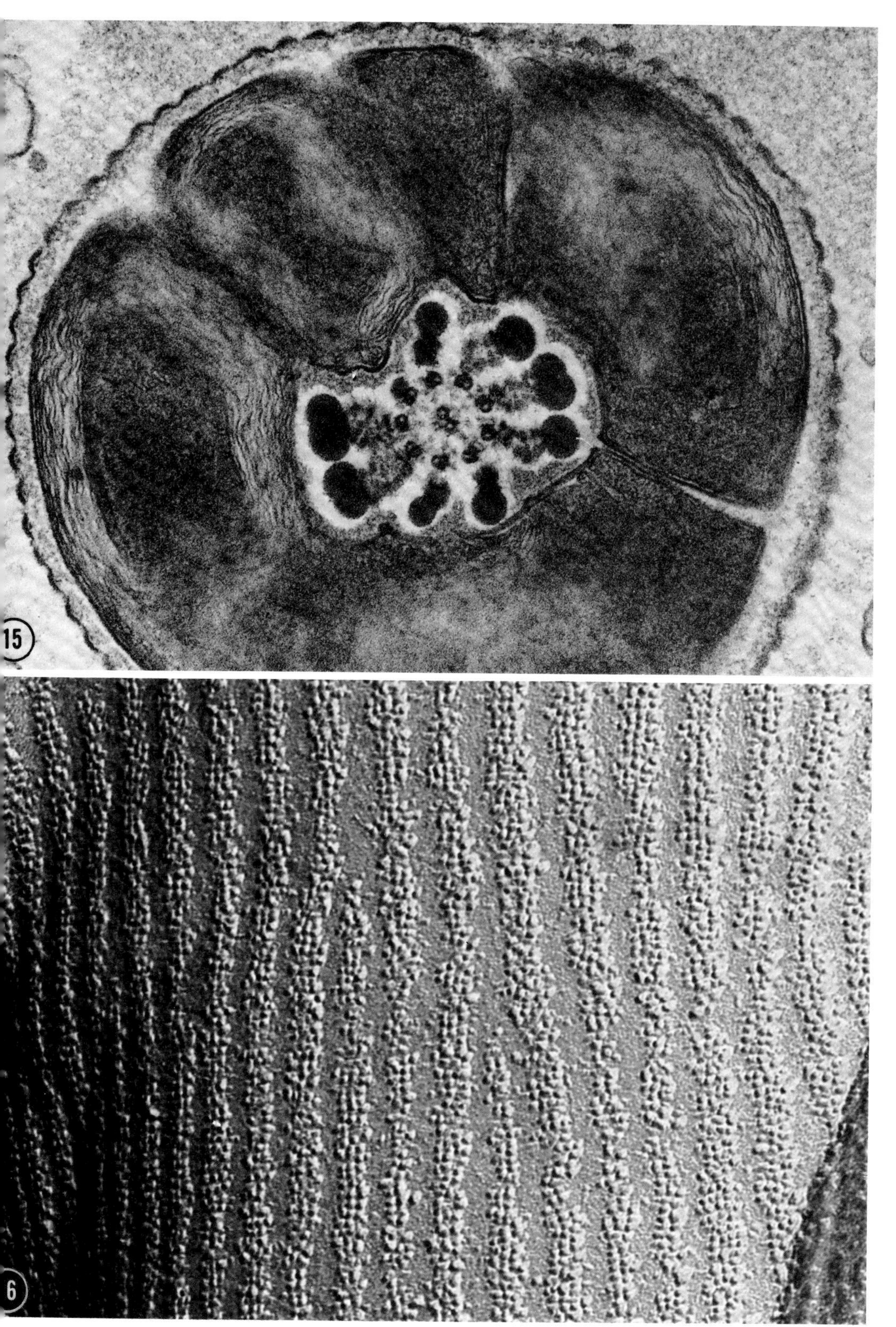

FIGURE 15 Transverse section through the midpiece of a *Didelphys* spermatozoon. Notice the regular scalloping of the surface membrane and the filamentous component associated with its cytoplasmic surface.

FIGURE 16 Freeze-fracture preparation of *Didelphys* sperm midpiece showing linear arrays of intramembrane particles separated by particle-free aisles. These correspond, respectively, to the ridges and grooves of the scalloped pattern seen in sections (micrograph by G. Olson).

by particle-free aisles of about the same width (Fig. 16). The particle-rich regions of the membrane correspond to the ridges of the scalloped membrane seen in thin section, and the particle-poor regions correspond to the grooves. Although this pattern bears no obvious relation to the arrangement of the underlying mitochondria, it may be significant in relation to the findings in the guinea pig, in that it is confined to that region of the spermatozoon where the plasmalemma is in close relationship to the mitochondrial sheath (Olson et al., 1977).

Integral membrane proteins are believed to be involved in phenomena such as ion and metabolite transport. Some particles may completely span the membrane and have hydrophilic peptide sequences on both external and internal surfaces of the membrane. The observation of linear particle arrays in the sperm midpiece of some species raises intriguing problems for both the reproductive biologist and the membrane biologist. Do these particle aggregations enhance the permeability of the membrane to energy-rich substrates for the mitochondrial enzymes? There is increasing evidence that the lateral mobility of certain classes of integral proteins within membranes is limited by cytoplasmic microfilaments that are linked to the intramembrane particles (Nicholson and Poste, 1976). Does the layer of material on the cytoplasmic surface of the opossum midpiece membrane represent peripheral membrane protein lending stability to the ordered arrangement of intramembrane particles? How does proximity to the underlying mitochondria control the linkage between the peripheral and the integral proteins of the membrane?

Remarkable progress has been made in our understanding of the morphogenesis and ultrastructure of the spermatozoon. I hope this brief paper has succeeded in drawing attention to some of the challenging, unanswered questions and fascinating structural problems that will test the skill and ingenuity of investigators between now and the next International Congress of Cell Biology.

ACKNOWLEDGMENTS

This work was supported in part by U.S. Public Health research grant HD-02134.

REFERENCES

Baccetti, B., V. Pallini, and A. G. Burrini. 1976. Accessory fibers of the sperm tail. II. Their role in binding zinc in mammals and cephalopods. *J. Ultrastruct. Res.* **54:**261–275.

Beams, H. W., and R. G. Kessel. 1974. The problem of germ cell determinants. *Int. Rev. Cytol.* **39:**413–479.

Burgos, M. H., and D. W. Fawcett. 1955. Studies on the structure of the mammalian testes. I. Differentiation of the spermatids in the cat. *J. Biophys. Biochem. Cytol.* **1:**287–300.

Comings, D. E., and T. A. Okada. 1972. The chromatoid body in mouse spermatogenesis: evidence that it may be formed by the extrusion of nuclear components. *J. Ultrastruct. Res.* **39:**15–23.

Fawcett, D. W. 1965. The anatomy of the mammalian spermatozoon with particular reference to the guinea pig. *Z. Zellforsch. Mikrosk. Anat.* **67:**279–296.

Fawcett, D. W. 1970. A comparative view of sperm ultrastructure. *Biol. Reprod.* **2**(suppl. 2):90–127.

Fawcett, D. W. 1971. Observations on cell differentiation and organelle continuity in spermatogenesis. *In* Genetics of the Spermatozoon. R. A. Beatty and S. Gluecksohn-Waelsch, editors. North-Holland Publishing Company, Amsterdam. 37–68.

Fawcett, D. W., W. A. Anderson, and D. M. Phillips. 1971. Morphogenetic factors influencing the shape of the sperm head. *Dev. Biol.* **26:**220–251.

Fawcett, D. W., and R. Hollenberg. 1963. Changes in the acrosome of guinea pig spermatozoa during passage through the epididymis. *Z. Zellforsch. Mikrosk. Anat.* **60:**276–292.

Fawcett, D. W., and D. M. Phillips. 1970. Recent observations on the ultrastructure and development of the mammalian spermatozoon. *In* Comparative Spermatology. B. Baccetti, editor. Academic Press, Inc., New York. 13–28.

Friend, D. S., and D. W. Fawcett. 1974. Membrane differentiations in freeze-fractured mammalian spermatozoa. *J. Cell Biol.* **63:**641–664.

Green, J. A., and G. L. Dryden. 1976. Ultrastructure of the epididymal spermatozoa of the Asiatic musk shrew, *Suncus murinus*. *Biol. Reprod.* **14:**327–331.

Koehler, J. K., and P. Gaddum-Rosse. 1975. Media induced alterations of the membrane associated particles of guinea pig sperm tail. *J. Ultrastruct. Res.* **51:**106–118.

Lindemann, C. B., and I. R. Gibbons. 1975. Adenosine triphosphate induced motility and sliding of filaments in mammalian sperm extracted with Triton X-100. *J. Cell Biol.* **65:**147–162.

Nicander, L., and L. Ploen. 1969. The fine structure of spermatogonia and primary spermatocytes in rabbits. *Z. Zellforsch. Mikrosk. Anat.* **99:**221–234.

Nicholson, G. L., and G. Poste. 1976. The cancer cell: dynamic aspects and modifications of cell surface organization (part 1). *N. Engl. J. Med.* **295:**197–203.

OLSON, G., M. LIFSICS, D. W. HAMILTON, and D. W. FAWCETT. 1977. Structural specializations in the flagellar plasma membrane of opossum spermatozoa. *J. Ultrastruct. Res.* In press.

SATIR, P. 1968. Studies on cilia. III. Further studies on the cilium tip and a "sliding filament" model of ciliary motility. *J. Cell Biol.* **39:**77–85.

SÖDERSTRÖM, K. O., and M. PARVINEN. 1976. Incorporation of tritiated uridine by the chromatoid body during rat spermatogenesis. *J. Cell Biol.* **70:**239–246.

SUMMERS, K. E., and I. R. GIBBONS. 1973. Effects of trypsin digestion on flagellar structures and their relationship to motility. *J. Cell Biol.* **58:**618–629.

Differentiation and Regulation of Photoreceptors

INTRODUCTORY REMARKS

RICHARD M. EAKIN

Considering the importance of terminology in biology, it may be helpful to introduce this symposium by defining a few terms.

First, *photoreceptors*. Photoreceptors are light-sensitive organelles, usually arrays of membranes which contain a photopigment composed of a glycoprotein combined with a chromophore that is invariably some form of vitamin A. These photoreceptoral organelles are of two morphological types, with few exceptions: (1) ciliary (Figs. 1–3), and (2) microvillar or rhabdomeric (Fig. 4), as I have called them (Eakin, 1972), to use the first word applied, by Grenacher, to photoreceptors of arthropods and mollusks. As both cilia and microvilli are specialized extensions of the cell membrane, photobiology is basically membrane biology. In this symposium we shall be concerned with both types of photoreceptors: Dr. Bok with ciliary photoreceptors of vertebrates, and Dr. Röhlich with an example of a rhabdomeric photoreceptor in an invertebrate. Incidentally, there is a misunderstanding in some quarters that invertebrates possess only rhabdomeric receptors. Several invertebrate phyla, however, have only ciliary photoreceptors, and they occur in certain subgroups of other major taxa (see Eakin, 1972, for examples).

Second, *differentiation*. This term literally means becoming different, and, as used by the developmental biologist, it includes both biochemical and morphological changes which transform an embryonic primordium or a regenerating blastema into an adult structure: organ, tissue, cell type, or organelle, such as cilium, microvillus, or cell junction. A century of study of eyes, ocelli, and eyespots, in recent times by electron microscopy, has given us a good understanding of morphological differentiation of several photoreceptors, particularly those of vertebrates and a few invertebrates. New structural features are discovered each year, even in previously well-studied eyes. And at the molecular level there is a large unknown. Indeed, chemodifferentiation is one of the major unsolved problems in biology. Why and how, for example, do two retinal cells, lying side by side, follow different paths of differentiation: one into a rod cell, the other into a cone cell? Moreover, to make the problem more complex, recall that there are different kinds of rods and various types of cones—all within a single retina. Or how do these cells become coupled for communication intraspecifically and extraspecifically? Or why do the microvilli in a planarian ocellus develop away from the light, whereas those in a snail's eye differentiate toward the light? Both authors will consider selected examples of differentiation: Dr. Röhlich, some morphological aspects of invertebrate photoreceptors, and Dr. Bok, chemodifferentiation of the rod membrane.

Third, *regulation*. The term "regulation" is used by both embryologist and physiologist. Thus, we speak of the regulation of presumptive epidermis into neural tissue when these districts are exchanged in the early amphibian embryo. Or, again, we speak of the regulation of blood sugar by the interplay of certain hormones, or the regulation of ion flux in a muscle fiber or in the outer segment of a rod. In both embryological and physiological contexts, the word implies readjustment, reorganization, and the emergence of a renewed or restored structure or state of equilibrium. In discussing the biochemical events which create or renew the photosensitive membrane of a vertebrate rod, Dr Bok will analyze not only an example of chemodifferentiation, but one of regulation, because the rod membrane is continuously reformed throughout the life of an organism from the moment the first disks are established in the embryo or newborn until the death of the seeing animal, and Dr. Röhlich, in discussing the regeneration of eyes in a flatworm, will be considering

---

RICHARD M. EAKIN Department of Zoology, University of California, Berkeley, California

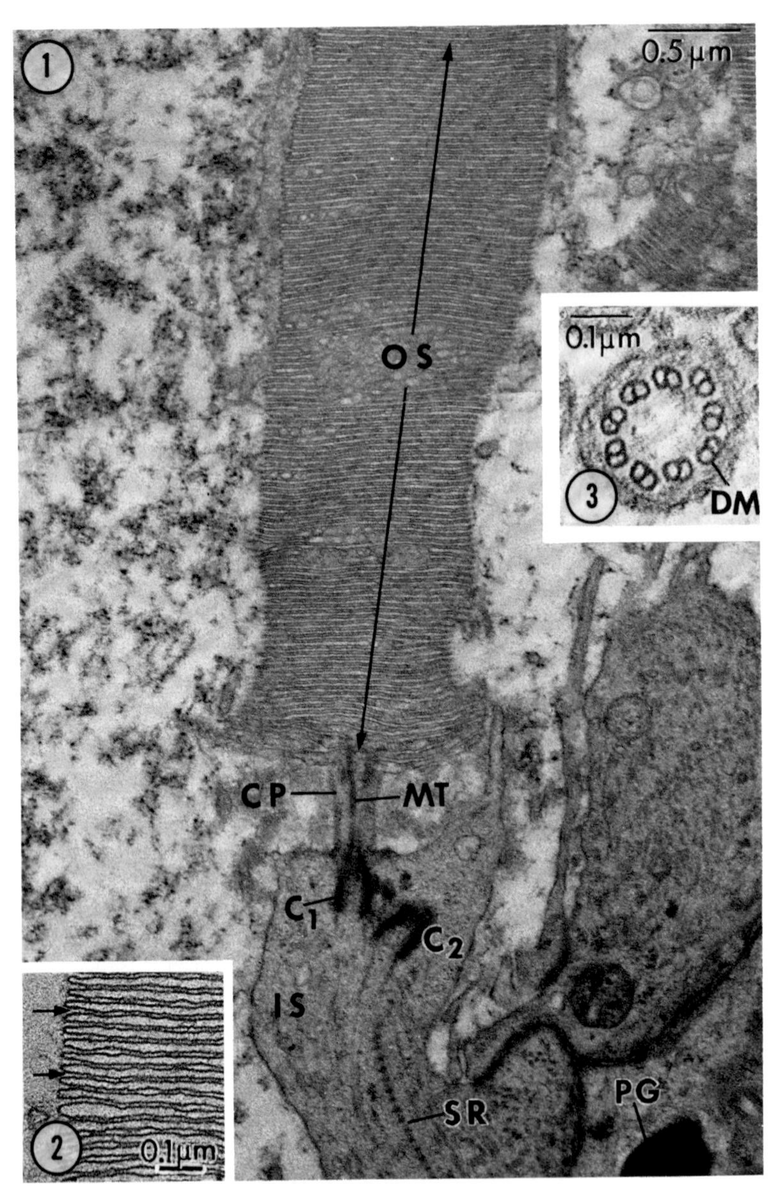

FIGURE 1 Electron micrograph of longitudinal section of ciliary photoreceptor in parietal eye of a lizard (*Sceloporus occidentalis*). C_1 and C_2, distal and proximal centrioles, respectively; *CP*, connecting piece; *IS*, inner segment; *MT*, one of the 18 ciliary microtubules; *OS*, outer segment composed of many disks; *PG*, pigment granule in adjacent supportive cells; *SR*, striated ciliary rootlet. × 23,000.

FIGURE 2 Higher magnification of edges of conelike disks in outer segment showing formation by infolding of ciliary membrane. × 57,000.

FIGURE 3 Cross section of connecting piece showing nine doublets (*DM*) of ciliary microtubules (9 × 2 + 0). × 86,000. (Figs. 1-3 from Eakin, 1973.)

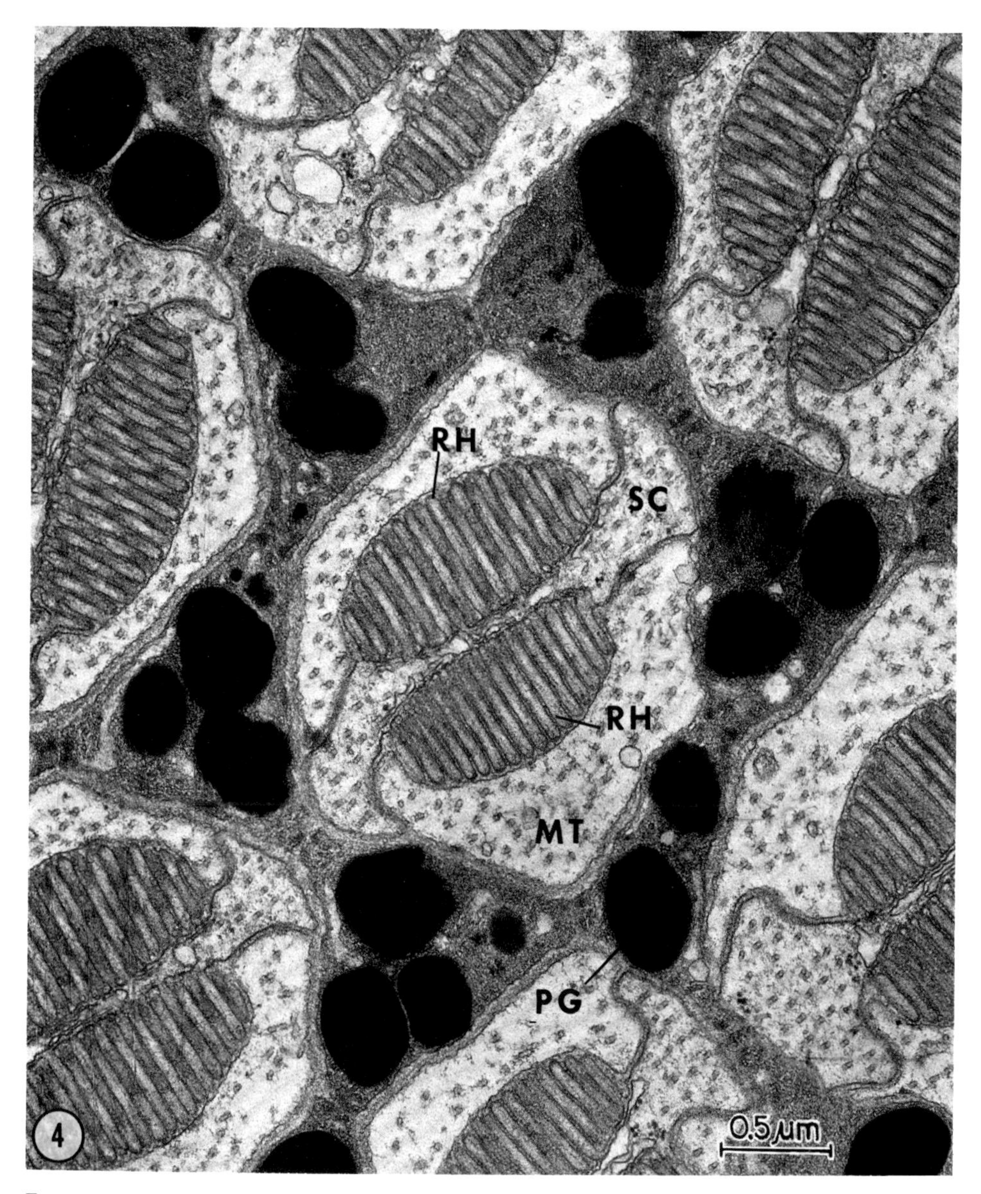

FIGURE 4 Electron micrograph of several rhabdomeric photoreceptors in anterolateral eye of a jumping spider (*Phidippus johnsoni*). Each sensory cell (*SC*) bears two rhabdomeres (*RH*) consisting of microvilli, and it is surrounded by two nonpigmented supportive cells containing many microtubules (*MT*) and by processes of pigmented supportive cells containing black granules (*PG*). × 31,000. (Eakin and Brandenburger, 1971).

an instance of regulation in the context of developmental biology.

REFERENCES

EAKIN, R. M. 1972. Structure of invertebrate photoreceptors. *In* Handbook of Sensory Physiology. H. J. A. Dartnall, editor. Springer-Verlag, New York. 625–684.

EAKIN, R. M. 1973. The Third Eye. University of California Press, Berkeley, California. 157.

EAKIN, R. M., and J. L. BRANDENBURGER. 1971. Fine structure of the eyes of jumping spiders. *J. Ultrastruct. Res.* **37:**618–663.

THE BIOSYNTHESIS OF RHODOPSIN AS STUDIED BY MEMBRANE RENEWAL IN ROD OUTER SEGMENTS

DEAN BOK, MICHAEL O. HALL, and PAUL O'BRIEN

The visual pigments are intrinsic membrane proteins present in high density in the light-sensitive membranes of retinal photoreceptors. In the vertebrate retina, these membranes consist of stacks of bimembranous, flattened saccules (disks) which are derived from the plasma membrane. The organelles of the photoreceptor are organized into separate cytoplasmic compartments that are sequentially arranged along the long axis of the cell. The light-sensitive disks are found within the outer segment, the most distal part of the cell. Mitochondria are segregated in the distal part of the inner segment, and the Golgi zone and endoplasmic reticulum are found in the proximal region of the inner segment. The nucleus is situated between the inner segment and the synaptic terminal, the most proximal (nearest to the brain) portion of the cell.

About a decade ago, it was discovered that the outer-segment disk membranes are continually assembled at the proximal end (base) of the vertebrate rod outer segment (Young, 1967) and intermittently shed from the distal end (Young and Bok, 1969). The discarded disks are then phagocytized by the retinal pigment epithelium (Young and Bok, 1969). It was quickly demonstrated that the rhodopsin polypeptide is renewed at a rate coupled to that of disk membrane renewal (Hall et al., 1969; Bargoot et al., 1969).

At about the same time, a method for the purification of rhodopsin became available (Heller, 1968). Although it was known from the early work of Wald (1935) that rhodopsin contains retinal (the oxidized form of vitamin A) as its chromophore, it was only after the successful removal of membrane lipids during purification that the rhodopsin polypeptide was found to be a glycoprotein (Heller and Lawrence, 1970). With knowledge of the chemical composition of rhodopsin in hand, we began a series of experiments that were designed to explore the sequence of cytological events which lead to the biosynthesis of the complete rhodopsin molecule and its subsequent assembly into rod disk membranes.

It was clear from our early studies in the late 1960s that the rhodopsin polypeptide is synthesized in the rough endoplasmic reticulum of the inner segment. Elucidation of the intracellular sites for glycosylation and addition of the chromophore did not come so easily. Intracellular sugar nucleotide pools in the photoreceptors interfered with the labeling kinetics of cell organelles that were presumed to be involved in glycosylation (Bok et al., 1974; O'Brien and Muellenberg, 1974). The fact that the retinal linkage is labile to conventional histological procedures thwarted our early attempts to study incorporation of the chromophore. We now believe that, in large measure, we have overcome the technical impediments to these studies and offer the following progress report.

MATERIALS AND METHODS

^{3}H-Amino Acid Incorporation into Photoreceptors

Leopard frogs (*Rana pipiens*) were injected intravenously with a 1:1 amino acid mixture containing 300 μCi/g of L-[4,5-^{3}H]leucine (29.1 Ci/mmol) and L-[4-^{3}H]phenylalanine (5.0 Ci/mmol). The solution contained equal amounts of radioactivity on account of the two amino acids. The animals were maintained at 23°C under

DEAN BOK Department of Anatomy and Jules Stein Eye Institute, University of California, Los Angeles, California

MICHAEL O. HALL Jules Stein Eye Institute, University of California, Los Angeles, California

PAUL O'BRIEN Laboratory of Vision Research, National Eye Institute, Bethesda, Maryland

normal laboratory illumination during the day and with the lights extinguished during the night. The frogs were killed at the following intervals after injection: 15 and 30 min; 1, 2, 6, 12, and 24 h; and 1, 2, 3, and 8.5 wk. The eyes were fixed by immersion overnight in buffered 4% formaldehyde (0.085 M sodium phosphate, pH 7.2) followed by 1-h fixation with 1% osmium tetroxide in the same buffer.

Effect of 6-Diazo-5-Oxo-L-Norleucine (DON) on the Intracellular Glucosamine Pools

Dark-adapted neural retinas (without pigment epithelium) were dissected from the eyes of *Rana pipiens*. Photoreceptors were labeled with radioactive sugars by incubation in an oxygenated Krebs-Ringer bicarbonate buffer as described by Basinger and Hall (1973). One-half of the retinas was preincubated for 0.5 h in the presence of 20 μg/ml of DON before the radioactive compounds were added. The other half was preincubated in the buffer alone. After preincubation, 5 μCi/ml of D-[6-^{3}H]glucosamine (7.3 Ci/mmol) and 1 μCi/ml of D-[1-^{14}C]mannose (50 mCi/mmol) were added, and incubation was continued for 3 h. When incubation was terminated, photoreceptor rod outer segments (ROS) were isolated in the dark, and their visual pigment extract was chromatographed as has been described (O'Brien and Muellenberg, 1974). After the first chromatographic step, the rhodopsin was rechromatographed on a 167 × 1.5-cm column of Agarose (Bio-Gel A, 1.5 m, 100–200 mesh; Bio-Rad Laboratories, Richmond, Calif., using 0.3% Ammonyx LO (Onyx Chemical Corp., Jersey City, N. J.) as the detergent (0.05 M Tris HCl, pH 7.8). Specific activities of the column fractions were measured by liquid scintillation spectrometry.

[^{3}H]Glucosamine Incorporation into Photoreceptors

Dark-adapted frog neural retinas were incubated as above, with the exception that [^{14}C]mannose was omitted and the radioactive concentration of [^{3}H]glucosamine was raised to 1.0 mCi/ml. After preincubation in Krebs-Ringer bicarbonate buffer alone or in buffer containing 20 μg/ml DON, the retinas were removed from the radioactive medium at 10, 20, and 30 min. They were fixed in buffered, 4% formaldehyde (0.085 M sodium phosphate, pH 7.2, plus 3% sucrose) overnight and further fixed for 1 h in 1% osmium tetroxide in the same phosphate buffer.

^{3}H-Vitamin A Incorporation into Photoreceptors

Light- and dark-adapted *R. pipiens* were injected in the dorsal lymph sac with 7 μCi/g of [11-^{3}H]retinol (1.24 Ci/mM), freshly prepared from [11-^{3}H]retinoic acid (Schwarzkopf et al., 1949) and dissolved in ethanol, Tween 80 (Atlas Chemical Industries, Inc., Wilmington, Del.), and 1% bovine serum albumin (1:1:14; Eakin and Brandenburger, 1968). The two groups of animals were maintained in their respective light- and dark-adapted states for the duration of the experiment. At selected times from 30 min to 3 wk after injection, one dark-adapted and one light-adapted animal were quickly killed in the dark, and the eyes were immersed in buffered 4% formaldehyde (0.085 M sodium phosphate, pH 7.2, plus 3% sucrose). Retinas from all animals were reduced in the dark with borane·dimethylamine (BDMA) at pH 1.5 for 3 h according to previously published methods (Hall and Bok, 1976). After reduction, the tissues were washed 5 times in 0.085 M sodium phosphate (pH 7.2). The retinas were cut in half, and lipids were extracted from half of each retina with chloroform methanol (Napolitano et al., 1967). The remaining half was left unextracted. Lipid-extracted retinas were osmicated with 4% osmium tetroxide in carbon tetrachloride. The unextracted retinas were treated as usual with phosphate-buffered 1% osmium tetroxide.

Autoradiography

After appropriate dehydration, all fixed tissues were embedded in Araldite 502 (Ciba-Geigy Corp., Ardsley, N.Y.). Light-microscope autoradiography was performed on 0.5-μm sections of all tissues according to previously published methods (Young and Bok, 1969). Quantitative electron-microscope autoradiography was performed on red rods labeled with ^{3}H-glucosamine according to a procedure described earlier (Basinger et al., 1976). Quantitative light-microscope autoradiography was performed in a similar manner on red rods labeled with ^{3}H-retinol. For all data, silver grain counts were expressed as the mean ± 1 SD.

RESULTS

^{3}H-Amino Acid Incorporation into Red Rods

Knowledge concerning rhodopsin polypeptide biosynthesis in the red rod inner segment has been available since the earliest experiments on this subject (Young, 1967; Young and Droz, 1968; Hall et al., 1969), and therefore we shall offer only a brief description of the events that follow an injection of radioactive amino acids.

When ^{3}H-amino acids were injected into living frogs, they were rapidly incorporated into the proximal part of the inner segment. Fifteen min after injection of a [^{3}H]leucine-[^{3}H]phenylalanine mixture, the radioactive label was observed exclusively in the rough endoplasmic reticulum (RER). The Golgi zone was not radioactive at this time (Fig. 1). However, the newly synthesized, radio-

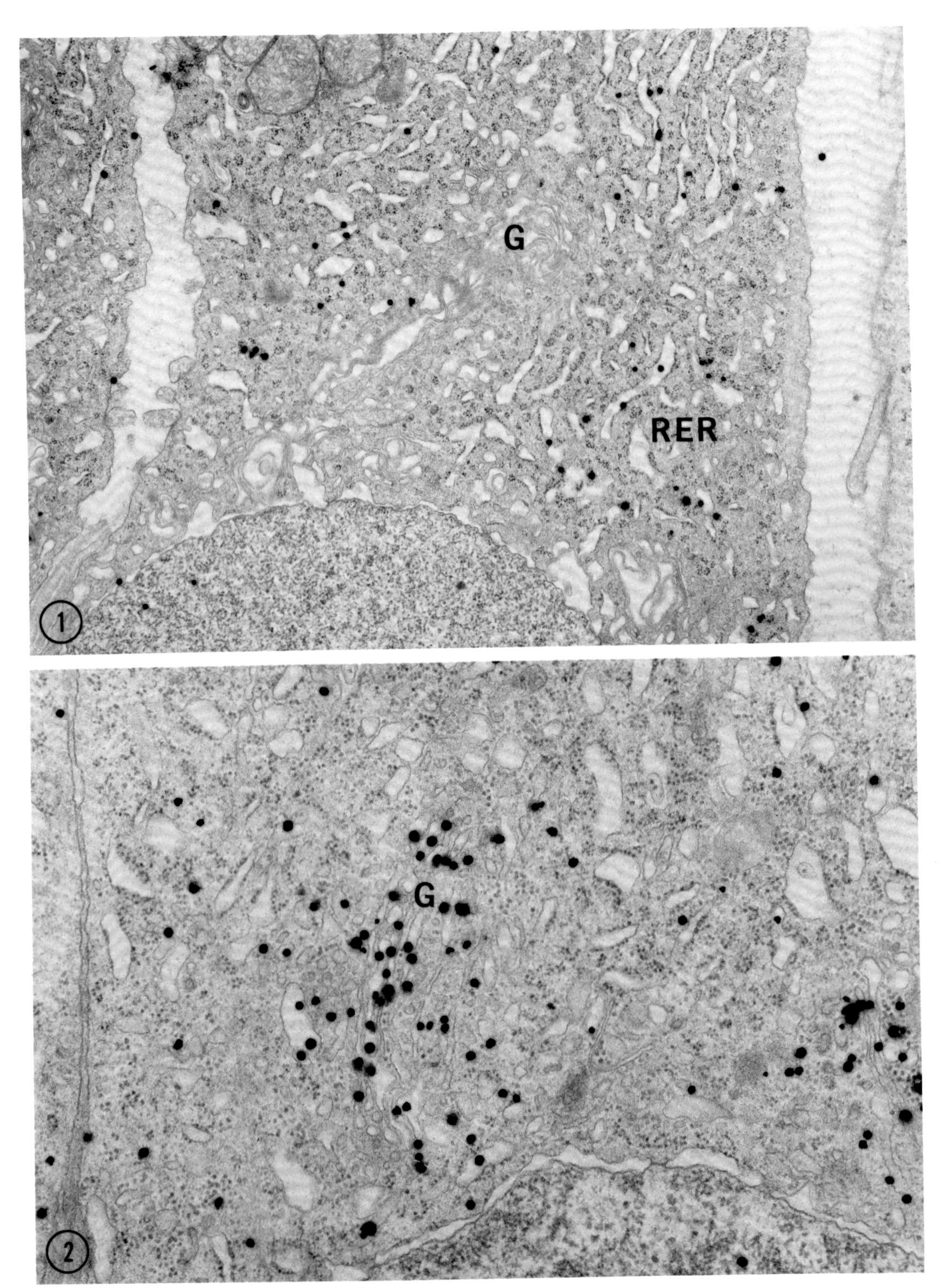

FIGURES 1–3 Electron microscope autoradiographs of frog red rods prepared at selected intervals after an intravascular injection of L-[^{3}H]leucine and L-[^{3}H]phenylalanine mixture. Autoradiographs were exposed for 60 days.

FIGURE 1 Rod inner segment 15 min after injection. The rough endoplasmic reticulum (*RER*) is radioactive but the Golgi zone (*G*) is not. × 13,000.

FIGURE 2 Rod inner segment 60 min after injection. The Golgi zone (*G*) is heavily radioactive, indicating that newly synthesized protein is transported from the RER to this organelle. × 24,400.

active protein soon began to migrate toward other organelles. By 30 min after injection there were perceptible levels of radioactivity in the Golgi zone, and by 1 h the Golgi zone was the most radioactive organelle in the cell (Fig. 2). Two hours after introduction of the radioisotope, the most proximal disks of the ROS became intensely radioactive. This was indicated by a band of silver grains that extended across the proximal end (base) of the outer segment (Fig. 3). At successive intervals after injection, this contingent of radioactive disks was displaced along the length of the outer segment until it reached the distal end and was actively shed and subsequently phagocytized by the retinal pigment epithelium.

Modulation of [^{3}H]Glucosamine and [^{14}C]Mannose Incorporation into Rhodopsin by DON

Both [^{3}H]glucosamine and [^{14}C]mannose were used as precursors for the oligosaccharide chain of rhodopsin because we had to rule out the possibility that the enzyme inhibitor, DON, might have an effect on glycoprotein synthesis per se. Because hexose metabolic pathways are not influenced by DON, alterations in the incorporation of mannose would indicate an interference with either the synthesis of the rhodopsin polypeptide or that of the oligosaccharide. DON blocks the synthesis of glucosamine 6-phosphate and all intermediates leading to and including UDP-N acetylglucosamine (UDP-GlcNAc) by inhibiting the enzyme L-glutamine D-fructose 6-phosphate amido-transferase (Mazlen et al., 1970). On the other hand, exogenous, [^{3}H]glucosamine can bypass the enzyme blockage because it is phosphorylated by hexokinase to glucosamine-6-phosphate (Kornfeld et al., 1964). The predicted result should be depletion of the endogenous UDP-GlcNAc pool. If [^{3}H]-glucosamine were to be provided under these circumstances, the rhodopsin polypeptide should be labeled more readily than it was in earlier studies (Bok et al., 1974).

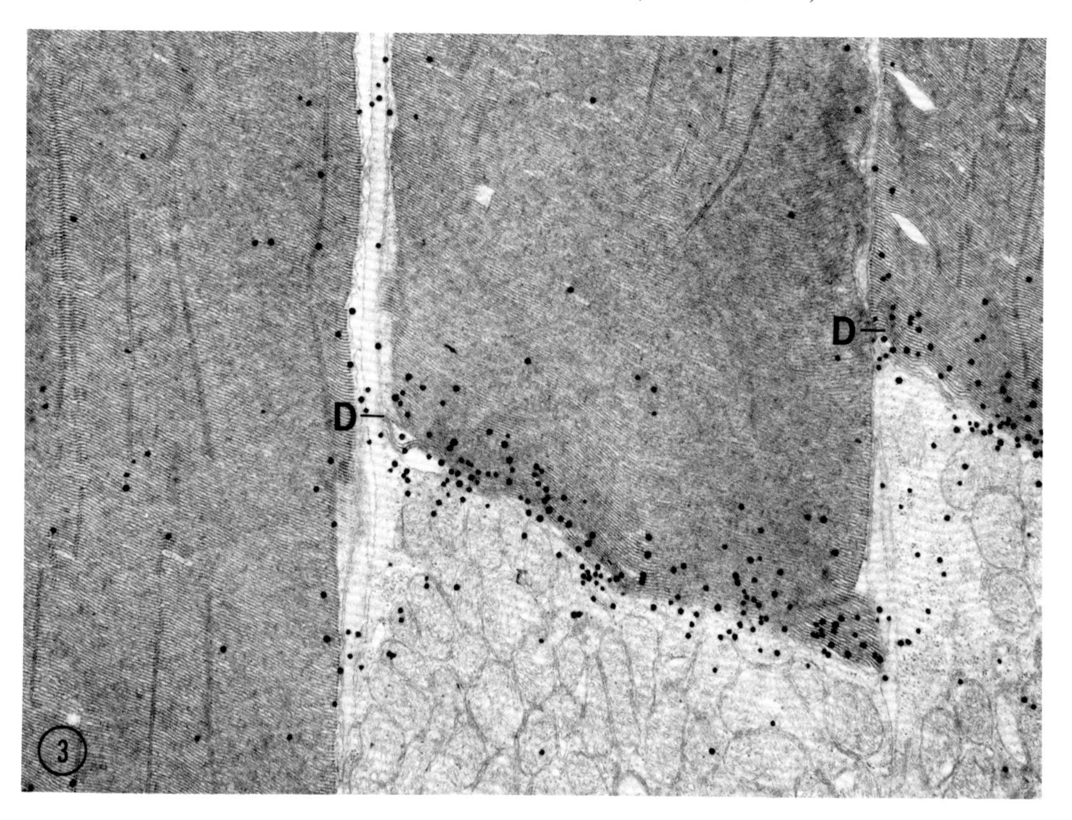

FIGURE 3 6 h after injection, basal disks (*D*) of the ROS are radioactive. Biochemical analysis has shown that most of this radioactivity is by reason of opsin that was synthesized in the inner segment and transported to the outer segment. × 13,000.

When frog retinas were first preincubated for 0.5 h in the presence of DON and further incubated in the presence of DON and the two radioactive sugars for 3 h, [^{3}H]glucosamine incorporation was dramatically increased, but [^{14}C]mannose incorporation remained almost unaltered. The specific activity of rhodopsin as a result of [^{3}H]glucosamine was 151% of the control, whereas rhodopsin specific activity as a result of [^{14}C]mannose was 88% of the control value (Table I). The specific activity of rhodopsin due to [^{14}C]mannose remained nearly the same as that of the control, so it was concluded that polypeptide and oligosaccharide syntheses were not altered significantly. However, because the specific activity of rhodopsin due to [^{3}H]glucosamine rose to 151% of the control, it was concluded that the addition of DON to the incubation medium did, in fact, result in the depletion of a significant portion of the preexisting sugar nucleotide pool.

[^{3}H]Glucosamine Incorporation into Red Rods

When isolated frog retinas were preincubated in DON and then incubated further in the presence of [^{3}H]glucosamine, both the RER and the Golgi zone were heavily labeled by 10 min (Fig. 4). By contrast, labeling of the Golgi zone and RER in the control retinas incubated without DON had detectable levels of radioactivity, but less than half that of the DON-treated retinas (Table II). As incubation in the presence of [^{3}H]glucosamine and DON continued, the differences in Golgi labeling between DON-treated and control retinas increased. After 20 min of incubation, Golgi labeling was six times higher in the DON-treated retinas than it was in the controls (Table II). Labeling of the RER remained about 2.3 times higher in the DON-treated rods. As was noted in our earlier work, Golgi labeling in rods that were not incubated in DON showed a lag during the first 20 min, after which the radioactive concentration increased sharply. As the incubation continued to 30 min, the Golgi labeling in control retinas began to rise, and the difference in the control and DON-treated retinas was no longer as great as it was at 20 min (Table II).

^{3}H-Vitamin A Incorporation into Red Rods

As noted in an earlier publication (Hall and Bok, 1974), the rate of incorporation of [^{3}H]retinol in both light- and dark-adapted photoreceptors is very slow in the frog retina. Although the outer segments of both light- and dark-adapted rods gradually became radioactive during the first 6 h after injection, there was no evidence for radioactivity above background levels in the rod inner segments. As time after injection progressed, the inner segments remained unlabeled, whereas the ROS became progressively more radioactive. From the earliest time after injection, ROS from dark-adapted animals were more radioactive than those from light-adapted animals. By 3 days, dark-adapted ROS were about 10 times more radioactive than the light-adapted ones (2.07 ± 0.064 counts/unit area for dark-adapted rods vs. 0.20 ± 0.013 counts/unit area for light-adapted rods). In addition, by 3 days after injection, the distribution of label differed for light- and dark-adapted rods. Whereas the radioactive label remained uniformly distributed throughout the length of light-adapted rods, dark-adapted rods were more heavily labeled at their proximal ends. This labeling pattern was even more evident in dark-adapted rods at 7 days after injection (Fig. 5). Since the ROS renewal rate, and hence the rate of disk production, is known for frog rods (36 disks/day at 22.5°C; Young and Droz, 1968), we were able to count, differentially, radioactivity in the part of the outer segment that was formed before and after the injection of [^{3}H]retinol. In animals sacrificed 3 days after injection, the counts over recently formed disks

TABLE I

Effect of DON on Simultaneous Incorporation of Sugars into Frog Rhodopsin

	Specific activity* after 3 h of incubation		
Preincubation medium	[^{3}H]glucosamine	[^{14}C]mannose	Isotope ratio
Krebs-Ringer bicarbonate (0.5 h)	31,020	3,714	8.35
Krebs-Ringer bicarbonate plus DON (0.5 h)	46,733 (151% of control)	3,267 (88% of control)	14.31

* DPM/ml/A_{498} nm.

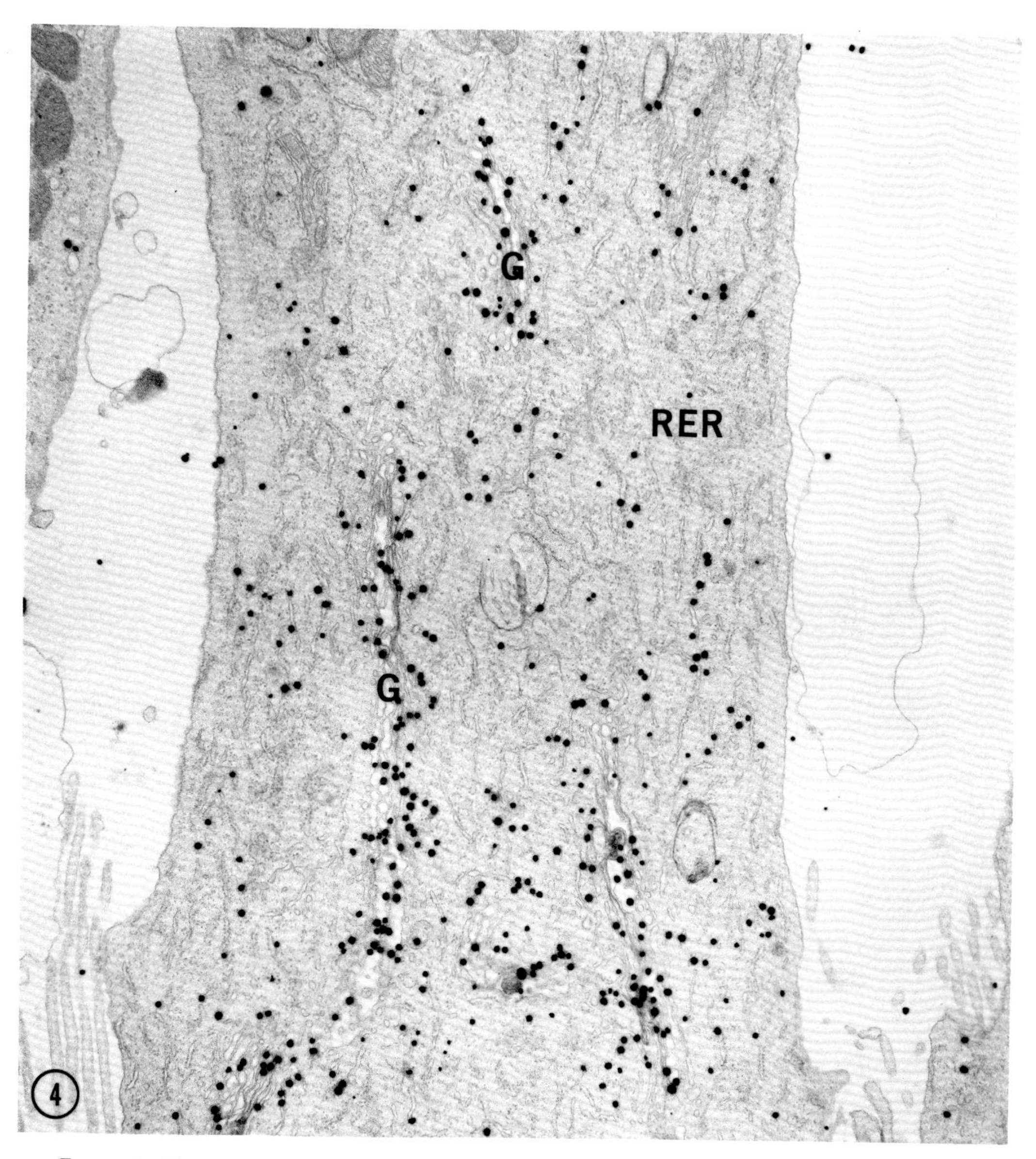

FIGURE 4 Electron microscope autoradiograph of frog red rod inner segment after 0.5-h preincubation in DON followed by 10-min incubation in D-[^{3}H]glucosamine plus DON. The rough endoplasmic reticulum (*RER*) and Golgi zone (*G*) are labeled simultaneously, indicating that glycosylation of protein occurs at both sites. Exposure 90 days, × 13,000.

TABLE II
Effect of DON on Incorporation of [^{3}H]glucosamine into the Golgi Zone and RER

Period of incubation in [^{3}H]glucosamine	Golgi zone (silver grains/unit area)		RER (silver grains/unit area)	
	Control	DON	Control	DON
10 min	0.563±0.031	1.502±0.067	0.158±0.004	0.363±0.008
20 min	0.503±0.031	2.998±0.099	0.256±0.006	0.574±0.011
30 min	1.322±0.063	3.148±0.101	0.392±0.009	0.752±0.015

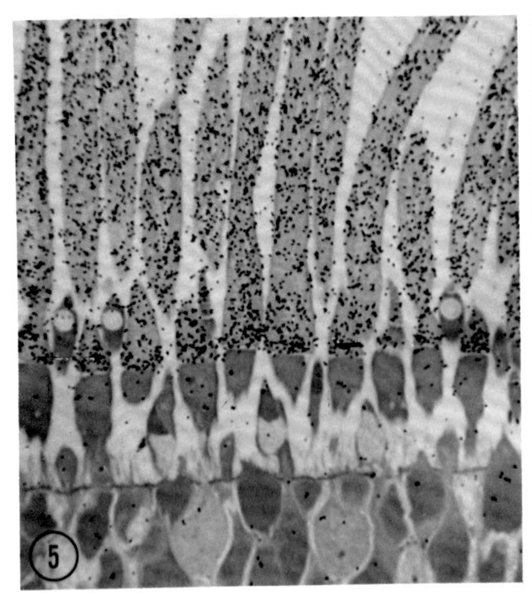

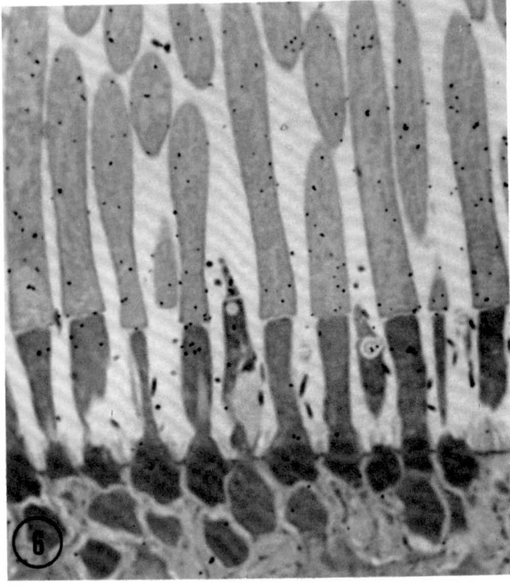

FIGURES 5 and 6 Light microscope autoradiographs of frog red rods 1 wk after injection of [11-^{3}H]retinol. The tissues were fixed in the dark and reduced with BDMA before osmication and embedding. Autoradiographs were exposed for 42 days. Toluidine blue stain, × 890.

FIGURE 5 Dark-adapted rods. Radioactivity is observed along the entire length of the outer segments. Quantitative silver grain analysis indicated that disks at the outer segment base (arrows), which were assembled after injection, were about twice as radioactive as preexisting disks. Radioactivity in other parts of the cell did not exceed background levels.

FIGURE 6 Light-adapted rods. Radioactivity/unit area is much reduced when compared to that of dark-adapted rods. Silver grains are evenly distributed with no preferential labeling of the outer segment base. As was the situation in dark-adapted rods, radioactivity in other portions of the cell did not exceed background levels.

were 4.61 ± 0.48/unit area, whereas counts over the preexisting disks were only 2.31 ± 0.06/unit area. In dark-adapted animals sacrificed 7 days after injection, recently formed disks had 5.24 ± 0.40 counts/unit area, whereas preexisting disks had 3.66 ± 0.9 counts/unit area. Thus it appeared that disks assembled after the injection of [^{3}H]retinol were significantly more radioactive than those present in the rod before injection. At no time in the course of the experiment was there any evidence in light-adapted rods for a concentration of radioactivity at the bases of the outer segments (Fig. 6).

Extraction of lipids from the formaldehyde-fixed, BDMA-reduced retinas did not significantly alter the number or distribution of silver grains over the outer segments of light- or dark-adapted rods. Because the method of lipid extraction employed is known to remove 93% of the phospholipids from formaldehyde-fixed retinas (Nir and Hall, 1974), it was concluded that nearly all of the silver grains were due to [^{3}H]retinal, which was irreversibly bound to protein by reduction of the aldimine linkage between retinal and the rhodopsin polypeptide.

DISCUSSION

Within 10 min after an injection of ^{3}H-amino acids into frogs, about 80% of the radioactivity in the cell is localized in the RER of the inner segment (Young and Droz, 1968). As is the case with many cell types, a large portion of this radioactive protein subsequently passes through the Golgi zone. Young and Droz (1968) have estimated that, by 1 h after injection of ^{3}H-amino acids, more than 15% of the total cell radioactivity is localized in the Golgi zone, although this organelle has less than 2% of the total cell volume. Opsin is a glycoprotein which contains glucosamine and mannose in its oligosaccharide (Heller and Lawrence, 1970). Because the Golgi zone is a known site for glycosylation, it has been tempting to speculate that opsin is one of the proteins that passes through the Golgi zone.

During the synthesis of some serum glycoproteins, the core sugars (glucosamine and mannose) are added to the protein while it is still in the RER (Molnar et al., 1965; Whur et al., 1969), although a number of investigators have shown that, in other instances, glucosamine and mannose can be added at multiple sites, namely, ribosomes and membranes of the RER and Golgi zone (Molnar et al., 1965; Sturgess et al., 1972; Moscarello et al., 1972). We now have provided unambiguous evidence that glucosamine residues are incorporated into macromolecules simultaneously in the RER and Golgi zone of frog rods. Unfortunately, autoradiography alone cannot tell us whether glucosamine residues are added to opsin at both sites. Nonetheless, when all of the available evidence is taken together, it appears likely that this is so. Heller and Lawrence (1970) have shown that the rhodopsin oligosaccharide is attached to opsin through an asparagine-glucosamine linkage. Therefore, since glucosamine is the first sugar to be added during synthesis of the rhodopsin oligosaccharide, it is most likely that this event occurs in the RER. Because a large portion of the protein synthesized in the inner segment passes through the Golgi zone on its way to the outer segment, and because rhodopsin is present in such high concentration in outer-segment disks, it is likely that partially glycosylated opsin molecules receive some of their more peripherally situated sugars in the Golgi zone. Some of the glucosamine residues are, in fact, peripherally located in the rhodopsin oligosaccharide (Heller and Lawrence, 1970).

The chromophore of rhodopsin, 11-*cis*-retinal, is linked to the glycoprotein through a retinyl-lysine Schiff base (Bownds, 1967). This linkage is labile to light and many other factors, including many of the organic solvents utilized in the processing of tissues for microscopic study. Any study that fails to take this into account is subject to question, as the label can be translocated in the handling of the tissue. During the current experiments, all [^{3}H]retinol-labeled tissues were processed in the dark and reduced with BDMA before dehydration and embedding. This reducing agent is capable of penetrating the hydrophobic region of the disk membrane where the retinyl-lysine Schiff base is situated (Hall and Bok, 1976). Reduction of the double bond produces a stable retinyl-opsin linkage and allows reliable autoradiography to be performed.

The present study indicates that the rhodopsin glycoprotein reaches the outer segment before the chromophore is added. This is demonstrated by the fact that silver grain counts never exceeded background levels in the rod inner segment at any time between 30 min and 3 wk after an injection of [^{3}H]retinol. Furthermore, labeling above background was never observed in any other part of the cell except the outer segment. The labeling of nuclei, inner segments, and synaptic terminals observed by others after an injection of [^{3}H]retinyl acetate (Sherman, 1970; Pourcho and Bernstein, 1975) is almost certainly a result of translocation of the label, as these investigators processed their tissues in the light and took no precautions for stabilization of the retinal-lysine Schiff base.

The absolute levels of radioactivity were always higher in dark-adapted rods. For reasons which are well documented but poorly understood, the rhodopsin chromophore migrates from the photoreceptor outer segment to the retinal pigment epithelium during light adaptation (Dowling, 1960). This, no doubt, accounts for the fact that light-adapted ROS contained less total radioactivity than their dark-adapted counterparts. On the other hand, not all of the radioactivity observed in dark-adapted rods is the result of labeled rhodopsin. Previous studies have shown that the specific activities of rhodopsin purified from dark- and light-adapted rods labeled with [^{3}H]retinal are similar (Bridges and Yoshikami, 1969; Hall and Bok, 1974). Furthermore, we have shown that about 20% of the total radioactivity can be extracted from dark-adapted, reduced outer segment membranes with chloroform:methanol or from unreduced, dark-adapted membranes with petroleum ether, a reagent which does not bleach rhodopsin (Hall and Bok, 1976).

Radioactivity in light-adapted rods was always evenly distributed along the length of the outer segment. At no time during the course of the experiment was there any evidence in light-adapted rods for a concentration of radioactivity at the outer segment base. In dark-adapted rods, the situation was quite different. Although there was a great deal of radioactivity throughout the length of the dark-adapted outer segment, there was also a statistically significant elevation of radioactivity at the outer segment base, which was discernible between 3 and 7 days after injection. Our interpretation of this observation is as follows: In the light-adapted rods, one never observes a concentration of radioactivity at the base of the outer segment because rhodopsin in all parts of the outer segment is constantly bleached by light. Thus, the label that is introduced in higher concentration at the outer segment base by disk assembly is quickly randomized throughout the outer segment. In the dark-adapted state, the randomization caused by bleaching does not occur. What, then, is the origin and nature of all of the randomly distributed radioactivity in the dark-adapted outer segment? As mentioned earlier, some of this is because of [^{3}H]retinol or its derivatives that are present in the dark-adapted outer segment but not linked to opsin. Bridges (1976) has recently reported that the frog outer segment contains enough of a "free" retinol pool to synthesize rhodopsin for 2 days. An additional possibility is that 11-*cis*-retinal is capable of exchange among opsin molecules in the dark. This is supported by our observation that BDMA-reduced outer segments that have undergone exhaustive lipid extraction retain most of their randomly distributed radioactivity. This treatment would be expected to remove all of the radioactivity except [^{3}H]retinyl opsin (formed by action of the reducing agent). Indeed, the pattern of labeling in dark-adapted [^{3}H]retinol-labeled rods is very similar to that obtained when ^{3}H-fatty acids are injected into frogs (Bibb and Young, 1974*a*). Fatty acids, as well as intact phospholipids (Bibb and Young, 1974*b*), are capable of exchange in the outer segment membrane. Since retinal is a lipid bound in a most tenuous manner to protein, it does not seem unreasonable to suggest that this molecule should also undergo exchange.

Rhodopsin biosynthesis begins on the ribosomes of the rod inner segment. The polypeptide is apparently glycosylated at multiple sites (RER, smooth ER, and Golgi) during its migraton from the inner segment to the outer segment. Upon arriving at the base of the outer segment, it is conjugated with its chromophore at about the time that it is assembled into proliferating disk membranes.

ACKNOWLEDGMENTS

We are most grateful for the technical assistance of Myunghee Chun, Marcia Lloyd, and Caryl Schechter. We also thank Dr. Richard Eakin for critical reading of the manuscript.

This work was supported by National Institutes of Health grants EY 00046, EY 00331, and EY 00444.

REFERENCES

Bargoot, F. G., T. P. Williams, and L. M. Beidler. 1969. The localization of radioactive amino acid taken up into the outer segments of frog (*Rana pipiens*) rods. *Vision Res.* **9:**385–391.

Basinger, S. F., D. Bok, and M. O. Hall. 1976. Rhodopsin in the rod outer segment plasma membrane. *J. Cell Biol.* **69:**29–42.

Basinger, S. F., and M. O. Hall. 1973. Rhodopsin biosynthesis *in vitro*. *Biochemistry*. **12:**1996–2003.

Bibb, C., and R. W. Young. 1974*a*. Renewal of fatty acids in the membranes of visual cell outer segments. *J. Cell Biol.* **61:**327–343.

Bibb, C., and R. W. Young. 1974*b*. Renewal of glycerol in the visual cells and pigment epithelium of the frog retina. *J. Cell Biol.* **62:**378–389.

Bok, D., S. F. Basinger, and M. O. Hall. 1974. Autoradiographic and radiobiochemical studies on the incorporation of [6-^{3}H]glucosamine into frog rhodopsin. *Exp. Eye Res.* **18:**225–240.

Bownds, D. 1967. Site of attachment of retinal in rhodopsin. *Nature (Lond.)*. **216:**1178–1181.

Bridges, C. D. B. 1976. 11-*cis*-vitamin A in dark-adapted rod outer segments is a probable source of prosthetic groups for rhodopsin biosynthesis. *Nature (Lond.)*. **259:**247–248.

Bridges, C. D. B., and S. Yoshikami. 1969. Uptake of tritiated retinaldehyde by the visual pigment of dark-adapted rats. *Nature (Lond.)*. **221:**275–276.

Dowling, J. E. 1960. Chemistry of visual adaptation in the rat. *Nature (Lond.)*. **188:**114–118.

Eakin, R. M., and J. W. Brandenburger. 1968. Localization of vitamin A in the eye of a pulmonate snail. *Proc. Natl. Acad. Sci. U. S. A.* **60:**140–145.

Hall, M. O., D. Bok, and A. D. E. Bacharach. 1969. Biosynthesis and assembly of the rod outer segment membrane system. Formation and fate of visual pigment in the frog retina. *J. Mol. Biol.* **45:**397–406.

Hall, M. O., and D. Bok. 1974. Incorporation of [^{3}H]vitamin A into rhodopsin in light and dark adapted frogs. *Exp. Eye Res.* **18:**105–117.

Hall, M. O., and D. Bok. 1976. Reduction of the retinal-opsin linkage in isolated frog retinas. *Exp. Eye Res.* **22:**595–609.

Heller, J. 1968. Structure of the visual pigments. I. Purification, molecular weight, and composition of bovine visual pigment. *Biochemistry*. **7:**2906–2913.

Heller, J., and M. A. Lawrence. 1970. Structure of the glycopeptide from bovine visual pigment$_{500}$. *Biochemistry*. **9:**864–869.

Kornfeld, S., R. Kornfeld, E. F. Neufeld, and P. J. O'Brien. 1964. The feedback control of sugar nucleotide biosynthesis in the liver. *Proc. Natl. Acad. Sci. U. S. A.* **52:**371–379.

Mazlen, R. G., C. G. Muellenberg, and P. J.

O'Brien. 1970. L-glutamine D-fructose 6-phosphate amidotransferase from bovine retina. *Exp. Eye Res.* **9:**1–11.

Molnar, J., G. B. Robinson, and R. Winzler. 1965. IV. The subcellular sites of incorporation of glucosamine-1-^{14}C into glycoprotein in the rat liver. *J. Biol. Chem.* **240:**1882–1888.

Moscarello, M. A., L. Kashuba, and J. M. Sturgess. 1972. The incorporation of [^{14}D]D-galactose and [^{3}H]D-mannose into Golgi fractions of rat liver and into serum. *FEBS (Fed. Eur. Biochem. Soc.) Lett.* **26:**87–91.

Napolitano, L., F. Lebaron, and J. Scaletti. 1967. Preservation of myelin lamellar structure in the absence of lipid: a correlated chemical and morphological study. *J. Cell Biol.* **34:**817–826.

Nir, I., and M. O. Hall. 1974. The ultrastructure of lipid-depleted rod photoreceptor membranes. *J. Cell Biol.* **63:**587–598.

O'Brien, P. J., and C. G. Mullenberg. 1974. The biosynthesis of rhodopsin *in vitro*. *Exp. Eye Res.* **18:**241–252.

Pourcho, R. G., and M. H. Bernstein. 1975. Localization of $^{3}H_2$-vitamin A in mouse retina. *Exp. Eye Res.* **21:**359–367.

Schwarzkopf, O., H. J. Cahnmann, H. J. Lewis, A. D. Swidins, and H. M. Wuest. 1949. Zur synthese des vitamins A. *Helv. Chim. Acta.* **32:**443–452.

Sherman, B. S. 1970. Autoradiographic localization of ^{3}H-retinol and derivatives in the rat retina. *Exp. Eye Res.* **10:**53–57.

Sturgess, J. M., M. Mitranic, and M. A. Moscarello. 1972. The incorporation of D-glucosamine-^{3}H into the Golgi complex from rat liver and into serum glycoproteins. *Biochem. Biophys. Res. Commun.* **46:**1270–1277.

Wald, G. 1935. Carotenoids and the visual cycle. *J. Gen. Physiol.* **19:**351–371.

Whur, P., A. Herscovics, and C. P. Leblond. 1969. Radioautographic visualization of the incorporation of galactose-^{3}H and mannose-^{3}H by rat thyroids in vitro and in relation to the stages of thyroglobulin synthesis. *J. Cell Biol.* **43:**289–311.

Young, R. W. 1967. The renewal of photoreceptor cell outer segments. *J. Cell Biol.* **33:**61–72.

Young, R. W., and D. Bok. 1969. Participation of the retinal pigment epithelium in the rod outer segment renewal process. *J. Cell Biol.* **42:**392–403.

Young, R. W., and B. Droz. 1968. The renewal of proteins in retinal rods and cones. *J. Cell Biol.* **39:**169–184.

DIFFERENTIATION AND REGULATION IN INVERTEBRATE PHOTORECEPTORS

PÁL RÖHLICH

If one reviews the many variations evolved by nature to produce cell types that can be stimulated by light (Eakin, 1972), one common structural feature becomes obvious: a specialized membranous array for light perception. This membranous system—the photoreceptor organelle—is a derivative of the cell membrane and contains the visual pigment necessary for the primary events of photoreception. In addition to this anatomical characteristic, a unique biological function also evolved. The primary role of this photoreceptoral structure is excitation by light. By necessity, it is dependent in many vegetative functions on other parts of the photoreceptor cell and, in vertebrates, on the auxiliary function of the pigment epithelium. Such a high degree of specialization is necessarily reflected in the fine structure of the photoreceptor cell, which is one of the highly differentiated cells that first appears in evolution.

Among the many questions related to the highly differentiated nature of the visual cell, there are two which I shall review in this paper: how an invertebrate photoreceptor differentiates and how the photoreceptor membranes are renewed and degraded. I shall discuss photoreceptoral differentiation, using the example of one species, a planarian. After describing the fine structure of the planarian visual cell, I shall deal with photoreceptor differentiation in regeneration and embryonic development and with the effect of light deprivation upon photoreceptor development. Finally, I shall summarize briefly findings on turnover of photoreceptoral membranes in invertebrates, which are similar to those in vertebrates.

PÁL RÖHLICH Laboratory I of Electron Microscopy, Semmelweis University of Medicine, Budapest, Hungary

The Planarian Photoreceptor Cell

The planarians have an important position in the evolution of lower invertebrates because, as representatives of the most ancient class (Turbellaria) of the phylum Platyhelminthes, they first exhibit all the basic features of higher animal organization: bilateral symmetry, cephalization, and a third germ layer (mesoderm). In addition, many planarian species have an amazing regenerative ability that makes them favorite experimental animals of developmental biologists.

The fine structure of the planarian eye has been studied in the following species: *Dugesia lugubris, Dugesia tigrina, Dugesia japonica, Dugesia dorotocephala,* and *Dendrocoelum lacteum* (Röhlich and Török, 1961; MacRae, 1964; Kishida, 1967*a*; Carpenter et al., 1974*a*). The eyes of these species are situated in the head above the two-lobed brain, and are slightly flattened vesicles consisting of *pigment cells* that surround a cluster of photoreceptor endings (Fig. 1). The processes of the photoreceptor cell, called clubs, enter the eye vesicle through a pigment-free part, the corneal membrane. The eye vesicle appears in the light microscope as a pigmented cup consisting of a layer of pigment cells in which the pigment granules are accumulated near the luminal side of each cell. Most of the cell organelles, such as Golgi complex, mitochondria, ergastoplasm, etc., are situated near the opposite side of each cell.

The perikarya of the *photoreceptor cells* lie outside the eye vesicle. Each cell is bipolar, with a characteristic sensory process (club) extending into the eye vesicle and an axon that connects the cell to the brain. The fine structure of the planarian photoreceptor cell is shown in Fig. 2. The specialized photoreceptor is situated at the end of the sensory process and consists of several

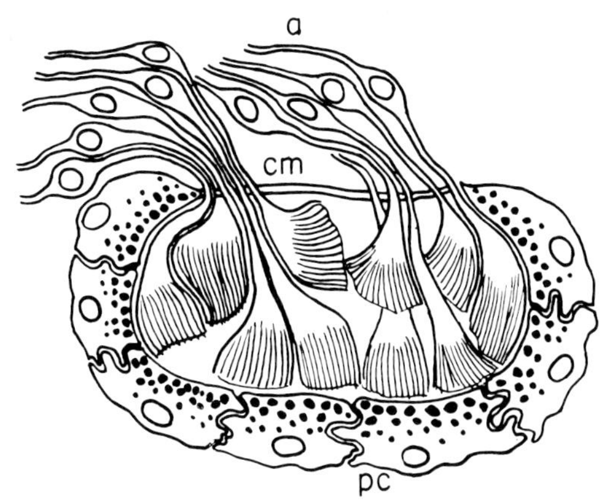

FIGURE 1 Microscopic structure of a planarian eye. Receptor endings (retinal clubs) of photoreceptor cells are embedded in eye vesicle consisting of pigment cells (*pc*). *a*, an axon; *cm*, corneal membrane.

hundred long *microvilli* (rhabdomeric photoreceptor) (Eakin, 1972), which radiate from a conical thickening of the sensory process, the conical body. The space between the microvilli is filled with a matrix of medium density. A similar intercellular humor in close contact with the photoreceptor organelle can be found in most retinas, including those of vertebrates (Röhlich, 1970). The assumption that the array of microvilli (rhabdomere) is the site of light perception is supported by its analogy with rhabdomeres of higher invertebrates, in which a number of data document their primary role in photoreception (Eakin, 1972). (1) Rhodopsinlike photopigments can be extracted from isolated rhabdomeres. (2) Microspectrophotometry and dichroic absorption of rhabdoms indicate the presence of an oriented photopigment in the microvilli. (3) Light sensitivity appears simultaneously with differentiation of the rhabdom in embryonic development. (4) Lack of vitamin A affects both rhabdom fine structure and visual perception. (5) Vitamin A and amino acids are incorporated into compounds that are transported from the site of synthesis to the rhabdom. (6) Many arthropods can analyze polarized light, which can best be explained by an orientation of photopigment molecules in microvillar membranes arranged perpendicularly to one another. (7) With freeze-fracturing, the cytoplasmic fracture face (face P) of the microvillar membrane shows a high number of intramembranous particles (Perrelet 1972; Fernandez and Nickel 1976; Brandenburger et al. 1976; Eguchi and Water-

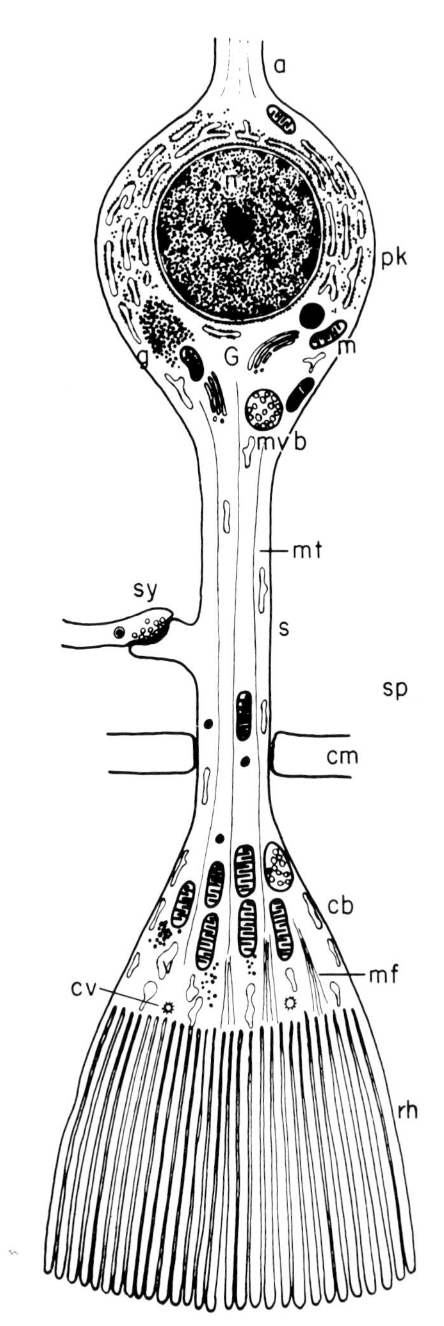

FIGURE 2 A planarian photoreceptor cell is bipolar with an axon (*a*), perikaryon (*pk*) and sensory process (*sp*); the latter consists of stalk (*s*), conical body (*cb*), and the microvillar rhabdomere (*rh*). *cm*, corneal membrane; *cv*, coated vesicle; *g*, glycogen particles; *G*, Golgi apparatus; *m*, mitochondrion; *mf*, bundle of thin microtubules; *mt*, microtubule; *mvb*, multivesicular body; *n*, nucleus, *sy*, axodendritic synapse; *v*, cistern of the vacuolar system.

man, 1976), which are assumed to represent rhodopsin or rhodopsin complexes because the number of particles is reduced by digitonin extraction or degradation in multivesicular bodies and also because they are affected by altering light conditions.

(8) Finally, there is indirect evidence from a planarian photoreceptor that the retinal clubs are involved in light perception (Röhlich and Tar, 1968). Planarians (*Dugesia tigrina*) exposed to complete darkness for several weeks lost the normal structure and size of their retinal clubs. The first sign of degeneration, after 1 wk of light deprivation, was the increasing sensitivity of the microvillar membranes to fixation in osmium tetroxide (Röhlich, 1966). By the end of the 3rd wk, the eyecup was almost empty because of the atrophy of the photoreceptor endings. At this time, the cross-sectional area of the retinal clubs was reduced to about one-fifth of the normal size. When the planarians were returned subsequently to normal illumination, the receptor endings rapidly regenerated and regained their insensitivity to osmium.

The conically shaped thickening of a retinal club, the *conical body,* contains a relatively high number of mitochondria, as does the inner segment of a vertebrate photoreceptor. Mitochondria in dark-adapted eyes swell when exposed to light (Carpenter et al., 1974*b*) and accumulate intramitochondrial calcium in the presence of actinomycin D and light (Röhlich, unpublished data). These findings support the idea that the clubs are stimulated by light and that their energy is used in some important step in photoreception. There is also a smooth-surfaced vacuolar system present in the conical body that also reacts to dark- and light-adaptation, dilating and proliferating in the dark and collapsing when illuminated (Röhlich and Török, 1962; Carpenter et al., 1974*b*). Unchanging constituents of a conical body are microtubules, 23 and 11 nm in thickness. The narrower tubules occur in bundles and radiate toward the microvilli.

The *stalk* of a sensory cell contains a large number of 23-nm microtubules, a few vacuoles, mitochondria, dense-cored vesicles, and multivesicular bodies. Two special features deserve attention. One is synaptic contacts with small nerve fibers, in which the photoreceptor cell is the postsynaptic element (Carpenter et al., 1974*a*). These junctions may function to feed back inhibition during short-term light adaptation. The other structure, coated vesicles, usually are present both at the base of the microvilli and along the whole length of the sensory process. As recent data on the turnover of the photoreceptor membrane indicate (see below), they may have important functions in membrane transport to or from the microvilli.

The *perikaryon,* with its rich ergastoplasm, is clearly the site of protein synthesis in the planarian photoreceptor cell. A peculiarity of this ergastoplasm is that its cisterns are devoid of ribosomes in many places. A few Golgi complexes, clusters of glycogen particles, and multivesicular bodies complete the picture.

Differentiation of the Planarian Photoreceptor in Regeneration and Embryonic Development

Regeneration in planarians, especially that of the head, has been studied extensively by experimental embryologists in the last two decades. If an animal is decapitated, a blastema is rapidly formed by migration and accumulation at the cut surface of totipotent cells, the neoblasts. The first important morphogenetic process in the blastema is the differentiation of a new head ganglion from neoblasts through homologous induction or by autodifferentiation. The new ganglion functions as a morphogenetic inductor of new eyes (Wolff and Lender, 1950; Lender, 1952; Török, 1958). It is still unclear what morphogenetic relation exists between the photoreceptoral and pigment cells in the eye. In other words, do the photoreceptoral cells (or the pigment cells) appear first and induce the formation of the other cell type?

There is good reason to assume that embryonic development of an eye proceeds as in regeneration. Therefore, we recently investigated the finestructural features of photoreceptoral differentiation during embryonic development (Röhlich and Török, unpublished data) and compared it with earlier observations by Kishida (1967*b*) and Röhlich (1967*a*) on the regenerating eye. Differentiation in both embryonic development and regeneration were found to be essentially similar, so they can be discussed together.

The eye rudiments of the planarian *Dugesia lugubris* can be first recognized under the dissecting microscope by the appearance of faint pigment spots on the *3rd day of regeneration and the 9th day of embryonic development,* respectively. The pigment spot is a small group of pigment cells differentiating from neoblasts. Regeneration and ontogenetic development seem to differ slightly in this early stage. In regeneration, the pigment cells very early form a narrow cavity which is filled with a substance of medium density and is closed by septate desmosomes (Fig. 3). In embryonic development, the cavity is formed somewhat later by the sensory processes of differentiating photoreceptor cells projecting into the group of cells at several places (Figs. 4 and 5). The pigment cell processes then enclose the photoreceptor endings and the two are joined by desmosomes. This leads to the formation of multiple cavities, each containing one or two sensory endings (Fig. 5). At this stage, pigment cells show the ultrastructural signs of a high synthetic activity, with many free and membrane-bound ribosomes, a well-developed Golgi system, prominent nucleoli, and a few forming pigment granules. The young photoreceptor cells can be differentiated from neoblasts only by their form and location. They grow a thick and relatively short, irregular sensory process which has a fine structure basically similar to that of the presumptive perikaryon. Many free ribosomes and a few microtubules characterize the cytoplasm of a sensory process at this stage. The first sign of a differentiating rhabdomere is the appearance of a few irregular and short microvilli at the ends of the photoreceptoral processes (Fig. 4).

During the next day (*4th day of regeneration and 10th day of embryonic development;* Fig. 5), differentiation progresses rapidly by a rich microvillar system developing at the ends of the sensory processes. Microvesicular bodies and coated vesicles appear in the photoreceptor processes, but their cytoplasm still contains a high number of ribosomes and a few Golgi complexes. There are no recognizable conical bodies at this stage.

The essential structural characteristics of the definitive eye can be easily detected on the following day (*5th day of regeneration and 11th–12th day of ontogenetic development;* Fig. 6). The multiple cavities have fused to form a large pigmented cup, but a few incomplete septa, consisting of narrow pigment cell processes, still persist. Some of them are connected to other pigment cell processes, forming the corneal membrane. The photoreceptor cells become more and more fully differentiated. The ends of the sensory processes thicken to form conical bodies into which mitochondria migrate from the perikaryon.

This young eye has essentially all the morphological characteristics of an adult eye, but its size is considerably smaller. It reaches its final size by the gradual addition of newly forming photoreceptor cells from neoblasts. Similarly, new pigment cells are added from the outside of the pigment cup. This final stage of development lasts until the end of the 2nd wk of regeneration and the conclusion of postembryonic development, respectively. However, if a single photoreceptor cell is considered, differentiation seems to be a rapid process which is completed in roughly 2–3 days.

Is Light Required for Photoreceptor Differentiation?

Prolonged darkness, as was shown (Röhlich and Tar, 1968), leads to involution of structures participating in light perception. In a planarian, this can lead to the eye cup becoming virtually empty, as a result of the strong atrophy of retinal clubs. If light is so essential for the maintenance of a differentiated eye, the question arises whether it is also required for the differentiation of the photoreceptor cell.

If planarians (*Dugesia tigrina*) are kept in complete darkness during eye regeneration (Röhlich 1967*a,b*), retinal clubs are fully differentiated. At the end of the 1st wk, such retinal clubs cannot be distinguished from the controls (differentiated in normal light conditions). However, if the experimental animals are continued in darkness for an additional period of 2 wk, the degenerating effect of darkness on fully differentiated receptors becomes increasingly prominent and, at the end of the 3rd wk, the retinal clubs are markedly atrophied.

A similar observation was made on planarians (*Dugesia lugubris*) which were reared in darkness from the beginning of embryonic development (Röhlich and Török, unpublished data). Retinal clubs in hatching animals were nearly normal, but if the animals were further deprived of light for additional weeks, the atrophic effect of light deprivation on differentiated photoreceptors gradually appeared.

As a consequence, it can be concluded that light is not required for photoreceptor differentiation and that light deprivation does not affect the genetically determined photoreceptoral development. On the other hand, a normally lighted environment is essential for the maintenance of well-differentiated photoreceptoral structures. A similar conclusion has been reached on differentiating crustacean photoreceptors (Roach and Wiersma, 1974).

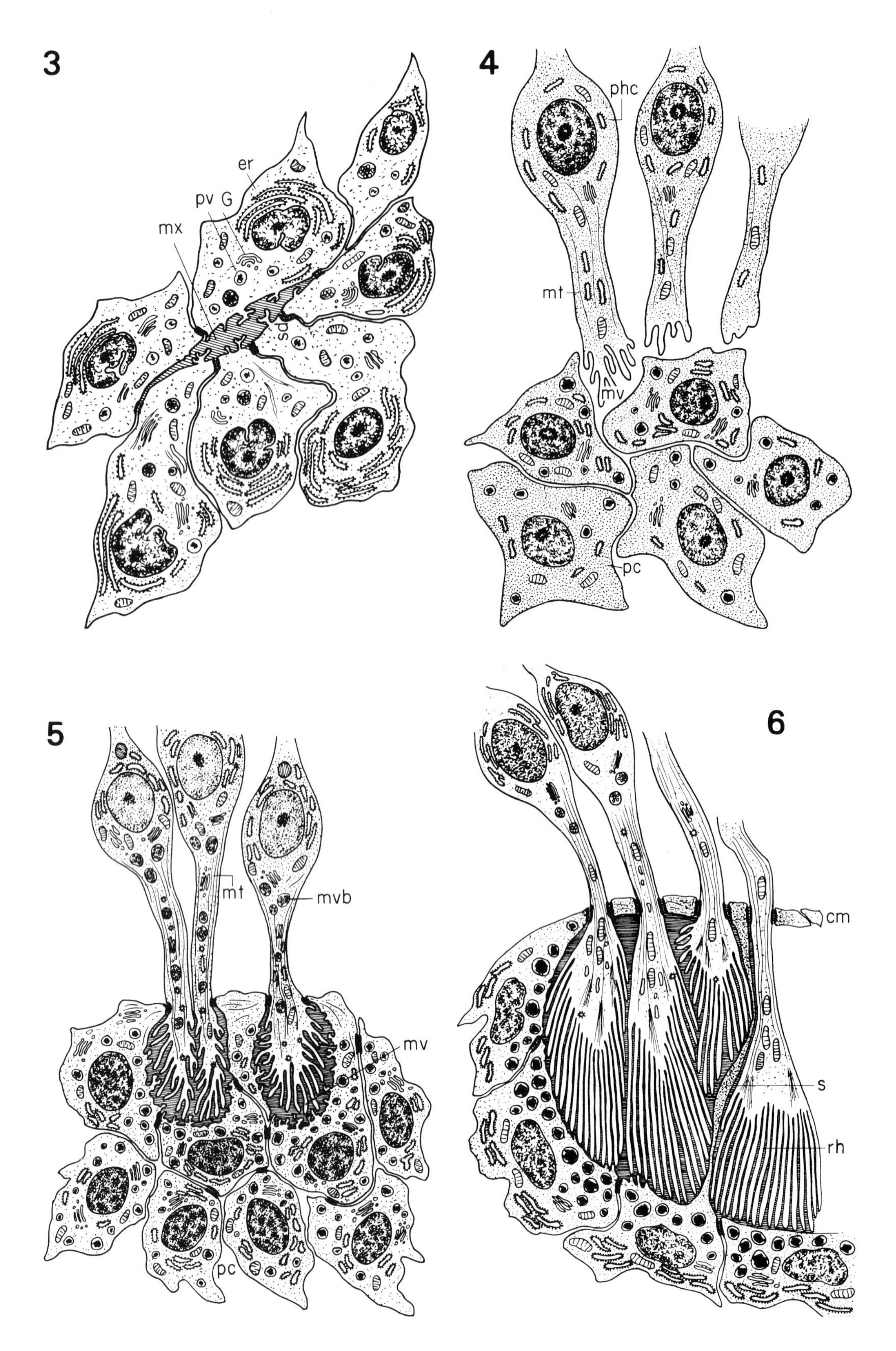
3
er
pv
G
mx
sd
4
phc
mt
mv
pc
5
mt
mvb
mv
pc
6
cm
s
rh

Photoreceptor Membrane Turnover in Invertebrates

In the last decade, much important data have accumulated concerning the assembly and turnover of photoreceptor membranes in vertebrate visual cells (see Bok, this volume). The photopigment is synthesized in the ergastoplasm of the inner segment of a rod or cone cell. Then it is transported via the Golgi region and the connecting cilium into the outer segment. In rod cells, rhodopsin is built into the membrane of newly formed disks at the base of the outer segment. These disks are moved by subsequently added disks toward the tip of the outer segment. Then the oldest disks are phagocytized and digested by the pigment epithelium. If a vertebrate photoreceptor cell is compared with those of invertebrates, one can recognize, in spite of the immense structural diversity (Eakin, 1972), one basic principle common to most photoreceptor cells: there is a specialized region of the cell where photopigment is incorporated into membranous structures, the photoreceptor organelles. Protein components, synthesized in the ergastoplasm of the visual cell, must therefore be transported to the photoreceptor organelle. What are the data supporting such an intracellular transport and a possible membrane turnover in invertebrate photoreceptor cells?

Electron microscope autoradiographic studies with labeled amino acids in insects (Perrelet, 1972) show that the newly synthesized protein is first found in the ribosome-containing cytoplasmic area and subsequently in the rhabdom (Fig. 8). Membrane turnover in the larval mosquito eye was calculated to be much faster than that in the vertebrate photoreceptor cell (White and Lord, 1975), and in the crayfish, synthesis and transport of the newly formed protein was found to be so fast that incorporation in the rhabdom occurred within a few minutes (Hafner and Bok, 1977). Evidence that the newly synthesized protein represents rhodopsinlike photopigments is still lacking. It is also unknown whether the photopigment is bound to any vesicular organelle during its transport to the rhabdom and, if so, whether these structures are identical with coated vesicles derived from the Golgi apparatus (White and Gifford, personal communication).

A detailed study of photoreceptor renewal has been carried out by Eakin and Brandenburger (1974) on a snail (*Helix aspersa*). They came to the conclusion that 80-nm vesicles, called photic vesicles, played a key role in the transport of photopigment from the soma of a cell to the microvilli (Fig. 7). This hypothesis is supported by the following findings. (1) Isotope-labeled vitamin A followed this cytoplasmic route:Golgi apparatus–photic vesicles–microvilli (Brandenburger and Eakin, 1970). (2) In light-tolerant slugs, the vesicles were found in the cytoplasmic hillocks, whereas in light-avoiding slugs they accumulated in the basal regions of the photoreceptor cells (Eakin and Brandenburger, 1975). (3) Reduction in number of photic vesicles and destruction by lysosomes were observed in snails after prolonged dark-adaptation (Eakin and Brandenburger,

FIGURES 3–6 Fine structure of planarian eyes in different stages of differentiation.

FIGURE 3 Third day of regeneration. Young pigment cells, differentiating from neoblasts, surround a narrow cavity filled with a substance of medium density (*mx*). *er*, ergastoplasm; *G*, Golgi apparatus; *pv*, propigment vesicle; *sd*, septate desmosome.

FIGURE 4 Ninth day of embryonic development. Differentiating photoreceptor cells (*phc*) send short processes toward a group of young pigment cells (*pc*). A few short and irregular microvilli (*mv*) and microtubules (*mt*) distinguish sensory processes from other parts of photoreceptor cells.

FIGURE 5 Tenth day of embryonic development and 4th day of regeneration. Endings of sensory processes are embedded in multiple cavities among differentiating pigment cells (*pc*). Microtubules (*mt*) increase in number, and multivesicular bodies (*mvb*) appear in sensory process. Microvilli (*mv*) are longer, but still irregular.

FIGURE 6 Eleventh day of embryonic development and 5th day of regeneration. One large cavity in pigment cell group is incompletely partitioned by thin pigment cell septa (*s*). Photoreceptor endings (retinal clubs) are differentiated into conical body with mitochondria and an ordered array of long microvilli, the rhabdomere (*rh*). *cm*, corneal membrane.

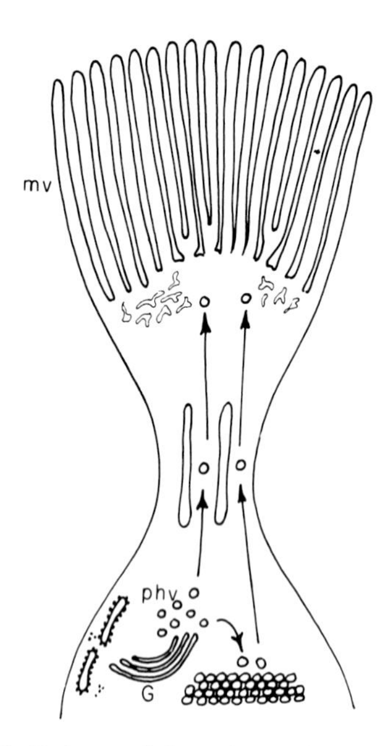

FIGURE 7 Pathway of photopigment in snail photoreceptor cell (Brandenburger and Eakin, 1970). Photic vesicles (*phv*) derived from Golgi apparatus (*G*) are assumed to carry photopigment to the rhabdomeric microvilli (*mv*) either directly or after intermediate storage.

1974). (4) Freeze-fracturing of snail eyes showed intramembranous particles in the vesicle membranes similar to those in the microvillar membranes (Brandenburger et al., 1976).

The other aspect of membrane turnover is degradation of old or damaged photoreceptor membranes. This task is accomplished in vertebrates by the pigment epithelial cells that phagocytize and digest the apical parts of rod outer segments. In arthropods, on which most invertebrate studies have been conducted, it seems that there is a retrieval of the photoreceptor membrane into the cytoplasm by endocytosis. When exposed to light, rhabdoms of the compound eye of *Daphnia* and of the larval mosquito eye diminish in size (Röhlich and Törő, 1965; White, 1967, 1968; White and Lord, 1975), accompanied by the appearance of many coated vesicles at the basal infoldings of the microvillar membranes (Röhlich and Törő, 1965; Eguchi and Waterman, 1967; White, 1967, 1968). The presence of coated vesicles, multivesicular bodies, and lamellar bodies in these photoreceptor cells led Eguchi and Waterman (1976) and White and Gifford (personal communication) to assume a light-dependent, receptor-membrane cycling. Parts of photoreceptor membrane would be internalized by pinocytosis to form coated vesicles, which, after losing their coat, would fuse and undergo reverse pinocytosis to produce multivesicular bodies (Fig. 8). The latter are regarded as secondary lysosomes in which degradation of the photoreceptor membrane would take place. Eguchi and Waterman support their hypothesis with freeze-etching of crayfish compound eyes: they found a gradual decrease in numbers of intramembranous particles, supposed to represent rhodopsin, as a result of transfer from microvillar membranes to multivesicular and lamellar bodies. Other indirect evidence for the proposed membrane degradation was the disappearance of these vesicular structures beginning at the onset of dark-adaptation of the larval mosquito eye, following the sequence: coated vesicle–multivesicular body–lamellar body (White and Gifford, personal communication). The vesicular organelles reappear in the same order in light-adaptation.

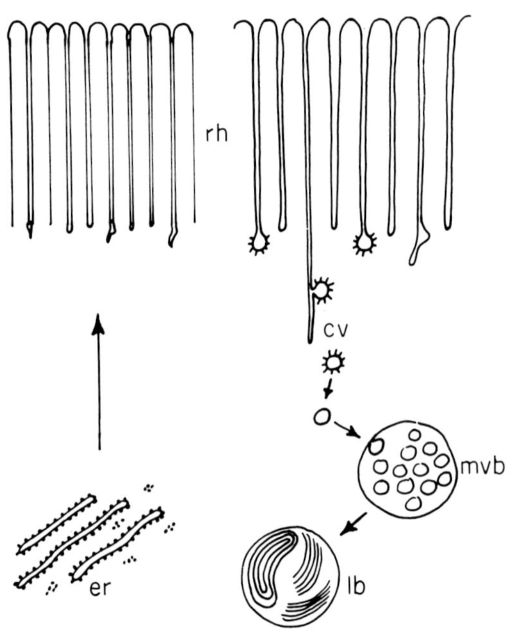

FIGURE 8 Renewal and degradation of photoreceptor membrane in arthropods. Diagram composed from the results of Perrelet (1972), Hafner and Bok (1977), Eguchi and Waterman (1976), and White and Gifford (1976, personal communication). Rhodopsin is transported from site of synthesis (*er*) to the microvillar membranes of rhabdomere (*rh*). Pieces of membrane are internalized as coated vesicles (*cv*), which, after losing their coat, form microvesicular (*mvb*) and lamellar bodies (*lb*) in which degradation takes place.

REFERENCES

BRANDENBURGER, J. L., and R. M. EAKIN. 1970. Pathway of incorporation of vitamin A-^{3}H into photoreceptors of a snail, *Helix aspersa*. *Vision Res.* **10:**639–653.

BRANDENBURGER, J. L., C. T. REED, and R. M. EAKIN. 1976. Effects of light- and dark-adaptation on the photic microvilli and photic vesicles of the pulmonate snail *Helix aspersa*. *Vision Res.* **16:**1205–1210.

CARPENTER, K. S., M. MORITA, and B. BEST. 1974*a*. Ultrastructure of the photoreceptor of the planarian *Dugesia dorotocephala*. I. Normal eye. *Cell Tissue Res.* **148:**143–158.

CARPENTER, K. S., M. MORITA, and B. BEST. 1974*b*. Ultrastructure of the photoreceptor of the planarian *Dugesia dorotocephala*. II. Changes induced by darkness and light. *Cytobiologie*. **8:**320–338.

EAKIN, R. M. 1972. Structure of invertebrate photoreceptors. *In* Handbook of Sensory Physiology. Vol. VII/1. H. J. A. Dartnall, editor. Springer-Verlag, Berlin. 623–684.

EAKIN, R. M., and J. L. BRANDENBURGER. 1974. Ultrastructural effects of dark-adaptation on eyes of a snail, *Helix aspersa*. *J. Exp. Zool.* **187:**127–133.

EAKIN, R. M., and J. L. BRANDENBURGER. 1975. Retinal differences between light-tolerant and light-avoiding slugs (Mollusca: Pulmonata). *J. Ultrastruct. Res.* **53:**382–394.

EGUCHI, E., and T. H. WATERMAN. 1967. Changes in retinal fine structure induced in the crab *Libinia* by light and dark adaptation. *Z. Zellforsch. Mikrosk. Anat.* **79:**209–229.

EGUCHI, E., and T. H. WATERMAN. 1976. Freeze-etch and histochemical evidence for cycling in crayfish photoreceptor membranes. *Cell Tissue Res.* **169:**419–434.

FERNANDEZ, H. R., and E. E. NICKEL. 1976. Ultrastructural and molecular characteristics of crayfish photoreceptor membranes. *J. Cell Biol.* **69:**721–732.

HAFNER, G. S., and D. BOK. 1977. The distribution of ^{3}H-leucine labeled protein in the retinula cells of the crayfish retina. *J. Comp. Neurol.* In press.

KISHIDA, Y. 1967*a*. Electron microscopic studies on the planarian eye. I. Fine structures of normal eye. *Sci. Rep. Kanazawa Univ.* **12:**75–110.

KISHIDA, Y. 1967*b*. Electron microscopic studies on the planarian eye. II. Fine structures of the regenerating eye. *Sci. Rep. Kanazawa Univ.* **12:**111–142.

LENDER, Th. 1952. Le rôle inducteur du cerveau dans la régénération des yeux d'une planaire d'eau douce. *Bull. Biol. Fr. Belg.* **86:**140–215.

MACRAE, E. K. 1964. Observations on the fine structure of photoreceptor cells in the planarian *Dugesia tigrina*. *J. Ultrastruct. Res.* **10:**334–349.

PERRELET, A. 1972. Protein synthesis in the visual cells of the honey-bee drone as studied with electron microscope autoradiography. *J. Cell Biol.* **55:**595–605.

PERRELET, A., H. BAUER, and V. FOYDER. 1972. Fracture faces of an insect rhabdomere. *J. Microsc.* (*Paris*). **13:**97–106.

ROACH, J. L. M., and C. A. G. WIERSMA. 1974. Differentiation and degeneration of crayfish photoreceptors in darkness. *Cell Tissue Res.* **153:**137–144.

RÖHLICH, P. 1966. Sensitivity of regenerating and degenerating planarian photoreceptors to osmium fixation. *Z. Zellforsch. Mikrosk. Anat.* **73:**165–173.

RÖHLICH, P. 1967*a*. Fine structure of photoreceptors in normal and experimental conditions. C. Sc. thesis, Budapest.

RÖHLICH, P. 1967*b*. Fine structural changes induced in photoreceptors by light and prolonged darkness. *In* Symposium on Neurobiology of Invertebrates. J. Salánki, editor. Hungarian Academy of Science, Budapest. 95–109.

RÖHLICH, P. 1970. The interphotoreceptor matrix: electron microscopic and histochemical observations on the vertebrate retina. *Exp. Eye Res.* **10:**80–96.

RÖHLICH, P., and E. TAR. 1968. The effect of prolonged light-deprivation on the fine structure of planarian photoreceptors. *Z. Zellforsch. Mikrosk. Anat.* **90:**507–518.

RÖHLICH, P., and I. TÖRŐ. 1965. Fine structure of the compound eye of *Daphnia* in normal, dark- and strongly light-adapted state. *In* Eye Structure. Vol. II. J. W. Rohen, editor. Schattauer-Verlag, Stuttgart. 175–186.

RÖHLICH, P., and L. J. Török. 1961. Elektronenmikroskopische Untersuchung des Auges von Planarien. *Z. Zellforsch. Mikrosk. Anat.* **54:**362–381.

RÖHLICH, P., and L. J. TÖRÖK. 1962. The effect of light and darkness on the fine structure of the retinal clubs of *Dendrocoelum lacteum*. *Q. J. Microsc. Sci.* **104:**543–548.

TÖRÖK, L. J. 1958. Experimental contributions to the regenerative capacity of *Dugesia lugubris* O. Schm. *Acta Biol. Acad. Sci.* (*Hung.*). **9:**79–98.

WHITE, R. H. 1967. The effect of light and light deprivation upon the ultrastructure of the larval mosquito eye. II. The rhabdom. *J. Exp. Zool.* **166:**405–426.

WHITE, R. H. 1968. The effect of light and light deprivation upon the ultrastructure of the larval mosquito eye. III. Multivesicular bodies and protein uptake. *J. Exp. Zool.* **169:**261–278.

WHITE, R. H., and E. LORD. 1975. Diminution and enlargement of the mosquito rhabdom in light and darkness. *J. Gen. Physiol.* **65:**583–598.

WOLFF, E., and Th. LENDER. 1950. Sur le rôle organisateur du cerveau dans la régénération des yeux chez une planaire d'eau douce. *C. R. Acad. Sci.* (*Paris*). **23:**2238–2239.

Cells of the Artery Wall and Atherosclerosis

THE CELLS OF THE ARTERY WALL IN THE STUDY OF ATHEROSCLEROSIS

RUSSELL ROSS, JOHN GLOMSET, BEVERLY KARIYA, ELAINE RAINES, and JACQUELINE BUNGENBERG DE JONG

The field of cell biology has reached a point where information gained from diverse approaches at the cellular and molecular levels can be applied to understand the etiology and pathogenesis of disease processes important to man. This is now being done in relation to atherosclerosis, an arterial disease that is the principal cause of death in the United States and Western Europe (Arteriosclerosis, 1971). The lesions of atherosclerosis are characterized by focal proliferation of smooth muscle cells, accumulation of extracellular connective tissue components, including collagen, elastic fiber proteins, and proteoglycans, and deposition of intracellular and extracellular lipid (Ross and Glomset, 1973, 1976). Recognition of the central importance of intimal smooth muscle proliferation in the formation of the lesions of atherosclerosis has permitted in depth studies utilizing approaches that include cell culture and in vivo stimulation of lesion formation. Studies in our laboratories have focused on the biology of arterial endothelium and smooth muscle in relation to a hypothesis that states that the lesions of atherosclerosis result from the reponse of the endothelium to different forms of "injury."

The Response to Injury Hypothesis

The response to injury hypothesis is diagrammatically presented in Fig. 1. This hypothesis suggests that endothelial cells are altered by some form of "injury" so that they desquamate at particular focal sites in the artery wall. The basis for endothelial desquamation is not clear. However, it may be related to factors important in maintaining cell-cell and cell-connective tissue (basement membrane) attachments responsible for maintaining the integrity of the endothelial lining. We have suggested (Ross and Glomset, 1976; Harker et al., 1976; Ross and Harker, 1976) that a threshold may exist below which the endothelial cells are capable of resisting the shearing stress of blood flow through the artery. Factors that may cause this threshold to be exceeded could be mechanical in nature, as at particular anatomic sites such as bifurcations, or in hypertensive individuals where marked elevation in blood pressure results in increased shear stress (Glagov, 1972; Fry, 1973). In individuals with normal blood pressure, factors such as chronic hypercholesterolemia (Ross and Harker, 1976) or chemical imbalances as found in chronic homocystinemia (Harker et al., 1976) may lower this threshhold so that the endothelium could become susceptible to the normal shearing forces. Such alterations in this threshold could lead to the detachment of the cells from their neighbors and from the underlying connective tissue, resulting in focal sites of desquamation and exposure of the underlying basement membrane and collagen fibrils. Platelets could then adhere to the exposed connective tissue and undergo the process of adherence, aggregation, and release of material (specifically mitogenic factors) into the underlying artery wall (Fig. 1). Plasma constituents such as lipoproteins would also have greater access to artery wall components.

The entry of such platelet and plasma factors into the artery could then lead to migration of smooth muscle cells from the media into the intima, to proliferation of these and preexisting intimal smooth muscle cells, and to formation of large amounts of connective tissue matrix by these cells. If the conditions that lead to injury of the endothelial cells are removed or prevented from acting, then the focal intimal proliferative response would cease and the lesions could regress. The sequence

RUSSELL ROSS and CO-WORKERS Departments of Pathology and Medicine, School of Medicine, University of Washington, Seattle, Washington

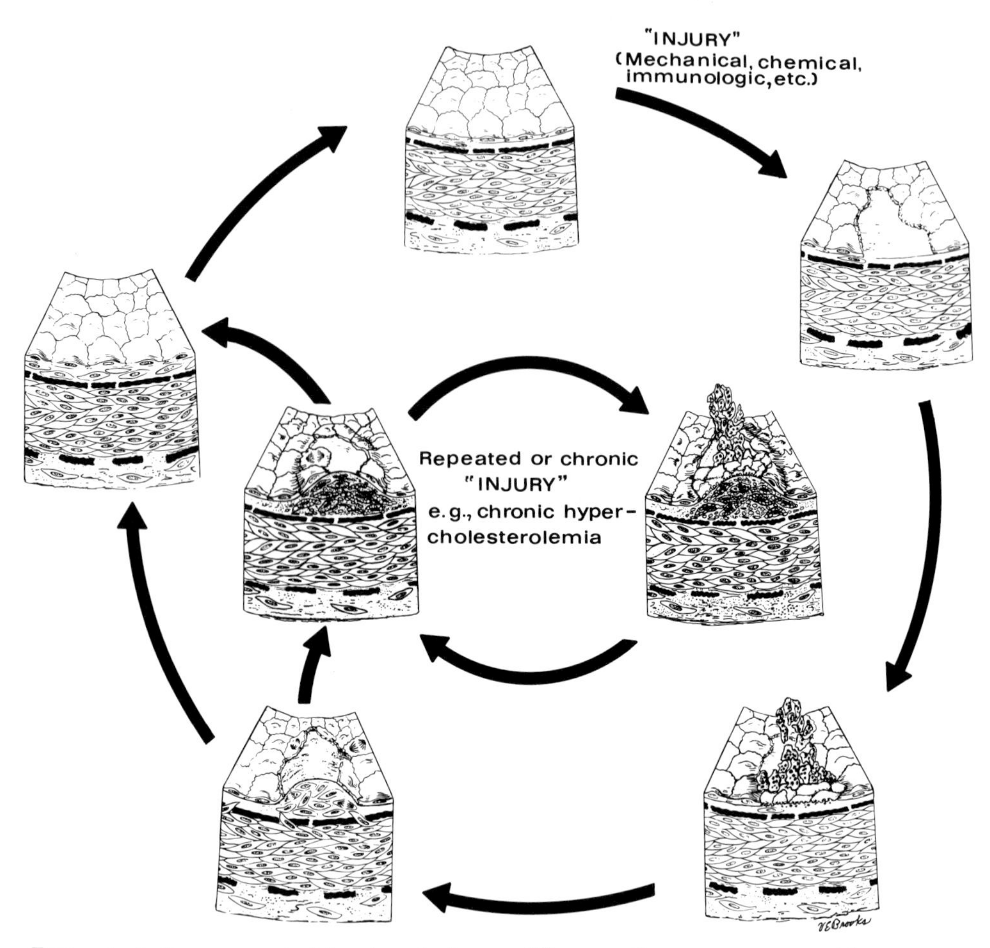

FIGURE 1 In the response to injury hypothesis, two different cyclic events may occur. The outer, or regression, cycle may represent common single occurrences in all individuals in which endothelial injury leads to desquamation, platelet adherence, aggregation, and release, followed by intimal smooth muscle proliferation and connective tissue formation. If the injury is a single event, the lesions may go on to heal and regression occur. The inner, or progression, cycle demonstrates the possible consequences of repeated or chronic endothelial injury as may occur in chronic hyperlipidemia. In this instance, lipid deposition as well as continued smooth muscle proliferation may occur after recurrent sequences of proliferation and regression, and these may lead to complicated lesions that calcify. Such lesions could go on to produce clinical sequelae such as thrombosis and infarction. (Reprinted with permission of *Science, Wash., D. C.*)

of events suggested by this hypothesis is presented in the outer, or regression, cycle in Fig. 1.

In contrast, if the conditions that lead to the initial injury continue on a long-term basis, then the repeated process of injury, proliferation, and repair could lead to lesion progression which at some point may become irreversible. These events are presented in the inner, or progression, cycle of Fig. 1.

We have been utilizing two principal approaches to examine the response to injury hypothesis. These include studies of smooth muscle cells in culture, and the in vivo response of these cells to several forms of mechanical and chemically induced injury. Correlation of the cell culture studies with the in vivo studies has greatly expanded our understanding of this disease process as induced in the subhuman primate. The results of some of these studies are presented in this report.

Cell Culture-Diploid Smooth Muscle Cells

Both endothelium and smooth muscle have been extensively studied in cell culture. The results of studies with endothelial cells are presented by Gimbrone in this volume. However, before discussing the growth response of arterial smooth muscle cells to various factors in culture, it is important to consider the limitations presented by diploid cells in culture as contrasted with cell lines derived from either normal or transformed cells.

LIMITATIONS AND POSSIBILITIES: Most studies of control of cell growth in culture have utilized cell lines that consist essentially of heteroploid or aneuploid cells which are "immortal" in culture. The use of such cells permits a certain reliance upon the fact that each time the cells are grown in culture they will behave in a fashion similar, if not identical, to that observed for their progenitors. In contrast, diploid cells derived from different donors cannot be relied upon for such predictably reproducible behavior as each strain of cells reflects the genetic variability of its donor. Such inherent genetic variability and the phenomenon of "senescence in culture" (Hayflick, 1965) necessitates internal standardization which can be accomplished with base-line studies of the growth response of each strain of cells and provide a basis for comparing the response of one donor strain of cells with another.

The smooth muscle cells used in our experiments are derived from explants of the inner one-third of the media of the thoracic aorta of 1- to 3-yr-old male *Macaca nemestrina* born and raised in the breeding colony of the Regional Primate Center of the University of Washington. The whole blood serum used for base-line studies was prepared from a large pool of blood from 20 or more male *M. nemestrina*. A large number of aliquots were stored at −70°C for use in each study.

These experiments assessed the growth response of the smooth muscle cells from each donor animal to increasing concentrations (0.5, 1, 2.5, 5, 10, 20%) of the same homologous whole blood serum. Figure 2 demonstrates typical examples of the wide range of variability in the response of different donor strains to increasing concentrations of the same lot of homologous serum. For example, the cells shown in Fig. 2*a* responded in an almost linear fashion to increasing doses of serum, whereas the cells shown in Fig. 2*c* grew as well in medium containing 1% serum as the cells shown in Fig. 1*a* did in 5% serum. Furthermore, the cells shown in Fig. 2*c* revealed little difference in growth response as the content of serum in the medium was increased from 2.5 to 20%. In general, the donor animals appeared to be divisible into three broad groups termed "hyporesponders," "normoresponders," and "hyperresponders."

This variability in donor response to the same serum lot raises important questions concerning the interpretation of experimental results with diploid cells in culture. This is particularly true with data from different laboratories where individual donor variability (as noted previously) has not been analyzed before studying the response of cells to specific growth factors. Such genetic variability also poses difficulties in making comparisons in experimental results between different laboratories. Clearly, this problem is decreased when studying cell lines such as 3T3 cells. However, it must be taken into consideration if one is going to analyze the growth response of diploid cells with a finite life span, such as arterial smooth muscle cells or endothelium.

Platelets Provide the Principal Mitogen in Serum

Many laboratories have attempted to identify and isolate the numerous growth factors known to be present in whole blood serum. Previous studies from our laboratory (Ross et al., 1974; Rutherford and Ross, 1976; Ross et al., 1977) and recent ones by Antoniades et al. (1975) have shown that the principal mitogen(s) in blood serum is a factor(s) derived from platelets during the physiological process of clotting. This was discovered as a result of the observation that the cells grown in medium containing 5% serum derived from cell-free plasma in which no platelet mitogens are present remain in a "quiescent" state for a prolonged period of time. Under these conditions (in contrast to 5% whole blood serum [Fig. 3]), 3% or less of the cells divide in any 24-h period. Addition of a partially purified platelet factor at a concentration of 100 ng/ml to medium containing cell-free plasma serum restores proliferative activity comparable to that of 5% whole blood serum (Fig. 4).

Cells such as arterial smooth muscle cells (Ross et al., 1974), dermal fibroblasts (Rutherford and Ross, 1976), 3T3 cells (Kohler and Lipton, 1974), and glial cells (Westermark and Wasteson, 1976; Busch et al., 1977) all respond in similar fashion to the thrombocyte-derived mitogen(s). The platelet factor(s) is a heat stable (56°C for 30 min), relatively low molecular weight (13,000–23,000 daltons), basic protein. Exposure of arterial smooth muscle cells or fibroblasts to the fac-

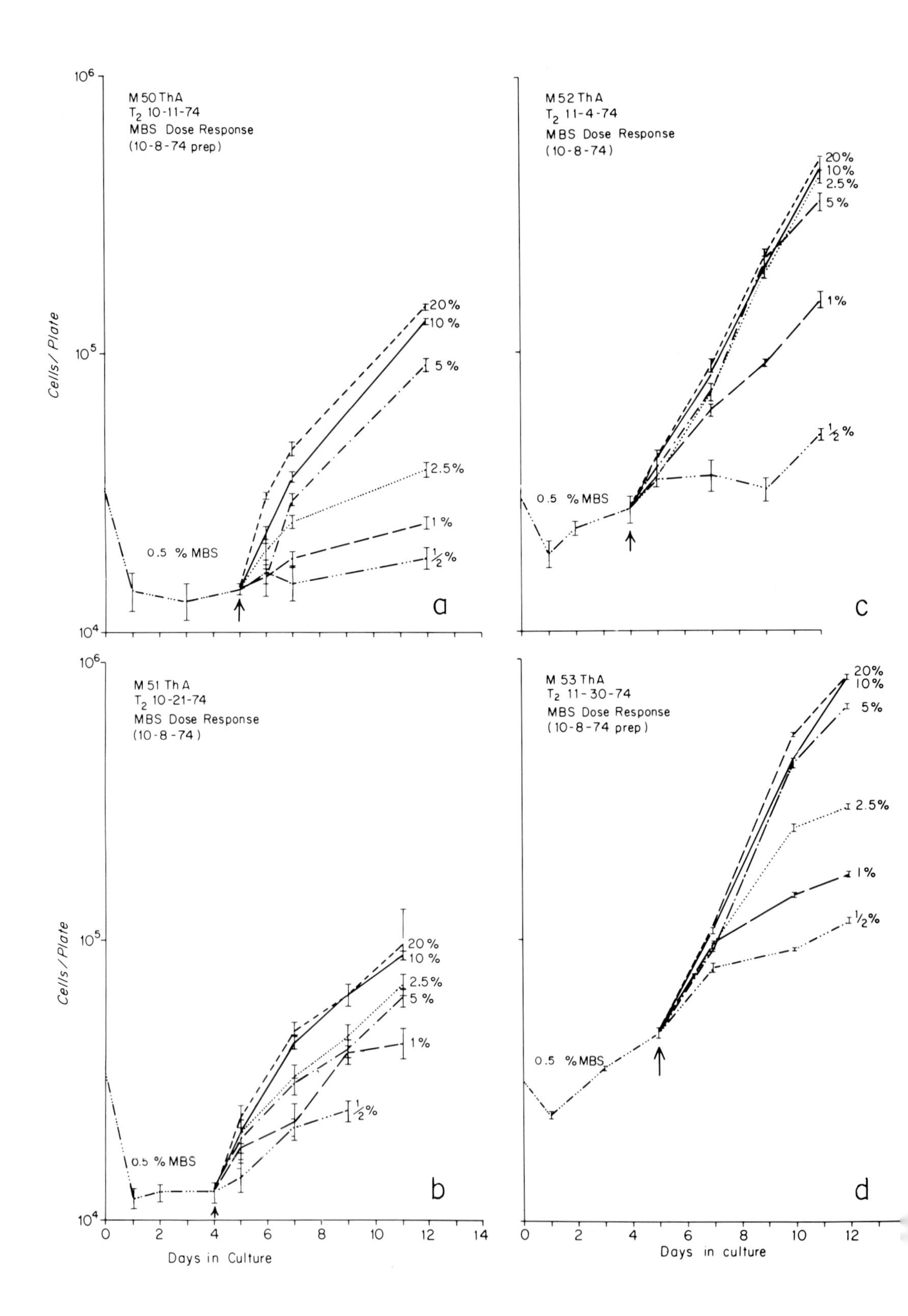

M 50 ThA
T2 10-11-74
MBS Dose Response
(10-8-74 prep)
20%
10 %
5 %
2.5%
1 %
½%
0.5 % MBS
a
M 52 ThA
T2 11-4-74
MBS Dose Response
(10-8-74)
20%
10%
2.5%
5 %
1%
½%
0.5 % MBS
c
M 51 ThA
T2 10-21-74
MBS Dose Response
(10-8-74)
20 %
10 %
2.5%
5 %
1%
½%
0.5 % MBS
b
M 53 ThA
T2 11-30-74
MBS Dose Response
(10-8-74 prep)
20%
10%
5%
2.5%
1%
½%
0.5 % MBS
d
Cells / Plate
Days in Culture
Days in culture

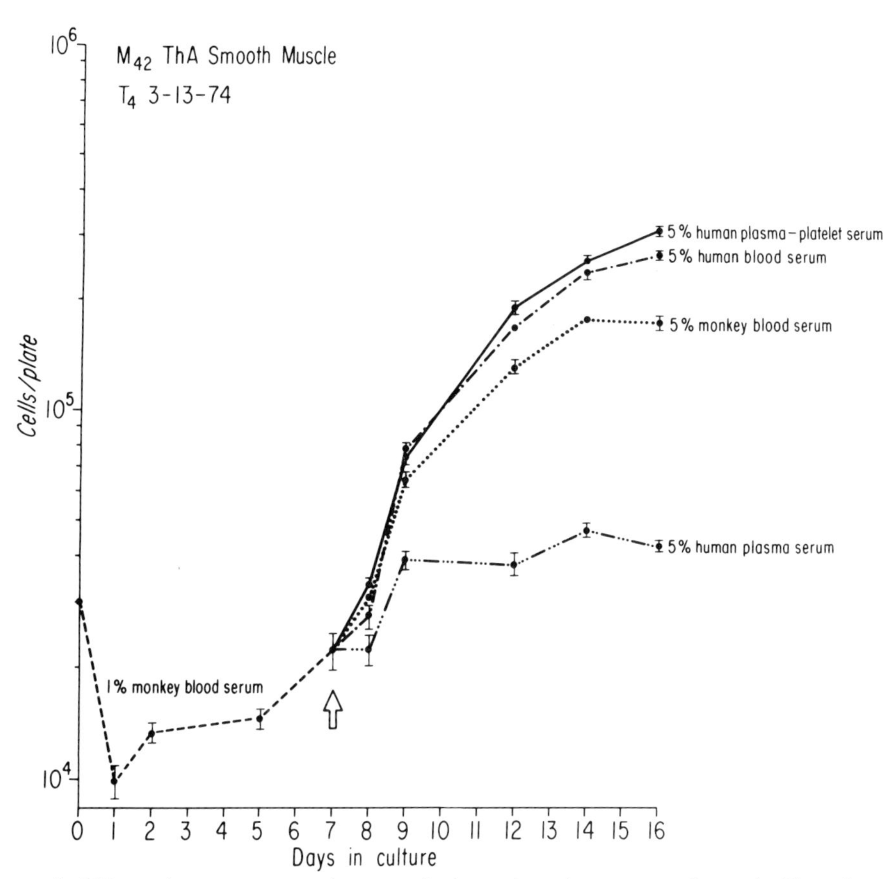

FIGURE 3 This graph represents growth curves of primate thoracic aorta smooth muscle. The cells were grown in the same medium containing the same pool of homologous serum. The differences in the growth characteristics are emphasized by these growth curves which compare the growth of randomly distributed groups of cells into four different media, containing either 5% monkey blood serum, 5% human platelet free plasma serum (human plasma serum), 5% human blood serum, or 5% human plasma serum to which platelet factors had been readded (human plasma-platelet serum). Similar effects were seen for fibroblasts in identical experiments. The smooth muscle cells grew logarithmically for 9 days or longer after exposure to whole blood serum or to platelet factors, whereas the dermal fibroblasts grew logarithmically for only two or three days before becoming quiescent under the same conditions. Part of this difference in response may be the result of the "hill and valley" pattern of growth demonstrated by smooth muscle cells and lacking in fibroblasts, and part may be the result of the difference in cell volume. Other as yet undiscovered factors may also be responsible for these characteristic differences in response.

FIGURE 2 These sets of four growth curves represent the response of aortic smooth muscle cells from the same trypsinization from approximately 1-yr-old male donor monkeys (*Macaca nemestrina*) to the same pool of homologous serum grown in the same culture medium. In sharp contrast, Fig. 2*c* demonstrates that cells from monkey no. 52 grew equally well in 1% serum as did cells from monkey no. 50 (2*a*) in 5% serum. Further, the cells from monkey no. 52 (Fig. 2*c*) grew equally well in 2.5–20% serum. The remaining growth curves in these two figures clearly demonstrate the marked variability from donor animal to donor animal that can be seen in response to the same level of the same pool of serum in the medium. This difference in response represents one of the principal problems in evaluating the growth response of diploid cells in culture and demonstrates the need for appropriate control experiments whenever such studies are performed with these cells.

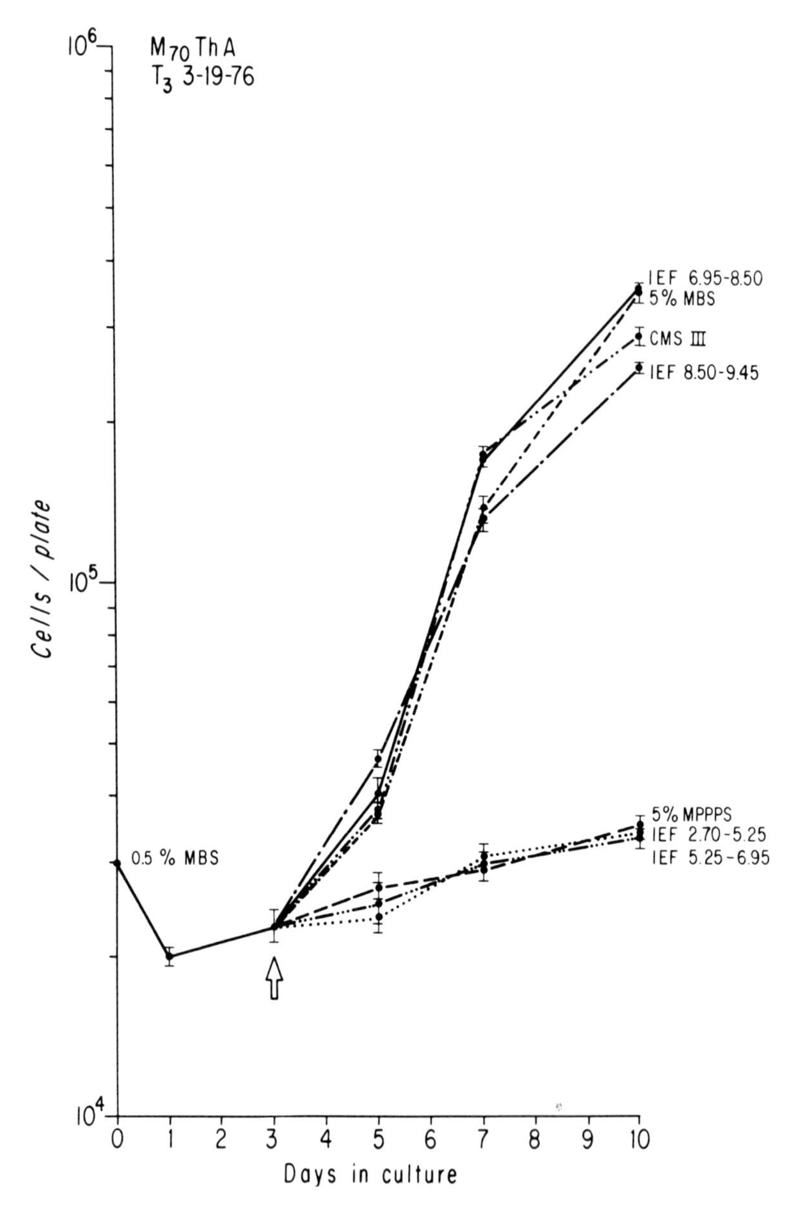

FIGURE 4 Response of arterial smooth muscle cells to fractions of platelet release material added to a 5% platelet-poor plasma serum base. A large series of 35-mm Falcon plastic petri dishes were inoculated with 3×10^4 smooth muscle cells in 0.5% pooled monkey whole blood serum. After 3 days in culture, the dishes were randomly divided into the seven groups indicated above. Fractions were prepared as follows: platelets were purified from platelet-rich plasma by gel filtration and specifically released by incubation with thrombin; release supernate was chromatographed on CM-Sephadex using a stepwise gradient: 0.01 M Tris-HCl, pH 7.4, 0.02 M benzamidine, 0.09–1.09 M NaCl; active fraction CMS III (0.01 M Tris-HCl, pH 7.4, 0.02 M benzamidine, 0.39 M NaCl) concentrated and passed over G-200 in 6 M GuHCl in 0.05 M Phos, pH 6.0; active component (from standard molecular-weight range 10,000–20,000) concentrated and further fractionated by isoelectric focusing in bed of Sephadex G-75SF containing 2% ampholine (pH 3.5–10) at +5°C; isoelectric focusing fractions scraped from plate and eluted with 6.0 M GuHCl in 0.05 M phosphate, pH 6.0, onto small column of G-50M equilibrated in same buffer. This procedure is necessary to separate the proteins and ampholines.

tor(s) results in recruitment of a large percentage of the cells into the cell cycle. If the factor(s) is presented to the cells for 1 h and then withdrawn, depending upon the state of synchrony of the culture, approximately 25% of the cells will be recruited into one round of DNA synthesis and division, and will then again become quiescent (Rutherford and Ross, 1976). When combined with the appropriate combination of other growth factors and hormones such as insulin, it may be possible, as suggested by Sato (unpublished data), to regulate the growth of different kinds of cells in culture in a totally defined medium. This has not yet been accomplished for diploid cells.

Other Plasma Factors

Other constituents derived from whole blood serum or cell-free plasma serum, including lipoproteins, have effects upon cells in culture. In earlier studies, we demonstrated that low-density lipoproteins, when coupled with lipoprotein-free serum, will also support the growth of smooth muscle cells in culture (Ross and Glomset, 1973). These studies suggested that the low-density lipoproteins may provide nutrients for membrane formation for the growth of cells in culture.

Connective Tissue Metabolism

Arterial smooth muscle cells are the principal connective tissue synthetic cell of the artery wall (Ross and Glomset, 1974). The determination of the factors that control the elaboration of each of the connective tissue components and the various amounts formed under different culture conditions will be important in helping to elucidate factors that control their production in vivo. In culture, arterial smooth muscle cells are capable of forming glycosaminoglycans (Wight and Ross, 1975), both types III and I collagen (J. Burke, unpublished observations), and both soluble and insoluble elastin (Narayanan et al., 1976; Abraham et al., 1974), as well as elastic fiber microfibrils (Ross, 1971).

In summary, it can be shown that arterial smooth muscle cells retain their phenotype in culture, have characteristic growth patterns and responses that differ from other connective tissue cells such as fibroblasts, and form all of the connective tissue matrix components found in vivo in the artery wall. The proliferative response evoked by serum in culture is the result of exposure of the cells to a mitogen(s) that is derived from platelet release. This platelet factor(s) is, by definition, present in all whole blood sera and has been shown to elicit a proliferative response from all mesodermally derived cells thus far examined.

In Vivo Studies of Atherogenesis

Two different types of in vivo studies have been performed to test the response to injury hypothesis. The first of these has utilized intra-arterial balloon catheters to mechanically remove the endothelium from selected segments of the iliac artery or the aorta (Baumgartner and Studer, 1966). These studies, and those using other forms of intra-arterial mechanical injury (Stemerman and Ross, 1972; Bjorkerud, 1969; Helin et al., 1971), have demonstrated that shortly after removal of the endothelium, platelets adhere to the exposed subendothelial connective tissue and undergo degranulation so that both platelet and plasma components have access to the injured segment of the artery wall. Within 1 wk, smooth muscle cells are observed in the process of migrating from the media into the intima at the sites of missing endothelium. During the period of endothelial regeneration, intimal smooth muscle proliferation continues so that 3 mo after injury normocholesterolemic animals have lesions containing 15–20 cell layers of smooth muscle cells surrounded by newly formed connective tissue. These lesions are reversible in normocholesterolemic monkeys (Ross and Glomset, 1973) but appear to be irreversible in hypercholesterolemic monkeys (Ross and Harker, 1976). In addition, hypercholesterolemic monkeys show a marked decrease in platelet survival together with large amounts of both intracellular and extracellular lipid deposits in the diet-induced lesions (Fig. 5).

These in vivo studies demonstrated that platelet adherence and release is associated with loss of endothelium and subsequent smooth muscle proliferation, but do not indicate whether the platelet factor(s) observed in vitro is required for smooth muscle cell proliferation in vivo. Studies of chronic homocystinemic baboons have provided data relevant to this question (Harker et al., 1976). After 6 days of chronic homocystinemia (induced by continuous intravenous infusion of homocystine at a rate that exceeded the animal's capability to clear the amino acid), endothelial loss of up to 10% was induced in the aorta together with a decrease in platelet survival of up to 50%. The decrease in platelet survival could be directly correlated to the amount of missing endothelium (Ross and Har-

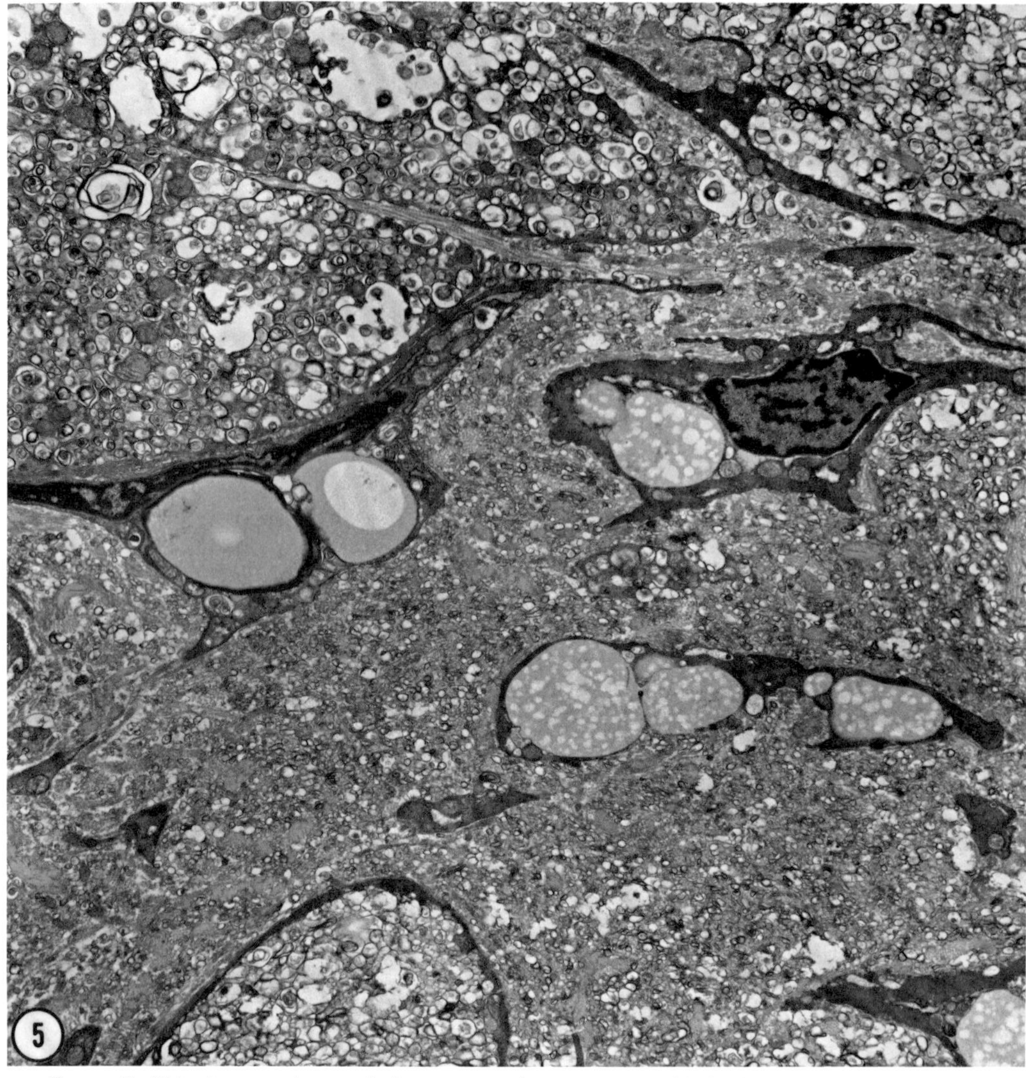

FIGURE 5 This electron micrograph demonstrates the appearance of intimal smooth muscle cells that have proliferated in an atherosclerotic lesion in a monkey fed a high fat diet for 1 yr. The smooth muscle cells contain large amorphous extracellular lipid deposits as well as numerous small membranous debris, similar in appearance to the membranous deposits found in the extracellular connective tissue. Such an appearance is characteristic of the altered smooth muscle cells in these lesions. × 7,200.

ker, 1976). Further studies with a large series of homocystinemic baboons demonstrated that after 90 days of induced chronic homocystinemia, there was not only loss of endothelium and a marked decrease in platelet survival, but an intimal smooth muscle proliferative response occurred identical to that observed after various forms of mechanical injury. However, it was possible to prevent the intimal proliferative response seen after 90 days by treating the chronic homocystinemic baboons with the inhibitor of platelet function, dipyridamole, for the entire period of the experiment. Appropriate doses of dipyridamole returned platelet survival to normal levels but had no effect on the endothelial desquamation induced by the chronic homocystinemia. This evidence, together with that from several other laboratories, supports the notion that platelets release a mitogen(s) that stimulates focal smooth muscle proliferation in vivo as well as in vitro. Moore et al. (1976) were able to completely suppress the in vivo smooth muscle mitogenic response normally produced by an indwelling catheter in rabbits made thrombocytopenic with an antiplatelet serum. In a different set of experiments, Bowie et al. (1975) were able to show that pigs with the platelet defect associated with von Willebrand's disease and fed a hypercholesterolemic diet developed no smooth muscle proliferative lesions in contrast to normal matched hypercholesterolemic pigs. Von Willebrand's disease is a genetic defect in which platelet aggregation is inhibited.

Thus it can be shown in vivo as well as in vitro that the smooth muscle proliferative response is dependent upon a factor released by platelets in their immediate environ. Inhibition of platelet release or absence of the factor(s) inhibits the proliferative response.

Summary

The response to injury hypothesis as an explanation for the pathogenesis of the lesions of atherosclerosis suggests that endothelial injury causes exposure of subendothelial connective tissues to whole blood, and promotes adherence of platelets to these tissues together with release of mitogenic factor(s) from the platelets into the underlying artery wall. This exposure to platelet mitogen(s) and plasma factors is thought to stimulate intimal proliferation of smooth muscle cells. Experiments in cell culture have supported this hypothesis by demonstrating that the principal mitogen(s) in whole blood serum is a platelet-derived factor. Experiments in vivo have provided evidence that platelet factors also play a role in the induction of both preatherosclerotic and atherosclerotic lesions. Further experiments concerning the isolation and identification of the platelet factor(s), modes of protecting the endothelium and inhibiting the activity of platelet mitogen(s), should provide important new data for understanding the etiology and pathogenesis of atherosclerosis. These studies show how the approaches of cell biology can be used to study an important disease process in man.

ACKNOWLEDGMENTS

These investigations were supported in part by grant HL14865 from the U. S. Public Health Service.

REFERENCES

Abraham, P. A., D. W. Smith, and W. H. Carnes. 1974. Synthesis of soluble elastin by aortic medial cells in culture. *Biochem. Biophys. Res. Commun.* **58:**597.

Antoniades, H. N., D. Stathakos, and C. D. Scher. 1975. Isolation of a cationic polypeptide from human serum that stimulates proliferation of 3T3 cells. *Proc. Natl. Acad. Sci. U. S. A.* **72:**2635–2639.

Arteriosclerosis: A Report by the National Heart and Lung Institute Task Force on Arteriosclerosis. 1971. Department of Health, Education, and Welfare Publication no. (National Institutes of Health) 72-219. Vol. 2.

Baumgartner, von, H. R., and A. Studer. 1966. Folgen des Gefasskatheterismus am normo- und hyperscholesterinaemischen Kaninchen. *Pathol. Microbiol.* **29:**393–405.

Bjorkerud, S. 1969. Reaction of the aortic wall of the rabbit after superficial, longitudinal, mechanical trauma. *Virchows Arch. Abt. A Pathol. Anat.* **347:** 197–210.

Bowie, E. J. W., V. Fuster, C. A. Owen, Jr., and A. L. Brown. 1975. Resistance to the development of spontaneous atherosclerosis in pigs with von Willebrand's disease. Fifth Congress of the International Society on Thrombosis and Haemostasis, Paris.

Busch, C., A. Watson, and B. Wasteson. 1977. Release of a cell growth promoting factor from human platelets. *Exp. Cell Res.* In press.

Fry, D. L. 1973. Responses of the arterial wall to certain physical factors. *Ciba Found. Symp.* **12:**93–125.

Glagov, S. 1972. Hemodynamic risk factors: mechanical stress, mural architecture, medial nutrition and the vulnerability of arteries to atherosclerosis. *In* The Pathogenesis of Atherosclerosis. R. W. Wissler and J. C. Geer, editors. Williams & Wilkins Co., Baltimore. 164–199.

Harker, L., R. Ross, S. Slichter, and C. Scott. 1976. Homocystine-induced arteriosclerosis: the role of endothelial cell injury and platelet response in its genesis. *J. Clin. Invest.* **58:**731–741.

Hayflick, L. 1965. The limited *in vitro* lifetime of human diploid cell strains. *Exp. Cell Res.* **37:**614–636.

Helin, P., I. Lorenzen, C. Garbarsch, et al. 1971. Arteriosclerosis in rabbit aorta induced by mechanical dilatation: biochemical and morphological studies. *Atherosclerosis.* **13:**319–331.

Kohler, N., and A. Lipton. 1974. Platelets as a source of fibroblast growth-promoting activity. *Exp. Cell Res.* **87:**297–301.

Moore, S., R. J. Friedman, D. P. Singal, J. Gauldie, and M. Blajchman. 1976. Inhibition of injury-induced thromboatherosclerotic lesions by antiplatelet serum in rabbits. *Thromb. Diath. Haemorrh.* **35:** 70–81.

Narayanan, A. S., L. B. Sandberg, R. Ross, and D. L. Layman. 1976. The smooth muscle cell. III. Elastin synthesis in arterial smooth muscle cell culture. *J. Cell Biol.* **68:**411–419.

Ross, R. 1971. The smooth muscle cell. II. Growth of smooth muscle in culture and formation of elastic fibers. *J. Cell Biol.* **50:**172–186.

Ross, R., and J. Glomset. 1973. Atherosclerosis and the arterial smooth muscle cell. *Science (Wash. D. C.).* **180:**1332–1339.

Ross, R., and J. Glomset. 1974. Studies of primate arterial smooth muscle cells in relation to atherosclerosis. *In* Arterial Mesenchyme and Arteriosclerosis. W. D. Wagner and T. B. Clarkson, editors. Plenum Publishing Co., New York-London. 265–279.

Ross, R., and J. Glomset. 1976. The pathogenesis of atherosclerosis. *N. Engl. J. Med.* **295**(pt. I):369–377; pt. II, **295:**420–425.

Ross, R., J. Glomset, B. Kariya, and L. Harker. 1974. A platelet-dependent serum factor that stimulates the proliferation of arterial smooth muscle cells *in vitro. Proc. Natl. Acad. Sci. U. S. A.* **71:**1207–1210.

Ross, R., J. Glomset, B. Kariya, and E. Raines. 1977. The role of platelet factors in the growth of cells in culture. *J. Natl. Cancer Inst.* In press.

Ross, R., and L. Harker. 1976. Hyperlipidemia and atherosclerosis. *Science (Wash. D. C.).* **193:**1094–1100.

Rutherford, R. B., and R. Ross. 1976. Platelet factors stimulate fibroblasts and smooth muscle cell quiescent in plasma-serum to proliferate. *J. Cell Biol.* **69:**196–203.

Stemerman, M. B., and R. Ross. 1972. Experimental arteriosclerosis. I. Fibrous plaque formation in primates: an electron microscope study. *J. Exp. Med.* **136:**769–789.

Westermark, B., and A. Wasteson. 1976. A platelet factor stimulating human normal glial cells. *Exp. Cell Res.* **98:**170–174.

Wight, T. N., and R. Ross. 1975. Proteoglycans in primate arteries. II. Synthesis and secretion of glycosaminoglycans by arterial smooth muscle cells in culture. *J. Cell Biol.* **67:**675–686.

THE LOW-DENSITY LIPOPROTEIN PATHWAY IN HUMAN FIBROBLASTS

Biochemical and Ultrastructural Correlations

JOSEPH L. GOLDSTEIN, MICHAEL S. BROWN, and RICHARD G. W. ANDERSON

In man, the development of atherosclerosis is potentiated by the deposition of massive amounts of esterified cholesterol within the substance of the artery wall. Perhaps the oldest, and the most straightforward, explanation for this cholesteryl ester deposition is the insudative theory (reviewed in Walton, 1975; Ross and Glomset, 1976; Goldstein and Brown, 1977). In its modern form, this theory states that the primary lesion in atherosclerosis involves an insult to the arterial endothelium. As a result, the normal barrier function of the endothelium breaks down and plasma constituents, including lipoproteins, penetrate through the endothelium into the deeper layers of the artery wall. Because of its insolubility, the cholesterol that is carried into the wall by the plasma lipoproteins can be removed only at a slow rate, and its progressive accumulation leads to the slow growth of the atheromatous plaque.

One way for the body to limit the penetration of cholesterol into the artery wall would be to keep the plasma lipoprotein cholesterol concentration low. On the other hand, it has recently been demonstrated that a variety of body cells require access to lipoproteins in order to supply themselves with cholesterol for membrane synthesis (Brown and Goldstein, 1976*a*). Thus, the body is faced with a dual problem of maintaining plasma lipoprotein concentrations high enough to supply its cells with cholesterol and yet not so high as to trigger atherosclerosis.

To solve this problem, human cells are equipped with a high-affinity membrane receptor that can bind a specific plasma lipoprotein, low density lipoprotein (LDL), when it is present at extremely low concentrations in the interstitial fluid (Goldstein and Brown, 1976, 1977). Binding of LDL to this membrane receptor initiates a process whereby the lipoprotein is internalized by endocytosis and delivered to lysosomes, where its protein and cholesteryl ester components are hydrolyzed and the resultant free cholesterol is made available for use by the cell in membrane synthesis. The liberated cholesterol, in turn, suppresses the cell's own cholesterol synthesis by reducing the activity of the rate-controlling enzyme, 3-hydroxy-3-methylglutaryl coenzyme A reductase. The liberated cholesterol also activates a microsomal acyl-CoA:cholesterol acyltransferase so that any excess cholesterol that enters the cell through the receptor pathway is reesterified and stored by the cell as cholesteryl ester. Finally, when cellular cholesterol stores have become adequate, the excess cholesterol acts to suppress the synthesis of LDL receptor molecules themselves, thus limiting the uptake of cholesterol and protecting the cell against an overaccumulation of cholesteryl esters (Brown and Goldstein, 1974, 1975, 1976*c*). The sequential steps in the pathway of LDL metabolism are shown schematically in Fig. 1.

The existence of the LDL pathway was deduced on the basis of extensive studies of cholesterol and lipoprotein metabolism in cultured fibroblasts derived from normal humans and from subjects with genetic blocks at specific sites in this pathway (Brown and Goldstein, 1976*a*). The LDL pathway has also been observed to operate in cultured human aortic smooth-muscle cells and human

JOSEPH L. GOLDSTEIN and CO-WORKERS Departments of Internal Medicine and Cell Biology, University of Texas Health Science Center at Dallas, Dallas, Texas

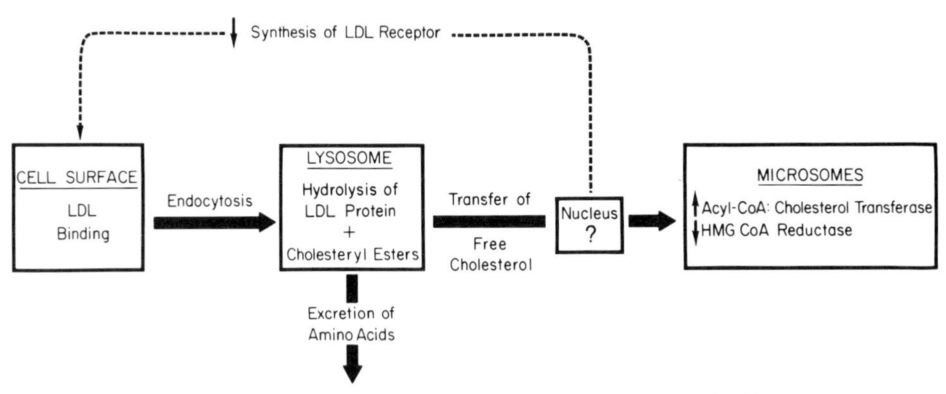

FIGURE 1 Sequential steps in the LDL pathway in human fibroblasts.

long-term lymphoid cell lines (Goldstein and Brown, 1975; Kayden et al., 1976; Ho et al., 1976*a*). Recently, its existence has been demonstrated in lymphocytes freshly isolated from human blood (Ho et al., 1976*b*).

We have postulated that most nonhepatic cells in man use the LDL pathway to obtain their cholesterol from LDL, which is derived ultimately from the liver (Brown and Goldstein, 1976*a*). The high affinity of the LDL receptor allows cells to take up LDL at a time when the plasma concentration of the lipoprotein is held low, and thus the LDL receptor serves as a major protective factor against atherosclerosis. When plasma LDL levels rise, either because of genetic defects in the LDL receptor (as in the disease Familial Hypercholesterolemia) or for other genetic or environmental reasons, deposition of the lipoprotein in the artery wall is enhanced and atheroclerosis ensues (Goldstein and Brown, 1977).

The LDL pathway, as described above, represents a classic example of adsorptive endocytosis, a process that was originally described in amoebae and in phagocytic cells from mammals (Korn, 1975; Jacques, 1975). In this article, we review recent biochemical and ultrastructural studies that deal with the coupling between the first two steps in the LDL pathway—namely, the binding of LDL to the cell-surface receptor and the subsequent endocytosis of the receptor-bound lipoprotein.

Biochemical Studies

The critical component of the LDL pathway is the specific receptor on the cell surface that binds plasma LDL. Indirect evidence suggests that the LDL receptor is a protein (or glycoprotein) molecule (Goldstein and Brown, 1976). In fibroblasts, the kinetics of LDL-receptor function have been studied by measurement of the binding of ^{125}I-labeled LDL to the receptor in intact monolayers *in situ*.

Binding of ^{125}I-LDL to the fibroblast receptor exhibits both high affinity (half-saturation achieved at 2 μg protein/ml at 4°C or 10–15 μg protein/ml at 37°C) and specificity (affinity for LDL more than 200-fold higher than for human high density lipoprotein or other plasma proteins). Under conditions of maximal derepression, actively growing fibroblasts exhibit about 15,000 high-affinity binding sites per cell (measured at 4°C). As shown in Fig. 2, binding of ^{125}I-LDL to the receptor shows an absolute requirement for divalent cations. When cells were incubated with ^{125}I-LDL at 4°C, binding was prevented by inclusion of 0.1 mM EDTA in the medium. This inhibition was completely overcome by the presence of low concentrations of Ca^{++} or Mn^{++}. Mg^{++} also restored the ^{125}I-LDL binding, but only at higher levels. Concentrations of Mn^{++} above 0.1 mM were inhibitory for binding.

In contrast to the direct study of ^{125}I-LDL binding to the cell-surface receptor at 4°C, studies of this reaction at 37°C in intact fibroblasts are complicated by the fact that internalization of the lipoprotein occurs and thus the total amount of ^{125}I-LDL measured as being bound to the cell represents the sum of the ^{125}I-LDL that is bound at the receptor site and that which has entered the cell through the receptor mechanism (Goldstein et al., 1976). To overcome this problem, we found it necessary to develop quantitative methods that could accurately distinguish between ^{125}I-LDL bound to the cell surface and ^{125}I-LDL that had been internalized. This was made possible through the use of heparin, a sulfated glycosaminoglycan

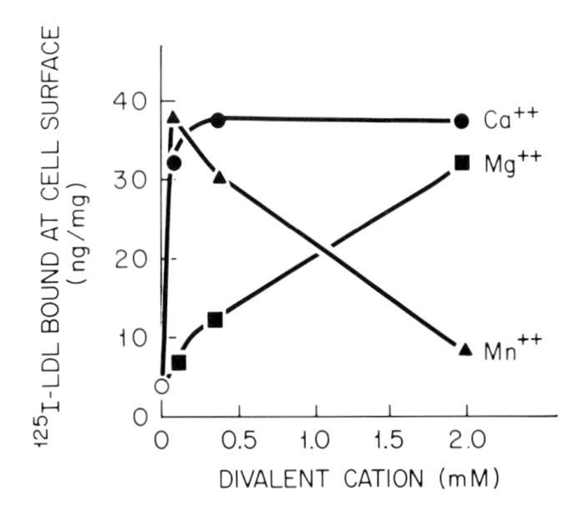

FIGURE 2 Effect of divalent cations on the binding of ^{125}I-LDL to the cell surface of normal fibroblasts at 4°C. Cells were grown in monolayer as previously described. On day 5 after seeding, when in the late logarithmic phase of growth, the cells were placed in medium containing lipoprotein-deficient serum (Goldstein et al., 1976). After incubation for 48 h (day 7), each monolayer was prechilled to 4°C for 30 min. Each monolayer was then washed with 3 ml of cold phosphate-buffered saline, after which each dish received 2 ml of phosphate-buffered saline containing 10 mg/ml of bovine serum albumin, 100 μM EDTA, pH 7.4, 5 μg protein/ml of ^{125}I-LDL (175 cpm/ng), and the indicated concentration of divalent cation: (○) none, (●) $CaCl_2$, (▲) $MnCl_2$, or (■) $MgCl_2$. The cells were incubated at 4°C for 2 h, after which they were washed extensively and the amount of ^{125}I-LDL that was releasable by heparin and hence bound at the cell surface was measured (Goldstein et al., 1976). The cells were dissolved in 1 ml of 0.1 N NaOH and an aliquot was used to measure the protein content. Each value represents the average of duplicate incubations.

that is known to form soluble ionic complexes with LDL (Iverius, 1972). Formation of such complexes allows heparin to release ^{125}I-LDL from its cell-surface receptor (Goldstein et al., 1976; Brown et al., 1976).

When normal fibroblasts have bound ^{125}I-LDL at the receptor site at 4°C, most of the bound ^{125}I-LDL can be released by heparin and hence is presumed to be on the cell surface. However, as shown by the data in Fig. 3, when cells were allowed to bind ^{125}I-LDL at 4°C, washed to remove unbound ^{125}I-LDL, and then warmed to 37°C, the amount of heparin-releasable ^{125}I-LDL on the cell surface declined by 70% within 12 min, and this fall was balanced by an equal rise in the amount of ^{125}I-LDL that remained associated with the cell after heparin treatment. This ^{125}I-LDL, which was resistant to release by heparin, represented receptor-bound ^{125}I-LDL that had become internalized by a temperature-dependent process that resembled adsorptive endocytosis. When the incubations were continued at 37°C, this internalized ^{125}I-LDL was delivered to the lysosomes where it was degraded, and after 12 min [^{125}I]monoiodotyrosine began to appear in the culture medium as TCA-soluble radioactivity. As the internal ^{125}I-LDL was degraded, the cellular content of internal ^{125}I-LDL reached a plateau and then declined.

With the amount of heparin-releasable ^{125}I-LDL as a measure of the amount of ^{125}I-LDL bound at the cell surface, we have been able to study the effect of temperature on each of the events in the process of LDL uptake and degradation (Fig. 4). The rate and extent of ^{125}I-LDL binding to the surface receptor (heparin-releasable

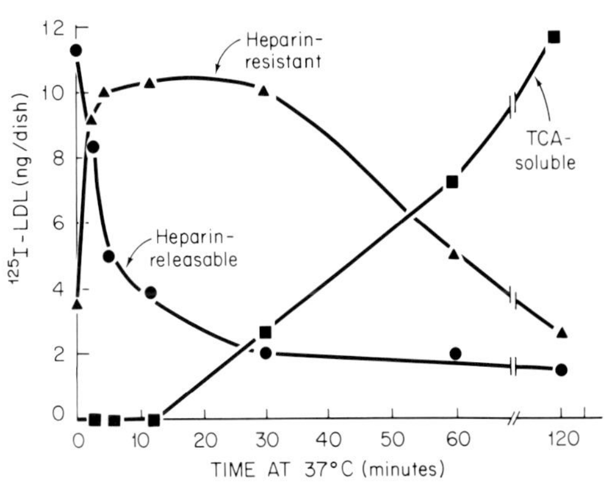

FIGURE 3 Internalization and degradation at 37°C of ^{125}I-LDL previously bound to the LDL receptor at 4°C. Nonconfluent monolayers of normal fibroblasts were prepared and incubated for 48 h in medium containing lipoprotein-deficient serum as described in the legend to Fig. 2. On day 7, the cells were prechilled to 4°C for 30 min, after which 10 μg protein/ml of ^{125}I-LDL (169 cpm/ng) was added. The ^{125}I-LDL was allowed to bind to the cells at 4°C for 2 h, after which each monolayer was washed extensively. Each dish then received 2 ml of growth medium containing 5% lipoprotein-deficient serum and 10 μg protein/ml of unlabeled LDL, and all the dishes were warmed to 37°C. After the indicated interval, groups of dishes were rapidly chilled to 4°C, the medium was removed, and its content of ^{125}I-labeled TCA-soluble material (■) was measured (Goldstein and Brown, 1974). The amount of ^{125}I-LDL bound to the cell surface and releasable by heparin (●) and the amount of ^{125}I-LDL that had entered the cell and was hence resistant to heparin release (▲) were determined as previously described (Goldstein et al., 1976). Each value represents the average of duplicate determinations. The protein content averaged 363 μg/dish.

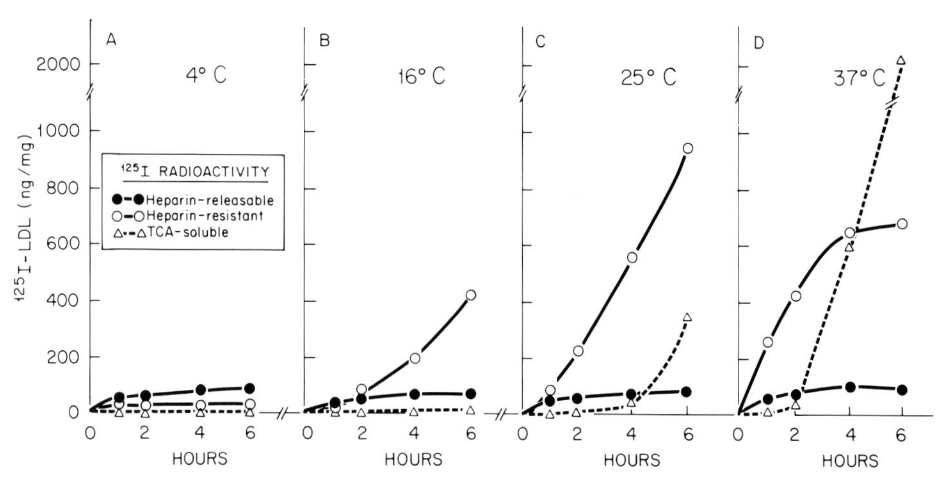

FIGURE 4 Effect of temperature on the binding of ^{125}I-LDL to the cell surface (●——●), its internalization (○——○), and its proteolytic degradation (△- -△) in monolayers of normal fibroblasts. Nonconfluent monolayers were prepared and incubated for 48 h in medium containing lipoprotein-deficient serum as described in the legend to Fig. 2. On day 7, the medium in each dish was replaced with 2 ml of medium that contained 5% lipoprotein-deficient serum and 10 μg protein/ml of ^{125}I-LDL (124 cpm/ng). The medium had been adjusted to the indicated temperature before its addition to the cells. After incubation at the indicated temperature for the indicated time, groups of monolayers were rapidly chilled to 4°C. The medium was immediately removed and its content of ^{125}I-labeled TCA-soluble material (△- -△) was measured (Goldstein and Brown, 1974). The cell monolayers were then extensively washed, and the amounts of heparin-releasable ^{125}I-LDL (●——●) and heparin-resistant ^{125}I-LDL (○——○) were determined as previously described (Goldstein et al., 1976). Each value represents the average of duplicate incubations.

^{125}I-LDL) were quantitatively similar at all temperatures. In contrast, the initial rate of internalization (i.e., the accumulation of heparin-resistant ^{125}I-LDL) was strongly temperature-dependent, increasing progressively with increasing temperature above 4°C. At 16 and 25°C the accumulation of heparin-resistant ^{125}I-LDL continued linearly for 6 h. However, at 37°C, a plateau in the cellular content of ^{125}I-LDL was reached after about 3 h (Fig. 4 *D*). The plateau at 37°C was by reason of the fact that, after 3 h, the rate of degradation of ^{125}I-LDL was equal to its rate of cellular uptake. Thus, a dynamic steady state was established in which the total cellular content of ^{125}I-LDL was constant, but degradation products continued to appear in the medium at a linear rate. On the other hand, at 16 and 25°C the rate of degradation of the internalized ^{125}I-LDL was much slower than at 37°C so that ^{125}I-LDL continued to accumulate in the cell after 3 h.

The internalization of receptor-bound ^{125}I-LDL at 37°C is relatively rapid, in that half of the receptor-bound particles are internalized through endocytosis every 5 min at 37°C (Fig. 3). During the endocytosis process, each internalized ^{125}I-LDL particle is replaced at the cell surface by a new particle of ^{125}I-LDL from the medium so that the amount of ^{125}I-LDL bound to the plasma membrane remains constant (Fig. 4 *D*). The attainment of such a dynamic steady state is because of the continuous regeneration of unoccupied receptors when receptor-bound LDL enters the cell at 37°C. Direct evidence for this regeneration of receptors is shown in the experiment in Fig. 5. In this experiment, one group of normal fibroblasts was incubated with a saturating concentration of unlabeled LDL at 4°C so as to occupy all functional receptors. A control group was not incubated with LDL. Both groups of cells were then warmed to 37°C in order to allow internalization of any previously bound LDL particles. At various intervals after warming to 37°C, both groups of cells were again chilled to 4°C to stop the internalization process, and the number of functional receptors on the cell surface was determined by measurement of the amount of ^{125}I-LDL that could be bound at 4°C. In the cells to which unlabeled LDL had been bound, the amount of ^{125}I-LDL that could initially be bound at 4°C was reduced from 37 to 8 ng/mg. However, warming

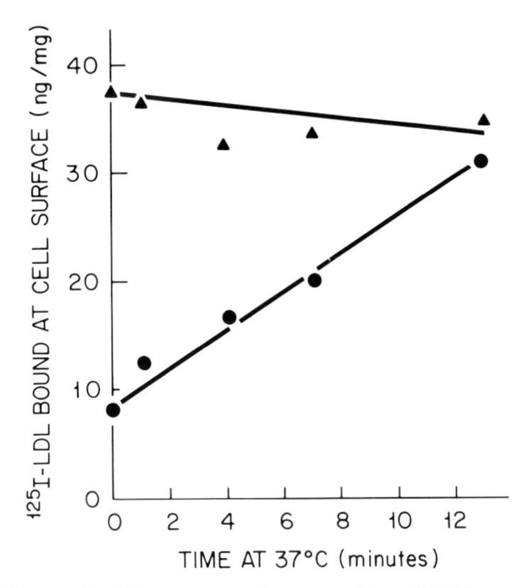

FIGURE 5 Time-course of regeneration of LDL-receptor sites at 37°C in fibroblasts previously incubated with unlabeled LDL at 4°C. Nonconfluent monolayers of normal fibroblasts were prepared and incubated for 48 h in medium containing lipoprotein-deficient serum as described in the legend to Fig. 2. On day 7, the cells were prechilled to 4°C for 30 min, after which they were incubated for 2 h at 4°C with 2 ml of fresh medium containing 10% lipoprotein-deficient serum and either no LDL (▲) or 50 μg protein/ml of unlabeled LDL (●). The monolayers were then washed extensively. Each dish then received 2 ml of growth medium containing 10% lipoprotein-deficient serum, and all of the dishes were warmed to 37°C. After the indicated interval at 37°C, groups of dishes were rapidly chilled to 4°C and the medium was replaced with 2 ml of ice-cold medium containing 10% lipoprotein-deficient serum and 10 μg protein/ml of ^{125}I-LDL (310 cpm/ng). The ^{125}I-LDL was allowed to bind to the cells for 30 min at 4°C, after which each monolayer was washed extensively and the amount of heparin-releasable ^{125}I-LDL bound to the cell surface was determined (Goldstein et al., 1976). Each value represents the average of duplicate incubations. (▲) Cells preincubated in the absence of LDL; (●) cells preincubated in the presence of 50 μg/ml of unlabeled LDL.

the cells to 37°C led to a progressive increase in the number of unoccupied receptor sites, so that by the end of 12 min the number of unoccupied receptors was the same in the cells that had been preincubated with unlabeled LDL as it was in those that had never been exposed to the unlabeled lipoprotein. The acquisition of new LDL receptors in Fig. 5 paralleled the internalization of ^{125}I-LDL (Fig. 3), indicating that as each particle of receptor-bound LDL is internalized, a fresh, unoccupied receptor molecule appears on the surface. The mechanism for this striking finding is currently under investigation.

Ultrastructural Studies

Recent ultrastructural studies have indicated that the functional LDL-receptor sites are not randomly distributed on the cell surface of normal fibroblasts (Anderson et al., 1976). With ferritin-labeled LDL as a probe for electron microscope studies, we have shown that more than 70% of the LDL-receptor sites are concentrated in short (0.5 μm) segments of plasma membrane where the membrane appears indented and coated on both of its sides by a fuzzy material. These so-called "coated regions," which constitute less than 2% of the total surface membrane of human fibroblasts, have been observed previously in a variety of other cell types, where they have been postulated to play a role in the specific uptake of proteins by adsorptive endocytosis (Roth and Porter, 1964; Fawcett, 1965; Friend and Farquhar, 1967; Bennett, 1969).

To obtain evidence that the LDL-ferritin complex was binding to the physiologic LDL receptor, we examined this binding with cells derived from a subject with the receptor-negative form of homozygous Familial Hypercholesterolemia (FH). Extensive biochemical studies have documented that the primary defect in this disease involves an absence of LDL receptors, so that fibroblasts from these subjects are unable to bind, internalize, or degrade LDL (Brown and Goldstein, 1976*b*). When fibroblasts from such an FH homozygote were exposed to the LDL-ferritin, no ferritin cores were seen on the cell surface. In all other respects, the ultrastructure of these cells was indistinguishable from that of normal cells. In particular, these mutant cells had the same number of coated regions per unit length of plasma membrane as did normal cells (Anderson et al., 1976).

Fig. 6 shows a series of electron micrographs of unstained sections of the cell surface of normal (*a*–*c*) and FH homozygote (*d*–*f*) fibroblasts that had been incubated at 4°C with LDL-ferritin. By using the criteria of indentation and localized thickening of the membrane, we were able to identify the coated regions in these unstained preparations. As in stained preparations, clusters of ferritin cores were localized in these coated regions in unstained normal cells but not in the FH homozygote cells. A detailed quantitative analysis of these sections has been published (Anderson et al., 1976).

When normal fibroblasts are first allowed to

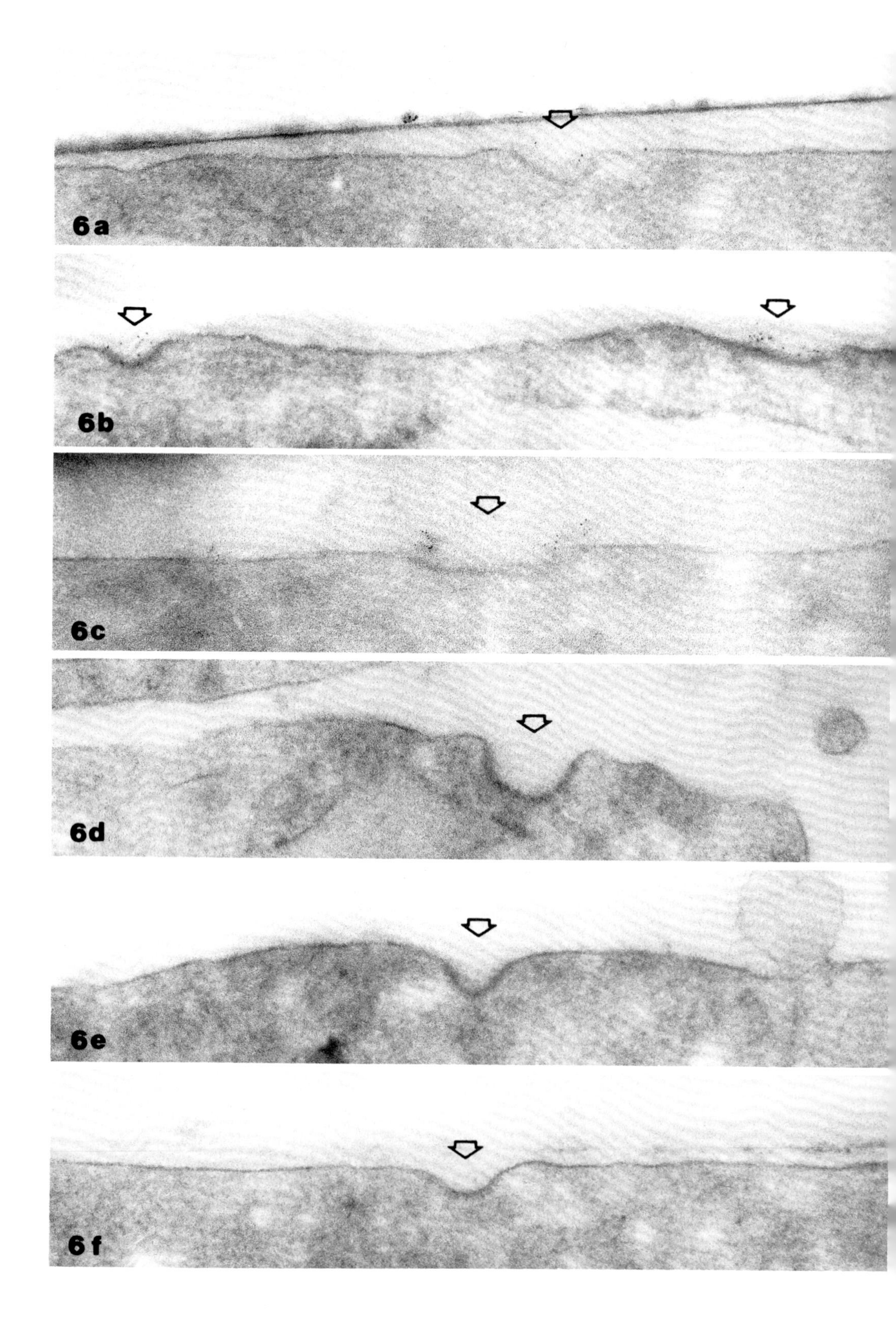
6a
6b
6c
6d
6e
6f

bind LDL-ferritin at 4°C and then warmed to 37°C, within 2 min the coated regions can be seen to invaginate and form coated endocytotic vesicles that carry the bound LDL-ferritin into the cell (Figs. 7 and 8). Within 5 min, much of the bound LDL-ferritin had reached the lysosome (Fig. 9). The time-course of internalization of LDL-ferritin as visualized with the electron microscope corresponds exactly to the time-course of internalization of ^{125}I-LDL studied biochemically (Fig. 3). Thus, within 12 min at 37°C nearly all of the

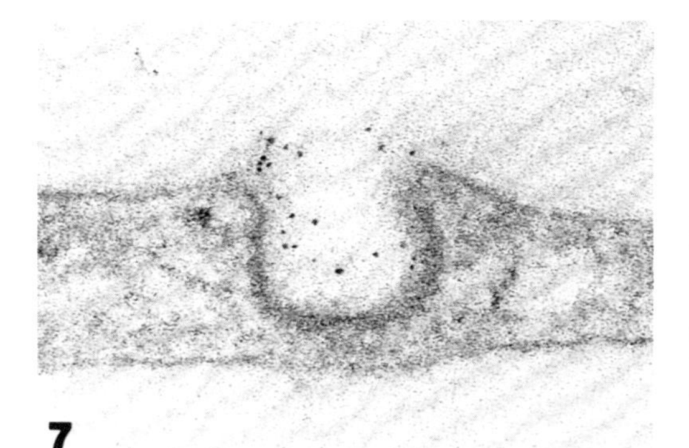

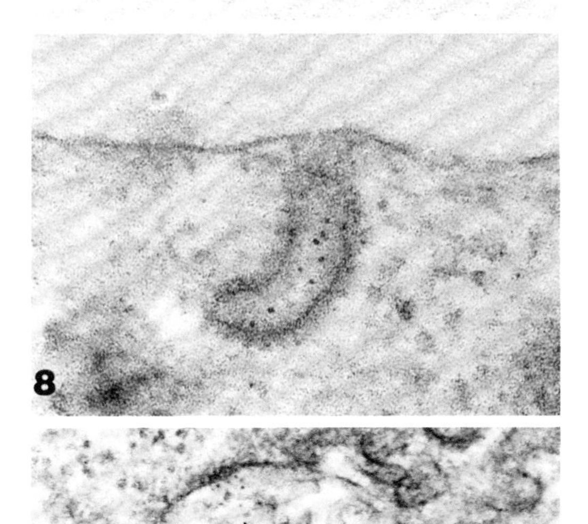

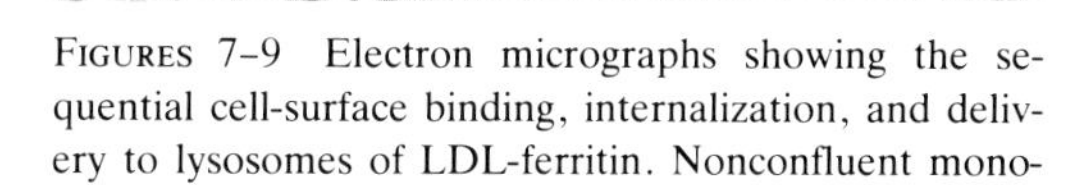

FIGURES 7–9 Electron micrographs showing the sequential cell-surface binding, internalization, and delivery to lysosomes of LDL-ferritin. Nonconfluent monolayers of normal fibroblasts were prepared as described in the legend to Fig. 3. On day 7, the cells were chilled to 4°C for 30 min, after which LDL-ferritin at a concentration corresponding to 47 μg/ml of LDL-protein was added to the growth medium. The LDL-ferritin was allowed to bind to the cells at 4°C for 2 h, after which each monolayer was washed extensively. Each dish then received 2 ml of growth medium containing 10% lipoprotein-deficient serum and all dishes were warmed to 37°C. After incubation at 37°C for the indicated time, the monolayers were fixed, embedded, and stained with uranyl acetate and lead citrate as previously described (Anderson et al., 1976).

FIGURE 7 An indented, coated region of the fibroblast membrane that was observed after the cells had been warmed for 1 min. This coated region contains LDL-ferritin and represents an intermediate step in the transformation of a coated membrane region into a coated vesicle. × 115,500.

FIGURE 8 A coated vesicle containing LDL-ferritin that was observed after the cells had been warmed for 2 min. At this time-point, many of the coated indentations of the plasma membrane had been transformed into coated vesicles. These coated vesicles were the only type of endocytotic vesicle that was observed to contain LDL-ferritin. × 94,000.

FIGURE 9 A lysosome containing LDL-ferritin that was observed after the cells had been warmed for 5 min. Ferritin cores within the lysosome are indicated by arrows. × 62,500.

FIGURE 6 Electron micrographs showing unstained sections of the cell surface of normal (*a–c*) and FH homozygote (*d–f*) fibroblasts incubated with LDL-ferritin. Nonconfluent monolayers of cells were grown, incubated with LDL-ferritin at a concentration corresponding to 33 μg/ml of LDL-protein at 4°C for 2 h, fixed, and embedded as previously described (Anderson et al., 1976). The indented regions of the cell membrane in these unstained sections corresponded to the coated regions and coated pits seen in stained sections. Whereas ferritin cores were associated with the indented regions of the cell surface of normal cells (*a–c*), no ferritin was found at any site on the cell membrane of the FH homozygote cells, including the indented regions (*d–f*). The arrows indicate the indented regions of the cell surface. × 64,000.

surface-bound LDL-ferritin had been internalized in coated vesicles.

We have previously shown that the lysosomal degradation of ^{125}I-LDL is blocked when fibroblasts are incubated with chloroquine, so that intact ^{125}I-LDL progressively accumulates within the lysosome (Goldstein et al., 1975*a, b*). Figs. 10 and 11 show the appearance of the lysosomes in chloroquine-treated fibroblasts that have been incubated with LDL-ferritin. In the normal cells, massive amounts of LDL-ferritin were seen within distended multivesicular bodies (Fig. 10). In the

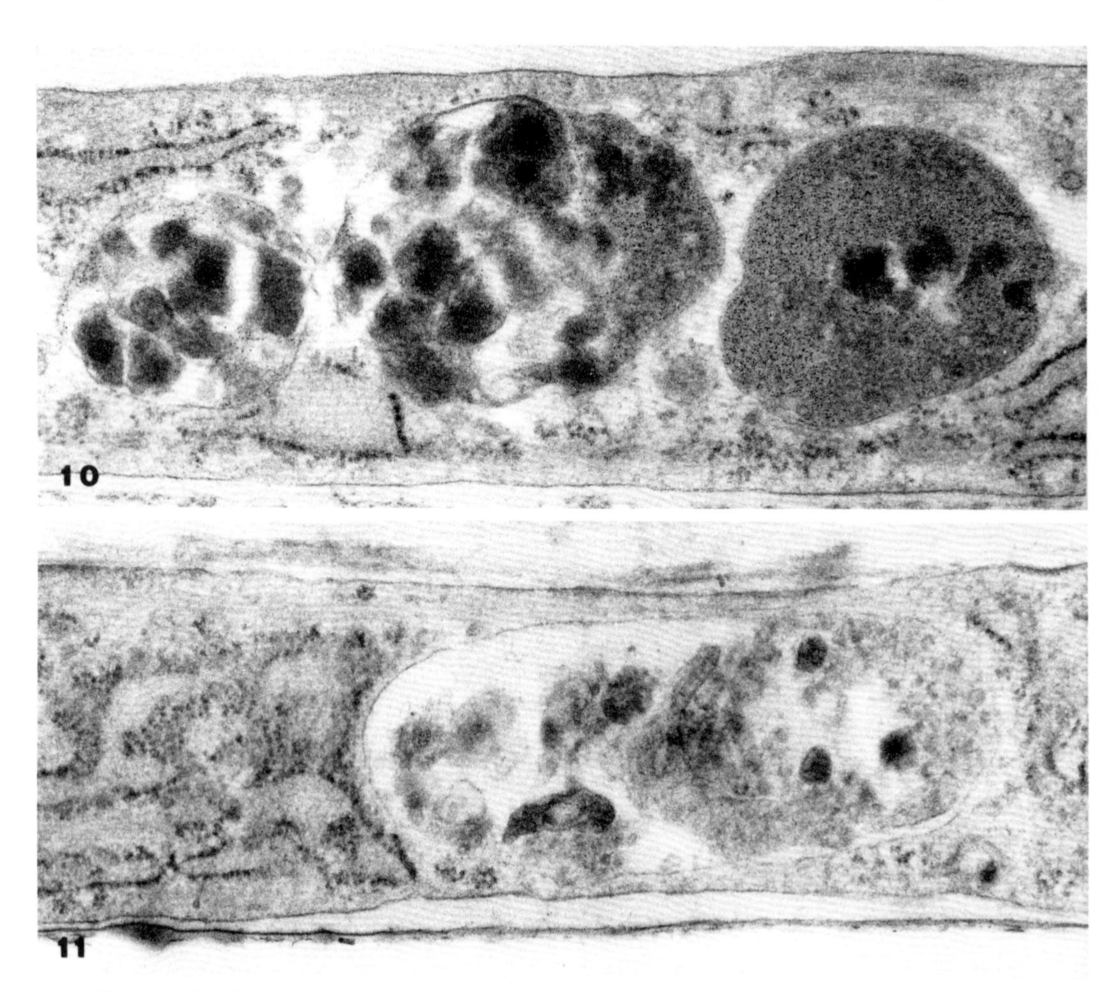

FIGURES 10 and 11 Electron micrographs of normal (Fig. 10) and FH homozygote (Fig. 11) fibroblasts that were incubated with LDL-ferritin in the presence of chloroquine. Nonconfluent monolayers were prepared as described in the legend to Fig. 3. On day 7, LDL-ferritin at a concentration corresponding to 120 μg/ml of LDL-protein was added to the growth medium in the presence of 75 μM chloroquine. After incubation at 37°C for 24 h, the monolayers were washed extensively, fixed, embedded, and stained with uranyl acetate and lead citrate as previously described (Anderson et al., 1976).

FIGURE 10 The chloroquine-treated normal fibroblasts typically showed large lysosomes that contained variable amounts of ferritin. Whereas the lysosome in the right of the micrograph is heavily labeled with ferritin, the other two lysosomes contain less ferritin. Except for the abnormally large lysosomes, other aspects of cell morphology were normal. × 39,000.

FIGURE 11 The chloroquine-treated FH homozygote fibroblast showed large lysosomes and other morphologic features that were indistinguishable from those of the normal fibroblasts. However, in contrast to normal cells, no ferritin was observed in the lysosomes of the homozygote cells. × 39,000.

FH homozygote cells, the same multivesicular bodies were seen, but they contained no LDL-ferritin (Fig. 11). These data further confirm the critical role of the coupling between the LDL receptor and the lysosome in allowing cells to take up the LDL particle and utilize its cholesterol.

SUMMARY

Recent studies indicate that human fibroblasts ingest plasma low density lipoprotein (LDL) by a process that kinetically and ultrastructurally resembles adsorptive endocytosis. The critical step is the binding of LDL to a high-affinity cell-surface receptor that is localized to coated regions of the plasma membrane. Uptake of LDL by this receptor-mediated process permits the cell to acquire cholesterol from the lipoprotein and this, in turn, not only provides sterol for membrane synthesis, but also provides sterol for cellular regulatory actions. The existence of this high-affinity uptake process allows nonhepatic cells in the body to take up LDL and utilize its cholesterol at a time when plasma LDL levels are held low. By allowing the body to maintain low LDL levels, the LDL pathway constitutes a powerful protective factor against atherosclerosis (Goldstein and Brown, 1977).

ACKNOWLEDGMENTS

Gloria Y. Brunschede, Mary Sobhani, and Margaret Wintersole provided excellent technical assistance. Jean Helgeson and Marian Eastman provided invaluable aid with the tissue culture.

This research was supported by grants from the American Heart Association and the National Institutes of Health (HL 16024, GM 21698, GM 19258). Dr. Goldstein is the recipient of a Research Career Development Award from the National Institute of General Medical Sciences. Dr. Brown is an Established Investigator of the American Heart Association.

REFERENCES

Anderson, R. G. W., J. L. Goldstein, and M. S. Brown. 1976. Localization of low density lipoprotein receptors on plasma membrane of normal fibroblasts and their absence in cells from a familial hypercholesterolemia homozygote. *Proc. Natl. Acad. Sci. U. S. A.* **73:**2434.

Bennett, H. S. 1969. The cell surface: movements and recombinants. *In* Handbook of Molecular Cytology. A. Lima-de Faria, editor. North-Holland Publishing Co., Amsterdam. 1295–1319.

Brown, M. S., and J. L. Goldstein. 1974. Familial hypercholesterolemia: defective binding of lipoproteins to cultured fibroblasts associated with impaired regulation of 3-hydroxy-3-methylglutaryl coenzyme A reductase activity. *Proc. Natl. Acad. Sci. U. S. A.* **71:**788.

Brown, M. S., and J. L. Goldstein. 1975. Regulation of the activity of the low density lipoprotein receptor in human fibroblasts. *Cell.* **6:**307.

Brown, M S., and J. L. Goldstein. 1976*a*. Receptor-mediated control of cholesterol metabolism. *Science (Wash. D.C.).* **191:**150.

Brown, M. S., and J. L. Goldstein. 1976*b*. Familial hypercholesterolemia: a genetic defect in the low density lipoprotein receptor. *N. Engl. J. Med.* **294:**1386.

Brown, M. S., and J. L. Goldstein. 1976*c*. The LDL pathway in human fibroblasts: a system for the study of receptor-mediated adsorptive endocytosis. *Trends in Biochemical Sciences.* **1:**193.

Brown, M. S., Y. K. Ho, and J. L. Goldstein. 1976. The low density lipoprotein pathway in human fibroblasts: relation between cell surface receptor binding and endocytosis of low density lipoprotein. *Ann. N.Y. Acad. Sci.* **275:**244.

Fawcett, D. W. 1965. Surface specializations of absorbing cells. *J. Histochem. Cytochem.* **13:**75.

Friend, D. S., and M. G. Farquhar. 1967. Functions of coated vesicles during protein absorption in the rat vas deferens. *J. Cell Biol.* **35:**357.

Goldstein, J. L., S. K. Basu, G. Y. Brunschede, and M. S. Brown. 1976. Release of low density lipoprotein from its cell surface receptor by sulfated glycosaminoglycans. *Cell.* **7:**85.

Goldstein, J. L., and M. S. Brown. 1974. Binding and degradation of low density lipoproteins by cultured human fibroblasts: comparison of cells from a normal subject and from a patient with homozygous familial hypercholesterolemia. *J. Biol. Chem.* **249:**5153.

Goldstein, J. L., and M. S. Brown. 1975. Lipoprotein receptors, cholesterol metabolism, and atherosclerosis. *Arch. Pathol.* **99:**181.

Goldstein, J. L., and M. S. Brown. 1976. The LDL pathway in human fibroblasts: a receptor-mediated mechanism for the regulation of cholesterol metabolism. *Curr. Top. Cell. Regul.* **11:**147.

Goldstein, J. L., and M. S. Brown. 1977. The low density lipoprotein pathway and its relation to atherosclerosis. *Annu. Rev. Biochem.* In press.

Goldstein, J. L., G. Y. Brunschede, and M. S. Brown. 1975*a*. Inhibition of the proteolytic degradation of low density lipoprotein in human fibroblasts by chloroquine, concanavalin A, and Triton WR 1339. *J. Biol. Chem.* **250:**7854.

Goldstein, J. L., S. E. Dana, J. R. Faust, A. L. Beaudet, and M. S. Brown. 1975*b*. Role of lysosomal acid lipase in the metabolism of plasma low density lipoprotein. *J. Biol. Chem.* **250:**8487.

Ho, Y. K., M. S. Brown, H. J. Kayden, and J. L. Goldstein. 1976*a*. Binding, internalization, and hydrolysis of low density lipoprotein in long-term lymphoid cell lines from a normal subject and a patient with homozygous familial hypercholesterolemia. *J. Exp. Med.* **144:**444.

Ho, Y. K., M. S. Brown, D. W. Bilheimer, and J. L. Goldstein. 1976*b*. Regulation of low density lipoprotein receptor activity in freshly isolated human lymphocytes. *J. Clin. Invest.* **58:**1465.

Iverius, P. H. 1972. The interaction between human plasma lipoproteins and connective tissue glycosaminoglycans. *J. Biol. Chem.* **247:**2607.

Jacques, P. J. 1975. The endocytic uptake of macromolecules. *In* Pathobiology of Cell Membranes. Vol. 1. B. F. Trump and A. V. Arstila, editors. Academic Press, Inc., New York. 255–282.

Kayden, H. J., L. Hatam, and N. G. Beratis. 1976. Regulation of 3-hydroxy-3-methylglutaryl coenzyme a reductase activity and the esterification of cholesterol in human long-term lymphoid cell lines. *Biochemistry.* **15:**521.

Korn, E. D. 1975. Biochemistry of endocytosis. *In* MTP International Review of Science. Vol. 2. C. F. Fox, editor. Butterworth & Co. Ltd., London. 1–26.

Ross, R., and J. A. Glomset. 1976. The pathogenesis of atherosclerosis. *N. Engl. J. Med.* **295:**369 and 420.

Roth, T. F., and K. R. Porter. 1964. Yolk protein uptake in the oocyte of the mosquito *Aedes Aegypti.* L. *J. Cell Biol.* **20:**313.

Walton, K. W. 1975. Pathogenetic mechanisms in atherosclerosis. *Am. J. Cardiol.* **35:**542.

CULTURE OF VASCULAR ENDOTHELIUM AND ATHEROSCLEROSIS

MICHAEL A. GIMBRONE, JR.

Vascular endothelium is a structurally simple but functionally complex tissue, the integrity of which is essential to the health of the arterial wall. Although involvement of endothelium in the pathogenesis of atherosclerosis has been recognized since the time of Virchow (1856), many relevant aspects of its biology have not been well studied. What does the endothelial cell synthesize and secrete? How does it interact with circulating hormones and plasma constituents, such as the lipoproteins? Which of its intrinsic properties are responsible for blood compatibility and selective permeability? What factors accelerate its aging or modify its response to injury? Answering such questions has been difficult because of the relative inaccessibility of the vascular lining to experimental manipulation in vivo. Recent advances in cell culture have changed this situation. Endothelial cells now can be harvested from the large vessels of animals and man, and maintained in vitro as homogeneous populations (see Gimbrone, 1976, for review). This approach permits direct study of metabolic, synthetic, and functional properties under controlled conditions. This report will discuss the basic methodology of endothelial culture, as well as two areas of current study relevant to the problem of atherosclerosis: growth control and regeneration, and the metabolism of vasoactive substances.

Methodological Considerations

Successful cultivation of vascular endothelium in vitro depends upon a selective isolation method, appropriate culture conditions, and reliable criteria for identifying the cultured cells.

ISOLATION OF ENDOTHELIAL CELLS: Endothelial cells can be harvested from large blood vessels by mechanical scraping of the luminal surface (Lewis et al., 1973) or dissociation of the intima with proteolytic enzymes (Maruyama, 1963; Jaffe et al., 1973*a*; Gimbrone et al., 1974; Booyse et al., 1975). In our experience, perfusing the lumen of intact vessel segments with bacterial collagenase consistently yields viable endothelial cells with a minimum of contamination by other cell types. The usefulness of collagenase may be the result, in part, of its selective digestion of the subendothelial basement membrane, releasing patches of the endothelial lining from the internal elastic lamina. In addition, unlike trypsin and other nonspecific proteases, collagenase does not attack the cell membrane and, therefore, tends to be less toxic to isolated cells. However, with this technique, control of enzyme concentration and digestion time, as well as gentle handling of the vessel, are important to avoid dislodging cells from deeper layers of the wall (Gimbrone and Cotran, 1975).

The following procedure is used routinely in our laboratory to obtain primary cultures of endothelium from umbilical cords, a readily available source of normal human vascular tissue (Gimbrone et al., 1974). This method can also be adapted to the aorta or inferior vena cava of experimental animals (Wechezak and Mansfield, 1973; Buonassisi, 1973; Booyse et al., 1975; Macarak et al., 1976).

Sterile umbilical cords are obtained immediately after normal vaginal deliveries or cesarean sections. Cord segments, at least 20–30 cm in length, are rinsed in Ringer's solution to remove adherent blood. Any crushed areas are excised, leaving a single length of untraumatized cord. The umbilical vein or artery is identified and cannulated with a siliconized glass cannula attached to a plastic three-way stopcock. The lumen is then

MICHAEL A. GIMBRONE, JR. Department of Pathology, Peter Bent Brigham Hospital and Harvard Medical School, Boston, Massachusetts gently perfused with 200 ml of Ringer's solution to remove any traces of blood. The opposite end of the vessel is similarly cannulated and the segment reflushed with 100 ml of Ringer's solution. The vessel lumen is filled with buffered collagenase solution (*Clostridium histolyticum,* Type I, Worthington Biochemical Corp., Freehold, N. J.; 1 mg = 125 U/ml, in Dulbecco's phosphate-buffered saline with magnesium and calcium, pH 7.2–7.4). The stopcocks are closed under slight pressure, and the cord segment is incubated for 10–20 min in a large jar containing Ringer's solution warmed to 37° C. The contents of the vessel are then gently flushed out with an equal volume of Hanks' balanced salt solution into a siliconized glass, conical centrifuge tube. Centrifugation (200 *g* for 5 min) yields a small white granular pellet. Contamination of the preparation can usually be detected at this point: a bulky, gelatinous pellet indicates the presence of connective tissue matrix and cellular elements from deeper areas of the vessel wall, whereas red cells reflect inadequate flushing of luminal blood.

PRIMARY CULTURES: After resuspension in culture medium and plating in plastic flasks, clumps of 5–10 rounded endothelial cells spread to form epitheloid clusters within the first few hours (Fig. 1*a*). These islands increase in size and gradually coalesce to form incomplete monolayers by 3–5 days (Fig. 1*b*). Cells within these early monolayers are uniform in appearance: elongated (30–40 μm wide, 60–80 μm long), with single ovoid nuclei which contain two to three prominent nucleoli, surrounded by a granular perinuclear region, and a broad, thin peripheral cytoplasm with indistinct borders (Fig. 1*c*). After 5–7 days, monolayers of closely packed, polygonal cells (35–50 μm) have usually formed (Fig. 1*d*). The tendency of overgrowth into multilayers is not observed in these cultures; cells retain their epitheloid appearance, close apposition, and attachment to the substrate for as long as 4–5 wk in primary culture.

Primary endothelial cultures thus obtained from umbilical vessels usually are morphologically uniform populations. However, when the collagenase digestion time is prolonged or the vessel has been traumatized, a second cell type appears, which is easily differentiated from endothelium by its large size and prominent cytoplasmic striations. The demonstration by electron microscopy of abundant myofilament bundles and the absence of endothelial-specific organelles (Weibel-Palade bodies) identifies this contaminant as smooth muscle. When each cell type is selectively cultured (Gimbrone and Cotran, 1975), a characteristic pattern of growth is observed. Endothelial cells remain closely apposed in simple monolayers (Fig. 2*a*), while smooth muscle cells become organized into multilayered hillocks and parallel arrays, interspersed with accumulations of extracellular material (Fig. 2b). These strikingly different organizational patterns are retained on serial passage, and mimic those expressed by each cell type in vivo. When mixed primary cultures are left undisturbed, however, endothelial cells invariably undergo degenerative changes first, and overgrowth by smooth muscle results. In practice, by avoiding trauma and controlling the extent of collagenase digestion, smooth muscle contamination in primary endothelial cultures can be held to a minimum.

Various commercially available nutrient media will support endothelial cell growth. In the experiments described here, the standard growth medium consisted of medium 199 (Hanks' salts with glutamine), supplemented with 15 mM HEPES buffer (pH 7.4), antibiotics, and 20–30% (vol/vol) fetal calf serum (heat-inactivated and mycoplasma-free). It is noteworthy that endothelial cells isolated from different vessels and species appear to require higher serum concentrations for optimal growth than do other diploid cells, including vascular smooth muscle (Gimbrone et al., 1974; Booyse et al., 1975; Blose and Chacko, 1975; Martin and Ogburn, 1976).

IDENTIFICATION OF CULTURED CELLS: Cultured endothelial cells exhibit certain morphological and functional features that help to differentiate them from other cells derived from the vascular wall, such as smooth muscle and fibroblasts (reviewed in Gimbrone, 1976). An especially useful identity marker in human umbilical endothelium is the Weibel-Palade body, an endothelial-specific organelle (Weibel and Palade, 1964), which has a characteristic ultrastructural appearance (Fig. 3) and is retained even after prolonged culture (Haudenschild et al., 1975). However, there is wide variation in the frequency and ease of demonstration of these endothelial organelles in different species and anatomical locations. A more universal marker may be antihemophilic factor (AHF, factor VIII) antigen which is localized to the vascular lining in vivo, and has been demonstrated by immunofluorescent microscopy in cultured endothelial cells from the human umbilical vein and bovine aorta (Jaffe et al.,

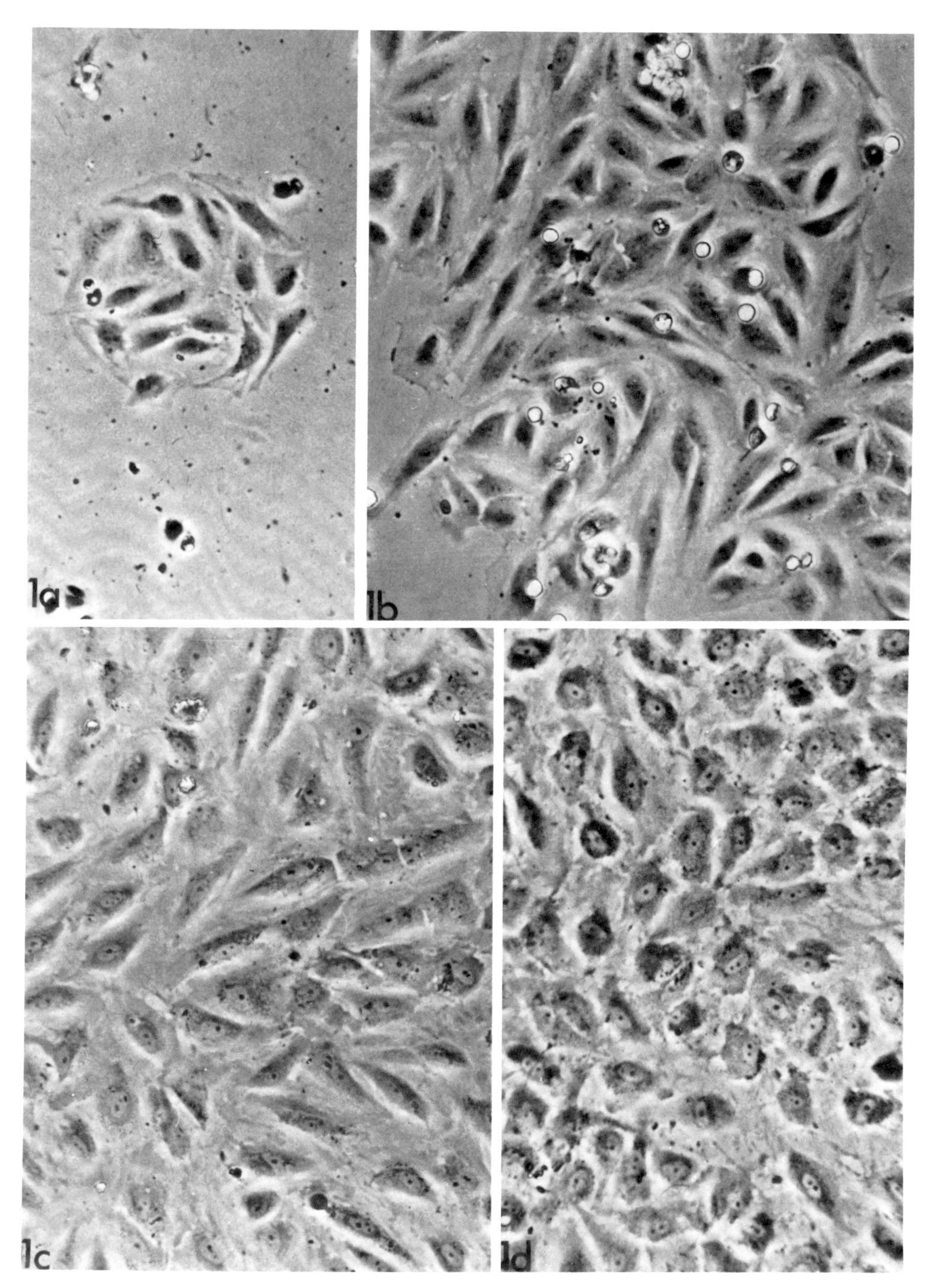

FIGURE 1 Primary cultures of human umbilical vein endothelium at different stages of growth (phase contrast, × 100). (*a*) 6 h, a small cluster of endothelial cells attaches to culture flask and spreads; (*b*) 24 h, adjacent clusters begin to coalesce; (*c*) 3 days, early monolayer; individual cells are elongated and have a broad, thin peripheral cytoplasm with indistinct borders; (*d*) 7 days, mature monolayer; cells are polygonal and tightly packed. (Reproduced from Gimbrone, 1976, by permission.)

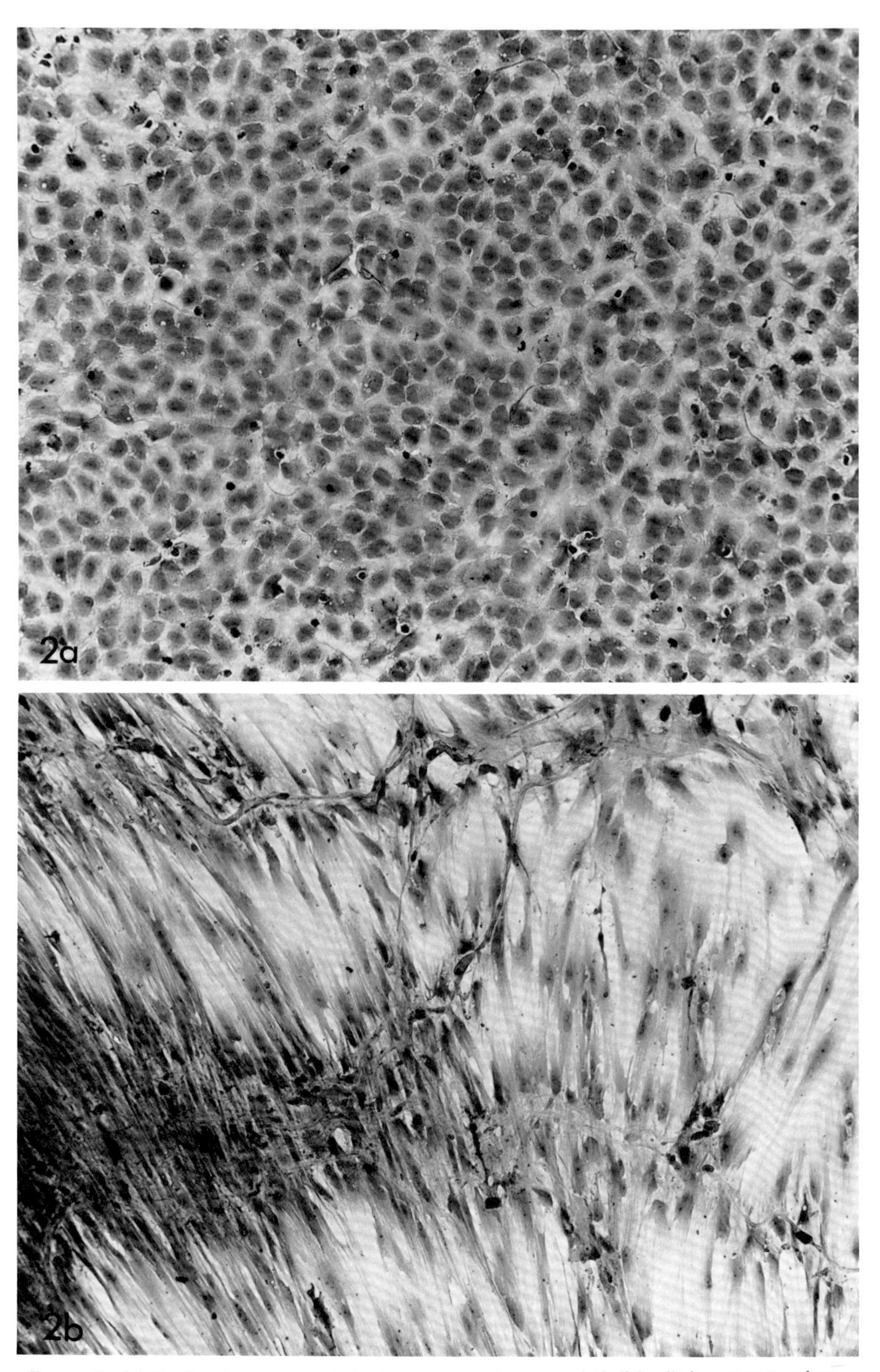

FIGURE 2 (*a*) Confluent monolayer of closely apposed polygonal endothelial cells in a mature culture from a human umbilical vein, × 50. (Reproduced from Gimbrone et al., 1974, by permission.) (*b*) Mature culture of human umbilical vein smooth muscle showing distinctive topographical pattern, with multilayered aggregates, parallel arrays, and accumulations of extracellular material, × 25.

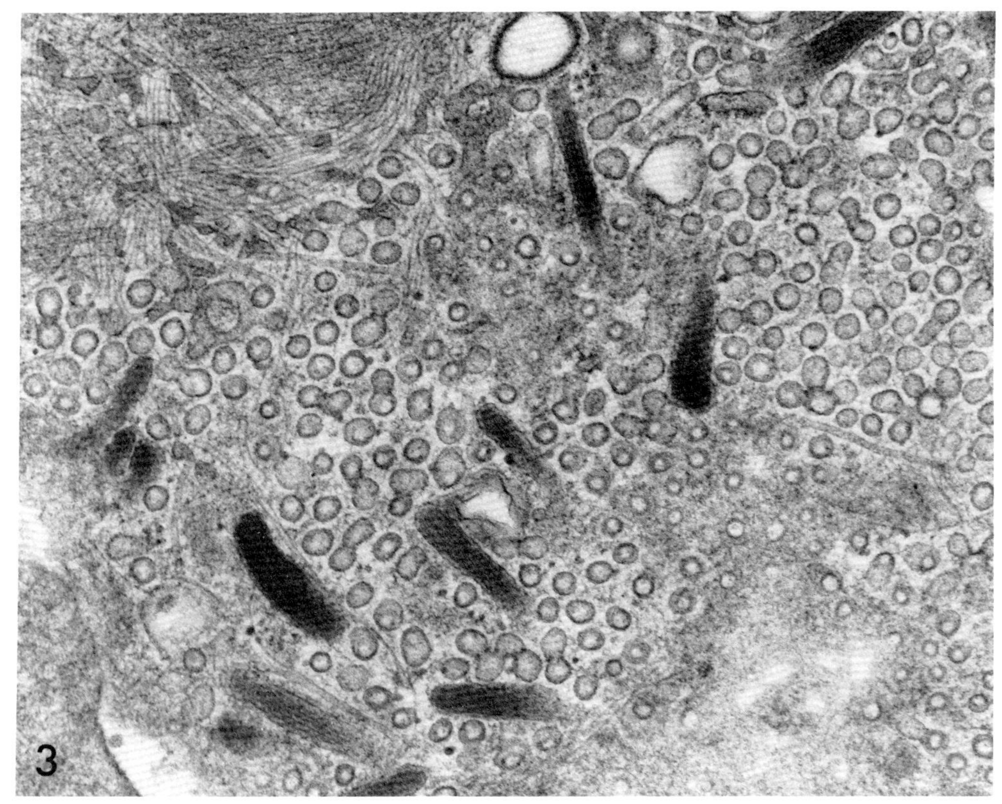

FIGURE 3 Electron micrograph of peripheral cytoplasm of cultured umbilical vein endothelial cell; note abundant pinocytotic vesicles and several rod-shaped, endothelial-specific organelles or Weibel-Palade bodies, × 36,000. (Micrograph courtesy of Dr. R. S. Cotran, Peter Bent Brigham Hospital.)

1973*b*; Booyse et al., 1975; Macarak et al., 1976). When more is known about the occurrence and properties of endothelial-specific enzymes, perhaps the identification of cultured endothelium by simple histochemical methods will become possible.

Growth Control and Regeneration

The endothelial lining of large blood vessels, such as the aorta, normally behaves as a slow renewal population: mitotic figures are rarely seen in histologic sections and in vivo autoradiographic studies with tritiated thymidine show low average labeling indices (Wright, 1972; Schwartz and Benditt, 1973). However, increased endothelial replication can be induced by a variety of pathologic stimuli, and this plays an important role in the reparative response of the vascular wall to mechanical, chemical, and immunologic injury. Recent studies have focused attention on endothelial injury and its sequelae, increased platelet-vessel wall reactivity, and abnormal intimal permeability, as initiating and promoting factors in the pathogenesis of atherosclerosis (Ross and Glomset, 1973, 1976; Mustard and Packham, 1975). Until the development of endothelial culture, no satisfactory in vitro system was available to study the mechanisms of endothelial growth, injury, and regeneration (Gimbrone et al., 1974).

When freshly isolated human umbilical endothelial cells are plated in medium supplemented with whole blood serum, after an initial attachment and spreading phase, they begin to synthesize DNA and proliferate to form a continuous monolayer (Gimbrone et al., 1974). As cell density increases, marked regional variations in replication become apparent. In central areas of the culture flask, where cells are in contact on all sides, the frequency of labeled nuclei in [^{3}H]thymidine autoradiographs is low. Simultaneously, around the periphery of the culture, where the monolayer is still incomplete, severalfold greater labeling is observed. These regional differences in replication are not influenced by the supply of fresh nutrient media and presumably reflect some form of growth control intrinsic to the cell population. Once a tightly packed monolayer has formed, cell density stabilizes and the overall labeling index drops to a basal level, reflecting a slow, fixed rate of cell attrition. A similar pattern of density-dependent DNA synthesis is seen when total TCA-precipitable [^{3}H]thymidine incorporation is determined at successive stages of culture growth (Fig. 4).

Although certain other cultured cells, such as primary diploid human fibroblasts or the BALB/c-3T3 line of mouse embryo cells, also exhibit density-dependent or "topo"-inhibition of division, endothelial monolayers appear to differ in a significant respect: they are refractory to serum growth factors in the postconfluent state. In other contact-inhibited culture systems, DNA synthesis and cell division can be stimulated by the addition of fresh medium containing increased concentrations of serum. However, basal levels of replication in confluent endothelial monolayers are not influenced by heterologous or homologous whole blood sera at concentrations ranging from 0–100% (Haudenschild et al., 1976). In contrast, vascular smooth muscle cells are especially sensitive, at all stages of culture growth, to mitogens which are present in whole blood serum and are thought to be derived from platelets (Ross et al., 1974; Wall et al., 1976). This major difference in the responsiveness of endothelium and smooth muscle may be important for growth control in the vascular wall. In vivo, the intact, quiescent endothelial lining may present a barrier to plasma constituents and platelet-release products, thus preventing stimulation of underlying smooth muscle cells. Factors that influence the ability of endothelial cells to restore this barrier after injury may be critically important in the pathogenesis of atherosclerosis.

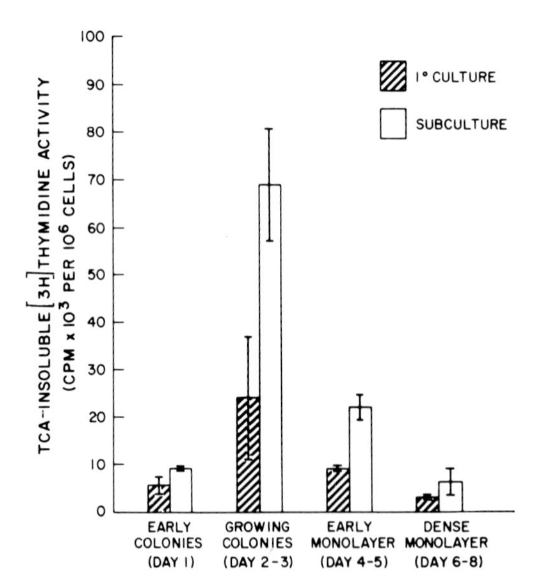

FIGURE 4 Tritiated thymidine incorporation into TCA-insoluble activity in primary and subcultured endothelial cells at successive stages of culture growth. Medium was exchanged daily and the stage of growth determined by phase-contrast examination. Each bar represents the average incorporation (±SD) of three separate cultures (10–20 replicate microcells per culture), after exposure to [methyl-^{3}H]thymidine for 12 h. (Reproduced from Gimbrone et al., 1974, by permission.)

In addition to their structural organization and slow turnover rate, confluent endothelial cultures also mimic the vascular intima in their response to mechanical injury. When small defects are created in confluent endothelial cultures, a regenerative response is elicited, which rapidly restores an intact monolayer (Gimbrone et al., 1974). By the use [^{3}H]thymidine autoradiography, the contribution of cell replication to this process can be assessed. Localized increases in DNA synthesis are detected in and around the denuded area during repair, but the healed defect soon returns to a serum-insensitive, topo-inhibited state. Phase contrast microscopy of the regenerating wound edge suggests that endothelial cell spreading and migration also may be involved. When cultures are preirradiated with 1,500 rads (a dose that prevents cell division but is not acutely cytotoxic), extensive healing of small mechanical defects still occurs (Sholley et al., 1976). It is possible that a similar process of endothelial cell migration and cytoplasmic spreading may help to rapidly restore intimal integrity following small localized injuries, in vivo. An irradiation-wounding model thus may prove useful in the in vitro assay of biological substances or drugs that promote or inhibit migrational and replicative aspects of endothelial regeneration.

Metabolism of Vasoactive Substances

In addition to functioning as a selective permeability barrier and nonthrombogenic lining for the vascular system, there is increasing evidence that endothelium actively participates in the metabolism of vasoactive substances. Endothelial cultures have proved especially useful for identifying metabolic products and characterizing the enzymatic activities involved.

PROSTAGLANDINS: Prostaglandins are a group of potent, vasoactive unsaturated C-20 fatty acids, which can be synthesized by many mammalian tissues and act primarily near their origins, as "tissue hormones" (Ferreira and Vane, 1967). Various lines of investigation have implicated certain prostaglandins, particularly prostaglandin E (PGE), in blood pressure regulation, the inflammatory process, and the function of leukocytes and platelets. Little is known, however, about the specific origins of endogenous prostaglandins, or the pathophysiologic factors controlling their biosynthesis, within the circulatory system. As an approach to this problem we have utilized endothelial cultures to study production of PGE under basal conditions and in response to certain vasoactive mediators (Gimbrone and Alexander, 1975, 1976).

Culture medium incubated with human umbilical vein endothelial cells accumulates immunoreactive PGE-like material (iPGE) at the rate of 56.0 ± 6.0 ng/mg cell protein per 24 h. When arachidonic acid is added to the culture medium, basal production is stimulated severalfold. To confirm that the extracellular accumulation of iPGE in this culture system reflects synthesis, rather than release of stored material, indomethacin, a potent inhibitor of prostaglandin cyclooxygenase, can be added to the medium. This results in a dose-related reduction in basal iPGE production (Fig. 5). Thus, cultured endothelial cells can generate PGE from arachidonic acid, presumably through the formation of an endoperoxide intermediate.

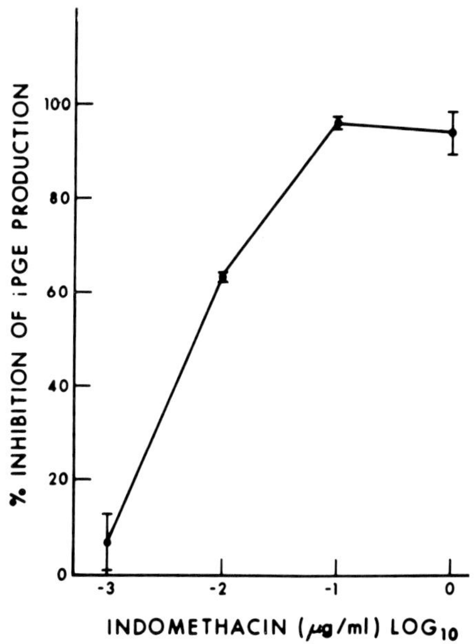

FIGURE 5 Inhibition of iPGE production in human endothelial cultures by indomethacin. Equal volumes of fresh medium containing no additions, or the indicated concentrations of indomethacin, were added to replicate confluent cultures in each experiment. After the cultures were incubated at 37°C for 24 h, the concentration of iPGE in the culture medium was determined by radioimmunoassay, and related to the protein content of the washed, sonicated cells in each culture well. Untreated cultures produced 0.129 ± 0.017 ng iPGE/mg cell protein per 24 h (mean ± SEM). Brackets indicate the SEM of mean values in two separate experiments. (Reproduced from Gimbrone and Alexander, 1975, by permission.)

There is indirect evidence, primarily from in vivo experiments, that vasoactive polypeptides such as angiotensin II can modulate PGE synthesis in vascular tissues. The role of endothelial cells in this phenomenon can be readily studied in cultures incubated with serum-free, chemically defined medium. This approach has several advantages: (1) enzymatic degradation and/or protein binding of test agonists and secreted prostaglandins are minimized; (2) unknown vasoactive substances, hormones, etc., present in serum or blood are not introduced; and (3) essential fatty acid precursors for prostaglandin synthesis are limited to those available from intracellular sources. When angiotensin II (hypertensin, Ciba-Geigy) is added to endothelial cultures plated in replicate, a dose-related stimulation of iPGE production is seen (Fig. 6). At maximal stimulation, the amount of iPGE in the culture medium is approximately 100 times greater than the intracellular content of unstimulated cells. This observation suggests that receptors for angiotensin II octapeptide, or a closely related metabolite (for example, des-Asp[1] Angiotensin II heptapeptide), are coupled to control points in the pathways for prostaglandin biosynthesis in vascular endothelium. It is an intriguing possibility that the effects of certain vasoactive hormones on the blood vessel wall may be mediated through receptor-coupled prostaglandin synthesis in endothelial cells.

ANGIOTENSIN METABOLISM: Angiotensin II, the principal effector hormone of the renin-angiotensin system (RAS), has been implicated in normal vascular homeostasis and hypertensive disease. This vasopressor octapeptide is generated from circulating renin substrate by the sequential action of two enzymes, renin, a protease secreted by the kidney, and angiotensin converting enzyme

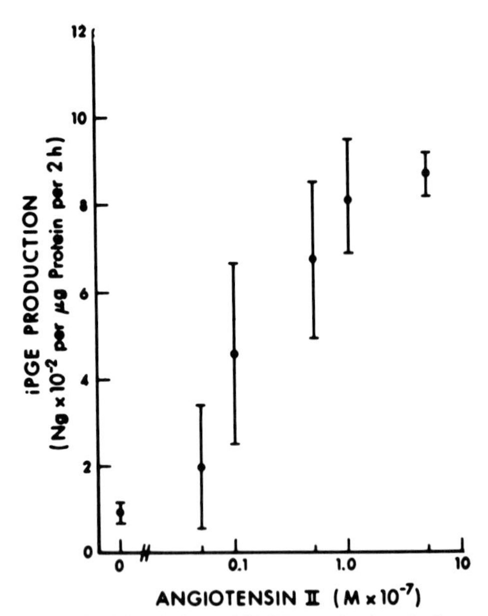

FIGURE 6 Stimulation of iPGE production in human endothelial cultures by angiotensin II. Angiotensin II (Hypertensin), freshly prepared in serum-free medium 199, contained negligible levels of iPGE before and after incubation in the absence of cells for 2 h at 37°C. In each experiment duplicate confluent cultures were incubated with equal volumes of serum-free medium, or medium containing the indicated concentrations of angiotensin II, for 2 h at 37°C. The concentration of iPGE in the culture medium was then determined and related to the protein content of the washed, sonicated cells from each culture well. Brackets indicate the SEM of mean values in three separate experiments. (Reproduced from Gimbrone and Alexander, 1975, by permission.)

(kininase II), a dipeptidyl-carboxypeptidase, found in high concentration in the lung. The situation of these key enzymes in separate organs, and the identification of a circulating prohormone, angiotensin I decapeptide, has indicated that the RAS functions as a systemic endocrine unit. However, a remarkable variety of other organs and tissues contain either angiotensin-converting or reninlike enzymes, though the specific cellular locations and physiological significances of these activities remain unclear. Recent studies with endothelial cultures have helped to clarify this problem (Johnson and Erdos, 1975; Hial et al., 1976; Ryan et al., 1976).

Endothelial cells cultured from human umbilical veins contain an enzyme, with an apparent molecular weight of 150,000 daltons, which generates biologically active angiotensin II (octapeptide) from angiotensin I (decapeptide), and inactivates bradykinin by cleaving a carboxyterminal dipeptide residue (phe-arg). Divalent cations and chloride are required for maximal converting activity. Converting and kininase activities are completely abolished by SQ20881 (bradykinin-potentiating factor, BPF9a), a competitive nonapeptide inhibitor. Smooth muscle cells isolated from the same vessel segments, and maintained under similar culture conditions, lack both activities. Thus, endothelial cells from extrapulmonary vessels contain an enzyme with chemical and biological properties similar to the angiotensin converting enzyme (kininase II) of lung. This suggests that the apparent ubiquity of this important component of the RAS may be a consequence of its presence in vascular endothelium throughout the body, and its high specific activity in the lung simply as a result of the marked density of capillary endothelium in that organ.

Further studies with cultured human umbilical endothelial cells provide evidence for another potentially important enzyme in the metabolism of angiotensin, a reninlike activity (J. Pisano, V. Hial, and M. Gimbrone, unpublished observations). Sonicated preparations of cultured endothelial cells, incubated with tetradecapeptide renin substrate, generate a material that contracts smooth muscle in both the guinea pig ileum and the gravid rat uterus. Furthermore, the contractile effect on rat uterus is reduced by inhibitors of angiotensin converting enzyme and augmented severalfold by preincubation with purified angiotensin converting enzyme. These bioassay data indicate that the product generated by the action of endothelial cells on tetradecapeptide renin substrate is angiotensin I (decapeptide). However, preliminary experiments on pH maximum and sensitivity to pepstatin suggest that this enzyme is different from plasma renin and may be related to a reninlike activity previously found in homogenates of blood vessel walls (Gould et al., 1964).

Data derived from studies on cultured endothelial cells thus indicate that the lining of systemic blood vessels may be a functional metabolic unit for certain vasoactive polypeptides. Endothelial cells have the capacity to (1) generate angiotensin I from a synthetic renin substrate, (2) convert this decapeptide to its active pressor form, angiotensin II, (3) degrade bradykinin, a potent depressor substance and inflammatory mediator, and (4) release "tissue hormones," such as prostaglandin E, in response to angiotensin stimulation. Further study of these complex interactions may shed light on the problem of hypertension and its role as a risk factor in atherosclerosis.

SUMMARY

Endothelial cells can be selectively harvested from the intimal lining of large blood vessels and cultivated in vitro as homogeneous populations which retain differentiated morphological and functional characteristics. Such cultures are especially useful for studying certain facets of endothelial cell biology, such as growth control and metabolic capabilities.

Confluent endothelial cultures appear to mimic naturally occurring endothelial tissues in their organization as simple monolayers, which have relatively stable cell densities and low cell turnover rates. Cultured endothelium differs significantly from vascular smooth muscle in its sensitivity to mitogenic factors found in whole blood serum: severalfold higher serum concentrations are required for optimal growth at subconfluent cell densities, and the postconfluent monolayer appears to become refractory to further stimulation. Endothelial monolayers also can reconstitute themselves after mechanical injury. This regenerative response involves both cell replication and migration, the relative contributions of which can be determined by [^{3}H]thymidine autoradiography in combination with cytostatic doses of X-radiation. Such culture models may aid in the analysis of factors that influence endothelial regeneration and intimal hyperplasia.

There is increasing evidence that endothelium plays an active role in the metabolism of vasoactive substances. Culture endothelial cells contain an indomethacin-sensitive prostaglandin synthetase system capable of generating E-type prostaglandins, which are known to influence vascular tone and permeability, as well as the function of leukocytes and platelets. Prostaglandin E production is stimulated by polypeptide hormones such as angiotensin II, perhaps through a receptor-coupled mechanism. In addition, at least two enzymes related to renin-angiotensin metabolism, a reninlike activity and angiotensin I converting enzyme (kininase II), are demonstrable in cultured endothelial cells. Further study of these metabolic capabilities may help unravel the interrelationships of complex disease processes such as hypertension and atherosclerosis, and the role of vascular endothelium in their pathogenesis.

ACKNOWLEDGMENTS

This work was supported in part by grants HL 08251 and HL 20054 from the National Heart, Lung and Blood Institute. The technical assistance of Ms. E. Shefton and Mr. G. Majeau is gratefully acknowledged.

REFERENCES

BLOSE, S. H., and S. CHACKO. 1975. *In vitro* behavior of guinea pig arterial and venous endothelial cells. *Dev. Growth Differ.* **17:**153–165.

BOOYSE, F. M., B. J. SEDLAK, and M. E. RAFELSON. 1975. Culture of arterial endothelial cells: characterization and growth of bovine aortic cells. *Thromb. Diath. Haemorrh.* **34:**825–839.

BUONASSISI, V. 1973. Sulfated mucopolysaccharide synthesis and secretion in endothelial cell cultures. *Exp. Cell Res.* **76:**363–368.

FERREIRA, S. H., and J. R. VANE. 1967. Prostaglandins: their disappearance from and release into the circulation. *Nature (Lond.).* **216:**868–873.

GIMBRONE, M. A., JR. 1976. Culture of vascular endothelium. *In* Progress in Hemostasis and Thrombosis. Vol. III. T. H. Spaet, editor. Grune & Stratton Inc., New York. 1–28.

GIMBRONE, M. A., JR., and R. W. ALEXANDER. 1975. Angiotensin II stimulation of prostaglandin production in cultured human vascular endothelium. *Science (Wash. D. C.).* **189:**219–220.

GIMBRONE, M. A., JR., and R. W. ALEXANDER. 1976. Prostaglandin production by human vascular endothelium and smooth muscle in culture. *In* Proceedings of the International Symposium on Prostaglandins in Hematology. M. J. Silver, editor. Spectrum Publications, New York. In press.

GIMBRONE, M. A., JR., and R. S. COTRAN. 1975. Human vascular smooth muscle in culture: growth and ultrastructure. *Lab. Invest.* **33:**16–27.

GIMBRONE, M. A., JR., R. S. COTRAN, and J. FOLKMAN. 1974. Human vascular endothelial cells in culture: growth and DNA synthesis. *J. Cell Biol.* **60:**673–684.

GOULD, A. B., L. T. SKEGGS, and J. R. KAHN. 1964. The presence of renin activity in blood vessel walls. *J. Exp. Med.* **119:**389–399.

HAUDENSCHILD, C., R. S. COTRAN, M. A. GIMBRONE, JR., and J. FOLKMAN. 1975. Fine structure of vascular endothelium in culture. *J. Ultrastruct. Res.* **50:**22–32.

HAUDENSCHILD, C., D. ZAHNISER, and M. KLAGSBRUN. 1976. Human vascular endothelial cells in culture: lack of response to serum growth factors. *Exp. Cell. Res.* **98:**175–183.

HIAL, V., M. A. GIMBRONE, JR., G. WILCOX, and J. J. PISANO. 1976. Human vascular endothelium contains angiotensin I converting enzyme and renin-like activity. *Fed. Proc.* **35**(3)**:**705.

JAFFE, E. A., R. L. NACHMAN, C. G. BECKER, and C. R. MINICK. 1973*a*. Culture of human endothelial cells derived from umbilical veins: identification by morphological and immunologic criteria. *J. Clin. Invest.* **52:**2745–2756.

Jaffe, E. A., L. W. Hoyer, and R. L. Nachman. 1973*b*. Synthesis of antihemophilic factor antigen by cultured human endothelial cells. *J. Clin. Invest.* **52:**2757–2764.

Johnson, A. R., and E. G. Erdos. 1975. Angiotensin I converting enzyme in human endothelial cells. *Circulation.* **51–52**(suppl. II)**:**II-59.

Lewis, L. J., J. C. Hoak, R. D. Maca, and G. L. Fry. 1973. Replication of human endothelial cells in culture. *Science* (*Wash. D. C.*). **181:**453–454.

Macarak, E. J., B. V. Howard, and N. A. Kefalides. 1976. Properties of calf endothelial cells in culture. *Lab. Invest.* In press.

Martin, G. M., and C. E. Ogburn. 1976. Cell, tissue and organoid cultures of blood vessels. *In* Growth, Nutrition and Metabolism of Cells in Culture. Vol. III. G. H. Rothblat and V. J. Cristofolo, editors. Academic Press, Inc., New York. In press.

Maruyama, Y. 1963. The human endothelial cell in tissue culture. *Z. Zellforsch. Mikrosk. Anat.* **60:**69–79.

Mustard, J. F., and M. A. Packham. 1975. The role of blood and platelets in atherosclerosis and the complications of atherosclerosis. *Thromb. Diath. Haemorrh.* **33:**444–456.

Ross, R., and J. A. Glomset. 1973. Atherosclerosis and the arterial smooth muscle cell: proliferation of smooth muscle is a key event in the genesis of the lesions of atherosclerosis. *Science* (*Wash. D. C.*). **180:**1332–1339.

Ross, R., and J. A. Glomset. 1976. The pathogenesis of atherosclerosis. *N. Engl. J. Med.* **295:**369–377, 420–425.

Ross, R., J. A. Glomset, B. Kariya, and L. A. Harker. 1974. A platelet-dependent serum factor that stimulates the proliferation of arterial smooth muscle cells in vitro. *Proc. Natl. Acad. Sci. U. S. A.* **71:**1207–1210.

Ryan, U. S., J. W. Ryan, C. Whitaker, and A. Chiu. 1976. Localization of angiotensin converting enzyme (Kininase II): immunocytochemistry and immunofluorescence. *Tissue Cell.* **8:**125–145.

Schwartz, S. M., and E. P. Benditt. 1973. Cell replication in the aortic endothelium: a new method for study of the problem. *Lab. Invest.* **28:**699–707.

Sholley, M. M., M. A. Gimbrone, Jr., and R. S. Cotran. 1976. Cellular migration and replication in endothelial regeneration: a study using irradiated endothelial cultures. *Lab. Invest.* In press.

Virchow, R. 1856. Phlogose und Thrombose im Gefasssystem, Gesammelte Abhandlungen zur Wissenschaftlichen Medicin. Frankfurt-am-Main, Meidinger Sohn and Company, p. 458.

Wall, R. T., L. A. Harker, and G. E. Striker. 1976. Differential response of human vessel wall cells to growth factors. *Circulation.* **53–54**(suppl. II)**:**II-54.

Wechezak, A. R., and P. B. Mansfield. 1973. Isolation and growth characteristics of cell lines from bovine venous endothelium. *In Vitro* (*Rockville*). **9:**39–45.

Weibel, E. R., and G. E. Palade. 1964. New cytoplasmic components in arterial endothelia. *J. Cell Biol.* **23:**101–112.

Wright, H. P. 1972. Mitosis patterns in aortic endothelium. *Atherosclerosis.* **15:**93.

Name Index

Subject Index

NAME INDEX

C

D

E

F

G

H

I

J

K

L

M

N

O

P

Q

R

S

T

U

V

W

Y

Z

SUBJECT INDEX

page references to figures and tables appear in italics

A

D

E

F

G

M

N

S

T